# FOOD BIOTECHNOLOGY
## Principles and Practices

# FOOD BIOTECHNOLOGY

## Principles and Practices


*Editors*

**V. K. JOSHI**

*Professor & Head*
Dr. Y.S. Parmar University of Horticulture & Forestry
Nauni, Solan, Himachal Pradesh, India

**R. S. SINGH**

*Professor & Head*
Department of Biotechnology
Punjab University, Patiala
Punjab, India



**I.K. International Publishing House Pvt. Ltd.**

NEW DELHI • BANGALORE

## DISCLAIMER

The views and ideas expressed are those of the contributors. The editors and publishers are not responsible for any damage or loss due to these views and ideas, though every care has been taken to present authentic and reliable information.

*Published by*

I.K. International Publishing House Pvt. Ltd.
S-25, Green Park Extension
Uphaar Cinema Market
New Delhi–110 016 (India)
E-mail:info@ikinternational.com
Website: www.ikbooks.com

ISBN 978-93-81141-49-6



Reprint 2013



Published by Krishan Makhijani for I.K. International Publishing House Pvt. Ltd., S-25, Green Park Extension, Uphaar Cinema Market, New Delhi–110 016 and Printed by Rekha Printers Pvt. Ltd., Okhla Industrial Area, Phase II, New Delhi–110 020.

*Dedicated*
*to our*
**Revered Parents**
*who*
*nurtured, motivated and enlightened our knowledge*
*and vision, which acted as catalyst*
*in transforming our lives*

# Preface

Food is the basic necessity of man, rather every living system. In fact, the whole life of human beings revolves around the food or the means to acquire it. Prehistorically speaking, the man has waged many wars and several civilizations had come and gone in the context of food directly or indirectly to acquire fertile land. History is witness to several incidents when the people perished simply because they could not afford food to fill their empty bellies in the conditions like famines, etc. Interestingly, the food still is used as a weapon by man. It is also not denying the fact that all over the world, efforts are made not only to enhance food production but also to enhance the quality and availability of the same.

Major challenges facing the world today are not just those of food production and food quality for meeting protein and calorie needs (basic nutritional needs) but also those related to better health of men. A significant challenge is the outbreak of oxidation-linked disease epidemics caused by excess calories. Diabetes is tied to other oxidation-linked diseases such as cardiovascular disease (CVD) and cancer or infectious diseases like AIDS, tuberculosis and food-borne illnesses among lower income populations. All these will inevitably place a tremendous burden on the emerging health-care systems of these countries. All major health challenges, be they of calorie sufficiency or calorie insufficiency-linked infectious diseases, are directly or indirectly related to diet and environmentally lined diseases. Therefore, technologies for disease chemoprevention through dietary modifications (reduced calorie intake with more fruits and vegetables and novel ingredients from other food-grade biological and microbial systems) will be very important to help in managing these emerging health-care challenges.

Biotechnology has become an important tool in recent years, and scientists across the world are investigating advanced and novel biological, cellular, molecular, and biochemical strategies for improving food production, processing qualities, enhancing food safety and quality, sensory and functional aspects of food and food ingredients for better human health. Precisely speaking, these aspects fall under the domain of food biotechnology, which is not a new discipline but certainly its subject area has enhanced so much that it can now stand on its own as an exciting field with immense practical implications in the field of human life. In addition, advances in food biotechnology have become more nutritionally relevant and must consider the environmental impacts and consequences of food production and consumption. Though the term 'food biotechnology' is far wider, it has been taken to mean in this book to the extent that how foods are processed, produced and/or modified, may be from plants, animals or microorganisms. The aspects associated with food additives, machinery and equipment, quality of products, etc. are an integral part of this book. The food biotechnology has imbibed a large number of researches conducted in different disciplines like genetics, biochemistry, engineering, microbiology and even biology. It has embraced the principles of nutrition to produce the product with enriched nutrition. The food biotechnologists have made the techniques to detoxify the food, made texteurised food and preserved the foods with methods that were developed by food technologists. New additives, biopreservatives, bioflavours and biocolours have certainly been the contribution of the food biotechnologists to the food industry.

The book covers the fundamental principles and the applied aspects of food biotechnology. It is slanted towards the latest research based information sandwitched with fundamentals, principles and practices. The advancement of our knowledge in different aspects of food biotechnology like solid state fermentation, protein engineering, waste management, etc. need a new text-cum-reference book. So publication of the book would need no apology. Rather it is a need of the hour.

The book is divided into four sections. Section-A deals with fundamentals of food biotechnology, Section-B deals with production of food and feed, Section-C is related with production of food additives and the last Section-D describes quality assurance and waste management. In all, there are 26 chapters in the book. The topics herein deal with bioconversion of food raw materials to processed products, improvement of food quality, food safety, designing of ingredients for functional foods, biochemical advances in traditional fermentation and most importantly, have provided an international perspective to the whole field.

All these chapters have been contributed by well-known experts in their respective fields. The text in each chapter is supported by numerous clear, informative tables and figures. The editors wish sincerely to thank all the contributors for their outstanding efforts to review and document the research and conceptual information about the subject matter of their respective chapters. Their efforts have certainly enriched our knowledge of food biotechnology. We place on record, deep sense of appreciation for the consistent hard work and generosity of Dr. Ranjeeta Bhari, who has ungrudgingly given the support for the completion of this book. We would also like to place on records our thanks to the publishers/authors of articles whose work have been cited/included in the MSS. Wherever such material has been included, proper reference/source has been acknowledged. Help received from the friends and well-wishers in the form of encouragement is gratefully acknowledged. Last, but not the least, we acknowledge the enthusiastic support and patience of our wives and children in bringing out this book. It is hoped that this book will provide information about the latest research and advances, especially the innovations in food biotechnology. It is hoped that the manuscript would be very useful to the academia and the industrialists.

**V.K. Joshi**
**R.S. Singh**

# Editors

**V.K. Joshi** (Ph.D., FBRSI), is Professor and Head, Department of Food Science and Technology, Dr. Y.S. Parmar University of Horticulture and Forestry, Nauni, Solan 173230 (HP), India, and former adviser of HPMC, Shimla (HP). He got various trainings viz., Food Processing & Nutrition from MAU, Parabhani; Sensory Analysis from CFTRI, Mysore; Postharvest Technology of Fruits and Vegetables from PAU, Ludhiana; Computer Applications, Networking and Internet from University Computer Centre, Nauni. He has varied professional experience of research, teaching and extension, including seven years in Quality Control in HP State Department of Horticulture, Shimla. His specific areas of work include fermentation technology, waste utilization, biocolours, quality control and sensory evaluation of food. His major contribution is in wine fermentation technology and he is a renowned wine consultant. He has to his credit, research, teaching and extension experience of more than 32 years including seven years in Quality Control. He has guided 10 M.Sc., and 5 Ph.D. students and has authored/edited eight books: *Fruits Wines; Biotechnology: Food Fermentation* (2 Vol.); *Postharvest Technology of Fruits and Vegetables* (2 Vol.); *Sensory Science: Principles and Applications in Food Evaluation; Hand Book of Enology: Principles, Practices, Recent Innovations* (3 Vol.); *Orientation in Quality Evaluation of Processed Foods; Food Processing Waste Management Technology; Food Processing and Preservation.* He was also guest editor of *Journal of Scientific & Industrial Research* (NISCAIRE, New Delhi, Spl. Issue on "Advances in Biotechnology") and *Natural Product Radiance* (NISCARE, New Delhi, Spl. Issue on "Production of Wines from Non-Grape Fruits"). Besides 120 research papers, he has published 57 book chapters/review/technical articles/popular articles. He has presented research papers in 40 seminars/conferences. He has handled many research projects funded by ICAR, NHB, DBT, MFP, etc. He was member of Board of Studies, HPU, Shimla, College of Horticulture, UHF, Nauni; Core Committee of SLITE, Longowal (Pb.), Board of Studies on Food and Nutrition, Punjabi University, Patiala. He has filed a patent on production of wines from plum, apricot and strawberry. He is a life member of Association of Food Scientists and Technologists (Mysore) and Trustee Biotechnological Research Society of India, Trivandrum. He is a member of scientific committees of Food Safety and Standards Authority of India on "Contaminants in Food, Alcoholic Beverages". He is recipient of N.N. Mohan Memorial Award (1982), Kejriwal Award (1995), N.A. Pandit Award (1996), N.N. Mohan Memorial Award (1998), Dr. J.S. Pruthi award by AIFPA, New Delhi (2009), Himachal Shree Award by Himotkrash Society, Himachal Pradesh (2003). He is editor-in-chief of *International Journal of Food & Fermentation Technology* and editor of *Indian Food Packer,* a reputed journal of the food industry. He is a referee of several presitgious journals including *Indian Food Packer; Journal of Applied Microbiology; Letters in Applied Microbiology; Bioresource Technology; Food Technology and Biotechnology; Indian Journal of Biotechnology* and *Journal of Scientific & Industrial Research.* He helped to set up/renovate the winery/fruit processing units, etc. and licensed commercialization of fruits and vegetable products. At present, he is actively engaged in lactic acid fermentation of fruits and vegetables, biocolours and tea fermentation.

**R.S. Singh** (Ph.D. Biotechnology) is Professor and Head, Department of Biotechnology, Punjabi University, Patiala 147 002 (Pb.), India. He is recipient of MASHAV-UNESCO Fellowship (2006) at Faculty of Agricultural, Food and Environmental Quality Sciences (Division for External Studies), Rehovot Campus, Hebrew University of Jerusalem, Israel. He was also INSA Visiting Scientist (2008) to the Department of Pharmaceutical Biotechnology, NIPER, Mohali (Pb.). He has more than 18 years of teaching and research experience in Fermentation Technology, Biochemical Engineering, Microbial Technology and Food Biotechnology. He has guided 75 M.Sc. and 8 Ph.D. students for their research work and presently, 8 Ph.D. and 6 M.Sc. students are pursuing research under his supervision in the areas of Food Biotechnology, Fermentation Technology and Microbial Biotechnology. He got training in strain improvement from CFTRI, Mysore and participated in many workshops on varied aspects, particulary "Patent Awareness and IPR in Biotechnology" organized by PSCST, Chandigarh and CIMAP, Lucknow as well as on "Protein Purification" oraganised by GE Healthcare Life Sciences, USA. Besides 80 research papers in reputed national/international journals, he has published 10 book chapters and 30 popular articles. He has filed two Indian patents on inulinase and gluconic acid production. He is a regular referee of many prestigious national/international journals. He has successfully completed 4 major research projects and 2 minor research projects funded by UGC and CSIR. He has presented 50 research papers in national/international conference/symposia as well as delivered many Plentary Lectures/Invited talks and also chaired the technical sessions. His current interest in research are high fructose syrups, maltose syrups, microbial polysaccharides and production & purification of industrial enzymes for use in food industry. He is Chairman of Graduate and Post-graduate Board of Studies in Biotechnology, Punjabi University, Patiala and DBT Nominee of Biosafety Committee, Department of Biotechnology & Environmental Sciences, Thapar University, Patiala. He is a member of Faculty of Life Sciences, Institutional Animal Ethical Committee, Institutional Clinical Ethical Committee, Institutional Biomedical Waste Management Committee and Academic Council of Punjabi University, Patiala. He is on advisory board of National Institute of Ayurvedic Pharmaceutical Research, Patiala and also on editorial advisory board of *International Journal of Food & Fermentation Technology*. He is life member of PASc, AMI, ISCA & BRSI and also member of Asian Federation of Biotechnology. He was executive member of Management Council (2009-11) of Biotech Research Society of India. Presently, he is President of BRSI Unit, Punjabi University, Patiala. He is honorary consultant to various regional food industries.

# List of Contributors

**Amit Pratush:** Department of Biotechnology, Himachal Pradesh University, Summer Hill, Shimla 171 005 (H.P.)

**Amit Seth:** Department of Biotechnology, Himachal Pradesh University, Summer Hill, Shimla 171 005 (H.P.)

**Aravindan Rajendran:** Department of Chemical Engineering, Annamalai University, Annamalai Nagar 608 002 (T.N.)

**Dev Raj:** Department of Food Science and Technology, Dr. Y.S. Parmar University of Horticulture & Forestry, Nauni, Solan 173 230 (H.P.)

**Devender Attri:** Department of Food Science and Technology, Dr. Y.S. Parmar University of Horticulture & Forestry, Nauni, Solan 173 230 (H.P.)

**G.K. Saini:** Department of Biotechnology, Punjabi University, Patiala 147 002 (Pb.)

**G.S. Abrol:** Department of Food Science and Technology, Dr. Y.S. Parmar University of Horticulture & Forestry, Nauni, Solan 173 230 (H.P.)

**Gurpreet Kaur:** Department of Food Engineering and Technology, Sant Longowal Institute of Engineering & Technology, Longowal 148 106 (Pb.)

**Himani Kansal:** Department of Pharmaceutical Technology, National Institute of Pharmaceutical Education & Research, S.A.S. Nagar 160 062 (Pb.)

**L.S. Devi:** Department of Food Science and Technology, Dr. Y.S. Parmar University of Horticulture & Forestry, Nauni, Solan 173 230 (H.P.)

**M.B. Bera:** Department of Food Technology, Sant Longowal Institute of Engineering & Technology, Longowal 148 106 (Pb.)

**M.P. Devi:** Department of Food Science and Technology, Dr. Y.S. Parmar University of Horticulture & Forestry, Nauni, Solan 173 230 (H.P.)

**M.S. Bahttacharyya:** Department of Pharmaceutical Technology, National Institute of Pharmaceutical Education & Research, S.A.S. Nagar 160 062 (Pb.)

**Manisha Kaushal:** Department of Food Science and Technology, Punjab Agricultural University, Ludhiana 141 001 (Pb.)

**Manpreet Singh:** Department of Pharmaceutical Technology, National Institute of Pharmaceutical Education & Research, S.A.S. Nagar 160 062 (Pb.)

**Munish Puri:** Department of Biotechnology, Punjabi University, Patiala 147 002 (Pb.)

**N.S. Rana:** Department of Vegetable Crops, Dr. Y.S. Parmar University of Horticulture & Forestry, Nauni, Solan 173 230 (H.P.)

**N.S. Thakur:** Department of Food Science and Technology, Dr. Y.S. Parmar University of Horticulture & Forestry, Nauni, Solan 173 230 (H.P.)
**Namita Bansal:** Department of Microbiology, Panjab University, Chandigarh 160 022
**Navdeep Thakur:** Department of Biotechnology, Himachal Pradesh University, Summer Hill, Shimla 171 005 (H.P.)
**Neha Gautam:** Department of Basic Sciences, Dr. Y.S. Parmar University of Horticulture & Forestry, Nauni, Solan 173 230 (H.P.)
**P.K. Mehta:** Department of Biotechnology, Himachal Pradesh University, Summer Hill, Shimla 171 005 (H.P.)
**P.S. Panesar:** Department of Food Engineering and Technology, Sant Longowal Institute of Engineering & Technology, Longowal 148 106 (Pb.)
**Pankaj Soni:** Department of Pharmaceutical Technology, National Institute of Pharmaceutical Education & Research, S.A.S. Nagar 160 062 (Pb.)
**Parveen Pahuja:** Department of Biotechnology, Punjabi University, Patiala 147 002 (Pb.)
**Praveen Kaul:** Department of Pharmaceutical Technology, National Institute of Pharmaceutical Education & Research, S.A.S. Nagar 160 062 (Pb.)
**R.C. Kasana:** Division of Biotechnology, Institute of Himalayan Bioresource Technology, Palampur 176 061 (H.P.)
**R.S. Singh:** Department of Biotechnology, Punjabi University, Patiala 147 002 (Pb.)
**Rajinder Kumar:** Department of Human Biology, Punjabi University, Patiala 147 002 (Pb.)
**Rakesh Sharma:** Department of Food Science and Technology, Dr. Y.S. Parmar University of Horticulture & Forestry, Nauni, Solan 173 230 (H.P.)
**Raman Soni:** Department of Biotechnology, DAV College, Chandigarh 160 022
**Ranjeeta Bhari:** Department of Biotechnology, Punjabi University, Patiala 147 002 (Pb.)
**Reeba Panesar:** Department of Food Engineering and Technology, Sant Longowal Institute of Engineering & Technology, Longowal 148 106 (Pb.)
**S.K. Bhatia:** Department of Biotechnology, Himachal Pradesh University, Summer Hill, Shimla 171 005 (H.P.)
**S.K. Sharma:** Department of Horticulture, G.B. Pant University of Agriculture & Technology, Hill Campus Ranichauri, Tehri-Garhwal 249 199 (Uttaranchal)
**S.K. Soni:** Department of Microbiology, Panjab University, Chandigarh 160 022
**Sachin Dubey:** Department of Pharmaceutical Technology, National Institute of Pharmaceutical Education & Research, S.A.S. Nagar 160 062 (Pb.)
**Sawinder Kaur:** Department of Food Technology, K.M. Vidyalaya, Jalandhar 144 004 (Pb.)
**Sawraj Singh:** Department of Pharmaceutical Technology, National Institute of Pharmaceutical Education & Research, S.A.S. Nagar 160 062 (Pb.)
**Shashi Bhushan:** Division of Biotechnology, Institute of Himalayan Bioresource Technology, Palampur 176 061 (H.P.)
**Somesh Sharma:** Department of Food Science and Technology, Dr. Y.S. Parmar University of Horticulture & Forestry, Nauni, Solan 173 230 (H.P.)
**Sumali Kapor:** Department of Pharmaceutical Technology, National Institute of Pharmaceutical Education & Research, S.A.S. Nagar 160 062 (Pb.)
**T.C. Bhalla:** Department of Biotechnology, Himachal Pradesh University, Summer Hill, Shimla 171 005 (H.P.)
**U.C. Banerjee:** Department of Pharmaceutical Technology, National Institute of Pharmaceutical Education & Research, S.A.S. Nagar 160 062 (Pb.)
**U.K. Narta:** Department of Biotechnology, Himachal Pradesh University, Summer Hill, Shimla 171 005 (H.P.)

**V.K. Joshi:** Department of Food Science and Technology, Dr. Y.S. Parmar University of Horticulture & Forestry, Nauni, Solan 173 230 (H.P.)
**V.S. Rana:** Department of Food Science and Technology, Dr. Y.S. Parmar University of Horticulture & Forestry, Nauni, Solan 173 230 (H.P.)
**Viruthagiri Thangavelu:** Department of Chemical Engineering, Annamalai University, Annamalai Nagar 608 002 (T.N.)
**Vivek Garg:** Department of Food Science and Technology, Dr. Y.S. Parmar University of Horticulture & Forestry, Nauni, Solan 173 230 (H.P.)
**Wamik Azmi:** Department of Biotechnology, Himachal Pradesh University, Summer Hill, Shimla 171 005 (H.P.)

# Contents

# Food Biotechnology: An Overview

# 1

**V.K. Joshi[1] and R.S. Singh[2]**

*[1]Department of Food Science and Technology, Dr. Y.S. Parmar University of Horticulture and Forestry, Nauni, Solan 173 230 (H.P.)*
*[2]Department of Biotechnology, Punjabi University, Patiala 147 002 (Pb.)*

## 1. INTRODUCTION

Biotechnology is the exploitation of the biochemical potential of microorganisms for medical, agricultural and industrial purposes, thus is not a new field. More appropriately, it is a multidisciplinary subject. A commercial biological process though based on biochemical or microbiological knowledge can be operated only by the application of chemical engineering [1]. The European Federation of Biotechnology has defined Biotechnology as "the integrated use of biochemistry, microbiology and chemical engineering to achieve the industrial application of microbes and tissue culture." It is rightly stated that the fruits of biotechnology are borne on a tree whose roots are the biological sciences (microbiology, genetics, molecular biology, etc.) and whose trunk is chemical engineering in a wider sense [2]. The Convention on Biological Diversity [3] defined Biotechnology as "any technological application that uses biological systems, living organisms or derivatives thereof to make or modify products or processes for specific use." It thus, includes activities such as traditional food fermentations, waste treatment, drug development, fish farming and crop development. Food biotechnology has been defined as "the application of biological techniques to food crops, animals and microorganisms to improve the quality, quantity, safety, ease of processing and production economics of food". It thus, includes the traditional manufacturing processes used for bread, beer, cheese and various fermented milk products [4].

The application of biological principles results in the production or conversion of food stuffs into more palatable, nutritious or stable food, which is more synonymous with definition and purpose of food biotechnology. Intentionally or naturally, microorganisms have been involved in the production of fermented foods since the ancient times, such as in bread making, cheese making and brewing wine, etc. [5, 6]. What was essentially a household item for the early man became one of the most important industrial enterprises. The production of alcoholic beverages constituted an important segment of this industry, witnessing a spectacular growth from the time of Pasteur. Since then, many biotechnological advances have made the food biotechnology an ever-expanding industry over the last century. In food processing, either the microorganisms are eliminated to prevent spoilage or they are encouraged to grow resulting in the production of fermented foods. The former can be called negative food biotechnology and latter as positive food biotechnology. The broad areas covered under food biotechnology are shown in Table 1. Often the food fermentation is referred as

*Food Biotechnology: Principles and Practices*
*V.K. Joshi & R.S. Singh (Eds.)*
*I.K. International Publishing House Pvt. Ltd., India*
*pp 1–37*

**Table 1:** Broad areas of food biotechnology.

| |
|---|
| • Microbial Technology |
| • Enzyme Technology |
| • Bioreactor Technology |
| • Plant Cell Technology |
| • Animal Cell Technology |
| • Genetically modified (GM) foods (Plant and animal products) |
| • Functional foods (Prebiotics, probiotics, synbiotics, etc.) |
| • Functional genomics and bioinformatics of food microorganisms |
| • Biotechnological waste management |

traditional or conventional food biotechnology, while that involving molecular biology or genetic engineering signifies modern food biotechnology. Both are playing a significant role in the organism, process and product improvement/development. The stress on latter, though appropriate, should not be at the expenses of basic microbiology and biochemistry. Who knows when somebody could isolate a microbe with the ability to ferment the substrate or carry out the reaction at present considered impossible. Perhaps, a super-non-engineered bug can be looked for in the environment or could be engineered with new capabilities by genetic engineering. Similarly, genetic modification of food raw materials offers several potential advantages and benefits compared to traditional selective breeding techniques [4, 7, 8]. Recent studies also indicated that major genes for meat quality could offer excellent opportunities not only for increasing the level of meat quality, but also for decreasing variability. Tremendous potential now exists for muscle development, fat content, tenderness, colour, flavour, etc. [9]. Advantages and benefits of GM foods to developing countries are likely to be considerable. GM technologies have the capability of reducing world hunger to 50% by 2015 [10]. As IFST [4] has stated "Food scientists and technologists can support the responsible introduction of GM techniques, provided that issues of product safety, environmental concerns, ethics and information are satisfactorily addressed so that the benefits of this technology become available both to improve the quality of the food supply and to help feed the world's escalating population in the coming decades."

Functional foods/nutraceuticals demonstrate various physiological benefits and reduced disease risk amongst the consumers. These foods can be easily incorporated into the diet because of their being similar in appearance to conventional foods. Nutraceuticals can be produced through chemical synthesis, fermentation and genetic engineering. They can also be incorporated into our diet in the form of fortificants, supplements or consumed directly as genetically enriched foods [11]. The market potential for nutraceutical/functional foods is steadily growing.

Bioinformatics provides software tools to analyze the enormous amount of biological data that gets generated from sequencing projects, transcription and metabolic profiling experiments, and investigations on functional analysis. The tools assist in the construction of complete catalogues of the genes, gene products, functional properties, identification of interaction amongst the gene products, functional properties, identification of interaction amongst the gene products, elucidation of pathways, comparison of genomes, gene and protein similarity searches, and identification of single nucleotide polymorphs (SNPs). Bioinformatics provides insights into the possibility of genetic modification of biological systems to result in new food products and new therapies of economical and clinical significance [12].

In this chapter, the various facets of food biotechnology have been discussed in a generalized manner citing specific examples to illustrate the point. Food fermentation, microbial biocatalysis, genetically modified foods, functional foods, impact of genomics and bioinformatics on food biotechnology and biotechnological waste management in food industry have been highlighted briefly.

## 2. FOOD BIOTECHNOLOGY SYSTEMS

Food biotechnology could be organized based on the products produced or the systems used in the food biotechnology industry as illustrated in Tables 2 and 3 [13]. Either of the structure underlines the magnitude of the food biotechnology industry, but the potential is endless and as varied as microorganisms themselves. What is, however, required is the sound knowledge of microorganisms, their ecology, genetics and manipulating their metabolism and control ability to handle them on a large

**Table 2:** Product oriented organization of foods produced by biotechnological processes.

| Mode of production | Food origin | |
|---|---|---|
| | **Plant** | **Animal** |
| Microbial | Fruits (Vinegar, Wine, Juices) | Dairy products (Cheese, Yoghurt) |
| | Vegetables (Olives, *Sauerkraut*) | Meat products (Fermented sausage, Uncooked ham) |
| | Cereal (Baked goods, Alcoholic beverages) | Fish products (Fermented fish) |
| | Seeds (*Miso, Tempeh, Oncom*) | Processing waste (Single cell proteins) |
| | Fungi (Mushrooms) | |
| | Luxury foods (Cocoa, Tea, Coffee) | |
| Plant | Sprouts (Malt, Bean sprouts) | |
| | Herbs and vegetables (Hydroponics, Plant cell culture) | |
| | Seeds (Soy protein hydrolysate) | |
| Enzymatic* | Fruits and vegetables (Juice using pectinases) | Meat products (using Papain) |
| | Cereals (High fructose corn syrup) | Dairy products (Hydrolysis of whey) |
| | Seeds (Soy protein hydrolysate) | Fish products (Hydrolysis of fish protein concentrate) |

*Partially purified enzyme preparations or immobilized whole cells.

*Source:* 13

**Table 3:** System oriented organization of biotechnology based food production and processing.

| Unit processes | Products | | | | |
|---|---|---|---|---|---|
| | **Raw material** | **Ingredients, production and processing aids** | **Food material** | **Waste material** | **Environment** |
| Production methods | Solid state fermentation | Enzymes, vitamins, flavours | Fermentation | - | - |
| Modification/ improvement | Bioconversion of polysaccharides | Hydrolysis of proteins | Meat tenderization | Vinegar, single cell proteins | Air/water purification |
| Preservation | Silage, coffee fermentation | Immobilization of biocatalysts | Lactic acid fermentation | Protein recovery | Soil preservation |
| Bioassays | Enzymatic/ microbial methods | Enzymatic/ microbial methods | Microbial/ biological methods, biosensors | Enzymatic/ microbial methods | Biosensors, biological methods |

*Source:* 13

scale and capacity to recover the products. Within the scope of the food biotechnology, these aspects have found a place in the subsequent chapters.

Biotechnology is a remarkable tool that can be used to improve our agriculture and food processing systems including fisheries, poultry and animal products. The technological innovations are one of the cornerstones with which the successful food biotechnology industries of 21$^{st}$ century can be assured of winning global influence and sustainable profitability. These include other market-oriented biotechnological innovations aimed at improving the efficiency of the production process, creating and realizing new strategically important opportunities. A lot of innovations have been reported in bioreactor technology [14-16]. High cell density reactors have become more popular in many processes [17]. For more details, see *Chapter 3*. Biochemical engineering is an important component of food biotechnology industry, for more details see *Chapter 2*. Enzymes are one of the biotechnology based products of commercial value. The applications of the genetic engineering would tremendously enhance the enzyme productions at industrial scale. For more details on enzyme technology, see *Chapter 5*. Solid-state fermentation technology is also one of the promising areas, for more details see *Chapter 4*. Genetically modified foods and functional foods have a lucrative market in the food industry, for details see *Chapter 9* and *10*. To ensure the safety and quality of food products, many new analytical tools have been developed, for more details see *Chapter 23* and *24*. Sensory evaluation of food products is one of the important criteria of quality assurance, for more details see *Chapter 25*.

## 3. FOOD FERMENTATION: THE CONVENTIONAL FOOD BIOTECHNOLOGY

Fermentation is the oldest form of biotechnology, essentially consisting of transformation of the simple raw materials into value added amazing range of products by utilizing the phenomemon of the growth of microorganisms or their activities on various substrates. Fermentation can be broadly defined as "the process in which the carbohydrate or like compounds are hydrolysed, under anaerobic or aerobic conditions" [18]. In ancient time, fermentations were mostly natural which are practiced even now in some foods, while controlled fermentations with defined microorganisms and process have replaced many on industrial scale. Microbes offer multitude of advantages like rapid growth using simple nutrient media. They adopt sophisticated methods for survival in nature, which permit them to interact with and metabolize a wide variety of substrates with variable physical forms, and convert them to cellular biomass during fermentation. Fermentation plays a very significant role as

**Table 4:** Why to conduct fermentation?

- Fermented foods add variety to the man's diet.
- Fermentation is a potent tool to preserve the food through the activity of microorganisms or their enzyme by producing metabolites inhibitory to pathogenic or spoilage causing microorganism.
- Some of the fermenting microorganisms synthesize vitamin, proteins, amino acids and enrich the food with these nutrients.
- The fermented foods are also made nutritious by the liberation of nutrients locked in the plant structure by the indigestible materials by fermentation.
- Fermentation processes are less energy consuming than their counterparts, and require less complex or costly equipment.
- Fermentation recycles the natural materials and can convert waste material into useful products.
- Fermentation produces metabolites or byproducts of immense value to human beings.
- Fermentation results in detoxification of the foods by inducing changes in plant tissue to make vacuole bound glucosides to diffuse membrane bound linamerase and hydrolysed products to leach or by increasing the β-glucosidase activity and thus, breaking down cyanogenic glucosides.

summarized in Table 4. Apart from its role in preservation, fermentation has a major role in the production of a variety of fermented food products/additives.

Depending upon the regional tastes and acceptability, there is quite a large variation in the methods of preparation of fermented foods. Compared to the technology involved in the production of fermented foods with that of processed foods, fermentation constitutes only a single unit operations while many other unit operations are common (Fig. 1). Thus, food fermentation technology would include many unit operations other than the microorganisms and fermentation. The food fermentation involves all those fermentation processes where either the ultimate product is used directly as a food, or as an additive to food or a basic ingredient to the food or the by-products formed during fermentation, their utilization, disposal or proper management. For any process to make fermented food, essential components are raw materials, microorganisms, fermentation vessel and associated controls, processing, recovery and packaging systems, but from fermentation technology point of view, the fermenter and microorganism involved are of prime importance. Since the end-product would serve as a food, it is essential to evaluate it from nutritional, toxicology and sensory qualities point of view, and thus, the knowledge of these aspects would form an integral part of food fermentation technology. The technological dimensions of food fermentation toward the input of engineering sciences in designing the fermenters and associated controls (Plate 1) for optimum fermentation and product recovery are immense. As the product on the commercial scale would be marketed, which can occur only if it is economical, the economic of such products and the associated marketing aspects, thus, cannot be ignored.

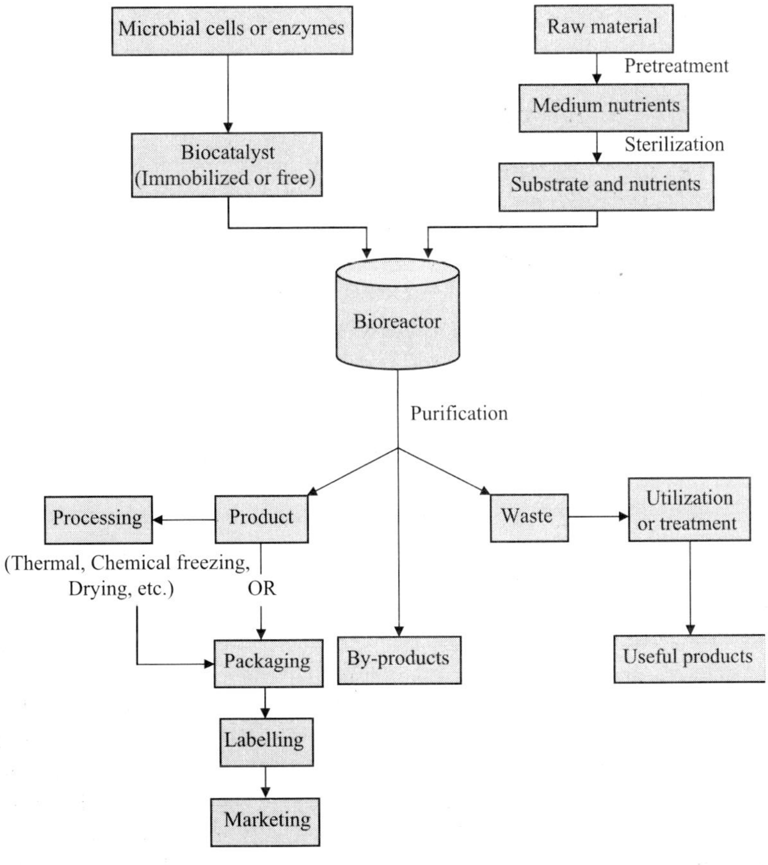


**Fig. 1:** A generalized scheme of various operations employed in food fermentation technology.

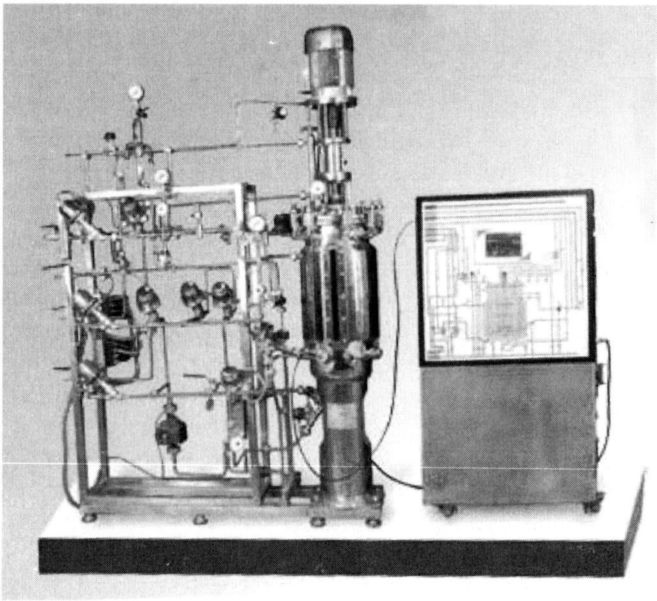

**Plate 1:** Overview of the top-driven 30 L fermenter with skid and computer control. (Courtesy: Scigenics India Pvt. Ltd.)

### 3.1 Food Fermenting Microorganisms

The whole gamut of food fermentation is dependent on the microorganisms and their activity in the food. While some convert milk to yoghurt, leaven bread or ferment fruit juice to wine, other cause food spoilage or even food poisoning. The latter group has received and continues to get attention from food safety considerations. The most important microorganism from the industrial standpoint is yeast, principally *Saccharomyces cerevisiae* var. *ellipsoideus*, and the oldest known process employing this microbe is the fermentation to produce wine and beer.

The yeasts are without doubt the most important group of microorganisms exploited commercially by man. Besides conventional products, the yeasts have been utilized in a number of other products (Table 5). Yeasts also play a prominent role in the fermentation process of many traditionally fermented foods (*Lafun* and *Garri*) including African fermented foods (Palm wine, *Burukutu, Agadagidi*) as such or in combination with lactic acid bacteria. In most of the instances, the development of desired sensory qualities of indigenous foods depends on the composite contribution of two or more types of microorganisms. But yeasts and molds are often present and contribute to the development of proper flavour and aroma of such foods. In natural fermentations of *Orient*, the important microbial group included *Lactobacillus plantarum, Streptococcus faecium, Leuconostoc* sp., etc., but the occurrence of *Staphylococcus haemolytica* and *Shigella* sp., the known pathogens, should be of concern [21].

With developments in science, microorganisms and pathways involved in different food fermentations as well as those associated with spoilage have been elucidated. But the food fermentations are so numerous and varied that still a considerable research is needed for their complete study in future. As food is the basic need of man, methods to improve the existing fermentation or develop new processes and products would be required. The genetic engineering encompasses the complete range of these high-tech microorganisms as a powerful tool in the process and product development.

**Table 5:** Some of the present and future uses of yeasts in food fermentation industry.

**(A) Proteins and peptides (heterologous molecular cloning in *Saccharomyces* sp.)**
- Antitrypsin
- Somatostatin
- Glucoamylase
- Amylase
- β-glucanase
- β-galactosidase
- Chymotrypsin and other proteinases
- Thaumatin

**(B) Biotransformations by yeast**
- Ethanol, glycerol, biomass and $CO_2$ by *Saccharomyces* and other yeasts
- α-carotene production by *Rhodotorula glutinis*
- L-phenylalanine production by *Rhodotorula rubra*
- Depolymerization of tannin extract by *Saccharomyces rouxii*
- Ascorbic acid production by *Candida norvegensis*

**(C) Flavour and other additives production (bioconversion) by yeast**
- Citric acid production by *Yarrowia lipolytica*
- Ethyl acetate and acetaldehyde by *Candida utilis*
- Monoterpene production by *Ambrosizyma monospora*
- Esters and other volatile production by *Kluyveromyces lactis*
- *Hansenula* produces ethyl acetate
- *Candida* produces lactones having fruity, coconut and buttery sweet aroma
- *Rhodotorula minuta* produces terpenes and lactones

*Source:* 19, 20

## 3.2 Microbial Diversity and Food Fermentation

Microbial diversity as a major source of biotechnological products and processes is significant from isolating a novel organism and secondly, to screen a collection of appropriate microorganisms with desirable attribute [22]. Presently, developments in recombinant DNA technology have dominated, providing unique opportunities for commercialization. But this approach may not be appropriate in all the biotechnological processes and products where a strong incentive to make use of natural organism to produce novel or natural product(s) to meet the growing consumer demands, or where agricultural products and clean-up operations in the context of increasing environmentally alert society are needed. Our present knowledge (12% of bacteria and 5% of fungi out of the estimated are being utilized), of the biological diversity including microorganisms is inadequate, indicating enormous potential for the biotechnological innovations in utilizing the diversity [22].

## 3.3 Role of Fermentation in Food Preservation

It is well established that fermentation and dehydration are amongst the oldest techniques used to preserve foods. While in all the other methods of preservation, microorganisms are killed, eliminated or suppressed, but in fermentation they are encouraged to grow and, multiply. The people of South East Asia have old customs of food storage by fermentation. The fermentation helps to preserve the perishables like milk, meat, fruits, vegetables, foods, etc. [6, 23, 24]. It is especially true in certain regions or countries of the world and in the earlier times, when bountiful harvest yielded larger quantity of fresh foods than the population could consume without the food getting bad or spoiled. For such foods, fermentation proved and is a potent tool to prevent the postharvest losses. Amongst their fermented products were rice wine, sugar cane wine, pickles, fish sauce fermented noodles, *tempeh, oncorm*, etc. [6]. The preservative effect of fermentation is due to the formation of ethanol, carbon dioxide, specific antibiotics or antimicrobial substances like bacteriocins, reutrin, etc. and the major effect is the lowering of pH such as by lactic or acetic acid bacteria, as illustrated by the bread making and alcoholic beverages. The yeast fermentation results in the formation of $CO_2$, thus lowering the pH which is complemented by the metabolites such as citric acid, ketoglutaric acid, etc. formed during fermentation. It is quite important in bread or beer production but not in wines, where the pH is already in the acidic range. The preservative effect is more pronounced in the sour rye or sour white bread where besides yeast, lactic and acetic acid bacteria produce the acid. Many fermented foods have good keeping qualities and may be kept for longer periods without refrigeration than the fresh unfermented foods. But like any other food, the fermented foods are also subjected to normal spoilage, and to prevent this one or combinations of the preservative technique are applied such as low temperature storage, thermal processing or addition of preservative.

## 3.4 Fermented Food Products/Additives

Fermented foods (liquid, solid or semisolid) are those that have been subjected to the action of microorganisms. Fermented products are so popular in many countries that out of four cassava based products surveyed in Africa, nearly three were fermented, employing submerged, mash and solid state or low moisture fermentation [25]. These delicious and nutritious foods are made and consumed both in the Western and Eastern countries. These are often associated with the social and cultural history of the people [6], and the trend is likely to remain so in the future, though these may be affected by the urbanization of the countries of this region [6]. The traditional fermented foods can be classified broadly into six groups: (1) Alcoholic beverages (2) Vinegar (3) Fermented milk products (4) Fermented pickles (5) Fermented fish/meat products (6) Fermented plant proteins.

From ancient times, people all over the world have developed traditional processes for the production of alcoholic beverages either from natural sugar (e.g., wine) or by degrading the starchy materials (e.g., beer). Fruit wines, referred to as *Soma* in our ancient scriptures, have been a part of food of man and were used as a medicine. These might have originated accidentally (transformation of fruit juice beverages having exhilarating or stimulating properties). Distillation was, however, a later development. These fall into two classes, differing in the nature of the raw materials used and the technique of preparation. Wines are made principally from fruits like grape or extract of other fruits [26, 27], while the beers are prepared from grains like barley, rice, corn, etc. The principal carbohydrate in grains is starch, which cannot be fermented directly. This process is brought about by starch splitting enzymes, by germinating the grains followed by drying and storage until needed. Distillation of beer produces a variety of liquors (whisky, rum, gin, vodka, etc.), while that of wine produces brandy. From condiments point of view, vinegar is the product made traditionally by alcoholic fermentation followed by acetic fermentation with *Acetobacter* sp. Vinegar includes cider vinegar, malt vinegar or other fruit vinegars and is used almost in every home. At industrial scale, it constitutes one of the important fermented products.

Another group of fermented products makes use of lactic acid bacteria (LAB) in which the sugars are converted into lactic acid. There are several advantages associated with this fermentation such as desired sensory qualities, extension of processing season, a little mechanical energy requirements and cheap in cost. It is the ancient mode of fermentation when the perishable produce is allowed to undergo natural process of fermentation resulting in the production of acid and change in the texture or other sensory quality of products. Among the fermented products include a large variety of fermented milk, cheese, *dahi, kumiss, kefir, sauerkraut* (from cabbage), *yoghurt, idli*, fermented meat, etc., produced all over the world. The fermentations are initiated in most of the cases by natural microflora or controlled fermentation. It is preferable to the acetic fermentation in preparation of pickles. Some of the fruits and vegetable based products and microflora involved can be seen in Table 6.

The fermentation uses minimum heat processing or such treatments which are in line with the present trend of using minimally processed food products. The microorganisms, especially LAB, are being added intentionally to serve as protective culture to the foods like bacon and chicken salads to preclude botulin toxin production with temperature abuse during the handling of bacon. A further development could be the addition of such cultures with antimicrobial activity due to bacteriocin production. For more details on lactic acid fermented foods, see *Chapter 12*.

The fermented foods from plants are prepared and consumed all over the world. Fruits, vegetables, cereals, roots, crops, legumes and oilseeds are all used for the purpose [28]. The fermentation contributes very much to the development of specific sensory qualities of these products, besides improving the digestibility and nutritional values. *Tempeh* can be cited to illustrate this aspect, though there are several other plant derived traditional fermented food products like Nigerian *Ogi*, Kenyan *Uji*, South African *Mageu*, Ghauaian *Kenkey* and *Mane* of Benin and *Togo*. More information on traditionally fermented foods can be seen in *Chapter 11*.

Production of microbial biomass as source of proteins has been advocated, and extensive research in various aspects has been conducted throughout the world. One approach to produce single cell proteins (SCP) was the use of petroleum products for the growth of such organisms as SCP, which has now been abandoned mainly because of the cost of petroleum products. It is also to some extent affected by the safety considerations, needing rigorous testing. Rather emphasis has been shifted towards animal feed than developing human food. The alternative sources like cellulosic material, food processing waste or such material for the production of single cell proteins is being encouraged. Techniques such as solid state fermentation have been used successfully in the production of protein enriched foods or enzymes. Extraction of protein from non-conventional sources e.g., wine lees, could become a better source of protein compared to that obtained from conventional sources like fish, oil seeds and leaves [30]. For more details on microorganisms for food and feed, see *Chapter 8*.

To meet the demands of fat from sources other than conventional, production of fat from microorganisms is certainly a novel idea. A number of microorganisms have been isolated and their fat characterized for its composition [31]. But it is still not commercialized due to economic considerations. However, production of polyunsaturated fatty acids (PUFA) from microbial origin has attained significance in the present age due to the medical benefits derived from them.

The colour of the foodstuff is one of the most important elements in determining its acceptance and enhancing the delicacy of food articles. The consumer expects to see his/her food looking 'natural', appetizing and enticing too. Now-a-days, some fermentative food-grade pigments are in the

**Table 6:** Bacteria used as starter cultures in fruits and vegetable products.

| Fruits/vegetables | Bacteria involved |
|---|---|
| *Pickled vegetables* | |
| Carrots | Mixture of *Lactobacillus plantarum, L. brevis, Leuconostoc mesenteroides* and *Pediococcus cerevisiae* |
| Cucumber | Mixture of *P. cerevisiae, L. plantarum* and *L. brevis* |
| Sliced/diced | Mixture of *P. cerevisiae* and *L. plantarum* |
| *Mixed vegetables* | |
| Green tomatoes | Mixture of *P. cerevisiae* and *L. plantarum* |
| Hot cherry peppers | *L. plantarum* |
| Olives | *L. plantarum* |
| Sauerkraut | Mixed fermentation by lactic acid bacteria in *L. plantarum* |

market. Microbial colours have been produced from a number of microorganisms, which are natural sources like *Rhodotorula, Phafia rhodozyma, Monoascus*, etc. [32]. Pigments producing microalgae are quite common in nature, and their exploitation for use in food industry is rightly prescribed [33]. The food industry is increasingly interested in these colours due to large concern shown to coaltar dyes having cancerous implications. The problem of economic yields is amongst the hindrance for its commercial production. The microbial production of food-grade pigments has been prescribed in detail by Dufosse [34]. For more details on microbial colours, see *Chapter 19.*

The present aversion of the consumers towards the chemicals added to the food, cosmetics and other such additives holds great potential for their production from natural sources like plants or microorganisms. Presently, most of the flavours used in the food industry are made chemically, while the preference of consumers is for natural flavours. Therefore, the microbial production of flavour compounds or bioconversion of such compounds is gaining importance. There is plethora of information on flavour production by microorganisms [20]. The filamentous fungi have been employed in food fermentation since ancient times, but their usefulness in modifying or enhancing the flavour or production of flavour metabolites, to produce enzymes either to generate or remove compounds which influence the flavour is quite recent and of great importance in modern food biotechnology [35]. The modification of flavour during fermentation is also accomplished by filamentous fungi. Filamentous fungal enzymes and metabolites influence the flavour, such as cheese. In Japan, China, Taiwan, Indonesia and other South East Asian countries, various oriental foods are fermented after treatment with preparation from *Aspergillus, Rhizopus, Neurospora, Monascus, Fusarium*, etc., to improve the flavour of meat, poultry, fish, pulses and cereals. Fungal amylases, lipases, cellulases, and hemicellulases play a key role in producing attractive aroma. Yeast and bacteria used in the subsequent fermentation, besides improving the flavour, also contribute to the nutrition and digestibility of food. The filamentous fungi also destroy undesirable flavours already present in the food. For more details on microbial flavours, see *Chapter 20.*

It is very significant contribution of food fermentation, wherein various microorganisms are employed to produce a variety of enzymes such as amylases, glucoamylases, cellulases, proteases, keratinases, pectinases, nucleases, lipases, etc. [29]. For their production various fungi and bacteria are employed. The enzyme production is an industry of large magnitude. Some of the applications of enzymes in food industry are listed in Table 7.

The enzymes have become almost indispensable tool in the hands of fermentation technologists for food production and modification. Pectinase is used extensively for juice and wine clarification and extraction of juices from pulpy fruits. Similarly, production of high fructose corn syrup by enzymatic conversion is of great commercial value. Many sugar syrups are produced enzymatically. Traditional high fructose corn syrup (HFCS) is produced from starch by enzymatic hydrolysis using α-amylase, glucoamylase and glucose isomerase. The production of high fructose syrup from inulin by inulinases is receiving great potential [36-38]. Generally, maltotriose rich syrups are being produced from maltose by cation exchange resin chromatography. Such syrups can be produced by the hydrolysis of pullulan using the debranching enzyme pullulanase [39, 40]. The application of genetic engineering is expected to increase the existing enzyme production besides commercial production of others, not being produced at present. Some of the recombinant enzymes marketed commercially for use in food industry are depicted in Table 8.

The production of biochemicals is an exciting field of fermentation having application in the food or allied industries. To name a few, vitamins, ergosterols, lipids, especially PUFA's, glycolipids, ascorbic acid, crude RNA, organic acids, polysaccharides, etc. are the products of microbial fermentation [41-43]. 'Bacteriocin', especially from lactic acid bacteria, has a potential to become preservative of natural origin due to antimicrobial activity associated with such compounds. The industrial production of such chemicals is of great significance due to the high price they fetch in the market. For more details on organic acids, see *Chapter 17.*

**Table 7:** Uses of enzymes in food industry.

| Market | Enzyme | Purpose/function |
|---|---|---|
| Dairy | Rennet (protease) | Coagulant in cheese production |
| | Lactase | Hydrolysis of lactose to give lactose-free milk products |
| | Protease | Hydrolysis of whey proteins |
| | Catalases | Removal of hydrogen peroxide |
| Brewing | Cellulases, β-glucanases, α-amylases, proteases, maltogenic amylases | For liquefaction, clarification and to supplement malt enzymes |
| Alcohol production | Amyloglucosidase | Conversion of starch to sugar |
| Baking | α-amylases | Breakdown of starch, maltose production |
| | Amyloglycosidases | Saccharification |
| | Maltogen amylase | Delays process by which bread becomes stale |
| | Protease | Breakdown of proteins |
| | Pentosanase | Breakdown of pentosan, leading to reduced gluten production |
| | Glucose oxidase | Stability of dough |
| Wine and fruit juice | Pectinase | Increase of yield and juice clarification |
| | Glucose oxidase | Oxygen removal |
| | β-glucanases | |
| Meat | Protease | Meat tenderizing |
| | Papain | |
| Protein | Proteases, trypsin, aminopeptidases | Breakdown of various components |
| Starch | α-amylases, glucoamylases, hemicellulases, maltogenic amylases, glucose isomerases, dextranases, β-glucanases | Modification and conversion (e.g., to dextrose or high fructose syrups) |
| Inulin | Inulinases | Production of fructose syrups |

**Table 8:** Recombinant enzymes and their applications in food industry.

| Recombinant enzyme | Application(s) |
|---|---|
| α-acetolactate decarboxylase | Brewing |
| α-amylase | Baking, brewing, distilling, starch, etc. |
| Catalase | Mayonnaise |
| Chymosin | Cheese |
| β-glucanase | Brewing |
| α-glucotransferase | Starch |
| Glucose isomerase | Starch |
| Glucose oxidase | Baking, egg mayonnaise, etc. |
| Hemicellulase | Baking |
| Lipase | Fats, oils |
| Maltogenic amylase | Baking, starch, etc. |
| Microbial rennet | Dairy |
| Phytase | Starch |
| Protease | Baking, brewing, dairy, distilling, fish, meat, starch, vegetable, etc. |
| Pullulanase | Brewing, starch, etc. |
| Xylanase | Baking, starch, etc. |

Functional foods have great potential in the market. In addition to health promising properties, many of them are useful for the modification of food flavour and physico-chemical characters. Probiotics, prebiotics and synbiotics are receiving great response in the category of functional foods. They are either produced in the food products during their production or supplemented as food ingredients with known health promoting functionalities. A number of probiotic microorganisms are produced by fermentation. For more details on concepts and applications of probiotics in food, see *Chapter 22*. Oligosaccharides are attracting increasing interest as prebiotic functional food ingredients [44-47]. Majority of the food-grade oligosaccharides are of microbial origin, which are either produced by fermentation or enzymatic synthesis.

## 3.5 Nutritional Benefits of Fermentation

Fermentation has various effects on the nutritional quality of food. Microbial fermentation leads to a decrease in the level of carbohydrates and indigestible oligosaccharides in legumes. A decrease in the levels of the latter reduces abdominal distention and flatulence. Moreover, certain amino acids may be synthesized, and the availability of B-group vitamins may be increased by fermentation. Fermentation techniques involving the addition of amylase-rich flour and a small amount of lactic acid starter culture provide a considerable increase in the nutrient density of the resulting food. The decrease in the viscosity of starchy foods mixed with water (porridges) after fermentation enables the desirable semi-liquid consistency.

Fermentation, by lowering pH, also provides more optimal conditions for enzymatic degradation of the phytate present in cereals in the form of complexes with proteins or polyvalent cations such as iron, zinc, calcium, and magnesium. This degradation occurs during the soaking of cereal grains in water for 12-14 hours. Such a reduction in phytate may increase the amount of soluble iron, zinc, and calcium several fold. Similarly, as a result of lactic acid fermentation, the tannin content may be reduced in some cereals, leading to the increased absorption of iron. Vitamin C is better preserved in lactic acid-fermented vegetables, which also facilitates the absorption of iron in a phytate rich diet. There are indications for the presence of other iron-promoting factors in fermented vegetables as well [48].

The microbes have the capacity to upgrade the nutritive value of the materials like those rich in starch by increasing their protein content. Nutritive value of many fermented foods is not altered appreciably during fermentation, rather it is enhanced in many cases. The microbes also degrade the complex carbohydrates into simple components, which are easily digestible. In wine and beer, the yeast synthesize vitamins, proteins and amino acids, and thus, are rich in these compounds besides sugar, acid, phenolics, etc. [49].

From nutritional contribution of fermentation point of view, citation of household product *Chhang*, a beer like traditional beverage prepared on cottage consumed extensively in the sub-Himalayan region can be made. It is safe from microbiological, biochemical and nutritional safety angles [50].

## 3.6 Therapeutic Value of Fermented Foods

The food habits of the Himalayas typically exclude additional medicines or supplementary products. Most of the ethnic foods, both fermented and non-fermented, have therapeutic value and are eaten for prevention of illness. Traditionally, the Himalayan ethnic people do not have the habit of taking drugs and medicines in the form of tablets, tonics, etc. This may be due to the therapeutic value of their low cholesterol ethnic foods, which contain antioxidants, antimicrobials, probiotics, essential amino acids, bionutrients, and some important bioactive or other beneficial compounds.

The ethnic people who consume these foods believe that they have medicinal properties. *Somar* is used to cure stomach upset, and *Sherpa* to control the diarrhea. *Gundruk* and *Sinki* are used to treat indigestion and are commonly eaten as an appetizer. *Soibum* has wound healing and anti-tumor properties for animals, particularly cows. *Kinema* is highly nutritive and is eaten to boost protein intake and to cure heart diseases. *Jaanr* and ethnic fermented beverages are high-calorie food beverages, rich in mineral content and vitamins, mostly given to post-natal women. *Bhaati jaanr* and *poko* promote good health, nourishing the body and giving good vigor and stamina. Ethnic Himalayan milk products have protective and probiotic properties that stimulate the immune systems and cure stomach-related diseases. *Raksi* is a stimulating alcoholic drink that has both social and medicinal importance in the food culture of the Himalayan people. Although clinical study of the Himalayan fermented foods is yet to be carried out, the people who consume these customary foods believe that they have therapeutic values, and these products have been used, both as foods and therapy for centuries. Such ethnic foods, if studied properly, may find a place in the global market [51].

These days there is renewed interest in the consumption of wine, but in limited quantities, sustained further by the recent findings on the decreased incidence of cardiovascular diseases amongst the consumers explaining the French Paradox [52]. Consumption of low levels of alcohol is associated with lowered mortality from coronary heart diseases as well as from all other causes [53], besides being an important adjunct to the diet. Wines also contain glucose-tolerance-factor (GTF), used in curing the diabetes. The beneficial effect is attributed to the components present in the wine such as anthocyanins, phenols including flavanone, even with the intake of fat in the diet.

## 4. MICROBIAL BIOCATALYSIS

Microbial biocatalysts (as both isolated enzyme and whole-cell system) are increasingly being exploited to assist in the synthetic routes to complex molecules of industrial interest [54, 55]. Currently, there is specific attention on the use of biocatalysis to create new routes to reduce the cost of production [56]. Apart from their use in the production of microbial flavours used in food industry, the biggest role for biocatalysis still remains in the pharmaceutical sector [57]. 'Green chemistry' is described as the design, development and application of chemical processes and products to reduce or eliminate the use and generation of substances hazardous to human health and the environment [58]. Biocatalysts constitute a greener alternative to traditional organic synthesis that offers appropriate tools for the industrial transformation of natural or synthetic materials under mild reaction conditions [59], low energy requirement and minimizing the problems of isomerisation and rearrangement [60]. Good examples of the replacement of traditional organic processes by a greener biocatalytic alternative include the industrial synthesis of semi-synthetic penicillins and cephalosporins [59], the transformation of natural and synthetic fibers [61], multi-step synthesis of polyketide and glycopeptide antibiotics [60, 62] and a number of microbial flavours [63]. Whole-cell processes are cheaper to implement upstream as there is no need for enzyme isolation [64]. Whole-cells can provide an efficient way to regenerate cofactors *in situ* and contain multiple enzymes enabling multistep chemistry. However, they generally do not tolerate higher concentrations of substrate or product (>1 g/l), have limited use with organic solvents and can generate unwanted side products as a result of the presence of multiple enzymes. However, the price to be paid for these benefits is higher; upstream costs and enzyme reuse is often essential to keep these costs manageable [65].

Microbial biocatalysis is employed in the commercial production of many flavours and fragrance chemicals and is one of the emerging fields of food biotechnology. Bulk flavoring chemicals such as citric acid, high fructose corn syrup, and glutamic acid worth millions of pounds are produced using microbial processes [63]. Macrocyclic musk compounds like ethylene brassylate, cyclopentadecanolide and hexadecanolide perform very well in fragrances, and now they are being produced commercially using microbial processes. With *Candida tropicalis*, α, ω alkanedioic acid is produced, which is chemically converted to these fragrance compounds. Palmitic acid is fermented to sophorolipid by *Torulopsis* sp. Another interesting flavour made by microbial biocatalysis is L-Rhamnose (D-deoxy-L-mannose) as a starting material for 2, 5 dimethyl hydroxy furan-3-one (DMHP) which is one of most intensely sweet aroma chemical. It is an active chemical for many fruit flavours like pineapple, and strawberry. L-rhamnose and proline produce this compound after heating. Two microorganisms *Klebsiella* sp. (strain K32) and *Acinetabacter calcoaceticus* are used for alternative microbial process. Biocatalysis certainly opens new opportunities of microbial flavours for the flavourists.

## 5. GENETICALLY MODIFIED (GM) FOODS OR BIOTECH FOODS

Biotechnology is acclaimed as the most important development in recent years with enormous promise in contribution to human welfare. It has a tremendous potential in the production of genetically modified (GM) foods. GM foods, also called transgenic or biotech foods are known to contain new proteins produced by the recombinant DNA molecule. With the growing world population, ensuring food security is a serious task, and biotech foods are the best suited and provide solution to relieve world from hunger. The development of recombinant DNA technology started early in 1980s, and is continuously being developed in view of its potential applications. Genetic modification has several useful applications in agriculture, food processing, plant and animal products, nutrition, health and microbiology. The genetic modification of food and/or raw materials offers several potential advantages (Table 9). However, many issues of concern about biotech foods have been raised by scientific community and public, some of which are listed in Table 10.

**Table 9:** Advantages of genetic modification of food raw materials.

- Provides more food, more economically
- Is faster in realising a desired trait and in a cheaper manner
- Is more precise in selecting particular desirable characteristics
- Allows more traits to be improved, e.g. herbicide and insecticide tolerance, insect resistance, stress (e.g. drought), temperature and virus resistance, salt tolerance, etc.
- Reduces the number of food-deficient regions of the world through the development of stress-resistant varieties more amenable to growing on poor soils
- Improves the shelf-life of fresh fruits and vegetables
- More easily reduces the levels of allergens, naturally-occurring toxicants and other undesirable constituents in food crops
- Increases the levels of desirable disease-resisting and health-promoting food constituents (i.e., functional foods and nutraceuticals)
- Improves the sensory and nutritional qualities of foods (e.g. levels of vitamins A, E, desirable fatty acids, iron, fibre, etc.)
- Produces a variety of ingredients and processing aids (e.g. enzymes, microorganisms)
- Enables optimum feed composition and quality for optimal conversion by animals selected for such feeds, with concomitant benefits for the environment (e.g. development of low-phytate seeds that reduce phosphorus and nitrogen levels in wastes)
- Reduces pesticide and herbicide usages, with environmental and cost reduction benefits
- Reduces tillage with consequent soil benefit and reduces hand-weedings
- Is used to improve desired features of farmed animals, including fish, such as disease resistance and food production with improved qualities (e.g. lower in fat or fat modified, higher yields of prime cuts, better feed conversion, etc.)

*Source:* 6-8

**Table 10:** Issues of concern about biotech foods.

- Safety of GM foods, especially in the long term and their unintended consequences
- Environmental impacts of GM organisms and crops
- Role of "big business" in patenting GM organisms and preventing public access to GM technologies, especially in developing countries
- Potential allergenicity of novel proteins in GM foods
- Ethical issues like transfer of gene from animal to a plant may lead to a controversy
- Religious concerns regarding animal based work
- Loss of biodiversity

*Source:* 10, 66

## 5.1 GM Plant Products

The first recombinant DNA-derived food plant was a tomato carrying an anti-senescence gene to reduce the level of the softening enzyme polygalacturonase to increase the shelf-life of the fruit. Many GM crops have now been developed including soybean, corn, rice, potato, canola, cotton, etc. To date more than 80 crops have been modified for agronomic performance and more than 25000 field trials were conducted [4]. The major countries cultivating GM crops are USA, Argentina, Canada, Brazil and China [67]. GM crops were first extensively commercialized in 1996. Since then, the area under the cultivation of GM crops is increasing rapidly (Table 11). These GM crops are processed for making many food products. The development in the GM food technology certainly ensures the food security to the masses. However, GM foods must be tested thoroughly for their antinutritional effects before they are permitted for consumption by the population. The scope for this technology is immense and it can be said that GM foods are the future foods. For more details, see *Chapter 9*.

**Table 11:** Million hectares planted to GM crops, 1996-2008.

| Country | 1996 | 1997 | 1998 | 1999 | 2000 | 2001 | 2002 | 2003 | 2004 | 2005 | 2006 | 2007 | 2008 |
|---|---|---|---|---|---|---|---|---|---|---|---|---|---|
| United States | 1.5 | 8.1 | 20.5 | 28.7 | 30.3 | 35.7 | 39.0 | 42.8 | 47.6 | 49.8 | 54.6 | 57.7 | 62.5 |
| Argentina | 0.1 | 1.4 | 4.3 | 6.7 | 10.1 | 11.8 | 13.5 | 13.9 | 16.2 | 17.1 | 18.0 | 19.1 | 21.0 |
| Brazil | - | - | - | - | - | - | 3.5 | 3.0 | 5.0 | 9.4 | 11.5 | 15.0 | 15.8 |
| Canada | 0.1 | 1.3 | 2.8 | 4.0 | 3.0 | 3.2 | - | 4.4 | 5.4 | 5.8 | 6.1 | 7.0 | 7.6 |
| India | - | - | - | - | - | - | - | 0.1 | - | 1.3 | 3.8 | 6.2 | 7.6 |
| China | 1.1 | 1.8 | - | 0.3 | 0.5 | 1.5 | 2.1 | 2.8 | 3.7 | 3.3 | 3.5 | 3.8 | 3.8 |
| Paraguay | - | - | - | - | - | - | - | - | - | - | 2.0 | 2.6 | 2.7 |
| South Africa | - | - | - | - | - | - | - | - | - | - | 1.4 | 1.8 | 1.8 |
| Uruguay | - | - | - | - | - | - | - | - | - | - | 0.4 | 0.5 | 0.7 |
| Bolivia | - | - | - | - | - | - | - | - | - | - | - | - | 0.6 |
| Philippines | - | - | - | - | - | - | - | - | - | - | 0.2 | 0.3 | 0.4 |
| Australia | - | 0.1 | 0.1 | - | 0.1 | 0.2 | - | 0.1 | 0.2 | 0.3 | 0.2 | 0.1 | 0.2 |
| Mexico | - | - | - | - | - | - | - | - | 0.1 | 0.1 | 0.1 | 0.1 | 0.1 |
| Spain | - | - | - | - | - | - | - | - | 0.1 | 0.1 | 0.1 | 0.1 | 0.1 |
| Romania | - | - | - | - | - | - | - | - | - | - | 0.1 | - | - |
| Others (estimated) | - | - | 0.1 | 0.2 | 0.0 | 0.0 | 0.2 | 0.2 | 0.6 | 0.2 | 0.4 | 0.5 | 0.3 |
| *Total* | 2.8 | 12.7 | 27.8 | 39.9 | 43.9 | 52.4 | 58.3 | 67.3 | 78.9 | 87.4 | 102.4 | 114.8 | 125.2 |

*Source:* 68

## 5.2 GM Animal Products

The genetic modification of animals can be applied to biomedical research, modeling of human disease, production of proteins or other substances for therapeutic aims, as an alternative source of cell tissues and organs for xenotransplantation, and to improve desired features of farm animals, poultry animals and fish. The enhancement of the quality of muscle foods traditionally involved selection of farm animals, selection of preferred traits by cross-breeding, the use of endogenous and exogenous growth factors and extensive post-harvest technologies to improve palatability attributes such as tenderness, flavour and juiciness [69]. Biotechnological techniques appear to hold great promise for improving the quality and yield attributes of animal products. A wide range of biotechnology strategies is available for altering the balance between the lean and adipose tissue growth and deposition in meat-producing animals. These include genetic selection and management strategies. More recently, the confirmation of the growth-promoting and nutrient repartitioning effects of somatotropin, somatomedin, β-adrenergic agonists, immunization of animals against target circulating hormones or releasing factors, myostatin mutations, polar overdominance (callipyge mutation) and gene manipulation techniques have given rise to a technological revolution for altering the growth and development in meat producing animals. In these animals, the genome contains recombinant DNA. The major difficulty is the inability to insert genes precisely into animal's DNA. Consequently, the animals develop severe health problems like ulcers, pneumonia, arthritis, dermatitis, etc. The successful use of such technology depends on the identification, isolation and modification of useful genes or group of genes for meat quality and quantity and controls the level of expression of the inserted genes in transgenic animals so that their health is not impaired. There is hardly any carcass data from the transgenic animals. Over the past decade, the progress has been very slow due to limited funding. Tremendous potential now exists for muscle development, fat content, tenderness, colour and flavour. Although rapid progress has been made, much needs to be accomplished from the point of view of eating quality and safety.

Biotechnological interventions indicated that we may create better performing starter culture by taking some useful properties from other organisms [9]. Thus useful properties of different organisms can be incorporated in one strain. Recently, the starter cultures have also been used for improving the hygienic status and shelf-life of foods. For example, introducing the catalase gene in the organism can prevent the adverse effects of hydrogen peroxide in the product. The transfer of lysostaphin gene of *S. staphylolyticus* into starter lactobacilli could prevent the growth of *S. aureus*.

# 6. FUNCTIONAL FOODS THROUGH BIOTECHNOLOGICAL APPROACHES

The concept of functional foods was introduced in 1980 by the Japanese Government to regulate the use of food for medicinal purposes, and since then, the concept still defies uniform definition. The first attempt to define them spoke of "any food that has a positive impact on an individual's health, physical performance or state of mind in addition to its nutritive value" [70]. Since then, several bodies have developed their own definitions, but to date neither Europe nor the US has issued a legal version. However, European scientists have made a thorough attempt by providing a preliminary definition "A food can be regarded as 'functional' if it is satisfactorily demonstrated to affect beneficially one or more target functions in the body, beyond adequate nutritional effects, in a way that is relevant to either an improved state of health and well being and/or reduction of risk of disease. Functional foods must remain foods and they must demonstrate their effects in amounts that can normally be expected to be consumed in the diet: they are not pills or capsules, but part of normal food pattern" [71]. Hence, functional foods are those foods which provide nutrition as well as additional health benefits.

Researchers began to isolate or purify some of the specific bioactive compounds, from food called 'nutraceuticals', which are generally sold in medicinal forms not associated with food, and demonstrated to have a physiological benefit or provide protection against chronic disease. Active ingredients effective in promoting human health include amino acids, fats, dietary fibers, antioxidants, pigments, vitamins and minerals. Nutraceuticals include a range of agri-food ingredients, or phytochemicals extracted from edible plants or animal products commonly found in diet. These bioactive compounds as a part of the food, consumed directly, constitute functional foods. Scientific community has just began to tap the large potential of biotechnology to produce varieties of plants that will have a whole range of traits including increased nutritional value, medicinal properties, industrial utility, novel taste and aesthetic appearance. Nutrients/bioactive compounds are produced in parts of plants (plants usually do not produce them) by biotechnological techniques, and their bioavailability is also improved. A wide range of biotechnological products including bioactive therapeutic proteins, blood proteins, animal health products and industrial enzymes could be produced through genetic engineering, which are viewed as environment-friendly, renewable and inexpensive resource of nutrients and chemicals.

Microbial systems related to functional foods include probiotics, prebiotics, synbiotics, lactic acid bacteria (LAB), etc. There is no doubt that dairy products are functional foods. They are one of the best sources of calcium, an essential nutrient which can prevent osteoporosis and possibly colon cancer [72]. In view of the former, the National Academy of Sciences of USA recently increased recommendations for this nutrient for most human age groups. In addition to calcium, however, recent research has focused specifically on other compounds in dairy products, particularly fermented dairy products, known as probiotics. LAB are widely used as starter cultures in the production of food and feed. Fermentation of milk, vegetables, sausages, beverages and bakery products result in products with changed composition, taste and prolonged shelf-life.

Oligosaccharides are attracting increasing interest as prebiotic functional food ingredients. They are relatively new functional food ingredients that have great potential to improve the quality of many foods [47]. They can be extracted from plant materials or synthesized by enzymatic reactions. In addition to providing useful modifications to food flavour and physicochemical characteristics, many of these possess bifidogenic effect and a number of health-promising properties [46, 73]. These include non-cariogenecity, low calorific value and the ability to stimulate the growth of beneficial bacteria in colon. Both, the production and applications of food-grade oligosaccharides are increasing rapidly. Research continues into the development of new oligosaccharides with a range of physiological properties and applications in food industry. Synbiotics, mixtures of prebiotics and probiotics, combine the effects of stimulating the body's own and new ones in the foods.

An interesting example of microorganisms producing nutrients is *Haematococcus* algae, nature's richest source of astaxanthan, a potent antioxidant that acts as a cell stabilizer [74]. Astaxanthan is available as a dietary supplement. The role of omega-3 (n-3) fatty acids in the prevention of cardiovascular disease (CVD) is also well known. N-3 PUFA enriched eggs can be produced by modifying hen's diet [75].

The global market of functional foods is estimated upto 33 billion US$. In US, the market is differentiated as functional foods with special health claims achieving a turnover of around 0.5 billion US$, and functional foods without claims with an annual turnover of atleast 15 billion US$ [73]. Japan is another important market of functional foods, with a specific health related food category called FOSHU (Food of Special Health Use). An estimated market value of FOSHU labeled products was 2 billion US$ in year 2000 with an estimated turnover of 14 billion US$. The European functional foods market contributes to 4-8 billion US$ that constitutes below 1% of the total foods and drinks market. Among the European countries, Germany, France, United Kingdom and Netherlands are the important ones with functional food market. The market place of foods is changing rapidly. Dietary supplements are being incorporated into food products. Plant foods are being genetically engineered to contain higher amounts of nutrients and health-promising constituents. Research on functional foods is changing the face of foods as we know them. We could select our produce and diet based on our individual genome, with the objective of reducing genetic risk factors for specific diseases. Other new products might be available on the shelves, such as breakfast cereal with added *Hypericum perforatum* for days when one is feeling low, or another cereal with added choline for extra energy on that morning before a tennis match.

## 7. SAFETY OF BIOTECHNOLOGICALLY PRODUCED PRODUCTS

The safety of food products is of utmost importance, both from consumer acceptance and marketing point of view. The microbial food safety is assured primarily through the control of microorganisms at the food source and in raw materials selection; product design and process control; applications of Good Hygiene/Manufacturing Practices (GHPs/GMPs); and implementation of the Hazard Analysis and Critical Control Point (HACCP) system throughout the food chain [76]. Microbial quality is important from a food safety point of view. The traditional approach has so far been the retrospective use of microbial results in order to improve the quality of the future lots of the products. A creative combination of DNA probes with other technologies could form the core of future monitoring systems. The safety of GM foods or food ingredients is a subject of heightened interest among consumers, public policy makers and within the scientific community. Unfortunately, a wide gap exists between what is offered and what is suitable for implementation due to various reasons.

### 7.1 Safety of Fermented Foods

The quality of raw material used in the production and fermentation conditions constitutes the major factor having an impact on physical, chemical, microbiological and sensory characteristics of fermented products. Proper understanding of the factors and their control is an integral part of fermented product production. Both production and quality control technologists need to have sound knowledge of analytical techniques normally employed.

Since fermentation is typically one part of a multi-step food-processing operation, assessing the safety of fermented foods and determining appropriate interventions to improve should be considered in the context of the entire process transforming food-stuffs from raw materials to end-products [48]. Traditional foods are accepted as safe being consumed from time immemorial. Such foods are not subjected to rigorous safety testing. But those made as new product or through the modification as a result of genetically manipulated microorganisms used in fermented food product would require mandatory safety testing. It is different from that used for additive testing where the safety is determined at specific dose level. The dose effect cannot be applied to the food being taken several times in a day [77]. Toxicological problems of traditionally fermented foods might be a possibility, as some of the fungi isolated from such foods are closely related to those which produce toxin (*A. flavus*), but the strains used in such fermentations did not produce toxin experimentally and the fermented foods tested remained free from mycotoxins and are safe to eat.

### 7.2 Safety of GM Foods

Recombinant DNA techniques are being used to develop and produce food crops. The questions regarding the safety of their consumption are often raised, especially the components which would be introduced in our diet such as proteins, carbohydrates, fats and oils. Variations in the structure of these components need scientific evaluation for safety, as some proteins are known to have toxicity, allergenicity, etc. [7], while the enzymes from different microorganisms could have different structure, hence catalytic activity, protein structure, etc., with inherent possibility for alterations in food.

Generally, the proteins/enzymes have structure similar to those consumed normally, thus, would not raise any concern for safety [77]. But in many cases, it is not known if the protein(s) involved such as bacterial or animal toxins, hemagglutinins, enzyme inhibitors, vitamin binding or those with selenium containing proteins has allergeniticity or is toxic to vertebrates. Only well designed serological tests, scientifically conducted, could throw light on such aspects. However, the present knowledge does not indicate any test which could be used to predict or assess the new proteins for allergenicity, but should include metabolic, toxicological or digestibility studies. However, if genetically modified plant used in food produces a substance that is indigestible, the question of nutrition rather than safety would emerge, besides, evaluation for their wholesomeness. With respect to food fermentations, this thinking is equally applicable rather it is more true with respect to the microorganisms after genetic engineering is used for fermentations.

### *7.2.1 Safety and safety evaluation*

GM crops and foods are now available and have been consumed for over 13 years, and there appears to be no credible scientific evidence to show that the ingestion of transgenic products is injurious to human health or the environment. GM foods and ingredients currently available in the international market have passed risk assessments conducted by appropriate national and international bodies. Risk is usually defined as "the probability of harm", while a hazard is anything that might conceivably go wrong. Risk is thus a combination of the hazard involved, the probability of its occurrence, and the consequences of that occurrence. Risk analysis generally follows the Codex Alimentarius principles [78]. Risk assessment is a scientific examination of the nature of the hazard involved (hazard identification), the extent of the hazard and its risk (hazard characterization), the likelihood of its effects on particular segments of the population (exposure assessment), and an overall assessment of the severity of the risk and its consequences (risk characterization). The safety assessment of GM foods compares its distinctiveness from conventional counterpart and involves assessment of factors listed in Table 12.

**Table 12:** Safety assessment factors of GM foods.

- Identity
- Source
- Composition
- Effects of processing/cooking
- Transformation processes
- The recombinant DNA (e.g., stability of insertion, potential for gene transfer)
- Protein expression product of the novel DNA (e.g., effects on function, potential toxicity and allergenicity)
- Possible secondary effects from gene expression or disruption of host DNA or metabolic pathways (e.g., macro/micronutrients, anti-nutrients, toxicants, allergens)
- Potential intake and dietary impact of the GM food

*Source:* 79

### *7.2.2 Allergenicity and naturally occurring toxicants of GM foods*

Food allergens produce abnormal immunological consequences in some consumers to some food constituents, usually particular proteins. The most frequent types of allergies are mediated by allergen-specific IgE antibodies, and the outcome can be mild to fatal. Although more than 160 foods have been associated with sporadic reactions, common food allergens (peanuts, soybeans, milk, eggs, some fish, crustaceans, wheat and tree nuts, etc.) account for about 90% of all the reactions [80].

The potential allergenicity of GM foods can be assessed using the decision-tree approach of the International Food Biotechnology Council of the International Life Sciences Institute [79, 81], now adopted and refined by the agricultural biotechnology industry. The assessment examines the source of the gene(s), amino acid sequence homology of the newly introduced protein(s) to known allergens, the immunoreactivity of the introduced protein with IgE antibodies from individuals known to suffer from allergic reactions, and the physicochemical properties (effects of digestion and heat on the proteins). The IFBC-ILSI strategy is continuously being refined to improve the detection of known and potentially allergenic material, or to test the allergenicity of proteins from foods with no history of allergenicity.

The enzyme introduced into glycophosphate-tolerant soybeans and the Bt proteins used in insect-resistant crops have no sequence homology to known allergens and are rapidly digested in mammalian systems. Reports that a Brazil nut gene had caused allergies in consumers eating GM soybean products have been shown to be incorrect, as the transfer of Brazil nut allergenicity to the GM soybean was identified during the development work, and the product was never released in the market.

Some foods or crops contain natural toxicants (e.g., solanine, enzyme inhibitors), although major human crops and animals contain few of such constituents through millennia of selective breeding. It is essential that adequate assessments be undertaken to maintain this status with GM foods.

### *7.2.3 Public attitudes towards GM foods*

The attitudes of consumers towards the GM foods are dependent on their assessment of risk and benefit of the foods produced and the technologies involved, and on the nature and extent of the communications provided to them. Although many national biotechnology or food standards organisations or agencies undertake surveys on consumer acceptance of GM foods and processes, few truly international studies have been undertaken. Most of the research on consumer attitudes has focused on the USA and Europe. In one extensive study, more than 35,000 respondents were asked whether the benefits of GM crops (with respect to reduced pesticide use) were greater than the risks. Over two-thirds of the respondents from USA, Colombia, Cuba, Dominican Republic, China, India, Indonesia and Thailand believed that the benefits outweighed the risks, while less than one-third of the respondents from France, Greece, Italy, Spain and Japan did so. The USA led the industrialized countries in support for biotechnology. Overall, people in developing countries are supportive of GM crops [82]. Support for GM foods or an ingredient is generally less than for other food biotechnology applications, where the benefit or involvement is less direct. Factors influencing the support for acceptance of GM products or technologies include, for example, whether the GM product is eaten or not; whether there is a direct benefit to consumers (e.g., golden rice); why the genetic modification was done, and how exotic the transformation was; the necessity for the genetic modification considering the food choices available; and the level of technical information obtained, especially on risk and safety assessment, and trust in the information. In some studies, government agencies and universities generally ranked high in credibility of the information provided, while activist groups and agribusiness companies ranked low. Some population groups have shown more concern for environmental and biodiversity issues than for food-related issues. Of considerable importance to many consumers is adequate labeling of GM products in order to make an informed decision. Labeling requirements for GM foods, ingredients and processes vary for different countries. Deciding what to label, and the level of GM in the food to trigger the labeling requirement, are issues that require debate and uniform implementation.

### *7.2.4 Ethical issues of GM foods*

Discussion on the ethical issues of GM foods by its opponents centres on "man is playing God" by transferring genes; or people who follow a religion that does not allow the consumption of a food such as pork should not be expected to eat a food containing pig genes, but do not address the ethics of denying to millions of starving or malnourished people that have opportunity of addressing these imbalances by access to GM foods or ingredients. The latter are, however, fully addressed by the two reports of the Nuffield Council on Bioethics [83, 84], the latter specifically directed at the needs of developing countries. Ethical considerations are also addressed in the TIFS report [8]. On the matter of choice for those who, for religious or other reasons, do not wish to consume GM foods, one key would be to adopt distinctive labeling as practiced in some countries; another would be to purchase organic foods in which GM ingredients are prohibited. The issues involved are complex, and the diametrically opposed views unlikely to be reconciled [85].

#### *7.2.5 International regulation of GM foods*

In July 2003, the 26th session of the Codex Alimentarius Commission approved principles for the risk analysis of foods derived from biotechnology, and guidelines for the conduct of food safety assessments of foods produced from recombinant plants and microorganisms [78, 86, 87]. These procedures now form the basis for the assessment of GM foods and ingredients used widely by national regulatory agencies. The Cartagena Protocol on Biosafety to the Convention on Biological Diversity was accepted on 29th January, 2000, and came into force on 11th September, 2003. To date, 111 countries are parties to the agreement. The definitions used in the Codex Principles are the same as those used in the Protocol so that the documents are compatible and supportable. The protocol is a legally binding instrument that regulates the international movement of living modified organisms (LMOs) resulting from modern biotechnology to protect the environment. On 27th February, 2004, the member states of the Cartagena Protocol adopted handling, transport, packaging and identification requirements for promoting the safety of LMOs or GMOs. The implementation of international agreements for the regulation of GM foods and ingredients and processes for genetic modification of plants, animals and microorganisms at the national or regional level is a complex issue and should be discussed in an elaborative way before its implementation.

#### *7.2.6 Global scenario*

In USA, USDFA has approved GM maize, soyabean, canola and cotton. The commodities grown in USA are Bt corn, Bt cotton, papaya and soyabeans. The first commercially safe and approved GM product in USA was in 1994 for the sale of delayed ripening tomato. Labeling of GM foods is mandatory in all the EU regions. Currently, two forms of GM foods are marketed in USA, EU and Japan: GM whole foods, and processed foods that contain products from GM plants. Labeling of GM soyabeans is done in Korea. European Union has adopted new rules to permit the development and sale of new GM products. Zambia has banned GM foods as of now. According to the UK Science Academy, improvements are required before clearing the use of GM plants. Although scientists in Japan claim to have created pigs implanted with spinach genes and the meat is said to be healthier to eat, it has to be got approved before consumption [66].

## 8. IMPACT OF GENOMICS AND BIOINFORMATICS ON FOOD BIOTECHNOLOGY

Food biotechnology, apart from the direct use of plants and animals, is mainly concerned with the applications of food-grade microorganisms in industrial processes. The aims of food biotechnology are not only directed towards improving food production, but are also defined by consumer demands for safe, natural, fresh, tasteful and convenient products. Earlier, the improvement of industrially important microorganisms was focused on strain selection after classical mutagenesis, but now-a-days more directed approaches involving genetic engineering are used. The main drawbacks of these new approaches are that they are time-consuming, side-effects occurring in the selected or constructed strains are hard to predict and assess, and full range of engineering possibilities cannot be exploited due to lack of knowledge of inter-related regulatory and metabolic processes going on in a cell. Furthermore, part of the consumer population has serious worries about the safety of these novel foods, and whether unwanted spread of genetically modified microorganisms in nature can be controlled. It is exactly in the last two fields that functional genomics can provide solutions. The implementation of functional genomics programs on food microorganisms will allow us to reach various industrial goals within the coming decade, e.g., to determine side-effects of genetic alterations on functionality in products, to create desired pleiotropic effects by specific regulatory mutations, to predict and improve stress responses, to develop novel antimicrobial systems, and to direct metabolic engineering efforts [88]. Furthermore, it will enable the identification of novel (enzyme) activities, the performance of high-throughput mutation analysis in strains of interest and the development of fast identification methods for food-spoilage or pathogenic microorganisms.

The recent burst in genome sequences of food-related microorganisms has opened the way for functional genomic approaches, including transcriptome, proteome and metabolome analysis, as well as structural genomics [88]. To exploit the information on genome sequences, novel high-throughput analytical methods are being developed, such as DNA microarrays (transcriptome), MALDI-ToF mass spectroscopic analysis (proteome), etc. DNA microarrays provide a powerful tool for analyzing transcription profiles of whole genome of any organism [89, 90]. These arrays have already been successfully used for bacterial transcriptome analysis of 100 genes of *Streptococcus pneumoniae* and 106 genes of *Haemophilus influenzae* [88].

The combination of transcriptome with proteome and metabolome research and the elucidation of structure-function relationships of biomolecules will eventually result in a true understanding of whole-cell functioning. An overview of approaches involved in functional genomics programs and their expected deliverables for food biotechnology are depicted in Fig. 2. The proteome can be regarded as the expressed protein complement of a genome. Essentially, two types of proteomic research can be distinguished: first expression proteomics which aims at determining global changes in protein production levels; and second, cell-map proteomics, which systematically studies the protein-protein interaction through the isolation of protein complexes [91]. Several pharmaceutical companies are heavily involved in proteomic research, with major goals to identify novel drug targets for use in diagnostics and therapeutics [92]. Food biotechnology faces similar needs, such as determining the right targets for metabolic engineering, studying resistance mechanisms of food pathogens and identification of probiotic strains and *in situ* activities of lactic acid bacteria. Computational genomics should also be integrated with proteomics for the detection of functional and structural evolutionary relationships between proteins with only limited sequence similarities [93].

The huge number of experimental data generated is made into large databases. Their interpretation greatly depends on novel bioinformatics methodologies, which should enable linking of the databases and facilitate classification and interpretation of the results. Fast developments are also taking place in bioinformatics. Systems for cluster analysis and display of genome-wise expression patterns have been developed that provide information on the status of cellular processes and the possible functions of genes without a known functionality. A nice overview of existing databases on microbial genomes, genomics and proteomics, including an extensive list of relevant websites is available [94]. Novel data-mining and -linking approaches get increasing attention, yielding methods for determining microbial genescapes. Examples are the determination of phyleit and functional patterns of open reading frame (ORF) distribution among prokaryote, providing a quantitative tool for comparative genomics.

An exciting development in bioinformatics is the building of integrative model called 'E-CELL' [95], for stimulating intracellular molecular processes to predict the dynamic behaviour of cells, and incorporating information on gene regulation signaling and metabolism. This model is extremely useful for simulation of food microbes with relatively small genomes, such as those of lactic acid bacteria. The system can be continuously improved by adding novel experimental data, which are currently generated at a high pace. Another interesting project is called 'the Virtual Cell' [96], which aims at

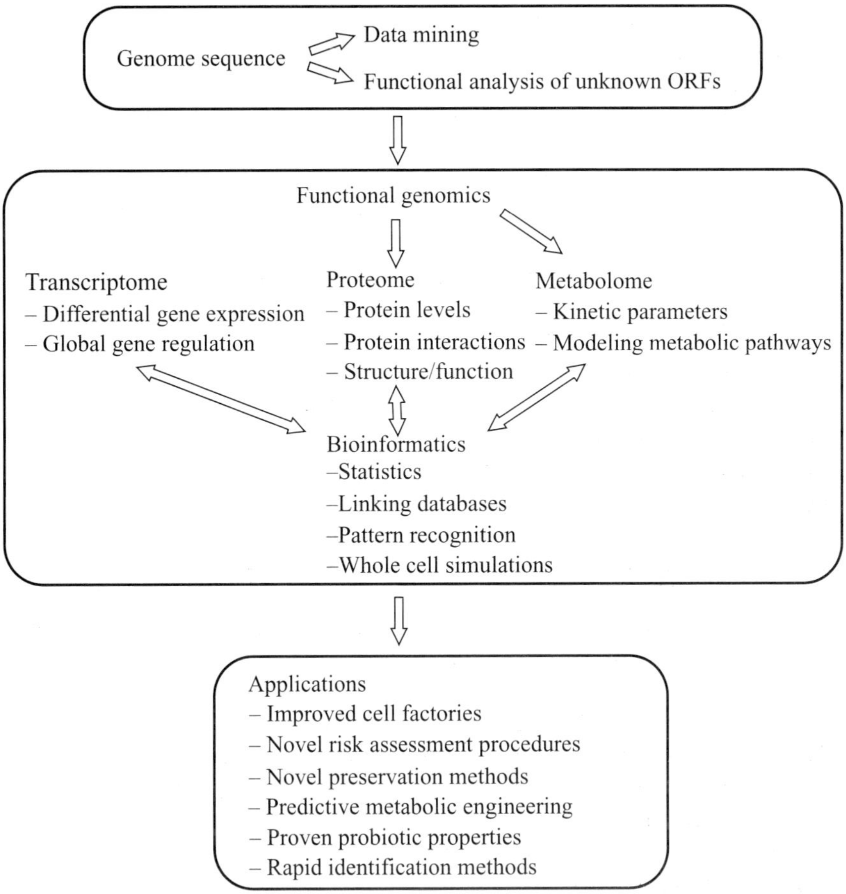


**Fig. 2:** Functional genomics of food microorganisms.
*Source:* 88

modeling of cellular processes to be used as a tool for experimental biologists, e.g., for metabolic engineering purposes. The powerful approach in functional genomics of using knowledge coming from metabolic engineering studies and gene-phenotype relationships for model building forms a major challenge in bioinformatics [97]. Further developments in bioinformatics are ongoing, especially in the fields of genome mining, including functional analysis of genes with known and unknown functions, linking databases with experimental data (transcriptome, proteome, metabolome, protein structure and function) and pattern recognition.

## 9. BIOTECHNOLOGICAL APPROACHES FOR WASTE MANAGEMENT IN FOOD INDUSTRY

The application of biotechnology in the waste management using microorganisms has received a considerable attention, where besides reducing pollution load, these produce metabolites of economic importance such as ethanol. In view of the dwindling oil resources, alternative fuel resources are being accorded priority. According to Carr [98], "*I believe that the process we shall need to fill this gap (oil depletion) will have to be microbiological, and applied microbiology*" and the present research on waste utilization and cellulose degradation for ethanol production as fuel testifies the correctness of his thoughts. The complete degradation of cellulose (a renewable source) to glucose could provide a major source of food, fuel and chemical feed stocks [99], besides reduction in the pollution level in the waste especially from the paper, lumber and textiles industry. Genetic engineering could provide new directions and solution to this problem.

Large amounts of fruit and vegetable processing wastes are produced from food industry, which may be disposed in various ways, including their immediate use for landfill or as animal feed or which, alternatively, may be processed biotechnologically in order to produce more valuable products [100, 101]. In recent times, food industry has continued to make progress in solving the waste problem through the recovery of by-products and waste materials using fermentation or other related biological processes.

Wastes like apple pomace from juice processing can be converted into many useful value added products [102, 103]. Readers would get a number of opportunities to find many more such illustrations on utilizing the waste from food processing industries in *Chapter 24*. From resource utilization, recycling and providing clean environment point of view, this aspect is likely to become almost a compulsive need of human race.

## 10. FOOD BIOTECHNOLOGY AND WORLD FEEDING PROBLEMS

Biotechnology is widely and deeply being applied in the food industry, driving a revolution in the food products. We have used biotechnology to manufacture food products for more than 8000 years. Bread, alcoholic beverages, vinegar, cheese, yoghurt and many other fermented foods owe their existence to enzymes found in various microorganisms. Even today, the development of healthy foods has shifted towards conventional fermentation and manufacturing, which has a lower technical level input but greater competitive pressure as well. This industry meets both high product yield and technical standards in order to be profitable.

Historically speaking, the fermented foods were prepared by the processes developed in homes by accidental observations of natural phenomenon of fermentation [5]. Until the last century, these processes have been more or less like an art without knowledge of the basic changes involved, and the information was passed on from generation-to-generation. For thousands of years, the people have relied upon the foods preserved by fermentation processes, being unaware of the microorganisms involved, and why or how the foods acquire their modified qualities or remain safe for consumption. Ultimately, the application of basic knowledge led to the production of foods of superior and uniform quality at the industrial scale.

Fermented foods are the result of the activity of a few specialized types of microorganisms among the thousands of species known to mankind. The success of such processes is dependent upon the control of environmental conditions ensuring the growth of the desired microbial species and inhibition of others, some of which may be food poisoning agents or spoilage causing. The control, inhibition, or elimination of undesirable species concomitant with favouring the growth and involved in fermentation by desired species, may be in sequence, must be the ultimate goal to produce fermented foods. With knowledge of food fermentation available now concerning the effects of physical and chemical environment, growth of various microbial species, desirable changes in fermented food products, the food fermentation technology has potential to feed the ever growing population of world.

Even though great strides have been made during the past century on different aspects of food fermentation, the advances in the industry revolved around improvement in methodology, product standardization and quality improvement rather than development of new products. Still many fermented foods have never been investigated, and research on such foods, which the people already consume, may improve their quality. Such products and practices are accepted by people instantly as no taste development and introduction of such foods is needed. Further, in the present world of information explosion, the technology moves globally, and in this context, the food fermentation can play a far greater role in feeding the human race.

The need today is not only for the production of more food, but nutritionally rich, better preserved food with safety for consumption, as it is useless to produce more food unless it can be harvested, preserved and made available to the people [5]. Many countries being rich in natural resources with excellent soil and abundant rainfall should not have problem in feeding their people, but the difficulties of proper postharvest infrastructure, preservation, transportation, etc. are a deterrent to nutritious feeding programme. Fermentation has proved itself as a cheap and effective means of food preservation, nutritional enrichment and taste improvement.

Biotechnology is already beginning to provide sustainable and life-sustaining assistance in the developing countries. Through the use of biotechnology, researchers are providing higher-yielding strains, foods with enhanced nutritional traits, GM foods and products that last longer and are resistant to spoilage. Food biotechnology, a dynamic new science that uses genetic engineering to enhance the output and value of many food products, may hold the key to prevent the world from hunger. It is estimated that GM technologies would reduce the problem of world hunger to 50% by 2015 [10]. Environmentalists argue that genetic engineering poses too many risks to human health and the environment, and that its use should be sharply curtailed or even banned together. However, an overwhelming number of scientists from around the world emphatically dismiss these objectives as unfounded. It would thus be a tragedy if misinformation spread by the environmental movement about GM foods is allowed to win the day and the world is deprived of its great potential to improve and save lives from hunger. If all the concerns of GM techniques are satisfactorily addressed, this technology may help to feed the world's escalating population in the coming decades [4]. At least 800 million people suffer from malnutrition. An estimated 1.3 billion people live on less than $1 per day, which is an insufficient income to purchase an adequate diet. One reason for so many nations suffering from hunger is that their climates are often inhospitable for efficiently producing agricultural products. Of the 42 highly indebted poor countries of the world, 39 are located in tropical or desert regions where growing conditions are less than optimal. Plant viruses, soil erosion, costly fertilizers and pesticides and inadequate storage are endemic problems in the developing countries. Agricultural biotechnology can play a major role in eliminating this human suffering. Through gene manipulation, scientists have been able to alter many of the staple food crops that developing nations depend on, such as cassava, rice, maize and potatoes, to make them more resistant to disease, more nutritional and more productive. Another major benefit of food biotechnology is that it can significantly improve the nutritional content of various foods. Researchers at the Swiss Federal Institute of Technology, for instance, have developed a new breed of rice that has a higher content of iron, thus helping to address an iron deficiency suffered by 3.7 billion people worldwide. Iron deficiency can lead to the development of anemia, a disease characterized by insufficient red blood cells. The rice also contains enough vitamin A. Besides causing blindness, lack of vitamin A has been linked to heart disease and some cancers. The United Nations children's fund estimates that 1-2 million deaths of children between 1 to 4 years of age, each year, could be prevented if these children get more vitamin A. Rice fortified with vitamin A would be especially welcomed by several Asian countries where 80% of daily caloric intake consists of rice. But rice is by no means the only food that can add this much-needed vitamin to the diets of the populations of developing countries. Just recently, an international team of scientists used genetic engineering to create a tomato with three times the normal levels of β-carotene, which the human body processes into vitamin A.

Biotech researchers have already developed and are field testing rice enhanced with β-carotene, a precursor of vitamin A, which is important because rice is a primary diet staple in the developing world. The United Nations Food and Agriculture Organization recently stated "The potential to create rice with an enhance micronutrient content illustrates one way in which genetic engineering can contribute to reducing malnutrition. Vitamin A deficiency, which is widespread in the developing world, can lead to morbidity and blindness and contribute to child mortality." Similarly, researchers at the Donald Danforth Plant Science Centre are developing high protein and vitamin enhanced cassava, the primary source of food calories in tropical regions of the developing world.

Biotech foods could one day reduce losses due to spoilage, especially in areas with limited transportation and refrigeration capability. According to a joint report issued by National Academy of Sciences of the United States, United Kingdom, Brazil, China, India and other developing countries, "it is possible that farmers in developing countries could benefit considerably from crops with delayed ripening or softening, as this may allow them much greater flexibility in distribution than they have at present. In many cases, small-scale farmers suffer heavy losses due to excessive or uncontrolled ripening or softening of fruit or vegetables." These approaches are strongly supportive of food biotechnology's potential to feed a hungry world and improve the environment at the same time.

## 11. FOOD BIOTECHNOLOGY INDUSTRY

The food biotech industry refers to all related industries of food processing that use biotechnological production methods, i.e. genetic engineering, cell engineering, enzyme engineering and conventional fermentation as well as fermentation engineering to improve the quality of food or food ingredients, enhance production or reduce costs, etc. For example, using genetic engineering to produce special enzymes or high value added products such as functional peptides; using cell engineering to produce special food flavour additives; using enzyme engineering to change the food's quality and nutritional values, e.g. protein that has undergone special enzyme hydrolysis, which enhances its digestion. In addition, hydrolysis of carbohydrates by amylases can produce glucose monosaccharide or low polysaccharides. Meanwhile, fermentation engineering uses modern day fermentation to re-create traditional fermented foods using microbial flora for fermenting to enhance the ingredient utilization and reduce the fermentation period needed for flavour improvement. World over the food biotech industry is expanding and is emerging as a diverse sector. The data on biotechnology based firms in various fields of some countries leading in biotechnology is given in Table 13.

**Table 13:** Biotechnology firms by application.

| Country | Health | Agriculture | Food & beverages processing | Natural resources | Environment | Industrial processing | Bioinformatics | Other | Total biotech firms |
|---|---|---|---|---|---|---|---|---|---|
| Australia | 353 | 185 | 0 | 22 | 85 | 90 | 84 | 48 | 527 |
| Austria | 101 | 7 | 9 | 1 | 2 | 12 | 5 | 0 | 121 |
| Belgium | 33 | 14 | X | 0 | 6 | 15 | 5 | 0 | 73 |
| Germany | 258 | 59 | 0 | 1 | 18 | 30 | 20 | 201 | 587 |
| Korea | 241 | 25 | 187 | X | 109 | 51 | 42 | 118 | 773 |
| New Zealand | 72 | 63 | 66 | NA | 33 | 48 | 39 | 27 | 135 |
| Poland | 8 | 1 | 0 | 0 | 1 | 0 | 1 | 0 | 11 |
| Philippines | 4 | 19 | 6 | 6 | 1 | 9 | 2 | 0 | 25 |
| South Africa | 76 | 0 | 0.08 | 1 | 0.16 | 0 | 0.08 | 0 | 78 |
| Canada | 310 | 107 | 39 | 21 | 39 | 39 | 5 | 11 | 532 |
| Sweden | 131 | 8 | 0 | 0 | 7 | 7 | 0 | 0 | 148 |
| Brazil | 25 | 16 | 0 | 3 | 10 | 10 | 0 | 17 | 71 |

X for Belgium (Flanders): "Agriculture" includes "Food and Beverage Processing" and for Korea: "Agriculture" includes "Natural Resources"; NA: Not Available

*Source:* 68

## 11.1 Magnitude

The food based industries are significant contributors to the economy of many countries of the world. The global market value of biotechnology based food industry is increasing day by day. As observed by Houwink [104] "Moneywise biotechnology is mainly beer and wine". Enzymes are the other biotechnology based products of commercial value. Many countries are producing enzymes for both food and feed. In 2002, the world wide enzyme market for food and animal feed use amounts to roughly 1.176 billion US$, representing an annual growth of about 11.4%. The world's main suppliers of enzymes are Denmark, US and Japan (Fig. 3). The quantities of enzymes produced in the world for use in food and feed also reflect the magnitude of the food biotechnology industry. Currently, the enzyme market has gradually saturated, e.g., the starch and sugar industry. In the baking industry, there has been an increasing trend for α-amylase, lipase and xylanase. Enzymes still have a huge development potential in health food market. The value of citric acid can be judged from its quantity produced in the world. The natural flavours produced by the microorganisms or those modified by them are emerging high value products. Another fermentation based product which made a great impact on the marketing is PUFA (Polyunsaturated fatty acids) which is highly valuable pharmaceutical product with health implications. Mushrooms as a flavoured and delicious food are relished universally and have established market world-over. The fermented foods having regional/local consumption/market in the past are now being consumed globally, thus expanding the market of food fermentation product enormously. *Idli, Dosa* or *Soy sauce* is not consumed in Asian countries, but is a delicacy in the European countries, adding greatly to the magnitude of the food fermentation.

The second-generation bioengineered foods, which provide direct consumer health benefits would be more easily accepted and have lucrative market. Functional foods/nutraceutical that demonstrate physiological benefits and reduced disease risk fall within this category. Nutraceutical sales (including vitamins, dietary supplements and herbs) in USA were valued at 86 billion US$ in 1996. By 2020, the nutritional market is anticipated to reach 500 billion US$ market [74]. Nowadays, a number of prebiotic, probiotic and synbiotic products/product supplements, as functional foods with health-promising functionalities, are available in the market. Today, the worldwide market value of health foods has exceeded 30 billion US$, with the US and Japan as the major countries. Digestion related health foods are most popular in Europe, especially probiotics. Other popular products include anti-

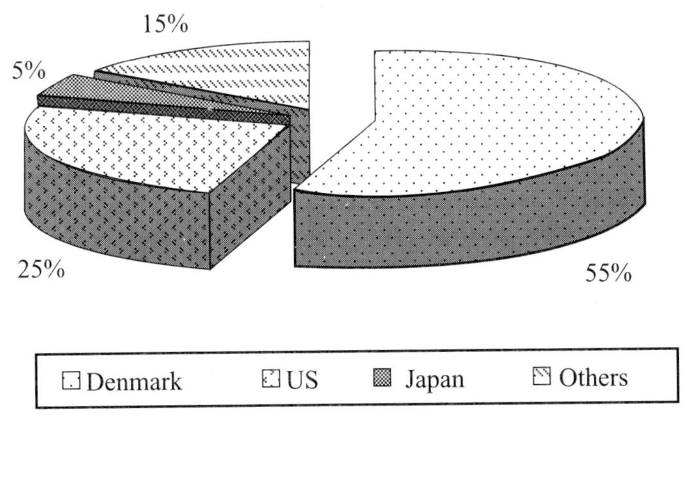


**Fig. 3:** World's main suppliers of enzymes.

oxidation related products and products that are good for the heart and blood vessels (such as soybean lecithin, DHA, EPA, anthocyanins, etc.). In the US, vitamins are the most popular, especially anti cancer products and botanical medicines. As for digestion related health products, the market for milk products made from probiotics still has a lot of room for development. Of Japan's certified specific usage health foods, 71% are related to improving digestion functions (lactic acid bacteria, fiber, oligosaccharide, etc.). In addition, there are also products related to blood pressure (polypeptide, glucoside), mineral related (hemoglobin iron, oligosaccharide, CCP, CCM, etc.), and cholesterol related (soybean protein, fiber, etc.). Japan has taken the leading role in the R&D and application of functional foods. As of 1999, 167 functional foods have already been approved. Of this, oligosaccharide group, lactic acid bacteria and dietary fiber account for 80%. There are also 13 protein and peptide products, which can help to reduce cholesterol and blood pressure. GM foods with well established functionalities with more understanding of ethics and other concerning issues would have a potent market in the food industry in future. The new products in future would add further to the list already available to the consumers, accordingly contributing significantly to world economy.

## 11.2 Research and Development Innovations

The food industry as a whole has a longer lag period between the invention and innovation. The maxim of chemical industry of America, "ten years from test tube to tank car", does not hold good for the innovations in food biotechnology. However, the industry is also labour oriented [105]. After 1960, many companies have cut dramatically the investment in the R&D expenditure due to increased regulations, higher rates of inflation and bigger apparent difficulties in obtaining technical breakthrough. But it depends upon the size of the company where the length of time it takes, before using a technique tends to be inversely related to its size and the profitability of its investment in the innovation. The food processing industry, though very conservative in the past regarding the investment in R&D, has recently started investing in biotechnology with corresponding increase in the number of potential products and applications, which include low energy products, improved enzymes and microorganisms, polysaccharides, flavour products and monoclonal antibody for food testing [106]. The biotech R&D expenditure of biotechnology firms in some countries is depicted in Table 14.

The main problem regarding the development of food biotech industry is the focus of research within academia has been concentrated on developing key technology and products for medical medicines, very little has been invested in food biotechnology. Therefore, there is very little original and innovative biotech technology or products that can be transferred to the industry. Moreover, the industry lacks investing in R&D as well while also in a rush to transform. As such, many business houses have invested in health food, creating overlap in products and vicious competition, which is not beneficial to the development of the food biotech industry. The industry lacks original and innovative technology and products to create a competitive edge, which can increase profits and enhance competitiveness against foreign manufacturers. There is also a serious lack of senior R&D talent in the industry, which makes it difficult to implement the research results from the academic field. Regarding the lack of senior R&D talent, in addition to strengthening the cultivation of new talent, the number of different research topics should also be reduced as well. Key technology and products that are original and innovative should be carefully selected and carried out through integration or by collaboration between the industry and academia. Academia should assist the industry to build up key platform technologies to support the industry's R&D capability. Related government departments should seek opportunities to work on international projects, encouraging foreign companies to set up R&D and production sites.

New innovations in technology improve the product quality or reduce the cost or both. To cite one example, adoption of Arnott's continuous fermentation produced crackers with reduction in the

**Table 14:** Biotechnology R&D in biotechnology firms by application.

| Country | Total biotech firms | Health | Agriculture | Food & beverages processing | Natural resources | Environment | Industrial processing | Bioinformatics | Other | Total biotech R&D expenditure |
|---|---|---|---|---|---|---|---|---|---|---|
| | | Million PPP$ | | | | | | | | Million PPP$ |
| Belgium | 73 | 517.1 | 80.2 | X | 0.0 | 0.5 | 23.0 | 2.0 | 0.0 | 622.6 |
| Korea | 773 | 374.7 | 18.2 | 153.7 | X | 39.1 | 20.4 | 28.9 | 74.1 | 709.3 |
| Poland | 11 | 6.0 | 2.6 | 0.0 | 0.0 | 0.1 | 0.0 | 0.1 | 0.0 | 8.8 |
| Canada | 532 | 1226.1 | 128.5 | 8.2 | 19.8 | 11.5 | 0.0 | 7.4 | 2.5 | 1404.0 |
| Germany | 496 | 894.9 | 56.2 | 0.0 | 0.1 | 21.2 | 30.7 | 7.1 | 188.0 | 1198.2 |

X for Belgium (Flanders): "Agriculture" includes "Food & beverages processing" and for Korea "Agriculture" includes "Natural resources"
*Source:* 68

processing cost without the loss of quality or the use of immobilized yeast in the form of a bioreactor made possible the beer production as a continuous process [107, 108]. A significant contribution to amino acid synthesis can be made by the fermentative production of methionine by the DNA recombinant technology, introducing such genes for large scale production similar to what genetic engineers have done to construct *S. cerevisiae* to produce 'thaumatin', a sweet taste plant protein. Companies which were producing enzymes since the early 1980s, now started using genetic engineering techniques to improve production efficiency and to develop new products. Companies can tailor their enzymes more precisely according to customer demands for products with specific properties. So clear advantages are here for both industry and customers with major improvements in enzyme production, giving better products and processes. Because of the importance of fermented foods to so many cultures, scientists are conducting a lot of work to improve the microorganisms that carry out food fermentation. The bacterium responsible for many of our fermented dairy products, such as cheese and yoghurt, is susceptible to infection by a virus that causes substantial economic losses to the food industry. Through recombinant technology, researchers have made strains of this bacterium and other important fermenters resistant to viral infection. It is known that some bacteria used in food fermentation produce compounds that have the ability to kill other contaminating bacteria, which cause food poisoning and spoilage. We are equipping many of our microbial fermenters with self-defense mechanism to decrease microbial contamination of fermented foods.

Several microbial flavour production processes have been developed, but the number of applications is limited. Due to very low concentration of the broth, the costs of microbial flavours are quite high. Recovery procedure especially absorbent could make the downstream processing cost effective. Currently, there is specific attention on the use of microbial biocatalysis to create new roots to reduce the cost of production. Biocatalysis constitute a greener alternative to traditional organic synthesis that offers appropriate tools for the industrial transformation of natural or synthetic materials. Another obstacle is the legal regulations, where new biotechnological product needs evaluation before it is awarded the status of natural for use in the food industry [20].

Innovative approaches in food fermentations, especially the production of biopreservative, have resulted in the discovery of an enzyme chitinase with antimicrobial activity from germinating soyabean seeds. Another emerging and exciting area is the enzyme cocktail consisting of lactoperoxidase, glucose oxidase and lysozyme, as an affective antimicrobial food system.

Regarding the non-commercialization of technology for the production of fat despite considerable research, the simple apparent reason is the inability of biotechnologically produced fat to compete against very low cost of agricultural seed oils [31]. However, the production of PUFA (gamma linolenic) is being undertaken commercially in both United Kingdom and Japan, mainly because of higher cost of production from agricultural sources than microbial sources.

Some of the crops, particularly utilized by the developed countries like cassava could be greatly benefited by the application of food fermentation principles, but such countries are unfortunately unable to either support the research programme or implement the strategies already developed [21]. Another problem that hinders the benefit of development in biotechnology to be fully exploited by these countries is the poor mechanisms of promoting and facilitating technology transfer coupled with existing economic, legal and social barriers preventing university industry cooperation [109]. Even simple and easy to use techniques could not be accomplished in the places where fermented foods are produced as a routine matter due to several reasons such as expensive equipment, their maintenance, and inability to adopt the technology by the villagers, and more so the lack of interaction between the research institutes and the users.

The technology to produce fermented foods must be economically effective, but should also preserve the wealth of generation of world's largest industry. Though economical parameters would continue to strive for more profit, the basic fact to keep in view is that the food is eaten and must be wholesome though in conformity with changing pattern of society. The food must, therefore, satisfy both the biological requirements as well as providing palatability to it.

Presently, R&D is head towards high functionality and high technology segmentation. Functional and nutritional peptides belong to this category of products. In the recent years, the market growth rate for peptide products has grown swiftly. Appropriate incentives to the food industry would greatly enhance the development of functional foods. Food companies have traditionally funded research for new food product formulations, but regarding functional foods the stakes are higher, both for the food companies as well as the consumers. The research required for a functional food to meet scientific standards for efficacy is a substantial investment, but the returns on the investment is not exclusively to that company. As soon as the health claim is adequately documented, competing companies can use the claim. Incentives such as period of exclusively or tax incentives would encourage food companies to pursue functional food development as a profitable venture.

GM foods will certainly become popular when consumers see more of their direct benefits. Worldwide research is on to exploit rDNA technology to achieve numerous specific benefits like more and better food, process benefits, economic environmental benefits, diet and nutrition benefits, and also health and medical benefits. At present, GM foods are facing more legal problems than technical problems, and there is a need for change in mind set in using biotech foods. FAO also recommended great potential of GM foods and emphasized the need for a thorough check to develop public confidence. The potential allergenecity of GM foods can be assessed using the decision-free approach of the International Food Biotechnology Council of the International Life Sciences Institute, now adopted [110] and refined by biotech industry.

Functional genomics is currently the most effective approach for increasing the knowledge at the molecular level of metabolic and adaptive processes in whole cells. The implementation of functional genomics on food microorganisms will allow us to reach various industrial goals within the coming decade. Fast developments are taking place in bioinformatics which provide information on the status of cellular processes on the possible functions of genes without a known functionality.

## 12. FUTURE PERSPECTIVES

The future should see improvement in the existing knowledge of indigenous microbial strains relevant to food production, processing and preservation, efforts to understand the biochemical pathways of synthesis of metabolite or degradation and improvement in the fermentation processes including recovery of natural flavours, colours or such compounds of high added value. Optimizing of existing fermentation processes would include activities in solid state fermentation, as these processes are economical at low scale of operation due to higher production and do not require the large product demand, infrastructure and expertise necessary for submerged fermentation [111]. Moreover, solid state fermentation has been used to produce the traditional fermented foods of *Orient*.

Availability of microbial fat, microbial pigments, flavours or additives has potential to alleviate the short supply of these important agricultural commodities. With the advances made in the field of enzymology, bioreactor technology and genetic engineering, it would be elegant and novel approach to use the cellulosic waste after hydrolysis or other similar waste. The growing consumerism for the natural products would further accelerate the possibility of production of such compounds in future by fermentation.

In spite of all, the ability of the microorganisms to produce substances having applications in food industry has not been fully exploited. It is true both with the microorganisms in nature itself as well as genetically modified microorganisms. Because of large biodiversity, they remain still unexploited. The developments in genetic engineering have made food fermentation more attractive, whose application can be through two strategies — either to modify the existing catabolic pathway enabling utilization of newer substrates, or design a new pathway and look for the microbe either in nature or by manipulated DNA of existing one to carry out the fermentation.

It is well established that the traditional food fermentation technology is on sound footing. The applications of recently acquired knowledge in the genetic engineering DNA technology, biochemistry, new microorganisms and their metabolic capabilities, biosensor development, computer designing and monitoring of fermentation processing, bioreactor technology nutritional significance and role of fermented foods would accelerate its growth to provide food to millions of people.

Research is beginning to provide clear evidence of relationship between dietary components and health benefits, and consumers have enthusiastically embraced the evolving role of food in health. These consumers create a demand for food products with enhanced health benefits. Functional foods have a place in the health care continuum, providing low-cost approaches to disease prevention that give consumers a significant role in maintaining good health. Nutrigenomics, proteomics and metabolomics are three new disciplines that will contribute to the rapid development of functional foods. Bioinformatics is a new tool that uses computer database technology from multiple, and sometimes, disparate disciplines. These new disciplines and tools have improved a lot of understanding of food biotechnology. The recent burst in genome sequences of food-related microorganisms has opened a new era of functional genomics in food biotechnology.

Earlier, the improvement in quality of meat products was largely dependent on postharvest technologies. Biotechnology and alternative production methods appear to hold great potential for improving the quality and yield attributes of animal products. Food biotechnology has a great potential to feed the hungry world and to improve the environment at the same time. All the issues related to GM foods like safety, environmental concerns, ethics and information, etc. should be satisfactorily addressed before their launching in the market.

## REFERENCES


1. Angold R, Beech G & Taggart J (1989) *Food Biotechnology*, Cambridge University Press, Cambridge, UK, p 1.
2. Trevan MD, Boffey S, Goulding KH & Stantbrossy P (1987) What is biotechnology? In: *Biotechnology - The Biological Principles,* Trevan MD, Boffey S, Goulding KH & Stantbrossy P (eds), Open University Press, Milton Keynes, UK, p 1.
3. CBD (2000) *Cartagena Protocol on Biosafety to the Convention on Biological Diversity,* United Nations Environment Programme, January 2000, p 24.
4. IFST (2004) *IFST Current Hot Topics: Genetic Modification and Food,* Institute of Food Science and Technology, UK.
5. Pederson CS (1971) *Microbiology of Food Fermentation*, AVI Publ Co, Inc, Westport, USA
6. Lim G (1991) Indigenous fermented foods in South East Asia. *ASEAN Food J* 6: 83-101.
7. IFT (2000) IFT Expert Report on Biotechnology and Foods (4 parts). *Food Technol* 54(8): 124-136; 54(9): 53-61; 54(9): 62-74; 54(10): 61-80.
8. TIFS (2003) Genetically modified foods for human health and nutrition: The scientific basis for benefit/risk assessment. *Trends Food Sci Technol* 14: 169-338.
9. Murthy TRK (2004) Biotechnological approaches for enhancing meat quality. *Indian Food Ind* 23: 24-26.
10. IUFoST (2005) Biotechnology and food. Scientific Information Bulletin of International Union of Food Science & Technology, February 2005, UK.
11. Ebbertt J (1998) The impacts of biotechnology on the grain processing industry. *AgBioForum* 1: 78-80.

12. Jagadish MN & Manjunath TM (2004) Impact of genomics and bioinformatics on crop and food biotechnology. *Indian Food Ind* 23: 30-32.
13. Knorr D (1987) Biotechnological processes in food production. *Food Biotechnol* 1: 225-248.
14. Deshusses MA, Chen W, Mulchandani A & Dunn IJ (1997) Innovative bioreactors. *Curr Opin Biotechnol* 8: 165-168.
15. Singh RS, Dhaliwal R & Puri M (2008) Development of a stable continuous flow immobilized enzyme reactor for the hydrolysis of inulin. *J Ind Microbiol Biotechnol* 35: 777-782.
16. Singh RS, Sooch BS & Attri D (2010) Bioreactor technology in wine production. In: *Handbook of Enology: Principles, Practices and Recent Innovations*, Vol 2, Joshi VK (ed), Asiatech Publishers Co. Inc., New Delhi, India, pp 802-860.
17. Singh RS & Sooch BS (2009) High cell density reactors in production of fruit wines with special reference to cider - An overview. *Nat Prod Rad* 8: 323-333.
18. Potter NN (1978) *Food Science*, The AVI Publ Co, Inc, Westport, USA, p 354.
19. Crumplen R, D'Amore T, Panchal CJ, Russell I & Stewart GG (1989) Industrial uses of yeast: Present and future. *Yeast* 5: 3-9.
20. Janssen L, DePotter HL, Schamp NM & Vandamme (1992) Production of flavours by microorganisms. *Process Biochem* 27: 195-215.
21. Bokanga M (1995) Biotechnology and cassava processing in Africa. *Food Technol* 49: 86-90.
22. Bull TA, Goodfellow M & Slater JH (1992) Biodiversity as a source of innovation in biotechnology. *Ann Rev Microbiol* 46: 219-252.
23. Joshi VK & Bhutani VP (1995) Peach. In: *Handbook of Fruit Science and Technology - Cultivation, Storage and Processing*, Salunkhe DK & Kadam SS (eds), Marcel Dekker, New York, USA, p 243.
24. Bhutani VP & Joshi VK (1995) Plum. In: *Handbook of Fruit Science and Technology– Cultivation, Storage and Processing*, Salunkhe DK & Kadam SS (eds), Marcel Dekker, New York, USA.
25. Westby A (1994) Importance of fermentation in cassava processing. In: *Proceedings of the 9$^{th}$ Symposium of the International Society for Tropical Root Crops,* October 20-26, Accra, Ghana.
26. Amerine MA, Berg HW, Kunkee RE, Ough CS, Singleton VL & Webb AD (1980) *The Technology of Wine Making*, AVI Publishing Co, Westport, USA.
27. Joshi VK (1997) *Fruit Wines*, Directorate Extension Education, Dr YS Parmar University of Horticulture & Forestry, Nauni, Solan, India.
28. Roumbouts FM & Nout MRJ (1995) Microbial fermentation in the production of plant foods. *J Appl Bacteriol Symp Suppl* 79: 108.
29. Pandey A (1994) *Solid State Fermentation*, New Age International (P) Limited, Publishers, New Delhi, India.
30. Sciancalepore V (1989) Extraction of proteins from wine lees. *Am J Enol Vitic* 40: 65-66.
31. Ratledge C (1991) Microorganisms for lipids. *Acta Biotechnol* 11: 429-437.
32. Joshi VK, Attri D, Bala A & Bhushan S (2003) Microbial pigments. *Indian J Biotechnol* 2: 362-369.
33. Dufosse L, Galaup P, Yaron A, Arad SM, Blanc P, Murthy KNC & Ravishankar GA (2005) Microorganisms and microalgae as sources of pigments for food use: A scientific oddity or an industrial reality. *Trends Food Sci Technol* 16: 389-406.

34. Dufosse L (2006) Microbial production of food grade pigments. *Food Technol Biotechnol* 44: 313-321.
35. Bigelis R (1992) Flavour metabolite and enzymes from filamentous fungi. *Food Technol* 46: 151-161.
36. Singh RS, Dhaliwal R & Puri M (2007) Production of high fructose syrup from *Asparagus* inulin using immobilized exoinulinase from *Kluyveromyces marxianus* YS-1. *J Ind Microbiol Biotechnol* 34: 649-655.
37. Singh RS, Dhaliwal R & Puri M (2007) Partial purification and characterization of exoinulinase from *Kluyveromyces marxianus* YS-1 for the preparation of high fructose syrup. *J Microbiol Biotechnol* 17: 733-738.
38. Ricca E, Calabro V, Curcio S & Lorio G (2007) State of the art in the production of fructose from inulin enzymatic hydrolysis. *Crit Rev Biotechnol* 27: 129-145.
39. Singh RS, Saini GK & Kennedy JF (2010) Maltotriose syrup preparation from pullulan using pullulanase. *Carb Poly* 80: 402-408.
40. Singh RS, Saini GK & Kennedy JF (2010) Covalent immobilization and thermodynamic characterization of pullulanase for the hydrolysis of pullulan in batch system. *Carb Poly* 81: 252–259.
41. Singh RS, Saini GK & Kennedy JF (2008) Pullulan: Microbial sources, production and applications. *Carb Poly* 73: 515-531.
42. Joshi VK & Pandey A (1999) Biotechnology: Food fermentation. In: *Biotechnology: Food Fermentation, Microbiology, Biochemistry and Technology,* Joshi VK & Pandey A (eds), Educational Publishers & Distributors, New Delhi, India, pp 1-24.
43. Soni SK & Sandhu DK (1999) Microbiology of food fermentation. In: *Biotechnology: Food Fermentation, Microbiology, Biochemistry and Technology,* Joshi VK & Pandey A (eds), Educational Publishers & Distributors, New Delhi, India, pp 25-86.
44. Crittenden RG & Playne MJ (1996) Production, properties and applications of food-grade oligosaccharides. *Trends Food Sci Technol* 7: 353-361.
45. Meyer D & Stasse-Wolthuis M (2009) The bifidogenic effect of inulin and oligofructose and its consequences for gut health. *Eur J Clin Nutr* 3: 1277-1289.
46. Rastall RA (2010) Functional oligosaccharides: Applications and manufacture. *Ann Food Sci Technol* 1: 305-339.
47. Singh RS & Singh RP (2010) Production of fructooligosaccharides from inulin by endoinulinase and their prebiotic potential. *Food Technol Biotechnol* 48: 435-450.
48. Motarjemi Y & Nout MJR (1996) Food fermentation: a safety and nutritional assessment. *Bull World Health Org* 74: 553-559.
49. Gasteineau FC, Darby JW & Turner TB (1979) *Fermented Food Beverages in Nutrition*, Academic Press, London, UK.
50. Basappa SC, Someshekar D, Agarwal R, Suma K & Bharathi K (1997) Nutritional composition of fermented ragi (*Chhang*) by phals and defined starter cultures as compared to unfermented ragi (*Eleucine coracana* G.). *Int J Food Sci Nutr* 48: 313-319.
51. Tamang JP (2009) *Himalayan Fermented Foods: Microbiology, Nutrition and Ethnic Values*, CRC Press, Boca Raton, USA, pp 247-248.
52. Muller CJ (1995) Wine and health. Is it more than alcohol. In: *Wine Analysis and Production*, Zocklein BW, Fungelsang KC, Gump BH & Nury FS (eds), Chapman and Hall, London, UK, p 3.
53. Delin CR & Lee TH (1991) The J shaped curve revisited: Wine and cardiovascular health update. *Aus New Zealand Wine Ind J* 6: 15-16.

54. Aertsen A, Meersman F, Hendrickx MEG, Vogel RF & Michiels CW (2009) Biotechnology under high pressure: applications and implications. *Trends Biotechnol* 27: 434-441.
55. Chi Z, Chi Z, Liu G, Wang F, Ju L & Zhang T (2009) *Saccharomycopsis fibuligera* and its applications in biotechnology. *Biotechnol Adv* 27: 423-431.
56. Schoemaker HE, Mink D & Wubbolts MG (2003) Dispelling the myths-biocatalysis in industrial synthesis. *Science* 299: 1694-1697.
57. Patel RN, Chu L & Mueller R (2003) Diastereoselective microbial reduction of (S)-[3-chloro-2-oxo-1-(phenylmethyl)propyl]carbamic acid, -1,1-dimethylethyl ester. *Tetrahed Asym* 14: 3105-3109.
58. Armor JN (1999) Striving for catalytically green processes in the 21st century. *Appl Cat A: Gen* 189: 153-162.
59. Sheldon RA & van Rantwijk F (2004) Biocatalysis for sustainable organic synthesis. *Aus J Chem* 57: 281-289.
60. Bojarová P & Kren V (2009) Glycosidases: a key to tailored carbohydrates. *Trends Biotechnol* 27: 199-209.
61. Gubitz G & Schmid MG (2004) Chiral separation principles. In: *Chiral Separations: Methods and Protocols*, Gubitz G & Schmid MG (eds), Humana Press Inc, New Jersey, USA, pp 1-28.
62. Bode HB & Muller R (2005) The impact of bacterial genomics on natural product research. *Angew Chem Int Edit* 44: 6828-6846.
63. Hagedorn S & Kaphamer B (1994) Microbial biocatalysis in the generation of flavour and fragrance chemicals. *Ann Rev Microbiol* 48: 773-800.
64. Zheng G, Yuan Q, Yang L, Zhang X, Wang J & Sun W (2006) Preparation of (2S, 3R)-methyl-3-phenylglycidate using whole cells of *Pseudomonas putida*. *J Mol Cat B: Enz* 43: 133-136.
65. Cao L (2005) Immobilised enzymes: Science or art? *Curr Opin Chem Biol* 9: 217-226.
66. Rao NM (2004) Genetically modified foods (GMF): An overview. *Indian Food Ind* 23: 14-16.
67. James C (2004) *Global status of commercialized biotech/GM crops: 2004.* ISAAA Briefs 32. International Service for the Acquisition of Agri-biotech Applications, New York, USA.
68. van Beuzekom B & Arundel A (2009) OECD biotechnology statistics 2009, Organization for Economic Co-operation and Development, Paris, France.
69. Solomon MB (2001) Meat biotechnology. In: *Meat Science and Technology*, Hui YH, Hui YH & Nip W-K (eds), Marcel Dekker, Inc, New York, USA, pp 127-148.
70. Goldberg I (1994) *Functional Foods, Designer Foods, Pharma Foods, Nutraceuticals*, Chapman & Hall, London, UK.
71. Diplock AT, Aggett PJ, Aswell M, Bornet F, Fern EB & Roberfroid MB (1999) Scientific concepts of functional foods in Europe: Consensus document. *Brit J Nutr*: Supplement 1.
72. Mestre Prates JA & Mateus CMRP (2002) Functional foods from animal sources and their physiological active components. *Revue Med Vet* 153: 155-160.
73. Sangeetha PT, Ramesh MN & Prapulla SG (2005) Recent trends in microbial production, analysis and application of fructooligosaccharides. *Trends Food Sci Technol* 16: 442-457.
74. Jeevaratnam K & Bawa AS (2004) Functional foods through biotechnological approach. *Indian Food Ind* 23: 21-24.
75. Lewis NM, Seburg S & Flanagan NL (2000) Enriched eggs as a source of n-3 polyunsaturated fatty acids for humans. *Poult Sci* 79: 971-974.

76. FAO/WHO (2001) Principles for the establishment and application of microbiological criteria for foods. General requirements. CAC/GL21-1997. Joint FAO/WHO Food Standards Programme, Codex Alimentarius Commission, Food and Agricultural Organization of the United Nations/World Health Organization, Rome, Italy.
77. Kessler DA, Taylor MR, Maryanski JH, Flamn EL & Kahl LS (1992) The safety of food developed by biotechnology. *Science* 256: 1747-1749.
78. FAO/WHO (2003) Principles of risk analysis of foods derived from modern biotechnology. CAC/GL 44-2003, Rome, Italy.
79. FAO/WHO (2000) Safety aspects of genetically modified foods of plant origin. Report on a joint FAO/WHO expert consultation on "Foods Derived from Biotechnology", 29 May-2 June 2000, Geneva, Switzerland.
80. Hefle SL, Nordlee JA & Taylor SL (1996) Allergenic foods. *Crit Rev Food Sci Nutr* 36: S69-S89.
81. Metcalfe DD, Astwood JD, Townsend R, Sampson HA, Taylor SL & Fuchs RL (1996) Assessment of the allergenic potential of foods derived from genetically engineered crop plants. *Crit Rev Food Sci Nutr* 36: S165-S186.
82. Hoban, TJ (2004) Public attitudes towards agricultural biotechnology. ESA Working Paper No. 04-09. FAO, Rome, Italy.
83. Nuffield Council on Bioethics (1999) Report on "genetically modified crops: the ethical and social issues". http://www.nuffieldbioethics.org/go/ourwork/gmcrops/introduction
84. Nuffield Council on Bioethics (2003) The use of genetically modified crops in developing countries. http://www.nuffieldbioethics.org/go/ourwork/gmcrops/publication_313.html
85. IUFoST (2010) Biotechnology and food. Scientific Information Bulletin of International Union of Food Science & Technology, March, 2010, UK.
86. FAO/WHO (2003) Guideline for the conduct of food safety assessment of foods derived from recombinant-DNA plants. CAC/GL 45-2003, Rome, Italy.
87. FAO/WHO (2003) Guideline for the conduct of food safety assessment of foods produced using recombinant-DNA microorganisms. CAC/GL 46-2003, Rome, Italy.
88. Kuipers OP (1999) Genomics for food biotechnology: Prospects of the use of high-throughput technologies for the improvement of food microorganisms. *Curr Opin Biotechnol* 10: 511-516.
89. Schena M, Heller RA, Theriault TP, Konrad K, Lachenmeier E & Davis RW (1998) Microarrays: Biotechnology's discovery platform for functional genomics. *Trends Biotechnol* 16: 301-306.
90. Ekins RE & Chu FW (1999) Microarrays: Their origins and applications. *Trends Biotechnol* 17: 217 218.
91. Blackstock WP & Weir MP (1999) Proteomics: Quantitative and physical mapping of cellular proteins. *Trends Biotechnol* 17: 121-127.
92. Dove A (1999) Proteomics: Translating genomics into products? *Nat Biotechnol* 17: 233-236.
93. Khalak H & Francesco VD (1999) Analysing biological molecules: Onward to function. *Trends Biotechnol* 17: 262-264.
94. Burks C (1999) Molecular biology database list. *Nucleic Acids Res* 27: 1-9.
95. Tomita M, Hashimoto K, Takahashi K, Shimizu T, Matsuzaki Y, Miyoshi F, Saito K, Tanida S, Yugi K, Venter JC & Hutchison C (1999) E-CELL: Software environment for whole cell simulation. *Bioinformatics* 15: 72-84.
96. Schaff J & Loew LM (1999) The virtual cell. *Pac Symp Biocomput* 4: 228-239.

97. Bailey JE (1999) Lessons from metabolic engineering for functional genomics and drug discovery. *Nat Biotechnol* 17: 616-618.
98. Carr JG (1983) Microbes I have known: A study of those associated with fermented products. *J Appl Bacteriol* 55: 383-402.
99. Coughlan MP (1985) The properties of fungal and bacterial cellulases with comment on their production and application. In: *Biotechnology and Genetic Engineering Reviews*, Vol 3, Russel GE (ed), Intercept Ltd, Newcastle upon Tyne, UK, p 39.
100. Joshi VK, Pandey A & Sandhu A (1999) Fermentation technology for food industry waste utilization. In: *Biotechnology: Food Fermentation, Microbiology, Biochemistry and Technology*, Joshi VK & Pandey A (eds), Educational Publishers & Distributors, New Delhi, India, pp 1291-1348.
101. Gupta K & Joshi VK (2000) Fermentative utilization of waste from food processing industry. In: *Post Harvest Biotechnology of Fruits and Vegetables*, Verma LR & Joshi VK (eds), Indus Publishing Company, New Delhi, India, pp 1171-1193.
102. Vendruscolo F, Albuquerque PM, Streit F, Esposito E & Ninow JL (2008) Apple pomace: A versatile substrate for biotechnological applications. *Crit Rev Biotechnol* 28: 1-12.
103. Bhushan S, Kalia K, Sharma M, Singh B & Ahuja PS (2008) Processing of apple pomace for bioactive molecules. *Crit Rev Biotechnol* 28: 285-296.
104. Houwink EH (1984) *A Realistic View on Biotechnology*, Decherna, Frankfort, USA, (Published on behalf of the European Federation of Biotechnology).
105. Mansfield E, Romeo A, Schwertz M, Teeca D, Wagner S & Brach P (1982) *Technology Transfer, Productivity and Economic Policy*, W.W. Norton and Co, New York, USA.
106. Newell N (1986) Large US food firms will take a wait and see approach to application of biotechnology. *Genet Eng News* 6: 4-5.
107. Fox D, Andrade M, Depalo M, Harrison S, Hill N, Kenett A, McKenzie J, Maiorana M, Marsh S & Annuk D (1990) Arnott's continuous fermentation process. *Food Australia* 42: 46-48.
108. Lommi H, Gronqvist A & Paujunen E (1990) Immobilized yeast reactor speeds beer production. *Food Technol* 44: 128-133.
109. Knorr D (1995) Improving food biotechnology resources and strategies in developing countries. *Food Technol* 49: 91-93.
110. FAO/WHO (2009) Annual Report on "The State of Food Insecurity in the World: Economic Crises – Impacts and Lessons Learned", Food and Agriculture Organization, USA.
111. Krishna C (2005) Solid-state fermentation systems — an overview. *Crit Rev Biotechnol* 25: 1-30.

# Biochemical Engineering in Food Biotechnology

# 2

**M.B. Bera**

*Department of Food Engineering and Technology, Sant Longowal Institute of Engineering & Technology, Longowal 148 106 (Pb.)*

## 1. INTRODUCTION

Biochemical engineering has developed as a branch of chemical engineering in which one applies principles of chemical process system on biological system. The differences between biochemical engineering and chemical engineering lie not in the principles of unit operations and unit processes, but in the nature of living systems [1]. When one deals with food biotechnology i.e., application of biotechnology in food processing and technology, it requires theoretical understanding of the biochemical engineering as a subject. The subject matter in this chapter deals with the basics of biochemical engineering starting with the unit's dimensions and their application in studying the homogeneity of equations. The later section involves the understanding of mass transfer and agitation. Material balance with its relevance in quantitative analysis of system provides an approach in prediction as well as determination of metabolites produced. Finally the bioreactors in which all these biochemical reactions take place are discussed with their classifications and theoretical consideration.

## 2. UNITS AND DIMENSIONS

All quantities in terms of which the laws of physics can be expressed and which can be measured directly or indirectly are known as physical quantities [2, 3]. Dimension is a physical entity, which can be recognized by our sensory perception. It is measured to define the concepts of length, time and mass, etc. For example, the concept of time and length is easy to visualize but difficult to define. Unit in turn quantifies the magnitude, amount or the size of dimension under consideration. Some common units used in SI system are given in Table 1. Units are further classified into two types: (i) primary units, which are used to express the magnitude of single dimension, mentioned by international system (SI) of units and (ii) secondary units derived from primary units [5]. Primary units are also known as base units, while Secondary units are also known as derived units (Table 2).

### 2.1 Dimensional Analysis and its Applications

It is a method of algebraic treatment of the symbols for units considered independently of magnitude. It is useful in checking the consistency of the units in equations, converting units, and in the scale-up

*Food Biotechnology: Principles and Practices*
*V.K. Joshi & R.S. Singh (Eds.)*
*I.K. International Publishing House Pvt. Ltd., India*
*pp 39–70*

**Table 1:** SI system (Systeme International d' Unites).

| Quantity | Symbol* | Name |
|---|---|---|
| Amount of substance | mol | Mole |
| Electric current | A | Ampere |
| Length | m | meter |
| Mass | kg | Kilogram |
| Luminous | cd | Candela |
| Temperature | K | Kelvin |
| Time | s | Second |
| Plane angle | rad | Radian |
| Solid angle | sr | Steradian |

*Symbols are written in lowercase letters unless they are from the name of the person.

*Source:* 4

**Table 2:** Derived SI units.

| Quantity | Unit | Symbol |
|---|---|---|
| Angular speed | radian per second | rad/s |
| Area | meter squared | $m^2$ |
| Density | kilogram per cubic meter | $kg/m^3$ |
| Diffusivity | meter square per second | $m^2/s$ |
| Dynamic viscosity | Pascal second | Pa.s |
| Energy, work, heat quantity | Newton meter | J |
| Force | kilogram meter per second squared | N |
| Kinematic viscosity | meter square per second | $m^2/s$ |
| Linear acceleration | meter per second squared | $m/s^2$ |
| Linear speed | meter per second | m/s |
| Momentum | kilogram meter per second | Kg.m/s |
| Pressure, stress | Newton per meter squared | Pa |
| Specific heat capacity | Joule per kilogram per Kelvin | J/kg.K |
| Thermal conductivity | Watt per square meter Kelvin | $W/m^2K$ |
| Volume | cubic meter | $m^3$ |

*Source:* 4

of data obtained in model test units to predict the performance of full scale equipment. Conversion of physical quantities from one system of units is possible through conversion factors that can be obtained based on Fourier's law of homogeneity using dimensional analysis [6]. A simple example to obtain conversion factor is discussed in the following section

## 2.2 Conversion Factor from Dimensional Analysis

It can be used to obtain conversion factor for conversion of units from one system to another, when the value of physical quantity in one system is given. Mathematically, the above statements can be written in a simple expression to obtain the desired conversion factor as follows:

Let the measure of a physical quantity given by

$$X = NU \tag{1}$$

where, $U$ is the size of the unit and $N$ is the numerical value of the physical quantity for the unit chosen. If $U_1$ and $U_2$ are the units for measurement of the physical quantity in two systems and $N_1$ and $N_2$ are the numerical values for the two units, then:

$$N_1U_1 = N_2U_2 \tag{2}$$

If the dimension of the physical quantity in terms of mass, length, time, and temperature is given as $M^aL^bT^c\theta^d$, then Eq. (2) can be written as

$$N_1\ [M_1{}^aL_1{}^bT_1{}^c\theta_1{}^d] = N_2\ [M_2{}^aL_2{}^bT_2{}^c\theta_2{}^d] \tag{3}$$

where, $M_1$, $L_1$, $T_1$ and $\theta_1$ and $M_2$, $L_2$, $T_2$, and $\theta_2$ are the units of mass, length, time and temperature in the two systems. Therefore, the conversion factor for converting one system of units to another can be obtained by rearranging Eq. (3)

$$\text{Conversion factor} = \frac{N_2}{N_1} = \frac{\left[M_1{}^aL_1{}^bT_1{}^c\theta_1{}^d\right]}{\left[M_2{}^aL_2{}^bT_2{}^c\theta_2{}^d\right]} \tag{4}$$

Equation (4) can be used to find out the value of the conversion factor to convert physical quantity from one system to another [7].

**Example 1:**
Universal gas constant '*R*' has an experimental value of 8.31447J/K.mol. Obtain the value of *R* in ergs/K.mol.

***Solution:***
Using Eq. (4), one can write

$$\text{Conversion factor} = \frac{N_2}{N_1} = \frac{\left[M_1{}^aL_1{}^bT_1{}^c\theta_1{}^d\right]}{\left[M_2{}^aL_2{}^bT_2{}^c\theta_2{}^d\right]} \tag{5}$$

By comparing the units of the gas constant '*R*' in different systems, one can observe that all other units are similar except units of energy, which are respectively Joule and erg. It is known that both erg and Joule are units of energy in centimeter gram second (CGS) system and SI system respectively, with the following dimensions:

$$\text{Energy} = [M_1L_1{}^2T_1{}^{-2}\theta_1{}^0] \tag{6}$$

Substituting the value of dimension from Eq. (6) in Eq. (5)

$$\text{Conversion factor} = \frac{N_2}{N_1} = \frac{\left[M_1L_1{}^2T_1{}^{-2}\theta_1{}^0\right]}{\left[M_2L_2{}^2T_2{}^{-2}\theta_2{}^0\right]} \tag{7}$$

Now substituting the values of units of mass, length, time, etc in CGS system as gram, centimeter and second, and SI system as kilogram, meter and second, respectively, one can obtain the conversion factor to obtain the value of *R* in terms of erg.

$$\text{Conversion factor} = \frac{N_2}{N_1} = \frac{\left[(\text{kg})(\text{m})^2(s)^{-2}\right]}{\left[(\text{g})(\text{cm})^2(s)^{-2}\right]} \tag{8}$$

$$\text{or Conversion factor} = \frac{N_2}{N_1} = \frac{\left[(1000\ \text{g})(100\ \text{cm})^2(s)^{-2}\right]}{\left[(\text{g})(\text{cm})^2(s)^{-2}\right]}$$

Cancelling the units in numerator and denominator

$$\text{or Conversion factor } = \frac{N_2}{N_1} = \left[1000\,(100)^2\right]$$

$$\text{or Conversion factor } = \frac{N_2}{N_1} = 1 \times 10^7$$

Therefore value of $R$ = Conversion factor x 8.31447 ergs/K.mol

$$\text{or } R = 1 \times 10^7 \times 8.31447 \text{ ergs/K.mol}$$

$$\text{or } R = 8.31447 \times 10^7 \text{ ergs/K.mol}$$

## 2.3 Checking Consistency of Equations

If a theoretical equation exists among the variables affecting a physical process, that equation must be dimensionally homogeneous [8]. Simple steps required in checking the consistency of equation are:

1. Tabulate the dimensions of the quantities on left and right side of the given equation.
2. Add and subtract the power coefficients of the dimensions on either side of the equations.
3. Check whether the power coefficients of the dimensions on both sides of the equations are same or not?
4. If power coefficients on both sides of equations are same, then the given equation is dimensionally consistent.

One can have a better idea about the dimensional homogeneity of an equation system by the following example.

**Example 2:**
The pressure ($P$) inside a bioreactor due to the liquid column of height ($h$) having a density ($d$) is given by $P = hdg$, where '$g$' is acceleration due to gravity. Check whether the equation is dimensionally consistent or not?

***Solution:***
Here the physical quantities involved are pressure ($P$), height ($h$), density ($d$) and acceleration due to gravity ($g$). The dimensions of these quantities $P$, $h$, $d$, and g are respectively $[ML^{-1}S^{-2}]$, $[L]$, $[ML^{-3}]$, $[LS^{-2}]$. Tabulating the dimensions of the quantities on both side of the equation:

| **Left hand side** | | **Right hand side** |
|---|---|---|
| P | = | hdg |
| $[ML^{-1}S^{-2}]$ | = | $[L]\,[ML^{-3}]\,[LS^{-2}]$ |

or rearranging the above, one can write

| | | |
|---|---|---|
| $[ML^{-1}S^{-2}]$ | = | $[M]\,[L\,L\,L^{-3}]\,[S^{-2}]$ |
| or $[ML^{-1}S^{-2}]$ | = | $[M]\,[L^1\,L^1L^{-3}]\,[S^{-2}]$ |

Now adding and subtracting the coefficients of similar dimensions one gets

| | | |
|---|---|---|
| $[ML^{-1}S^{-2}]$ | = | $[M][L^{1+1-3}]\,[S^{-2}]$ |
| or $[ML^{-1}S^{-2}]$ | = | $[M][L^{-1}]\,[S^{-2}]$ |
| or $[ML^{-1}S^{-2}]$ | = | $[ML^{-1}S^{-2}]$ |

Now comparing both sides of the equation, one can observe that coefficients of dimensions $M$, $L$ and $S$ are respectively 1, -1 and -2 on both sides of the equation. Therefore, the given equation is dimensionally consistent.

## 3. DIMENSIONLESS GROUPS AND THEIR APPLICATIONS

The study of problems in fluid dynamics and heat and mass transfer is made difficult by many parameters which appear to affect them. By grouping the variables into dimensionless groups, it is possible to reduce the effective number of variables. Thus the experimenter can greatly reduce the number of observations to be made by varying the dimensionless groups, rather than observing the effect of varying each dimensional property [9, 10]. Various dimensionless groups commonly known as dimensionless number used in flow ($N_{Re}$, $N_Q$), mixing and agitation ($N_{Fr}$, $N_P$), heat ($N_{Pr}$, $N_{Nu}$), and mass transfer ($N_{Sc}$, $N_{Sh}$) are described in the following section.

### 3.1 Reynolds Number

Osbrone Reynolds studied the conditions under which one type of flow changes into the other, and in 1883 he found that various factors, which affect the critical velocity of flow are: diameter of tube, viscosity, density and average linear velocity of the fluid. He combined these factors into a dimensionless group known as Reynolds number, such that changes in the kind of flow occur at a particular value of the number.

$$N_{Re} = \frac{DV\rho}{\mu} \tag{9}$$

where, $D$ is the diameter of tube or pipe; $V$ is the average linear velocity of flow; $\mu$ is the viscosity of liquid; and $\rho$ is the density of liquid. The above equation for Reynolds number is true only for straight circular pipes. Reynolds number describes the ratio of inertial forces to viscous forces (or convective momentum transport to molecular momentum transport) in a flowing fluid. Therefore it serves to indicate the degree of turbulence [4].

### 3.2 Flow Number

It is the ratio of volumetric flow rate to the total flow leaving the impeller, as measured at the tips of the blades.

$$N_Q = \frac{q}{n(D_a)^3} \tag{10}$$

where, $D_a$ is the diameter of the impeller; $n$ is the number of rotation; and $q$ is the volumetric flow rate ($m^3/s$). The values of flow number ($N_Q$) are important in design consideration and may vary from 0.5 to 1.3 for baffled agitated vessels.

### 3.3 Froude Number

It is the ratio of inertia force to gravitational force per unit area acting on the fluid, and is important when there is a significant wave motion on a liquid surface (i.e., vortex formation occurs). It is not important when baffles are used or $N_{Re} < 300$. It can be obtained as:

$$N_{Fr} = \frac{u^2}{gl} \tag{11}$$

where, $u$ is the linear velocity of the stirrer; $g$ is acceleration due to gravity; and $l$ is the characteristic length.

### 3.4 Power Number

The power number, also known as Newton number is analogous to friction factor or drag coefficient. It is proportional to the ratio of the drag force acting on a unit area of the impeller and the inertial stress. The power number has different specifications according to the field of application. For stirrers, the power number is defined as:

$$N_p = \frac{P}{\rho n^3 d^5} \tag{12}$$

where, $P$ is the impeller power input; $\rho$ is the fluid density; $n$ is the rotational speed and $d$ is the diameter of the stirrer.

### 3.5 Prandtl Number

It is the ratio of kinematic viscosity to thermal diffusivity. It is actually the ratio of velocity boundary layer to thermal boundary layer, and hence one can relate it to the thickness of the thermal and velocity boundary layers. If $N_{Pr} = 1$, the boundary layer coincides with thermal layer. Mathematically, Prandtl number is:

$$N_{Pr} = \frac{\nu}{\alpha} \quad \frac{\mu C_p}{K} \tag{13}$$

### 3.6 Nusselt Number

It is the dimensionless heat transfer coefficient and appears when one deals with convection. It may be viewed as the ratio of convection to conduction for a layer of fluid. If $N_{Nu} = 1$, pure conduction is observed. It provides a measure of the convection heat transfer at the surface. It is expressed as:

$$N_{Re} = \frac{h\,L}{K} \tag{14}$$

where, $h$ is the heat transfer coefficient; $L$ is a characteristic length and $k$ is the thermal conductivity.

### 3.7 Schmidt Number

It describes the ratio between kinematic viscosity of air and molecular diffusivity and is used in mass transfer in general and diffusion in flowing systems. It is analogous to Prandtl number in heat transfer. It physically relates the relative thickness of the hydrodynamic layer and mass transfer boundary layer and can be expressed as:

$$N_{Sc} = \frac{\mu}{\rho D_v} \tag{15}$$

where, $\mu$ is the kinematic viscosity of the fluid; $\rho$ is the density of the fluid; and $D_v$ is the diffusivity.

### 3.8 Sherwood Number

It is proportional to the ratio of mass diffusivity to molecular diffusivity; in other words, it represents the ratio of length scale to diffusive boundary layer. It is a convenient parameter for the analysis of transfer processes occurring in a stationary medium in laminar flow.

$$N_{Sh} = \frac{K_c L}{D_v} \tag{16}$$

where, $K_c$ is the diffusion rate or overall mass transfer coefficient; $D_v$ is the diffusivity; and $L$ is the characteristic length.

## 4. CONCEPT OF MASS BALANCE

Although biological systems are quite different from chemical process systems, the material balance principles of chemical process systems apply equally well to biological systems [11]. One of the basic laws of physical science is the law of conservation of mass. This law, stated simply, says that "mass cannot be created or destroyed (excluding nuclear or atomic reactions)". Hence the total mass of all materials leaving plus the mass of any material accumulating or left in the process is:

$$\text{Input} = \text{Output} + \text{Accumulation} \tag{17}$$

As in majority of cases, the materials entering a process undergo chemical as well as biological reaction, so that materials leaving are different from those entering. In these cases, it is usually convenient to make a molar and not weight balance on individual components.

### 4.1 Mass Balances for Cell Growth and Product Formulation

Bioprocessing usually comprises a series of processing operations viz., nutrient mixing, growth, aeration, separation etc., which occurs in a sequence. The individual operations are often referred as "unit operations". The simplest form of representation of a process is by means of a *block diagram*, in which each unit operation or processing unit is represented by a rectangle (Fig. 1). The flows into and out of the unit operations are represented by single lines with arrows to show the direction of flow.

Although, biochemical processes are complex than other chemical processes, they obey the law of conservation of matter. It involves the consumption of substrates, which provides energy and raw materials required for the synthesis of additional cell mass. As the cell requires substrate for its growth, it utilizes atoms of carbon, hydrogen, nitrogen, oxygen present in the environment of the reactor for their metabolic activities and produces the desired metabolite.

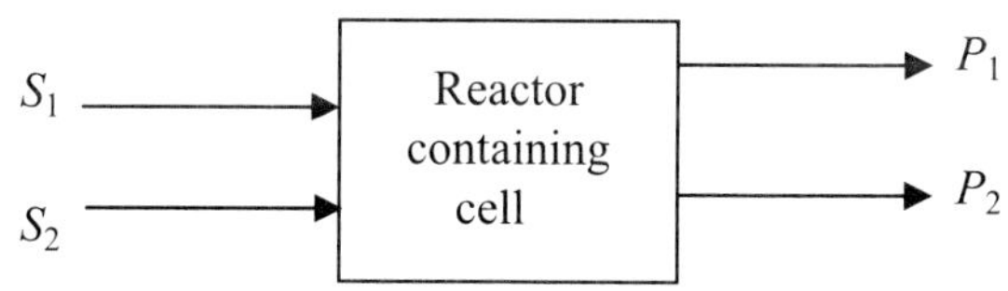


**Fig. 1:** Schematic diagram for mass balance over a bioreactor.

where, $S_1$ and $S_2$ are the substrate provided for the growth of cell, and $P_1$ and $P_2$ are the metabolites or product formed by the cell during a biological reaction.

In principle, mass balances are straightforward. Simply add up the materials entering the system and the materials leaving the system. If the system is in a steady state, these two totals should be the same. Any unknown streams may then be found by difference. One way to do this is to represent the mass balances by equations relating input and output. Each processing operation will have a mass balance equation associated with it. In addition, there will be an equation relating to the inputs and outputs for the whole process.

Mass balance over a reactor can be carried out either by developing algebraic equations and solving them for unknown values, or by using Pearson's method (generally used to determine the relative quantity of the materials involved in blending and mixing). The common steps involved in solving simple mass balance problems [4] are:

1. Draw a block diagram of the given process.
2. Select suitable time basis (operation hour) or weight basis (product output or feed input).

3. Draw up a mass balance table and enter all the available information into the table.
4. Write required mass balance equations for overall mass balance and component balance.
5. Identify and calculate the unknown values by solving the mass balance equations and enter them into the table for ease of remembrance, calculation.

## 4.2 Stoichiometry in Biochemical Engineering

Cell growth and related metabolic activities can be represented in the same way as used for simple chemical reactions, but for that it is important to know the chemical formula or elemental composition of dry cells [12]. One can understand the application of simple stoichiometric balance approach in a better way from the following example:

**Example 3:**
Discuss the steps involved in the determination of empirical formula $C_aH_bO_cN_d$ of the cell, considering it as an aerobic reaction and assuming that a single molecule is obtained. The sources provided for the growth of cells are $CH_xO_y$ and $H_pO_qN_r$, and the respiratory quotients for the reaction is $RQ$.

***Solution:***
It is given that sources provided are $CH_xO_y$ and $H_pO_qN_r$. It is an aerobic reaction, therefore $O_2$ will be consumed by the organisms for growth. Thus, the stoichiometric relation for growth can be written as

$$\alpha CH_xO_y + \beta H_pO_qN_r + \gamma O_2 \longrightarrow C_aH_bO_cN_d + \theta CO_2 + \nu H_2O \quad (18)$$

Now considering elemental balances and comparing the stoichiometric coefficient on both sides of the equation:

For carbon ($C$) balancing stoichiometric coefficient results

$$\alpha = a + \theta \quad (19)$$

For hydrogen ($H$) balancing stoichiometric coefficient results

$$\alpha x + \beta p = b + 2\nu \quad (20)$$

For oxygen ($O$) balancing stoichiometric coefficient results

$$\alpha y + \beta q + 2\gamma = c + 2\theta + \nu \quad (21)$$

For nitrogen ($N$) balancing stoichiometric coefficient results

$$\beta r = d \quad (22)$$

Also respiratory coefficient for the reaction is $RQ$, which is ratio of moles of $CO_2$ formed and moles of $O_2$ consumed, i.e.

$$RQ = \theta/\gamma \quad (23)$$

Solving Eqs. (19), (20), (21), (22) and (23), one can obtain the values of $\delta$, $a$, $b$, $c$, $d$, and finally the empirical formula of the dry cell can be obtained.

## 4.3 Mass Transfer

It can be defined as the migration of substance through a mixture under the influence of a concentration gradient in order to reach chemical equilibrium. They are broadly classified into three categories: (i) mass transfer operations involving direct contact of two immiscible phases; (ii) mass transfer operations involving phases separated by a membrane; and (iii) direct contact of miscible phases. A number of biochemical engineering operations such as absorption, distillation, extraction, membrane separation, and mixing involve mass transfer principles. The major components involved in these mass transfer processes are water, sugars, salt, oils, proteins, acids, flavour/aroma substances, oxygen (in aerobic fermentation), carbon dioxide, and various other compounds produced by microorganisms [13, 14]. This section deals with the different types of mass transfer, concept of boundary layer and effects of agitation on the rate of transfer. The general equation for transfer at molecular level is given as:

$$\text{Rate of a transfer process} = \frac{\text{Driving force}}{\text{Resistance}} \tag{24}$$

Two common modes of heat transfer are conduction and convective heat transfer; analogous to it in mass transfer one can come across molecular diffusion and convective mass transfer.

*4.3.1 Gas-liquid mass transfer*

It is important when considering the transfer of oxygen from a gaseous to a dissolved phase. One of the common examples of processes involving gas-liquid mass transfer is carbon dioxide removal from the fermentation broth and distillation of volatile products during downstream processing.

*4.3.2 Liquid-liquid mass transfer*

It is of importance in bioprocessing of different products such as (i) steroid transformation; (ii) breakdown and biological conversion of oils and alkanes; and (iii) in downstream processing operations involving solvent extraction.

*4.3.3 Solid-liquid mass transfer*

It must be considered when dealing with microbial pellets and immobilized cell and enzyme systems. It is also important in downstream processing operations such as de-ionization, adsorption and crystallization.

## 4.4 Diffusion

It can be defined as the transfer or movement of individual molecules through a fluid by the means of random, individual movements of the molecule. Diffusion of molecule is mainly due to concentration gradient. The equation for mass transfer by diffusion in a binary mixture of two components $A$ and $B$ is given by Fick's law.

$$J_{AX} = \frac{-D_{AB}\, dC_A}{dx} \tag{25}$$

where, $J_{AX}$ is the molar flux of the component $A$ in the $x$ direction due to molecular diffusion of component $A$ in kg.mol/s.m$^2$, $D_{AB}$ is the molecular diffusivity of the molecule $A$ in $B$ in m$^2$/s, $C_A$ is the concentration of $A$ in kg.mol/m$^3$, and $x$ is the distance of diffusion in m.

*4.4.1 Diffusion in gases*

There are theoretical and empirical correlations for estimating the binary diffusion coefficient $D_{AB}$ in gases, liquids, and solids. The steady state diffusion of $A$ through non-diffusing $B$ (Fig. 2) is given by:

$$J_A = \frac{D_{AB}\, P_t \left(p_{A1} - p_{A2}\right)}{RT\, z\, p_{BM}} \tag{26}$$

where, $J_A$ is the molar flux of $A$ (kg.moles/m$^2$.sec, $D_{AB}$ is the diffusivity of molecule $A$ in $B$ (m$^2$/sec); $P_t$ is the total pressure (N/m$^2$); $p_{A1}$ is the partial pressure of component $A$ (N/m$^2$); $p_B$ is the partial pressure of component $B$ (N/m$^2$); $R$ is universal gas constant; $T$ is the absolute temperature; $z$ is the column height of the diffusing component '$A$'; and $p_{BM}$ is analogous to log mean temperature difference and is given as:

$$P_{BM} = \frac{(p_{A1} - p_{A2})}{\ln[(P - p_{A2})/(P - p_{A1})]} \tag{27}$$

### 4.4.2 Diffusion in liquids

Analogous to steady state diffusion in gases, steady state diffusion in liquid $A$ through non-diffusing $B$ can be obtained as:

$$J_A = \frac{D_{AB}\, P_t (x_{A1} - x_{A2})}{RT\; z\; x_{BM}} \tag{28}$$

where, $x_{A1}$ is the mole fraction of component $A$; $x_B$ is the mole fraction of component $B$ (N/m$^2$); $R$ is universal gas constant; $T$ is the absolute temperature; $z$ is the column height of the diffusing component '$A$'; and $x_{BM}$ is analogous to log mean temperature difference and is given as:

$$x_{BM} = \frac{(x_{A1} - x_{A2})}{\ln[(1 - x_{A2})/(1 - x_{A1})]} \tag{29}$$

### 4.4.3 Diffusion in solids

The diffusion of gases and liquids through porous solid materials may occur by a combination of Fick's diffusion and Knudsen diffusion. The mass transfer within the gas or liquid contained in the pores is governed by Fick's law if the pores are large. Knudsen diffusion occurs when the pores are small and their size approaches the mean free path of the gas molecule. The flux in such case is described in terms of Knudsen diffusivity, which is independent of the pressure and is calculated as

$$D_{KD} = \tfrac{2}{3}\,(r_{avg} v_{avg}^{\,2}) \tag{30}$$

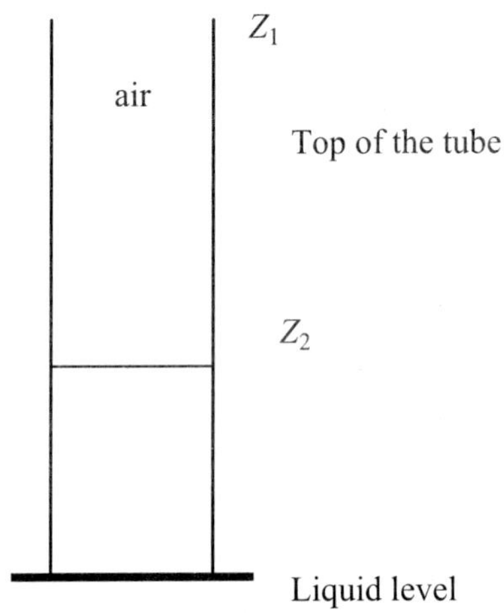


**Fig. 2:** Diffusivity of carbon tetrachloride ($CCl_4$) through air.
*Source:* 14

where, $D_{KD}$ is Knudsen diffusivity in $m^2/s$ and $r_{avg}$ is the average molecular velocity for component $A$ in m/s. The molar flux for gas and liquids through a porous solid can be described in terms of $D_{KD}$ by

$$N_A = \frac{D_{KD}\left(p_{A1} - p_{A2}\right)}{RTz} \tag{31}$$

## 4.5 Convection

Convective mass transport, which occurs in liquids and gases, results from bulk motion of the fluid [15, 16]. Two major types of convective mass transfer include (i) free convective mass transfer occurring naturally due to the concentration difference or density variations and (ii) forced convective mass transfer imposed by external forces.

### *4.5.1 Natural convection*

In natural convection, the movement of fluids is due to temperature differences. These processes are too slow to provide a sufficient rate of mass transfer in most reaction systems.

### *4.5.2 Forced convection*

In forced convection, the movement of fluids is resulted from the input of a power source. It is induced by an external power source to increase the rate of mass transfer. The power source can be an agitator, pump or compressor.

When a fluid is flowing outside a solid surface in forced convection motion, one can express the rate of convective mass transfer from the surface to the fluid, or vice versa, by the following equation:

$$N_A = k_c\,(C_{L1} - C_{Li}) \tag{32}$$

where, $k_c$ is mass transfer coefficient in m/s; $C_{L1}$ the bulk liquid concentration in kg mol/m$^3$; and $C_{Li}$ is the concentration in the fluid next to the surface of the solid.

# 5. CONCEPT OF BOUNDARY LAYER

Convective mass transfer problems in turbulent flow are not always amenable to analytical methods of solution. Therefore, the concept of boundary layer theory was introduced in one of the models proposed for better understanding of turbulent regime. Boundary layer is nothing but a hypothetical stagnant film of thickness '$x$', and it is assumed that all the concentration changes occur in this layer. It is central to studying interfacial mass transfer process.

A boundary layer can be described as a thin layer of fluid surrounding a body, which is immersed in or is in contact with a fluid (Fig. 3). The body can be a solid object (e.g., a pipe wall, a reactor tank wall, an immobilized enzyme particle, a cellular aggregate, or a solid substrate such as wood) or a fluid of a different phase than the medium (e.g., an oxygen bubble or an oil globule in water). The boundary layer concept can be cleared by considering the profile of a fluid flowing in a pipeline (Fig. 4).

The liquid velocity will be highest in the centre of the pipe and will be slowest at the walls. In fact, the flow rate at the walls must be zero (since the walls are not moving). Thus, surrounding the walls will be a region of very slow flow. This region is known as a boundary layer. Since fluid movement in a boundary layer is slow, the movement of solutes across the boundary layer will be determined by diffusion forces rather than by forced convection. Thus in assessing mass transfer rates between the phases, it is assumed that the movement of solutes through boundary layers is the rate limiting step.

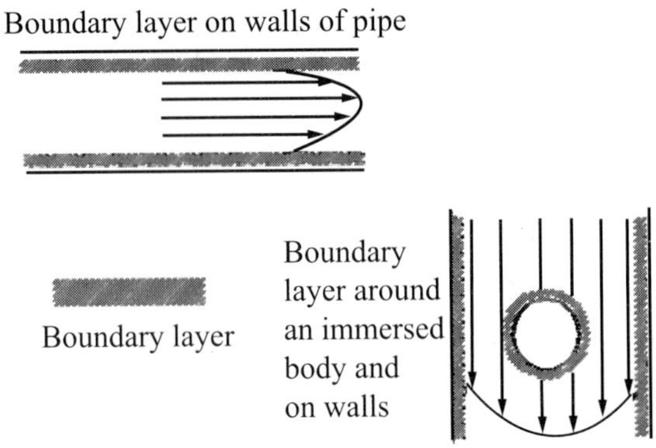


**Fig. 3:** Boundary layers around the different bodies.
*Source:* 17

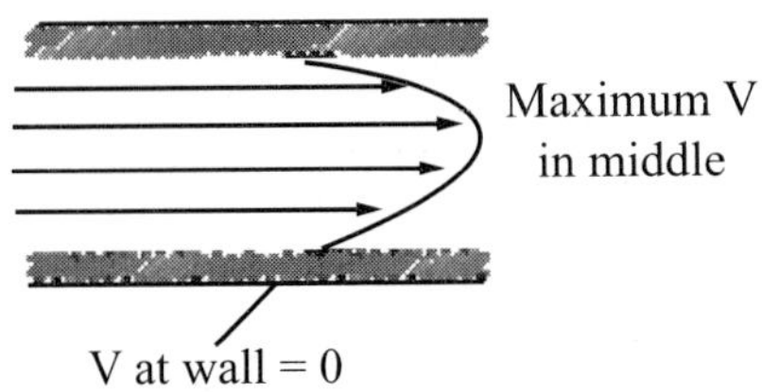


**Fig. 4:** Profile of liquid flowing in a pipeline.
*Source:* 17

## 5.1 Mass Transfer in Boundary Layer

The rate of mass transfer between the phases is largely determined by the rate of movement across the boundary layer. The rate of solute movement through a boundary layer is determined by diffusion forces alone, and the rate of movement across the boundary layer can be increased by:

(i) Increasing the interfacial area between the phases
(ii) Increasing the concentration gradient across the boundary layer
(iii) Decreasing the size of the boundary layer
(iv) Increasing the rate of movement of molecules through the boundary layer

### *5.1.1 Agitation and its effect*

In biochemical engineering, many operations are dependent to a great extent on effective agitation and mixing of fluids. In general, agitation refers to forcing a fluid by mechanical means to flow in a circulatory or other pattern inside a vessel. Mixing usually implies the taking of two or more separate phases, such as a fluid and a powdered solid, or two fluids, and causing them to be randomly distributed through one another. Agitation and mixing play an important role in mass transfer rates between the phases, and high agitation conditions have the following effects:

(i) Creates high shear conditions and promotes the breakdown of bubbles and globules.
(ii) Removes solutes from the boundary layer maintaining concentration gradients.

Other than these effects, agitation also helps in dispersing a gas in a liquid as fine bubbles (e.g., oxygen from air in a suspension of microorganisms for fermentation). It also increases the heat transfer between the fluid and a coil or jacket in the vessel wall. The size of the velocity gradient is increased by increasing the mixing conditions, and thus agitation decreases the size of the boundary layer (Fig. 5).

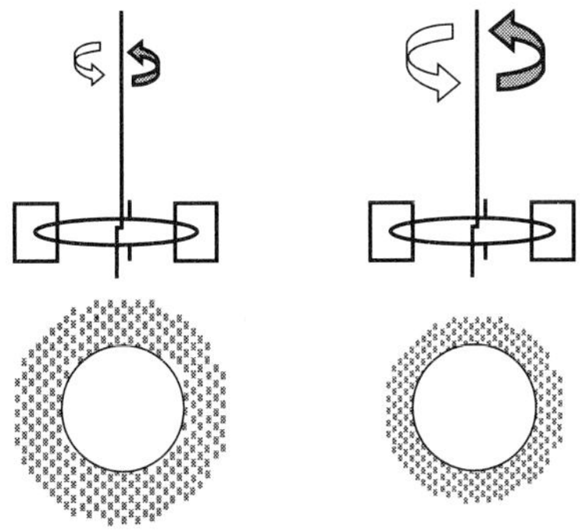

Increasing the mixing conditions in a reactor reduces the size of the boundary layer.

**Fig. 5:** Effect of high mixing on boundary layer.
*Source:* 18

Smaller boundary layers lead to reduced time required for solutes to diffuse into bulk liquid, thus increasing mass transfer rates. Increasing the interfacial area (A) will increase the mass transfer rate. The value of A is dependent on (i) the presence of coalescence promoting agents and (ii) agitation and shear conditions.

5.1.1.1 Impeller types and power requirement: Impellers/agitators are divided into two classes: those which generate currents parallel with the axis of the impeller shaft, and those which generate currents in a radial or tangential direction. The first are called axial flow impellers and the second are tangential flow impellers. The three major types of impellers are propellers, paddles and turbines, and axial flow turbines. Each type includes many variations and subtypes. Various other impellers are also reported, which are useful in certain situations, but the above three major types can solve nearly 95 percent of the liquid agitated problems. The radial flow disc turbine is most common for aerobic fermentation.

The power, $P$, drawn by mixing impellers in the reactors with low Reynolds number and medium viscosity range is proportional to:

$$P \,\alpha\, N^3 D^5$$

or

$$P = \rho N_P n^3 (D_a)^5 \tag{33}$$

where, $\rho$ is the density of the fluid; $N_P$ is power number; $n$ is impeller speed and $D_a$ is the impeller diameter. For highly turbulent flow, it is constant. In case when the fluid is non-Newtonian, then the power number consumption is calculated using apparent viscosity and its corresponding Reynolds number.

## 5.2 Rheology

It is the science of flow and deformation. The relationship between the shear rate and shear stress in a real fluid is part of rheology. In general, food materials exhibit both fluid and solid behaviour and are called viscoelastic. Viscosity is one of the important parameters in characterizing fluid flow. The rheological properties of the materials being processed in bioreactors will influence the power consumption and hence the heat and mass transfer rates [19].

The rheological characteristics and types of fluid encountered in a bioreactor are normally divided into two groups: Newtonian fluids and non-Newtonian fluids, according to their flow behaviour.

### *5.2.1 Newtonian fluids*

These fluids exhibit a linear increase in the shear stress with the rate of shear. The viscosity of a Newtonian fluid is dependent only on temperature but not on shear rate and time. Examples of Newtonian fluid include water, milk, sugar solution, mineral oil, etc. The behaviour of these fluids can be expressed as:

$$\tau = -\mu\left(\frac{dv}{dr}\right) \tag{34}$$

where, $\tau$ is the shear stress, $\mu$ is the viscosity of the given fluid and $(dv/dr)$ is the velocity gradient.

### *5.2.2 Non-Newtonian fluids*

These do not exhibit linear relationship like Newtonian fluids. They are further classified into time independent and time dependent fluids. Their behaviour can be expressed as:

$$\tau = -\mu\left(\frac{dv}{dr}\right)^n \tag{35}$$

where, $n$ is the flow behaviour index. Equation (35) is known as power law. Newtonian fluids are special case of power law fluids for which flow behaviour index ($n$) is 1.

5.2.2.1 Time independent non-Newtonian fluids: They can be characterized as:

5.2.2.1.1 Shear thinning fluids: They are also called pseudoplastic. The viscosity of these fluids decreases with increased shear rate. The flow behaviour index for these fluids is less than unity but greater than 0. Examples of shear thinning fluids include fruit juice concentrates, vegetable soups, puree, etc.

5.2.2.1.2 Shear thickening fluids: They are known as dilatant fluids. The viscosity of these fluids increases with increased shear rate. Their flow behaviour index is greater than 1. Examples of these fluids are concentrated starch suspensions, sausage slurry, homogenized peanut butter, etc.

5.2.2.1.3 Binghamplastic fluids: These fluids exhibit a so-called yield value, i.e. a certain shear stress must be applied before the flow occurs. Examples of binghamplastic fluids are tomato paste, french dressing, fudge sauce, tomato ketchup, etc. The power law Eq. (35) for these fluids is modified as:

$$\tau = \tau_0 - \mu\left(\frac{dv}{dr}\right)^n \tag{36}$$

where, $\tau_0$ is the shear stress that must be applied before the flow occurs.

5.2.2.2 Time dependent non-Newtonian fluids: The viscosity of these fluids is dependent on temperature, shear rate and time. They are further classified as:

5.2.2.2.1 Thixotropic: These fluids are quite common in chemical as well as food industry. They are also known as time thinning fluids, i.e. their viscosity decreases with time. An example of thixotropic fluid is yoghurt.

5.2.2.2.2 Rheopectic: These fluids are also known as time thickening, i.e. their viscosity increases with time. Rheopectic fluids are rare in food industry.

The viscosity of the non-Newtonian fluid is determined at a particular instant, and this instantaneously viscosity is called apparent viscosity. Mathematically, apparent viscosity can be expressed as:

$$\mu_a = (2\ \omega)^{n-1}\ K(1/n)^n \tag{37}$$

wherc, $\mu$ is the viscosity of fluid; $\omega$, is the angular velocity of rotation; $K$, is the consistency index and $n$ is the flow behaviour index. The apparent viscosity has the same unit as viscosity, but the value varies with the rate of shear. Thus, when reporting an apparent viscosity, the shear rate under which it was determined must also be specified.

## 5.3 Characterization of Flow

At low stirring speeds, a phenomenon known as circulation or swirling occurs. During circulation, the fluid follows the laminar flow lines, moving in circles around the tank; the result being ineffective mixing. Fluid mixing is described as existing one of three types:

(i) Laminar flow
(ii) Turbulent flow
(iii) Transient flow

In laminar flow, the fluid tends to flow in one direction as visualized as "flow lines". There is little mixing between these flow lines and fluid follows an ordered trend. In turbulent flow, the flow lines break down to form eddies (Fig. 6). These eddies carry the fluids away from the flow lines leading to mixing of materials through the reactor.

In between laminar and turbulent flow, there is a state which is called transient flow. In this state, the fluid moves randomly between the laminar and turbulent states. Normally turbulent flow conditions are required for good heat and mass transfer and are often desirable [20]. In an unbaffled tank, as the stirrer speed is increased, the liquid will continue to circulate, although at a faster speed. Although high stirring speeds are used, but mixing is still poor as the bulk of the fluid is simply circulating throughout the tank. Eventually, the liquid and air will be drawn into the impeller leading to a phenomenon known as vortexing, which is generally undesirable. It can be prevented with the use of (i) off-centre impellers and (ii) baffles.

(a) (b)

**Fig. 6:** (a) Laminar flow (b) Turbulent flow.

## 5.4 Effect of Temperature on Viscosity

Temperature has a strong influence on the resistance of flow of a fluid. It is very important that temperatures be maintained constant when making rheological measurements. The temperature dependence of the viscosity and the consistency index can be expressed in terms of Arrhenius equation:

$$\ln\left(\frac{\mu_1}{\mu_2}\right)=\frac{E_a}{R}\left(\frac{1}{T_1}-\frac{1}{T_2}\right) \tag{38}$$

where, $\mu_1$ is the viscosity at absolute temperature $T_1$ (273 K) and $\mu_2$ is the viscosity at a temperature $T_2$, $Ea$ is the activation energy and $R$ is the universal gas constant (8.314 J/g.mole.K).

### *5.4.1 Viscosity of liquids*

Viscosity of liquids, in general, decreases with increasing temperature. When sufficient data are available, Eq. (38) for liquids can be modified to:

$$\ln \mu = a + b/T \tag{39}$$

where, '$a$' and '$b$' are constants and can be obtained from the slope and intercept of the graph obtained by plotting temperature versus viscosity data, and $T$ is the absolute temperature of the fluid.

### *5.4.2 Viscosity of gases*

Viscosity of gases increases with an increase in temperature. The viscosity ($\mu$) of many gases is approximated by the formula:

$$\mu = \mu_o(T/T_o)^n \tag{40}$$

where, $T$ is the absolute temperature, $\mu_0$ is the viscosity at an absolute reference temperature $T_o$ (273 K), and $n$ is an empirical exponent that best fits the experimental data. The viscosity of an ideal gas is independent of pressure, but the viscosities of real gases and liquids usually increase with pressure. Viscosities of liquids are generally two orders of magnitude greater than the gases at atmospheric pressure.

## 6. INTRODUCTION TO BIOREACTORS

Bioreactors are vessels in which raw materials are biologically converted into specific products, using microorganisms, plant, animal or human cells or individual enzymes. A bioreactor supports the natural process of cells by trying to maintain their environment to provide optimum growth conditions by providing suitable conditions — temperature, pH, substrate, salts, vitamins and oxygen [21-26].

Bioreactor is also referred to as fermenter. Although the two terms are used more or less interchangeably to describe vessels in which controlled biological processes occur, there is minor difference between them. Fermenters are usually used for growing microbial cells, while bioreactors are used for growing mammalian cells. They vary in size and complexity from a 10 ml volume in a test tube to computer controlled fermenters with liquid volumes greater than 100 $m^3$. The performance of any bioreactor mainly depends on:

(i) Biomass concentration which must remain high.
(ii) Sterile conditions being maintained.
(iii) Effective agitation so that the distribution of substances and microorganisms in the reaction is uniform.
(iv) Heat removal.
(v) Creation of the correct shear conditions — high may damage cells, low may lead to flocculation or growth on wall and stirrer.

There are different modes of classification of bioreactors. They can be classified as follows:

(i) On the basis of air requirement — aerobic and anaerobic
(ii) On the basis of material of construction — glass, plastic, steel and wood
(iii) Mode of operation — batch, fed-batch and continuous
(iv) On the basis of agitator location — bottom entry, top entry, centrally stirred, off-line

The simplest bioreactor is a standing culture in which little or no power is used for aeration, and it is dependent on the transfer of oxygen through the still surface of the culture. It means that even a simple T-flask or Erlenmeyer flask can also be used to work as a bioreactor. Shake flasks are next category of bioreactors in which agitation can be achieved. More advanced bioreactors include bubble column, airlift reactors, stirred tank reactors, and fluidized bed reactors. One of the most common classifications of the reactors on the basis of mode of operation [19] is as follows:

## 6.1 Batch Reactor

A bioreactor is filled with fresh medium and then inoculated with the desired organism to produce the desired metabolic product or organism inside the reactor (Fig. 7). At the end of the fermentation, the contents are removed for downstream processing. The reactor is then cleaned, sterilized and refilled for the next fermentation.

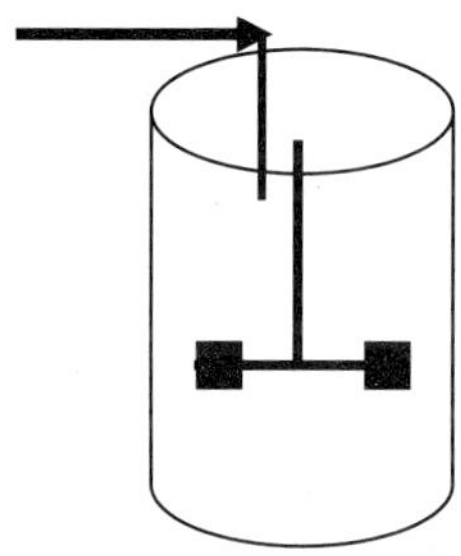

**Fig. 7:** Typical batch reactor.

Biochemical processes involve batch growth of cell populations where after seeding a liquid medium with an inoculum of living cells, nothing is added to the culture as growth proceeds. Typically in such a reactor, the concentrations of nutrients, cells and products vary with time as the growth proceeds. In a batch reactor (ideal), the change of the cell concentration is equal to the growth rate and can be obtained from material balance for component '*i*' in the reactor, as

Rate of accumulation of component '*i*' = Rate of formation of '*i*' due to chemical reaction

Or in other words

Rate of change of culture volume × molar concentration of component '*i*' =

$$\text{Culture volume} \times \left(\frac{\text{moles of } i \text{ formed by reaction}}{\text{culture volume} \times \text{time}}\right)$$

$$\text{or} \quad \frac{d(Vx_i)}{dt} = VR_{fi} \tag{41}$$

where $V$ is the culture volume and $x_i$ is the moles of '*i*' per unit culture volume and $R_{fi}$ is the moles of '*i*' formed by reaction per unit volume per unit time. As the volume of the reactor is maintained constant, one can write

$$\frac{dx_i}{dt} = R_{fi} \tag{42}$$

## 6.2 Fed-Batch Reactor

In a fed-batch reactor, the fresh media is added continuously without removal until the maximum liquid reactor volume is reached (Fig. 8). The reactor may then be allowed to continue or be partially or completely emptied depending on the process. These are the most common bioreactors used to produce biological products. Fed-batch reactors can be used to maintain cells under low substrate or nutrient conditions without washout occurring.

With both fed-batch and continuous fermenters, it is possible to set and maintain the specific growth rate and substrate concentration at optimal levels. The major physical difference between a fed-batch and continuous fermenter is that the effluent is not continuously removed and thus washout does not occur.

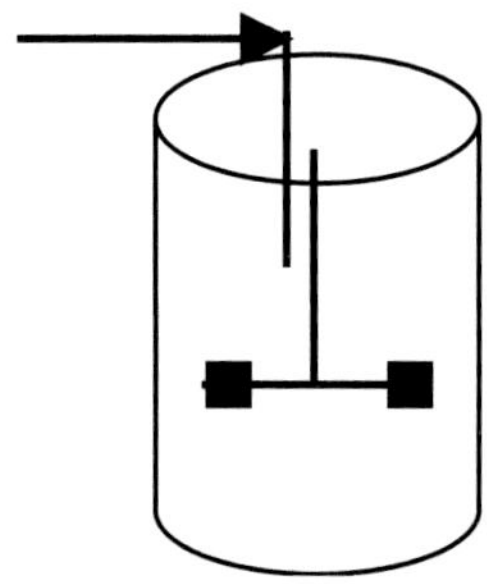

**Fig. 8:** Typical fed-batch reactor.

### *6.2.1 Advantages of fed-batch reactor over continuous reactors*

(i) Since cells are not removed during the fermentation, fed-batch fermenters are well suited for the production of compounds produced during very slow or zero growth.

(ii) Unlike a continuous fermenter, the feed does not need to contain all the nutrients needed to sustain growth. The feed may contain only a nitrogen source or a metabolic precursor.

(iii) A fed-batch reactor can be operated in a variety of ways.

## 6.3 Continuous Reactor

In continuous reactors, fresh media is added continuously and bioreactor fluid is removed continuously (Fig. 9). As a result, cells continuously receive fresh medium, and products and waste products and cells are continuously removed for processing. The reactor can thus be operated for long periods of time without having to be shut down. Continuous reactors can be many times more productive than batch reactors. This is partly due to the fact that the reactor does not have to be shut down as regularly, and also due to the fact that the growth rate of the bacteria in the reactor can be more easily controlled and optimized.

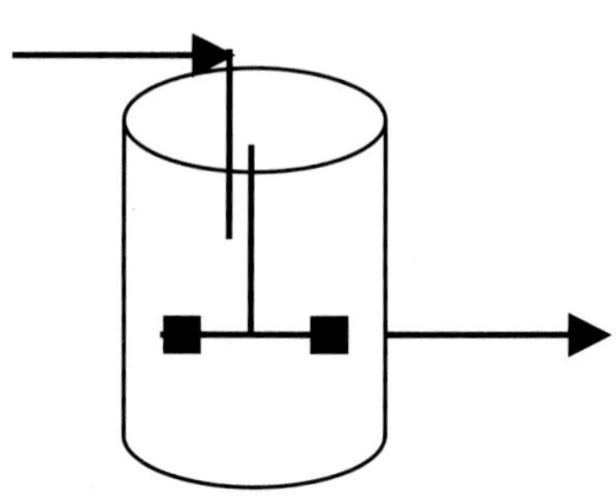

**Fig. 9:** Typical continuous reactor.

## 6.4 Some Common Industrial Bioreactors

Other than the above-mentioned bioreactors, some of the common bioreactors used in industries include stirred tank reactor, plug flow reactor, bubble column reactor, air lift reactors, fluidized bed reactor, packed bed reactor, etc. [7] and are subsequently described in the following section.

In stirred tank reactors, mechanical stirrers are used to mix the reactor to distribute heat and materials (Fig. 10). These are the most widely used design in industrial fermentation. It can be employed for both aerobic and anaerobic fermentation of a wide range of cells including microbial, animal and plant cells.

In plug flow reactors, fluid enters at one end of a cylindrical tube and cells grow while they pass through (Fig. 11). Long tube and absence of stirring device prevents racial mixing. There is no or very less radial mixing between the fluid components, although longitudinal mixing is there.

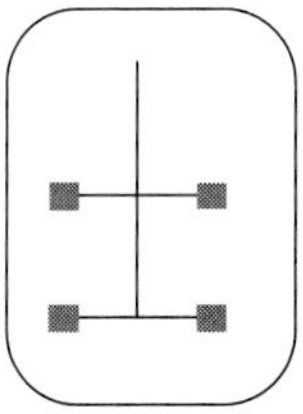

**Fig. 10:** Stirred tank reactor.

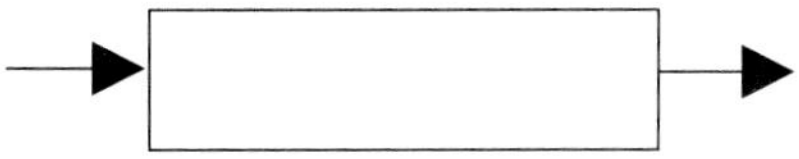

**Fig. 11:** Plug flow reactor.

By bubble column bioreactors, one means the reactors with large aspect ratio (height to diameter ratio). They take the form of columns instead of more squat tanks typical of agitated vessels. These are tall reactors in which mixing is done entirely by forcing compressed gas into the reactor vessel (Fig. 12). They have been used for the cultivation of microorganisms for use as animal feed e.g., single cell protein (SCP).

These reactors are similar to bubble column reactors, but differ by the fact that they contain a draft tube. The draft tube is typically an inner tube which improves circulation and oxygen transfer and equalizes shear forces in the reactor. The need for highly cost-efficient oxygen transfer in fermentations such as those with hydrocarbon feed stocks has led to air-lift fermenters as shown in Fig. 13.

In fluidized bed reactors, liquid flows upward through a long vertical cylinder (Fig. 14). In this biocatalyst particles are suspended by drag forces exerted by rising liquid. The small particles create a large surface area for cells to stick to and enable a high rate of transfer of oxygen and nutrients to the cells. These are considerably more complex than constant stirred tank reactor and plug flow reactors.

In packed bed reactors, particles do not move with the liquid (Fig. 15). Packed bed reactors are simple to construct and operate, but can suffer from blockages and poor oxygen transfer. In many of

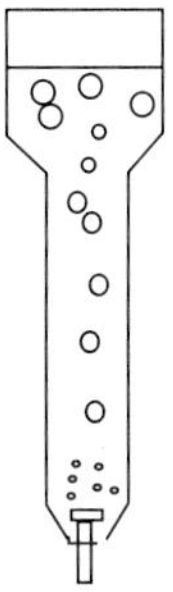

**Fig. 12:** Bubble column reactor.

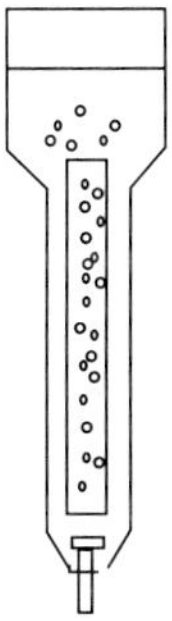

**Fig. 13:** Air lift reactor.

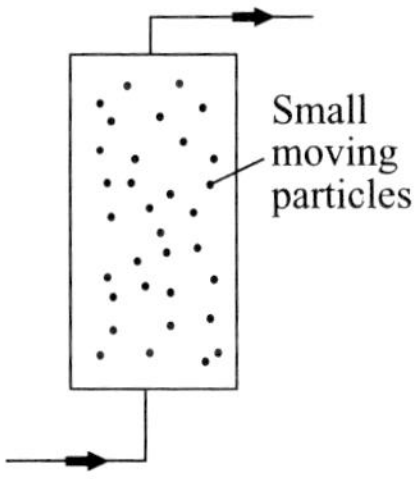


**Fig. 14:** Fluidized bed reactor.

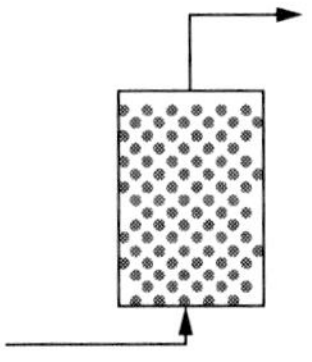

**Fig. 15:** Packed bed reactor.

these systems, immobilized enzymes are used for glucose isomerization, selective penicillin hydrolysis and selective separation of racemeic mixtures of amino acids.

## 6.5 Comparison of the Advantages and Disadvantages of Continuous Reactors System with Batch Fermentation

Continuous cultures have following advantages over batch reactors system [27]:

(i) In a chemostat, the cells can be maintained at a constant physiological state and growth rate. The growth rate can be adjusted by changing the feed flow rate. Consequently, it is easier to optimize the productivity.

(ii) It is not necessary to shut the continuous fermenter down as frequently as a batch fermenter. At the end of a batch fermentation, the reactor must be emptied, cleaned, sterilized and re-filled. The time required for these operations is known as turnaround time. Theoretically, a continuous fermenter could operate indefinitely without having to be shut down. In practice, however this is not possible.

(iii) A biological phenomenon observed in batch cultures is the lag phase. The lag phase occurs at the beginning of the fermentation and represents a physiological adaption of the culture to the new environment. Growth during this phase is very slow and the lag phase represents a period of very low productivity. Because continuous cultures are shut down with less frequency as compared to batch reactors, there is less loss of productivity during lag phases.

(iv) When referring to continuous culture systems, the terms lag phase, exponential phase, stationary phase and death phase have no meaning. This is because the system is operating continuously and growth cannot be segregated into phases.

(v) Most downstream processing operations are most productive when operated in a continuous manner. Using a continuous culture allows the fermentation to be in-tune with other operations in the plant. Thus, overall plant productivity is easier to optimize.

(vi) Continuous cultures thus offer the potential of higher productivities. As a consequence, it is possible to have smaller reactors and associated equipment and thus lower capital costs. This should lead to higher profits.

Despite many potential benefits, pure culture applications of continuous cultures are not well established in industry. Following are some disadvantages of continuous culture:

(i) The batch fermentation process is easy to understand and relatively well established in industry. To switch from a batch to a continuous process represents a large risk because complex phenomenon occurring in continuous system is not well understood.

(ii) In the processing of pharmaceutical, all products must be able to be segregated into batches. For this reason, unit operations in pharmaceutical processing are performed in a batch manner. Therefore, the use of continuous cultures to produce many therapeutic products is not accepted by the US Food and Drug Administration as a Current Good Manufacturing Practice.

(iii) Many commercially important products are produced only when cell growth stops. These products are not amenable to production in a conventional continuous flow system, as new biomass will not be produced to replace that removed in the effluent. Eventually the culture will be washed out.

(iv) Not all products are produced well in a continuous flow system. For example, in full flavour development, some fermented foods and beverages require cellular products released from different phases of batch culture growth. As continuous fermenters maintain the cells at a single physiological condition, the resultant product is generally inferior. Beer and soya sauce, for example, cannot be successfully produced by continuous culture.

## 6.6 Standard Geometry Stirred Tank Bioreactor

A stirred tank reactor normally has a cylindrical or curved base (Fig. 16). A curved base assists in the mixing of the reactor contents. Stirred tank bioreactors are generally constructed according to standard dimensions as those published by the International Standards Organization and the British Standards Institution. These dimensions take into account both mixing effectiveness and structural considerations. Typical dimensions of a mechanically stirred tank bioreactor fitted with a sparger and a turbine have been provided in Table 3.

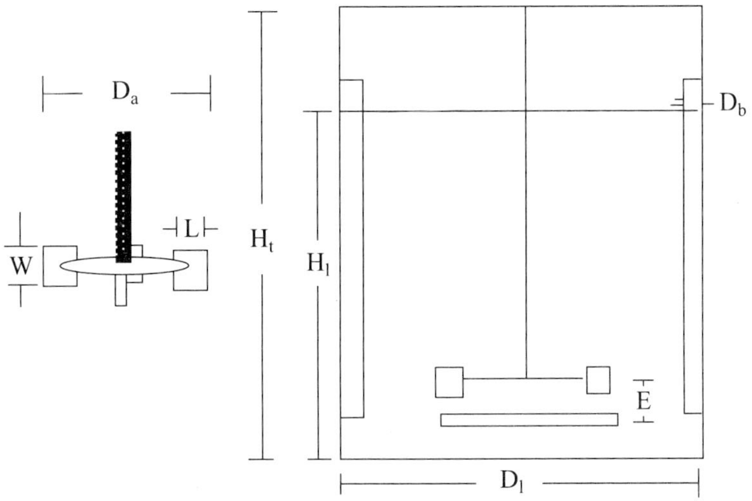


**Fig. 16:** Schematic line diagram of a typical stirred tank reactor with major dimensions.

**Table 3:** Standard dimensions for a stirred tank reactor.

| **Ratio** | | **Typical values** |
|---|---|---|
| Height of liquid in reactor to height of reactor | $H_L/H_t$ | ~0.7-0.8 |
| Height of reactor to diameter of tank | $H_t/D_t$ | ~1-2 |
| Diameter of impeller to diameter of tank | $D_a/D_t$ | 1/3-1/2 |
| Diameter of baffles to diameter of tank | $D_b/D_t$ | ~0.08-0.1 |
| Impeller blade height to diameter of impeller | $W/D_a$ | 0.2 |
| Impeller blade width to diameter of impeller | $L/D_a$ | 0.25 |
| Distance between middle of impeller blade and impeller blade height | $E/W$ | 1 |

($D_a$ = diameter of impeller; $W$ = length of the turbine blade; $L$ = width of turbine blade; $H_t$ = height of the reactor; $H_L$ = height of liquid in the reactor; $D_b$ = diameter of baffle, $D_t$ = tank diameter; $E$ = distance between middle of impeller and impeller blade).

**Example 4:**

A stirred tank bioreactor is approximately cylindrical in shape. It has a total volume ($V_t$) of 100,000 litres. Calculate the dimensions of the reactor if the geometry of the reactor is defined by the following ratios:

$$D_t : H_t \quad 0.50$$
$$D_a : D_t \quad 0.35$$
$$D_b : D_t \quad 0.1$$

***Solution:***

Given the volume of tank = 100,000 litres = 100 $m^3$

One can find out the volume of the cylinder ($V_t$) using the following formula:

$$V_t = \frac{\pi D_t^2}{4} H_t \tag{43}$$

Where $D_t$ is the tank diameter and $H_t$ is height of the tank. Since it is given that

$$H_t = 2 \times D_t \tag{44}$$

Thus one can write

$$V_t = \frac{\pi D_t^2}{4} 2D_t \tag{45}$$

$$\text{or } V_t = \frac{\pi D_t^3}{2} \tag{46}$$

Therefore one can find out the diameter of the tank using

$$D_t = \left(\frac{2V_t}{\pi}\right)^{\frac{1}{3}} \tag{47}$$

Substituting the values of volume of the tank in Eq. (47) one gets

$$D_t = (2 \times 100/\pi)^{0.33} = 3.99 \text{ m} \sim 4.0 \text{ m}$$

Now substituting the value of $D_t$, one can obtain the remaining dimensions by using given geometric ratios and hence:

$$H_t = 2 \times D_t = 2 \times 4 = 8 \text{ m}$$

$$D_a = 0.35 \times D_t = 0.35 \times 4 = 1.4 \text{ m}$$

$$D_b = Dt/0.1 = 4/0.1 = 40 \text{ m}$$

## 6.7 Dilution Rate and Residence Time

The liquid flow rate into a reactor can be expressed in terms which relate it to the volume of the reactor. Two terms that are commonly used are dilution rate ($D$) and residence or retention time. The dilution rate and residence time have a major influence on the productivity of the bioreactor [28]. These terms are defined in term of feed rate ($F$) as:

$$\text{Dilution Rate} = D = \frac{\text{Flow rate}}{\text{Reactor volume}} = \frac{F}{V} \tag{48}$$

where, $F$ is flow rate of feed in litre/h or $m^3$/h, and $V$ is the reactor volume in litre or $m^3$. The unit of dilution rate is $h^{-1}$.

Residence time is reciprocal of dilution rate and can be expressed as:

$$\text{Residence time} = \tau = \frac{\text{Reactor volume}}{\text{Flow rate}} = \frac{V}{F} \tag{49}$$

Normally the flow rate term ($F$) used in Eqs. (48) and (49) refers to the volumetric flow rate and not the mass flow rate. The volumetric flow rate is equal to the volume of liquid delivered per unit time and can be expressed as:

$$F = \frac{\text{Volume delivered } (l)}{\text{time } (h)} \tag{50}$$

## 6.8 Continuous Stirred Tank Reactor and Plug Flow Reactors

Reactions are carried out as batches or with continuous streams through a vessel. Flow reactors are distinguished by the degree of mixing of successive inputs. The limiting cases are: (1) with complete mixing, called an *ideal continuous stirred tank reactor* (CSTR), and (2) with no axial mixing, called a *plug flow reactor* (PFR). Real reactors deviate more or less from these ideal behaviours. Deviations may be detected with *residence time distributions* (RTD) obtained with the aid of tracer tests. In other cases, a mechanism may be postulated and its parameters are checked against the test data. The commonest models are combinations of CSTRs and PFRs in series and/or parallel. Thus, a stirred tank may be assumed completely mixed in the vicinity of the impeller and in plug flow near the outlet.

## 6.9 Theoretical Consideration for Fed-Batch Reactor

It is often desirable to add liquid streams to add precursors for desired products in a batch reactor as the reaction process occurs, or to add nutrients in stationary phase to obtain additional products. When a liquid feed enters the reactor, the culture volume is also altered, and this must be taken into account in the equations used to describe the reactor.

Let $F$ denotes the volumetric flow rate of the entering feed stream at time '$t$' and '$x_{if}$' is the initial concentration of component '$i$' in this entering stream, the material balance on component '$i$' yields

$$\frac{d(Vx_i)}{dt} = VR_{fi} + Fx_{if} \tag{51}$$

If densities of both the entering liquid stream and the culture fluid are equal to '$\rho$', a total mass balance on the reactor contents can be written as:

$$\frac{d(\rho V)}{dt} = \rho F \tag{52}$$

Assuming that density does not change substantially with time during batch operation, one can write

$$\frac{dV}{dt} = F \tag{53}$$

Differentiating Eq. (51) with respect to time as now the volume of reactor is function of time, one can obtain

$$x_i \frac{dV}{dt} + V\frac{dx_i}{dt} = VR_{fi} + Fx_{if} \tag{54}$$

Now substituting the value of $dv/dt$ from Eq. (53) in Eq. (54) one can write

$$\frac{dx_i}{dt} = \frac{F}{V}\left[x_{if} - x_i\right] + R_{fi} \tag{55}$$

which gives the rate of change of concentration of component '$i$' in the fed-batch reactor.

## 6.10 Theoretical Consideration in a Batch Reactor

Let $X$ be the amount of biomass of a microorganism present at any time, $t$ in a bioreactor with a specific growth rate '$\mu$'. Then the rate of biomass formation can be simply obtained from Monod model given as:

$$\frac{dX}{dt} = \mu X \tag{56}$$

where '$\mu$' is specific growth rate and describes how fast the cells are producing. The higher the value of specific growth rate, the faster the cells are growing. When the cells are not growing, their specific growth rate is zero. It is constant during the exponential phase of growth. Specific growth rate ($\mu$) can be obtained from the following relation, known as Monod model.

$$\mu = \frac{\mu_m S}{K_s + S} \tag{57}$$

where $K_s$ is saturation or Monod constant; $\mu_m$ is the maximum specific growth rate and $S$ is the concentration of the growth limiting substrate. Equation (56) can be modified by the substitution of value of $\mu$, from Eq. (57) to give:

$$\frac{dX}{dt} = \frac{\mu_m S}{K_s + S} X \tag{58}$$

If biomass formation and product formation are assumed to be directly linked to substrate utilization by yield coefficients, then one can write for biomass formation:

$$\frac{dX}{dt} = -Y_{XS} \frac{dS}{dt} \tag{59}$$

and for product formation:

$$\frac{dP}{dt} = -Y_{PS} \frac{dS}{dt} \tag{60}$$

Or from Eqs. (58) and (59), one can obtain the relation for substrate utilization or consumption:

$$\frac{dS}{dt} = -\frac{1}{Y_{xs}} \frac{\mu_m S}{K_s + S} X \tag{61}$$

and from Eqs. (60) and (61), one can obtain the relation for product formation:

$$\frac{dP}{dt} = \frac{Y_{PS}}{Y_{xs}} \frac{\mu_m S}{K_s + S} X \tag{62}$$

where, $Y_{PS}$ and $Y_{XS}$ are called stoichiometric yield coefficients for product and biomass formation. These give the values of product or biomass formed for each unit of reactant or substrate utilized.

## 6.11 Theoretical Consideration in Continuous Stirred Tank Reactors (CSTR)

The liquid entering the stirred tank reactor (Fig. 17) is called the feed or the influent, and the liquid which flows out of the reactor is called the effluent. It is an open system and nutrients are fed continuously to the reactor and the output is taken continuously. It may be either (i) chemostat or (ii) turbidostat.

### *6.11.1 Chemostat*

It can be defined as a continuous culture device controlled by the concentration of limiting nutrients and dilution rate.

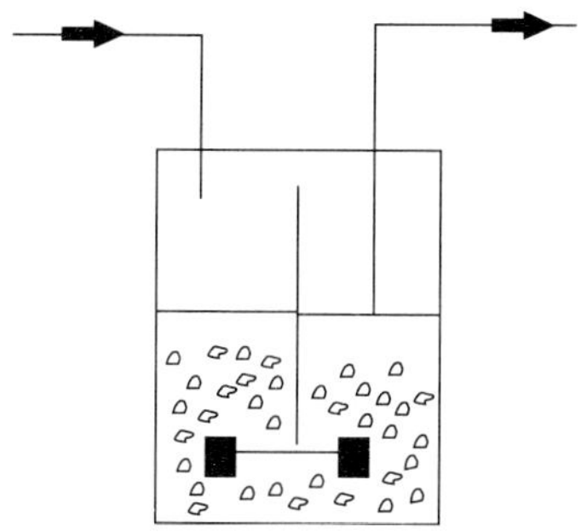

**Fig. 17:** Schematic diagram of continuous stirred tank reactor.

### 6.11.2 Turbidostat

It can be defined as the device which maintains a constant cell density by continuous harvesting and dilution of the remaining biomass and constant addition of nutrient chemicals, etc. The flow rate of media through the vessel is automatically regulated to maintain a predetermined turbidity or cell density. If the growth medium after fermentation has become highly turbid, then increase the flow rate; if it is not turbid, then decrease the flow rate.

### 6.11.3 Assumptions in the model

An important objective of continuous culture operation is to control cell growth at a level at which the productivity is optimum. There are several ways in which this can be achieved. One is to maintain a constant fermenter volume and to use a flow rate that gives an appropriate productivity. In this mode of operation, the fermenter system is known as a chemostat, and analyzing the CSTR here chemostatic approach is followed. Several assumptions which have been considered while obtaining mathematical analysis for the CSTR are as follows:

(i) The feed is assumed to be sterile.
(ii) The reactor is assumed to be perfectly mixed such that the composition of the effluent is the same as that of the reactor.
(iii) In addition, product formation is assumed to be growth associated.
(iv) Growth is assumed to follow the Monod model.

As with our batch culture model, the equations can be modified to consider non-growth associated product formation and variations of the Monod model, such as inhibition.

For a batch reactor, Eqs. (58), (61) and (62) can be modified to describe biomass, substrate and biomass accumulation in a chemostat:

Biomass (X)
$$\frac{dX}{dt} = \frac{\mu_m S}{K_s + S} X - DX \tag{63}$$

and for growth limiting nutrient (S)
$$\frac{dS}{dt} = -\frac{1}{Y_{xs}} \frac{\mu_m S}{K_s + S} X + D(S_0 - S) \tag{64}$$

and for Product (P)
$$\frac{dP}{dt} = \frac{Y_{PS}}{Y_{xs}} \frac{\mu_m S}{K_s + S} X - DP \tag{65}$$

where $S_0$ is the concentration of the growth limiting substrate in the feed and $D$ is the dilution rate. Equations (63) and (64) can be solved using numerical techniques or by steady state analysis. These equations (63) and (64) predict that under stable constant operating conditions, biomass, substrate and product concentrations become constant (Fig. 18). The model has been developed from a mass balance on the biomass, substrate and product components of the system, since at steady state, biomass, substrate and product concentrations do not vary. Thus, the following equations hold:

$$\frac{dX}{dt} = 0 \tag{66}$$

$$\frac{dS}{dt} = 0 \tag{67}$$

$$\frac{dP}{dt} = 0 \tag{68}$$

Equations (66) to (68) can be used to predict the steady state concentrations of $X$, $S$ and $P$ at any dilution rate. In practice, a true steady state condition is almost impossible to achieve.

**6.11.3.1 Steady state analysis of the biomass equation:** Steady state biomass concentration with respect to time in a CSTR is shown in Fig. 19. A biomass material balance obtained from Eq. (63) yielded the following equation:

$$\frac{dX}{dt} = \mu X - DX \tag{69}$$

But at steady state from Eq. (66) $\frac{dX}{dt} = 0$

then, substituting the value of $dx/dt$ from Eq. (66) into Eq. (68) one gets

$$\mu X = DX$$

and finally

$$\mu = D \tag{70}$$

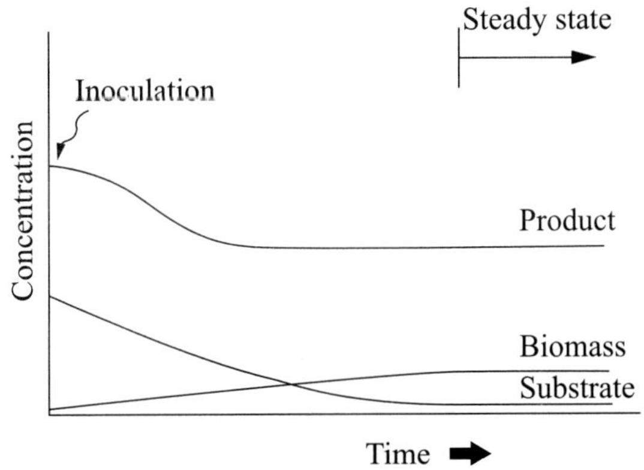


**Fig. 18:** Change in the concentration of biomass, product and substrate with time.

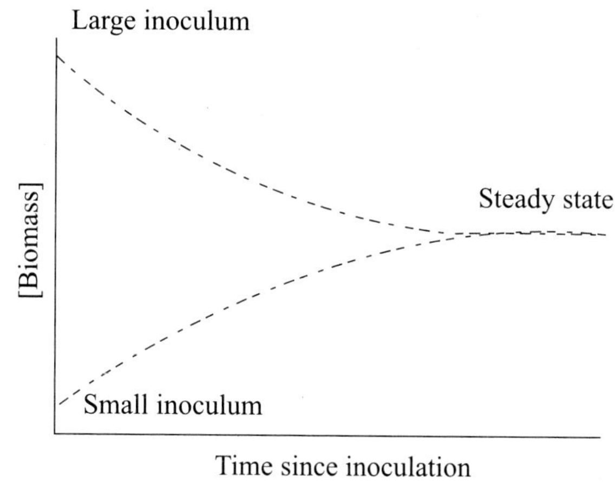


**Fig. 19:** Steady state biomass concentration in a CSTR.

Thus in a chemostat, the steady state specific growth rate of an organism is set by the dilution rate. This means that by simply setting the dilution rate, it is possible to set the growth rate of the organism. If one wants the organisms to grow at a faster rate, increase the dilution rate and vice versa. Typical biomass and substrate concentration variations with dilution rate are shown in Fig. 20. Optimization of product and biomass formation can therefore be performed by setting the correct dilution rate. For example, if an organism produces a product maximally only when it grows slowly, then a low dilution rate should be used.

6.11.3.2 Steady state relationship between the dilution rate and substrate concentration: In a chemostat, the specific growth rate is equivalent to the dilution rate

$$\mu = D$$

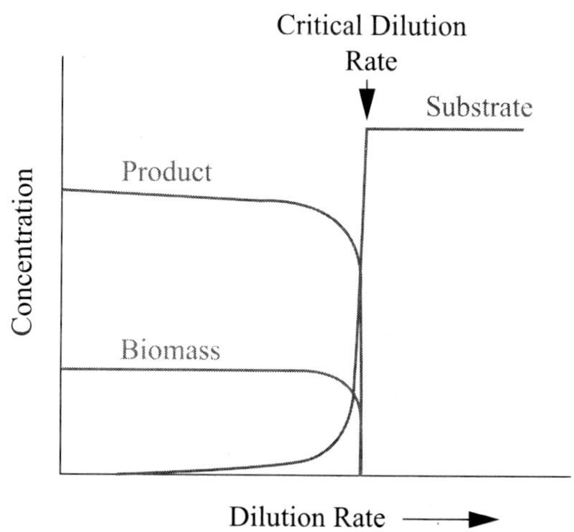


**Fig. 20:** Variation in biomass and substrate concentration with dilution rate.

Now assuming that Monod Model holds true, substituting the value of μ from Eq. (57) in Eq. (70)

$$D = \frac{\mu_m S}{K_s + S} \tag{71}$$

or, $D\ (K_s + S) = \mu_m\ S$

or $D\ K_s + D\ S = \mu_m\ S$

or $D\ K_s = \mu_m\ S - D\ S$

$$D\ K_s = (\mu_m - D)\ S \tag{72}$$

Finally one can have a relationship between the dilution rate and the concentration of the growth limiting nutrient.

$$S = \frac{DK_S}{\mu_m - D} \tag{73}$$

If Monod model describes cell growth, then the steady substrate concentration in the fermenter can be correlated to the dilution rate.

6.11.3.3 Steady state biomass concentration equation: Equation (64) obtained from a mass balance for the growth limiting substrate across the reactor gives:

$$\frac{dS}{dt} = -\frac{1}{Y_{xs}} \frac{\mu_m S}{K_s + S} X + D(S_0 - S)$$

or one can write it as

$$\frac{dS}{dt} = -\frac{1}{Y_{xs}} \mu X + D(S_0 - S) \tag{74}$$

But, at steady state from Eq. (67)

$$\frac{dS}{dt} = 0$$

and therefore

$$0 = -\frac{1}{Y_{xs}} \mu X + D(S_0 - S)$$

or

$$D(S_0 - S) = \frac{1}{Y_{xs}} \mu X \tag{75}$$

or at steady state from Eq. (70) $\mu = D$, therefore

$$(S_0 - S) = \frac{1}{Y_{xs}} X \tag{76}$$

And finally the steady state equation for the biomass concentration in a chemostat can be expressed as:

$$X = Y_{xs}\ (S_0 - S) \tag{77}$$

As the relationship between the specific growth rate and the substrate concentration is described by the Monod model, the equation can be described by using Eq. (73), as:

$$X = Y_{XS}\left(S_0 - \frac{D.K_S}{\mu_m - D}\right) \tag{78}$$

For achieving maximum utilization of substrate concentration ($S_0$) and maximum population of $X$, optimum $D$ value for a fermenter to be run can be given as:

$$D_{opt} = \mu_m \left(1 - \sqrt{\frac{K_s}{K_s + S_0}}\right) \tag{79}$$

## 6.12 Theoretical Consideration in a Plug Flow Reactor (PFR)

In a plug flow reactor there is no variation in axial velocity over the cross section. When the fluid moves through a large pipe or channel with sufficiently large Reynolds number ($N_{Re}$ >2100 in a pipe), it approximates the plug flow. In reality, the plug flow reactor (PFR) is hard to be found, however, the packed bed fermentor can be approximated as PFR. In a PFR (Fig. 21) at steady state condition, considering the section $t = 0$ and $t = \tau$ the mass balance yields:

$$\frac{dX}{d\tau} = \mu X \tag{80}$$

**Fig. 21:** Schematic diagram of plug flow reactor.

where, $\tau$ is the residence time for the PFR. The above equation (80) is similar to that of a batch reactor balance, except in this case $t = \tau$ (residence time of PFR). Further derivation to obtain biomass and substrate are similar to that of a batch reactor. In contrast to a CSTR, sterile feed to a PFR means that there is zero biomass concentration in effluent. PFR provides greater substrate conversion and higher product concentration than CSTR of equal volume. CSTRs are often used in series to decrease the average residence time required for a given conversion (relative to a single, large CSTR). PFR typically maximizes the effluent product concentration for microbial processes [29, 30].

## 6.13 Applications of CSTR and PFR in Industry

The most widespread large scale application of continuous reactors is in wastewater treatment [31]. Activated sludge plants, trickle bed filters, anaerobic digesters and ponds all operate in continuous manner. Cell immobilization is also often employed to improve the efficiency of the process. These are well established in the wastewater industry for several reasons:

(i) Contamination is not a consideration, as the wastewater feed will always contain microorganisms.

(ii) Use of batch cultures is simply not economically feasible, as the flow of wastewater is often measured in mega litres per hour and batch reactors simply could not cope with the load.

Applications of continuous reactor in other food industries are limited. A one million-litre continuous reactor is reported to make single cell protein from animal feed (Pruteen). In this process, methanol is continuously converted to microbial biomass for sale as single cell protein. Some Japanese companies use immobilized non-growing cells for the continuous production of certain amides. Despite

**Table 4:** Industrial applications of CSTR and PFR.

| Product | Reactor Type | Applications |
|---|---|---|
| Alkylate | CSTR | Gasoline component |
| Butadiene | CSTR | Rubber monomer/polymer |
| Caprolactum | CSTR | Plastic or fibers |
| Polyethylene | PFR | Plastics |
| Nylon-6 | PFR | Fiber optics |
| Ethyl acetate | PFR | Soaps and detergents |

the limited applications, considerable effort is being put into research and development for continuous reactors. The future widespread use of continuous reactors in industry should not be ignored.

There has been a wide range of applications of CSTR and PFR in chemical process industries, and some of them are listed in Table 4.

## 7. FUTURE TRENDS IN BIOCHEMICAL ENGINEERING

The extraordinarily rapid pace of biotechnology research has fuelled the industry growth and produced a glamour image that sometimes prove hard to live up to. The early 1980's brought rash properties that biological products and processes would provide cheap and easy solutions to a vast array of industrial problems, but these have gone by the way side, replaced by more realistic evaluations of biotechnology's limited but valuable role in creating novel, high-value products for specialist's niches in a spectrum of industrial applications. Research has already produced far more successes than failures, and the industry is now mature enough and secures enough to focus its technological expertise on realistic targets that have a reasonable probability of producing useful products.

Biochemical engineering includes those fermentation and enzymatic conversions that yield food stuffs. Biochemicals used directly for human consumption include individual amino acids, fructose and citric acid. It has many possibilities in the long run, as it can revolutionize medical, agricultural, and industrial biotechnology [32-34]. However, the time span required for the revolution is decades.

The scope of biochemical engineering is beginning to include the complex structures of biological origin, for e.g., monoclonal plants for agriculture just as chemical engineer now participate in manufacture of fabricated plastic structures and microelectronic components, topics far a field from the old chemical process industries [35]. Therefore, there is a serious need to transform complex technological processes into the engineering applications. Concentrated efforts from the engineers can reflect better growth of the industry in the field.

## REFERENCES

1. Anonymous (1988) *The International Biotechnology Handbook,* Euromonitor Publication Ltd, 87-88, Turmoil Street, London, UK.
2. Batty JC & Folkman SL (1983) *Food Engineering Fundamentals,* Wiley, New York, USA.
3. Charm SE (1978) *Fundamentals of Food Engineering,* AVI Publishing Co, Westport, USA.
4. Bailey JE & Ollis DF (1986) *Biochemical Engineering Fundamentals,* Mc Graw-Hill Book Co, New York, USA.

5. Earle RL (1983) *Unit Operations in Food Processing*, Pergamon Press, Elmsford, New York, USA.
6. Brennan JG, Butters JR, Cowell ND & Lilly AEV (1976) *Food Engineering Operations,* Applied Science London, UK.
7. Atkins B (1974) *Biochemical Reactors,* Pion Limited, London, UK.
8. Leniger HA & Beverloo WA (1975) *Food Process engineering*, D. Reidel Publishing Company, Dordrecht, Holland.
9. Mc Cabe WL, Smith JL & Harriott P (1993) *Unit Operations of Chemical Engineering*, McGraw-Hill, Inc., New Delhi, India.
10. Ibarz A & Barbosa-Canovas GV (2003) *Unit Operations in Food Engineering,* CRC Press, Boca Raton, USA.
11 Nag PK (1985) *Engineering Thermodynamics*, Tata McGraw Hill Publishing Co., New Delhi, India.
12. Pandey H, Sharma HK, Chauhan RC, Bera MB & Sarkar BC (2004) *Experiments in Food Process Engineering,* CBS Publishers and Distributers, New Delhi, India.
13. Toledo RT (1980) *Fundamentals of Food Process Engineering,* AVI Publishing Co., Westport, USA.
14. Treybal RE (1981) *Mass Transfer Operations,* Mc Graw Hill International, Singapore.
15. Rao MA & Rizvi SSH (1986) *Engineering Properties of Foods*, Marcel Dekker, New York, USA.
16. Rizvi SSH & Mittal GS (1997) *Experimental Food Engineering*, CBS Publishers, New Delhi, India.
17. Rajput RK (1999) *Heat and Mass Transfer,* S. Chand and Company Ltd., New Delhi, India.
18. Lee JM (1991) *Biochemical Engineering*, Prentice Hall, Engelwood Cliffs, New Jersey, USA.
19. Stanbury PF, Whitaker A & Hall JS (1997) *Principles of Fermentation Technology,* Aditya Books (P) Ltd., New Delhi, India.
20. Watson EL & Harper JC (1988) *Elements of Food Process Engineering,* AVI Publishing Co., Westport, USA.
21. Bungay HR & Belfort G (1987) *Advanced Biochemical Engineering,* John Wiley and Sons, New York, USA.
22. Farrall AW (1976) *Food Engineering Systems,* Vol 1, *Operations,* AVI Publishing Co., Westport, USA.
23. Harper JC (1976) *Elements of Food Engineering*, AVI Publishing Co., WestPort, USA.
24. Heldman DR & Lund DB (1992) *Handbook of Food Engineering,* Marcel Dekker, Inc., New York, USA.
25. Heldman DR & Singh RP (1975) *Fundamentals of Food Process Engineering,* AVI Publishing Co., Westport, USA.
26. Heldman DR & Singh RP (1981) *Food Process Engineering,* AVI Publishing Co., Westport, USA.
27. Merkel JA (1983) *Basic Engineering Principles*, AVI Publishing Co., WestPort, USA.
28. Bu'lock J & Kristiansen B (1987) *Basic Biotechnology*, Academic Press, London, UK.
29. Loncin M & Merson RL (1979) *Food Engineering, Principles and Selected Applications,* Academic Press, New York, USA.
30. Levenspiel O (1972) *Chemical Reaction Engineering,* John Wiley & Sons, New York, USA.

31. Lee BH (1996) *Fundamentals of Food Biotechnology*, VCH Publishers Inc., New York USA.
32. Solomons GL (1969) *Materials and Methods in Fermentation,* Academic Press, London, UK.
33. Webb FC (1964) *Biochemical Engineering*, D Van Nostrand Company Ltd., London, UK.
34. Crueger W & Crueger A (1984) *Biotechnology: A Text Book of Industrial Microbiology*, Sinaeur Associates, Inc., Sunderland, MA 01375 and Science Tech, Inc., Madison, WI 53705, USA.
35. Henderson SM & Perry RL (1979) *Agricultural Process Engineering,* AVI Publishing Co., WestPort, USA.

# Industrial Fermentation Technology

# 3

**Mani S. Bhattacharyya, Sachin Dubey, Praveen Kaul, Sumali Kapor and Uttam C. Banerjee**

*Department of Pharmaceutical Technology, National Institute of Pharmaceutical Education & Research, S.A.S. Nagar 160 062 (Pb.)*

## 1. INTRODUCTION

The use of natural biological processes to obtain usable products is certainly not new. Since recorded history, microbes have been involved in the preparation and processing of items in man's daily diet. Long before the term "biotechnology" was coined for the process of using living organisms to produce improved commodities, people were utilizing living microorganisms to produce valuable products (Table 1). Usually the actions of these microbes ultimately made the food unacceptable, either by altering the appearance or odour of the food to a point which was no longer appetizing or by producing poisonous toxins, some of which were lethal. Occasionally, however, microbial infections of food materials made it appear more appetizing and the taste enhanced. Ultimately, microbial infections of these foods were exploited, and now the fermented foods and beverages form a large and important sector of the food industry. Today the main groups of microbes involved in the industry include yeasts, moulds and bacteria.

**Table 1:** List of ancient biotechnology applications.

| Milestone | Period or year |
|---|---|
| Proving bread with leaven | Prehistoric period |
| Fermentation of juices to alcoholic beverages | Prehistoric period |
| Knowledge of vinegar formation from fermented juices | Prehistoric period |
| Cultivation of vine | Before 2000 BC |
| Manufacture of beer in Babylonia and Egypt | 3rd Century BC |
| Wine growing promoted by Roman Emperor Marcus Aurelius Probus | 3rd Century AD |
| Production of spirits of wine (ethanol) | 1150 |
| Vinegar manufacturing industry | 14th Century |
| Discovery of the fermentation properties of yeast by Erxleben | 1818 |
| Description of lactic acid fermentation by Pasteur | 1857 |
| Detection of fermentation enzymes in yeast by Buchner | 1897 |
| Discovery of penicillin by Fleming | 1928/29 |
| Discovery of many other antibiotics | From about 1945 |

*Food Biotechnology: Principles and Practices*
*V.K. Joshi & R.S. Singh (Eds.)*
*I.K. International Publishing House Pvt. Ltd., India*
*pp 71–94*

Since then "biotechnology" has rapidly progressed and expanded. In the mid-forties, scale-up and commercial production of antibiotics such as penicillin occurred. The techniques used were: (a) isolation of an organism producing the metabolites of interest using screening/selection procedures, and (b) improvement of production yields via mutagenesis of the organism or optimization of media and fermentation conditions. This type of "antique" biotechnology is limited to metabolites produced in nature. It is also limited by its trial-and-error approach, and requires a lengthy timeframe (years or even decades) for yield improvement.

Fermentation is the term used by microbiologists to describe any process for the production of a product by means of the mass culture of a microorganism. The product can either be:

1. *The cell itself:* referred to as biomass production.
2. *A microorganism's own metabolite:* referred to as a product from a natural or genetically improved strain (Table 2).
3. *A foreign product produced in a microorganism:* referred to as a product from recombinant DNA technology or genetically engineered strain, i.e. recombinant strain.

For each biotechnological process, the most suitable containment system must be designed and then monitored and controlled. The environment in which the desired biocomponents can interact with the environment and material supply is known as the fermenter. Fermenters range from simple stirred tanks to complex integrated systems involving varying levels of computer input. Fermenters occur in two distinct types. The first are non-aseptic systems where it is not necessary to operate with strictly pure cultures of microorganisms. The other type of fermenter is used for production of such compounds as antibiotics, amino acids and polysaccharides. Thus, this type of fermenter involves considerable challenges on the part of the engineering team. The purpose of any fermenter is to optimize the growth of the organism or of a product produced by the organism. To obtain such conditions, the factors taken into consideration are an energy source, other essential nutrients to satisfy the needs of the organism, absence of inhibiting compounds in the medium, a reliable inoculum, and the most advantageous physico-chemical conditions. There are two types of fermentation systems: closed or open. A closed system implies that all the nutrient components are added at the beginning of the fermentation process and, as a result, the growth rate of the contained organisms will eventually proceed to zero either due to diminishing nutrients or accumulation of toxic waste products. A modification of the batch process is the fed batch system. Here, volumes of nutrients may be added to augment depletion of nutrients. Overall, the system remains closed and there is no continuous flow. In contrast to the above types, in the open system, feeding of nutrients is carried out continuously in the fermenter and spent medium is removed at the same rate from the growing culture.

**Table 2:** Products produced by microbial fermentation.

| Products | Examples |
|---|---|
| Antibiotics | Penicillin, Cephalosporins, Tetracyclines |
| Commodity enzymes | Amylases, Proteases, Lipases |
| Specialty enzymes | Restriction endonucleases, Streptokinases |
| Catalytic enzymes (Biocatalysts) | Lipases, Nitrilases, Oxidoreductases |
| Biomass | Yeast, Cheese starter cultures, Bioremediation starter cultures, Algal biomass |
| Biopolymers | Xanthan gum, Polyhydroxy alkanoates, Dextran |
| Biopesticides | *Bacillus thuringiensis* preparations, Baculovirus preparations |
| Foods | Soya sauce, Yoghurt, Vinegar, Cheese |
| Pigments | β-carotene, astaxanthine |
| Therapeutics | Interleukin 2, Tumour necrosis factor, Interferons, Vaccines, Growth Hormones, Antibodies |

To achieve optimization of the fermenter system, the following guidelines must be closely followed: the fermenter should be designed to exclude entrance of contaminating organisms as well as to prevent the outgo of the desired organism, the culture volume should remain constant, the dissolved oxygen level must be maintained above critical levels of aeration and culture agitation for aerobic organisms, parameters such as temperature and pH must be controlled, and the culture volume must be well mixed. The following types of materials should be used in fermenter construction:

1. All materials coming into contact with the solutions entering the fermenter or the actual organism culture must be corrosion resistant to prevent trace metal contamination of the process.
2. The materials must be non-toxic so that slight dissolution of the material or components does not inhibit culture growth.
3. The materials of the fermenter must withstand repeated sterilization with high pressure steam.
4. The fermenter stirrer system, entry ports and end plates must be sufficiently rigid not to be deformed or broken under mechanical stress.
5. The fermenter system must be compatible with utility systems like gas analyzer, online biochemical analyzer, data recording and processing softwares, etc.
6. Visual inspection of the medium and culture is advantageous, transparent materials should be used where ever possible.

## 2. SCALE-UP OF FERMENTATION PROCESS

One aspect of the fermentation development program is scale-up, which is the ability to transfer procedures developed in the laboratory to protocols used in the plant. This can often involve an increase in volume of one thousand-fold or more. This requires knowledge of plant operations, the differences between large and small scale equipment, as well as "non-scientific issues" such as utility costs, availability of materials and supplies, and labour practices.

Following development at the laboratory scale, the cell culture and fermentation processes are scaled up and the operation conditions and target ranges are tested at the pilot scale. Process modifications are made as necessary to facilitate operation at the manufacturing scale before transfer to the cGMP facility. Having chosen a purification stratergy from the laboratory scale evaluation, the process is scaled-up and run in the pilot plant. This allows assessment of process for its ability to meet the target specification for purity and product characteristics and to make modifications as necessary before transferring the process to manufacturing. The product produced at this stage is often used for pre-clinical studies. Scale-up scientists work closely with the development, manufacturing and quality specialists to document all product specific processes and transfer them into the manufacturing facility. This intimate interaction between teams continues throughout the project to ensure that during the early full-scale manufacturing batches, the process operates in a reliable and predictable manner. Template documentation systems permit rapid transfer of new processes into the cGMP system. The scale-up group is also responsible for pilot scale studies on the process as part of the development program examining key parameters to ensure successful operation in the cGMP environment. The optimization of the fermentation process is included in the strategic analysis of the viability of a project. Optimization takes place once the feasibility of production with the selected organism has been demonstrated. This implies that an expression system has been constructed and that, at least from a theoretical viewpoint, the system should be regarded as the optimum. Before starting long and expensive optimization work, it is important that the stability of the strain is established, at least for the number of generations necessary for cell banking and largest-scale fermentation, including the pre-cultures. Expression systems based on plasmids are sometimes unstable. Several parameters affect the segregational stability of plasmids. The stability is different for each plasmid and depends on the host strain [1]; high instabilities are correlated with low copy number plasmids [2]. The size of the inserted DNA also affects the stability; the larger the plasmid, the less stable it is [3]. Culture conditions, such as temperature [4], media composition and growth rate [5] can also modify the plasmid stability. Antibiotics are often used to stabilize plasmids. The compensation of an auxotrophic defect of the host coded on the plasmid is more suitable than using antibiotics for regulatory reasons [6]. However, the preparation of media for auxotrophic strains is laborious. The choice of the producing strain and the expression vector will depend on the constitutive or inducible expression into the cytoplasmic compartment or the secretion into the outer medium. If the strain has been acquired outside the laboratory, the quality of documentation of source of original strain and the cloning procedure must be evaluated and approved by a competent quality assurance service. The codon usage of the gene of interest should be optimized to facilitate the expression in the chosen microorganism. A screening of expression in a shake-flask must be performed immediately to find a clone that produces the recombinant protein at a high level of expression.

## 2.1 Optimization of Culture Conditions

The main aim of optimization is to maximize the production, a process can be initiated only once at a laboratory-scale and a minimum set of quality control tools are available to quantify and assess the quality of the product. The fermentation protocol affects the impurity profile and thus impacts strongly the efficacy of the downstream processing (DSP). Similarly, the conditions of fermentation can determine whether the product of interest will be in its soluble or insoluble form, which will deeply affect the DSP and consequently the quality, and yield of the purified product. Overproduction at high yield can also lead to the depletion of certain factors that are necessary for good conformation of the product. The fermentation must thus be developed concomitantly with the purification because they will affect each other. The quality of the communication between the developers of the fermentation and the DSP is pivotal to the success of the scaling-up process (Fig. 1).

Optimization of the fermentation process can be conducted either by changing one factor at a time or by varying several factors at the same time and looking for interactions using statistical analysis. The statistical design of experiments is an organized approach that gives more reliable information per experiment than unplanned approaches. Statistical data analysis enables the visualization of the interactions among experimental variables, leading to predictions of the response data in areas not directly covered by experimentation [7]. One of the most important operations for the investigator is

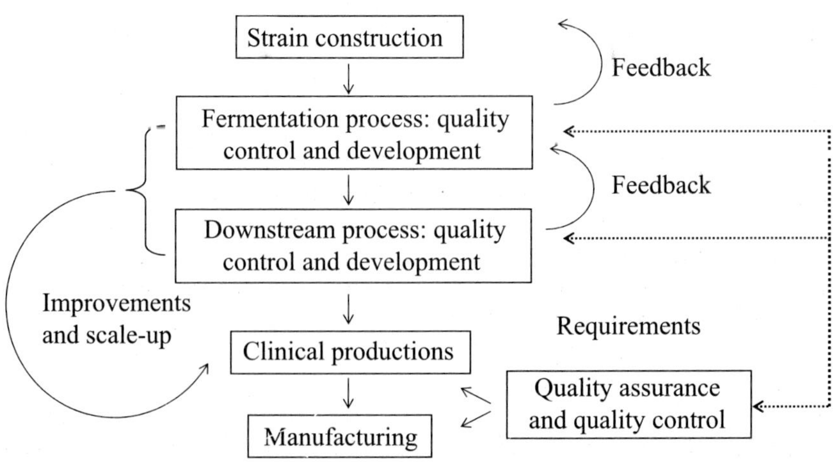


**Fig. 1:** Interactions between various divisions for the development of a bioprocess.

to define the variables that could affect the productivity or the quality of the product. Fermentation can be performed in batch, fed-batch or continuous mode. Continuous culture is not common in the pharmaceutical industry because the probability of mutation and contamination is higher and the definition of a batch size is not evident. Batch processes are simple and robust, but the only way to reach a high cell density is the fed-batch mode, which is more complex but allows the metabolism of the strain to be controlled. For *S. cerevisiae*, high concentration of glucose induces the Crabtree effect: the yeast enters in a fermentative metabolism even if there is no limitation of oxygen [8]. The metabolism can be controlled by using a fed-batch method. Feed-limiting the glucose concentration in the broth allows the cells to be grown to high density in a respiratory metabolism for *S. cerevisiae* [9] and avoids the production of acetate by *E. coli* [10]. The comparison of the productivity of the recombinant product per day is used to choose the most economical process. The medium can be either completely defined or complex when it includes peptones and growth promoters. Complex media are easy to prepare, cheap and fast growing, but are difficult to analyze and variation can occur between batches. Defined media gives reproducible growth and the purity in the supernatant for secreted product is better than with complex media. In both cases, the choice of the carbon source is crucial. High concentration of glucose will lead to the accumulation of acetate, which reduces biomass yield and expression levels in *E. coli*. Glycerol should be preferred to glucose for *E. coli* [10]. The use of raw materials of animal origin must be avoided when possible. When this is not feasible, the raw materials must be produced in a country free from transmissible spongiform encephalopathies (TSEs). Optimization of the medium pH is crucial especially for yeast-secreting protein, because they can grow in a wide range of pH. The choice of pH depends on the stability of the recombinant protein expressed. Working with *Pichia pastoris* at low pH allows protease degradation to be avoided. This yeast cannot grow at pH below 2.2. Temperature has an important role in the solubility of the expressed protein and also on the productivity. Lowering the temperature from 30 to 25°C during the methanol induction phase results in a four-fold increase in the yield of galactose oxidase cloned in *Pichia pastoris* [11]. The factors affected by scale-up are the number of generations, the mutation probability, media sterilization, the quality of temperature and pH regulations, agitation, aeration and pressure. The best way to prepare the scaling-up of a process is to first scale-down to the pilot scale of the conditions of culture that will be used at the final scale of production [12]. When the scale is increased, the broth will become more and more heterogeneous. In large fermenters, oxygen can be depleted in some area of the reactor [13]. The process must be characterized and validated by some controls to demonstrate the robustness of the procedure and thus its reliability and reproducibility. All the equipment must be validated for good manufacturing practices, including the qualification of the installation, validation of cleaning in place (CIP) and sterilizing in place (SIP), maintenance and calibration plan [14]. The fermentation protocol can be improved until clinical phase III, which must be performed at the final scale of production. The optimization of a heterologous protein production starts with the construction of a well-designed recombinant strain. The fermentation development stratergy involves the knowledge of the behaviour of strain and the recombinant protein. The fermentation process is developed taking into account the feedback from the DSP development, the final scale of production and the good manufacturing practice requirements. Even though there is a trend to standardize as much as possible of the fermentation processes, it has been shown that even subtle changes would powerfully increase the productivity.

## 2.2 Scale-up of Aeration and Agitation

Aeration and agitation affect a large list of environmental parameters, and it is highly unlikely that the conditions of small-scale fermentation are replicated precisely on the large-scale. Suitable conditions of mixing and oxygen transfer can be maintained with a range of aeration/agitation combinations, i.e., the agitation rate must fall between a maximum and minimum value. Mixing is inadequate below minimum level, and shear-damage to the cells is too great above the maximum value. There are mainly three ways to trouble-shoot scale-up problems:

1. Identify principal environmental parameters affected by aeration and agitation.
2. Identify the process variables that affect the identified environmental domain.
3. Calculate value of the process variable to be used on the large-scale for exact replication of same environmental conditions on both scales.

The process variables that affect mixing and mass transfer are summarized in Table 3. If DO concentration is perceived as the over-riding parameter, then power consumption per unit volume should be maintained constant on scale-up. The most important environmental domains affected by aeration and agitation for majority of fermentations are oxygen concentration and shear. Thus, the mostly used scale-up criteria are maintenance of constant $K_La$ or constant shear conditions.

**Table 3:** Effect of process variables on mass transfer or mixing.

| Process variable | Affected mass transfer or mixing characteristic |
|---|---|
| Power consumption per unit volume | Oxygen transfer rate |
| Volumetric air flow rate | Oxygen transfer rate |
| Impeller tip speed | Shear rate |
| Pumping rate | Mixing time |
| Reynolds number | Heat transfer |

## 2.3 Scale-down Methods

Scale-down is the situation where laboratory or pilot-scale experiments are conducted under conditions, which mimic the industrial-scale conditions. This approach is utilized for the development of a new product and the improvement of a current full-scale fermentation [15]. The aspects that must be considered in design of laboratory or pilot plant experiments in the context of scale-down may be regarded as:

1. *Medium design:* Only the relevant media should be used.
2. *Medium sterilization:* In case of large-scale batch sterilization, exposure time at high temperature will be much greater than the pilot plant. Thus sterilization time for small-scale should be increased. Alternatively, medium sterilized in production fermenter may be utilized on pilot-scale.
3. *Inoculation procedure:* It is not always possible to inoculate the production fermenter under optimal conditions. The scale-down approach may be used to predict the consequences of such events by mimicking these situations in the laboratory.
4. *Number of generations:* The requirement for greater number of generations in case of industrial-scale as compared to pilot plant demands greater stability criteria for the strain. The industrial situation may be modelled in the laboratory by using serial subculture to ensure that the strain is sufficiently stable.
5. *Mixing:* An inevitable decrease in degree of mixing will occur with increase of scale.
6. *Oxygen transfer:* The oxygen transfer rate (OTR) decreases drastically as the scale is increased. Thus, unrealistic demands may arise for a fermentation plant if the development work has been done at very high OTR.

## 3. BASIC FUNCTIONS OF A FERMENTER

The main function of a fermenter is to provide a controlled environment for the growth of micro-organisms or animal cells to obtain a desired product. The following points must be considered in the design:

1. Should be operable aseptically for a number of days.
2. Adequate aeration and agitation must be provided without any shear damage to the cells.
3. Minimum power consumption.
4. Provisions for temperature and pH control.
5. Sampling facilities.
6. Evaporation losses should be minimized.
7. Should be designed to require minimum labour in operation, harvesting, cleaning and maintenance.
8. Should have smooth internal surfaces using welds instead of flange joints.
9. Vessel should be of similar geometry to both smaller and larger vessels in pilot plant or industrial plant to facilitate scale-up.

The first two points are probably the most critical, and it is obvious that the design of a fermenter requires collaboration from specialists of different fields. Fig. 2 represents the diagram of a typical mechanically agitated and aerated fermenter having one multi-bladed impeller.

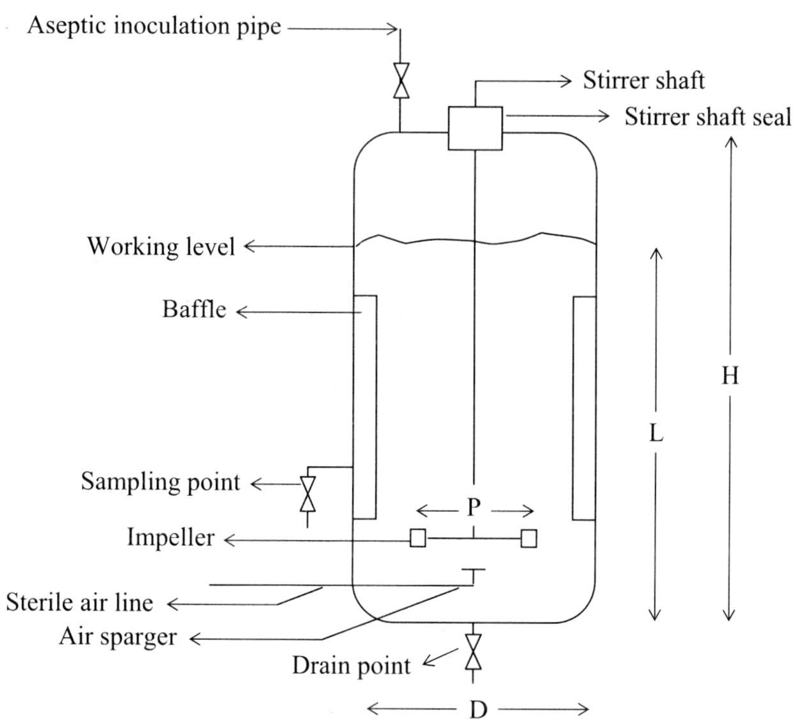


**Fig. 2:** Diagram of a fermenter with one multi-bladed impeller.

## 4. MASS TRANSFER

### 4.1 Boundary Layers

Boundary layer theory is central to the study of interfacial mass transfer process. A boundary layer is described as a thin layer of fluid surrounding a body which is immersed in or is in contact with a fluid. The body can be a solid object, e.g. a pipe wall, a reactor tank wall, an immobilized enzyme particle, a cellular aggregate, or a solid substrate such as wood or a fluid of a different phase than the medium, e.g. an oxygen bubble or an oil globule in water (Fig. 3). The following example illustrates the boundary layer concept. Consider the profile of a fluid flowing in a pipeline. The liquid velocity will be highest in the centre of the pipe and will be slowest at the walls. Thus, surrounding the walls will be a region of very slow flow. This region is known as a boundary layer. Extending this concept, we see that boundary layers also exist around walls of tanks and other solid surfaces as well as around objects immersed in a fluid. Because fluid movement in a boundary layer is slow, the movement of solutes across the boundary layer will be determined by diffusion forces rather than by forced convection. Thus in assessing mass transfer rates between phases, it is assumed that movement of solutes through boundary layers is the rate limiting step.

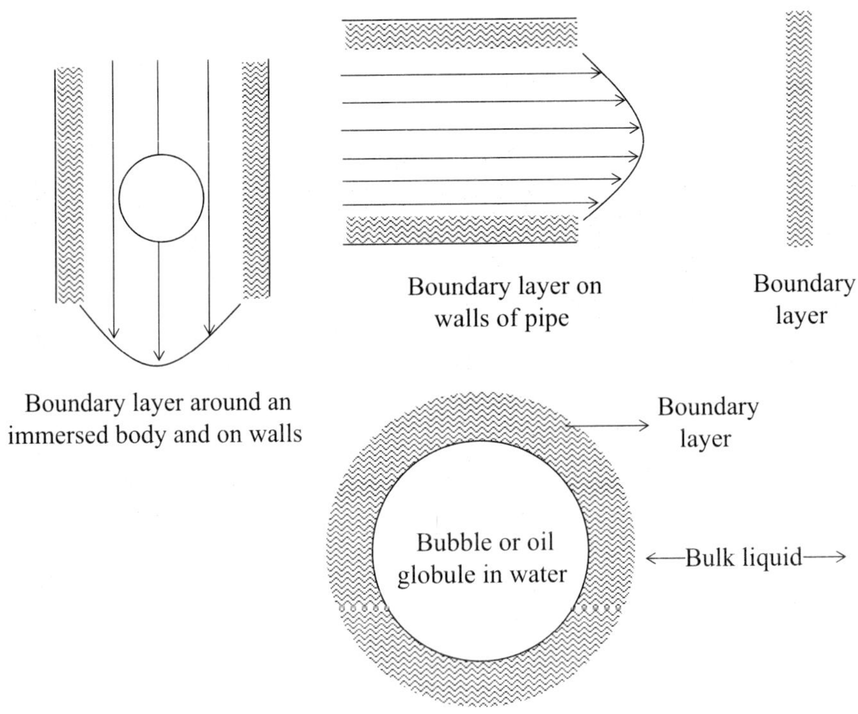


**Fig. 3:** Various examples of boundary layer.

### 4.2 Aeration and Agitation

Fermenters usually require very good agitation, and that takes considerable power. Hence, a strong agitator shaft is required, and entry of the shaft into the fermenter could provide a path for contamination. Also, the fermenter is held above the atmospheric pressure, so the place where the shaft enters should not be a place where air escapes. If the shaft enters the top of the vessel and extends to near

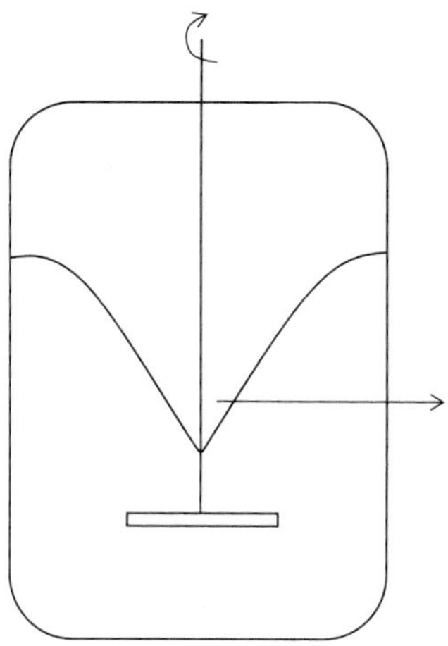


**Fig. 4:** Diagrammatic explanation for power loss due to absence of baffles in a fermenter.

the bottom, some support is required. Without baffles, the contents would swirl, and power would be wasted (Fig. 4). With no baffles, centrifugal force drives the swirling fluid towards the walls. Its level is higher if there is no mixing, so it falls into the open space (the vortex). This simple sliding or falling does not create much turbulence or good mixing. Baffles minimize swirling. They are moving relative to the fluid, and it is like having more paddles for mixing. Majority of fermentations are aerobic in nature and therefore require provisions for oxygen. The oxygen demand of an industrial fermentation is normally satisfied by agitating and aerating the broth. However, the productivity of many fermentations is limited by oxygen availability, and therefore it is important to consider the factors which will influence the fermenter's efficiency in supplying oxygen to the cells. Oxygen is usually supplied in the form of air, the cheapest available gaseous source. The transfer of oxygen from air to the cell can be described in a number of steps [16]:

1. The transfer of oxygen from an air bubble into solution.
2. Transfer through the fermentation medium to the cell.
3. Uptake of dissolved oxygen by the cell.

The rate of oxygen transfer from air bubble to the liquid phase may be described by the following equation:

$$dC_L/dt = K_La(C^*-C_L) \quad (1)$$

where $dC_L/dt$ : the oxygen transfer rate $(g/l/h)$,
$K_L$ : mass transfer coefficient $(h^{-1})$,
a : interface area $(m)$,
$C^*$ : saturated oxygen concentration $(g/l)$,
$C_L$ : DO concentration $(g/l)$,
$t$ : time $(h)$.

$K_La$ may be considered as the sum of reciprocals of the resistances to the transfer of oxygen from gas to liquid and $(C^* - C_L)$ may be considered as the driving force. Since it is extremely difficult to measure '$K_L$' and '$a$' separately, the two terms are combined as $K_La$, the volumetric mass transfer coefficient and is regarded as the measure of aeration capacity of the fermenter. The greater the $K_La$, the greater is the aeration capacity. Thus, it is very essential to measure $K_La$ of a fermenter.

## 5. $K_La$ DETERMINATION

### 5.1 Sulphite Oxidation Technique

It measures the rate of conversion of 0.5 M solution of sodium sulphite to sodium sulphate in the presence of copper or cobalt catalyst:

$$Na_2SO_3 + 0.5\ O_2 \rightarrow Na_2SO_4$$

As soon as oxygen enters the solution, it is immediately consumed in the oxidation of sulphite, so that oxidation rate equals oxygen transfer rate (OTR).

$K_La$ may be calculated from the equation:

$$\text{OTR} = K_La.\ C^* \tag{2}$$

This method is simple and represents the bulk state inside the fermenter, however, it is time consuming and very inaccurate. Also, the rheology of sodium sulphite solution is completely different from any fermentation broth.

### 5.2 Gassing-out Techniques

It relies upon the estimation of increase in DO of a solution during aeration and agitation. The OTR will decrease during the period of aeration as $C_L$ approaches $C^*$ due to decline in driving force $(C^* - C_L)$. In order to monitor the increase in DO over an adequate range, it is necessary to decrease the oxygen level first. Two methods have largely been employed for this purpose i.e., static and dynamic method.

#### *5.2.1 Static gassing out method*

In this technique, the DO concentration of the liquid is lowered by scrubbing it with nitrogen [17]. The deoxygenated media is now aerated and the DO level is monitored by a DO probe. Integrating equation 1, we get

$$\ln\ [1\text{-}(C_L/C^*)] = -\ K_La.\ t \tag{3}$$

Thus a plot of ln $(C^* - C_L)$ with respect to time will yield a straight line, with a slope of - $K_La$ as shown in Fig. 5. The advantage of this method is that it is very rapid and utilizes the fermentation media.

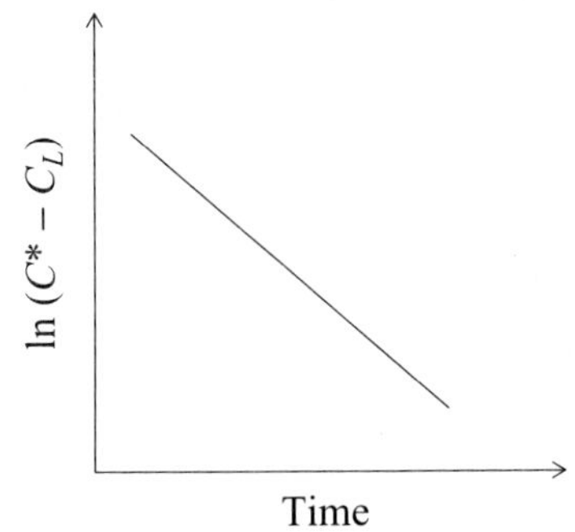


**Fig. 5:** A plot of ln $(C^* - C_L)$ against time.

This method is suitable mainly for smaller vessels. In large fermenters, the gas residence time is very high. As the larger vessel is oxygenated, $C^*$ will no longer be constant.

### 5.2.2 Dynamic gassing out method

The method utilizes respiratory activity of a growing culture in the fermenter to lower the oxygen [18]. Hence the estimation is carried out in conditions that mimic the actual fermentation. However, it requires the use of membrane type covered DO probes and hence the use of response correction factors. The procedure involves stopping the supply of air to fermenter, resulting in a linear decline in DO and then subsequently allowed to increase (Fig. 6 a, b).

The slope AB in Fig. 6a represents the respiration rate of the culture. Aeration is resumed at point B until DO is restored. Over the period BC, the increase in DO is the difference between the transfer of oxygen into the solution and its uptake by the cells which is expressed by the following equation:

$$dC_L/dt = K_La(C^* - C_L) - xQO_2 \qquad (4)$$

Rearranging above equation

$$C_L = -1/K_La\ \{(dC_L/dt) + xQO_2\} + C^* \qquad (5)$$

where $x$ = cell mass concentration & $QO_2$ = specific oxygen uptake rate.

The term $xQO_2$ is given by the slope of line $AB$. Thus a plot of $C_L$ versus $dC_L/dt + xQO_2$ will yield a straight line with slope equivalent to $-1/K_La$ (Fig. 6b). The technique is convenient because it utilizes DO tension rather than concentration, since it is the rates of transfer and uptake that are monitored so that the percentage saturation values generated by the probe may be directly utilized. The advantage of the technique is that it can determine $K_La$ during the actual fermentation run. It is also very rapid and requires just a DO probe. A major limitation is the range over which the increase in DO may be measured. It is important not to allow DO to drop below $C_{critical}$ during the deoxygenation step as the specific oxygen uptake rate will then be limited and the term $xQO_2$ would not be constant over resumption of aeration.

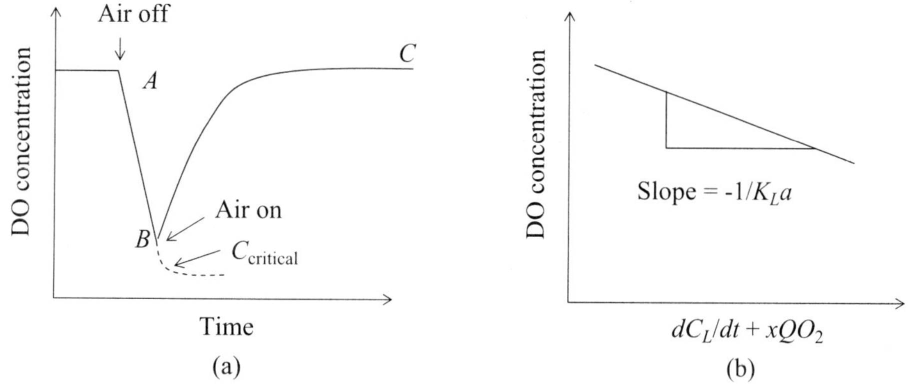


**Fig. 6:** Dynamic gassing out method for estimation of $K_La$.

## 6. RHEOLOGY OF FERMENTATION FLUIDS

Fluids may be described as Newtonian or non-Newtonian depending on whether their rheology (flow characteristics) obey Newton's law of viscous flow or not. Newton's law of viscous flow states that the viscous force, $F$, opposing motion at the interface between the liquid layers, flowing with a velocity gradient of $dv/dx$, is given by equation

$$F = \mu A(dv/dx) \tag{6}$$

where $\mu$ is the fluid viscosity, which may be regarded as the resistance to the fluid flow,
Equation 6 may be rewritten as $F/A = \mu(dv/dx)$
$F/A$ is termed as shear stress ($\tau$) and is the applied force per unit area, $dv/dx$ is the shear rate ($\gamma$) and is the velocity gradient. Thus:

$$\tau = \mu\gamma \tag{7}$$

Equation 7 confirms to the general equation

$$\tau = K\gamma^n \tag{8}$$

where $K$ = consistency coefficient, $n$ = flow behaviour index or power law index.

For a Newtonian flow $n$ is 1 and the consistency coefficient is the viscosity which is the ratio of shear stress and shear rate. Thus a plot of shear stress against shear rate, for a Newtonian fluid would produce a straight line, slope of which would be equal to viscosity. Such a plot is termed as a rheogram (Fig. 7). Thus, a Newtonian fluid has constant viscosity regardless of shear, so that its viscosity will not vary with agitation rate. However, a non-Newtonian liquid does not obey Newton's law of viscous flow and its viscosity is therefore not constant. The value of $n$ (equation 8) for such liquids deviates from 1 and its behaviour is said to follow power law model. Thus, the viscosity of a non-Newtonian fermentation broth will vary with agitation and is described as an apparent viscosity ($\mu_a$). The various types of non-Newtonian fluids are discussed below.

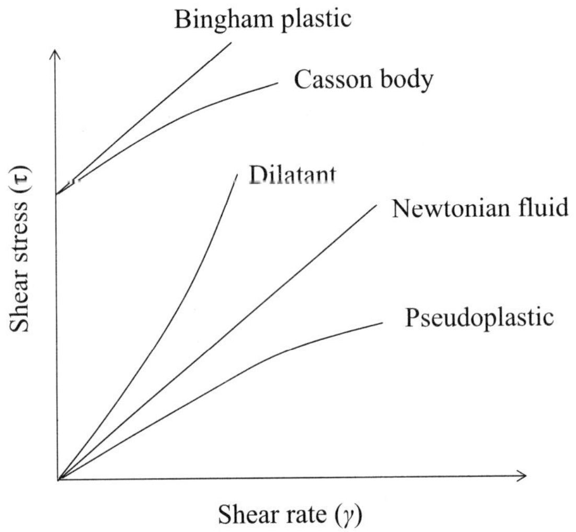


**Fig. 7:** Rheogram of fluids of different properties.

### 6.1 Bingham Plastic Rheology

These are similar to Newtonian fluids except that shear rate does not increase until a threshold shear stress is exceeded. The threshold shear stress is termed the yield stress or yield value, $\tau_0$. A linear relation of shear stress and shear rate is observed once the yield stress is exceeded and the slope of this line is termed as coefficient of rigidity or the plastic viscosity. Thus, the flow of a Bingham plastic is described by the following equation:

$$\tau = \tau_0 + n\gamma \tag{9}$$

where $n$ = coefficient of rigidity, $\tau_0$ = yield stress.

Mycelial fermentation broths, toothpaste and clay are the known Bingham plastics. $K$ has the same units as viscosity and may be taken as the apparent viscosity. The flow behaviour index is less than unity for a pseudoplastic fluid, the smaller the value of $n$, the greater the flow characteristic of the fluid deviate from that of a Newtonian fluid. Equation 8 may be converted to the logarithmic form as:

$$\log \tau = \log K + n \log \gamma \tag{10}$$

Thus, a plot of log shear stress versus log shear rate will produce a straight line, the slope of which will equal the flow behaviour index, and the intercept on the shear stress axis will be equal to the logarithm of consistency coefficient. Many mycelial fermentations have been reported to behave like pseudoplastics.

### 6.2 Dilatant Rheology

The apparent viscosity of a Dilatant liquid increases with shear rate. The flow of a dilatant liquid may also be described by equation 8. But in this case the value of the flow behaviour index is greater than 1. The greater the value, the greater the flow characteristics deviate from those of a Newtonian fluid. Thus, the values of $K$ and $n$ may be obtained from a log plot of shear stress against shear rate. Fortunately, this type of behaviour is not exhibited by fermentation broths. An everyday example is liquid cement slurry.

### 6.3 Casson Body Rheology

Another type of non-Newtonian fluids have also been described [19] termed as Casson body, which behave as a pseudoplastic, that is, the apparent viscosity decreases with increase in shear rate, but they display a high yield stress and therefore also resemble a Bingham plastic. The flow characteristic of Casson body may be described by the following equation:

$$\sqrt{\tau} = \sqrt{\tau_0} + K_c\sqrt{\gamma} \tag{11}$$

where $K_c$ = Casson viscosity.

A plot of $\sqrt{\tau}$ versus $\sqrt{\gamma}$ will give a straight line, the slope of which will equal Casson viscosity and the intercept of the $\sqrt{\tau}$ will be $\sqrt{\tau_0}$. Rheology of penicillin broth has been described to be a Casson body [20]. Hence, in order to determine the rheological nature of a fluid, it is necessary to construct a rheogram which requires the use of a viscometer. Furthermore, mycelial suspensions present greater difficulty and therefore must be assessed critically [21].

## 7. MEDIA

For the production of a particular compound, biochemical processes are highly dependent on the medium constituents used for the growth of the organisms. A stable source of carbon, nitrogen, mineral elements, oxygen (in case of aerobic culture) and water has to be supplied for the cultivation of the microorganisms through the medium of cultivation. Addition of vitamins and sometimes supplementary nutrients is essential for the production of a particular product. Therefore, a detailed investigation is obvious for finding out a suitable medium for an industrial process. Formulation of a suitable media is affected by the scale of production. The same medium though showing suitability in the small scale may be unsuitable for the large scale process due to its economic consideration and process feasibility. It may also be unsuitable for its lower gas transfer potential, high viscosity causing more power consumption, etc. Therefore, medium used for the large scale process must meet the following criteria:

1. The media should give a maximum growth or yield of the product per gram of substrate used.
2. It will produce the maximum concentration of biomass and/or product which ever is important.
3. Undesired product formation must be minimum.
4. Must facilitate the product formation at a higher rate.
5. The substrate that will be used must be of consistent quality.
6. The substrate must be available throughout the year without any variation in quality.
7. Should not interfere with the downstreaming processes.
8. It must not cause problems during the media preparation and also in sterilization.
9. Problems during the process running particularly in terms of aeration and agitation, extraction and purification, and also waste treatment must be minimized.

For a particular process, the media selection will influence the design of the fermenter to be used for the process. On the other hand, if a fermenter is already available, then during the designing of the media, the influence of the type of fermenter has to be taken into consideration. Other important considerations are influence of pH variation on the medium, foam formation, oxidation and reduction potential and morphology of the organism that will be cultivated. For some processes inducers play an important role, and therefore, during the media design, along with all above said considerations, effect of inducer has to be analyzed.

Generally for fermentation process complex sources of medium components are used, as they are cheaper than that of the defined components, thereby lowering the production cost. As these sources contain undefined components, therefore, the control of batch to batch variation of the product quality becomes difficult. These small variations are also difficult to rectify. On the other hand, due to its undefined nature, product recovery and effluent treatment also become a trouble-some task. Often all the components are not utilized by the microorganisms and cause environmental problems in terms of increased BOD of the effluent. On the other hand, for the production of low volume - high cost material, batch to batch variation is not affordable and thus the medium used should give highly precise productivity. The medium used for these kinds of processes generally contain defined components. The ultimate choice of the media components depends on the detailed analysis of various associated parameters and judicious application of the knowledge.

Designing of a medium for the microbial cell culture is somewhat easier taking advantage of their unique *de novo* synthetic potential of various essential compounds. Various aspects of microorganism generating medium have been widely studied and a handful number of literature is available in this aspect. Medium for the animal cell culture contains various complex components and needs more extensive studies as compared to the microbes. These media are generally used for the production of vaccines, monoclonal antibodies and therapeutic proteins. Generally, foetal calf serum (FCS) is used as an essential component of mammalian cell culture medium. But due to the undefined nature of FCS and its variability, it is being replaced by designed components that can mimic the effect of FCS and the medium is known as serum free medium (SFM).

## 7.1 Medium Formulation

The medium for the culture of microorganisms must supply the essential components for the growth and metabolite production and therefore, a constant supply of energy must be maintained for the biosynthesis and cell maintenance. Therefore, media formulation is essential in the laboratory, pilot scale and also actual production scale. For the formulation of medium, the most essential step is to consider the stoichiometry for the growth and product formation. For an aerobic process:

$$\underset{\text{energy source}}{\text{Carbon \&}} + \underset{\text{source}}{\text{Nitrogen}} + O_2 + \underset{\text{requirements}}{\text{Other}} \longrightarrow \text{Biomass} + \text{Product} + CO_2 + H_2O + \text{Heat}\uparrow$$

For the economical designing of media, it is important to express all the parameters of this equation in quantitative terms. By the proper application and evaluation of these parameters, actual quantification of the amount of product or biomass is possible. But the calculation of these parameters is not always an easy task, and in practical situation, it is not possible to calculate these parameters precisely. Knowledge of the elemental composition of the medium and micronutrients is also required and may require similar quantification. It is important to identify the specific components that are essential for microbial growth and specific product formation. These compounds may be vitamins, amino acids, nucleotides or some other nutrients.

Often carbon sources used in the media play dual role i.e., in the biosynthesis and also energy generation. In case of aerobic organisms, carbon requirement for the biomass production is estimated from the growth yield coefficient ($Y$) and is defined as:

$$Y = \frac{\text{Quantity of cell dry matter produced}}{\text{Quantity of carbon substrate utilized}}$$

Other crucial parameters such as oxygen requirement, $CO_2$ release, maintenance of pH, etc. needs special attention while designing the medium and its optimization.

## 7.2 Water

Water is essential for the over all processing of whole fermentation process. Good quality of water with uninterrupted supply is essential for the media preparation, sterilization, and maintenance of temperature while running the process apart from cleaning and rinsing of the vessel and utility components. Assessment of suitability of water is done generally for its effluent load, bacterial content and pH. In laboratory scale, it is possible to use distilled water, but in large scale production units, generally water is deionized first and then pH is adjusted and additives are added according to the need. In industrial scale, recycling or semi-recycling of water is economical; various workers have discussed the proper use of water in detail [22, 23].

## 7.3 Carbon Sources

Carbon sources are the primary substrates which when metabolized influence the growth and metabolism of microorganisms, and thus the formation of primary and secondary metabolites. Faster growth rate due to rapid metabolism of sugars, often reduces the secondary metabolite production. But the problem can be overcome by the addition of non-readily metabolizing sugars like lactose. Recent trend is to use readily metabolizing sugars in semi-continuous or fed batch mode. Other problems like catabolite repression may be overcome by genetic modifications. The main product of the fermentation process usually determines the choice of sugars. This is also associated with the market price of the product. In case of high volume, low purity and low price product, cheaper carbon sources are used; while when the product needs high purity and high quality, the sugar used must be of high purity and high quality. Purity of the carbon sources also affects the choice of substrate. The presence of metallic ions in the carbon sources reduces the production of citric acid and therefore the removal of metallic ions from the carbon sources is essential [24, 25, 26]. Special care should be taken during the sterilization of medium. Heating often reduces the quality and availability of the sugars to the microorganisms. Therefore, it is recommended to sterilize the sugar separately and then to add it to the sterilized medium. Sugars often react with ammonium ions to form a black nitrogen containing product, which partially inhibits the growth of the microorganisms. Starch, when heated for sterilization gelatinizes to produce a viscous solution, and it has been seen that only up to 2% concentration of starch can be used without modification [27]. In case of conventional fermentation, malt extract, beef extract, and cane molasses are most frequently used carbohydrates. Barley malt contains at least 58-60% starch (dry weight) and the rest are different types of sugars [28]. Analysis of beet and cane molasses has shown that they contain almost 48.5% and 33.4% of sucrose, respectively. In addition, cane molasses also contain 21.2% invert sugar and the rest are non-sugar components. Apart from that, for different processes, as per the requirement of the process, oils and fats and different hydrocarbons are used for the biomass production [29, 30].

### 7.4 Nitrogen Sources

In most industrial bioprocesses, microorganisms used can utilize both the organic and inorganic nitrogen sources. Inorganic nitrogen sources can be supplied as ammonia gas. Organic nitrogen may be supplied as amino acid, proteins or urea. Growth is generally faster with the organic nitrogen. In some cases, microorganisms show specific requirements of amino acids as they may be deficient in producing the amino acids. However, the use of pure amino acids makes the process far expensive and therefore, cheaper and readily available sources are used. Other pertinacious nitrogenous compounds that are used as the ready sources of nitrogen are corn-steep liquor, soya meal, peanut meal, cotton seed meal and yeast extract.

## 8. STERILIZATION

Sterilization is the first and the most important step of any fermentation. Improper sterilization will hamper the product quality as the contaminating organisms will deplete the medium components needed for the growth of the desired organisms. On the other hand, a fermentation process must be started with a pure culture inoculum for the product quality and process economy. Therefore, not only the medium but also the fermenting vessels and materials to be used for the process have to be sterilized. Sometimes sterility is maintained at the area where the process is being operated. Sterilization can be done by radiation, filtration, ultrasonic treatment and heat. Heat sterilization is the preferred method of sterilization for the media and liquid components, whereas air is sterilized by using various filters (primary and secondary).

## 9. INOCULUM DEVELOPMENT

It is essential that the culture used to inoculate fermentation satisfies the following criteria:

1. It must be in a healthy condition.
2. The composition of medium used for the growth of inoculum plays a very important role in extension of lag phase during fermentation.
3. Inoculum volume is added to a large volume of medium. This can cause outward diffusion.

4. Inoculum size should be such that it helps in minimizing the length of the lag phase in the fermentation.
5. It must be in a suitable morphological form.
6. It must be free from contamination.
7 It must retain its product-forming capabilities.

The preparation of a population of microorganisms from a dormant stock culture to a state useful for inoculating a final productive stage is called inoculum development. Once the fermentation has been started, it can be made worse but not better [31]. A critical factor in obtaining a suitable inoculum is the choice of the culture medium. The design of the production medium is determined not only by the nutritional requirement for the growth of organism but also by the requirement of maximum product formation. The formation of product in the seed culture is not an objective during inoculum development, so that the seed medium may be of a different composition from the production medium. The quantity of inoculum normally used is between 3 and 10% of the medium [32, 34-33]. A relatively large volume of inoculum is used to minimize the length of the lag phase and this will generate maximum biomass in the production fermenter in a short time and subsequently increase the productivity. Starting from a stock culture, the inoculum must be built up in a number of stages to produce sufficient biomass to inoculate the production-stage fermenter. This may involve two or three stages in fermenters, depending on the size of the ultimate vessel. The greater the number of stages between the master plate culture and the production fermenter, more is the risk of contamination and strain degeneration. Therefore, an appropriate size of the inoculum should be used in fermentation vessels.

A detailed procedure for the development of inoculum for bacterial fermentation has been suggested which with minor modifications, is applicable to any type of culture [32]. The procedure involves the use of one sub-master culture to develop a bulk inoculum which was sub-divided, stored in a frozen state and used as a inocula for several months. A single colony, derived from a sub-master culture, was inoculated into liquid medium and grown to maximum log phase. This culture was then transferred into nineteen times its volume of medium and incubated again to the maximum log phase, at which point it was dispensed in 20 $cm^3$ volume, plug frozen and stored at below –20°C. At least 3% of the sample is tested for purity and productivity for each subsequent fermentation, provided it is found suitable for the process. The remaining sample can be used as initial inocula for preparation of starter culture for subsequent fermentation.

## 9.1 Criteria for the Transfer of Inoculum

The physiological condition of the inoculum have a major effect on the performance of the fermentation. The optimum time of the transfer must be determined experimentally and then the procedures should be established so that inoculation with an ideal culture may be achieved routinely. Criteria which may be monitored on line are the most convenient parameters to use as indicators of inoculum quality and these would include dissolved oxygen, pH and oxygen or carbon dioxide in the effluent gas. Real time expert computer system is a good option, which can be used to predict the time of inoculum transfer for industrial-scale fermentation. The system involves the comparison of on-line fermentation data with detailed historical data of the process.

Smith and Calam [35] have compared the quantity and enzymatic profile of differently prepared inocula of *Penicillium patulum* (producing griseofulvin) and demonstrated that a low level of glucose 6-phosphate dehydrogenase is indicative of a good quality inoculum. The enzyme profile of good quality inoculum was established early in the growth of the seed culture. Thus, this approach could be used to assess the cultural condition giving rise to satisfactory inoculum, but would be of less value in determining the time of transfer.

## 10. PRESERVATION OF INDUSTRIALLY IMPORTANT MICROORGANISMS

Microorganisms require special preservation methods in order to ensure optimal long-term viability and genetic stability. The isolation of organism for a commercial process may be a long and a very expensive procedure. The culture used to initiate an industrial fermentation must be viable and free from contamination. Microorganisms require special preservation methods in order to ensure optimal long-term viability and genetic stability. In general, each preservation method can be assigned to one of the following groups:

### 10.1 Metabolically Inactive Preservation Techniques

*10.1.1 Cryopreservation*

- Freezing and low temperature storage in liquid nitrogen
- Freezing and low temperature storage at 60–70°C

*10.2.2 Drying*

- Preservation by shelf freeze-drying
- Preservation by spin freeze-drying
- Preservation by vacuum drying over calcium chloride

### 10.2 Metabolically Active Methods

1. Periodic transfer on agar slopes
2. Keeping agar cultures under mineral oil

Maintenance of strains by metabolically active methods should be used only in case a strain cannot be preserved by one of the metabolically inactive methods.

### 10.3 Quality Control after Preservation

Quality control of microbial strain after each preservation cycle is extremely necessary, at least viability and purity, and where appropriate, the identity of the preserved culture has to be checked immediately after preservation. Details of these controls are inserted in the protocols of the preservation. If available, a registration form for the freeze-drying process per batch e.g. vacuum, product temperature, shelf temperature, condenser temperature, time, etc. has to be filled. Any remark on the viability or properties of a batch has to be archived and should remain available to compare with future controls.

## 11. FOOD-BASED FERMENTATIONS

Fermentation technology is one of the oldest food technology applications that was developed and utilized for survival since ancient times. The origin of Asiatic fermentation technology evolved as early as the littoral forager's period of the Primitive Pottery Age (8,000-3,000 B.C.), which led the Neolithic culture of agriculture in the Northeast Asia. Fermentation technology has developed indigenously all around the world by using natural products from the respective regions to produce required food materials, from which the characteristic taste and aroma of each cultural society has been made. Steinkraus have classified fermentation technology of the world, which is responsible for man's survival into six groups: alcoholic, lactic acid, leavened breads, meat substitutes, meat flavoured sauces/pastes and protein/flavouring agent's fermentation [36]. The meat eating habit of Western culture enforces the need of food preservation technology, so as to keep their perishable meat and milk edible for longer period. Meat sausage, cheese and acid fermented milk making were important food preservation technology methods until the time when refrigerators were not available in homes. On the other hand, people who ate cereals as staple food in the East wanted to have meaty flavoured and salty condiments, which make the bland tasting cereal foods more palatable. This demand led Asian people to develop soybean sauce and fish sauce through fermentation technology. However, consumer perception of food has now changed. Until the 20$^{th}$ century, the food situation of the world was always problematic. Food shortage could occur any time even in the affluent societies. Therefore, man mostly lived for food, and we may call this earlier period the "survival food age." In the 20$^{th}$ century, access to food was no longer a problem in most of the affluent societies, but time was short for work and leisure. People wanted to save time cooking and obtaining food. Consequently, convenience food became the major item in the food market. Now, we all notice that the 21$^{st}$ century is the "functional food era." Health oriented food and nutraceuticals are major concerns of today's consumer.

Fermentation technology has adapted itself to social demands. During the survival food age, fermentation was used mainly for food preservation and condiments production. In the convenience food age, it was used for flavour production and other ingredients for industrial mass production of food. The 21$^{st}$ century is called the era of tailor-made goods satisfying personal demands, and together with the health benefit demands, fermentation technology finds new challenges in the market. Many of the traditional fermented foods are receiving new attention for their health promoting or disease preventing/curing effects. Scientific evidences for their physiological functions are accumulating and the technologies enhancing the beneficial effects are developing rapidly by using modern biotechnological and genetic engineering techniques. Some of the recent developments and future prospects in the fermentation food industry are discussed below.

## 11.1 Lactic Acid Fermentation for Probiotics

Yoghurt and sour milk products are well known in Western society as probiotic food. Similar products made with rather inexpensive raw materials are found in many indigenous fermented foods around the world. Korean *Kimchi* is one of the examples. It is made from vegetables, lightly salted and fermented by lactic acid bacteria, mainly *Leuconostoc mesenteroides*. The lactic acid bacteria grown in *kimchi* can survive in gastric acid and bile juice and reach the large intestine [37]. More importantly, *kimchi* contains various functional components either originating from the ingredients or formed during fermentation, as shown in Table 4. In addition to the probiotic effect, these vegetable products have excellent prebiotic function.

Combining Western milk fermentation technology and Eastern cereal processing skills led to the creation of fusion foods like soybean yoghurt and rice yoghurt. *Risogurt* is an example of lactic acid fermented vegetable drink, made from the mixture of rice and soybean protein and studied in Korea [38]. Many of the vegetable yoghurt products are found in markets of USA and Japan.

The antimicrobial activity of some lactic acid bacteria against *Helicobacter pylori*, which is known to cause stomach cancer, is well applied to acid fermented milk beverages and highly accepted by the consumers in Korea [39]. Anti-diarrhea effect of lactic acid bacteria like *Lactobacillus acidophilus* is enhanced by protein coating [40]. The freeze dried bacterial cells coated

**Table 4:** Biologically-active compounds in *kimchi*.

| Chemical compounds | Occurrence | Possible effect |
|---|---|---|
| Benzylisothiocyanate, indole compound, thiocyanate, flavonoid | Chinese cabbage, allium vegetable, red pepper | Antibiotic, anticarcinogenic, immune stimulant |
| Sistosterol | Chinese cabbage | Reduces the cholesterol level |
| Diallylsulfide, diallyltrisulfide, diallylmethylsulfide | Allium vegetable | Anticarcinogenic, antioxidant, fibrinolytic |
| Gingerrol, gingerin | Ginger | Antibiotic, fibrinolytic |
| Capsaicin | Red pepper | Laxative, secretion of neuropeptides |
| Lactic acid bacteria | *Kimchi* | Antagonistic |
| Bacteriocin | *Kimchi* | Antibiotic |
| L-(+)-lactic acid | *Kimchi* | Modulation of T-cell function |
| Acethylcholine | *Kimchi* | Laxative |
| Dextran | *Kimchi* | Laxative |
| γ-Aminobutyric acid | *Kimchi* | Laxative |

*Source:* 46

with protein are added to food for preventing diarrhea and other health benefits, and this market is increasing rapidly.

## 11.2 Soybean Fermentation for Cancer and Degenerative Disease Prevention

The physiological function and health benefit of soybean is widely recognized today. The blood cholesterol level reduction activity of soybean protein and its hydrolysate and the pseudoestrogenic effect of soybean isoflavones are experimentally proved. In addition to flavour formation, soybean fermentation increases the digestibility and the nutritional value, and also produces health beneficial functional compounds [41]. Soybean fermentation for soybean sauce and paste, Korean *chongkukjang* and Japanese *natto* production involves the process of enzymatic hydrolysis of protein to make peptides and amino acids. Some peptides in soybean sauce and paste are known to have ACE inhibition, antithrombotic and anti-cancer effects [42, 43-44]. Fermentation with moulds, *Aspergillus* and *Rhizopus*, converts glycoside form of isoflavones, daidzin and genistin, to aglycones, which have higher potency of estrogenic effect. Production of mucous polysaccharides and enzyme kinase during bacterial (*Bacillus*) fermentation of soybean is related to the fibrinolytic and immune modulating activity of *chongkukjang* and *natto* [45].

## 11.3 Mixed Culture Fermentation

Industrial yoghurt production in the West and soybean sauce production in the East were achieved by purely isolated starter culture techniques. However, many of the traditional fermentation starters in the East are naturally fermented mixed culture system. The consumer often prefers the traditional fermented products made by the naturally fermented starter culture for their natural taste and aroma. The deep and bountiful aroma of Korean rice wine, *chongju*, and the sharp, strong flavour of Korean soybean sauce, *kanjang*, is more preferred by Korean people than those of industrial counterparts. The anti-cancer activity of *doenjang* made from Korean traditional mixed culture *meju* was reported to be higher than that of the industrial product made from *koji*. Studies on the physiological interactions for growth and biosynthesis of functional compounds by different species in the traditional fermentation system will reveal the actual control of typical aroma and flavour production. Computer aided analysis of microbial interaction will make this sort of study possible and manageable.

## 11.4 Additional Information on Food Fermentations

### *11.4.1 Sausage*

#### 11.4.1.1 Semi-dry fermented sausages

- Summer sausage: *Pediococcus cerevesiae* and *Lactobacillus plantarum*

#### 11.4.1.2 Dry fermented sausages

- Pepperoni: *Pediococcus cerevesiae* and *Lactobacillus plantarum*
- Genoa salami: *Micrococcus* spp. mixed with either *Pediococcus cerevesia*e or *Lactobacillus plantarum*
- Snack sticks: *Pediococcus cerevesiae* and *Lactobacillus plantarum*

Lactic acid is produced, lowering the pH of the sausage, which in turn helps in preservation as well as a flavouring agent.

### *11.4.2 Enzymes*

11.4.2.1 Chymosin: Replaces rennet in over 80% of the cheese produced. Rennet is extracted from the stomach of milk-fed veal calves. The supply is linked to the veal market, causing wide shift in availability and price of the rennet. Recombinant chymosin is produced by an *E. coli.* The gene for chymosin was transferred from a calf to *E. coli.*
11.4.2.2 Amylase: Amylase is the enzyme that breaks starch into its separate glucose components. It is used in the brewing industry for malting and also used in baking.
11.4.2.3 Glucose isomerase: The enzyme glucose isomerase converts or isomerizes glucose (an aldehyde) to fructose (a ketone), which is a sweeter product. The fructose is 1.8 times sweeter than glucose.
11.4.2.4 Pectinase: Pectinase is the enzyme that breaks down pectin, a polysaccharide found in fruits. It is used to remove particulate matter or clarify fruit juices.
11.4.2.5 Glucose oxidase: Glucose oxidase is an enzyme used in the production of dried egg whites. On drying egg whites, the glucose present in the white portion can react with amines in a reaction known as Millard browning. This will cause the dried egg whites to turn brown. Addition of glucose oxidase will breakdown the glucose and prevent glucose from reacting and causing the off colour of the egg whites. To produce these recombinant enzymes, the gene is first transferred into bacteria. The bacteria are then grown in fermenter, enzyme produced is purified and sold.

## 12. SAFETY ISSUES OF FERMENTED PRODUCTS

Since fermentation makes raw food materials edible without cooking, the risks of hazardous microbial contamination always exist in fermented food, especially naturally fermented traditional foods. The uneven distribution of salt in lactic acid fermented fish products and contamination of *Aspergillus flavus* in traditional starter cultures for rice wine and soybean sauce may result in severe food poisoning incidences [47, 48].

On the other hand, most of the traditional fermentation methods have their own inbuilt safeguard mechanisms. The food poisoning microorganisms contaminating the vegetables and other raw materials of *kimchi* are killed within one week of fermentation period mainly due to the acid formation and bacteriocin production [49]. Also, large amounts of nitrate and secondary amines in vegetable products are reduced by fermentation [50].

## 13. FUTURE TRENDS

Fermentation technology has confronted new challenge in the era of functional food with its efficient biosynthesis potential. Research for useful strains from traditional fermented foods continues worldwide and relevant information is accumulating. Exchange of knowledge and skills of fermentation technology of the West and the East will accelerate the technology innovation and new product development. Considering the importance of fermentation technology in the era of functional food, the public sectors and international bodies should pay attention to the relevant R&D efforts of academic institutions and industrial research groups. The R&D support of national and international funding organizations should include the following research priorities:

1. Nationwide search for useful microbial strains from indigenous fermented products and characterization of their physiological properties.
2. Establishment of a global network on fermentation technology and microbial strains and starter culture information.
3. Studies on the mixed culture fermentation and the interaction of microorganisms in fermentation system.
4. Evaluation of physiological function and health benefit mechanism of the metabolites obtained during fermentation process.
5. Evaluation of safety of fermented products, the microorganisms involved and the metabolites.
6. Evaluation and dissemination of fermentation technology for the improvement of hygienic condition of developing countries.

## REFERENCES


1. Posno M, Leer RJ, van Luijk N, van Giezen MJ, Heuvelmans PT, Lokman BC & Pouwels PH (1991) Incompatibility of *Lactobacillus* vectors with replicons derived from small cryptic *Lactobacillus* plasmids and segregational instability of the introduced vectors. *Appl Environ Microbiol* 57: 1822-1828.
2. Futcher AB & Cox BS (1984) Copy number and the stability of 2-μm circle based artificial plasmids of *Saccharomyces cerevisiae. J Biotechnol* 157: 283-290.
3. Leers, RJ, van Luijk N, Posno M & Pouwels PH (1992) Structural and functional analysis of two cryptic plasmids from *Lactobacillus pentosus* MD353 and *Lactobacillus plantarum* ATCC 8014. *Mol Gen Genet* 234: 265-274.
4. Koizumi, J-I, Monden Y & Aiba S (1985) Effects of the temperature and the dilution rate on the copy number of recombinant plasmid in continuous culture of *Bacillus staerothermophilus* (pLP11). *Biotechnol Bioeng* 27: 721-728.
5. Reinikainen P & Virkajärvi I (1989) *Escherichia coli* growth and plasmid copy umbers in continuous cultivations. *Biotechnol Lett* 11: 225-230.
6. Kumar PKR, Maschke HE, Friehs K & Schuegerl K (1991) Strategies for improving plasmid stability in genetically modified bacteria in bioreactors. *Trends Biotechnol* 9: 279-284.
7. Murphy TD (1977) Design and analysis of industrial experiments. *Chem Eng* 6: 168-182.
8. Käpelli O (1986) Regulation of carbon metabolism in *Saccharomyces cerevisiae* and related yeasts. *Adv Microb Physiol* 28: 181-208.
9. Mendoza-Vega O, Sabatie J & Brown SW (1994) Industrial production of heterologous proteins by fed-batch cultures of the yeast *Saccharomyces cerevisiae. FEMS Microbiol Rev* 15: 369-410.

10. Lee SY (1996) High cell-density culture of *Escherichia coli. Trends Biotechnol* 14: 98-105.
11. Whittaker MM & Whittaker JW. (2000) Expression of recombinant galactose oxidase by *Pichia pastoris. Protein Expr Purif* 20: 105-111.
12. Kwanmin JJ (1989) Scale-down techniques for fermentation. *Biopharmacology* 2: 30-39.
13. Enfors S, Jahic M, Rozkov A, Xu B, Hecker M, Jürgen B, Krüger E, Schweder T, Hamer G, O'Beirne D, Noisommit-Rizzi N, Reuss M, Boone L, Hewitt C, McFarlane C, Nienow A, Kovacs T, Trägårdh C, Fuchs L, Revstedt J, Friberg PC, Hjertager BH, Blomsten G, Skogman H, Hjort S, Hoeks F, Lin HY, Neubauer P, van der Lans R, Luyben K, Vrabel P & Manelius A (2001) Physiological responses to mixing in large scale bioreactors. *J Biotechnol* 85: 175-185.
14. Lubiniecki AS, Gardner AR, Smith TM, Wang WK, McAllister PR & Federici MM (2003) Validation of fermentation processes. *Dev Biol (Basel)* 113: 37-44.
15. Jem KJ (1989) Scale-down techniques in fermentation. *Biopharmacology* 2: 30-39.
16. Bartholomew WH, Karrow EO, Sfat MR & Wilhelm RH (1950) Oxygen transfer and agitation in submerged fermentations. Mass transfer of oxygen in submerged fermentations of *Streptomyces griseus. Ind Eng Chem* 42: 1801-1809.
17. Wise WS (1951) The measurement of aeration of culture media. *J Gen Microbiol* 16: 1886- 1891.
18. Taguchi H & Humphery AE (1966) Dynamic measurement of volumetric oxygen transfer coefficient in fermentation systems. *J Ferm Technol* 44: 881-889.
19. Casson N (1959) *Rheology of Disperse Systems*, Pergamon Press, Oxford, UK.
20. Roels JA, Van Denberg J & Vonkem RM (1974) The rheology of mycelial broths. *Biotechnol Bioeng* 16: 181-208.
21. Vant Riet K & Tramper J (1991) *Basic Bioreactor Design*, Marcel Dekker, New York, USA.
22. Karrow EO & Waksman SA (1947) Production of citric acid in submerged culture. *Ind Eng Chem* 39: 821-825.
23. Woodward JC, Snell RL & Nicholls RS (1949) Condition molasses and the like for the production of citric acid. US Patent 2 492 673.
24. Smith JE, Nowakowska, Waszczuk K & Andreson JG (1974) Organic acid production by mycelial fungi. In: *Industrial Aspect of Biochemistry* Part A, Spencer B (ed), North Holland, Amsterdam, Netherlands, pp 297-317.
25. Solomons GL (1969) *Materials and Methods in Fermentation,* Academic Press, London, UK.
26. Stowell JD (1987) The application of oils and fats in antibiotic processes. In: *Carbon Substrate in Biotechnology*, Stowell JD, Beardsmore AJ, Keevil CW & Woodward JR (eds), IRL Press, Oxford, UK, pp 139-159.
27. Pan CH, Speth SV, McKillip E & Nash CH (1982) Methyl oleate-based medium for cephalosporins C production. *Dev Ind Microbiol* 23: 315-323.
28. Hammer G (1979) Biomass from natural gas. In: *Economic Microbiology*, Vol 4, Rose AH (ed), Academic Press, London, UK, pp 315-360.
29. Gaunt DM, Trinci APJ & Lynch JM (1984) Metal ion composition and physiology of *Trichoderma reesei* grown on a chemically defined medium prepared in two different ways. *Trans Br Mycol Soc* 83: 575-581.
30. Ramamoorthy S & Kushner DJ (1975) Binding of mercuric and other heavy metal ions by microbial growth media. *Microbial Ecol* 2: 162-176.
31. Calam CT (1976) Starting investigational and production culture process. *Biochemistry* 15: 7-12.
32. Ncoln RE (1960) Integrity of fermentation equipment contaminant and sterility. State of the Art Report 4, Warren Spring Laboratory, Department of Trade and Industry, Stevenage, UK.

33. Hunt GR & Stieber W (1986) Inoculation development. In: *Manual of Industrial Microbiology and Biotechnology*, Demain AL & Solomon NA (eds), American Society of Microbiology, Washington DC, USA, pp 32-40.
34. Meyrath J & Suchanek G (1972) Inoculation techniques-effect due to quality and quantity of inoculum. In: *Methods in Microbiology*, Vol 7B, Yamanda K, Kinoshita S, Tsunoda T & Aida KJ (eds), Halsted Press, New York, pp 159-209.
35. Smith GM & Calam CT (1980) Variations in inocula and their influence on the productivity of antibiotic fermentation. *Biotechnol Lett* 2: 261-266.
36. Steinkraus KH (1993) Comparision of fermented foods of the East and West. In: *Fish Fermentation Technology*, Lee CH, Steinkraus KH & Reilly PJA (eds), United Nations University Press, Tokyo, Japan, pp 1-12.
37. Lee CH (1997) Lactic acid fermented foods and their benefits in Asia. *Food Control* 8: 259-269.
38. Mok CK (1994) Lactic acid fermentation of rice and its quality improvement strategy. In: *Lactic Acid Fermentation of Non-dairy Food and Beverages*, Lee CH, Adler-Nissen J & Barwald G (eds), HarnLimWon, Seoul, Korea, pp 132-150.
39. Park MJ, Kim JS, Yim JY, Jung HC, Song IS, Yu E, Lee JJ, Huh CS, & Baek YJ (2001) The suppressive effect of a fermented milk containing *Lactobacilli* on *Helicobacter pylori* in human gastric mucosa. *Korea J Gastroenterol* 38: 233-240.
40. Chung MJ, Cho YC, Hong UP & Kim D (2001) Improvement of stability in protein-coated LAB (PROLAC) and dual coated LAB (DUOLAC) using soy protein and polysaccharides, Presented at the *First Asian Conference on Lactic Acid Bacteria for Industrial Application and New Technology*, Suwon, Korea, pp 15-22.
41. Lee CH (2001) *Fermentation Technology in Korea*, Korea University Press, Seoul, Korea.
42. Shin ZI, Ahn CW, Nam HS, Lee HJ, Lee HJ & Moon TH (1995) Fractionation of angiotensin converting enzyme (ACE) inhibitory peptides from soybean paste. *Korean J Food Sci Technol* 27: 230-234.
43. Shon DH, Lee KE, Ahn CW, Nam HS, Lee HJ & Shin JK (1996) Screening of antithrombotic peptides from soybean paste by the microplate method. *Korean J Food Sci Technol* 28: 684-688.
44. Kim SE (1995) Tumor suppressive peptides derived from traditional Korean soy sauce, MS Thesis, Seoul National University, Suwon, Korea.
45. Lee BY, Kim DM & Kim KH (1991) Physico-chemical properties of viscous substance extracted from *Chongkook-jang*. *Korean J Food Sci Technol* 23: 599-604.
46. Oh YJ, Hwang IJ & Leitzmann C (1994) Nutritional and physiological evaluation of *kimchi*. In: *The Science of Kimchi*, Korean Society of Food Science and Technology, Seoul, Korea, p 226.
47. Lee CH (1989) Fish fermentation technology. *Korean J Appl Microbiol Bioeng* 17: 645-654.
48. Lee CH & Lee SS (2002) Cereal fermentation by fungi. *Appl Mycol Biotechnol* 2: 151-170.
49. Lee CH (1997) Lactic acid fermented foods and their benefits in Asia. *Food Control* 8: 259-269.
50. Lee CH (1986) *Kimchi;* Korean fermented vegetable foods. *Korean J Dietary Culture* 1: 395-402.

# Solid State Fermentation Technology

4

**V.K. Joshi[1] and Neha Gautam[2]**

*[1]Department of Food Science and Technology,*
*[2]Department of Basic Sciences, Dr. Y.S. Parmar University of Horticulture and Forestry, Nauni, Solan 173 230 (H.P.)*

## 1. INTRODUCTION

Solid state (substrate) fermentation (SSF) generally refers to the growth of the microorganisms on moist solid substrate in the absence or near-absence of free water [1-3]. In recent years, SSF has proven more promising in the development of several bioprocesses and products. It seems that the two terms, solid state fermentation and solid substrate fermentation have often been indistinctly used, but can certainly be distinguished depending upon the nature of solid phase used [4]. Solid substrate fermentation is used to define only those processes in which the substrate itself acts as carbon/energy source, occurring in absence or near-absence of free water, while solid state fermentation is defined as the processes occurring in the absence or near-absence of free water, employing a natural substrate as above, or an inert substrate used as solid support. The understanding and modelling of microbial growth kinetics and transport phenomenon play an important role in SSF. In recent years, interest in SSF has been increasing because of its applications in producing enzymes, antibiotics, biopesticides, aroma, biopharmaceuticles, organic acids and other compounds. In this chapter, efforts have been made to summarize the information on solid state fermentation, historical developments, basis of SSF and its applications in various fields within the *ambit* of this book.

## 2. HISTORY AND DEVELOPMENT OF SSF

Discovery of solid state fermentation cannot be pin-pointed to a single date, but it can be said that SSF has its origin in food fermentation [5]. The recorded history of SSF has been described on cheese making by *Penicillium roqueforti* in Asia before the birth of Christ, on bread making in 2000 BC by Egyptians [6, 7] and on soy sauce by Koji process in China in the years 1000 BP and probably in the years 3000 BP. During the 18$^{th}$ and 19$^{th}$ century, SSF was used for the development of leather tanning process using gallic acid; production of vinegar from apple pomace; and development of composting and solid waste treatment, agroindustrial waste, etc. [8]. After this, in the beginning of the 20$^{th}$ century, SSF was witnessed for the production of primary metabolites such as enzymes and organic acids using microorganisms [9-11]. The period of 1940's was the remarkable period in the fermentation industry, which saw the significant discovery and development of the wonder drug, penicillin. Penicillin was produced using liquid as well as solid culturing techniques [12]. During 1950-60, reports

*Food Biotechnology: Principles and Practices*
*V.K. Joshi & R.S. Singh (Eds.)*
*I.K. International Publishing House Pvt. Ltd., India*
*pp 95–125*

were published describing steroid transformations in SSF using fungal cultures, which was yet another milestone achievement in the history of SSF. This was again followed by the period of 1970's when fungal cultures were successfully cultivated in SSF for the production of mycotoxins that has significant impact on cancer research. During this period, yet another important application oriented finding of SSF research was the production of protein enriched cattle feed (single cell proteins) [13]. Some other historical developments of solid state fermentation are summarized in Table 1.

During 1991-92, two reviews were published on SSF describing the general features of SSF and aspects of fermenter design in SSF (Table 2). During this period, maiden concepts appeared to

**Table 1:** History and development of solid state fermentation.

| Period | Development |
|---|---|
| 2000 BC | Bread making by Egyptians |
| Before birth of Christ in Asia (Recorded history 1000 BP) | Cheese making by *Penicillium roqueforti* |
| 2500 BP | Fish fermentation/preservation with sugar, starch, salts, etc. |
| 2500 BP | Koji process |
| 7$^{th}$ Century | Koji process from China to Japan |
| 18$^{th}$ Century | Vinegar from pomace |
| 18$^{th}$ Century | Use of gallic acid in tanning-printing, etc. |
| 1860-1900 | Sewage treatment |
| 1900-1920 | Production of fungal enzymes |
| 1920-1940 | Production of mycotoxins, protein enriched feed, gluconic acid, citric acid, development of drum-type fermenter |
| 1940-1950 | Fantastic development in fermentation industry, penicillin production |
| 1950-1960 | Steriod transformation |
| 1960-1980 | Production of mycotoxins, protein enriched feed |
| 1980-1990 | Production of various primary and secondary metabolites, development of column type of fermenter, work on kinetics and modelling aspects of SSF |
| 1990-present | Developments on fundamental aspects of SSF, bioprocesses/products developments:<br>***A. Bioprocess***<br>Bioremediation and biodegradation of hazardous compounds, biological detoxification of agro-industrial residues, biotransformation, biopulping, etc.<br>***B. Products***<br>*Bioactive compounds:* Aflatoxin, ochratoxin, bacterial endotoxins, gibberellic acid, zearalenone, ergot alkaloids, penicillin, cephalosporin, cephamycin C, tetracycline, chlorotetracycline, oxytetracycline, actinorhodin, methylenomycin, surfactin, cyclosporine A, ustiloxins, antifungal volatiles, clavulanic acid, mycophenolic acid<br>*Enzymes:* Cellulase, β-glucosidase, CMCase, laccase, xylanase, polygalacturonase, ligninase, xylanases, β-xylosidase, α-arbinofuranosidase, Li-peroxidase, proteases (acidic, neutral and alkaline), lipases, α-galactosidase, β-glucosidase, α-amylase, β-amylase, glucoamylase, glutaminase, inulinase, phytase, tannase, feruloyl para-counmaroyl esterase<br>Citric acid, fumaric acid, lactic acid, oxalic acid, gallic acid.<br>L-glutamic acid, pigments, carotenoid, xanthan gum, succinoglycan, ethanol, aroma compounds, vitamins $B_{12}$, $B_6$, riboflavin, thiamine, nicotinic acid, nicotinamide, gamma linolenic acid, biosurfactants, biopesticides/bioherbicides. |

*Source:* 5

**Table 2:** Developments in fermenter for solid state fermentation.

| Type of fermenter | Process/product | Microorganism |
|---|---|---|
| Wooden trays (1923) | Koji fermentation | *Aspergillus oryzae* |
| Rotating drum (1962) | Composting | Mixed cultures |
| Rotating drum (1964) | Koji fermentation | *A. oryzae* |
| Rotating drum (1969) | Enzymes | *A. oryzae* |
| Fermbach flask (1966, 1968, 1983) | Aflatoxins, spores | *A. oryzae* |
| Aluminium pot (1943) | Enzymes | *A. oryzae* |
| Horizontal vessel (1975) | Ochratoxins | *A. ochracus* |
| Horizontal drum (1976) | Continuous process feedlot | - |
| Cement mixer (1977) | Corn fermentation | - |
| Cement mixer (1975) | Straw fermentation | *Trichoderma, Candida* |
| Bread making blender (1979) | Protein enrichment | *A. niger* |
| Corn storage bin (1979) | Aflatoxins | *A. flavus* |
| Horizontal vessel (1979) | Koji | *A. oryzae* |
| Aluminium trays (1980) | Ligno-cellulose fermentation | - |
| Valmic bags (Porous films) (1985) | Mycotoxins | *A. flavus* |
| Fixed bed differential reactor (1989) | Spore production | *Penicilliun roqueforti* |
| Glass columns (1980, 1991, 1992) | Enzyme production | *A. niger* |
| Glass columns (1999, 2000) | Citric acid, gibberellic acid | Fungal strains |

*Source:* 15

develop fermenters (bioreactors) suitable for SSF processes, which led to the development of drum-type fermenters [14-18].

Current decade has witnessed an unprecedented spurt in SSF for the development of bioprocesses such as bioremediation and biodegradation of hazardous compounds, biological detoxification of agro-industrial residues, biotransformation of crops and crop-residues for nutritional enrichment, defouling, and production of value-added products such as biologically active secondary metabolites, including antibiotics, alkaloids, plant growth factors, etc., enzymes, organic acids, biopesticides, including myco-pesticides and bioherbicides, biosurfactants, biofuel, aroma-compounds, etc. With the advent of biotechnological innovations, mainly in the area of enzyme and fermentation technology [17, 19, 20], many new avenues have been opened for the application of SSF. Because of the advancement in the technology, more recent techniques have also been incorporated in SSF processes such as aroma sensing, infra-red spectrometry, artificial vision, tomographic techniques, magnetic resonance imaging or MRI, etc.

## 3. ADVANTAGES OF SSF

Solid state fermentation can be used advantageously in industrial and agricultural biotechnology processes, where the fermented product may be used directly or where the high product concentration in SSF allows minimum downstream processing [21]. In order to understand this, let us look into some fundamental questions. There is a lot of research going on into SSF. Why? SSF should be chosen if (a) particular economic conditions favoured it (Therefore, in some parts of the world some enzymes are produced by SSF, whereas in other parts, the same enzymes are produced by SmF); (b) the product is only produced in SSF—or if produced in both systems, the SSF product is far superior (for example, fungal spores for use as biopesticides tend to be much more robust when they are produced in SSF compared to when they are produced in SmF, and some fungi simply do not sporulate well in SmF); (c) use of solids becomes an imperative (government regulation in response to environmental pressures caused by dumping of organic solids).

It is possible that SSF processes could be routinely operated at large scale, following rational design rules, but we currently do not know enough to be really sure if this is the case or not. Certainly, there are a few quite successful SSF processes such as enzyme production processes, various biopesticide production processes, etc., even if many of these are at relatively small scales. There is also the koji industry. However, SSF should not be seen as a technology, which can simply replace SmF. In fact, SmF has many features, which would make it the preferred method in a case where SmF had similar economic performances. The greater homogeneity in a SmF system simply makes the process less problematic. However, there are a number of products for which SSF is the superior production technology, and a number of large-scale SSF processes have begun operating commercially. There is a continued need to develop SSF technology to allow such processes to operate at their full potential. SSF has marked advantages over submerged fermentation as described below:

- Solid state fermentation provides high yield as compared to submerged fermentation
- It requires lesser energy as compared to submerged fermentation
- It is a simple process and the product formation is found superior in it
- Simplicity in preparation of media
- Requirement of lesser fermentation space
- Simple machinery, equipment and control system
- Minimizes catabolite repression
- Easier scale-up of the process and superior yields
- Lower waste water output/less water used
- Absence of foam formation
- Absence of rigorous control of fermentation parameters
- Easier aeration
- Easier control of contamination
- Applicability of using fermented solids directly
- Storage of dried fermented matter
- The capital cost of overall production process is significantly lesser

On the basis of these advantages, it can be said that solid state fermentation holds tremendous potential over submerged fermentation in industrial processes in producing certain enzymes and metabolites [22]. All these advantages also justify the reason or the resurgence of SSF as a method for production of microbial products [23].

## 4. CHARACTERISTICS OF SSF AND SUBMERGED FERMENTATION

The major differences in the characteristics of solid and submerged fermentation are outlined in Table 3. Due to static nature of substrate and microorganism, the system poses difficulties for heat and mass transfer effects. Due to respiration, carbon dioxide is evolved and being heavier than oxygen remains accumulated in the substrate bed, which in turn results in increase in the temperature of the fermenting bed. Two other features of SSF, mainly low physical energy requirements and high human energy requirement led to an untrue assumption that SSF was a low technology. In fact, little knowledge or understanding of fundamental aspects of SSF is the main reason behind above mentioned aspects. Over a period of time, since the last decade, scenario is changing fast and it is hoped that generation of data-based knowledge on fundamental aspects of SSF would lead to the development of modern fermenters with automation as is the case of SmF.

**Table 3:** Comparison of characteristics of SSF and submerged fermentation.

| Characteristics | Solid state | Submerged |
|---|---|---|
| Microorganism, substrate | Static | Agitated |
| Water usage | Limited | Unlimited |
| Oxygen supply | Diffusion | Aeration |
| Volume of fermentation mash | Smaller | Larger |
| Liquid waste produced | Negligible | Significant volume |
| Physical energy requirement | Low | High |
| Human energy requirement | High | Low |
| Capital investment | Low | High |

*Source:* 5

## 5. BASIC PRINCIPLES OF SOLID STATE FERMENTATION

As already mentioned, SSF offers numerous advantages over submerged fermentation (SmF). These include high volumetric productivity, relatively higher concentration of the products, less effluent generation and simple fermentation equipment, etc. [24]. There are several factors, which influence the solid state fermentation. Among these selection of a suitable strain, substrate and selection of process parameters (physical, chemical and biochemical) are crucial [16, 25]. The major factors, which affect microbial growth and activity in SSF include selection of a suitable microorganism and substrate, pre-treatment of the substrate, particle size (inter-particle space and surface area) of the substrate, water content and $a_w$ (water activity) of the substrate, relative humidity, type and size of inoculum, control of temperature of fermenting matter/removal of metabolic heat, period of cultivation, maintenance of uniformity in environment of SSF and the gaseous atmosphere i.e., oxygen consumption rate and carbon dioxide evolution rate [26-31].

### 5.1. Selection of a Suitable Microorganism

Selection of a suitable strain is one of the most important criteria in SSF [32]. The vast majority of wild type microorganisms are incapable of producing commercially acceptable yields of the product. The most commonly used microorganisms in the production of secondary metabolites are fungi, yeast and actinomycetes. The mycelial morphology of such organisms is ideal for their invasive growth on solid and insoluble substrates. Solid state offers the greatest possibilities when the fungi are used [33]. Based upon the type of microorganisms involved, SSF process can be classified into two main groups, natural (indigenous) SSF and pure culture SSF using individual strains or a mixed-culture. Composting and ensiling are the two typical SSF processes, which involve natural micro-flora [34]. Bioprocesses involving pure culture SSF are also known since long. One such unique example is the Koji process utilizing a culture of *Aspergillus oryzae* [35]. Pure cultures are generally used in industrial SSF processes, as these help in optimum substrate utilization for the targeted product. Examples are the production of enzymes, organic acids, bioactive secondary metabolites, etc. Examples of SSF processes involving mixed (pure) cultures are bioconversion of agro-industrial residues such as wheat straw and sugarcane bagasse for protein enrichment in which a cellulolytic organism (e.g., a fungal strain of *Chaetomium cellulolyticum*) and an yeast strain (e.g. *Candida utilis*) are used [17, 20, 36]. In nature, SSF is mostly carried out by mixed cultures in which several microorganisms show development and symbiotic cooperation.

**Table 4:** Spectrum of enzymes obtainable from the strains of *Aspergillus niger*.

| | |
|---|---|
| • α amylase | • Xylanase |
| • β amylase | • Polygalactouronase |
| • Glucoamylase | • Amidase |
| • Lactase | • Cellulases |
| • Catalase | • Glucose oxidase |
| • Protease | • Glucose dehydrogenase |
| • Trehalase | • Urease |
| • Tannase | • Insulase |
| • Lipase | • Zymase |

*Source:* 5, 34

The importance of the microorganisms can be seen from the fact that the strains of *Aspergillus niger* alone can produce a large number of enzymes (Table 4). It is also important to ensure that the selected microbial strain is best suited not only for SSF but in other systems also. For example, strain of *Rhizopus formosaensis* has best potential to produce glucoamylase enzyme when grown as surface culture in liquid fermentation [37]. The yields of enzyme were less when the culture was grown in SmF or even as surface culture in SSF. Yet another crucial aspect involves the screening of the strain for its efficiency to produce the desired end-product as the sole product, although there could be other associated activities, but only in trace amounts. Thus, selection of the strain for any process development and so for SSF as well, is an extremely significant step and must be planned with utmost care. While doing so, it would be necessary to keep in mind all the necessary associated factors such as process technique, substrate, associated products formation, etc.

## 5.2 Selection of a Substrate

The selection of a substrate for SSF process depends upon several factors mainly related to the cost and availability and thus, may involve their screening. In SSF process, the solid substrate not only supplies the nutrients to the microbial culture growing on it, but also serves as an anchorage for the cells [38]. The substrate provides all the needed nutrients to the growing microorganisms. However, some of the nutrients may be available in sub-optimal concentrations, or even not present in the substrates. It would be necessary to supplement them externally [39, 40] with desired nutrients. It has also been a practice to pre-treat (chemically or mechanically) some substrates before use in SSF processes (e.g., Ligno-cellulosics), which makes them more easily accessible for microbial growth.

Agroindustrial residues including those of tropical origin offer potential advantages for the application of substrate such as cassava, soybean, sugar beet, sweet potato, potato and sweet sorghum [41, 42], crop residues [43] such as bran and straw of wheat rice [44], hull of soy, corn and rice, bagasses of sugarcane [45] and cassava [46], residues of coffee processing industry, residues of fruit-processing industries, waste of oil-processing mills, coconut cake, soybean cake, peanut cake, canola meal and palm oil mill waste, and others such as saw-dust, corn cobs, carob pods, tea waste, chicory roots, etc. [47]. In relation to SSF processes using inert substrates, two approaches have been adopted; one in which synthetic materials such as amberlite or polyurethane are used, and other which utilizes natural materials such as sugarcane bagasse as the inert solid support problems in the fermentation kinetics studies (due to their heterogeneous nature). Such studies could be performed better using synthetic inert solid substrates.

## 5.3 Particle Size of Substrate

Among the several other factors important for microbial growth and activity in a particular substrate, particle size and moisture level/water activity are the most critical. Generally, smaller substrate particles would provide larger surface area for microbial attack and thus, should be considered as desirable factor [15, 28, 34, 48]. Too small substrate particles may result in substrate agglomeration in most of the cases, which may interfere with microbial respiration/aeration and thus, may result in poor growth. At the same time, larger particles provide better respiration/aeration efficiency (due to increased inter-particle space) but provide limited surface for microbial attack. Thus, it would be necessary to arrive at a compromised particle size for a particular process. Typical example of particle size distribution in three different commercial samples of wheat bran, which is the most frequently used substrate in SSF processes, can illustrate the point. Samples A and B were obtained from a local market in India, while sample C was prepared by mixing A and B in 1:1 ratio (w/w). Best results on enzyme production were obtained by the substrate containing particles of 425-500 microns, followed by 500-600 microns (Table 5). Substrates with lower particles resulted in lower enzyme activity and the same was true for bigger particles substrates. The lowest enzyme activity was obtained with the substrate containing particles bigger than 1.4 mm. The specific enzyme activity of sample A was much better than sample B, and this was best in sample C, which showed a unique distribution of particles of different sizes. From the techno-economic feasibility point of view, it was most appropriate to use the mixed substrate.

**Table 5:** Particle size distribution in samples of wheat bran.

| | Per cent distribution | | |
|---|---|---|---|
| **Particle size** | **Sample A** | **Sample B** | **Sample C** |
| Bigger than 1.4 mm | 1.08 | 41.72 | 21.39 |
| 1.0-1.4 mm | 1.10 | 27.57 | 14.35 |
| 850 μm-1.00 mm | 2.38 | 11.69 | 7.04 |
| 600-850 μm | 10.04 | 4.90 | 7.47 |
| 500-600 μm | 8.41 | 3.04 | 5.73 |
| 425-400 μm | 17.03 | 3.35 | 10.19 |
| 300-425 μm | 13.83 | 2.76 | 8.30 |
| 180-300 μm | 21.98 | 2.02 | 13.00 |
| Smaller than 180 μm | 21.45 | 2.82 | 12.14 |

*Source:* 5, 34

## 5.4 Carbon Source and Carbon/Nitrogen Relationship (C/N)

Type, source and nature of carbon and nitrogen are among the most important factors for the success of any fermentation process. These represent the energy source that will be available for the growth of the microorganism. The rate at which the carbon source is metabolized can often influence the formation of biomass or production of primary or secondary metabolites. Type, source and nature of carbon and nitrogen are among the most important factors for any fermentation process. Carbon source could be as simple as a pure monosaccharide compound such as glucose or as complex as polymeric molecules such as cellulose or starch. As a matter of fact, in any fermentation process, when selecting a proper energetic or carbon source, there could be two points of view:

(a) Select the carbon source as a function of the microorganism to be employed and the products to obtain, which means in the widest sense, to formulate the proper medium.
(b) Select the adequate microorganism in order to employ a particular substrate component [49]. This point of view is related to the use of resources such as industrial residues, use of eventually harmful wastes to the environment or raw materials that could improve the economics of established technology. This of course, too means that the media need to be formulated.

Nitrogen is the next most crucial factor that determines the growth of microorganisms and needs to be calculated/determined to obtain maximum biomass. In media formulations, it is necessary to take into account the biomass composition. Cellular biomass presents average 40-50% carbon, 30-50% oxygen, 6-8% hydrogen and 3-12% nitrogen. Other elements such as P, S, Mg, K are important as part of the cell and are required in minimum quantities. Trace elements (Fe, Mn, Zn, Cu, Mo, B) may also be needed but in smaller quantities. From all the principal components of cellular biomass, carbon source is the most important one and should be given significant consideration. The yield of a product in kinetic studies is based on the consumption of carbon source ($Y_{x/s}$).

$$Y = \frac{\text{Quantity of cell dry matter produced}}{\text{Quantity of carbon substrate utilized}}$$

where, $Y$ is cellular yield coefficient.

Theoretically, this parameter has a value of 0.5 when the carbon sources are sugars. Nitrogen is the next most crucial factor that determines the growth of microorganisms and needs to be calculated/determined to obtain maximum biomass. This is based on the yield ($Y_{x/s}$) and the biomass composition. The other two components, oxygen and hydrogen are obtained through the organic carbon source. Generally, it is believed that the ratio between the mass of carbon and nitrogen ($C/N$) is most crucial for a particular process to obtain a specified product. This statement is even valid when the metabolite to be produced is non-growth associated or even partially associated. In the case when the metabolite is associated with growth, it becomes crucial from the beginning of the process to bring the balanced medium. Besides, when dealing with fungi, it is necessary to take into account how the $C/N$ ratio could induce or delay sporulation.

## 5.5 Moisture Content/Water Activity

Moisture is a factor that is intimately related with the definition of SSF and characteristics of the biological material. Substrate moisture and water activity ($a_w$) play very important role in SSF. The water activity of the medium is considered as a fundamental parameter for mass transfer of the water and solutes across the cell membrane [50]. Rather, the control of this parameter could be used to control and modify the metabolic activity of the microorganism [14]. In general, the types of microorganisms that can grow in SSF system are determined by the water activity ($a_w$). Water activity is defined as the ratio of the equilibrium vapour pressure of the medium to the equilibrium vapour pressure of pure water at the same temperature [46].

Pure water has $a_w$ = 1.00 and it decreases with the addition of solutes. Bacteria generally grow at higher $a_w$ values, while filamentous fungi and some yeasts can grow relatively at lower $a_w$ values. Due to this fact, generally filamentous fungi have been mostly preferred to other groups of microorganisms for SSF processes. In case of fungal growth in SSF, higher water activity generally results in sporulation, while low water activity with suitable range favours spore germination and mycelial growth. The importance of water in the system is due to the fact that the great majority of viable cells are characterized by moisture content of 70-80%. From this simple observation, it is easy to arrive at the conclusion that a certain quantity of water should be available for new cell synthesis. In general sense, it has been established that in the case of bacteria, the moisture of the solid matrix must be higher than 70%. For yeasts, the moisture range could be as wide as 70-90%. This certainty by itself could signify a great advantage for the development of a specific SSF process. In case where the optimum moisture level for the growth of employed microorganism is relatively low, it could be a natural barrier to the possible contaminants. In such a case, the need of sterilization treatment could be less rigorous or even absent [5, 51].

## 5.6 Temperature

Biological process is characterized by a relatively narrow range of temperature. High temperature limits for development could be found in many cases at values not higher than 60-80°C, and with particular strains at utmost 120°C [5]. The significance of temperature in the development of a biological process is such that it could determine protein denaturation, enzymatic inhibition, promotion or inhibition of the production of a particular metabolite, cells death, etc. Considering the temperature or the temperature range in which the microorganism grows, following classification has been made:

- **Extreme thermophiles:** These are the microorganisms which survive at temperatures as high as 400°C (for example, those found on sea beds in sulphur-chimneys). These may have an optimum activity between 100-125°C.
- **Thermophiles:** These microorganisms present capacity of growth at such temperatures as 80-95°C with optimum growth temperature in a range between 45-70°C.
- **Mesophiles:** The capacity of growth for these microorganisms does not exceed a temperature of 50°C. Optimum temperature for growth is between 30-45°C.
- **Psychrophiles:** These microorganisms can grow at temperatures below 0°C with optimum for growth between 10-20°C.

Thus depending upon the kind of microorganism, adequacy for a particular process can be determined. Normally, the fermentation processes (SSF as well as SmF) develop with mesophilic microbial strains. However, the trend is to look for thermophilic ones, which in some sense could allow solving the problems of heat evaluation during the process. However, mesophilic strains are the most existent. Temperature rise due to exothermic characteristics of fermentation processes is one of the main difficulties faced in SSF. Temperature control is more difficult in SSF processes compared to SmF processes. In the case of SSF, the solid characteristics of the substrate lacks homogeneity in the reactor and therefore, results in the existence of temperature gradients. Such facts make the heat exchange difficult in the system [24].

A typical example of the temperature effect in SSF can be observed in the process of composting. This process is developed spontaneously through the growth of natural microflora present in it. As temperature rises due to heat evolution, the activity of microorganisms belonging to mesophilic temperature growth range under environmental conditions tends to diminish and start to impose higher temperature range (thermophilic) microorganisms [52]. Undoubtedly, control of one of the factors that decide the good development of SSF must be taken into account when any SSF process is designed (Fig. 1).

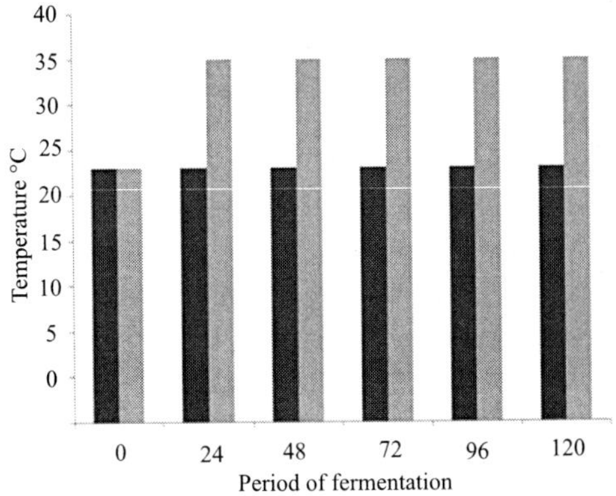


**Fig. 1:** Temperature variation during SSF.

## 5.7 pH

pH is the most important factor for any fermentation process. The substrate and medium acidity has great importance in the bioprocess not only because the pH partly screens the microbial growth from a large number of competitors, but also because it partly gives a right habitat for the microorganisms [53]. An attempt to overcome the problem of pH variability during the SSF process is obtained by the substrate formulation considering the buffering capacity of the different components employed, or by the use of buffer formulation with components that have no deleterious influence on the biological activity. In general, it can be observed that fungi have a pH range for growth between 3.5-6.0, yeast between 4.5-7.0, and bacteria a little higher than the last ones.

## 5.8 Aeration and Agitation

The aeration and agitation have a determinant influence on SSF due to two fundamental aspects:

(1) $O_2$ demand in the aerobic processes.
(2) Heat and mass transport phenomenon in heterogenous system [54].

Oxygen does not present the difficulties from the physiological point of view as observed in the submerged fermentation. It is not necessary to have mechanisms, physical or chemical, which allow that the $O_2$ be dissolved previously in a liquid phase before it could be employed for the metabolic process. In this sense, $O_2$ demand in SSF can be satisfied with relatively low aeration levels [55].

Considering the aeration levels, it is very important that they can be expressed through intensive and not extensive values in order to make independent levels from the scale in which the process is developing. A criteria to express the aeration in a particular system is obtained by the airflow intensity or VKgM. This value is the airflow expressed by unit substrate weight, commonly as $l$air $kg^{-1}$ $m^{-1}$. It must be made clear that the weight considered for the calculation is on dry or wet basis. Once the level of aeration is found as an intensive factor, this value could serve as aeration criteria for higher scales of the process. This indeed does not mean that the value can be taken up as scale-up criteria, but could be used as a starting point at the new levels.

Other point to take into account in the aeration is the quality at the entrance to be fermented, in particular, the air moisture [56]. Use of saturated air is a common practice to avoid substrate drying or change in water activity. This procedure is not always advisable. Airflow is the main tool to transfer the heat evolved during the process. Water evaporation and heat transfer with the surrounding are the processes that keep the process temperature in proper limits. Obviously, a contradiction exists between the heat to eliminate and the moisture in the system. A variable flow of air and its quality could improve temperature control.

As has been pointed out earlier in heat transfer, there occurs the existence of gradients owing to the heterogeneous nature of the system. Substrate heterogeneity causes too possible air canalization. When air canalization occurs, there will be non-aerated zones that could produce non-desired cell catabolism changes. A possible solution to solve heterogeneity problems is agitation. Agitation could improve the homogeneity and disrupt gradients. However, if it is right that agitation improves homogeneity, there may exist problems related to the integrity of the microbial cell. This may be particularly true and related to fungal mycelium.

## 5.9 Fermenters for SSF

The design of fermenter for SSF processes is an important aspect. In any fermentation process, the bioreactor (fermenter) provides proper environment for the growth and activity of the microorganisms [15, 16] to carry out biological reactions. An ideal fermenter should have several characteristics:

- Material used for construction should be non-toxic so that slight dissolution of the material or component does not inhibit culture growth.
- The material used in bioreactor must withstand repeated sterilization with high pressure steam.
- There should be proper arrangements for aeration/agitation with sampling, charging and discharging ports and cooling mechanism to control the generated metabolic heat energy.
- All the materials coming into contact with the solutions entering the bioreactor or the actual organism culture must be corrosion resistant to prevent trace metal contamination of the process.
- The bioreactor stirrer system, entry ports and end plates must be easily machinable and sufficiently rigid not to be deformed or broken under mechanical stress.
- Visual inspection of the medium and culture is advantageous; transparent material should be used, wherever possible.
- During the period of fermentation, it should be capable of preventing the release of internal biomass/media into the environment as well as preventing the entry of foreign substances into the reaction media.
- Bioreactor should maintain the environmental parameters such as temperature, pH, oxygen, etc.
- Bioreactor should be capable of operating under aseptic conditions, although some of the processes e.g., composting may be treated as non-sterile natural inocula [25].

Several types of fermenters have been used for solid state fermentation in batch or continuous mode. One common feature in almost all is their simplicity in operation that has been carried out in flasks, beakers, Roux bottles, petri dishes, glass jars and columns. Wide mouth Erlenmeyer's flasks offer ease in handling and have been most commonly employed for SSF processes without any agitation and aeration. Inoculum is added after substrate autoclaving (unlike glass columns in which pre-inoculated substrate is packed). Glass columns have been considered useful for the processes with forced aeration (moist and cool air) so as to control the fermenting bed's temperature and to remove accumulated carbon dioxide from it. Columns can be placed in a temperature controlled water bath so as to control the temperature of the fermenting substrate. The size of glass column could vary from 2.0-7.5 cm diameter and 20-35 cm length, depending upon the specific SSF process.

### 5.9.1 Tray fermenter

Tray fermenters are the simplest of all types of fermenters used in SSF. These reactors are very simple, without forced aeration and usually with no mixing. Trays could be constructed using wood, metals (iron or aluminium) or plastic material. The material of construction could be stainless steel or should be painted to avoid corrosion, though stainless steel may lead to higher capital investment than aluminium. The bottom of the tray is perforated in such a way that it holds the thin layer of substrate that allows aeration and avoids overheating (to the under surface of the bed and to some extent to the inner layers in the bed) [57-60].

Usually, these trays are arranged one above the other with a suitable gap between them and are placed in a temperature and relative humidity controlled room. Temperature is generally controlled by circulating the air (cold or warm as the case may be). Koji fermentation has traditionally been carried out in tray fermenters since its inception at commercial scale. Several modifications in design in trays have been proposed and used to improve the process efficiency and ease in handling. Swinging type tray fermenter are useful for loading and unloading the substrate. Tray fermenters have been used in studies at laboratory scale, pilot plant scale and commercial scale. However, this type of fermenter requires a large operational area and is labour intensive. Their design does not lead readily to mechanical handling. The substrate requires separate sterilization. For these reasons, it is not so attractive to the users.

### 5.9.2 Drum fermenter

Drum (type) fermenters basically consist of drum-type vessel usually equipped with a rotation device and arrangements for air circulation i.e., an air inlet port and outlet pipe [61, 62]. The inlet pipe may run parallel to the bottom or centre, or it may branch at several points over the whole length of the drum to facilitate air distribution, which is normally attained by forced aeration. These could be subjected to rotation and thereby, achieving the mixing of the fermenting substrate. Growth of the microorganisms in this type of fermenter is considered to be better and more uniform than the tray fermenters [63]. Like tray fermenters, several modifications have been proposed to improve process efficiency in drum fermenters also. These include baffling of the drum, fixing a paddle to achieve better mixing of the substrate, fitting of a perforated bottom, etc. Fig. 2 shows a schematic version of a drum loaded with solid substrate.

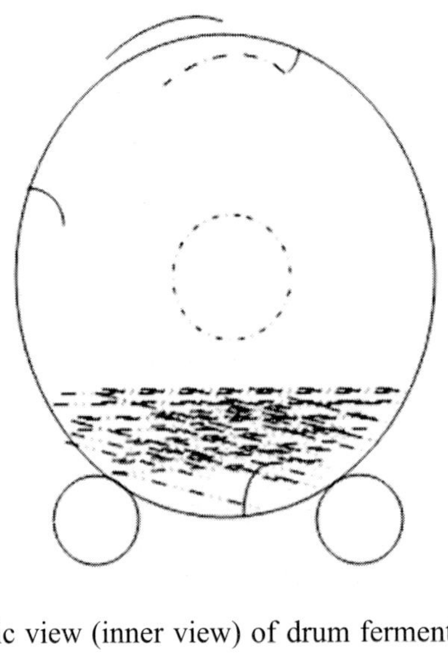

**Fig. 2:** Schematic view (inner view) of drum fermenter with substrate.

The extent of the mixing or rotation speed of the drum fermenter would depend upon the type of SSF process. Usually, low rotation (1-15 rpm) has been recommended, although high speeds of rotation have also been applied. In some continuous composting processes, long drums mounted on a gentle incline with continuous rotation have been considered very useful. However, there are contradictions on the application of rotation and its extent in various SSF processes, in particular when fungal cultures are used, as mixing could lead to damage of the mycelia. Table 6 shows the effect of rotation (and mixing) on glucoamylase production in solid state fermentation by *A. niger* using wheat bran as the substrate. Evidently, the mixing of substrate at any rotation speed was not found useful, rather it resulted in reduced growth of the fungal culture consequently leading to lower enzyme yields. The difficulty faced with drum fermenters having mixing was the aggregation of the substrate, thus harming the fermentation. Large-scale drum fermenters also pose handling difficulties.

**Table 6:** Effect of rotation speed on glucoamylase synthesis by a strain of *Aspergillus niger* in a drum-type fermenter.[a]

| Rotation speed (rpm) | Glucoamylase (IU/gds[b]) | Glucosamine (mg/gds) |
|---|---|---|
| 12 | 255 | 10.2 |
| 20 | 222 | 9.3 |
| 50 | 211 | 7.1 |
| 100 | 199 | 6.0 |
| Control[c] | 360 | 18.0 |

[a]All the fermentations were carried out at 30°C for 48 h; [b]gds-gram dry fermented substrate; [c]control-no mixing.

*Source:* 5

### *5.9.3 Column fermenter*

Column fermenters consist of a glass or plastic column with lids at both the ends. It may be fitted with a jacket for the circulation of water to control the temperature of the fermenting substrate; alternatively the whole column may be placed in a temperature controlled water bath [64]. Column fermenters have most commonly been used in laboratory studies for the production of enzymes, organic acids, biologically active secondary metabolites, spores, etc. Usually, air is circulated from bottom to top. A typical set-up of column fermenter, vertical and horizontal columns for laboratory studies, with estimation of growth by respirometry analysis of exit gas for oxygen and carbon dioxide is shown (Figs. 3-5).

### *5.9.4 Miscellaneous designs of SSF*

Several other types of fermenter designs have been applied for various purposes, although the basic concept still remained around tray type or drum type bioreactor. Other concepts involved in the application of fermenter design is the action of conveyor belts as the base to support the solid substrate. There have been some efforts to use different materials for the construction of fermenter, for example, a microspores film containment system was developed for laboratory scale SSF studies. The bags were made from polypropylene material with micropores of less than 0.4 μm, which allowed efficient air and moisture vapour transmission without any cross contamination. A unique system developed for comparing SSF and SmF in combination consisted of a cylindrical vessel. In its upper part SSF was conducted, while in the bottom part SmF was carried out. The fermenter was suitable for direct comparison of SSF and SmF. Deep trough type design was also introduced for SSF process and has been claimed to offer potential advantages in some cases i.e., industrial enzyme production.

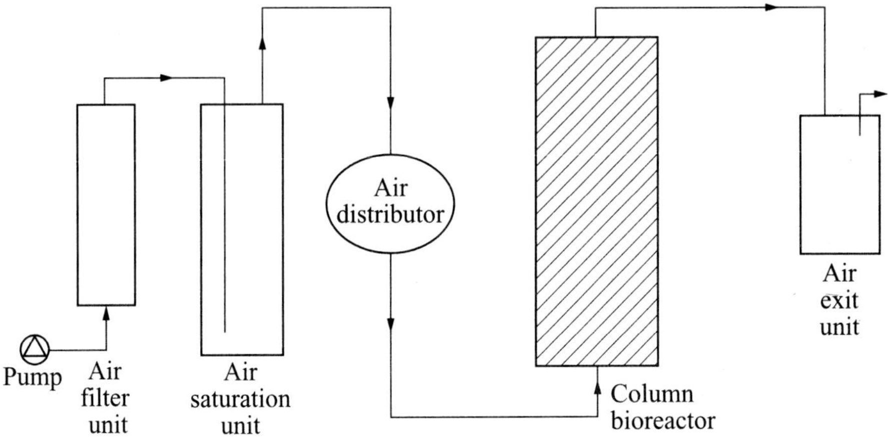


**Fig. 3:** Schematic set-up of column bioreactor for enzyme production in SSF.
*Source:* 57

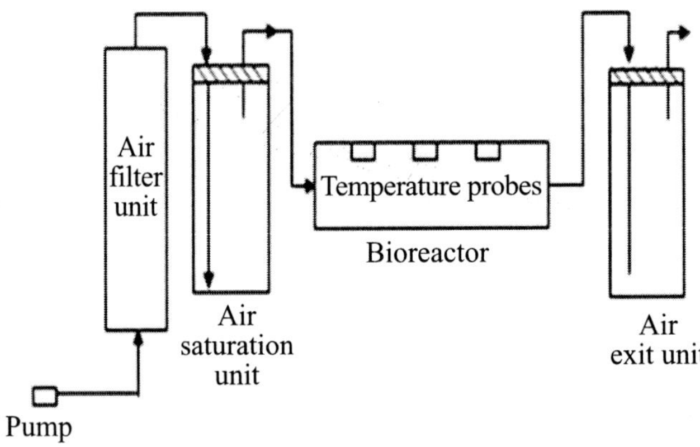


**Fig. 4:** Schematic set-up of horizontal column fermenter for SSF.
*Source:* 57

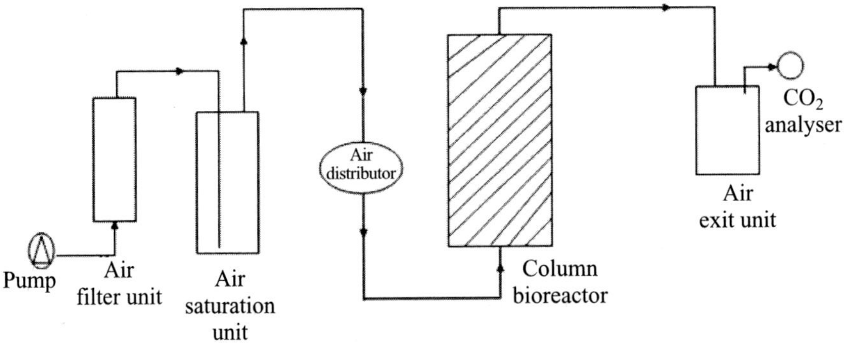


**Fig. 5:** Schematic set-up of vertical column fermenter for SSF.

In this method (which is also called deep bed culture method or pile cultivator method), the fermenter is in the form of a bed upto 5 m in diameter and 12-40 cm deep with a perforated bottom. These types of fermenters have been widely used for traditional food fermentation in East and South-East Asian countries.

Critical evaluation of the available information on design aspects of fermenters for SSF processes shows that not enough details are available and mostly these are still empirical. There is a continuous increase in availability of information on mechanization and automation of SSF bioreactors. It is hoped that enormous efforts being made all over the world would certainly lead to the development of suitable fermenters for SSF process.

Based on the need for agitation, solid state fermentation can be divided in two groups: (a) fermentation without agitation, and (b) fermentation with occasional or continuous agitation. Second group can be further subdivided into:

- Fermentation with occasional agitation, without forced aeration
- Fermentation with slow continuous agitation, without forced aeration
- Fermentation with occasional agitation, with forced aeration
- Fermentation with slow continuous agitation, with forced aeration

The choice of a particular design of fermenter in SSF would depend upon the type of substrate being used, as the structure of the solid matrix would plan an important role in heat and mass transfer effects. Indeed equally important is the type of microorganism being employed, and the end-application of the research would be equally important to select a type of fermenter.

## 6. APPLICATIONS OF SOLID STATE FERMENTATION

### 6.1 Waste and By-product Utilization

Agricultural and industrial wastes are among the causes of environmental pollution. Their conversion into useful products may ameliorate the problems they cause. Plate 1 shows pile of waste materials viz., peel, seed, pomace, rags, kernels, etc. that is biodegradable. These wastes have the potential to support the growth of microorganisms because of the availability of many nutrients like carbohydrates, dietary fibres and minerals. Research carried out on SSF of a waste from apple processing industry revealed the possibility of production of several products [65]. Solid state fermentation of apple pomace followed by drying increases the nitrogen and fat content in the fermented and dried apple pomace. SSF of apple pomace with *Saccharomyces cerevisiae,* in sequential interactive culture, improved the soluble protein contents of the fermented apple pomace. It has proved to be a suitable substrate for the production of ethanol and animal feed concomitantly when the apple pomace was fermented with *Saccharomyces cerevisiae,* using SSF and recovering ethanol. The product (animal feed) could successfully be fed to poultry after mixing with standard feed in the ratio of 1:1. SSF of apple pomace using *Aspergillus niger* has yielded pectin esterase enzyme more than submerged fermentation. Using apple pomace as a base, solid state fermentation processes for the production of biocolours have been developed for *Chromobacter* sp., *Sarcina* sp., *Rhodotorula* sp. and *Micrococcus* sp. Citric acid is another product developed by SSF having a large commercial value. The process of SSF is easy to perform and economical. The studies truly reflect the magnitude of benefit, which could be reaped from bioprocessing of waste from food industry that otherwise would go waste and is a serious loss of natural resources. Though these studies are useful at lab scale, yet these have to be validated at pilot scale level for imbibation by the industry. Nevertheless, the example of apple pomace waste utilization could become a model for the utilization of similar wastes. Development of solid state fermenter and associated equipment for harvesting of products can go a long way in transfer of technology from laboratory to industry. Production of different value added products by using SSF has been described in sections subsequent.

### 6.1.1 Production of ethanol and high crude protein

Apple pomace was fermented using solid state fermentation (Fig. 6). After solid state fermentation of apple pomace, removal of ethanol and drying of fermented apple pomace was done. There was a considerable enhancement of nutritive value of fermented apple pomace. There was increase in fermentability and ethanol production when apple pomace was blended with 10 per cent molasses and 1.8 per cent of nitrogen source (ammonium sulphate). The same concentration also increased the crude protein in the apple pomace.

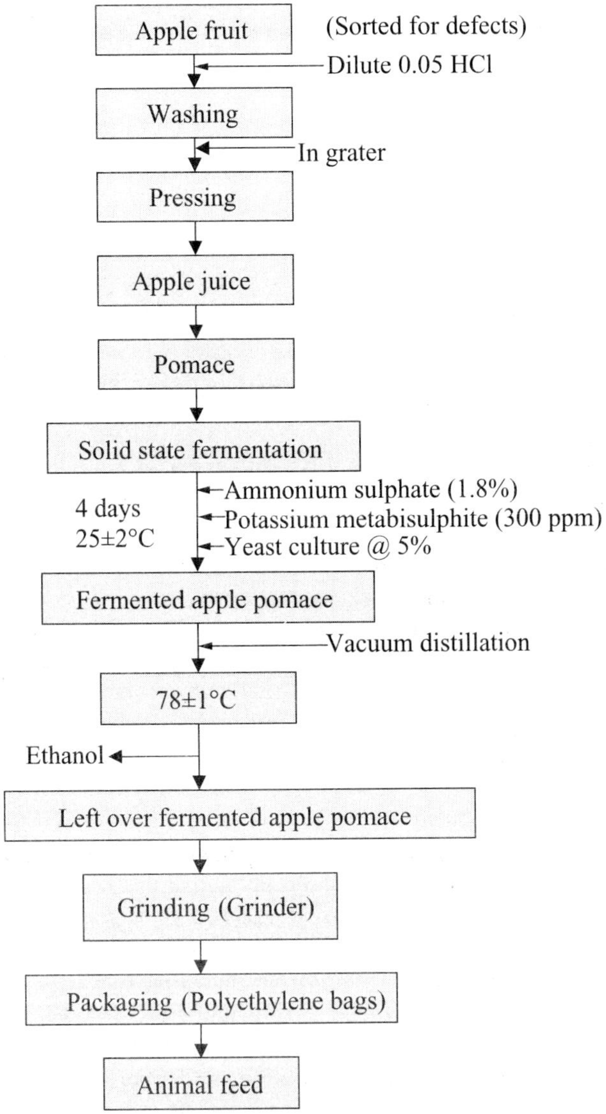


**Fig. 6:** Production of animal feed by using fermented apple pomace.
*Source:* 67

**Plate 1:** Waste and by-product utilization.

### *6.1.2 Production of pectin esterase enzyme*

Pectinase finds extensive applications in fruit processing industries for clarification of fruit juices and wines [68], extraction of fruit juice [69], manufacture of pectin free starch, curring of toffees, cocoa and tobacco, and scouring as an analytical tool for the production of plant products. Pectinase is produced by several fungi including *Aspergillus* sp., *Fusarium moniliforme, Rhizoctonia solani* and *Trichoderma* sp., etc. The presence of pectin in the apple pomace substrate (SSF) induces the production of pectin esterase, so such a substrate is preferred (Fig. 7). Studies carried out on apple pomace to produce pectinases revealed that dilution of 1:3, pH 4.0, temperature 25°C for 96 hours gives the highest pectin esterase production in solid state fermentation using *Aspergillus niger*. Among the different nitrogen sources, ammonium sulphate @ 0.2 per cent and sodium chloride @ 0.5 percent gave the highest PME activity. The pectinases produced by *Aspergillus niger* by SSF of apple pomace shows better extraction and juice recovery. Production of bulk enzyme using apple pomace substrate in solid state fermentation and submerged fermentation shows higher PME production in SSF than SmF [70]. The difference in pectinase production in SSF and SmF is apparently related to its mode of synthesis in two fermentations.

### *6.1.3 Production of microbial colours in solid state fermentation*

Colour is the most important attribute of any article especially food, rather it would not be exaggeration to say 'we all eat with our eyes'. It is an important element in determining the acceptance and enhancing the acceptability of a food product. No matter how much the food is rich in nutrients, flavoured or well textured, it is unlikely to be eaten unless it has the right colour. Microbes are important source of natural colours [71, 72]. Apple pomace has been utilized for the production of microbial colours using solid state fermentation (Plate 2). Apple pomace (10-50 g/l) has been incorporated to the basic medium for the production of *Rhodotorula* (pink colour), *Sarcina* sp. (dark yellow), *Chromobacter* sp. (dark red) and *Micrococcus* sp. (light yellow) [73]. The production of pigment in apple pomace based medium using SSF gives the better yield of biomass and carotenoids. The flow sheet for the production of microbial colours is given in Fig. 8. Similarly, *Rhodotorula* sp. grows best at a temperature of 30°C with an incubation period of 72 hours at pH 5.5 in the apple pomace based medium [74], whereas *Chromobacter* sp. producing dark red colour grows best at a temperature of 35°C with pH of 6.0 for 48 hours of incubation. Different conditions for the growth of pigment producing microorganisms are shown in Table 7.

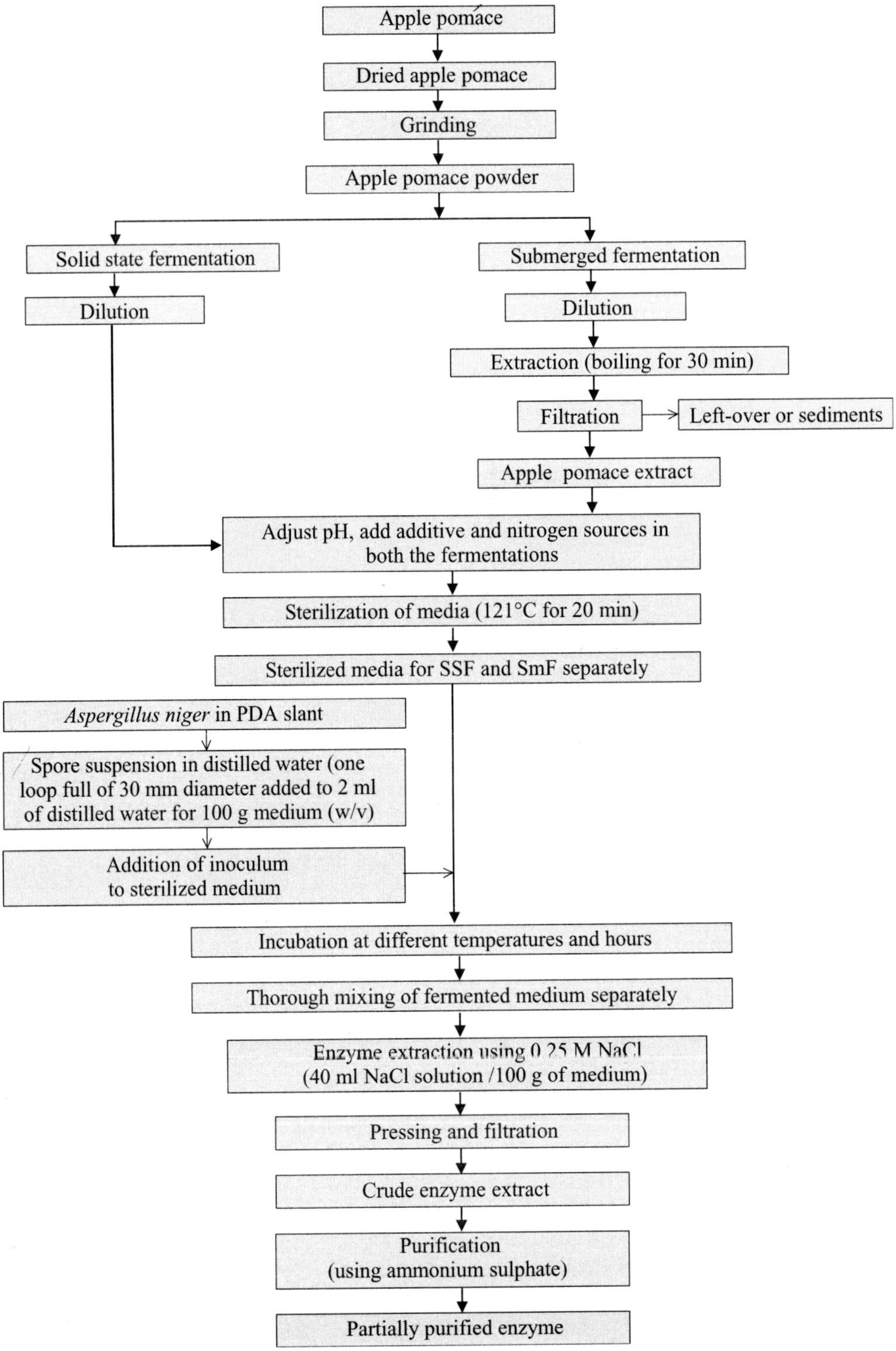


**Fig. 7:** Production of pectin esterase enzyme by using apple pomace.
*Source:* 67

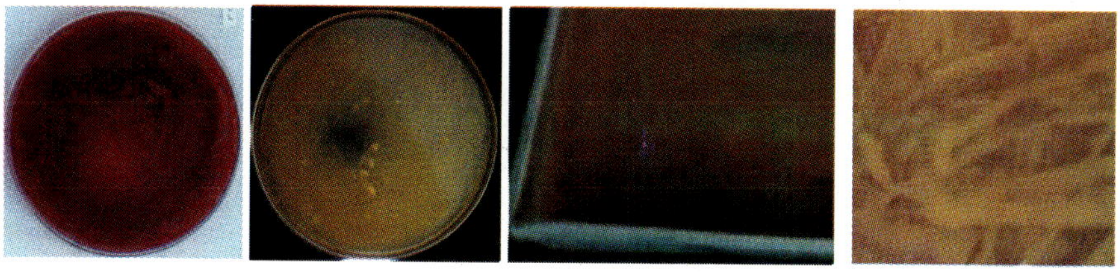

A B C D

**Plate 2:** Colour producing microorganisms grown on apple pomace-based medium. A = *Chromobacter* sp. on a petri plate; B = *Sarcina* sp. on a petri plate; C = *Rhodotorula* sp. on tray; D = *Micrococcus* sp. on tray.

**Table 7:** Incubation temperature, time and pH for different pigment producing microorganisms.

| Microorganism | Temperature (°C) | Time | pH |
|---|---|---|---|
| *Rhodotorula* sp. | 30±1 | 72 hrs | 4.5 |
| *Monascus purpureus* | 28-35 | 4 - 14 days | 5.5-6.5 |
| *Yarrowia lipolytica* | 25 | 3 days | 5.5 |
| *Phaffia rhodozyma* | 20-22 | 3 days | 5.0 |
| *Micrococcus* sp. | 35 | 4 days | 6.0 |
| *Sarcina* sp. | 35 | 3 days | 5.5 |
| *Chromobacter* sp. | 35 | 2 days | 6.0 |

*Source:* 71, 74, 76

### *6.1.4 Production of citric acid*

Citric acid is an important additive used as an acidulant in food processing industries. Apple pomace has also proved to be a good substrate for citric acid production [77]. SSF of apple pomace with five different citric acid producing strains of *Aspergillus niger* was carried out. Addition of methanol at the rate of 4 per cent enhanced the citric acid production by all the strains. *Aspergillus niger* strain gave the maximum citric acid production when grown on apple pomace at different temperatures in the presence of 4% methanol (Fig. 8). Methanol is not assimilated by *A. niger,* so it enhances the citric acid production. It might be due to the effect of methanol on cell permeability, thus effecting the excretion of citric acid [78]. Applications of citric acid in industry are enlisted in Table 8.

## 6.2 Enzyme Production

Microorganisms have been employed extensively to produce various industrial enzymes. Apple pomace has proved to be a good substrate for the production of many enzymes. Enzymes like amylase, xylanase and cellulose have been produced from apple pomace as a substrate. Pectinase is produced by several fungi including *Aspergillus* sp., *Botrytis cinerea, Fusarium moniliformis, Rhizoctonia solani, Trichoderma* and *Neurospora crassa,* but *Aspergillus* is a major source. Production of various industrial enzymes by using SSF has been enlisted in Table 9.

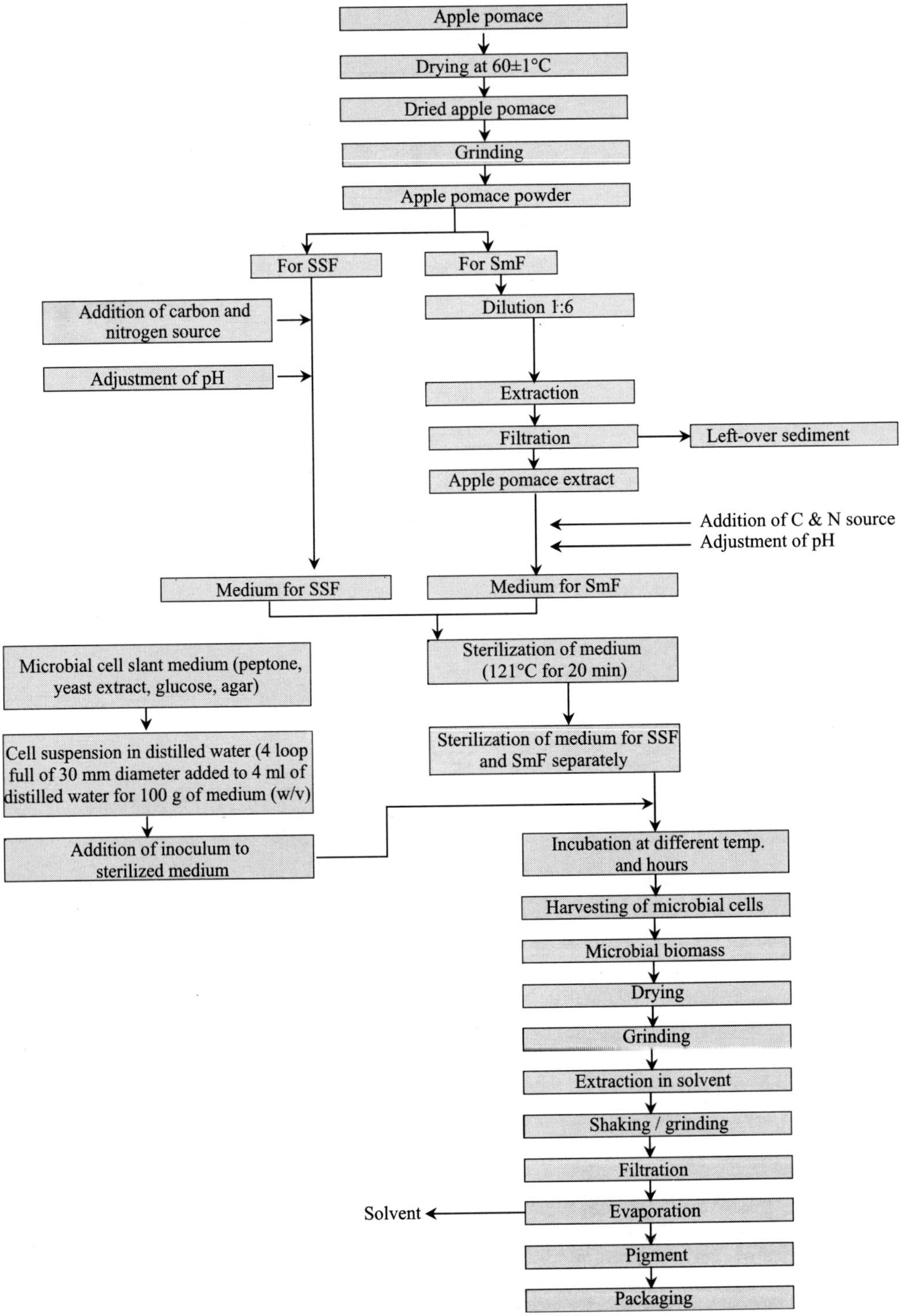


**Fig. 8:** General flow sheet for microbial colour production from apple pomace based medium.
*Source:* 75

**Table 8:** Uses of citric acid.

| Industry | Applications |
|---|---|
| Beverages | Provides tartness and complements fruit and berry flavours. Increases the effectiveness of antimicrobial preservatives. Used in pH adjustment to provide uniform acidity |
| Candy | Provides tartness. Minimizes sucrose inversion. Produces dark colour in hard candies. Acts as acidulant |
| Cosmetic and toiletries | pH adjustment, antioxidant as a metallic-ion chelator, buffering agent |
| Dairy products | As emulsifier in ice creams and processed cheese; acidifying agent in many cheese products and as an antioxidant |
| Fats and oils | Synergist for other antioxidants, as sequestrant |
| Frozen fruit | Lowers pH to inactivate oxidative enzymes. Protects ascorbic acid by inactivating trace metals |
| Industrial applications | Sequestrate metal ion, neutralizant, buffer agent |
| Jellies, jams and preserves | Provides tartness, pH adjustment |
| Metal cleaning | Removes metal oxides from the surface of ferrous and non-ferrous metals, for preparational and operational cleaning of iron and copper oxides |
| Pharmaceuticals | As effervescent in powders and tablets in combination with bicarbonates Provides rapid dissolution of active ingredients. Acidulant in mild astringent formulation. Anticoagulant |
| Others | In electroplating, copper plating, metal cleaning, leather tanning, printing inks, bottle washing compounds, floor cement, textiles, photographic reagents, concrete, plaster, refractories and moulds, adhesives, paper, polymers, tobacco, waste treatment, etc. |

*Source:* 34

**Table 9:** Solid state fermentation for the production of enzymes.

| Enzyme | Microorganisms | Substrate |
|---|---|---|
| Cellulases | *Trichoderma reesei* + *Aspergillus phoenics* | Bagasse |
| Cellulases, β-glucosidase | *A. niger* | Coconut coir pith |
| Cellulases, xylanase, ligninases | *Cerrena unicolor* | Grapevine trimming dust |
| Cellulases | *Penicillium citrinum* | Rice husk |
| Cellulases (FP) | Mesophilic fungal strains | Rice husk |
| CMCase, xylanase, laccase | *Cerrena unicolor, Coriolus hirsutus, Pleurotus ostreatus* | Tea production waste |
| Cellulase, xylanase, polygalacturonase | *A. niger* | Wheat bran |
| Cellulase, amylase | *T. viride, A. niger* | Cellulose, starch |
| Various enzymes | *Lentinula edodes* | Lignocellulosic materials |
| Cellulase, ligninase | Strains of Basidiomycetes | Bagasse |
| Cellulase | *T. resei* | Cellulosics |
| Cellulase, xylanase | *Gliocladium* sp. | Sweet sorghum silage |
| | *Trichoderma* sp., *Penicillium* sp. | Wheat straw |
| Cellulase, β-glucosidase, | *A. niger* | Agro-wastes |
| Cellulase | *T. resei* | Agro-wastes |
| Polysaccharide degrading | *P. capsulatum* | Sugar beet pulp |
| Cellulase, β-glucosidase | *Trichoderma* sp., *Botritis* sp. | Wheat bran + rice straw, spent wheat bran |
| Xylanase | *A. ustus, Sporotrichum plverulentum* | Spent wheat bran |

Contd...

Contd...

| Enzyme | Microorganisms | Substrate |
|---|---|---|
| Cellulase, ligninase | *Polyporus* sp. | Bagasse |
| CMCase, β-glucosidase | *Neurospora crassa* | Wheat straw |
| β-glucosidase, xylanase | *Botritis* sp., *A. ustus, Sporotrichum pulverulentum* | Rice straw + wheat bran |
| Cellulase | *T. resei, Sporotrichum pulverulentum* | Wheat bran |
| Cellulase | *T. harzianum* | Wheat straw + wheat bran |
| Cellulase | *T. resei* | Cellulosic wastes |
| CMCase | *Spiecellum roseum* | Agro-wastes |
| Cellulase, β-glucosidase | *Aspergillus* sp. | Bagasse, wheat bran, rice bran |
| Cellulase | *Streptomyces* sp. | Lignocellulosic substrates |
| Cellulase | *Pestaloptiopsis versicolor* | Agro-wastes |
| Cellulase | *T. resei* | Wheat bran |
| Cellulase | *T. viride* | Cellulosic wastes |
| Cellulase | *Trichoderma* sp. | Agro-wastes |
| Cellulase | *T. resei* | Wheat bran |
| Cellulase | *T. koningii* | Saw dust + wheat bran |
| Cellulase, xylanase, laccase | *C. unicolor* | Grape wine waste |
| Cellulase, xylanase | *A. niger* | Palm oil mill waste |
| Cellulases | *T. viride* | Cellulosics |
| Cellulase, β-glucosidase | *A. ellipticus, A. fumigatus* | Bagasse |
| Cellulase | *T. resei* | Wheat bran |
| Cellulase, xylanase, laccase | *P. sajor-caju* | Sago hampas |
| Cellulase, xylanase | Lignolytic fungal cultures | Sweet sorghum pulp, wheat straw |
| Cellulase, xylanase | *T. harzianum* | Cassava waste |
| Cellulase | *T. resei* | Steam pre-treated willow |
| CMCase | *T. resei* | Wheat bran |
| Cellulase, xylanase | *Gliocladium* sp. | Sweet sorghum |
| Cellulase | *T. resei* | Wheat straw |
| Cellulase | *Gliocladium* sp. | Soyhull |
| Cellulase | *T. resei* | Paddy straw |
| Cellulase | *Lentinus edodes* | Wheat straw |
| Cellulase, xylanase | *T. resei, A. niger* | Sweet sorghum silage |
| Cellulase, β-glucosidase | *A. oryzae* | Cellulosic wastes |
| Xylanase | *T. resei, A. niger, A. phoenicis* | Bagasse |
| Xylanase, laccase, aryl-alcohol oxidase | *Phelibia radiate, P. eryngii* | Wheat straw |
| Xylanase | *Bacillus licheniformis* | Wheat bran |
| Xylanase | *A. niger* | Wheat bran + bagasse |
| Xylanase, β-xylanase | *Phanerochaete chrysosporium* | Wheat bran, defined media |
| Xylanase, cellulase, | *P. sajor-caju* | Wheat straw |
| β-glucosidase | Bacterial isolate B 698 | |
| Xylanase | | Lignocellulosics |
| Xylanase, α-arabinofuranosidase | *A. sajae* | Rice straw, soy bean hull, wheat bran, oat spelt, kraft pulp, beet pulp |
| Xylanase, acetoesterase | *Melanocarpus albomyces* | Wheat/rice straw, bagasse, rice husk |
| Xylanase | *Chaetomium globosum, A. niger* | Wheat bran, apple pomace, sugar beet |
| Pulp, wheat straw xylanase, xylosidase | *A. fumigatus* | Rice straw, corn hull, corncobs |
| Bagasse, wheat bran, xylanase, CMCase, catecholoxidase | *P. sanguineous* | Wheat straw |
| xylanase | *Thermomyces lanuginosus (Humicola lanuginosa)* | Wheat bran |

Contd...

Contd...

| Enzyme | Microorganisms | Substrate |
|---|---|---|
| *Thermascus aurantiacus* Xylanase | *Humicola* sp. | |
| Xylanase, CMCase | *A. niger, A. fumigatus, T. viride* | |
| Xylanase, cellulase, α-arabinofuranosidase β-xylosidase | *Thermomonospora* sp. | |
| Xylanase | *Talaromyces emersonii* | Wheat bran, barley straw, oat straw, wheat straw, wheat bran |
| β-xylosidase | *A. awamori* | Wheat straw |
| β-xylosidase | *T. harzianum* | Wheat bran |
| β-xylosidase | *A. tereus* | Wheat bran |
| β-xylosidase, β-glucosidase | *A. fumigatus* | Wheat bran |
| Laccase | *P. ostreatus* | Wheat bran |
| Laccase, Mn-peroxidase | *Pleurotus* sp. | 14C-lignin (perlite) |
| Laccase, Mn-peroxidase | *Pleurotus* sp. | Wheat straw |
| Laccase, Li-peroxidase, Mn-peroxidase Alcohol-oxidase | *P. floridae* | Synthetic media |
| Mn-peroxidase, laccase, catalase | *P. ostreatus* | Synthetic media |
| Mn-peroxidase, laccase, catalase | *Phlebia radiata* | Wheat straw |
| Laccase, Li-peroxidase, Mn-peroxidase | *Trametes versicolor* | Bagasse |
| Phenol oxidase | *Flammulina velutipes* | Bagasse |
| Laccase, Aryl-alcohol oxidase | *Phlebia radiata* | Wheat straw |
| Laccase, Li and Mn-peroxidase | *P. chrysosporium* | Wood chips |
| Laccase, Li-peroxidase. Mn-peroxidase | *Panus tigrinus, Phlebia radiata* | Wheat straw |
| Laccase | *P. chrysosporium, P. ostreatus* | Wood chips |
| Mn-peroxidase | *Rigidoporus lignosus* | Wheat straw |
| Laccase, Mn-peroxidase | *Panus tigrinus* | Cotton stalks |
| Laccase, Li-peroxidase | *T. versicolor, P. ostreatus* | Saw dust |
| Mn-peroxidase | | Wheat straw |
| Laccase | *Polyporus* sp. | Wheat straw, kraft lignin, cellulose-powder Bagasse |
| Protease | *Penicillium citrinum* | Soybean flour, sunflower flour, Coffee husk |
| Alkaline protease | *Bacillus amyloliquefaciens* | Soybean meal |
| Acid protease | *A. niger* | Wheat bran |
| Neutral protease, α-amylase | *Bacillus licheniformis* | Wheat bran |
| Acid protease | *Rhizopus oligosporus* | Rice bran |
| Neutral protease, α-amylase | *Bacillus licheniformis* | Wheat bran |
| Acid protease | *Rhizopus oligosporus* | Rice bran |

*Source:* 5

## 6.3 Production of Aroma

The current technical knowledge and new developments are exploited to provide insight into the fermentation bases of flavours, aroma and fragrances (Tables 10 and 11) and it has become apparent how biotechnology is providing new opportunities for industry with product quality [5]. The manufacture of aroma compounds from apple pomace are special case, being one of the few high value products that can be made. Flavouring compounds after extracting with liquid $CO_2$ fractioned at different temperatures gave flavourless and intensely flavoured fractions. The flavour had broad spectrum than those obtained by distillation [79]. Solid state fermentations were also carried out to test the efficacy of *Ceratocystis fimbriata* to grow on different agro-industrial substrates and for aroma production [80], seven media including apple pomace supported fungal growth, but that with apple pomace produced a strong fruity aroma, was growth dependent and the maximum aroma intensity was detected a few hours before or after the maximum respirometric activity. Gas chromatographic analysis of the components present in the headspace identified several compounds and 15 of them were found as acid (1), alcohols (6), aldehyde (1), ketones (2) and esters (5).

**Table 10:** Some important flavours synthesized by fungal cultures.

| Flavour | Compounds | Microorganism |
|---|---|---|
| Organ blossom | Methyl anthranilate | *Pycnoporous cinnabarinus* |
| Cherry | Benzaldehyde | *Bjerkandera adusta* |
| Jasmonates | | *Botryodipoldia theobromae*<br>*Gibberella fujikuroi* |
| Aniseed | 4-methoxy benzaldehyde | *Ischnoderma benzoinum* |
| Blue cheese | Methyl ketones | *Aspergillus* and *Penicillium* sp. |
| Vanilla pod | Vanillin | *Pycnoporous cinnabarinus* |
| Raspberry | 4-(4-hydroxyphenyl)-2-butanone | *Nidula niveo-tomentosa* |
| Woodruff | Coumarins | *Pleurotus euosmus* |
| Wintergreen | Methylsalicylate | *Phellinus* sp. |
| Mushroom | $C_8$-compounds | *Pleurotus pulmonarious, Agaricus bisporus, Rhizopogan rubescens, Morchella* sp., *Lentinus edodes, Grifola frondosa* |
| Shiitake | Lenthionin | *Lentinus edodes* |

**Table 11:** Production of aroma in solid substrate fermentation of cereal grains.

| Microbes | Substrate | Aroma group | Product | Aroma and intensity |
|---|---|---|---|---|
| *Rhizopus oryzae* | Amaranth grain | Aldehyde, alcohol esters | Acetaldehyde, ethanol, 1-propanol, ethyl acetate, ethyl propionate, 3-methyl butanol acetaldehyde, ethanol, propanol, ethyl acetate, ethyl propionate, 3-methyl butanol | Slightly alcoholic |
| | Amaranth grain + salt solution | Aldehyde, alcohol esters | Acetaldehyde, ethanol, isopropanol, ethyl acetate, isobutyrate, isobutyl acetate, ethyl butyrate, isoamyl acetate, 2-octanone | Slightly alcoholic |
| *Penicillium aurantiogriseum* | Oat meal agar | Alcohols, ketones | Ethanol, 1-propanol, 2-methyl-1-propanol, 3-methyl-1-butanol, sesquiterpenes | |
| | Barley meal agar | Alcohols, ketones | Ethanol, 1-propanol, 2-methyl-1-propanol, 3-methyl-1-butanol, sesquiterpenes | |
| | Wheat meal agar | Alcohols, ketones | Ethanol, 1-propanol, 2-methyl-1-propanol, 3-methyl-1-butanol, sesquiterpenes | |
| | Oat grain | Alcohols | 3-methylfuran, 1-propanol, 2-methyl-1-octen-2ol, 1-octen-1-ol, acitic acid, sesquiterpenes | |

Contd...

Contd...

| Microbes | Substrate | Aroma group | Product | Aroma and intensity |
|---|---|---|---|---|
| *Aspergillus niger* | Coarse wheat meal (CWM) | Alcohols | 3-methylbutanol, 3-octanone, 3-octanol, 1-octen-3-ol, 1-octanol, 2-octen-1-ol | |
| *Aspergillus ochraceus* | CWM | Alcohols | 3-methylbutanol, 3-octanone, 3-octanol, 1-octen-3-ol, 1-octanol, 2-octen-1-ol | |
| *Aspergillus oryzae* | CWM | Alcohols | 3-methylbutanol, 3-octanone, 3-octanol, 1-octen-3-ol, 1-octanol, 2-octen-1-ol | |
| *Aspergillus parasiticus* | CWM | Alcohols | 3-methylbutanol, 3-octanone, 3-octanol, 1-octen-3-ol, 1-octanol, 2-octen-1-ol | |
| *Penicillium chrysogenum* | CWM | Alcohols | 3-methylbutanol, 3-octanone, 3-octanol, 1-octen-3-ol, 1-octanol, 2-octen-1-ol | |
| *P. citrinum* | CWM | Alcohols | 3-methylbutanol, 3-octanone, 3-octanol, 1-octen-3-ol, 1-octanol, 2-octen-1-ol | |
| *P. funiculosum* | CWM | Alcohols | 3-methylbutanol, 3-octanone, 3-octanol, 1-octen-3-ol, 1-octanol, 2-octen-1-ol | |
| *P. raistricki* | CWM | Alcohols | 3-methylbutanol, 3-octanone, 3-octanol, 1-octen-3-ol, 1-octanol, 2-octen-1-ol | |
| *P. viridicatum* | CWM | Alcohols | 3-methylbutanol, 3-octanone, 3-octanol, 1-octen-3-ol, 1-octanol, 2-octen-1-ol | |
| Fungi imper-fecti *Alternaria Cephalosporium* | CWM | Alcohols | 3-methylbutanol, 3-octanone, 3-octanol, 1-octen-3-ol, 1-octanol, 2-octen-1-ol | |
| Fungal odour *Fusarium* | CWM | Alcohols | 3-methylbutanol, 3-octanone, 3-octanol, 1-octen-3-ol, 1-octanol, 2-octen-1-ol | Fungal odour |
| *Bacillus subtilis* | Soybean+ acetoin + L-threonine | Pyrazine | 2,5dimethylpyrazine, trimethylpyrazine, ethylpyrazine | Roasted flavours |

*Source:* 5

## 6.4 Production of Biosurfactant

Biosurfactants are important compounds produced in SSF. These are amphiphilic compounds which contain hydrophobic and hydrophilic moieties that reduce the surface tension between the individual molecules at that surface and interface, respectively [81]. Considerable attention has been given to the production of surface active molecules of biological origin because of their potential utilization in food processing, pharmacology and oil industry. Biosurfactants increase the microbial degradation of pollutants. Different types of microbial surfactant are produced by different microbes (Table 12).

**Table 12:** Structural types of microbial surfactants.

| Surfactant | Source |
|---|---|
| Trehalose dimycolates | *Mycobacterium* sp. |
| Trehalose dicornemycolates | *Nocardia* sp., *Rhodococcus* sp., *Corynebacterium* sp. |
| Rhamnolipids | *Pseudomonas* |
| Sophorolipids, Amino acid-lipids | *Torulopsis* sp. |
| Lipopeptides | *Bacillus* sp., *Streptomyces* sp., *Corynebacterium* sp., *Mycobacterium* sp. |
| Ornithine-lipid | *Pseudomonas* sp., *Thiobacillus* sp., *Agrobacterium* sp., *Gluconobacter* sp. |
| Phospholipids | *Candida* sp., *Corynebacterium* sp., *Micrococcus* sp., *Thiobacillus* sp. |
| Fatty acids/natural lipids | *Acinetobacter* sp., *Pseudomonas* sp., *Micrococcus* sp., *Mycococcus* sp., *Candida* sp., *Penicillium* sp., *Aspergillus* sp. |

*Source:* 81

## 6.5 Production of Vitamins

The role of several strains of *Rhizopus oligosporous, R. arrhizus, R. stolonifer* and several bacteria has been observed in vitamin synthesis. The formation of water-soluble vitamins, vitamin $B_{12}$, $B_6$, riboflavin, thiamine, nicotinic acid and nicotinamide has been achieved in the SSF. The *Rhizopus* strain formed riboflavin, nicotinic acid, nicotinamide and vitamin $B_6$. The isolates of *Rhizopus oligosporous* are generally the best vitamin producers. The best formation capabilities have been shown by *Citrobacter froundii* and *Klebsiella pneumoniae* [9].

## 6.6 Mushroom Production

Mushrooms have been recognized worldwide as a source of food due to high biological value of their proteins, apart from the presence of carbohydrates, fats, vitamins and minerals [82]. It represents a group of highly specialized fungi, producing conspicuous fruiting bodies of pleasant and characteristic biting texture [83]. The folic acid content, which is essential in blood formation, is higher in mushrooms than that in liver and spinach [84].

### *6.6.1 Technology of mushroom production*

6.6.1.1 Compost preparation: This is the first and important stage in mushroom cultivation. The principal aim of composting is to produce selective growth medium, which renders successful mycelial colonization and fruiting body formation. Natural compost is prepared from agro-industrial waste products and residues like wheat straw, paddy, barley straw, rice bran, saw dust, banana, maize stover, wool waste, sugarcane bagasse [85, 86].

6.6.1.2 Spawn preparation: The production of spawn or inoculum for SSF is the most important step in the process of mushroom cultivation. The spawn production involves the growth of mycelia of mushroom on cereal grains. Different cereal grains such as rye, wheat, sorghum are used as carriers for fungal mycelium.

6.6.1.3 Mushroom development: The requirement for fruiting body formation differs from that for mycelia development. The fruiting body requires less temperature compared to mycelia cultures [85]. Accumulation of $CO_2$ favours good mycelial growth, but it is detrimental for fruit body formation. Hence, proper ventilation of accumulated $CO_2$ should be taken care of. Once the fruiting bodies are formed, harvesting is based upon the species and to some extent on market demand [86].

## 7. LIMITATIONS OF SSF

In spite of the benefits, there are a few limitations of SSF in comparison to SmF, which have to be solved to make the process workable at industrial scale. Some of the limitations of SSF for future investigations are summarized here.

- Heat buildup
- Occasional bacterial and fungal contamination
- Limitation of microbial types that can be used
- Need for use of indirect methods for biomass estimation
- Need for production of large scale inoculum
- Non-availability of well defined scale-up criteria
- Problem in product recovery and harvest

## 8. CONCLUSION

An overview of the various aspects shows that the solid state fermentation has large economic potential for the conversion of waste material into several value added products. SSF is a simple, high yielding and economically by feasible process. It is now technologically and economically possible to produce low bulk, high value products like pectinases, microbial colours and animal feed. In future, solid state fermentation could become a potential tool for solid waste management of food processing plant and to prevent environment pollution as well. More in-depth research on the various problems could help in commercializing the various technologies.

## REFERENCES

1. Chahal DS (1985) Solid state fermentation with *Trichoderma ressei* for cellulase production. *Appl Environ Microbiol* 49: 205-210.
2. Pandey A, Francis F, Sabu A & Soccol CR (2004) General aspects of solid state fermentation. In: *Concise Encyclopedia of Bioresource Technology*, Pandey A (ed), Haworth Press, New York, USA, pp 702-709.
3. Sharma N, Bansal KL & Neopaney B (2006) Effect of moisture level on bio-degradation of forest waste under SSF. *J Sci Ind Res* 65: 675-679.
4. Bhalla T & Joshi M (1994) Protein enrichment of apple pomace by co-culture of cellulolytic moulds and yeasts. *World J Microbiol Biotechnol* 10: 116-117.
5. Pandey A, Carlos RS, Leon JAR & Nigam P (2001) History and development of solid state fermentation. In: *Solid State Fermentation in Biotechnology*, Asiatech Publishers Inc, New Delhi, India, pp 3-7.
6. Aidoo KE, Hendry R &Wood JB (1982) Solid state fermentations. *Adv Appl Microbiol* 28: 201-237.
7. Hesseltine CW (1972) Solid state fermentation part I. *Biotechnol Bioeng* 14: 517-532.
8. Pandey A, Ashakumary L, Selvakumar P & Vijaylakshmi KS (1994) Influence of water activity on growth and activity of *Aspergillus niger* NCIM 1245 for enzyme production in solid state fermentation. *World J Microbiol Biotechnol* 10: 488-486.
9. Zheng Z & Shetty K (2000) Solid state production of polygalacturonase *Lentinus edodes* using fruit processing wastes. *Proc Biochem* 35: 825-830.

10. Reddy GV, Babu R, Komaraiah P, Roy K & Kithari I (2003) Utilization of banana waste for the production of lignolytic and cellulolytic enzymes by solid substrate fermentation using two *Pleurotus* species (*P. ostreatus* and *P. sajor-caju*). *Process Biochem* 38: 1457-1462.
11. Oboh G & Akindahunsi A (2003) Biochemical changes in cassava products (flour and gari) subjected to *Saccharomyces cerevisiae* solid media fermentation. *Food Chem* 82: 599-602.
12. Balakrishnan K & Pandey A (1998) Production of biologically active secondary metabolities in solid state fermentation. *J Sci Ind Res* 55: 365-372.
13. Krishna C & Young MM (2004) Single cell protein. In: *Concise Encyclopedia of Bioresource Technology*, Pandey A (ed), Haworth Press, New York, USA, pp 293-295.
14. Pandey A (1992) Recent process developments in solid state fermentation. *Process Biochem* 27: 109-117.
15. Pandey A (1991) Effect of particle size of substrate on enzyme production in solid state fermentation. *Biores Technol* 37: 169-172.
16. Mitchell DA, Pandey A, Sangsurasak & Kreiger N (2000) Scale up strategies for packed bed bioreactors for solid state fermentation. *Process Biochem* 35: 1211-1225.
17. Nigam P & Prabho KA (1987) A note on utilization of bagasse for the production of proteinaceous cattle feed. *Biol Waste* 19: 275-280.
18. Grajek W (1988) Cooling aspects of solid state cultures of mesophilic and thermophilic fungi. *J Ferment Technol* 66: 675-679.
19. Berry DF, Tomkinson RA, Hetzel GH, Mullins DE & Young RW (1993) Evaluation of solid state fermentation techniques to dispose off atrazine and carbofuran. *J Environ Quality* 22: 366-374.
20. Zadrazil F & Brunnert H (1982) Solid state fermentation of lignocelluloses containing plant residues with *Sporotrichum pulverulentum* Nov. and *Dichomitus squalens* (Karst.) Reid. *Eur J Appl Microbiol Biotechnol* 16: 45-51.
21. Tengerdy RP (1996) Cellulase production by SSF. *J Sci Ind Res* 55: 313-316.
22. Sharma N & Bansal KL (2005) Composition of forest waste degradation under solid state fermentation and submerged fermentation by co-culture of *Bacillus licheniformis & Bacillus coagulans. Asian J Microbiol Biotech Env Sci* 7: 387-390.
23. Kimura P & Satyanarayna (2004) Xylanase production by the thermophilic mold *Humicola lanuginose* in solid state fermentation. *Appl Biochem Biotechnol* 119: 145-151.
24. Ramanamurthy MVR, Karanth NG & Raghavrao KSMS (1993) Biochemical engineering aspect of solid-state fermentation. *Adv Appl Microbiol* 38: 99-147.
25. Durand A (2003) Bioreactor designs for SSF. *Biochem Eng J* 13: 113-125.
26. Pandey A, Selva Kumar O, Soccol CR & Nigam P (1999) Solid state fermentation for production of industrial enzymes. *Curr Sci* 77: 149-162.
27. Balakrishnan K & Pandey A (1998) Production of biologically active secondary metabolities in solid state fermentation. *J Sci Ind Res* 55: 365-372.
28. Pandey (1992) Production of glucoamylase in solid state fermentation. In: *Industrial Microbiology*, Malik VS & Sridhar P (eds), IBH & Oxford Publishing Co., New Delhi, India, pp 525-537.
29. Nigam P & Singh D (1994) Solid state (substrate) fermentation systems and their applications in biotechnology. *J Basic Microbiol* 34: 406-423.
30. Ramesh MV & Lonsane BK (1991) Ability of a solid state fermentation technique to significantly minimize catabolite repression of alpha-amylase production by *Bacillus licheniformis M27. Appl Microbiol Biotechnol* 11: 649-652.

31. Gonzalez JB & Tomasini (1996) Production of aflatoxins in solid state fermentation. *J Sci Ind Res* 55: 424-430.
32. Pandey A, Francis F, Sabu A & Soccol CR (2004) General aspects of solid state fermentation. In: *Concise Encyclopedia of Bioresource Technology*, Pandey A (ed), Haworth Press, New York, USA, pp 702-708.
33. Mitra P, Chakraverty R & Chandra AL (1996) Production of proteolytic enzymes by solid state fermentation—An overview. *J Sci Ind Res* 56: 439-442.
34. Rodreguez-Leon JA, Soccol CR & Pandey A (2007) Factors affecting solid state fermentation. In: *Current Developments in Solid State Fermentation*, Pandey A, Soccol CR & Larroche C (eds), Asia Tech Publishers, Delhi, India, pp 27-47.
35. Joshi VK, Pandey A & Sandhu DK (1999) Food factory waste management technology. In: *Biotechnology-Food Fermentation: Microbiology, Biochemistry & Technology*, Vol 2, Joshi VK & Pandey A (eds), Educational Publishers and Distributors, New Delhi, India, pp 1291-1348.
36. Singhania RR, Soccol CR & Pandey A (2007) Application of Tropical agro-industrial residues as substrate for solid state fermentation processes. In: *Current Developments in Solid State Fermentation*, Pandey A, Soccol CR & Larroche C (eds), Asiatech Publishers, New Delhi, India, pp 412-442.
37. Fujio Y & Morito H (1996) Improved glycoamylase production by *Rhizopus* sp. A-11 using metal ion supplemented liquid medium. *J Ferment Bioeng* 82: 554-557.
38. Pandey A, Wamik A, Jasvir Singh & Banerjee UC (1999) Types of fermentation and factors affecting it. In: *Biotechnology: Food Fermentation,* Joshi VK & Pandey A (eds), Asiatech Publishers, New Delhi, India, pp 383-425.
39. Pandey A, Soccol CR, Nigam P & Soccol VT (2000) Biotechnological potential of agro industrial residues. I sugarcane bagasse. *Biores Technol* 74: 69-80.
40. Villas-Boas S, Esposito E & Mendonça M (2003) Bioconversion of apple pomace into a nutritionally enriched substrate by *Candida utilis* and *Pleurotus ostreatus*. *World J Microbiol Biotechnol* 19: 461-467.
41. Tengerdy RP (1998) Solid substrate fermentation for enzyme production. In: *Advances in Biotechnology*, Pandey A (ed), Educational Publishers & Distributors, New Delhi, India, pp 13-16.
42. Gunasekaran P & Chandra Raj K (1999) Ethanol fermentation technology – *Zymomonas mobilis*. Curr Sci 77: 56-68.
43. Silman RW (1980) Enzyme formation during solid state fermentation in rotating vessel. *Biotechnol Bioeng* 22: 411- 420.
44. Shamla TR & Sreekantiah KR (1987) Successive cultivation of selected cellulolytic fungi on rice straw and wheat bran for economic production of cellulases and D-xylanase. *Enz Microb Technol* 9: 97-101.
45. Kurakake M, Ooshima H & Harano Y (1991) Pretreatment of bagasse by UCT-solvent for the enzymatic hydrolysis. *Appl Biochem Biotechnol* 27: 111-121.
46. Soccol CR (1998) Citric acid and glucoamylase production from cassava by-products in solid state fermentation. In: *Solid State Fermentation*, Raimbault M, Soccol CR & Chuzel G (eds), OSTROM Publishers, Montpellier, France, pp 125-136.
47. Tran C & Mitchell D (1995) Pineapple waste—a novel substrate for citric acid production by solid state fermentation. *Biotechnol Lett* 17: 1107-1110.
48. Zadrazil F, Ostermann D & Compare GD (1992) *Solid Substrate Cultivation*, Elsevier Applied Science, London, UK.

49. Stanbury, PP, Whitakar A & Hall SJ (1977) Media for industrial fermentations. In: *Principles of Fermentation Technology*, Aditya Books (P) Ltd, New Delhi, India, pp 93-94.
50. Gervias P, Marechal PA & Molin P (1996) Water relation of solid state fermentation. *J Sci Ind Res* 55: 343-357.
51. Ranganna S (1986) *Handbook of Analysis and Quality Control for Vegetable Products*, Tata Mcgraw-Hill Publ Co, New Delhi, India.
52. Subbarao NS (1999) Organic matter decomposition. In: *Soil Microbiology IV Edition of Soil Microorganisms and Plant Growth*, Oxford & IBH Publishing Co., pp 252-270.
53. Giovannozzi SG & Porri A (1989) The potentiality of solid-state biotransformation of lignocellulosic materials. *Chim Oggi* 7: 15-19.
54. Rathbun BL & Shuler ML (1983) Heat and mass transfer effect in static solid substrate fermentation. Design of fermenter chambers. *Biotechnol Bioeng* 25: 929-938.
55. Auria R, Palacios J & Revah S (1992) Determination of the interparticular effective diffusion coefficient for $CO_2$ & $O_2$ in SSF. *Biotechnol Bioeng* 39: 898-902.
56. Sarrette M, Nont MJR, Gervais P & Rombents FM (1992) Effect of water activity on production and activity of *Rhizopus oligosporus* polysaccharide. *Appl Microbiol Biotechnol* 61: 623-629.
57. Ashakumari L, Selvakumar P & Pandey A (1994) Column fermenter for solid state fermentation. In: *Solid State Fermentation*, Pandey A (ed), Wiley Eastern, New age International Publishers, New Delhi, India, pp 33-37.
58. Raghavrao KSMS, Ranganathan TV & Karanth NG (2003) Some engineering aspects of solid state fermentation. *Biochem Eng J* 13: 127-135.
59. Schutyser MAI, Weber FJ, Briels WJ, Rinzema A & Boom RM (2002) Heat and water transfer in rotating drum containing solid substrate particles. *Biotechnol Bioeng* 82: 552-563.
60. Durand A, Renavd R, Maratray J, Almanza S & Diez M (1996) INRA–Dijon reactors for solid state fermentation design and applications. *J Sci Ind Res* 55: 317-332
61. Lindenfelser LA & Ciegler (1975) Solid state fermenter for ochratoxin A production. *Appl Microbiol* 29: 323-327.
62. Giovannozi–Sermanni G, D'Annibale A, Peranic C, Porri A, Pastina F, Minelli V, Vitale NS & Gelsomino A (1994) Characteristics of paper handsheets after combined biological pretreatments & conventional pulping of wheat straw. *Tappi J* 77: 151-157.
63. Prado FC, Vandenberghe LPS, Woiciechowski A L, Rodrigues-Leon JA & Soccol CR (2005) Citric acid production by solid state fermentation on a semi-pilot scale using different percentages of treated cassava bagasse. *Braz J Chemical Eng* 22: 547-555.
64. Medeiros ABP, Pandey A, Christen P, Fontoura PSG, de Freistas RJS & Soccol CR (2001) Aroma compounds produced by *Kluyveromyces marxianus* in solid state fermentation on a packed bed column bioreactor. *World J Microbiol Biotechnol* 17: 767-771.
65. Nigam P & Singh D (1996) Processing of agricultural wastes in solid state fermentation for cellulose production. *J Sci Ind Res* 55: 373-380.
66. Joshi VK & Sandhu DK (1996) Preparation and evaluation of animal feed using solid state fermentation of apple pomace. *Biores Technol* 56: 251-255.
67. Joshi VK & Attri D (2008) Ethanol recovery from solid state fermented apple pomace and evaluation of physico-chemical characteristics of the residue. *Nat Prod Rad* 7: 127-132.
68. Joshi VK & Bhutani VP (1991) The influence of enzymatic clarification on the fermentation behaviour, composition and sensory qualities of apple wine. *Sciences Des Aliments* 11: 491-496.
69. Joshi VK, Chauhan, SK & Lal BB (1991) Extraction of juices from plum, peach and apricot by the pectolytic enzyme treatment. *J Food Sci Technol* 28: 64-65.

70. Parmar M, Joshi VK, Lal BB & Rana N (2004) Effect of fermentation parameters for pectin methyl esterase production in solid state fermentation of apple pomace. In: National Symposium on Advances in Industrial Biotechnology, Punjabi University, Patiala, p 36.
71. Joshi VK, Attri D, Bala A & Bhushan S (2003) Microbial pigments. *Ind J Biotechnol* 2: 362-369.
72. Joshi VK, Parmar M & Rana NS (2006) Pectin esterase production from apple pomace in solid state fermentations. *Food Technol Biotechnol* 44: 253-256
73. Attri D (2004) *Production and Evaluation of Microbial Colours using Apple Pomace*, PhD Thesis, Dr YS Parmar University of Horticulture and Forestry, Nauni, Solan, HP, India.
74. Sandhu DK & Joshi VK (1997) Development of apple pomace based medium, optimizing pigment production by *Rhodotorula* and its characterization. *Adv Food Sci* 19: 31-34.
75. Joshi VK & Attri D (2006) Solid state fermentation of apple pomace for the production of value added products. *Nat Prod Rad* 5: 289-296.
76. Attri D & Joshi VK (2005) Optimization of apple pomace based medium and fermentation conditions for pigment production by *Micrococcus* species. *J Sci Ind Res* 64: 598-601.
77. Shankaranand VS & Lonsane BK (1994) Citric acid by solid state fermentation: A case study for commercial exploitation. In: *Solid State Fermentation*, Pandey A (ed), Willey Eastern, New Delhi, India, pp 149-152.
78. Kumar J, Jain V, Shanker G & Srivastava A (2003) Utilization of fruit waste for citric acid production by solid state fermentation. *Process Biochem* 38: 1725-1729.
79. Sharma RC & Joshi VK (2005) Apple processing technology. In: *The Apple*, Chadda KL & Awasthi RP (eds), Malhotra Publishing House, New Delhi, India, p 445.
80. Karanth NGK, Deo PG & Veenanad NK (1999) Microbial production of biosurfactants and their importance. *Curr Sci* 77: 116-123
81. Purkayastha RP & Chandra A (1985) *Manual of Indian Edible Mushrooms*, Jagmander Book Agency, New Delhi, India.
82. Bano Z, Ghosh PK, Rajarathnam S & Shaskirekha MN (1996) Biotransformation efficiencies of lignocellulose wastes by mushroom in solid state fermentation. *J Sci Ind Res* 55: 400-407.
83. Kohli MS (1990) *The Hindu Survey of Indian Agriculture*, National Press, Madras, India.
84. Chang ST (1993) Mushroom biology: The impact on mushroom production and mushroom product. In: *Mushroom Biology and Mushroom Products*, Chang ST, Buswell JA & Chin SW (eds), The Chinese University Press, Hongkong, pp 3-20.
85. Srinivas MRS, Nagin CH & Lonsane, K (1994) Use Plackett-Burman design for rapid screening of several nitrogen source, growth/product promoters, minerals and enzyme inducers for the production of alpha-galactosidase by *Aspergillus niger* MRSS 234 in solid state fermentation system. *Bioprocess Engg* 10: 139-144.
86. Levanon D, Danai O & Masaphy SD (1988) Chemical and physical parameters in recycling organic wastes for mushroom production. *Biol Wastes* 26: 341-348.

# Enzyme Technology

# 5

**Manpreet Singh, Himani Kansal, Sawraj Singh,
Pankaj Soni and Uttam C. Banerjee**

*Department of Pharmaceutical Technology, National Institute of Pharmaceutical Education & Research, S.A.S. Nagar 160 062 (Pb.)*

## 1. INTRODUCTION

The existence of enzymes has been known for nearly two centuries, and has been associated with the history of ancient Greece where they used enzymes from microorganisms in baking, brewing, alcohol production, cheese making, etc. The Sumerians, for instance, produced at least 19 different types of beers. Some of the earliest studies were performed in 1835 by the Swedish chemist Jon Jakob Berzelius who termed their chemical action as catalytic. This practical art was the fuse for the explosion in understanding of organic and biological chemistry that took place in the $19^{th}$ century. Coining the term 'catalysis', Berzelius proposed that it plays a central role in various life processes. In the living plants and animals, thousands of catalytic processes go on between the tissues and the fluids, and produce the amount of dissimilar chemical syntheses from the same raw materials. Studies of fermentation led to key insights into life's chemistry by Liebig, Pasteur and Emil Fischer, among others, culminating in the identification of enzymes (in yeast) as nature's catalytic molecules and Fischer's intuitive leap of the 'lock and key' mechanism for their specificity. It is the specificity that drew the interest of chemists seeking selective catalytic agents. But the trials of putting biocatalysts to industrial use are amply illustrated by attempts in 1941 to produce fungal penicillin, which basically was whole cell process. The isolation of a large number of beneficial thermophilic microorganisms from different exotic ecological zones of the earth and the subsequent extraction of useful enzymes from them had paved the way for the commercialization of enzymes [1-4].

Over the last 20 years, many reservations like limited substrate specificity of enzymes, their availability, existence, protein stability, etc. with respect to biocatalysis have been voiced. About ten years ago, with 3196 different enzymes already listed in enzyme nomenclature [5], only about 50 enzymes were fully characterized and only about a dozen enzymes were available commercially on a regular basis. However, recombinant DNA technology, discovered in 1978 by Cohen and Boyer (Stanford University, Palo Alto, CA, USA), allowed the enzymes to be produced much more efficiently in higher purity, and in a more economical way [6]. But, the protein stability is the major concern, as they commonly require ambient temperature conditions, neutral pH and aqueous media.

Enzymes are used in numerous new applications in the food, feed, agriculture, paper, leather and textile industries, resulting in significant cost reduction [7]. With better knowledge and purification of enzymes, the number of applications has increased manifold. Enzymes have been used for thousands of years for a wide variety of biotechnological processes, which include the

*Food Biotechnology: Principles and Practices*
*V.K. Joshi & R.S. Singh (Eds.)*
*I.K. International Publishing House Pvt. Ltd., India*
*pp 127–144*

production of beer, wine, vinegar and cheese and commodity products such as leather and linen. The concept for the development of clean technology marks a paradigm shift from the management of waste to the prevention of pollution at source. Biotechnology represents a powerful and versatile tool enabling technology deliver, clean industrial products and processes. Recent evidences show that this technology is competitive with other technologies in terms of cost and risks, is not limited by scale of operation, and has a major impact on greening of industry.

Only during the past few decades, researchers had started to purify and characterize a large variety of different enzymes, cloned the respective genes and determined the three-dimensional structures of the enzymes. "White biotechnology" is the term, which has been coined to describe the biotechnological production of chemical compounds using enzymes and microorganisms. The most important property of a catalyst for application in a process is not its activity but rather its selectivity, followed by its stability. Enzymes are destined for selective synthesis of molecules with several similar functional groups or chiral centers. A growing emphasis is laid on the synthesis of enantiomerically pure compounds. The interest in these compounds shown by all areas of the life science industries (e.g., pharmaceuticals, food and agriculture) stems from the challenge to develop structurally optimized inhibitors, almost always containing chiral centers. They are capable of catalyzing reactions exhibiting strict *enantio- and regio*-selectivity, or both, under benign condition, thus producing a pure isomer as compared to racemic mixture. With the use of enzymes, one can omit not only the tedious blocking and deblocking steps that are common in enantioselective and regioselective organic synthesis, but also reduce the wastage of fossil fuels required for preparing high-temperature and high-pressure condition [8]. Thus, they are considered more satisfactory catalysts than conventional chemical ones from commercial and environmental points of view. They catalyze chemical reactions with great specificity and rate enhancements. The reactions are the basis of the metabolism for the industry to carry out elegant, efficient and economical biocatalytic conversion. The enzymes can be used widely in industrial synthetic chemistry as excellent catalysts. The importance of the drive for enantiomeric purity for biocatalysis can be discerned from the fact that fully 88% of all biocatalytic processes above a scale of 100 kg involve chiral targets [9]. Of the new drugs introduced in 2000, 76% were single enantiomers compared with 21% in 1991 [10]. Thus, in short, it can be said that green chemistry has an overarching approach that is applicable to all aspects of chemistry [11]. Catalysis offers numerous advantages i.e., novel, high-yield, shorter process routes, increased selectivity, lower temperature and pressure, etc. However, there are three essential drawbacks of today's biocatalysts:

1. Biocatalysts are often not sufficiently stable in the desired media.
2. Very few biocatalysts exist for the desired reactions, from the available substrates to the targeted products.
3. Development cycles are too long for new and improved biocatalysts.

## 1.1 Enzyme Classification and Nomenclature

Classification of enzymes has been subdivided in six different classes according to their catalytic properties:

1. Oxidoreductases: oxidation/reduction
2. Transferases: transfer of functional groups
3. Hydrolases: hydrolysis
4. Lyases: cleavage of molecules
5. Isomerases: isomerization
6. Ligases: formation of molecules

Of the various enzyme classes, hydrolases, which catalyze the cleavage or synthesis of ester and amide-bonds, have the widest range of applications. They are used in laundry detergents (lipases, proteases) and in the food industry (amylases, proteases, lipases, esterases), and also as biocatalysts for the synthesis of chemicals. In addition, these enzymes catalyze slightly exergonic equilibrium reaction and do not require cofactors or complex reaction conditions. Also the direction of the reaction can be reversed by using appropriate reaction conditions (product removal, low water concentration, etc.). Isomerases are also easy to utilize, but have found few applications, the most important of which is the production of high fructose corn syrup from glucose using glucose isomerase. They are also used to racemize unwanted stereoisomers in amino acid resolutions. In general, lyases are hardly ever applied, as these enzymes generally exhibit narrow substrate ranges. Each enzyme has a unique systematic name, which describes the action of that enzyme, and an associated four-number code. The first number indicates the type of reaction that is catalyzed, the next two numbers indicate the sub-class and sub-subclass of reaction and the fourth number identifies the enzyme. This classification scheme for enzymes was produced by the International Commission on Enzymes in 1961 [12] and is regularly updated (the commission was established in 1956 by the International Union of Biochemistry in consultation with IUPAC.) In addition each enzyme has a working (trivial) name which is more commonly used, e.g., the enzyme commonly referred to as lipase has the systematic name glycerol ester hydrolase and the associated code number is 3.2.1.1. While specifying the properties of the original enzyme, its working name as well as its systematic name and associated code number must be stated.

### 1.2 Enzyme Activity

Catalytic activity is the most important enzyme parameter from an analytical standpoint because it has a direct bearing on sensitivity. The most widely used unit of activity is the International unit (IU); one IU is the amount of enzyme activity that catalyzes the transformation of one micro-mole of substrate per minute under the optimal experimental conditions. However, the katal is the recommended unit for enzyme activity and is defined as the amount of enzyme activity that catalyzes the transformation of one mole of substrate per second under optimal experimental conditions [13]. The specific activity is the number of units per milligram of protein. The molar activity (turnover number) is the number of substrate molecules transformed per minute by one μmole of enzyme.

## 2. SOURCES OF ENZYMES

There are three primary sources of enzymes, i.e., animal tissue, plants and microbes (Table 1). Enzymes may be extracted from a given source (organism) by a number of different methods [14]. Nature is an almost inextinguishable source of biocatalysts. Even though, the systematic exploitation of selected natural niches as sources of enzymes is rather recent. These organisms include the moulds *Aspergillus niger, Rhizopus oryzae, Rhizomucor meihei*, blights such as *Endothia parasitica* and yeasts such as *Candida* sp. and *Saccharomyces* sp. with a part from bacteria and the remainder divided between animals (8%) and plant (4%) sources. Non-microbial sources have become a good source in recent times. Most of the strains used have either been employed by the food industry for many years or have been derived from such strains by mutation and selection including bacteria, fungi, higher plants, and animals [15]. Research on extremophiles, i.e. organisms able to survive and thrive in extreme environmental conditions, as promising sources for highly stable enzymes, is a subject of present decade [16-17] and certainly, a very active research area at present [18].

**Table 1:** Some important industrial enzymes and their sources.

| Enzymes | Sources | Industrial uses |
|---|---|---|
| *(a) Microbial sources (Bacteria, Fungi and Yeast)* | | |
| α-Amylase | *Bacillus* | Starch |
| Asparaginase | *Escherichia coli* | Health |
| Glucose isomerase | *Bacillus* | Fructose syrup |
| Lipase | *Pseudomonas aeruginosa* | Pharmaceutical |
| Protease | *Bacillus* | Detergent |
| Pullulanase | *Klebsiella* | Starch |
| α-Amylase | *Aspergillus* | Baking |
| Aminoacylase | *Aspergillus* | Pharmaceutical |
| Glucoamylase | *Aspergillus* | Starch |
| Cellulase | *Trichoderma* | Paper industry |
| Lipase | *Rhizopus* | Food |
| Raffinase | *Mortierella* | Food |
| Oxidoreductase | *Candida vishwanathii* | Pharmaceutical |
| Invertase | *Saccharomyces* | Confectionery |
| Lipase | *Candida* | Food |
| *(b) Animal and plant sources* | | |
| Catalase | Liver | Food |
| Chymotrypsin | Pancreas | Leather |
| Lipase | Pancreas | Food |
| Rennet | Abomasum | Cheese |
| Trypsin | Pancreas | Leather |
| α-Amylase | Malted barley | Brewing |
| β-Amylase | Malted barley | Brewing |
| Lipoxygenase | Soybeans | Food |

Enzyme manufacturers produce enzymes in accordance with all applicable governmental regulations, including the appropriate federal agencies (e.g., Food and Drug Administration, United States Department of Agriculture, Environmental Protection Agency, etc.).

## 3. ISOLATION OF ENZYMES

Isolation of enzyme from intracellular source begins by physically separating the cells from the production medium. The cell wall of the organisms is then disrupted through a mechanical process of homogenization (French press), sonication (sonicator), or mechanical shearing (dynomill). In case of extracellular enzymes, the organism excretes the enzyme that does not involve breaking of the cell. Further impurities are removed by various extraction, purification and standardization techniques. Specific enzymes are precipitated or adsorbed by the use of ion exchange columns [19]. Various chemical extractants including organic solvents (acetone), polymers (methylcellulose or polyvinyl alcohol) or salts (sodium phosphate) are also used [15]. Final purification removes the extractants by further centrifugation, adsorption to a suitable adsorbent, and subsequent elution [20]. Recent technological advances in genetic engineering have made it possible to alter cellular genetic content, resulting in new production capabilities of the cell.

## 4. ENZYME STORAGE AND USE

In spite of the impressive number of applications, enzymes do not exhibit satisfying performance or characteristics in terms of activity, stability and above all selectivity because of the problem of long term stability. The purified proteins often need to be stored for an extended period of time while retaining their original structural integrity and/or activity. Three approaches are possible for the longer stability of enzymes:

1. Use of additives
2. Controlled use of covalent modification
3. Enzyme immobilization

The extent of storage 'shelf-life' can vary from a few days to more than a year and is dependent on the nature of the protein and the storage conditions used. Storage at room temperature often leads to protein degradation and/or inactivity, commonly as a result of microbial growth. Optimal conditions for storage are distinctive to each protein. Generally, proteins are best stored at ±4°C in a simple buffer. For long term storage from 1 month to 1 year, storing them in cryovials under liquid nitrogen is desirable [21]. Freezing at –20 or –80°C is the more common form of protein storage, but once thawed, protein has to be refrozen again. This problem can be solved by the addition of additives such as 50% glycerol or ethylene glycol, which will prevent solution from freezing at –20°C, enabling repeated use from a single stock without thawing. But the common practice of freezing proteins in a glycerol solution is also problematic because the cryoprotectant can possibly interfere with crystallization and may need to be removed by dialysis after thawing. Different methods used to store proteins are illustrated in Table 2.

Enzymes are versatile biocatalysts and find increasing applications in many areas, including organic synthesis [4, 22-24]. Enzymologists have always strived to identify methods by which they can link an enzyme to a carrier or modify the enzymes to suit them for the required application [25]. Even after an enzyme has been identified as being useful for a certain reaction, its application is often hampered by the lack of long-term stability under process condition and also by difficulties in recovery and recycling. Some of the problems can also be overcome by modifying enzymes

**Table 2:** Different methods for storage of proteins at or below 4°C.

| Characteristics | Solution at 4°C | Solution in 25-50% glycerol or ethylene glycol at –20°C | Frozen at –20° to –80°C | Lyophilized (usually also frozen) |
|---|---|---|---|---|
| Typical shelf-life | 1 month | 1 year | Years | Years |
| Requires sterile condition or addition of antibacterial agent | Yes | Usually | No | No |
| Number of times a sample may be removed for use | Many | Many | Once: repeated freezing thawing cycles generally degrade proteins | Once: it is impractical to lyophilize the samples a number of times |

(e.g., immobilization). Researchers have also overcome some of the other shortcomings by altering substrates (substrate engineering), modifying reaction system (medium engineering) or by enzyme engineering.

## 5. PRINCIPLES OF ENZYME AND WHOLE CELL TECHNOLOGY

Both isolated enzymes and whole cells have been used industrially as catalysts. Among the biocatalysts, choice is between commercially available enzymes of different purity and whole cell catalysts. Compared with isolated enzymes, whole cell catalysts can be much more readily and inexpensively prepared [22, 24, 26]. Whole cell catalysts offer a number of advantages over the use of isolated enzymes in biocatalysis. They are convenient and stable source of enzymes since they are protected from the external environment and are synthesized by the cell in response to the presence of substrate. One feature that makes whole cell biocatalysts so attractive is that they generate cofactors together with the actual biocatalysts, therefore they are practically convenient in application in which their reducing power is employed. The most important application is reduction of carbonyl groups to secondary alcohols. Also the reaction exploiting the oxidation and hydroxylation activity of biocatalysts is carried out with whole cell biocatalysts due to limited stability of the isolated enzymes. In particular, microbial cells have been most commonly used for industrial purposes because of their diversity and ease of handling. Rapid advances in the life sciences such as recombinant DNA technology have greatly increased the availability of whole cell catalysts. It has enabled the overproduction of a desired enzyme in various heterologous hosts (microorganism, plant, and animal cells). This technique has even made it possible to modify metabolic pathways in the host cells, such as cofactor regeneration system, biosynthesis of complex metabolites, etc.

## 6. BIOCONVERSION

The term bioconversion is the use of biological processes to transform biomass materials from one form to another. The ethanol production was probably the first bioconversion process discovered by man. A critical consideration in any bioconversion application is the attainment and development of a suitable biological catalyst. This may require application of various microbiological, biochemical engineering and, particularly, molecular technologies. Thus, exploration of biodiversity and/or molecular engineering are the important steps in the gaining of novel enzyme activities. The majority of enzymes currently being used or developed for biotransformation are microbial in origin, and the field of molecular genetics has provided much scope for the isolation and manipulation of enzymes to produce biocatalysts with predetermined properties. In most bioconversion processes, immobilization or stabilization of the biological agent provides significant advantages, such as facilitating reuse of an enzyme, containment of cells and support of a growing biofilm for use in a continuous system, etc. In bioconversion, biocatalysts are being used extensively in the industrial production of bulk chemicals, pharmaceutical and agrochemical intermediates, active pharmaceuticals and food ingredients [22-23]. The demand for chiral drugs is increasing, and the pressure is also on the development of environmentally cleaner processes [27-29]. Bioconversion can be explained by taking examples of lipases from *Pseudomonas aeruginosa* MTCC 5113, isolated from the soil to enantiospecifically hydrolyse the (±)-methyl *trans*-3(4-methoxyphenyl) glycidate (MPGM), an intermediate in the synthesis of cardiovascular drug, diltiazem [30] and carbonyl reductase, which was produced by *Candida tropicalis* PBR-2, isolated from soil, which is capable of carrying out the enantioselective reduction of N, N-dimethyl-3-keto-3-(2-thienyl)-1-propanamine to (S)- N, N-dimethyl-3-hydroxy-3-(2-thienyl)-1-propanamine, a key intermediate in the synthesis of the chiral drug (S)-Duloxetine [31] as illustrated in Fig. 1.

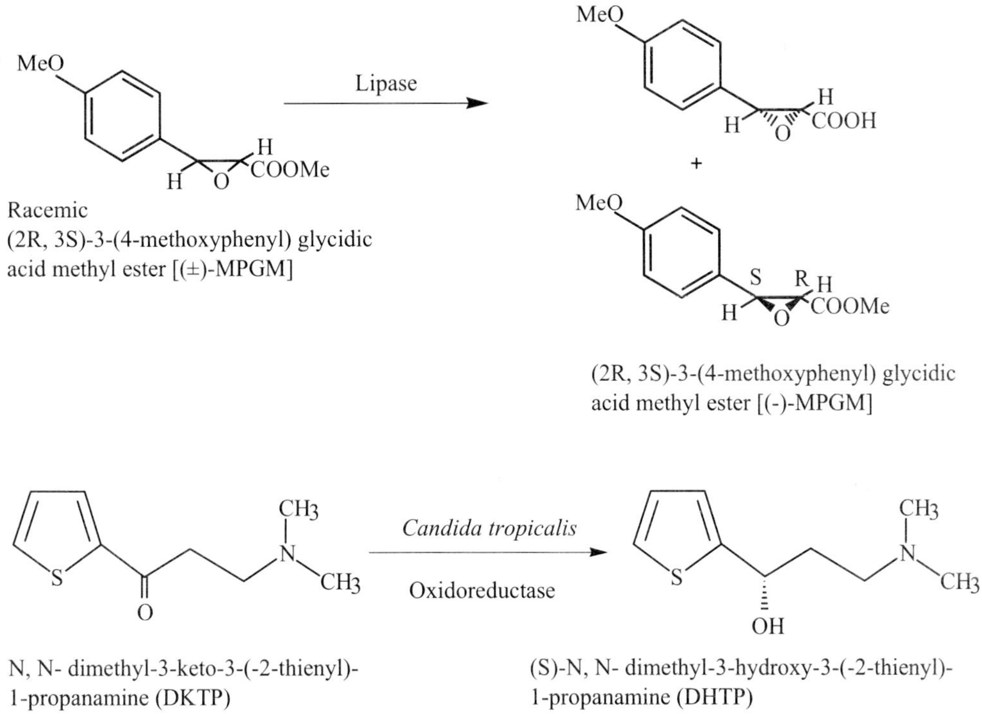


**Fig. 1:** Lipase and oxidoreductase-mediated biotransformations.

The ideal biotransformation system must include [32]:

- Efficient enzyme production system, using a readily culturable microbial source which should be genetically stable and non-pathogenic.
- Efficient biocatalyst, used as resting cells or purified enzymes with high *stereo*-selectivity and high activity in the desired reaction with flexible substrate selectivity and minimal side reactions.
- A biocatalyst stable under optimal reaction conditions, amenable to immobilization/stabilization and bioreactor constraints.

## 7. BIOREACTORS

Bioreactors are the key components for enzyme production, providing a controlled environment to direct cellular responses toward specific enzyme production. Bioreactors offer a possibility to provide a controlled environment for microbial fermentation processes, which is required for optimal yield [33]. Ideally, a bioreactor must provide:

1. An appropriate environment e.g., temperature, humidity, the maintenance of biochemical conditions (e.g., pH, $pO_2$), concentrations of nutrients and growth factors to support cell proliferation.
2. Sufficient metabolite transport to and from the cells.
3. Structurally defined support for cell attachment.

There are several important factors that determine the choice of reactor for a particular process. The choice depends on:

1. The cost, which is inclusive of the costs associated with substrate(s), downstream processing, labour, depreciation, overheads and process development, in addition costs concerned with building and running the enzyme reactor.
2. The form of the enzyme of choice (i.e., free or immobilized).
3. The kinetics of the reaction and the chemical and physical properties of an immobilization support including whether it is particulate, membranous or fibrous, and its density, compressibility, robustness, particle size and regenerability.
4. The scale of operation, the possible need for pH and temperature control, the supply and removal of gases and the stability of the enzyme, substrate and product.

In addition, the enzyme systems are characterized biochemically to provide information, which is required in modification, adaptation, and scale-up of the bioreactors. Industrial enzyme reactions are carried out in a variety of bioreactors with different properties.

## 7.1 Stirred Tank Reactor

The basic stirred tank reactor (STR) consists of a tank equipped with a stirring device. It will also have a heating/cooling system and baffles to ensure adequate mixing. The stirred tank has many advantages. It is simple to use and so is very versatile. In this kind of reactor, the enzyme is either added to the solution in the tank or an immobilized enzyme is dispersed in the reaction solution. A variation is to retain the immobilized enzyme in baskets attached to the impeller shaft, leading to low mass-transfer resistance. This kind of reactor can be used for substrate-inhibited enzymes, where substrate is continuously added at a low concentration (fed-batch operation).

The main disadvantage of stirred tank reactor is that the tank has to be emptied and refilled at the end of the batch run. This leads to longer downtime and a loss of productivity. In addition, stirred tanks can suffer from batch to batch variations. The stirred tank is very good for small-scale production of high-cost products. Productivity of a stirred tank reactor is given by:

$$V_{max}t = [\mathrm{So}]\,F - K_M \ln(1\text{-}F)$$

Where $t$ = reaction time

$$F = \text{fractional conversion} = \frac{[\mathrm{So}]-[\mathrm{S}]}{[\mathrm{So}]}$$

[So] = substrate concentration at $t = 0$
[S] = final substrate concentration

This equation allows the calculation of the amount of enzyme needed to give a desired conversion of substrate to product in a given time, or the time taken to reach a desired conversion with a given amount of enzyme.

## 7.2 Continuous Flow Stirred Tank Reactor

The continuous flow stirred tank reactor (CSTR) is essentially a stirred tank with a continuous feed of substrate and an outflow of product. In a CSTR, the enzyme is subjected to low substrate concentration and high product concentration. This is desirable for substrate-inhibited enzymes, but will cause problems with product inhibited enzymes. The low substrate concentration also reduces the effectiveness factor, which can lead to very inefficient use of the enzyme. One advantage of CSTRs is the ease of monitoring parameters such as temperature and pH.

### 7.3 Packed Bed Reactor

In a packed bed reactor (PBR), the immobilized enzyme particles are held in a column and substrate is pumped under plug flow conditions. In a PBR, the enzyme at the bottom of the column is exposed to substrate, whereas the enzyme at the top is exposed largely to product. This type of reactor is not well suited for substrate-inhibited enzymes. The main disadvantage with PBRs is the difficulty of maintaining even flow and constant temperature.

### 7.4 Fluidized Bed Reactor

Fluidized bed reactors are intermediate in behaviour between PBRs and CSTRs. The immobilized enzyme bed is fluidized by upward motion of the substrate stream. Fluidized beds are particularly useful if a reaction utilizes or releases a gas. The gas stream can then be used to fluidize the bed. The principle disadvantage of fluidized beds is that they are difficult to scale-up and their use is generally restricted to small scale high priced products.

### 7.5 Membrane Reactor

In a membrane reactor, the enzyme is contained within a semi-permeable membrane immersed in a stirred tank. The membranes are usually in the form of hollow fibres around 0.2 mm diameter. Such reactors can be operated in either batch or continuous mode. The principal advantage of membrane reactors is that they are easy to use and involve no enzyme immobilization. They are often used for small scale production and applications involving multi-enzyme systems and co-factor recycling. The principal disadvantage is the high cost of the membranes.

## 8. IMMOBILIZATION OF ENZYMES

Immobilization is the act of limiting movement. Immobilized enzymes have been defined as "The enzymes that are physically confined or localized, with retention of their catalytic activity and which can be used repeatedly and continuously." This helps in their economic reuse and in the development of continuous bioprocesses. There are a variety of methods by which enzymes can be localized. The methods used for the immobilization of enzymes broadly fall into four main categories [34]:

1. Physical adsorption onto an inert carrier.
2. Inclusion in the lattices of a polymerized gel.
3. Cross-linking of the protein with a bifunctional reagent.
4. Covalent binding to a reactive insoluble support.

The attractions of immobilized enzymes from an analytical standpoint are primarily their reusability, greater efficiency and control of their catalytic activity (e.g., potentially longer half-lives, predictable decay rates and more efficient multi-step reactions) and thus, cost saving.

## 8.1 Choice of Support Material for Immobilization

A variety of techniques and supports have been investigated for the immobilization [35]. The support material can have a critical effect on the stability of the enzyme and the efficiency of enzyme immobilization. The choice of the support as well as the technique depends on the nature of the enzyme, nature of the substrate and its ultimate application. The most important requirements for a support material are:

- Insolubility in water
- High enzyme binding capacity
- Chemical inertness
- Mechanical stability

An activated support is defined as "A material having an enzyme with a reactive functional group covalently attached to an otherwise inert surface." The percentage of enzyme immobilized is usually calculated by measuring the amount of enzyme remaining in the supernatant after immobilization and subtracting this from the amount originally present. The absolute enzyme activity remaining on the support after immobilization is more difficult to determine, and an apparent activity is usually measured which takes into account mass transfer and diffusion restrictions in the experimental procedure. The other critical performance indicator is the stability of the immobilized enzyme with respect to time, temperature and other storage conditions and experimental variables. The effect of storage conditions, e.g., pH, temperature and ionic strength and impurities incorporated during the immobilization step must also be considered. Other quantifiable parameters for an immobilized enzyme are its pH optimum (and working range), oxidation-reduction potential working range and the apparent Michaelis constants (kM) for appropriate substrates (which also give an indication of immobilized enzyme selectivity). These parameters can be affected by immobilization and experimental conditions, and therefore, comprehensive experimental details must be provided.

## 8.2 Techniques for Immobilization of Enzymes

Techniques for immobilization have been broadly classified into four categories:

1. Adsorption
2. Covalent binding
3. Entrapment
4. Membrane confinement

A general comparison of methods of immobilization is depicted in Table 3.

**Table 3:** Generalized comparison of different enzyme immobilization techniques.

| Characteristics | Adsorption | Covalent binding | Entrapment |
|---|---|---|---|
| Preparation | Simple | Difficult | Difficult |
| Cost | Low | High | Moderate |
| Binding force | Variable | Strong | Weak |
| Enzyme leakage | Yes | No | Yes |
| Applicability | Wide | Selective | Wide |
| Matrix effects | Yes | Yes | Yes |
| Large diffusional barriers | No | No | Yes |

### 8.2.1 Adsorption

Adsorption of enzymes onto insoluble supports is a very simple method of wide applicability and capable of high enzyme loading (about one gram per gram of matrix). Simply mixing the enzyme with a suitable adsorbent, under appropriate conditions of pH and ionic strength, followed by washing off loosely bound and unbound enzyme after certain period will produce the immobilized enzyme in a directly usable form. The driving force causing this binding is usually due to a combination of hydrophobic effects and the formation of several salt links per enzyme molecule. The particular choice of adsorbent depends principally upon minimizing leakage of the enzyme during use. Although the physical links between the enzyme molecules and the support are often very strong, they may be reduced by many factors including the introduction of the substrate, pH or ionic strength. Moreover, a dynamic equilibrium is normally observed between the adsorbed enzyme and the support, which is often affected by pH as well as the ionic strength of the surrounding medium. This property of reversibility of binding has often been used for the economic recovery of the support. This has also been successfully adapted in industry for the resolution of racemic mixtures of amino acids, using amino acid acylase [35]. Examples of suitable adsorbents are ion-exchange matrices, porous carbon, clays, hydrous metal oxides, glasses and polymeric aromatic resins. Ion-exchange matrices, although more expensive than other supports, may be used economically due to the ease with which they may be regenerated when the bound enzyme has come to the end of its active life. Regeneration of matrix simply involves washing off the used enzyme with concentrated salt solutions and re-suspending the ion exchanger in a solution of active enzyme.

### 8.2.2 Covalent binding

Covalent binding is an extensively used technique for the immobilization of enzymes, though it is not a good technique for the immobilization of cells. The functional groups extensively investigated are the amino, carboxyl and the phenolic group of tyrosine [36]. Lysine residues are found to be the most generally useful groups for covalent bonding of enzymes to insoluble supports due to their widespread surface exposure and high reactivity, especially in slightly alkaline solutions. They also appear to be only very rarely involved in the active sites of enzymes. Enzymes are covalently-linked to the support through the functional groups in the enzymes, which are not essential for the catalytic activity. Enzymes like glucose oxidase, peroxidase, invertase, etc. have been immobilized using this technique. Covalent binding has been extensively investigated using inorganic supports. Enzymes covalently bound to inorganic supports have been used in the industry. Enzymes have also been bound to synthetic membranes, thus integrating biconversion and downstream processing. Large-scale processes using such an approach have been demonstrated for the preparation of invert sugar using invertase [37].

### 8.2.3 Cross-linking

Biocatalysts can also be immobilized through chemical cross-linking using homo- as well as hetero-bifunctional cross-linking agents [35]. Among these, glutaraldehyde (Fig. 2) has been extensively used in view of its GRAS status, low cost, high efficiency and stability. The enzymes or the cells have been normally cross-linked in the presence of an inert protein like gelatine, albumin and collagen [38].

### 8.2.4 Entrapment

Entrapment has been extensively used for the immobilization of cells, but not for enzymes. The major limitation of this technique for the immobilization of enzymes is the possible slow leakage during continuous use in view of the small molecular size compared to the cells [38]. Entrapment of enzymes within gels

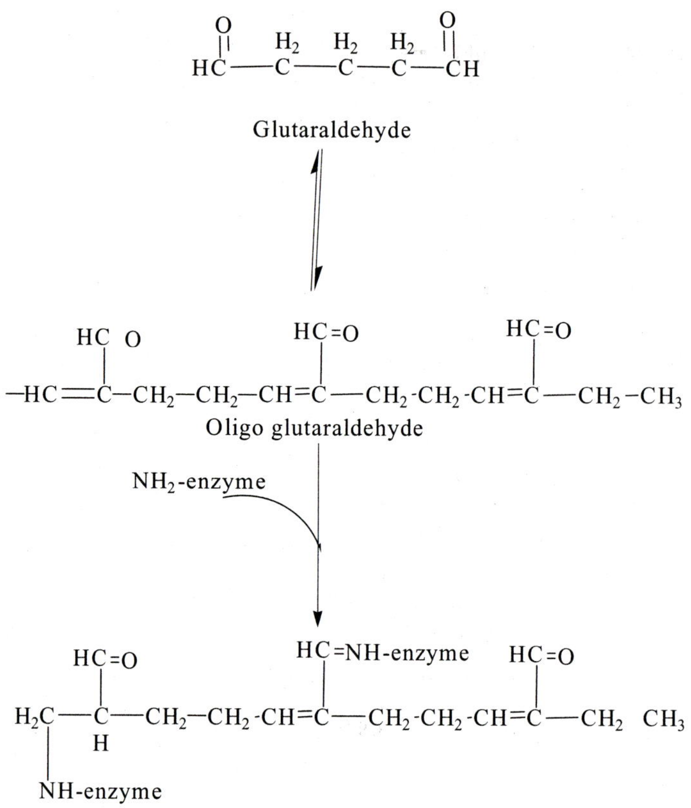

**Fig. 2:** Illustration of cross linking of glutaraldehyde with enzyme.

or fibres is a convenient method for use in processes involving low molecular weight substrates and products. Amounts in excess of 1 g of enzyme per gram of gel or fibre may be entrapped. Biocatalysts have been entrapped in natural polymers like agar, agarose and gelatin through thermo-reversal polymerization [36]. A number of synthetic polymers have also been investigated like photo-crosslinkable resins, polyurethane pre-polymers and acrylic polymers like polyacrylamide [39]. Among these, the most widespread matrix made from monomeric precursors is the polyacrylamide gel. One of the major limitations of entrapment technique is the diffusional limitation as well as the steric hindrance, especially when the macromolecular substrates like starch and proteins are used [37]. Diffusional problems can be minimized by entrapment in fine fibres of cellulose acetate or other synthetic materials or by using an open pore matrix. Recently, the development of so-called hydrogels and thermoreactive water-soluble polymers, like the albumin-poly (ethylene glycol) hydrogel have attracted the attention of biotechnologists.

## 8.3 Benefits of Enzyme Immobilization

Immobilization of enzymes have several advantages such as:

1. Multiple and repetitive use of an enzyme or in a continuous process.
2. The ability to stop the reaction rapidly by removing the enzyme from the reaction system.
3. Enzymes are usually stabilized by binding.
4. Product is not contaminated with the enzyme (especially in the food and pharmaceutical industry).

5. Easy separation of the enzyme from the product.
6. Allows development of a multienzyme reaction system.
7. Reduces effluent disposal problem.

## 9. RECENT PROGRESS IN BIOPROCESS TECHNOLOGY

Until recently, scientists wishing to utilize biotransformation needed to rely upon the availability of particular organism and isolated enzyme to catalyze the desired reaction for the synthesis of a particular product. New technologies developed have made an impact on the ability to utilize tailor made organisms and develop new biotransformation processes. These technologies include isolation, characterization and utilization of microorganisms and their enzymes from the extreme conditions, manipulation and alteration of the enzymes and their metabolic pathways, etc. Since many of these environments are similar to those found in chemical processes, the potential of performing extremophiles based biotransformations under harsh conditions has attracted the attention of various researchers. From biotechnological perspective, there are many potential advantages of employing enzymes in organic solvents opposed to aqueous media. These include higher substrate solubility, reversal of hydrolytic reactions, modified enzyme specificity and improved thermostability. The recent advancement in this field is the development of active and stable homogenous (soluble) biocatalysts, since native enzymes are insoluble in nearly all the organic solvents. The tools of molecular biology have also played a crucial role in generating new enzyme catalysts. One particular technique developed over the past few years is gene shuffling in which rounds of *in-vitro* recombinations and mutations of pool of related gene sequences are performed and selection of the desired enzymatic activity is made *in vivo*. This process is known as molecular evolution. An example of this is the conversion of β-galactosidase into a β-fucosidase. Similar to rapid generation of improved enzymes, improved metabolic pathways can be obtained by this method. Enzyme catalyzed transformations are increasingly being considered and developed for the manufacture of a wide range of intermediates in the pharmaceuticals, agrochemicals, fine chemicals and food industries. For obtaining non-racemic chiral molecules, the reactions should proceed with high yield and the product should be recovered with high chemical and optical purity. Various parameters such as solvent, pH, immobilization, chemical modification and temperature have effect on the enantioselectivity of enzyme-catalyzed reactions. Subtle modification of the substrate to change the enantioselectivity of the reaction is one of the simplest methods for obtaining changes in optical purity, which is known as substrate engineering. For the oxidative kinetic resolution of sec-alcohols using *Rhodococcus ruber* DSM 44541, it was found that the introduction of C=C bond units into the side chain improved the enantioselectivity of the reaction for the lipases from *Pseudomonas* sp. and *Alcaligenes* sp. The enantioselectivity of acylation of 1-phenylethanol using vinyl acetate as acyl donor was found to be improved when the reaction was carried out in $[BMIM](CF_3SO_2)_2N$ as solvent. Penicillin G acylase isolated from *E.coli* immobilized on glyoxyl agarose also showed the change in catalytic activity and enantioselectivity towards racemic esters and amides. The enantioselective deacylation of racemic amino acid esters has also been studied using the chemically modified enzymes decanoyl-α-chymotrypsin, which showed lower activity relative to native enzyme but showed an increase in enantioselectivity. Some of the functions of different enzymes in industrial processes are given in Table 4.

Generally, the enzymatic reactions are carried out by passing a liquid stream of substrate over the immobilized enzyme. The reaction would proceed and products are then carried through the reactor and collected while the enzyme remained in place. This proved to be a very important development for the starch-sugar industry because it was a much more economical method for processing high glucose

**Table 4:** Applications of enzymes in industrial processes.

| Market | Enzyme | Purpose/function |
|---|---|---|
| Dairy | Rennet (protease) | Coagulant in cheese production |
| | Lactase | Hydrolysis of lactose to give lactose-free milk products |
| | Protease | Hydrolysis of whey proteins |
| Brewing | Cellulases, beta-glucanases, alpha amylases, proteases, maltogenic | For liquefaction, clarification and to supplement malt enzymes |
| Alcohol production | Amyloglucosidase | Conversion of starch to sugar |
| Baking | Alpha-amylases | Breakdown of starch, maltose production |
| | Amyloglycosidases | Saccharification |
| | Protease | Breakdown of proteins |
| | Glucose oxidase | Stability of dough |
| Wine and fruit juice | Pectinase | Increase of yield and juice clarification |
| | Glucose oxidase | Oxygen removal |
| Meat | Protease | Meat tenderising |
| Protein | Proteases, trypsin, aminopeptidases | Breakdown of various components |
| Starch | Alpha amylase, glucoamylases, hemi-cellulases, maltogenic amylases | Modification and conversion (to dextrose or high fructose syrups) |

syrups into HFCS. Most HFCS production is now accomplished using continuous fixed-bed reactors in which the glucose (xylose) isomerase has been immobilized. The high glucose syrup is passed through the reactors, resulting in a much more rapid and efficient isomerization of glucose to fructose. The rate of conversion depends on many factors, including temperature and the length of time the glucose syrup is in contact with the enzyme (i.e., the fluid flow rate). The saccharization step can be done in a similar way, using columns of immobilized glucoamylase. Both high glucose and high fructose syrups find application in the food industry, primarily as sweeteners, flavour enhancers and fermentation substrates. HFCS is used extensively in baked products, beverages and dairy products.

A canning process has been developed that uses lower-than-normal cooking temperatures, a brief holding period and naturally-occurring plant tissue enzyme, pectin methylesterase, to reverse the softening effects of cooking. The enzyme reacts with cholesterol and converts it to coprostanol, a sterol that passes through the body when consumed. Three enzymes have been isolated from maize that digest insoluble plant products composed of complex xylans, feraxan or ferulated arabinoxylan. Applications include biomass degradation and modification of insoluble plant products to soluble substances. Blends of enzymes are finding growing use as bromate substitutes in baking applications. Companies are starting to market these enzymes and are looking to work with others to tailor them to specific applications.

The use of enzymes has the potential to increase productivity, efficiency and quality output in agro-industrial processing operations in many developing countries. Enzyme-catalyzed processes generally have requirements for a simple manufacturing base, low capital investment and consume relatively small amounts of energy, when compared to other methods of food processing.

## 9.1 Ultra High Fructose Syrup

One of the major discoveries in the field of nutritive sweeteners was the development of a process for the isomerization of glucose to fructose. It became possible to produce starch derived syrup which was as sweet as sucrose (sugar) and known as 'High fructose syrup' (HFS) which has become a major sweetener. High Fructose (Corn) Syrup - HFS or HFCS is a nutritive sweetener with high commercial potential. It is a product in which a large percentage of the glucose derived from starch hydrolysis has been converted into its sweeter-tasting isomer fructose by the use of enzymes. The crystal clear syrup has sweetness and calories approximately equal to that of a sugar solution. It performs many of the same functions as sugar, chiefly the "clean" sweetening of beverages, ketchup, dairy products, baked goods and so on. They are produced from a variety of starch raw materials including corn, rice, tapioca, wheat, cassava and sugar beet pulp. A portion of glucose in ordinary corn syrup is converted to fructose through incubation with the enzyme glucose isomerase. Since fructose is a much sweeter monosaccharide than glucose, the sweetness of the syrup increases relative to corn syrup. Fructose is also more desirable than glucose as it has increased solubility at low temperature. In addition, HFCS makes foods such as bread and breakfast cereal "brown" better when baked and gives chewy cookies and soft textured snack bars. Through further processing, the fructose content can be increased to 55% (yielding a product that has the same sweetness as sucrose) or any desired higher amount. Common commercial grades of high fructose corn syrup include grades having 42%, 55% or 90% fructose. When compared to table sugar (sucrose), HFCS is not at all "high" in fructose. In fact, HFCS is nearly identical in composition to table sugar (sucrose), which is composed of 50% fructose and 50% glucose. HFCS is composed of either 42 or 55% fructose, with the remaining sugars being glucose and higher sugars.

The corn wet milling industry makes HFCS from corn starch using a series of unit processes that include steeping corn to soften the hard kernel; physical separation of the kernel into its separate components - starch, corn hull, protein and oil; breakdown of the starch to glucose; use of enzymes to invert glucose to fructose; removal of impurities and finally blending of glucose and fructose to make HFCS-42 and HFCS-55. The sweeter HFCS-55 has a predominant role in carbonated colas and soft drinks. HFCS-42 is popular in canned fruits, condiments, baked goods and other processed foods for its mild sweetness that would not mask natural flavours. The dairy industry uses HFCS-42 in fluid products such as flavoured milks and eggnog yoghurt, ice cream and other frozen desserts and novelties.

The conversion of starch to sugar is generally described by referring to a unit called "dextrose equivalent" (DE). (The terms dextrose and glucose both refer to the same sugar molecule). A "DE" is a measure of the degree of hydrolysis of the syrup product. Starch itself has no free dextrose and so has a DE of zero. Dextrose has a DE of 100, i.e., it is 100% dextrose. All DE values are therefore between 0 and 100.

The conversion of starch to sugars is basically a two-step process. First comes the liquefaction step, followed by a saccharification step. Initially, the starch is gelatinized by heating, producing a thick and viscous mixture. The temperature required to completely gelatinize the starch depends on the source. Waxy maize starch, a corn starch, can be gelatinized effectively at about 110ºC (230ºF). The liquefaction step is the initial thinning of this starch slurry. The subsequent addition of a heat-stable bacterial α-amylase decreases the viscosity by converting the gelatinized starch into a mixture of linear and branched dextrins along with some sugars, mostly glucose and maltose. Sugar syrups of up to about 40 DE can be achieved using bacterial α-amylases alone. But, for very high rates of conversion, a saccharification step is required. This is the step that results in syrup composed almost entirely of sugars. In this step, a glucoamylase is added to the mixture of dextrins produced by the α-amylase. Depending on the glucoamylase used, a pullulanase enzyme may also be added more efficiently to hydrolyze the α-1,6 bonds at the branch points in amylopectin. The result of this saccharification step is a sugar syrup with a DE in the range of about 96–98.

The next step is isomerisation of glucose to fructose. The isomerization of glucose to fructose is part of the glycolysis cycle that converts glucose to pyruvate and is catalyzed by the enzyme glucose isomerase (D-glucose ketoisomerase). One of the triumphs of enzyme technology, so far, has been the development of 'glucose isomerase', which in turn led to the commercialization of high fructose corn syrups. Now, it is known that several types of bacteria can produce such glucose isomerases.

They are resistant to thermal denaturation and will act at very high substrate concentrations, which have the additional benefit of substantially stabilizing the enzymes at higher operational temperatures. The vast majority of glucose isomerases are retained within the cells that produce them, but need not be separated and purified before use. The degree of enzymatic isomerization depends on several operating parameters. While isomerization levels in excess of 50% fructose are possible, conversion to 42% fructose content was selected as having the best balance between increased sweetness and process economics. After isomerisation, the pH of the syrup is lowered to 4 – 5 and it is purified by ion-exchange chromatography and treatment with activated carbon. Then, it is normally concentrated by evaporation to about 70% dry solids.

## 9.2 Applications in Food and Food Ingredients

Enzymes can be used to modify raw materials and aid in the processing or cooking stages. The roles of enzymes include enhancement of flavour and aroma, removal of unwanted flavours and taints, enhancement of digestibility, modification of texture to aid processing and final product appearance, upgrading raw materials, etc. The main enzyme activity utilized in food processing applications is protease. However, applications utilizing lipases [40] and carbohydrate degrading activities are also becoming widespread. Some well-known specific applications include:

- Cheese making. Rennet is responsible for curd production from milk; many flavours in cheeses are a result of protease action.
- Production of infant milk formulas from cow's milk.
- Soy sauce production.
- Production of hydrolyzed vegetable protein.
- Gelatin hydrolysis.
- Production of meat extracts.
- Pet food manufacture.

## REFERENCES

1. Groboillot A (1994) Immobilization of cells for application in the food industry. *Crit Rev Biotechnol* 14: 75-107.
2. Bharat B & Hoondal G (1998) Isolation, purification and properties of thermostable chitinase from an alkalophilic *Bacillus* sp. BG-11. *Biotechnol Lett* 20: 157-159.
3. Bauer M, Driskil L, Callen W, Snead M, Mathur E & Kelly R (1999) An endoglucanase EglA, from the hyperthermophilic archaeon *Pyrococcus furiosus* hydrolyzes α-1,4 bonds in mixed linkage (1-3), (1-4)-β-D-glucans and cellulose. *J Bacteriol* 181: 284-290.
4. Kohilu U, Nigam P, Singh D & Chaudhary K (2001) Thermostable, alkaliphilic and cellulase free xylanases production by *Thermoactinomyces thalophilus* subgroup C. *Enzyme Microb Technol* 28: 606-610.
5. Moss GP (1992) *Enzyme Nomenclature*, Academic Press, San Diego, USA.
6. Baneyx F (1999) Recombinant protein expression in *Escherichia coli. Curr Opin Biotechnol* 10: 411-421.
7. Godfrey T & West SI (1996) Introduction to industrial enzymology. In: *Industrial Enzymology*, Godfrey T & West S (eds), Macmillan Press, London, UK.
8. Schmidt A, Dordick JS, Hauer B, Kiener A, Wubbolts M & Witholt B (2001) Industrial biocatalysis today and tomorrow. *Nature* 409: 258-268.

9. Straathof JJ, Panke S & Schmid A (2002) The production of fine chemicals by biotransformations. *Curr Opin Biotechnol* 13: 548-556.
10. Agranat HC & Caldwell J (2002) Putting chirality to work: the strategy of chiral switches. *Nat Rev Drug Discov* 1: 753-768.
11. Anastas PT & Kirchhoff MM (2002) Origins, current status, and future challenges of green chemistry. *Acc Chem Re* 35: 686-694.
12. International Union of Biochemistry (1984) *Enzyme Nomenclature*, Academic Press, Orlando, USA.
13. Commission on Enzyme Nomenclature (1973) *Enzyme Nomenclature*, Elsevier, New York, USA.
14. Nielsen PH (1992) Enzyme applications (Industrial and Therapeutic). In: *Kirk-Othmer Encyclopedia of Chemical Technology*, Vol 9, Kroschwitz JI & Seidel A (eds), John Wiley & Sons Inc, New Jersey, USA, pp 567-620.
15. White JS & White C (1997) *Source Book of Enzymes*, CRC Press, Boca Raton, USA.
16. Herbert R (1992) A perspective on the biotechnological potential of extremophiles. *Trends Biotechnol* 7: 349-353.
17. Haard N (1998) Specialty enzymes from marine microorganisms. *Food Technol* 52: 64-67.
18. Davis M (1998) Making a living at the extremes. *Trends in Biotechnol* 16: 102-104.
19. Lilly MD (1979) Intracellular microbial enzyme production. In: *Enzyme Technology,* Wingard LB, Katchalsi-Katzin E & Goldstein L (eds), Academic Press, New York, USA, pp 219-258.
20. Kula MR (1979) Extraction and purification of enzymes using aqueous two-phase systems. In: *Enzyme Technology,* Wingard LB, Katchalsi-Katzin E & Goldstein L (eds), Academic Press, New York, pp 71-95.
21. Gulick AM, Song H, Endow SA & Rayment I (1998) X-ray crystal structure of the yeast Kar3 motor domain complexed with Mg.ADP to 2.3Å resolution. *Biochemistry* 37: 1769-1776.
22. Drauz K & Waldmann H (2002) *Enzyme Catalysts in Organic Synthesis*, VCH, New York, USA.
23. Schoemaker HE, Mink D & Wubbolts MG (2003) Dispelling the myths–biocatalysis in industrial synthesis. *Science* 299: 1694-1697.
24. Faber K (2000) *Biotransformations in Organic Chemistry*, Springer Verlag, Berlin, Germany.
25. Gupta MN & Roy I (2004) Enzymes in organic media: Forms, functions and applications. *Eur J Biochem* 271: 2575-2583.
26. Breuer M, Ditrich K, Habicher T, Hauer B, Kesseler M, Stuermer R & Zelinski T (2004) Industrial methods for the production of optically active intermediates. *Angew Chem Int Ed* 43: 788-824.
27. Huisman GW & Gray D (2002) Towards novel processes for the fine-chemical and pharmaceutical industries. *Curr Opin Biotechnol* 13: 352-358.
28. Galperin MY & Baker AJM (2004) Environmental biotechnology: From biofouling to bioremediation: The good, the bad and the vague. *Curr Opin Biotechnol* 15: 167–169.
29. Parales RE & Haddock JD (2004) Biocatalytic degradation of pollutants. *Curr Opin Biotechnol* 15: 374-379.
30. Singh S & Banerjee UC (2005) Enantioselective hydrolysis of methoxyphenyl glycidic acid methyl ester [(±)-MPGM] by a thermostable and alkalostable lipase from *Pseudomonas aeruginosa*. *J Mol Catal: B Enzym* 36: 30-35.

31. Soni P & Banerjee UC (2005) Biotransformation for the production of the chiral drug (S) - Duloxetine catalyzed by a novel isolate of *Candida tropicalis. Appl Microbiol Biotechnol* 67: 771-777.
32. Stephanie GB (2001) Development of bioreactors for application of biocatalysts in biotransformations and bioremediation. *Pure Appl Chem* 73: 77-83.
33. Williams JA (2002). Keys to bioreactor selection. *Chem Eng Mag* 10: 34-46.
34. Carr PW & Bowers LD (1980) *Immobilized Enzymes in Analytical and Clinical Chemistry. Fundamentals and Applications*, Wiley, New York, USA.
35. Hartmier W (1988) *Immobilized Biocatalysts: An Introduction*, Springer Verlag, Berlin, Germany.
36. Tampion J & Tampion MD (1987) *Immobilized Cells: Principles and Applications*, Cambridge University Press, Cambridge, UK.
37. Kolot FB (1981) Microbial carriers. Strategy for selection. *Process Biochem* 16: 30-46.
38. Ulbricht T & Papra A (1997) Polyacrylonitrile enzyme ultrafiltration membranes prepared by adsorption, cross-linking, and covalent binding. *Enzyme Microb Technol* 20: 61-68.
39. D'Souza SF & Nadkarni GB (1980) Continuous inversion of sucrose by gel-entrapped yeast cells. *Enzyme Microb Technol* 2: 217-222.
40. Sharma R, Chisti Y & Banerjee UC (2001) Production, purification, characterization and applications of lipases. *Biotech Adv* 19: 627-662.

# Protein Structure, Rheology and Texturization

6

**Munish Puri and Parveen Pahuja**

*Department of Biotechnology, Punjabi University, Patiala 147 002 (Pb.),*

## 1. INTRODUCTION

Proteins constitute an important group of biomolecules having a great significance in human nutrition. They play a significant role in metabolism, synthesis of membrane, enzymes, immune system, etc. in living beings. In foods, the proteins beside nutrition have sensory properties, especially the texture. The chewiness in various foods is imparted by proteins. Texturization of proteins, the subject matter of this chapter, assumes immense importance. All relevant aspects are discussed here.

## 2. PROTEIN STRUCTURE

Proteins, like all other biomolecules, are made up of carbon, hydrogen and oxygen. The basic structural unit of proteins is L-α-amino acids which are 20 in number. The amino acids contain α-carbon atom covalently attached to a hydrogen atom, a carboxyl (–COOH) group, an amino group ($-NH_2$) and a side chain R-group (Fig. 1a). The amino acids differ in their chemical nature with respect to R-group only. They join in different sequences to give different structures of proteins. The wide variety of functions which proteins perform are attributed to their three-dimensional structures [1].
Proteins can structurally be divided into four categories:

- Primary structure
- Secondary structure
- Tertiary structure
- Quaternary structure

### 2.1 Primary Structure

Primary structure defines the sequence of amino acid residues. Amino acids are joined by the peptide bond, which is formed by condensation reaction between the carboxyl group of one amino acid and amino group of the other amino acid. A protein is a chain of many amino acids joined in this manner, thus referred as a polypeptide. The peptide bond is planar and very rigid. Therefore the polypeptide

*Food Biotechnology: Principles and Practices*
*V.K. Joshi & R.S. Singh (Eds.)*
*I.K. International Publishing House Pvt. Ltd., India*
*pp 145–156*

chain has rotational freedom about two bonds namely phi (Φ) and psi (Ψ) (Fig. 1b); Φ is bond between $C\alpha$ and $N$ and Ψ between $C\alpha$ and $C_1$`. Still there are only few allowed configurations due to steric hindrance.

## 2.2 Secondary Structure

Secondary structure may be termed as the local conformation of the polypeptide. There are two main structural conformations which are regular and most prevalent. These are alpha (α) helix and beta (β) sheets. Besides, few other structural arrangements are also found in some proteins. These are dependent upon the allowed conformations of Ψ and Φ angles.

Helix conformation is obtained by curving the polypeptide backbone so as to result into the formation of a coil. The helices may have right-hand or left-hand orientation, but the right hand α-helix are more favoured, being more stable sterically. Among the right handed helices, α-helix is the most prevalent e.g. wool keratin. α-helices (Fig. 1c) have axis length of 5.4Å and 3.6 residues per turn. Besides α-helices, there are other forms which are found in some proteins like $3_{10}$. New $3_{10}$ (Fig. 1c) helices

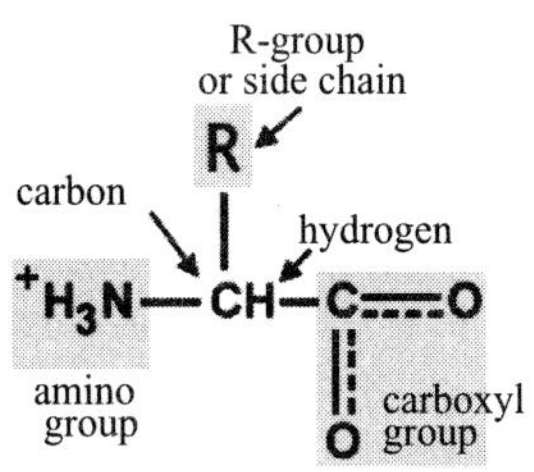


(a) *Source:* www.biology.arizona.edu/biochemistry

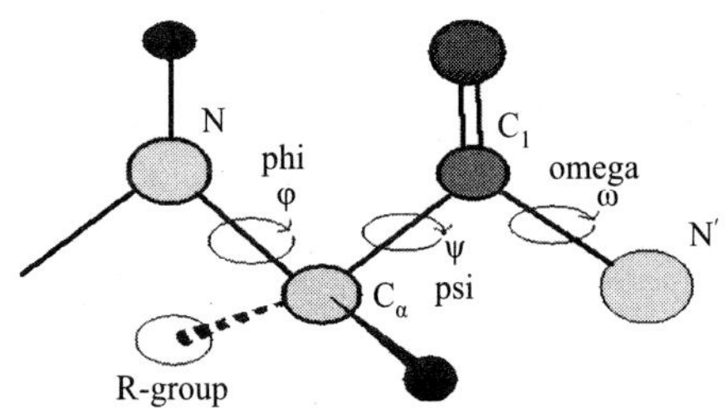


(b) *Source:* www.bioinfo.witz.ac.za

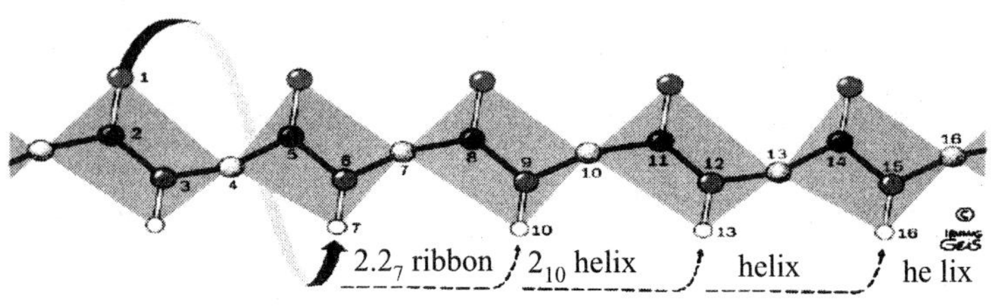


(c) *Source:* 2

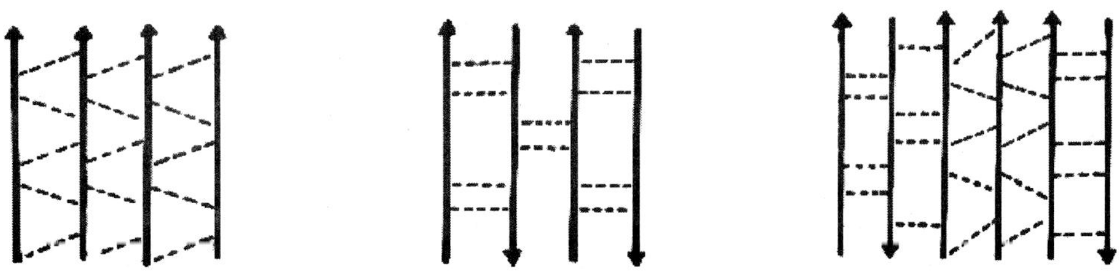

(d) *Source:* www.cryst.bbk.ac.uk (e) *Source:* www.cryst.bbk.ac.uk (f) *Source:* www.cryst.bbk.ac.uk

**Fig. 1:** Fundamentals of protein structures: (a) fundamental structure of an amino acid; (b) a dipeptide showing Ψ and φ angles; (c) showing different types of helices in single diagram; (d) parallel β sheet; (e) anti-parallel β sheet and (f) mixed β sheet.

have 3 residues per turn. Turns of this type are seen in short segments like the end of a helix and do not exist as an independent structure. Another secondary conformation which is comparatively rare is π-helix (Fig. 1c). It has 4.4 residues per turn. π-helix is a theoretical structure. The helices have most of the hydrogen bonds satisfied within the structure which provides extra stability to the polypeptide chain.

Beta (β) sheets are formed unlike α-helices by hydrogen bonding between two adjacent polypeptide chains (e.g. silk fibroin). In proteins, representation of β-sheet structures can be of three types. Parallel β-sheets are those where the alternating strands have opposite orientation (Fig. 1d). In case of anti-parallel β-sheets (Fig. 1e), two adjacent polypeptide chains have same terminal orientation. While in mixed β-sheets (Fig. 1f) the adjacent polypeptide chains have alternating terminals.

## 2.3 Tertiary Structure

Tertiary structure of a protein molecule is the uniquely folded three-dimensional spatial arrangement of its polypeptide chain, which is also known as its 'Fold', obtained by a complex process termed protein folding. Thus, tertiary structure is the global three-dimensional structure of its polypeptide chain. In tertiary structures, the hydrogen bonding between its side-chain compounds plays a very important role in providing stability to it. Besides a very characteristic type of interaction found in tertiary structure is the hydrophobic interaction. The non-polar residues of the polypeptide chain have the tendency to form a hydrophobic core such that all hydrophobic amino acids face away from water and form a sort of core. This hydrophobic core takes part in providing high stability to a folded protein molecule. Also some polar residues participate in this core whose polar groups form hydrogen bonding with each other to provide extra stability (e.g. amylase, wool keratin).

## 2.4 Quaternary Structure

Quaternary structures are formed by the non-covalent interactions between two or more independently folded polypeptide chains. This is true that most of the proteins in their functional state are the multimeric forms that are formed of more than one polypeptide chain. These may be attached by hydrophobic interaction or electrostatic interaction, etc. Formation of quaternary structures becomes a necessity for those polypeptide chains whose hydrophobic residue content is more than 30%. To accommodate high levels of hydrophobicity, formation of quaternary structure is encouraged (e.g. catalase and urease) [3].

# 3. FUNCTIONAL PROPERTIES OF PROTEINS

Those physical properties of proteins which participate in overall behaviour (sensory effects like texture, smell, etc.) of the food stuffs are called the functional properties of proteins. In order to exhibit functional properties, proteins must interact with other food components which reflect different types of interactions like protein-protein, protein-lipid, protein-carbohydrate and protein-water interactions. These interactions are non-spontaneous, and rather may require change in free energy or the change in structure of protein molecule. Thus, entropy and free energy play vital roles in protein functionality. These may sometimes get influenced by the presence of other molecules and sometimes may accompany changes in the microscopic environment of the system like pH, temperature, dielectric constant, ionic strength, etc. In fact, proteins have some bonds which have freedom of energy that provides protein molecules slight possibility of attaining various three-dimensional conformations. At different microscopic environment, it always attains different structure, but there is always a native state which is at the lowest free energy that is the most stable configuration. In such a case, while binding with other components to obtain functional properties, a protein molecule has to adjust for a change in its free energy and entropy [4]. The major functional properties of the proteins are water binding, foaming, emulsification, flavour binding, viscosity, gelation and dough formation.

## 3.1 Water Binding

Water binding is the interaction of water molecules with amino acids of the protein. Such an interaction affects both the structure and functionality of the proteins. Water binds in different ways with a protein molecule. *Structural* water is that which is intimately associated with the final structure of the protein. *Monolayer* is that which has dipole-dipole or ion-dipole type of interaction and forms a kind of layer around the protein molecule. *Hydrophobic* is that which affects the ordering of water molecules around hydrophobic residues. All the above type of water molecules do not behave like bulk water and also do not freeze at normal freezing temperatures. So this is called *Un-freezable* water. This is used as the underlying principle in many techniques used to measure water molecules bound to proteins. Finally *capillary* water is that which freezes at normal temperatures and behaves like the bulk water. This water is available as solvent for small molecules. Different factors which affect interaction of water molecules with the protein are amino acid composition, protein conformation, surface polarity/hydrophobicity, ionic concentration, ion species, pH and temperature.

Proteins contain mainly three types of amino acids, polar charged, polar uncharged and non-polar [5]. Polar charged amino acids have very high interaction with water molecules and have free energy of 5 Kcal/mole associated with it, and there are 6-7 water moieties that can attach to such a residue. Polar charged residues are those that are involved in dipole-dipole interaction with water molecules. There can be two water moieties attached with one such amino acid and the energy associated with this type of interaction is 2-6 Kcal/mole. The third type is non-polar amino acids and these have a different type of interaction with water. This is called hydrophobic hydration. There is a kind of ordering of water molecules around the non-polar amino acid, which results in lowering of entropy, and the energy involved in this case is 4 Kcal/mole.

A denatured protein will bind to more water upon denaturation. During the same process, the amino acids are exposed more to the aqueous media, thus becoming more available to hydration.

In some cases, proteins may require electrolytes to dissolve completely. This enhances the water binding potential of the protein and the added electrolytes combined. However, if more electrolytes are added, then electrolyte may even compete with the protein for binding to water molecules, which decreases the hydration.

## 3.2 Viscosity

Viscosity or the consistency of the food products is a very important factor contributing to their general acceptability. Ideal solutions tend to have constant viscosity coefficient, given below as a mathematical expression:

$$F/A - \mu dv/dr$$

where, $F/A$ is shear stress, $\mu$ is the viscosity coefficient, $dv/dr$ is shear rate.

In non-Newtonian fluids, velocity gradient is not directly proportional to the shear stress. In such solutions, viscosity decreases with increase in shear stress. The solutions which show such behaviour are known as *pseudoplastic*. Some solutions can regain their original nature after the withdrawal of shear stress, and are called *thixotropic*.

Viscosity can be related to the volume occupied by the solute to the volume occupied by the solvent. The protein solutions which occupy more volume have greater chance of interacting with other molecules besides protein-protein interactions. Thus, high molecular weight proteins (glucose oxidase, hexokinase) greatly increase the viscosity even at low concentrations. There are many factors that affect the viscosity of solutions namely size, shape, flexibility and hydration of the solute, shear rate, pH, temperature, ionic strength, etc.

## 3.3 Gelation

A gel is an intermediate phase between solid and liquid substance. It is a system lacking the steady state flow. The formation of gels may be achieved by enzymes, heat or divalent cations, though most protein gels are formed by the heating process.

A gel is a substance which has a high level of networking that may consist of covalent or non-covalent bonding. When a protein is heated, the formation of the 'pro-gel' occurs. Networking starts in this state itself. It exposes many functional groups of the protein to undergo more interactions, such as H-bonding, hydrophobic and electrostatic interactions, The relative contribution of each depends upon the type of protein, environment, heating conditions, etc.

Proteins form two types of gels: coagulum and translucent. Proteins that contain more non-polar residues upon denaturation undergo hydrophobic aggregation. This interaction between them is very high, thus, the rate of formation of aggregates is also fast. A coagulum is irreversibly formed even when the heating is still continuing. On the other hand, the proteins which contain more polar residues form soluble complexes upon denaturation. These soluble complexes show low levels of interaction, thus, the rate of formation is slow and generally requires cooling. It finally leads to the formation of translucent type of gels.

The hydration in case of protein gels is very high, even up to 98%. But like other hydrated systems, this water does not contribute to the flowability, the reason for which is poorly understood. The translucent gels are more hydrated than coagulum type gels.

There are two critical factors responsible for the thermal and mechanical stability of the gels. First is the number of cross-links formed per monomer chain and the second is protein concentration. For a stable gel, a 'least concentration endpoint (LCE)' is required. For example, for soy proteins, LCE is 8%.

In addition to these two very important factors, several other factors like pH, salts and other additives affect the gelation. At extreme pH, weak gels are formed. For most of the protein gels, the optimum pH is between 7 and 8. Besides heating, other methods of gelation are also employed like enzymes (e.g. transglutaminase) and use of divalent cations like $Ca^{2+}$ and $Mg^{2+}$ [6].

## 3.4 Emulsification

An emulsion is a thermodynamically unstable mixture of immiscible liquids. Several natural and processed foods are emulsions like egg yolk, coconut milk, soy milk, margarine, mayonnaise, sausage, etc. These systems may be stabilized by the addition of substances which are partially soluble in both the phases and are commonly known as *emulsifiers*. The system may coalesce if an energy barrier to phase separation is not maintained. There are many small and large molecules that play the role of emulsifiers, and proteins are one of those.

Proteins are larger and more complex molecules than many small emulsifiers and are often used to aid in the formation of emulsions which also result in their increased stability. The first thing for protein to perform this function is to reach at interface and unfold. As non-polar amino acids are mostly located in the interior of the molecule, these have to be exposed so to get inserted into the hydrophobic region. As a general case, around 6 to 8 amino acids are inserted, and if more fall into the flexible portion, the unfolding continues.

## 3.5 Flavour Binding

Flavour plays an important role in the acceptance of any food consumed in any part of the world. Flavour is such a compound which either interacts with any of the four basic taste receptors or is volatile enough to interact with odour receptors in olfactory organs. However, the proteins are odourless compounds. They are only able to exhibit flavouring properties if any flavouring component is bound to them or through any break-down product. Several proteins are known to exhibit certain off-flavours like oil seed proteins and whey protein concentrates. These bad flavours are due to some compounds (aldehydes, ketones and alcohols) generated upon the oxidation of unsaturated fatty acids. The hydrophobic regions of the proteins interact with lipids, which generally results into bad odours. Thus a thorough knowledge is of high importance for removal of these off flavours from protein components. Proteins actually can be used as flavour carriers or flavour modifiers in fabricated foods. For example, meat analogues contain plant proteins where the creation of meat like flavour becomes a necessity for the success of the product. For a protein to be a good flavour carrier, it should tighten the flavour tightly, retain them during processing and release after mastication. Most of the flavouring components would impart their effect only after they are released from the carrier.

Most flavour compounds are volatile and lipid soluble, and they make strong interactions with the hydrophobic regions of the protein molecules. The effect of absorbed flavouring components in such a way depends only on the type of the absorbed molecule, which may be positive or negative. Degradation products of proteins may also lead to flavour generation. Proteolysis often results in a bitter flavour, which is due to creation of hydrophobic peptides. Further degradation may lead to decarboxylation of free amino acids. For example, cadaverine and putrescine, the results of lysine and ornithine fermentation respectively, are known to have a very unpleasant smell. However, interaction of proteins with reducing sugars leads to formation of pleasant flavour compounds. One of the examples is freshly baked bread.

### 3.6 Dough Formation

Dough can be defined as an extensible, visco-elastic protein network formed upon the mixing of an appropriate amount of water to cereal proteins, e.g. wheat, rye and barley. The proteins involved in dough formation are gliaden and glutenin. The gliadins are a mixture of proteins that are soluble in 70% alcohol. They have molecular weights ranging from about 25,000 to 100,000. They are usually associated with small amounts of glycolipids. The molecules are held together with disulfide linkages and are fairly extensible, but are not very elastic. The glutenins average about 100,000 in molecular weight. They are elastic molecules with low extensibilities. Glutenins are soluble in solutions of dilute acid or alkali and are associated with non-polar lipids. These molecules contain both inter and intra molecular disulfide bonds. Glutenin and gliadin have very few charged amino acids and thus are not very water-soluble. They do contain large amounts of glutamine (around 35%), and thus, are excellent at immobilizing water. They also contain a large number of hydrophobic groups that are buried in their interiors.

When glutenin and gliadin are suspended together in an aqueous medium, with the input of mechanical energy, both molecules start to unfold. This unfolding exposes hydrophobic groups which cause the molecules to aggregate. Also during the mixing process, the glutenin and gliadin are covalently cross-linked through the formation of inter-molecular disulfide bonds. The cross-linked protein is known as gluten. If enough cross-linking occurs, a visco-elastic structure results. This structure is able to expand when gas is produced either through the metabolism of yeast or through the expansion of water vapours during heating. As the temperature increases, the proteins begin to denature. The denaturation of gluten along with gelatinization of starch gives structure to products like bread [7, 8].

## 4. CHANGES IN 3-D STRUCTURE OF PROTEINS

Proteins, as explained before, are organized in four levels of structural hierarchy. The three-dimensional conformation of protein at which it exhibits its functional state is called 'native state'. It is commonly observed in laboratory as well as in nature that some proteins stop performing their functions, and the reason in many cases can be attributed to the change in three-dimensional conformation, i.e., native state of the protein. These changes are brought about by different features like change in pH, temperature, addition of organic solvents and other denaturing agents [9]. There can be two types of changes:

(1) Chain-chain interaction leading to association, aggregation, flocculation, coagulation and precipitation.
(2) Chain-solvent interaction which leads to solubilization, dissociation, swelling and denaturation.

### 4.1 Temperature

High temperatures lead to unfolding of the proteins and the equilibrium of this change of state is related to the enthalpy change. For most proteins, the unfolding enthalpy $\Delta H^0$ is in the range of 50-150 kcal/mole. The non-covalent interactions play a key role in stabilizing the protein structure and heating provides the energy to break those interactions. Proteins generally show sharp transition from the native state to unfolded state and this happens as a result of co-operativity. Many non-covalent interactions break off together. Once the first bond is broken, it helps adjacent bond to break with a relative ease. Thus a small change in temperature leads to a large change in structure.

### 4.2 pH

Change in pH alters the distribution of cationic, anionic and non-ionic polar sites on the protein molecule, which in turn affects water-protein and protein-protein interactions. Protein exhibits high interaction among the charged groups and the molecule becomes shrunken. These charged groups are least available to water, and very low level of interaction occurs with water that results into minimal hydration, swelling and swellability. At extreme pH, the proteins tend to unfold.

### 4.3 Organic Solvents

Organic solvents generally have lower dielectric constants than the water, which cause high levels of interaction among the opposite charges, leading to aggregation. Besides, there are many salts like guanidium salt, which lead to denaturation of the proteins. However, for most proteins, low salt concentrations tend to increase the solubility of the protein which is known as "salting in" process. Actually the salt molecules bind to the counter charges present on protein and affect the electrostatic interactions.

## 5. TEXTURIZATION

In affluent countries or at places where food is available in abundance, people tend to choose food based on quality characteristics. Quality includes sensory properties like taste and appearance, and nutritional value. Quality determines the general acceptability of the food product, though acceptability is subjective. But, in general, quality of any product is one of the most important factors in selling of a product, the other being the price. When we eat food, we use all of our senses e.g., when we eat a potato chip, besides its taste we also determine its quality with the sound of its snap. This indicates that there are more than one factors that determine the food quality, thus, quality can be sub divided into three categories i.e appearance, texture and flavour. Of these presently, we shall be discussing the textural factors only (for more details, see the chapter on sensory evaluation). Texture can be regarded as the degree of softness, chewiness, sponginess, crunchiness, plasticity and many more terms (adjectives) galore. It is defined by the International Organization of Standardization (ISO) as "all of the theological and structural (geometric and surface) attributes of a food product perceptible by means of mechanical, tactual and, where appropriate, visual and auditory receptors." In general, texture can be explained as mouthfeel, handfeel, feel at the palate, e.g., while chewing a gum one expects it to be chewy and the steak to be compressible [10].

In the recent past, the food market has seen many innovations. With more wealth creation, people have been looking for many food alternatives as well as altogether newer food items. This has led to many food scientists to work on different aspects of the foods and especially in texturization process. It has made possible for great sources of nutrition, which were earlier dumped as waste, now being utilized to make many new products. e.g., whey proteins, which are good source of proteins, a well acknowledged fact. However, more than 50% of the whey proteins are not utilized in foods or are wasted. But many efforts are continuously being made to increase the utilization of this rich source of proteins. One of the ways is to use different forms of whey products like whey protein concentrates (WPC), whey protein isolates (WPI) and whey protein lactalbulins (WLAC) [11]. These may be added to many food items such as extruded snacks and meats or result into the formation of all new products like highly stable acidified drinks and many more.

Many religious sentiments have led to the creation of meat analogues. These are the food products which have the textural characteristics like meat. These are used both as meat mimickers and meat substitutes [12]. Meat mimickers are those that duplicate the feeling as that of eating a meat product. On the other side, meat substitutes are those food products that not only aim to feel like meat while eating, but also have the comparable nutritional value. The market for meat analogues was reported to grow at 17% annually in 2002, and the size for such products was pegged at US$ 4 billion during 2004 for the US alone [12]. Thus, a lot of attention has been given to such products. The major chunk of the market was shared by soy and whey proteins. The usage of these proteins has been popularized mainly after processing them for texturization and with some added flavors and other supplements. Whey and soy proteins have been the main leaders of this industry, but other protein sources have also caught up with the trend of textured proteins like wheat, egg white and vegetables.

The first processed soy protein products were mainly flours or powders which were added to foods such as bread, pasta or beverages. The objective was to suppress their feel of presence to an extent that renders these powders sufficiently flavourless and white, and to counteract any change in the accepted characteristics of the "host" food caused by the incorporation of soy protein products at nutritionally and economically significant levels. During 1960s, such products became commercially available widely.

## 5.1 Technique of Texturization

In texturization, the intramolecular binding forces are broken and the protein gets unfolded. They are stabilized by interaction with neighbouring chains. The starting materials for texturization vary from sesame, soy, wheat, groundnut, corn, yeast and blood plasma. The molecular weight of the starting protein material should be between 10-50 kDa. Applied to soy protein products, the terms "texturization or texturing" mean the development of a physical structure which will provide, when eaten, a sensation of eating meat. Meat "texture" is a complex concept comprising visual aspect (visible fibres), chewiness, elasticity, tenderness and juiciness. The principal physical elements of meat which create the texture complex are the muscle fibres and the connective tissue.

The concept of producing meat analogs is not a new one. References can be traced to as back as 1000 years. However, in modern day science the first patents were granted in late 1950s. These inventions produced homogeneous, isotropic (un-oriented, of equal structure in all directions) gels, which had only one of the elements of meat texture i.e., chewiness. But they were not accepted very well commercially. Now-a-days, the production of textured proteins has been established well and can be categorized into two categories:

(1) **Spinning:** Spinning is procedurally somewhat similar to production techniques, which are applied at textile industries in which fibres (protein) are assembled into a binding material.
(2) **Extrusion:** In this technique, the input material is converted into a hydratable, laminar, chewy mass without true fibres.

### *5.1.1 Spun-fibre based texturization*

The process for the production of textured soy products containing spun protein fibres was first described in a 1954 patent [13]. Since then, many additions to and modifications of the basic concept have been suggested. The basic flow diagram of the process is shown in Fig. 2. The first requirement of the process is the ISP (isolated soy protein), which can either be produced in plant or can be procured commercially. A concentrated protein solution is prepared by adding alkali to the ISP slurry. The solution, containing approximately 20% protein at pH 12 to 13 is "aged" (to permit unfolding of the protein molecules) until the formation of dope. "Dope" is a highly viscous solution that has viscosity equivalent to that of honey (50,000 to 100,000 centipoise). This viscous concentrated protein solution is technically known as "dope".

The next step is the transformation of the dope into distinct, stretched fibres (spinning) by coagulating fine jets of the solution in an acid bath. The "dope" is pumped into the coagulating bath through a spinneret, which is a plate with thousands of fine holes (about 75 microns in diameter). The bath contains a solution of phosphoric acid and salt, maintained at pH of about 2.5. As the jet of "dope" contacts the acid medium, the oriented protein molecules are suddenly coagulated and form a fibre. The fibres are picked up as a "tow" and stretched to enhance the molecular orientation and increase fibre strength. Stretching reduces the diameter of the fibre well below that of the holes on the spinneret. The tows of fibre pass

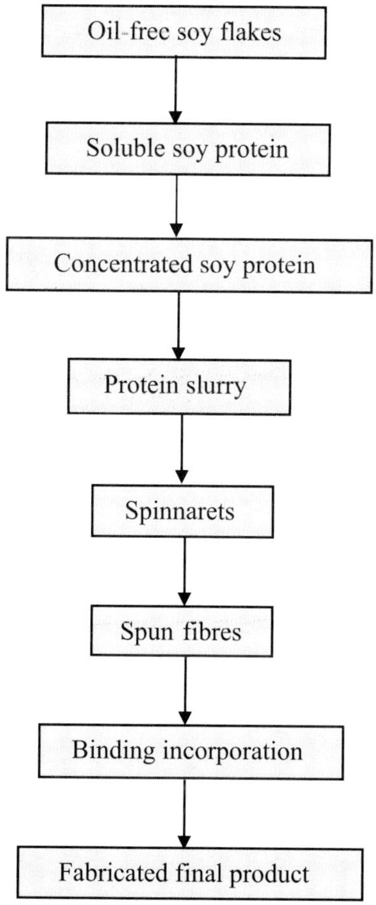


**Fig. 2:** Flow diagram of spun fibre texturization process.

through a step of washing to remove excess acidity and salt. The subsequent operations depend on the final product. Soy protein fibres are only one ingredient of the meat-like structure. The other ingredients include fat, binders, colouring and flavouring additives, etc. The nature of these ingredients, the proportion of fibres and their orientation in the binder matrix depend on the type of flesh food to be imitated. The binder matrix contains heat-coagulable components, commonly egg albumen and the final structure is usually stabilized by thermal setting.

Spun fibre-based textured soy products have been used as "total" meat analogs (i.e. to replace meat totally) and as meat extenders (i.e. to replace part of the meat in ground meat, patties, etc.). The visco-elastic properties of the dope affect the fibre formation during spinning. The amount of presence of sodium hydroxide has a marked effect on the physical properties of the fibres. The visco-elastic properties also change as per the molecular weight of the protein, which ultimately affect its spinnability [14]. The spun fibre production of textured protein has not been very successful over the years. This has been because of the cost involved at every step. The process in itself is very cost intensive, both in initial capital investment and running expenses. The process also requires an expensive starting material i.e., isolated soybean protein. Today, there are very few producers of spun soy protein fibres and textured products containing them. The most successful spun fibre based meat analog has been the imitation bacon chip.

### *5.1.2 Extrusion texturization*

Mc. Anelly described a process for the production of spongy, elastic particles from soy flour [15]. A mixture of defatted flour and water was extruded through a food grinder. The extruded strands were heat-set in an autoclave, chopped, leached with hot water and dried. Although Mc. Annelly's invention has been considered as the frontrunner in this field of extrusion texturization processes, the breakthrough, however, was the invention of continuous cooking-extrusion process, for which a patent was awarded to Atkinson in 1970 [16]. Defatted soy flour containing a certain amount of moisture is used in this process which is passed through a high-pressure extruder-cooker to result into an expanded, porous, somewhat oriented structure described as "pleximellar". Although devoid of true fibres, the product possessed the textural characteristics of chewiness and elasticity, and was deemed to imitate meat in this respect. Extrusion texturized soy flour became popularly known as TVP (Textured Vegetable Protein) or TSP (Textured Soy Protein).

The design of the extruder is such that it contains a rotating screw and a barrel (Fig. 3). The barrel can be smooth or grooved. The screw configuration is such that the free volume delimited by one screw flight and the inside surface of the barrel decreases gradually as one goes from one end of the screw shaft to the other [17].

This configuration helps to compress the material as it is conveyed forward by the rotating screw. Screws may be of different compression ratios, which can be used for different applications. Also the barrel is, generally, equipped with a number of sections of steam heated jackets or induction heating elements or cooling jackets. At last at the exit end of the barrel from where the textured product has to come out, a "die" is fitted. Different dies have different shapes of small orifice, which determine the shape of the extruded product.

Extrusion texturization uses low fat soybean flour or meal (sometimes). However to obtain the better product quality and uniformity in the quality, it is important to condition the flour with steam before it is fed into the extruder. This conditioning helps to adjust the protein solubility index. One more important factor at this initial stage of the process is the moisture content. Generally, moisture content of 20-25% is considered enough for most of the extrusion processes, but there have been works carried out at high moisture content. This is also called wet extrusion, i.e., moisture content >40%. There is a great difference in the terms of the rheological properties, energy requirements between the low moisture extrusion and the high moisture extrusion [18].

Now, the flour is ready to be fed to the extruder. When it enters the extruders, it first meets the rotating screw. This rotating screw propels the flour forward along the barrel. As it passes down the barrel, the energy is supplied from outside by the heating elements around the barrel. Also the heat is generated by the friction with the screw device. Finally, the temperature reaches upto 150-180°C. This rapid "pressure cooking" process transforms the mass into a thermoplastic "melt", hence the name of "thermoplastic extrusion".

Thermoplastic extrusion imparts alignment to many high molecular weight components while causing proteins to undergo thermal denaturation. Sudden release of pressure causes instant evaporation of some of the water and "puffing". Puffing and, therefore, porosity can be controlled by monitoring melt temperature at the die. The extrudate has to be brought to desired shape and size so that it could be sold as shelf stable product; it is cut continuously with a knife. Then, many different flavors, other ingredients for better marketing etc. can be added. While texturizing the soy material, extrusion cooking also provides the heat treatment necessary to reduce the microbial load and to inactivate the trypsin inhibitor. It should be noted that despite the high temperatures in the extruder, trypsin inhibitor inactivation may be incomplete due to the relatively short processing time. Besides thermoplastic extrusion, one more method of extrusion was described in a patent. They described a process wherein moistened soy flour is fed into the pressurized reactor where it meets high pressure steam. However, it is reported that this procedure is not in use nowadays. Shelf stable dry products are usually marketed at a moisture level of 8%.

## 6. APPLICATIONS

### 6.1 Meat Extenders

The principal use of texturized soy protein products is as a meat extender in comminuted meat products such as patties, fillings, meat sauces, meat balls, etc. In such products, as much as 30% of the meat can be replaced by hydrated texturized soy products without loss of eating quality. The cost of textured soy flour is approximately 0.60 U.S. Dollars per kilogram. About 3.5 kilograms of hydrated base is obtained from each kilogram of textured flour. Thus, the cost of meat replacement is only 17 cents of a dollar for each kilogram of meat saved. Furthermore, textured soy products offer not only economic savings but also certain types of product improvement. Their ability to absorb water and fat results in increased product juiciness and permits the use of meat with higher fat content. Ground beef extended with TVP has been used extensively in school lunch programs with good results. The property of TVP to withstand cooking in a retort (retortability, retort stability) is relevant to its use in canned luncheon meat, meat loaf and similar products.

### 6.2 Meat Analogues

Chunks of extrusion texturized soy protein products and spun fibre based preparations are marketed as "imitation meat" or "meat analogues". The market for these products was, at first, limited to the relatively small sector of vegetarians. Recently, there is a marked trend to reduce the consumption of red meat, associated with the demand for low-cholesterol foods. At the same time, the industry has been successful in developing more attractive meat analogues made from

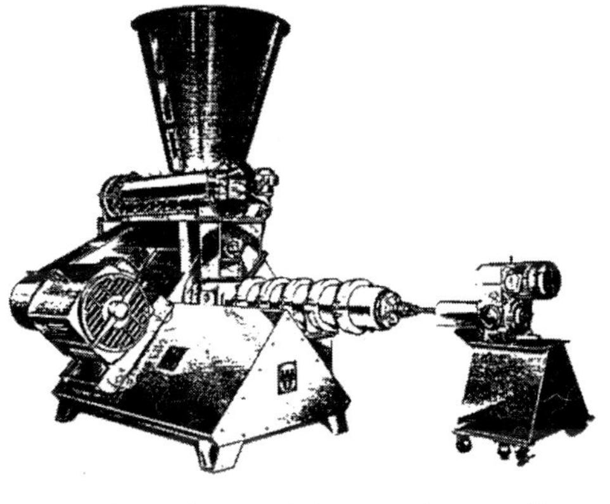

**Fig. 3:** Extruder used for texturing soy flour. (Courtesy: Wenger Manufacturing Co.) *Source:* www.fao.org

rehydrated textured soy proteins, alone or in combination with wheat gluten. These products are marketed as flavoured, fully prepared, frozen ready-to eat entrées. The present marketing strategy for meat analogs is to present them to the public as new, high quality products, and not as inexpensive substitutes for meat. So far, this strategy seems to be successful. The market for these sophisticated (and by no means inexpensive) products is rapidly expanding, particularly in Western Europe.

## 6.3 Other Applications

Imitation bacon bits based on texturized soy protein products have been mentioned earlier. The price range for this product is 1.50 to 2 U.S. Dollars per kilogram. A pasta product containing texturized soy protein granules is being offered on the retail market, in addition to its use in institutional feeding.

## REFERENCES


1. Nelson DL & Cox MM (2005) The three-dimensional structure of proteins. In: *Lehninger Principles of Biochemistry*, WH Freeman & Co, New York, USA, pp 159-202.
2. Voet D & Voet J (1995) Three-dimensional structure of proteins. In: *Biochemistry,* John Wiley and Sons, New York, USA, pp 225-245.
3. Crieghton TE (1983) *Proteins: Structures and Molecular Properties*, WH Freeman & Co, New York, USA.
4. Hudson BJF (1982) *Developments in Food Proteins-1*, Applied Science Publ, London, UK.
5. Walsh G (2002) *Proteins: Biochemistry and Biotechnology*, John Wiley & Sons, New York, USA.
6. Kinsella JE (1976) Functional properties of proteins in foods: A survey. *Crit. Rev Food Sci Nutr* 7: 219-280.
7. Cherry JP (1981) Protein functionality in foods. In: *ACS Symposium Series* 147, American Chemical Society: Washington, USA, pp 1-6.
8. Belitz HD & Grosch W (2007) *Food Chemistry*, Springer-Verlag, New York, USA.
9. Fersht A (1999) *Structure and Mechanism in Protein Science*, WH Freeman & Company, New York, USA.
10. Potter NN & Hotchkiss JH (2005) Quality factors in foods. In: *Food Science*, Heldman DR (ed), Chapman & Hall, New York, USA, pp 90-112.
11. Onwulata CI, Konstance RP, Cooke PH & Farrell Jr HM (2003) Functionality of extrusion-texturized whey proteins. *J Dairy Sci* 86: 3775-3782.
12. Border RE & Cheryl (2006) Achieving success with meat analogs. *Food Technol* 60: 28-34.
13. Boyer RA (1954) High protein product and process for its preparation. US Patent 2,682,466.
14. Hayakawa I, Katsuta K & Kawasak SI (1982) Effects of rheological properties and molecular weight distribution on the spinnability of soy protein. *Agric Biol Chem* 46: 1997-2001.
15. McAnelly JK (1964) Method for producing a soybean protein product and the resulting product. US Patent 3,142,571.
16. Atkinson WT (1970) Meat like protein food product. US Patent 3,488,770.
17. Clark, JP (1978) Texturization by extrusion. *J Texture Studies* 9: 109-123.
18. Akdogan H (1999) High moisture food extrusion. *Int J Food Sci Technol* 34: 195-207.

# Genetic Manipulation of Food Microorganisms

# 7

**S.K. Soni[1], Raman Soni[2] and Namita Bansal[1]**

*[1]Department of Microbiology, Punjab University, Chandigarh 160 022*
*[2]Department of Biotechnology, DAV College, Chandigarh 160 022*

## 1. INTRODUCTION

Microorganisms are associated with all the foods that we eat and are responsible for the formation of certain foods through the process of fermentation. They may also be directly used as a source of food in the form of single cell proteins and food supplements in the form of pigments, amino acids, vitamins, organic acids, enzymes and lipidic substances, and also inhibit other microbes, which may produce the toxins. The exploitation of microorganisms in food industry is as old as the history of civilization. The traditional application has been in the production of various indigenous fermented foods by involving the natural microflora coming from the environment. The current applications involve the production of various foods and food additives by the process of fermentation with genetically modified microbes.

Fermentation is one of the oldest forms of food preservation technologies in the world. Indigenous fermented foods such as bread, cheese and wine have been prepared and consumed for thousands of years and are strongly linked to culture and tradition, especially in rural households and village communities. The development of fermentation technology is lost in the mists of history. Anthropologists have suggested that it was the production of alcohol that motivated primitive people to settle down and become agriculturists. Some even think that the consumption of fermented foods is pre-human [1]. The first fermented food consumed probably was fermented fruits. Hunter-gatherers would have consumed fresh fruits, but at the times of scarcity would have eaten rotten and fermented fruits. Repeated consumption would have led to the development of taste for fermented fruits. There is a reliable information that fermented drinks were being produced over 7,000 years ago in Babylon (now Iraq), 5,000 years ago in Egypt, 4,000 years ago in Mexico and 3,500 years ago in Sudan [2, 3].

Bread-making probably originated in Egypt as early as 10000 BC [4]. Several triangular loaves of bread have been found in ancient tombs. Fermentation of milk started in many places with evidence of fermented products in use in Babylon over 5,000 years ago. There is also an evidence of fermented meat products being produced for King Nebuchadnezer of Babylon. China is thought to be the birthplace of fermented vegetables and the use of *Aspergillus* and *Rhizopus* moulds to make food. The book called "Shu-Ching" written in the Chou dynasty in China (1121-256 BC) refers to the use of "chu", a fermented grain product [5].

*Food Biotechnology: Principles and Practices*
*V.K. Joshi & R.S. Singh (Eds.)*
*I.K. International Publishing House Pvt. Ltd., India*
*pp 157–232*

The history of winemaking is as old as the civilization. Wine fermentation was probably the first fermentation known to the mankind. Different techniques were tried for the production of various types of wines since ages, but the basic principle remained the same. During the last 150 years, the scientific basis of winemaking has gradually become clearer, and many new varieties of wine have evolved [6]. In 1863, Louis Pasteur established the microbial activity on the basis of wine fermentation, and demonstrated that the fermentation of grape musts and production of good quality wines is a complex biochemical process involving the sequential appearance of different microbial species whose development is affected by various intrinsic and extrinsic factors. The process includes the combinatorial action of various yeasts, lactic acid and acetic acid bacteria with the yeast belonging to *Saccharomyces cerevisiae* as the primary organism in wine fermentations.

There are several varieties of wines available around the world and are generally, classified on the basis of the type of fruit used, source of yeast inocula, colour, alcohol content, presence or absence of carbon dioxide, processing of the fruit prior to fermentation, etc. For the production of wines, various raw materials including fruits, sugar, acid, nitrogen source, clarifying enzyme, filter aid, etc. are required. The fruit constitutes the most important raw material. Most of the wines are generally prepared from over-ripened grapes, but commercial wines are also prepared from several fruits other than grapes. In some cases, the higher level of acidity in the fruit requires dilution, consequently lowering the sugar level which is compensated by the exogenous addition of table sugar to the must. Unlike grapes, which are a good culture medium for the growth of wine, yeast needs no sugar supplementation, however the juice/pulp from other fruits need sugar supplementation for successful alcoholic fermentation.

Indigenous fermented legume and cereal products form an important part of human diet in South-East Asia including India, the middle East and Africa. They are also becoming popular in the developed world due to their high nutritional value and organoleptic characteristics. People of Asia are the pioneers in the development of means of fermenting vegetable proteins to produce meat-like flavours. Indonesians developed the fermentation methods for introducing meat-like texture into vegetable products. Koreans are credited for the production of acid fermented vegetables. People of Egypt developed wheat breads leavened with yeasts, while Indians discovered the methods for souring and leavening cereal-legume batters. Generally, the traditional methods of preparing indigenous fermented foods are simple and inexpensive. However, the ancient methods of making such foods are changing rapidly through modern microbial technology. Some of the fermented products including dairy products, coffee, cocoa and bread have been studied extensively for many decades and their production methods have been modernized keeping pace with the advancement in modern science. During the last few years, much interest has been generated in the fermented foods of Asian and African countries including India where such foods are still being manufactured according to the traditional, technologically less advanced methods using simple equipment, and are produced at cottage industry scale by means of natural microflora from the staples and the surroundings [7-9].

Fermented dairy products also constitute a vital part of human diet in many parts of the world. The preservation and concomitant transformation of flavour and texture of milk by fermentation has been known since centuries before the role of microorganisms in the process was appreciated. Historically, fermentation processes were based upon spontaneous souring of milk caused by inherent microflora. Modern processes affect milk fermentation under predictable, controllable and extracting conditions to yield cultured dairy products of high nutritional and sanitary standards. The origin of the fermented dairy products dates back to the dawn of civilization. The ancient Sanskrit scriptures of India, the Vedas, document the food value of dahi, a fermented milk product similar to modern yoghurt. Further evidence for the existence of soured milk as a food in early times is corroborated by the Bible. Milk is fermented as such or in combination of some additives to produce cultured butter milk, cultured cream, yoghurt, acidophilus milk, kefir, koumiss and cheese. Cultured butter milk is obtained from pasteurized skim milk or part skim milk cultured with lactic acid and aroma producing organisms. The term buttermilk is also used for the phospholipids-rich fluid fraction obtained as a by-product during the churning of cream in butter manufacture. However, cultured butter milk is a viscous, cultured, fluid milk, having a characteristic pleasing aroma and flavour. Cultured cream or sour cream is manufactured by ripening pasteurized cream having 18% fat with lactic and aroma-producing bacteria. This product resembles cultured buttermilk in terms of culturing procedure. However, in consistency it is an acid gel containing butter-like aromatic flavour. Cultured cream is used as a toppling on vegetables, salads, fish, meats and fruits, as a filling in cakes and in soups and cookery in place of buttermilk or sweet cream. Yoghurt, a fermented curd product, is prepared from a mix standardized from whole, partially defatted milk, condensed skim milk, cream and non-fat dry milk. Alternatively, milk may be partly concentrated by the removal of 15-20% water in a vacuum pan. Kefir and Koumiss belong to the class of acid and alcoholic fermented milks, which are very popular in Russia. Pickles from vegetables and fruits like mango, gooseberry and lemons are prepared and used throughout India. More than their nutritive value, they act as food adjuncts and appetizers as they add palatability to Indian foods. Indian pickles is a very big industry now and the products are popular throughout the world [8].

## 2. MICROORGANISMS AS THE AGENTS OF FERMENTATION IN FOODS

Beer, wine and bread making by yeasts, the processing of milk to dairy products by lactic acid bacteria and the fermentation of vinegar by acetic acid bacteria show that these microbes are among the oldest domesticated microorganisms. The most common groups of microorganisms involved in food fermentations include bacteria, yeasts and moulds.

### 2.1 Bacteria

Several bacterial families are present in foods, majority of which are concerned with food spoilage. As a result, the important role of bacteria in the fermentation of foods is often overlooked. The most important bacteria in desirable food fermentations belong to lactobacillaceae, which have the ability to produce lactic acid from carbohydrates. Other important bacteria, especially in the fermentation of fruits and vegetables, are the acetic acid producing *Acetobacter* species.

#### *2.1.1 Lactic acid bacteria*

The lactic acid bacteria (LAB) represent a group of Gram positive bacteria, non-respiring, non-spore forming, cocci or rods, which are functionally related by their ability to produce lactic acid during homo- or hetero-fermentative metabolism. The acidification and enzymatic processes accompanying the growth of LAB impart the key flavour, texture, and preservative qualities to a variety of fermented foods. Industrial applications of the LAB rely on six key beneficial and non-pathogenic species: *Lactococcus* (milk), *Lactobacillus* (milk, meat, vegetables, cereal), *Leuconostoc* (vegetables, milk), *Pediococcus* (vegetables, meat), *Oenococcus oeni* (wine) and *Streptococcus thermophilus* (milk). Other members of the LAB, notably lactobacilli, occupy important niches in the gastrointestinal tracts of humans and animals and are considered to offer a number of probiotic benefits to general health and well being. These benefits include a positive influence on the normal microflora, competitive exclusion of pathogens, and stimulation/modulation of mucosal immunity. More recently, LAB are being used in the production of industrial chemical and biological products including biopolymers (*Leuconostoc* sp.), bulk enzymes (*Lactobacillus brevis*), ethanol and lactic acid (*Lactobacillus casei, L. lactis, delbrueckii, L. brevis*) [10]. The LAB are also strong candidates for development as oral delivery vehicles for digestive enzymes and vaccine antigens [11, 12] Their innate acid tolerance, ability to survive gastric passage, and safety record during human consumption are key features that can be exploited effectively by the biologics to targeted locations and tissues.

Lactic acid bacteria (LAB) have been essential in food and feed fermentation for centuries, because they considerably contribute to the nutritional value of products. Owing to some of their metabolic properties, LAB play an important role in the food industry because they significantly contribute to the flavour, texture and in many cases to the nutritional value of the food products [13]. The interest in the applications of microorganisms and their metabolites in the prevention of food spoilage and extension of shelf-life of foods has increased during the last decade [14]. In addition, the data are rapidly accumulating that some strains of LAB could be probiotics. Therefore, an increasing interest exists for dairy products containing specific bacterial species with potential health-improving properties [15]. The fermentation of milk with intestinal species of *Lactobacillus acidophilus* and/or *Bifidobacterium bifidum* is being applied for a number of products [16]. These species are more and more connected with health-promoting effects in human and animal intestinal tract [17], and that is why probiotics are considered as starter cultures for the production of functional foods.

Lactic acid bacteria also produce a number of antimicrobial substances that might be of importance for food and feed fermentation and preservation for which they are economically significant. Besides metabolic end products, some LAB strains also secrete antimicrobial proteinaceous compounds, termed bacteriocins, which kill closely related bacteria. Bacteriocins produced by LAB were intensively investigated [18]. On the basis of their chemical, structural and functional properties, bacteriocins could be classified into three main classes, class I, the lantibiotics; class II, the small heat stable non-lantibiotics and class III, large heat labile bacteriocins [19]. A new class of bacteriocins has been proposed [20] representing the bacteriocins such as circularin A, which are ribosomally synthesized, non-modified cyclic peptides with antimicrobial activity. Since LAB are generally regarded as safe (GRAS), they or their bacteriocins could be safely used in food production and preservation. Typical LAB bacteriocins have a narrow antibacterial spectrum, but LAB also produces bacteriocins having broad antibacterial spectrum. Thus, some LAB bacteriocins can inhibit the growth of Gram-positive pathogenic and spoilage bacteria as well as yeasts [21]. Besides, it has been reported that bacteriocins also inhibit the growth of some Gram-negative species [22, 23].

Lactic acid bacteria carry out their reactions i.e., the conversion of carbohydrate to lactic acid, carbon dioxide and other organic acids, without the need for oxygen. They are described as microaerophilic as they do not utilise oxygen. Because of this, the changes that they effect do not cause drastic changes in the composition of the food. Some of the families are homofermentative, i.e., they only produce lactic acid, while others are heterofermentative and produce lactic acid plus other volatile compounds and small amounts of alcohol. *Lactobacillus acidophilus, L. bulgaricus, L. plantarum, L. caret, L. pentoaceticus, L. brevis* and *L. thermophilus* are examples of lactic acid producing bacteria involved in food fermentations. All species of lactic acid bacteria have their own particular reactions and niches, but overall *L. plantarum,* a homofermenter, produces high acidity in all vegetable fermentations and plays the major role. All lactic acid producers are non-motile Gram positive rods that need complex carbohydrate substrates as a source of energy. The lactic acid they produce is effective in inhibiting the growth of other bacteria that may decompose or spoil the food. Since the whole group is referred to as 'lactic acid bacteria', it might appear that the reactions they carry out are very simple, with the production of one substrate. This is far from the truth. The lactic acid bacteria are a diverse group of organisms with a diverse metabolic capacity. This diversity makes them very adaptable to a range of conditions and is largely responsible for their success in acid food fermentations [21-23].

Despite their complexity, the whole basis of lactic acid fermentation centers on the ability of lactic acid bacteria to produce acid, which then inhibits the growth of other non-desirable organisms. Species of the genera *Streptococcus* and *Leuconostoc* produce the least acid. Next are the heterofermentative species of *Lactobacillus*, which produce intermediate amounts of acid, followed by *Pediococcus* and lastly the homofermenters of *Lactobacillus* species, which produce the most acid. Homofermenters convert sugars primarily to lactic acid, while heterofermenters produce about 50% lactic acid, 25% acetic acid and ethyl alcohol, and 25% carbon dioxide. These other compounds are important as they impart particular tastes and aromas to the final product. The heterofermentative lactobacilli produce mannitol and some species also produce dextran [1, 3].

*Leuconostoc mesenteroides* is a bacterium associated with the sauerkraut and pickle fermentations. This organism initiates the desirable lactic acid fermentation in these products. It differs from other lactic acid species in that it can tolerate fairly high concentrations of salt and sugar. *L. mesenteroides* initiates growth in vegetables more rapidly over a range of temperatures and salt concentrations than any other lactic acid bacteria. It produces carbon dioxide and acids which rapidly lower the pH and inhibit development of undesirable microorganisms. The carbon dioxide produced replaces oxygen, making the environment anaerobic and suitable for the growth of subsequent species of lactobacillus. Removal of oxygen also helps to preserve the colour of vegetables and stabilises any ascorbic acid that is present.

Organisms from the Gram positive *Propionibacteriaceae* family are responsible for the flavour and texture of some fermented foods, especially Swiss cheese, where they are responsible for the formation of 'eyes' or holes in the cheese. These bacteria break down lactic acid into acetic and propionic acids and carbon dioxide. Several other bacteria for e.g., *Leuconostoc citrovorum, L. dextranicum, Streptococcus lactis, S. cremis, S. liquefaciens* and *Brevibacterium* species are important in the fermentation of dairy products [1, 3].

All the milk products are generally produced by intense activity of lactic cultures. In fact milk is a natural habitat of a number of lactic acid bacteria, which may cause spontaneous souring. For desired fermentation processes, specially prepared cultures of harmless lactic acid bacteria are used which are called "starter culture" or "starters" and yield desirable flavour and texture to a particular fermented milk product. The lactic acid bacteria generally used in milk fermentation are from two groups viz., Gram +ve cocci like *Streptococcus* sp. *(S. thermophilus, S. lactis, S. lactis* subsp. *diacetylactis, S. cremoris)* and Gram +ve asporogenous rod-shaped bacteria viz., *Lactobacillus* sp. *(L. bulgaricus, L. acidophilus).*

The important microorganisms playing an important role in the preservation of vegetables and fruits for the production of pickles include salt-resistant lactic acid bacteria, initially heterofermentative *Leuconostoc* sp. and *Lactobacillus brevis*, are generally replaced by homofermentative *L. plantarum* and *Pediococcus* sp. [1, 3]. As acids are produced and the pH drops, yeasts become more important.

### *2.1.2 Acetic acid bacteria*

A second group of bacteria of importance in food fermentations are the acetic acid producers from the *Acetobacter* species. *Acetobacter* are important in the production of vinegar (acetic acid) from fruit juices and alcohols. The same reaction also occurs in wines, oxygen permitting, where the *Acetobacter* can cause undesirable changes – the oxidation of alcohol to acetic acid. This produces a vinegary off-taste in the wine. The most desirable action of acetic acid bacteria is the production of vinegar. This is essentially a two stage process, where yeasts convert sugars into alcohol, followed by *Acetobacter*, which oxidise alcohol to acetic acid. Vinegar, literally translated as sour wine, is one of the oldest products of fermentation used by man. It can be made from almost any fermentable carbohydrate source, for example, fruits, vegetables, syrups and wine. Whatever the raw material used, the fermentation process follows a definite sequence. The yeasts and bacteria exist together in a form known as commensalism. The *Acetobacter* are dependent upon the yeasts to produce an easily oxidisable substance (ethyl alcohol). It is not possible to produce vinegar by the action of one type of microorganism alone. For a good fermentation, it is essential to have an alcohol concentration of 10 to 13%. If the alcohol content is much higher, the alcohol is incompletely oxidised to acetic acid. If it is lower than 13%, there is a loss of vinegar because the esters and acetic acid are oxidised. In addition to acetic acid, other organic acids are formed during the fermentation which become esterified and contribute to the characteristic odour, flavour and colour of the vinegar. Acetaldehyde is an intermediate product in the transformation of reducing sugar in fruit juice to acetic acid or vinegar. Oxygen is required for the conversion of acetaldehyde to acetic acid. In general, the yield of acetic acid from glucose is approximately 60%, i.e., three parts of glucose yield two parts of acetic acid [1, 3].

The organisms involved in vinegar production usually grow at the top of the substrate, forming a jelly-like mass. This mass is known as 'mother of vinegar'. The mother is composed of both *Acetobacter* and yeasts, which work together. The principal bacteria are *Acetobacter aceti, A. xylinum* and *A. ascendens*. The main yeasts are *Saccharomyces ellipsoideus* and *S. cerevisiae.* It is important to maintain an acidic environment to suppress the growth of undesirable organisms and to encourage the presence of desirable acetic acid producing bacteria. It is a common practice to add 10 to 25% by volume of strong vinegar to the alcoholic substrate in order to attain a desirable fermentation. The alcoholic fermentation of sugars should be completed before the solution is acidified because any remaining sugar will not be converted to alcohol after the acetic acid is added. Incomplete fermentation of the juice results in a "weak" product. The acetic acid strength of good vinegar should be approximately 6%.

### *2.1.3 Bacteria of alkaline fermentations*

A third group of bacteria are those which bring about alkaline fermentations — the *Bacillus* species; particularly *B. subtilis, B. licheniformis* and *B. pumilius. B. subtilis* are the dominant species, causing the hydrolysis of proteins to amino acids and peptides, and releasing ammonia, which increases the alkalinity and makes the substrate unsuitable for the growth of spoilage organisms. Alkaline fermentations are more common in protein rich foods such as soybeans and other legumes, although there are a few examples utilising plant seeds. For example, water melon seeds (Ogiri in Nigeria), sesame seeds (Ogiri-saro in Sierra Leone) and others where coconut and leaf proteins are the substrates (Indonesian semayi and Sudanese kawal, respectively). Although the range of products of alkaline fermentation do not match those brought about by acid fermentations, they are important in that they provide protein rich, low cost condiments from leaves, seeds and beans, which contribute to the diet of millions of people in Africa and Asia [1, 3].

## 2.2 Yeasts

Yeasts and yeast-like fungi are widely distributed in nature. They are present in orchards and vineyards, in the air, the soil and in the intestinal tract of animals. Like bacteria and moulds, yeasts can produce beneficial and non-beneficial effects in foods. The most beneficial yeasts in terms of desirable food fermentation are from the *Saccharomyces* family, especially *S. cerevisiae*. Yeasts play an important role in the food industry as they produce enzymes that favor desirable chemical reactions such as the leavening of bread and the production of alcohol and invert sugar. Some yeasts are chromogenic and produce a variety of pigments, including green, yellow and black. Others are capable of synthesizing essential B group vitamins [1, 3].

Although there is a large diversity of yeasts and yeast-like fungi (about 500 species), only a few are commonly associated with the production of fermented foods. They are all either ascomycetous yeasts or members of the genus *Candida*. Varieties of *Saccharomyces cerevisiae* genus are the most common yeasts in fermented foods and beverages based on fruit and vegetables. All strains of this genus ferment glucose and many ferment other plant derived carbohydrates such as sucrose, maltose and raffinose. In the tropics, *Saccharomyces pombe* is the dominant yeast in the production of traditional fermented beverages, especially those derived from maize and millet.

Of the various yeast genera, *Brettanomyces, Candida, Cryptococcus, Debaryomyces, Dekkera, Hanseniaspora, Kloeckera, Kluyveromyces, Metschnikowia, Pichia, Rhodotorula, Saccharomyces, Saccharomycodes, Schizosaccharomyces* and *Zygosaccharomyces* are associated with wine-making. Yeasts participating in the fermentations generally come from the grapes and their existence depends on the sum of various physico-chemical and biotic factors that pertain to vineyards and wineries [24]. *Kloeckera apiculata* and *Hanseniaspora uvarum* are the predominant yeasts intimately associated on the surface of grape berries, accounting for roughly 50-75% of the total yeast population [25]. Fermentative species of *Saccharomyces cerevisiae* occur at extremely low populations on healthy, undamaged grapes and are rarely isolated from intact berries and vineyard soils [26]. In contrast to its low occurrence in natural habitats such as vineyards, *S. cerevisiae* is abundant in the grape juice and must-coated surfaces of winery equipment, forming an important component of a so-called 'residential' or 'winery' yeast flora [27]. In fact, *S. cerevisiae* is by far the most dominant yeast species colonizing surfaces in wineries, demonstrating the selective effects of grape juice and wine as the growth substrates. Preeminent of the role of *S. cerevisiae* in winemaking been firmly established, there is an ever-growing demand for new and improved wine yeast strains. The genetic engineering of yeasts has as its targets more efficient fermentation processing and control of microbial spoilage, as well as the improvement of 'the sensory quality of wine' and 'the wholesomeness of wine'. Yeast strains may be developed which enhance the health benefits and reducing the risks. Wine yeast strains currently are being genetically programmed with surgical precision for the cost-competitive production of high quality grapes and wine with relatively minimal resource inputs and a low environmental impact [6]. This entails the establishment of yeast strains with increased productivity, efficiency, sustainability and ecofriendly. With regard to the wine yeast biotechnology, the emphasis is on the development of *Saccharomyces cerevisiae* strains with improved fermentation, processing and biopreservation abilities, and capabilities for an increase in the wholesomeness and sensory quality of wine [28].

### 2.3 Moulds

Moulds are also important organisms in the food industry, both as spoilers and preservers of foods. Certain moulds produce undesirable toxins and contribute to the spoilage of foods. The *Aspergillus* species are often responsible for undesirable changes in foods. These moulds are frequently found in foods and can tolerate high concentrations of salt and sugar. However, others impart characteristic flavours to foods and some others produce enzymes, such as amylase for bread making. Moulds from the genus *Penicillium* are associated with the ripening and flavour of cheeses. Moulds are aerobic and therefore require oxygen for growth. They also have the greatest array of enzymes, and can colonize and grow on most types of food. Moulds do not play a significant role in the desirable fermentation of fruit and vegetable products [1, 3].

## 3. MICROORGANISMS AS SOURCE OF ENZYMES IN FOOD FERMENTATIONS

The changes that occur during the fermentation of foods are the result of enzymatic activity. Enzymes are complex proteins produced by living cells to carry out specific biochemical reactions. They are known as catalysts since their role is to initiate and control the reactions, rather than being used in a reaction. Because they are proteinaceous in nature, they are sensitive to fluctuations in temperature, pH, moisture content, ionic strength and concentrations of substrate and inhibitors. Each enzyme has requirements at which it can operate most efficiently. Extremes of temperature and pH can denature the protein and destroy enzyme activity. Since they are so sensitive, reactions can easily be controlled by slight adjustments to temperature, pH or other reaction conditions. In the food industry, enzymes have several roles—the liquefaction and saccharification of starch, the conversion of sugars and the modification of proteins. Microbial enzymes play a role in the fermentation of fruits and vegetables. A wide range of microbial enzymes are also widely used in food processing (Table 1) which help to control the texture, appearance and nutritive value of food as well as to generate desirable flavours and aromas [29].

**Table 1:** Applications of microbial enzymes in food industry.

| Enzymes | Source: Bacteria | Source: Fungi | Functions |
|---|---|---|---|
| Proteases | *Bacillus subtilis, Lactobacillus acidophilus, L. plantarum, L. lactis, L. delbrueckii, L. helveticus* | *Aspergillus oryzae* | Baking, brewing, cheese production, meat tenderization, flavourants |
| Lipases | *Bacillus* sp. | *Aspergillus oryzae, A. niger, Rhizopus* sp., *Rhizomucor miehei, Rhodotorula* sp. | Flavorants, in dairy industry |
| α-amylases | *Bacillus* sp. | *Aspergillus* sp. *(Aspergillus oryzae)* | Baking, brewing, starch processing |
| β-amylases | *Bacillus* sp. | *Rhizopus* sp. | Brewing, maltose syrup production |
| Amyloglucosidases (Glucoamylase) | - | *Aspergillus niger, Rhizopus niveus* | Brewing, baking, glucose syrup production |
| β-galactosidases (Lactases) | *Lactobacillus* sp., *Bacillus* sp., *E. coli, Streptococcus lactis* | *Kluyveromyces* sp., *Candida* sp. | Processing of dairy products, whey-fructose syrup production |
| Glucose isomerases | *Actinoplane* sp., *Arthrobacter* sp., *Streptomyces* sp. | - | High-fructose syrup production |
| Glucose oxidases | - | *Aspergillus niger, Penicillium notatum* | As antioxidant, used along with catalase to remove oxygen from food products |
| Pullulanases | *Klebsiella pneumoniae, Bacillus acidopullulyticus* | - | Brewing, debranching of starch in sugar syrup manufacture |
| Pectinases | - | *Aspergillus niger, Penicillium* sp. | Extraction of plant and fruit juices, coffee bean fermentation |
| Hemicellulases & β-glucanases/ Cellulases | *Bacillus* sp. | *Trichoderma reesei, Penicillium funiculosum, P. emersonii, Aureobasidium pullulans, Cryptococcus* sp., *Trichosporon* sp. | Juice production, alcoholic beverages, baking, brewing, nutraceutics |
| Urease | *Lactobacillus fermentum* | - | Wine production |
| Invertase | - | *Saccharomyces cerevisiae* | Variously in food processings |

*Source:* 29

## 4. MICROORGANISMS AS A SOURCE OF FOOD SUPPLEMENTS

A food supplement, mycoprotein, is the processed cellular mass that is obtained from the filamentous fungus *Fusarium venenatum*. The filamentous nature of the cells gives the cellular mass a meat-like texture that makes it suitable for a variety of applications in food, including use as a muscle replacer in meat-alternative products. Mycoprotein can also be used as a fat replacer in certain dairy products and as a cereal replacer in products such as breakfast cereals or puffed snacks. The protein content of mycoprotein typically ranges from 42-50% percent on a dry weight basis. Mycoprotein is a high quality protein because it contains all of the essential amino acids. The fat content of mycoprotein typically ranges from 12-14% on a dry weight basis, and this fat content is more like vegetable fat than animal fat because it has a low proportion of saturated fatty acids and a high proportion of mono- and polyunsaturated fatty acids. The fungus is grown aerobically under steady-state conditions maintained by a continuous feed of nutrients and simultaneous removal of the culture. Quorn is the trademark of a fungus-based food product, sold (largely in Europe) as a meat substitute or imitation meat. It is marketed at the health-conscious, and to vegetarians. Quorn is the first commercial mycoprotein to be used in general food supplies [29]. A wide range of metabolites, produced by microorganisms, in the form of organic acids, amino acids, vitamins, lipids, proteins, etc. are also used as food supplements for the purpose of preservation, flavour enhancement, stabilization and improvement in digestibility and nutritive value (Table 2).

Bacteria and fungi are producers of a variety of organic acids including lactic acid, citric acid, acetic acid, gluconic acid and malic acid, which are important in food industry. The traditional organisms for citric acid fermentation are selected strains of *Aspergillus niger*, though citric acid has also

**Table 2:** Food additives of microbial origin.

| Additive function | Product | Producing microorganisms |
|---|---|---|
| Acidulants | Citric acid | *Aspergillus niger, Candida guilliermondii, Candida lipolytica* |
| | Acetic acid | *Acetobacter aceti, Clostridia* |
| | Lactic acid | *Lactobacillus delbrueckii, L. bulgaricus* |
| | Malic acid | *Brevibacterium flavum* |
| Amino acids | Glutamic acid | *Corynebacterium glutamicum, Brevibacterium flavum* |
| | Lysine | *Corynebacterium glutamicum, Brevibacterium flavum* |
| | Tryptophan | *Klebsiella aerogenes* |
| Antimicrobial agents | Nisin | *Lactobacillus lactis* |
| | Natamycin | *Streptomyces natalensis, S. chattanoogensis* |
| Antioxidants | Ascorbic acid | *Acetobacter suboxidans* |
| Colours | β-Carotene | *Blakeslea trispora, Dunaliella salina* |
| | Lycopene | *Blakeslea trispora, Streptomyces chattanoogensis* |
| | Zeaxanthin | *Flavobacterium* sp. |
| | Lutein | *Spongiococcum excentricum, Chorella pyrenoidosa* |
| | Cantaxanthin | *Cantharellus cinnabarinus, Rhodococcus maris* |
| | Astaxanthin | *Mycobacterium lacticola, Xanthophyllomyces dendrorbous* |
| Emulsifiers | Various polysaccharides | Many organisms |
| | Glycolipids | *Rhodococcus erythropolis* |
| Flavours and flavour enhancer | Vanillin | *Saccharomyces* sp. |
| | Nucleosides | *Bacillus subtilis, Brevibacterium ammoniagenes* |
| | Monosodium glutamate | *Corynebacterium glutamicum, Brevibacterium flavum* |
| Humectants | Glycols | *Bacillus* sp. |

Contd...

**Contd...**

| Additive function | Product | Producing microorganisms |
|---|---|---|
| Lipids and fatty acids | Glycolipids | *Bacillus subtilis* |
| | γ-linoleic acid | *Mucor* sp. |
| Non nutritive sweetners | Aspartame | From microbial production of aspartic acid and phenylalanine |
| Stablizers and thickners | Glycan | *Saccharomyces* sp. |
| | Pullulan | *Aureobasidium pullulans* |
| | Xanthan | *Xanthomonas campestris* |
| Vitamins | Ascorbic acid (C) | *Acetobacter suboxidans* |
| | Cyanocobalamin ($B_{12}$) | *Propionibacteria* sp., *Pseudomonas denitrificans, Streptomyces griseus* |
| | Riboflavin ($B_2$) | *Ashbya gossypii, Ermothecium ashbyii,* recombinant *Bacillus subtilis* |

*Source:* 29

been produced with several yeasts, mainly species of *Candida,* especially, *Candida lipolytica.* Citric acid provides a pleasant tartness and complements fruit, berry and other flavours. Blends of citric acid combined with malic acid, lactic acid or fumaric acid are also used in beverage formulations. Other functions of citric acid in these applications rely on its ability to prevent metal-catalyzed off-flavours and deterioration of colour. This important food additive also serves as an antimicrobial preservative, retarding the growth of spoilage organisms. Acetic acid fermentation is applied in the aerobic, two-step process for the production of vinegar. However, in the last decade, a significant amount of research has been devoted to the anaerobic production from biomass and its potential to compete with chemical synthesis. Vinegar is a liquid condiment used as a food flavouring and preservative. It is produced by the action of aerobic bacteria of the genus *Acetobacter* on dilute solutions of ethanol such as cider, wine or diluted distilled alcohol solutions. The ethanol solutions utilized are also produced microbiologically by the ethanol fermentation of sugars by *Saccharomyces cerevisiae.* When starch materials are used, the enzymatic saccharification of starch is required prior to the alcohol fermentation. Lactic acid has been produced from various lactic acid bacteria in indigenous fermented foods including dairy products and is mainly responsible for the precipitation of milk proteins for yoghurt and cheese production. *Aspergillus niger* is used to produce gluconic acid at large scale. A portion of gluconic acid is converted to δ-gluconolactone, a slow acting acidulant which is used with sodium bicarbonate for controlled leavening of baked goods. The lactone is also used in the dairy industry for cheese curd formation and for the improvement of heat stability of milk. In addition, it serves as an additive in cured meat products, luncheon meats, tofu, and packet desert mixes. L-malic acid obtained with fungal fermentations plays a role similar to that of citric acid as a food ingredient. As an acidulant, it has a smooth, long-lasting flavour profile. Malic acid exhibits a synergy with aspartame permitting a 10% reduction in the usage of this sweetner in beverages [29].

Amino acids include an important class of compounds required by living cells as growth factors. The production of amino acids by fermentation processes was initiated in 1956 when L-glutamic acid was discovered as a product in the spent medium of *Corynebacterium glutamicum.* About 66% of amino acids are used in food industry and 31% as feed additives. In food industry, amino acids in the form of sodium glutamate, sodium aspartate, D, L-alanine and glycine are used to enhance flavours. L-cysteine improves the quality of bread during the baking process and acts as antioxidant in fruit juices. L-tryptophan and L-histidine act as antioxidants in powdered milk. L-phenylalanine and L-aspartic acid are used as the starting molecules for aspartame (L-aspartyl-L-phenylalanine methyl ester), which is used as a low calorie sweetener. Lysine is added in bread in Japan, and in some countries soya products are enhanced by the addition of methionine and L-lysine. D, L-methionine is used to upgrade animal feeds.

The principal microorganisms capable of producing glutamic acid include *Corynebacterium, Brevibacterium, Microbacterium and Arthrobacter,* however, *Corynebacterium glutamicum* is mainly used commercially. L-methionine is synthesized with immobilized *Aspergillus oryzae* aminoacylase. L-Lysine is produced with L-amino-caprolactum hydrolase from *Cryptococcus laurentii*, and L-phenylalanine is produced with immobilized *Rhodotorula rubra* cells bearing phenylalanine ammonia lyase. Phenyl alanine is in demand for the synthesis of aspartame, a low-calorie, artificial sweetner used in soft drinks and desserts. Tryptophan, lysine and methionine, all essential amino acids in the human diet, are used to enrich foods. Certain fungi are able to produce substantial amounts of these amino acids, but are surpassed by improved industrial strains of bacteria. Strains of *Pichia anomala* are prolific producers of tryptophan. Strains of the yeasts *Candida utilis* and *Saccharomyces cerevisiae* and the molds *Ustilago maydis* and *Gliocladium* sp. produce significant quantities of lysine. Mutants of the yeasts *Candida tropicalis, Y. lipolytica, Pichia pastoris* and *C. utilis* have an increased methionine content and may have potential as food supplements [29].

Vitamins are essential mediators of metabolism and play an important role in the human diet. They are often supplements in foods and animal feeds. Many microorganisms are the potential producers of various vitamins. Many fungi and bacteria can produce riboflavin. The two ascomycetes *Ermothecium ashbyii* and *Ashbya gossypii* are the potential producers of riboflavin. Industrial fermentation with *A. gossypii* is now preferred and generates about 30% of world output. Sterols are widely distributed among the fungi. Ergosterol, a commercially important lipid, extracted from dehydrated yeast can be converted into vitamin D by irradiation with ultraviolet light. Such yeast strains may serve as alternative sources of vitamin D in future and can be used in various dairy products, often milk and margarine and in breakfast cereals and baby foods. Certain fungal species, such as *Blakeslea trispora*, a heterothallic fungus of the Mucorales group and *Phycomyces blakesleeanus* are extremely rich in mycelial β-carotene, which are a potential source of carotenoides that act as precursors for the production of vitamin A.

Microbial polysaccharides in the form of pullulan, scleroglucan and agar have attracted considerable interest as industrial fungal polymers with unique properties in food application. Pullulan is a neutral, linear homopolysaccharide composed of maltotriose units and is produced by the dimorphic black yeast *Aureobasidium* (*Pullularia*). Its ability to form strong transparent films with low oxygen permeability enhances its usefulness as a food coating and packaging agent. Scleroglucan, a highly branched, neutral homopolysaccharide with a backbone of about 90 D-glucose units is produced by *Sclerotium glucanicum* or *S. rolfsii*. Scleroglucan exhibits pseudo-plasticity over a broad range of temperature, pH, and salt conditions, making it suitable for use as a food additive. Agar, a linear galactan sulphate produced by red sea weed (*Gelidium*), having an ability to absorb water forming gels, is used as a stabilizer in foods [29].

About 70% of the world's supply of fats and oils is derived from plants and currently no food lipids are produced by microbiological means. It is notable that the processes for fat production with yeasts *Rhodotorula gracilis, Lipomyces lipofer, Lipomyces starkeyi, Endomyces vernalis, Cryptococcus terricolus* and *Candida* species, which accumulate significant levels of fat, have been developed. The molds *Mortierella vinacea, Mucor circinelloides, Aspergillus ochraceous, A. terreus*, and *Penicillium lilacinum* are also good producers of fat, as are certain fungi of genera *Hansenula, Chaetomium, Cladosporium, Malbranchea, Rhizopus* and *Pythium*. Fatty acids for the food, medical and cosmetic industries could be standard products of fungal fermentation in coming years [29].

## 5. STRAIN DEVELOPMENT TECHNOLOGIES FOR FOOD MICROORGANISMS

The desired microbes can be modified genetically in many ways. Some techniques alter limited regions of the genome, whereas others are used to recombine or rearrange the entire genome [6, 30-34]. Techniques having the greatest potential in genetic programming include selection of variants, mutation and selection, hybridization, rare-mating, protoplast fusion, gene cloning and transformation. The combined use of replica plating, mutagenesis, hybridization and recombinant DNA methods have dramatically increased the genetic diversity that can be introduced into microbial cells. The purpose of the genetic manipulation is to create a new strain with precisely defined genetic properties. The techniques used for the manipulations may be random or specific and are discussed below.

## 5.1 Random Mutagenesis and Selection

Mutation of microbial strains can lead to improvement of certain traits with the simultaneous debilitation of other characteristics. Through mutations, the genetic constitution of microorganisms can be altered. The process of mutation for the production of beneficial products began at the end of 1920's when Alexander Fleming, by using ultraviolet rays, increased the production of penicillin by the mold, *Penicillium chrysogenum*, by about 1000-fold [6]. Mutations are permanent, sometimes transmissible changes in the genetic material of a cell which are caused by copying errors in the genetic material during cell division and by exposure to radiation, chemicals or viruses. Although mutations are probably induced with the same frequency in haploid, diploid or polyploids, they are not easily detected in diploid and polyploid cells because of the presence of non-mutated alleles. Only if the mutation is dominant and is a phenotypic effect, it is detected without the need for additional alterations. Therefore, haploid strains of microorganisms are preferred, though not essential, when inducing mutations. Successful mutation and breeding is usually associated with mutations in meiotic segregants where the two mating parents of a genetically stable hybrid provide a good basis for the introduction of recessive mutations. Mutagenesis has the potential to disrupt or eliminate undesirable characteristics and enhance favourable properties of microbes. Though the use of mutagens for directed strain development is limited, the method could be applied to isolate new variants of the microorganisms prior to further genetic manipulation. In multicellular organisms, mutations can be subdivided into germ line mutations, which can be passed onto progeny, and somatic mutations, which often lead to the malfunction or death of a cell and can cause cancer. Mutations are considered the driving force of evolution, where less favorable (or deleterious) mutations are removed from the gene pool by natural selection, while more favorable (or beneficial) ones tend to accumulate. Neutral mutations do not affect the organism's chances of survival in its natural environment and can accumulate over time, which might result in what is known as punctuated equilibrium, the modern interpretation of classic evolutionary theory. Structurally, mutations can be classified as small-scale mutations and large-scale mutations in chromosomal structure [35, 36].

## 5.2 Small-Scale Mutations

These affect one or a few nucleotides and may include point and frameshift mutations. Point mutations, often caused by chemicals or malfunction of DNA replication, exchange a single nucleotide for another. Most common is the transition that exchanges a purine for a purine or a pyrimidine for a pyrimidine (A ↔ G, C ↔ T). A transition can be caused by nitrous acid, base mispairing or mutagenic base analogs such as 5-bromo-2-deoxyuridine (BrdU). Less common is a transversion, which exchanges a purine for a pyrimidine or a pyrimidine for a purine (C/T ↔ A/G). A point mutation can be reversed by another point mutation, in which the nucleotide is changed back to its original state (true reversion) or by second-site reversion (a complementary mutation elsewhere that results in regained gene functionality). These changes are classified as transitions or transversions. An example of a transversion is adenine being converted into a cytosine. There are many other examples also that can be found. There are three kinds of point mutations, depending upon what the erroneous codon codes for: (i) silent mutations: codes for the same amino acid, so has no effect, (ii) missense mutations: codes for a different amino acid, (iii) nonsense mutations: codes for a stop codon, which can truncate the protein. Frameshift mutations involve the insertion or deletion of one or few nucleotides in the DNA molecule which shifts the reading frame of nucelotide sequences. These are usually caused by transposable elements, or errors during the replication of repeating elements (e.g., AT repeats). Most insertions or deletions in a gene can either alter splicing of the mRNA, or cause a shift in the reading frame, both of which can significantly alter the gene product [35, 36].

## 5.3 Large-Scale Mutations

These bring about a change in chromosomal structure and include (i) amplifications or gene duplications, leading to multiple copies of chromosomal regions, increasing the dosage of the genes located within them, (ii) deletions of large chromosomal regions, leading to loss of the genes within those regions, (iii) chromosomal translocations attaching DNA from separate chromosomes, (iv) interstitial deletions removing the regions of DNA from a single chromosome, thereby apposing previously distant genes, (v) chromosomal inversions switching the orientation of a segment of a chromosome, thereby apposing its ends to previously distant genes, and (vi) loss of heterozygosity with loss of one allele, either by a deletion or recombination event, in organisms which previously had two.

Mutations can take place spontaneously or can be induced intentionally in various microorganisms. In nature, mutation appears random or chance events with no regard to environmental conditions. Spontaneous mutations occur in the absence of any agent known to alter the genetic material. Induced mutations are produced by chemical and physical agents, termed as mutagens including base analogs, alkylating agents, deaminating agents, acridine derivatives, X-rays, gamma rays and UV rays [35, 36].

The average spontaneous mutation frequency in microbes at any particular locus is approximately $10^{-6}$ per generation. The use of mutagen greatly increases the frequency of mutations in a wine yeast population. Mutation and selection (often through replica-plating on selective agar media) appear to be a rational approach to strain development when a large number of performance parameters are to be kept constant while only one is to be changed.

The phenotypic results of mutation can vary dramatically. If the changes occur in a non-functional area of DNA or, because of the redundant nature of the genetic code, do not change a gene at the level of the protein sequence, the mutation will be silent. Such a mutation will not have any phenotype, although it can accumulate and change the genetic structure of subsequent generations. Even if a mutation causes a change in the amino acid sequence of a protein, the consequences to the organism may be negligible. For example, conservative amino acid changes may not affect the activity of a mutated enzyme, or the enzyme may not be required in the environmental niche in which the organism is growing. In the former case, such a change is an evolutionary step affecting the protein and may have consequences on protein activity at a latter stage of evolution change. In the latter case, if the environment changes, the lack of the enzyme may become critical. In general, changes at the amino acid level in a protein will have a consequential effect on the protein's activity, usually detrimental. Thus, an important metabolic pathway may be blocked by the lack of a functional enzyme resulting in identifiable phenotypes. These phenotypes can be classified into four main groups: auxotrophic mutations, substrate utilization mutations, resistance mutations and essential mutations.

Auxotrophic mutations are mutations that knock out the genes that are involved in anabolic pathways inside the bacterial cells. The metabolic pathways within the cells synthesize a wide range of amino acids, nucleic acids, lipids, cofactors, etc. that are central to the normal growth of the cell. Obviously, all bacterial species do not synthesize all such cellular requirements, and for a particular species, some additions to a basic medium which provides a carbon, nitrogen and energy source are needed. Such a medium is called a defined minimal medium, and it is the simplest medium on which bacteria can grow. An auxotrophic mutation, because it interrupts a central anabolic function, would now require an additional chemical added to its minimal medium. Thus, a mutation that inactivates an enzyme in the pathway that synthesizes histidine will result in a requirement for histidine to be added to the minimal medium for bacterial growth. Similarly, an interruption of the purine biosynthetic pathway would need the addition of purines. Such mutations are easily identified by the failure of such a strain to group on the minimal medium for that species, but the strain will grow in the presence of the added factor.

A mutation in either an enzyme involved in the breakdown or involved in the uptake will result in the loss of the ability to utilize such a resource. For example, a mutation in β-galactosidase in *E. coli* stops the utilization of lactose as a carbon and energy resource. Similarly, a mutation that affects a cellulase gene will reduce to some extent, the degradation of cellulase by that organism. Interestingly in this case, since most cellulase utilizing bacteria contain a number of cellulase genes, the utilization will probably still continue, but at a lower rate. Resistance mutations are a class of mutations which can be considered artificial for the most part, although they can occur in nature. The presence of antibiotics, heavy metals and other toxic substances in the environment means that such mutations do occur in specific environments. In general, they can be divided into three types: firstly, mutations that inhibit the uptake of toxic substance, usually by inactivating a gene involved in uptake; secondly, mutations that change the biological target of the toxic substance so that it is less affected by the substance; and finally, by changing an enzyme so that it can now degrade the toxic substance, either by switching the gene on, if it is already present or by natural selection of a new gene activity [35, 36].

## 5.4 DNA Damage in Mutations

Although DNA is naturally quite resistant to change because of its structure, certain aspects of its structure can lead to limited change. Fig. 1 shows the tautomeric changes that can occur in certain bases, which allow mispairing during DNA replication. Such mispairing results in a change in the base of one strand of the daughter DNA molecule. A second round of replication will result in the fixing of such a mutation. Usually, the mispairing is picked up by the DNA repair systems of the organisms. At that point, there is a 50% chance that the repair mechanisms will fix the mutation [35, 36]. Another

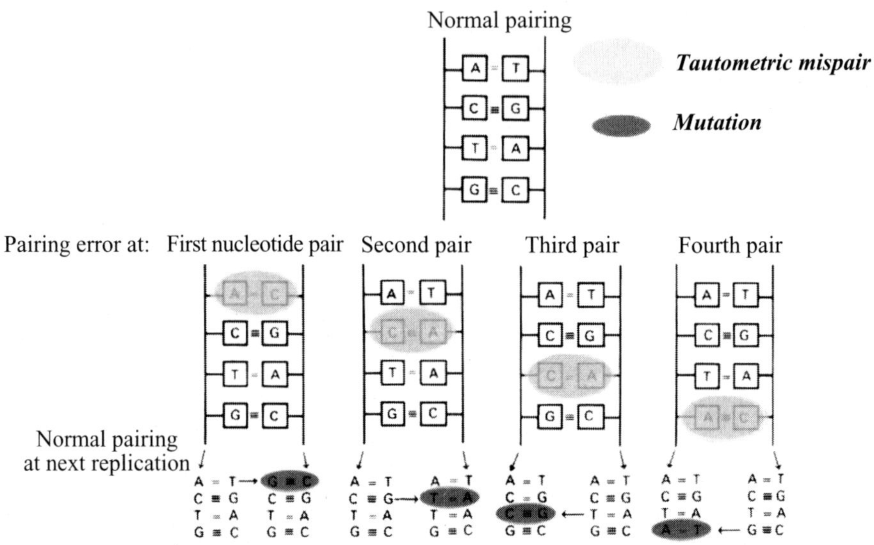


**Fig. 1:** Mutation by tautomeric mispairing.
*Source:* 35

form of DNA damage at the base level is the formation of thymidine dimers by ultraviolet light. As shown in Fig. 2, two adjacent thymidines can be cross-linked and such a complex stops DNA replication. Again, DNA repair comes in to remove the problem. Finally, at the base level, a variety of chemicals can damage DNA by modification of the base, as shown in Fig. 3. The resulting change may alter DNA pairing and result in a base sequence change.

Large scale changes can be induced by environmental factors such as ultraviolet light, nuclear radiation and chemicals. These can result in DNA strand breakage and rejoining, which causes change in the order of genes and the control of gene expression. The end point can be so major that normal sexual reproduction becomes impossible. The latter is exploited when insects are sterilized by irradiation *en mass* and then released as a control measure for insect-borne disease.

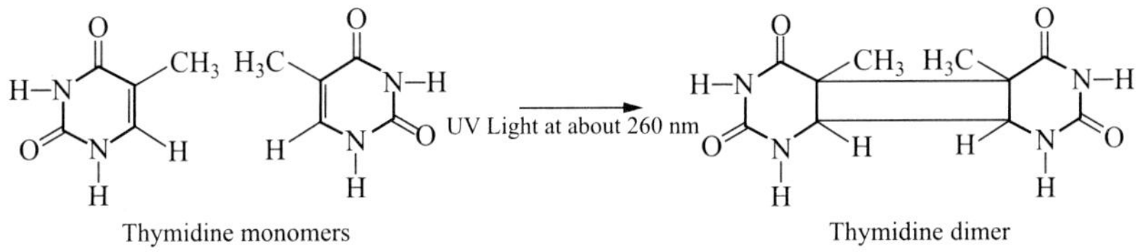


**Fig. 2:** Formation of a thymidine dimer.
*Source:* 35

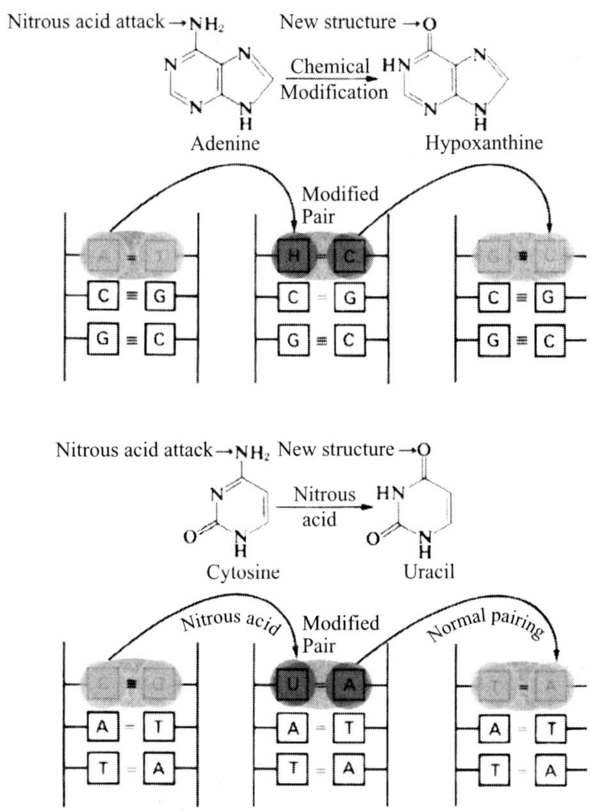


**Fig. 3:** Base pair modification caused by nitrous acid.
*Source:* 35

## 5.5 DNA Repair

The most significant agent for mutation in microorganisms is DNA repair. This paradox is due to the fact that DNA repair is an induced emergency mechanism, which has evolved to remove mutation that may cause the death of the cell. Because of this, some DNA repair mechanisms are error prone and as such cause further but less lethal DNA changes. An example of a DNA repair mechanism that is error free, is the removal of thymidine dimers by photoreaction. Here, the enzyme photolyase uses visible light to split the linked bases and restore the ability of the DNA to replicate (Fig. 4). However, error-prone repair in prokaryotes involves the enzyme DNA polymerase 1, which has a higher error rate than the DNA polymerase, which the cell uses for normal DNA replication. Thus, when DNA polymerase fills in the area where the mutation has been removed, as shown in Fig. 5, the rate at which the wrong base is inserted is 100 to 1000 fold higher than in normal DNA replication. Since prokaryotes are haploid, such a mutation immediately becomes part of the genetic make-up of the cell unlike eukaryotic diploids, where a normal second copy of the gene usually protects the cell from the potentially lethal effect of such a mutation [35, 36].

## 5.6 Conjugation

Conjugation is a genetic exchange process that requires the physical connection of two bacterial cells. Although not absolutely essential, conjugation usually involves a plasmid as the means of establishing the connection and causing the genetic exchange. In many cases, conjugation involves only plasmid exchange, but chromosomal exchange also occurs at a significant rate. Conjugation, whereby chromosomal DNA is exchanged between bacteria, seems to occur as a by-product of the presence of plasmids in bacteria, although plasmid independent chromosome transfer is not unknown. Whether it is a circular or a linear plasmid, if the plasmid is capable of interacting with the main chromosome, it seems to be able to mobilize that chromosome [35].

## 5.7 Transduction

Transduction is genetic exchange involving bacterial viruses as intermediate vectors that allow DNA transfer from cell to cell. Unlike transformation and conjugation, the extent of natural genetic exchange by transduction is not really known. There are two broad classes of bacterial viruses or bac-

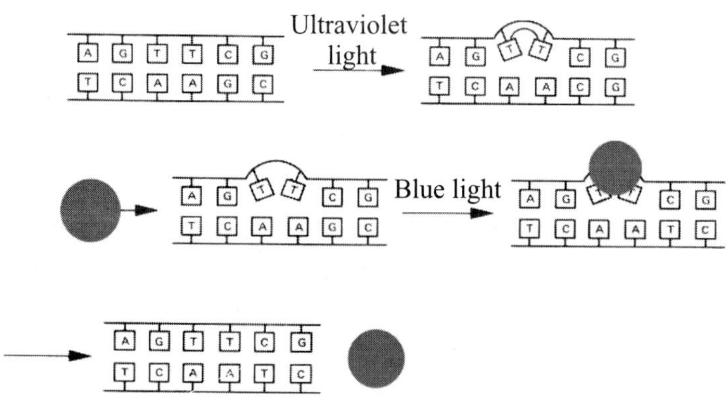


**Fig. 4:** Removal of thymidine dimers by photoreaction.
*Source:* 35

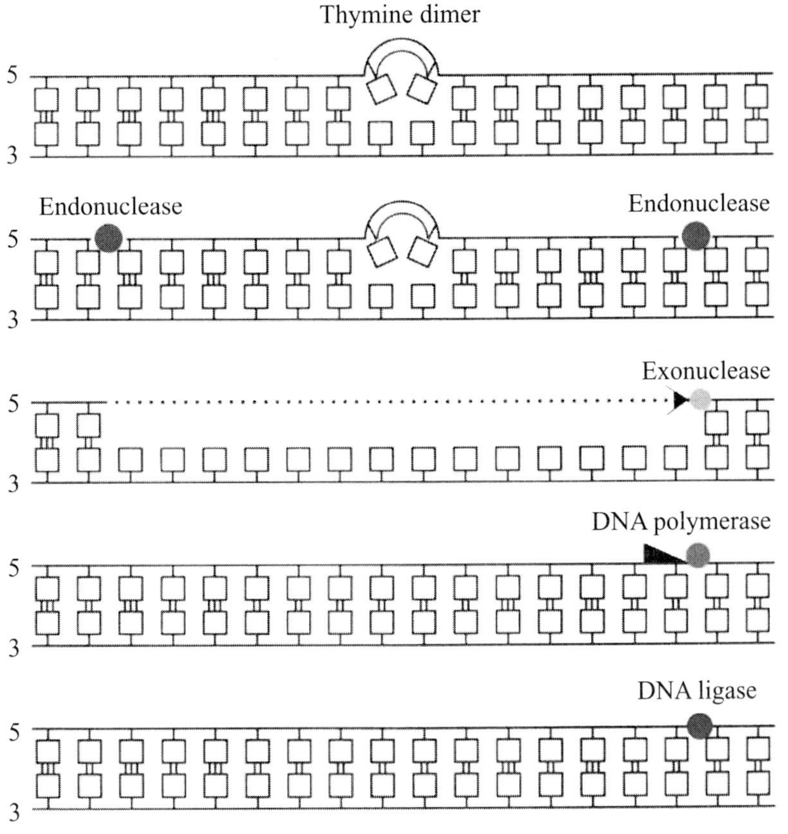


**Fig. 5:** Error-prone DNA repair in prokaryotes.
*Source:* 35

teriophages. The first are virulent bacteriophages that do not interact with the bacterial chromosome as part of their life cycle and replicate only inside the bacterial cell. The second are temperate bacteriophages, which as well as undergoing a virulent life cycle, can integrate their genome stably into the bacterial chromosome (Fig. 6). Virulent and temperate bacteriophages give rise to two different types of transduction, generalized and specialized.

### 5.7.1 *Generalized transduction*

In this type of transduction, random pieces of chromosome DNA are packaged by mistake into a bacteriophage head and can thus be transferred on infection into the new host DNA (Fig. 7).

### 5.7.2 *Specialized transduction*

Specialized transduction (Fig. 8) involves the incorrect excision of DNA when lysogeny ends, with the result that the bacteriophage picks up a collinear piece of DNA from beside its insertion site. When lysogeny is re-established, the additional DNA is also inserted into the new host.

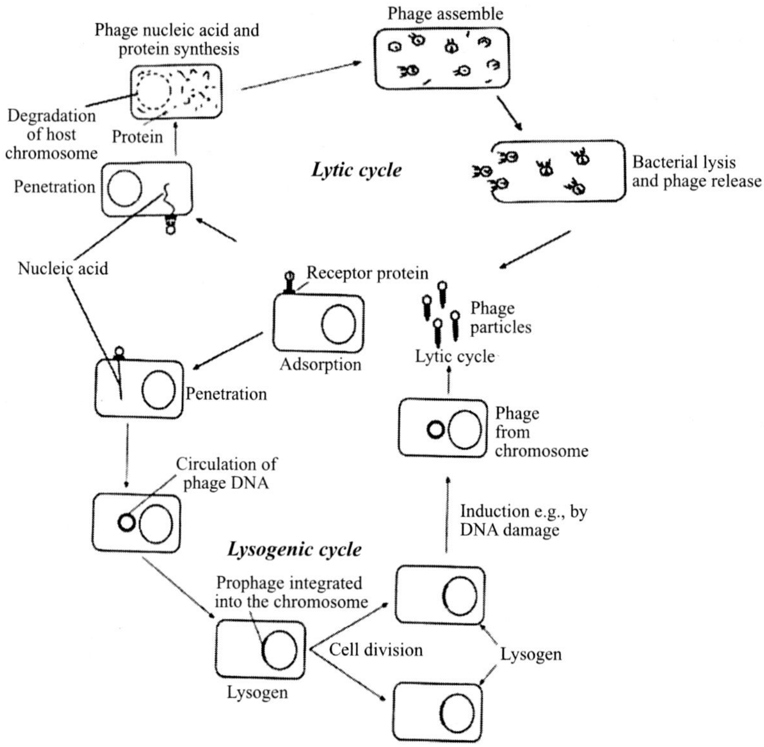


**Fig. 6:** Transduction by virulent and temperate bacteriophages.
*Source:* 35

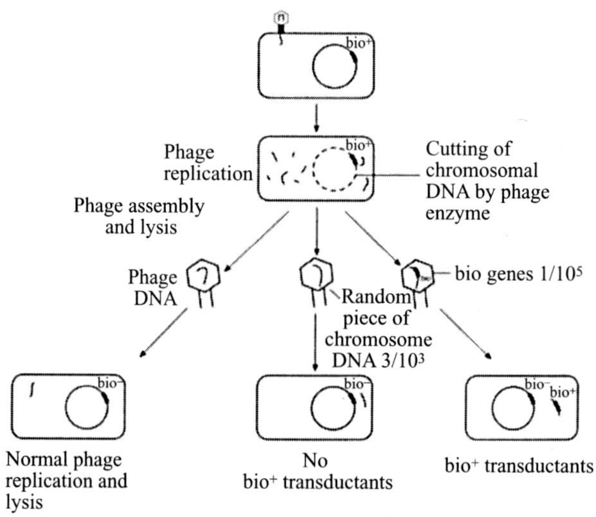


**Fig. 7:** Generalized transduction.
*Source:* 35

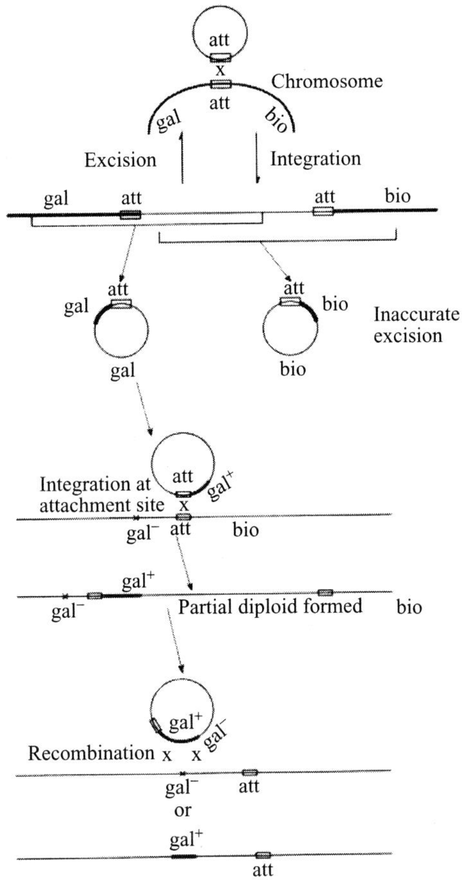


**Fig. 8:** Specialized transduction.
*Source:* 35

## 5.8 Transposition

Transposition is a mechanism by which the pieces of DNA can move from one place to another on the DNA molecules within a cell. There are effectively two types of genetic units that can cause this type of move: insertion sequences and transposons. Both use specific enzymes called transposases to allow the movement. Insertion sequences are the smaller of the two and do not encode any function other than the ability to move. The structure of an insertion sequence is shown in Fig. 9. The transposase and the invert repeats are the essential requirements for transposition. As insertion is relatively random, insertion sequences can insert into the middle of function genes. The result of this is that the gene will cease to function and this process is called insertional inactivation [35].

Three mechanisms of transposition have been postulated — replicative transposition, non-replicative transposition and conservative transposition (Fig. 10). Various entities may use one or more of these mechanisms to effect transposition. One particular example is the TnA family of transposons which does not seem to be closely related to any particular insertion sequence family. It can only transpose by replicative transposition. It contains a second enzyme, a resolvase, which is essential for separation of the cointegrated intermediates produced by replicative transposition.

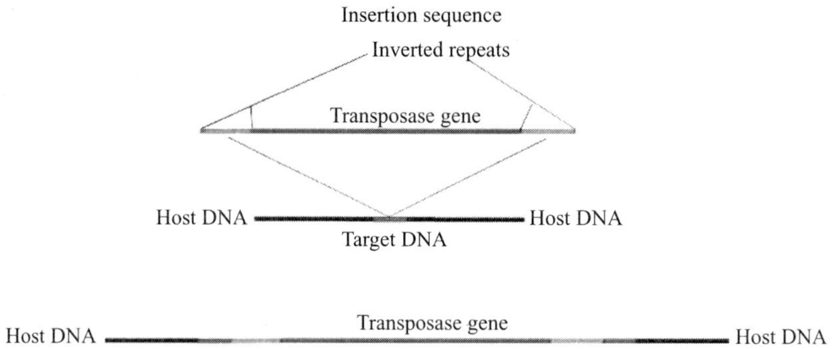


**Fig. 9:** Transposition by insertion sequence.
*Source:* 35

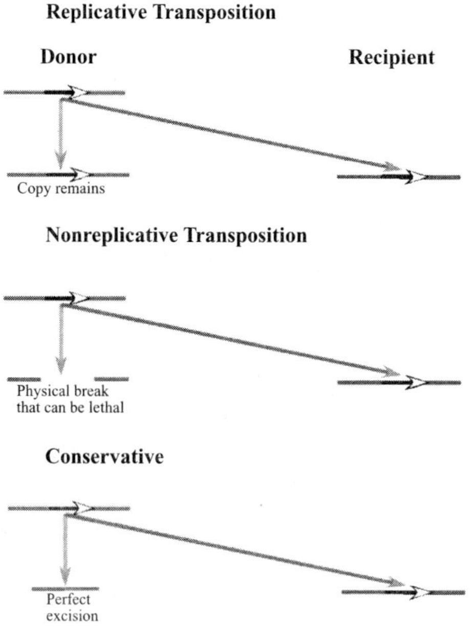


**Fig. 10:** Three mechanisms of transposition.
*Source:* 35

## 5.9 Hybridization

This is the first method, which is put into consideration when the desired property, discovered in another strain, is under multigenic control. Its main disadvantage is its imprecision and consequent disregard for the genome of the original strain. There are of course procedures, such as backcrossing, that can potentially regenerate the original genome. The regeneration is often only partial due to meiotic and mitotic recombination phenomena. Hybridization is usually only feasible between two haploid cells of opposite mating types belonging to the same species. Wing and Lausten [37] were the first to produce new yeast types by recombination of existing variation between commercial strains.

Others also attempted the production of new hybrids by conventional cross breeding, but encountered problems related to the life cycle of commercial strains [38]. Because of this, repeated backcrossing of a derived strain to its original parent to remove undesirable characteristics is both tedious and inefficient. The hybrids obtained by crossing the mating derivatives of commercial yeasts displayed intermediate properties when compared to the parental strains, a result which would be expected for characteristics under polygenic control e.g., fermentation rate in yeast.

Intra-species hybridization involves the mating of haploids of opposite mating-types to yield a heterozygous diploid (Fig. 11). Recombinant progeny are recovered by sporulating the diploid, recovering individual haploid ascospores and repeating the mating/sporulation cycle as required [31]. Haploid strains from different parental diploids, possessing different genotypes, can be mated to form a diploid strain with properties different from that of either parental strain. Thus in theory, crossbreeding can permit the selection of desirable characteristics and the elimination of undesirable ones [39].

Generally, hybridization provides a solution for precise objectives, based on a correctly, defined genetic context. Non-$H_2S$ producing flocculant hybrids by conjugating spores of two homothallic commercial strains were constructed, which on analysis had shown to be homozygous for the desired character. This combination is rarely found in indigenous populations. By conjugating spores, Thornton [40] introduced the flocculant property of a heterothallic haploid laboratory strain into a non-flocculant homothallic commercial strain. To regenerate the original genome of the commercial strain,

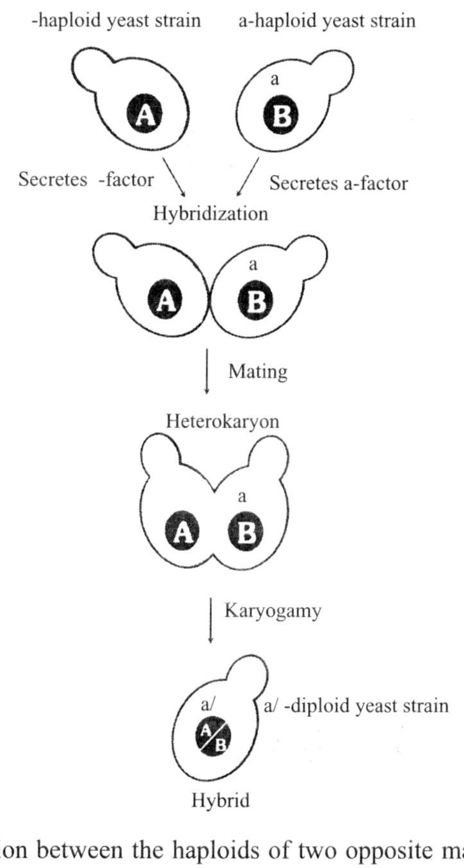


**Fig. 11:** Hybridization between the haploids of two opposite mating-types of yeast.
*Source:* 6

five successive back crosses were then attempted, thereby obtaining clones that did not differ from the original strain with regard to fermentation kinetics and organoleptic properties. The modified strain containing the flocculant property allowed better separation of yeast cells from the wine, thereby facilitating the process of clarification.

Unfortunately, many wine yeasts are homothallic and the use of hybridization techniques for the development of wine yeast strains has proved difficult. However, this problem can be circumvented by direct spore-cell mating using a micromanipulator, in which four homothallic ascospores from the same ascus are placed into direct contact with heterothallic haploid cells (Fig. 12). Mating takes place between compatible ascospores and cells. The resulting diploid is sporulated. Since two spores in each ascus are homothallic and two are heterothallic, stable haploids can be isolated from the sporulated diploids. Elimination or inclusion of a specific property can thus, be achieved relatively quickly by hybridization, provided that it has a simple genetic basis, for example one or two genes [32]. But, many desirable wine yeast characteristics are specified by several genes or are the result of several gene systems interacting with one another. Four ascospores from the same ascus are micromanipulated into direct contact with heterothallic haploid yeast cells. Mating takes place between compatible spores and cells. The resulting diploid is sporulated. Since two spores in each ascus are homothallic and two spores are heterothallic, stable haploids can be isolated from the sporulated diploids.

## 5.10 Rare Mating

Yeast strains that fail to express a mating-type can be force-mated (rare-mating) with haploid *MATa* and *MAT*α strains (Fig. 13). Typically, a large number of cells of the parental strains are mixed and a strong positive selection procedure is applied to obtain the rare hybrids formed. For instance, industrial strains that have a defective form of, or lack, mtDNA (respiratory deficient mutants) can be force-mated with auxotrophic haploid strains having normal respiratory characteristics [31]. Mixing of these non-mating strains at high cell density will generate only a few respiratory-sufficient prototrophs. These true hybrids with fused nuclei can then be induced to sporulate for further genetic analysis and crossbreeding.

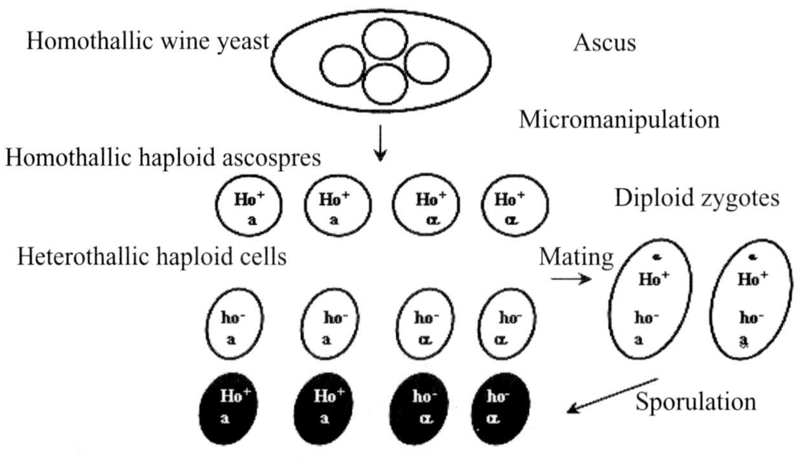


**Fig. 12:** The isolation of haploid strains from homothallic yeast by spore-cell mating.
*Source:* 6

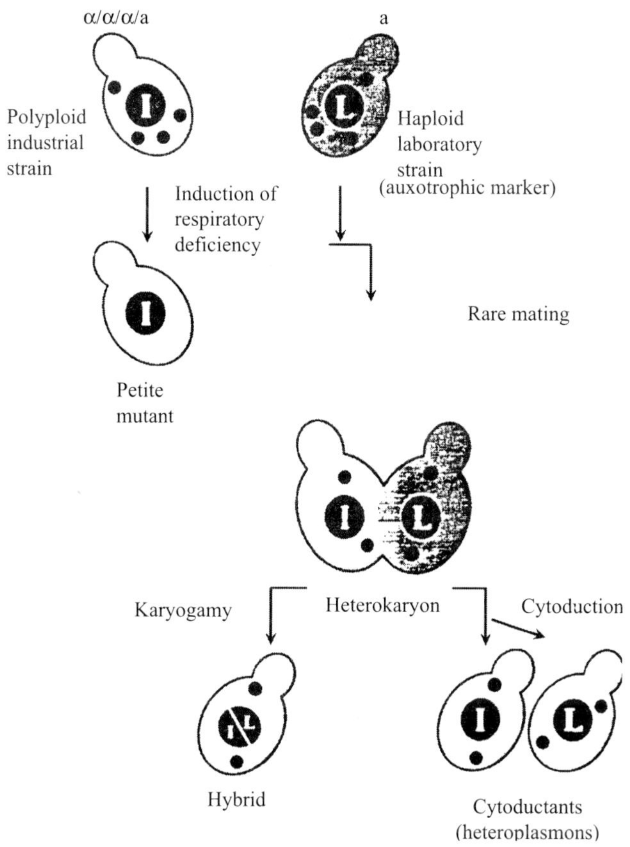


**Fig. 13:** Rare-mating between industrial and laboratory strains of yeast.
*Source:* 6

Industrial strains that fail to show a mating type are force-mated with haploid strains, exhibiting a or α mating type. A large number of cells of the parental strains are mixed and the rare hybrids or cytoductants are selected as respiratory-sufficient prototrophs from crosses between a respiratory-deficient mutant of the industrial strains and an auxotrophic haploid laboratory strain. Rare mating is based on the fact that in a population of diploid or polyploid cells, some individuals become homozygous for the mating type and can therefore, be crossed with a haploid cell of the opposite type. This cross is a rare event that can only be detected with a highly effective screen. The procedure has great advantage of allowing hybridization between the strains of different ploidy and avoids the need for sporulation, which often does not work with commercial strains and runs the risk of genome alteration. Respiratory-deficient auxotroph rare mating [41] involves the crossing of commercial strains lacking the ability to grow on glycerol or ethanol (because of a defective form or lack of mitochondrial DNA), with auxotrophic strains having normal respiratory characteristics. The hybridization product, although rare, is easily detected by its ability to grow on medium lacking amino acids but containing glycerol or ethanol as the sole carbon source. The hybrid so obtained can then be encouraged to sporulate and the segregants tested for their wine making performance. Tubb *et al.* [41] used this method to construct dextrin fermenting derivatives of *Saccharomyces uvarum/ Saccharomyces diastaticus*. The genes responsible for dextrin utilization and phenolic off-flavours from *Saccharomyces diastaticus* were found to segregate independently in a Mendelian manner and so were separated during meiosis. A detailed description of rare mating is explained during the transfer of the killer property to a brewing strain [42].

## 5.11 Cytoduction

Rare-mating is also used to introduce cytoplasmic genetic elements into yeasts without the transfer of nuclear genes from the yeast parent. This method of strain development is termed as cytoduction. Cytoductants receive cytoplasmic contributions from both the parents, but retain the nuclear integrity of only one [31]. Cytoduction requires that a haploid mating strain carry the *kar1* mutation that impedes karyogamy (nuclear fusion) after mating. This more specific form of strain construction can for example, be used to introduce the dsRNA determinants for the $K_2$ zymocin and associated immunity into a particular wine yeast.

Cytoduction can also be used to substitute the mitochondrial genome of yeast or introduce a plasmid encoding desirable genetic characteristics into specific yeast strains. Mating between strains, one of which carries the *kar1* allele, occasionally generates progeny that contain the nuclear genotype of one parent together with an additional chromosome from the other parent [32]. The donation of a single chromosome from an industrial strain to a haploid *kar1* recipient is termed as 'single-chromosome transfer', and is used to examine individual chromosomes of industrial yeast strains in detail [31]. Conde *et al.* [43] isolated Kar mutants in which nuclear fusion does not occur after cells fusion. These mutants can be used in the following procedures: (1) transfer of cytoplasmic material without the transfer of nuclear material; and (2) simultaneous transfer of one or two chromosomes at a frequency depending on the size of the chromosome involved [44]. The latter characteristic can be used for the partial and selective transfer of nuclear material. Loiseau [45] used the first procedure for marking purpose, introducing mitochondria from a haploid Kar laboratory strain, with mutations conferring resistance to specific inhibitors into a commercial strain. However, the results in technological validation trials were less favourable than with strains obtained by direct mutagenesis with $MnCl_2$ or by selection of spontaneous mutants. Vezinhet *et al.* [46] have used the second procedure to introduce a single chromosome 1, containing gene *FLO5,* which confers the flocculent characteristic into a strain used in producing sparkling wine by method champagnoise.

## 5.12 Protoplast Fusion

Protoplast fusion is a direct, asexual technique employed in crossbreeding as a supplement to mating. Like rare-mating, protoplast fusion can be used to produce either hybrids or cytoductants (Fig. 14). Both these procedures overcome the requirement for opposite mating types to be crossed, thereby extending the number of crosses that can be done [31]. Cell walls of yeasts can be removed by lytic enzymes in the presence of an osmotic stabilizer to prevent osmolysis of the resulting protoplasts. Spheroplasts from the different parental strains are mixed together in the presence of a fusing agent, polyethylene glycol and calcium ions, and then allowed to regenerate their cell walls in an osmotically-stabilized selective-agar medium.

Protoplast fusion of non-sporulating industrial yeast strains removes the natural barriers to hybridization. The desirable (and undesirable) characteristics of both the parental strains will recombine in the offspring. Cells of different levels of polyploidy can be fused. For instance, a diploid wine yeast strain can be fused to a haploid strain to generate triploid strains. Alternatively, two diploid wine yeasts with complementing desirable characteristics can be fused to generate a tetraploid strain with all the genetic backgrounds of the two parental wine yeasts [32].

Protoplast fusion can be used to obtain interspecific and even intergeneric hybrids via a mechanism that has not been completely elucidated but is similar to somatic cell fusion. The final result is

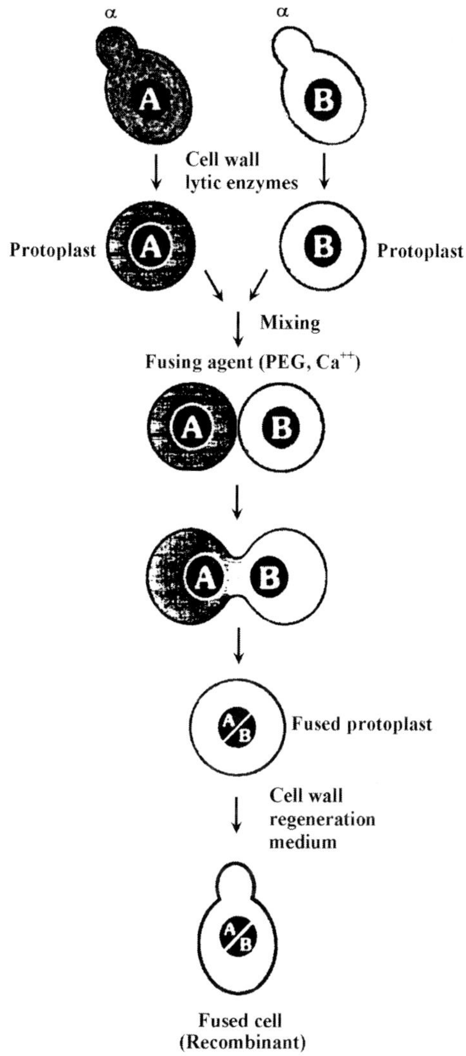


**Fig. 14:** Protoplast fusion between two different yeast cells is a direct asexual technique to produce either hybrids or cytoductants.
*Source:* 6

often a large predominance of the genome of one of the strains plus several genes of the other strain. However, considerable difficulties have been encountered in developing a coherent screen and stabilizing the fusion products. The incomplete enzymatic removal of yeast cell walls and the polyethylene glycol-induced fusion of the resulting protoplast have been used to produce hybrids between what was previously non-mating commercial yeast strains. This technique allows the mating between the strains of the same species or across the species that normally do not mate. In general, protoplast fusion serves only to remove any natural barriers to hybridization, the products are still likely to combine both desirable and undesirable characteristics of the parents. Therefore, the parents must have some unique easily traceable traits.

## 5.13 Transformation

Transformation involves the genetic exchange by the uptake of naked DNA. In other words, the microbial cell either naturally or under artificial conditions moves a foreign DNA molecule across its membrane(s) and then introduces it into its chromosome by recombination. Inherently, the naked DNA is highly vulnerable while floating free in the natural or artificial environment. The export of nucleases into the environment and the presence of restriction endonucleases in the bacterial cytoplasm restrict transformation. Not withstanding this, transformation is probably the most important method of horizontal genetic exchange between highly divergent microorganisms.

Selection of variants, mutagenesis, conjugation, transduction, hybridization, rare-mating and protoplast fusion have value in strain development programmes, but these methods lack the specificity required to modify the organisms in a well-controlled fashion [32]. It may not be possible to define precisely the change required using these genetic techniques, and a new strain may bring an improvement in some aspects, while compromising other desired characteristics. Transformation offers the possibility of altering the characteristics of the microorganisms with surgical precision: the modification of an existing property, the introduction of a new characteristic without adversely affecting other desirable properties, or the elimination of an unwanted trait. By using such procedures, it is possible to transform microbial cells with DNA fragments or purified genes from the same or other organisms. The transformation requires (i) cell permeability to donor DNA, (ii) an appropriate DNA vector, (iii) an easily selectable gene marker, and (iv) a suitable recipient strain.

The exogenous DNA normally does not enter the microbial cells due to the rigid cell wall, which is generally partially removed by lytic enzymes which are commercially available as mixtures of α, β-glucanases (Helicase, Zymolyase, Glusulase) for yeast cells and lysozyme for bacterial cells. The resulting spheroplasts are suspended in osmotically stabilized hypertonic solution along with donor DNA in the presence of fusing agents, polyethylene glycol (PEG) and calcium chloride. The spheroplasts are then embedded in 3% agar to allow the regeneration of cell wall and placed on selective medium that supports the growth of transformed cells (Fig. 15).

## 5.14 Gene Cloning and Transformation (Recombinant DNA Technology)

Recombinant DNA consists of a DNA sequence that has been artificially produced by joining pieces of DNA from different organisms. This technique can also be used to develop "useful" microorganisms that have the ability to degrade toxic wastes. New strains of agriculturally essential plants such as wheat, corn and rice represent by-products of rDNA technology. Recombinant DNA technology can be used in a wide range of industrial enterprises to develop: (i) microorganisms that produce new products, (ii) more efficient methods of producing currently available products, (iii) large quantities of products that are scarce in the natural environment.

Recombinant DNA technology comprises a mixture of techniques, some new and some borrowed from other fields, such as microbial genetics. These techniques consist of: (i) specific cleavage of DNA by restriction nuclease, (ii) DNA cloning, whereby a specific DNA fragment is integrated into rapidly replicating vectors, so that it can be amplified in bacteria or yeast cells (Fig. 16), (iii) nucleic acid hybridization, which makes it possible to identify specific sequences of DNA or RNA with great accuracy and sensitivity by their ability to bind a complementary nucleic acid sequence, and (iv) DNA sequencing of the nucleotides in a cloned DNA fragment.

## 5.15 Specific Enzymes Used in Recombinant DNA Technology

The specific cleavage of DNA is carried out by the use of naturally occurring enzymes, most of which are derived from bacteria or viruses (Fig. 17). These include: (i) restriction endonucleases produced

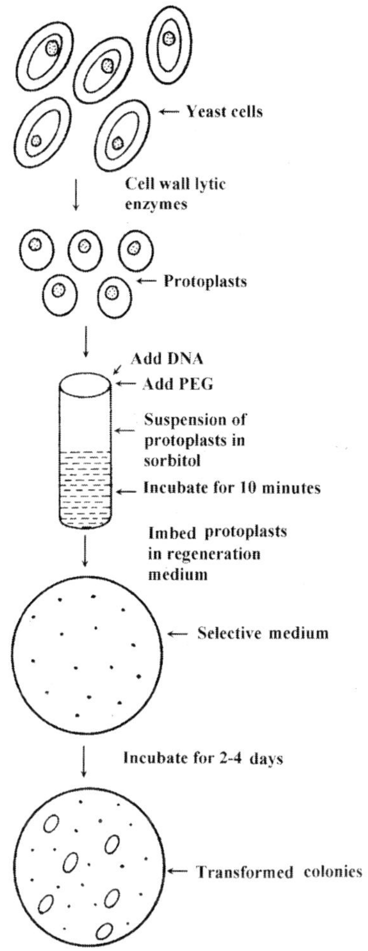


**Fig. 15:** The method of yeast transformation.
*Source:* 34

by bacteria in natural environments where they are continually exposed to a dilute 'soup' of foreign DNA released from other organisms, which have died or lysed as well as the bacteriophage DNA. They protect themselves from becoming infected or killed by producing many different restriction-modification systems, called restriction enzymes which are named after the organisms from which they were originally isolated e.g., EcoRI from *Escherichia coli* and BamHI from *Bacillus amyloliquefaciens*. These systems operate by enzymes which recognize specific short regions of DNA sequence, which are usually palindromic e.g., 5' GGATCC 3', 3' CCTAGG 5'. Molecular biologists use highly purified preparations of restriction endonucleases to cut DNA into specific fragments. (ii) DNA ligase which provides the ability to rejoin cut fragments of DNA and form artificial recombinant molecules. The enzyme utilizes ATP to create new phosphodiester bonds, covalently linking together the ends of DNA strands. The ligase encoded by bacteriophage T4 is particularly valuable since it has very good properties *in vitro,* such as the ability to join together 'blunt-ended' DNA fragments, which allow the joining of otherwise non-compatible fragments e.g., different restriction enzymes, (iii) Other modifying enzymes (phosphatases, kinases, single-strand specific nucleases, etc.) that allow precise modifications to pieces of DNA to be made *in vitro* in order to add, remove or alter the structure of DNA, (iv) RNA

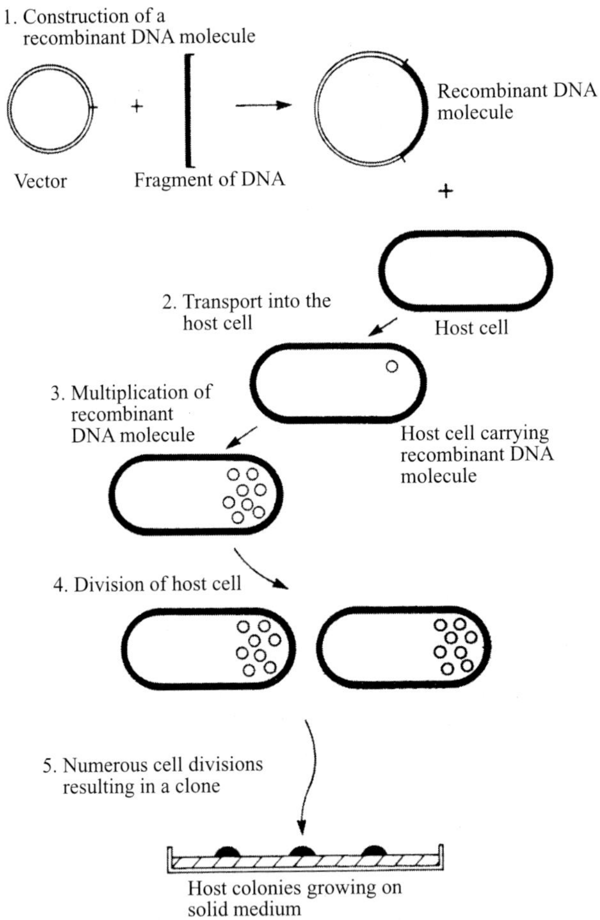


**Fig. 16:** The basic steps in gene cloning.
*Source:* 35

modifying enzymes e.g., exonucleases, RNA ligase and reverse transcriptase. RNA is much more difficult to work with *in vitro* because the enzymes available are generally not as sophisticated as the set available which modify DNA, (v) DNA polymerases that allow the synthesis of DNA from a pre-existing template *in vitro*, which can then be used for a variety of purposes, including sequence analysis by chain-termination methods.

Thermostable polymerases like Taq DNA polymerase from *Thermus aquaticus* have become important in rDNA technology. This bacterium has evolved to grow in hot springs at temperatures which kill most other species. These enzymes allow the amplification of as little as one molecule of DNA into a large amount by means of repeated cycles of melting, primer annealing and extension by the enzyme, which is not destroyed by the high temperatures used in this process. This is known as the polymerase chain reaction (Fig. 18).

## 5.16 Vectors Used in DNA Cloning

As DNA fragments obtained after the action of restriction endonucleases do not generally replicate on their own when introduced into a bacterium, they need to be joined to a DNA molecule that can

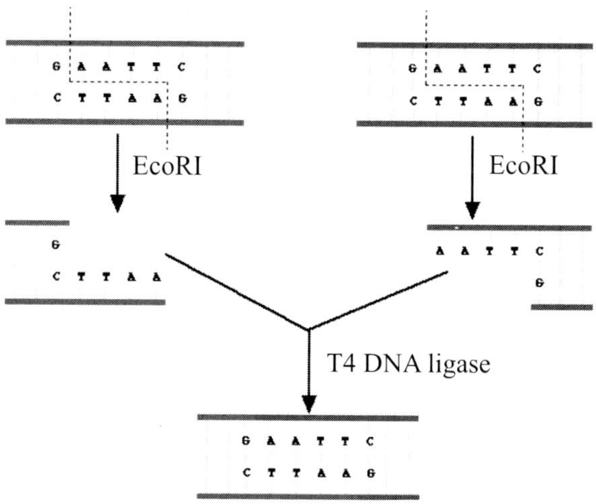


**Fig. 17:** Restriction digestion of DNA by restriction endonucleases and rejoining of cut fragments of DNA by DNA ligase.
*Source:* 35

replicate—a vector. Vectors need to be small enough to be manipulated conveniently and also to allow the joined DNA to be analysed easily. Vectors are derived either from plasmids or bacterial viruses—bacteriophage [35, 47].

### *5.16.1 Plasmid vectors*

A plasmid is a DNA element that can replicate autonomously within a bacterium. Plasmids can be linear molecules, but the ones which are most commonly exploited as vectors for genetic manipulation are circular, which often carry antibiotic resistance genes. Apart from anything else, circularity means that if the DNA molecule is cut once, all the plasmid functions remain together on the linear fragment formed so that only one vector segment needs to be joined to the foreign DNA segment to allow it to replicate (Fig. 19). Many naturally occurring plasmids are either "self-transmissible", being able to spread to other bacteria via a plasmid-encoded mating or conjugation bridge, or "mobilisable", being able to spread by conjugation so long as there is an appropriate self-transmissible plasmid present. With the concern about escape of the products of genetic engineering, it is generally desirable that genes promoting such transfer are deleted or inactivated before plasmids are exploited as vectors. However in some applications, mobilisability is important to allow the vector to be transferred back to the host to be manipulated. Plasmid vectors have been manipulated to make it easier to detect or even select for the acquisition of an inserted fragment of DNA. The most common method is to use a gene which confers on bacteria a property that can be detected by a colour reaction when that bacteria are placed on a specific medium or are sprayed with a particular chemical. The best example is the *lacZ* gene whose product β-galactosidase will hydrolyse Xgal (the chromogenic substrate) to give a blue compound that is detected easily. However, *lacZ* is a large gene (about 3 kb) and therefore not the best choice to insert into a small vector. Fortunately β-galactosidase can be split into two domains—the very small a-fragment and the much larger b-fragment, neither of which exhibit enzyme activity on their own but associate *in vivo* to generate a functional enzyme. With a host strain encoding the b-fragment and the plasmid encoding the a-fragment, one can develop plasmid derivatives with cloning sites within the region coding for the a-fragment so that insertions will disrupt the orf and prevent functional a-fragment production and so give a plasmid-host combination which is colourless [47]. Specific cloning vectors e.g., yeast episomal plasmid (YEp), yeast

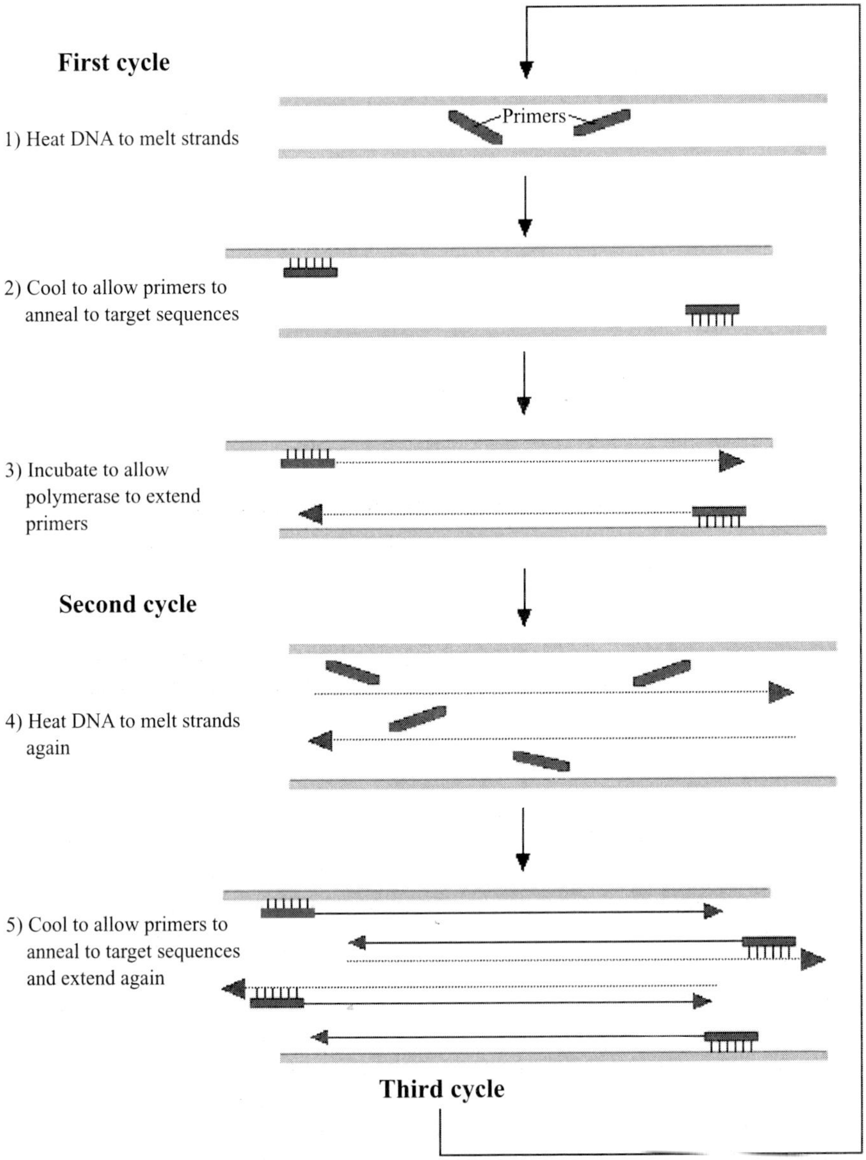


**Fig. 18:** A schematic representation of polymerase chain reaction.
*Source:* 35

replicating plasmid (YRp), yeast centromere plasmid (YCp), yeast integrating plasmid (YIp), yeast artificial chromosomes, plasmids based on killer factor determinants, etc. are also used for manipulation involving yeasts and filamentous fungi.

### *5.16.2 Phage vectors*

Bacterial viruses called bacteriophages (eaters of bacteria) are generally classified into virulent and temperate phages. Virulent phages have a single "lytic" life cycle, which involves infection,

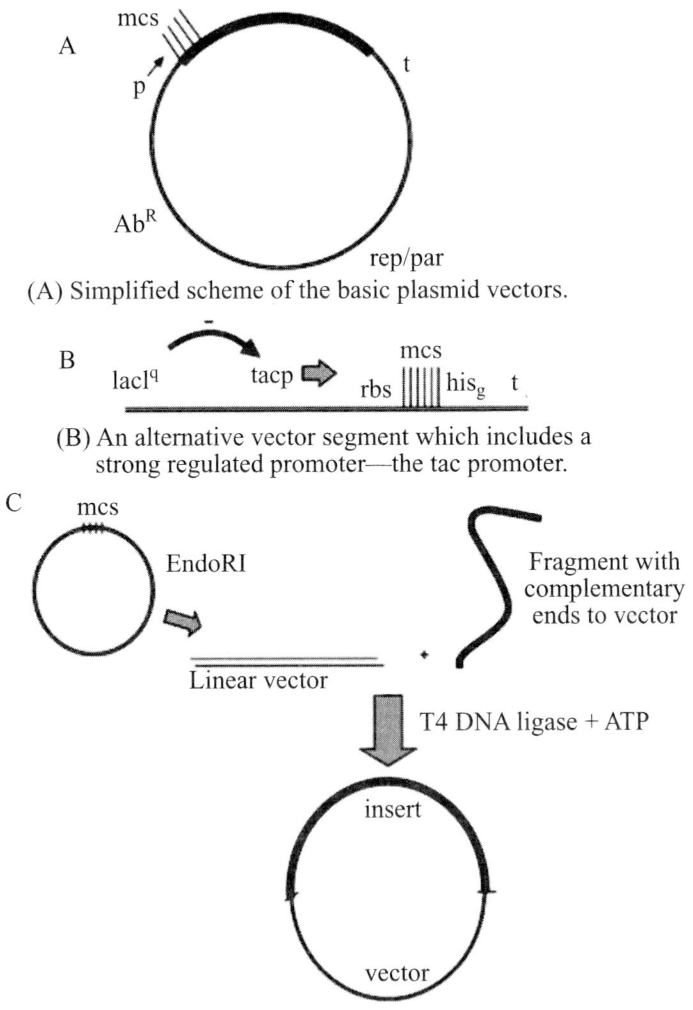


(A) Simplified scheme of the basic plasmid vectors.

(B) An alternative vector segment which includes a strong regulated promoter—the tac promoter.

(C) Simple scheme showing the ligation of a foreign fragment into the vector.

**Fig. 19:** A simplified scheme of the basic plasmid vectors.
*Source:* 47

multiplication and host death (lysis) with the release of new viruses. In general, such phages have not been used to develop vectors. However, the M13 phage that forms the basis of vectors for generating single-stranded DNA for DNA sequencing uses a slight variant of this cycle in which the infected bacteria become factories for producing phage particles, which bud out from the bacterial surface in a continuous stream so that the host is debilitated and grows slowly but never dies. Temperate phage can go through the lytic cycle but also have an alternative, the "lysogenic" cycle, in which a phage-encoded repressor system is switched on shortly after infection and shuts down the lytic cycle so that the phage enters a quiescent state called a prophage that replicates in step with its host. For bacteriophage lambda, the prophage integrates into the chromosome, whereas for P1 the prophage multiplies like a plasmid. The prophage can be activated to re-enter the lytic cycle by inactivation of the phage repressor. The natural cause of this activation is normally stress for example the lambda repressor is degraded by the RecA co-protease activity which is activated by DNA damage. Bacteriophage lambda is the most developed phage cloning system. Lambda DNA in phage particles exists as a linear molecule with 12 base cohesive ends which are generated by the cleavage process at the *cos* site that occurs simultaneously with the packaging process. This molecule circularises on injection into a bacterium and this is necessary to generate the series of genes that are needed for the production of phage head and tail that assemble to form the phage particle [35, 47]. The essential components are a selectable marker ($Ab^R$ - conferring resistance to antibiotics such as ampicillin, tetracycline, kanamycin or chloramphenicol); a replication and stability system providing a high or a low copy number; one or more sites that can be cut by restriction endonucleases to allow insertion of foreign DNA fragments—a cluster of such sites is called a multiple cloning site—mcs; and a gene running across the mcs whose disruption can be detected easily. Inclusion of a strong transcriptional terminator (t) downstream of the mcs can prevent strong transcription from interfering with plasmid replication.

In fig.19B, an alternative vector segment is shown which includes a strong regulated promoter — the tac promoter — that can be used to direct controlled expression of the coding region for any gene inserted into the mcs. If the gene lacks its own translation signals then the presence of a ribosome binding site (rbs) can direct translation on the mRNA produced so long as the distance between the rbs and the start codon for the coding region is approximately 7-8 nucleotides in length. The *$lacI^q$* gene encodes a repressor protein that binds to the tac promoter and prevents transcription except when an "inducer" such as isopropyl-β-D-thiogalactoside (IPTG), which binds to the repressor and alters its conformation so that it no longer binds to the DNA, is present.

C. Simple scheme showing the ligation of a foreign fragment into the vector by cutting with a restriction enzyme and using T4 DNA ligase to join the fragments together before introducing them into bacteria by transformation or electroporation.

## 5.17 *In Vitro* Recombination

The basic techniques of *in vitro* genetic manipulation were developed in bacteria: isolated plasmid or phage DNA, cut with a restriction endonuclease is joined to DNA fragments from any source so long as the fragment ends are identical to those of the vector—either flush ends or over-hanging single-stranded ends. Restriction endonucleases are derived from bacteria in which they appear to act as a system to cut foreign DNA into pieces after entry and thus protect the bacteria from completely alien DNA. They may also serve to prepare related but not identical DNA for homologous recombination. Self DNA is protected from cutting by methylation of the DNA at the sites which would otherwise be recognised by the endonuclease. Restriction systems must be considered during genetic manipulation because they can be a reason why the transfer of DNA from one bacterial strain to another is not successful: if unmodified vector or "insert" DNA is used with a strain carrying a restriction system, then the efficiency can be significantly reduced. One strategy that has been used to reduce the efficiency of the restriction barrier to DNA entry is to grow the bacteria at high temperature for a number of generations. The original source of DNA ligase is the phage T4 where it plays a key role in phage DNA replication. Normally, the enzyme is derived from bacteria that have been manipulated to over-produce this enzyme [35, 47].

Restriction enzymes can be used to generate fragments either by complete or partial digestion of the DNA. Complete digestion is the simpler strategy in which DNA is digested with excess enzyme (1 unit = enough to cut all sites in 1 mg DNA at 37°C) and individual fragments which can be as small as 100 bp (enzymes with a 4 bp recognition sequence) to >10 kb (enzymes with a 6 bp recognition sequence) are joined to vector. Partial digestion involves limited amount of enzyme for a limited amount of time and gives random segments running from one restriction site to another but generating fragments consisting of two or more "fragments" still joined together. A given gene may thus end up on more than one type of fragment. The advantages of this are that if the genetic region of interest has a target site for the restriction endonuclease to be used within it, then the only way that the region can be obtained intact is by partial digestion. The other advantage is that a library of random fragments produced by partial digestion will contain fragments that overlap with each other. Therefore, a long DNA segment can be built up as a series of different cloned segments and identified as such by hybridization of the fragments with each other, even though they are not identical fragments. Subsequent mapping of restriction sites along the DNA of each related clone can identify those segments that are identical segments in different clones.

### 5.18 Introduction of DNA into Bacteria

Transformation, mixing naked DNA with bacteria that are "competent" to take up DNA, is the commonest and most straightforward way of introducing DNA into bacteria. Some bacteria become competent naturally towards the end of exponential phase or upon starvation (for example, *Bacillus subtilis, Streptococcus pneumoniae* and *Haemophilus influenzae*), but the active uptake process often converts the DNA during uptake into single-stranded fragments that do not regenerate the desired combination of vector and cloned fragment. For uptaking the products of *in vitro* recombination, bacteria are normally treated with a mixture of divalent cations, the most basic requirement being calcium ($Ca^{2+}$), that allow DNA to bind to the negatively charged outer surface of the bacteria in such a way that during a short period at 42°C after holding on ice, the DNA is taken up into the cytoplasm. When naked phage DNA is taken up, this is called transfection. Not all bacteria can be easily made competent for transformation, presumably due to the structure of the outer surfaces of the bacterial cell. Electroporation has provided an alternative, more general method in which bacteria and naked DNA are placed in a small chamber between two electrodes and then subject to a short but intense electric surge which temporarily creates pores through which DNA can enter the cell. A variety of application notes and publications are available providing details for the optimisation of this process for most bacterial genera currently studied. A third way of introducing DNA is by injection after *in vitro* packaging into phage particles, but this is more specialised and only applies to vectors that carry the DNA sequence signals that allow the DNA to be recognised by the proteins that initiate the process. In the case of bacteriophage lambda, this is the cos site recognised by gene product A, an endonuclease that introduces nicks 12 bp apart to create an end that start the wrapping of the DNA into the protein head of the phage. Extracts from different bacterial strains carrying mutant phage that have accumulated different components of the packaging process, can be mixed *in vitro* so as to provide all that is necessary, and ligated DNA added to this mix will be packaged. Production of stable phage particles depends on the DNA being approximately the same size as the normal phage genome, so it provides a selection for clones that have substantial inserts [35, 47].

## 6. CURRENT STATUS AND FUTURE PROSPECTS OF GENETICALLY MODIFIED MICROBES

Theoretically, almost any gene can be transferred from one organism to another and carry out its function, provided that the proper signals are present to enable the host recognize the gene. The strain improvement involves the re-routing of metabolic pathways, and engineering of the flux of a certain route in industrial microorganisms. Some examples of genetic manipulation in microorganisms for developing novel strains are discussed below:

## 6.1 Re-routing of Carbon Metabolism in *Lactobacillus lactis*

Lactic acid bacteria (LAB) are used worldwide in the industrial manufacture of fermented food products. Their most important application in this respect is undoubtedly in the dairy industry, where these microorganisms are used to convert milk or milk-derived products to an enormous variety of fermented dairy products. However, other unprocessed food materials, also like meat and vegetables, are subjected to fermentation by LAB on an industrial scale. In milk, these bacteria encounter lactose as the major carbon source, but they have the capacity to use a number of other mono- and disaccharide substrates. The major product of these fermentations is lactic acid, which plays a crucial role in protection of the final fermented product against spoilage. Besides this acidification that acts as a natural preservative effect, LAB metabolism is essential for development of desired product properties like flavour, shelf-life and texture. The metabolic conversions that generate these end-products vary widely, depending on the lactic acid bacterium. Genetic engineering strategies that involve inactivation of undesired genes and/or over-expression of existing or novel ones have been used to create re-routing of the metabolic fluxes by changing the energy metabolism or the concentrations of metabolic intermediates or of existing or completely new end-products. To generate a rational approach for these strategies, it is essential to understand the pathways that are manipulated, their fluxes, control factors and the genes involved. Knowledge of the traditional pathways recognized in LAB and the characterization of new ones, in combination with physiological studies, has led to an extensive insight in LAB metabolism. Proper appreciation of the application potential of LAB as cell factories is related to several features that are shared by these microorganisms. In general, they have a long history of safe use and grow rapidly on media that are derived from milk and contain lactose as the major carbon source, which is converted to a number of interesting metabolic end-products during fermentation. Moreover, LAB are applied worldwide as starter cultures in industrial food fermentations, implying that LAB with modified metabolic pathways cannot only be used in bioreactors but also for *in situ* production of flavour, texture, or health promoting compounds in food products or even in the gastrointestinal tract of consumers of food. Several other features of LAB, like their relatively simple metabolism, limited biosynthetic capacity, and apparent lack of gene multiplicity, illustrate their advantages for genetic engineering purposes compared to other microorganisms like yeast or fungi. Among lactobacilli, the best studied is *Lactobacillus lactis*, and in the following text we will highlight some important examples of efficient re-routing of carbon metabolism [35, 49].

### 6.1.1 *Sugar transport and glycolysis in L. lactis*

*L. lactis* can use a variety of mono- and disaccharides as carbon source, including glucose, fructose, galactose, lactose, maltose, mannose and cellobiose. Three different systems for sugar import are known in *L. lactis*. Primary transport systems display a direct coupling of sugar import to ATP hydrolysis via a transport specific ATPase [48]. In secondary sugar transport systems, the import of sugar is coupled to transport of ions or other solutes [49]. Finally, sugars that are imported via the phosphoenolpyruvate (PEP)-dependent phosphotransferase system (PTS) are phosphorylated during transport. This system transfers the phosphate group from PEP, resulting in the formation of pyruvate, to the incoming sugar via two cytoplasmic phosphocarrier proteins: enzyme I and HPr [50]. In lactic acid bacteria, many sugars are imported via this PTS system, which is an energetically efficient system for sugar transport because sugar import is directly coupled to sugar phosphorylation at the expense of one ATP that would otherwise be generated by PEP conversion to pyruvate by pyruvate kinase (pyk). The ptsHI operon of *L. lactis*, encoding the general PTS components HPr and enzyme I, respectively, has been cloned and characterized. HPr plays an important role in regulation of sugar utilization, both at the level of transcription of genes involved in use of a specific sugar, and at the level of sugar transport.

This phosphocarrier protein contains two residues that are target for phosphorylation. A histidine residue in the N-terminal part of the protein, that is a target for enzyme I mediated phosphorylation and the resulting HPr-(His-P) primarily plays a role in PTS sugar transport, but can also influence the use of non-PTS sugars [51, 52]. The second HPr-residue that is a target of phosphorylation is a conserved serine residue in the C-terminal part of HPr. The HPr-(Ser-P), like HPr-(His-P), plays a complex role in sugar use, where it can: (i) allosterically control PTS and non-PTS permease activity, thereby regulating sugar import, (ii) control the activity of a sugar-phosphate phosphatase, thereby playing a role in the phenomenon called inducer exclusion, and (iii) interact with the catabolite control protein CcpA that negatively controls the expression of a large number of genes or operons that are subject to catabolite repression [51]. Moreover, it has recently been shown that the lactococcal CcpA, besides its role in catabolite repression, also regulates sugar metabolism by affecting the glycolytic flux through activation of the *lac* operon [53].

In dairy fermentations, lactose is the main carbon source encountered by *L. lactis* and the pathway used to convert this sugar is relatively well understood. In many lactococcal strains, the genes involved in lactose use (*lac* operon) are plasmid encoded, and the best described is the lactose mini-plasmid pMG820 [54, 55]. This plasmid contains genes involved in lactose-PTS transport (*lacEF*), lactose-6P hydrolysis (*lacG*), and the tagatose-6P pathway (*lacABCD*). Expression of the *lac*-operon genes is controlled by the transcriptional repressor encoded by *lacR* and repression of the *lac* gene expression can be relieved by the inducer tagatose-6P, which is an intermediate of the tagatose-6P pathway [57]. An elegant example of engineering of lactose utilization in *L. lactis* was described in 1985 by Thompson *et al.* By classical mutagenesis, a *L. lactis* double mutant was constructed that lacked both the glucokinase (*glk*) activity and the mannose PTS system involved in glucose import. As a result, when this strain was grown on lactose, the galactose-6P moiety of this disaccharide can be converted via the tagatose-6P pathway, whereas the glucose moiety remains unphosphorylated and is secreted. The lack of a mannose PTS in these cells prevents the subsequent uptake of the expelled glucose. The overall result is an apparent quantitative conversion of lactose to glucose by these cells, which shows that galactose and glucose metabolism in lactose grown *L. lactis* can be completely separated [57].

Many of the genes encoding glycolytic enzymes in *L. lactis* have been cloned and characterized. Moreover, a predictive metabolic flux model of lactococcal glycolysis was recently developed on basis of *in vivo* NMR measurements, allowing qualitative prediction of metabolic shifts upon changes in the environmental conditions [59]. Interestingly, the transcription of the *las* operon [59], encoding three key-enzymes involved in glycolysis and production of lactate, phosphofructokinase (*pfk*), pyruvate-kinase (*pyk*) and lactate dehydrogenase (*ldh*), has recently been shown to be activated by CcpA [53]. By inactivation of the lactococcal *ccpA* gene, strains were generated that displayed a strongly reduced expression level of the *lac* operon genes, which resulted in a shift from homo-lactate to mixed acid fermentation [53]. This study is an elegant example of the possibility of metabolic engineering by manipulation at the level of global regulation, and the availability of the key regulatory components that exert this control in *L. lactis* (HPr, CcpA and Enzyme I) [52, 53]. This will probably allow further fine-tuning of metabolic engineering strategies in the future [35].

### 6.1.2 *Engineering of the pyruvate metabolism*

Pyruvate is the key metabolic intermediate; it is the end product of the substrate level phosphorylation reactions of the glycolytic pathway and is also produced through the phosphorylation cascade involved in PTS-mediated sugar import. Pyruvate can be converted to various end-products by *L. lactis*, including acetate, ethanol, diacetyl, acetoin and 2,3-butanediol. However, the main endproduct is evidently lactate, which is formed by the lactate dehydrogenase reaction. This reaction acts as main electron sink under anaerobic conditions, leading to homolactic fermentation. However, a shift towards mixed acid fermentation has been observed under certain fermentation conditions, such as carbohydrate limitation [60], diminished rates of sugar metabolism [61] and aerobic conditions [62]. Several genetic engineering strategies have been used to create mutant lactococcal strains that display a significant shift in their metabolism and that produce either one of the endogenous end products more efficiently, or produce an end product that is normally not produced by *L. lactis*. Here we will describe examples of efficient re-routing of pyruvate metabolism toward the end products, diacetyl and alanine [35].

**6.1.2.1 Engineering diacetyl production:** One of the end products that can be generated from pyruvate is diacetyl, which is an important flavour compound in dairy products generating the typical butter-aroma. This compound is naturally produced by *L. lactis*, especially by *L. lactis* subsp. *lactis* biovar. *diacetylactis*, from citrate in co-metabolic fermentation with lactose [63]. During this fermentation, the intermediate α-acetolactate is accumulated, which is chemically (non-enzymatically) converted to diacetyl through an oxidative decarboxylation reaction. Based on its aroma value, efficient diacetyl (or its precursor α-acetolactate) production from lactose, rather than from citrate by *L. lactis* has been one of the targets of various metabolic engineering strategies. An obvious target for engineering is the lactate dehydrogenase (LDH), and an LDH deficient strain was created by *ldh* gene disruption. The mutant displayed a mixed acid fermentation under anaerobic conditions, producing high amounts of formate and ethanol. In contrast upon aeration, a dramatic shift in pyruvate metabolism was observed, leading to increased acetate production and acetoin as the main end product [64]. An alternative approach is to increase the amount of enzyme involved in conversion of pyruvate to α-acetolactate. Two enzymes are known in *L. lactis* that catalyse this reaction: α-acetolactate synthase (ALS; encoded by the *als* gene) [65] and acetohydroxy acid synthase (ILVBN; encoded by the *ilvBN* gene) [66].

**6.1.2.2 Efficient diacetyl production by cofactor engineering:** The intracellular redox balance, most importantly reflected by the NADH/NAD 1 ratio, plays a predominant role in control of the fermentation pattern displayed by *L. lactis* [62, 67]. Pyruvate reduction by LDH effectively regenerates NAD 1 and thereby effectively transfers reduction equivalents formed during glycolysis to the metabolic end product. Interestingly, and illustrative for the energetically unfavorable pathways that will be used by these cells to maintain their redox balance, it has recently been shown that the rate of sugar fermentation by LDH-deficient *L. lactis* strains can be increased by co-metabolization of acetate. The acetate consumed under these conditions is exclusively converted into ethanol at the expense of NADH and ATP, thereby creating a redox rescue pathway for these cells that lack an effective reductive pathway [68].

**6.1.2.3 Homo-L-alanine fermentation by modified *L. lactis*:** Most metabolic engineering strategies that have been applied in *L. lactis* were aimed at increasing the production of one of the natural lactococcal metabolic end products. As an alternative, introduction of a heterologous pathway in *L. lactis* could lead to lactococcal variants that produce interesting new products from lactose (or another carbon source). An example of this type of approach is the overexpression of the *Zymomonas mobilis* genes encoding pyruvate decarboxylase (*pdc*) and alcohol dehydrogenase (*adh*). Expression of this ethanol forming pathway in LDH-deficient *L. lactis* cells led to a significant amount of ethanol production (2%). Extremely efficient re-routing of the pyruvate flux has been achieved by overexpression of L-alanine dehydrogenase (ALAD), encoded by the *alaD* gene derived from *Bacillus sphaericus* [69].

## 6.2 Complex Pathway Engineering

A major challenge in current and future engineering approaches is the modulation of complex pathways, leading towards, interesting end products. A good example of such a product is given by the polysaccharides that are naturally produced by some so-called ropy strains of LAB. Recent years have shown increasing interest in the production of exopolysaccharides (EPS) by LAB, caused by their relevance in dairy-product properties like texture and mouth-feel. Based on the food-grade status of LAB, these EPSs could readily be applied as food additives, where they can contribute to several product properties (i.e., viscosity, emulsion stability). Moreover, it has been suggested that they can be active as prebiotic [70], cholesterol lowering neutraceutical [43] or immunomodulant [71, 72].

# 7. GENETIC ENGINEERING FOR YEASTS

*Saccharomyces cerevisiae* is the most thoroughly investigated eukaryotic microorganism, which aids our understanding of the biology of the eukaryotic cell and hence ultimately, human biology. For several centuries, *S. cerevisiae* has been used in the production of food and alcoholic beverages, and today this organism is also used in a number of different processes within the pharmaceutical industry. *S. cerevisiae* is a very attractive organism to work with since it is nonpathogenic, and due to its long history of application in the production of consumable products such as ethanol and baker's yeast, it has been classified as a GRAS organism (generally regarded as safe). Also, the well-established fermentation and process technology for large-scale production with *S. cerevisiae,* make this organism attractive for several biotechnological purposes. Another important reason for the applicability of *S. cerevisiae* within the field of biotechnology is its susceptibility to genetic modifications by recombinant DNA technology, which has been even further facilitated by the availability of the complete genome sequence of *S. cerevisiae*, published in 1996 [73]. Yeasts have been used to produce food and beverages since the Neolithic age. Their use in fermentation was recognized in 1836-1838; and Louis Pasteur demonstrated their unequivocal role in the conversion of sugar to ethanol and carbon dioxide in 1861. The genetic improvement of industrial strains traditionally relied on classical genetic techniques (mutagenesis, hybridization, protoplast fusion, cytoduction, etc.), followed by selection for broad traits such as fermentation capacity, ethanol tolerance, absence of off-flavours (e.g., $H_2S$ for wine strains), fast dough fermentation, osmotolerance, rehydration tolerance, organic acid resistance (baker's strains), flocculation and carbohydrate utilization (brewer's strains). Despite considerable work, a major limitation of these classical genetic techniques was the difficulties of adding or removing features from a strain without altering its performance. One major advantage of gene technology over classical genetic techniques is that just one characteristic can be precisely modified, without affecting other desirable properties. In addition, molecular biology approaches have introduced a new dimension. The expression of heterologous genes has substantially increased the possibilities. In the past 20 years, impressive progress has been made in the development of molecular techniques for *Saccharomyces cerevisiae.* These advances have been successfully applied to industrial strains in the past decade, allowing the development of a new generation of specialized industrial yeast strains. The principal targets for strain development fall into two broad categories: (1) improvement of fermentation performance and simplification of the process and (2) improvement of product quality e.g., organoleptic and hygienic characteristics [74].

The use of pure culture strains in brewing had been widespread since the end of the nineteenth century. Top-fermenting (ale-brewing) yeasts form a diverse group of polyploid yeasts that are closely related to the laboratory strains of *S. cerevisiae* [75]. Bottom-fermenting (lager-brewing) yeasts were initially classified as *S. carlsbergensis*, later included in *S. cerevisiae* [76] and then renamed *S. pastorianus* [77]. These strains are allo-tetraploid and exhibit poor sporulation and low spore viability. Extensive studies, mainly by Carlsberg, indicate that this species contains part of two divergent genomes, probably derived from *S. cerevisiae* and *S. monacencis* [75). Conversely, other studies favour the formation of hybrids between *S. cerevisiae* and *S. bayanus* [78]. Despite this complex genetic constitution, brewers took an active interest in yeast genetics as early as the beginning of the 1980s, leading to the production of hybridized and genetically engineered yeasts with improved characteristics [75, 79, 80]. Compared to brewer's yeast, the study of wine yeast genetics is relatively recent. The majority of commercial wine yeasts are strains of *S. cerevisiae*, including those described by enologists as *S. bayanus*, which are in fact *S. cerevisiae* [81]. These strains are predominantly homothallic, diploid/aneuploid and display low sporulation ability. Compared to laboratory strains, they exhibit chromosomal-length polymorphisms and possess rearranged chromosomes with multiple translocations [82].

Classical genetic approaches were first applied to wine yeast strains in the middle of the 1980s, in response to increasing demand for new characteristics resulting from the development of pure culture strains. Strains with new properties e.g., with mitochondrial markers, flocculation properties or expressing the killer toxin, have been generated by mutagenesis, hybridization or cytoduction [30]. Only a few of these strains have been commercialized, including mitochondrial mutants, which have been very useful for implantation studies. This is mainly due to the lack of specificity of the methods, making it difficult to modify a characteristic precisely without altering other properties. In the past 10 years, the demand for more specialized wine yeasts has been growing and the number of commercialized, selected wine yeast strains has increased from about 20 to over 100. Substantial amounts of work took place during the 1990s to develop new strains, which mainly involved recombinant DNA approaches [6, 30, 32, 33, 83, 84]. Compared to the huge amount of work performed to engineer brewer's yeast and wine-making yeast, the improvement of baker's yeast has received less attention.

Attempts to construct baker's yeast with improved broad traits, such as fermentative performance, osmo- annd acid-tolerance and tolerance to rehydration [85], were performed in the 1980s, using breeding procedures. These approaches were successful. For example, they generated new strains with rapid fermentative properties. Other traits, especially tolerance to stresses (drying, freezing, osmotic stress), which involves complex multigenic responses, are more difficult to improve by classical breeding programs. Our limited knowledge of the genetic basis of these important commercial properties has also limited their manipulation by recombinant DNA techniques. Therefore, much of the effort in genetic engineering of baker's yeast has gone into improving sugar degradation and fermentation performance [86]. Many targets for yeast improvement are relevant to several of these fields and have been presented in the following section. Conversely, each of these fields has specific demands, which will be reviewed in separate sections.

## 7.1 Extension of Substrate Range: Starch, Lactose, Melibiose and Xylose Consumption

The extension of the substrate range of *S. cerevisiae* serves as an important option in utilization of new raw materials e.g., agricultural waste products, for large-scale production of several biotechnological products as well as in decreasing the biological content in the waste stream of an industrial process. Many examples of substrate range extension are described in the literature, and some of these fall into more than a single category of metabolic engineering as defined in the introduction, since redirection of the flux to obtain an improved yield of a given product e.g., ethanol, becomes important immediately after the substrate range has been extended to include the desired substrate. The utilization of starch, lactose, melibiose, and xylose is reviewed in the following sections.

### *7.1.1 Starch utilization*

Starch is the storage carbohydrate in plants, and it serves as an important energy and carbon source in biotechnological processes. Starch is made up of long chains of glucose units joined by α-1,4 linkages and joined at branch points by α-1,6 linkages. Many microorganisms, including *S. cerevisiae*, are not able to degrade starch since they do not produce starch-decomposing enzymes such as α-amylase (which cleaves α-1,4-glycosidic bonds), β-amylase (which cleaves maltose units from the non-reducing end of starch), pullulanase or isoamylase (debranching enzymes that hydrolyze α-1,6-glycosidic bonds), and glucoamylase (which hydrolyzes glucose units from the non-reducing end of starch). Hence, it is necessary either to add starch-decomposing enzymes to the starch before fermentation or to use a recombinant strain that produces starch-decomposing enzymes in order to utilize this carbon source. In the baking industry, it is advantageous to use a certain recombinant strain that does not require α-amylase-enriched flour. For production of low-calorie beer in the brewing industry, it is of interest to use a recombinant strain of *S. cerevisiae* that secretes a glucoamylase whereby the larger oligomers (dextrins), which are formed from the partial hydrolysis of barley starch, are decomposed. Expression of a glucoamylase gene from *Aspergillus awamori* and secretion of the heterologous protein from *S. cerevisiae* has been successfully demonstrated, but the transformed strains grow on dextrins at a lower rate than observed when glucoamylase is added externally to the medium [87]. Co-expression of the *STA2* gene of *Saccharomyces diastaticus* encoding a glucoamylase and an *AMY1* gene encoding an α-amylase of *Bacillus amyloliquefaciens* was demonstrated to synergistically enhance starch degradation [88]. When the *pulA* gene of *Klebsiella pneumoniae*, encoding a pullulanase, was introduced into this transformant strain containing the *STA2* gene and the *AMY1* gene and secreted using the yeast mating pheromone α-factor, an even higher starch hydrolysis rate was obtained compared with the parental strain and almost complete assimilation (99%) of starch was achieved [89]. This illustrates that it may be necessary to introduce several different starch-degrading enzymes in order to obtain an efficient starch utilization. Hollenberg and Strasser [90] also succeeded in constructing a *S. cerevisiae* strain that is able to hydrolyze starch by introduction of both the *GAM1* gene and the *AMY1* gene, which encode a glucoamylase and an α-amylase, respectively, both originating from the yeast *Schwanniomyces occidentalis*. The transformed strain of *S. cerevisiae* showed ethanol productivity equal to that observed when a conventional strain was used together with decomposing enzymes added prior to the fermentation.

### *7.1.2 Lactose utilization*

Lactose is abundant in milk and is a major constituent of cheese whey, which is a by-product of cheese production. Only half of the cheese whey production is used for different purposes, including production of biogas, ethanol and single-cell proteins, whereas the other half is discarded as effluent [91]. Thus, cheese whey causes a major environmental problem due to its high biological oxygen demand and the high chemical oxygen demand, primarily as a result of its high lactose content [92]. *S. cerevisiae* does not have a lactose permease system, and hence lactose cannot be transported across the cell membrane; it also does not encode a β-galactosidase that cleaves the β-1,4 bond of lactose into a glucose residue and a galactose residue. Since *S. cerevisiae* is a good choice for ethanol production, it may be advantageous to introduce heterologous genes that enable *S. cerevisiae* to utilize lactose. For this reason, the genes that code for lactose permease and β-galactosidase of *Kluyveromyces lactis* were introduced into *S. cerevisiae* [93] and protoplast fusions resulting in hybrids of *K. lactis* and *S. cerevisiae* were also employed [94]. Unfortunately, these attempts resulted in low specific growth rates on lactose, genetic instability, and low ethanol yields when fermented in an immobilized yeast membrane bioreactor [93]. A flocculating *S. cerevisiae* strain containing genes encoding the lactose permease and the β-galactosidase of *Kluyveromyces marxianus* have been constructed [95]. An improved flocculation capacity and strain stability, as well as complete lactose consumption resulting in an ethanol yield of 80% of the theoretical value were obtained when this strain was fermented in a continuous bioreactor setup [96]. Hence, the possibility is presented of ethanol production from whey, which also contributes to solving the environmental problem referred to above. Another strategy to metabolically engineer *S. cerevisiae* in order to enable it to utilize whey has been successfully demonstrated by introduction of an extracellular β-galactosidase [97]. A thermostable β-galactosidase gene of *Aspergillus niger* (*lacA*) was expressed from an *ADH1* promoter in *S. cerevisiae* and 40% of the recombinant protein was secreted to the extracellular medium, whereby the cells grew with a specific growth rate of 0.43 $h^{-1}$ (equal to a doubling time of 1.6 h) on whey permeate containing 4% (w/v) lactose.

### *7.1.3 Melibiose utilization*

Molasses, which is used for ethanol and baker's yeast production, contains mainly glucose, fructose and sucrose, but the trisaccharide raffinose is also present to some extent. Invertase, encoded by the *SUC2* gene, decomposes raffinose into fructose and the disaccharide melibiose, which also contains a glucose residue and galactose residue similar to lactose, but the glucose and galactose moieties of melibiose are linked by an α-1,6 bond. Only a few strains of *S. cerevisiae* are able to assimilate melibiose, and many strains are not able to cleave the α-1,6 linkage of melibiose since they are missing one of the *MEL1* to *MEL11* genes, all encoding melibiase (α-galactosidase), which is secreted to the medium outside the cell [98]. Baker's yeast strains able to utilize melibiose by expressing the *MEL1* gene have been constructed by genetic engineering and classical breeding [99]. Thus, industrial use of these strains makes it possible to reduce the biological oxygen demand of an industrial waste stream and improve the overall yield of either biomass or ethanol produced from molasses. Since both glucose and galactose are formed from hydrolysis of melibiose, melibiase-producing strains partly relieved of glucose control over galactose metabolism have been constructed [99].

### *7.1.4 Xylose utilization*

D-Xylose is the second most abundant sugar in nature (after D-glucose) and is present in plant material as part of the heterogeneous polymer hemicellulose, which is one of the main constituents of the low-cost carbohydrate lignocellulose. Lignocellulose is renewable, and due to its abundance in nature in some areas of the world, this carbohydrate is potentially attractive for production of value-added products. Production of ethanol to be used as a liquid fuel is a well-described example of this [100]. This is an attractive application from an environmental point of view, since no net carbon dioxide is released from combustion of this carbon source [101]. To design an economically viable process, the ethanol yield should be close to the maximum theoretical yield (0.51 g of ethanol per g of xylose or glucose). To maximize the ethanol yield, the production organism should be capable of utilizing all pentoses and hexoses present in the lignocellulose, and furthermore, it is important that the production organism has a high inhibitor tolerance, since the lignocellulose hydrolysates contain inhibitory substances. Dependent on the nature of the inhibitors, the toxicity may be removed by the use of a variety of detoxification methods, but doing so will greatly increase the cost of operation [102]. Furthermore, for ethanol production, a high product concentration diminishes the cost involved in ethanol recovery, which means that a high ethanol tolerance is preferred. Among yeasts and bacteria, *S. cerevisiae* fulfils the above criteria the best and has the ability to produce high ethanol concentrations rapidly but with the disadvantage of not being able to utilize xylose. Expansion of the substrate spectrum to enable xylose consumption by *S. cerevisiae* seems to be a good approach in the development of a process for large-scale ethanol production from hemicellulose.

## 7.2 Improved Viability and Vitality of Active Dried Yeast Starter Cultures

Both the genetic and physiological stability of stock cultures of seed yeast and wine yeast starter cultures are essential for optimal fermentation performance. Physiological stability and 'fitness' of active dried wine yeast cultures relate to the maintenance of cell viability and vitality during the process of yeast manufacturing, including desiccation and storage. The manufacturers of active dried yeast starter cultures can positively influence the degree of viability and vitality, as well as the subsequent fermentation performance of their cultures by the way they cultivate their yeasts [103]. Industrial cultivation of yeasts can have profound effect on the microbiological quality, fermentation rate, production of hydrogen sulphide, ethanol yield and tolerance, resistance to sulphur dioxide as well as tolerance to drying and rehydration. Due to the roles that trehalose and glycogen play in a yeast cell's response to variations in environmental conditions, it is generally recommended that the manufacturers of active dried wine yeast starter cultures cultivate their yeast in such a way that the maximum amount of these storage carbohydrates is accumulated in the yeast cells.

In *S. cerevisiae,* trehalose is synthesized from glucose-6-phosphate and UDP-glucose by the *TPS1*-encoded trehalose-6-phosphate synthetase and converted to trehalose by the *TPS2*-encoded trehalose-6-phosphate phosphatase [104]. The regulation of trehalose synthesis and degradation is mediated by cAMP-dependent phosphorylation mechanisms [105]. Trehalose is associated with nutrient-induced control of cell cycle progression, control of glucose sensing, transport and initial stages of glucose metabolism, as well as stress protection against dehydration, freezing, heating and osmo-stress, toxic chemicals, such as ethanol, oxygen radicals and heavy metals [106]. This storage carbohydrate plays an important role during sporulation, nutrient starvation, growth resumption and growth rate. Trehalose content in the yeast cell is probably one of the most important factors affecting the resistance of yeasts to drying and subsequent rehydration [103]. The accumulation of this disaccharide on both sides of the plasma membrane is thought to confer stress protection by stabilizing the yeast's membrane structure.

Glycogen, another carbohydrate reserve whose accumulation by yeast propagated for drying has been linked to enhanced viability and vitality upon reactivation, provides a readily mobilisable carbon and energy source during the adaptation phase. The biosynthesis of glycogen is effected by glycogen synthase, which catalyzes the sequential addition of glucose from UDP-glucose to a polysaccharide acceptor in a linear α-1,4 linkage, while branching enzymes are responsible for the formation of α-1,6 branches [106]. There are two forms of glycogen synthase in S. *cerevisae, Gsy1p* and *Gsy2p*. The *GSY1* gene is expressed constitutively at a low level along with growth on glucose, while the level of the *GSY2*-encoded glycogen synthase increases at the end of the exponential phase of growth when glycogen accumulates [104]. Glycogen breakdown quickly following depletion of nutrients at the end of fermentation is accompanied by sterol formation. Since sterol is essential for yeast vitality, low levels of accumulated glycogen in active dried wine yeast starter cultures may result in insufficient yeast sterols which in turn, may impair yeast performance upon inoculation into grape juice [106]. In this regard, it is important to note that the over expression of the *SUT1* and *SUT2* genes has been shown to promote the uptake of sterol from the medium under fermentative conditions.

Owing to their multiple roles in increasing survival of *S. cerevisiae* cells exposed to several physical and chemical stresses, trehalose and glycogen have important implications for the viability, vitality and physiological activity of active dried wine yeast starter cultures upon reactivation emphasizing the need to develop wine yeast strains with a proficient trehalose and glycogen accumulation capacity. However, due to the complexity of yeast viability, vitality and physiological activity, it is unclear at this stage whether the modification of the expression levels of the *TPS1, TPS2, GSY1, GSY2, SUT1* and/or *SUT2* genes would contribute to yeast fitness and fermentation performance of starter culture strains.

## 7.3 Improved Nitrogen Assimilation

Of all the nutrients assimilated by yeast during fermentations, nitrogen is quantitatively second only to carbon [107]. Carbon-nitrogen imbalances and more specifically, deficiencies in the supply of assimilable nitrogenous compounds, remain the most common causes of poor fermentative performance and sluggish or struck fermentations [108, 109]. Such problematic and incomplete fermentations occur because nitrogen depletion irreversibly arrests hexose transport. Other problems related to the nitrogen composition of grape must include the formation of reduced-sulphur compounds, in particular, hydrogen sulphide and the potential formation of ethyl carbamate from metabolically produced urea [107]. Unlike grape, sugars that are usually present in large amount (often exceeding 20%, w/v) to that needed for maximal yeast growth, the total nitrogen content of grape juices ranges from 60 to 2400 mg/l and can, therefore, be growth-limiting [107].

There are two basic strategies to circumvent problems linked to nitrogen deficiency: prevention of nitrogen deficiency in grape juice by optimizing vineyard fertility, and more commonly, supplementation with ammonium salts such as diammonium phosphate (DAP). However, the injudicious use of DAP supplements often contravenes the wine industry's desire to minimize its use of additives while producing wines of high quality. Moreover, excessive addition of inorganic nitrogen often results in excessive levels of residual nitrogen, leading to microbial instability and ethyl carbamate (and phosphate in case of DAP) accumulation in wine [108-110]. The degree of supplementation of inorganic nitrogen in grape juice is therefore often regulated. This implies that knowledge of the nitrogen content of grape juice and the requirement for nitrogen by yeast are important considerations for optimal fermentation performance and the production of wines that comply with the demands of regulatory authorities and the consumers.

The major nitrogenous compounds in the average grape must are proline, arginine, alanine, glutamate, glutamine, serine and threonine [111]; the ammonium ion levels may also be high, depending on grape variety and vineyard management. Proline and arginine account for about 30-65% of the total amino acid content of grape juices. *S. cerevisiae* is incapable of adequately hydrolyzing grape proteins to supplement nitrogen-deficient musts and therefore, relies on the ammonium and amino acids present in the juice. Wine yeasts differ from each other in terms of their ability to use various nitrogen sources. Ammonium is the preferred nitrogen source [112], and when a readily used nitrogen source, such as ammonium, glutamine or asparagine is present, genes involved in the uptake and catabolism of poorly utilized nitrogen sources (including proline) are repressed. This nitrogen catabolite repression exerted upon non-preferred nitrogenous compounds damages the assimilation of proline as well as arginine since both the amino acids depend on the proline utilization pathway [112]. Since the proline content of wine is generally not less than grape juice, it appears that proline is not taken up by wine yeast under anaerobic fermentative conditions [111]. Proline is transported into *S. cerevisiae* by the general amino acid permease and the *PUT4*-encoded proline-specific permease. Once inside the yeast cell, proline is converted to glutamate in mitochondria by the *PUT1*-encoded proline oxidase and *PUT2*-encoded pyrroline-5-carboxylate dehydrogenase. The expression of both *PUT1* and *PUT2* is regulated by the *PUT3*-encoded activator and the *URE2*-encoded repressor. Ure2p represses transcription of *PUT1* and *PUT2* under nitrogen-limiting conditions, while the *GLN3* encoded regulator has no effect on these genes [112].

Since yeast strains vary widely in their nitrogen requirement, an obvious target for strain improvement is to select or develop starter strains that are more nitrogen efficient for use in low nitrogen musts [108-110]. To achieve this, a thorough understanding of the regulation of nitrogen assimilation by yeast under fermentative conditions is required. In an effort to develop wine yeast strains that are relieved from nitrogen catabolite repression and that are capable of utilizing proline more efficiently under winemaking conditions, a mutant containing a *ure2* recessive allele was constructed [112].

It was demonstrated that this mutation strongly deregulates the proline utilization pathway, thereby improving the overall fermentation performance of the ure2-carrying yeast. This may be the first step towards the development of wine yeasts that are able to efficiently assimilate the abundant supply of proline in grape juice under fermentative conditions.

## 8. RELEVANT TARGETS FOR BREWER'S AND WINE-MAKING YEASTS

### 8.1 Flocculation

Yeast flocculation, the asexual aggregation of cells into flocs and their subsequent removal from the fermentation medium by sedimentation, is of major interest for some fermentation processes. In brewing, especially bottom fermentation, good flocculation towards the end of primary fermentation is essential for a bright beer with sufficient aroma. The onset of cell sedimentation is critical. If it occurs too late, the beer will be cloudy and hard to filter; and if it occurs too early, maturation will be inefficient. Flocculation is also of interest for the elaboration of sparkling wines. In the classic Champagne method, a secondary fermentation is conducted in the bottle to develop specific organoleptic characteristics. After 1 year, the yeast cells are removed from the bottle by a time-consuming and expensive procedure. This process could be advantageously simplified by the sedimentation of flocculent yeast cells [74].

Yeast flocculation is calcium-dependent, reversible aggregation of cells into flocs, involving a protein-sugar interaction between a specific cell-surface lectin and cell-wall mannan [113]. Several attempts have been made to construct flocculent industrial strains by classical genctic approaches. Electrofusion was used to convert a non-flocculent brewer's yeast into a flocculent one with adequate brewing performance [114, 115]. Flocculation was also introduced into wine yeast strains by hybridization or cytoduction. Extensive work in the past decade has led to a better understanding of the molecular and biochemical basis of flocculation. Several dominant flocculation genes, *FLO1, FLO5* and *FLO8*, are involved in flocculation [6]. The dominant *FLO1* gene has been isolated, characterized and shown to encode a cell-wall protein containing a lectin domain. Transfer of the flocculation characteristics into a wine yeast strain was achieved by transforming the strains with a multicopy vector containing *FLO1,* isolated from a flocculent *S. cerevisiae* strain [116].

A strong flocculation phenotype was achieved in a non-flocculent brewer's yeast by integration of the *ADH1*-regulated *FLO1* into the *ADH1* locus [117]. However, the onset of flocculation still cannot be precisely controlled: flocculation of the recombinant strain occurred too early in brewing trials. A major difficulty in the development of precise control of gene expression is the lack of regulated promoters that can be used under industrial conditions. The most well known inducible promoters for yeasts cannot be used, due to the composition of the industrial medium and regulatory constraints preventing major modifications to it. An example of a promoter that is induced by environment signals is Hsp30 (heat-shock protein 30). This gene is induced by factors that occur late or towards the end of fermentation (high ethanol concentration, depletion of sugars and nitrogen). When the native *FLO1* promoter was replaced by the Hsp30 promoter, flocculation occurred towards the end of fermentation, under laboratory conditions [118].

### 8.2 Glycerol Overproduction

Glycerol is the most abundant by-product of alcoholic fermentation after ethanol and carbon dioxide. Due to its non-volatile nature, glycerol has no direct impact on the aromatic characteristics of wine. However, this triol imparts certain other sensory qualities. It has a slightly sweet taste, and owing to its viscous nature, also contributes to the smoothness, consistency and overall body of wine [119]. In particular, higher levels of glycerol may contribute to the perceived sweetness of wine. The amount of glycerol in wines depends on several factors: grape variety, nitrogen composition, degree of ripeness, mould infection, sulphite levels, pH of grape must, fermentation temperature, aeration, choice of starter culture strain and inoculation level [119, 120]. Typically, under controlled conditions, glycerol concentrations are higher in red wines than in white wines, varying from 1 to 15 g/l. The threshold taste level of glycerol is observed at 5.2 g/l, whereas a change in the viscosity is only perceived at a level of 25 g/l in wine [119]. Wine yeast strains overproducing glycerol could therefore, be of considerable value in improving the organoleptic quality of wine [120]. Many attempts have been made to increase the glycerol yield during fermentation. However, interest for re-routing the carbon flux towards glycerol pathway is expected at the cost of ethanol production. This may be an alternative approach to the current methods for removing ethanol from beverages (e.g., dialysis), which are expensive and detrimental for aroma compounds. In addition, the overproduction of glycerol at the expense of ethanol could fulfill a growing need for table wine with lower levels of ethanol. About 4-10% of the carbon source is usually converted to glycerol resulting in glycerol levels of 7-10% of that of ethanol [119].

The physiological functions of glycerol synthesis are related to redox balancing, resistance to hyperosmotic and oxidative stress, recycling of cytosolic inorganic phosphate and nitrogen metabolism [121]. Futhermore, glycerol-3-phosphate, the precursor of glycerol, is an essential intermediate in the biosynthesis of membrane lipids. It is also noteworthy that glycerol is not only produced by yeasts, but can also serve as carbon source in aerobically grown cultures. The utilization of glycerol is coupled to respiration via a glycerol kinase. This *GUT1*-encoded glycerol kinase converts glycerol to glycerol-3-phosphate, which is then oxidized to dihydroxyacetone phosphate by the *GUT2*-encoded flavin-dependent and membrane-bound mitochondrial glycerol-3-phosphate dehydrogenase [119]. *GUT2* is strongly repressed in the presence of glucose. *FPS1* that encodes a channel protein was shown to act as a glycerol transport facilitator controlling both glycerol influx and efflux [119].

It has been reported that the over expression of *GPD1*, together with constitutive expression of *FPS1*, successfully redirected the carbon flux towards glycerol and the extracellular accumulation of glycerol. Depending on the genetic background in these engineered strains, 1.5 to 4-fold increase in glycerol levels were obtained [122]. As a result of redox imbalances resulting from glycerol overproduction, ethanol formation was decreased and the metabolite pattern of these recombinant wine yeasts was considerably changed. These yeasts offer new prospects to improve the quality of wines lacking in smoothness and body, and production of low-alcohol wines [120].

## 8.3 Enhanced Liberation of Grape Terpenoids

The varietal flavour of grapes is due to the accumulation of volatile secondary metabolites in *V. vinifera* [123]. However, a high percentage of these metabolites occur as their respective non-volatile o-glycosides. Several studies have shown that increased enzymatic hydrolysis of aroma precursors present in grape juice can intensify the varietal character of wines for instance, terpenes such as geraniol and nerol can be released from terpenyl-glycosides by the β-D-glucosidase activities present in grape juice. However, these enzymatic activities are inhibited by glucose and exhibit poor stability at the low pH and high ethanol levels of wine [123].

Attempts are underway to replace less efficient grape glucosidases by aroma liberating β-glucosidases from *Aspergillus* and other fungal species, which have been developed as components of commercial enzyme preparations, added to fermented juice (as soon as the glucose has been consumed by the yeast) or young wine [124]. The addition of exogenous enzyme preparations to wine, however, is an expensive practice. This has led to renewed interest in the more active β-glucosidase produced by certain strains of *S. cerevisiae* and other wine-associated yeasts. Like the grape β-glucosidases, yeast β-glucosidases are not inhibited by glucose, and the liberation of terpenols during fermentation can be ascribed to their action on the terpenyl-glycoside precursors [124]. Since these β-glucosidases are absent in most *S. cerevisiae* starter culture strains, researchers have functionally expressed the β-glucosidase gene *(BGL1)* of the yeast *Saccharomycopsis fibuligera* in *S. cerevisiae* [125, 126]. When the β-l,4-glucanase gene from *Trichoderma longibrachiatum* was expressed in wine yeast, the aroma intensity of wine increased, presumably due to the hydrolysis of glycosylated flavour precursors [127]. Likewise, the *S. cerevisiae* exo-β-1,3-glucanase gene *(EXG1)* has been over-expressed and introduced the endo-β-1,4-glucanase gene *(EVD1)* from *Butyrivibrio fibrisolvens,* the endo-β-l,3 and l,4-glucanases *(BEG1)* from *Bacillus subtilis* and the α-arabinofuranosidase *(ABF2)* in *S. cerevisiae* [125, 126].

## 8.4 Enhanced Production of Desirable Volatile Esters

During the primary or alcoholic fermentation of grape sugars, wine yeast produces ethanol, carbon dioxide and a number of by-products including esters, of which alcohol acetates and $C_4$ - $C_{10}$ fatty acid ethyl esters are found in the highest concentration in wine [6]. Although these compounds are ubiquitous to all wines, the level of esters formed varies significantly. Apart from these factors, such as grape cultivar, rootstock and grape maturity, the ester concentration produced during fermentation is dependent on the yeast strain, fermentation temperature, insoluble material in the grape must, its pH, the amount of sulphur dioxide, amino acids present in the must and malolactic fermentation [128].

The characteristic fruity odours of wine are primarily due to a mixture of hexyl acetate, ethyl caproate, caprylate, isoamyl acetate, and 2-phenylethyl acetate. The synthesis of acetate esters such as iso-amyl acetate and ethyl acetate in *S. cerevisiae* is a result of at least three acetyltransferase activities: alcohol acetyltransferase (*AAT*), ethanol acetyl transferase (*EAT*) and iso-amyl alcohol acetyl transferase (*IAT*) [129]. Isoamyl acetate (banana-like aroma) is an important determinant for beer, sake and young wines. This ester is produced by yeast from isoamyl alcohol, which is itself a by-product of leucine synthesis. Overexpression of one of the genes responsible for leucine synthesis (*LEU4* encoding-isopropylmalate synthase) in a sake yeast strain results in a very slight increase in isoamyl alcohol concentrations and the corresponding ester [74].

Another strategy in a brewing strain was to increase the formation of acetate esters by overexpression of the *ATF1* gene, which encodes the alcohol acetyltransferase catalyzing the formation of esters from acetyl CoA and relevant alcohols. The *ATF1*-encoded alcohol acetyltransferase activity (*AAT*) is the best-studied acetyl transferase of *S. cerevisiae.* It has been reported that the 61 kDa *ATF1* gene product (Atflp) is located within the vacuoles and plays a major role in the production of iso-amyl acetate and to a lesser extent ethyl acetate during beer fermentation. To investigate the role of *AAT* in wine, the *ATF1* gene from a widely used commercial wine yeast strain (VIN13) has been cloned and placed under control of the constitutive yeast phosphoglycerate kinase gene (*PGK1*) promoter and terminator [130]. Integration of this modified copy of *ATFI* into the genomes of three commercial wine yeast strains (VIN7, VIN13 and WE28) resulted in the over expression of AAT activity and increased levels of ethyl acetate, iso-amyl acetate and 2-phenylethyl acetate in wine. The concentration of ethyl caporate, ethyl caprylate and hexyl acetate showed only minor changes, whereas the acetic acid concentration decreased by more than half. This study established the concept that the over expression of acetyltransferase genes such as *ATF1* could profoundly affect the flavour profiles of wines deficient in aroma, thereby paving the way for the production of products maintaining a fruitier character for longer periods after bottling. However, a major disadvantage of this approach is that it concomitantly increases the formation of esters e.g., ethyl acetate, undesirable in wine above a critical amount. Another approach was developed on sake yeast strains to disrupt the *EST2* gene, which codes for the major esterase hydrolyzing isoamyl acetate. However, this only resulted in a limited increase (approximately 2-fold) in isoamyl acetate production.

## 8.5 Increased Production of Acetate Esters

Isoamyl acetate (banana-like aroma) is an important determinant for beer, sake and young wines. This ester is produced by yeast from isoamyl alcohol, which is itself a by-product of leucine synthesis. Over expression of one of the genes responsible for leucine synthesis (*LEU4* encoding—isopropylmalate synthase) in a sake yeast strain resulted in a very slight increase in isoamyl alcohol concentrations and the corresponding ester [131]. Another strategy developed in a brewing strain was to increase the formation of acetate esters by overexpression of the *ATF1* gene, which encodes the alcohol acetyltransferase catalyzing the formation of esters from acetyl CoA and the relevant alcohols. This led to a 27-fold increase in the production of isoamyl-acetate during fermentation on a laboratory scale [132]. The strategy has been also successfully applied to wine yeast strains [130]. However, a major disadvantage of this approach is that it generates a simultaneous increase in the formation of esters e.g., ethyl acetate, which can be undesirable in wine above a critical amount. Another approach developed on sake yeast strains was to disrupt the *EST2* gene, which codes for the major esterase hydrolyzing isoamyl acetate [133]. However, this only resulted in a limited increase (approximately 2-fold) in isoamyl acetate production.

## 8.6 Elimination of Ethylcarbamate

Ethylcarbamate is a suspected carcinogen found in many fermented foods and beverages e.g., wine, sherry, brandy and sake [134]. It is mainly formed by the spontaneous chemical reaction of ethanol and urea at elevated temperatures in acidic media. Urea is produced mainly from the cleavage of arginine by arginase. To reduce the formation of urea in sake, two copies of the *CAR1* gene, coding for arginase, were disrupted in industrial sake yeast. This resulted in the elimination of urea and ethylcarbamate formation during sake brewing [135]. It may also be of interest to reduce the formation of ethylcarbamate in wines. Arginine is one of the most abundant amino acids in grape must and is quickly assimilated by yeast. Therefore, wines made from arginine-rich grape musts may sometimes contain amounts of ethylcarbamate that exceed the authorized concentrations. Although the approach described above might be conceivable for wine yeast, it has the major drawback of reducing the amount of nitrogen available during fermentation. Another possibility is to express an acidic urease that degrades urea into ammonia and carbon dioxide in yeast [136]. The acidic urease from *Lactobacillus fermentum* was recently expressed in *S. cerevisiae*, with the aim of degrading urea during wine fermentation. However, this approach was ineffective because *S. cerevisiae* lacks several auxiliary proteins and appropriate cofactors required for the correct fold ing of urease [137].

## 8.7 Antimicrobial Properties

In brewing, sake-brewing and wine-making, there is always a risk that wild yeast becomes predominant. Many indigenous yeasts secrete zymocin, a toxin that kills sensitive strains. Zymocin production and immunity to this toxin are determined by a cytosolic, double-stranded RNA. There are three major types of killer toxins (K1-K3). As most of the yeast strains found in grape musts produce the K2 zymocin, the K1 dsRNA has been integrated into the genome of K2 wine yeast [138]. A brewing strain with increased resistance to contamination was also constructed by expressing the genes coding for zymocin production and immunity. The modified yeast was resistant to the zymocin, which can be produced by contaminating yeast and can kill yeast flora that is sensitive to the toxin [42]. Although the idea of achieving a better control of yeast implantation is relevant, the utilization of starter strains with a selective advantage is questionable. Industrial yeasts, for example wine yeasts, are classically used without containment and are released into the environment. Therefore, great attention must be paid to the potential impact of yeast strains possessing a competitive advantage over the natural microflora.

The fermentation media are also susceptible to bacterial contamination. A brewer's strain carrying both the killer factor and antibacterial properties has been constructed [139]. This strain was obtained by mating a respiratory-deficient strain with antimicrobial properties and a killer strain, followed by fusion of the resulting hybrid with a brewer's strain. Attempts to develop bactericidal wine yeast strains have been recently described. Two bacteriocin genes, encoding a pediocin and a leucocin gene from *Pediococcus acidilactici* and *Leuconostoc carnosum*, respectively have been expressed in *S. cerevisiae* [140]. As these bacteriocins have a rather narrow spectrum, this approach should be extended to other toxins to warrant a satisfying stability of wines. The use of bactericidal yeasts would be useful for the production of wine with reduced levels of sulfur dioxide and other chemical preservatives. However, as sulfur dioxide has other interesting properties as an antioxidant, its addition to wine will probably remain necessary.

## 9. SPECIFIC TARGETS FOR BREWER'S YEAST

### 9.1 Fermentation of Dextrins

Brewer's yeast strains cannot utilize dextrins, which represent about 25% of malt wort sugars and contribute significantly to the calorific content of beer. Crude commercial preparations containing glucoamylases (usually from *Aspergillus*) are used to produce light beers. An early target for yeast genetic improvement was the development of amylolytic yeast strains that hydrolyze residual starch or dextrin in wort, producing low carbohydrate beers. Although classical genetic approaches have succeeded in transferring the amyloglucosidase activity of *S. diastaticus* (which hydrolyzes the α-1,4 linkages of dextrins) to brewer's strains, a major problem was the co-transfer of the *POF1* gene, causing phenolic off-flavour. This side-effect was overcome by the expression of *S. diastaticus STA2* gene encoding amyloglucosidase [141]. To overcome plasmid instability problems, an all-yeast, multicopy plasmid carrying *STA2* under the control of the PGK promoter, was used to transform a commercial lager-brewing yeast strain [142]. The recombinant brewer's yeast secreting amyloglucosidase exhibited improved carbohydrate degradation and produced approximately 1% more ethanol, without altering the fermentation performance or the quality of the beer. The degree of dextrin degradation was further increased by expressing amyloglucosidases able to hydrolyze α-1,6 glycosidic links. The corresponding genes from *A. niger* [143] or *A. awamori* [144] have been integrated into the genome of brewer's yeast strains to give recombinant strains with high levels of secreted enzymes and dextrin degradation. Good quality, superattenuated beers have been produced on a pilot scale by brewer's yeast secreting the *A. niger* enzyme [145]. Increased dextrin degradation has also been obtained with a brewer's yeast expressing the gene coding for the amyloglucosidase of *Schwanniomyces occidentalis*. This enzyme has both α-1,4 and α-1,6 activities and has the additional advantage of being thermolabile, thus being inactivated by pasteurization. This ensures that the beer does not become sweet during storage. To increase starch degradation further, various *Saccharomyces cerevisiae* yeasts coexpressing the *STA2* gene from *S. distaticus*, coding for a glucoamylase and the *AMY* gene from *Bacillus amyloliquefaciens* were constructed. These strains were very efficient at starch degradation in laboratory conditions [88].

## 9.2 Beer Viscosity and Filtration

The β-1,4 and β-1,6 linkages of β-glucans, a polysaccharide found in barley cell walls, are cleaved by a specific endo-β-glucanase during malting. This thermolabile enzyme is present only in small amounts in the wort due to the elevated temperature during malt drying. Consequently, β-glucan degradation is often insufficient, which leads to increased viscosity, reduced filterability and the formation of gels and hazes in the beer. The filtration performance of beer can be significantly improved by the addition of a commercial preparation of bacterial β-glucanase to the fermenter. A cheaper solution is the use of yeast strains secreting glucanase. The secretion of *B. subtillis* β-glucanase in brewer's yeast strains was obtained by expressing the appropriate gene fused to the promoter and signal sequence of factor in a multicopy plasmid (Lancashire and Wilde, 1987). More efficient β-glucan degradation was subsequently achieved by expressing a β-glucanase of *Trichoderma reesei* with a lower optimum pH (4-5), instead of 6.7 for the *B. subtilis* enzyme [146-148]. Brewer's strains carrying one copy of the *T. reesei* EG1 gene integrated into the chromosome (ADH1 or PGK1 locus) and devoid of bacterial sequences have been constructed and tested on a pilot scale. Almost all β-glucans were digested by some of the integrants during fermentation, resulting in a significant reduction of beer viscosity. The growth and brewing properties of the strains were unaltered [149].

## 9.3 Diacetyl Elimination

Brewer's yeasts positively or negatively influence the flavour of the beer by producing aroma compounds during primary fermentation. Vicinal diketones, pentanedione and diacetyl, in particular are considered unpleasant. The main reason for lagering (secondary fermentation) is to decrease the concentration of diacetyl, which can be tasted at concentrations below 0.02-0.10 mg/l. Diacetyl is formed outside the cell by the chemical oxidation of acetolactate, an intermediate in the biosynthesis of valine, which diffuses into the fermenting wort. Once diacetyl has been formed, it is enzymatically reduced inside the yeast cell to acetoin and finally to 2,3-butanediol, which has no impact on the flavour of the beer. The complete removal of diacetyl sometimes requires lengthy maturation (1-3 weeks). Various approaches based on the engineering of enzymes of the valine biosynthetic pathway have been tested to reduce yeast diacetyl formation. Diacetyl formation was reduced in *ILV2* mutants and in strains in which the expression of *ILV5*, which codes for acetohydroxy acid reductoisomerase, was increased [151]. Another strategy, which has the advantage of not interfering with the biosynthesis of valine or isoleucine, consisted of expressing the acetolactate decarboxylase gene from *Enterobacter aerogenes* in brewer's yeast. Brewer's strains with the ADHI-controlled *ALDC* gene integrated into the chromosome produced considerably less diacetyl than the parent yeast in fermentation tests [152].

## 9.4 Reduction of Hydrogen Sulfide Production

Low levels of hydrogen sulfide are produced by *S. cerevisiae* by the reduction of sulfate during methionine synthesis. Due to its very low sensory threshold, trace amounts of hydrogen sulfide can alter the organoleptic characteristics of beer. This compound must be eliminated during beer maturation. An alternative strategy is to develop brewing yeasts with reduced hydrogen sulfide formation. Increased expression of *NHS5*, encoding the cystathionine-synthase in brewing yeast, has been shown to suppress the formation of hydrogen sulfide in beer produced on the laboratory scale, without affecting other fermentation characteristics [152]. Another approach, based on the overexpression of *MET25*, which encodes o-acetylhomoserine and o-acetylserine sulfhydrylase, resulted in a 10-fold decrease in hydrogen sulfide production in beer in pilot-scale experiments [153]. This can be explained by increased consumption of the substrate by the overproduced enzyme and by a decrease in sulfate uptake. An alternative approach was to partially or fully eliminate *MET10*, which encodes a putative sulfite reductase subunit. This led to a substantial reduction in hydrogen sulfide and the accumulation of sulfite. The beer produced showed increased flavour stability [154].

Another sulfur compound that may cause organoleptic problems, especially for some lager beers, is dimethyl sulfide (DMS) which is produced during fermentation by reduction of dimethyl sulfoxide (DMSO). The *MXR1* gene was recently shown to encode a methionine sufoxide reductase. A mxr1 disruption mutant was reported to be unable to reduce DMSO in laboratory conditions. This work opens up the construction of brewing strains that do not produce DMS.

### 9.5 Increased Production of Sulfur Dioxide

Sulfite is an antioxidant and a key compound for the flavour stability of beer, because it combines with aldehydes. There has been extensive work to control the accumulation of sulfite by brewer's yeast. One approach, based on the over expression of *MET3* and *MET14* in brewer's yeast, led to increased sulfite production [155]. Alternatively, brewer's yeast with several of the four copies of *MET2* inactivated was shown to produce higher amounts of sulfite. However, hydrogen sulfide production was also increased. As mentioned above, the inactivation of *MET10* was shown to be an efficient strategy for increasing sulfite formation and decreasing hydrogen sulfide production [154].

## 10. SPECIFIC TARGETS FOR WINE YEAST IMPROVEMENT

### 10.1 Malolactic Fermentation

Alcoholic fermentation by pure-cultured wine yeast strains has become an increasingly well controlled process. In contrast, malolactic fermentation remains unreliable in numerous situations, due to the poor development of lactic acid bacteria in wine. This secondary fermentation (decarboxylation of malate to lactate by the malolactic enzyme) has an essential role for the deacidification and stabilization of wine. Delayed or stuck malolactic fermentation leads to scheduling problems in cellars and increases the risks of wine alteration. To circumvent the problems of unreliability of malolactic fermentation, *S. cerevisiae* strains able to degrade malic acid completely into lactic acid and carbon dioxide have been constructed. The malolactic gene from *Lactococcus lactis* has been cloned and efficiently expressed in *S. cerevisiae* [156-158]. Due to the inability of *S. cerevisiae* to transport malate efficiently, the malate permease from *Schizosaccharomyces pombe* was coexpressed with the malolactic enzyme in *S. cerevisiae* [159, 160]. The recombinant strains fully degraded up to 7 g of malate/l in 4 days, simultaneously with alcoholic fermentation and without affecting the growth properties and fermentation rate [159]. A careful comparison of wines, obtained on a pilot scale with engineered industrial strains and traditional malolactic fermentation, will be necessary. Attention will have to be paid to the impact on flavours, because bacterial malolactic fermentation has been described in some instances to improve the organoleptic complexity of wine. Similarly, it will be interesting to compare the advantages of both yeast and bacterial malolactic fermentation in terms of wine stabilization, because the composition of residual micronutrients might differ in the wines obtained by the two systems. Although the use of engineered yeast strains for malolactic fermentation to produce safer wines of high quality might compete with bacterial starter cultures in the future, the use of malolactic-engineered yeast would provide an additional advantage. Indeed, the elimination of secondary fermentation limits the risk of developing bacterial flora responsible for the alteration of wine and/or the production of undesirable metabolites.

## 10.2 Production of Lactic Acid

Achieving the correct balance between sugar and acidity is a major requirement for wine quality. In some hot regions, grape musts are often insufficiently acidic and the balance must be corrected. Wine acidification is usually accomplished by adding tartaric acid, which is authorized in certain specific situations and the efficiency of which is limited by the instability of the K-tartarate. Biological acidification using a lactic acid-producing *S. cerevisiae* strain is a promising alternative due to the high stability and organoleptic properties of this organic acid. Moreover, lactic acid is naturally found in wines after malolactic fermentation. Therefore, *S. cerevisiae* strains performing a dual fermentation (ethanol and lactate) have been constructed by expressing a lactate dehydrogenase (LDH) gene from *Lactobacillus casei* [161-162]. These strains can efficiently adjust the acidity of grape must: production of lactate at around 5 g/l increased total acidity by 0.2-0.3 pH units [163]. Consistent with the diversion of sugars towards lactate, the ethanol content of the wines obtained is slightly lower [approximately 0.25% (v/v) for 5 g lactate/l]. It is technically possible to construct a set of wine yeast strains producing different amounts of lactic acid across a range suitable to correct grape musts lacking in acidity. Lactate-producing yeasts are one of the most promising examples of wine-yeast improvement and may also be of interest in other fermentation fields (production of acidic doughs for baking, industrial lactate production).

## 10.3 Decrease of Volatile Acidity

The level of acetic acid, the main component of volatile acidity, is critical for the quality of wines. The concentration of acetic acid in wines is usually approximately 0.5 g/l and must remain below 0.8 g/l. Yeasts sometimes produce excessive acetic acid, either due to the genetic background of the yeast or the wine-making processes (e.g., excessive clarification). Despite the importance of this organic acid, the genetic basis of acetate production during alcohol fermentation is still largely unknown. It has recently been shown that the level of acetate produced by *S. cerevisiae* can be effectively controlled by increasing or decreasing the level of expression of the cytosolic acetaldehyde dehydrogenase encoded by *ALD6* [164]. Inactivation of the two alleles of this gene in a wine yeast leads to a 2-fold reduction in the amount of acetate produced during wine fermentation. As a consequence of the resulting redox imbalance, glycerol, succinate and 2,3-butanediol production is slightly increased.

## 10.4 Improved Ethanol Tolerance

The physiological basis of ethanol toxicity is complex and not well understood, but it appears that ethanol mainly impacts upon membrane structural integrity and membrane permeability [111]. The chief sites of ethanol action include the yeast cell's plasma membrane, hydrophobic proteins of mitochondrial membranes, nuclear membrane, vacuolar membrane, endoplasmic reticulum and cytosolic hydrophilic proteins [106]. Several intrinsic and external factors are known which synergistically aggravate the inhibitory effects of ethanol. These factors include high fermentation temperatures, nutrient limitation (especially oxygen, nitrogen, lipids, magnesium ions and metabolic by-products, such as other alcohols, aldehydes, esters, organic acids especially octanoic and decanoic acids), certain fatty acids and carbonyl and phenolic compounds [166]. By manipulating the physicochemical environment during the cultivation and manufacturing of active dried wine yeast starter cultures and during the actual vinification process, the yeast cells acquired protective adaptations can be promoted. Prior exposure of yeast cells to ethanol (pre-conditioning) elicits adaptive stress responses that confer a degree of resistance to subsequent exposure to high levels of ethanol. Most of the so-called survival factors including oleic and linoleic acids, and ergosterol are formed only in the presence of molecular oxygen, which in part explains the great success in the use of commercial starter cultures that are cultivated under highly aerobic conditions and in low glucose concentrations. These starter yeast cells contain high levels of survival factors that can be passed onto the progeny cells during the six or seven generations of growth in a typical wine fermentation [111].

Wine yeast strains usually contain higher levels of survival factors than non-wine *Saccharomyces* strains [111]. The physiological response of wine yeast to ethanol challenge is also greater than in the case of non-wine strains. These defensive adaptations of wine yeasts conferring enhanced ethanol tolerance range from alterations in membrane fluidity to synthesis of detoxification enzymes. Responses include a decrease in membrane saturated fatty acids (e.g., palmitic acid); an increase in membrane unsaturated long-chain fatty acids (e.g., oleic acid); phosphatidylinositol biosynthesis (thereby increasing the phospholipid: protein ratio in the membrane); elevated levels of cellular trehalose that neutralize the membrane-damaging effects of ethanol; stimulation of stress protein biosynthesis; enhanced mitochondrial superoxide dismutase activity that counter effects ethanol-induced free radical synthesis; increased synthesis of cytochrome P450, alcohol dehydrogenase activity and ethanol metabolism [32, 111]. It is clear that the genetics of ethanol tolerance is multifarious and a very complex process. It is estimated that more than 250 genes are involved in the control of ethanol tolerance in yeast [111]. Nevertheless, some reports claim that continuous culture of yeasts in a feedback system, in which the ethanol was controlled by the rate of carbon dioxide release, enabled the selection of viable mutants with improved ethanol tolerance and fermentative ability [165].

## 10.5 Improved Polysaccharide Degradation

The use of pectinases and glucanases in winemaking has several advantages. These enzymes can be used to facilitate wine clarification and to improve liquefaction of grapes, thereby increasing the juice yield. Polysaccharide-degrading enzymes may also enhance the liberation of various compounds trapped in grape skins, thereby improving the bouquet and colour of the wine. As these commercial enzyme preparations are expensive, much effort has been made to develop yeast strains that secrete heterologous pectinases, glucanases, xylanases or a combination of these enzymes [6]. For example, polypectate degradation was increased by wine yeast co-expressing the *Erwinia chrysanthemi* pectate lysate gene and the *E. carotovora* polygalacturonase gene [167]. Another approach was based on the expression of the *Fusarium solani* pectate lyase gene [168]. Glucanolytic wine yeast strains have been constructed by the expression of fungal and bacterial endo-β-1,4-glucanases [127, 169], or a combination of endo- and exoglucanases [170].

## 10.6 Improvement of Wine Flavour and Other Sensory Qualities

The single most important factor in winemaking is the organoleptic quality of the final product which is due to the presence of desirable flavour compounds and metabolites in a well-balanced ratio and the absence of undesirable ones. Many variables contribute to the distinctive flavours of wine. Oenological practices including the yeast and fermentation conditions have a prominent impact on the primary flavours of *V. vinifera* wines. The volatile profile of wines is dominated by those components that are formed during fermentation, since these compounds are present in the highest concentrations. The flavour of wine immediately after fermentation only approximates that of the finished product. After the sudden and dramatic changes in composition during fermentation, chemical constituents generally react slowly during aging to move to their equilibria, resulting in gradual changes in flavour. The harmonious complexity of wine can be subsequently increased by volatile extraction during oak barrel aging [171].

Despite the extensive information published on flavour chemistry, odour thresholds and aroma descriptions, the flavour of complex products such as wine cannot be predicted. With a few exceptions (e.g., terpenes in the aromatic varieties and alkoxypyrazines in the vegetative or herbaceous cultivars), perceived flavour is the result of specific ratios of many compounds rather than being attributable to a single impact compound [128]. In wines, the major product of yeast fermentation, esters and alcohols contribute to a generic background flavour, whereas subtle combinations of trace components derived from the grapes usually elicit the characteristic aroma notes of these complex beverages.

## 10.7 Enhanced Liberation of Grape Terpenoids

The varietal flavour of grapes is due to the accumulation of volatile secondary metabolites in *V. vinifera* [123]. However, a high percentage of these metabolites occur as their respective non-volatile o-glycosides. Several studies have shown that increased enzymatic hydrolysis of aroma precursors present in grape juice can intensify the varietal character of wines, for instance, terpenes such as geraniol and nerol can be released from terpenyl-glycosides by the β-D-glucosidases activities present in grape juice. However, these enzymatic activities are inhibited by glucose and exhibit poor stability at the low pH and high ethanol levels of wine [123].

Attempts are underway to replace less efficient grape glucosidases by aroma liberating β-glucosidases from *Aspergillus* and other fungal species, which have been developed as components of commercial enzyme preparations added to fermented juice (as soon as the glucose has been consumed by the yeast) or young wine [124]. The addition of exogenous enzyme preparations to wine, however, is an expensive practice. This has led to renewed interest in more active β-glucosidase produced by certain strains of *S. cerevisiae* and other wine-associated yeasts. Like the grape β-glucosidases, yeast β-glucosidases are not inhibited by glucose, and the liberation of terpenols during fermentation, and can be ascribed to their action on the terpenyl-glycoside precursors [124]. Since β-glucosidases are absent in most *S. cerevisiae* starter culture strains, researchers have functionally expressed the β-glucosidase gene *(BGL1) of* the yeast *Saccharomycopsis fibuligera* in *S. cerevisiae* [125]. When the β-1,4-glucanase gene *from Trichoderma longibrachiatum* was expressed in wine yeast, the aroma intensity of wine increased, presumably due to the hydrolysis of glycosylated flavour precursors [127]. Likewise, the *S. cerevisiae* exo-β-1,3-glucanase gene *(EXG1)* has been overexpressed and introduced the endo-β-1,4-glucanase gene *(EVD1)* from *Butyrivibrio fibrisolvens,* the endo-β-1,3 and 1,4-glucanases *(BEG1)* from *Bacillus subtilis* and the α-arabinofuranosidase *(ABF2)* in *S. cerevisiae* [125, 126].

## 10.8 Enhanced Production of Desirable Volatile Esters

During the primary or alcoholic fermentation of grape sugars, wine yeast produces ethanol, carbon dioxide and a number of by-products including esters, of which alcohol acetates and $C_4$ - $C_{10}$ fatty acid ethyl esters are found in the highest concentration in wine [6]. Although these compounds are ubiquitous to all wines, the level of esters formed varies significantly. Apart from these factors, such as grape cultivar, rootstock and grape maturity, the ester concentration produced during fermentation is dependent on the yeast strain, fermentation temperature, insoluble material in the grape must, its pH, the amount of sulphur dioxide, amino acids present in the must and malolactic fermentation [172].

The characteristic fruity odours of wine are primarily due to a mixture of hexyl acetate, ethyl caproate, caprylate, isoamyl acetate, and 2-phenylethyl acetate. The synthesis of acetate esters such as isoamyl acetate and ethyl acetate in *S. cerevisiae* is as a result of at least three acetyltransferase activities: alcohol acetyltransfersase (*AAT*), ethanol acetyl transferase (*EAT*) and iso-amyl alcohol acetyl transferase (*IAT*) [173]. Isoamyl acetate (banana-like aroma) is an important determinant for beer, sake and young wines. This ester is produced by yeast from isoamyl alcohol, which is itself a by-product of leucine synthesis. Over expression of one of the genes responsible for leucine synthesis (*LEU4* encoding and isopropylmalate synthase) in sake yeast strain results in a very slight increase in isoamyl alcohol concentrations and the corresponding ester [74].

Another strategy in a brewing strain was to increase the formation of acetate esters by overexpression of the *ATF1* gene, which encodes the alcohol acetyltransferase catalyzing the formation of esters from acetyl CoA and relevant alcohols. The *ATF1*-encoded alcohol acetyltransferase activity (*AAT*) is the best-studied acetyl transferase of *S. cerevisiae*. It has been reported that 61 kDa *ATF1* gene product (Atf1p) is located within the vacuoles and it plays a major role in the production of iso-amyl acetate and to a lesser extent ethyl acetate during beer fermentation. To investigate the role of *AAT* in wine, the *ATF1* gene from a widely used commercial wine yeast strain (VIN13) has been cloned and placed under the control of constitutive yeast phosphoglycerate kinase gene (*PGK1*) promoter and terminator [130]. Integration of this modified copy of *ATFI* into the genomes of three commercial wine yeast strains (VIN7, VIN13 and WE28) resulted in the over expression of AAT activity and increased levels of ethyl acetate, iso-amyl acetate and 2-phenylethyl acetate in wine. The concentration of ethyl caproate, ethyl caprylate and hexyl acetate showed only minor changes, whereas the acetic acid concentration decreased by more than half. This study established the concept that the over-expression of acetyltransferase genes such as *ATF1* could profoundly affect the flavour profiles of wines deficient in aroma, thereby paving the way for the production of products maintaining a fruitier character for longer periods after bottling. However, a major disadvantage of this approach is that it concomitantly increases the formation of esters e.g., ethyl acetate, undesirable in wine above a critical amount. Another approach was developed on sake yeast strains to disrupt the *EST2* gene, which codes for the major esterase hydrolyzing isoamyl acetate. However, this only resulted in a limited increase (approximately 2-fold) in isoamyl acetate production.

## 10.9 Optimized Fusel Oil Production

Fusel oils or higher alcohol are produced by wine yeasts during fermentation from intermediates in the branched chain amino acids biosynthetic pathways for the production of Ile, Leu and Val by decarboxylation, transamination and reduction [173]. Higher alcohols do not affect the wine flavour and aroma when their concentration is low. In some cases, they even contribute to wine quality [174]. Initial attempts to use Ile$^-$, Leu$^-$ and Val$^-$ mutants succeeded in lowering the levels of isobutanol, active amyl alcohol and isoamyl alcohol production in fermentations, but these mutants were of no commercial use as their growth and fermentation rates were compromised [175]. A Leu$^-$ mutant derived from the widely used Montrachet wine yeast has been reported to produce more than 50% less isoamyl alcohol during fermentation than the prototrophic parent [176]. It will be of great interest to confirm further by disruption of specific *ILE, LEU* and *VAL* genes of yeasts in lowering the levels of fusel oil in wine.

## 10.10 Improved Liberation of Varietal Aroma

The bouquet of a wine is determined by a combination of many compounds, the production of which is influenced by grape variety, terroir, viticulture practices, enological practices and yeast strain fermentation conditions. Due to this complexity, only a small number of suitable targets can be defined. A well identified target for aroma produced by yeast metabolism is isoamyl acetate (described above). In addition to producing aromatic compounds, yeast can increase the liberation of varietal aroma by producing enzymes that hydrolyze non-aromatic precursors in the grape must. For example, various grape monoterpenols (e.g., linalol, geraniol) are linked to diglycosides, which can be converted to monoglucosides by cleaving of the 1,6-glycosidic linkage. The flavour compound is then liberated by the action of a β-glucosidase. However, grape and yeast β-glucosidases are inhibited by glucose and are unstable at low pH, therefore they poorly hydrolyze monoterpenyl-glucosides during wine fermentation. The addition of commercial preparations of fungal β-glucosidases after fermentation has been shown to improve the hydrolysis of glyco-conjugated aroma compounds. Progress in this area relies on the isolation of new glucose-tolerant β-glucosidases that are stable at low pH. A highly glucose-tolerant β-glucosidase has been purified from *Candida peltata* [177]. A β-glucosidase from *A. oryzae* that is highly resistant to inhibition by glucose and is stable at low pH was recently described and the corresponding gene cloned [178]. This enzyme has a broad specificity, because it can hydrolyze 1,3-, 1,4-, and 1,6-β-diglycosidase and can release flavour compounds such as geraniol, nerol and linalool from the corresponding monoglucosides in a rich-glucose medium at pH 2.9. An alternative approach for increasing the flavour of wine involves the modification of existing *S. cerevisiae* metabolic pathways associated with the production of aromatic compounds. Interestingly, mutants of the ergosterol biosynthetic pathways have been shown to produce monoterpenes (geraniol, citronelol, linalool) similar to those of the floral grape cultivars [179].

## 10.11 Reduced Sulfide

Hydrogen sulfide is a highly undesirable compound in wine and is a product of the yeast's sulfur metabolism. Hydrogen sulfide is produced during wine fermentation, mainly in response to the depletion of nitrogen and possibly certain vitamins, and its production is influenced by many environmental factors and by the yeast strain. In conditions of nitrogen starvation, hydrogen sulfide may accumulate and diffuse out of the cells [108]. Wine yeast strains with low hydrogen sulfide production have been obtained by hybridization [180]. Alternatively, reduced sulfide production might be achieved by manipulating the sulfate metabolic pathway, as described above for brewer's yeast. However, strategies such as elimination of *MET10* cannot be used for wine yeast, because they result in an increased sulfite production. The final amount of sulfur dioxide in wine is regulated, and sulfur dioxide is currently added to grape must and wine as a preservative. However, due to the current unfavourable public opinion, the tendency is to reduce its content in the wine.

## 10.12 Maltose Utilization

A high fermentation rate is a prerequisite for bakers' yeast strains. The free sugars present in the flour (sucrose, glucose, fructose and maltose) are sequentially consumed. Fermentation continues due to the action of amylases present in the dough, which release maltose from starch. Maltose utilization by yeast requires a maltose permease and a maltase, and both are induced in the presence of maltose. A major factor limiting the dough fermentation rate is the repression of the synthesis of maltose-utilizing enzymes and the inactivation of the maltase enzyme by glucose [181]. Low concentrations (1-2%) of free sugars (mainly glucose and fructose) in the dough repress maltose utilization, causing a lag phase in carbon dioxide production. To avoid this lag phase, maltose-utilizing enzymes have been de-repressed by replacing the native promoters of the maltase and maltose permease with constitutive promoters [182]. Baker's yeast strains display intrinsic differences in their rates of maltose utilization. Non-lagging bakers' strains are characterized by rapid maltose fermentation, unlike lagging strains, which exhibit a lag phase. It has been reported that the non-lagging phenotype depends on a high level of maltase expression in the presence of glucose [183]. It has been suggested that non-lagging strains display higher constitutive basal levels of expression of *MAL* genes than lagging strains under non-inducing and non-repressing conditions [184]. It has been proposed that these differences are due to divergences in the genetic structure of the *MALX3* gene, which codes for an activator of the other

*MAL* genes. Therefore, expression of a *MALX3* gene isolated from a non-lagging strain in a lagging strain might improve maltose metabolism in unsweetened dough [184].

## 10.13 Other Approaches to Increase the Fermentative Capacity

Several attempts have been made to increase the glycolytic flux. Various control steps have been suggested, including sugar uptake, the hexokinase, the phosphofructokinase and the pyruvate kinase. However, overexpression of a combination of these enzymes and other glycolytic enzymes failed to increase the glycolytic flux [185]. An alternative strategy was based on the hypothesis that a decrease in the intracellular ATP concentration would result in an increased carbon dioxide production rate. This was tested by de-repression of the ATP consuming gluconeogenic enzymes, fructose 1,6-diphosphatase and phosphoenolpyruvate carboxykinase [186]. Although no evidence could be obtained of functional futile cycling, this resulted in a 25% increase in the rate of glucose consumption. This strain also displayed a significant reduction in biomass yield. Recently, recombinant strains, simultaneously overexpressing a set of seven enzymes involved in the lower part of glycolysis, were constructed. The recombinant strain exhibited increased glycolytic flux, but only under conditions of increased ATP demands [187].

# 11. FOOD PROCESSING ENZYMES FROM RECOMBINANT MICROORGANISMS

Enzymes are commonly used in food processing and in the production of food ingredients. Enzymes traditionally isolated from culturable microorganisms, plants, and mammalian tissues are often not well-adapted to the conditions used in modern food production methods. The use of recombinant DNA technology has made it possible to manufacture novel enzymes suitable for specific food-processing conditions. Such enzymes may be discovered by screening microorganisms from diverse environments or developed by modification of known enzymes using modern methods of protein engineering or molecular evolution. As a result, several important food-processing enzymes such as amylases and lipases with properties tailored to particular food applications have become available. Another important achievement is improvement of microbial production strains. For example, several microbial strains recently developed for enzyme production have been engineered to increase enzyme yield by deleting native genes encoding extracellular proteases. Moreover, certain fungal production strains have been modified to reduce or eliminate their potential for production of toxic secondary metabolites [178].

The industrial production of enzymes for use in food processing dates back to 1874, when Danish scientist Christian Hansen extracted rennin (chymosin) from calve's stomach for use in cheese manufacturing [189]. Chymosin is now produced from microorganisms that contain the bovine prochymosin gene introduced through recombinant deoxyribonucleic acid (rDNA) techniques. Bovine chymosin expressed in *Escherichia coli* K-12 became the first recombinant enzyme approved for use in food by the US Food and Drug Administration (FDA) [27].

Many enzymes currently used in food processing are derived from recombinant microorganisms. Enzyme manufacturers take advantage of new genetic techniques to develop and manufacture enzymes with improved properties. Such enzymes often originate from microorganisms that cannot be easily cultured under laboratory or industrial conditions. By judicious selection of host microorganisms, recombinant production strains can be constructed to allow efficient production of enzymes that are substantially free of undesirable enzymes or other microbial metabolites.

The increasing sophistication of food processing created a demand for a broad variety of food processing enzymes with characteristics compatible with food processing conditions. For example, commonly used sweeteners such as glucose or fructose syrups are typically produced from corn starch using hydrolytic enzymes. In the first step of starch hydrolysis, starch is liquefied with α-amylase by heating at 105°C for 2-5 min followed by 1-2 h at 90-100°C. With the advent of rDNA technology, it became possible to engineer α-amylases with increased heat stability and improved compatibility with other parameters of the liquefaction process. These improvements were accomplished by introducing changes in the α-amylase amino acid sequences through DNA sequence modifications of the α-amylase genes. Other enzymes currently used in food processing have also been improved using rDNA techniques.

Enzymes used in food processing are sold as enzyme preparations. An enzyme preparation typically contains the enzyme of interest and several added substances such as diluents, preservatives, and stabilizers. The added materials are usually well-known substances suitable for use in food. Enzyme preparations may also contain other enzymes and metabolites from the production organism and the residues of raw materials used in fermentation media and during isolation and purification of the enzyme. All these materials are expected to be of appropriate purity consistent with current good manufacturing practice (cGMP). The safety evaluation of food processing enzymes from recombinant microorganisms has been extensively discussed in the literature [190-192] and in guidance documents issued by regulatory authorities and international organizations, for example, by the Scientific Committee for Food [193]. In principle, the same safety considerations apply to enzymes derived from native and recombinant microorganisms. The key component in evaluating enzyme safety is the safety assessment of the production strain, in particular, its pathogenic and toxigenic potential [191]. Although neither pathogenic nor toxigenic microorganisms are intentionally used in the production of food processing enzymes, certain fungi traditionally used as sources of enzymes have been found to produce low levels of toxic secondary metabolites under fermentation conditions conducive to the synthesis of these compounds. Some of these microorganisms are now used as sources of recombinant enzymes.

In the last two decades, FDA has received a number of petitions and GRAS notices on food-processing enzymes derived from recombinant microorganisms. These enzymes are listed in Table 3 with their source microorganisms and references to FDA regulations, GRAS affirmation petitions, and GRAS notices. Lists of commercial enzymes used in food processing can be found at the web sites of Enzyme Technical Association and the Association of Manufacturers and Formulators of Enzyme Products.

## 11.1 Steps in the Development of Recombinant Production Strains

Industrial production of recombinant enzymes is preceded by an extensive research and development phase that culminates in the construction of a successful production strain. This process typically involves the following stages: (1) development of the host (recipient) strain; (2) construction of the expression vector; (3) transformation of the host strain; (4) identification of the best recombinant strain; (5) additional improvements; and (6) characterization of the production strain. Each of these steps is dictated by specific circumstances related to the identity and properties of the host organism and availability of genetic methods suitable for its modification and transformation.

## 11.2 Characteristics of Host Strains

As shown in Table 3, most host strains used to develop production strains for food-processing enzymes have been derived from a relatively small number of bacterial and fungal species, primarily *B. subtilis, B. licheniformis, A. niger*, or *A. oryzae*. These microorganisms have a long history of use as safe sources of native enzymes and a proven record of efficient growth under industrial production conditions. They are also amenable to genetic manipulations and known for their ability to secrete ample quantities of enzymes into fermentation media. These characteristics make these microorganisms particularly desirable for use as hosts for a variety of heterologous enzymes. Several microorganisms with no history of use in the industrial production of native enzymes, such as *E. coli* K-12, *F. venenatum*, and *P. fluorescens* have also been successfully utilized as hosts for expression of food-processing enzymes. Although majority of host microorganisms used to date secrete enzymes in the fermentation media, the Gram negative bacteria *E. coli* K-12 (source of chymosin) and *P. fluorescens* Biovar I (source of α-amylase) accumulate rather than secrete heterologous enzymes. Isolation and purification of the accumulated enzymes generally involve more steps during the production process than purification of the secreted enzymes.

The wild-type strains of several host microorganisms produce a variety of extracellular enzymes. Such enzymes may be carried over to the final enzyme preparation and catalyze undesirable reactions in food. Extracellular proteases are particularly troublesome because they tend to degrade heterologous enzymes. To increase enzyme yield, protease-deficient host strains have been constructed. Other modifications include developing sporulation-deficient mutant strains. Microorganisms used as hosts for recombinant enzymes listed in Table 3 are recognized as non-pathogenic (i.e., they do not cause disease in healthy humans and animals). It is less clear whether all of these microorganisms should also be described as non-toxigenic. Some of these microorganisms, most notably *A. niger*, *A. oryzae*, and *F. venenatum*, may produce low levels of toxic secondary metabolites under certain cultivation conditions. Pariza and Foster [193] defined the non-toxigenic organism as one which does not produce injurious substances at levels that are detectable or demonstrably harmful under ordinary conditions of use or exposure. Under this definition, *A. niger*, *A. oryzae*, and *F. venenatum* production strains are sometimes described as non-toxigenic in submissions to FDA.

**Table 3:** Enzymes from recombinant microorganisms.

| Source | Enzyme |
|---|---|
| *Aspergillus niger* | Phytase |
| | Chymosin |
| | Lipase |
| *Aspergillus oryzae* | Esterase–lipase |
| | Aspartic proteinase |
| | Glucose oxidase |
| | Laccase |
| | Lipase |
| | Pectin esterase |
| | Phospholipase A1 |
| *Bacillus licheniformis* | α-Amylase |
| | Pullulanase |
| *Bacillus subtilis* | α-Acetolactate dccarboxylase |
| | α-Amylase |
| | Maltogenic amylase |
| | Pullulanase |
| *Escherichia coli* K-12 | Chymosin |
| *Fusarium venenatum* | Xylanase |
| *Kluyveromyces marxianus* var. *lactis* | Chymosin |
| *Pseudomonas fluorescens* Biovar I | α-Amylase |
| *Trichoderma reesei* | Pectin lyase |

*Source:* 188

## 12. ENGINEERING STABILITY OF BIOCATALYSIS

Biocatalysts are the catalysts of cell metabolism i.e., the enzymes. However, this concept has expanded beyond its physiological meaning, a biocatalyst being, in broader terms, any biological entity capable of catalyzing the conversion of a substrate into a product. Accordingly, biocatalysts can be divided in to cellular (whether growing, resting or non-living cells) and non-cellular (enzymes that have been removed from the cell system that produce them). Ribozymes, abzymes and peptide mimics can also be considered as biocatalysts [195]. Biocatalysts have competed and will compete with conventional chemical catalysts. Potential advantages of biocatalysts are their high specificity, high activity under mild environmental conditions and high turnover number; their biodegradable nature and their label as a natural product have now also become very important assets of biocatalysts [196]. Drawbacks are inherent to their complex molecular structure that makes them costly to produce and also intrinsically unstable. Different agents like temperature and chemicals promote enzyme inactivation. Inactivation by chemicals can be avoided rather easily by keeping them out of the reaction medium. Temperature, however, produces opposed effects on enzyme activity and stability, and is therefore a key variable in any biocatalytic process [197]. In fact, both enzymes active at low temperatures and stable at high temperatures are of great technological potential [198]. Biocatalyst stability i.e., the capacity to retain activity through time, is undoubtedly the limiting factor in most bioprocesses, biocatalyst stabilization being then a central issue of biotechnology. In fact, biocatalyst operational stability will determine to a large extent the viability of the process, be it new or faced to compete with an already existing technology. It is reasonable then that a significant effort in R&D in the field of biocatalysis is devoted, from different perspectives, to enzyme stabilization [199].

### 12.1 Strategies for the Production of Stable Biocatalysts

Enzyme stability is dictated by its three-dimensional configuration, which in turn is determined by genetic (primary structure) and environmental (interaction with the surroundings) factors. Screening for intrinsically stable biocatalysts is a prominent area of research in biocatalysis. Nature biodiversity is an almost inextinguishable source of biocatalysts. Even though, the systematic exploitation of selected natural niches as sources of enzymes is rather recent. Research on extremophiles i.e., organisms able to survive and thrive in extreme environmental conditions, as promising sources for highly stable enzymes, is a subject of the present decade [200] and certainly a very active research area at present [201]. Those conditions (extreme temperatures, extreme pH and aggressive chemicals) may be present in the reaction medium, being then fruitful to produce enzymes suited to express and retain their activities in such conditions. Biocatalyst thermostability allows a higher operation temperature, which is clearly advantageous because of a higher reactivity (higher reaction rate, lower diffusional restrictions), higher stability, higher process yield (increased solubility of substrates and products and favourable equilibrium displacement in endothermic reactions), lower viscosity and fewer contamination problems [202]. These advantages surmount certain drawbacks arising from more stringent requirements for materials, harder post-reaction inactivation, and restrictions in the case of labile substrates or products.

Thermostable biocatalysts are therefore highly attractive. Thermostable enzymes can be obtained from mesophilic and thermophilic organisms; even psycrophiles have some thermostable enzymes. Thermophiles represent an obvious source of thermostable enzymes, being reasonable to assume that such character will confer their proteins a high thermal stability. This is certainly so, as can be appreciated in the case of several biotechnologically relevant enzymes from the hyperthermophilic archaebacteria *Pyrococcus furiosus* and *Thermotoga* sp. (Table 4) [203, 204]. In fact, it has been claimed that enzymes from thermophiles are stable at temperatures higher than 20°C to the temperature

**Table 4:** Thermostable enzymes from *Pyrococcus furiosus* and *Thermotoga* sp.

| Enzyme | Temperature optimum (°C) *Pyrococcus furiosus* | Temperature optimum (°C) *Thermotoga* sp. |
|---|---|---|
| α-Amylase | 100 | 90 |
| Cellulase | | 105 |
| α-Glucosidase | 115 | |
| α-Galactosidase | | 105 |
| β-Glucosidase | 105 | 105 |
| β-Galactosidase | | 80 |
| β-Glucanase | | 95 |
| Glucose isomerase | | 95 |
| Protease | 110 | |
| Pullulanase | 110 | |
| Xylanase | | 105 |

*Source:* 200

**Table 5:** Industrial thermostable enzymes, commercial enzymes from thermophiles and thermophilic genes cloned in mesophilic hosts.

| Industrial thermostable mesophilic enzyme | | | Commercial enzymes from thermophiles | | Thermophilic genes cloned in mesophilic host | |
|---|---|---|---|---|---|---|
| **Thermostable enzyme** | **Mesophile producer** | **Temp °C** | **Thermostable enzyme** | **Thermophile producer** | **Cloned enzyme** | **Thermophilic donor** |
| α-Amylase | *Bacillus* | 95 | O-malate dehydrogenase | *T. thermophilus* | Thermolysin | *B. thermoprot.* |
| Glucoamylase | *Aspergillus* | 60 | β-Amylase | *C. thermosulphuricum* | Acetyl quinase | *B. stearotherm.* |
| Pullulanase | *Aerobacter* | 60 | α-Galactosidase | *B. stearothermophilus* | Alanine dehydrogenase | *B. stearotherm.* |
| Glucose isomerase | *Actinoplanes* | 60 | DNA polymerase | *P. furiosus* | Leucine dehydrogenase | *B. stearotherm.* |
| Pectinase | *Aspergillus* | 60 | α-Amylase | *P. furiosus* | DNA Polymerase | *T. aquaticus* |
| Alcalase | *Bacillus* | 60 | Glutamate dehydrogenase | *P. furiosus* | endonuclease | *T. thermophilus.* |
| Lipase | *Aspergillus* | 60 | Cellulase | *R. marinus* | Malate dehydrogenase | *T. flavus* |
| Acid protease | *Mucor* | 50 | | | β-Glucanase | *T. cellulolytic* |
| Lactase | *Aspergillus* | 50 | | | β-Glucosidase | *C. saccahrol* |
| | | | | | Pullulanase | *T. brockii* |
| | | | | | Lipase | *B. thermocat* |

*Source:* 200

optimum for growth of such organisms [205]. Interestingly, enzymes from mesophiles exhibit the same pattern. By now, a high number of thermostable enzymes from thermophiles has been reported; most of them belong to eubacterial and archaebacterial kingdoms [206]. However, the technological use of thermophiles still faces several challenges since knowledge on physiology and genetics of such organisms is poor, they are fastidious, grow slowly and are not recognized as safe. Therefore, even though thermal stability can be considered a rare event in mesophilic organisms, thermostable enzymes used by industry are still produced from mesophiles and commercial enzymes from thermophiles are still scarce, as can be appreciated in Table 5 [206, 207]. Opportunities for thermophilic enzymes in industrial processes have been recently highlighted [208].

It follows then the convenience of cloning thermophilic genes into more suitable mesophilic hosts. Those systems will be highly productive and the enzymes produced will retain their original thermostability. In fact, in a number of cases, thermophilic genes have been cloned and expressed in mesophilic hosts, producing enzymes highly active and stable at high temperatures. Some examples are in Table 2 [207, 209-211]. *E. coli* and other bacterial hosts pose some problems in expressing genes from archaebacteria, because of misreading of intervened genes. This is not the case with eubacterial genes, being therefore better candidates for cloning into bacterial hosts [207].

In some cases, remarkable similarities are observed between thermophilic enzymes and their mesophilic counterparts, homology being as high as 85% [212]. Thermostability is the result of differences in specific amino acid sequences, and it has been ascribed to a more rigid configuration and high number of hydrophobic interactions. By examining primary sequences of thermophilic enzymes and mesophilic counterparts, the non-conserved regions, as those possibly linked to the thermostable phenotype, can be identified. This opens up the possibility of using protein engineering techniques [213] to produce point mutations in the mesophilic structural gene, which will result in the corresponding amino acid substitution in the primary structure of the encoded protein [205]. Good results have been obtained in several cases while replacing amino acids for those corresponding to the thermophilic protein [212]. However, homology between mesophilic enzymes and their thermophilic counterparts is usually between 30 and 50%, and no general strategy for converting mesophilic into thermophilic enzymes has emerged yet, making thermophiles or the genes derived from them the preferred source for thermostable enzymes in the foreseeable future [204]. Genetic engineering and protein engineering are modern techniques already in use for the commercial production of biocatalysts of improved stability, not only to high temperatures, but also to extremes of pH, oxidizing agents and organic solvents. Cloning and expression in suitable hosts is being used routinely by major enzyme production companies to produce improved biocatalysts; this certainly applies to the cloning of thermostable enzyme genes. Protein engineering is also being used to obtain improved biocatalysts, the case of alkaline protease being a paradigm. Already in the market, thermostable proteases capable to withstand harsh washing conditions (high pH, high concentration of strong oxidants) are products of protein engineering produced by point amino acid substitutions in the most labile region of the molecule. In the production of syrups from cornstarch, thermostability of α-amylase is severely reduced below pH 6, which poses the inconvenience of pH adjustment before and after starch liquefaction. A thermostable α-amylase from *Bacillus licheniformis*, active at low pH and low $Ca^{++}$ concentration has been patented [214] A thermostable glucose isomerase is a major challenge in the production of high-fructose corn syrup. Equilibrium is favoured at high temperature, so that at 110°C, 55% HFCS could be produced at the enzyme reactor stage, without the cumbersome process of sugar fractionation now used [215]. It was shown that specific substitution of a surface arginine residue for lysine, obtained by site-directed mutagenesis, produced a substantial thermal stabilization in the glucose isomerase from *Actinoplanes missouriensis* [216]. Protein engineering is a powerful tool for the design of robust biocatalysts and probably most future biocatalysts will be produced by engineered organisms.

## 12.2 Strategies for the Operational Stabilization of Biocatalysts

Biocatalysts are required to perform in an environment quite different from their natural habitat. Most enzymatic reactions are performed in aqueous media, which favours inactivation. Water acts as a reactant in inactivation reactions and also as a lubricant in conformational changes associated with protein unfolding [202]. Therefore, biocatalyst stabilization under operation conditions is a key issue of biocatalysis. Several strategies are at hand to increase operational stability: the use of stabilizing additives, chemical modification of enzyme structure, derivatization, immobilization, crystallization and medium engineering. Use of stabilizing additives is a customary practice in enzyme technology and shelf-life of enzyme products very much relies upon such additives. However, its use as operational stabilizers has little significance and poor predictability [217], although in the case of enzymes performing in non-conventional media, the use of additives has proved to enhance enzyme activity and stability [218].

Stabilization by chemical modification of the protein molecule is attractive, but has not received much attention [219, 220]. Increased stability has been obtained by the introduction of hydrophilic groups on the surface of the enzyme molecule that reduce the contact of hydrophobic regions with water, thereby preventing incorrect refolding after reversible denaturation [202]. Derivatization with polymers is being increasingly proposed for the stabilization of soluble enzymes. Modification of proteases with carbohydrate polymers, like polymeric sucrose and dextran, has proven to stabilize them against inactivation induced by temperature and chaotropic agents [221]. Horseradish peroxidase has been recently stabilized with several methoxypolyethylene glycols [222].

Immobilization to solid carriers is perhaps the most used strategy to improve the operational stability of biocatalysts, other benefits being obtained as well, like better control of operation, flexibility of reactor design, and facilitated product recovery without catalyst contamination [223]. Thermal stability upon immobilization is the result of molecular rigidity and the creation of a protected microenvironment. Among immobilization methods available, multipoint covalent attachment is the most effective in terms of thermal stabilization [224], although thermal stabilization has also been reported for gel-entrapped enzymes [225]. We have observed dramatic increase in thermal stability by immobilizing different enzymes to glutaraldehyde-activated chitin matrices, where multiple Schiff-base linkages are established between free amino groups in the protein and the aldehyde group in the glutaraldehyde linker [226]. Despite its great technological potential, few large-scale processes utilize immobilized enzymes. Severe restrictions may arise because of additional costs, activity losses and diffusional restrictions. In the last few years, improvement in carrier and immobilization techniques are opening new options for process development. In general, immobilized biocatalysts will compete advantageously when the cost of the catalyst is a major component of the processing cost (which is not always the case) and substrates and products are readily soluble and of low molecular weight.

Cross-linked enzyme crystals (CLEC) are highly stable novel type biocatalysts. They are produced by stepwise crystallization and molecular cross-linking to preserve the crystalline structure. Uniform size crystals can be obtained in the range from 1 to 100 mm; under 5 mm, diffusional restrictions are insignificant. CLEC are extremely stable, not only with respect to temperature, but also to other inactivating agents like organic solvents. In CLEC, enzyme molecules are compacted almost to the theoretical limit, stabilization being a consequence of intense polar and hydrophobic interactions. Again, molecular rigidity is responsible for thermal stability. Besides, CLEC are highly resistant to proteolysis, proteases being excluded from the tight crystal matrix. Specific activity is significantly higher for CLEC than for immobilized enzymes, although activity can be severely restricted by the low molecular flexibility of the enzyme and substrate size exclusion. Stability of CLEC in organic hydrophobic solvents and water-miscible co-solvents is remarkably high [227], therefore most applications of CLEC are being developed in connection with enzyme synthesis in organic media. Some examples are the production of chiral compounds, peptides and esters. Several CLEC are now in the market. Some of them are lipases, thermolysin, glucose isomerase and penicillin acylase, the last two of paramount commercial significance [228]. A completely different approach for biocatalyst stabilization is medium engineering i.e., the manipulation of reaction medium [229]. Since water is involved in enzyme inactivation, partial or almost total substitution of water might be beneficial for biocatalyst stability [230]. In fact, numerous cases have been reported where remarkable enzyme stability has been obtained in such media. Until recently, the use of enzymes in non-aqueous media seemed unfeasible because of the very low activities obtained. However, its tremendous technological potential has been a powerful driving force for research and development in that area of biocatalysis, and recent advances are outstanding [231]. Organic solvents, reputedly incompatible with enzyme activity, have notable exceptions like polyglycols [232] and glymes and recently, guidelines for the proper selection of solvents have emerged [233]. Even though the main purpose of medium engineering in biocatalysis is associated with the utilization of robust commercial hydrolytic enzymes in organic synthesis [234], thermostability in organic media is an additional bonus of great significance in process economy. In recent years, proteases, lipases, acylases and glycosidases in organic media have been studied in the synthesis of peptides [235, 236], esters [237], oligosaccharides [238] and glycosides [239]. Products have considerable pharmacological and industrial relevance, and in all those cases, thermal stability of the biocatalyst was a key issue. The use of protein engineering to design biocatalysts especially suited to perform in organic solvents is also being studied [240].

## 13. CONCLUSIONS

Lactic acid bacteria and yeasts have been playing a major role in the food and beverage industries since long and are mankind's oldest domesticated organisms, generally regarded as safe. Their metabolism has made possible various biotechnological processes like beer brewing, leavening bread doughs, wine making, cheese making, yoghurt production and many indigenous fermented foods. *S. cerevisiae* was also the first genetically modified organism (GMO) to be cleared for food use, as a baking and brewing strain. Although considerable progress has been made in genetic manipulation of wine yeasts for specific wine making processes, these strains have not yet been cleared for commercial use probably due to the public fear and perception of risk with regard to genetically modified food. It has been generally agreed that these fears are baseless as no scientific test has shown any of the GM foods currently being used to be toxic.

Fears about food and environmental safety have evoked a plethora of strict legislation and regulatory guidelines. Such regulations, although differing in detail, are broadly similar in most countries. Guidelines for approval of GM products and the release of GMOs usually require a number of obvious guarantees. These include a complete definition of the DNA sequence introduced, and the elimination of any sequence that is not indispensable for expression of the desired property, the absence of any selective advantage conferred on the transgenic organism that could allow it to become dominant in natural habitats, no danger to human health and/or the environment from the transformed DNA, and a clear advantage to both the producer and the consumer.

The regulatory authorities appear more willing to approve the use of GMOs than the public as the latter suspects that GM foods will prove unhealthy in the long term and that the escape of GMOs with transplanted genes will damage the environment and result in the loss of biodiversity. As the fears of the general pubic are baseless, consumer education is essential to remove their fear of the unknown. Scientists must consistently inform the public and remain open about experiments, research and products. The consumer should be reassured of transparent regulatory systems and the meticulous implementation of biosafety legislation with clear technical standards and definitions with respect to GM products.

Successful application of recombinant DNA technology in the wine industry will depend on assuring commercial users of genetically modified wine yeasts that existing desirable characteristics have not been damaged, that the requirements of beverage legislation are met, and that the engineered strain will be stable in practice. Enzymes found in nature have also been used in the production of fermented foods for millennia. The production of enzyme preparations isolated from natural sources dates back to the late $19^{th}$ century. The developments in molecular genetics and cell biology in the last four decades have re-shaped enzyme production. It became possible to clone genes encoding enzymes and express them in host microorganisms that are well-adapted to large-scale industrial fermentation. Enzyme yield could be substantially increased by using efficient promoters and introducing multiple copies of the enzyme-encoding gene. It also became possible to tailor enzyme properties to food-processing conditions such as temperature or pH. This has been accomplished by modifying the amino acid sequence of the enzyme using either rational design or molecular evolution. As examples, we described α-amylases that were modified for increased thermostability to match the conditions of starch hydrolysis during production of sweeteners from corn.

The safety of enzyme production strains continues to be a focus of attention. The concept of developing safe strain lineages using well-characterized non-pathogenic, non-toxigenic microbial strains, particularly those with a history of safe use in food enzyme manufacture has been embraced by the enzyme industry. Industrial strains of microorganisms used as sources of native enzymes are now often used as hosts for heterologous enzymes. It has been recognized that certain fungi traditionally used as sources of food-processing enzymes and considered to be non-pathogenic and non-toxigenic, for example, *A. oryzae* or *A. niger*, are capable of producing low levels of certain mycotoxins under fermentation conditions conducive to the synthesis of these compounds. Mycotoxin production by these fungi can be minimized or avoided by controlling the fermentation conditions used in the production of enzymes. Advances in genetics and molecular biology enriched the knowledge about these microorganisms and provided new tools for reducing their toxigenic potential. Some fungal host strains described above have been altered using classical mutagenesis and/or rDNA methods to inactivate or impair their mycotoxin synthetic pathways. Most effective is the inactivation of the entire pathway by deletion or disruption of genes encoding key enzymes involved in mycotoxin synthesis. For example, the *F. venenatum* strain, from which the trichodiene synthase gene was deleted, lost the capacity to produce trichothecene mycotoxins.

Another interesting approach relies on avoiding potential unintended effects of DNA insertion, such as an increase in the levels of secondary metabolites, by targeting the cloned genes into designated chromosomal loci in the host genome. Once it has been shown that the DNA insertion does not affect secondary metabolite pathways, it is assumed that other cloned genes can be safely inserted into the same locus without triggering unintended effects. This strategy is useful for host strains that naturally produce secondary metabolites at toxicologically insignificant levels. Microbial host strains have also been genetically modified to improve enzyme production. For example, several bacterial and fungal species used as sources of food-processing enzymes naturally produce extracellular proteases that may degrade the target enzyme. Over the years, classical mutagenesis has been effectively employed to generate protease-deficient mutants of these microorganisms. However, upon the identification of genes encoding proteases, it became possible to generate strains deficient in specific proteases. Current strain improvement strategies have already contributed to creating more efficient and safer enzyme production strains. This trend will undoubtedly continue as the knowledge about the genetic make-up of microorganisms used for enzyme production expands and new genetic techniques emerge.

## REFERENCES

1. Stanton RW (1998) Food fermentation in the tropics. In: *Microbiology of Fermented Foods*, Vol 2, Wood BJB (ed), Elsevier Applied Science Publishers, London, UK, pp 696-712.
2. Dirar H (1993) *The Indigenous Fermented Foods of the Sudan: A Study in African Food and Nutrition*, CAB International, Wallingford, USA.
3. Pedersen CS (1979) *Microbiology of Food Fermentations*, AVI Publishers, Westport, USA.

4. Wood E (1989) *World Sourdoughs from Antiquity*, Sinclair Publishing, Idaho, USA.
5. Yokotsuka T & Sasaki M (1998) Fermented protein foods in the orient, with emphasis on *shoyu* and *miso* in Japan. In: *Microbiology of Fermented Foods*, Vol 1, Wood BJB (ed), Elsevier Applied Science Publishers, London, UK, pp 351-415.
6. Pretorius IS (2000) Tailoring wine yeasts for the new millennium: novel approaches to the ancient art of winemaking. *Yeast* 16: 675-729.
7. Soni SK & Sandhu DK (1999) Fermented cereal products. In: *Biotechnology: Food Fermentation Microbiology, Biochemistry and Technology*, Vol 2, Joshi VK & Pandey A (eds), Educational Publishers and Distributors, New Delhi, India, pp 895-950.
8. Soni SK & Arora JK (2000) Indian fermented foods: biotechnological approaches. In: *Food Processing: Biotechnological Applications*, Marwaha SS & Arora JK (eds), Asiatech Publishers Pvt Ltd, New Delhi, India, pp 143-190.
9. Soni SK & Marwaha SS (2003) Cereal products: Biotechnological approaches for their production. In: *Biotechnology: Strategies in Agro-Processing*, Marwaha SS & Arora JK (eds), Asiatech Publishers Pvt Ltd, New Delhi, India, pp 236-266.
10. Hofvendahl K & Hahn-Hagerdal B (2000) Factors affecting the fermentative lactic acid production from renewable resources. *Enzyme Microb Technol* 26: 87-107.
11. Wells JM, Robinson K, Chamberlain LM, Schofield KM & Le Page RW (1996) Lactic acid bacteria as vaccine delivery vehicles. *Antonie van Leeuwenhoek* 70: 317-330.
12. Pouwels PH, Leer RJ, Shaw M, Heijne den Bak-Glashouwer MJ, Tielen FD, Smit E, Martinez B, Jore J & Conway PL (1998) Lactic acid bacteria as antigen delivery vehicles for oral immunization purposes. *Int J Food Microbiol* 41: 155-167.
13. McKay LL & Baldwin KA (1990) Applications for biotechnology: present and future improvements in lactic acid bacteria. *FEMS Microbiol Rev* 87: 3-14.
14. Stiles ME (1996) Biopreservation by lactic acid bacteria. *Antonie van Leeuwenhoek* 70: 331-345.
15. Portier A, Boyaka P, Bougoudoga F, Dubarry M, Huneau F, Tome D, Dodin A & Coste M (1993) Fermented milks and increased antibody responses against cholera in mice. *Int J Immunotherapy* 94: 217-224.
16. Driessen F & Boer R (1989) Fermented milks with selected intestinal bacteria a healthy trend in new products. *Netherlands Milk Dairy J* 43: 369-382.
17. Holzapfel WH, Haberer P, Snel J, Schillinger U & Huis In't Veld JHJ (1998) Overview of gut flora and probiotics. *Int J Food Microbiol* 41: 85-101.
18. Klaenhammer TR (1988) Bacteriocins of lactic acid bacteria. *Biochimie* 70: 337-349.
19. Nes IF, Diep DB, Havårstein LS, Brurberg MB, Eijsink V & Holo H (1996) Biosynthesis of bacteriocins in lactic acid bacteria. In: *Lactic Acid Bacteria: Genetics, Metabolism and Applications*, Venema G, Huis in't veld JHJ & Hugenholtz J (eds), Kluwer Academic Publishers, Dordrecht, The Netherlands, pp 17-32.
20. Kemperman R, Kuipers A, Karsens H, Nauta A, Kuipers O & Kok J (2003) Identification and characterization of two novel clostridial bacteriocins, circularin A and closticin 574. *Appl Environ Microbiol* 69: 1589-1597.
21. Atanassova M, Choiset Y, Dalgalarrondo M, Chobert JM, Dousset X, Ivanova I & Haertlé T (2003) Isolation and partial characterisation of a proteinaceous anti-bacteria and anti-yeast compound produced by *Lactobacillus paracasei* sub sp. *paracasei* M3. *Int J Food Microbiol* 87: 63-73.
22. Arihara K, Ogihara S, Mukai T, Itoh M & Kondo Y (1996) Salivacin 140, a novel bacteriocin from *Lactobacillus salivarius* sub sp. *salicinus* T140 active against pathogenic bacteria. *Lett Appl Microbiol* 22: 420-424.

23. Cardi A (2002) Selection of *Escherichia coli*-inhibiting strains of *Lactobacillus paracasei* subsp. *paracasei. J Ind Microbiol Biotechnol* 29: 303-308.
24. Lachance MA & Starmer WT (1998) Ecology and yeasts. In: *The Yeasts, A Taxonomic Study*, Kurtzman CP & Fell JE (eds), Elsevier Science, Amsterdam, The Netherlands, pp 21-30.
25. Fleet GH (1998) The microbiology of alcoholic beverages. In: *Microbiology of Fermented Foods*, Wood BJB (ed), Vol 1, Blackie Academic and Professional, Glassgow, UK, pp 217-262.
26. Martini A (1993) The origin and domestication of the wine yeast *Saccharomyces cerevisiae. J Wine Res* 4: 165-176.
27. Flamm EL (1991) How FDA approved chymosin: a case history. *Biotechnology* 9: 349-351.
28. Pretorius IS & Bauer FF (2002) Meeting the consumer challenge through genetically customized wine yeast strains. *Trends Biotechnol* 20: 426-432.
29. Rahi DK & Soni SK (2007) Applications and commercial uses of microbes. In: *Microbes: A Source of Energy for 21st Century*, Soni SK (ed), New India Publishing House, New Delhi, India, pp 69-126.
30. Barre P, Vezinhet F, Dequin S & Blondin B (1993) Genetic improvement of wine yeasts. In: *Wine Microbiology and Biotechnology*, Fleet GH (ed), Harwood Academic Publishers, Chur, Switzerland, pp 265-287.
31. Hammond JRM (1996) Yeast genetics. In: *Brewing Microbiology*, Priest FG & Campbell I (eds), Chapman and Hall, London, UK, pp 43-82.
32. Pretorius IS & van der Westhuizen TJ (1991) The impact of yeast genetics and recombinant DNA technology on the wine yeast industry. A review. *S Afr J Enol Vitic* 12: 3-31.
33. Querol A & Ramon D (1996) The application of molecular techniques in wine microbiology. *Trends Food Sci Technol* 7: 73-78.
34. Srivastava DK & Raverkar KP (1999) Genetic manipulation of industrially important microorganisms. In: *Biotechnology: Food Fermentation Microbiology, Biochemistry and Technology*, Vol 1, Joshi VK & Pandey A (eds), Educational Publishers and Distributors, New Delhi, India, pp 173-258.
35. Soni R, Soni SK & Goyal N (2007) Biotechnology for developing novel microbes. In: *Microbes: A Source of Energy for 21st Century*, Soni SK (ed), New India Publishing House, New Delhi, India, pp 495-538.
36. Dubey RC & Maheshwari DK (2004) Mutation and mutagenesis. In: *A Textbook of Microbiology*, S Chand & Company Ltd, New Delhi, India, pp 193-213.
37. Wing Q & Laustsen O (1938) Artificial species hybridization in yeast. Comptes rendus des travaux du Laboratoire Carlsberg. *Series Physiologique* 22: 235-240.
38. Johnston JR (1965) Breeding yeasts for brewing; Il Production of hybrid strains. *J Inst Brew* 71: 135-137.
39. Barre P, Vezinhet F, Dequin S & Blondin B (1993) Genetic improvement of wine yeast. In: *Wine Microbiology and Biotechnology*, Fleet GH (ed), Harwood Academic Publishers, Chur, Switzerland, pp 421-447.
40. Thornton RJ (1985) The introduction of flocculation into a homothallic wine yeast. A practical example of the modification of wine making properties by the use of genetic technique. *Am J Enol Vitic* 36: 47-49.
41. Tubb RS, Seárle BA, Goody AR & Brown AJP (1981) Rare mating and transformation for the construction of novel brewing yeasts. In: Proccedings of 18th European Brewery Convention Congress, Copenhagen, IRL Press, London, UK, pp 487-496.

42. Hammond JRM & Eckersley KW (1984) Fermentation properties of brewing yeast with killer character. *J Brew* 90: 167-177.
43. Conde J & Fink GR (1976) A mutant of *Saccharomyces cerevisiae* defective for nuclear fusion. *Proc Natl Acad Sci USA* 73: 3651-3655.
44. Dutcher SK (1981) Internuclear transfer of genetic information in Kar 1-1/Kar-1 heterokaryons in *Saccharomyces cerevisiae. Mol Cell Biol* 1: 245-253.
45. Loiseau G (1986) Obtention dun marquage mitochondrial de levures oenologigsues selectionnees en Champagne. Application au controle microbiologique en oenologie ScD Thesis, Universite des, Sciences et Tecniques du Languedoc, Montpellier, France.
46. Vezinhet F, Blondin B & Barre P (1991) Mapping of the FLO5 gene of *Saccharomyces cerevisiae* by tansfer of a chromosome during cytoduction. *Biotechnol Lett* 13: 47-52.
47. Thomas CM (2007) Genetic manipulation of bacteria. In: *Biotechnology, Encyclopedia of Life Support Systems* (EOLSS), Doelle HW & Rokem JS (eds), Eolss Publishers, Oxford, UK, pp 20-25.
48. Cocaign-Bousquet M, Garrigues C, Loubiere P & Lindley N (1997) Physiology of pyruvate metabolism in *Lactococcus lactis. Antonie van Leeuwenhoek* 70: 253-267.
49. De Vos WM, Boerrigter I, van Rooijen RJ, Reiche B & Hengstenberg W (1990) Characterization of the lactose-specific enzymes of the phosphotransferase system in *Lactococcus lactis. J Biol Chem* 265: 22554-22560.
50. Deutscher J, Kuster E, Bergstedt U, Charrier V & Hillen W (1995) Proteinkinase dependent HPr/CcpA interaction links glycolytic activity to carbon catabolite repression in Gram-positive bacteria. *Mol Microbiol* 15: 1049-1053.
51. Fath MJ & Kolter R (1993) ABC exporter. *Microbiol Rev* 57: 995-1017.
52. Garrigues C, Loubiere P, Lindley ND & Cocaign-Bouquet M (1997) Control of the shift from homolactic acid to mixed acid fermentation in *Lactococcus lactis*: predominant role of NADH/NAD1 ratio. *J Bacteriol* 179: 5282-5287.
53. Gibson GR & Roberfroid MB (1995) Dietary modulation of the human colonic microbiota: introducing the concept of prebiotics. *J Nutr* 124: 1401-1412.
54. Godon JJ, Delorme C, Bardowski J, Chopin MC, Ehrlich SD & Renault P (1993) Gene inactivation in *Lactococcus lactis* branched chain amino acid biosynthesis. *J Bacteriol* 175: 4383-4390.
55. Hols P, Kleerebezem M, Schank AN, Ferain T, Hugenholtz J, Delcour J & de Vos WM (1999) Conversion of *Lactococcus lactis* from homolactic to homoalanine fermentation through metabolic engineering. *Nat Biotechnol* 17: 588-592.
56. Hols P, Ramos A, Hugenholtz J, Delcour J, de Vos WM, Santos H & Kleerebezem M (1999) Acetate utilization in *Lactococcus lactis* deficient in lactate dehydrogenase: A rescue pathway for maintaining redox balance. *J Bacteriol* 181: 5521-5226.
57. Hosono A, Lee J, Ametani A, Natsume M, Hirayama M, Adachi T & Kaminogawa S (1997) Characterization of a water-soluble polysaccharide fraction with immunopotentiating activity from *Bifidobacterium adolescentis* M101-4. *Biosci Biotech Biochem* 61: 312-316.
58. Hugenholtz J (1993) Citrate metabolism in lactic acid bacteria. *FEMS Microbiol Rev* 12: 165-178.
59. Kitazawa H, Yamaguchi T, Miura M, Saito T & Itoh H (1993) B-cell mitogen produced by slime-forming, encapsulated *Lactococcus lactis* subspecies *cremoris* isolated from ropy sour milk, viili. *J Dairy Sci* 76: 1514 -1519.
60. Llanos RM, Marin CJ, Hillier AJ & Davidson BE (1993) Identification of a novel operon in *Lactococcus lactis* encoding enzymes for lactic acid synthesis: phosphofructokinase, pyruvate kinase and lactate dehydrogenase. *J Bacteriol* 175: 254-255.

61. Lopez de Felipe F, Starrenburg MJC & Hugenholtz J (1997) The role of NADH-oxidation in acetoin and diacetyl production from glucose in *Lactococcus lactis*. *FEMS Microbiol Lett* 156: 15-19.
62. Luesink EJ, Beumer CM, Kuipers OP & de Vos WM (1999) Molecular characterization of the *Lactococcus lactis ptsHI* operon and analysis of the regulatory role of HPr. *J Bacteriol* 181: 764-771.
63. Luesink EJ, van Herpen RE, Grossiord BP, Kuipers OP, de Vos WM (1998) Transcriptional activation of the glycolytic *lac* operon and catabolite repression of the *gal* operon in *Lactococcus lactis* are mediated by the catabolite control protein CcpA. *Mol Microbiol* 30: 789-798.
64. Marugg JD, Goelling D, Stahl U, Ledeboer AM, Toonen MY, Verhue WM & Verrips CT (1994) Identification and characterization of the α-acetolactate synthase gene from *Lactococcus lactis* sub sp. *liacetylactis*. *Appl Environ Microbiol* 60: 1390-1394.
65. Neves AR, Ramos A, Nunes MC, Kleerebezem M, Hugenholtz J, de Vos WM, Almeida J & Santos H (1999) *In vivo* NMR studies of glycolytic kinetics in *Lactococcus lactis*. *Biotechnol Bioeng* 64: 200-212.
66. Platteeuw C, Hugenholtz J, Starrenburg MJC, van Alen-Boerrigter I & de Vos WM (1995) Metabolic engineering of *Lactococcus lactis*: influence of the overproduction of α-acetolactate synthase in strains deficient in lactate dehydrogenase as a function of culture conditions. *Appl Environ Microbiol* 61: 3967-3971.
67. Poolman B (1993) Energy transduction in lactic acid bacteria. *FEMS Microbiol Rev* 12: 125-148.
68. Postma PW, Lengeler JW & Jacobsen GR (1993) Phosphoenolpyruvate dependent carbohydrate phosphotransferase systems of bacteria. *Microbiol Rev* 57: 543-94.
69. Thomas TD, Ellwood DC & Longyear VMC (1979) Change from homo- to heterolactic fermentation by *Streptococcus lactis* resulting from glucose limitation in anaerobic chemostat cultures. *J Bacteriol* 138: 109-117.
70. Thompson J, Chassy BM & Egan W (1985) Lactose metabolism in *Streptococcus lactis*: studies with a mutant lacking glucokinase and mannose-transferase activities. *J Bacteriol* 162: 217-233.
71. Van Rooijen RJ, Dechering KJ, Wilmink CMJ, de Vos WM (1993) Lysines 72, 80, 213, and aspartic acid 210 of the *Lactococcus lactis* LacR repressor are involved in the response to the inducer tagatose-6-phosphate leading to induction of lac operon expression. *Protein Eng* 6: 208-215.
72. Van Rooijen RJ, van Schalkwijk S & de Vos WM (1991) Molecular cloning, characterization, and nucleotide sequence of the tagatose 6-phosphate pathway gene cluster of the lactose operon of *Lactococcus lactis*. *J Biol Chem* 266: 7176-7181.
73. Goffeau A, Barrell BG, Bussey H, Davis RW, Dujon B, Feldmann H, Galibert F, Hoheisel JD, Jacq C, Johnston M, Louis EJ, Mewes HW, Murakami Y, Philippsen P, Tettelin H &, Oliver SG (1996) Life with 6000 genes. *Science* 274: 546-567.
74. Dequin S (2001) The potential of genetic engineering for improving brewing, wine making and baking yeasts. *Appl Microbiol Biotechnol* 56: 577-588.
75. Hansen J & Kielland-Brandt M (1997) Brewer's yeast. In: *Yeast Sugar Metabolism, Biochemistry, Genetics, Biotechnology and Applications*, Zimmerman FK & Entian KD (eds), Technomic Publishing, New York, USA, pp 503-526.
76. Yarrow D (1984) *Saccharomyces*. In: *The Yeasts, A Taxonomic Study*, Kreger-Van Rij NJW (ed), Elsevier, Amsterdam, The Netherlands, pp 379-395.
77. Vaughan-Martini A & Martini A (1987) Three newly delimited species of *Saccharomyces* sensu stricto. *Antonie Van Leeuwenhoek* 53: 77-84.

78. Yamagashi H & Ogata T (1999) Chromosomal structures of bottom fermenting yeasts. *Syst Appl Microbiol* 22: 341-353.
79. Hammond JRM (1995) Genetically-modified brewing yeasts for the 21[st] century, progress to date. *Yeast* 11: 1613-1627.
80. Benitez T, Gasent-Ramirez JM, Castrejon F & Codon AC (1996) Development of new strains for the food industry. *Biotechnol Prog* 12: 149-163.
81. Masneuf I, Aigle M & Dubourdieu D (1996) Development of a polymerase chain reaction/restriction fragment length polymorphism method for *Saccharomyces cerevisiae* and *Saccharomyces bayanus* identification in enology. *FEMS Microbiol Lett* 138: 239-244.
82. Rachidi N, Barre P & Blondin B (2000) Examination of the transcriptional specificity of an enological yeast. A pilot experiment on the chromosome-III right arm. *Curr Genet* 37: 1-11.
83. Henschke PA (1997) Wine yeast. In: *Yeast Sugar Metabolism, Biochemistry, Genetics, Biotechnology and Applications,* Zimmermann FK & Entian KD (eds), Technomic Publishing, New York, USA, pp 527-559.
84. Butzke CE & Bisson, LF (1996) Genetic engineering of yeast for wine production. *Agro-Food Ind Hi-Tech* (Jul/Aug): 26-30.
85. Evans IH (1990) Yeast strains for baking: Recent developments. In: *Yeast Technology*, Spencer JFT & Spencer DM (eds), Springer, Berlin Heidelberg, New York, USA, pp 13-45.
86. Randez-Gil F, Sanz P & Prieto JA (1999) Engineering baker's yeast: room for improvement. *Trends Biotechnol* 17: 237-244.
87. Innis M A, Holland MJ, McCabe PC, Cole GE, Wittman VP, Tal R, Watt KWK, Gelfand DH, Holland JP & Meade JH (1985) Expression, glycosylation, and secretion of an *Aspergillus* glucoamylase by *Saccharomyces cerevisiae. Science* 228: 21-26.
88. Steyn AJ & Pretorius IS (1991) Co-expression of a *Saccharomyces diastaticus* glucoamylase-encoding gene and a *Bacillus amyloliquefaciens* alpha-amylase-encoding gene in *Saccharomyces cerevisiae. Gene* 100: 85-93.
89. Janse BJ & Pretorius IS (1995) One-step enzymatic hydrolysis of starch using a recombinant strain of *Saccharomyces cerevisiae* producing alpha-amylase, glucoamylase, and pullulanase. *Appl Microbiol Biotechnol* 42: 878-883.
90. Hollenberg CP & Strasser AWM (1990) Improvement of baker's and brewer's yeast by gene technology. *Food Biotechnol* 4: 527-534.
91. Siso MGI (1996) The biotechnological utilization of cheese whey: a review. *Biores Technol* 57: 1-11.
92. Mawson AJ (1994) Bioconversions for whey utilization and waste abatement. *Biores Technol* 47: 195-203.
93. Jeong YS & Vieth WR (1991) Fermentation of lactose to ethanol with recombinant yeast in an immobilized yeast membrane bioreactor. *Biotechnol Bioeng* 37: 587-590.
94. Farahnak F, Seki T, Ryu DY & Ogrydziak D (1986) Construction of lactose-assimilating and high ethanol producing yeasts by protoplast fussion. *Appl Environ Microbiol* 51: 362-367.
95. Dominques L, Teixeira JA & Lima N (1999) Construction of a flocculent *Saccharomyces cerevisiae* fermenting lactose. *Appl Microbiol Biotechnol* 51: 621-626.
96. Dominques L, Dantas MM, Lima N & Teixeira JA (1999) Continuous ethanol fermentation of lactose by a recombinant flocculating *Saccharomyces cerevisiae* strain. *Biotechnol Bioeng* 64: 692-697.
97. Kumar V, Ramakrishnan S, Teeri TT, Knowles JKC & Hartley BS (1992) *Saccharomyces cerevisiae* cells secreting an *Aspergillus niger* β-galactosidase grow on whey permeate. *Bio/Technology* 10: 82-85.

98. Naumov GI, Naumova ES, Turakainen H & Korhola M (1996) Identification of the β-galactosidase *MEL* genes in some populations of *Saccharomyces cerevisiae*: a new gene *MEL11*. *Genet Res Camb* 67: 101-108.
99. Rønnow B, Olsson L, Nielsen J & Mikkelsen JD (1999) Derepression of galactose metabolism in melibiase producing bakers' and distillers' yeast. *J Biotechnol* 72: 213-228.
100. Olsson L & Hahn-Hägerdal B (1996) Fermentation of lignocellulosic hydrolysates for ethanol production. *Enzyme Microb Technol* 18: 312-331.
101. Lynd LR, Cushman JH, Nichols RJ & Wyman CE (1991) Fuel ethanol from cellulosic biomass. *Science* 251: 1318-1323.
102. Buchert J, Niemelä K, Puls J & Poutanen K (1990) Improvement in the fermentability of steamed hemicellulose hydrolysate by ion exclusion. *Proc Biochem* 25: 176-180.
103. Degre R (1993) Selection and commercial cultivation of wine yeast and bacteria. In: *Wine Microbiology and Biotechnology*, Fleet GH (ed), Harwood Academic Publishers, Chur, Switzerland, pp 421-447.
104. Francois J, Blazquez MA, Arino J & Gancedo C (1997) Storage carbohydrates in the yeast *Saccharomyces cerevisiae*. In: *Yeast Sugar Metabolism*, Zimmermann FK & Entian KD (eds), Technomic Publication, Lancaster, UK, pp 285-311
105. Thevelein JM (1996) Regulation of trehalose metabolism and its relevance to cell growth and function. In: *The Mycota III*, Brambl R & Marzluf GA (eds), Springer-Verlag, Berlin and Heidelberg, pp 395-420.
106. Walker GM (1998) *Yeast Physiology and Biotechnology*, Wiley, New York, USA.
107. Henschke PA & Jiranek V (1993) Yeasts-metabolism of nitrogen compounds. In: *Wine Microbiology and Biotechnology*, Fleet GH (ed), Harwood Academic Publishers, Chur, Switzerland, pp 183-223.
108. Jiranek V, Langridge P & Henschke PA (1995) Regulation of hydrogen sulphide liberation in wine-producing *Saccharomyces cerevisiae* strains by assimilable nitrogen. *Appl Environ Microbiol* 61: 461-467.
109. Jiranek V, Langridge P & Henschke PA (1995) Amino acid and ammonium utilization by *Saccharomyces cerevisiae* wine yeasts from a chemically defined medium. *Am J Enol Vitic* 46: 75-83.
110. Jiranek V, Langridge P & Henschke PA (1995) Validation of bismuth containing indicator media for predicting $H_2S$-producing potential of *Saccharomyces cerevisiae* wine yeast under enological conditions. *Am J Enol Vitic* 46: 269-273.
111. Boulton B, Singleton VL, Bisson LF & Kunkee RF (1996) Yeast and biochemistry of ethanol fermentation. In: *Principles and Practices of Winemaking*, Boulton B, Singleton VL, Bisson LF & Kunkee RE (eds), Chapman and Hall, New York, USA, pp 139-172.
112. Salmon JM & Barre P (1998) Improvement of nitrogen assimilation and fermentation kinetics under ecological conditions by derepression of alternative nitrogen-assimilatory pathways in an industrial *Saccharomyces cerevisiae* strain. *Appl Environ Microbiol* 64: 3831-3837.
113. Stratford M (1992) Yeast flocculation: reconciliation of physiological and genetic viewpoints. *Yeast* 8: 25-38.
114. Urano N, Sato M, Sahara H & Koshino S (1993) Conversion of a non-flocculent brewer's yeast to flocculent ones by electrofusion. 1. Identification and characterization of the fusants by pulse field gel electrophoresis. *J Biotechnol* 28: 237-247.
115. Urano N, Sato M, Sahara H & Koshino S (1993) Conversion of a non-flocculent brewer's yeast to flocculent ones by electrofusion. II. Small-scale brewing by fusants. *J Biotechnol* 28: 249-261.

116. Bidard F, Blondin B, Dequin S, Vezinhet F & Barre P (1994) Cloning and analysis of a *FLO5* flocculation gene from *S. cerevisiae. Curr Genet* 25: 196-201.
117. Watari J, Nomura M, Sahara H & Koshino S (1994) Construction of flocculent brewer's yeast by chromosomal integration of the yeast flocculation gene *FLO1. J Inst Brew* London 100: 73-77.
118. Verstrepen K, Bauer FF, Michiels G, Derdelinckx F, Delvaux F & Pretorius IS (2000) Controlled expression of *FLO1*in *Saccharomyces cerevisiae. Eur Brew Conv Monogr* 28: 30-42.
119. Scanes KT, Hohmann S & Prior BA (1998) Glycerol production by the yeast *Saccharomyces cerevisiae* and its relevance to wine: a review. *S Afr J Enol Vitic* 19: 17-23.
120. Remize F, Roustan JL, Sablarrolles JM, Barre P & Dequin S (1999) Glycerol overproduction by engineered *Saccharomyces cerevisiae* wine yeast strains leads to substantial changes in by-product formation and to a stimulation of fermentation rate in stationary phase. *Appl Environ Microbiol* 65: 143-149.
121. Prior BA & Hohmann S (1997) Glycerol production and osmoregulation. In: *Yeast Sugar Metabolism*, Zimmermann FK & Entian KD (eds), Technomic Publication, Lancaster, UK, pp 313-337.
122. Michnick S, Roustan JL, Remize F, Barre P & Dequin S (1997) Modulation of glycerol and ethanol yields during alcoholic fermentation in *Saccharomyces cerevisiae* strains overexpressed or disrupted for *GDP1* encoding glycerol 3-phosphate dehydrogenase. *Yeast* 13: 783-793.
123. Henschke PA (1997) Wine Yeast. In: *Yeast Sugar Metabolism*, Zimmermann FK & Entian KD (eds), Technomic Publication, Lancaster, UK, pp 527-560.
124. Canal-Llaubers RM (1993) Enzymes in winemaking. In: *Wine Microbiology and Biotechnology*, Fleet GH (ed), Harwood Academic Publishers, Chur, Switzerland, pp 477-506.
125. Van Rensburg P, Van Zyl WH & Pretorius IS (1998) Engineering yeast for efficient cellulose degradation. *Yeast* 14: 67-76.
126. Van Rensburg P, Van Zyl WH & Pretorius IS (1998) Over-expression of the *Saccharomyces cerevisiae* exo-β-1,3-glucanase gene together with the *Bacillus subtilis* endo-β-1,3-1,4-glucanase gene and the *Butyrivibrio fibrisolvens* endo-β-1,4-glucanase gene in yeast. *J Bacteriol* 55: 43-53.
127. Perez-Gonzalez JA, Gonzalez R, Querol A, Sendra J & Ramon D (1993). Construction of a recombinant wine yeast strain expressing β-(1,4)-endoglucanase and its use in microvinification processes. *Appl Environ Microbiol* 59: 2801-2806.
128. Cole VC & Noble AC (1995) Flavour chemistry and assessment. In: *Fermented Beverage Production*, Lea AGH & Piggott JR (eds), Blackie Academic and Professional, London, UK, pp 361-385.
129. Malcorps P, Cheval JM & Dufour JP (1991) A new model for the regulation of ester synthesis by alcohol acetyltransferase in *Saccharomyces cerevisiae* during fermentation. *J Am Soc Brew Chem* 49: 47-53.
130. Lilly M, Lambrechts MG & Pretorius IS (2000) The effect of increased yeast alcohol acetyltransferase activity on the sensorial quality of wine and brandy. *Appl Environ Microbiol* 66: 774-753.
131. Hirata D, Aoki S, Watanabe K, Tsukioka, M & Suzuki T (1992) Stable overproduction of isoamyl alcohol by *Saccharomyces cerevisiae* with chromosome-integrated multicopy *LEU4* genes. *Biosci Biotechnol Biochem* 56: 1682-1683.
132. Fuji T, Nagazawa N, Iwamatsu A, Bogaki T, Tamai Y & Hamachi M (1994) Molecular cloning, sequence analysis and expression of the yeast alcohol acetyltransferase gene. *Appl Env Microbiol* 60: 2786-2792.

133. Fukuda K, Yamamoto N, Kiyokawa Y, Yanagiuchi T, Wakai Y, Kitamoto K, Inoue Y & Kimura A (1998) Brewing properties of sake yeast whose *EST2* gene encoding isoamyl acetate-hydrolysing esterase was disrupted. *J Ferm Bioeng* 85: 101-106.
134. Ough CS (1976) Ethylcarbamate in fermented beverages and foods. I. Naturally occurring ethylcarbamate. *J Agric Food Chem* 24: 323-328.
135. Kitamoto K, Oda K, Gomi K & Takashi K (1991) Genetic engineering of a sake yeast producing no urea by successive disruption of arginase gene. *Appl Environ Microbiol* 57: 2568-2575.
136. Ough CS & Trioli G (1988) Urea removal from wine by an acid urease. *Am J Enol Vitic* 39: 303-307.
137. Visser JJ, Coton EPN, Bauer F, Viljoen M, Vuuren HJJ van (1999) Engineering an acid urease for heterologous expression in *Saccharomyces cerevisiae. Curr Genet* 35: 321-326.
138. Boone C, Sdicu AM, Wagner J, Degré R, Sanchez C & Bussey H (1990) Integration of the yeast K1 killer toxin gene into the genome of marked wine yeasts and its effect on vinification. *Am J Enol Vitic* 41: 37-42.
139. Sasaki T, Watari J, Kohgo M, Nishikawa N & Matsui Y (1984) Breeding of a brewer's yeast possessing anticontaminant properties. *J Am Soc Brew Chem* 42: 164-167.
140. Schoeman H, Vivier MA, Du Toit M, Dicks LM & Pretorius IS (1999) The development of bactericidal yeast strains by expressing the *Pediococcus acidilactici* pediocin gene (*pedA*) in *Saccharomyces cerevisiae. Yeast* 15: 647-656.
141. Perry C & Meaden P (1988) Properties of a genetically-engineered dextrin-fermenting strain of brewers' yeast. *J Inst Brew London* 94: 64-67.
142. Vakeria D & Hinchliffe E (1989) Amylolytic brewing yeast: their commercial and legislative acceptability. In: *European Brewing Convention, Proc 22nd Congr Eur Brew Conv,* IRL Press, Eynsham, England, UK, pp 475-482.
143. Yocum RR (1986) Genetic engineering of industrial yeasts. *Proc Bio Expo* 86: 171-180.
144. Cole GE, McCabe PC, Inlow D, Gelfand DH, Ben-Bassat A Innis MA (1988) Stable expression of *Aspergillus awamori* glucoamylase in distiller's yeast. *Biotechnology* 6: 417-421.
145. Gopal CV & Hammond JRM (1992) Use of genetically-modified yeasts for beer production. In: *European Brewing Convention, Proc 5th Int Brew Technol Conf,* Brewing Technology Services, London, UK, pp 297-306.
146. Enari TM, Nikkola M, Suihko ML, Penttilä M & Knowles J (1987) Yeast strains suitable for accelerated brewing. US Patent 7,044,244.
147. Penttilä M, Suihko ML, Lehtinen U, Nikkola M & Knowles JKC (1987) Construction of brewer's yeasts secreting fungal endo-β-glucanases. *Curr Genet* 12: 413-420.
148. Penttilä M, André L, Saloheimo M, Lehtovaara P & Knowles JKC (1987) Expression of two *Trichoderma reesei* endoglucanases in the yeast *Saccharomyces cerevisiae. Yeast* 3: 175-185.
149. Suihko ML, Lehtinen U, Zurbriggen B, Vilpoa A, Knowles J, Penttilä M (1991) Construction and analysis of recombinant glucanolytic brewer's yeast strains. *Appl Microbiol Biotechnol* 35: 781-787.
150. Gjermansen C, Nilsson-Tillgren T, Petersen JG, Kielland-Brandt MC, Sigsgaard P & Holmberg S. (1988) Towards diacetyl-less brewers' yeast. Influence of *ilv2* and *ilv5* mutations. *J Basic Microbiol* 28: 175-183.
151. Fuji T, Nagazawa N, Iwamatsu A, Bogaki T, Tamai Y & Hamachi M (1994) Molecular cloning, sequence analysis and expression of the yeast alcohol acetyltransferase gene. *Appl Env Microbiol* 60: 2786-2792.
152. Tezuka H, Mori T, Okumura Y, Kitabatake K & Tsumura Y (1992) Cloning of a gene suppressing hydrogen sulfide production by *Saccharomyces cerevisiae* and its expression in a brewing yeast. *J Am Soc Brew Chem* 50: 130-133.

153. Omura F, Shibano Y, Fukui N & Nakatani K (1995) Reduction of hydrogen sulfide production in brewing yeast by constitutive expression of *MET25* gene. *J Am Soc Brew Chem* 53: 58-62.
154. Hansen J & Kielland-Brandt M (1996) Inactivation of *MET10* in brewer's yeast specifically increases $SO_2$ formation during beer production. *Nat Biotechnol* 14: 1587-1589.
155. Korch C, Moutain HA, Gyllang H, Winge M & Brehmer P (1991) A mechanism for sulphite production in beer and how to increase sulphite levels by recombinant genetics. In: *European Brewing Convention, Proc 23rd Eur Brew Conv,* IRL Press, Eynsham, England, UK, pp 201-208.
156. Ansanay V, Dequin S & Barre P (1993) Clonage et expression du gène de l'enzyme malolactique de *Lactococcus lactis*. Eur Patent 0656056.
157. Ansanay V, Dequin S, Blondin B & Barre P (1993). Cloning, sequence and expression of the gene encoding the malolactic enzyme from *Lactococcus lactis. FEBS Lett* 332: 74-80.
158. Ansanay V, Dequin S, Camarasa C, Schaeffer V, Grivet JP, Blondin B, Salmon JM & Barre P (1996) Malolactic fermentation by engineered *Saccharomyces cerevisiae* as compared with engineered *Schizosaccharomyces pombe. Yeast* 12: 215-225.
159. Bony M, Bidard F, Camarasa C, Dulau L, Barre P & Dequin S (1997) Complete malolactic fermentation by *S. cerevisiae* strains expressing heterologous genes for malolactic enzyme and malate transport. *FEBS Lett* 410: 452-456.
160. Volschenk H, Viljeon M, Grobler J, Bauer F, Lonvaud-Funel A, Denayrolles M, Subden RE & Vuuren HJJ van (1997) Malolactic fermentation in grape musts by a genetically engineered strain of *Saccharomyces cerevisiae. Am J Enol Vitic* 48: 193-197.
161. Dequin S & Barre P (1992) Souches de levures exprimant le gène de la LDH lactique et vecteurs utilisables pour l'obtention des dites souches. Eur Patent 0651784.
162. Dequin S & Barre P (1994) Mixed lactic acid-alcoholic fermentation by *Saccharomyces cerevisiae* expressing the *Lactobacillus casei* L(+) LDH. *Bio/Technology* 12: 173-177.
163. Dequin S, Baptista E & Barre P (1999) Acidification of grape musts by *S. cerevisiae* wine yeast strains genetically engineered to produce lactate. *Am J Enol Vitic* 50: 45-50.
164. Remize F, Andrieu E & Dequin S (2000) Engineering of the pyruvate dehydrogenase by-pass in *S. cerevisiae* – role of the cytosolic $Mg^{2+}$ and mitochondrial $K^+$ acetaldehyde dehydrogenases Ald6p and Ald4p in acetate formation during alcoholic fermentation. *Appl Environ Microbiol* 66: 3151-3159.
165. Brown SW & Oliver SG (1982) Isolation of ethanol-tolerant mutants of yeast by continuous selection. *Eur J Appl Microbiol Biotechnol* 16: 119-122.
166. Edwards CG, Beelman RB, Bartley CE & McConnel AL (1990) Production of decanoic acid and other volatile compounds and the growth of yeast and malolactic bacteria during vinification. *Am J Enol Vitic* 41: 48-56.
167. Laing E & Pretorius IS (1993) Co-expression of an *Erwinia chrysanthemi* pectate lyase-encoding gene (pelE) and an *E. caratorova* polygalacturonase-encoding gene (peh1) in *Saccharomyces cerevisiae. Appl Microbiol Biotechnol* 39: 181-188.
168. Gonzalez-Candelas L & Cortell A, Ramon D (1995) Construction of a recombinant wine yeast strain expressing a fungal pectate lyase gene. *FEMS Microbiol Lett* 126: 263-270.
169. Van Rensburg P, Van Zyl WH & Pretorius IS (1994) Expression of the *Butyrivibrio fibrisolvens* endo-β-1,4-glucanase gene together with the *Erwinia* pectate lyase and polygalacturonase gene in *S. cerevisiae. Curr Genet* 27: 17-22.
170. Van Rensburg P, Van Zyl WH & Pretorius IS (1997) Over-expression of the *S. cerevisiae* exo-β-1,3-glucanase gene and the *Bacillus subtilis* endo-β-1,3-1,4-glucanase gene and the *Butyrivibrio fibrisolvens* endo-β-1,4-glucanase gene in yeast. *J Biotechnol* 55: 43-53.

171. Cantagrel R, Lurton L, Vidal JP & Galy B (1995) From vine to Cognac. In: *Understanding Natural Flavours*, Piggott JR & Patterson A (eds), Blackie Academic and Professional, Glasgow, UK, pp 208-228.
172. Houtman AC, Marais J & Du Plessis CS (1980) Factors affecting the reproducibility of fermentation of grape juice and of the aroma composition of wine. I. Grape maturity, sugar, inoculum concentration, aeration, juice turbidity and ergosterol. *Vitis* 19: 37-54.
173. Malcorps P & Dufour JP (1987) Ester synthesis by *Saccharomyces cerevisiae*: localization of acetyl-CoA: iso-amylalcohol acetyltransferase (AT). *Proc Eur Brew Conv* 21: 377-384.
174. Kunkee RE & Amerine MA (1970) Yeast in wine-making. In: *The Yeast*, Vol 3, Yeast Technology, Rose AH & Harrison JS (eds), Academic Press, New York, USA, pp 5-71.
175. Ingraham JL, Guymon JF & Crowell EA (1961) The pathway of formation of n-butyl and n-amyl alcohols by a mutant strain of *Saccharomyces cerevisiae. Arch Biochem Biophys* 95: 169-175.
176. Snow R (1983) Genetic improvement of wine yeasts. In: *Yeast Genetics – Fundamental and Applied Aspects,* Spencer JFT, Spencer DM & Smith ARW (eds), Springer-Verlag, New York, USA, pp 439-459.
177. Saha BC & Bothast RJ (1996) Production, purification and characterization of a highly glucose-tolerant novel β-glucosidase from *Candida peltata. Appl Environ Microbiol* 62: 3165-3170.
178. Riou C, Salmon JM, Vallier MJ, Günata Z & Barre P (1998) Purification, characterization, and substrate specificity of a novel glucose-tolerant β-glucosidase from *Aspergillus oryzae. Appl Environ Microbiol* 64: 3607-3614.
179. Chambon C, Ladeveze V, Oulmouden A, Servouse M & Karst F (1990) Isolation and properties of yeast mutants affected in farnesyl diphosphate synthetase. *Curr Genet* 18: 41-46.
180. Romano P, Soli MG, Suzzi G, Grazia L & Zambonelli C (1985) Improvement of a wine *Saccharomyces cerevisiae* strain by a breeding program. *Appl Environ Microbiol* 57: 830-835.
181. Needleman R (1991) Control of maltase synthesis in yeast. *Mol Microbiol* 5: 2079-2084.
182. Osinga KA, Beudeker RF, Plaat JB & van der Hollander JA de (1988) New yeast strains providing for an enhanced rate of the fermentation of sugars, a process to obtain such strains and the use of these strains. Eur Patent 0306107A2.
183. Oda Y & Ouchi K (1990) Role of yeast maltose fermentation genes in $CO_2$ production rate from sponge dough. *Food Microbiol* 7: 43-47.
184. Higgins VJ, Braidwood M, Bell P, Bissinger P, Dawes IW & Attfield PV (1999) Genetic evidence that high noninduced maltase and maltose permease activities, governed by Malx 3-encoded transcriptional regulators, determine efficiency of gas production by baker's yeast in unsugared dough. *Appl Environ Microbiol* 65: 680-685.
185. Schaaff I, Heinisch J & Zimmermann FK (1989) Overproduction of glycolytic enzymes in yeast. *Yeast* 5: 285-290.
186. Navas MA, Cerdan S & Gancedo JM (1993) Futile cycles in *Saccharomyces cerevisiae* strains expressing the gluconeogenic enzymes during growth on glucose. *Proc Natl Acad Sci USA* 90: 120-124.
187. Smits HP, Hauf J, Müller S, Hobley TJ, Zimmermann FK, Hahn-Hägerdal B, Nielsen J & Olsson L (2000) Simultaneous overexpression of enzymes of the lower part of glycolysis can enhance the fermentative capacity of *Saccharomyces cerevisiae. Yeast* 16: 1325-1334.
188. Olempska-Beer ZS, Merker RI, Ditto MD & DiNovi MJ (2006) Food-processing enzymes from recombinant microorganisms – a review. *Regul Toxicol Pharmacol* 45: 144-158.

189. Nielsen PH, Malmos H, Damhus T, Diderichsen B, Nielsen HK, Simonsen M, SchiV HE, Oestergaard A, Olsen HS, Eigtved, P & Nielsen TK (1994) Enzyme applications (industrial). In: *Encyclopedia of Chemical Technology*, Vol 9, Othmer (ed), John Wiley & Sons Inc, New York, USA, pp 567-620.
190. International Food Biotechnology Council (IFBC) (1990) Safety evaluation of foods and food ingredients derived from microorganisms. *Regul Toxicol Pharmacol* 12: S114-S128.
191. Pariza MW & Johnson EA (2001) Evaluating the safety of microbial enzyme preparations used in food processing: update for a new century. *Regul Toxicol Pharmacol* 33: 173-186.
192. Jonas DA, Antignac E, Antoine JM, Classen HG, Huggett A, Knudsen I, Mahler J, Ockhuizen, T, Smith M, Teuber M, Walker R & de Vogel P (1996) The safety assessment of novel foods. Guidelines prepared by ILSI Europe Novel Food Task Force. *Food Chem Toxicol* 34: 931-940.
193. Pariza MW & Foster EM (1983) Determining the safety of enzymes used in food processing. *J Food Protect* 46: 453-468.
194. Scientific Committee for Food (SCF) (1992) Guidelines for the Presentation of Data on Food Enzymes, 27th series, Ref. No. EUR 14181 EN, pp. 13-22. (<Http:/europa.eu.int/comm/food/fs/sc/scf/reports_en.html>).
195. Benkovic S & Ballesteros A (1997) Biocatalysts - the next generation. *Trends Biotechnol* 15: 385-386.
196. Polastro E (1989) Enzymes in the fine-chemicals industry: Dreams and realities. *Bio/Technology* 7: 1238-1241.
197. Wasserman B (1984) Thermostable enzyme production. *Food Technol* 38: 78-88.
198. Somkuti G & Holsinger V (1997) Microbial technologies in the production of low lactose in dairy foods. *Food Sci Technol Int* 3: 163-169.
199. Illanes A (1999) Stability of biocatalysts. *Elec J Biotechnol* 2: 7-15
200. Haard N (1998) Specialty enzymes from marine microorganisms. *Food Technol* 52: 64-67.
201. Davis M (1998) Making a living at the extremes. *Trends Biotechnol* 16: 102-104.
202. Mozhaev V (1993) Mechanism-based strategies for protein thermostabilization. *Trends Biotechnol* 11: 88-95.
203. Fischer L, Bromann R, Kengen S, de Vos W & Wagner F (1996) Catalytic potency of β-glucosidase from the extremophile *Pyrococcus furiosus* in glucoconjugate synthesis. *Bio/Technology* 14: 88-91.
204. Adams M & Kelly R (1998) Finding and using hyperthermophilic enzymes. *Trends Biotechnol* 16: 329-332.
205. Daniel R (1996) The upper limits of enzyme stability. *Enzyme Microbial Technol* 19: 74-79.
206. Coolbear T, Daniel R & Morgan H (1992) The enzymes from extreme thermophiles: bacterial sources, thermostability and industrial relevance. *Adv Biochem Eng Biotechnol* 45: 57-98.
207. Kristjansson J (1989) Thermophilic organisms as sources of thermostable enzymes. *Trends Biotechnol* 10: 395-401.
208. Caruana C (1997) Enzymes tackle tough processing. *Chemical Eng Progr* 11: 13-20.
209. Adams M, Perler F, & Kelly R (1995) Extremozymes: expanding the limits of biocatalysis. *Bio/Technology* 13: 662-668.
210. Halldorsdottir S, Thorolfsdottir E, Spilliaert R, Johansson M, Thorbjarnardottir S, Palsdottir A, Hreggvidsson G, Kristjonsson J, Holst O & Eggertsson G (1998) Cloning, sequencing and overexpression of a *Rhodothermus marinus* gene encoding a thermostable cellulase of glycosilhydrolase family 12. *Appl Microbiol Biotechnol* 49: 277-284.

211. Rua M, Atomi H, Schmidt-Dannert C & Schmid R (1998) High-level expression of the thermoalkalophilic lipase from *Bacillus thermocatenulatus* in *Escherichia coli. Appl Microbiol Biotechnol* 49: 405-410.
212. Vieille C & Zeikus J (1996) Thermoenzymes: Identifying molecular determinants of protein structural and functional stability. *Trends Biotechnol* 14: 183-190.
213. Imanaka T, Shibazaki M & Takagi M (1988) A new way of enhancing the thermostability of proteases. *Nature* 324: 695-687.
214. Crabb W & Mitchinson C (1997) Enzymes involved in the processing of starch and sugars. *Trends Biotechnol* 15: 349-352.
215. Pedersen S (1993) Industrial aspects of immobilized glucose isomerase. In: *Industrial Application of Immobilized Biocatalysts,* Tanaka A, Tosa T & Kobayashi T (eds), Marcel Dekker, New York, USA, pp 185-208.
216. Quax W, Mrabet N, Liiten R, Schuurhuizen P, Stanssens P & Lasters I (1991) Enhancing the thermostability of glucose isomerase by protein engineering. *Bio/Technology* 9: 738-742.
217. Ye W, Combes D & Monsan P (1988) Influence of additives on the thermostability of glucose oxidase. *Enzyme Microbial Technol* 10: 498-502.
218. Triantafyllou A, Wehtje E, Adlercreutz P & Mattiasson B (1997) How do additives affect enzyme activity and stability in non-aqueous media. *Biotechnol Bioeng* 54: 67-76.
219. Besson C, Favre-Bonvin G, O'Fagain C & Wallach J (1995) Chemical derivatives of *Pseudomonas aeruginosa* elastase showing increased stability. *Enzyme Microbial Technol* 17: 877-881.
220. Erarslan A & Ertan H (1995) Thermostabilization of penicillin G acylase obtained from a mutant of *E. coli* ATCC 11105 by bisimidoesters as homobifunctional cross-linking agents. *Enzyme Microbial Technol* 17: 629-635.
221. Sundaram P & Venkatesh R (1998) Retardation of thermal and urea induced inactivation of a-chymotrypsin by modification with carbohydrate polymers. *Protein Eng* 11: 699-705.
222. Garcia D & Marty J (1998) Chemical modification of horseradish peroxidase with several methoxypolyethylene glycols. *Appl Biochem Bioeng* 73: 173-184.
223. Katchalsky-Katzir E (1993) Immobilized enzymes Ð learning from past successes and failures. *Trends Biotechnol* 11: 471-478.
224. Guisan J, Fernandez-Lafuente R, Rodraguez V, Bastida A & Alvaro G (1993) Enzyme stabilization by multipoint covalent attachment to activated pre-existing supports. In: *Stability and Stabilization of Enzymes,* van den Tweel W, Harder A & Buitelaar R (eds), Elsevier, Amsterdam, The Netherlands, pp 55- 62.
225. Gianfreda L, Modafferi M & Greco Jr G (1985) Enzyme stabilization towards thermal, chemical and proteolytic deactivation. *Enzyme Microbial Technol* 7: 78-82.
226. Illanes A, Zuniga M, Chamy R & Marchese M (1988) Immobilization of lactase and invertase on crosslinked chitin. In: *Bioreactor Immobilized Enzymes and Cells*, Moo-Young M (ed), Elsevier Applied Science, London, UK, pp 233-249.
227. Noritomi H, Koyama K, Kato S & Nagahama K (1998) Increased thermostability of cross-linked enzyme crystals of subtilisin in organic solvents. *Biotechnol Tech* 12: 467-469.
228. Margolin A (1996) Novel crystalline catalysts. *Trends Biotechnol* 14: 223-230.
229. Gupta M (1992) Enzyme function in organic solvents. *Eur J Biochem* 203: 25-32.
230. Koskinen A & Klibanov A (1996) *Enzymatic Reactions in Organic Media*, Blackie Academic & Professional, London, UK.
231. Klibanov A (1997) Why are enzymes less active in organic solvents than in water? *Trends Biotechnol* 15: 97-101.

232. Khmelnitsky Y, Levashov A, Klyachko N & Martinek K (1988) Engineering biocatalytic systems in organic media with low water content. *Enzyme Microbial Technol* 10: 710-724.
233. Rosell C, Terreni M, Fernandez-Lafuente R & Guisan J (1998) A criterion for the selection of monophasic solvents for enzymatic synthesis. *Enzyme Microbial Technol* 23: 64-69.
234. Halling P (1984) Effects of water on equilibria catalysed by hydrolytic enzymes in biphasic reaction systems. *Enzyme Microbial Technol* 6: 513-516.
235. Clapes P, Pera E & Torres J (1997) Peptide bond formation by the industrial protease, neutrase, in aqueous media. *Biotechnol Lett* 19: 1023-1026.
236. Van Unen D, Engbersen J & Reinhoudt D (1998) Large acceleration of a-chymotrypsin-catalyzed dipeptide formation by 18-Crown-6in organic solvents. *Biotechnol Bioeng* 59: 553-556.
237. Coulon D & Ghoul M (1998) The enzymatic synthesis of non-ionic surfactants: the sugar esters. *Agro Food Industry Hi-Tech* 9: 22-26.
238. Bucke C (1996) Oligosaccharide synthesis using glycosidases. *J Chem Technol Biotechnol* 67: 217-220.
239. Scheckermann C, Wagner F & Fischer L (1997) Galactosylation of antibiotics using the β-galactosidase from *Aspergillus oryzae. Enzyme Microbial Technol* 20: 629-634.
240. Chen K, Robinson A, Van Dam M, Martinez P, Economou C & Arnold H (1991) Enzymatic engineering for nonaqueous solvents II. Additive effects of mutation on the stabilization and activity of subtilisin E in polar organic media. *Biotechnol Progr* 7: 125-129.

# Microorganisms for Food and Feed

8

**T.C. Bhalla, Praveen Kumar Mehta, Shashi Kant Bhatia, Navdeep Thakur and Amit Pratush**

*Department of Biotechnology, Himachal Pradesh University, Summer Hill, Shimla 171 005 (H.P.)*

## 1. INTRODUCTION

Increase in human population, rapid industrialization and dwindling natural resources on earth have posed serious threats to food and energy supplies [1]. Inspite of the advancements in traditional crop growing methods coupled with improved crop varieties, these have not been able to meet the world's demand for food grade proteins and vitamins. Agriculture, animal husbandry and fisheries are vulnerable to climatic changes and are prone to a number of diseases, so human population can not exclusively depend on these for meeting protein supply. The global shortage of dietary protein has been particularly noticed since the 1950's. Animal proteins are expensive to produce and this, together with other practical problems such as transportation, the need for refrigeration and the decrease in fish stocks, make these largely unavailable in many developing countries. The fauna and flora of sea are immensely affected by the changes in water and climate. The supply of plant protein is also problematic, especially in view of soil impoverishment, pollution, drought and other adverse climatic imbalances. These problems have led the scientists to formulate strategies for the production of additional and alternative sources of protein for both human and animal consumption.

It was felt in 1950's that traditional sources of protein foods would no longer be able to meet the ever increasing demand and the possibilities of using microbial protein might be assessed to meet future protein demand. The incidental consumption of microbes by human in fermented food and that of waste products of brewery and distillery is quite old. However, the use of microbes for human consumption started in Germany with the drying of spent brewers yeast in 1910 [2]. The use of dried yeast for animal feed by 1914 was about 10,000 tonnes [3].

Although the use of microorganisms as food and feed has been familiar for centuries, but it was for the first time in 1967 at a conference held at Massachusetts Institute of Technology (MIT), USA that the term "Single Cell Protein" was coined to include all forms of microbial cell biomass grown for its protein content, which in turn may be of bacterial, yeast, algal or fungal origin. SCP production technologies arose as a promising way to solve the problem of worldwide protein shortage. They evolved as bioconversion processes, which turned low value by-products, often wastes, into products with added nutritional and market value. Intensive research into fermentation science and technology for biomass production, as well as feeding, has resulted in a profound body of knowledge, the

*Food Biotechnology: Principles and Practices*
*V.K. Joshi & R.S. Singh (Eds.)*
*I.K. International Publishing House Pvt. Ltd., India*
*pp 233–267*

benefits of which now span far beyond the field of SCP production. The system for large-scale production of yeast was developed in Germany particularly after the First World War. This was in Berlin for cultivation of *Saccharomyces cerevisiae*, which replaced as much as 60% of the foodstuffs that Germany had been importing after the war [4]. During World War II, the other yeasts *Candida arborea* and *Candida utilis* were also used as important source of protein in Germany.

In the 1950s, some oil industries became interested in growing microorganisms on alkanes. Many companies including British Petroleum (BP), Kanegafuichi (Japan) and Liquichimica (Italy) entered in this venture of microbial biomass production. In the USSR, 12 out of 86 plants making SCP were said to rely on hydrocarbons as a source of carbon and energy for the microorganisms. Other potential substrates for the production of microbial biomass as protein include bagasse, citrus wastes, sulphite waste liquor, molasses, animal manure, whey, starch and sewage, etc. The microorganisms are now not only explored for their protein content, but are also being investigated as health promoting/protecting agents and are being used as nutraceuticals or probiotics. In this chapter, a broader view of microorganisms as a source of single cell protein, single cell oils and mycoproteins is presented.

## 2. MICROORGANISMS USED FOR SINGLE CELL PROTEINS

A large number of different species of algae, fungi, yeasts and bacteria are used for single cell protein production. Some microbial sources of SCP are listed in Table 1. The microorganisms used for SCP production must have a reasonable protein content, and should be non-toxic and non-pathogenic to human and animals. Microorganisms possess the ability to utilize a range of inexpensive carbon and nitrogen sources, and require moderate growth conditions to convert it into valuable products, which is a prerequisite for the purpose of SCP production. Some biochemical and cultural features of various groups of microorganisms for the production of SCP have been summarized in Table 2. One of the main advantages of SCP compared to other types of proteins is the small doubling time of the microbial cells as shown in Table 3. A temporal comparison of protein production among some microorganisms (yeast and bacteria) to that of well known animal (beef cattle) and plant (soya beans) has been presented in Table 4. Due to shorter doubling time, microbes have higher productivities of biomass than the animal and plant systems.

**Table 1:** List of various bacteria, yeasts, algae and fungi used for SCP production.

| Bacteria | Yeasts | Fungi | Algae |
|---|---|---|---|
| *Acromonas hydrophila* | *Candida guilliermondii* | *Aspergillus fumigatus* | *Chlorella* sp. |
| *Acinetobacter calcoaceticus* | *Candida lipolytica* | *Aspergillus niger* | *C. sorokiniana* |
| *Alcaligenes eutrophus* | *Candida tropicalis* | *Aspergillus oryzae* | *Porphyrum* sp. |
| *Bacillus* sp. | *Candida utilis* | *Chaetomium cellulolyticum* | *Scenedesmus* sp. |
| *Cellulomonas* sp. | *Debaromyces kloekeri* | *Fusarium graminearum* | *Spirulina* sp. |
| *Methylomonas* sp. | *Hansenula polymorpha* | *Paecilomyces varioti* | |
| *Mycobacterium* sp. | *Kluyveromyces fragilis* | *Rhizopus chienensis* | |
| *Nocardia* sp. | *Pichia* sp. | *Trichoderma viride* | |
| *Pseudomonas* sp. | *Rhodotorula* sp. | *Thermoleophilum album* | |
| *Rhodopseudomonas* sp. | *Saccharomyces* sp. | | |
| | *Torulopsis candida* | | |
| | *Torulopsis methanosorbosa* | | |

**Table 2:** Comparison of some cultural and biochemical characteristics of various groups of microbes for SCP production.

| Feature | Bacteria | Yeast | Filamentous fungi | Algae |
|---|---|---|---|---|
| Growth rate | Highest | Quite high | Lower than bacteria and yeast | Low |
| Substrate | A wide range of substrates | Most substrates except hydrocarbons and $CO_2$ | Limited substrates (mostly starchy and cellulosic materials) | Light, inorganic carbon sources e.g., $CO_2$ (preferably) |
| pH range | 5-7 | 5-7 | 3-8 | Up to 2 |
| Cultivation system | Bioreactors | Bioreactors | Bioreactors | Open ponds, tanks in sunlight |
| Risk of contamination | High; precautions necessary | Low | Low if grown below pH 5 | High |
| Biomass recovery | Sometimes problematic; new improved methods are needed | Easy by centrifugation | Easy for filamentous or pellet forms | Difficult and costly with unicellular algae |
| Protein content (crude) | 80% or more | 55-60% | 50-55% | Up to 60% |
| Amino acid profile | Generally good, a small deficit in S-containing acids | Generally good, deficit in S-containing amino acids | Low in S-containing amino acids | Generally good, low in S-containing amino acids |
| Nucleic acid content | High (8-14%) | High (5-12%) | High (3-10%) | Low (4-6%) |
| Removal of nucleic acids | Necessary | Necessary | Necessary | Necessary |
| Toxins | Gram–ve bacteria may produce endotoxins | - | Many species produce mycotoxins | Three types of toxin: Endotoxin, Neurotoxin, Hepatotoxin |
| Other features | - | High vitamin B content | Chitin may contain a significant proportion of N content, which is unavailable | Low yield (1-2 g dry wt/l). High chlorophyll content, unsuitable for humans |

**Table 3:** A comparison of mass doubling time of various groups of organisms.

| Organism | Mass doubling time |
|---|---|
| Bacteria | 10-120 min |
| Algae/moulds | 2-6 hr |
| Yeasts | 10-120 min |
| Plants | 1-2 weeks |
| Chickens | 2-4 weeks |
| Pigs | 4-6 weeks |

**Table 4:** Comparison of protein production efficiency of some organisms in 24 hours.

| Organism (1,000 kg) | Amount of protein produced in 24 h (kg) |
|---|---|
| Beef cattle | $10^{-1}$ |
| Soyabeans | $10^{1}$ |
| Yeast | $10^{2}$ |
| Bacteria | $10^{12}$ |

## 2.1 Advantages of SCP over Conventional Protein Sources

(i) The production of SCP is based upon the availability of renewable and non-renewable raw materials.

(ii) The production of SCP is independent of land, climate and is carried out in a continuous culture.

(iii) Microorganisms can be genetically modified and easily controlled.

However, there are some other factors that impair the usefulness of SCP which includes:

(i) Unicellular biomass contains non-digestible cell wall and fatty acid glycerides, which never occur in traditional foodstuffs.

(ii) The content of nucleic acids in unicellular biomass is higher than the permissible levels and cause disorder of purine metabolism in the human body.

(iii) Some biomass has a disagreeable flavour, unacceptable colouration and taste, which make them unpalatable even for animal consumption.
In order to overcome some of the limitations of SCP mentioned above, methods have been developed to lyse the cell.

## 2.2 Nutritional Value of Single Cell Proteins

The nutritional value of SCPs can be compared and assessed on the basis of their nutrient composition, amino acid profile, vitamin and nucleic acid content. In addition to these aspects, palatability, allergies and gastrointestinal effects should be taken into consideration [5]. For toxicological effects and carcinogenicity, long term feeding trials are undertaken. Average biochemical composition of the major groups of microorganisms is shown in Table 5.

Bacterial protein is similar to fish protein, yeast's protein is like soya, and the fungal protein is somewhat lower than the yeast's. Most of the microbiological proteins have comparatively lower content of sulphur containing amino acids cysteine and methionine, and require supplementation, while they have higher levels of lysine (Table 6).

The vitamins of microorganisms are primarily of the B type. The vitamin $B_{12}$ is mostly found in bacteria, while vitamin A is usually present in algae. Other nutritional parameters which evaluate the quality of a given SCP are: (i) digestibility, (ii) biological value, (iii) protein efficiency ratio, and (iv) net protein utilization.

It is important to note that digestibility (D) of SCP (microbial cells) is low especially with algal cells because of cell walls. Although the protein efficiency ratio (PER) of several SCP is comparable with the PER of conventional protein sources, yet acceptability and palatability results on nutritional trials are not very encouraging [8]. Moreover, we cannot as such accept or extend the outcome of animal feeding trials of SCP with regard to their acceptability and tolerance to human beings.

### *2.2.1 The problem of nucleic acids*

About 70-80% of the total cell nitrogen is represented by amino acids, while the rest occurs in nucleic acids. The concentration of nucleic acids in SCP is higher than other conventional proteins and it is the characteristic of all fast growing organisms. The problem that results from the consumption of

**Table 5:** Biochemical composition of the main groups of microorganisms (% dry weight).

| Biochemical parameter | Fungi | Algae | Yeasts | Bacteria |
|---|---|---|---|---|
| Protein | 30–45 | 40–60 | 45–55 | 50–65 |
| Fat | 2–8 | 7–20 | 2–6 | 1.5–3.0 |
| Ash | 9–14 | 8–10 | 5–9.5 | 3–7 |
| Nucleic acid | 7–10 | 3–8 | 6–12 | 8–12 |

*Source:* 6

**Table 6:** Essential amino acid content of some single cell proteins in comparison with several reference proteins (grams of amino acid per 100 grams of protein).

| Amino acid | *Saccharomyces cerevisiae* | *Spirulina maxima* | *Cellulomonas* | BP* SCP | Wheat | Egg | Cow milk |
|---|---|---|---|---|---|---|---|
| Lysine | 7.7 | 4.6 | 7.6 | 7.0 | 2.8 | 6.3 | 7.8 |
| Threonine | 4.8 | 4.6 | 5.4 | 4.9 | 2.9 | 5.0 | 4.6 |
| Methionine | 1.7 | 1.4 | 2.0 | 1.8 | 1.5 | 3.2 | 2.4 |
| Cysteine | - | 0.4 | - | - | 2.5 | 2.4 | - |
| Tryptophan | 1.0 | 1.4 | - | - | 1.1 | 1.6 | - |
| Isoleucine | 4.6 | 6.0 | 5.3 | 4.5 | 3.3 | 6.8 | 6.4 |
| Leucine | 7.0 | 8.0 | 7.3 | 7.0 | 6.7 | 9.0 | 9.9 |
| Valine | 5.3 | 6.5 | 7.1 | 5.4 | 4.4 | 7.4 | 6.9 |
| Phenylalanine | 4.1 | 5.0 | 4.6 | 4.4 | 4.5 | 6.3 | 4.9 |
| Histidine | 2.7 | - | 7.8 | 2.0 | - | - | - |
| Arginine | 2.4 | - | 6.4 | 4.8 | - | - | - |

* British petroleum (BP)

*Source:* 7

proteins with high level of nucleic acids (7.8-25 g/100 g protein dry weight) is the high concentration of uric acid in the blood, which sometimes causes the disease gout [9]. Most mammals, reptiles and molluscs possess the enzyme uricase that further converts uric acid to allantoin. Uricase is absent in man, birds and some reptiles, therefore, the end product of purine degradation is uric acid. Nucleic acid content of various SCP's can be removed either by chemical treatment with NaOH, treatment with 10% NaCl or thermal shock [10]. These methods aim to reduce the RNA content from about 7% to 1%, which is considered within the acceptable limits.

## 2.3 Substrates for SCP Production

The key factor in the economic viability of any microbial process is the choice of substrates. The substrate to be used should be inexpensive, abundant and in proximity to the production plant. The most widespread and commonly used substrates for SCP production have been mostly carbohydrates, which are widely available renewable resource. Traditional substrates for biomass production can be grouped in three categories as shown in Fig. 1:

(i) Carbohydrates either in the form of assimilable sugar or polymer form (e.g., starch, cellulose etc.), which can readily yield assimilable sugars by their enzymatic hydrolysis have been the traditional substrate. These sources include molasses, starchy grain and wood sugar from spent sulphite liquor or wood hydrolysates and whey. The yeast of the genera *Saccharomyces cerevisiae, S. uvarum, C. utilis* and *Kluyveromyces fragilis* have been grown on these substrates [2].

(ii) During the 1960's, there was an upsurge of development of SCP processes based on substrates from non-renewable sources such as methane, petroleum hydrocarbons and petrochemicals. These substrates were chosen for their low cost and availability in large quantities in mediterranean regions and Venezuela [11]. Many processes were developed

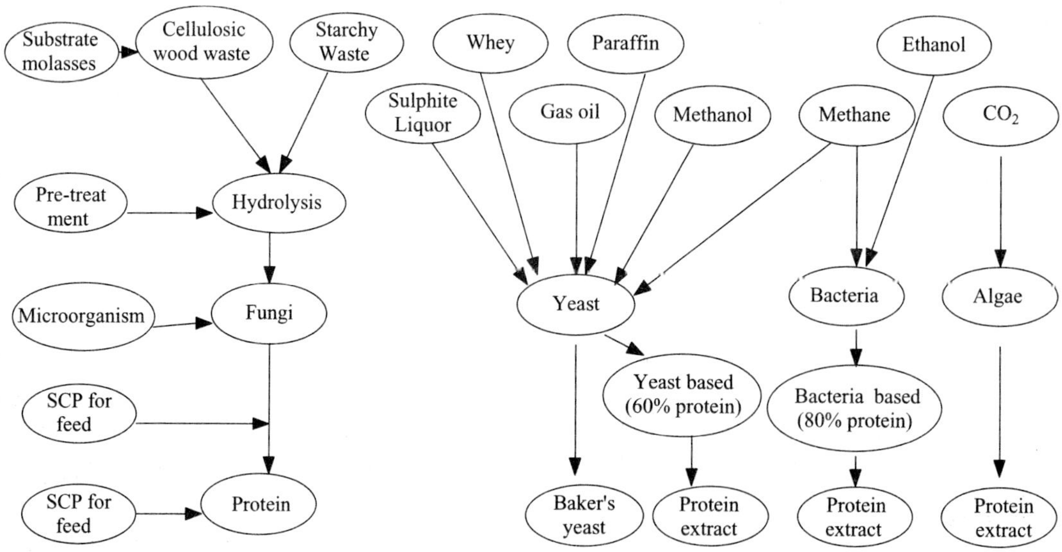


**Fig. 1:** An overall view of the substrates and microorganisms used for the production of single cell proteins.

for biomass production using bacteria (*Brevibacterim, Acinetobacter, Bacillus, Pseudomonas, Methylobacter, Methylococcus, Methylomonas, Methylophilus, Corynebacterium*) and yeasts (*Candida, Debaromyces, Endomycopsis, Hansenula, Saccharomyces* and *Rhodotorula*, etc). The last two decades have witnessed a continuous rise in the prices of hydrocarbon sources due to their low availability, and this has resulted in drastic shift towards the use of agricultural and food processing industry wastes for the production of microbial biomass. These raw materials require the use of microorganisms (bacteria, yeasts and fungi as well as algae and photosynthetic/lithotrophic bacteria) other than those mentioned above [12].

(iii) Third category of substrate is $CO_2$ used to culture photosynthetic microbes, especially algae (*Chlorella, Spirulina*) for SCP purpose.

### *2.3.1 SCP from n-alkanes*

It was only in 1960's that the large number of microorganisms and substrates were screened and evaluated for single cell protein production. British Petroleum (BP) Co. Ltd. constructed a commercial plant for cultivation of yeast cells on n-alkane for the production of SCP. Some crude oils contain upto 15% wax and these waxes must be removed since they make oil more viscous, precipitate out at low temperatures and sometimes block plant pipes. *Candida lipolytica* and *C. tropicalis* are used as source microorganism and 16,000 tons SCP/year in France and 4,000 tons SCP/year in England were manufactured. The product produced was called "TOPRINA" for which $NH_3$ was used as N-source and $Mg^{+2}$ ions to increase yield. Later Toprina$^{TM}$ was reported to be carcinogenic and toxic, which led to its opposition in many countries by environmental groups, and its marketing as a human food supplement was prevented in several countries.

### *2.3.2 SCP from methane*

The production of SCP from methane is cost effective as the substrate is abundant and does not have the toxicity problem that alkanes pose. In addition to being a constituent of North Sea gas, it is also produced by anaerobic digestion. Methane is the highly reduced form of carbon, thus it gives high cell yields relative to the amount of gas consumed. The two important genera *Methylomonas* and *Methylococcus* have been identified for SCP production on methane as they efficiently utilize this as carbon source. Nitrates or ammonium salts serve as good nitrogen source for the growth of these bacteria. The most interesting work was carried out by Shell in the England involving the methane utilizing gram-negative rods such as *Hyphomicrobium* sp., *Acinetobacter* and *Flavobacterium* for the production of SCP [4]. All developmental activities and commercialization of the product were indefinitely postponed by Shell because of the availability of soyabean and maize at low cost vis-à-vis difficulty in extending its sophisticated process at that point of time.

### *2.3.3 SCP from methanol*

Methanol is available in large quantities and high purity. It is directly produced from methane or methane related raw materials. The use of methanol in the production of single cell protein (SCP) is of great importance since the substrate is a common product of industrial operation. The most advanced process developed for the production of SCP from methanol using *Methylophilus methylotropha* belongs to Imperial Chemical Industries (ICI). Ammonia was used as a nitrogen source and SCP so produced was marketed as PRUTEEN$^{TM}$ that contained 72% crude protein. It was mainly recommended for animal feed and served as a good source of energy, vitamins and minerals as well as highly balanced protein. Methanol is a preferred substrate because of its solubility in water (unlike methane and n-paraffin).

For the production of SCP on methanol, source feed stock is mixed with a heat labile micronutrient source filtered and introduced into a fermenter (Fig. 2). A portion of the cell rich ferment is removed periodically from fermenter, and passed through a separator to separate SCP product and effluent. A portion of effluent is held in hold tank where it is maintained at a temperature, which is effective to inhibit the growth of microorganism by a steam jacket. A part of effluent is removed from the held tank and combined with enhanced medium containing concentrated fresh media. The enhanced medium is then sterilized to produce a 'sterilized enhanced medium' which is then cooled to produce 'cooled sterilized medium'. Sterilized cooled medium is introduced into the second fermenter along with complimentary nutrients. SCP product from first and second fermenter is introduced into the pasteurizer to produce pasteurized SCP product, which is further dried to produce final SCP product that can be further processed as per the requirement for animal or human feed supplement.

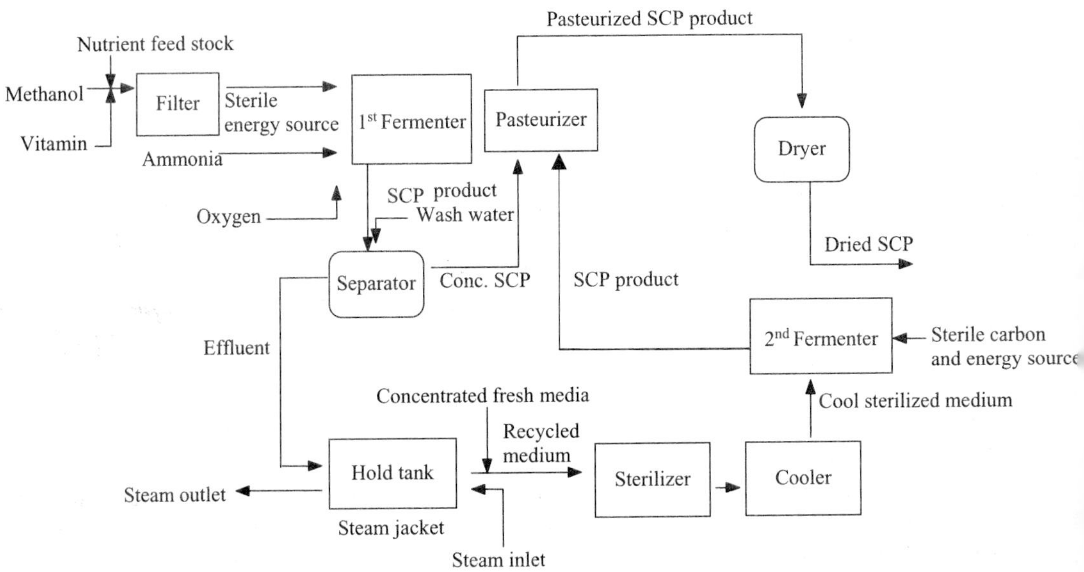


**Fig. 2:** Flow diagram of SCP production.

### *2.3.4 SCP from ethanol*

Ethanol was used as a substrate by Amoco Co. in the US for the production of food grade yeast *Torula* as SCP and the product was sold by the name TORUTEIN™. The TORUTEIN had 52% protein and was marketed as a flavour enhancer of high nutritional value, a substitute of meat, milk and egg protein. However, it could not compete with soya in United States, since the later is cheaper, plentiful and can serve to replace meat and egg diets.

### *2.3.5 SCP from lignocellulose*

The lignocellulosic wastes mainly from agriculture, forestry and horticulture constitute the most abundant substrate for SCP, but the necessity of pretreatment to breakdown the complex polysaccharide weighs against the production cost. Approximately six hundred million tonnes of wheat/paddy/ragi, etc. is produced in the world annually. In Greece, the availability of straw from wheat and rye, the two most important cereals, is estimated to be 1.5 million tonnes per year [14]. Pretreatment with some physical or chemical methods becomes essential because the lignin, hemicellulose and cellulose (LHC) complex is tough to unravel and does not allow easy accessibility to hydrolytic enzymes for saccharification of lignocelluloses. University of Waterloo, Ontario, Canada developed a process based on the cellulolytic fungus *Chaetomium cellulolyticum* and utilized agricultural wastes like crop residues of straw, corn stover and bagasse, animal manure, and forestry residues such as wood sawdust and pulp mill sludge. At present, the only economical utilization of cellulosic waste is in the production of mushroom.

### *2.3.6 SCP from other renewable resources*

Many renewable substrates generated by some food/paper/agro-processing industries as wastes (such as sugarcane bagasse, waste from paper pulp industry, starch waste, wheat bran, fruit processing wastes, sulphite liquor waste, and waste from activated sewage sludges, etc.) are used in SCP production. Microbial biomass has been commercially produced from whey since 1940's in the United States and European countries. Among the several processes that have been described [15], the Pekilo, Vienna and the Bel processes stand out to be quite attractive for SCP production based on industrial wastes.

## 2.4 Production of SCP

Microorganisms are endowed with metabolic capabilities of using a wide range of various substrates both from renewable and non-renewable sources, but all of them require carbon, nitrogen and phosphorus sources as well as other minerals and vitamins. Yeast cells were probably the first to be harvested on commercial scale since 1905. It was only in 1960 that a large number of microorganisms and substrates were screened for single cell protein production. British Petroleum Co. Ltd. was first to establish a commercial plant for the production of SCP by cultivating yeast cells on n-alkanes. Pruteen$^{TM}$, a SCP for animal feed, was also produced by growing *Methylophilus methylotrophus* on methanol by Imperial Chemical Industries (ICI), but increasing oil prices forced the closure of these plants. Large-scale production of SCP is economically feasible if it is produced with cheaper feed stocks [16].

Over the last 30 years, many pilot plants have been developed that utilize a range of substrates and microorganisms. The various types of substrates used for SCP include petroleum based hydrocarbons, methane or methanol, carbohydrates (starch crops), agricultural and forestry wastes such as corn stover, bagasse, wood pulp, saw dust, wheat bran, paper mill wastes, sulphite waste liquor, mesquite wood, cane and beet sugar molasses, fruit processing wastes, dairy industry wastes such as cheese whey, whey permeate and other miscellaneous wastes [1].

Yeast biomass grown on protein enriched maize flour has resulted in a novel protein called INCAPARINA$^{TM}$ for human consumption and has been successfully introduced in some African countries to combat the protein deficiency. The Phillips Petroleum Corporation in UK developed a process for production of SCP from yeast called Phillips/Provesta process by which Provesteen P is produced from methanol, while two products, Provesteen T and K have been produced from sugar feed stocks of crop byproducts and whey permeate, respectively [1]. The main stages of single cell protein production are medium preparation, fermentation and downstream processing. Few processes used for SCP production by different commercial plants are described below:

### *2.4.1 The Symba process*

The Symba process was developed in Sweden to produce SCP for animal feed from starchy wastes combining two yeasts in a sequential mixed culture: the amylase producing *Endomycopsis fibuligera* and the fast growing *Candida utilis*. This process is operated in two stages (Fig. 3). In the first stage, *E. fibuligera* is grown in a small reactor containing potato waste, which is supplemented with

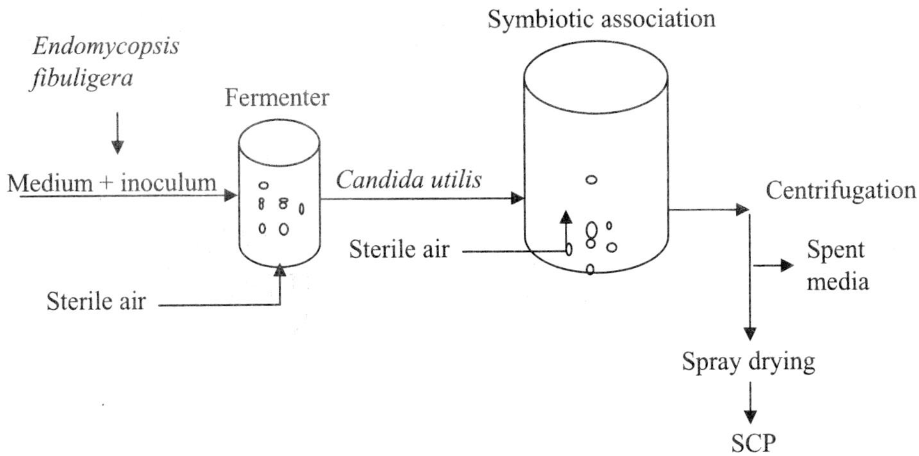


**Fig. 3:** Diagrammatic representation of Symba process used for SCP production.

nitrogen and phosphorus sources. Starch is hydrolysed at this stage. In the second stage, the broth is pumped into next reactor where both the organisms are present. After few days of fermentation, biomass is recovered by centrifugation and dried by spray or drum drying. This process can be operated continuously and after 10 days up to 90% reduction in pollution load of waste is recorded. In this process, *C. utilis* dominates the final product and constitutes 90% of the SCP. Resultant protein-rich biomass (45% protein) is concentrated by centrifugation and finally spray or drum dried. The final product called 'Symba yeast' contains about 45% protein besides vitamins.

### *2.4.2 The Bel process*

This is the most popular process for the production of SCP from dairy industry waste using *Kluyveromyces lactis* or *Kluyveromyces marxianus* by Bel Industry (France). Whey invariably contains about 5% lactose, 0.8% protein, 0.2-0.6% lactic acid and is used as a substrate. The SCP produced by this process is marketed as Proteibel, which is used for consumption by human and animal. These processes initially involve whey pasteurization, during which 75% of whey proteins are precipitated. The lactose concentration is adjusted to 34 g/l and mineral salts are also added. This supplemented whey is introduced into a 22 $m^3$ continuous fermenter, maintained at 38°C and pH 3.5, with an aeration rate of 1700 $m^3$/h. The yeast utilizes the lactose and attains a biomass concentration of 25 g/l, with a biomass yield of 0.45-0.55 g/g lactose. Yeast cells are recovered by centrifugation, then resuspended in water, recentrifuged and finally roller-dried to 95% solids. Levels of residual sugar remaining in the spent medium are less than 1 g/l [17].

### *2.4.3 The Bioprotein process*

During late 1970's when the prices of conventional feed protein were high and oil prices were low, single cell protein (SCP) were produced using alkanes (methane) and straight chain hydrocarbon by several oil companies. However, due to their immiscibility in water and explosiveness when mixed with oxygen especially methane, use of these compounds posed some problems.

In 1990, a company named Norferm produced SCP by growing *Methylococcus capsulatus* in a medium fed with methane rich natural gas as the sole carbon and energy source, and this process is known as Bioprotein process. Fermentation is carried out in continuous loop fermenters containing medium with ammonia, minerals and methane. Downstream processing comprises centrifugation, ultra filtration and spray drying to harvest the biomass. The final product obtained is marketed as Pronin [17].

Similar technique had been developed by ICI for the production of single cell proteins from methanol and called 'Pruteen Process'. However, microorganisms can only tolerate a low concentration (0.1-1.0%, v/v) of methanol. The advantages of using methanol over methane and many other carbon sources are the complete miscibility with water and its availability in a very pure form. The Pruteen Process in UK is the most technically accepted process that is being used for SCP production since 1980 [17].

## 2.5 Problems with SCP

SCP is normally considered as a source of protein. However, like any other biological material, it also contains cell wall, nucleic acids, carbohydrate, lipids, minerals and vitamins. The major problem in the use of single cell protein as a human food is the presence of high concentration of nucleic acids. About 10-15% of the total nitrogen in fungi and yeasts is in the form of nucleic acids. These are not metabolized in the same way as proteins but follow a different route. The consumption of proteins with high concentration of nucleic acids (8-25 g/100 g of protein dry weight) causes an increase in uric acid level in blood resulting in kidney stones and gout. The problems associated with the SCP from hydrocarbons are the presence of residual alkanes, polycyclic aromatic hydrocarbons and fatty acids. The accumulation of these compounds in the adipose tissues of animals causes serious health hazards [16]. Consumption of SCPs may also cause skin reactions or allergies and gastrointestinal reactions resulting in nausea and vomiting. The assimilation of some heavy metals, microbial toxins and chemical residues from nutrient media can also cause serious health hazards [16]. SCP has an unpleasant colour, odour and taste, which make them unpalatable even for animal consumption. It is therefore necessary to reduce nucleic acid content in SCP to an acceptable low level.

# 3. SINGLE CELL OILS (SCO)

The ability of some oil accumulating yeast, moulds, photosynthetic microflora and bacteria suggests that they might compete economically as a source of fats and oils with plants and animals, and the idea of using them as an alternative to plant oils was conceived in Germany during the world wars, (1914-18) and (1939-45). However, no large–scale process was developed for SCO production and earlier the lipid-rich microbial mass that was produced was fed only to army horses. The oils derived from microbial sources are called single cell oils, microbial oils or unicellular oils; the terms used are parallel to single cell proteins to denote oils of microbial origin. Oleaginous microorganisms are those microbes that accumulate more than 20-25% of their biomass as oil and this oil is called single cell oil [18]. In 1980, cocoa butter was the first single cell oil preparation, which was used as a substitute of cheese [19]. Due to the ability of microorganisms to synthesize high quality fats, their potential roles have been further explored.

To increase the accumulation of oil in oleaginous microorganisms, a culture medium with highly assimilable carbon and low nitrogen content is preferred so that cells exhaust nitrogen quickly but continue carbon utilization. Not all microorganisms have equally good potential for oil production. Bacteria are relatively poor producers of triacylglycerols, whereas a few yeasts and a larger group of fungi have extremely good capabilities for synthesis of lipids and oils. The yield does not depend alone on high oil content but many other factors such as growth characteristics of organisms, their stability in the culture medium, efficient substrate utilization and extractability of oil, etc. Moulds and yeast have received comparatively more attention for single cell oil production. The bacteria do not find preference for oil production because of their unfavourable growth characteristics and presence of allergen and toxin with their cell mass. The requirement of large surface area and specific conditions for growth also make algae unsuitable for oil production at commercial scale. Hence, only yeasts and moulds are actively considered for oil production.

The oleaginous yeasts accumulate lipids in the form of triacylglycerol. The distribution pattern of fatty acyl groups on glycerol moiety has revealed that sn-2 position is exclusively occupied by an unsaturated acyl group. In this respect, the yeast triacylglycerol are similar to plant oils, but differ from animal fats which usually have saturated fatty acid at the corresponding position. In most of the yeasts, the relative abundance of individual fatty acid is oleic acid > palmitic acid > linoleic acid and stearic acid [18]. The polyunsaturated fatty acids (PUFA) like γ-linoleic acid (GLA), dihomo-γ-

**Table 7:** Some oleaginous microorganisms, their oil content and carbon source.

| Species | Oil content (%) | Substrate |
|---|---|---|
| ***Yeasts*** | | |
| *Candida curvatus* | 58 | - |
| *Candida tropicalis* | 32 | n-alkane |
| *Candida guilliermondii* | 30 | n-alkane |
| *Cryptococcus albidus* | 65 | Glucose |
| *Lipomyces lipofer* | 67 | Glucose |
| *Rhodotorula glutinis* | 64 | Glucose |
| *Tricosporon cutaneum* | 45 | Glucose |
| ***Moulds*** | | |
| *Aspergillus ficheri* | 53 | Sucrose |
| *Aspergillus terreus* | 57 | Sucrose |
| *Chaetomium globosum* | 54 | Glucose |
| *Cunninghamella japonica* | 37 | On soaked rice |
| *Humicola lanuginose* | 75 | Glucose |
| *Mucor circinelloides* | 65 | Glucose |
| *Penicillium spinulosum* | 64 | Molasses |
| *Penicillium lilacinum* | 56 | Glucose |
| ***Algae*** | | |
| *Botryococcus braunii* | 40 | - |
| *Chlorella pyrenoidosa* | , | - |
| *Dunaliella salina* | , | - |
| *Monalanthus salina* | , | - |
| *Nitzschia loinssima* | 34 | - |
| ***Bacteria*** | | |
| *Arthrobacter* sp. | 80 | - |
| *Corynebacterium* sp. | 30-40 | - |
| *Mycobacterium* sp. | 30-40 | - |
| *Rhodococcus* sp. | 30-40 | - |

linolenic acid (DHGLA), arachidonic acid (AA), docosahexanoic acid (DHA) and eicosahexaenoic acid (EPA), which have dietary significance, are not present in large amounts.

Yeast system is amenable to genetic manipulations, so it can be easily modified to produce oil of desired qualities. The fats synthesized by moulds are generally characterized by high proportion of palmitic acid, oleic acid and linoleic acid. The ability of many moulds to synthesize and accumulate some nutritionally and neutraceutically important PUFA belonging to both n-3 (EPA, DHA) and n-6 (GLA, DHGLA and AA) series make moulds better resource for fat and oil production.

However, microbial fermentations have not been able to compete economically with vegetable and animal sources, chiefly because of the capital and recurring cost involved in purchase and running of fermenter. Some oleaginous microorganisms are listed in Table 7.

The various advantages of microbial oil production over animals and plants are given below:

(i) Fat production in plant depends upon climatic factors, but microbial fat production is independent of climatic condition.
(ii) Microbial fats are rich in polyunsaturated fatty acids, which have nutritional and therapeutic value [20].
(iii) Certain sterols and biosurfactants can also be produced by using microbes, which cannot be produced by plants.
(iv) Antioxidants, vitamins and trace elements are also present in microbial fat.
(v) Genetic engineering of microbes for desirable characteristics and higher production of lipid is an easy task.
(vi) Single cell oil can also form substitute for fuel.

## 3.1 Composition of Single Cell Oils

Single cell oils are derivatives of triglyceride oil e.g., docosahexaenoic acid rich single cell oil (DHASCO) and arachidonic acids rich single cell oil (ARASCO). These oils are rich in long chain polyunsaturated acid (PUFA). Single cell oils are free of cyclic and trans fatty acids and have low amount of eicosapentanenoic acid (EPA).

### *3.1.1 Docosahexaenoic acid rich single cell oils*

Docosahexaenoic acid rich single cell oil (DHASCO) is a dienosterol unique to algae. It is rich in DHA fatty acid i.e., 40-45%. Other fatty acids like oleic acid (10-30%), palmitic acid (10-20%) and myristic acid (10-20%) are also present in large amount but less than DHA. Palmitoleic, stearic, linoleic and behnic acid contents in SCO are low i.e., 0-2%. Silicon, sulphur, arsenic, cadmium, chromium, lead, mercury and molybdenum are also present in traces. Dinosterol is the most commonly found sterol in DHASCO i.e., 31.5%. Other sterols like dehydrocholesterol (9.6%), dehydrodinosterol (8.2%), lathosterol (7.5%), dinosterone (6.0%) are present in medium amount, while denosterol (2.4%) and cholesterol (1.7%) content are low in this SCO.

### *3.1.2 Arachidonic acid rich single cell oils*

ARA is the most abundant fatty acid in ARASCO i.e., 38-44%. Oleic acid (5-38%), stearic acid (5-20%) and linoleic acid (4-15%) are present in lower amounts than ARA. Other fatty acids palmitoleic acid (0-2%), linolenic acid (1-5%) and behnic acid (0-3%) are present in very low amount. Silicon content may range from 280-350 ppm, while level of other elements arsenic, cadmium, chromium, iron, lead, mercury and molybdenum is low. Desmosterol is the most abundant sterol (67.3%) and lanosterol is present in very small amounts (1.1%). These sterols are most commonly found in plants and edible fungi.

### 3.2 Production of DHASCO

*Crypthecodinium cohnii* is a member of Dinophyta. Some species of photosynthetic Dinophyta produce toxin, but none of the heterotrophic species produce toxin and are pathogenic to humans or other mammalian species [21]. Fermentation process is used for the DHASCO production. Seed culture is prepared in liquid medium and transferred to large vessels. Cells are centrifuged and spray dried when fatty acid and cell density reach at a certain level. Blending process with hexane is used for the extraction of oil from biomass. Deoiled solids are separated from extracted oil and desolventization is done under vacuum and winterized condition to remove the more saturated oil fractions. Standard procedures of bleaching and deodorization are then used for the refining of extracted oil. High oleic sunflower oil is used to dilute the DHASCO to a standard 40% DHA concentration and mixed with 0.025% antioxidant-tocopherol and ascorbyl palmitate.

### 3.3 Production of ARASCO

*Mortierella alpina*, a member of Phycomycetes group of fungi, is used for the ARASCO production. For the production of ARASCO, a specific strain of *M. alpina* is selected so that only oil rich in arachidonic acid with no eicosapentaenoic acid is produced [22]. *M. alpina* requires a reduced carbon source for growth because it is a psychotropic and non-photosynthetic organism. Seed culture is grown in shake flask and then added to production medium in larger vessel of desired volume. Fungal biomass is then harvested, dried and subjected to oil extraction process as mentioned in the preceding section.

## 4. FUNGI AS FOOD

Since antiquity, fungi are influencing human affairs as a direct source of food or medicine. Fungi have 20-30% dry matter as a crude protein and thus serve as an ideal food for human beings. The properties of fungal food are that it is rich in dietary fibre, free from cholesterol and has high protein content with essential amino acids.

Many of the mushrooms grow wildly in forest and constitute important component of forest minor produce. Mushrooms are cultivated all over the world and their global annual production is about 8 million metric tonnes. *Agaricus* sp. (button mushroom) alone accounts for 30% of the total production. About 60% of the world's mushrooms are grown in China. *Lentinus edodes* and *Volvariella volvacea* are the oldest cultivated species of edible fungi. Fungal food is used as a substitute to meat, while some species have necrotic and toxic effects [17]. The use of fungi in production of edible biomass has several advantages as they most efficiently convert the plant wastes (lignocelluloses) into edible foods [23]. The agricultural waste (straw, bagasse, and saw dust) and solid liquid waste of paper and agricultural industry can be easily converted to consumable biomass by growing mushroom. Mushrooms (single cell/fungal food) have wide acceptability and consumption due to their flavour and texture. Traditional method and solid state fermentations are used for the production of mushrooms and other fungal food products. Applications of some important fungi as food are given in Table 8.

### 4.1 Mushrooms

The people of various regions of world identified, selected and used the wildly growing mushrooms much before the practice of cultivation of mushroom started. The Japanese were first to grow

**Table 8:** Application of fungi in the production of some food.

| Application | Species |
|---|---|
| ***Cheeses*** | |
| Roquefort, 'blue' stilton | *Penicillium roquefortii* |
| Camembert, brie, soft ripened cheeses | *Penicillium camembertii* |
| ***Cultivated edible macrofungi*** | |
| Champignon, button mushroom | *Agaricus bisporus (A. brunnescens), A. bitorquis* |
| Winter mushroom, enoki | *Flammulina velutipes* |
| Shimeji | *Hypsizygus tessulatus* |
| Shiitake or shiang-gu | *Lentinula edodes* |
| Oyster mushroom | *Pleurotus* sp. |
| Truffle | *Tuber melanosporum* |
| Chinese or straw mushroom | *Volvariella volvacea* |
| ***Oriental food fermentations*** | |
| *Hamanatto* | *Aspergillus oryzae* |
| *Shoyu (soy sauce)* | *Aspergillus oryzae, A. sojae* |
| *Miso* | *Aspergillus oryzae, A. sojae* |
| *Wine* | *Botrytis cinerea* |
| *Ang-kak* | *Monascus purpurea* |
| *Ontjom* | *Neurospora intermedia* |
| *Tempeh* | *Rhizopus oligosporus* |
| ***Commercially harvested wild edibles*** | |
| Fungo parcino | *Boletus edulis* |
| Yellow chanterelle | *Cantharellus cibaricus* |
| Blewit | *Clitocybe nuda* |
| Gypsy mushroom | *Cortinarius caperatus* |
| Maitake | *Grifola frondosa* |
| Tooth fungus | *Hericium erinaceus* |
| Sweet tooth fungus | *Hydrum repandum* |
| ***Conditionally edible species*** | |
| Fly agaric | *Amantia muscoria* |
| Tippler's Bane | *Caprinopsi atramentaria* |
| Beefsteak morel | *Gyromitra esculenta* |

mushroom about two thousand years ago. The word 'mushroom' is thought to have its origin from the French 'mousseron', a term that included both edible as well as poisonous mushrooms. Today, the word refers only to edible fungi, that even to its portion above the ground, the fruiting body. The cultivation of mushrooms in Europe started nearly 300 years ago. Mushrooms are now being produced in over eighty countries around the world. The total annual production today is estimated to be over eight million tonnes. Saprophytic edible mushrooms can be divided into two groups on the basis of their natural substrate: (i) primary rot, (ii) secondary rot.

## 4.2 Primary Rot Fungi

The group of fungi that grow on wood are called primary rot. Each kind of species requires special type of wood and cultivation conditions. Trees used for cultivation should be felled in winter season because trees felled in spring or summer have soft bark and are prone to infection by competitive microorganisms and get damaged easily. Trees are cut down into small logs of 1 meter length and 5-15 cm diameter and holes are drilled into the wood log. Inoculum is prepared on grains or pieces of wood. When pieces are well colonized with mycelium, they are used to inoculate the holes and covered with wax to avoid evaporation and infection. Inoculated logs are incubated at high humidity and desirable temperature that ranges from 20-30°C for the vegetative growth.

## 4.3 Secondary Rot Fungi

The fungi that require microbially modified or partially decomposed organic matter for growth are called secondary rot fungi. For high yield of fruiting bodies and colonization of substrate by the fungus, humic acid and organic nitrogen are needed, which get accumulated into the substrate due to microbial decomposition. Some carbon and nitrogen containing residues from other sources can also be added into the substrate followed by microbial treatment for high yield of fruiting bodies. There are several advantages and limitations of secondary rot over primary rot which are listed below:

**Advantages**

- For production of fruiting bodies, plant waste materials or mixture of them can be used.
- Yield of fruiting body per unit time is higher.
- Cost of per unit fruiting production is lower.

**Limitations**

- To achieve reproducible production, highly sophisticated technology is used.
- High investment is required for machinery and air conditioned houses.
- High input of manpower and energy is needed if machines are not used.

Out of hundred species of edible mushrooms that grow wildly, only a few species are cultivated [24]. The major species of cultivated mushrooms are listed in Table 9.

## 4.4 Cultivation of Mushrooms

*Agaricus bisporus* (replaced by *A. bitorquis* in regions with Mediterranean climates) is by far the most commonly cultivated mushroom in the world. *Agaricus bisporus* (button mushroom) grows on garden waste, manure heaps and on roadsides. In the beginning of its domestication, a suitably prepared bed of horse manure in gardens was used for mushroom cultivation [25]. Later it was cultivated in caves and old mines, which showed that mushroom production no longer depends on seasons and weather. Now it can be cultivated in properly built mushroom houses under the guidance of specialists. The flow sheet for mushroom cultivation is outlined in Fig. 4. The steps involved in the cultivation process of typical mushroom are: (i) preservation of fungal cultures and production of spawn (inoculum) on sterilized cereal grains, (ii) preparation of compost and inoculation with spawn, (iii) colonization of substrate, (iv) initiation of fruiting body formation, and (v) harvesting and post harvest management.

**Table 9:** List of some primary and secondary rot fungi.

| Species | Common name | Substrate | Area | Production × $10^3$ tonnes |
|---|---|---|---|---|
| ***Primary rot fungi*** | | | | |
| *Auricularia* sp. | Wood-ear | Wood, sawdust, rice bran, waste sugar cane | China, Taiwan | 485 |
| *Flammulina velutipes* | Winter mushroom | Sawdust and rice bran | Europe and Eastern Asia | 285 |
| *Grifola frondosa* | Siting-hen mushroom or Limuo, Maitaka | - | - | 33 |
| *Hypsizygus marmoreus* | Shimeji | - | - | 74 |
| *Lentinus edodes* | Shiitake or oak mushroom | Logs of broad leaves trees, sawdust blocks | Temperate Eastern Asia, USA and Europe | 1564 |
| *Pholiota nameko* | Nameko or viscid mushroom | Logs of broad leaves trees, sawdust, rice bran | Eastern Asia | 56 |
| *Pleurotus* sp. | Oyster mushroom | Logs, sawdust, cereal straw, newspaper, banana stems, cotton cloth, etc. | Eastern Asia, India, USA and Europe | 876 |
| *Tremella fuciformis* | White jelly fungus or silver ear | Logs of broad leaves trees | Japan, China and Taiwan | 130 |
| ***Secondary rot fungi*** | | | | |
| *Agaricus bisporus* | Button mushroom | Composted straw | Temperate areas | 1956 |
| *Volvariella volvacea* | Straw mushroom or Chinese mushroom | Composted rice straw, cotton waste | Africa and South-Eastern Asia | 181 |

### *4.4.1 Preservation of fungal cultures and preparation of spawn*

The production of spawn or inoculum is the most important and crucial step in the cultivation of microorganisms. Fungal culture medium impregnated with mycelial fragments of mushroom is called spawn, which serves as an inoculum for mushroom cultivation. By $17^{th}$ century, France had started and standardized the production of spawn for cultivation of button mushroom (*Agaricus bisporus*) by mixing colonized substrates. Different cereal grains such as rye, wheat, sorghum, millet and sawdust/bran mixture are generally used for spawn production. The reason for using cereal grains for spawn production is that these allow fast mycelia development and easy handling during sterilization. The epidermis and endosperm of cereal grains have enough nutrients to support rapid growth and colonization of mycelium. The mycelia grow on the seed surface. Some inorganic carriers i.e., perlite can also be employed for spawn production. Perlite spawn comprises of 1450g perlite, 1650g wheat bran, 200g $CaSO_4$, 50g $CaCO_3$ and 6650 ml

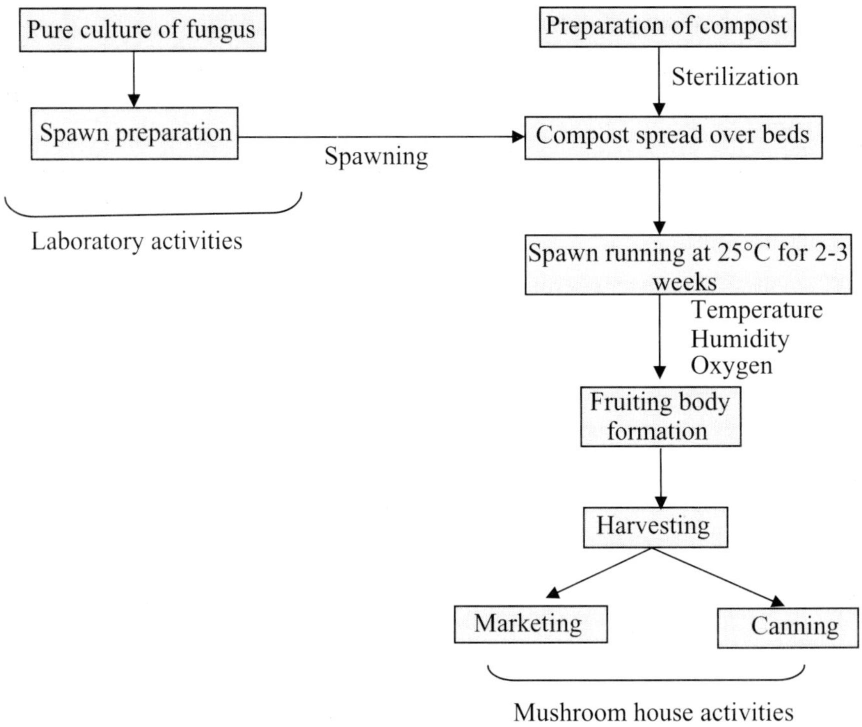


**Fig. 4:** Flow diagram for mushroom cultivation.

water. For grain spawn formation, the cereal grains are cooked in water until they swell, water is decanted and 2% lime (calcium carbonate) is mixed to the grains and transferred to test tubes/ flasks. These are autoclaved at 121°C for 30 min, cooled down to 30-40°C, inoculated with pure culture of mushrooms and incubated at suitable temperature for the growth of mycelium. The strains used for spawn production are maintained on agar slants and preserved by lyophilization and cryopreservation. Spawn is generally stored at 2-5°C, and during storage it remains viable but in latent state.

#### *4.4.2 Pretreatment of compost*

4.4.2.1 Pretreatment of substrate: The pretreatment is given to the substrate to provide an optimum moisture level and porosity for proper aeration. This significantly enhances the mycelial colonization of substrates and results in better yield of mushrooms.

4.4.2.2 Composting: The main aim of composting is to prepare medium for rapid growth of mycelium, and it takes about 10-14 days. For the preparation of compost, the material used must contain a cheap source of carbon and nitrogen e.g., horse manure and wheat straw which should not include substances that inhibit the mycelial growth and must be degradable by microorganisms. For composting, the substrates are mixed, stacks are made and immersed in water so as to make these moist but not waterlogged. The compost is kept for sterilization at 60-70°C for 1-2 days and it is also known as 'peak heating' of compost. To avoid the excessive heating and anaerobic conditions during sterilization of compost, periodic turning or mixing is done [25]. A complete sterilization of substrates generally gives better yields of fruiting bodies. After sterilization, the compost is cooled till the temperature comes down to 25°C, which is favourable for mycelial growth. Inoculum or spawn is mixed with compost and put into trays and covered with a thin layer (3-5 cm) of soil, peat or chalk to prevent drying of the compost. It also maintains temperature and gives support to the fruiting bodies. The spreading of a small amount of spawn all over the bed of sterilized compost is called spawning or run. The quantity of inoculum may vary from 0.5% (*Agaricus bisporus*) to 5% (*Agrocybe aegerita*) of the substrate weight. Temperature and humidity is maintained for optimum mycelial growth, as these immensely affect spawn running and also the formation of fruiting body.

### *4.4.3 Colonization of substrates*

The availability of free nutrients and water greatly influence the colonization of sterile substrates. Mycelia colonize more rapidly when the substrates are supplemented with desired nutrients. Different physical, chemical and biological parameters affect the growth of mycelia on solid substrates. Among the physical factors, variation in temperature can be tolerated, but the moisture content of the substrate and the humidity in the environment are comparatively more important for the growth of mycelium. About 80-90% humidity in environment and 70-80% moisture content of substrate is desirable for mycelial growth. The pH of the substrate has a minor affect on fungal growth, but the gas exchange between substrate and environment during fungal growth in solid substrates has more impact on mushroom yield. Some species of fungus e.g., *Pleurotus* sp. and *Stropharia* sp. suppress the growth of competitive microorganisms by secreting organic acids and antibiotics. However, some competitive microorganisms are also suppressed due to the accumulation of $CO_2$ in the medium.

### *4.4.4 Initiation and formation of fruiting bodies*

Mushroom crop matures at different intervals producing flushes. The first flush of fruiting bodies comes at about 3 weeks, and after 5-6 flushes the culture has to be renewed [26]. The factors which affect the fruiting body formation include substrate composition, temperature, air, substrate humidity, light, oxygen and the presence of other microorganisms [27].

### *4.4.5 Harvest and post harvest management*

The harvesting and post harvest management are very important as mushrooms have a high rate of metabolism and will destroy themselves in a short time. Harvesting starts on each crop after about 65 days from the beginning of the composting process. The first crop of mushrooms is ready to pick in 20 to 26 days after casing. Once the first mushroom is picked up, many more mushrooms mature and can be harvested within three to four days. This production peak is referred to as "flush". A new flush occurs every week. Generally after each flush, the level of usable nutrients and water in the compost depletes. By the end of the third flush, the culture is renewed. Mushroom has a short shelf life, especially in tropical countries. Therefore after harvesting, the mushrooms are immediately transported to a pre-cooler place to reduce the pulp temperature to 35°C or below from a pre-harvested, 68°C as quickly as possible. Some of the mushrooms such as Shiitake and Dingri can be air-dried or freeze dried and stored for a longer period without the loss of nutritional value.

## 4.5 Nutritional Value of Mushrooms

The nutrient content of mushrooms is also related to moisture content, which in turn depends on the cultivation conditions. The button mushroom (*A. bisporus*) generally has a high moisture if harvested at an immature stage, and thus has a different nutrient profile as compared to the fully mature mushrooms. The protein content in mushroom typically varies between 20-30% on dry mass basis.

However, there is extreme variation in protein content among various mushroom species e.g., 3.5% in *Cantharellus cibarius* and 44% in *Agaricus bisporus*. High protein content makes them an ideal food because they contain all the essential amino acids. Mushrooms can be an important dietary source of these amino acids. These contain very high percentage of water (85-90%). However, the temperature and humidity during growth and storage greatly influence the moisture content of mushrooms.

After moisture, carbohydrates are the major component of mushrooms (average of 4.2%), particularly pentoses, hexoses, disaccharides and polysaccharides, β-glucan, a soluble fibre [28]. Soluble fibre has been shown to help prevent and manage cardiovascular disease by lowering total and LDL cholesterol levels. It also helps to regulate blood sugar levels [29]. There is some evidence that beta-glucans offer anti-cancer potential [30, 31]. Mushrooms are low in fat (comprising 2-8% dry weight) and there is no cholesterol in this food. Some mushrooms in Basidiomycotina have the ability to lower the level of serum cholesterol. The fruit body of a mushroom is rich in B-complex vitamins including riboflavin ($B_2$), niacin, pantothenic acid, thiamin ($B_1$), biotin, folate and vitamin $B_{12}$. Since mushrooms do not contain starch, it makes them suitable for the diet of diabetic people.

Outila *et al.* [32] found that ergocalciferol in mushrooms increased the serum 25-hydroxy vitamin D concentrations and can be reliably recommended as a natural source of vitamin D. Mushrooms also contain some calcium, iron, copper, potassium, zinc and selenium. However, only copper, selenium and potassium are present in nutritionally significant amount. These provide more selenium than other fruits and vegetables. Selenium in food is considered to have anticancer effect.

## 4.6 Mycoproteins

Microorganisms have long been used in the elaboration of food for human consumption. In addition to the macroscopic fungi (e.g., mushrooms), the microscopic fungi have also been cultured in fermenters and used as a part of food. Among the microfungi, brewer's yeast is the first organism recognized as supplement for animal proteins and it was produced at industrial scale for food during the First World War. *Fusarium ventenatum* (a filamentous fungus) has been investigated in the last two decades as a primary source of protein. This fungus is grown in continuous flow culture to produce mycoprotein, which is sold under the trade name Quorn™ in USA, UK and some other European countries. The first mycoprotein product was marketed in 1985. The efforts to develop a meat substitute based on fungal mycelium were made by the British company Rank Hovis McDougall (RHM) during the 1960s.

Mycoproteins are rich in dietary fibre, polyunsaturated fatty acids, and have a low degree of allergenicity. The nutritional quality of mushroom is comparable with milk protein [33]. As compared to most bacteria and yeasts, the growth rate of filamentous fungi is low, but the harvesting from culture medium is easier. The fungal biomass contains comparatively less nucleic acids. Advantages of using fungi over bacteria, yeasts and algae for protein production are as follows:

(i) Mycoproteins contain lesser nucleic acid as compared to that of single cell protein from other microorganisms.

(ii) The final product can be converted into various textured foods because of its fibrous nature.

(iii) The penetration power of fungal mycelium into insoluble substrates is more than other microbes, which makes them suitable for cultivation under solid substrate fermentation using lignocellulosic materials.

(iv) The culture of filamentous fungi for protein production is more economical and beneficial.

(v) Digestibility and net protein utilization from fungal biomass is higher, as it has greater retention time in digestive system.
(vi) These are acceptable for human consumption due to their mushroom like odour and taste.
(vii) The fungal proteins do not require further processing before use.

### *4.6.1 Fungi used for mycoprotein production*

In 1970's, many industries used various fungi (*Aspergillus niger, Trichoderma harzianum, Fusarium graminarium,* etc.) for the large scale production of mycoprotein using various renewable substrates or industrial waste material. However, relatively few have operated commercially due to obstacles encountered on scale-up or for economic reasons. The physiological problems that are often encountered on scale-up include difficulties with:

(i) Oxygen requirements and oxygen transfer rates
(ii) Nutrient and temperature gradients
(iii) Effects of $CO_2$, as high levels may inhibit respiration in certain microorganisms
(iv) Hydraulic pressure in deep fermenters

In some instances, the economics of production can be improved by either increasing the value of the product or reducing the production costs through:

(i) Use of cheaper substrates
(ii) Improvements in the efficiency of the organism
(iii) Enhanced nutritional value/composition of the microbial protein
(iv) Marketing the protein as a premium product for human rather than animal consumption
(v) Development of a multiproduct process
(vi) Lowering downstream processing costs e.g., by reducing endogenous RNA levels

The various fungi and processes used for mycoprotein production are given in Table 10. The mycoproteins were cultured in continuous stirred tank reactors and final product is harvested by centrifugation.

### *4.6.2 Method of mycoprotein production*

The process for mycoprotein production involves the continuous stirred tank reactor containing modified starch or glucose medium. The carbon to nitrogen ratio (C:N) is maintained at 5:1 to 15:1 depending upon the fungus used. In continuous system, ammonium salts are used as a source of nitrogen, whereas phosphoric acid is generally employed as a source of phosphorus. The preculture containing fungal biomass is inoculated in the fermenters and the nutrient medium supplemented with carbon

**Table 10:** List of some fungi and processes used for mycoprotein production.

| Fungus | Process | Substrate | Reference |
|---|---|---|---|
| *Aspergillus niger* | Tate and Lyle, U.K. | Carob bean extract | 34 |
| *Chaetomium cellulolyticum* | Waterloo Process, Canada | Corn starch, cellulosic wastes, wood sawdust | 35 |
| *Fusarium graminearum* | Ranks Hovis McDougall, U.K. | Starch/glucose syrup | 36 |
| *Gliocladium deliquescens* | Impro Systems, Iowa | Corn wet milling waste | 37 |
| *Paecilomyces variotii* | Pekilo Process, Finland | Sulphite waste liquor | 38 |
| *Trichoderma harzianum* | ICAITI (EI-Salvador) | Coffee processing waste | 39 |

and nitrogen sources are added continuously. The temperature is maintained at 30°C and the cells are harvested continuously. The fungal mycelium is harvested by centrifugation, and the final product containing about 30% total solids is obtained.

4.6.2.1 The Pekilo process: This process began its operation in 1975 and was the first commercial continuous process for the production of filamentous fungus. It had to overcome the special problems caused by the pseudoplastic rheological behaviour of submerged cultures of fungal mycelium, which particularly affect oxygen transfer rates. The process was developed in Finland for the production of SCP using *Paecilomyces variotii* on the medium comprising spent sulphite liquor derived from wood processing that contains monosaccharides and acetic acid. This continuous process is operated aseptically and produces over 10000 tonnes of SCP a year from two 360 $m^3$ fermenters. The protein obtained by culture of this organism is called Pekilo protein and is mainly used as an animal feed. The Pekilo protein contains upto 59% crude protein on dry weight basis. The production of mycoproteins in fermenter is shown in Fig. 5. Mycoprotein production is an example of fermentation process that involves an efficient use of a cheap and readily available energy source. The harvested product absorbs a wide variety of flavourings and colouring compound, while its fibrous texture (arising from its mycelial nature) makes it easy to formulate products that imitates the feel of meat and fish.

### *4.6.3 Nutritional value of mycoproteins*

The composition of Quorn® mycoprotein is presented in Fig. 6. Crude protein values based on total nitrogen (TN × 6.25) comes to be about 60% of the total fungal biomass. However, these values

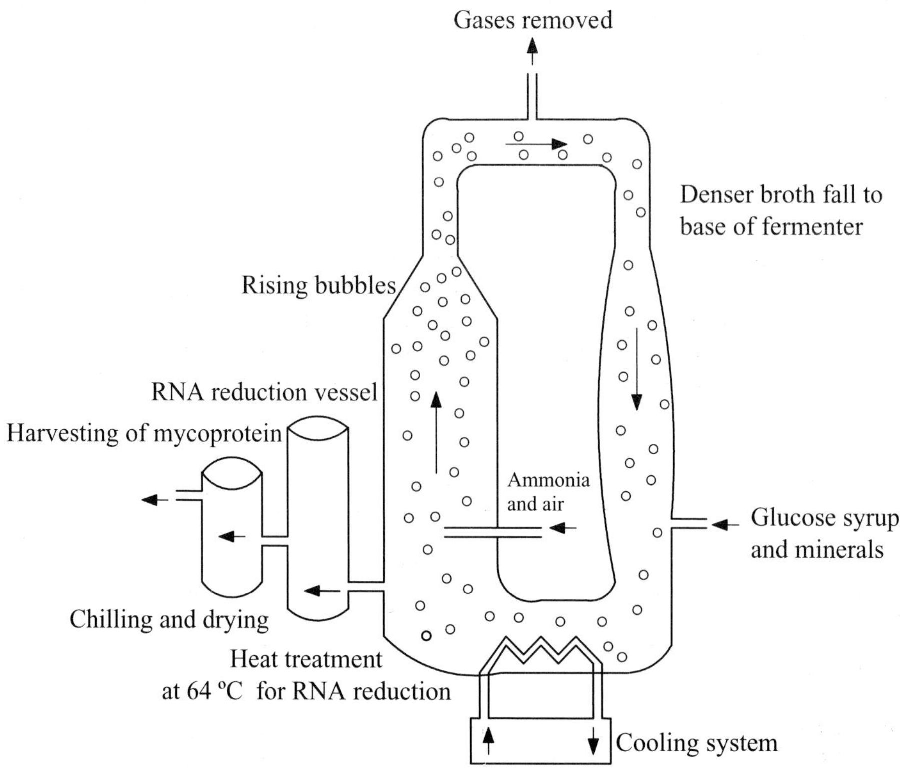


**Fig. 5:** Schematic diagram of production of mycoproteins in a fermenter.

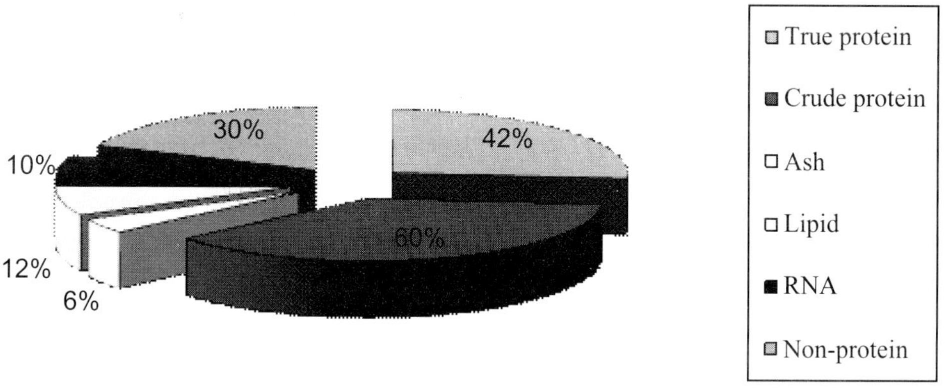


**Fig. 6:** Proximate composition of *F. graminaerum* (% dry matter).

are meaningless since fungal biomass contains substantial amount of non-protein nitrogen such as n-acetyl glucosamine in the chitin, which is the major component in the fungal cell wall as well as high RNA content. True protein content in the fungal cell is about 15%, and the chitin in the cell wall is 10% [40].

Average non-protein nitrogen content for thirteen fungi grown on different media has been shown to be 35%. Distribution of nitrogen in *Fusarium graminearum* grown in continuous culture at a growth rate of 0.1 $hr^{-1}$ has been studied [36] (Table 11). The protein content of this mycoprotein is comparable to that of milk and cottage cheese [41]. It contains all the eight essential amino acids required in human nutrition [40, 42]. Its amino acid composition is close to the one recommended by FAO, though methinone and cysteine are the limiting amino acids in this product.

The minerals in mycoprotein are comparable with meat, and its iron content (1.3 mg/100g) is equivalent to pork and higher than chicken. It is very rich in zinc and has the same dietary fibre content (5.1%, w/w) as in fresh green vegetables such as peas, but more than whole meal bread. Incorporation of Quorn® mycoprotein in food makes it a low-calorie (80 kcal/100g) food that lacks animal fats, cholesterol and is low in saturated fats. Trinci [41] has presented a comparison of the composition of Quorn® mycoprotein and braised beef as given in Table 12.

**Table 11:** Distribution of nitrogen of *Fusarium graminearum* grown in continuous culture.

| Component | Nitrogen (%) | Total nitrogen (%) |
|---|---|---|
| Free amino acid | 0.71 | 7.35 |
| Protein | 6.34 | 65.63 |
| Nucleotide | 0.15 | 1.55 |
| RNA | 1.46 | 15.11 |
| N-acetyl glucosamine | 1.00 | 10.35 |
| Total | 9.66 | - |
| Total nitrogen (by Kjeldahl) | 9.72 | - |

*Source:* 36

**Table 12:** Comparison of some biochemical components in braised beef and mycoprotein.

| | Component as % of total (DW) | |
|---|---|---|
| **Component** | **Braised beef** | **Mycoprotein** |
| Protein | 30.9 | 12.2 |
| Dietary fibre | 0 | 5.1 |
| Fat (total) | 11.0 | 2.9 |
| Fat (saturated) | 4.6 | 0.4 |
| Polyunsaturated:saturated fatty acid | 0.1 | 2.5 |
| Carbohydrate | 0 | 1.3 |
| Cholesterol | 0.08 | 0 |

*Source:* 41

**Table 13:** Comparison of some of Quorn (a mycoprotein) with various meats (value per 100g).

| Biochemical parameter (Per 100 gm) | Quorn™ Pieces | Skinless Chicken Pieces | Quorn™ Southern Burgers | Breaded Chicken burger | Quorn™ Burgers | Beef Burger | Quorn™ Sausages | Low Fat Pork Sausages |
|---|---|---|---|---|---|---|---|---|
| Protein | 12.3 | 24.8 | 10.7 | 12.9 | 12.8 | 15.0 | 13.5 | 13.0 |
| Carbohydrate | 1.8 | 0.0 | 12.3 | 17.1 | 5.8 | 3.5 | 5.2 | 9.6 |
| Oil/ Fat | 3.2 | 5.4 | 9.6 | 16.7 | 4.6 | 23.8 | 4.7 | 9.3 |
| Fibre | 4.8 | 0.0 | 3.1 | 0.6 | 4.1 | 0.4 | 3.4 | 1.6 |

*Source:* Quorn Foods Inc. product data; Ground beef information from USDA Nutrient Database for Standard (http://www.quorn.us).

#### *4.6.4 Problems associated with mycoproteins*

Some of the problems associated with mycoproteins include:

(i) Slow growth in culture media as compared to bacteria and yeast
(ii) Presence of nucleic acid
(iii) Some fungi may produce mycotoxin too
(iv) Utilization of limited substrates (mostly starchy and cellulosic materials) for growth of fungi
(v) The environmental conditions used for the growth of fungi also favour contamination by yeasts.

To decrease nucleic acid content, heat shock at 68°C for about 20 min is given to fungal biomass to allow the endogenous ribonucleases to reduce the RNA content to 1% with minimum loss of protein and fibrous texture.

## 5. ALGAE AS FOOD

The use of algae as food and feed is known since antiquity, as they form part of the diets in East Asia and Central Africa. Out of the 40,000 species of prokaryotic and eukaryotic algae, only a few species are grown commercially for use as food or food supplements [43]. The most common macroalgae used as food include the red algae *Porphyra* (mori, kim, laver), *Asparagopsis taxiformis* (limu), *Gracilaria, Chondrus crispus* (Irish moss), *Palmaria palmate* (dulse), the kelps *Laminaria,*

*Undaria* (wakame), *Macrocystis, Caulerpa racemosa, Codium* and *Ulva* [44-46]. These algae are either harvested from wild populations or cultured in farm and ponds. Macroalgae can be consumed either fresh or in dried form and are also used in the preparation of pickles [47]. The use of micro-algae, especially blue-green algae, has been reported for long time as a nutritional supplement or food source [48]. The application of microalgae as food supplement is becoming popular in the recent time due to their nutraceutical value. Algae, being autotrophic organisms, can utilize energy from sunlight, carbon source from $CO_2$ and nutrients present in water for synthesizing their food.

## 5.1 Use of Blue-Green Algae as Human Food

Blue-green algae (cyanobacteria) are very good source of proteins and other nutrients. These algae are invariably free from heavy metals or harmful bacteria and contain the most complete range of essential and non-essential amino acids, vitamins and minerals present in any single food [49]. These are rich in the antioxidant beta-carotene, which the human body converts into vitamin A. These also have high level of B vitamins, providing even more $B_{12}$ than liver. In fact, the only vitamins that blue-green algae lack are vitamins D and E. Another benefit of blue-green algae is that they have a plenty of minerals, which are important because of their synergistic relationship with vitamins. Vitamins and minerals promote the assimilation of one another. Minerals are also necessary to form body fluids, blood and bone and to maintain healthy nerve function. Blue-green algae contain almost all organic minerals in chelated form, meaning that each is attached to a protein molecule that transports it to the bloodstream and enhances their absorption [50].

### *5.1.1 Nutritional value of micro-algae*

The nutritional value of any algal species depends on its cell size, digestibility, production of toxic compounds and biochemical composition. The gross composition of 16 species of micro-algae is presented in Table 14. Although there are differences in the biochemical composition among the micro-algal species, yet proteins are the major organic component followed by lipids and carbohydrates in algae considered for use in human food and animal feed. The level of protein, lipid, and carbohydrate in blue green algae ranges 12-35%, 7.2-23% and 4.6-23% on dry weight basis, respectively.

The content of highly unsaturated fatty acids (HUFA), particularly eicosapentaenoic acid (20:5 n-3, EPA), arachidonic acid (20:4 n-6, ARA), and docosahexaenoic acid (22:6 n-3, DHA) is of major importance in the evaluation of the nutritional value of an algal species. Significant concentrations of EPA are present in the diatoms (*Chaetoceros calcitrans, Centruroides gracilis, Skeletonema costatum, Thalassiosira pseudonana*) and *Platymonas* sp., whereas high concentrations of DHA are found in the prymnesiophytes (*P. lutheri, Isochrysis* sp.) and *Chroomonas salina.*

Micro-algae can also be considered as a rich source of vitamin C i.e., ascorbic acid (0.11-1.62% of dry weight). The nutritional value of micro-algae can vary considerably according to the culture conditions. For example, the effect of the composition of the culture medium on the proximate composition of various species of micro-algae is depicted in Table 15. The protein content per cell, which is considered as one of the most important parameter assessing nutritional value of micro-algae as feed in aquaculture, was found to be more susceptible to variation than the other cellular constituents. Moreover, the growth of animals fed on a mixture of several algal species is often superior to that obtained when feeding only one algal species. A particular alga may lack a nutrient, while another alga may contain that nutrient and lack a different one. In this way, a mixture of both algal species supplies the animals with an adequate amount of both nutrients. The nutritional aspects of micro-algae used in marine culture of bivalve molluscs, crustaceans and fish have been extensively reviewed [53].

**Table 14:** Concentrations (pg/cell) of chlorophyll a, protein, carbohydrates and lipids in some species of micro-algae used in aquaculture.

| Algal Class & Species | Dry wt. | Chl a | Protein | Carbo-hydrate | Lipid |
|---|---|---|---|---|---|
| ***Bacillariophyceae*** | | | | | |
| *Chaetoceros calcitrans* | 11.3 | 0.34 | 3.8 | 0.68 | 1.8 |
| *Chaetoceros gracilis* | 74.8 | 0.78 | 9.0 | 2.0 | 5.2 |
| *Nitzschia closterium* | - | - | - | - | - |
| *Phaeodactylum tricornutum* | 76.7 | 0.41 | 23.0 | 6.4 | 10.7 |
| *Skeletonema costatum* | 52.2 | 0.63 | 13.1 | 2.4 | 5.0 |
| *Thalassiosira pseudonana* | 28.4 | 0.27 | 9.7 | 2.5 | 5.5 |
| ***Chlorophyceae*** | | | | | |
| *Dunaliella tertiolecta* | 99.9 | 1.73 | 20.0 | 12.2 | 15.0 |
| *Nannochloris atomus* | 21.4 | 0.080 | 6.4 | 5.0 | 4.5 |
| ***Cryptophyceae*** | | | | | |
| *Chroomonas salina* | 122.5 | 0.98 | 35.5 | 11.0 | 14.5 |
| ***Eustigmatophyceae*** | | | | | |
| *Nannochloropsis oculata* | 6.1 | 0.054 | 2.1 | 0.48 | 1.1 |
| ***Prasinophyceae*** | | | | | |
| *Tetraselmis chui* | 269.0 | 3.83 | 83.4 | 32.5 | 45.7 |
| *Tetraselmis suecica* | 168.2 | 1.63 | 52.1 | 20.2 | 16.8 |
| ***Prymnesiophyceae*** | | | | | |
| *Isochrysis galbana* | 30.5 | 0.30 | 8.8 | 3.9 | 7.0 |
| *Isochrysis aff. Galbana (T-iso)* | 29.7 | 0.29 | 6.8 | 1.8 | 5.9 |
| *Pavlova lutheri* | 102.3 | 0.86 | 29.7 | 9.1 | 12.3 |
| *Pavlova salina* | 93.1 | 0.34 | 24.2 | 6.9 | 11.2 |

*Source:* FAO, Fisheries Technical Paper, 361, 1996 (http://www.fao.org); 51

### *5.1.2 Production*

In some countries, algae are cultivated in large trenches e.g., sewage oxidation ponds by using sunlight [54]. In Japan, the mass cultivation of mixed culture of algae *Chlorella ellipsoids* and *Scenedesmus obliguus* is done in open pond systems. Oxidation ponds are prepared for the cultivation of algae on sewage wastes, where the sewage is allowed to accumulate (Fig. 7). Algal cultures are inoculated in the ponds and allowed to grow. Various factors that affect the yield of algae are illumination, time, light intensity, $CO_2$, agitation and nitrogen source. Nitrogen can be supplemented in the form of ammonium salt, which enhances the biomass yield. Harvesting of algal biomass is generally done by concentration, dewatering and drying.

### *5.1.3 Spirulina*

*Spirulina platensis* is classified as *Arthrospira (Spirulina) platensis* (Nordstedt) Gomont, and it is photosynthetic, filamentous, spiral-shaped, multicellular and blue-green micro-alga which has many kinds of vitamins in large quantities, high protein content (60-70%), varieties of minerals, 5-6% fat of good quality and also kinds of high effective biological active substances. A microscopic view of *Spirulina* is shown in Fig. 8(a-c). The two most important species for SCP production and nutraceutical point of view are *Spirulina maxima* and *Spirulina platensis*. *Spirulina platensis* has been commercially exploited for the production of human food supplements, animal feed and

**Table 15:** Cellular density and proximate composition of four marine micro-algae grown in different culture media.

| Organism<br>Culture medium | Cellular density ($10^6$ cells/ml) | Protein (pg/cell) | Carbohydrates (pg/cell) | Lipids (pg/cell) |
|---|---|---|---|---|
| *T. suecica* | | | | |
| Walne | 2.29 | 13.31 | 6.20 | 7.04 |
| ES | 2.58 | 16.98 | 6.93 | 7.22 |
| F/2 | 2.38 | 21.75 | 8.37 | 7.92 |
| Algal-1 | 4.11 | 32.22 | 8.83 | 8.65 |
| *D. tertiolecta* | | | | |
| Walne | 4.04 | 13.37 | 13.22 | 22.28 |
| ES | 4.24 | 14.88 | 15.73 | 23.94 |
| F/2 | 4.97 | 13.26 | 17.91 | 23.67 |
| Algal-1 | 8.45 | 18.82 | 11.08 | 18.18 |
| *I. galbana* | | | | |
| Walne | 10.11 | 5.17 | 4.28 | 25.95 |
| ES | 12.09 | 7.23 | 5.21 | 28.38 |
| F/2 | 10.81 | 8.13 | 5.59 | 26.82 |
| Algal-1 | 16.15 | 9.57 | 4.28 | 20.68 |
| *P. tricornutum* | | | | |
| Walne | 19.01 | 2.65 | 6.42 | 6.51 |
| ES | 16.23 | 5.21 | 9.20 | 6.45 |
| F/2 | 24.65 | 3.34 | 6.90 | 5.52 |
| Algal-1 | 39.04 | 4.20 | 5.98 | 5.79 |

*Source:* 52

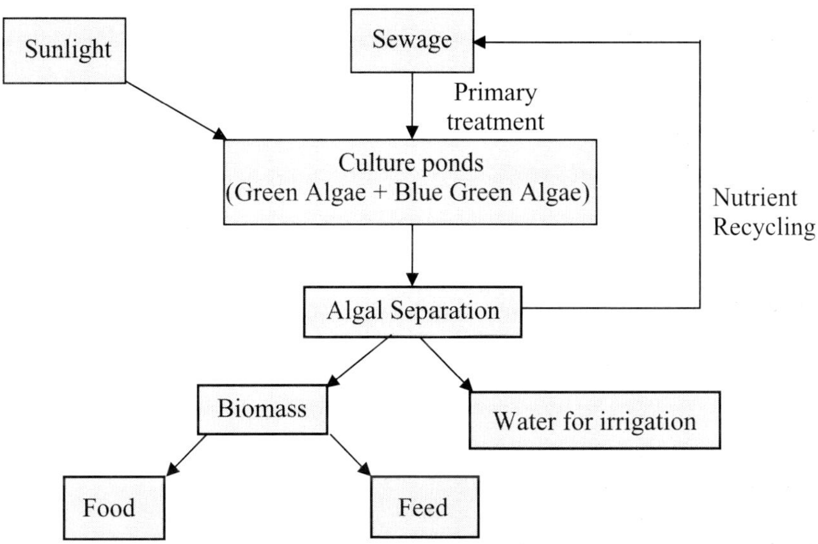


**Fig. 7:** Flow diagram for the cultivation of algae on sewage.

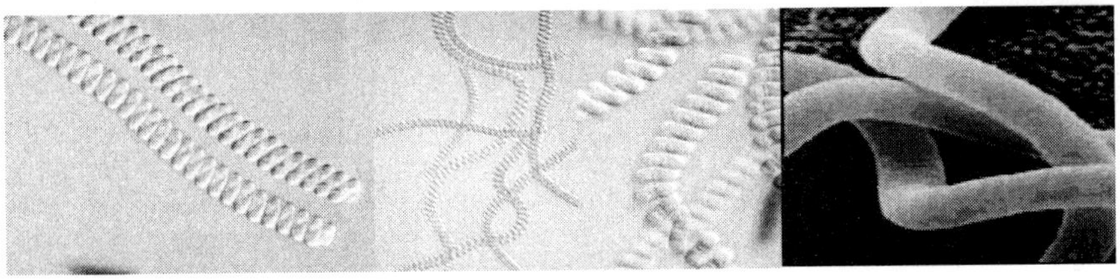

(a) Long filament (b) Perfect spiral coils (c) Electron micrograph

**Fig. 8:** Microscopic view of *Spirulina.*

pharmaceuticals as it has relatively high concentration of valuable products like phycocyanin and polyunsaturated fatty acids.

*Spirulina* is one of the most promising micro-alga for the production of cyanocobalamine (vitamin $B_{12}$), antioxidant pigment like β-carotene, tocopherols, α-linolenic acid and single cell protein (SCP). The blue coloured pigment phycocyanin and other pigments including myxoxanthophyl and zeaxanthin of *Spirulina* are widely used as a natural colourant for food additive purposes. Phycocyanin also has anticancer, antioxidant, antiviral and anti-inflammatory properties.

Some of the best worldwide known *Spirulina* producing companies are Earthrise Farms (USA), Cyanotech (USA), Hainan DIC Microalgae Co. Ltd. (China), Murugappa Chettiar Research Center (India), Genix (Cuba) and Solarium Biotechnology (Chile) [55-57].

5.1.3.1 Protein: It unusually contains high amount of protein between 60-70% by dry weight [48]. *Spirulina* is useful in human nutrition due to its high quality and quantity of protein. The nutritive value of a protein is related to the quality of amino acids, digestibility coefficient as well as its biological value [58, 59]. It is a complete protein containing all essential amino acids, though with lower level of methionine, cysteine and lysine as compared to meat, eggs or milk. However, it is superior to all standard plant proteins from legumes [60].

5.1.3.2 Essential fatty acids: It is rich in γ-linoleic acid (GLA) and also provides α-linoleic acid (ALA), linoleic acid (LA), stearidonic acid (SDA), eicosapentaenoic acid (EPA), docosahexaenoic acid (DHA) and arachidonic acid (AA) [48, 60]. *Spirulina* contains 4-7% lipids and has essential fatty acids: linoleic acid (LA) $(C_{18:2})\Delta^{9,\,12}$ and γ-linoleic acid $(C_{18:3})\Delta^{9,\,12,\,15}$ (GLA, Table 3). The later is claimed to have medicinal properties and is required for arachidonic acid and prostaglandin synthesis. GLA decreases low-density lipoprotein, since it is 170-fold more effective than LA [60].

5.1.3.3 Vitamins: It has significantly high concentration of vitamins $B_1$ (thiamine), $B_2$ (riboflavin), $B_3$ (nicotinamide), $B_6$ (pyridoxine), $B_9$ (folic acid), $B_{12}$ (cynocobalamin), C, D and E [48, 60]. *Spirulina* is a very rich source of vitamin $B_{12}$ and that is why these cyanobacteria are of great value for people needing supplements in the treatment of pernicious anemia [57, 59, 61].

5.1.3.4 Minerals: It a very good source of potassium and also contains calcium, chromium, copper, iron, magnesium, manganese, phosphorus, selenium, sodium and zinc [60, 62]. Iron in some nutritional complements is not appropriately absorbed. Iron from *Spirulina* is better absorbed than ferrous sulphate and other complements. Thus, it represents an adequate source of iron for anemic pregnant women [63].

**5.1.3.5 Carbohydrates:** *Spirulina platensis* has about 13.6% carbohydrates, and some of these are glucose, rhamnose, mannose, xylose and galactose [64]. It does not have cellulose in its cell wall, and therefore, it is an interesting foodstuff for people with problems of poor intestinal absorption and geriatric patients [59]. A new high molecular weight polysaccharide with immunostimulatory activity has been isolated from *Spirulina* and called "Immulina". This highly water-soluble polysaccharide represents between 0.5% and 2.0% (w/w) of the dry micro-algae [65].

**5.1.3.6 Nucleic acids:** One of the major concerns about the consumption of microorganisms is their high content of nucleic acids that may cause problem like gout. *Spirulina* contains 2.2%-3.5% of RNA and 0.6 %-1% of DNA in total which is less than 5% of dry weight. These values are smaller than those of other microalgae like *Chlorella* and *Scenedesmus* [48].

**5.1.3.7 Photosynthetic pigments:** *Spirulina* contains many pigments including chlorophyll-a, xanthophyll, beta-carotene, echinenone, myxoxanthophyll, zeaxanthin, canthaxanthin, diatoxanthin, 3'-hydroxyechinenone, beta-cryptoxanthin, oscillaxanthin plus the phycobiliproteins c-phycocyanin and allophycocyanin [66].

These pigments are responsible for the characteristic colours of certain flamingo species that consume these cyanobacteria in some African countries. This knowledge has promoted the use of these algae as a source of pigmentation for fish eggs [48, 67, 68] and chicken. *Spirulina* also enhances the yellowness and redness of broiled chicken due to accumulation of zeaxanthin [69].

**5.1.3.8 Consumption benefits:** Studies have revealed that *Spirulina* intake for 4 weeks reduces serum cholesterol levels in human beings by 4.5% [68], and significantly reduces body weight by 1.4 ± 0.4 kg [70]. The β-carotene is one of the most effective components to counteract those free radicals that alter cells causing cancer [71, 72]. *Spirulina* extract induces the tumour necrosis factor in macrophages, suggesting a possible mechanism for the destruction of tumour [73]. An extract of sulfated polysaccharides called Calcium-Spirulan (Ca-SP), made of rhamnose, ribose, mannose, fructose, galactose, xylose, glucose, glucuronic acid, galacturonic acid and calcium sulphate, from *Spirulina* has exhibited activity against HIV, Herpes Simplex virus, Human Cytomegalovirus, Influenza A virus, Mumps virus and Measles virus [68, 74].

Cell free extract of *S. maxima* has shown antimicrobial activity against *Bacillus subtilis, Streptococcus aureus, Saccharomyces cerevisiae* and *Candida albicans*. The presence of high quantities of acrylic acid in *Spirulina* was substantiated at the end of the seventies. This substance shows antimicrobial activity in a 2 mg/l of biomass concentration. Propionic, benzoic and mandelic acids are also found in this alga [75]. *Lactobacillus* population in human gastrointestinal tract is increased by *Spirulina* consumption, which in turn leads to better digestion and absorption of food and intestinal protection against bacterial infections and immune system stimulation [68].

Other benefits attributed to *Spirulina* include anti-arthritic effect due to the anti-inflammatory and antioxidative properties of phycocyanin [76], anti-atherogenic property [77], tumor inhibition [78], chemoprotective and radio-protective effect [79] and antioxidant activity on lead-induced toxicity in rats [80].

**5.1.3.9 Toxicology:** Studies in Mexico showed that the administration of *S. platensis* to mice does not cause embryonic or fetal damages [81, 82]. Absence of phycotoxins in *Spirulina* is an advantage with respect to *Microcystis, Anabaena* and *Aphanizomenon*. However, some fresh water cyanobacteria have caused death in livestock and allergy or gastrointestinal reactions in human beings [83].

5.1.3.10 Cultivation: Most of the commercially available *Spirulina* is cultivated in open-channel raceway ponds with paddle-wheels to agitate the water. *Spirulina* has been proposed by both National Aeronautics and Space Administration (NASA) [84], and the European Space agency [85] as one of the primary foods to be cultivated during long-term space missions.

The main commercial large-scale culture of microalgae started in early 1960s in Japan with the culture of *Chlorella* followed by *Spirulina* in the early 1970s at Lake Texcoco, in México. The first plant in USA (Earthrise Farms) for the exploitation of *Spirulina* was built in 1981 in California. Zarrouk medium [86, 87] is frequently used during the isolation process and the SSM medium is preferred in the industrial production stage [88]. The nitrogen of human urine mainly exists in urea; studies proved that *Spirulina platensis* could grow well using urea as nitrogen source [89-93]. Feng and Wu [94] cultivated *Spirulina platensis* with pre-treated urine, studied the $O_2$ evolution and discovered that *Spirulina platensis* cultivated in the urine could release $O_2$ quickly, while at the same time much biomass could be obtained. The batch cultures of *Spirulina platensis* can treat human urine effectively; the consumption of Total Phosphorus (TP) could reach to more than 99.99%; Total Nitrogen (TN) finally stabilised at 0.06-0.32 mg/l; and the consumption of TN could reach to more than 99%, and 1.05 g dry biomass could be obtained by treating 12.5 ml Synthetic Human Urine (SHU). This not only makes sewage purification, but also obtains much fresh *Spirulina platensis*, which can be used as a feed, fertilizer and biofuel.

*Spirulina* production may be carried out in closed and open systems. The first one involves laboratory scale and this is not used in industrial production. The open system denominated raceway due to its low production cost, easy handling and high production of biomass. That is why this system is frequently chosen for industrial production. This method uses a pond with a central islet, and a motor operating a paddle wheel, which allows continuous displacement of liquid culture in the peripheral channel. Paddle wheels speed in the order of 20 cm/s has been recommended. If necessary, plates may be introduced to avoid dead point formation. Production process has five stages:

(i) **Filtration and Cleaning:** A nylon filter at the entrance of the water pond is needed
(ii) **Concentration:** To obtain algal biomass, which is washed to reduce salts content
(iii) **Neutralization:** To neutralize the biomass with the addition of acid solution
(iv) **Disintegration:** To break down trichomes by a grinder
(v) **Dehydration:** By spray-drying. This operation has great economic importance since it involves about 20-30% of the production cost [95].

Packing is done in sealed plastic bags to avoid hygroscopic action on the dry *Spirulina,* and storage is done in corrugate cardboard boxes and fresh, dry, dim, pest-free and clean store room, preventing *Spirulina* pigments from deteriorating [96]. Quality control for *Spirulina* as a food includes microbiological standard tests, chemical composition test and test for heavy metals, pesticides and extraneous materials (insect fragments, rodent hair and feather fragments) [57].

## 6. CONCLUSIONS

There has been a tremendous progress to identify and evaluate the potential of bacteria, yeast, microalgae, micro and macrofungi as single cell protein on one hand, and to find out cheaper and sustainable substrate for the mass cultivation of microbes for use as single cell proteins. Some of these findings have been translated in industrial ventures. The contribution of single cell proteins, mycoproteins and mushroom in comparison to plant and animal protein sources to meet the protein demand of human population is little. We need to address issues related to nucleic acid content, toxicity, palatability, consumer acceptability of single cell protein (SCP), and efforts are also needed to find out still better and inexpensive substrates rather than biodegradable waste from industries and agriculture, and municipal activities should be explored for the production of SCP. Moreover, these microorganisms need to be selected for SCP production, which in addition to protein, have some health promoting biomolecules. The research and development in these lines will really lead to harness the potential of microbes to significantly supplement protein, oil and vitamin supply for human food and domestic animal feed.

## REFERENCES


1. Puniya AK, Singh S, Kumar CG & Singh K (1995) Single cell protein: A promising dietary substitute. *Indian J Exp Biol* 33: 545-551.
2. Reed G (1982) Microbial biomass, single cell protein and other microbial products. In: *Prescott and Dunn's Industrial Microbiology*, AVI Publishers Inc, USA, pp 541-592.
3. Bulschek G (1962) Food and feed yeasts. In: *The Yeasts,* Reiff F, Kautzmann R, Luers H & Lindemann M (eds), Verlag Hans Carl, Nurnberg, Germany, pp 123-125.
4. Rose AH (1981) The microbiological production of food and drink. *Sci Am* 6: 95-104.
5. Lichtfield JH (1968) The production of fungi. In: *Single-Cell Proteins*, Mateles I & Tannenbaum (eds), The MIT Press, Cambridge, MA, UK, pp 309-329.
6. Brinton MM & Warren L (1976) *Single Cell Proteins in Industrial Microbiology*, McGrow-Hill Book Company, New York, USA.
7. Han YW, Duhlap CE & Callihan CD (1971) Single cell protein from cellulosic wastes. *Food Technol* 25: 130-154.
8. Ware SA (1977) Single cell protein and other food recovery technologies from waste. Municipal Office of R&D, USEPA, Cincinnati, Ohio, USA.
9. Wood DA (1986) Useful microbial processes for upgrading lignocellulosics. In: *XIV International Congress of Microbiology*, Manchester, England.
10. Zee JA & Simard RE (1974) Simple process for the reduction in the nucleic acid content in yeast. *Appl Microbiol* 29: 59-62.
11. Legrys GA (1978) Power demand and mass transfer capability of mechanically agitiated gas liquid contractors and their relationship to air lift fermenters. *Chem Engg Sci* 43: 33-83.
12. Litchfield JH (1987) Production of foods, and feeds from biomass by microbiological processes. In: *Biomass Conversion Technology*, Moo-Young M (ed), Pergamon Press, New York, USA, pp 113-145.
13. Malick EA, Vanderveen JW, Hitzman DO & Wegner EH (1984) Production of single cell material. US Patent 4: 439.
14. Israelidis CJ & Coduonis MI (1982) Utilisation of agricultural wastes for animal feed and energy. *Agri Res* 6: 243-253.
15. Sico MIG (1996) The biotechnological utilization of cheese whey: A review. *Biores Technol* 57: 1-11.
16. Bhalla TC, Gajju H & Agrawal HO (1999) Production of single-cell proteins. In: *Biotechnology: Food Fermentation, Microbiology, Biochemistry and Technology*, Joshi VK & Pandey A (eds), Educational Publishers and Distributors, New Delhi, India, pp 1003-1022.
17. Waites MJ, Morgan NL, Rockey JS & Higton G (2002) Microbial biomass production. In: *Industrial Microbiology: An Introduction*, Waites MJ, Morgan NL, Rockey JS & Higton G (eds), Blackwell Science, Oxford, UK, pp 218-228.
18. Ratledge C & Tan KH (1990) Oils and fats: Production, degradation and utilization by yeasts. In: *Yeast Biotechnology and Biocatalysis*, Verachtert H & Mot RD (eds), Marcel Dekker Inc, New York, USA, pp 223-248.

19. Ward PO & Singh A (2005) Omega-3/6 fatty acid: Alternative source of production. *Process Biochem* 40: 3627-3652.
20. Colquhoun DM (2001) Nutraceuticals: Vitamins and other nutrients in coronary heart diseases. *Curr Opin Lipidol* 12: 639-646.
21. Hoek C, Van den, Mann DG & Jahns HM (1995) *Algae: An Introduction to Modern Phycology*, Cambridge University Press, Cambridge, UK.
22. Kyle DJ (1997) Production and use of single cell oil highly enriched in arachidonic acid. *Lipid Tech* 9: 116-121.
23. Stamets P (1993) *Growing Gourment and Medicinal Mushroom*, Ten Spreed Press, Berkeley, California, USA.
24. Campbell-Platt G & Cook PE (1989) Fungi in the production of foods and food ingredients. *J Appl Bacteriol* 117S-131S.
25. Carlile MJ, Watkinson SC & Gooday GW (2001) The fungi. In: *Fungi and Biotechnology*, Carlile MJ, Watkinson SC & Gooday GW (eds), Academic Press, New York, USA.
26. Dubey RC (2002) Single cell protein (SCP) and mycoprotein. *A Text Book of Biotechnology*, S Chand & Sons, New Delhi, India, pp 300-327.
27. Zadrazil F & Kamra DN (1997) Edible mushrooms. In: *Fungal Biotechnology*, Hamlyn PF (ed), Chapman and Hall, Weinheim, Deutschland, Germany, pp 14-25.
28. Sadler M (2003) Nutritional properties of edible fungi. *Br Nutr Found Nutr Bull* 28: 305-308.
29. Beelman RD, Royse D & Chikthimmah N (2003) Bioactive components in button mushroom *Agaricus bisporus* (J. Lge) Imbach (Agaricomycetideae) of nutritional, medicinal, and biological importance. *Int J Med Mushrooms* 5: 321-337.
30. Kidd PM (2000) The use of mushroom glucans and proteoglycans in cancer treatment. *Altern Med Rev* 5: 4-27.
31. Zaidman BZ, Yassin M, Mahajna J & Wasser SP *(2005*) Medicinal mushroom modulators of molecular targets as cancer therapeutics. *Appl Microbiol Biotechnol* 67: 453-468.
32. Outila TA, Mattila PH, Piironen VI & Lamberg-Allardt CJE (1999) Bioavailability of vitamin D from wild mushrooms (*Cantharellus tubaeformis*) as measured with a human bioassay. *Am J Clin Nutr* 69: 95-98.
33. Edwards G (1986) Mycoprotein—The development of a new food. *Food Lab Newslett* 4: 21-24.
34. Imric FKE & Vlitos AJ (1975) Production of fungal protein from carob (*Ceratonia siliqua L.*). In: *Single Cell Proteins*, Tannenbaum SR & Wang DIC (eds), MIT Press, USA, pp 223.
35. Moo-Young M, Daugulis AJ, Chahal DS & MacDonald DG (1979) The waterloo process for SCP production from waste biomass. *Process Biochem* 14: 38-45.
36. Anderson C, Longton J, Maddix C, Seammell GW & Solomons GL (1975) The growth of microfungi on carbohydrate. In: *Single Cell Proteins*, Tannenbaum SR & Wang DIC (eds), MIT Press, USA, pp 314-322.
37. Church BD, Widmer CM & Boon WC (1977) Water reuse and animal feed production for continuous fungal bioconversion of agro-industrial food processing waste. In: *Proceedings of Second International Biochemistry Symposium*, Toronto, Canada, pp 1-15.
38. Romantschuk H & Lehtomaki M (1978) Operational experiences of first full scale Pekilo SCP-mill application. *Process Biochem* 13: 16-29.
39. Aguirre F, Maldonado O, Rotz C, Menchu JF, Espinosa R & de Cubrera S (1976) Protein from waste. Growing fungi on coffee waste. *Chem Technol* 6: 636-640.
40. Solomons GL (1985) Production of biomass by filamanteous fungi. In: *Comprehensive Biotechnology*, Vol 3, Moo-Young M (ed), Pergamon Press, Oxford, UK, pp 483-495.

41. Trinci APJ (1992) Myco-protein: A twenty-year overnight success story. *Mycol Res* 96: 1-8.
42. Sadler M (1988) Quron®. *Nutr Food Sci* 112: 9-12.
43. Borowotzka M (1998) Commercial production of microalgae: Ponds, tanks, tubes and fermenters. *J Biotechnol* 70: 313-321.
44. Tseng CK (1981) Commercial cultivation. In: *The Biology of Seaweeds*, Lobban CS & Wynne MJ (eds), Blackwell Scientific, Oxford, UK, pp 680-725.
45. Druehl LD (1988) Cultivated edible kelp. In: *Algae and Human Affairs*, Lembi CA & Waaland JR (eds), Cambridge University Press, Cambridge, UK, pp 119-134.
46. Mumford TF & Miura A (1988) Porphyra as food: Cultivation and economics. In: *Algae and Human Affairs*, Lembi CA & Waaland JR (eds), Cambridge University Press, Cambridge, UK, pp 87-117.
47. Abbott IA (1988) Food and food production from seaweeds. In: *Algae and Human Affairs*, Lembi CA & Waaland JR (eds), Cambridge University Press, Cambridge, UK, pp 135-147.
48. Ciferri O (1983) *Spirulina*, the edible microorganism. *Microbiol Rev* 47: 551-578.
49. Batch JF & Batch PA (1990) *Prescription for Nutritional Healing*, Avery Publishing Group, Garden City Park, New York, USA.
50. Weiner MA & Weiner JA (1994) *Herbs That Heal: Prescription for Herbal Healing*, Quantum Books, Mill Valley, CA, USA, pp 176-178.
51. Brown MR (1991) The amino acid and sugar composition of 16 species of microalgae used in mariculture. *Aquaculture* 145: 79-99.
52. Herrero C, Angeles C, Fabregas J & Abalde J (1991) Yields in biomass and chemical constituents of four commercially important marine microalgae with different culture media. *Aqua Eng* 2: 99-110.
53. Brown LR (1968) World food problems. In: *Single-Cell Protein*, Mateles RI & Tannenbaum SR (eds), The MIT Press, Massachusetts, USA, pp 1-26.
54. Oswald WJ (1988) Large-scale algal culture systems (engineering aspects). In: *Microalgal Biotechnology*, Borowitzka MA & Borowitzka LJ (eds), Cambridge University Press, Cambridge, UK, pp 357-394.
55. Ayala F, Vargas T & Cardenas A (1988) Chilean experiences on microalgae culture. In: *Algal Biotechnology*, Stadler T, Mollion J, Verdus MC, Karamanos Y, Morvan H & Christiasen D (eds), *Proceedings of the Fourth International Meeting of the SAA*, Elsevier Applied Science, London, UK, pp 229-236.
56. Jourdan JP (1993) Solarium *Spirulina* farm in the Atacama Desert (North Chile). In: *Spiruline algue de vie. Bulletin de l'Institut Océanographique Monaco. Musée Océanographique*, Doumenge F, Durand-Chastel H & Toulemont, A (eds), Numéro spécial, 12: 191-194.
57. Belay A (1997) Mass culture of *Spirulina* outdoors. The earthrise farms experience. In: *Spirulina platensis (Arthrospira): Physiology, Cell-biology and Biotechnology*, Vonshak A (ed), Taylor and Francis, London, UK, pp 131-158.
58. Dillon JC & Phan PA (1993) *Spirulina* as a source of proteins in human nutrition. In: *Spiruline algue de vie. Bulletin de l'Institut Océanographique Monaco. Musée Océanographique*, Doumenge F, Durand-Chastel H & Toulemont, A (eds), Numéro spécial, 12: 103-107
59. Richmond A (1992) Mass culture of cyanobacteria. In: *Photosynthetic Prokaryotes*, Mann N & Carr N (eds), Plenum Press, New York, USA, pp 181-210.
60. Babadzhanov AS, Abdusamatova N, Yasupava FN & Faizullaeva N (2004) Chemical composition of *Spirulina platensis* cultivated in Uzbekistan. *Chem Nat Comp* 40: 276-279.

61. Becker EW (1984) Nutritional properties of microalgal potentials and constraints. In: *Handbook of Microalgal Mass Culture*, Richmond A (ed), CRC Press Inc, Boca Ratón, USA, pp 339-408.
62. Una BC & Tokusoglu O (2003) Unal, i profili nutrienti della biomassa di M.K. "di tre microalghe: *Spirulina platensis, clorella vulgaris* e galbana di Isochrisis. *Giornale di scienza dell'alimentazione* 68: 4-12.
63. Puyfoulhoux G, Rouanet JM, Besancon P, Baroux B, Baccou JC & Caporiccio B (2001) Iron availability from iron-fortified *Spirulina* by an *in vitro* digestion/Caco-2 cell culture model. *J Agric Food Chem* 49: 1625-1629.
64. Shekharam K, Ventakaraman L & Salimath P (1987) Carbohydrate composition and characterization of two unusual sugars from the blue-green algae *Spirulina platensis. Phytochemistry* 26: 2267-2269.
65. Pugh N, Ross SA, Elsohly HN, Elsohly MA & Pasco DS (2001) Isolation of three weight polysaccharide preparations with potent immunostimulatory activity from *Spirulina platensis, Aphanizomenon flos-aguae* and *Chlorella pyrenoidosa. Planta Med* 67: 737-742.
66. Onshak A (1997) *Spirulina platensis* (Arthrospira): *Physiology, Cell-biology and Biotechnology*, Taylor & Francis, London, UK.
67. Saxena PN, Ahmad MR, Shyan R & Amla DV (1983) Cultivation of *Spirulina* in sewage for poultry feed. *Experientia* 39: 1077-1083.
68. Henerikson R (1994) Microalga *Spirulina*, superalimento del futuro. In: *Ronore Enterprises*, Ediciones U (ed), España, Barcelona, Spain, pp 222-237.
69. Toyomizu M, Sato K, Taroda H, Kato T & Akiba Y (2001) Effects of dietary of *Spirulina* on meat color muscle of broiler chickens. *Br Poul Sci* 42: 197-202.
70. Becker EW, Jakober B & Luft D (1986) Clinical and biochemical evaluations of the alga *Spirulina* with regard to its application in the treatment of obesity. A double-blind crossover study. *Nutr Rep Int* 33: 565-574.
71. Fedkovic Y, Astre C, Pinguet F, Gerber M, Ychou M & Pujol H (1993) Spiruline and cancer. *Bulletin de l'Institut Oceanographique*12: 117-120.
72. Schwartz J, Flynn E & Shklar G (1990) The effect of carotenoids on the antitumor immune response *in vivo* and *in vitro* with hamster and mouse immune effectors. In: *Micronutrients and Immune Functions–Cytokines and Metabolism,* Bendich A, Chandra R, Gerard K, Cerami A & Takaku F (eds), New York Academy of Sciences, USA, pp 92-109.
73. Shklar G & Schwartz J (1988) Tumor necrosis factor in experimental cancer regression with alphatocopherol, beta-carotene, canthaxanthin and algae extract. *Eur J Cancer Clin Oncol* 24: 839-850.
74. Hayashi K, Hayashi T & Kojima I (1996) A natural sulfated polysaccharide, Calcium-Spirulan, isolated from *Spirulina platensis: in vitro* and *ex vivo* evaluation of anti-herpes simples virus and anti-human immunodeficiency virus activities. *AIDS Res Hum Retroviruses* 12: 1463-1471.
75. Balloni W, Tomaselli L, Giovannwtti L & Margheri MC (1980) Biologia fondamentale del genere *Spirulina.* In: Progetto finalizzato "*Ricerca di nuove fonti proteiche e di nuove formulazioni alimentari". Atti del Convegno: Prospettive della coltura di Spirulina in Italia. Consiglio Nazionale delle Richerche,* Cantarelli C, Ciferri O, Florenzano G, Kapsiotis G, Materassi R & Treccani U (eds), Firenze-Academia dei Georgofili, CNR, Tipografia Coppini, pp 49-82.
76. Ramírez D, Gonzalez R, Merino N, Rodríguez S & Ancheta O (2002) Inhibitory effects of *Spirulina* in zymozan-induced arthritis in mice. *Mediators Inflamm* 11: 75-79.
77. Kaji T, Fujiwara Y, Inomata Y, Hamada C, Yamamoto C, Shimada S, Lee JB & Hayashi T (2002) Repair of wounded monolayers of cultures bovine aortic endothelial cells is

inhibited by calcium spirulan, a novel sulfated polysaccharide isolated from *Spirulina platensis*. *Life Sci* 70: 1841-1848.

78. Dassgupta T, Banerjee S, Yadav PK & Rao AR (2001) Chemonodulation of carcinogen metabolizing enzymes, antioxidant profiles and skin and fore stomach papillomagenesis by *Spirulina platensis*. *Mol Cell Biochem* 226: 27-38.
79. Zhang HQ, Lin AP, Sun Y & Deng YM (2001) Chemo-and radio-protective effects of polysaccharide of *Spirulina platensis* on hemopoietic system of mice and dogs. *Acta Pharmacol Sin* 22: 112-114.
80. Upasani CD, Khera A & Balaraman R (2001) Effect of lead with vitamin E, C or *Spirulina* on malondialdehyde, conjugated dienes and hydroperoxides in rats. *Indian J Exp Biol* 39: 70-74.
81. Chamorro G, Salazar M & Salazar S (1989) Estudio teratogénico de *Spirulina* en rata. *Arch Latin Nutr* 39: 641-649.
82. Chamorro G & Salazar M (1990) Estudio teratogénico de *Spirulina* en ratón. *Arch Latin Nutr* 40: 66-94.
83. Chamorro G, Salazar M, Favila L & Boureges H (1996) Farmacología y toxicología del alga *Spirulina*. *Rev Invest Clin* 48: 389-399.
84. Alabama A (1988) Characterization of *Spirulina* biomass for CELSS diet potential. Normal: Al, University of Manchester, UK.
85. Cornet JF & Dubertret G (1990) The cyanobacterium *Spirulina* in the photosynthetic compartment of tthe MELISSA artificial ecosystem. In: *Workshop on Artificial Ecological Systems,* DARA-CNES, Marseille, France, pp 24-26.
86. Zarrouk C (1966) Contribution a letudevdune cynobphycee. Influence de divers facteurs physiques et chimiques surla croissance *et al* photosynthe'se de *Spirulina* maxima, Ph.D. Thesis, Universite de Paris, Paris, France.
87. Borowitzka M (1992) Algal growth media and sources of algal cultures. In: *Microalgal Biotechnology*, Borowitzka M & Borowitzka L (eds), Cambridge University Press, UK, pp 456-465.
88. Ayala F (2000) *Solarium Biotechnology*, Huayca La & Región I, Comunicación Personal, Chile, South America.
89. Cao S (2000) Comparative study on the two types of nitrogen source of *Spirulina*. *Sea-Lake Salt Chem Ind* 5: 30-32.
90. Danesi EDG, Rangel-Yagui CO, Carvalho JCM & Sato S (2002) An investigation of effect of replacing nitrate by urea in the growth and production of chlorophyll by *Spirulina platensis*. *Biomass Bioenergy* 23: 261-269.
91. Torre P & Carlos EN (2003) Fed-batch addition of urea for *Spirulina platensis* cultivation thermodynamics and material and energy balances. *Enz Microb Tech* 33: 698-707.
92. Luis DS, Attilio C & Gabriela CT (2004) Continuous and pulse feedings of urea as a nitrogen source in fed-batch cultivation of *Spirulina platensis*. *Aqua Eng* 3: 237-245.
93. Soletto D & Binaghi L (2005) Batch and fed-batch cultivations of *Spirulina platensis* using ammonium sulphate and urea as nitrogen sources. *Aqua Eng* 243: 217-224.
94. Feng D-l & Wu Z-c (2006) Culture of *Spirulina platensis* in human urine for biomass production and $O_2$ evolution. *J Zhejiang Univ Science B* 7: 34-37.
95. Ayala F & Laing I (1990) Commercial mass culture techniques for producing microalgae. In: *Introduction to Applied Phycology*, Akatsuka I (ed), SPB Academic Publishing, The Hague, Netherlands, pp 447-477.
96. Ayala F (1998) Guía sobre el cultivo de Spirulina. In: *Biotecnología de Microorganisms Fotoautótrofos*, Motril, Granada, España, pp 3-20.

# Genetically Modified Foods 9

**Rajinder Kumar**

*Department of Human Biology, Punjabi University, Patiala 147 002 (Pb.)*

## 1. INTRODUCTION

Mother nature was the first genetic engineer. For as long as ten thousand years, humans have been purposefully choosing the genetic make up of the crops they are growing. Genetic selection for features such as faster growth, larger seeds, or sweeter fruits has dramatically changed domesticated plant species as compared to their wild relatives. The world now sits at the cusp of a second potential agricultural revolution, the 'Gene Revolution' in which modern biotechnology enables the production of genetically modified (GM) crops that may be tailored to address on going agricultural problems in specific regions of the world. The GM crop movement has the potential to do enormous good, but also presents novel risks and has significant challenges to overcome before it can truly be considered revolutionary. It is hoped that this technology can illuminate opportunities to increase farm production, rural income and food security in developing countries.

By the year 2050, there are likely to be 9 billion people on the earth, an increase of 50 per cent over the present day. Most of the increase will occur in the cities of developing countries, primarily in Asia. If present economic development continues, this population increase will require a doubling in food production. Only a fraction of the food that all these people will need can be produced in the breadbaskets of the world. Most of this food has to be grown locally. Not only this, the problem of feeding all the people is worsened by the uneven distribution of cropland. For example, China has a quarter of the human population, but only seven percent of the world's farmland.

During the last doubling of the human population from 3 billion in 1960 to 6 billion in 2000, food production increase kept up with the population growth because we created and adopted multiple new technologies. Better techniques to cultivate the soil, new irrigation technologies, more advanced pesticides which are biodegradable, better genetic strains, machinery that harvests more of the crop, synthetic fertilizers, and green manures that restore the nutrients to the soil have helped to raise food production.

Food security encompasses sufficient physical, social and economic access to safe, nutritious and culturally acceptable food at the household level. However, the potential of genetic engineering in agriculture has remained controversial. There is considerable uncertainty about the impact on human and environmental health and also whether these products will provide a sustainable solution to food problems. The fundamental aim of GM foods is to increase crop production and conquer world hunger. The risks and benefits associated with genetically modified (GM) foods are difficult to quantity. Here in this chapter, all these aspects would be discussed.

*Food Biotechnology: Principles and Practices*
*V.K. Joshi & R.S. Singh (Eds.)*
*I.K. International Publishing House Pvt. Ltd., India*
*pp 269–291*

## 2. WHAT ARE GENES?

All organisms, from viruses to humans, contain a unique set of instructions which set down how they develop, grow and live. These instructions are found inside the cells on a long molecule called DNA, which is divided into small sections that control different aspects of the organism's growth and characteristics, and these sections are called genes. Very simple organisms such as bacteria may have a few thousand genes [1], while more complicated organisms have many more. For example, it has been estimated that maize has around 50,000 genes, while in man theses are 30-40,000 in number.

What we are today is because of genes. It is because our genes code for human characteristics like a common body plan of two eyes, ears, etc. Thus all humans are related because of similar genes. In fact man may think that he is better than a monkey, but the genetic difference between us is less than three percent. The minor differences of big eyes or small eyes are determined by the sub-groups of the genes containing the information of that part. Genes have the blueprint for the nectar of flowers, venom of cobra, colour of skin, intelligence, aptitude, behaviour and everything that living things have [2].

In nature, you cannot cross a fish with a vegetable. Animals and plants have long been separate in evolution. But scientists have produced a "frost-resistant" tomato by splicing its genetic code, a gene that protects a flounder from the cold. The first transgenic plant is said to have been created in the early eighties, when a gene from a bacterium was spliced into a petunia. Subsequently, oilseed rape has a bay tree gene spliced into it to improve its oil, and a potato has been given a disease-resistant chicken gene. The first commercially grown GM food crop was a tomato callled 'FlavrSavr' created by a California company 'Calgene'. This tomato was made more resistant to rotting by adding an antisense gene which interferes with the production of the enzyme polygalacturonase. Calgene submitted it to the US Food and Drug Administration (FDA) for assessment in 1992. The agency determined that the FlavrSavr was in fact a tomato, did not constitute a health hazard, and did not need special labelling. Calgene was allowed to release it into the market in 1994, where it was welcomed by consumers who purchased the fruit at two to five times the price of standard tomatoes. However, production problems, and competition from a conventionally bred long-shelf-life (LSL) variety prevented the product from becoming profitable and Calgene was bought by Monsanto in 1995. A variant of the FlavrSavr was used by Zeneca to produce tomato paste which was sold in Europe during the summer in 1996. Its labelling and pricing were designed as a marketing experiment, which proved at that time, European consumers would accept genetically engineered foods (GM Foods). This attitude drastically changed after outbreaks of mad cow disease weakened consumer trust in government regulators, and protesters rallied against the introduction of Monsanto's Roundup-Ready soybeans.

At the same time, to feed the growing population, scientists had no means to increase the cultivable land and so they resorted to increasing the yield from the same area. They modified using technology the seeds, fertilizers and pesticides and they succeeded. As it always happens, any solution is 'genetically' entangled with new problems. New solutions to these new problems have again the same basic characteristics. The vicious cycle continues, so long there is life. The side effects of GM foods on our health represent one such point in this cycle. Since we cannot go back in technology, we will have to again use the very technology to neutralize its own ills. You cannot cross-breed a carrot and a tree, because they are sexually incompatible. But with molecular techniques you can take a gene from a tree and put it directly into a carrot without having to drag all the other tree genes along with it.

## 3. GENETIC ENGINEERING

This technology has the power to break down fundamental genetic barriers — not only between species, but also between humans, animals, and plants. By randomly inserting together the genes of non-related species — utilizing viruses, antibiotic-resistant genes and bacteria as vectors, markers, and promoters - and permanently altering their genetic codes, genetically modified organisms are created that pass these genetic changes onto their offspring through heredity. Gene engineers all over the world are now snipping, inserting, recombining, rearranging, editing, and programming genetic material. Animal genes and even human genes are randomly inserted into the chromosomes of plants, fish, and animals, creating heretofore unimaginable transgenic life forms. For the first time in history, transnational biotechnology corporations are becoming the architects and "owners" of life [3]. As we enter a new generation, the demand for food increases. With the staggering advancements in technology, science has been able to reach at new levels. Scientists are now able to produce foods cheaper, faster, and with better features by altering the genetic information contained in the foods. For example, all the seedless grapes we eat today are genetically-modified crops as the scientists have altered the genetic information in the grapes. The scientists inserted a gene into the grape that prevents the seed from forming an outer cover. Subsequently, the seed dissolves away, leaving the final product — seedless grapes.

## 4. GENETICALLY MODIFIED FOODS

The term GM foods or GMOs (genetically-modified organisms) is most commonly used to refer to crop plants created for human or animal consumption using the latest molecular biology techniques. These plants have been modified in the laboratory to enhance desired traits such as increased resistance to herbicides or improved nutritional content can create plants with the exact desired trait very rapidly and with great accuracy. For example, plant geneticists can isolate a gene responsible for drought tolerance and insert that gene into a different plant. The new genetically-modified plant will gain drought tolerance as well. Not only genes can be transferred from one plant to another, but genes from non-plant organisms also can be used. The best known example of this is the use of *Bacillus thuringiensis* genes in corn and other crops. *Bacillus thuringiensis* is a naturally occurring bacterium that produces crystal proteins that are lethal to insect larvae [4]. Bt. crystal protein genes have been transferred into corn, enabling the corn to produce its own pesticides against insects such as the European corn borer.

Most of the food we eat today comes from plants and animals that have been grown and bred by humans for countless generations, undergoing substantial genetic changes over thousand of years. Traditionally, plants or animals with the most desirable characteristics were chosen for food and for breeding the next generation. The desirable characteristics arose from naturally occurring variations in the genetic make-up of individual plants or animals. Thus, genetic modification, in this sense, occurs naturally and forms the fundamental basis of evolution and breeding. Today's techniques of genetic modification provide new ways to identify particular characteristics and transfer them between the living organisms. For example, it is now possible to make a copy of a particular gene from the cells of a plant, animal or microbe and insert the copy into the cells of another organism to give it a desired characteristic. Because the resulting plants or animals had their genetic material altered in some way, they are commonly referred to as 'genetically modified' or 'GM' organisms. Consequently, foods derived from GM plants or animals are often called 'GM foods'. The term 'GM Food' is also sometimes applied to foods that contain GM ingredients, and to food additives or processing aids produced using genetic modification.

The introduction of GM foods into our food supply has stimulated much discussion about the nature of these foods, their safety and their place in global food production. There has been much concern about GM foods, some of which are being tested and some of which are already used as ingredients in the food we eat. GM stands for 'genetically modified', that describes the food has been made by the genetic manipulation. To some extent, humans have been involved in genetic modification for centuries. For example, larger cattle which gave more milk were bred to produce even larger offsprings. Seeds from cereals and other crops that were hardier and grew better were selected for planting to produce better yields. With GMO, however, the modifications involved are often of a kind that could not possibly occur naturally. For example, adding cow growth hormone to the embryo of a broiler chicken to produce larger, faster growing chicken, or adding genes from a virus to a plant to allow it to become resistant to the virus.

## 5. FOOD AND GENES — WHAT HAPPENS WHEN WE EAT?

Firstly, it is important to dispel some misconceptions about the nature of food, digestion and existing practices, as well as about genetics. Humans, like all other animals, obtain the basic materials for bodily growth and maintenance from complex chemical materials like proteins, fats, carbohydrates, in other words, food. These chemicals are broken down by normal digestive processes into relatively simple components, which are re-synthesized into complex molecules within our bodies. Some of our food contains the DNA of plants, animals and microorganisms, but as far as we are aware, this DNA is not directly integrated into our own human genetic make-up [5]. Eating GM food is thus no more likely to change our own genes than eating irradiated food is going to irradiate us. Concern has been expressed that DNA might, in some circumstances, be taken up by the bacteria of the human gut, but this seems highly improbable. Moreover, in many processed foods, DNA is already broken down during the manufacturing process. Our diet comes from a relatively small number of organisms available on earth. As much as 93 percent of the human diet is provided by only 29 basic crop species.

## 6. HOW TO MAKE A TRANSGENIC PLANT?

Plant cells are 'totipotent' — meaning that any one cell can be cultured to grow into a whole new plant. Scientists can easily add desired genes (DNA) to a plant cell by either using a common soil bacteria that naturally brings DNA into plant cells (called *Agrobacterium*) or by literally 'shooting' DNA into plant cells using a high pressure helium 'gun', and by these techniques the incoming DNA is 'spliced into' plant's genome, resulting in a transgenic plant carrying one or more new genes that produce desirable proteins.

## 7. ADVANTAGES OF GM CROPS

There are many reasons why GM foods could be advantageous. For example, a crop could be made to grow quicker, with increased protein and vitamin levels, or with less fat [6]. An often-used argument in favour of GM crops is that drought-resistant crops could help to alleviate famine in developing countries, where low rainfall often leads to food shortage. Techniques have also been developed to make fresh produce to last longer, so that it can ripen on the plant and be transported more easily with less wastage.

The first GM food products, tomato puree and vegetarian cheese, appeared in British supermarkets in 1996. The puree was made from tomatoes which were designed to stay firmer for longer, leading to less waste in harvesting. The tomatoes also held less water, meaning that less water was required to grow them and less energy was used removing water from them to process them into puree. This in turn made the puree cheaper for the consumer [7]. The first GM soya was planted in the US in the same year. Monanto, a major GM manufacturer, has developed a strain of GM soya which is resistant to Roundup, its own brand of herbicide. This allows weeds to be controlled even after the soya has started to grow, saving an estimated 33% on the amount of herbicide used. GM foods have been largely accepted by the Americans, with nearly 70% of them saying that they would buy GM foods even if they were simply engineered to stay fresh for longer. Even more would like to purchase foods modified to resist insects, pests, resulting in less use of pesticides.

## 7.1 Insect Resistance

Insect resistant GM crops have raised particular concerns because of potential adverse effects on population of non-target species. The soil bacterium *Bacillus thuringiensis* produces specific proteins (collectively known as Bt) active against many agricultural pests such as beetles, moths and flies. Several thousand tonnes of these proteinaceous bacterial toxins have been in use for a number of years as biopesticides with no detrimental effects on the environment. However, because of their poor stability and high cost, they have not achieved extensive market penetration. To get around this problem, several genes encoding for insecticidal Bt proteins have been inserted into food crops, most notably corn. Corn is particularly vulnerable to attack by the European corn borer, which causes millions of dollars worth of damage every year. The effect of insect infestation on Bt and non-Bt cotton bolls is shown in Plate 1.

**Plate 1:** Results of insect infestation on Bt (right) and non-Bt (left) cotton bolls.
*Source:* **31**

## 7.2 Herbicide Tolerance

Herbicide tolerance weed control is one of the farmer's biggest challenges in crop production, because poorly controlled weeds drastically reduce crop yields and quality. Many herbicides in the market control only certain types of weeds, and are approved for use only on certain crops at specific growth stages. Residues of some herbicides remain in the soil for a year or more so that farmers must pay close attention to the herbicide history of a field when planning what to plant there. For some crops, it is not cost-effective to remove weeds by physical means such as tilling, so farmers often spray large quantities of different herbicides (weed-killer) to destroy weeds, a time-consuming and expensive process that requires care so that the herbicide does not harm the crop plant or the environment. Crop plants genetically engineered (GE) to be resistant to one very powerful herbicide could prevent environment damage by reducing the amount of herbicides needed. For example, Monsanto has created a strain of soybeans GM to be not affected by one of their herbicide product roundup [8]. Farmer grow these soybeans, which requires only one application of weed-killer. Herbicide tolerant crops resolve many of those problems because they include transgenes providing tolerance to the herbicides roundup (chemical name: glyphosate) or liberty (glufosinate). These herbicides are broad-spectrum, meaning that they kill nearly all kinds of plants except those that have the tolerant gene. Thus, a farmer can apply a single herbicide to his fields of herbicide tolerant crops. Another important benefit is that this class of herbicides breaks down quickly in the soil, eliminating residue carry-over problems, thus reducing environmental impact.

## 7.3 Disease Resistance

There are many viruses, fungi and bacteria that cause plant diseases. Plant biologists are working to create plants with genetically-engineered resistance to these diseases [9]. Unexpected frost can destroy sensitive seedlings. An antifreeze gene from cold water fish has been introduced into the plants such as tobacco and potato. With this antifreeze gene, these plants are able to tolerate cold temperature that normally would kill unmodified seedlings [10].

### 7.4 Drought Tolerance/Salinity Tolerance

As the world population grows and more land is utilized for housing instead of food production, farmers will need to grow crops in locations previously unsuited for plant cultivation. Creating plants that can withstand long period of drought or high salt content in soil and groundwater will help people to grow crops in formerly inhospitable places [11].

### 7.5 Pharmaceutical Medicines

Pharmaceutical medicines and vaccines are often costly to produce and sometimes require special storage conditions not readily available in third world countries. Researchers are working to develop edible vaccines in tomatoes and potatoes. These vaccines will be much easier to ship, store and administer than traditional injectable vaccines.

### 7.6 Phytoremediation

Not all GM plants are grown as crops. Soil and groundwater pollution continues to be a problem in all parts of the world. Plants such as popular trees (*Populus* sp.) have been genetically engineered to clean up heavy metal pollution from contaminated soil. Plants can be engineered to grow in harsher climates with more efficiency. The world's supply of arable land is not growing, but demands for food are continuosly increasing. GM crops could be developed to grow in colder climates with less water, in saltier soils, or with more cycles in a given year. Once again rice is being looked, this time as a way to increase the amount of food that can be grown from a given plot of land.

## 8. GM FOODS AND ERADICATION OF WORLD HUNGER

Whilst the genetic modification of agronomic traits has already produced many new crop varieties, improvements in food quality brought about by the genetic route have been slower to emerge for both technical and regulatory reasons. Rice is the staple food for nearly one-half of the world's population, and in Asia, it provides 50-80% of the average daily calorie intake. Although thousands of varieties of rice are available, all are poor in Vitamin A content. It has been calculated that 124 million children worldwide are deficient in Vitamin A intake leading to 14 million cases of poor vision and 5 lakh cases of irreversible blindness. Increased intake of green vegetables and animal products as well as vitamin supplements are all possible conventional solutions to the problem of Vitamin A deficiency, but are too costly for long-term implementation in countries, where the majority of people are very poor. The genetic solution to this problem has been to introduce genes coding for phytoene syntheses and desaturase, carotene desaturase and lycopene cyclase from the daffodil has been cloned into rice to produce a GM cultivar high in carotene, a precursor of Vitamin A. The new rice has a yellow hue and has been named "golden rice". It has been calculated that 300 g of cooked "golden rice" could provide an adult's daily requirement of Vitamin A, although these estimates need to be confirmed experimentally [12].

In an unprecedented gesture, the scientists, research funding bodies, companies and technology licence owners involved in the development of "golden rice" gave up their intellectual property rights in 2000 for the benefit of resource-poor farmers in the developing world. Samples of the rice were delivered to the International Rice Research Institute (IRRI) in Los Banos, Philippines, in early 2001 for further safety and utility testing. It remains to be seen whether "golden rice" fulfils its promise, but if successful, it will be given free-of-charge to subsistence farmers in developing countries. Golden Rice is a wonderful product, and just the type of GM crop that appeals to the public, but there are a few drawbacks:

- A person would have to consume pounds of rice a day just to get the recommended daily allowance of β-carotene/vitamin A currently made in golden rice.

- β-carotene is fat-soluble vitamin, and starving people usually do not have enough body fat to effectively bring the β-carotene into their bodies to turn it into vitamin A.
- Alleviating one vitamin deficiency is a beginning, but a multivitamin's worth (and more) of nutritional deficiencies need to be corrected in people living on a cup of rice a day.

## 9. GLOBAL AREA (GM CROPS)

In 2006, the first year of the second decade of commercialization of GE crops 2006-2015, the global area of biotech crops continued to climb for the tenth consecutive year at a sustained double-digit growth rate of 13 percent or 12 million hectares (30 million acres), reaching 102 million hectares (252 million acres). Biotech crops area grew 12 percent or 12.3 million hectares to reach 114.3 million hectares, the second highest increase in the past five years. GE crops achieved several milestones in 2006-2007, annual hectare of biotech crops exceeded 100 million hectares (250 million acres). For the first time, the number of farmers grown GE crops (10.3 million) exceeded 10 million; the accumulated hectare from 1996 to 2006 exceeded half a billion hectares at 577 million hectares (1.4 billion acres), with an unprecedented 60-fold increase, making it the fastest adopted crop technology in recent history (Fig. 1). It is notable that the year-to-year increase of 12 million hectares in 2006 is the second highest in the last 5 years in absolute area, despite the fact that adoption rates in the US, the principal grower of GE crops, are already over 80 percent for soybean and cotton. It is also noteworthy that in 2006, India, the largest cotton growing country in the world, registered the highest proportional increase with an impressive gain that almost tripled its Bt cotton area to 3.8 million hectares. In 2006, the number of countries planting GE crops increased from 21 to 22 with the EU country Slovakia, planting Bt maize for the first time and bringing the total number of countries planting GE crops in the EU to six out of 25. Spain continued to be the lead country in Europe planting 60,000 hectares in 2006.

In 2006, 10.3 million farmers from 22 countries planted biotech crops, up from 8.5 million farmers in 2005 (Table 1). Of the 10.3 million, 90 percent or 9.3 million (up significantly from 7.7 million in 2005) were small, resource-poor farmers from developing countries whose increased income from

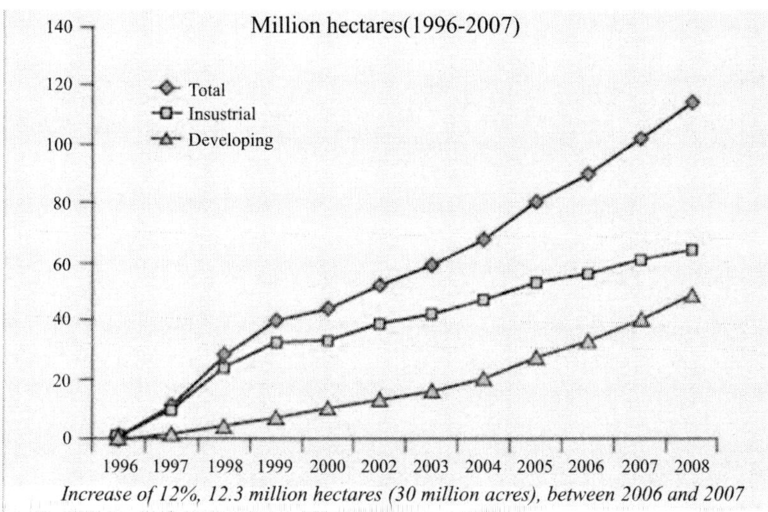


**Fig. 1:** Global area of biotech crops (1996-2007).
*Source:* **30**

**Table 1:** Global area of biotech crops in 2006-2007.

| | | Area (Million hectare) | | |
|---|---|---|---|---|
| **Rank** | **Country** | **2006** | **2007** | **Biotech crops** |
| 1 | USA | 54.6 | 57.7 | Soybean, maize, cotton, canola, squash, papaya, alfalfa |
| 2 | Argentina | 18.0 | 19.7 | Soybean, maize, cotton |
| 3 | Brazil | 11.5 | 15.0 | Soybean, cotton |
| 4 | Canada | 6.1 | 7.0 | Canola, maize, soybean |
| 5 | India | 3.8 | 6.2 | Cotton |
| 6 | China | 3.8 | 3.8 | Cotton, tomato, poplar, petnnia, sweet pepper, papaya |
| 7 | Paraguay | 2.0 | 2.6 | Soybean |
| 8 | South Africa | 1.4 | 1.8 | Maize, soybean, cotton |
| 9 | Uruguay | 0.4 | 0.5 | Soybean, maize |
| 10 | Philippine | 0.2 | 0.3 | Maize |
| 11 | Australia | 0.1 | 0.1 | Soybean/cotton |
| 12 | Romania | 0.1 | <0.1 | Soybean |
| 13 | Mexico | 0.1 | <0.1 | Cotton, soybean |
| 14 | Spain | 0.1 | <0.1 | Maize |
| 15 | Columbia | <0.1 | <0.1 | Cotton/cotton carnation |
| 16 | France | <0.1 | <0.1 | Maize |
| 17 | Iran/Poland | <0.1 | <0.1 | Rice/maize |
| 18 | Honduras | <0.1 | <0.1 | Maize |
| 19 | Czech Republic | <0.1 | <0.1 | Maize |
| 20 | Portugal | <0.1 | <0.1 | Maize |
| 21 | Germany | <0.1 | <0.1 | Maize |
| 22 | Slovakia | >0.1 | <0.1 | Maize |
| 23 | Chile | >0.1 | <0.1 | Maize, soybean, canola |

*Source:* 30

biotech crops contributed to their poverty alleviation. Of the 9.3 million small farmers, most of them were Bt cotton farmers, 6.8 million were in China, 2.3 million in India, one lakh in the Philippines, several thousands in South Africa, with the balance in the other seven developing countries which grew GE crops in 2006. This initial modest contribution of biotech crops to the Millennium Development Goal of reducing poverty by 50 percent by 2015 is an important development, which has enormous potential in the second decade of commercialization from 2006 to 2015.

## 10. GLOBAL MARKET VALUE OF GM CROPS

In 2005, the global market value of biotech crops, estimated by Cropnosis, was \$5.25 billion. The 5.25% billion biotech crop market comprised of \$2.42 billion for biotech soybean, \$1.91 billion for biotech maize (36%), \$0.72 billion for biotech cotton (14%), and \$0.21 billion for biotech canola (4%). In 2007, the global market value of biotech crops, estimated by Cropnosis, was US \$6.9 billion. The US \$6.9 million biotech crop market comprised of US \$ billion for (equivalent to 47% of global biotech crop market, up from 39% in 2006), US \$2.6 billion for biotech soybean (37%, down from 44% in 2006), US \$0.9 billion for biotech cotton (13%), and US \$0.2 billion for biotech canola (3%). Of the US \$6.9 billion biotech crop market, US \$5.2 billion (76%) was in the industrial countries and US \$1.6 billion (24%) in the developing countries. The accumulated global value for the eleven year period, since biotech crops were first commercialized in 1996, is estimated at US \$42.4 billion.

The global area of biotech crops from 1995-2007 is depicted in Fig. 2. The projection for 2009 is approximately US $7.5 billion. The country wise global area of biotech crops in 2006-07 is shown in Table 1. Biotech crop countries and mega countries growing 50,000 hectares or more of biotech crops are shown in Fig. 2.

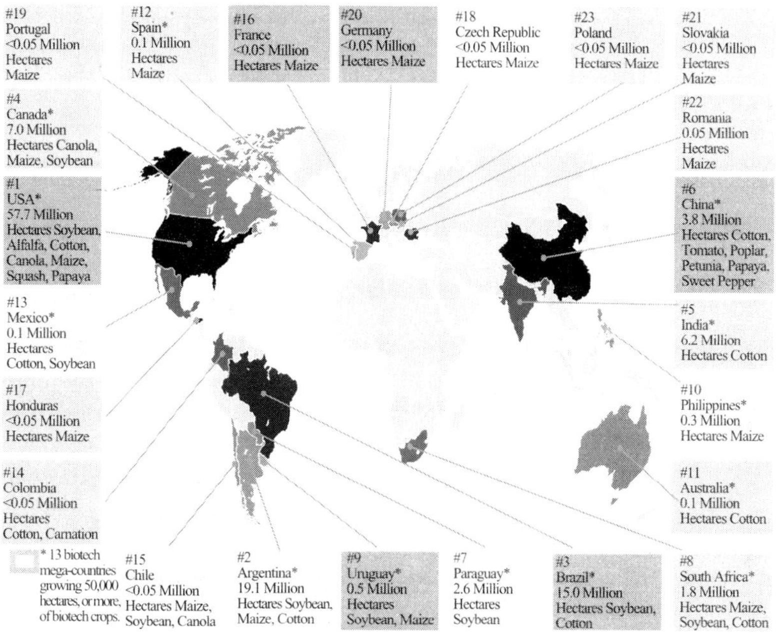


**Fig. 2:** Biotech crop countries and mega countries growing 50,000 hectares or more of biotech crops.
*Source:* 30

## 11. INDIAN SCENARIO OF GM FOODS

In India, research on transgenic plants has been in progress in several government and private laboratories, but none has been commercialized so far. Several Indian institutes and organizations claim that they have developed transgenic plants which are ready for green house/screen house/poly house evaluation and some for field evaluation as well. The Government of India through the Department of Biotechnology (DBT) has been mainly responsible for the propagation of research in transgenic plants and development of transgenic cultivars in the country. There had also been major support from ICAR, DST, CSIR and certain other agencies including private sector units. Some important institutes where research work on transgenic development for mustard cotton, brinjal, chickpea, lentil, pigeon pea, tomato, rice, cauliflower, potato is in progress include National Research Centre on Plant Biotechnology, New Delhi; Jawaharlal Nehru University, New Delhi.

India, the largest democracy in the world, is highly dependent on agriculture which generates almost one quarter of its GDP and provides two-thirds of its people with their means of survival. India is a nation of small resource-poor farmers, most of whom do not make enough income to cover their meagre basic needs and expenditures. The National Sample Survey (NSS) last conducted in 2003 reported that 60.4% of rural households were engaged in farming, indicating that there are 89.4 million farmer households in India. Sixty percent of the farming households own less than one hectare of land, and only 5% own more than 4 hectares. Only the 5 own farming households (5% of 90 million) have an income that is greater than their expenditures. The average income of farm households in India (based on Rs. 45 per US Dollar) was $46 per month and the average consumption expenditures was $62. Thus, of the 90 million farmer households in India, approximately 85 million, which represent about 95 percent of all farmers, are small resource-poor farmers who do not make enough money from the land to make both ends meet. In the past, these included the vast majority of 5 million or more Indian cotton farmers. India has a larger area of cotton than any country in the world; about 9 million hectares cultivated by approximately 5 to 5.5 million farmers. Whereas India's cotton area represents 25% of the global area of cotton, in the past it produced only 12 percent of world production [14].

Bt cotton, which confers resistance to important insect pests of cotton, was first adopted in India as hybrids in 2002 and doubled its Bt cotton area to approximately 100,000 hectares in 2003. Bt cotton area again increased four-fold in 2004 to reach over half a million hectares. In 2005, the area planted to Bt cotton in India continued to climb reaching 1.3 million hectares, an increase of 160 percent over 2004.

In 2006, the record increase in adoption in India continued with almost a tripling of area of Bt cotton from 1.3 million hectares to 3.8 million hectares. In 2006, this increase in area was the highest, year after year growth for any country in the world. Of the 6.3 million hectares of hybrid cotton in India in 2006, which represents 70 percent of all the cotton area in India, 60 per cent or 3.8 million hectares was Bt cotton — a remarkably high proportion in a fairly short period of five years. The distribution of Bt cotton in the major growing states in 2004, 2005 and 2006 is shown in Table 2. The major states growing Bt cotton in 2006, listed in order of hectare, are Maharashtra (1.840 million hectares or 22 percent), Gujarat (470,000 hectares or 12 percent), Madhya Pradesh (310,000 hectares or 6 percent) in the northern zone and the balance in Karnataka and Tamil Nadu and other states.

It is estimated that, in India, approximately 2.3 million small farmers planted on an average 1.65 million hectares of Bt cotton in 2006. The number of farmers growing Bt cotton hybrids in India has increased from 300,000 small farmers in 2004 to one million in 2005, with over a two-fold increase in 2006 to 20.3 million farmers, who are reaping significant benefits from the technology. Coincidental with the steep increased adoption of Bt cotton between 2002 and 2005, the average yield of cotton in India, which had one of the lowest yield in the world, increased from 308 kg per hectare in 2001-02 to 450 kg per hectare in 2005-2006, with most of the increase in yield of up to 50 percent or more attributed to Bt cotton.

The Union Government of India enacted the Environmental Protection Act (EPA) in 1986 to preserve the environment and minimize the risks from pollutants and contaminants that can be of risk to human and animals. The Department of Biotechnology (DBT) has recently issued guidelines for

**Table 2:** BT cotton cultivation in India (2004-07).

| | Area × 1000 hectares | | | |
|---|---|---|---|---|
| **State** | **2004** | **2005** | **2006** | **2007** |
| Maharashtra | 200 | 607 | 1,840 | 2880 |
| Andhra Pradesh | 75 | 280 | 830 | 1190 |
| Gujarat | 122 | 150 | 470 | 818 |
| Madhya Pradesh | 80 | 146 | 310 | - |
| Northern Zone* | - | 60 | 215 | - |
| Karnataka | 18 | 30 | 85 | - |
| Tamil Nadu | 5 | 27 | 45 | - |
| Others | - | - | 5 | - |
| Total | 500 | 1,300 | 3,800 | - |

*Punjab, Haryana, Rajasthan

*Source:* 30

research in transgenic plants in India. Under these guidelines, the rules and procedures for handling GMO are implemented and monitored by the Institutional Biosafety Committee (IBSC) and the Review Committee on Genetic Manipulation (RCGM). Any research institute/private entrepreneur who wishes to develop DNA products is required to take approval from the above committees. Keeping in view the important role that GMOs can play in the economic upliftment of the country, the Government of India realized that there could be unintended hazards and risks from the use of GMOs and products thereof, if the new technology was not properly assessed and utilized. GM foods are not yet approved for commercialization in India. Under the current global scenario on the regulation of GM foods, the India Government has to set out rules, regulations, guidelines and other legal provisions concerning the safety assessment of GM foods.

## 12. IMPORT OF GM FOODS

The rules governing WTO's Agreement on Sanitary and Phytosanitary (SPS) measures prevent countries from taking action to restrict the import of such food after their harmful effects come to light. Thus, the government must take necessary steps to establish guidelines for GM foods from other countries. Such guidelines should take into account the following factors:

- The type of GM food imported, whether it is whole or processed GM food.
- The country of origin.
- Certificate of safety assessment of the GM food complying with the guidelines of the regulatory agencies in both the exporting country as well in the importing country.
- Labelling of GM foods, including the type of genetic information, the percentage of GM material present, possible health effects associated with the organism modified (such as allergenicity), alteration in the chemical/nutritional composition.
- Provision for recourse to legal redressal in cases of unexpected adverse effects. This calls for establishment of infrastructure and expertise to test GM foods.

Safety assessment protocols for GM foods have to be established by generating the following information:

- The amount of GM food that people are likely to consume.
- A detailed description of what the food is and how it is produced.
- A history of, if any, possible health effects linked to the organism being modified.
- A detailed description of the genetic modification process.
- An evaluation, if any, of possible changes in the nutritional composition of the modified food.
- Evaluation of the community consuming the modified foods under controlled conditions.

GM foods can survive in the world if there are proper cautionary steps take to ensure that consumers are aware of their existence. Labelling is a very basic form of control every consumer has the right to know what they are buying. In 1998, the European Union (EU) passed a law, making labelling mandatory.

## 13. ENVIRONMENTAL IMPACT OF GM CROPS

The second most important concern about GM crops is their environmental impact. Many questions arise. For example, could herbicide tolerant GM crops become "super weeds"? The results of recent studies have shown that GM crops do not become self-seeding or spread into neighbouring areas. Pesticide-tolerant corn, sugar beet and rapeseed and insect-resistant potatoes together with controls were planted at 12 sites in the UK in 1990 and the sites were monitored for 10 years. All the plants, whether modified or not, died out within 4 years of sowing except one unmodified potato, which lasted for 10 years. Of greater concern is the risk of a herbicide-tolerant gene spreading through pollination to other plants which are already weeds. Field studies have shown that pollen from GM plants is rarely carried over very long distances. For example, the percentage of cross-pollination in a field of GM rapeseed has been determined experimentally as 0.1 percent, 0.0003 percent, and zero at distances from the test site of 6, 47 and 70 metres, respectively. For GM potato, cross-pollination was 0.02 percent at 10 metres and zero at 20 m from the site [15, 16]. A problem that remains is in predicting the effects of extremely rare hybridization events that may not be evident in relatively small plots of one or two hectare, but may occur when GM crops are grown on thousands of hectares of land. There are historical examples of genes moving from one organism to another by unknown mechanisms over several million years, and such events may eventually result in the transfer of the GM trait to a weed. However, these occurrences are so rare that it would be impossible to monitor them in practice.

Two main concerns about the effects of GM plants on the environment are that the new plants will become pernicious weeds or they will transfer their new genes to wild relatives or similar crops growing nearby with unforeseen effects. Evidence from thousands of field trials suggests that the new plants will behave just like the varieties currently in cultivation. There is also evidence to suggest that the transfer of genetic material from transgenic crops to their wild relatives or unmodified plants occurs, although the frequency of such transfer and its significance is still debated. The ecologists involved in such work have emphasized the need for caution and the importance of case-by-case analysis (in other words, it is difficult to generalize about the impact of GM crops).

A further concern is whether plants with introduced genes that enable them to resist insect attack will quickly lead to the establishment of resistant populations of pests. Because of the intense selection pressure (favouring naturally-resistant individuals) that crops carrying for example, Bt genes will exert, refugia of susceptible plants are usually grown alongside transgenic crops. Indeed, this has been a legal or voluntary requirement in the USA and Australia where transgenic, insect-resistant cotton has been grown. To date, there have been no confirmed cases of resistant populations developing, but it is generally accepted that without measures such as the limited use of a range of pesticides and the use of refugia, resistant pest populations will certainly develop. With this in mind, in January, 1999, four major producers of Bt maize plants *(Monsanto, Pioneer, Hi-Bred, Novartis,* and *Mycogen-Dow Agro Science)* proposed that 20 percent of farmland should be set aside for non-transgenic crops when Bt maize is grown.

An article in Nature claimed that only 56% of monarch butterfly larvae survived being fed milkweed covered with GM corn pollen. The study claimed that the insect toxic in the corn pollen poisoned the larvae. The results were unconfirmed in further studies, and even showed no effect on monarch survival rates at low doses. Monarch survival rates are affected when the milkweed has more than 1000 grains of GM corn pollen per square centimetre of the milkweed plant. In nature, typical figures do not exceed 600 grains per square centimetre and average around 170. The study raised awareness about the effects of GM pollen spreading with the wind. Other species may be similarly harmed at lower doses, and the effects of GM crop pollen on the ecosystem are not well understood [17].

## 14. CONTAMINATION

One of the main difficulties which farmers will encounter when growing GM crops is that there is no way to restrict pollen movement. In the case of oilseed rape, researchers have found that its pollen can travel up to 4 km and can escape from fields even when they are surrounded by barrier crops to prevent them. The US Government has separation distance requirements between GM and non-GM crops. These only require farmers to leave a distance of 50-20 meters between GM and non-GM oilseed rape, 6-60 m for sugar/fodder beet and 50-20 m for maize. But the evidence shows that this is clearly not enough to protect farmers and consumers from GM contamination. There have already been several serious incidents of GM contamination, despite the fact that GM crops are only grown by a minority of farmers worldwide. In spring 2000, the seed company Advanta announced that they had discovered that the batches of oilseed rape seeds they sold to farmers in France, Germany, Sweden and the UK were contaminated with a GM oilseed rape variety. The GM oilseed rape, produced by Monsanto, was not authorized for cultivation in the European Union. Advanta blamed the contamination on cross pollination of their seed crop with a GM crop that had been at least 4 km away. Over 5,000 hectares of contaminated seeds were grown in the UK and farmers affected could not sell their crops or were forced to destroy them.

Later in 2000, it was found that GM maize called Star Link, which was not approved for human consumption, had contaminated foods across the USA. Suspected allergic reactions were reported and more than 300 brands of taco shells, crisps and other maize products had to be withdrawn from shops. The US Government was forced to buy up stocks and Kellogg's closed production lines for two weeks. The loss to the US economy had been estimated billions of dollars. In November 2002, it was reported that half a million bushels of soya was contaminated by GM maize plants used to produce a pharmaceutical or industrial chemical. The soya was planted on the same site used to grow GM "pharmaceutical." The maize grew and contaminated the new crop. The GM oilseed rape supplied by Aventis for the farm scale trials in England and Scotland was found to be contaminated with an unauthorized GM variety containing antibiotic resistance markers.

Mycotoxins are chemicals produced by moulds that are detrimental to human health. Many different mycotoxins are produced by various fungi such as *Aspergillus* or *Fusarium* species that grow on plants. Some of these chemicals cause liver damage or cancer. In the case of the chemical called fumonisin, which is a mycotoxin produced by certain *Fusarium* fungal species that are natural colonizers of maize plants, the fungal toxin is known to cause (i) severe human birth defects when pregnant women ingest food such as torillas made from mouldy maize, and (ii) cancer in adults when either men or women drink maize based alcoholic beverages fermented from mouldy maize. These food safety problems are serious health issues in regions where maize is a staple food in Central America, South Africa and China. Fungal growth on maize is promoted by moisture, climatic factors, and most notably, insect predation. Several reports demonstrate that insect protected GM maize can have lower mycotoxin levels due to reduced insect damage.

## 15. TOXINS AND POISONS

GE products clearly have the potential to be toxic and a threat to human health. In 1989, a genetically engineered brand of L-tryptophan, a common dietary supplement, killed 37 Americans and permanently disabled or afflicted more than 5,000 others with a potentially fatal and painful blood disorder, eosinophilia myalgia syndrome (EMS), before it was recalled by the (FDA). The manufacturer, Showa Denko, Japan's third largest chemical company, had for the first time in 1988-89 used GE bacteria to produce the over-counter supplement. It is believed that the bacteria somehow became contaminated during the recombinant DNA process. Showa Denko had to pay over compensation of $2 billion to EMS victims.

In 1999, front-page headline stories in the British press revealed Rowett Institute scientist Dr. Arpad Pusztai's explosive research findings that GE potatoes, spliced with DNA from the snowdrop plant and a commonly used viral promoter, the Cauliflower Mosaic Virus (CaMv), are poisonous to mammals. GE-snowdrop potatoes, found to be significantly different in chemical composition from regular potatoes, damaged the vital organs and immune systems of laboratory rats fed with GE potatoes. Most alarming of all was the damage to the rats' stomach linings — apparently a severe viral infection — which was most likely caused by the CaMv viral promoter, a promoter spliced into nearly all GE foods and crops.

Animals fed GMO corn, for example, developed problems of blood (Leukaemia), blood pressure, kidney, liver (including significant changes in kidney weight as well as lesions on the kidney and liver), allergies to infection, increased blood sugar levels, diseases, and even cancer. These are not minor problems, but major changes in the body. These are toxic reactions to a poison. (Even if the same effect does not occur in humans, think of the repercussions of us eating the meat of animals that ate the GMO corn and therefore unhealthy animals). Corn and Soy are the top two largest GM crops. You can find corn and soy in practically every processed food. Chickens and cows eat corn and soy on feedlots (another reason to choose only grass-fed meats and organic chicken), so the problem is multiplied in every way. Then there is nature's way of creating crops naturally with cross-pollination and it becomes clear that our entire food line is in jeopardy. Many claims have been made that there is a lack of proper and long term testing for these foods, and the danger inherent in the lack of labelling these foods for the consumer. But in the USA both these have failed miserably. In a country where 65 percent of farming area grows GE crops, the pesticide use has not come down even by a gram in the last ten years, whereas the pesticide use in the rest of the world (where hardly less than 5 percent of farmers grow GE crops) is coming down steadily during the same period. The yields have not increased in USA. It is not just a simple case of there being more people, so more food should be grown. There is more than enough food to feed everyone very well at the moment, yet hundreds of millions of people go hungry and nearly two billion are malnourished. For example, in 1998, it was estimated that 36 million people, including 14 million children, were hungry or on the brink of hunger in the USA, one of the richest countries in the world.

Most GM crops being grown at the moment are destined for markets in rich countries. Soya and maize are used mainly for animal feed and for adding to processed food in rich countries. Such products will not help to feed the poor and hungry people of the world. The majority of GM crops being grown around the world at the moment are herbicide tolerant. These crops are designed for use in intensive farming systems, with single crops in large fields requiring heavy use of chemical inputs. Majority of countries are lacking biosafety regulations. Scientists have already raised similar concerns about the feeding of GM in high quantities to both farm animals and babies. Scientists working for the Zambian Government which rejected GM food aid stated that GM maize is consumed by millions of Americans, it is eaten in highly processed form and is not a staple food in the USA. In Zambia, maize is the staple food and is usually the only carbohydrate source. This argument is commonly used to advocate the development and use of GM crops, but it is not currently supported by either the consensus of scientists or any comprehensive comparison of agricultural alternatives. Over long periods of time, these communities have constructed complex systems of knowledge about their environment.

## 16. EFFECTS ON FARMING STRUCTURE, BIODIVERSITY AND SOIL ECOSYSTEM

GE has the potential to affect world agriculture dramatically. Although great benefits may come, it has been suggested that there might also be accompanying disadvantages. Several of these disadvantages are no different to existing trends in world agriculture, such as the shift towards larger farms and more capital-intensive farming systems. This tends to favour only wealthy farmers who can invest in new technologies rather. In the developed world, there are concerns about over-production of food, although these worries are unlikely to be shared by those countries where the growth in population far outstrips the capacity of farmers to provide sufficient food alongside other changes, and technologies offer a realistic prospect of long-term sustainable agriculture to farmers in the Third World. Plant breeding methods have produced plants with greatly improved characteristics compared with the old cultivars. Biotechnology might encourage the production of a far wider variety of new crops, increasing biodiversity. Modern agriculture has been so successful in increasing the yield of food that farmers in the USA and Europe are paid to take land out of cultivation or to grow new crops, again increasing diversity. Because it has the potential to reduce waste, biotechnology may accelerate this trend. The opposing argument is that plant breeding will increasingly fall into the hands of just a few companies, and that the plant varieties they offer to the farmer will be correspondingly reduced. This could make crops more susceptible to attack by pests and diseases and lead to a reduction in the use of important old cultivars and their wild relatives [18]. If the numerous surveys that have been undertaken in Western Europe are to be believed (and there is no reason to think that they are inaccurate), the vast majority of the population here does not wish to consume GM food. The first GM product in Europe (the Zeneca tomato puree) presented no problem in this respect, as the cans were both clearly labelled and always offered alongside a similar non-GM product. The problem arose when GM food ingredients that are traded as bulk commodities entered the market. First GM soya, then maize started to be grown in the USA and traded internationally. This presented UK retailers, who had planned carefully for the introduction of the tomato puree, with the unforeseen problem. The presence of the toxin in exudates from Bt corn verified that it was active in an insecticidal bioassay using larvae of the tobacco hornworm. If transgenic crops substantially alter soil biota and affect processes such as soil organic matter decomposition and mineralization, this would be of serious concern to organic farmers and most poor farmers in the developing world. These farmers cannot purchase or do not want to use expensive chemical fertilizers. They rely instead on local residues, organic matter and especially soil organisms for soil fertility (e.g., key invertebrate, fungal or bacterial species), which can be affected by the soil bound toxin. Soil fertility could be dramatically reduced if crop leachates inhibit the activity of the soil biota and slow down natural rates of decomposition and nutrient series. Laboratory tests have shown that pollen from GM maize in the US damaged the caterpillars of the Monarch butterfly. This is a cause of damage to a single species, but it does show that GMO could have the potential to do unexpected harm to other plants and animals. In the end, this could lead to a loss of biodiversity, with certain animals and wild plant species effectively being rendered extinct.

## 17. FOOD ALLERGIES

One of the public health concerns over human consumption of GM foods is that some percentage of the public may be allergic to a protein coded by the gene introduced into the GM food. Allergic reactions can be extremely mild, or deadly. The past decade has seen a 300 percent increase in allergies around the world. Allergies may also develop only after repeated exposure to a substance. Exposure to new foods can reveal new allergies, and the combination of genetic material can lead to "cross-allergenicity" where people allergic to one thing may be allergic to the genetic material that has been added to in item they once ate safely [19].

In 1996, a major GE food disaster was narrowly averted when Nebraska researchers learned that a Brazil nut gene spliced into soybeans could induce potentially fatal allergies in people sensitive to Brazil nuts. Animal tests of these Brazil nut-spliced soybeans had turned up negative. People with food allergies (which currently afflicts 8 percent of all American children), whose symptoms can range from mild unpleasantness to sudden death, may likely be harmed by exposure to foreign proteins now being gene-spliced into foods. Mandatory labelling is also necessary so that those suffering from food allergies can avoid hazardous GE foods, so that public health officials can trace allergens back to their source when GE-induced food allergies break out. Unfortunately, the FDA and other global regulatory agencies do not routinely require pre-market animal and human studies to ascertain whether new allergies or toxins or increased levels of human allergens or toxins we already know about, are present in GE foods.

Scientists in favour of GM food argue that there has never been any public concern about the fact that, for the past 50-60 years, plant breeders have used chemical and radiation mutagenesis to create new varieties, involving major and quite unknown genetic modifications followed by selection of useful outcomes. So why is there so much opposition now? It is if GM is a lightning rod for a whole series of consumer concerns, already present but unexpressed. Consumers assess risk in a different way from the scientists. The technical definition of risk triggers so much more alarm, anxiety or outrage than others, seemingly regardless of scientific estimates of their seriousness. Research over many years has identified several 'fright factors', which affect the way, the public sees risk. People are generally more worried if they perceive a risk to be involuntary, for example, exposure to pollution, rather than voluntary, such as smoking. Risks are also seen as more serious if they are inequitably distributed, inescapable by taking personal precautions, if they arise from an unfamiliar or novel source or if they cause hidden and irreversible damage. Danger to small children, pregnant women or future generations also arouses particular dread, as does the possibility of certain forms of death, injury or illness. Risks are perceived to be more threatening if they damage identifiable rather than anonymous victims. The public fears also increase if the issue seems to be poorly understood by science and is subject to contradictory statements from responsible sources.

## 18. CONSUMER RELATED ISSUES

Consumer often ask, are GM foods safe? Can you guarantee that GM foods are 100 percent safe? Unfortunately, science cannot guarantee absolute safety because "absence of evidence" is not the same as evidence of the harm, so the question we should be asking is "How safe are GM foods? The safety testing of GM foods is based on the concept of' substantial equivalence.' This is the idea that if a GM food can be shown to be 'substantially' the same as a non-GM food, then it is considered to be safe. It was developed because of the difficulties and cost of conducting traditional safety tests (like those used for new drugs) on GM foods. But it has been severely criticized by some scientists because it is not clear what level of similarity makes something 'substantially' equivalent [20, 21]. Attitudes towards GM foods vary considerably at one end of the spectrum, are those who see this as the technology to feed the world, and at the other end are groups who are opposed to such techniques in principle. Surveys suggest that many European consumers do not reject GM foods out of hand, rather they weigh the perceived benefits to themselves. Is the product cheaper, tastier, and healthier against the potential risks (e.g., could it harm the environment, human health or animal welfare). In practice, this means that GM products are those where the perceived benefits are least steadily accepted.

GM foods are not necessarily bad, but permitting the expansion of GM crop planting and use in our food without proper knowledge as to the effects, both short- and long-term is at best unwise and at worst highly dangerous. It is not just we humans who could suffer, but ultimately many animal and plant species as well. Can biotech companies really be sure that their products will have no undue effects on biodiversity and food chains, and indeed on people's health? Until they are, the mass introduction of GM crops should not be permitted. We must proceed down the route of genetic modification with extreme caution, without denying that in some cases there could be great benefits to be bad. We have potentially more to lose by the introduction of GM crops. In America, farming takes place on an industrial scale, with millions of acres used exclusively for growing crops. Intensive use of pesticides has virtually wiped out wild animals and plants in the huge crop fields. The use of herbicide or insect resistant crops could potentially have severe effects on our biodiversity by virtually wiping out wild flowers and consequently the insects that feed on them, and further up the food chain, the predators that eat the insects. Each new genetic modification needs to be extensively tested for its safety, not only for humans but also for animals and plants. No single test taken in isolation can either legitimate or condemn all GM food, or indeed any single GMO.

## 19. ANTIBIOTIC RESISTANCE

Researchers also worry about the possible health impacts of the antibiotic 'marker gene' that are commonly used to trace gene transfers. All GE products have antibiotic marker genes. After GM food is absorbed into the human digestive tract, these antibiotic genes could move from what we have eaten into the blood stream or into the bacteria in the intestines. Such transfers might alter our health directly or change the beneficial symbiosis between people (the bacteria that inhibit our intestines). It would also help further spur resistance to antibiotic pharmaceuticals — one of the major emerging public health threats. While the BMA and consumer's Union both recommend a ban on the use of antibiotic markers in GM foods, the FDA said there is a 'vanishingly small' risk that antibiotic marker genes will be transferred to disease causing bacteria. All of the GE foods, up until now, contain antibiotic resistant material. This is because of the haphazard nature of gene splicing. Scientists cannot tell whether they have successfully spliced a foreign protein into a common food unless they also insert an antibiotic resistant marker gene. The antibiotic resistant material in GE foods in getting into the gut of animals and humans and can cause bacterial diseases that are resistant to antibiotic treatment. The BMA and the WHO have said that these antibiotic resistant marker genes should not be in foods and should be phased out as quickly as possible [22].

Seventy percent of all antibiotics in America are fed to animals to make them grow faster and to enable them to survive intensive confinement. It has been found in the tests of chicken and poultry that are sold in supermarkets — the monoorganic ones. The most of these animals are riddles not only with bacteria but also with bacteria that resist commonly used antibiotics. That is why Europeans have banned the use of putting antibiotic in animal feed and same is going to happen in the US. Livestock in USA are being injected with antibiotics as a growth hormone and human run the risk of developing immunity to antibiotics.

Fears that genes are antibiotic resistant, and could jump from bacteria in the gut may be fuelled by new research from the Netherlands. The results show that DNA lingers in the intestine and confirm that GM bacteria can transfer their antibiotic resistance genes to bacteria in the gut. Using an "artificial gut", researchers showed that DNA remains intact for several minutes in the large intestine. One concern about some GM crops, such as maize, used as animal fodder, is that they include a gene for antibiotic resistance. The resistance genes are used to track the uptake of modified genes, and are not expressed in the crops. While some scientists fear that these genes could jump into bacteria in the guts of livestock and create antibiotic resistant pathogens, others have said there is no such risk because the modified DNA breaks down quickly. The Dutch results cast doubt on these assurances. In the modified bacteria were a type normally found in the gut, such as *Enterococcus*. The experiment showed each had one in 10 million chance of passing DNA containing antibiotic resistance genes to an indigenous gut bacterium when they came in contact. There are normally around a thousand billion gut bacteria, suggesting many would be transformed. If some normal gut inhabitants were killed off — as in the guts of people or animals on antibiotics — the transfer rate from gut-type bacteria increased ten fold. The patenting, exclusive use and pricing of seed crops carry with it a range of issues that may be far more significant [23]. It is more about profits than looking after the people of this world. Everyone should have the right to grow their own food using seed that will reproduce after its own kind. GM producers take no responsibility for contamination of organic crops, and it is an outrage that this is permitted to happen. Then the Monsantos of this world want to sue the organic farmers for the use of their patent. Shoppers should have the right to know whether a product contains GM. It is not impossible to have food GM free. Most people believe that the use of genetic modification should be approached with caution. They want strong regulation and monitoring and in addition, some want a framework of rules for coexistence of GM and non-GM crops, and customers want a clear regime for traceability and labeling so that they can make their own choices.

## 20. SAFETY OF GM FOODS

Foods are complex mixtures of compounds characterized by a wide variation in composition and nutritional value. Although priorities vary, food safety is a concern among consumers in all countries. They would like assurances that GM products reaching the market have been adequately tested and that these products are being monitored to ensure safety and to identify problems as soon as they emerge. Because of the complexity of food products, research on the safety of GM foods is still thought to be more difficult to carry out than studies on components such as pesticides, pharmaceuticals, industrial chemicals and food additives [24]. Through the Codex Ailmentarius Commission (CAC), countries discuss standards of GMOs and ways to ensure their safety. One approach, which is being used in assessing the risks of GMOs, derives from the concept of substantial equivalence. Substantial equivalence acknowledges that the goal of the assessment is not to establish absolute safety, but to consider whether the GM food is as safe as its traditional counterpart, where such a counterpart exists. It is generally agreed that such an assessment requires an integrated and stepwise, case-by-case approach. Factors taken into account when comparing GM food with its conventional counterpart include identity, source and composition; effects of processing and cooking; the transformation process, the DNA itself and protein expression products of the introduced DNA; effects on function; potential toxicity, potential allergencity and possible secondary effects; potential intake and dietary impact of the introduction of the GM food. If the GMO-derived food is judged to be substantially equivalent to its conventional counterpart, then it is considered to be as safe as the counterpart. If it is not, further tests are conducted. By 2007, more than 150 scientific publications had reported on the nutritional value of animals feed. The safety of GM foods depends on government assessment and tests conducted by the GM companies themselves. Most of these tests have never been published or subjected to independent peer review. For example, a Spanish researcher who investigated this concern in 2000 could find only eight published safety studies on food from different GM crops, while there were over 40 GM crops approved for sale around the world at that time. We are aware of the widespread use of GM technology in the production of enzymes and other processing aids, which has so far attracted little concern.

The potential benefits of GM technology in human and animal health care are immense and there is a real danger that the crisis of consumer confidence over GM food and food ingredients will act to the detriment of advances in health care and medicine. This would be a tragedy. The medical benefits of GM technology must be fully supported, rather than undermined by hasty commercial GM food developments. Most of us would defend the use, for example, of GM-derived enzymes as delivering real customer benefit. GM-derived cheese coagulant for instance, delivers both animal welfare benefits in the eyes of the customer, and hygiene benefits in the eyes of the food safety professional. In short, it is very important in this debate to distinguish between the different applications of GM-derived food materials to arrive at a healthy conclusion for the welfare of the society [25]. The key issue is that any new food product/ingredient/processing aid can only hope to be commercially successful in today's market if it is seen to deliver a tangible customer benefit. GM insect resistance and GM herbicide resistance have failed to be seen to deliver this. There may be tangible customer benefits associated with them, but if so, they have not been seen to be so which means they have not been communicated.

## 21. EUROPEAN MORATORIUM

GM foods were first allowed in Europe in the early 1990s. However, in 1998 Environment Ministry in Europe agreed to improve the rules governing the release of GMOs into the environment. Whilst this was happening, a number of European countries agreed not to approve any new GMOs until the public and the environment were better protected. This was called the '*de facto*' moratorium, and since October 1998, no new GMOs were approved in Europe. Since then, new laws for releasing GMOs were agreed and came into force in October 2002. They are a marked improvement on the previous legislation and require companies to carry out risk assessments to judge the long-term impacts of growing GM crops and the indirect effects of GMOs. In addition, GM crops that have been modified to contain antibiotic resistant genes will be banned. In contrast, the approval process in the United States relies heavily on voluntary testing by the biotechnology companies [26]. Europe has also developed new laws for the labelling and traceability of GM foods and animal feeds to provide better information for the public. These came into force on April 18$^{th}$ 2004 and are probably the most comprehensive GM labelling laws in the world. Foods that either contain GMOs or have been derived from GM crops (even if they don't contain any modified DNA in the final product, like oil) have to be labelled if their genetic modification exceeds 0.9%. Animal feed also has to be labelled for the first time. It took us 60 years to realize that DDT might have oestrogenic activities and affect humans, but we are now being asked to believe that everything is OK with GM foods because we have not seen any dead bodies yet. It would be wise to avoid GE foods if at all possible. There may be nothing wrong but it would seem the better side of prudence to limit your exposure to them [27]. They were not designed with your health in mind.

Concern at environmental safety is only half of the GMO problem in public eyes. GM foods have created doubts as to whether a pizza containing GM Soya, tomato and cheese is as safe to eat as conventionally produced foods. It should however be noted that "many products from GM plants, such as sugar, as sugar prepared from GM sugar beet, are absolutely identical to conventional products. Additionally, the majority of GM foods will not contain viable genes or DNA, although the issue of whether these might be transferred to human has vexed the public. It is acknowledged that DNA can survive in human saliva for up to twenty minutes and in the stomach for up to eight seconds. Similarly, there is evidence for uptake of DNA microorganisms and human cells. Few researchers think it as a perfectly natural event which has been occurring throughout humankind. There was no evidence however for the incorporation of such DNA into human cells genetic material. It is worth remembering that the medical profession has been trying to develop ways to insert genes into the body cells of human some time, with so far rather limited success. We are not aware of any evidence for transfer of intact genes to humans for some time with so far rather limited success as potatoes, wheat or chickens, despite daily consumption of DNA in the diet. The transfer of genes between higher organisms depends on the probability of sexual crossing. Natural sexual barriers mean that it would be extremely rare for any gene (including transgenes) to get transferred between non-sexually compatible plants, let alone between plants and animals. The fear that transgenes may be transferred to human by eating GM foods seems to be unfounded.

Plant varieties are already selected for particular industrial purposes. Genetic modification can be used to increase the efficiency of production of chemicals already extracted from plants by causing their "over expression." The high starch potato, engineered by the University of Oxford, is grown not for food use but for its starch. Zeneca is developing a paper pulp tree with modified lignin to help produce paper using less energy and bleach. Plants (or more likely, microorganisms used in containment, as with insulin) can be modified to produce chemicals which they did not previously. Agriculture will change significantly as it responds to the new products which can be engineered into plants. Crops grown primarily for fuel for the manufacture of biodegradable plastics, pharmaceuticals and chemical raw materials are all possible. Monsanto has developed blue cotton which much reduces the need to dye the product genes, which produce proteins able to break down major pollutants into their component; non-toxic units may be useful in heavily polluted land [28]. Plants that assist environmental clean-up are already in production, such as those which extract metals from the soil, are useful for the reclamation of land contaminated by heavy industry.

## 22. FUTURE DEVELOPMENTS

- Continued development of herbicide-tolerant, virus - and pest-resistant crops.
- Methods to speed up traditional plant breeding.
- Further development of fruit and vegetables in which the production of ethylene is suppressed so that they take longer to ripen.
- Modification of oils, fats and starches to improve processing or dietary characteristics.
- Improvement of flavour, texture, bioabsorbability, nutritional content and elimination of genes for toxic substances and allergens.
- Identification of genes controlling salt tolerance, resistance to drought, flood and extreme temperatures, and response to day length.
- GM crops that fix nitrogen with greater efficiency, thereby reducing the need for fertilizers.
- GM plants that produce vaccines or therapeutic agents.
- GM biodegradable plastic grown in plants such as oilseed rape could being to replace plastics from fossil fuels within a decade.
- GM plants for bioremediation, i.e. removing toxic chemicals and agrochemical residues from the soil.
- Heart disease prevention (soybean and canola).
- Better digestibility (potatoes).
- Allergy prevention (rice, wheat, rye and pat).
- Increased nutritional value (cassava, rice and soybean).

Attempts are being made in various laboratories throughout the world to raise GM crops resistant to various diseases like cholera (tobacco, tomato, banana and potato), malaria (tobacco), foot-and mouth virus (alfalfa), hepatitis-B (tobacco), rabies (spinach, muskmelon and tobacco), diarrhoea (potato), influenza (tobacco), colon cancer (tobacco and soybean) and HIV/AIDS (tomato and tobacco) [29]. Recombinant DNA technology faces our society with problems unprecedented not only in the history of science, but also of life on earth. It places in human hands the capacity to redesign loving organisms, the products of three billion years of evolution. Such intervention must not be confused with previous intrusions upon the natural order of living organisms: animal and plant breeding. All the earlier procedures worked within single or closely related species. Our morality up to now is to go ahead without restrictions to learn its impact on nature. Restructuring nature is not a part of the bargain. This direction may not be only unwise, but also dangerous. Potentially, it could breed new animal and plant diseases, new sources of cancer, and novel epidemics.

Scientists warn that genetic engineers cannot always accurately predict the outcome of their experiments. Many scientists argue that the genetic engineering process is inherently unpredictable and that genetic engineers are operating with incomplete knowledge about the genes interacting with each other and with their external environment. While genetic engineers can with some precision locate and isolate a trait or gene to be inserted, they cannot control with any precision where that gene will be inserted into the host plant or how it will interact with other genes in the host plant. The new gene may disrupt the function or regulation of a plant's existing genes. GM foodstuffs should be segregated at source to enable identification and traceability of GM products. There should be a ban on the use of antibiotic resistance marker genes in GM food, as the risk to human health from antibiotic resistance developing in microorganisms is one of the major public health threats that will be faced in the 21$^{st}$ century. The risk that GM crops may increase the use of herbicides and pesticides in the environment needs to be comprehensively assessed to determine their full environmental impacts. We are at the early stages of this technology and there needs to be the proper safeguards built into it, particularly the regulatory framework to make sure we do not make mistakes. Public should not go against the technology. Find and weigh the pros and cons. Let us apply the technology carefully and precisely to get the best out of it. Responsible scientists recognize that there are genuine concerns regarding the introduction of genetically modified organisms on the planet. The application of every new technology involves some risk and may produce unforeseen problems. As with any human endeavour, mistakes will be made when applying GM technology to food production. The challenge will be to use scientific tools and knowledge to attempt to predict problems and solve them before they happen. There is a need for public concerns to be addressed by continuous research, communication, and appropriate legislative measures.

We cannot turn back with the clock on agriculture and only use methods that were developed to feed a much smaller population. It took some 10,000 years to expand food production to be current level of about 5 billion tonnes per year. We will have to nearly double current production again. This increase cannot be accomplished unless farmers across the world have an access to current high yielding crop production methods as well as new biotechnological breakthrough that can increase the yields, dependability and nutritional quality of our basic foods crops. We need to bring common sense into the debate on agricultural science and technology, and the sooner the better. James D. Watson, co-discoverer of the structure of DNA said, "Never put off doing something useful for the fear of evil that may never arrive."

## 23. ABBREVIATIONS

| | |
|---|---|
| BMA | British Medical Association |
| FDA | Food and Drug Administration |
| WTO | World Trade Organisation |
| GM | Genetically Modified |
| GMF | Genetically Modified Food |
| GMC | Genetically Modified Crop |
| GE | Genetic Engineering |
| EC | European Commission |
| EU | European Union |
| ISAAA | International Service for Acquisition of Agri-Biotech Applications |
| FAO | Food and Agriculture Organisation |
| GMO | Genetically Modified Organism |
| EPA | Environmental Protection Agency |
| GEAC | Genetic Engineering Approval Committee |
| RCGM | Review Committee on Genetic Manipulation |
| MEC | Monitoring and Evaluation Committee |
| IRRI | International Rice Research Institute |
| ICAR | Indian Council of Agriculture Research |

| | |
|---|---|
| IARI | Indian Agricultural Research Institute |
| WHO | World Health Organisation |
| DST | Department of Science and Technology |
| CSIR | Council of Scientific and Industrial Research |
| BRI | Botanical Research Institute |
| IBC | International Biosafety Commission |
| SPS | Sanitary and Phyto Sanitary |
| DBT | Department of Biotechnology |
| IBC | Institutional Biosafety Committee |
| EMS | Eosinophilia Myalgia Syndrome |
| CAC | Codex Ailmentarius Commission |

## 24. IMPORTANT WEBSITES

- www.agbio.org
- www.wto.org
- www.isaaa.org
- www.pubmed.org
- www.grain.org
- www.cialexpress.com
- www.colostate.edu/transgeniccrops
- www.sciencedirect.com
- www.finanwww.purefood.org
- www.tribuneindia.com
- www.indianfarmers.org

## REFERENCES

1. Steinbrecher RA (1996) From green to gene revolution: The environmental genetically engineered crops. *The Ecologist* 26: 273-282.
2. Shcmidt T & Heslop-Harrison JS (1998) Genomes, genes and junk: The large scale organisation of plant chromosomes. *Trends Plant Sci* 3: 195-199.
3. Gelvin SB (1998) The introduction and expression of transgenes in plants. *Curr Opin Biotech* 9: 227-232.
4. Trevavus A (1999) Much food many problems. *Nature* 402: 231-232.
5. Duke SO (1996) *Herbicide Resistant Crops. Agricultural, Environment, Economic Regulatory and Technical Aspects*, Lewis Publishers, Boca Raton, USA.
6. Feitelson JS, Payne J & Kain L (1992) *Bacillus thuringensis*: Insects and beyond. *Bio/ Technology* 10: 271-275.
7. Stephen N (2003) *Eat Your Genes: How GM Food is Entering Our Diet*, Zed books Ltd, London, UK.
8. Rissler J & Mellon M (1996) *The Ecological Risks of Engineered Crops*, MIT Press, Cambridge, UK.
9. Roller S (2001) Genetically modified foods. *Food Technol Biotechnol* 39: 259-263.
10. Domingo JL (2000) Health risk of GM foods: Many opinions but few data. *Science* 228: 1748-1749.
11. Smith JM (2003) *Seeds of Deception: Exposing Industry Government Lies About the Safety of the GE Foods You are Eating*, Yes! Books, Fairfield, France.
12. Millstone E, Brunner E & Mayor S (1999) Beyond substantial equivalence. *Nature* 401: 625-626.
13. Financial Express, India, 5 June 2006, *The Times of India*, 1 December, 2005, India News, Hyderabad, May 6, 2005.

14. Losey JE, Rayor LS & Cater ME (1999) Transgenic pollen harms Monarch larvae. *Nature* 399: 241.
15. Altieri MA (1996) *Agro Ecology, The Science of Sustainable Agriculture*, West View Press, Boulder, USA.
16. Anderson JA (1996) Allergic reactions to food. *Crit Rev Food Sci Nutr* 36(Suppl): S19-S38.
17. Freese W & Schubert D (2004) Safety testing and regulation of GE foods. *Biotech Genetic Eng Rev* 21: 299-325.
18. Martin T (2001) *Genetically Engineered Foods: Changing the Nature of Nature*, Park Street Press, Rochester, USA.
19. Laura T & Robin T (1998) *Genetically Engineered Foods, Are They Safe? You Decide*, Keats Publication, Illinois, USA.
20. Millstone E, Brunner E & Mayar S (1999) Beyond substantial equivalence. *Nature* 401: 625-626.
21. Antoniou M (1997) Substantial equivalence: A license to kill? *Nutr Ther Today* 7: 5-6.
22. Jones L (1999) Science, medicine and the future: Genetically modified foods. *BMJ* 318: 581-584.
23. Benett P (1999) Understanding response to risk: Some basic findings. In: *Risk Communication and Public Health,* Bennet P & Calman K (eds), Oxford University Press, Oxford, UK, pp 3-32.
24. Ghosh PK (1997) Transgenic plants and biosafety concerns in India. *Curr Sci* 72: 172-179.
25. Kakralya BL & Ahuja I (2001) *Transgenic Plants, Promise and Danger*, Agrobios, Jodhpur, India.
26. Das HK (2006) *Text Book of Biotechnology*, Wiley Precise, New Delhi, India.
27. The Economist (2002,) Better Dead than GM Fed? March 13, p 31.
28. Dawkins K (2000) Battle royale of the 21$^{st}$ century. *Seedling* 17: 2-8.
29. Talukdar K (2002) Genetic manipulation for nutritional value in improvement of crop plants. *Everyman Sci* 37: 21-25.
30. ISAAA (2008) Global status of commercialized biotech/GM crops: 2008, International Service for Acquisition of Agri-Biotech Services, USA.
31. USDA (2008) United States Department of Agriculture, Washington, USA.

# Functional Foods

# 10

**R.S. Singh and G.K. Saini**

*Department of Biotechnology, Punjabi University, Patiala-147 002 (Pb.)*

## 1. INTRODUCTION

The combination of consumer desires, advances in food technology and new evidence-based science linking diet to disease, and disease prevention has created an unprecedented opportunity to address public health issues through diet and lifestyle. Widespread interest in select foods that might promote health has resulted in the use of the term "functional foods." The term "functional food" is arbitrary, but it is nonetheless useful since it conveys to the consumer both the unique characteristics of the food and the associated health benefits. In the last decade of the $20^{th}$ century, consumers began to view their diets from a radically different vantage point. Food is no longer viewed merely as a means to satisfy hunger, prevent diet-deficiency diseases or to provide the essential building blocks of nutrition (e.g., water, protein, carbohydrate, fat, vitamin, minerals, etc.) for the maintenance and/or repair of body tissues; it has become the primary vehicle to transport us along the road to optimal health and wellness. We have begun to view our diets as first lines of defence in the prevention of various chronic diseases of aging, including cancer, heart disease, osteoporosis, arthritis and age-related macular degeneration [1]. This 'changing face' of food has given way to the most exciting area today—the functional foods [2].

## 2. FUNCTIONAL FOODS

Japanese researchers introduced the term "functional food" more than twenty years ago, but the concept still defies uniform definition. The first attempt to define them was "any food that has a positive impact on an individual's health, physical performance or state of mind in addition to its nutritive value" [3]. Since then, several bodies have proposed their own definitions, but to date neither Europe nor the US have issued a legal version. However, European scientists have made a thorough attempt by providing the following preliminary definition [4]:

> "A food can be regarded as 'functional' if it is satisfactorily demonstrated to effect beneficially one or more target functions in the body, beyond adequate nutritional effects, in a way that is relevant to either an improved state of health and well-being and/or reduction of risk of disease. Functional foods must remain foods and that must demonstrate their effects in amounts that can normally be expected to be consumed in the diet: they are not pills or capsules, but part of a normal food pattern."

*Food Biotechnology: Principles and Practices*
*V.K. Joshi & R.S. Singh (Eds.)*
*I.K. International Publishing House Pvt. Ltd., India*
*pp 293–336*

Functional foods have been defined as the foods that by virtue of the presence of physiologically-active components provide a health benefit [5]. The International Food Information Council (IFIC), is supported primarily by food, beverage and agricultural industries, defines functional foods as "foods that provide health benefits beyond basic nutrition." The conundrum that such definitions create is illustrated by a position paper on Functional Foods of the American Dietic Association [6], which appears to accept the IFIC definition, but also includes "whole foods" with beneficial health effects such as fruits and vegetables, low fat cheese and low fat snack foods. Such definitions are unable to clear the difference between functional foods and other foods, because almost any food can have some beneficial effect on some bodily functions. By the virtue of such definitions, even tap water can also be called a functional food, because a liberal intake of water prevents cystitis, kidney and bladder stones, and possibly bladder cancer. Thus, the definition of the type "providing a health benefit beyond basic nutrition" need to be avoided.

Institute of Medicine of the US National Academy of Sciences gave a concrete definition, which defined functional foods as, "those foods in which the concentrations of one or more ingredients have been manipulated or modified to enhance their contribution to a healthful diet" [7]. However, this definition also lacks such aspects of the intimate connection between the term "functional food" and the marketing of foods with a health claim. The term "functional foods" in the reality of the market place is exclusively attached to branded products that tend to claim or suggest improving health. Thus, functional food may be defined as follows: A functional food is a branded food, which claims explicity or implicity to improve health or well-being. This is similar to the definition given by Nestle [8], who described functional foods as, "products created just so that they can be marketed using health claims." Most, but not all, of such foods will also meet the definition of the Institute of Medicine, "foods in which the concentration of one or more ingredients has been manipulated or modified to enhance their contribution to a healthful diet."

## 3. NEED AND SIGNIFICANCE

Belief in the medicinal power of foods is not a recent event, but has been a widely accepted philosophy for generations. Although, Hippocrates may not have started the functional foods movement, he stated almost 2,500 years ago, "Let food be thy medicine and medicine be thy food." Today, consumer beliefs in the health benefits of selected foods and their components appear to be increasing at an unprecedented pace [9]. Health-conscious consumers are seeking out functional foods (designer foods and sometimes considered as nutraceuticals) to control their own health and well-being.

One of the most compelling reasons for the widespread interest in functional foods comes from the rather consistent findings that increased fruit and vegetable consumption is accompanied by a reduction in the risk of heat disease and cancer [10, 11] Below listed are some factors which explain why a functional food is one of the fastest growing categories of foods:

- Advances in nutritional science, agricultural technologies, and processing techniques e.g., biotechnology, genetic engineering [12-15].
- Ready access to nutrition and medical information through widespread media coverage, the internet, and other advances.
- Escalating health care costs which are placing more emphasis on disease prevention [12, 16].
- Recent legislative events such as government regulations, which have changed how foods are marketed and labelled. The more flexible the regulatory environment, the easier it is to market functional foods [16].
- The rise in the aging population, which is increasing the demand for healthier foods or food ingredients to improve health [16-18].

Many functional foods contain added vitamins, minerals and other essential nutrients. Some of these added nutrients indeed promote health: folic acid reduces the risk of neural tube defects, table salt with potassium reduces blood pressure, etc. Functional foods may also contain non-nutritive ingredients. Examples of effective non-nutritive ingredients are sugar alcohols in chewing gum, which reduce the risk of dental caries; plant stanols and sterols, which lowers low density lipoprotein cholesterol; and probiotic bacteria, which may diminish rotavirus diarrhoea in infants [19]. The best-known example of a functional food is probably margarine enriched with phytosterols in order to reduce cholesterol levels in the consumers.

Functional foods tend to be more expensive than their regular counterparts, but that expense is justified if the product promotes health. There is an analogy here: with the pharmaceutical drugs includes a premium that provides the money needed for the research that leads to new drugs. In the same way, the food industry could see the profits from functional foods to fund nutrition research that leads to healthy new foods, and industry also knows how to produce foods that appeal to consumers. Functional foods have been engineered to fit consumer's taste. This engineering is crucial; drugs will sell whatever their look and taste, but foods need taste, convenience, and appeal, or else they will not be consumed [20].

From a market perspective, functional foods are difficult to quantify because of different definitions being used. On the basis of viewing functional foods as foods that make specific health claims, the market in USA, Japan and Europe is estimated to be worth about € 7 billion. The functional food market can, however, be seen as part of a broader health-based/driven foods market. The functional foods found in the market today are often based on general discoveries in nutritional science, and less on a deliberate research strategy to develop functional foods. Often, research was aimed at identifying positive aspects of existing products and functional foods were then developed by optimizing these positive aspects [21].

With regard to biological benefits in functional foods, the International Life Sciences Institute's concreted action on Functional Food Science in Europe [4] has proposed six broad groups that are considered relevant from a scientific perspective. These include: (1) growth, development and differentiation; (2) substrate metabolism; (3) defence against reactive oxidative species; (4) the cardiovascular system; (5) gastrointestinal physiology and function; and (6) behaviour and psychological functions.

Although the future for functional foods appears promising, it ultimately depends on scientific evidence of their efficacy, safety and organoleptic quality. Biomarkers are needed to adequately assess the physiological impact of functional foods. Importantly, consumers must become aware of the beneficial health effects of functional foods. As our knowledge of functional foods is in its infancy, it is especially important that these foods be considered in the context of an overall healthful diet and lifestyle and not as a "magic bullet" to improve health and prevent disease [22].

## 4. FUNCTIONAL OAT PRODUCTS

Today's fast life with over-use of refined food products and decreased interest in natural foods has created a need for a new generation of food products i.e., functional foods [23-25]. Recent research suggests that oats, particularly their soluble fibre fraction i.e., beta glucan, plays a vital role in human physiological processes such as moderating the effect of hypertension, regulating blood glucose as well as insulin levels, weight management and promoting gastrointestinal health [26, 27]. The use of oats in functional food is receiving increased attention. In recent years, many health claims have been made [28]. Oat beta-glucans are found in various breakfast cereals as well as snacks. Products such as "oat milk"® [29, 30], "oatrim"® [31], calorie-reduced yoghurts fortified with oat fibre [32], etc.

containing fibre are offered to the market. "Proviva"® is an oatmeal product produced by the Swedish Company Probi AB using *Lactobacillus plantarum* for fermentation. It is produced by replacing milk with oat medium [33, 34].

Oat products are a widely studied dietary source of cholesterol-lowering soluble fibre β-glucan. Since 1997, the Food and Drug Administration (FDA) has allowed a health claim to appear on food labels linking consumption of whole oat products to reduced risk of heart disease [35]. This was in response to a petition submitted by the Quaker Oats Company. In health claim petition, 37 human clinical intervention trials conducted between 1980 and 1995 were summarized. The daily amount of oat bran or oat meal consumed in the above studies ranged from 34 to 123g. Quaker oats also determined that 3g of β-glucan would be required to achieve a 55 reduction in serum cholesterol [36]. The FDA has allowed that the main active ingredient is the soluble fibre (1→3) (1→4)-β-D-glucan, or β-glucan. Although β-glucan has been accepted as the main active ingredient, but this does not imply that any source of β-glucan is allowed. The health claim is specifically for oat bran, rolled oats (oat meal), and oat flour.

## 4.1 Oat Products

### *4.1.1 Whole oats*

Oats are a whole-grain food capable of providing nutritious sustenance in a more general sense i.e., they provide protein, energy, vitamins, minerals and a good amino acid balance [37]. The botanical name of commonly cultivated food and feed oat is *Avena sativa*. Oats used for food are first "dehulled." The highly lignified and fibrous covering is not suitable for human consumption without processing, though it is readily digested by ruminants. Detection of β-glucan is done by selective dye-binding characteristics [38]. Oat β-glucan is found throughout the endosperm and is located in the endosperm cell walls. Microscopic evaluation reveals that β-glucan is also present in the aluerone walls [39] in lesser amounts than the endosperm. Chemical analysis indicated that β-glucan constitutes about 75% of the endosperm cell walls [40].

Oat β-glucan is involved in carbohydrate and lipid metabolism (Table 1). Besides β-glucan, oats have a multitude of other components namely, phytates, phenolics, vitamins, etc. Oat β-glucan is also being produced and marketed to the pharmaceutical as well as cosmetics industries as a new product that moisturizes and smoothes the skin [41].

### *4.1.2 Oat bran*

The component of oat bran, thought to be responsible for lowering of cholesterol, is its soluble, non-starch polysaccharide, β-glucan. Characteristics of oat bran are summarized in Table 2. Oat bran is 19.5% total dietary fibre, of which 9.6% is insoluble and 6.3% is soluble [43].

**Table 1:** Physiological effects of oat β-glucan.

| Metabolism | Effects |
|---|---|
| Carbohydrate | Attenuates post blood glucose and insulin response |
| | Lowers glycemic index |
| | Delays gastric emptying |
| Lipid | Lowers serum cholesterol |
| | Reduces LDL-cholesterol |
| | Reduces rate of fat absorption |
| | May bind or inactivate pancreatic lipase |

*Source:* 42

**Table 2:** Characteristics of oat bran β-glucan.

| Component | Oat bran |
|---|---|
| Dietary % fiber (% β-glucan) | 15.9 |
| Nonfibre carbohydrate % | 50 |
| Fat % | 7.0 |
| Protein % | 17.3 |
| Simple sugar % | 3.0 |
| Moisture % | 9 |
| Ash % | 3 |
| β-glucan linkage | (1→3) (1→4) |
| Solubility | High |
| Viscosity | Moderate to high |
| 3g of β-glucan is found in | 44 g |

*Source:* 44

The micro structural characteristics of oat bran have been described in a review by Fulcher and Miller [45]. Anatomically, oat bran is considered to be the outer layer of the kernel, which includes a number of morphologically distinct layers ending in going from the outside into the kernel, with the aleurone cells (Fig. 1). American Association of Cereal Chemists (AACC) have defined oat bran as "a food which is produced by grinding clean oat groats or rolled oats and separating the resulting oat flour by sieving, bolting and/or other suitable means into fractions such that the oat bran fraction is not more than 50% of the starting material and has a total β-glucan content of atleast 5.5% (dry weight basis) and a total dietary fibre content of atleast 16.0% (dry weight basis), and such that atleast one-third of the total dietary fibre is soluble fibre" [46]. This definition was also accepted by FDA for the health claim ruling. A simplified flow process for extraction and purification of β-glucan from oat bran is depicted in Fig. 2 [47].

### *4.1.3 Food applications of oats*

A number of oat-derived ingredients for use in food are known as they are more important for their functionality than their food value. A product known as Oatrim was developed by Inglett [51], which

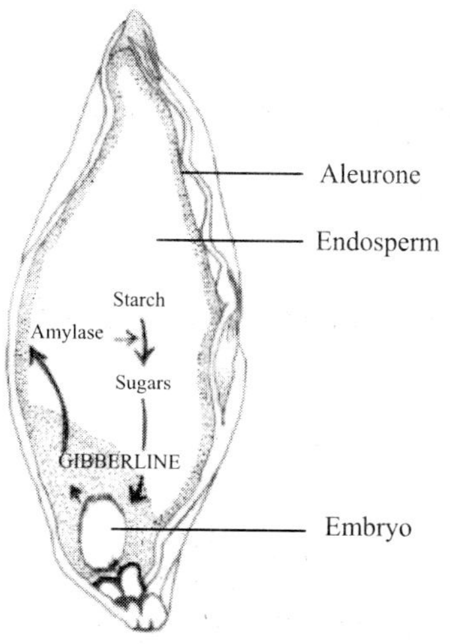


**Fig. 1:** Oat kernel.

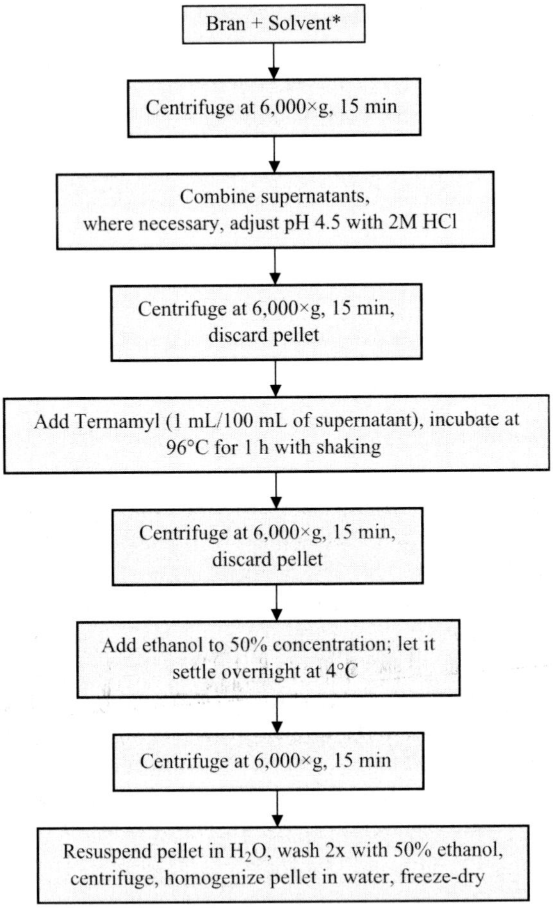


**Fig. 2:** Flow diagram of the extraction and purification process for oat bran β-glucan.
*Source:* **47**

*Solvents:

1. Distilled water adjusted to pH 10 with 20% sodium carbonate.
   Bran: Solvent→1:10
   Extraction time→ 30 min at 45 °C [48].
2. Distilled water adjusted to pH 7.
   Bran: Solvent→1:10
   Extraction performed sequentially at 40, 65 and 95 °C and 3 supernatant fractions combined [49].
3. 4% sodium hydroxide solution.
   Bran: Solvent→1:50
   Extraction time→ 18 h at room temperature [50].

is presently marketed by Quaker Oats-Rhône Poulenc and ConAgra-Mountain Lake Manufacturing. For preparing Oatrim, oats or oat bran is extracted with hot water containing heat-stable α-amylase to give low dextrose-equivalent maltodextrin with 1-10% of β-glucan by weight. Oatrim may be used as a fat replacer. Nurture, Inc. has produced Oat Vantage™, an oat based ingredient that manufacturers can use in food, beverages and supplements to make the FDA heart health claim.

#### *4.1.4 Structure of oat β-glucan*

The most remarkable constituent of the oat is its unique soluble fibre. There is a very large, and steadily expanding, body of clinical evidence attesting to the extraordinary therapeutic significance of oat soluble fibre, which is a unique mixed linkage beta-glucan comprised of (1→3), (1→4) beta-D-glucan. Beta-glucan is the principal constituent of cell walls in the oat seed and, accordingly, it is present in relatively low concentrations (~2-4%). The outer, branny regions of the oat have the thickest cell walls and represent a richer source of beta-glucan (~6-7%).

#### *4.1.5 Significance of oat β-glucan*

4.1.5.1 Cardiovascular health: Daily ingestion of modest quantities of oat β-glucan can cause a significant reduction in serum cholesterol, notably harmful LDL cholesterol [52]. Clinically observed outcomes for LDL reduction arising from oat β-glucan can exceed 20% for carefully controlled studies utilizing the prepared diets [53, 54]. Oat β-glucan usually has no effect on triglycerides or beneficial HDL cholesterol.

Cholesterol reduction with oat β-glucan is dose responsive, which have been affirmed by a metaanalysis, including the finding that the response is greater for subjects with higher initial levels of cholesterol [55]. On the basis of these studies and many more in 1997, the Food and Drug Administration enacted a health claim specifically addressing the cardiovascular health benefits of oat β-glucan [56]. The claim provides for a minimum effective daily dose of 3g, which may be divided into sub-doses as small as 0.75g. β-glucan being a hydrocolloid, has the capacity to form highly viscous solutions at low concentrations *in vivo*. This can lead to binding and increased excretion of bile acids, reduced absorption of fat and cholesterol, and a reduced absorption rate of available carbohydrates. This suggests that β-glucan is likely to be more effective when taken with meals.

4.1.5.2 Carbohydrate metabolism: Studies have revealed that oat β-glucan blunts the glycemic response, which occurs subsequent to ingestion of carbohydrates [57, 58]. This phenomenon is related to the viscosity of β-glucan as is the case with serum lipids, which can increase the duration of intestinal transit and result in delayed digestion of starches and other available carbohydrates, leading to reduction of glycemic response. Significant blunting of the glycemic response naturally leads to a significant blunting of the insulinic response. Oat β-glucan is emerging as a tool for diabetes and serum glucose management. As the glycemic and insulenic responses for β-glucan are driven by viscosity, which is concentration-dependent, there is an inherent dose relationship. Significant glycemic and insulenic responses have been observed for doses as low as 1.4g β-glucan [59].

4.1.5.3 Hypertension: There is a mounting evidence that dietary intervention with oats can reduce blood pressure. Reductions in systolic and diastolic pressure of 7.5 mm Hg and 5.5 mm Hg, respectively, have been observed [60]. Epidemiologic evidence suggests an inverse relationship between oat consumption and blood pressure [61]. Oat-derived hypotensive effects are due to β-glucan. The reduced insulin levels associated with oat β-glucan are related to reduced insulin resistance which, in turn, has been linked to reduced blood pressure. There is also preliminary evidence that reduced blood cholesterol may contribute to improved blood pressure [62].

4.1.5.4 Weight loss: Blunting of the postprandial glucose and insulin responses has been linked to reduction in the rate of return of hunger and, also, reductions in subsequent energy intake [63]. The increased viscosity of the alimentary bolus caused by solublized β-glucan can delay nutrient absorption and result in reduced hunger and/or increased satiety. This phenomenon may be mediated by increased secretion of gut hormones, such as cholecystokinin, which enhances satiety [64]. The soluble fibre in oats i.e., β-glucan readily induces very high levels of satiety. In a study, oats were found to have the third highest satiety rating among all the foods tested [65]. Satiety can lead to weight loss, although few studies are available for a significant weight loss demonstration for diets containing oat β-glucan [66, 67].

## 5. FUNCTIONAL WHEAT BRAN PRODUCTS

In the last decade, preventive medicine has undergone a great advance. Research has demonstrated that nutrition plays a crucial role in the prevention of chronic diseases [68]. In recent years, cereals and their ingredients are accepted as functional foods and nutraceuticals because of providing dietary fibre, proteins, energy, minerals, vitamins and antioxidants required for human health [69]. In North America, the food industry plays a pivotal role in shaping the diet. Currently in the United States, a health claim for a food or food component is based on "significant scientific agreement" totally on the publicly available evidence. Much of the research in this area has focused on specific nutrients of particular foods. Wheat and wheat bran provide an excellent example of this research process. Wheat is one of the major grains in the diet of a vast number of the world's population, and, therefore plays an important role in the nutritional quality of diet and human health. Wheat bran is widely available as a food ingredient.

### 5.1 Structure of Wheat Kernels

Structure of wheat kernels (caryopses) shows three main anatomical parts: an embryo (germ), an endosperm, and a pericarp (fruit coat, containing layers making up the bran) that covers the endosperm (Fig. 3) like a typical cereal grain, and the major milled fractions are summarized in Table 3. Bran layers are the outermost fibre rich layers. Endosperm is rich in starch and protein, while the germ is high in fat. The bran and germ fractions also have high vitamins and minerals content. Commercial bran fractions always have some endosperm part associated with them. The germ is typically removed in the milling process as a final product.

### 5.2 Milling of Grains

Milling of grains can be categorized into: wet milling or dry milling. Specialized milling technology has been developed to allow fractionation of grains according to very specific needs. However,

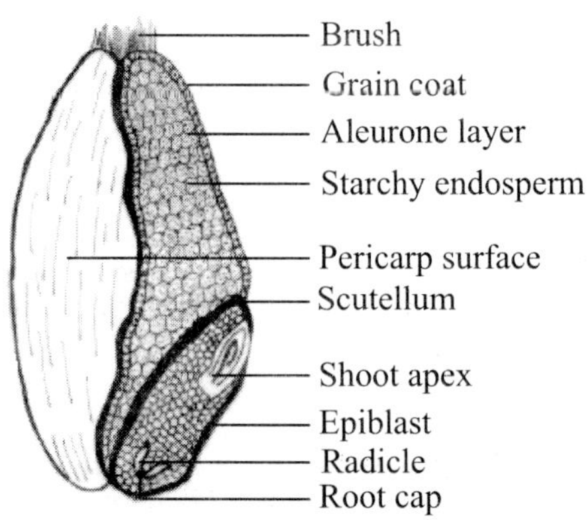


**Fig. 3:** Structure of wheat kernel (Caryopses).

**Table 3:** Structural relationship between botanical constituents and major mill fractions.

| Grain Component | Mill Fractions | |
|---|---|---|
| **Grain (Caryopsis)** | ↑ BEESWING | ↑ BRAN |
| 1. Pericarp (fruit coat) | | |
| (a) Outer pericarp | | |
| Outer epidermis (epicarp) | | |
| Hypodermis | | |
| Thin-walled cell remnants over most of grain, but cell walls remain in crease & attachment region; includes vascular tissue in crease | ↓ | |
| (b) Inner pericarp | | |
| Intermediate cells | | |
| Cross cells | | |
| Tube cells (inner epidermis) | | |
| 2. Seed | | |
| (a) Seed coat (testa) and pigment strand | | |
| (b) Nucellar epidermis (hyaline layer) | | |
| (c) Endosperm | | |
| Aleurone layer | | **WHITE FLOUR** |
| Starchy endosperm | | ↓ |
| (d) Embryo | | ↑ **GERM** |
| Embryonic axis | | |
| Scutellum | | ↓ |

*Source:* 70

wheat is typically dry milled. The main quality of dry milling is to separate the anatomical parts of the grain i.e., bran, endosperm and germ as effectively as possible. Wheat milling is done through a series of paired rolls and sifters that facilitate separation of the major components of the grain. Flour, bran and germ are the primary products. Terms used for milled bran fractions include "heavy bran," "light bran," "fine bran," "coarse bran," and "ground bran." For simplification, whole wheat, high-density (higher content of starchy endosperm material) bran and low-density (lower content of starchy endosperm material) bran (Table 4). Bran with approximately 35-40% of dietary fibre is a common milled product used as a raw material for breakfast cereal production. It is obtained in the "break" section of the mill where corrugated rolls allow gross separation of the fibrous branny layers from the endosperm and germ. This bran has endosperm material associated with it.

**Table 4:** Composition of wheat and bran fractions.

| Material[1] | Size[2] | Protein (%) | Fat (%) | Starch (%) | Dietary Fibre (%) |
|---|---|---|---|---|---|
| Whole wheat | 2000 | 10.2 | 1.5 | 65.0 | 12.0 |
| High density bran | 2200 | 12.5 | 1.7 | 61.1 | 13.0 |
| Low density bran | 1400 | 15.5 | 3.1 | 21.1 | 37.0 |

[1]All materials were analyzed and results reported on "as is" moisture basis, 14.5%, 13.1% and 11.7% for whole wheat, high-density bran & low-density bran, respectively.
[2]Size is mean particle size in microns

*Source:* Kellogg Company Laboratory values.

## 5.3 Breakfast Cereal Products

A large number of process technologies have been developed to produce hot or cold breakfast cereals from grain-based ingredients, including whole grains along with their fractions as bran and flour. Breakfast cereal manufacturing involves three basic steps: addition of flavouring, conditioning of tempering, cooking, drying, shaping/forming, toasting, vitamin spray/coating (optional), and packaging. Continuous processing in an extruder is one of the widely practiced techniques for cooking cereal manufacture. Many technologies are used to process pre-conditioned or cooked grain materials. These include flaking grains or grain pieces by roller milling, puffing that pressure differentiates as the basis for expansion, shredding of cooked grains and the formation of "mats" from shreds in various orientations, blending of grains and slurry components (containing a syrup mixture which binds ingredients together) for granola-type products, and drum drying of cooked flours to produce farina and typical baby cereals.

## 5.4 Role of Wheat Bran Fibre

Generally accepted role of wheat bran is in preventing constipation. Growing research has focused on its preventive effect against colon and breast cancer. Both soluble and insoluble fibres have value in promoting regularity in colonic function. Insoluble fibres as from wheat bran are found to be resistant to fermentation by colonic bacteria and increase faecal bulk by retaining water [71, 72]. Among the different fibre sources compared by Cummings [73], wheat bran ranked among the highest in faecal bulking, exceeding fibres from fruits and vegetables, gum and mucilages, cellulose, oats, corn, legumes and pectin [70]. It has been shown that the faecal bulking effect of wheat bran is reduced as particle size decreases [74, 75]. Particle size of bran is apparently not affected by mastication and digestion in the small intestine [76]. Particle size can affect the faecal bulking effect of wheat bran by altering water-holding capacity and resistance to colonic bacterial fermentation [74].

### *5.4.1 Gastrointestinal disorders*

5.4.1.1 Constipation: Controlled studies have revealed the effectiveness of fibre and wheat bran in relieving constipation [77]. After reviewing the available data, it is suggested that constipated patients have lower stool output and slower transit time with or without bran supplementation [78]. Constipation is more common in women than men [79]. That is why women need more fibre for normal laxation as compared to men. The effect of gender on stool weight may be explained by differences in transit [80]. In addition, hormonal changes contribute to occurrence of constipation more in women than men [81].

5.4.1.2 Diverticulosis: Diverticulosis refers to abnormalities of the colon characterized by the presence of sacs protruding through the colonic wall. Coarse wheat bran (unprocessed) has shown best results in preventing diverticulosis [82]. The mechanism behind the role of dietary fibre in the prevention or management of diverticulosis, however, is unclear. The hypothesis of Painter that diverticular disease being a result of colonic under filling would suggest that correcting this condition by maintaining certain faecal mass with high-fibre diet could potentially reduce the risk of developing diverticulosis [83].

### *5.4.2 Colon cancer risk reduction*

In North America, colon cancer is the second leading cause of cancer death. It has been associated with dietary risk factors that may be controlled [84]. As many as 40 epidemiologic studies have been reviewed that proved an association between dietary fibre and colon cancer risk [85]. The fibre intake from cereal grains including wheat bran or whole wheat has reported significant protective effects in several ecological studies [86-88].

### *5.4.3 Breast cancer risk reduction*

Epidemiological association between increased dietary fibre and a reduced risk of breast cancer is confounded by the fact that high fibre diets are low in fat and high in fruit and/or vegetable consumption [89, 90]. Elevations in bioavailable estrogens have been linked to an increased risk of developing breast cancer [91-93]. Recent data show that women with breast cancer have higher levels of circulating estrogens and higher levels of bioavailable estrogens than do women who are free from breast cancer [94]. In the case of colon cancer, some researchers have described a unique role in estrogen metabolism to the consumption of wheat products in comparison to other grains. Wheat bran in the form of ready-to-eat breakfast cereals favourably influences the breast cancer risk in a study, which examined the effects of three levels of wheat bran supplementation in the form of Kellogg's® All-Bran® cereal.

## 6. FUNCTIONAL PRODUCTS FROM RICE

Rice (*Oryza sativa* L.) is the staple food of half of the world's population. It is known to be a rich source of calories in the form of starch besides providing protein with higher nutritional quality in comparison to other cereal grains. Rice has versatile functional properties and is easily digestible.

### 6.1 Structure of Rice Kernel

The structure of rice kernel is shown in Fig. 4. It is composed of a hull, consisting of the lemma, palea, sterile lemmas and rudimentary glumes, the caryopsis or brown rice and an embryo. Further, characterization of the components of the brown rice and embryo are depicted in cross-section of the kernel (Fig. 5). Pericarp, tegmen and aleurone layers constitute the bran, and during commercial milling operations, bran is usually removed with much of the embryo. Hull makes 20% approx. of the rice kernel, the bran and embryo about 8-12% and the endosperm or milled rice about 70-72%.

Three forms of rice are known from market point of view i.e., rough, brown and milled. Rough rice (paddy) is the harvested unshelled rice as it comes from the field. Brown rice, also termed as shelled rice, husked rice, cargo rice or loozian, is the rice from which the hull has been removed via shelling or hulling. Milled rice or white rice is the rice from which most or all of the bran has been removed by milling. This is also known as scouring or whitening.

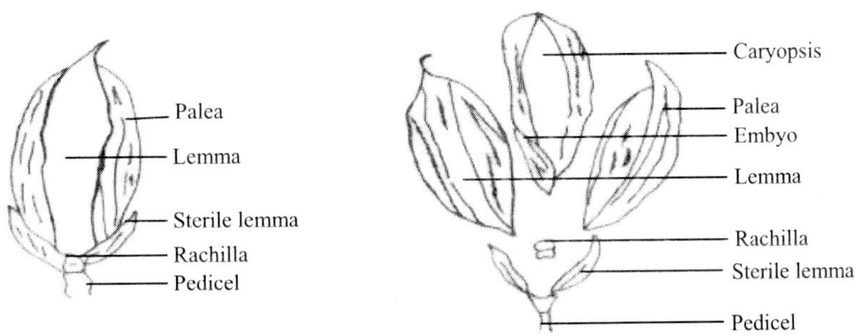


**Fig. 4:** Structure of a rice grain.

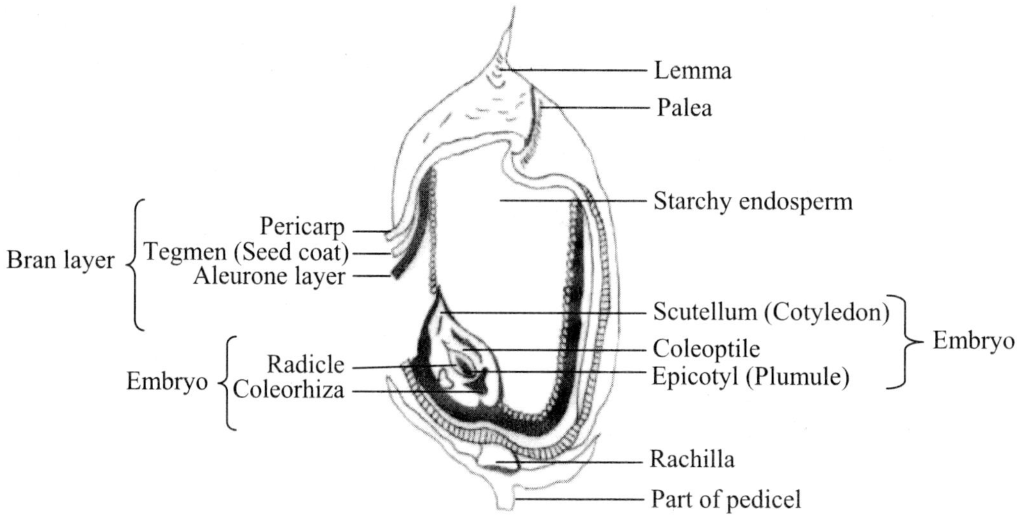


**Fig. 5:** Cross section of a rice kernel.

## 6.2 Rice Bran

Rice bran is the most abundant product of the U.S. rice milling industry. In an unstabilized form, the rice bran is used as an animal feed ingredient. With the improvement in methods of rice bran in past years, there are currently many arenas for the use of rice bran. Highest concentration of nutrients is found in bran layers of the rice kernel. Rice bran is used as a source of protein and is also employed for extraction of oil. Composition of commercial rice bran is depicted in Table 5 [95-97]. Phytate, a form of phosphorus, is a major mineral constituent of rice bran. Potassium, magnesium and silicon are also present in high levels, while sodium and calcium are in low levels. The rice bran possesses abundant B-vitamins as well as tocopherols, but is low in Vitamin A and C [97].

Rice bran has both insoluble and soluble fibres. Insoluble fibres belong to cellulose, while soluble fibres are hemicellulose in nature. Insoluble fibres add bulk to the gastrointestinal (GI) tract, thus causing more frequent stools to pass through the system and that too more quickly. Hemicellulose consists of several carbohydrate components, including hexoses, pentoses, hexuronic acids and amino acids [98]. The amount of sugars i.e., arabinose, glucose, galactose, xylose, etc. in hemicellulose vary according to the rice cultivar tested and the environment in which the rice is cultivated. The amount of the protein/amino acids also varies widely ranging from 9.6% for Louisiana-grown Starbonnet, a long-grain cultivar to 26.4% for California-grown Calrose, a medium-grain type [99].

**Table 5:** Composition of commercial rice bran.

| Component | Amount |
|---|---|
| Protein | 11.5–17.2 |
| Fat | 12.8–29.6 |
| Fiber (with soluble dietary fiber→1.9-2.4%) | 6.2–31.5 |
| Ash | 8.0–17.7 |
| Starch (Depending upon the degree of milling) | 10–55 |

Rice bran has phytonutrients, which have shown promising disease-preventing and health-related benefits [100]. Brown rice and rice bran contain compounds with putative cancer chemopreventive properties, like phenols (tricin), that are present at much lower levels in white than in brown rice. Thus, the consumption of rice bran or brown rice may be advantageous with respect to cancer prevention [101]. But consumption of rice bran has been limited as it gets rancid readily and microbial activity is also generally associated with raw rice bran [102-104]. Nutritional qualities of rice bran may be improved by heat treatment, which is involved in the stabilization of the bran [105].

Rice bran is a potentially valuable source of natural antioxidants such as tocopherols, tocotrienols and oryzanols [106]. An increased percentage of fibres ranging from 35-48% is found in the defatted rice bran [107]. So, defatted rice bran can be used in specialty high-fibre products and baked goods. Rice bran also possesses emulsifying [108] as well as foaming properties for the baked products [109, 110]. Defatted rice bran consists of about 15-20% protein [111]. It can be extracted with dilute alkali to produce a bran protein concentrate i.e., 50-60% proteins [108]. Bran protein has a good nutritional quality for human consumption. Rice bran also has high lysine content in comparison to most other cereal proteins [112].

## 6.3 Rice Bran Oil

Another co-product of the rice milling industry is rice bran oil. The rice bran (with germ) fraction contains the majority of the oil in the rice kernel. The oil content of clean rice bran is similar to that of soybeans and cottonseed i.e., 20-22% [113, 114]. The factors contributing to the yield of refined rice bran oil include the age and storage conditions of the rice, milling practices, bran stabilization, conditions used for extraction, and the method of refining the oil. Extraction can be carried out at either high or low temperatures. Many organic solvents can be used for rice bran oil extraction, but the most commonly used solvent is n-hexane [115]. Hexane extraction is carried out at 60°C yielding a greater quantity of crude oil (95-99%), but only 80% yield of refined oil [115, 116]. In contrast, low temperature extraction is carried out at 18°C removing neutral oil, with minimal quantities of gums and waxes, and 98% yield of the refined oil [116]. Alkali and physical refining are used for rice bran oil. Alkali refining gives better results with oils that contain relatively high amounts of free fatty acids, though at the expense of greater loss of neutral oil. Physical refining depends on molecular distillation from a thin film, and refining losses can approach the actual free fatty acid content [114, 116].

Rice bran oil is excellent for frying foods giving low peroxide, foam, free fatty acid and polymer formation [115, 117]. Under frying conditions, rice bran oil is more stable in comparison to the other common vegetable oils because of the more even balance between linoleic and oleic acid [114, 116]. Rice bran oil has been reported to have the beneficial effect of lowering serum cholesterol [118-124]. High temperature hexane extraction may result in 3-4% wax in the crude oil. These waxes can be isolated and purified by crystallizing and precipitating at low temperature [115]. Removal of residual oil may be done by centrifugation or filtering off and then washing with acetone or ethanol. Rice bran wax is composed of esters of long chain fatty acids, $C^{16}$-$C^{26}$, and fatty alcohols, $C^{22}$-$C^{30}$ in length [114, 116]. This wax is FDA approved and is used as a coating for fresh fruits and vegetables to prevent moisture loss. It is also used in confectionary, lipstick and other cosmetics [115].

## 6.4 Rice Starch

Milling of brown rice results in white rice, which is the endosperm of the kernel and is composed of starch. Like other starches, rice starch consists of two polymeric forms of glucose namely, amylose and amylopectin. Amylose has linear chains with only 3-4 branched chains on an average [125] and comprises 0-33% of the rice starch content [126]. Amylopectin is a highly branched polymer of glucose with 96% α-(1→4) bonds and 4% α-(1→6) bonds. Amylose and amylopectin molecules are organized into a radially anisotropic, semi-crystalline structure in the starch granule [127, 128]. In comparison to other cereal grains, rice starch granules are very small ranging in size from 3 to 5 microns. Isolated, native rice starch is a very fine, white powder providing applications as a cosmetic dusting powder, laundry-stiffening agent, paper and photographic paper powder, sugar coating in confectionary, and excipient for pharmaceutical tablets. Gelatinized rice starch has a bland taste and is smooth, creamy and spreadable. Rice starch provides functional as well as nutraceutical roles when isolated and *in situ* [129]. Waxy rice tends to lose its shape when cooked and is very sticky. U.S. long-grain rice has intermediate to high amylose contents and is marketed primarily as milled rice, parboiled rice, quick-cooking rice, etc. The medium- and short-grain cultivars have low amylose contents and are preferred for ready-to-eat (RTE) breakfast cereals, baby foods and brewing [130].

### 6.5 Health Benefits of Rice

Rice appears to be quickly and completely digested, which results in a high glycemic response [131] and low colonic fermentation [132]. In general, rice with high amylose content has a lower glycemic index (digests at slower rate). Current applications use waxy rice starch because of its innate heat and freeze-thaw stability. Development of rice starch-based foods with low glycemic index would be beneficial to diabetics and athletes. A starch-based sports beverage or food bar that slowly digests would be beneficial in maintaining blood glucose concentration and carbohydrate oxidation during the latter stage of exercise for the athlete. A rice bran oil diet reduces total serum cholesterol and LDL cholesterol, and Apo B as compared to other fat sources [119].

## 7. FLAXSEED PRODUCTS FOR DISEASE PREVENTION

Flax (*Linum vistatissimum* L.) and certain perennial species have been used in food in Europe and Asia since 5000-8000 BCE, and the fibre of flax stem has been used for linen cloth and many other uses. Approximately 200 species of *Linum* (flaxseed or linseed) are known, and it is an economically important oilseed crop, especially in Canada [133]. Flaxseed is an abundant source of α-linolenic acid, viscous fibre components and phytochemicals, such as lignans and protein. These components of flaxseed are of great interest both for the food and pharmaceutical industries. In addition, flaxseed incorporation into the diet is particularly attractive from the perspective of development of foods with specific health advantages. The demonstration of clinical activity associated with the consumption of flaxseed led the US National Cancer Institute (NCI) to target flax as one of the six plant materials for study as cancer preventive foods [134]. The physiological aspects of flaxseed components responsible for disease prevention have been reviewed earlier [135]. Flax is fairly easy to genetically engineer traits of importance for the production of nutrient-tailored foods or ingredients [136].

### 7.1 Structure of Flaxseed

Flaxseeds are flat, oval, and pointed at one end. The seed consists of a seed coat, an embryo consisting of two large, flattened cotyledons, a short hypocotyl and a radicle [137]. Cotyledons form 55%, the seed coat and the endosperm 36%, and the embryo axis 4%, respectively, of the total weight of hand-dissected flaxseed [138]. Abrasive dehulling removes the seed coat, which makes up most of the hull [139]. The hull obtained comprises approximately 22% of the seed. Structural features of the seed are illustrated in Fig. 6.

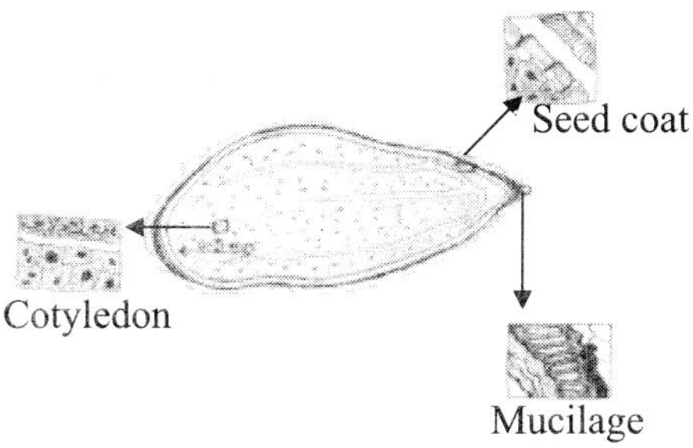


**Fig. 6:** Structure of flaxseed.

## 7.2 Biologically Active Components

### 7.2.1 Lipids

Flaxseed oil has been consumed for centuries in India, China and Europe [140], but it is traditionally considered non-edible oil in many western countries due to its rapid oxidation and polymerization and it does not have GRAS (Generally Recognized as Safe) status [141]. In 1996, flaxseed grown in the Canadian prairie provinces contained 5% palmitic acid (16:0); 3% stearic acid (18:0); 17% oleic acid (18:1n-9); 15% linoleic acid (18:2n-6) and 59% α-linoleic acid (ALA; 18:3n-3) [142, 143]. The reported biological properties of ALA demonstrate a broad spectrum of potential health benefits associated with the consumption of this fatty acid. These include reduced production of "4 series" leukotrienes [144, 145]; protection against cardiovascular disease and myocardial infarction [146, 147]; reduced mortality due to cancer; retared tumour growth [148] and antiparasitic and antimalarial activities [149]. Flaxseed also contains phospholipids, out of which 2.9% of sterol glycosides can be extracted [150]. The sterols can be further fractioned into cholesterol (2% of sterol fraction), campesterol (26%), stigmasterol (7%), sitosterol (41%), $\Delta^5$-avenasterol (13%), cycloartenol (9%) and 24-methylenecycloartanlo (2%) [151].

### 7.2.2 Proteins

Flaxseed, as vegetable protein, is commercially available in the form of seed, full-fat flour and meal. Flaxseed and flour are known to be least refined form of flax protein ingredients. There are two commercial grades of flaxseed meal i.e., one with 30-32% protein and second with 40% protein. Genetic and environmental factors attribute to the varied protein content of flaxseed. Cool growing conditions usually result in lower protein but higher oil content [152]. The protein fraction contains a favourable ratio of amino acids with lysine, threonine and tyrosine as the limiting amino acids. Flaxseed meals possess an essential amino acid index of 69. Flaxseed protein concentrate contains 66% protein with some lipids. Commercial flaxseed meal contains only 39% protein and up to 14% lipids.

Flaxseed proteins are structurally more lipophilic than soybean proteins. Their hydrophilic properties are influenced by the presence of polysaccharide gums in flaxseed protein preparations [153]. A class of native seed proteins called "oleosins," or oil-body proteins, are structural proteins found in the seeds. They are highly lipophilic with a unique secondary structure [154].

### 7.2.3 Polysaccharides and other compounds

Flaxseed hydrocolloidal gum, also referred to as mucilage, comprises about 8% of the seed weight [155]. Its yield depends upon the extraction regimes and can range from 3.5 to 9.4% [156, 157]. Fibre content of flax has also been associated with the seed coat colour. Meal obtained from yellow-seeded flax has less crude fibre (8.7 vs. 11.7%) compared to meal from brown-seeded flax [158].

Lignans are the phenolic compounds formed by the union of two cinnamic acid residues and are comprised of a 2,3-dibenzylbutane nucleus [159]. The biological role of lignans in plants is not yet clear, but their antifungal and insecticidal properties contribute to their role in plant host defence systems, while in some plants they may function as growth regulators [160]. Some bioactive functions i.e., anti-oxidant, antimicrobial and anticancer, have been attributed to phenolic acids. Dehulled, defatted flaxseed meal possesses 81 mg/100g of total phenolic acids and 73.9 mg/100g of esterified phenolic acids [161]. Trans-ferulic (46%), trans-sinapic (36%), f-coumaric (7.5%) and trans-caffeic (6.5%) are the major phenolic acids for dehulled, defatted flaxseed for both total as well as esterified phenolic acids [162].

Flavonoids have also been reported to exhibit antibacterial, antiviral, anti-inflammatory, antiallergic actions. Flavonoids are defined as a group of polyphenolic compounds that contain 15 carbon atoms as their basic skeleton arranged in a $C_6$-$C_3$-$C_6$ configuration [163]. Flavone C- and O-glycosides are the major flavonoids found in flax cotyledons [164]. The content of flavonoids varies with the cultivar and environmental conditions and ranges from 35 to 71 mg/100g [165].

## 7.3 Flax as a Food and Health Benefit

Flaxseed and flaxseed oil have been used for food for centuries in Asia, Europe, and Africa. Recently flaxseed has come in for use in North America. The three major components making flaxseed beneficial in human and animal nutrition are [166]:

(i) A very high content of α-linoleic acid (omega-3 fatty acid) essential for humans.
(ii) A high percentage of dietary fibre, both soluble and insoluble.
(iii) The highest content of plant "lignans" of all the plant or seed products used for human food. Lignans appear to be anticarcinogenic compounds.

Scientific literature over the last 30 years supports the many expected health benefits of consuming flaxseed and cold-pressed flaxseed oil. The National Cancer Institute has evaluated flaxseed, along with a number of other potential food ingredients, as a component of one or more ingredients [167]. Designer foods are composed of one or more ingredients that contribute essential nutrients for health along with protection against certain diseases as cancer and coronary heart disease. The average composition of 11 flax cultivars is shown in Table 6 [168].

**Table 6:** Composition of whole flaxseed (Produced in North Dakota, average of 11 cultivars).

| Component | Content (%) |
|---|---|
| Moisture | 7.1 – 8.3 |
| Lipids (DW basis) | 31.9 – 37.8 |
| Protein | 26.9 – 31.6 |
| Total dietary fiber | 36.7 – 46.8 |
| Insoluble | 30 ± S.E.* |
| Soluble | 10 ± S.E. |
| Fatty acid composition | |
| Palmitic, $C_{16:0}$ | 4.6 – 6.3 |
| Stearic, $C_{18:0}$ | 3.3 – 6.1 |
| Oleic, $C_{18:1}$ | 19.3 – 29.4 |
| Linoleic, $C_{18:2}$ | 14.0 – 18.2 |
| Linolenic, $C_{18:3}$ (omega 3) | 44.6 – 51.5 |

*Standard error

The North Dakota flaxseed cultivar 'omega', a yellow or golden flaxseed is preferred in Europe and also sold in Japan, Korea and United States food markets. Ground or whole flaxseed can be added to almost any baked products, imparting a nutty flavour to bread, waffles, and pancakes. Some other food uses of flaxseed are ready-to-eat breakfast cereals, breakfast drinks, salad dressings made with cold-pressed flaxseed oil, salad toppings, meat extenders, soups, bagels and cakes. Flaxseed flour is used commercially in breads in the United States by one or more large bakeries [169].

Flaxseed is an abundant source of α-linolenic acid (ALA), viscous fibre components and phytochemicals i.e., lignans and protein. These constituents are of great interest both for the food and the pharmaceutical industries. The physiological benefit of flaxseed oil is attributed to the high content of ALA. But the functional properties of ALA are not the same as those of flaxseed oil, because flaxseed oil contains other fatty acids ranging from 40 to 60%. So functional properties of flaxseed oil as a food ingredient go beyond the effects of ALA. The application of flaxseed gums in food systems with their beneficial health effects on carbohydrate and lipid metabolism is also known. Improvement in the physiological functionality of flaxseed gum can be made by enzymatically modifying its physiochemical and rheological properties to extend its application in food and beverages with longer shelf life, improved flow, consistent viscosity and a pleasing appearance.

Flaxseed polysaccharide gum or mucilage may have nutritional value as dietary fibre, which may appear to play a role in diabetes and coronary heart disease risk, preventing colon and rectal cancer and the incidence of obesity [170]. Flaxseed mucilage may lower blood glucose. It has been observed that flaxseed protein may influence blood glucose in two ways: by stimulating insulin secretion, which could result in reduced glycemic response [171] and by interaction with the polysaccharides.

## 8. FUNCTIONAL GRAPE, CITRUS AND VEGETABLE PRODUCTS

### 8.1 Grapes

Grapes are known to be the world's largest temperate fruit crop with approximately 67.2 million tonnes produced annually [172]. European grapes belong to the species *Vitis vinifera* L. and over 95% of all the grapes produced are of this species [173, 174]. North American grapes belong to two main species i.e., *Vitis labrusca* and *V. rotundiflolia* [174].

#### *8.1.1 Grape products*

8.1.1.1 Wine: Wine is a moderately alcoholic drink prepared from juice extracted from fresh, ripe grapes by fermentation process. Six main classes of wines are namely, red table wines, white table wines, rosé table wines, dessert wines, sparkling wines and appetizer wines. White wines are usually more delicate in flavour than red, and because of the lack of anthocyanins and very low levels of tannins [175], defects in taste and appearance are more apparent in them [176]. The main difference in preparing the two types of wines is in their pre-fermentation operations. White wine is produced by the fermentation of pressed grape juice, and red wine is produced by the fermentation of the juice on the skins in order to extract the red colour. Flow diagram for the production of white and red wine is given in Fig. 7.

Sparkling wines are produced by carbonation under pressure or by allowing continuing fermentation (to produce $CO_2$) in the bottle. Dessert wines are produced by fortification with alcohol so as to retain some of the sugars present in the grapes [177]. For more details, see chapter on "Wine and Brandy."

8.1.1.2 Juice: Grape juice is the liquid extracted from suitable ripened grapes. It differs a little in composition from the grapes, except for the content of crude fibre and the oils known to be present primarily

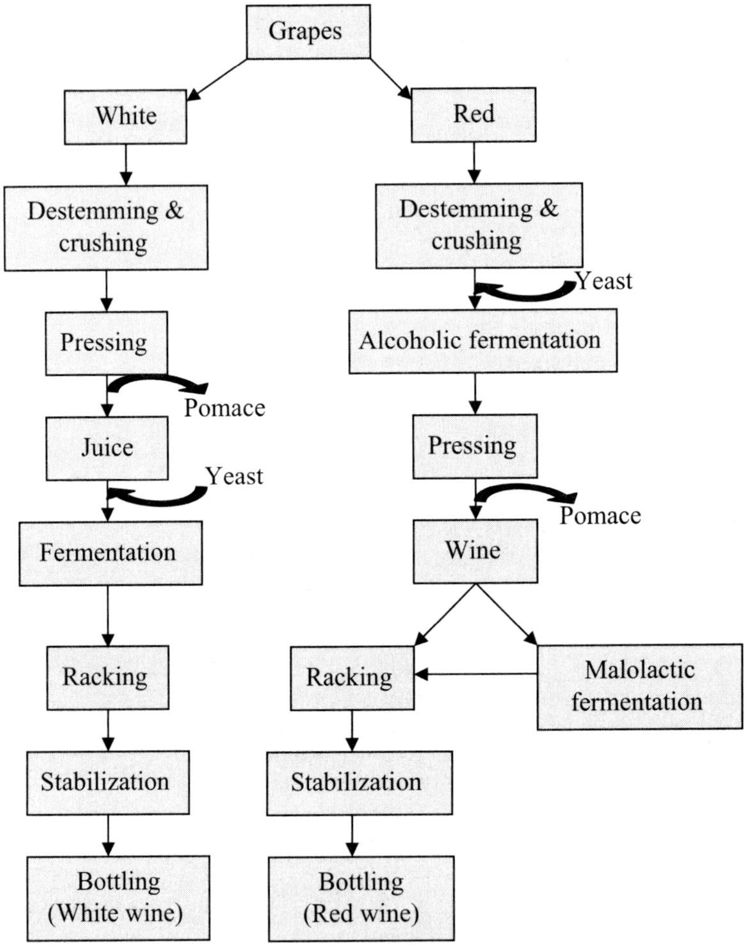


**Fig. 7:** Flowchart for the production of white and red wine from grapes.

in the seed. *Vinifera* grapes are much higher in sugar, lower in acidity and not flavourful as *labrusca* grapes [178]. Mature grapes yield juice with superior sensory quality as well as more desirable colour than unripe grapes. Juice prepared from Concord grapes (*V. labrusca*) is depicted in Fig. 8.

8.1.1.3 Grape pomace: Pressed skins, seeds and stems constitute grape pomace or marc and it represents as much as 20% of the weight of the grapes processed into wine and juice [177, 179]. Grape pomace can be used for producing a variety of products including ethanol, tartrates, citric acid, grape seed oil, grape seed tannins, hydrocolloids and anthocyanins. Fig. 9 represents the production of anthocyanin pigments, tannins and oils.

### *8.1.2 Physiological role of grapes and grape products*

Numerous studies in recent years have shown that flavonoids and phenolics present in grapes and grape products possess anticarcinogenic, anti-inflammatory, antihepatoxic, antibacterial, antiviral, antiallergic, antithrombiotic and antioxidant effects [180, 181]. The components of grapes and grape products

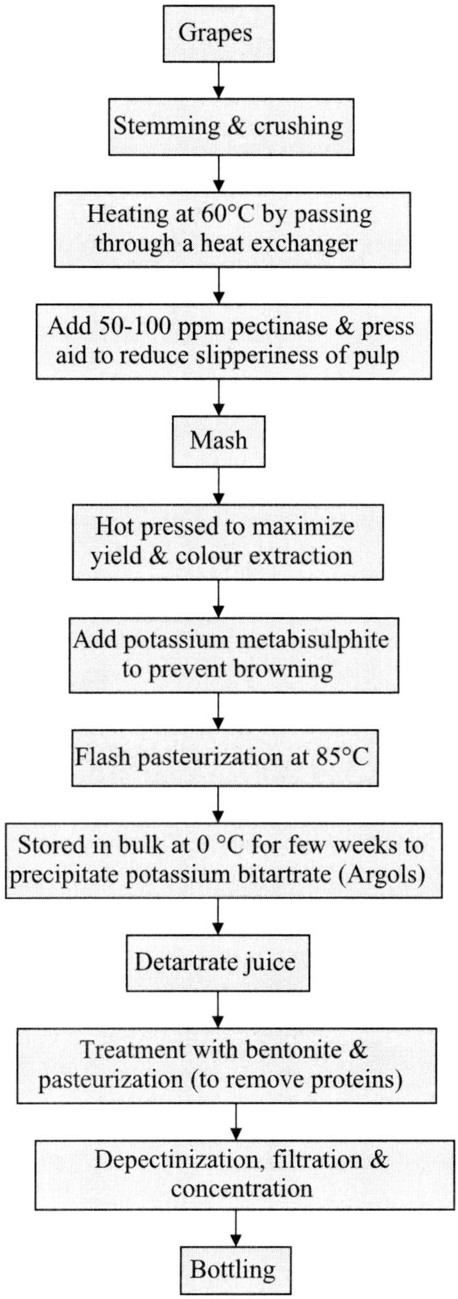


**Fig. 8:** Flow chart for the production of concord juice.

are believed to play a significant role in preventing or delaying the onset of the diseases including cancer and cardiovascular diseases. These components include phenolic compounds [182, 183]. The flavonoids including anthocyanins, flavonols and flavan-3-ols are powerful antioxidants [183-185]. Anthocyanins possess known pharmacological properties and are used for therapeutic purposes.

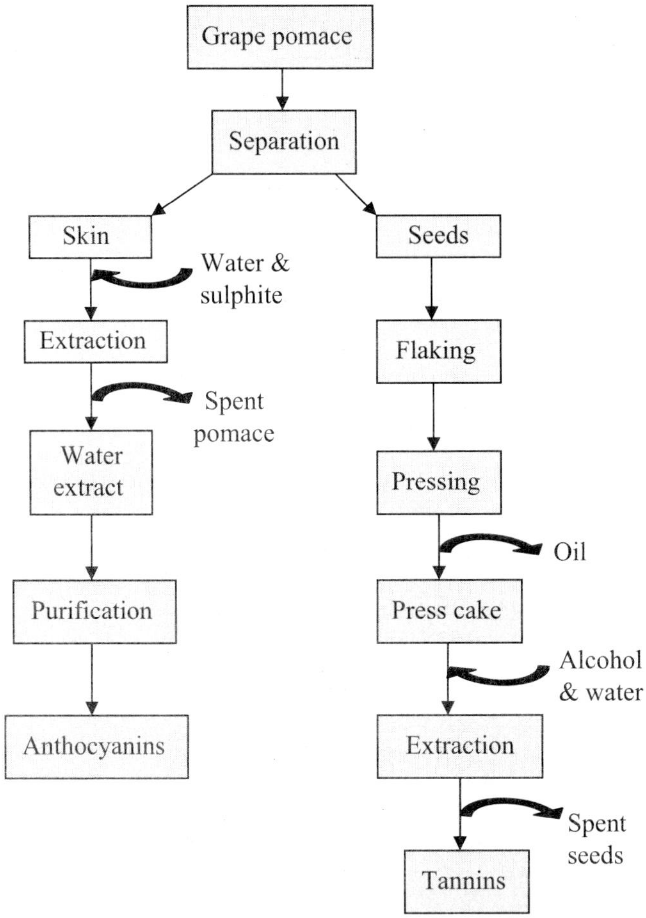


**Fig. 9:** Flow diagram representing the production of anthocyanin, tannins and oils.

Pharmaceutical preparations of *Vaccinium myrtillus* anthocyanins (VMA), applied orally or intravenal or intramuscular injection reduce capillary permeability and fragility [186]. Anti-ulcer activity of anthocyanins has also been reported [187]. It also provides protection against UV radiation [188].

Ellagic acid and resveratrol are the two components found in grapes that reduce the risk of cancer and coronary heart diseases [189]. Ellagic acid ($C_{14}H_6O_8$) is an acid hydrolysis product of ellagitannin found in grape juice and wine [190]. Resveratrol (3,5,4'-trihydroxystilbene) is a stilbene phytoalexin produced in the leaves and fruit of grape in response to UV radiation [191]. Thus, it is evident that grapes and grape products are rich source of phenolic compounds, especially flavonoids, demonstrating a wide range of biochemical and pharmacological effects. In brief, the regular consumption of available grape products has long-term health benefits.

## 8.2 Citrus Fruits

Citrus are the world's largest sub-tropical and tropical fruit crops, and their production is approximately over 105 million tonnes/year. Oranges account for 70% of this total production. Anatomically, citrus fruits i.e., orange (*Citrus sinensis*), tangerine (*Citrus reticulata*), lemon (*Citrus limon*), etc., contain an outer peel or rind including epidermis, flavedo and albedo. Flavedo in the sub-epidermal region contains chromoplasts that impart green, yellow or orange colour to the fruit. It also contains numerous oil glands filled with aromatic essential oils. The mesocarp of 'albedo' consists of an inner white spongy portion of the rind (peel) and is made up of elongated branched cells with large inter-cellular spaces. It is rich in sugars, pectin, cellulose and vitamin C. Below this lays the edible portion, consisting of several closely packed carpel segments arranged in a soft pithy core or axis of the fruit. Transverse section of citrus fruit or hesperidium is shown in Fig. 10.

### *8.2.1 Components of citrus fruits*

Two classes of compounds that have been extensively investigated are citrus flavonoids and limonoids due to their flavour impact on citrus juices besides their pharmacological activity and use as taxonomic markers. Four types of flavonoids occur in citrus fruits: flavanones, flavones, flavonols and anthocyanins. Flavanones form the predominating category among the citrus flavonoids; flavones and flavonols are present in small amounts. Naringin is the main flavanone glycoside found in grapefruit; narirutin and hesperidium in sweet orange; and eriocitrin and hesperidin in lemon [192, 193].

Limonoids are a group of chemically related triterpene derivatives [present in orange, grapefruit, lemon and lime]. Citrus limonoids have biological activities that may be used as chemopreventive agents [194]. Volatile components attribute to the flavour of fruits and fruit products. In citrus fruits, the water-soluble and oil-soluble volatile components present in juice sacs are released during juice extraction. Citrus volatiles contain 95-96% terpene hydrocarbon (mostly D-limonene), 1.6% aldehydes (octanal and decanal), 0.8% alcohols (linalool mostly) and 0.3% esters (octyl and neryl acetates) [195].

Citrus fruits constitute one of the richest sources of the high-quality pectin. Pectin is classified as a dietary fibre due to its resistance to hydrolysis by enzymes of the human small intestine. Extracted pectins are a group of heterogenous polysaccharides with high molecular weight consisting of linear chains of (1→4) linked α-D-anhydrogalacturonic acid units with portions rich in L-rhamnose and with variable side chains consisting of arabinose, galactose and xylose [196]. Oranges, tangerines and their juices contain moderate amounts of folic acid. Folic acid and its derivatives are water-soluble compounds collectively termed as folate or folacin. The principal form of folate in citrus is the reduced

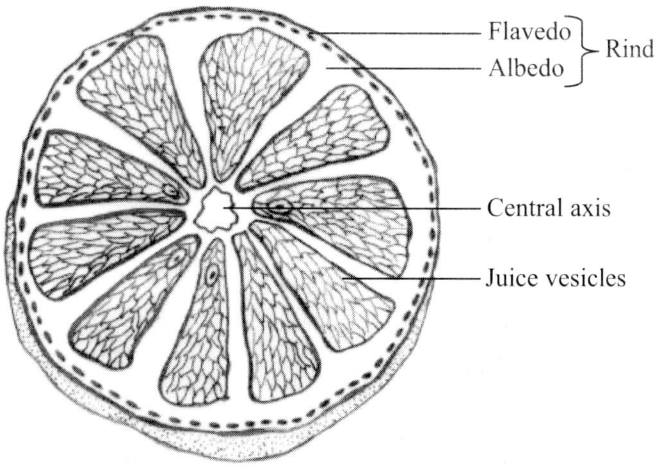


**Fig. 10:** T.S. of citrus fruit (hesperidium).

5-methyl tetrahydrofolate (monoglutamate form) [197]. Folic acid is light and oxygen sensitive and heat labile [198]. Vitamin C is found in greatest amounts in fruits and vegetables, especially citrus fruits and juices. The most active antiscurvy form of vitamin C is L-ascorbic acid (the reduced form). The greatest proportion of vitamin C (over 60%) in citrus is located in the flavedo and albedo [199].

### *8.2.2 Citrus products*

Citrus fruits are widely marketed as fresh commodities. Citrus fruits are produced for processing either as single-strength juice or concentrate along with other products such as essence and oil, pectin, flavonoids and limonoids such as naringin and limonin.

Large quantities of fruits are converted into juice concentrates. Orange juice is defined as an unfermented juice obtained from mature oranges of the species *Citrus sinensis* [200]. Weighing, grading and washing are the front-end operations, which also include material selection, preparation, handling and testing. Pectin is found in peel, membranes, juice sacs and cores. The pectin content in orange juice is distributed between the pulp and serum. Levels of pectin in peeled orange fruit as well as in juice are normally too low. Citrus fruit peel, particularly the albedo portion, serves as the major raw material for pectin manufacture. On dry weight basis, the pectin content in the peels of citrus fruits ranges from 30 to 33%. Fig.11 depicts the process for the production of pectins.

### *8.2.3 Health benefits of citrus components*

The recognition of physiologically active components in citrus fruits and their contribution to human health has been well established. A large number of citrus product constituents are capable of preventing or alleviating diseases and promoting health. Vitamin C, E and carotenoids are thought to play a significant role in preventing or delaying the onset of diseases like cancer, cardiovascular disease, and cataracts by counteracting processes.

Grapefruit juice is known to increase the system bioavailability of certain drugs as felodipine [201]. Selected citrus and other naturally occurring flavonoids possess anticancer activities [202]. Quercetin inhibits the activity of tumour promoters [203]. Limonene has been shown to block and suppress carcinogenic events. Dietary incorporation of pectin appears to affect several metabolic and digestive processes including the effects on lowering glucose absorption and cholesterol levels.

Vitamin A plays a vital role in many essential biological processes. It functions as a chromophore in the visual process, and is also involved in foetal development as well as in the regulation of proliferation and differentiation of many cell types throughout life. Carotenoids are considered to be vitamin A precursors, β-carotene being the most prevalent. β-carotene is approved for the general treatment of erythropoietic protoporphyria, a genetically inherited, light-sensitive skin disease [204]. Carotenoids without provitamin A activity e.g., canthaxanthin possess antimutagenic and anticarcinogenic properties [205]. β-carotene has been found to reduce the incidence of cardiovascular events [206].

The principal biochemical function of vitamin C in humans is to maintain a reducing environment for enzyme systems. Vitamin C has been categorized as a potential cancer-inhibiting agent by preventing the formation of carcinogens from the precursor compounds [207].

## 8.3 Vegetable Products

### *8.3.1 Cruciferous vegetables*

The Cruciferae family consisting of over 2000 species includes several important food crops such as *Brassica oleracea* L. var. *capita* L. (white cabbage), broccoli, cauliflower, turnip, radish, etc. Cruciferae may be used in the preparation of condiments or garnishes, but they do not contribute significant nutrients to the diet. The crucifers that are eaten as vegetables, in comparison, provide many nutrients including vitamins, minerals and fibres. Crucifers have long been recognized as an important

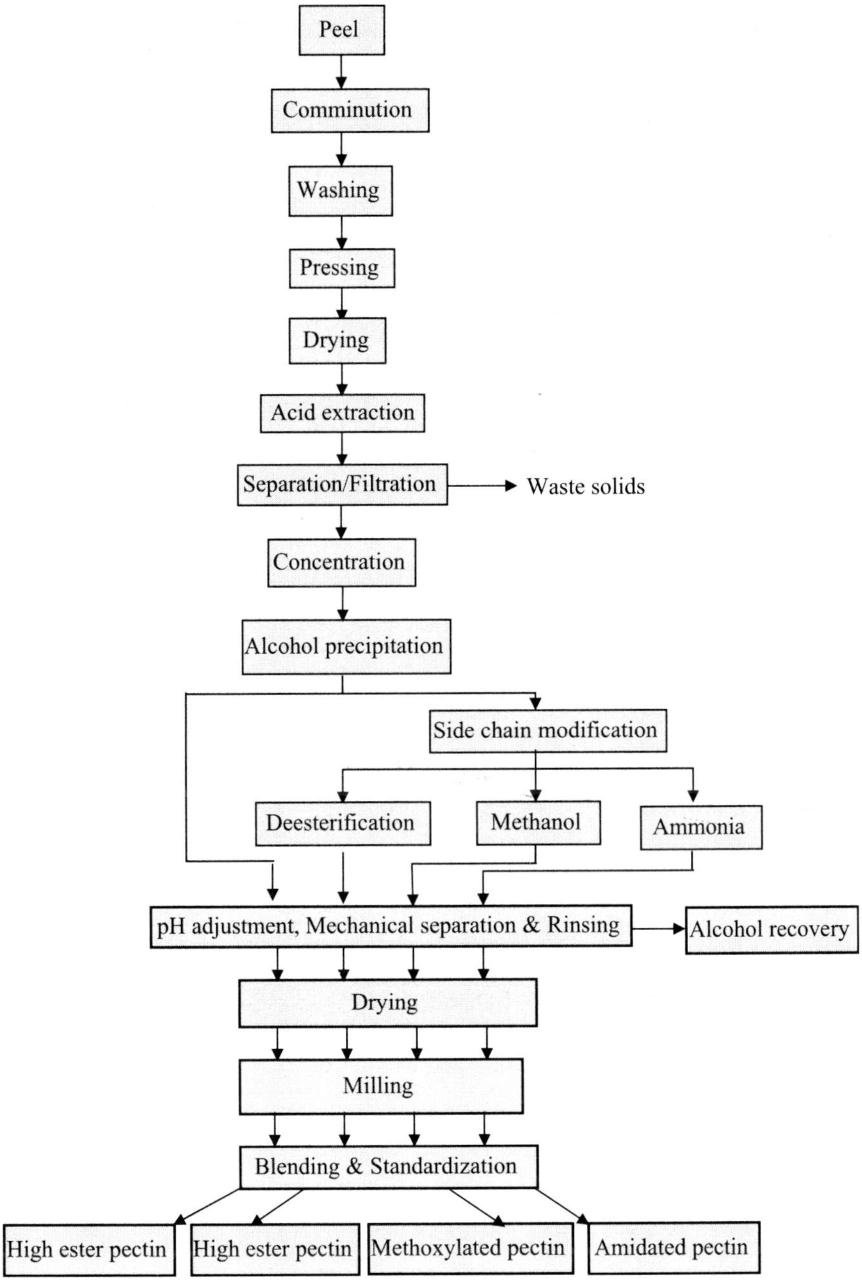


**Fig. 11:** Schematic presentation of pectin production process.

source of vitamin C, other water-soluble vitamins such as riboflavin, niacin and thiamine and vitamin A. Vitamin $K_1$ (phytonadione), the fat-soluble vitamin is also found in Crucifers.

**8.3.1.1 Biologically active compounds of crucifers:** Glucosinolates are vacuole-bound glycosides consisting of a β-thioglucose group, R side chain and oxime moiety and are found in all the species of cruciferae. They are categorized according to the structure of the side chain as aryl, alkyl, aromatic or indole glucosinolates. Though the content of glucosinolate is species specific, the most concentrated source of glucosinolates in diet appears to be brussel sprouts [208]. Glucosinolate content may increase during storage [209]. Atleast one substituted sulphoxide, S-methyl cysteine sulphoxide (SMCSO) occurs in cruciferae [210].

8.3.1.2 Products from crucifers: Characteristic pungent, sulphuraceous odours and flavours are produced during the preparation by cutting or slicing of the fresh crucifers. This is due to complex enzymic and non-enzymic reactions that usually involve physiologically active sulphur-containing compounds [211].

Majority of crucifers are consumed raw as salads, or immediately after cooking, some are dehydrated or canned while others are preserved by freezing. Cabbage is used for the production of a fermented product, sauerkraut, which is occasionally pasteurized for longer shelf life. Fermentation of salted cabbage is initiated by several species of homo- and heterofermentative lactic acid bacteria. Pickled cabbage and cauliflower are also prepared that are heat-processed after being mixed with vinegar brines. Cabbage is also used commercially for the production of coleslaw, a prepared salad made by mixing raw cabbage with a vinegar-based dressing, either mayonnaise or vinaigrette.

8.3.1.3 Health benefits of crucifers: Epidemiological evidence has associated the frequent consumption of cruciferous vegetables with decreased cancer risk. An inverse relationship between the consumption of total *Brassica* vegetables and cancer risk has been demonstrated [212]. Several isothiocyanates have been shown to inhibit or block tumours by chemical carcinogens. A potent isothiocyanate, 1-isothiocyanate-4(methylsulfinyl) butane (sulforaphane), an inhibitor of mammary tumours, was isolated from broccoli [213]. Vitamin C and A have been found to protect against cancer in laboratory studies, though their mode of action is not clear [214]. Vitamin C is also required for the development of cartilage, bone and teeth, wound healing, haemoglobin and erythrocytes formation, while vitamin A plays a vital role in vision along with the synthesis of glycoproteins. Vitamin $K_1$ is necessary for the synthesis of blood-clotting factors in the liver.

### *8.3.2 Onion and garlic*

The onion family i.e., genus *Allium* consists of more than 250 members. Traditionally, two members i.e., onions (*A. cepa*) and garlic (*A. sativum*) have been used as medicine for over 4,000 years [215].

8.3.2.1 Biologically active compounds of Alliums: Sulphur compounds are not as such found in the intact cells, rather they are formed as a result of reaction between the enzyme allinase and volatile precursors i.e., S-alk(en)yl cysteine sulphoxide and sulphonic acid. This takes place when the cells are ruptured, resulting in the formation of thiosulphinates and sulphonic acid derived compounds [215, 216]. Bulbs and leaves of alliums possess flavonoids. The flavonoids found in onion include eight quercetin glucosides, the 4'-glucoside, the 7,4'-diglucoside, the 3,4'-diglucoside, the 3-glucoside, the 7-glucoside, the 3,7-diglucoside, the 3-rutinoside (rutin), the 3-rhanmoside (quercetrin), the 7,4'-, and 3-glucosides of kaempferol, isorhamnetin 4'-glucoside and eight anthocyanins [217]. Onion is also a potential source of prostaglandins [218]. Various alliums have sterols and steroidal glycosides. In garlic, 0.021% and in onion, 0.095%, saponin level has been reported [219]. Fructans in high concentration are found in onions bulbs (35-40% dry weight), constituting a major portion of the water-soluble carbohydrates [220]. Fructans are fructosyl polymers that consist of D-fructose molecules joined by $\beta$ (2$\rightarrow$1) linkages, and this chain is terminated by a D-glucose molecule linked to fructose by an $\alpha$ (1$\rightarrow$2) bond.

8.3.2.2 Products from Alliums: Currently, a variety of products from alliums are marketed. Onion products include dehydrated onion pieces, onion powder, oleoresins, onion salts, pickled onions, canned, frozen and packaged onions, etc. Dehydrated garlic powder, garlic juice, garlic salt, garlic flavouring are some of the advantageous garlic products being marketed. Essential oil of garlic is extracted by distillation in boiling water in a still or by introducing live steam generated in a separate steam boiler. Due to the therapeutic properties of alliums, onion and garlic capsules, tablets have also been made.

8.3.2.3 Health benefits of Alliums: Alliums are used as medicine for many major and minor disorders. Alliums have anticarcinogenic activity, but garlic is known to be more effective [221] due to the interaction of the sulphur components of garlic with the tumour cell metabolism [222, 223]. Garlic and onion suppress/inhibit the platelet aggregation i.e., induced by collagen, arachidonic acid, ADP, adrenaline and the thrombin and the effect is dose dependent [224]. Alliums enhance the blood's fibrinolytic activity, suppress blood fibrinogen concentration and increase the blood coagulation time [225]. Garlic and onion also play a role in lowering serum cholesterol levels both in animals and humans. Garlic and onion bulbs can also reduce the level of blood sugar, thus they are beneficial to diabetes [226], besides reduction of hypertension [227] as well as stimulation of the body's immune system [215]. Flavonoids present in the bulbs as well as the leaves of alliums are potent antioxidants with a wide array of biochemical functions. They are involved in immune function, gene expression, liver function, enzyme activity, platelet aggregation, histamine metabolism, etc. [217]. Oligofructans are helpful in proliferation of bifidobacteria and in reduction of detrimental bacteria in the colon, reduction of toxic metabolites, prevention of constipation, protection of liver function, reduction of serum cholesterol, reduction of blood pressure and various anticancer effects [228].

## 9. FUNCTIONAL MILK AND DAIRY PRODUCTS

There is no doubt that dairy products are functional foods. They are known to be one of the best sources of calcium, an essential nutrient which can prevent osteoporosis and possibly colon cancer. Dairy products, particularly fermented dairy products are well known as probiotics. Probiotics are defined as "live microbial feed supplements which beneficially affect the host animal by improving its intestinal microbial balance [229]." Besides these, dairy products like cheese, fluid milk, yogurt, etc. have been recognized as an excellent source of several important vitamins and minerals [230].

### 9.1 Techniques for Physiological Functionality

Traditional milk and dairy products contain large amounts of components or precursors of agents of biological significance. The traditional processing techniques though have served the needs of manufacturing dairy foods with optimal sensory quality failed to meet their optimal physiological functionality, but with the present interest in functional foods, new processing techniques based on membrane technology are used to produce modified products. Currently, industry uses ultrafiltration (UF) for the production of quark and other cheeses in Germany and many other countries [231].

The process of microfiltration (MF) can be advantageously used either to augment the traditional separation processes, such as centrifugation for removal of fat or pathogenic or spoilage bacteria or to replace the heating processes as pasteurization [232, 233]. Development of functional dairy products as bioactive peptides utilizes membrane separation processes and is based on enzymatic or microbial modifications of traditional dairy components. The downstream processing of functional food ingredients includes water removal by concentration followed by drying. Processes such as reverse osmosis, nanofiltration, freeze-concentration or freeze-drying are now used in the production of dairy products.

Genetic engineering may be employed to produce milk with higher whey protein content, or with higher content of only the most desirable whey protein components. Technologically, the whey protein content of dairy products by ultrafiltration of milk or the addition of an isolated whey protein fraction can be increased [234]. Cultured dairy products such as buttermilk, kefir, yogurt, etc., have been in focus due to their physiological functionality. The health-promoting effects of cultured dairy products are related to the biological activity of the bacteria employed in the production of such products during the fermentation processes.

## 9.2 Fermented Dairy Products, Probiotics and Prebiotics

In Europe, Asia and Africa, traditional dairy products have been consumed for many generations. Development of commercial products as acidophilus milk (*Lactobacillus*), yoghurt (*Lactobacillus* and *Streptococcus*), etc., is a result of identification of the microorganisms responsible for fermentation and ability to prepare pure starter cultures.

The bacterial genera most often used as probiotics are Lactobacilli and Bifidobacteria. At present, probiotics are almost extensively consumed as fermented dairy products such as yoghurt or freeze-dried cultures [235, 236]. Lactic acid bacteria constitute a diverse group of organisms providing considerable benefits to mankind, some as natural inhabitants of the intestinal tract and others as fermentative lactic acid bacteria used in food industry, imparting flavour, texture and processing preservative properties. Some selected strains of *Lactobacillus, Bifidobacterium, Streptococcus, Lactococcus* and *Saccharomyces* have been promoted in food products because of their reputed health benefits [237, 238]. For safety, the probiotic microorganisms should not be pathogenic, have no connection with diarrhoegenic bacteria and no ability to transfer antibiotic resistance gene, as well as should be able to maintain genetic stability. To be recognized as functional food components, they should have following properties: acid- and bile-stability, resistance to digestive enzymes, adhesion to intestine surface, anti-carcinogenic and anti-mutagenic activity, cholesterol-lowering effects, enhancement of bowel motility and production of vitamins and enzymes [237].

Prebiotics are defined as non-digestible or low digestible food ingredients that benefit the host organism by selectively stimulating the growth or activity of one or a limited number of probiotic bacteria in colon [239]. This role is played by fermentable carbohydrates, which are not digested or poorly digested in the small intestine and stimulate the growth of bifidobacteria and some gram-positive bacteria. Lactulose, galactooligosaccharides, fructooligosaccharides, inulin and its hydrolysates, maltooligosaccharides, and resistant starch are prebiotics commonly used in human nutrition.

## 9.3 Physiologically Functional Food and Bioactive Compounds

Casein forms about 80% of the total milk protein fraction. It is known as a source of amino acids needed for the growth of the neonate. The most important casein-related bioactive peptides are casomorphins with opioid-like activity, immunopeptides with immunostimulating activity, peptides with antihypertensive activity and phosphopeptides having the ability to sequester calcium and possibly other minerals, thus acting as biocarriers [240].

Whey proteins form approximately 20% of the total milk protein fraction. Whey proteins in comparison to casein are sensitive to acid coagulation as well as to the action of chymosin. The whey protein is a heterogenous group of several dissimilar proteins, many of which are having physiological functionality. Use of whey proteins as specific ingredients in dairy or non-dairy functional foods has been increasing with improvements in the technological capabilities of the industry to produce commercially attractive whey protein concentrates (WPC) or whey protein fractions. The high nutritional quality of whey proteins has been recognized for a long time [241]. The two major whey proteins, α-lactalbumin and β-lactoglobulin, are being produced commercially as isolated protein fractions of relatively high purity using a variety of patented processes [242]. Calcium as a nutritionally important bioavailable milk component [230] is involved in bone development and blood pressure regulation [243], besides preventive effects in carcinogenesis, etc. [244]. Lipid-based bioactive compounds of milk include mainly fatty acids [245]. The most abundant milk solid component, lactose, has been considered as an enhancer of calcium absorption [246]. Lactose is an important starting material for the production of lactulose, and this is its novel nutraceutical application. This is added to infant formulas and other dairy products produced in Japan [247]. It is also used as a factor promoting probiotic growth. Lactose-free or lactose-reduced milk products have been available in the markets in many countries for some time, and may be regarded as examples of modern dairy products with improved physiological functionality.

### 9.4 Health Benefits

A major avenue for increasing the beneficial physiological effects of dairy products is likely to be "probiotic lines" of various traditional dairy foods. The hypocholesterolemic effect of fermented milk was discovered more than 30 years ago during studies conducted in Maasai Tribesmen in Africa [248]. Evidences support the role of probiotics in cancer risk reduction, particularly colon cancer [249]. Relatively less attention has been focused on the consumption of fermented milk products and breast cancer risk, although an inverse relationship has been observed in some studies [250]. It was found that whey proteins have a putative anti-cancer activity [251]. These proteins are effective in retardation of colon cancer in young rats, as compared with other dietary proteins i.e., meat and soy [252].

## 10. FUNCTIONAL SEAFOOD LIPIDS AND PROTEINS

### 10.1 Seafoods

Seafood plays an important role in human diet, representing the largest stock available of ω-3 polyunsaturated fatty acids, especially eicosapentaenoic (EPA) and docosahexenaenoic (DHA). Omega-3 (n-3) fatty acids are an essential class of polyunsaturated fatty acids (PUFA) derived primarily from fish oil. These foods are the greatest source of several fatty acids, especially the ω-3 series ones [253]. Fish oil has been used in food products such as margarine, for long time [254] because of its considerable biochemical, metabolic, nutritional and pharmaceutical importance [253]. A large number of benefits have been attributed to fish oil, especially its EPA and DHA fatty acids [255].

### 10.2 ω-6 and ω-3 Polyunsaturated Fatty Acids

Polyunsaturated fatty acids are produced from monosaturated fatty acids by the action of specific desaturases to the double-bond position in the chain. The monosaturated fatty acids are formed from saturated ones. This takes place in liver through a reaction catalyzed by microsomal desaturases. Linoleic acid (C18:2 ω-6), the precursor of arachidonic acid (C20:4 ω-6) is representative of the ω-6 family. The ω-3 family is represented by the α-linolenic acid (C18:3 ω-3). It forms the eicosapentaenoic (EPA, 20C:5 ω-3) and docosahexenoic acid (DHA, 22C:6 ω-3) via prolongation and desaturation [256]. Marine lipids possess high content of omega-3 PUFA, remain liquid, and oppose any tendency to crystallize [257].

### 10.3 Seafood Proteins and Marine Oils

Seafood proteins are a good source of amino acids and are digestible. Additionally, non-protein nitrogen (NPN) compounds are also available, depending upon the species. The dark muscles of fish generally contain NPN in higher amount. Pelagic fish in comparison to demersal fish have higher content of sarcoplasmic proteins. In some species, dark muscles contain less sarcoplasmic proteins than their white muscles. The mycofibrillar proteins of muscles are composed of myosin, actin, tropomyosin and troponins C, I and T [258].

Processing for the production of marine oils depends on the source of the raw material used, and the processing of marine oils is similar to that of vegetable oils [259]. PUFA (Polyunsaturated fatty acids) rich oils are generally prone to oxidation producing off-flavours and off-odours [260]. Use of methods like addition of antioxidants, microencapsulation and hydrogenation can be used for stabilizing marine oils.

### 10.4 Health Benefits

Epidemiological studies have shown that a regular consumption of fish has favourable effects on the triacylglycerol levels, blood pressure, coagulation mechanism and heart rhythm [261]. Physiological effects of omega-3 fatty acids are in the areas of heart and circulatory, immune response and cancer. These are useful in prevention or treatment of atherosclerosis [262], thrombosis [263] and high blood pressure [264]. Immune responses include treatment of asthma, arthritis [265], migraine headache and nephritis [263]. Cancers area involves cancer of the breast [266], prostrate and colon [264]. Marine oils may retard artherogenesis through their effects on platelet function, platelet-endothelial interactions and inflammatory response [267].

## 11. REGULATORY ASPECTS OF FUNCTIONAL PRODUCTS

Functional foods lie at the interface between food and drugs. Earlier food products were developed for taste, appearance and convenience for the consumer. But in past few years, the development of products focused on their health benefits. Most researches in the "functional" and ascribed specific heath benefits should receive scientific scrutiny before health claims are permitted. Generally, regarding the regulations to functional foods and nutraceuticals, the issue is the ability of developed products to deliver with health benefits.

### 11.1 Foods and Drugs

Food and Drugs act and Regulations in Canada or the Federal Food, Drug, and Cosmetic Act in the United States are for controlling both food and drugs. In other countries as Sweden has a separate law on Food Act and an Act on Medicinal Products [268]. With the growing interest in foods and components of foods, natural products, herbal products and dietary supplements for promoting health and reducing the risk of disease, the definitions of food and drug are being questioned. According to the Dietary Supplement Health and Education Act (DSHEA) in the United States in 1994, the dietary supplement term is defined as [269, 270]:

- containing one or more nutrients, herbs, botanicals, or a concentrate, metabolite or constituent extract from the ingredients i.e. vitamins and minerals;
- in the form of a supplement (meaning tablet, capsule or powder);
- is not represented as a food or sole item of a meal or the diet; and
- includes a similar new drug or biologic approved under previous legislation and not currently being investigated.

## 11.2 Regulations of Functional Foods

### *11.2.1 European countries*

According to European EC Directive 65/65/Eec, "any substance or combination of substances presented for treating or preventing disease or which may be administered with a view to making a medical diagnosis or to restoring, correcting or modifying physiological function" as a medical product requiring a license or marketing authorization. A few provisions are made for dietary supplements by current EU regulations. These regulations are more concerned with the protection of consumers from the unsafe products. Dietary supplements that seek acceptance under current dietetic regulations must provide evidence to inspecting authorities that supports the statements claiming the purpose, which it claims [271].

### *11.2.2 Japan*

Japan has developed a process for approval of functional foods to enable health claims to be made. In Japan, where consumers demand an increasing emphasis on safety and health, functional foods with proven clinical efficacy (officially now termed as Foods for Specific Health Use (FOSHU)) are distinguished by their beneficial physiological effects. Such foods are designed to be consumed as a constituent part of a regular diet taken daily to promote and maintain health by regulating bodily functions and to protect against a wide range of diseases including cancer, heart diseases, diabetes, osteoporosis and hypertension.

FOSHU regulations are the most advanced in the world in a number of ways. These are primarily designed to allow established and accepted ingredients to be used in food rather than to encourage the development of new ingredients. The requirements for achieving a FOSHU license are as follows [271]:

- the product must contribute to the improvement of dietary habits and enhance health
- the health benefits of the food or ingredient should have a clear medical basis
- the food/ingredient must have definable levels of appropriate consumption based on medical knowledge
- the food/ingredient should be safe as judged from experience
- relevant information should be defined in terms of physiochemical properties
- nutritional composition of the product should not differ greatly from that of ordinary foods
- the product must be consumed regularly rather that occasionally
- the product must be in the form of ordinary food rather than pills or capsules

### *11.2.3 The United States*

An important dilemma persisting with respect to the regulation of functional foods is that they exist at the interface between foods and drugs [272]. There is no provision for foods consumed with the intention of preventing disease in the existing US food regulations. The Dietary Supplement Health and Education Act [269] allows many functional foods to be considered as dietary supplements which are exempted from regulation as drugs. The Nutritional Labelling Education Act [273] permits health and disease prevention claims on a food label.

When dietary supplement manufacturers plan to market a new ingredient, information must be submitted 75 days in advance to FDA. FDA concludes whether the new ingredient can be considered to be safe. This information should be in the public domain 90 days after receipt by the FDA [274]. Having complied with DSHEA requirements, once a dietary supplement is marketed, the FDA has the responsibility for showing that a dietary supplement is unsafe before it can take action to restrict the product's use.

In the United States, medical foods are defined as "Food formulated to be consumed or administered under the supervision of a physician and which are intended for the specific dietary management or a disease or condition for which distinctive nutritional requirements, based on recognized scientific principle, are established by medical evaluation." In the United States, health claims for foods permitted under the Nutritional Labelling and Education Act are generic, not product-specific. Generic health claims are those that relate a nutrient, in the context of a total diet, to a particular disease or condition. At present, functional foods are regulated under these guidelines.

### *11.2.4 Canada*

Foods in Canada are regulated under the Food and Drugs Act and Regulations, but there is no separate category in the Food and Drug Act in Canada for functional foods or nutraceuticals [275]. Due to growing interest in functional foods and nutraceuticals and limitations of the Food and Drugs Act, the Health Protection Branch has taken initiatives to address concerns of food and pharmaceutical companies, health professionals and consumers. In Canada, no health claims can be made for foods, unless a submission is made through the 'Therapeutic Products Programme' for the food to be regulated as a drug [276].

## 12. FUTURE TRENDS

Formulated to be significant sources of specific nutrients or other components of nutritional significance, functional foods have the potential to make a major impact on the food supply and public health. The potential success of functional foods in developed countries is often attributed to two trends. Increasingly affluent and ageing populations have become more concerned with protecting their health through diet. Also, longer working hours and interest in fulfilling leisure activities have left younger individuals 'time-challenged,' difficulty in purchasing and preparing foods to meet dietary recommendations [277]. On a short-term basis, functional foods do have a future. There is a plenty of demand. In 1999, United States consumers spent about $ 16 billion on functional foods [278].

The major difficulty encountered in assessing the potential of functional foods is the lack of a universally accepted understanding or legal interpretation of the concept [279-282]. As consumers have no prior experience of many novel functional foods, marketing is required to build and maintain favourable attitudes to achieve profit margins for the commercial sector [283]. It is hard to predict the pre-requisites for success of a functional food, but the factors undoubtedly include consumer need and awareness, consumer acceptance of a food solution, powerful communication of health benefits to the consumer, and uncompromised taste, optimal convenience, proven safety, acceptable price level, a clear regulatory framework for making claims, providing a level of playing ground for all companies. The marketed functional foods are often based on general discoveries in nutritional science, and less on a deliberate research strategy to develop functional foods [284].

## REFERENCES

1. Sloan AE (2000) The top 10 functional food trends. *Food Technol* 54: 33-62.
2. Kent A & Mellentin J (1999) Eggs aim to crack healthy eating market. *New Nutr Bus* 4: 25-30.
3. Goldberg I (1995) *Functional Foods, Designer Food, Pharmafoods, Nutraceuticals,* Chapman & Hall, London, UK.
4. Dipock AT, Aggett PJ, Ashwell M, Bornet F, Fern EB & Roberfroid MB (1999) Scientific concepts of functional foods in Europe: Consensus document. *Brit J Nutr 81* (Suppl.1.): S1-S27.

5. Clydesdale FM (1999) ILSI North America food component reports. *Crit Rev Food Sci Nutr* 39: 203-316.
6. Thomson C, Block AS & Hasler CN (1999) Position of the American dietetic association: Functional foods. *J Am Diet Assoc* 99: 1278-1285.
7. Committee on Opportunities in the Nutrition and Food Sciences FaNBIoM (1994) *Opportunities in the Nutrition and Food Sciences: Research Challenges for the Next Generation of Investigators*, National Academy Press, Washington DC, USA.
8. Nestle M (2002) *Food Politics: How the Food Industry Influences Nutrition and Health*, University of California Press Ltd, London, UK.
9. Meyer A (1998) The 1998 top R & D survey. *Food Process* 58: 32-40.
10. Craig WJ (1997) Phytochemicals: Guardians of our health. *J Am Diet Assoc* 97: S199-S204.
11. deLorgeril M, Salen P, Martin JL, Monjaud I, Boucher P & Mamelle N (1998) Mediterranean dietary pattern in a randomized trial: prolonged survival and possible reduced cancer rate. *Arch Intern Med* 158: 1181-1187.
12. Milner JA (1999) Functional foods and health promotion. *J Nutr* 129 (suppl 7): 1395S-1397S.
13. Witwer RS (1999) Marketing bioactive ingredients in food products. *Food Technol* 50: 50-53.
14. Jelen P & Lutz S (1998) Functional milk and dairy products. In: *Functional Foods: Biochemical & Processing Aspects*, Mazza G (ed), Technomic Publishing Co, Pennsylvania, USA, pp 365-367.
15. Bldlack WR & Wang W (1998) Designing functional foods. In: *Modern Nutrition in Health and Disease*, Shils ME, Olson JA, Shike M & Ross AC (eds), Lipincott Williams & Wilkins, Philadelphia, USA, pp 1823-1833.
16. Milner JA (1998) Do "functional foods" offer opportunities to optimize nutrition and health? *Food Technol* 52: 24-29.
17. Hasler CM (1998) Functional foods: their role in disease prevention and health promotion. *Food Technol* 52: 63-70.
18. Hollingsworth P (1999) Food priorities for an aging America. *Food Technol* 53: 38-40.
19. Katan MB (2004) Health claims for functional foods. S*tudents BMJ* 12: 91-92.
20. Katan MB & Roos NM (2004) Promises and problems of functional foods. *Crit Rev Food Sci Nutr* 44: 369-377.
21. Weststrate JA, van Poppel G & Verschuren PM (2002) Functional foods, trends and future. *Brit J Nutr* 88 (suppl. 2): S233-S235.
22. McBean LD (1999) Functional foods: an overview. *Dairy Council Digest* 70: 31-36.
23. Goldberg I (1994) *Functional Foods*, Chapman and Hall, New York, USA.
24. Knorr D (1998) Technology aspects related to microorganisms in functional food. *Trends Food Sci Technol* 9: 295-306.
25. Sanders ME (1998) Development of consumer probiotics for the US market. *Brit J Nutr* 80: 213-218.
26. Geiger CJ & Ink SL (1996) The future of oat and barley as functional foods. In: *Proc V International Oat Conference & VII International Barley Genetics Symposium*, Scoles G & Rossnagel B (eds), Canada, pp 45-49.
27. Wood P (1997) Functional foods for health. In: *Cereals-Novel Uses and Processes*, Campbell GM (ed), Plenum Press, New York, USA, pp 233-239.

28. Sungsoo CS & Prosky L (1999) Application of complex carbohydrates to food product fat mimetics. In: *Complex Carbohydrates in Foods*, Sungsoo CS (ed), Marcel Dekker Inc, New York, USA, pp 411-429.
29. Frolich W (1996) Novel food—is there a market for barley and oat? In: *Proc V International Oat Conference & VII International Barley Genetics Symposium,* Scoles G & Rossnagel B (eds), Canada, pp 65-71.
30. Onming G, Akesson B, Oste R & Lundquist I (1998) Effects of consumption of oat milk, soya milk, or cow's milk on plasma lipids and antioxidative capacity in healthy subjects. *Ann Nutr Metab* 42: 211-220.
31. Pszczola DE (1996) Oatrim finds application in fat-free cholesterol-free milk. *Food Technol* 50: 80-81.
32. Fernandez-Garcia E, McGregor JU & Traylor S (1998) The addition of oat fiber and natural alternative sweetners in the manufacture of plain yogurt. *J Dairy Sci* 81: 655-663.
33. Bukowska H, Pieczul-Mroz J, Jastrezebsk K, Chelstowski K & Naruszewicz M (1998) Significant decrease in fibrinogen and LDL-cholesterol levels upon supplementation of the diet with *Lactobacillus plantarum* (Pro Viva) in subjects with moderately elevated cholesterol concentrations. *Atherosclerosis* 137: 437-438.
34. Documentation of *Lactobacillus plantarum* 299V, http://www.probi.se.
35. DHHS/FDA (1997) Food labeling: Health claims: Oats and coronary heart disease. Department of Health and Human Services/Food and Drug Administration. *Fed Reg* 62: 3584-3601.
36. Hasler CM (1998) A new look at an ancient concept. *Chem Industry* 2: 84-89.
37. Welch RW (1995) Oats in human nutrition and health. In: *The Oat Crop*, Welch RW (ed), Chapman and Hall, London, UK, pp 433-479.
38. Wood PJ, Fulcher RG & Stone BA (1983) A studies on the specificity of interaction of cereal cell wall components with congo red and colcoflour. Specific detection and histochemistry of (1→3) (1→4)-β-D-glucan. *J Cereal Sci* 1: 95-110.
39. Fulcher RG (1986) Morphological and chemical organization of the oat kernel. In: *Oats-Chemistry and Technology*, Webster FH (ed), American Association of Cereal Chemists, St Paul MN, USA, pp 47-74.
40. Miller SM, Fulcher RG, Sen A & Arnason JT (1995) Oat endosperm cell walls: I. Isolation, composition and comparison with other tissues. *Cereal Chem* 72: 421-427.
41. Oomah BD & Mazza G (1999) Health benefits of phytochemicals from selected Canadian crops. *Trends Food Sci Technol* 10: 193-198.
42. Wood PJ & Beer MU (1998) Functional oat products. In: *Functional Foods, Biochemical and Processing Aspects*, Mazza G (ed), Technomic Publishing Co Inc, Lancaster, USA, pp 1-37.
43. On File (1997) Alpha-beta Technology Inc, Worcester, Mass, UK, Personal Communication.
44. Inks S & Mathews R (1997) Oatmeal and oat-bran: heart healthy benefits and more. In: *New Technologies for Healthy Foods and Nutraceuticals*, Yalpani M (ed), ATL Press, Shrewsbury, USA.
45. Fulcher RG & Miller SS (1993) Structure of oat bran and distribution of dietary fibre components. In: *Oat Bran*, Wood PJ (ed), American Association of Cereal Chemists, St Paul MN, USA, pp 1-24.
46. Marlett JA (1993) Comparison of dietary fibre and selected nutrient compositions of oat and other grain fractions. In: *Oat Bran*, Wood PJ (ed), American Association of Cereal Chemists, St Paul MN, USA, pp 49-82.

47. Bhatty RS (1993) Extraction and enrichment of (1→3), (1→4)-β-D-glucan from barley and oat brans. *Cereal Chem* 70: 73-77.
48. Wood P, Weisz J, Fedec P & Burrows VD (1989) Large-scale preparation and properties of oat fractions enriched in (1→3) (1→4)-β-D-glucan. *Cereal Chem* 66: 97-103.
49. McCleary BV (1988) Solubility properties of barley beta-glucan. In: *Alternative End Uses of Barley*, Sparrow DHB, Lance RCM & Henry RJ (eds), Waite Agricultural Research Institute: Glen Osmond, Australia, pp 117-127.
50. Carr JM, Glatter S, Jeraci JL & Lewis BA (1990) Enzymatic determination of β-glucan in cereal-based food products. *Cereal Chem* 67: 226-229.
51. Inglett GE (1991) Method for making a soluble dietary fibre composition from oats. US Patent 4,996,063.
52. Katz DL (2001) http:// www.quakeroatmeal.com/HealthPros/LIB/RA_Articles.cfm.
53. Anderson JW, Story L, Seiling B, Chen W-JL, Petero MS & Story J (1984) Hypocholesterolemic effects of oat bran or bean intake for hyper cholesterolemic men. *Am J Clin Nutr* 40: 1146-1155.
54. Behall KM, Scholfield DJ & Hallfrisch J (1997) Effect of beta-glucan level in oat fiber extracts on blood lipids in men and women. *J Am Coll Nutr* 16: 46-51.
55. Ripsin CM, Keenan JM, Jacobs DR *et al.* (1992) Oat products and lipid lowering: A meta-analysis. *JAMA* 267: 3317-3325.
56. U.S. Food & Drug Administration (1997) Food labeling: health claims, oats and coronary heart disease. Final rule. *Federal Register* 62: 3583-3601.
57. Braaten JT, Wood PJ, Scott FW, Riedel KD, Poste Lm & Collins MW (1991) Oat gum lowers glucose and insulin after an oral glucose load. *Am J Clin Nutr* 53: 1425-1430.
58. Hallfreisch J, Scholfield DJ & Behall KM (1995) Diets containing soluble oat extracts improve glucose and insulin responses of moderately hypercholesterolemic men and women. *Am J Clin Nutr* 61: 379-384.
59. Wood PJ, Braaten JT, Scott FW, Riedel KD, Wolynetz MS & Collins MW (1994) Effect of dose and modification of viscous properties of oat gum on plasma glucose and insulin following an oral glucose load. *Brit J Nutr* 72: 731-734.
60. Keenan JM, Pins JJ, Frazel C, Moran A & Turnquist L (2002) Oat ingestion reduces systolic and diastolic blood pressure in patients with mild or borderline hypertension: a piloy trial. *J Fam Pract* 51: 369-374.
61. He J, Klag MJ, Whelton Pk, Mo J-P, Chen J-Y, Qian M-C, Mo P-S & He G-Q (1995) Oats and buckwheat intakes and cardiovascular disease risks in an ethnic minority of China. *Am J Clin Nutr* 61: 366-372.
62. Crago M, West S, Hoeprich K, Michaelis K & McKenzie J (1998) Effects of hyperlipidemia on blood pressure and coronary blood flow in swine. *FASEB J* 12: A238-A241.
63. Roberts SB (2000) High-glyccmic index foods: is there a connection. *Nutr Rev* 58: 63-70.
64. Mossner J, Grumann M, Zeeh J & Fischbach W (1992) Influence of various nutrients and their mode of application on plasma cholescystokinin (CCK) bioactivity. *Clin Investigator* 70: 125-129.
65. Holt SHA, Brand-Miller JC, Petocz P & Farmakalidis E (1995) A satiety index of common foods. *Eur J Clin Nutr* 49: 675-690.
66. Krotkiewski M (1985) Use of fibres in different weight reduction programs. In: *Dietary Fibre and Obesity-Current Topics in Nutrition and Disease*, Bjoerntorp P, Vahouny GV & Kritchevsky D (eds), Allan R Liss, New York, USA, pp 85-109.
67. Rytter E, Erlanson-Albertsson C, Lindahl, L, Lundquist I, Viberg U, Akesson B & Oste R (1996) Changes in plasma insulin, enterostatin, and lipoprotein levels during an

energy-restricted dietary regimen including a new oat-based liquid food. *Ann Nutr Metab* 40: 212-220.

68. López-Varela S, González-Gross M & Macros A (2002) Functional foods and the immune system: A review. *Eur J Clin Nutr* 56: 29-33.
69. Charalampopoulos D, Wang R, Pandiella SS & Webb C (2002) Application of cereals and cereal components in functional foods: A review. *Int J Food Microbiol* 79: 131-141.
70. Evers AD & Bechtel DB (1988) Microscopic structure of the wheat grain. In: *Wheat Chemistry and Technology*, Vol 1, Pomeranz Y (ed), American Association of Cereal Chemists Inc, St. Paul MN, USA, pp 1-68.
71. Cummings JH, Hill MJ, Jenkin DJA, Pearson JR & Wiggins HS (1976) Changes in fecal composition and colonic function due to cereal fiber. *Am J Clin Nutr* 29: 1468-1473.
72. Stephens AM & Cummings JH (1980) Mechanism of action of dietary fibre in the human colon. *Nature* 284: 283-284.
73. Cummings JH (1993) The effect of dietary fiber on fecal weight and composition. In: *Handbook of Dietary Fiber in Human Nutrition*, Spiller GA (ed), CRC Press Inc, Boca Raton, USA, pp 263-349.
74. Brodribb AJM & Groves C (1978) Effect of bran particle size on stool weight. *Gut* 19: 60-63.
75. Heller SN, Hckler LR, Rivers JM, Van Soest PJ, Roe DA, Lewis BA & Robertson J (1980) Dietary fibre: The effect of particle size of wheat bran on colonic function in young adult men. *Am J Clin Nutr* 39: 1734-1744.
76. Holland B, Unwin ID & Buss DH (1988) *Cereals and Cereal Products*. The Third Supplement of McCance & Widdowson's The Composition of Foods, The Royal Society of Chemistry, Cambridge, UK.
77. Badiali D, Corazziari E, Habib FI, Tomei E, Bausano GM, Margrini P, Anzini F & Torsoli A (1995) Effect of wheat bran in treatment of chronic nonorganic constipation: A double-blind controlled trial. *Dig Dis Sci* 40: 349-356.
78. Muller-Lissner SA (1988) Effect of wheat bran on weight of stool and gastrointestinal transit time: a meta analysis. *Br Med J* 296: 615-617.
79. Taylor R (1990) Management of constipation. *Br Med J* 300: 1063-1064.
80. Stephen AM, Wiggins HS, Englyst HN, Cole TJ, Wayman BJ & Cummings JH (1986) The effect of age, sex and level of intake of dietary fibre from wheat on large-bowel function in thirty healthy subjects. *Br J Nutr* 56: 349-361.
81. Lawson K, Kern F & Everson GT (1985) Gastrointestinal transit time in human pregnancy: prolongation in the second and third trimester followed by postpartum normalization. *Gastroenterol* 89: 996-999.
82. Naitove A & Almy TP (1993) Diverticular disease of the colon. In: *Gastrointestinal Disease-Pathophysiology, Diagnosis, Management*, Sleisenger MH, Fordtran JS & Saunders WB (eds), American College of Physicians, Philadelphia, USA, pp 1423-1424.
83. Painter NS (1967) Diverticulosis of the colon. *Am J Dig Dis* 12: 222-227.
84. Vargas PA & Alberts DS (1992) Primary prevention of colorectal cancer through dietary modification. *Cancer* 70: 1229-1235.
85. Greenwald P, Lanza E & Eddy GA (1987) Dietary fiber in the reduction of colon cancer risk. *J Am Diet Assoc* 87: 1178-1188.
86. Englyst HN, Bingham SA, Wiggin HS, Southgate DAT, Seppänen R, Helms P, Anderson V, Day KC, Choolun R, Collinson E & Cummings JH (1982) Nonstarch polysaccharide consumption in four Scandinavian populations. *Nutr Cancer* 4: 50-60.

87. McKeown-Eyssen GE & Bright-See E (1985) Dietary factors in colon cancer: international relationships: An update. *Nutr Cancer* 7: 251-253.
88. McKeown-Eyssen GE & Bright-See E (1984) Dietary factors in colon cancer: International relationships. *Nutr Cancer* 6: 160-170.
89. Rose DP (1986) Dietary factors and breast cancer. *Cancer Surv* 5: 671-687.
90. Weisburger JH & Kroes R (1994) Mechanisms in nutrition and cancer. *Eur J Cancer Prevent* 3: 293-298.
91. Jones LA, Ota DM, Jackson PM, Kemp K, Anderson DE, McCamant SK & Bauman DH (1987) Bioavailability of estradiol as a marker for breast cancer risk assessment. *Cancer Res* 47: 5224-5229.
92. Langley MS, Hammon GL, Bardsley A, Sellwood RA & Anderson DC (1985) Serum steroid binding proteins and the bioavailability of estradiol in relation to breast diseases. *J Natl Cancer Inst* 75: 823-829.
93. Ota DM, Jones LA, Jackson GL, Jackson PM, Kemp K & Bauman D (1986) Obesity, non-protein-bound estradiol levels, and distribution of estradiol in the sera of breast cancer patients. *Cancer* 57: 558-562.
94. Toniolo PG, Levitz M, Zeleniuch-Jacquotte A, Banerjee S, Koenig KL, Shore RE, Starx P & Pasternack BS (1995) A prospective study of endogenous estrogens and breast cancer in postmenopausal women. *J Natl Cancer Inst* 87: 190-197.
95. Kahlon TS, Saunders RM, Chow FI, Chiu MM & Betschart AA (1990) Influence of rice bran, oat bran and wheat bran on cholesterol and triglycerides in hamsters. *Cereal Chem* 67: 439-443.
96. Kahlon TS, Chow FI & Sayre RN (1994) Cholesterol-lowering properties of rice bran. *Cereal Foods World* 39: 99-103.
97. Wells JH (1993) Utilization of rice bran and oil in human diets. *Louisiana Agriculture* 36: 5-8.
98. Normand FL, Ory RL & Mod RR (1987) Binding of bile acids and trace minerals by soluble hemicelluloses of rice. *Food Technol* 41: 86-99.
99. Mod RR, Conkerton EJ, Ory RL & Normand FL (1978) Hemicellulose composition of dietary fiber of milled rice and rice bran. *J Agric Food Chem* 26: 1031-1035.
100. Jariwalla RJ (2001) Rice-bran products: phytonutrients with potential applications in preventive and clinical medicine. *Drugs Exp Clinical Res* 27: 17-26.
101. Hudson EA, Dinh PA, Kokubun T, SImmonds MS & Gescher A (2000) Characterization of potentially chemopreventive phenols in extracts of brown rice that inhibit the growth of human breast and colon cancer cells. *Cancer Res* 9: 1163-1170.
102. Gobder JS, Martin D, Shin TS, Setlhako G, Tricon C & Gervais M (1993) Optimizing rice parameters important in rice bran for human consumption. *Louisiana Agriculture* 36: 9-12.
103. Martin D, Godber JS, Setlhako G, Verma L & Wells JH (1993) Optimizing rice bran stabilization by extrusion cooking. *Louisiana Agriculture* 36: 13-15.
104. McCaskill DR & Orthoefer FT (1993) Capturing the value of rice phytonutrients. In: *Proceedings of Rice Utilization Workshop Developing Innovative, Non-Conventional Uses for Rice*, New Orleans, USA, pp 58-67.
105. Barber S & Barber CBde (1980) Rice bran: chemistry and technology. In: *Rice - Production and Utilization*, Luh BS (ed), AVI Publishing Co Inc, Westport, USA, pp 790-862.
106. Godber JS & Wells JH (1994) Rice bran as a viable source of high value chemicals. *Louisiana Agriculture* 37: 13-17.

107. Babcock D (1987) Rice bran as a source of dietary fiber. *Cereal Foods World* 32: 538-539.
108. Marshall WE (1993) Utilization of rice bran/hulls in value-added products. In: *Proceedings of Rice Unitlization Workshop Developing Innovative, Non-Conventional Uses for Rice,* New Orleans, USA, pp 68-76.
109. James C & Sloans S (1984) Functional properties of edible rice bran in model systems. *J Food Sci* 49: 310-311.
110. Hettiararchchy NS, Kalapathy U & Griffin K (1994) Natural antioxidants from fenugreek and rice bran. In: *Arkansas Rice Research Studies 1993,* Wells BR (ed), Fayettevile AR, Arkansas Agricultural Experiment Station Research Series 439, pp 218-221.
111. Randall JM, Satre RN, Schltz WG, Fong RY, Mossman AP, Tribelhorn RE & Saunders RM (1985) Rice bran stabilization by extension cooking for extraction of edible oil. *J Food Sci* 50: 361-368.
112. Juliano BO (1985) Rice bran. In: *Rice-Chemistry and Technology*, Juliano BO (ed), American Association of Cereal Chemistry, St. Paul MN, USA, pp 647-687.
113. Enochain RV, Saunders RM, Schultz WG, Beagle EC & Crowley PR (1981) Stabilization of rice bran with extrusion cookers and recovery of edible oil: A preliminary analysis of operations and financial feasibility. In: *Marketing Research Report 1120*, USDA-ARS, Washington DC, pp 18-25.
114. Sayre RN & Saunders RM (1990) Rice bran and rice bran oil. *Lipid Technol* 2: 72-76.
115. Kao C & Luh BS (1991) Rice oil. In: *Rice Stabilization*, Vol 2, Luh BS (ed), AVI Book Publs, New York, USA, pp 295-311.
116. Sayre RN (1988) Rice bran as source of edible oil and higher value chemicals. Presented at the American Association of Cereal Chemists 73rd Annual Meeting, San Diego, USA, pp 1-11.
117. Lynn L, Steen BJ & Anderson RM (1968) New rice oil. *Food Technol* 22: 1250-1252.
118. Nicolosi R (1990) Unsaponifiable in rice bran oil under study. *INFORM* 1: 831-835.
119. Nicolosi RJ, Ausman LM & Hegsted DM (1991) Rice bran oil lowers serum total and low density lipoprotein cholesterol and apoB levels in nonhuman primates. *Atherosclerosis* 88: 133-142.
120. Nicolosi RJ (1993) Nutritional aspects of rice oil. In: *Proceedings of Rice Ultilization Workshop-Developing Innovative, Non-Cnoventional Uses For Rice*, New Orleans, USA, pp 87-91.
121. Rukmini C & Raghuran CT (1991) Nutritional and biochemical aspects of the hypolipidemic action of rice bran oil: a review. *J Am College Nutr* 10: 3661-3669.
122. Hegsted M & Koushik CS (1994) Rice bran and rice bran oil may lower heart disease risk by decreasing cholesterol synthesis in the body. *Louisiana Agriculture* 37: 16-17.
123. Linchtenstein AH, Ausman LM, Carrasco W, Gualteiri Lj, Jenner JL, Ordovas JM, Nicolosi RJ, Glodin BR & Shaefer EJ (1994) Rice bran oil consumption and plasma lipid levels in moderately cholesterolemic humans. *Artheriosclerosis and Thrombosis* 14: 549-556.
124. Sharma RD & Rukmini C (1986) Rice bran oil and hypocholesterolemia in rats. *Lipids* 21: 715-717.
125. Hizukuri S, Takeda Y, Maruta N & Juliano BO (1989) Molecular structures of rice starch. *Carbohydr Res* 189: 227-235.
126. Juliano BO (1993) Improving food quality of rice. *Int Crop Sci* 1: 677-681.
127. Nikuni Z (1978) Studies on starch granules. *Starch/Staerke* 30: 105-111.

128. Lineback DR (1984) The starch granule: Organization and properties. *Bakers Digest* 58: 18-21.
129. Webb BD U.S. (1985) Criteria of quality. In: *Rice Chemistry and Technology*, Juliano Bo (ed), American Association of Cereal Chemists, St. Paul MN, USA, pp 425-427.
130. Webb BD (1991) Rice quality and grades. In: *Rice Utilization*, Vol 2, Luh BS (ed), AVI Book Publs, New York, USA, pp 89-119.
131. Wolever TMS, Jenkins DJA, Vuksan VRG, Wong GS & Jenkins AL (1990) Glycemic index of foods in individual subjects. *Diabetes Care* 13: 126-132.
132. Kerlin P, Wong L, Harris B & Capra S (1984) Rice flour, breath hydrogen and malabsorption. *Gastroenterology* 87: 578-585.
133. Carter JF (1993) Omega 3!! Report for the North Dakota Legislature, USA.
134. Caragay AB (1992) Cancer preventive foods and ingredients. *Food Technol* 46: 65-68.
135. Oomah BD & Mazza G (1998) Flaxseed products for disease prevention. In: *Functional Foods - Biochemical and Processing Aspects*, Mazza G (ed), Technomic Publ Co Inc, Lancaster, UK, pp 91-138.
136. Cullis CA & Cleary W (1985) Fluidity of the flax genome. In: *Plant Genetics,* Freeling M (ed), Alan R Liss, London, UK, pp 303-310.
137. Freeman TP (1995) Structure of flaxseed. In: *Flaxseed in Human Nutrition,* Cunnane SC & Thompson LU (eds), AOCS Press, Champaign, USA, pp 1-10.
138. Dorell DG (1970) Distribution of fatty acids within the seed of flax. *Can J Plant Sci* 50: 71-75.
139. Oomah BD, Mazza G & Kenaschuk EO (1996) Dehulling characteristics of flaxseed. *Lebensm Wiss u Technol* 29: 245-250.
140. Schiefer HB (1992) In pursuits of GRAS status for flaxseed: some practical and hypothetical considerations. In: *Proceedings of the 54$^{th}$ Flax Institute of the US*, Fargo, USA, pp 67-72.
141. Carter JF (1993) Potential of flaxseed and flaxseed oil in baked goods and other products in human nutrition. *Cereal Food World* 38: 753-759.
142. Declercq DR (1996) Preliminary Flaxseed Quality estimate-Report No. 2, Canadian Grain Commission, Winninpeg, MB, Canada, pp 1-3.
143. Anonymous (1997) Prairies Registration Recommending Committee for Grain, oilseeds Subcommittee Report 1997. Report of Flax Co-Operative Test 1996, pp 1-26.
144. Fritsche KL & Johnston PV (1990) Effect of dietary alpha-linolenic acid on growth, metastasis, fatty acid profile and prostaglandin production of two murine mammary adenocarcinomas. *J Nutr* 120: 1601-1609.
145. Henry MM, Moore JN, Feldman EB, Fischer JK & Russell B (1990) Effect of dietary alpha-linolenic acid on equine monocyte procoagulant activity and eicosanoid synthesis. *Circ Shock* 32: 173-188.
146. de Lorgeril M, Renaud S, Mamelle N, Salen P, Martin Jl, Monjaud I, Guidollet J, Touboul P & Delaye J (1994) Mediterranean alpha-linolenic acid-rich diet in secondary prevention of coronary heart disease. *Lancet* 343: 1454-1459.
147. Singer P (1992) Alpha-linolenic acid vs. long chain n-3 fatty acids in hypertension and hyperlipidemia. *Nutrition* 8: 133-135.
148. Johnston PV (1995) Flaxseed oil and cancer:,α-linolenic acid and carcinogenesis. In: *Flaxseed in Human Nutrition*, Cunnane SC & Thompson LU (eds), AOCS Press, Champaign, USA, pp 207-218.

149. Levander AO & Ager AL Jr (1995) Antimalarial effects of flaxseed and flaxseed oil. In: *Flaxseed in Human Nutrition*, Cunnane SC & Thompson LU (eds), AOCS Press, Champaign, USA, pp 237-243.
150. Aylward F & Nichols BW (1962) Plant lipids. II - free and combined sterols (sterol esters and glycosides) in commercial oilseed phospholipids. *J Sci Food Agric* 13: 86-91.
151. Middleditch BS & Knights BA (1972) Sterols of *Linum usitatissimum* seed. *Phytochemistry* 11: 1183-1184.
152. Declercq RD, Daun JK & Tipples KM (1995) Quality of western Canadian flaxseed. *Crop Bulletin* 216: 10-17.
153. Oomah BD & Mazza G (1995) Functional properties, uses of flaxseed protein. *INFORM* 6: 1246-1252.
154. Huang AHC (1992) Oil bodies and oleosins in seeds. *Ann Rev Plant Physiol Plant Mol Biol* 43: 177-200.
155. Bhatty RS & Cherdkiatgumchai P (1990) Compositional analysis of laboratory-prepared and commercial samples of linseed meal and of hull isolated from flax. *J Am Oil Chem Soc* 67: 79-84.
156. Cui W, Mazza G & Biladeris CG (1994) Chemical structure, molecular size distribution and rheological properties of flaxseed gum. *J Agric Food Chem* 42: 1891-1895.
157. Fedenuik RW & Biliaderis CG (1994) Composition and physiochemical properties of linseed (*Linum usitatissimum* L.) mucilage. *J Agric Food Chem* 42: 240-247.
158. Bell JM & Keith MO (1993) Nutritional evaluation of linseed meals from flax with yellow or brown hulls, using mice and pigs. *Ani Feed Sci Technol* 43: 1-18.
159. Setchell KDR (1995) Discovery and potential clinical importance of mammalian lignans. In: *Flaxseed in Human Nutrition*, Cuccane SC & Thompson LU (eds), AOCS Press, Champaign, USA, pp 82-98.
160. Ayres DC & Loike JD (1990) Lignans: Chemical, biological and clinical properties. In: *Chemistry and Pharmacology of Natural Products*, Phillipson JD, Ayres Dc & Baxter H (eds), Cambridge University Press, Cambridge, UK, pp 402-420.
161. Dabrowski KJ & Sosulski FW (1984) Composition of free and hydrolizable phenolic acids in defaated flours of ten oilseeds. *J Agric Food Chem* 32: 128-130.
162. Oomah BD, Kenaschuk EO & Mazza G (1995) Phenolic acids in flaxseed. *J Agric Food Chem* 43: 2016-2019.
163. Cook NC & Samman S (1996) Flavonoids - chemistry, metabolism, cardioprotective effects and dietary sources. *J Nutr Biochem* 7: 66-76.
164. Ibraham RK & Shaw M (1970) Phenolic constituents of the oil flax (*Linum usitatissimum*). *Phytochemistry* 9: 1855-1858.
165. Oomah BD, Mazza G & Kenaschuk EO (1996) Flavonoid content of flaxseed. Influence of cultivar and environment. *Euphytica* 90: 163-167.
166. Lay CL & Dybing DD (1989) Linseed. In: *Oil Crops of the World,* Robbelen G, Downey RK & Ashri A (eds), McGraw Hill, New York, USA, pp 416-430.
167. Stitt P (1990) Flax, the ideal "designer food" ingredient. In: *Proceedings of 53rd Annual Meeting of Flax Institute of the United States*, USA, pp 20-24.
168. Hettiararchchy N, Hareland G, Ostenson A & Balder-Shank G (1990) Composition of eleven flaxseed varieties grown in North Dakota. In: *Proceedings of 53rd Annual Meeting of Flax Institute of the United States*, USA, pp 36-40.
169. Burckhardt A (1989) Flax bread is a rising star of Wisconsin bakery. Minneapolis Star Tribune, St Paul, USA, 1/18/89, p 45.

170. Mazza G & Oomah BD (1995) Flaxseed, dietary fiber and cyanogens. In: *Flaxseed in Human Nutrition*, Cunnane SC & Thompson Lu (eds), AOCS Press, Champaign, USA, pp 56-81.

171. Nuttall FQ, Mooradian AD, Gannon MC, Billington C & Krezowski P (1984) Effect of protein ingestion on the glucose and insulin response to a standardized oral glucose load. *Diabetes Care* 7: 465-470.

172. Wine Institute (2003) World Vineyard, Grape and Wine Report, San Francisco, USA.

173. Nelson KE (1991) The grape. In: *Quality and Preservation of Fruits*, Eskin NAM (ed), CRC Press Inc, Boca Raton FL, pp 125-167.

174. Childers NF (1976) *Modern Fruit Science*, Horticultural Publications, Rutgers University, New Brunswick, USA, pp 666-713.

175. Singleton VL (1980) *Grape and Wine Phenolics, Background and Prospects*, Coll CR (ed), Centenaire, Paris, pp 215-227.

176. Singleton VL & Trousdale E (1983) White wine phenolics: Varietal and processing differences as shown by HPLC. *Am J Enol Vitic* 34: 27-34.

177. Amerine MA, Berg HW & Cruess WV (1967) *The Technology of Wine Making*, AVI Publs, Westport, CT, USA, pp 288, 367, 658-659.

178. McLelland MR & Rau EJ (1990) Grape juice processing. In: *Production and Packaging of Non-Carbonated Fruit Juices and Beverages*, Hicks D (ed), Blackie and Sons, Glasgow, UK, pp 226-242.

179. Hang YD (1988) recovery of food ingredients from grape pomace. *Process Biochem* 23: 2-4.

180. Macheix JJ, Flueriet A & Billot J (1991) *Fruit Phenolics*, CRC Press Inc, Boca Raton, USA.

181. Middleton E & Kandaswami C (1992) Effects of flavonoids on immune and inflammatory cell functions. *Biochem Pharmacol* 43: 1167-1179.

182. Hertog MGL, Hollman PCH, Katan MB & Kromhout D (1993a) Intake of potentially anticarcinogenic flavonoids and their determinants in adults in the Netherlands. *Nutr Cancer* 20: 21-29.

183. Hertog MGL, Feskens EJM, Hollman PCH, Katan MB & Kromhout D (1993b) Dietary antioxidant flavonoids and risk of coronary heart disease: the Zutphen elderly study. *Lancet* 342: 1007-1011.

184. Kanner J, Frankel E, Granit R, German B & Kinsella JE (1994) Natural antioxidants in grapes and wines. *J Agric Food Chem* 42: 64-69.

185. Maxwell JJ, Cruckshank A & Thorpe G (1994) Red wine and the antioxidant activity of serum. *Lancet* 334: 193-196.

186. Wagner H (1985) New plant phenolics of pharmaceutical interest. In: *Annual Proceedings of Phytochemistry Society of Europe,* Vol 15, Van Sumere CF & Lea PJ (eds), Clarendon Press, Oxford, UK, pp 409-425.

187. Cristoni A & Magistretti MJ (1987) Antiulcer and healing activities of vaccinium myrtillud anthocyanosides. *Farmaco Pract* 42: 29-40.

188. Kano E & Miyakoshi J (1976) UV protection effect of keramycin an anthocyanin derivative on cultured mouse fibroblast L ceels. *J Radiat Res* 17: 55-65.

189. Jang M, Cai L, Udeani GO, Slowing KV, Thomas CF, Beecher CWW, Fong HS, Farnsworth NR, Kinghorn AD, Mehta RG, Moon RC & Pezzuto JM (1997) Cancer chemopreventive activity of resveratrol, a natural product derived from grapes. *Science* 275: 218-220.

190. Boyle JA & Hsu L (1990) Identification and quantiifcation of ellagic acid in muscadine grape juice. *Amer J Enol Vitic* 41: 43-47.
191. Okeda T & Yokotsuka K (1996) Trans-resveratrol concentrations in berry skins and wines from grapes grown in Japan. *Amer J Enol Vitic* 47: 93-99.
192. Rouseff RL (1988) Liquid chromatographic determination of naringin and hesperidin as a detector of grapefruit juice in orange juice. *J Assoc Anal Chem* 71: 798-802.
193. Park GL, Avery SM, Byers JL & Nelson DB (1983) Identification of bioflavonoid from citrus. *Food Technol* 37: 98-105.
194. Dreyer DL (1968) Limnoid bitter principles. *Fortsch Chem Org Nat* 26: 190-244.
195. Redd JB & Hendrix CM Jr (1993) Processing of natural citrus oils and flavols. In: *Fruit Juice Processing Technology,* Nagy S, Chen CS & Shaw PE (eds), AgScience, Auburndale, USA, pp 83-109.
196. Briton G & Goodwin TW (1982) *Carotenoid Chemistry and Biochemistry*, Pergamon Press, Elmsford, New York, USA.
197. Abad AR & Gregory JF (1987) Determination of folate bioavailability with a rat bioassay. *J Nutr* 117: 866-873.
198. White DR, Lee HS & Kruger RE (1991) Reversed-phase HPLC/EC determination of folate in citrus juice by direct injection with column switching. *J Agric Food Chem* 39: 714-717.
199. Eaks IL (1964) Ascorbic acid content of citrus during growth and development. *Bot Gaz* 125: 186-191.
200. Chen CS, Shaw PE & Parish ME (1993) Orange and tangerine juices. In: *Fruit Juice Processing Technology*. Nagy S, Chen CS & Shaw PE (eds), AgScience, Auburndale, USA, pp 111-165.
201. Bailey DG, Malcolm J, Arnold O, Strong A, Munoz C & Spence JD (1993) Effect of grapefruit juice and naringin on nisoldipine pharmacokinetics. *Clin Pharmacol Therap* 54: 589-594.
202. Verma AK, Johnson JA, Gould MN & Tanner MA (1988) Inhibition of 7,12-dimethylbenz [a] anthracene and N-nitrosolmethylurea-induced rat mammary cancer by dietary flavonol quercetin. *Cancer Res* 48: 5754-5758.
203. Fujiki H, Horiuchi T, Yamashita K, Hakii H, Hirata Y & Sugimura T (1986) Inhibition of tumor promotion by flavonoids. In: *Plant Flavonoids in Biology and Medicine: Biochemical, Pharmacological and Structure-Activity Relationship*, Cody V, Middleton E & Harborne JB (eds), Alan R Liss Inc, New York, USA, pp 429-440.
204. Mathews-Roth MM, Pathak MA, Fitzpatrick TB, Harber LC & Kass EH (1970) Beta-carotene as a photoprotective agent in erythropoetic protopophyria. *New Engl J Med* 282: 1231-1234.
205. Zhang LX, Cooney RV & Bertram JS (1991) Carotenoids enhance gap junctional communication and inhibit lipid peroxidation in C3H/10T1/2 cells: Relationship to their cancer chemoprotective action. *Carcinogenesis* 12: 2109-2114.
206. Gaziano Jm, Manson JE, Ridker PM, Buring JE & Hennekens CH (1990) Beta-carotene therapy for chronic stable angina. *Circulation* 82 (Suppl. III): 111-201.
207. Mirvish SS, Cardesa A, Wallcave L & Shibik P (1975) Induction of mouse lung adenomas by amines or ureas plus nitrite and N-nitroso compounds: Effect of ascorbate, gallic acid, thiocyanate and caffeine. *J Natl Cancer Inst* 55: 633-636.
208. Haeney RK & Fenwick GR (1980) Glucosinolates in *Brassica* vegetables: Analysis of 22 varieties of brussel sprouts. *J Sci Food Agric* 31: 785-793.

209. Guffy SK & Hicks JR (1985) Effect of cultivar, maturity and storage on respiration, dry weight and glucosinolate content of cabbage. *Acta Hortic* 157: 211-218.
210. Stoeward GS (1995) Bioactive organosulfur phytochemicals in *Brassica oleracea* vegetables a review. *Food Chem Toxicol* 33: 537-543.
211. Dateo GP, Clapp RC, Mackey DAM, Hewitt EJ & Hasselstorm T (1957) Identification of the volatile sulfur components of cooked cabbage and the nature of the precursors in the fresh vegetable. *Food Res* 22: 440-446.
212. Verhoeven DTH, Goldbohm RA, van Poppel G, Verhagen H & vanden Brandt PA (1996) Epidemiological studies on *Brassica* vegetables and cancer risk. *Cancer Epidemiol Bomarkers Prev* 5: 733-748.
213. Zhang Y, Kensler TW, Cho C-G, Posner GH & Talalay P (1994) Anticarcinogenic activities of sulforaphane and structurally related synthetic norbornyl isothiocyanates. *Proc Natl Acad Sci* 91: 3147-3150.
214. Greenwald P (1996) Chemoprevention of cancer. *Sci Am* 275: 96-99.
215. Srivastava KC, Bordiov A & Verma SK (1995) Garlic (*Allium sativum*) for disease prevention. *South African J Sci* 91: 68-77.
216. Nakamura YK, Matsuo T, Shimoi K, Yakamura Y & Tomita I (1996) S-methyl methanethiosulfonate, bio-antimutagen in homogenates of *Cruciferae* and *Liliaceae* vegetables. *Biosci Biotech Biochem* 60: 1439-1443.
217. Mazza G & Miniati E (1993) *Anthocyanins in Fruits, Vegetables and Grains*, CRC Press Inc, Boca Raton, USA, pp 269-270.
218. Attrep KA, Bellman WP Sr, Attrep M Jr, Lee JP & Braselton WE Jr (1980) Separation and identification of prostaglandin $A_1$ in onions. *Lipids* 15: 292-296.
219. Smoczkiewicz MA, Nitschke D & Wieladek H (1982) Microdetermination of steroid and triterpene saponin glycosides in various plant materials. I. *Allium* species. *Mikrochim Acta* 11: 42-50.
220. Suzuki M & Cutcliffe JA (1989) Fructans in onion bulbs in relation to storage life. *Can J Plant Sci* 69: 1327-1333.
221. Belman S (1983) Onion and garlic oils inhibit tumour promotion. *Carcinogenesis* 4: 1063-1066.
222. Sparnins VL, Mott AW & Wattenberg LW (1986) Effects of allyl methyl trisulfide on glutathione S-transferase activity and PB-induced neoplasia in the mouse. *Nutr Cancer* 8: 211-215.
223. Sparnins VL, Barany G & Wattenberg LW (1988) Effcets of organosulfur compounds from garlic and onion on benzo (a) pyrene-induced neoplasis and glutathione S-transferase activity in the mouse. *Carcinogenesis* 9: 131-134.
224. Apitz-Castro R, Escalante J, Vargas R & Jain MK (1986) Ajoene, the antiplatelet principle of garlic, synergistically potentiates the antiaggregatory action of prostacylcin, forskolin, indomethacin and dipyridamole on human platelets. *Thromb Res* 42: 303-311.
225. Bordia A & Verma SK (1980) Effect of garlic feeding on regression of experimental atherosclerosis in rabbits. *Artery* 7: 428-437.
226. Jain RC & Vyas CR (1975) Garlic in alloxan-induced diabetic rabbits. *Am J Clin Nutr* 28: 684-685.
227. Foushee DB, Ruffin J & Banerjee U (1982) Garlic as a natural agent for the treatment of hypertension: a preliminary report. *Cytobios* 34: 145-152.
228. Tomomatsu H (1994) Health effects of oligosaccharides. *Food Technol* 48: 61-65.
229. Fuller R (1994) History and development of probiotics. In: *Probiotics*, Fuller R (ed), Chapman & Hall, New York, USA, pp 1-8.

230. Flynn A & Cashman K (1997) Nutritional aspects of minerals in bovine and human milks. In: *Advanced Dairy Chemistry*, Vol 3, Fox PF (ed), Chapman & Hall, London, UK, pp 257-302.
231. Maubois JL (1990) Nouvelles applications des technologies à membrane dans I' industrie laitière. In: *Proceedings, 23rd Int Dairy Congress*, Montreal, Brussels, Belgium: IDF, pp 1775-1790.
232. Jost R & Jelen P (1997) Cross-flow microfiltation - an extension of membrane processing of milk and whey. In: *Implications of Microfiltration on Hygiene and Identity of Dairy Products*, Bullletin No. 320, Brussels, Belgium: IDF, pp 9-15.
233. Kelly PM & Touhy JL (1997) The effectiveness of microfiltration for the removal of microorganisms. In: *Implications of Microfiltration on Hygiene and Identity of Dairy Products,* Bullletin No. 320, Brussels, Belgium: IDF, pp 9-15.
234. Peter S, Rattray W & Jelen P (1996) Heat stability and sensory quality of protein-standardized 2% fat milk. *Milchwissenscaft* 51: 611-616.
235. Salminen S, Isolauri E & Salimen F (1996) Clinical uses of probiotics for stabilizing the gut mucosal barrier: successful strains and future challenges. *Antonie Van Leeuwenhoek* 70: 347-358.
236. Mortomi M (1996) Properties of *Lactobacillus casei: Shirota* strains as probiotics. *Asia Pac J Clin Nutr* 5: 29-30.
237. Ouwehand AC, Kirjavainen PV, Shortt C & Salminen S (1999) Probiotics: mechanisms and established effects. *Int Dairy J* 9: 43-52.
238. Puupponen-Pimia R, Aura A-M, Oksman-Caldentey K-M, Myllaerinen P, Saarela M, Mattila-Sandholm T & Poutanen K (2002) Development of functional ingredients for gut health. *Trends Food Sci Technol* 13: 3-11.
239. Manning TS & Gibson GR (2004) Prebiotics. *Best Pract Res Clin Gastroenterol* 18: 287-298.
240. West DW (1986) Structure and function of the phosphorylated residues of casein. *J Dairy Res* 53: 333-352.
241. Renner E (1983) *Milk and Dairy Products in Human Nutrition*, Volk-swirtschaftlicher Verlag, Munich, Germany.
242. Pearce RJ (1988) Whey Protein Fractions. Patent Application PCT/AU88/00141.
243. McCarron DA, Morris CD, Henry HJ & Stanton JL (1984) Blood pressure and nutrient intake in the United States. *Science* 224: 1392-1398.
244. Sorenson W, Slattery ML & Ford M (1988) Calcium and colon cancer: a review. *Nutr Cancer* 11: 135-145.
245. Clandinin MT (1995) Infant nutrition: effects of lipid on later life. *Current Opinion Lipidology* 6: 28-31.
246. Mustapha A, Hertzler SR & Saviano DA (1997) Lactose: nutritional significance. In: *Advanced Dairy Chemistry*, Vol. 3, Fox PF (ed), Chapman & Hall, London, UK, pp 127-154.
247. Regester GO, Smithers GW, Mitchell IR, McIntosh GH & Dionysius DA (1997) Bioactive factors in milk: Natural and induced. In: *Milk Composition, Production and Biotechnology*, Welch RAS, Burns DJW, Davis SR, Popay AI & Prosser CG (eds), CAB International, New York, USA, pp 243-264.
248. Mann GV, Schaffer RD, Anderson RD & Sandstead HH (1964) Cardiovascular disease in the Maasai. *J Atheroscler Res* 4: 289-312.
249. Mital BK & Garg SK (1995) Anticarcinogenic, hypocholesterolemic, and antagonistic activities of *Lactobacillus acidophilus. Crit Rev Microbiol* 21: 175-214.

250. Van't Veer P, Dekker JM, Lamers JWJ, Kok FJ, Schouten EG, Brants HAM, Sturmans F & Hermus RJJ (1989) Consumption of fermented milk products and breast cancer: A case-control study in the Netherlands. *Cancer Res* 49: 4020-4023.
251. McIntosh GH, Royle PJ, Leu RK, Regester GO, Johnson MA, Grinsted RL, Kenward RS & Smithers GW (1998) Whey proteins as functional food ingredients. *Int Dairy J* 8: 425-434.
252. Barth CA & Behnke U (1997) Nutritional physiology of whey and whey components. *Nahrung* 41: 2-12.
253. Stansby ME (1969) Nutritional properties of fish oils. *World Rev Nutr Dietetics* 11: 46-105.
254. Labomar-Mun Cida Program (1989) Ciência e technologia de produtos pesqueiros. St. John's, New foundland, MUN Printing Services. *Ciência e technologia de organismos aquàticos* 1A: 1162.
255. Wang YJ, Miller LA, Perren M & Addis PB (1990) Omega-3 fatty acids in lake superior fish. *J Food Sci* 55: 71-73.
256. Belda MCR & Pourchet-Campos MA (1991) Ácidos graxos essenciais em nutrição: uma visão atualizada. *Ciência e technologia deAlimentos* 11: 5-35.
257. Ackman RC (1988) The year of fish oil. *Chem Ind* 3: 139-145.
258. Suzuki T (1981) *Fish and Krill Processing Technology*, Applied Science Publishers, London, UK.
259. Bimbo AP & Crowther JB (1991) Fish oils: processing beyond crude oil. *Infofish Int* 6: 20-25.
260. Shahidi F, Synowiecki J & Balejko J (1994) Proteolytic hydrolysis of muscle proteins of harp seal (*Phoca groenlandica*). *J Agric Food Chem* 42: 2634-2638.
261. Burr ML (1991) Is fish good for the heart. *Trends Food Sci Technol* 2: 17-20.
262. Mehta J, Lopez LM, Lowton D & Wargovich T (1980) Dietary supplementation with omega-3 polyunsaturated fatty acids in patients with stable coronary disease: Effects on indices of platelet and neurophil function and exercise performance. *Am J Med* 84: 45-52.
263. Kinsella JE (1986) Food components with potential therapeutic benefits. The n-3 polyunsaturated fatty acids on fish oils. *Food Technol* 40: 89-97.
264. Dyerberg J (1986) Linolenate-derived polyunsaturated fatty acids and prevention of atherosclerosis. *J Nutr Rev* 44: 125-134.
265. Singh G & Chandra RK (1988) Biochemical and cellular effects of fish and fish oils. *Prog Food Nutr Sci* 12: 371-419.
266. Branden LM & Carroll KK (1986) Dietary polyunsaturated fats in relation to mammary carcinogenesis in rats. *Lipids* 21: 285-288.
267. Fischer S (1989) Dietary polyunsaturated fatty acids and eicosanoid formation in humans. *Adv lipid Res* 23: 169-198.
268. Swedish Nutrition Foundation (1996) *Health claims in the Labelling and Marketing of Food Products. The Food industry's rules (Self-Regulating programme)*, Swedish Nutrition Foundation, Lund, Sweden.
269. Dietary Supplememts Health and Education Act (1994) Public Law 103-417, 25 October 1994. Codified at 42USC 287C-11.
270. Department of Health and Human Services, Food and Drug Administration (1994) Food Labelling: general requirements for health claims for dietary supplements. Final rule. *Federal register* 59: 395-426.
271. Ehrnreich H (2000) Foods as medicine. *World Food Ingr* June/July: 18-28.

272. Kottke MK (1998) Scientific and regulatory aspects of nutraceutical products in the United States. *Drug Dev Ind Pharmacy* 24: 1177-1195.
273. Nutrition Labelling and Education Act (1990) Pub L No. 101-525, 104 Stat 2353.
274. Wasser SP, Nevo E, Sokolov D, Reshetnikov S & Timor-Tismenetsky H (2000) Dietary supplements from medicinal mushrooms: diversity of types and variety of regulations. *Int J Medicinal Mushrooms* 2: 1-20.
275. Smith BL, Marcotte M & Harrison G (1996) *A comparative Analysis of the Regulatory Framework Affecting Functional Food Development and Commercialization in Canada, Japan, the European Union and the United States of America*, Inter/sect Alliance Inc, Ottawa, Canada.
276. Health Canada. Food Directorate and Drugs/Medical Devices Programme (1997) Discussion Document. Functional Foods and Nutraceuticals. Consultation Workshop, Ottawa, Canada.
277. Mogelonsky M (1999) Functional foods. *American Demographics* 21: 15-19.
278. United States General Accounting Office (2000) Food Safety: Improvements needed in overseeing the safety of dietary supplements and "functional foods." GAO/RCED-00-156, pp 1-30.
279. Blades M (2000) Functional foods or nutraceuticals. *Nutr Food Sci* 30: 73-75.
280. Heasman M & Mellentin J (2001) *The Functional Foods Revolution: Healthy People, Healthy Profits*, Earthscan, London, UK.
281. Luccina L (2003) Improving the success of functional foods. *Food Technol* 57: 42-47.
282. Roberfroid M (2002) Global view on functional foods: European perspectives. *Brit J Nutr* 88 (suppl. 2): S133-S138.
283. Gray J, Armstrong G & Farley H (2003) Opportunities and constraints in the functional food market. *Nutr Food Sci* 33: 213-218.
284. Weststrate JA, van Poppel G & Verschuren PM (2002) Functional foods, trends and future. *Brit J Nutr* 88 (suppl. 2): S233-S235.

# Indigenous Fermented Foods

# 11

**V.K. Joshi, V. Garg and Ghan Shyam Abrol**

*Department of Food Science and Technology, Dr Y.S. Parmar University of Horticulture & Forestry, Nauni, Solan 173 230 (H.P.)*

## 1. INTRODUCTION

Production of foods such as wine, beer, bread, yoghurt, cheese and pickles by fermentation constitutes most ancient technologies developed by man to improve the quality and storage life of some food materials from time immemorial [1, 2]. But it remained more or less an art till the dawn of microbiology in the $18^{th}$ and $19^{th}$ century, when it was proved beyond doubt that microorganisms are the agents responsible for fermentation. The first fermented foods consumed probably were fermented fruits. There is a reliable information that fermented drinks were produced over 7,000 years ago in Babylon (now in Iraq), 5,000 years ago in Egypt, 4,000 years ago in Mexico and 3500 years ago in Sudan [3]. The art of fermentation practiced by common man continued in spite of scientific and technological revolution, but largely remained confined to have scale, rural and tribal areas due to high cost of inaccessibility of the industry-made products in remote areas, taste of the people for the traditional fermented products and finally, their socio-cultural linkages with such products. In Himachal Pradesh, the traditional fermented foods and beverages form important constituents of staple diet of the people belonging to the tribal belts of Lahaul and Spiti, Kinnaur, Chamba and rural area of Kullu, Shimla, Mandi and Kangra districts [4].

Generally, the traditional methods of preparing indigenous fermented foods are simple and inexpensive. However, the ancient methods of making such foods are changing rapidly through modern microbial technology. During the last few years, much interest has been generated in the fermented foods of Asian and African countries including India where such foods are still being manufactured according to the traditional but technologically less advanced methods using simple equipment at cottage industry scale by means of natural microflora. An attempt has been made to present the available information on different aspects of the traditional fermented foods.

## 2. TRADITIONAL FERMENTED FOODS

India has a rich knowledge of fermented foods prepared from, milk, cereals, pulses, vegetables, fruits fish, etc. Fermented milks like *dahi*, butter milk (*chhash*), *lassi* are popular all over the country. The fermented cereal legume based products like *dhokla, khaman* are very popular in western part of India, and *idli* and *dosa* are popular in southern India. It is equally true with northern part of India.

*Food Biotechnology: Principles and Practices*
*V.K. Joshi & R.S. Singh (Eds.)*
*I.K. International Publishing House Pvt. Ltd., India*
*pp 337–374*

Some of the common indigenous fermented foods consumed in Northern Indian plains are listed in Table 1.

Indigenous fermented products constitute an important part of human diet in South-East Asia including India, the Middle East and Africa. They are also becoming popular in the developed world due to their high nutritional value and organoleptic characteristics. There are several fermented foods, which are prepared and consumed such as *Bhalle, Warries, Kulche,* and *Jalebies* [2, 3]. Of course, a variety of alcoholic beverages are produced and consumed in Punjab and adjoining area of Northern India (Tables 2 and 3).

Fermented milk *dahi* is a universal fermented product prepared and consumed in Northern India, similar to the people elsewhere in the world. Another product *Lassi* is a common fermented product of rural India, and Northern India is not an exception. Himachal Pradesh has several products common with Northern India, may be with some modifications depending upon the availability of raw material and liking of the consumers. At the same time there are several fermented products

**Table 1:** Some important indigenous fermented foods of India.

| Fermented food | Ingredient(s) | Nature of product | Produce use |
|---|---|---|---|
| *Bhallae* | Black gram | Deep fried patties | Snack after soaking in curd, water |
| *Vadia* | Black gram | Deep fried patties | Snack |
| *Warries* | Black gram and spices | Ball like hollow, brittle | Spicy condiment, eaten with vegetables, legumes, rice |
| *Bhatura* | White wheat flour | Deep fried bread | Breakfast |
| *Jalebie* | White wheat flour | Deep fried crispy | Confection food |
| *Kulcha* | White wheat flour | Flat bread | Staple food |
| *Nan* | White wheat flour | Flat bread | Staple food |

*Source:* 5

**Table 2:** Examples of acid-leavened bread and noodles used in Asia-Pacific region.

| Product name | Country of use | Major ingredients | Microorganisms | Appearance & usage |
|---|---|---|---|---|
| *Idli* | South India; Sri Lanka | Rice grits; Black gram powder | *L. mesenteroides; S. fecalis; T. candida; T. pullulans* | Steamed cake |
| *Dosa* | India | Rice flour; Black gram powder | *L. mesenteroides; S. faecalis; T. candida; T. pullulans* | Griddled pancake |
| *Dhokla* | India | Rice; Bengal gram | *L. mesenteroides; S. faecalis; T. candida; T. pullulans* | Steamed cake |
| *Jalebies* | India; Nepal; Pakistan | Wheat flour | *S. bayanus* | Pretzel-like confection |
| *Mantou* | China | Wheat flour | *Saccharomyces* sp. | Steamed cake |
| *Kichudok* | Korea | Rice, Takju | *Saccharomyces* sp. | Steamed cake |
| *Puto* | Philippines | Rice, Sugar | *L. mesenteroides; S. faecalis;* yeast | Steamed cake |
| *Brem* | Indonesia | Glutinous rice | | Cake |
| | China; Thailand; Korea; Japan | Mungbean starch | *L. mesenteroides; L. casei; L. cellobiosus; L. fermenti* | Noodle |
| *Khanomjeen* | Thailand | Rice | *Lactobacillus* sp.; *Streptococcus* sp. | Noodle |
| *Me* | Vietnam | Rice | Lactic acid bacteria | Sour food ingredient |

*Source*: 6

**Table 3:** Examples of cereal alcoholic beverages prepared in the Asia-Pacific Region.

| Product name | Country | Major ingredients | Microorganisms | Appearance & usage |
|---|---|---|---|---|
| | | **RICE WINE** | | |
| *Shaosinghjiu* | China | Rice | *S. cerevisiae* | Clear liquid |
| *Chongju* | Korea | Rice | *S. cerevisiae* | Clear liquid |
| *Sake* | Japan | Rice | *S. sake* | Clear liquid |
| | | **RICE BEER** | | |
| *Takju* | Korea | Rice, wheat | Lactic acid bacteria, *S. cerevisiae* | Turbid drink |
| *Tapuy* | Philippines | Rice, glutinous rice | *Saccharomyces, Mucor, Rhizopus, Aspergillus, Leuconostoc, L. plantarum* | Sour, sweet, liquid, paste |
| *Brembali* | Indonesia | Glutinous rice | *Mucor indicus, Candida* | Dark brown liquid, sour, alcoholic |
| *Jaanr* | India Himalaya | Finger millet | *Hansenula anomala, Mucor rouxianus* | Sweet-sour alcoholic paste; mix with water |
| | | **ALCOHOLIC RICE PASTE** | | |
| *Khaomak* | Thailand | Glutinous rice | *Rhizopus, Mucor, Saccharomyces, Hansenula* | Semisolid, sweet alcoholic |
| *Tapai pulut* | Malaysia | Glutinous rice | *Chlamydomucor, Endomycopsis, Hansenula* | Semisolid, sweet alcoholic |
| *Tape-ketan* | Indonesia | Glutinous rice, rice | *A. rouxii, E. burtonii, E. fibulinger* | Sweet-sour alcoholic paste |
| *Lao-chao* | China | Rice | *Rhizopus, A. rouxii* | Paste |
| *Bhattejaanr* | India Sikkim | Glutinous rice | *Hansenula anomala, Mucor rouxianus* | Sweet-sour alcoholic paste |
| | | **ALCOHOLIC RICE SEASONING** | | |
| *Mirin* | Japan | Rice, alcohol | *A. oryzae, A. usamii* | Clear liquid seasoning |

*Source:* 7

which are unique to Himalayan region, while others are specific to Himachal Pradesh, especially tribal areas. Some of the products are listed in Table 4. Different regions of Himachal Pradesh are known for their typical products like *Angoori* of Kinnaur district, while *Chhang* is known throughout the Himalayan region. Many of the fermented products are well known as house-hold items, while a few are prepared at cottage scale. Industrialization of these products has not taken place to any great extent, though some of these products have already found a place in the menu of several restaurants and hotels.

**Table 4:** Some of the traditional fermented foods of Himachal Pradesh.

| Fermented food | Substrate | Area |
|---|---|---|
| *Aska* | Rice flour | Lahaul |
| *Bari* | Black gram | Kullu, Kangra, Mandi, Bilaspur |
| *Bedvin roti* | Wheat floor, opium seeds, linseeds | Kullu, Kangra, Mandi, Chamba, Shimla |
| *Bhaturu* | Wheat flour | Kullu, Kangra, Mandi, Chamba, Shimla |
| *Borhe* | Black gram | Spiti |
| *Babru/Suhala* | Wheat flour | Kullu, Kangra, Mandi |
| *Chilra (Lwar)* | Rice/wheat/buck-wheat/barley | Lahaul, Kangra |
| *Dosha* | Wheat flour | Lahaul |
| *Gulgule* | Wheat flour | Kullu, Kangra, Mandi, Chamba, Spiti |
| *Jute* | Buck wheat flour | Kinnour |
| *Khoblu* | Wheet Flour, butter milk, lassi | Mandi, Suket |
| *Manna* | Wheat flour | Lahaul |
| *Marchu* | Wheat flour | Lahaul |
| *Pakk* | Barley, butter and lassi | Kullu, Kangra, Mandi, Bilaspur |
| *Pinni/Bagpinni* | Barley flour | Kullu, Kangra, Mandi, Chamba and Shimla |
| *Seera/Kheera* | Wheat grains | Kullu, Kangra, Mandi, Chamba, Spiti |
| *Sepubari* | Black gram | Kullu, Kangra, Mandi, Bilaspur |
| *Siddu* | Wheat flour | Kullu, Kangra, Shimla |
| *Thuktal* | Barley | Bilaspur |

*Source:* 4, 8

## 3. FERMENTED FOODS OF HIMACHAL PRADESH

Some of the traditional fermented foods, commonly prepared in Himachal Pradesh, are listed in Table 4. These foods are mostly cereal-based (wheat/barley/buckwheat ragi), but some legume (black gram) and milk-based fermented foods are also common. Some of the products like *Bhaturu, Siddu, Chilra, Marchu, Manna, Dosha, Pinni/Bagpinni, Seera,* etc. are unique to Himachal Pradesh. A large variety of fermented foods are prepared either daily, during special occasions or for consumption during journey. Traditional starter cultures like '*Malera*' and '*Treh*' are used as inocula in making these fermented foods. However, the natural fermentation (without the addition of inoculum, as micro-organisms present in the raw materials carry out fermentation) is used in the production of *Seera, Sepubari, Bohre*, etc.

A number of cereal (rice/barley), fruit (grapes/wild apricot/apple), jaggery and millet-based indigenously made fermented beverages are popular among the people of Himachal Pradesh. Some of the very common traditional fermented beverages of Himachal Pradesh are listed in Table 5. These fermented beverages are prepared and consumed regularly and also during special occasions. In the preparation of these fermented beverages, *'Phab'*/Dheli are the inoculum used for fermentation (Plate 1).

**Table 5:** Some traditional fermented beverages of Himachal Pradesh.

| Fermented food | Substrate | Area |
|---|---|---|
| *Arak/Ara/Ghanti* | Apple, chulli, barley, pear, etc. | Kinnaur |
| *Angoori* | Grapes | Kinnaur |
| *Chakti/Chhang/Lugri* | Rice/barley | Kangra, Lahaul & Spiti, Kullu & Kinnaur |
| *Daru/Ghanti* | Jaggery | Shimla, Kullu |
| *Sur* | Millet (Ragi) flour | Kullu, Kangra |

*Source:* 4, 8

**Plate 1:** Natural source of fermentation used in tribal areas of HP for fermentation of alcoholic beverages.
*Source:* 10

## 4. PREPARATION TECHNOLOGY

A large variation in production techniques of various fermented foods exists from region to region. The knowledge collected from various sources such as published literature, personal communications and the results of survey carried out has been documented [4, 8, 9]. Like other regions preparing traditional foods, in Himachal Pradesh, traditional fermented products are prepared and the know-how of the traditional processes and technologies involved in the production of fermented products has been transferred from one generation to another [4]. To prepare these fermented products, the raw materials needed are locally available and equipment used are easy to maintain and affordable by the folk people.

## 5. FERMENTED FOODS

### 5.1 *Warries* and *Papadam*

*Warries* are spicy, hollow, brittle, friable balls 5-8 cm in diameter, very popular in Northern India, especially in Amritsar city (Punjab) and are used as condiment in cooking with vegetables, legumes or rice (Plate 2). Dehulled black gram (*Phaseolus mungo*) grains after being washed and soaked overnight in water are ground to a past. The paste is then spiced with asafoetida (*Ferula foetida*, 0.5-1.0%), caraway (*Carum carvi*, 0.5-1.0%), cardamom (*Elettaria cardamomum*, 1.0%), cloves (*Syzygium*

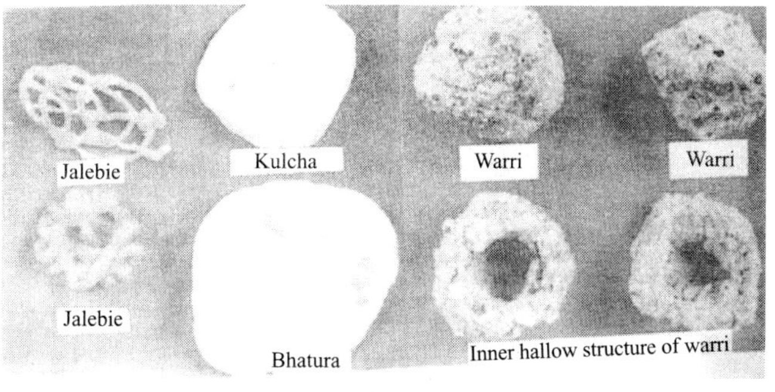


**Plate 2:** Some of the cereal based fermented foods.
*Source:* 5

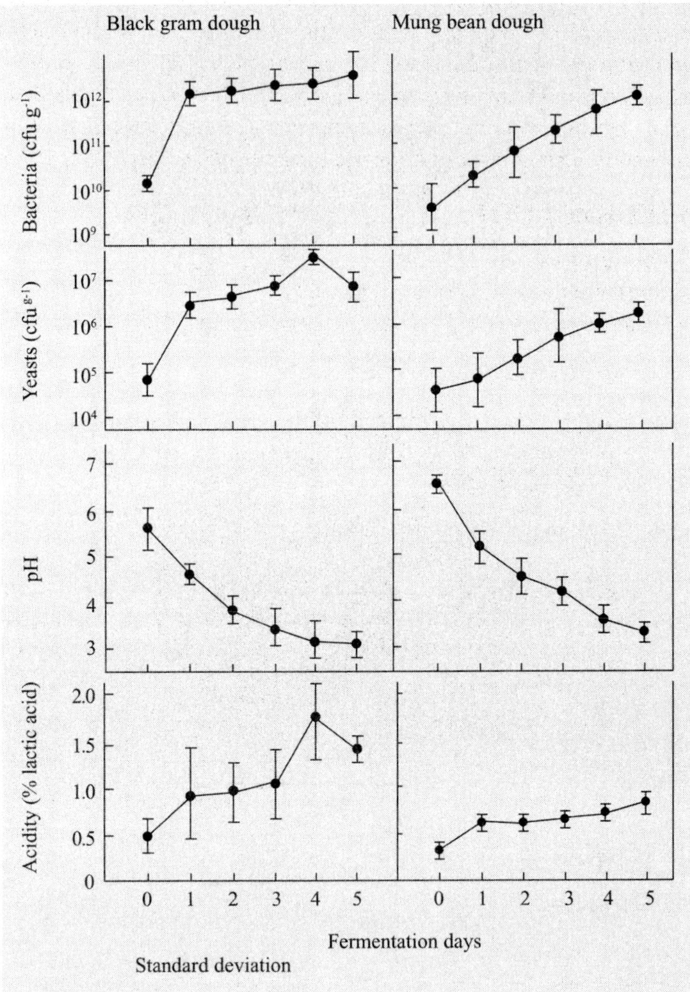


**Fig. 1:** Various changes accompanying warri fermentation.
*Source:* 5

*aromaticum*, 0.4-0.5%), fenugreek (*Trigonella foenumgraecum*, 1.0 %), ginger (*Zingiber officinale*, 8.0%) and red pepper (*Capsicum annum*, 1.0%) and moulded into small balls, which are allowed to undergo fermentation, and drying is carried out in open air for 4-8 days [11]. Changes accompanying the formation of warries are given in Fig.1.

*Papadam*, another condiment, is similarly prepared from black gram paste but without fenugreek and ginger. It is allowed to ferment for 4-6 hours and rolled into tortilla like wafers, which are served roasted or deep fried [5, 11].

## 5.2 *Bhatura*

*Bhatura* is leavened, deep fried bread (Plate 2) made from white wheat flour, consumed as a breakfast food and snacks with *masala chana* in Northern India. Dough is prepared by mixing white wheat flour with water and kneading for 20 min at room temperature. The dough is supplemented with 1% common salt and inoculated with a portion of previous batch ripe batter and kept in open to undergo fermentation for 8-14 h. The dough is then made into balls, which are then flattened by hand using

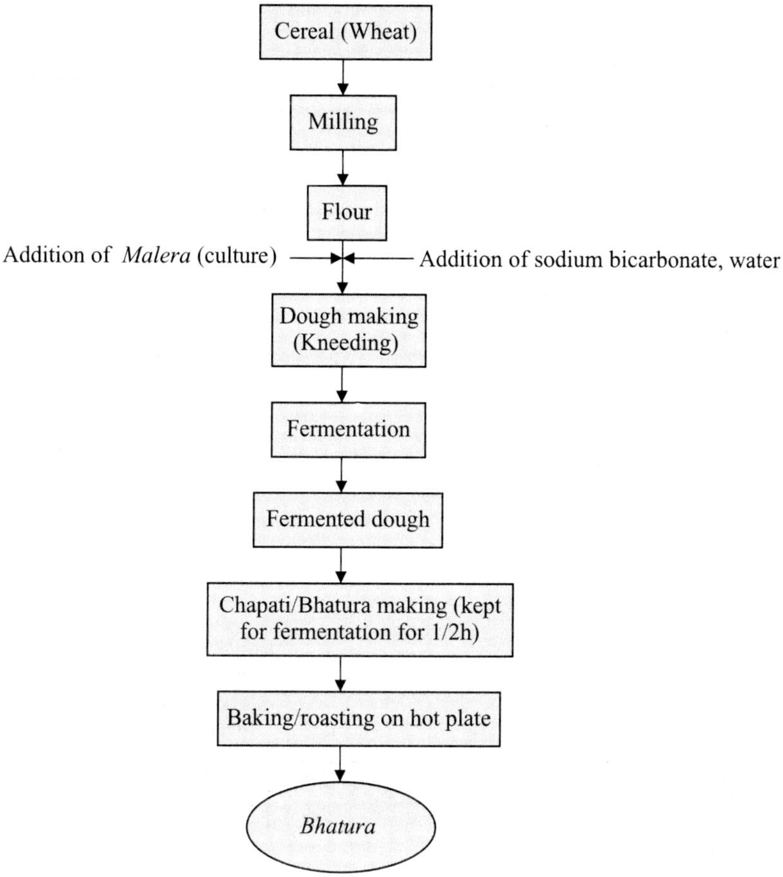


**Fig. 2:** Flowsheet for the *bhatura* preparation.

vegetable oil to 12-18 cm in diameter before deep frying in oil [5]. Formation of *Bhatura* is schematized in Fig. 2.

## 5.3 *Jalebie*

This is prepared by mixing equal quantities of white wheat flour and water and inoculating with some portion of a previous batch to raise the batter. It is then allowed to ferment at room temperature for 8-12 h. The fermented soft and silky batter is squeezed through an embroidered hole, about 4 mm in diameter, deposited as continuous spiral into hot vegetable oil and deep fried. The light brown crispy pretzels of 5-7 cm in diameter are removed from oil with the help of a sieve-like spatula, and immediately immersed in 70% sucrose syrup while hot. Quite often, rose (*Rosa india* L.) or *kewda* water (*Pandanus tectarius*) and orange food colour is added to the syrup. The crispy pretzels are finally removed from the syrup with spatula and served hot as a confection food in India and Pakistan [5].

## 5.4 *Kulcha*

It is a popular North Indian bread eaten with vegetables and meat and is generally used as a staple diet. Dough preparation is usually done by the straight dough method. White wheat flour, sugar, portion of the dough from previously fermented batch and water are mixed (100:1:10:50) at room temperature, kneaded by hand and allowed to ferment overnight in earthen pots. Sometimes gram flour (*Cicer aerietinum* L.) is also added at 20 per cent level. After fermentation, the dough is worked into balls, placed on a smooth surface, lightly sprinkled with flour and flattened by hand to 15-20 cm in diameter. Smoothly flattened dough is transferred onto a metallic baking tray or scooped in a special oven, made of clay. The baking is carried out for 30-40 min to get the final product [5].

## 5.5 *Nan*

*Nan* is generally consumed as a staple food by the people of India, Pakistan, Afghanistan and Iran. Its dough is raised by using white wheat flour, salt, inoculum from previous fermented batch and water (100:2:10:50), which is then kneaded with hand at room temperature and allowed to ferment in open for 6-8 hours. Sometimes, the dough is prepared by sponge dough method when yoghurt is added as an inoculum along with baker's yeast or some chemical leaveners. The flattened dough is then transferred onto a circular pad of cotton cloth and plastered onto the inner wall of a clay oven called tandoor. Before baking, some spices like caraway seeds (*Carum carvic*), coriander (*Coriandrum sativum*) and *kalaunji (Bochanania lauzan*) are sprinkled on it. The finished product is obtained by baking at 120-150°C for 10-15 min [5].

## 5.6 *Chilra*

*Chilra*, called *Lwar* in Lahaul, is more or less like *'Dosa'* but differs from it in terms of ingredients and shape. In its preparation, the major ingredients used are wheat/barley/rice and buckwheat flour. Inoculum used for its preparation is called *'Treh'* (previously fermented and left-over wheat/rice flour slurry). The traditional bucket-shaped wooden vessel used for fermentation is called *'Lwarenza'* [4]. The process for the preparation of *Chilra* is presented in Fig. 3.

*Chilra* is served with coriander chutney, potato and mutton soup in Lahaul valley. It is a staple food of the people of the area, and is a popular snack/staple food of the people in Lahaul/Kangra besides

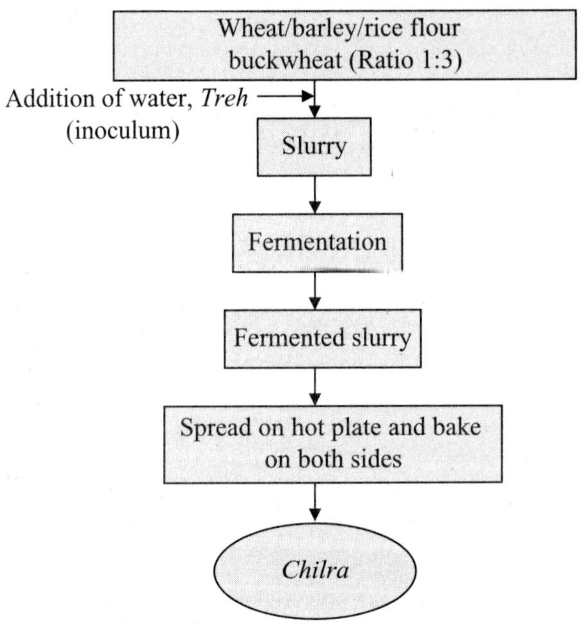


**Fig. 3:** The process for the preparation of *chilra*.

its preparation and consumption during marriage ceremonies and local festivals. *Chilra* is also used with milk in Mandi district, while sweet *Chilra* is made and consumed in Kangra [8].

## 5.7 *Siddu*

*Siddu* is a traditionally fermented steam cooked, oval or disc-shaped dish prepared as a delicacy in rural areas of Kullu and Shimla districts of Himachal Pradesh. In *Siddu* preparation, wheat flour, spices, mixed paste of opium seeds/walnut/black gram are the main ingredients. Inoculum used in its preparation is *'Malera'* (previously natural fermented left over dough). At present, yeast powder is also used for fermentation [4]. The process of *Siddu* preparation is illustrated in Fig. 4.

## 5.8 *Marchu*

Major ingredient for *Marchu*, prepared during the local festivals (*phagli, halda*) in Lahaul valley, is wheat flour. The inoculum used for fermentation is called *'Malera'*, which is previously natural fermented

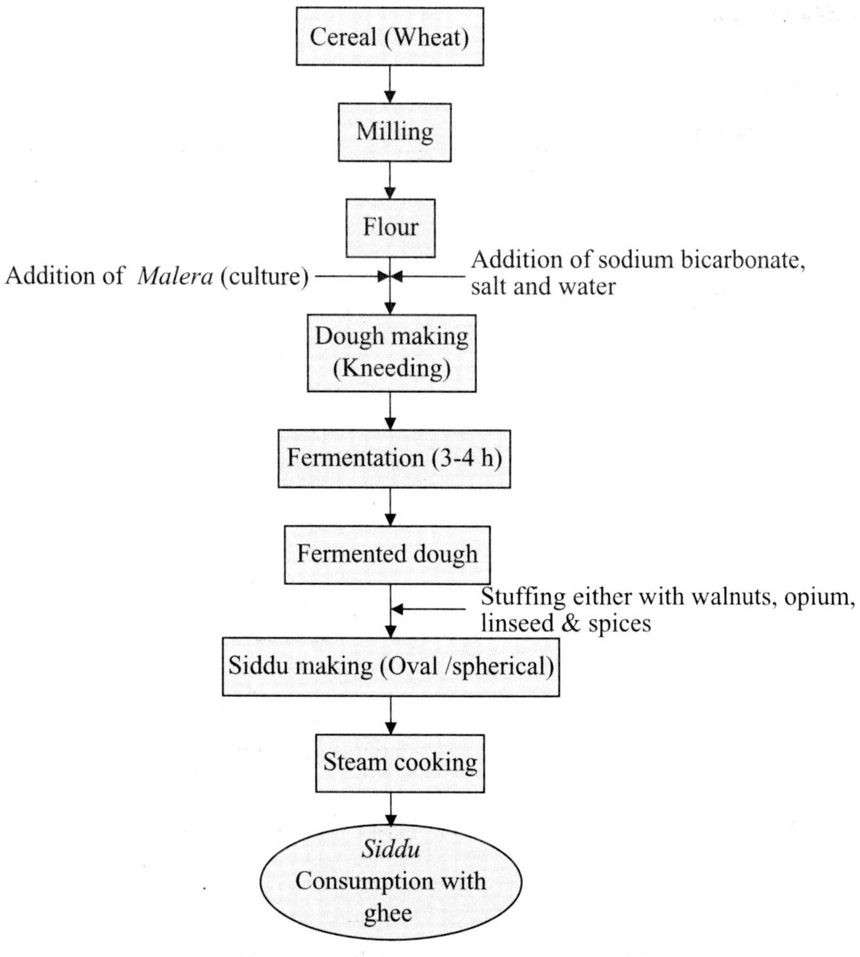


**Fig. 4:** The process of *siddu* preparation from wheat.

left-over dough [4]. The *Marchu* are roties made on a wooden base with carving to give designed imprints on *Marchu*. The method of making *Marchu* is illustrated in Fig. 5.

These are prepared during festivals, religious and marriage ceremonies as a snack/breakfast food to be taken with tea. It is also customary for daughter to take *Marchu* with her, whenever she visits her maternal home from in-laws or vice-versa [4]. When it is shaped like folded spirals, it is called *Doshu*. It can be sweet or salty.

## 5.9 *Sumkeshi Roti/Bhaturu*

*Bhaturu*, also known as *Sumkeshi roti* in Lahaul, is an 'indigenous bread' or *roties* and constitutes staple diet of the Himachalis living in rural areas of Kullu, Mandi, Kangra, Chamba and Shimla districts of Himachal Pradesh (India). *Bhaturu* serves as a staple diet for rural people who even take it during both meals along with vegetables or curries. This item is especially prepared by old people who have difficulty in chewing [8]. Rural migrants in urban area also prepare *Bhaturu*. The starting material is wheat flour and inoculum used is called '*Malera*'. Normal fermentation time is 2-3 hours in summer and 4-5 hours in winter. People knead flour using more inoculum and warm water to reduce the time [4]. After fermentation, one can see the rise in dough. The process of making *Bhaturu* is illustrated in Fig. 6.

## 5.10 *Suhalu/Babroo*

When *bhaturu* is fried in oil, it is *Suhalu/Babroo*. The product is specially prepared during special ceremony. This item is popular in Mandi Suket [8].

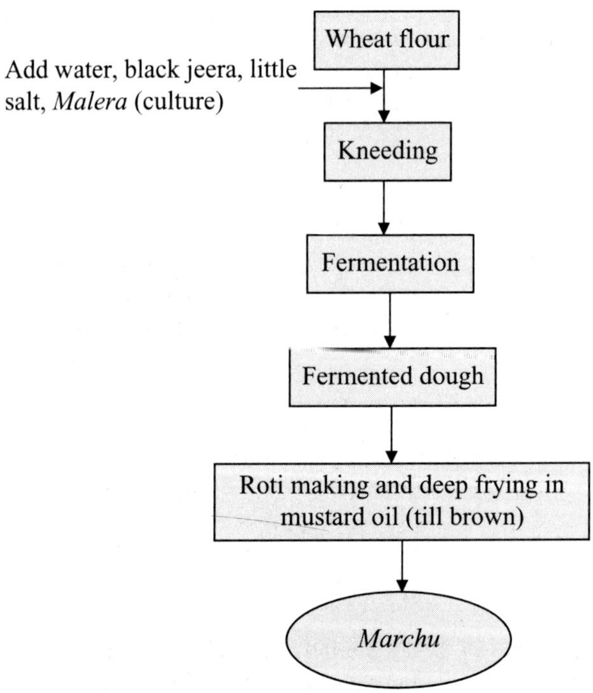


**Fig. 5:** The method of making *marchu*.

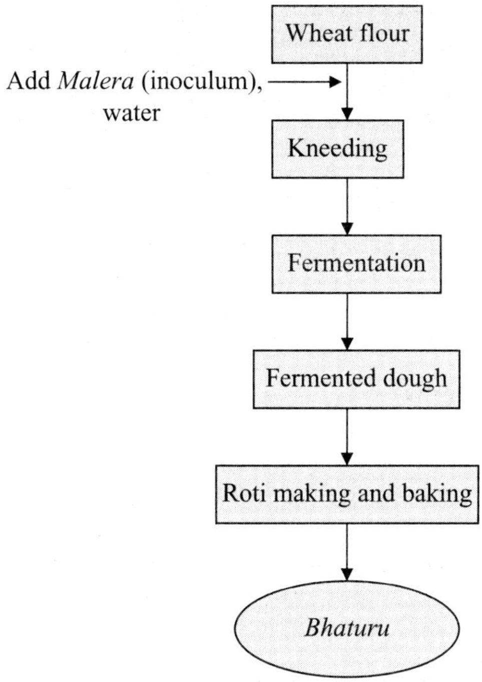


**Fig. 6:** Flowsheet for the preparation of *bhaturu* in Lahaul (H.P.).

## 5.11 *Bedvin Roti*

It is a modified form of *Bhatura* as these are stuffed with spices mixed paste of dal (black gram), opium seeds or walnut. These are either simple baked or deep-fried. The fried ones are called '*Kachoris*' in Mandi district [8]. There are taken as breakfast or snack food with tea.

## 5.12 *Gulgule*

*Gulgule* are prepared from wheat flour, made into viscous slurry by adding water, sugar and *'Malera'* (inoculum). This sweetened slurry is given small oval shape to be deep fried in oil until brown. These are prepared during religious and social ceremonies by both urban and rural population.

## 5.13 *Pinni/Bagpinni*

It is prepared in the tribal district of Lahal and Spiti. *Sattu* (roasted barley flour) is mixed with *Lassi* (buttermilk) or *Chhang* and kneaded in the form of dough, which is finally given the shape of a ball with a cavity in the centre [4] that is filled with ghee/butter.

## 5.14 *Seera/Kheera*

In its preparation, wheat grains are soaked in water for 2-3 days so as to allow fermentation to occur by natural microflora. The grains are then ground, steeping is done to allow the starch grains and some proteins to seetle down, and then bran is separated. The settled material is removed, sun-dried and is called *'Seera'/Kheera*. The dried material is made into slurry by soaking in water, which is then poured into hot ghee. To this sugar is added, cooked, and it serves as a sweet dish/snack. *Seera* is considered to be nutritious, easily digestible and fast snack food. *Seera* is prepared occasionally or offered to the guests as a sweet dish in the rural/urban areas of Kullu, Kangra, Mandi and Chamba districts.

### 5.15 *Khoblu*

It is a traditional fermented food of Mandi Suket area. Cooked rice is further cooked in butter milk. During cooking process, wheat flour *pedas* are also put in butter milk mixtures and cooking is continued till the *pedas* are cooked properly. Cooked *pedas* are consumed along with butter milk rice mixture [4, 8].

## 6. FERMENTED BEVERAGES

### 6.1 *Sura/Sur*

It is a millet-based (*Eleucine coracana*) fermented beverage prepared mostly in *Lug* valley of Kullu district and Chhota Bhangal area of Kangra District. Millet, locally called *Kodra,* has a long storage life and because of this it was popular as the 'famine grain'. *Phaff/Dheli* is the inoculum used for its preparation. Natural mixed microflora in *Phaff* or *deli* carry out starch hydrolyzing (saccharolytic) and ethanol forming (fermentative) activities. Herbal mix or *Dheli* preparation is an annual community effort, in which elderly people go to forests on the 20$^{th}$ day of *Bhadrapada* month (usually 5$^{th}$ or 6$^{th}$ September) and collect approximately 36 fresh herbs. Next day, the herbs are crushed in *'Ukhal'* (stone with a large conical cavity) using *'Mussal'* (a wooden bar) and the extract as well as the plant biomass are added into the *Sattu* and are roughly kneaded. This is poured into a wooden mould to give the shape of a brick and dried, which is called *Bhehli*. It is divided among the villagers, and it used whenever *Sura* is to be prepared. The *Dheli* provides bioactive compounds as well as stimulatory effect.

Some of the important barks used in *Dheli* preparation are *Pistacia integerima* (Kkakar shinga), *Solanum xanthocarpum* (Katati), *Clitoria ternatea* (Kkayal), *Aegel marmelos* (Bhel), *Viola cinerea* (Banaksa), *Cannabis sativa* (Bhang), *Trachyspermum copticum* (Ajwain), *Micromeria biflora* (Chharbara), *Spiranthes australis* (Bakarshingha), *Saussurea* sp. (Bbacha), *Bupleurum lanceolatum* (Nimla), *Drosera lunata* (Oshtori), *Salvia* sp. (Kotugha), *Arisaema helleborifolium* (Chidiri chum), *Fragaria* sp. (Dudlukori) and a number of other herbs. *Sur* is consumed during local festivals and marriage ceremonies. The process for the preparation of *Sura* is given in Figs. 7 and 8.

### 6.2 *Chhang/Lugri Chakhti/Jhol*

This is an indigenous rice beer made in the tribal belt of Lahaul and Spiti. Kullu and Kangra Huskless Sherokh variety of barley, locally called Grim, is used in its preparation [12]. The preparation of *Chhang* involves solid state fermentation as no additional water is added to the ingredients i.e., cooked rice and *'Phab'* (the traditional inoculum) or phapus. The quantity of water is used that is sufficient for absorption. *Chhang* is also prepared from barley; however, it takes longer time (one week) to ferment and the beverage is called *Lugri* in Lahaul. The traditional vessel, made of metal or stone used to store *Chhang,* is called *'Uthi'* in Lahaul. The *Chhang/Lugri* is a very popular fermented beverage which is served during *Phagli* (traditional new year of Lahulis) and marriage ceremonies to guests. *Chhang* is also called *Jhol, Chakti* in Kullu. Distilled form is known as *Sera* in Lahaul valley. This beverage has a religious significance, as it is sprinkled on guests as *shagun* (tribal custom). *Chhang* serves as a

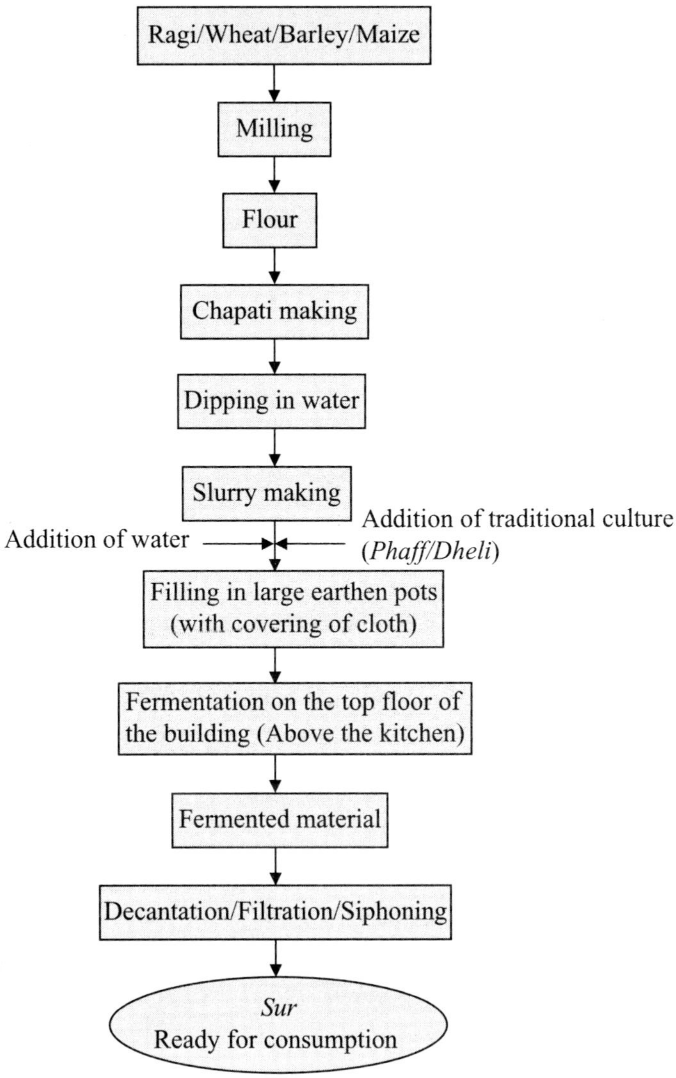


**Fig. 7:** Flowsheet for the preparation of *sur* (fermented drink).

tonic in winters as it contains vitamins, amino acids, and sugars besides alcohol and is considered to provide protection against cold. The process of *Chhang* preparation is outlined in Fig. 9.

## 6.3 *Daru/Ghanti*/Liquor

*Daru* is a jaggery-based traditionally fermented beverage prepared and consumed in rural areas of Shimla, Kullu and other regions of Himachal Pradesh. It is one of the popular alcoholic beverages prepared in the rural areas especially for entertaining guests during local festivals and marriage ceremonies. Babool wood locally called *kikar* (*Acacia nilotica*) is added to give taste and aroma to Daru. The process of *Daru* (distilled alcoholic beverage) preparation is depicted in Plate 3 and Fig. 10.

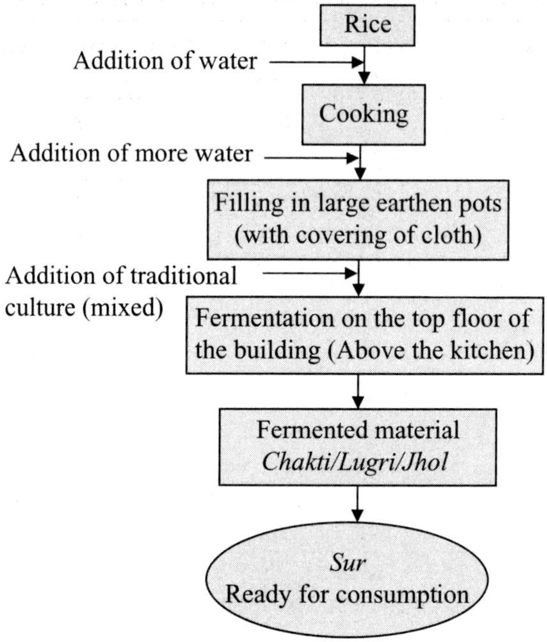


**Fig. 8:** Flowsheet for the *Sur* preparartion from the rice.

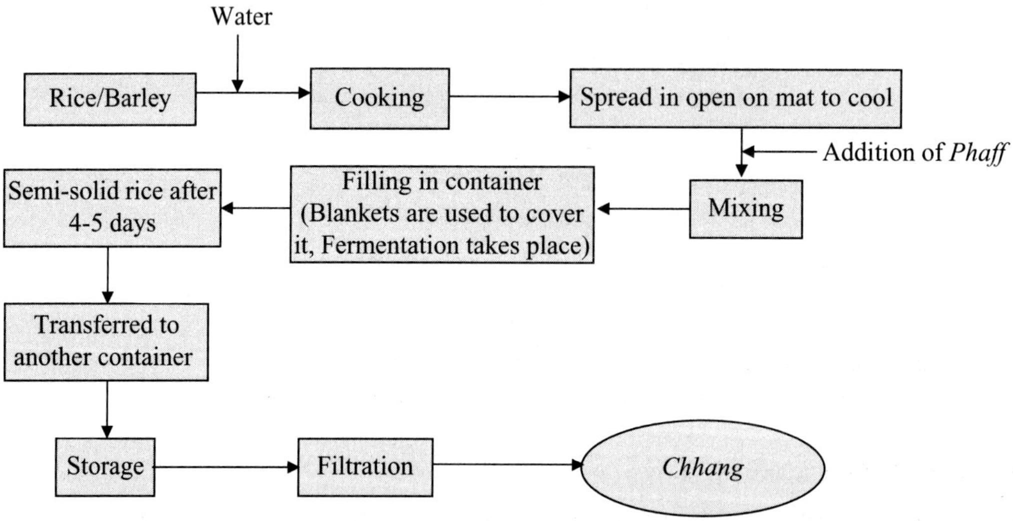


**Fig. 9:** Flowsheet for the *Chhang* preparation.

**Plate 3:** A view of distillation of alcoholic beverages in a tribal village of Himachal Pradesh, India. *Source:* 10

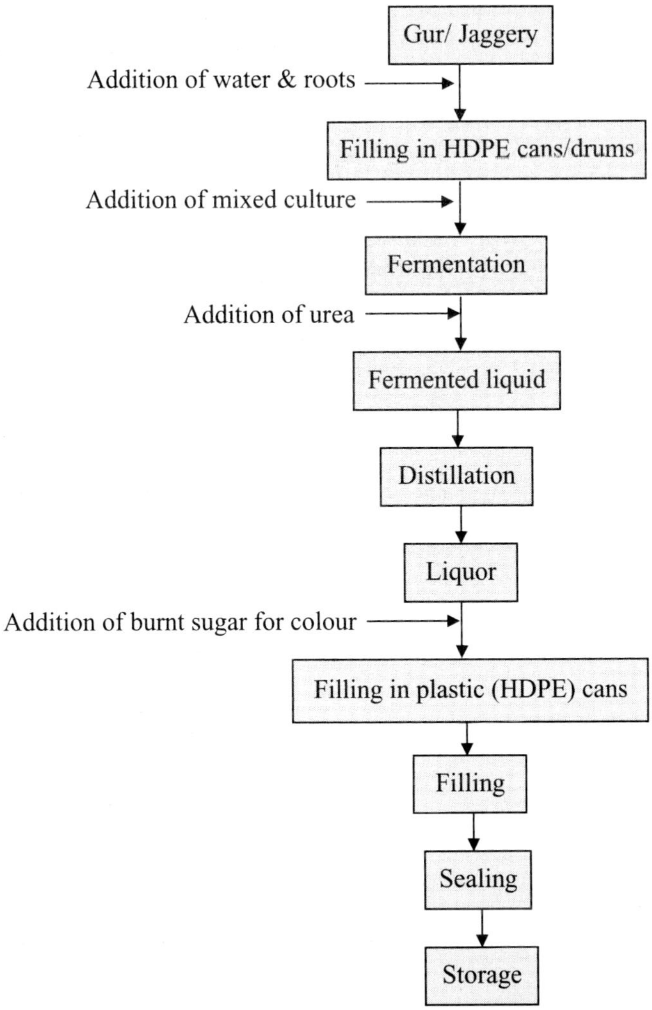


**Fig. 10:** Flowsheet for *Ghanti/Daru* preparation from *gur*/jaggery.

### 6.4 *Angoori*/Fruit Brandy

In the preparation of *Angoori*, Chholtu red and Chholtu white variety of grapes as used. A number of grape varieties are grown in the Kinnaur district and their produce is used to make this local alcoholic beverage (wine) as shown in Fig. 11. Angoori is a grape liquor. This is one of the very popular traditional fermented beverages consumed during local festivals and marriage ceremonies by the tribal people of Kinnaur. The *chulli* (wild apricot) is also employed to make a similar product after fermentation of fruit pulp and distillation of wine [13, 14].

## 7. FERMENTED MILK PRODUCTS

It is estimated that about 9.1% of the total milk produced in India is converted into various fermented milk products excluding cheese [15]. In all milk based products, the main biochemical change is the production of lactic acid from lactose by a variety of lactic acid bacteria (LAB) belonging to *Lactococci*, *Streptococci, Leuconostoc, Lactobacilli* and few others. The metabolism of citrate by some of these bacteria leads to production of diaceyl that gives typical pleasant flavours. Growth of LAB also makes the product microbiologically safe by controlling the unwanted spoilage causing microorganisms and pathogens. Due to antimicrobial properties of LAB towards intestinal pathogens, these foods have been used since very early times for the control of gastrointestinal disease, especially diarrhoea.

A method was optimized to prepare *lassi* using mixed culture of *Streptococcus thermophilus*, and *Lactobacillus acidophilus* using different levels of stabilizer, sugar, etc. The fresh *lassi* contained $20.4 \times 10^7$/ml live cells of *Lactobacillus acidophilus,* which remained almost constant upto storage period of 27 days. The product was highly acceptable, showing sensory scope of 17.49 out of 20 when fresh and 15.90 after 15 days of storage in pouch [16].

Acidophilus milk was made by process using *Lactobacillus acidophilus* adapted to human strain, was blended with banana paste, tomato juice and sugar and, finally spray dried. The powdered product was organoleptically acceptable and had higher nutritive value with respect to its vitamin content and protein quality [17]. The product showed as high as 14% survival of *Lactobacillus* culture ($8 \times 10^7$ cfu/g), which reduced to about one log cycle during 2 months of ambient storage [18].

Surplus milk in the house is collected preferably in pots and allowed to ferment naturally. It develops natural culture of lactic acid bacteria that carry out continuous fermentation of milk added everyday. The fermented clotted milk (gel) is called *dar* and is consumed with rice or chapatti. It

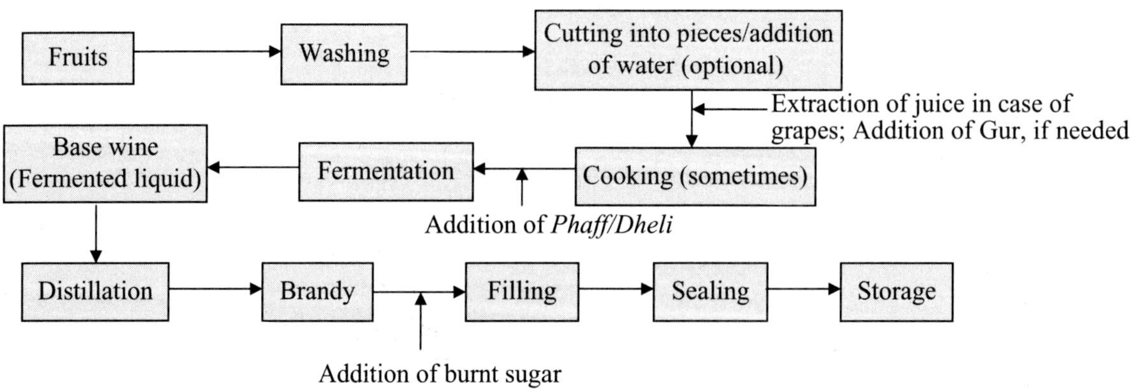


**Fig. 11:** Flowsheet for the preparation of *Anguri* in the Kinnaur Distt. of Himachal Pradesh.

is agitated, salt or sugar is added and relished as *lassi*. The *chhash* may also be spiced and cooked along with a small portion of flour (*besan*), and the preparation is called *kadhi*.

In the eastern part of India, there is a tradition of making *dahi* from concentrated milk and rich amount of sugar (upto 15.25%) with caramel, which looks like strong brown junket. Traditionally, this was made in small earthen pots, but is now available in plastic containers.

In western part of India, there is a tradition of removing excess whey from the curd by hanging it in a muslin cloth. The concentrated curd mass is called *Chhakka*. This is kneaded usually with equal amount of sugar and added with different flavours, colours, fruits, nuts etc. and relished as sweetmeat called *Shrikhand*. It is a common sweet dish in many social functions too.

## 8. FERMENTED ALCOHOLIC PRODUCTS

In the eastern and hilly regions of India, rice is the main crop. Fermented rice is used to make traditional beer called *Opo*. Products like *Chhcespi, Chhang*, Sher, *Shosim, Kenima* etc. are also popular in the region. Traditional alcoholic brewing is a home-based industry in the Himalayas, practiced mostly by rural women using their indigenous knowledge [19]. The people in the Eastern Himalayas eat cooked rice as a staple food. *Bhaati jaanr*, a traditional fermented rice paste, is also commonly consumed as a mild-alcoholic sweet beverage in the rural areas of Darjeeling hills, Sikkim and north-east states in India, Nepal and Bhutan. Drinking alcohol is the traditional custom of the ethnic people of the Eastern Himalayas, which has been documented in the historical literature [20]. *Bhaati jaanr* is a Nepali word for fermented rice beverage. During the production of *Bhaati jaanr*, glutinous rice is cooked for about 15 min in an open cooker, excess water is drained off, and cooked rice is spread on a bamboo mat for cooling (-40°C). A dry mixed starter culture called *marcha* is powdered, and about 2% of it is sprinkled over the cooked rice, mixed well and kept in a vessel or an earthen pot for 1-2 days at room temperature for saccharification [21, 22]. A sweet aroma is emitted during saccharification. After saccharification, the vessel is made airtight and fermented for 2-3 days in the summer and 7-8 days in the winter. *Bhaati jaanr* is made into a thick paste by stirring the fermented mass with the help of a hand-driven wooden or bamboo-made stirrer (Fig. 12).

*Kodo ko jaanr* is the most common fermented alcoholic beverage prepared from dry seeds of finger millet [*Eleusine coracana* (L) Gaertn.], locally called '*kodo*' in the Eastern Himalayan regions of the Darjeeling hills and Sikkim in India, Nepal and Bhutan. *Jaanr* is the common name for all alcoholic

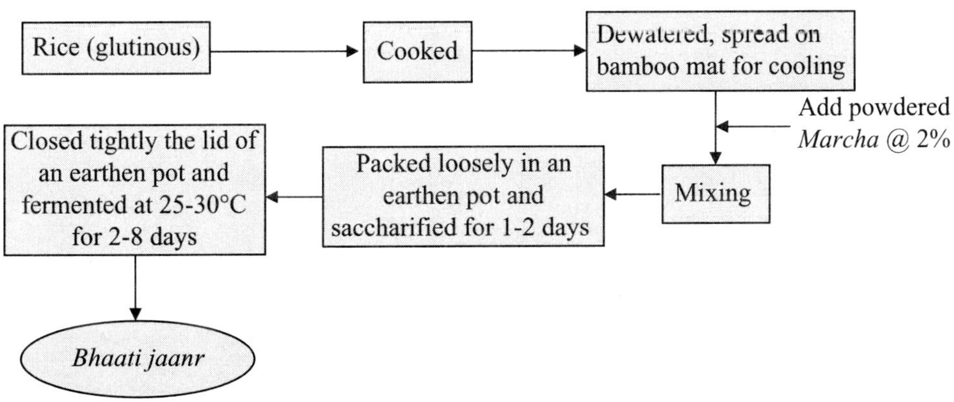


**Fig. 12:** Flow sheet of *Bhaati jaanr* preparation in South Sikkim.

beverages in Nepal; Tibetans call it '*minchaa chhyaang*', and the Lepchas calls it '*mong chee*'. Description of alcohol-drinking custom in the Sikkim Himalayas has been cited in some historical documents [20]. Similar product called *chhang* has been reported from Ladakh region in India [12]. During the traditional method of *kodo ko jaanr* preparation, dry seeds of finger millet are cleaned, washed and cooked for about 30 min in an open cooker, excess water is drained off and the cooked seeds are spread on a mat made up of bamboo locally called '*mandro*' for cooling. About 2% of dry powdered *marcha*, a mixed starter culture [21] is sprinkled over cooked seeds, mixed thoroughly and packed in a bamboo basket lined with fresh fern, locally called '*thadre unioon*'(*Thelypteris erubescens* Well ex Hook.) or banana leaves, then covered with sack clothes, and kept for 2-4 days at room temperature for saccharification. During saccharification, sweet aroma is emitted out and the saccharified mass of finger millets is transferred into an earthen pot or specially made bamboo basket called '*septu*' and made it air-tight and fermented for 3-4 days during summer and 5-7 days in winter at room temperature (Fig. 13). Good quality of *Kodo ko jaanr* has a sweet taste with mild alcoholic flavour. *Kodo ko jaanr* is consumed in an unique way in the Eastern Himalayan regions of Nepal, India, Bhutan and China (Tibet). About 200-500 g of fermented seeds of finger millets (*kodo ko jaanr*) are put into a vessel called '*toongbaa*' and lukewarm water is added up to the edge of the '*toongbaa*'. After 10-15 min, milky white extract of *kodo ko jaanr* is sipped through a narrow bamboo straw called 'pipsing' having a hole in a side near the bottom to avoid passing of grits. Water can be added 2-3 times after sipping up the extract. After consumption, grits of *kodo ko jaanr* are used as a fodder for pigs and cattle. This is a good example of total utilization of substrate as food and fodder [22].

The proximate composition of finger millet and *kodo ko jaanr* show that it has pH, acidity expressed as lactic acid, and the alcohol content of the product ranged from 3.7-4.5, 0.23-0.5% and 1.8-8.7%, respectively. Moisture content of the product was slightly higher in fermented product than unfermented cooked millet. No remarkable change was observed in ash, fat and protein contents of *kodo ko jaanr* over the substrate. Crude fibre content increased during fermentation. Calorie content of the unfermented millet and fermented product was almost the same. Increase in calcium, magnesium, manganese, iron and phosphorus was also observed in *kodo ko jaanr*, which is a mild-alcoholic sweet-

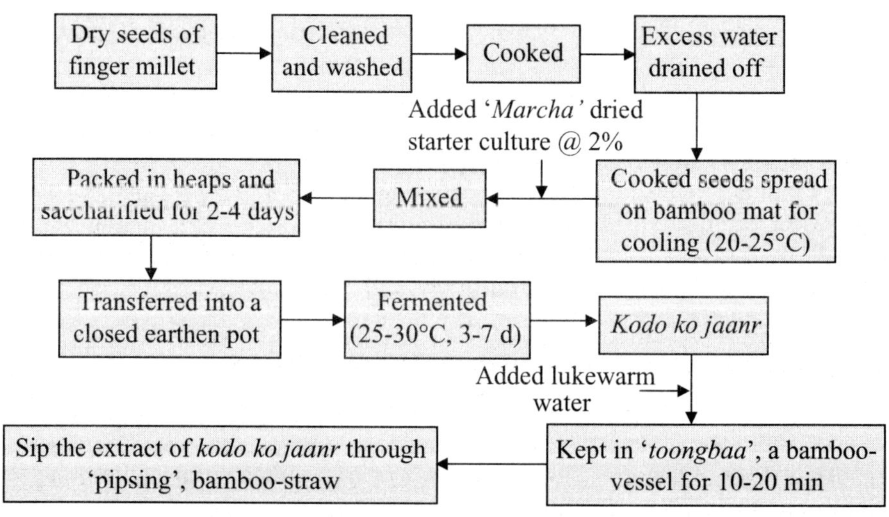


**Fig. 13:** Flow sheet of *kodo ko jaanr* preparation in Sikkim.

flavoured beverage. Because of high calorie, ailing persons and post-natal women consume the extract of *kodo ko jaanr* to regain the strength. Increase in minerals could be due to the recipients used (e.g., septum, *toongbaa*) or the material used (fresh fern) to line the bamboo basket during production. Finger millet, also called 'ragi' in South India, has been reported to be a good source of iron, calcium, magnesium and phosphorus [23].

India grows a variety of legumes like red gram (*tur*), black gram (*urad*), Bengal gram (*gram*), green gram (*moong*), cow pea (*chora*), moth beans (*muth*), massor, peas, etc. The most common food items *idli* and *dosa,* are prepared from rice and curd dal mixture, which is wet ground and fermented for 16-48 h with 60-75% moisture under atmospheric conditions. The fermented paste is then, steam cooked (*idli*) or fixed as pan-cake (*dosa*). The mixture of rice and various types of pulses depending upon the availability and liking are mixed, wet ground, fermented with additions of butter milk and *dahi*, and the legumed paste is spiced and steamed cooked to have *khamon* and *dhokla*, which are very popular in Gujarat State. *Handwa* is the spiced *rake* made from such mixtures.

*Bhalle* or *vada* are deep fried patties, usually prepared from black gram and served off soaking in *dahi*. All these products involve fermentation of carbohydrates by lactic acid bacteria (LAB) and yeasts. LAB produces acids to make it sour, and gives proper texture to the products.

*Warries* are spicy, hollow, brittle balls made from fermented black gram spiced with asafoetida, caraway, cardamom, cloves, fenugreek, ginger and seed pepper.

*Roti* or flattened bread is the most common staple food in India. The baked staple food made from raised dough of white wheat flour with salt and previous day dough or *dahi* as inoculum is called *nan*, which is very popular throughout India. These are baked in *tandur* by direct fire. There are several varieties of this category of food called *kullcha, rumali roti, bhatura*, etc. This food is more digestible and palatable. *Jalebees* and *Imaarti* are confectionary prepared from fermented syrup wheat flour, which are deep fried and dipped in sugar syrup.

Pickles find an important place in Indian dish as spicy taste enhancer. Mangoes are the main fruits used for the purpose, while main vegetables include carrots, chillies etc. Fermented vegetables like sauerkraut have not found place in India diet, yet, but there is an ample scope to expose that possibilities for fermented vegetables to prevent their spoilage and minimize wastage.

## 9. FERMENTED SOY AND PEANUT MILK

Both soy and peanut milk have been developed to combat the short supply of animal protein in many developing countries. The aqueous extracts of soybean and peanut marked as milk in many Asian countries with a characteristic beany flavour. This off flavour can be reduced by fermenting the milk using lactic acid bacteria to yield yoghurt like textured products having highly acceptable quality. A soybean beverage has been prepared [24], while a lipid protein concentrate in which lipoxygenase was inactivated and acid pH was developed [25] which was free from off flavours. *Miltone* is a toned peanut milk having acceptable flavour developed by (CFTRI), India [26]. Fermentation of soybean and peanut milks is the best means of masking the undesirable flavour [27]. Fermentation of soy milk with lactic acid bacteria reported a product with gelatinous consistency and acceptable flavour [28, 29].

## 10. FERMENTED FOODS OF SRI LANKA AND MALDIVES

In Sri Lanka, the closest neighbour of India, fermented foods have been used as meals or part of a meal for a long time. The most well known is the curd made from buffalo milk, and has been a delicacy since ancient times. However, *string hoppers, roti, thosai, idli* together with one or two side dishes are few dishes that make the breakfast and dinner, and some are fermented products. Especially *thosai* and *idli* together with *wadai* are common foods in Tamil community. *Hoppers, thosai, idli wadai, curd, toddy, jaadi* (caused fish) lime pickle dry fish, based are common fermented foods that have been consumed for a long time. However, yoghurt arrack, condiments like vinegar and mouldier fish, and newly developed vegetable pickles *salami, pepperoni* and *tempeh* are also commonly available fermented products, while curd made from buffalo milk is the most popular form of fermented milk in Sri Lanka [30]. Just before the preparation of hoppers, coconut milk is added and cooked in a hot griddle. Hoppers are much like sour dough pancakes or muffins [31].

*Tempeh*, produced by incubating soya with a fungus, is also used in population. When making *tempeh*, soya seeds are boiled (24 h), soaked (24 h), skin removed and kernels steamed and mashed kernels are incubated with the mould and wrapped in banana leaves and kept for 24 h [30]. *Tempeh* is mainly used for the preparation of snack as a textured vegetable portions, a meat substitute has been introduced, recently.

*Toddy* is a common fermented alcoholic beverage popular in the villages made by the fermentation of the soup from the coconut (commonest), palmrah (northern parts) and Kithul (wet-zone) palm. The soup is collected by slicing off the tip of an unopened flower. The soup oozes out and collected twice daily in a small pot tied underneath the flower. It is white and sweet with a characteristics flavour (4-6% alcohol) and has a shelf life of about 24 h. *Toddy,* when fermented, becomes arrack, which comes in varying degrees of strength. *Toddy* of alcohol content, 6.7%, is distilled in copper or stainless steel plants to obtain arrack. It is then stored in halmilla (*Bersya cordifolis*) vats for maintaining flavour component and colour. The matured spirit is blended with water to a final alcohol strength of 33% and bottled.

*Salami* and *Pepperoni* produced by using starter cultures on meat are also getting popularity though not traditional products. Meat is cultured with starter cultures, frozen, and kept at room temperature for 3 days. Then it is put in a steam chamber (34-40°C, 3-4 h) and in a controlled atmospheric room (3 weeks) where the temperature and relative humidity are gradually decreased. Depending on temperature, relative humidity and time, *Salami* and *Pepperoni* are produced.

*Jaddi* (Pickle curse) is prepared by the columbo curse or pickle cure method practical in Sri Lanka. Gutted fish are mixed with salt in a ratio of 3 parts of fish to 1 part of salt in large cement tables and dried pods of *Goraka* or tamarind are placed on the body. *Goraka* improves the process by lowering the pH below 4.0. A significant improvement in the quality of the product is obtained by replacing tamarind with 5% acetic acid in the ratio of 1 part of acid to 20 parts of fresh. It can be kept for one year [31]. The curry of *Ambulthiyal,* the traditional tuna fish in Sri Lanka has a shelf life of around 7 days at ambient temperature. Chemical composition of cured and fermented fish products in Sri Lanka is given in Table 6. Changes associated with fermentation are described in Table 7.

**Table 6:** Chemical composition of cured and fermented fish products in Sri Lanka.

| **Parameters** | **Rawfish** | ***Ambulthiyal*** | **Dried fish** | **Maldive fish** | **Jaadi pH 4.0** | **Fish sauce pH 5.9** |
|---|---|---|---|---|---|---|
| Moisture (%) | 70.80 | 58.52 | 28.0 | 15.73 | 48.8 | 69.12 |
| Ash (%) | 2 | 5.9 | 3.16 | 6.40 | 22.8 | 31.5 |
| Crude Protein (%) | 15.24 | 28.93 | 64.51 | 68.11 | 25.71 | 16 |
| Fat (%) | 0.1-2 | 1.78 | 3.5 | 2.46 | 2.24 | 1.5 |
| Salt (%) | 0.2 | 4.77 | 17.13 | 5.98 | 18.3 | 29.8 |
| Shelf life (days) | 1 | 7.180 | 80 | 120 | 360 | 44 |
| Water activity | 0.96 | 0.95 | 0.56 | 0.76 | 0,764 | - |

*Source:* 31

**Table 7:** Microbial, chemical and sensory changes during fermentation.

| Parameter | Maldive fish | *Ambulthiyal* | Dried fish | Jaadi | Fish sauce |
|---|---|---|---|---|---|
| TVN (total volatile nitrogen mg/wg) | 29.58 | 24.56 | 62.3 | 52.3 | 240 |
| Peroxide value (%) | 246.51 | 28.5 | 109.5 | 2050 | - |
| Free fatty acids (%) | 52.59 | 56.7 | 54.7 | 18.81 | - |
| Total bacterial count/g | $6.30 \times 10^5$ | $6.72 \times 10^3$ | $1.97 \times 10^5$ | $1.986 \times 10^3$ | $2.5 \times 10^3$ |
| Mould count/g | $2.99 \times 10^2$ | $1.52 \times 10^2$ | $2.4 \times 10^5$ | $1.13 \times 10^3$ | $1.6 \times 10^3$ |

*Source:* 31

## 11. FERMENTED FOODS OF INDONESIA

There are many kinds of indigenous fermented foods in Indonesia that require the use of soybeans, cassava, and glutinous rice. The main ones are *tempeh, kecap* (soy sauce), *tauco* (fermented whole soybean, similar to *miso*), *tape ketela* (cassava), *tape ketan* (glutinous rice), and brem liquid or solid. Indonesian fermented foods are produced by small- and medium-sized traditional enterprises. Government policy protects the small manufacturers. A production monopoly by large enterprises has to be avoided. On the other hand, considering the cultural and economic characteristics of fermented products, it would be self-defeating to modernize the traditional processes.

Traditional *tempeh* is a fermented food in which fungi, particularly *Rhizopus oligosporus*, play an essential role. Fresh *tempeh* is defined as a compact and sliceable mass of hydrated pre-cooked bean cotyledons bound together by mold mycelium. The major desirable characteristics of *tempeh* are its flavour, texture, and nutritional properties. In Indonesia, *tempeh* is consumed as a protein-rich meat substitute by all economic groups. Yellow soybeans are most commonly used, but black-coated varieties are used in some areas. The best-quality *tempeh* is made with soybeans, but lower-cost, lower-quality *tempeh* is made with other kinds of beans, such as wing, pigeon, and velvet beans. Production of *tempeh* is shown in Fig. 14.

*Tempeh* is a good source of protein, essential fatty acids, vitamin B, especially $B_{12}$, and minerals, including iron. *Tempeh* also contains antibacterial substances that can protect against infectious diseases, such as diarrhoea. *Tempeh* contains potent antioxidants due to the conversion of isoflavonoid compounds present in soybeans into the 6,7,4 trihydroxy isoflavon known as Factor 2, and a more active antioxidant enzyme superoxide dismutase. The antioxidant activity of *tempeh* may be related to its potential activity to retard the aging process. *Tempeh* has been assumed to have hypocholesterolemic properties. To obtain the maximum benefit from *tempeh* as a nutritious and functional food, consumption should be increased. However, the quality of *tempeh* in terms of the content of essential nutrients and functionality can only be assured by correct processing. Even though *tempeh* is being promoted, it is not widely consumed by group.

*Kecap* (soy sauce) is a condiment used to add flavour and colour to other foods, such as soybean curd, rice, fish, meat, and vegetables. It is consumed in several grades or levels of quality by all societal levels in urban and rural areas. The quality consumed depends on the consumer's income level. Most of the manufacturers are small-scale producers with limited capital who cannot afford to invest in modern equipment and facilities. However, *kecap* is part of the Indonesian fermented food industry, and demand is on the rise. There are two types of Indonesian *kecap*: *Kecap manis* is sweetened soy sauce, and *kecap asin* is salty soy sauce. The sweetened variety is most popular. It is thicker and dark brown because of the addition of palm sugar. The dominant flavour of *kecap* is derived from spices, such as *Alpinia galanga, Carum roxburghianum*, leaves of *Citrus hystrix, Eugena polyantha, Foeniculum volagare* (fennel), or aniseeds, *Andropogen ceriferus, Coriandrum sativum* (coriander), garlic and *Polianthes tuberosa*.

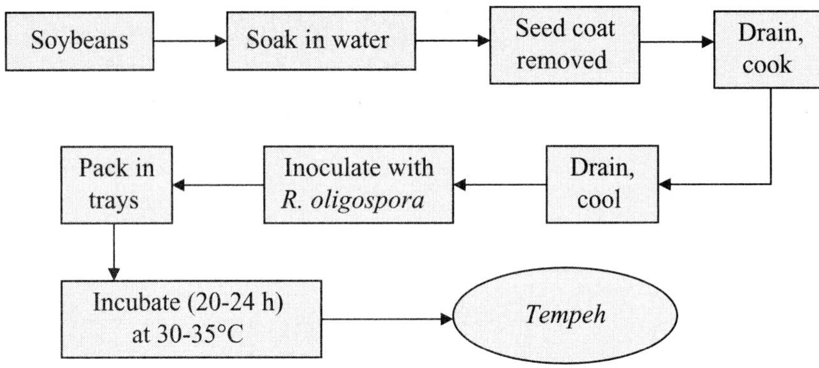


**Fig. 14:** Flow diagram for the production of *tempeh*.
*Source:* 32

Indonesian *tauco* is a yellow, saline paste with the consistency of porridge and a meat-like flavour. *Tauco* is a product of mold, bacteria, and yeast fermentation of soybeans, closely related to Philippine *taosi* and Japanese *miso*. The first stage of the fermentation, the overgrowth of soybean cotyledon with moulds, is very similar to the manufacture of *tempeh*. There are two types of *tauco*: one is a moist, slurried *tauco*, and the other, a dry paste *tauco*. The dry paste, which is sun-dried, is generally rehydrated before use.

Indonesian *tape ketan* is a sweet/sour rice-based alcohol in which an amylolytic mould of the *Amylomyces rouxii* type and at least one yeast of *Endomycopsis burtonii* type hydrolyse steamed rice (starch) to maltose and glucose and then ferment the sugar to ethanol and organic acids, which produces an attractive flavour and aroma. The most common *tape* is *tape ketela* (cassava) or *peuyem* (West Java). By fermenting cassava to *tape ketela*, it is possible to raise the protein content to 4% or higher. When consumed in this form, it imparts important protein and amino acids. During the processing of *tape ketela*, 40-70% of cyanide is lost during steaming; on the other hand, 9 to 35% is lost during 3 days of fermentation [33]. Manufacture of Indonesian *tape ketan* and *Indian Bhattejaanr* is given in Fig. 15. The juice of *tape ketan* can be used to produce *brem*. There are three types of *brem*, of which two are solid. The first, *brem* Madiun, is yellowish-white, sweet-sour, and prepared in blocks of 0.5 × 5 to 7 cm; the second, *brem* Wonogiri, is white, sweet, very soluble and is prepared in thin circular blocks of about 5 cm in diameter; the third type, brem Bali, is a famous liquor produced in Bali.

## 12. FERMENTED FOODS OF NEPAL, BHUTAN AND BANGLADESH

The most popular fermented foods of Nepal are *Kinema* (fermented soyabean), *Gundruk* (*Brassica compestris*) leaves, *Sinki (Raphanus sativus)*, *Tama* (succulent bamboo shoot), *Dahi* (similar to yoghurt), *Mohi* (butter milk), *Sher, Shergum* (soft cheese from buttermilk), *Chhurpi* (Dried very hard cheese from buttermilk), *Selroti* (deep fried preparation from rice flour), *Jand* (Local beer from rice/ maize millet) and *Tumba* (fermented millet drink).

Fermented milk products are other popular food products in this mountainous region. Very popular fermented foods of milk are *Dahi, Mohi, Sher, Sher ghum, chhurpi, Yak Cherso* and other cheeses. *Sher, Shergum, Shosis, Churtsi Chlauga or Chhurpi* are produced from Nak (Yak) milk in the higher altitude region of Bhutan [34]. Fermented foods in Bangladesh means some food items manufactured by the meat industry. In Bangladesh e.g., *Dahi* (sweet and sour Yoghurt), *matha* (Medo from

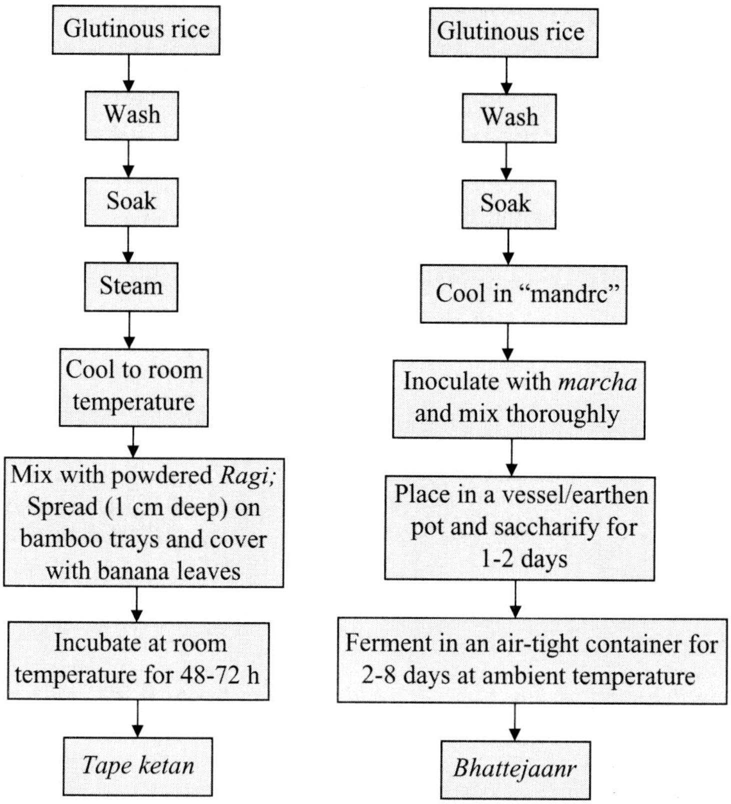


**Fig. 15:** Flow chart for the manufacture of Indonesian *tape ketan* and *Indian Bhattejaanr. Sources:* 21, 7

buttermilk) and fermented whey dried are some of the fermented products. Fermented whey dried is prepared from channa whey by adding 8% sugar, pH is adjusted to 3.8 and inoculated with 0.5% mixed culture, and incubated for 24 h at 40°C [35]. The drink is heated at 85°C for five minutes before filling in the bottles.

Soft-variety *Chhurpi* is a traditional cheese-like product which is commonly consumed by different ethnic groups of people in the Himalayan regions of the Darjeeling hills, Sikkim and Ladakh, Nepal, Bhutan and Tibet [36]. This type of *Chhurpi* has a white soft texture with a mild to strong flavoured taste and is consumed as curry mix with wild edible ferns (*Diplazium* sp.), pickles and condiments, along with boiled rice in meals [37]. The traditional production method of soft-variety *Chhurpi* is illustrated in Fig. 16 [38].

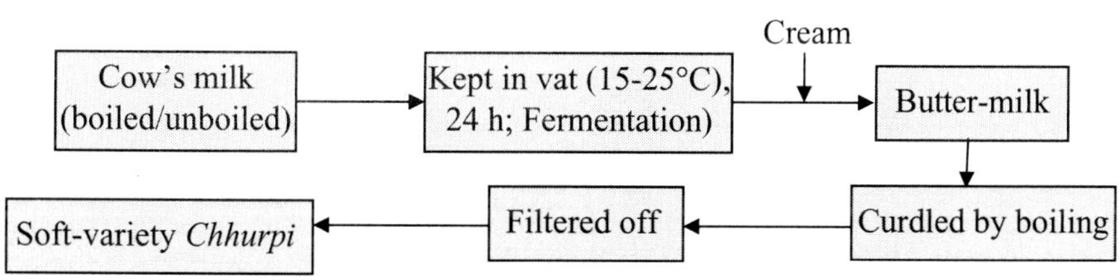


**Fig. 16:** Flow sheet of traditional preparation of soft-variety *Chhurpi.*

Soybean and its products have served as the principal protein components of million of people in the Orient for many centuries [39]. One of the products which is highly relished as a food condiment is the soy sauce. It is obtained by a complex fermentation of a mixture of soybeans and wheat using a mould in the initial aerobic stage followed by yeast and bacterial fermentation in the second anaerobic stage. Soy sauce is a liquid food condiment used to add flavour and colour to the blend, basic Orient diet including rice, raw fish, bean curd, fermented beans, boiled vegetables, etc. [40]. There are two major types of soy sauce in the Orient: the Chinese type and the Japanese type. In China, *Shoyu* is mainly made from soybeans alone or from a mixture of soybeans and wheat with a higher proportion of soybeans. Japanese soy sauce is made from equal amounts of soybeans and wheat. The four major characteristics that distinguish Japanese soy sauce from others are aroma, flavour, colour and stability [41]. Legume-based fermented foods are enlisted in Table 8.

*Kinema* is a non-salted, solid-substrate fermented, flavoursome, alkaline food, traditionally consumed mainly by the Nepalis, Lepchas and Bhutanese. It is a low-cost protein complement to rice. In the traditional method of *kinema* preparation, yellow seeded soybeans are cleaned, washed, soaked overnight (12-20 h) at ambient temperature (10-25°C), cooked by boiling (90-95°C) in spring water for about 90 min, crushed lightly to grit, wrapped in fern banana leaves and sackcloth, and left to stand (25-35°C) for 1-3 days. The desired state of fermentation is indicated by the formation of a typical *kinema* flavour dominated by ammonia as well as a white viscous fluid on the beans. Fresh

**Table 8:** Legume-based fermented foods.

| Product | Country specific | Substrate | Microorganisms involved | Nature use |
|---|---|---|---|---|
| *Chee-fan* | China | Soybean whey curd | *Mucor* sp., *A. glaucus* | Solid, eaten fresh like cheese |
| *Danadawa* | West Africa, Nigeria | African locust bean | Lactic acid bacteria, yeast | Solid, eaten fresh |
| *Hamanatto* | Japan | Soybeans, wheat flour | *Aspergillus ozyzae, Streptococcus* sp., *Pediococcus* sp. | Raisin like, flowering agent for meat or fish or eaten as a snack |
| *Kecap* | Indonesia & nearby region | Soybeans, wheat | *A. oryzae, Lactobacillus* sp., *Hansenula* sp., *Saccharomyces* sp. | Liquid, condiment & seasoning agent |
| *Kinema* | Nepal, Sikkim & Darjeeling | Soybeans | - | Solid snack food |
| *Ketjap* | Indonesia | Black soybeans | *A. oryzae* | Syrup, seasoning agent |
| *Khaman* | India | Bengal gram | - | Solid cake like breakfast food |
| *Meju* | Korea | Soybeans | *A. oryzae, Rhizopus* sp | Paste, seasoning agent |
| *Natto* | Northern Japan | Soybeans | *Bacillus natto* | Solid cake, as a meat substitute |
| *Ontjom* | Indonesia | Peanut cake inter media | *Neurospora* | Solid, roasted or fried |
| Soybean milk | China, Taiwan | Soybeans | Lactic acid bacteria | Liquid drink |
| *Sufu* | | Soybean, whey, curd | *A. elegans, Mhiemalis silvaticus, M. subtilissimus* | Solid, soybean cake, condiment |
| *Tempeh* | Indonesia & nearby system | Soybean | *Rhizopus* sp. | Frizzed in oil, coasted as a meat substitute |
| *Warries* | India | Black gram flour | *Candida* sp., *Saccharomyces* sp. | Spongy, spray condition out |

*Source:* 42

*kinema* is briefly fried in oil and added with vegetables, spices, salt and water to prepare a thick curry [36, 43, 44, 45]. *Bacillus subtilis* was the most dominant bacterium on raw soybeans. *Kinema* contained *B. subtilis* as well as *Enterococcus faecium.* In addition to these two types of bacteria, *Candida parapsilosis* and *Geotrichum candidum* occurred in many of the samples of *kinema* [45]. *Bacillus* is the sole microorganism carrying out this fermentation; the members of the accompanying flora are merely opportunists [46]. As natural fermentation results in products of variable quality, development of controlled fermentation is essential for the manufacture of products of reproducible quality. This requires the knowledge of type and impact of the microorganisms involved in the fermentation. Fermentation starters used in different countries are given in Table 9.

*Kinema* was prepared under optimized process conditions [46]. Soybean [*Glycine max* (L.) Merrill cultivar 'local yellow'] seeds are cleaned, washed, soaked in tap water for 8 h, dewatered and autoclaved with fresh water at 110°C for 10 min. The cooked beans are drained and crushed lightly by a pestle to dehull and break the seeds to mainly half-cotyledons. The whole soybean grits are wrapped loosely in perforated polythene bags, covered with cloths to check over-drying and incubated at 37°C for 48 h. For the aseptic preparation of *kinema*, the method described above has been modified [44]. Soybeans (250 g) are washed thoroughly and soaked in distilled water for 8 h. After decanting the water, 125 g (wet weight) of soaked beans are put into each of several 500 ml screw-capped bottles, and 250 ml distilled water is added in each. The bottles are autoclaved at 121°C for 15 min to obtain sterile cooked beans. The beans are drained, collected in a sterile polythene bag and a pestle is applied to the outside of the bag to dehull and break the seeds to mainly half-cotyledons. *B. subtilis,* previously isolated from kinema from a Darjeeling market [45], is grown on nutrient agar (M087, HiMedia, Bombay, India) at 37°C for 16 h. The culture is suspended in sterile water and 2 ml of the suspension of predominantly vegetative cells are mixed with the cooked beans to give approximately one million total cells/g wet weight. The inoculated beans are distributed in 50 g (wet weight) amounts to sterile 250 ml wide-mouthed bottles plugged with cotton wool and incubated at 45°C. Studies on changes in microbial and proximate composition are carried out by removing five bottles at every specific time interval.

**Table 9:** Fermentation starters used in different countries.

| Country | Name | Ingredients | Shape | Microorganism |
|---|---|---|---|---|
| China | *Chu* | Wheat, barley, millet, rice (whole grain, grits or flour) | Granular or cake | *Rhizopus, Amylomyces* |
| Korea | *Nuruk* | Wheat, rice, barley (whole grain, grits or flour) | Large cake | *Aspergillus, Rhizopus,* yeasts |
| | *Meju* | Soybean (whole seed) | Large ball | *Aspergillus, Bacillus* |
| Japan | *Koji* | Wheat, rice (whole grain, grits or flour) | Granular | *Aspergillus* |
| Indonesia | *Ragi* | Rice (flour) | | *Amylomyces* |
| Malaysia | *Ragi* | Rice (flour) | Small cake | *Endomycopsis* |
| Philippines | *Bubod* | Rice, glutinous rice (flour) | Small cake | *Mucor, Rhizopus, Saccharomyces* |
| Thailand | *Loopang* | Bran | Powder | *Amylomyces, Aspergillus* |
| India | *Marchaa* | Rice | Flat cake | *Hansenula anomala, Mucor fragilis, Rhizopus arrhizus* |

*Source:* 47

*Mesu*, a non-salted fermented bamboo shoot product, has traditionally been consumed by people in the bamboo-growing regions of the Darjeeling hills of West Bengal and Sikkim in India [36]. History of the art of pickling this vegetable is lost in antiquity. *Mesu* is prepared only during the months of June to September when bamboo shoots sprout. In the traditional method of its preparation, young edible shoots of bamboo *(Dendrocalamus hamiltonii Nees* and Arnott, *Bambusa tulda* Roxb. and *Dendrocalamus sikkimensis* Gamble, locally known as "choya bans", "karati bans" and "bhalu bans", respectively) are defoliated, chopped, pressed tightly into a hollow bamboo stem, covered tightly with leaves of bamboo or other wild plants and left at ambient temperature (20-25°C) to ferment for 7-15 days. Completion of fermentation is indicated by the typical *mesu* flavour and taste. *Mesu* is commonly used as a pickle by mixing it with salt, mustard oil and green chillies. It is also used for preparing curry by frying and mixing it with cooked meat. While fresh *mesu* has a shelf-life of only about a week, its pickle can be stored for a year or more.

*Marcha* or *murcha* is a traditional amylolytic starter used to produce sweet–sour alcoholic drinks, commonly called *jaanr* in the Himalayan regions of India, Nepal, Bhutan, and Tibet (China). These starter preparations have a variety of names such as *marcha* or *murcha* in the Himalayan regions of India, Nepal and Bhutan; ragi in Indonesia; loogpang in Thailand; bubod in the Philippines; Chinese yeast in Taiwan [48]; and *nuruk* in Korea [49]. *Marcha* is a mixed dough inocula used as a starter culture for production of various indigenous alcoholic beverages such as *jaanr* in the Darjeeling hills and Sikkim in India, Nepal, and Bhutan [21]. It is a dry, round-to-flattened, creamy white to dusty white, solid ball-like starter ranging from 1.9 to 11.8 cm in diameter and from 2.3 to 21.2 g in weight. *Marcha* or *murcha* is a Nepali word. Different ethnic communities of the region call it by their own dialect such as *khesung* by Limboo, *bharama* by Tamang, *bopkha* or *khabed* by Rai, *phab* by Bhutia and Tibetans, and *buth/thanbum* by the Lepcha. During preparation, glutinous rice is soaked in water for 6-8 h and crushed in a foot-driven heavy wooden mortar and pestle. In 1 kg of grained rice, ingredients added include roots of *Plumbago zeylanica* L., 2.5 g; leaves of *Buddleja asiatica* Lour, 1.2 g; flowers of *Vernonia cinerea* (L.) Less, 1.2 g; ginger, 5.0 g; red dry chili, 1.2 g; and previously prepared *marcha* as a mother culture, 10.0 g. The mixture is kneaded into flat cakes of varying sizes and shapes and placed individually on a platform suspended below the ceiling that is made up of bamboo strips above the kitchen, bedded with fresh fronds of ferns [*Glaphylopteriolopsis erubescens* (Wall ex Hook.) Ching], and covered with dry ferns and jute bags. These are left to ferment for 1-3 days. Finally, cakes of marcha are sun-dried for 2-3 days and stored in a dry place for more than a year [50]. *Marcha* is produced at home exclusively by women. This art is practiced as a hereditary trade that passes from mother to daughters. The microflora of *marcha* consists of filamentous moulds such as *Mucor circinelloides* and *Rhizopus chinensis*, yeasts such as *Saccharomycopsis fibuligera* and *Pichia anomala*, and bacteria such as *Pediococcus pentosaceus* [51]. Among these microbes, saccharifying yeast species belonging to *Saccharomycopsis* predominate, suggesting that they are involved in the amylolytic fermentation [51, 52] although alcohol producing microbes have not yet been recovered.

*Gundruk* is a non-salted, fermented and acidic vegetable product indigenous to the Himalayas. During fermentation of *gundruk*, fresh leaves of local vegetables known as drayo-sagT [*Brassica rapa campestris* (L.), *Clapham* variety *cuneifolia* Roxb.], mustard leaves [*Brassica juncea* (L.) Czern], cauliflower leaves (*Brassica oleracea* L. variety *botrytis*) and cabbages (*Brassica* sp.) are wilted for 1-2 days. Wilted leaves are crushed mildly and pressed into a container or earthenware pot, made airtight and fermented naturally for about 15-22 days. After desirable fermentation, products are removed and sun dried for 2-4 days. *Gundruk* is consumed as pickle or soup, and has some resemblance with other fermented acidic vegetable products such as *kimchi* of Korea, *sauerkraut* of Germany and *sunki* of Japan. *Sinki*, an indigenous fermented radish tap-root food, is traditionally prepared by pit fermentation, which is a unique type of biopreservation of foods by lactic acid fermentation in the Sikkim Himalayas. During *sinki* production, a 2-3 ft pit of same diameter is dug in a dry place. The pit is cleaned, plastered with mud and warmed by burning. After removing the ashes, the pit is lined with bamboo sheaths and paddy straw. Radish tap-roots are wilted for 2-3 days, crushed, dipped in lukewarm water, squeezed and pressed tightly into the pit, covered with dry leaves and weighted down by heavy planks or stones. The top of the pit is plastered with mud and left to ferment for 22-30 days. After completion of fermentation, fresh *sinki* is removed, cut into small pieces, sun-dried for 2-3 days, and stored at room temperature for future consumption. *Sinki* is consumed as soup and pickle in a meal. *Khalpi* or *khaipi* is a fermented cucumber (*Cucumis sativus* L.) product, commonly consumed by the Brahmin Nepalis in Sikkim. During *khalpi* preparation, matured and ripened cucumber is cut into suitable pieces, sun dried for 2 days and then put into a bamboo vessel, locally called *ddhungroo*T, and made air-tight. It is left at room temperature for 3-7 days. It is consumed as pickle by adding mustard oil, salt and powdered chillies. *Inziangsang* is a traditional fermented leafy vegetable product of Nagaland and Manipur in North-East India, mostly consumed by the Naga. It is prepared from mustard leaves and is very similar to *gundruk*. The major microflora of *gundruk* from Nepal is represented by species of lactic acid bacteria (LAB) including *Lactobacillus plantarum, L. casei* subsp. *casei, L. casei* subsp. *pseudoplantarum, L. cellobiosus* (*L. fermentum*) and *Pediococcus pentosaceus* [53].

## 13. FERMENTED FOODS OF DIFFERENT COUNTRIES OF THE ASIA-PACIFIC REGION

### 13.1 *Natto*

*Natto* is a popular fermented whole soybean Japanese food. Generally, *natto* products are dark in colour, having a pungent aroma. *Natto* is consumed with boiled rice or is used as a seasoning agent with cooked meats, seafoods and vegetable. Three types of *natto* are traditionally prepared in Japan, each of which differs in the manufacturing method [54]. *Itohiki-natto* is the main *natto* product of Eastern Japan and is often referred to as *natto* only. It is produced with *Bacillus natto*. In this method, the soybeans are first washed and soaked in water to increase its volume by twice the original. The soaked beans are then steamed for 15 minutes and cooled and inoculated with *B. natto*. The inoculated beans are packaged in 150 g quantities and allowed to ferment at 40-45°C for 18-20 h [55]. After fermentation, the surface of soybean is covered with viscous polymeric substances. High quality *natto* has a lot of viscous substances around the beans. *Yukimari-natto* is made by mixing *Itohiki-natto* with rice *Koji* and salt and then aged at 25-30°C for about 15 days. *Hama-natto* is the third type of *natto* made in Japan and is produced mainly in the region of Hamanatsu.

### 13.2 *Takju*

*Takju* is a popular turbid drink of Korea. The major ingredients for the preparation of *Takju* are rice and wheat, and microorganisms used are lactic acid bacteria and *S. cerevisiae*. Approximate chemical composition of *Takju* is given in Table 10, and alcohol (%), pH and total acidity (%) during fermentation are shown in Fig. 17.

### 13.3 *Poi*

In Hawaii and the islands of South Pacific, taro (*Colocasia esculenta*) corms are used to prepare a product called *poi*. Corms are first cooled, peeled and ground to a fine consistency. Water is added to the material, mixed and used as fresh *poi* for preparing fermented *poi*. The pounded material is fermented at an ambient temperature for 1 to 3 days. As fermentation proceeds, the thick consistency of *poi* changes to a product having a more watery and fluffy consistency.

**Table 10:** Approximate chemical composition of *takju.*

| Nutrient | Amount |
|---|---|
| Energy (kcal/100g) | 55.0 |
| Moisture (%) | 90.7 |
| Ethanol (%) | 7.0 |
| Protein (g/100g) | 1.9 |
| Fat (g/100g) | 0.1 |
| Carbohydrates (g/100g) | 1.2 |
| Ash (g/100g) | 0.1 |
| Calcium (mg/100g) | 14.0 |
| Phosphorus (mg/100g) | 28.0 |
| Iron (mg/100g) | 0.8 |
| Thiamine (mg/100g) | 0.01 |
| Riboflavin (mg/100g) | 0.03 |

*Source:* 7

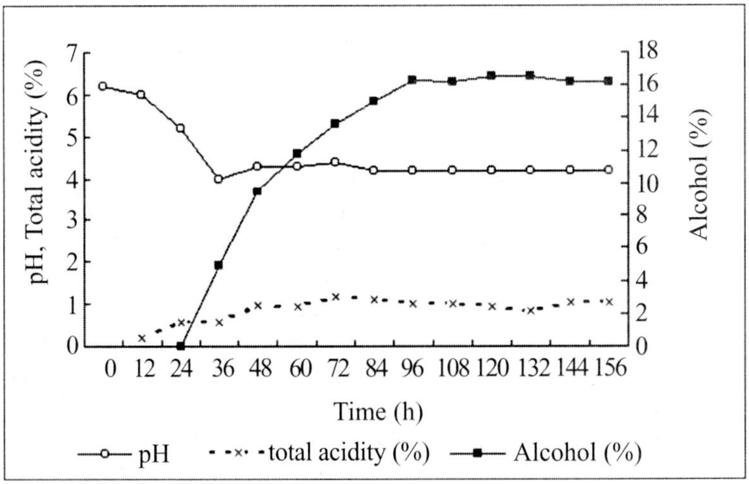


**Fig. 17:** Changes in alcohol (%), pH and total acidity (%) during *takju* fermentation.
*Source*: 56

Figure 18 compares the preparation processes of Korean *nuruk*, Japanese *koji*, Indonesian *ragi*, and Philippine *bubod* [57]. *Nuruk, ragi* and *bubod* are similar in that they are prepared by the natural fermentation of raw cereal powders, which are moulded into the shape of a cake or ball. *Koji,* on the other hand, is prepared by controlled fermentation of cooked cereals in a granular form, which are commonly inoculated with the mold, *Aspergillus oryzae*. Numerous types of microorganisms, moulds, bacteria and yeasts are found in these naturally fermented products. *Aspergillus oryzae* ($1 \times 10^7$ cfu/g), *Aspergillus niger* ($1 \times 10^7$ cfu/g), *Rhizopus* ($1 \times 10^6$ cfu/g), bacteria ($1 \times 10^7$ cfu/g) and yeasts ($1 \times 10^5$ cfu/g) were identified in *nuruk* [55]. The number of moulds ($1 \times 10^3$-$10^7$/g), yeasts ($1 \times 10^5$-$10^7$/g) and lactic acid bacteria ($1 \times 10^5$-$10^7$/g) were observed to vary with the source and district of collection of *bubod* [58]. Important microorganisms in *ragi* fermentations were mould of (*Amylomyces rouxii*) and yeast (*Endomycopsis burtonii*) [59], while those in *loog-pang* were *Amylomyces, Aspergillus, Rhizopus, Mucor* and *Absidia* [60].

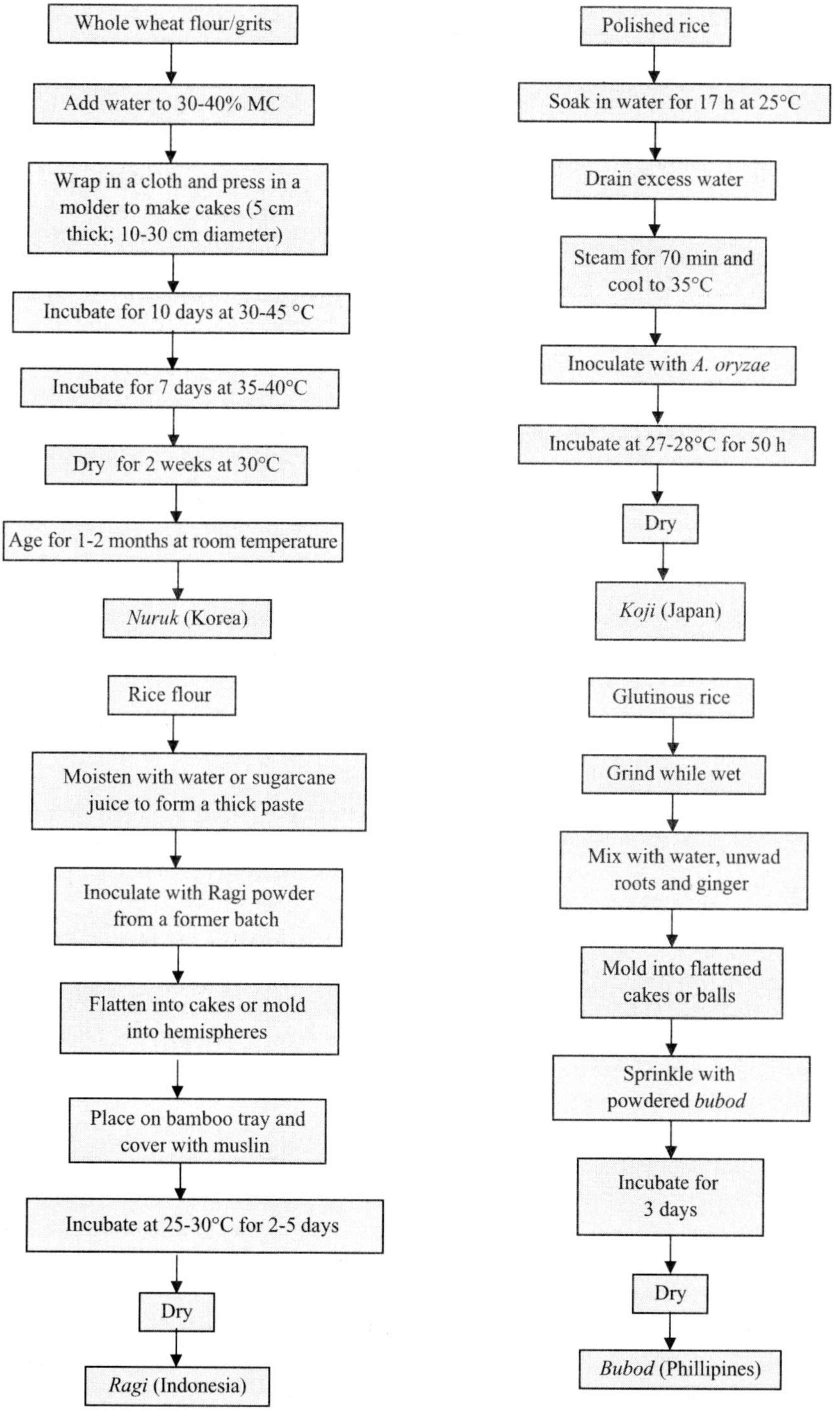


**Fig. 18:** Flow charts for the preparation of solid-fermented starters in different countries of the Asia-Pacific region.

**Table 11:** Main enzyme activities in *koji* and *nuruk*.

| Enzyme | *Koji* | *Nuruk* |
|---|---|---|
| α-Glucosidase (Wohlgemuth value $D^{40}$ 30/g) | 1225 | 256 |
| Glucoamylase (mg glucose produced/h/g) | 201 | 260 |
| Acid protease (mg tyrosine produced/h/g) | 3674 | 181 |

*Sources:* 56, 61

Table 11 compares the enzyme activities of Japanese *koji* and Korean *nuruk* [56, 61]. Enzyme activities in *koji* are generally higher than those in *nuruk*. This may be due to the fact that the pure culture of *Aspergillus oryzae* on loose cereal granules allows maximum growth during the preparation of *koji*, while in *nuruk* manufacture, mold growth is confined mainly to the surface of the cake or ball, thus allowing yeasts and lactic acid bacteria to grow simultaneously, and contribute to the deeper flavour notes of Korean rice-wine in later alcoholic fermentation stages.

## 14. QUALITY, SAFETY AND NUTRITIVE VALUE OF FERMENTED FOODS

Preparation of customary alcoholic beverages (naturally fermented beverages) and their consumption is permitted at home using natural source of microflora as fermenting agent, using locally available raw materials (ragi, apricot, apple, etc.) in tribal areas of Himachal Pradesh. Due to lack of proper scientific technology, many a times, the quality of such beverages is uncertain and their consumption leads to higher incidence of fatal diseases among the consumers [62]. Presence of some compounds besides affecting the quality can make the natural fermented beverages toxic. Among the toxic compounds in alcoholic beverages is methanol, which occurs in wines and brandies also, ranging from 32 to 238 mg/l in grape wine, 410-700 mg/l in French Cognac, 180-500 mg/l in Spanish distillates, 390-920 mg/l in German distillates, 1200 gh/l in plum, 59 gh/l cognac, 700 gh/l in pear [63-65].

In a study, samples (22) of naturally fermented alcoholic (NFA) beverages produced in the tribal district Kinnaur of Himachal Pradesh (India) were analysed [9]. Appreciable differences among various types of beverages for methanol contents were observed (Table 12) and different type of predominant microorganisms in natural source of fermentation are shown in Table 13. A very large variation can be observed as indicated by very large value of SD. In comparison to the experimental wines, NFA beverages had 14 times more methanol, while the distilled liquors prepared from the NFA beverages in the tribal areas contained about 40 times more methanol, than their experimental counterparts. The cereal based NFA beverages, irrespective of being distilled or not, had considerably lower methyl alcohol contents. The higher methanol contents recorded in NFA beverages has been attributed to the activity of mixed natural microflora including fungi and bacteria involved in such fermentations and pectolytic enzyme activity of the source of fermentation or the raw materials and the method of distillation [1, 66-68]. The higher methanol content in NSF beverages or home brewed beverages is of great concern due to its toxicity causing suspected bowel ulceration, blindness or death although the lethal properties of some of these country made wines are legendary [69]. It would underline the need for caution but not undue fears as the human body can metabolize 340 mg/kg of body weight of methanol compared to 1400 mg/kg body of ethanol [70]. Nevertheless, the wine should have methanol content as low as possible from total safety considerations.

The microbiological analysis of fermented products or their inocula revealed *Saccharomyces cerevisiae* as the main fermenting yeast. Few species of *Candida* have also been isolated, along with

**Table 12:** Methanol content (μl/l) in different types of alcoholic beverages.

| Type of alcoholic beverage | Mean ± S.D. (n–1) |
|---|---|
| Naturally fermented alcoholic beverages | |
| Naturally fermented wines | 772 ± 156.4 |
| Naturally fermented distillates | 4414 ± 2463 |
| Naturally fermented cereal products | 206 ± 223.6 |
| Naturally fermented cereal distillates | 448 ± 599.7 |
| | 1460 |

*Source:* 9; S.D. = n–1 Standard Deviation.

**Table 13:** Types of predominant microorganisms in natural source of fermentation.

| Microbial group | Predominant types |
|---|---|
| Fungi | *Aspergillus flavus* |
| | *Aspergillus oryzae* |
| | *Mucor* sp. |
| | *Rhizopus* sp. |
| Yeast | *Saccharomyces* sp. |
| | *Scytalidium thermophilus* (*Torula* sp.) |
| | *Rhodotorula* sp. |
| Bacteria | *Pediococcus* sp. |
| | *Leuconostoc* sp. |
| | *Pseudomonas flourescens* |

*Source:* 71

the species *Zygosaccharomyces bisporus* and *Kluyveromyces thermotolerance*. The bacterial species of *Lactobacillus* and *Leuconostoc* are prominently involved in fermentation, causing acidification and leavening. A large variation among the samples of NFA beverages for ethanol, volatile acidity and pH was noted (Table 14). The very low ethanol content in non-distilled NFA beverages is expected but not in the distilled NFA beverages, where it indicates the faulty method of fermentation and distillation. About 80 per cent of the total samples of NFA had volatile acidity of more than 0.04 per cent indicating the occurrence of acetification after the fermentation [9]. In another study [4], the fermented beverage samples were analysed for ethanol content, which exhibited a large variation. The highest ethanol content was recorded in *Sura* (a millet-based naturally fermented beverage), followed by *Chhang* and *Lugri. Behmi, Chulli* and *Angoori* had a low content of alcohol, while jaggery and red grapes based beverage contained 5% ethanol [4].

Out of 22 samples of NFA beverages collected, 4 (18.2%) were found to be positive for microbial contamination showing standard plate count (SPC) ranging from 10 to $3 \times 10^4$ cfu/ml showing the presence of bacteria in all the samples except one which had yeast almost in pure culture. The bacteria belonged to coli form group, while the yeast was identified as belonging to *Brettanomyces* sp. [9].

Fermented foods and beverages are nutritionally beneficial and easy to preserve. A number of fermented foods (e.g., *Siddu, Chilra, Marchu, Manna, Dosha, Bhaturu, Bedvin roti, Bagpinni* and *Churpa*) and fermented beverages (e.g., *Sura, Chhang, Lugri, Daru* and *Angoori*) were identified, which have become an integral part of the food and nutrition of the rural and tribal people of Himachal

Pradesh. *Chhang* - a beer like traditional beverage prepared on cottage consumed extensively in the sub-Himalayan region was found to be safe from microbiological, biochemical and nutritional safety angles. A comparison of nutritional composition of fermented (*Chhang*) with that of unfermented ragi (Table 15) clearly revealed the contribution of fermentation.

Low alcoholic beverage like wine (made from fruits) has a considerable role to play in nutrition of the consumers [14]. A typical wine contains ethyl alcohol, sugars, acid, higher alcohols, tannins, aldehydes, esters, amino acids, minerals, vitamins, anthocyanins, fatty acids and minor constituents like fla-

**Table 14:** Range of ethanol, volatile acidity and pH of naturally fermented alcoholic beverages.

| Parameters | No. of Samples* |
|---|---|
| **Ethanol (%, v/v)** | |
| Below 10 | 1 (4.5) |
| 10-20 | 7 (31.8) |
| 20-30 | 6 (27.2) |
| 40-60 | 8 (36.6) |
| **Volatile acidity (%, Acetic acid)** | |
| Below 0.04 | 4 (18.2) |
| 0.04 to 0.10 | 8 (36.6) |
| 0.10 to 0.40 | 10 (45.5) |
| **pH** | |
| Below 4 | 8 (36.6) |
| 4 to 4.5 | 5 (22.2) |
| 4.5 to 5.0 | 5 (22.2) |
| Above 5 | 4 (19.2) |

* Total number of samples = 22; Values in parentheses indicate percentage.

**Table 15:** Nutritional composition of unfermented and fermented ragi (*Chhang*).

| Components | Unfermented ragi (g/100 g) | Fermented ragi by starter culture (g/100 g) |
|---|---|---|
| Carbohydrates (Starch & sugar) | 78.0 ± 3.4 | 35.0 ± 2.5 |
| Proteins | 7.6 ± 0.34 | 10.5 ± 0.50 |
| Fat | 1.1 ± 0.05 | 1.0 ± 0.05 |
| Phosphorus | 0.3 ± 0.05 | 0.3 ± 0.05 |
| Calcium | 0.4 ± 0.06 | 0.4 ± 0.08 |
| Iron | 0.004 ± 0.009 | 0.004 ± 0.0004 |
| Ethanol | --- | 16.0 ± 0.68 |
| Total acidity (lactic acid) | --- | 3.0 ± 0.27 |
| Microbial biomass | --- | 2.7 ± 0.36 |
| Volatile acidity | --- | 0.02 ± 0.004 |
| Thiamin | 0.48 ± 0.08 | 0.50 ± 0.09 |
| Riboflavin | 0.20 ± 0.01 | 0.60 ± 0.10 |
| Niacin | 1.00 ± 0.05 | 4.20 ± 0.45 |
| Pantothenic acid | 0.40 ± 0.07 | 1.60 ± 0.08 |
| Folic acid | 0.006 ± 0.001 | 0.010 ± 0.001 |
| Cyganocobalamin | --- | 0.04 ± 0.004 |

*Source:* 72; All values are Mean ± Standard deviation (SD).

vouring compounds. The nutritional composition of alcoholic beverage varies with the wine type. Since the wines and beers are not distilled, they have more nutrients like vitamins, minerals and sugars than distilled beverages like brandy and whisky. Ethyl alcohol is an odourless liquid and solvent obtained through the fermentation of sugar containing liquid. It is a source of energy like other food components.

Other components of wine like tannins have considerable importance in health. On dry weight basis, consumption of wine has been positively correlated with the reduction of cardiovascular diseases. The relationship of alcohol and mortality has been described as a J-shaped curve, which successfully explained the French Paradox, where the people consume a large quantity of fat as well as wine, but have reduced incidence of heart diseases. The term 'French Paradox' was first coined by Dr Serge Renaud, the French Epidemiologist. Rather, co-existence of alcohol and antioxidant provides the answer to the French Paradox. The major beneficiary components are broadly called tannins or polyphenolic components. The major ingredient in wine is ethyl alcohol, whose excessive consumption leads to several problems.

## 15. NUTRITION AND PUBLIC HEALTH ASPECTS

Fermentation is a relatively efficient, low energy preservation process which increases the shelf life and decreases the need for refrigeration or other form of food preservation technology. It is therefore, a highly appropriate technique for use in developing countries and remote assess where access to sophisticated equipment is limited. Fermented foods are popular throughout the world and in some regions make a significant contribution to the diet of millions of individuals. Approximately 30% of women consume less than their daily requirements of energy, and at least 40% of women worldwide, suffer from iron-deficiency anaemia. Fermentation thus can enhance the nutritional value of a food product through increased vitamin level and improved digestibility.

*Saccharomyces cerevisiae* is able to concentrate large quantities of nicotinic acid and biotins, and thus forms enriched products. Sorghum bees in Southern Africa contain selectively high levels of riboflavins and nicotinic acid, which are important for people consuming a high maize diet Pellagra (a vitamin deficiency disease associated with high maize diets) is unusual in communities in which sorghum beer is consumed. Even children consume the drugs which contain selectively little alcohol but are rich in vitamins.

Plum wine in West Africa is high in vitamin $B_{12}$, which is very important for people with low meal intake, and who subsist primarily on a vegetarian diet. *Pulque* (a fermented plant sop) is an important source of vitamin for the economically weaked people in Mexico. The fermentation process involved in *Pulque* production increases its vitamin content, for instance, the vitamin content (Milligrams of vitamins per 100 g of product) of *pulque* increases from 5 to 29 for thiamine, 54 to 515 for niacin and 10 to 33 for riboflavin [73] during fermentation. *Idli* (a lactic acid bacteria fermented product consumed in India) is high in thiamine and riboflavin. Fermented foods are often more easily digestible than unfermented foods [74]. Certain lactic acid bacteria (e.g., *Lactobacillus acidophilus*) and moulds have been found to produce antibiotics and bacteriocin [75]. Despite this, the safety of fermented foods resulting from the probable production of toxic factors by microorganisms is a matter of public health concern. The aflatoxin forming ability of 238 strains of *Aspergillus* used in the Japanese food industry was studied [76], and it was found that 52 strains produced fluorescent compounds but none produced aflatoxins. Analysis of *miso* and rice-*Koji* samples of Japan also did not show the presence of aflatoxins. Contrary to the above reports, there are also reports on the toxic substances produced by fermenting micro-organisms. The presence of aflatoxin in soy sauce collected from Hong Kong was observed [77]. Bong Krek (a fermented food similar to *Ontjan*, produced from coconut press cake) is susceptible to form two very potent toxing under certain fermentation conditions unfavourable to the mold. Many deaths have been reported in Indonesia from Borgkrek poisoning [78].

Use of pure cultures in food fermentation atmospheres can avoid the possibility of contamination from toxic microbial species and enhance the utilization potential of fermented foods. Owing to their specific nutritional superiority, many of the fermented foods have now crossed the boundaries of the countries of the orient and built a safe niche in the occidental cuisine.

## 16. CONCLUSION

Fermented foods and beverages are nutritionally beneficial and easy to preserve. It is apparent that indigenous fermentation improves the digestibility and nutritional value of various foods and contributes towards the nutritional well-being of the world population. A number of fermented foods and fermented beverages were found to be an integral part of the food and nutrition of the rural and tribal people of Himachal Pradesh. The microbiological analysis of fermented products or their inocula revealed *Saccharomyces cerevisiae* as the main fermenting yeast, though other species of yeast have also been isolated. The bacterial species of *Lactobacillus* and *Leuconostoc* are prominently involved in fermentation, causing acidification and leavening.

Large variations in the ethanol contents of the NFA beverages alongwith far greater volatile acidities than prescribed indicate lack of standardized technology in the production of alcoholic beverages. Presence of coliform bacteria in the NFA beverages also point out the unhygienic conditions practised in the preparation of such beverages in the tribal areas. The low pH of NFA beverages, however, might have prevented the growth of any pathogenic microorganisms in wines. Naturally fermented alcoholic beverages have some other undesirable quality characteristics including the high level of methanol. To improve the quality, the use of pure culture of wine yeast rather than mixed bacterial and yeast fermentation needs to be encouraged. Addition of preservative like potassium metabisulphite, distillation practice where head and tail are cut-off, and manipulation of conditions of wine fermentation can be done. The traditional methods of preparing indigenous foods are simple and the fermentations are brought about by the microflora that comes from the staples and the environment. Though the microflora brings about several changes leading to improved digestibility and nutritional value, there is a possibility of development of various undesirable microorganisms and the production of toxic substances by certain species, which is a point of serious concern.

Traditional fermentations have low cost of production, need less labour input and raw materials needed for preparation are locally or easily available. These fermented products have a potential to grow into a small or medium size industry, if scientific and technological support is extended to the existing folk technologies/practices. There is much to be learnt about the role of individual microorganisms, standardization of their levels by artificially inoculating the ingredients with different microbes along with the optimization of various physico-chemical factors to improve the quality of the product.

## REFERENCES


1. Amerine MA, Berg HW, Kunkee RE, Ough ES, Singleton VL and Webb AD (1980) *The Technology of Wine Making*, AVI Publishing Co Inc, Westport, USA.
2. Joshi VK, Sandhu DK and Thakur NS (1999) Fruit based alcoholic beverages. In: *Biotechnology Food Fermentation (Microbiology, Biochemistry and Technology),* Joshi VK & Pandey A (eds), Educational Publishers and Distributors, New Delhi, India, pp 647-744.
3. Pedersen CS (1979) *Microbiology of Food Fermentation*, AVI Publishing Co., Inc. Westport, USA.
4. Thakur N, Savitri and Bhalla TC (2004) Characterisation of some traditional foods and beverages of Himachal Pradesh. *Indian J Trad Know* 3: 325-335.

5. Soni SK and Sandhu DK (1999) Fermented cereal products in food fermentation starter for amylolytic fermentation. *Int J Food Microbiol* 99: 135-146.
6. Lee CH (1989) Fish fermentation technology. *Korean J Appl Microbiol Bioengg* 17: 645-654.
7. Korea Rural Nutrition Institute (1991) *Food Composition Table*, RDA, Seoul, Korea.
8. Joshi VK (2005) Traditional fermented foods and beverages of northern India: An overview. In: *First Workshop and Strategic Meeting of the Swidish South Asia Network on Fermented Foods*, Anand, India, pp 59-72.
9. Joshi VK and Sandhu DK (2000) Quality evaluation of naturally fermented alcoholic beverages: microbiological examination of source of fermentation and ethanol productivity of the isolate. *Acta Alimentaria* 29: 323-326.
10. Joshi VK (1993) Studies on alcoholic fermentation of apple juice and pomace utilization. Ph.D. Thesis, Guru Nank Dev University, Amritsar, India.
11. Soni SK and Sandhu DK (1989) Biochemical and nutritional changes associated with Indian Punjabi warri fermentation. *J Food Sci Technol* 27: 82-85.
12. Bhatia AK, Singh RP and Atal CK (1977) Chhang - fermented beverage of Himalayan folk. *Indian food Packer* 31: 1-4.
13. Joshi VK, Bhutani VP and Sharma RC (1990) Effect of dilution and addition of nitrogen source on chemical, mineral and sensory qualities of wild apricot wine. *Am J Enol Vitic* 41: 229-232.
14. Joshi VK (1997) *Fruit Wines,* Directorate of Extension Education, Dr YS Parmar University of Horticulture & Forestry, Nauni, Solan, India.
15. Prajapati JB (2003) Fermented foods of India In: *Fermented Foods, Health Status and Social Well-Being,* International Seminar and workshop, Anand, India, Nov. 13-14, pp 1-4.
16. Patidar SK and Prjapati JB (1998) Standardization and evaluation of lassi prepared using *Lactibacillus acidophilus* and *Streptococcus thermophilus. J Food Sci Technol* 35: 428-431.
17. Prajapathi JB, Shah RK and Dave JM (1986) Nutritional and therapeutic benefits of a blended spray dried acidophilus prepovation. *Cult Dairy Prod J* 21: 16-21.
18. Prajapati JB, Shah RK and Save JM (1987) Survival of *Lactobacillus acidophilus* in blended spray dried *acidophilus* preparations. *Aust J Dairy Technol* 42: 17-21.
19. Tamang JP (2001) Food culture in the Eastern Himalayas. *J Himalayan Res Cultural Foundation* 5: 107-118.
20. Hooker JD (1987) *Himalayan Journals: Notes of a Naturalist in Bengal, the Sikkim and Nepal Himalyas, the Khasia Mountains and Sea (Microform),* Today and Tommarow's Printers and Publishers, New Delhi, India.
21. Tamang JP, Thapa S, Tamang N and Rai B (1996) Indigenous fermented food beverages of Darjeeling Hills and Sikkim: Process and product characterization. *J Hill Research* 9: 401-411.
22. Thapa S (2001) Microbiological and biochemical studies of indigenous fermented cereal-based beverages of the Sikkim Himalyas, Ph.D. Thesis, Food Microbiology Laboratory, Sikkim Govt., India.
23. Samantray GT, Misra PK and Patnaik KK (1989) Mineral composition of ragi. *Indian J Nutr Diet* 26: 113-116.
24. Sugimoto H, Van Busen JP and Robinson WB (1971) An enzymatic process for protein containing beverages based on soyabeen protein and lemon juice. *J Food Sci* 36: 729-734.
25. Mustakas GC (1974) A new soya lipid protein concentrate beverage. *Cereal Sci Today* 19: 62-73.

26. Swaminathan M and Parpia HAB (1967) Milk substitute based on oil seeds and nuts. *World Rev Nutr Diet* 8: 184-190.
27. Wang HL and Hesseltine CW (1970) Sofu and Laochao. *J Agric Food Chem* 18: 572-575.
28. Mital BK, Shallenberger RS and Stein Kraus KH (1973) L-Galactosidase activity of *Lactobacilli. Appl Environ Microbiol* 26: 783-788.
29. Mital BK and Stein Kraus KH (1976) *J M K Food Technol* 39: 342 (Cited in: Padmaja G and Mathew G (1999) Oriental fermented foods. In: *Biotechnology Food Fermentation (Microbiology, Biochemistry and Technology),* Educational Publishers & Distributors, New Delhi, India, pp 523-581).
30. Wikramanayake TW (1996) *Food and Nutrition,* Hector Kobbekaduwa Agrarian Research and Training Institute, Sri Lanka.
31. Sagarika E and Pradeepa J (2003) Fermented foods of Sri Lanka and Maldives: In: *Proceedings of Seminar on "Fermented Foods, Health Status and Social Well Being",* Solivenir, Anand, India, p 11.
32. Kuswanto Kapti R (2004) Industrialization of *Tempe* fermentation, In: *Industrialization of Indigenous Fermented Foods,* Steinkraus KH (ed), Marcel Dekker, Inc., New York, USA, pp 587-635.
33. Arihantana MB and Buckle KA (1987) Cassava detoxication during tape fermentation with traditional inoculum. *Int J Food Sci Technol* 22: 41-48.
34. FAO (1990) *The Technology of Traditional Milk Products in Developing Countries,* FAO Animal Production and Health Paper - 85.
35. Rahman MM and Hassan MN (1998) Utilization of chana-whey in the form of fermented and non-fermented drinks. *Bangladesh J Ani Sci* 122: 57-63.
36. Tamang JP, Sarkar PK and Hesseltine CW (1988) Traditional fermented foods and beverages of Darjeeling and Sikkim - a Review. *J Sci Food Agric* 44: 375-385.
37. Tamang JP and Tamang N (1998) *Traditional Food Recipes of the Sikkim Himalaya. Report,* Sikkim Biodiversty and Ecotourisms Project, GBPIHED, Gangtok, India.
38. Tamang JP (1992) Studies on the microflora of some traditional fermented foods of the Darjeeling hills and Sikkim, Ph.D Thesis, North Bengal University, Darjeeling, India.
39. Shibasaki K and Hesseltine CK (1962) Meso fermentation. *Econ Bot* 16: 180-195.
40. Yong FM and Wood BJB (1974) *Adv Applied Microbiol* 17: 15 (Cited in: Padmaja G and Mathew G (1999) Oriental fermented foods. In: *Biotechnology Food Fermentation (Microbiology, Biochemistry and Technology),* Educational Publishers & Distributors, New Delhi, India, pp 523-581).
41. Yokotsuka T (1985) Fermented protein foods in the Orient with emphasis on *shoyu* and *meso* in Japan. In: *Microbiology of Fermented Foods,* Vol 1, Wood BJB (ed), Elsevier, London, UK, pp 197-247.
42. Padmaja G and Mathew G (1999) Oriental fermented foods. In: *Biotechnology Food Fermentation (Microbiology, Biochemistry and Technology)*, Educational Publishers & Distributors, New Delhi, India, pp 523-581.
43. Beuchat LR (1983) Indigenous fermented foods. In: *Biotechnology,* Vol 5, Rehm HJ & Reed SG (eds), Verlag Chemie, Weinhiem, Germany, pp 477-528.
44. Sarkar PK, Cook PE and Owens JD (1993) *Bacillus* fermentation of soybeans. *World J Microbiol Biotechnol* 9: 295-299.
45. Sarkar PK, Tamang JP, Cook PE and Owens JD (1994) Kinema a traditional soybean fermented food: Proximate composition and microflora. *Food Microbiol* 11: 47-55.
46. Sarkar PK and Tamang JP (1994) The influence of process variables and inoculum composition on the sensory quality of kinema. *Food Microbiol* 11: 317-325.

47. Yokotsuka T (1985) Fermented protein foods in the Orient, with emphasis on *shoyu* and *miso* in Japan. In: *Microbiology of Fermented Foods,* Vol 1, Wood BB (ed), Elsevier Applied Science, London, UK, pp 197-247.
48. Hesseltine CW, Rogers R and Winarno FG (1988) Microbiological studies on amylolytic oriental fermentation starters. *Mycopathologia* 101: 141-155.
49. Park KI, Mheen TI, Lee KH, Chang CH, Lee SR and Kwon TW (1977) Korean *yakju* and *takju*. In: *The Proceeding of Symposium on Indigenous Fermented Foods*, November 21-27, GIAMI, Bangkok.
50. Saono S, Hull RR and Dhamcharee B (1986) *A Concise Handbook of Indigenous Fermented Foods in the ASCA Countries,* Indonesian Institute of Sciences, Jakarta, Indonesia.
51. Tamang JP and Sarkar PK (1996) Microflora of marcha: An amylolytic fermentation starter. *Microbios* 81: 115-122.
52. Hesseltine CW and Kurtzman CP (1990) Yeast in amylolytic food starters. Anales del Instituto de Biología, Universidad Nacional Autónoma de México. *Serie Botánica* 60: 1 -7.
53. Karki T, Okada S, Baba T, Itoh H and Kozaki M (1983) Studies on the microflora of Nepalese pickles gundruk. *Nippon Shokuhin Kogyo Gakkaishi* 30: 357-367.
54. Kiuchi K, Ohta T, Itrh H, Takabayashi T and Ebine H (1976) *J Agric Food Chem* 24: 404-407.
55. Hesseltine CW (1965) A millennium of fungi, food and fermentation. *Mycologia* 57: 149-197.
56. Kim CJ (1968) Microbiological and enzymological studies on Takju brewing. *J Korean Agricul Chem Soc* 60: 69-99.
57. Steinkraus KH (1983) *Handbook of Indigenous Fermented Foods,* Marcel Dekker Inc, New York, USA.
58. Tanimura W, Sanches PC and Kozaki M (1978) *The Fermented Food in the Philippines*, Part 1, pp 105-120.
59. Ko SD (1972) Tape fermentation. *Appl Microbiol* 23: 976-978.
60. Pichyangkura S and Kulprecha S (1977) Survey of mycelial molds in loogpang from various sources in Thailand. In: *Symposium on Indigenous Fermented Foods*, Bangkok, Thailand, pp 75-92.
61. Nunikawa Y and Ouchi K (1973) Kagaku to Seibutsu, 11: 216-220.
62. Sharma PC (1986) *Home wine making,* Govt of H.P. Publication, Shimla, India.
63. Lisle DB, Richards CP and Wardleworth DF (1978) The identification of distilled alcoholic beverages. *J Inst Brew* 84: 93-96.
64. Postel W and Adam L (1979) Gas chromatographic characterization of brandy cognac and Armagnac. I: Study of methods. *Dia Brand Wein Wirtschaft* 2: 404-409.
65. Postel W and Adam L (1980) Gas chromatographic characterization of brandy, Cognac and Armagnac II: Content of volatile substances. *Die Brand Wein Wistschoft* 2: 154-164.
66. Lee CY, Robinson WR, Van buren JP, Acree TE and Stuews GS (1975) Methanol in wines in relation to processing and variety. *J Enol Vitic* 26: 184-187.
67. Ali MZ and Dirar HA (1984) Microbiological study of Sudanese Date Wines. *J Food Sci* 49: 459-460.
68. Frazier WC (1971) *Food Microbiology*, Tata McGraw Hill Publ Co, New Delhi, India.
69. Fowles G (1989) The complete home wine maker. *New Scientist* 9: 38-42.
70. Christensen HE (1973) *The Toxic Substances List*, US Dept Health, Education, and Welfare, Public Health Service, National Institute for Occupational Safety and Health, Washington, USA.

71. Joshi VK and Sandhu DK (2000) Quality evaluation of naturally fermented alcoholic beverages: Microbiological examination of source of fermentation and ethanol productivity of the isclate. *Acta Alimentaria* 29: 323-324.
72. Basappa SC, Someshekar D, Agarwal R, Suma D and Bharathi K (1997) Nutritional composition of fermented ragi (*chhang*) by *Phab* and defined starter culture as compared to unfermented ragi (*Eleucine coracana* G.) *Int J Food Sci Nutri* 48: 313-319.
73. Steinkraus K (1992) Lactic acid fermentations. In: *Applications of Biotechnology to Traditional Fermented Foods,* National Academy Press, Washington, USA, pp 43-51.
74. Rusmini S & Ko SD (1974) Rice-grown *Rhizopus oligosporus* inoculum for *Tempeh* fermentation. *Appl Microbiol* 28: 347-350.
75. Matsusaki H, Endo N, Sonomoto K & Ishizaki A (1996) Lantibiotic nisin Z fermentative production by *Lactococcus lactis* IO-1: Relationship between production of the lantibiotic and lactate and cell growth. *Appl Microbiol Biotechnol* 45: 36-40.
76. Matsuura S, Manabe M and Sato T (1970) Surveillance for aflatoxins of rice and fermented rice products in Japan. In: *Proceedings of 1st US-Japan Conference on Toxic Micro-Organisms*, US-Japan Natural Resources Program, Honolulu, USA, pp 17-27.
77. Shank RC, Wogan GN, Gibson JB and Nodesutta A (1972) *Food Cosmet Toxicol* 10: 6 (Cited in: Padmaja G and Mathew G (1999) Oriental fermented foods. In: *Biotechnology Food Fermentation (Microbiology, Biochemistry and Technology)*, Educational Publishers & Distributors, New Delhi, India, pp 523-581).
78. Van Veen AG and Stein Kraus KH (1970) Nutritive value and wholesomeness of fermented foods. *J Afr Food Chem* 18: 576-579.

# Lactic Acid Fermented Foods

# 12

**Somesh Sharma, V.K. Joshi and N.S. Thakur**

*Department of Food Science and Technology, Dr. Y. S. Parmar University of Horticulture and Forestry, Nauni, Solan 173 230 (H.P.)*

## 1. INTRODUCTION

Fermentation is one of the oldest forms of food preservation in the world. Earlier the term 'Fermentation' was used for the production of alcoholic beverages like wine, but it encompasses all the foods produced by the application of microorganisms including lactic acid bacteria (LAB) or their enzymes. Lactic acid fermentation using LAB cultures is used throughout the world, in conjunction with chemical preservation, using salt and acid to preserve fruits and vegetables. Until the development of curing and freezing during the past 180 years, lactic acid fermentation was one of the most important methods for preserving foods. In consistence with the changing life style of consumers, attitude towards food products with fresh and convenience food, lactic acid fermentation provides food with diversified taste and flavour characteristics. The Chinese were the first to ferment the vegetables, and there is evidence to show that they had prepared such vegetables at the time of building the Great Wall of China [1]. Today, because of the development of efficient heat sterilizing and refrigeration systems, lactic acid fermentation to some extent has lost its importance as a food preservation method in industrialized countries, but is being used as a tool for taste diversification. It is extensively used in developing countries, but is gaining more importance in developed countries as a food with therapeutic value. It is true especially for vegetables, where either natural fermentation or fermentation by pure culture of lactic acid bacteria is practiced even in the developed countries [1-4]. In this chapter, lactic acid fermentation, its role and microorganisms involved along with the various products that can be made, are discussed.

## 2. ROLE OF LACTIC ACID FERMENTATION AND ITS ADVANTAGES

Fermented foods play an important role in providing food security, enhancing livelihoods and improving the nutrition and social well being of millions of people around the world, particularly the marginalized and vulnerable (Table 1). This is achieved through improved food preservation, increasing the range of raw materials that can be used to produce edible food products and removing anti-nutritional factors to make food safe to eat. Fermented foods are known to contain a plenty of polysaccharides and functional vitamins. Some earlier studies show that the technique of fermentation may create unique metabolites by yeasts symbiosis, the end result of which is a mechanism enhancing human health [5, 6]. Further, fermentation is a cheap and energy efficient means of preserving perishable

*Food Biotechnology: Principles and Practices*
*V.K. Joshi & R.S. Singh (Eds.)*
*I.K. International Publishing House Pvt. Ltd., India*
*pp 375–415*

**Table 1:** Role of lactic acid fermentation.

- Preserve food (fruit, vegetable, milk, meat) and increase the shelf stability
- Destroy naturally occurring toxins and undesirable components in the raw material
- Develop characteristic sensory properties i.e., flavour, aroma and texture
- Improves digestibility, especially of some legumes
- Enriches products with desirable microbial metabolites e.g., lactic acid or amino acids
- Create new products for new markets
- Enhance dietary value

*Source:* 7

raw materials. Although there are several options for preserving fresh fruits and vegetables including drying, freezing, canning and pickling, many of these are inappropriate for use on the small-scale in developing countries. For instance, the canning of vegetables at the small-scale has serious food safety implications and contamination with botulism is a possibility. Freezing of fruits and vegetables is not economically viable at the small-scale. Fermentation requires very little sophisticated equipment, either to carry out the fermentation or for the subsequent storage of fermented product. There are examples from around the world of the role fermented foods have played in preserving food to enhance food security. Lactic acid fermentation removes fermentable sugars, prevents the growth of pathogenic microorganisms, stabilizes the product and enhances the flavour [8]. Besides being a simple and cheap technique, it is the method of food preservation that combines salting to selectively control the microorganisms and fermentation to stabilize the treated materials. Lactic acid fermentation is used for commercial bulk storage, to increase the availability of seasonal vegetables and to obtain desired sensory product quality. Although most vegetables can be fermented by lactic acid bacteria, however, cabbage, olives and cucumber are the only vegetables that are fermented commercially in large volumes for human consumption [9]. These contain sugars and nutritionally adequate substances as substrate for the growth of lactic acid bacteria and other organisms. Fermentation can salvage waste food, which otherwise would not be usable as food by changing the consistency of the product and making it digestible. This increases the range of raw materials available as food. Even a wide range of waste products are fermented to edible foods such as dodery, which is produced by chopping bones.

Many fruits and vegetables contain naturally occurring toxins and anti-nutritional compounds. These can be removed or detoxified by the action of microorganisms during fermentation. For instance, the fermentation process that produces the Sudanese product *Kawal*, removes the toxins from the leaves of *Cassia obtusifolia,* and fermentation is an important step in ensuring that cassava is safe to eat. The production of fermented fruit and vegetable products provides income and employment to millions of people around the world. Recently, lactic acid fermentation has been used in vegetables such as lupins, peas, lentils, cassava, carrot and radish to achieve technological or nutritional advantages [7, 10]. In Korea, fermented vegetables known as *kimchi* is an almost ubiquitous accompaniment to meals [11, 12]. Chinese cabbage or radish is generally used as a source vegetable for *kimchi,* adding various ingredients such as red pepper, garlic, onion, ginger, jeotkal (salted and fermented sea foods) [13]. Lactic acid fermentation reduces the pungency of radish in *kimchi* [14]. Lactic fermentation was once the only method for successful preservation of cucumber pickling in brine. Fermentation can enhance the nutritional value of a food product, through increased vitamin levels and improved digestibility. Fermentation can improve the flavour and appearance of food. One important area is the creation of meat-like flavour. Fermented vegetables such as pickles, *gundruk* and sauerkraut are used as condiments to enhance the overall flavour of the meal. A small amount of pickle can make a bland starchy diet (like dahi and rice in Asia) much more appealing [15].

In Africa, fermented cassava products (like *Gari* and *Fufu)* are a major component of the diet of more than 800 million people, and in some parts of Africa they constitute over 50% of the diet. In Asia, the preparation of fermented foods is a widespread tradition. *Kimchi* (a fermented cabbage product) is the major food product of Korea. In Latin America, fermented cereal products, alcoholic drinks and fermented milk products are three of the most important sectors of the economy. Even the lactic acid bacteria are used commercially today as starter cultures to produce a variety of fermented meat and milk products such as dry salami, pepperoni, summer sausages, fermented meat snacks and milk products such as cheese, yoghurt and sour cream. The optimum health and nutrition of individuals is dependent upon a regular supply of food and a balanced diet. When diets are sub-optimal, the individual's capacity for work and achievements is greatly reduced. The most vulnerable groups are women, children and weaning infants. Availability of food, dietary restrictions and taboos, misconceptions, limited time available for feeding or eating compound to create a group of individuals who are nutritionally disadvantaged.

During the past 40-50 years, there has been a considerable improvement in the vegetable fermentation. Initially, it was mixed microbial fermentation and the preservative effect was by chance. Apart from contributing certain desirable characteristics, it also prolongs the availability and processing period of the product at relatively low cost.

## 3. LACTIC ACID FERMENTATION AND PRESERVATION

### 3.1 Lactic Acid

Lactic acid produced through lactic acid fermentation has been used for centuries in many foods i.e., yoghurt, cheese, sauerkraut to achieve a relative stability and safety. Now-a-days, lactic acid in a concentrated commercial form is added to several low acid and acid food products to achieve the same effect. Lactic acid is an organic acid like citric or acetic acid, and has two optically active forms, the L(+) and D(-). L(+) lactic acid is made in the body and can be converted into energy for muscles or in building blocks for body, while D(-) form is not produced in the body.

### 3.2 Preservative Action

#### *3.2.1 pH drop*

Early research had confirmed that fermented food preserved well because of the souring effect that certain bacteria imparted to the food by the conversion of sugars to organic acid. Reduction of pH and removal of large amounts of carbohydrates by fermentation are the primary preserving actions that these bacteria provide to a fermented food. Acid production decreases the pH of a food product, thereby inhibiting the growth of microorganisms. The lower the pH, the longer is the antimicrobial effect.

#### *3.2.2 Concentration of undissociated acid*

Undissociated weak acids are 10-600 times as effective in growth inhibition and death of microorganisms as the dissociated forms [16]. Further, organic acids have stronger antimicrobial effect than inorganic acids.

#### *3.2.3 Specific acid effect*

The difference in the antimicrobial activity of various organic acids under the same conditions of pH and acid dissociation is called specific acid effect. Buffering of an acid increases the effectiveness of retarding microbial growth. More undissociated molecules are present in buffered solutions giving a stronger inhibition. The inhibitory action of *Lactobacillus* sp. upon other microorganisms owes much to their ability to produce acid. *L. acidophilus*, being homofermentative organism, produces over 85%

lactic acid from the fermentation of glucose. It is the production of lactic acid which establishes an acidic condition in the intestines of most of the animals. Since many enteropathogenic microorganisms prefer a neutral or slightly alkaline environment, so acidity alone is quite inhibitory to many of them.

### 3.3 Salts of Lactic Acid

Natural sodium or potassium lactate are salts of L(+) lactic acid. Because of their neutral pH, these salts can be used in non-acid foods like meat and poultry products, fish products, ready-to-eat meals and fresh noodles. The preservative action of lactate salts is based on two principles: (i) addition of sodium or potassium lactate lowers the water activity ($a_w$) of a product resulting in an inhibition of the growth of microorganisms and secondly; (ii) at the same water activity, microorganisms are more inhibited by sodium lactate than by sodium chloride. Thus, lactate ions have antimicrobial effect besides the water activity decreasing effect.

## 4. ANTIMICROBIAL SUBSTANCES PRODUCED BY LACTIC ACID FERMENTATION

Lactic acid bacteria (LAB) are capable of producing inhibitory substances other than organic acids, which are antagonistic to other microorganisms. These substances are produced in much smaller amounts and include hydrogen peroxide, diacetyl, bacteriocins and secondary reaction products such as hypothiocyanate.

### 4.1 Toxicity to Pathogenic Microorganism

Lactic acid bacteria like *L. bulgaricus* possess antagonism towards *E. coli*, besides their metabolites which demonstrate cell-free enterotoxin activity. Other lactic acid bacteria have also been found to produce antimicrobial cell-free metabolites e.g., *Streptococcus faecium* and *S. diacetylactis* against many enteric pathogenic microorganisms such as *E. coli* and *Salmonella* sp.

### 4.2 Lactolin and Acidophilin

The presence of antibiotic substance called Lactolin in lactic acid bacteria has been recorded [17]. An antimicrobial agent in the cultures of *L. acidophilus* known as acidolin has also been identified. Low molecular weight nitrogenous compound called Acidophilin from the cultures of *L. acidophilus* has also been purified.

### 4.3 Bacteriocin

Bacteriocin is produced by both Gram positive and Gram negative bacteria as an 'essential biologically active protein moiety' possessing a bactericidal mode of action i.e., bacteriocins are small bacterial proteins or peptides with bactericidal properties [18]. Bacteriocins vary widely in their molecular weight, mode of action and host range. Bacteriocins produced by lactic acid bacteria inhibit the growth of pathogenic Gram positive organisms such as *Clostridia, Staphylococci, Bacilli* and *Listeria* in various food stuffs. It is added directly to the food as a purified product or alternatively, bacteria used in the production of the food may produce sufficient amount of bacteriocin to prevent the growth of spoilage microorganisms. Antimicrobial activity of the partially purified bacteriocin

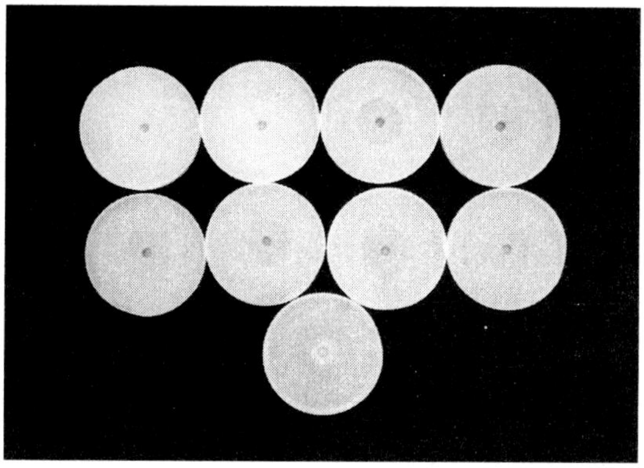

**(a) Antimicrobial zone against *Bacillus cereus***

**(b) Antimicrobial zone against *E. coli***

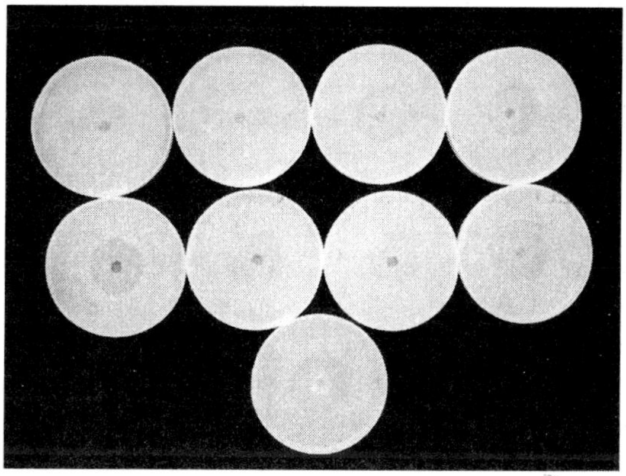

**(c) Antimicrobial zone against *Staphylococcus aureus***

**Plate 1:** Antimicrobial activity of isolated strain from natural lactic acid fermentation of carrot.

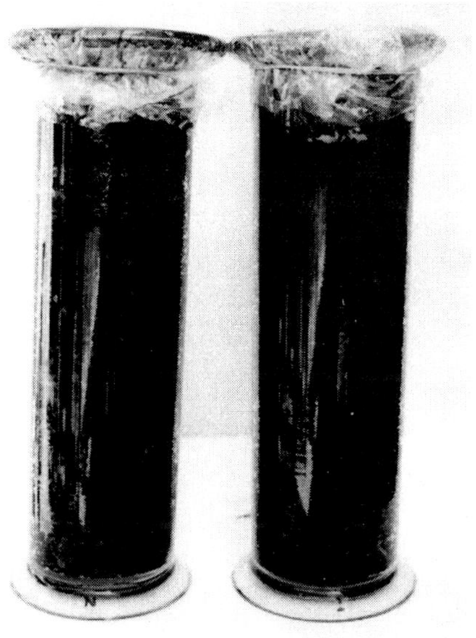

**Plate 2:** Lactic acid fermentation of carrot in jars.

**Plate 3:** Lactic acid fermentation of cucumber in brine.

from isolated strain CA 44 of *Lactobacillus* genus from natural lactic acid fermentation of carrot was reported against *Escherichia coli, Staphylococcus aureus* and *Bacillus cereus* (Plate 1) [19, 20].

## 4.4 Bacteriocins Produced by *Lactobacilli*

Bacteriocins that have been characterized from *Lactobacilli* exhibit only a narrow range of activity against closely related species within the Lactobacillaceae, and are therefore at this stage, unlikely to be of much use as preservatives in the food industry.

## 4.5 Nisin

Nisin is a polypeptide antibacterial substance or bacteriocin produced by lactic acid bacteria. The mature nisin molecule is a peptide of 34 amino acids, which is produced by some strains of *Lactococcus lactis* sub sp. *lactis* (formerly known as *Streptococcus lactis*). It is synthesized in the cell as a protein containing an additional 23 amino acids. The mature nisin molecule also contains a number of unusual amino acids not normally found in proteins. The unusual amino acids are lanthionine, methyl lenthionine and dehydroalanine dehydrobutyrine. It has been suggested that at least a part of antibacterial activity of nisin is due to the interaction of the dehydro residues of dehydroalanine and dehydrobutyrine with sulphydryl residues on the bacterial surface [21, 22].

Nisin is active against a wide range of Gram positive bacteria including *Streptococci, Staphylococci, Micrococcus lysodeikticus, Pneumococci, Bacilli, Clostridia, Corynebacteria, Mycobacteria, Lactobacilli* and *Actinomyces*. Its use as a food preservative is the ability to inhibit the outgrowth of *Clostridium* and *Bacillus* spores and its activity against *Listeria monocytogenes*. Nisin producing strains of *L. lactis* sub sp. *lactis* have also been used to prevent the *Clostridial* blowing of Swiss type cheese [23, 24]. Nisin is generally considered to be non-toxic to humans and is degraded by α-chymotrypsin, an enzyme produced in the pancreas and released into the small intestine and could not be detected in the saliva of humans 10 min after consuming liquid containing nisin. Moreover, there is no evidence of sensitization to nisin. Therefore, there is an extensive use of nisin as a food preservative. A number of bacteriocins, termed lactostrepcins, which inhibited the growth of other strains of *Lactococci, Streptococci* and selected strains of *Lactobacillus helveticus, L. citrovorum* and *L. paracitrovorum* have been identified [25].

## 4.6 Hydrogen Peroxide

The *Lactobacilli* have the ability to generate hydrogen peroxide during growth by several different mechanisms.The formation and accumulation of hydrogen peroxide by *Lactobacilli* in growth media with a subsequent antagonistic effect was shown with *Staphylococcus aureus* [26] and *Pseudomonas* sp. [27]. Hydrogen peroxide can also react with other components to form inhibitory substances.

## 4.7 Diacetyl

Diacetyl (2, 3-butanedione) is a metabolic end product of LAB that is synthesized from the intermediary metabolites, pyruvate [28]. It was shown that diacetyl was inhibitory at a level of 200 μg/ml for yeasts and Gram negative bacteria and 300 μg/ml for non-lactic Gram positive bacteria [29]. Although diacetyl is generally recognized as safe (GRAS), its utility as a food preservative is limited because of the relatively large amounts needed to provide preservation. The intense aroma of diacetyl (buttery) however, would preclude its use in many foods.

## 4.8 Reuterin

It is a low molecular weight, non-proteinaceous, highly soluble, pH neutral product produced by the heterofermentative species, *Lactobacillus reuterii*. It has been described as a broad spectrum antimicrobial agent active against certain Gram negative and positive bacteria, yeasts, fungi and protozoa [30].

## 5. CHANGES IN FOOD COMPOSITION

Fermented foods are more nutritious than unfermented ones due to synthesis or release of nutrients by some of the microorganisms employed [31]. The benefits of food fermentation are shown in Table 2. Generally, a significant increase in the soluble fraction of a food is observed during fermentation. The quantity as well as quality of the food proteins and the content of water soluble vitamins is generally increased, while the antinutritional factors show a decline during fermentation. Fermentation results in a lower proportion of dry matter in the food and the concentrations of vitamins, minerals and protein appear to increase when measured on a dry weight basis [32]. The applications of microorganisms in vegetable processing to achieve technological or nutritional advantages are given in Table 3. As there is scanty information on the effect of lactic acid fermentation on nutritional value of fermented vegetable, this aspect is illustrated by taking example of other foods. Combination of cooking and fermentation improved the nutritive quality of all the tested sorghum seeds and reduced the content of antinutritional factors to a safe level in comparison with other methods of processing [33]. The digestibility of starch in Bengal gram, cowpea and green gram was increased by fermentation. The protein efficiency ratio (PER) of wheat was found to increase on fermentation, partly due to an increase in the availability of lysine. A mixture of wheat and soybeans in equal amounts would provide an improved pattern of amino acids. The fermentation process raised the PER value of the mixture to a level, which was comparable to that of casein [34]. Fermentation may not increase the content of protein and amino acids unless ammonia or urea is added as a nitrogen source to the fermentation media [35].

During fermentation, certain microorganisms produce vitamins at a rate higher than others do. The content of thiamin and riboflavin in *dhokla* and *ambali* was about 50% higher after fermentation. Fermented milk products, in general, showed an increase in folic acid content and a slight decrease in vitamin $B_{12}$, while other B-vitamins were affected only slightly [36] in comparison to raw milk. The levels of vitamin $B_{12}$, riboflavin and folacin were increased by lactic acid fermentation of maize flour, but the level of pyridoxine was decreased [37]. Fermented whole onion plant retained 97% of vitamin A activity, while fermented egg plant only retained 34% of the vitamin A activity [38]. The mineral content is not affected by fermentation unless some salts are added to the product during fermentation or by leaching when the liquid portion is separated from the fermented food. Sometimes, when fermentation is carried out in metal containers, some minerals are solubilized by the fermented product, which may cause an increase in mineral content [39].

**Table 2:** Benefits of food fermentation.

| Raw material | Stability | Safety | Nutritive value | Acceptability |
|---|---|---|---|---|
| Meat | ++ | + | - | (+) |
| Fish | + | + | - | (+) |
| Milk | ++ | + | (+) | (+) |
| Vegetables | + | (+) | - | (+) |
| Fruits | + | - | - | ++ |
| Legumes | - | (+) | (+) | + |
| Cereals | - | - | (+) | + |

++: definite improvement; +: usually some improvement; (+): some cases of improvement; -: no improvement

*Source:* 32

**Table 3:** Application of microorganisms in vegetable processing to achieve technological or nutritional advantages.

| Application | Species | Effect |
|---|---|---|
| Sauerkraut | Starter cultures *Leuconostoc mesenteroides, Lactococcus lactis* | Aroma formation, nisin resistant |
| Fermented vegetables | Various species of lactic acid bacteria | Increase bioavailability of iron by reducing the phytate content |
| Lupins, peas, lentils and other legumes | Various species: *Lactobacillus acidophilus, Lactobacillus buchneri, Lactobacillus fermenti* | Improved digestibility |
| Cassava | *Lactobacillus plantarum* | Reduction in linamarine |
| Delicacy salads, potato, salads, ready-to-eat salads | Various species | Extension of shelf-life |
| Salads, baby foods, vegetable juices | Various species | Reduction of nitrate content |

*Source:* 7

# 6. TECHNOLOGY OF LACTIC ACID FERMENTED PRODUCTS

## 6.1 Factors effecting Lactic Acid Fermentation

Various factors such as microorganisms, salt concentration, temperature, chemical additives, amount of fermentable carbohydrates in the vegetables and the availability of nutrients in the brine effect the lactic acid fermentation [1, 40].

### *6.1.1 Microorganisms*

There are several bacterial classes responsible for food fermentation. Lactic acid bacteria (LAB) are one of the important microorganisms in food fermentation. The term lactic acid bacteria (LAB) has no strict taxonomic significance, although the LAB has been shown by serological techniques and 16S ribosomal RNA cataloging to be phylogenetically related and share a number of common features [11]. These are a group of Gram positive bacteria, non-respiring, non-spore forming, cocci or rods, which produce lactic acid as the major end product of the fermentation of carbohydrates [15]. LAB are responsible for the natural fermentation of fruits and vegetables and are within the genera of *Lactobacillus, Streptococcus, Pediococcus* and *Leuconostoc.* The microorganisms, which predominate the inhibition process and conduct primary fermentation include *Lactobacillus brevis, Lactobacillus plantarum, Leuconostoc mesenteroides, Streptococcus faecalis* and *Pediococcus cerevisiae* [41]. Lactic acid bacteria predominantly associated with plants are enumerated in Table 4. So the fermentation of vegetables may be affected by various groups of microorganisms. Owning to lack of oxygen, non-lactic acid bacteria are suppressed and overgrown by the facultative anaerobic lactic acid bacteria. Lactic acid fermentation is initiated by heterofermentative lactic acid bacteria, namely *Leuconostoc mesenteroides,* followed by homolactic bacteria e.g., *Lactobacillus brevis*. Next stage of fermentation is dominated by homofermentative lactic acid bacteria, which are selectively favoured by the complete lack of oxygen, lowered pH and an elevated salt content. Approximately 90 per cent of homofermentative bacteria consist of Streptobacteria, whereas *Streptococci* and *Pediococci* represent usually less than 10 per cent of the total number of lactic acid bacteria.

Recent investigations have revealed that only 30-80 per cent of the Streptobacteria in sauerkraut are truly *Lactobacillus plantarum*. Besides this, two other closely related species are of importance,

**Table 4:** Lactic acid bacteria associated with plants.

| *Lactococcus* | *Lactobacillus* | *Leuconostoc* | *Pediococcus* | *Enterococcus* |
|---|---|---|---|---|
| *L. lactis* | *L. arabinosus* | *L. mesenteroides* | *P. cerevisiae* | *E. faecalis* |
| | *L. brevis* | | *P. acidilactici* | *E. faecium* |
| | *L. buchneri* | | *P. acidophilus* | *E. faecium* |
| | *L. casei* | | | ssp. *liquefaciens* |
| | *L. curvatus* | | | |
| | *L. fermentum* | | | |
| | *L. plantarum* | | | |
| | *L. sakei* | | | |

*Source:* 42

namely *Lactobacillus sakei* and *Lactobacillus curvatus*. Of the special interest is the occurrence of *Lactobacillus bavaricus*, which is characterized by the exclusive formation of L-(+)-lactic acid [7, 43]. The initial population and the growth rate of microorganisms, as well as salt and acid tolerance, are important factors that influence the sequential development of various lactic acid bacteria in most of the vegetable fermentations. In cucumber fermentation, excessive growth of heterofermentative *Leuconostoc mesenteroides* is particularly undesirable, since the production of $CO_2$ may contribute to gaseous spoilage [1, 7, 11, 41, 42]. Lactic acid bacteria carry out their reactions i.e., the conversion of carbohydrate to lactic acid, $CO_2$ and other organic acids in absence of oxygen and therefore, described as micro-aerophilic [15]. These bacteria are highly specialized and are unable to grow in circumstances that permit the growth of many other microorganisms [1]. The lactic acid produced is effective in inhibiting the growth of other bacteria that may decompose or spoil the food [15]. Reduction of pH and removal of carbohydrates by fermentation are the primary preserving actions that these bacteria provide. But they are also capable of producing inhibitory substances such as bacteriocin, nisin, that are antagonistic towards other microorganisms [11, 20, 44-47].

Fresh plant material harbours numerous and varied types of microorganisms. A particular type of vegetable provides a unique environment in terms of type of vegetable, availability and concentration of substrate, buffering capacity, competing microorganisms and perhaps, natural plant antagonists [42]. Although an extremely small population of lactic acid bacteria is present, it is assumed that plants are a natural habitat for some species. An analysis of 30 different samples of white cabbage from 4 growing seasons has shown that the microflora normally is dominated by aerobic bacteria and yeasts, while lactic acid bacteria represent 0.15 to 1.5 per cent of the total bacterial population [48].

The lactic fermentation is a natural process brought about by the lactic acid bacteria (LAB) present in the raw food or those derived from a starter culture [49]. The interest in probiotic bacteria has been increasing throughout the last couple of decade. Probiotic bacteria belong to the group of lactic acid bacteria (LAB). *Bifidobacterium* and *Lactobacillus* species are the most common species applied to probiotic bacteria [50]. In lactic acid fermented products, salt tolerant lactic acid bacteria, initially heterofermentative *Leuconostoc* sp. and *Lactobacillus brevis* are generally replaced with homofermentative *Lactobacillus plantarum* and *Pediococcus* sp. [51]. All lactic acid producers are microaerophilic i.e., they require small amount of oxygen to function. Species of genera *Streptococcus* and *Leuconostoc* produce the least acid. Next are the heterofermentative species of *Lactobacillus,* which produce intermediate amounts of acid, followed by the *Pediococcus* and lastly, the homofermentors of *Lactobacillus* species, which produce most of the acid [15].

Fermentation of a majority of vegetables is done by naturally occurring lactic acid microflora rather than by defined starter cultures [1]. Starter cultures are used to ensure consistency between batches, inhibition of undesirable microorganisms, enhancing the speed of fermentation process as there is no lag time [15]. Sauerkraut is directly fermented using homofermentative *Lactobacillus helveticus*, which mainly produces L-(+)-lactic acid as a starter culture. The *Lactobacillus helveticus* strain used is highly competitive so that the initial *Leuconostoc* population is suppressed and even the homofermentative lactic acid bacteria are not able to compete. The suppression of the homofermenters leads to less developed flavour, which however, is accepted by some consumers [43]. Sequential culture fermentation with *Lactobacillus plantarum, Pediococcus cerevisiae* and *Streptococcus lactis* var *diacetylactis* was better than natural fermentation of carrot, radish and cucumber [19, 52, 53].

Most fermented vegetable juices are manufactured according to the "Lactoferment Process" [54]. In contrast to other vegetable products, mash or raw juice is pasteurized before fermentation to achieve pure culture fermentation through the addition of a starter culture. Lactic acid bacteria used for this purpose include *Lactobacillus* species (*Lactobacillus acidophilus, Lactobacillus delbrueckii, Lactobacillus helveticus, Lactobacillus plantarum, Lactobacillus salivarius, Lactobacillus xylosus, Lactobacillus bifidus, Lactobacillus brevis* and *Lactobacillus casei*), *Lactococcus lactis* and *Lactococcus mesenteroides* [55]. The species most often used are *Lactobacillus plantarum* and sometimes *Lactobacillus casei.*

The fermentation of vegetables is difficult to control, primarily because of variation in shape, large number and types of naturally occurring microorganisms, and variability in nutrient content. Since, it is not possible to eliminate the natural microflora by appropriate methods, consistent products made by pure culture fermentation cannot be achieved realistically. One promising approach might be the application of defined starter cultures, which are capable of growing rapidly and are highly competitive under environment conditions used to ferment products [7].

In sauerkraut fermentation, different kinds of bacteria begin to grow, but the acid forming types soon predominate. Prominent among the bacteria that attain appreciable numbers early in the fermentation are the coliform bacteria for example, *Enterobacter cloacae*, which produces gas and volatile acids, as well as some amount of lactic acid. Soon however, *Leuconostoc mesenteroides* begin to outgrow all other organisms and continue lactic acid production. Next, non-gas forming *Lactobacilli*, chiefly of the species *Lactobacillus plantarum*, continue the production of acid and raise acidity to 1.5-2.0 per cent. These bacteria complete the desired fermentation in the production of sauerkraut [56-58].

*Pediococcus cerevisiae* plays a major role in vegetable fermentation, along with *Lactobacillus plantarum*, particularly those in brine [1]. In olive pickling, the spoilage bacteria present in the primary stage are eliminated, due to acid production by LAB mostly *Streptococcus, Pediococcus* and *Leuconostoc* [59]. The intermediate stage has *Lactobacillus* as a dominating microflora. In the last stage, as the acid content is increased (0.8 to 1.0 per cent), the pH decreased. However, the use of starter cultures for Spanish type green olives was made [60], which increased the rate of acid production and reduced gas pocket formation and other types of spoilage [61]. In order to produce sauerkraut of consistent quality, starter culture (similar to those used in dairy industry) has been recommended. Starters traditionally used for milk fermentation, such as *Streptococcus lactis* can possibly be used without affecting the final quality as these organisms survive only for a short time in the kraut medium and do not disturb the natural sequence of microorganisms. However, if *Leuconostoc mesenteroides* is added in the early stages, it gives good flavour, though alters the sequence and results in a product that is incompletely fermented [15]. Extensive investigations of lactic acid bacteria used in sauerkraut fermentation have shown that strains of *Lactobacillus plantarum* caused a rapid decrease in pH but led to products with poor flavour. Sauerkraut with excellent sensory characteristics has also been obtained using selected strains of *Leuconostoc mesenteroides,* which simultaneously improved the uniformity of sauerkraut from different batches [62]. *Leuconostoc mesenteroides* and acidifying microorganism *Lactobacillus plantarum* are the main microorganisms responsible for the fermentation of *kimchi*, a product made in Korea with cabbage or radish as the main ingredient [63].

Influence of Gram-negative bacteria on the sauerkraut fermentation was studied and *Aerobacter cloacae* and *Flavobacterium rhamanus* were identified as members of Gram negative group that increased in numbers during initial stages of normal sauerkraut fermentation. *Aerobacter cloacae* was the most prevalent of the two strains [64]. The measurement of oxidation reduction potentials during normal fermentations showed that the pH decreased with the growth of Gram negative bacteria, but when a large inoculum of *Leuconostoc mesenteroides* or any other acid forming bacteria was present, the formation of acid was retarded. With a small inoculum, *Aerobacter cloacae* apparently produced favourable conditions for the growth of *Leuconostoc mesenteroides* as indicated by greater acid formation in a mixed culture. Batches of pickled cucumbers by using the pure culture of *Lactobacillus plantarum* and /or *Pediococcus cerevisiae* were prepared and the best results were obtained by mixed culture [65].

Effect of inoculation (*Lactobacillus plantarum* alone and mixed with *Lactobacillus brevis*), brine purge with $N_2$ and $CaCl_2$ addition on cucumber fermentation and on product quality was studied [66], where large size fresh cucumber fruits (8-11cm) in low salt brine (4.7% NaCl) at 17°C were fermented. In inoculated samples, the fermentation was completed in 11-21 days, however, non-inoculated samples took 35 days and non-inoculated samples with 10.7% NaCl had hardly began fermentation on the 87$^{th}$ day. Brine purge helped to avoid bloater formation. Model studies carried out to evaluate the fermentative bacteria (*Bifidobacterium bifidium, Lactobacillus casei, Lactobacillus plantarum, Lactococcus diacetylactis, Leuconostoc mesenteroides, Leuconostoc oenos, Pediococcus pentosaceus,* and a mixed culture of *Propionibacterium shermanii* and *Pediococcus pentosaceus*) in low salt (2.5% NaCl) cucumber juice brine at 22-26°C for 39 days resulted in formation of products like acetic acid, acetoin, ethanol, lactic acid, mannitol and propionic acid [67]. Each fermentation differed quantitatively in substrates and products formed, suggesting the use of these bacteria as cultures to ferment low salt brined cucumbers and generate a variety of unique organoleptic properties.

The lactic acid bacteria tolerate high salt concentrations which gives them an advantage over other less tolerant species and thus, allowing the lactic acid bacteria to produce acid. The latter inhibits the growth of undesirable organisms. *Leuconostoc* is noted for its high salt tolerance and for this reason, initiates majority of lactic acid fermentations [15].

Many researchers have studied the changing patterns of microflora involved in fermentation of *kimchi* alongwith the taste and odour produced during fermentation [68, 69]. Major microflora isolated from *kimchi* were *Lactobacillus plantarum, Lactobacillus brevis, Streptococcus faecalis, Leuconostoc mesenteroides, Pediococcus cerevisiae* and *Achromobacter, Flavobacterium* species as aerobic microorganisms. The major microorganisms involved in various fermented fruits and vegetable products are given in Table 5.

Mushroom-in-brine inoculated with lactic acid bacteria produced more acid than uninoculated mushroom. *Lactobacillus plantarum* produced the highest amount of acid followed by *Streptococcus lactis* [77]. In cucumber and green olive fermentation also, *Lactobacillus plantarum* produced the highest acid [78, 79].

### *6.1.2 Temperature*

Temperature affects the growth and activity of all the living cells and microbial cells are not an exception. At high temperature, microorganisms are destroyed while at low temperature, their rate of activity is decreased or suspended. Microorganisms are classified into three distinct categories on the basis of their temperature preference and these classes are psychrophiles, mesophiles and thermophiles. Lactic acid bacteria come under the mesophilic class and work best at a temperature range of 18-22°C [15]. The temperature is a critical factor for producing high quality fermented vegetables where it affects the vegetable acidification rate and promotes the growth of single microbial species, giving it a competitive edge over other species.

**Table 5:** New fruits and vegetables fermented products developed along with the microorganisms involved.

| Fermented Vegetables | Microorganisms involved | Reference(s) |
|---|---|---|
| Fermented fruit juices | *Lactobacillus casei* | 70 |
| Fermented vegetable juices | *Lactobacillus* strains | 71 |
| Pickled leafy vegetables | | |
| *Pak-Gard-Dong* | *Lactobacillus brevis*<br>*Pediococcus cerevisiae*<br>*Lactobacillus plantarum* | 72, 73 |
| *Gundruk* | *Lactobacillus plantarum*<br>*Pediococcus pentosaceus* | 74 |
| *Bossam kimchi*<br>(Pickled cabbage kimchi) | *Lactobacillus plantarum*<br>*Lactobacillus brevis*<br>*Streptococcus faecalis*<br>*Leuconostoc mesenteroides*<br>*Pediococcus pentosaceus* | 73 |
| *Sunki* (Otaki turnip) | *Lactobacillus plantarum*<br>*Lactobacillus brevis*<br>*Bacillus coagulans*<br>*Pediococcus pentosaceus* | 75 |
| *Kawal* (Sickle pod plant) | *Bacillus subtilis*<br>*Lactobacillus plantarum* | 15 |
| Sogurt (Yoghurt like product from soya proteins) | *Streptococcus salivarius*<br>*Lactobacillus delbrueckii* | 76 |
| Camembert cheese-like product from soya proteins | *Lactococcus lactis* sub sp. *lactis* and *cremoris*<br>*Leuconostoc mesenteroides* | 76 |

### *6.1.3 Water activity*

In general, bacteria require a fairly high water activity (0.9 or higher) to survive. There are a few species which can tolerate water activities lower than this, but usually the yeasts and fungi will predominate on foods with a lower water activity.

### *6.1.4 pH*

The optimum pH for most bacteria is near the neutral point (pH 7.0). Certain bacteria are acid tolerant and will survive at reduced pH levels. Notable acid-tolerant bacteria include the *Lactobacillus* and *Streptococcus* species, which play a role in the fermentation of dairy and vegetable products.

### *6.1.5 Oxygen availability*

Some of the fermentative bacteria are anaerobes, while others require oxygen for their metabolic activities. Some, *Lactobacilli* in particular, are microaerophilic. That is, they grow in the presence of reduced amounts of atmospheric oxygen. In aerobic fermentations, the amount of oxygen present is one of the limiting factors. It determines the type and amount of biological product obtained, the amount of substrate consumed and the energy released from the reaction. *Acetobacter* requires oxygen for the oxidation of alcohol to acetic acid.

### *6.1.6 Additives*

6.1.6.1 Salt concentration: Lactic acid fermentation of fruits and vegetables is often carried out in salted medium where it plays an important role in initiating fermentation. From the beginning, there were three methods i.e., high salt brine salting, low salt brine salting and dry salting [1]. Salting is done by adding dry salt to vegetables with high water content or by soaking others in brine. Dry salt must be added regularly to successive layers of the product in order to avoid high local salt concentrations that could promote yeast development. Any variety of common salt is suitable as long as it is pure; impurities or additives cause problems. Salt with chemicals to reduce caking should not be used as it makes the brine cloudy, while with lime impurities can reduce the acidity and the shelf-life of the final product. Impurities of iron and magnesium could result in the blackening of vegetables and impart bitter taste. Carbonates in pickles can result in soft texture [15, 80]. In principle, the vegetables can be fermented without the addition of sodium chloride [81]. But it is often a very important ingredient for many reasons, a major one being its contribution to flavour besides its role in supporting the development of anaerobic conditions in fermentation vessels. Its amount to be used depends on the particular vegetable and on the consumer demand [82].

In the pickling industry, salt (sodium chloride - NaCl) has historically been used for directing the fermentation of cucumber, radish and carrot [19, 83-86]. The complex changes that occur in vegetable fermentation are produced by the growth of a sequence of lactic acid bacteria. The growth of each species depends upon its initial presence on the vegetable, the sugar and salt concentration and the temperature. Salt concentration in vegetable fermentation can range from 20-80 g/l during fermentation and upto 160 g/l in some stored vegetables. A high salt concentration inhibits the growth of unwanted microorganisms and induces plasmolysis of plant cells, thus promoting anaerobiosis in the medium. Anaerobiosis is more effective in finally cut and shredded plant material. Whole or unjuicy vegetables are therefore fermented in highly salted brine. Plasmolysis also promotes the growth of lactic acid bacteria by releasing nutrients contained in the plant cells. The dominant bacteria at the outset of fermentation are resistant to high salt concentration. Due to progressive acidification of the medium, acid tolerant bacteria are present by the final stages of the process. A dominant flora sequence over the course of lactic acid fermentation has been identified i.e., successively *Pediococcus, Leuconostoc* and *Lactobacillus* [87]. Most vegetables can be fermented at 12.5-20° salometer salt. If so, the microbial sequence of lactic acid bacteria generally follows the classical sauerkraut fermentation described earlier [1]. At higher salt levels of about 40° salometer, the sequence is skewed towards the development of homofermentation, dominated by *Lactobacillus plantarum*. But at higher salt concentration of about 60° salometer, the lactic acid fermentation ceases to function [15, 88].

In sauerkraut fermentation, salt withdraws water and nutrients from the vegetable tissue and the latter acts as the substrate for growth of lactic acid bacteria [1]. Since the fermentation of sauerkraut involves several types of microorganisms, their relative rates of growth will depend on environmental factors. In a brine with 5% salt, the fermentation takes place appropriately, but at 10% salt concentration, it is slow and at 20% fermentation stops altogether. No doubt, lactic acid bacteria are salt tolerant, but at higher concentration they are inhibited [15, 57]. Salt content of fermenting sauerkraut not only influences the bacterial flora but also affects the quality and chemical composition of the product [89]. A high local salt concentration (above 2.5%) would lead to pink kraut [87, 90]. It exerts a selective effect on microorganisms naturally present on vegetables, and increasing the amount of salt inhibits the growth of undesirable bacteria and fungi but selects lactic acid bacteria. In sauerkraut fermentation, heterofermentative lactic acid bacteria are favoured by low concentration (1%) and are greatly inhibited at 3%. Higher amount of salt favours homofermentative species, resulting in accelerated fermentation. It is generally agreed that salt content below 0.8% often results in undesirable fermentation as well as in soft kraut. The latter quality defect is reportedly contributed by the salt, which has an adverse effect on the texture of final products and several plant and microbial enzyme systems [7].

Three different concentrations of salt i.e., 1, 2.25 and 3% for sauerkraut preparation were tried [91], and it was found that salt concentration of 2.25% produced the highest acidity (1.60% as lactic acid). However for consumer acceptability, the highest score for colour was given to the product of 2.25% salt concentration and lowest for 1.0% salt concentration. The sauerkraut with 2.25% salt content was awarded the highest score for flavour and texture. The differences among different treatments had been attributed to the quantity of salt, acid development and production of flavouring constituents during fermentation. It has been found that too low salt concentration results in poor flavour of sauerkraut [1]. Generally, salt is added to a final concentration of 2.0 - 2.5% as in sauerkraut process, whereas *Lactobacilli* are slightly inhibited, but cocii are not affected. Although, this concentration of salt has a great inhibitory effect against the desirable organisms than against those responsible for spoilage. So it is not the salt concentration that inhibits the spoilage organisms, but the acidic environment created by the *Lactobacilli* that keeps the spoilage bacteria at bay. The use of salt brines is not recommended in sauerkraut making, but is common in vegetables that have a low water content [15].

Sodium chloride is an important ingredient in cucumber fermentation and is also used to provide selective environment for lactic acid bacteria responsible for desirable changes in the quality of the product [3, 9]. The effect of sodium chloride is most likely due to the fact that it either inhibits degradative enzymes or prevents the growth of microorganisms that produce degradative enzymes [83]. In cucumbers, lactic acid bacteria may use brined cucumbers as a fermentation substrate. Both the concentration and chemical composition of the brine play a governing role in the sequence of microorganisms that occur in natural cucumber fermentations. Proper salt concentration exerts a selective effect on the natural flora, resulting in the growth of LAB. Salt concentration $< 5\%$ favours the growth of Enterobacteriaceae members along with the lactic acid bacteria and allows the development of salt tolerant fermentative yeasts [63]. A few examples of salt concentration applied during lactic acid fermentation of most common vegetables are summarized in Table 6.

Several types of pickles are fermented in low salt concentration such as dill pickles and may be consumed after fermentation without further processing [1]. Olive pickle of cultivar Carotina in 4% brine and 0.1% citric acid was also made and had better flavour, taste, acidity and overall acceptability than the pickle prepared with other cultivars: Lacella and Frantio [94].

They are fermented in brine of 5% or even less salt, where the fermentation is rapid. The cucumbers absorb salt until equilibrium of acid and salt between the brine and cucumbers is attained. Initiation of cucumber fermentation is carried out in many instances by *Leuconostoc mesenteroides*, particularly at lower salt concentrations and temperatures [95, 96]. But at higher temperature and salt concentrations, fermentation was often initiated by *Streptococcus faecalis, Pediococcus cerevisiae, Lactobacillus plantarum* and to a lesser extent by *Lactobacillus brevis*. The latter two species are

**Table 6:** Salt concentration used during lactic acid fermentation of vegetables.

| Vegetable | Salt concentration (%, w/w) | Type of salting | Reference(s) |
|---|---|---|---|
| Cabbages sauerkraut | 2-3 | Dry salt | 9 |
| Korean *kimchi* | 3 | Dry salt | 92 |
| Cucumbers | 3, 5-8 | Brine | 9, 19 |
| Green olives | 4-7 | Brine | 9 |
| Carrots | 1.5-3, 2.5 | Brine, salt | 53, 93 |
| Radish | 2.5 | Salt | 52 |
| Black olives | 2-5 | Brine | 15 |
| Over-night dill pickles | 5.3 | Brine | 56 |
| Genuine dill pickles | 7.5-8.5 | Brine | 56 |
| Sauerruben (turnip) | 2.2 | Dry salt | 56 |
| Pickled leafy vegetables | 2.5 | Dry salt | 56 |

responsible for the high acidity. However, large sized cucumbers and those packed in high concentration brines may soften because of enzymatic action produced by yeasts during natural fermentation or as a result of the native enzymes of the cucumber [1].

An initial salt concentration of 5.0 to 7.5% at equilibrated level was optimum for fermentation of pickling cucumbers [97], and a good quality brine stock pickling cucumber could be obtained by fermentation in 7.5% equilibrated concentration of sodium chloride solution for 4-5 days initially and raising the salt level to 14%. Sewon [98] patented a method for conserving cucumbers in bulk brine, which was low in NaCl and described a method for conserving cucumbers making fermentation unnecessary [99]. The effect of salt concentration on first order softening rates of cucumber pickle during acid storage was determined [100], wherein softening rates of unfermented tissues increased, as sodium chloride concentration was increased from 0-1.5 M. But controlled fermentation of cucumbers reduced the salt requirements in the brine, formerly used to enable the naturally low count of lactic acid forming bacteria to develop, and reduced the incidence of bloater formation [101].

Chavasit *et al.* [67] carried out model studies to evaluate the effect of fermentative bacteria in low salt cucumber brine (2.5% NaCl). The per cent sugar fermented ranged from 16.2-87.7, and substrates were citric acid, fructose, glucose and malic acid. Thus the use of pure or mixed cultures to ferment low salt brined cucumbers was suggested.

Carrots were fermented in brine with different salt concentrations ranging from 2-10% and it was found that fermentation of carrot in 2% brine at 30±2°C for six days yielded an acceptable product [102] in which 60% carotenoids were retained at the end of four months of storage at room temperature of 30°C. The brine containing 2 and 4% salt had faster drop in pH compared to 6, 8 and 10% and the fermentation was completed in 6-7 days, whereas in case of 6, 8 and 10% brine, it took 3-4 days more to develop maximum acidity. On the basis of organoleptic evaluation (flavour, taste and overall acceptability), the product with 2% brine, 0.78 % acid, pH 3.38 and final count of $67 \times 10^8$ cfu/ml was the best.

**6.1.6.2 Other additives:** Many ingredients apart from salt can be used in the preparation of lactic acid fermented fruits and vegetables. They have three main functions: as a source of nutrients such as sugars, mineral salts and vitamins for fermentation causing microorganisms, help to restrict the growth of unwanted bacteria either through a regulatory effect on pH or by producing inhibitory substances and spices that have a final flavour determining role in fermented vegetables [63].

6.1.6.2.1 Growth stimulators: Many foods contain the ingredients essential for the growth of even the most fastidious true lactic acid bacteria. The biochemical changes that occur in food as a result of microbial action are in their simplest form, a reflection of the changes required by the microorganisms to obtain certain components for growth and to obtain energy for the necessary conversions. The nutritional requirements of the lactic acid bacteria are particularly complex and exacting. Besides an energy source, a variety of essential growth factors must be made available [1]. Some of the ingredients, added more or less empirically to lactic acid fermented vegetables or fruits, seem to enhance the development of lactic flora, which generally requires very special conditions to proliferate. It was reported that lactic acid bacteria depends essentially on plant sugars for the growth [63]. For some vegetables with low nutrient contents, such as turnip and cucumber, the addition of sugar promotes bacterial growth, thereby accelerating fermentation as jaggery (*gur*) is added during lactic acid fermentation of sweet turnip. Adding 1-2 g of sucrose or jaggery to 100 g of hand-shredded turnip (mixed with 6 g of salt and 6 g of mustard seed) increased lactic acid production during fermentation [103]. Further acidification was however, much slower at high sucrose concentration (25 g), which is attributed to the partial inhibition of lactic bacterial growth.

6.1.6.2.2 A substrate - whey: Whey is often added in the traditional lactic acid vegetable fermentation processes, as it has high lactose content, a potential energy substrate for lactic bacteria. It is highly recommended for the preparation of only slightly salted sauerkraut (0.2-0.3% salt, w/w of brine) [104], or for fermenting low nutrient vegetables [105]. Dry salting fermentation of mushroom with or without blanching was not found possible, and even addition of lactose did not stimulate the fermentation and led to the chemical and microbiological spoilage [77]. Japanese rice bran is used in "*tsukemono*" (a lactic acid fermented mixture of vegetables and leaves) [104], while Russian wheat bran is used for lactic acid fermentation of apples and pears.

6.1.6.2.3 Spices: Spices or aromatic herbs are added to most of the lactic acid fruit and vegetable fermentation preparations to improve the flavour of finished products. Aromatic compounds in these spices (including mainly terpenes and polyphenols) often have an antimicrobial effect, which means that they can have a selective role in the development of bacteria during fermentation [106].

6.1.6.2.4 Mustard seed: Mustard seeds are also of interest, as they contain allyl isothiocynate, a volatile aromatic compound with antibacterial and antifungal properties. Ground mustard seeds or oil are widely used in traditional fermented vegetable preparations in India. Adding 6-10 g of mustard seed powder to turnip (100 g) and salt (6 g) mixture increased lactic acid levels during fermentation [103].

Method for the preparation and preservation of a nutritious and colourful beverage from fermented black carrots has been standardized [71]. Among the treatments, addition of 3% salt, 1.0% mustard, 0.015% sodium benzoate and 0.01% potassium metabisulphite to 1:1 carrot water mixture was recommended for preparation of fermented beverage. The shelf-life of fermented juice could be increased either by addition of 1.0% mustard or by the use of chemical preservatives. So the addition of mustard along with preservatives had complementary role for extending the shelf-life of fermented juice upto 20 weeks. The juice preserved with 3% salt alone got spoiled within 3 weeks, while those preserved with mustard were spoiled after 10 weeks. The rate of development of acidity as well as microbial population was also faster in 1% mustard than in 0.5% mustard samples. The retention of colour, flavour and taste was the best in treatment fermented with 1% mustard and 3% salt.

It was found that increasing the proportion of mustard from 1-2% with different salt concentrations (2-12%) progressively increased the rate of lactic acid production and suppressed the spoilage by mycoderma in cauliflower slurry [107]. The maximum acidity was developed in 4% salt when used either alone or along with 20% mustard. Mustard showed a selective preservative action, allowing LAB to grow and suppress the surface yeast. Among the different varieties of mustards tried, *Brassica nigra* (Banarsi Rye) was the best followed by *B. juncea* (Rye), *B. campestris* var *sarson* or *B. campstris* var *Toria*. Mustard, being an oil source, is an important condiment and used in various food products. Mustard, like other condiments and spices, not only imparts flavour and taste to the food products, but also possesses preservative action [108, 109]. Addition of mustard powder to pickle of turnip, cabbage and cucumber increased the rate of lactic fermentation [103, 110, 111].

## 6.2 Lactic Acid Fermented Vegetables

Preparation of fermented foods using LAB has been suggested to improve their acceptability [112]. It is established that more than one species of lactic acid bacteria (LAB) are responsible for vegetable fermentation [113, 114]. Moreover, controlled fermentation using starter cultures are being used now, whereas traditional fermentations depended upon the growth of naturally occurring LAB. Wallace [115] has described a controlled fermentation process, which prevented the formation of bloaters in cucumber. Incoming cucumbers were treated in a chlorinated tank to destroy the natural microflora of which LAB constitute a small proportion as desirable lactic acid bacteria. A pure culture of LAB was therefore, added to give a controlled fermentation. Anders [101] carried out controlled fermentation of cucumbers involving removal of majority of the natural microorganisms from the raw cucumber stock and introduction of desirable lactic acid bacteria. Efforts to improve cultures include development of non-malate decarboxylating strains of *Lactobacillus* for use in cucumber fermentation [4]. In cucumber fermentation, two processes (chlorination and acidification) effectively suppress the growth of undesirable bacteria during the initial salt addition and equilibrium of the brine and cucumbers. After about 10 to 24 hours following equilibration of the brine and cucumbers, they are inoculated with special cultures of *Pediococcus cerevisiae* and *Lactobacillus plantarum* [56]. A general flow sheet for the production of fermented vegetables is given in Fig. 1. Such fermentation has been induced in ripe olives from *Hojiblanka* cultivar [116] as the growth of lactic acid bacteria in olive is rare because of the effect of oleoperine and other polyphenols from the fruit.

A special fermented beverage traditionally known as *Kanji* has been prepared from crimson coloured carrots supplemented with salt and mustard powder [71]. Lactic acid fermentation is employed to prepare various products such as fermented grape juice [117], fermented nut milk [118], sogurt from soyabeans [119], fermented cucumber extract [120], fermented beverage from wheat and maize [121] and many more products.

Fermented foods have been a part of staple foods in the mountainous region of Bhutan and Nepal as well as for the Nepalese origin ethnic community living in different parts of the world. *Kinema* (fermented soyabean), *Gundruk (Brassica campestris* leaves), *Sinki (Raphnus sativus*), *Tama* (Succulent bamboo shoots), *Dahi, Mohi, Sher, Shergum, Chhurpi* and *Selroti* are some of the popular fermented vegetables from the hills and mountain of Nepal and Bhutan [122]. Lactic acid fermentation is used to stabilize beverages by acidification. The process renders the lactose more digestible for people with lactose intolerance. Traditional East-African beverages such as "*Muratina*" (Sauce tree) or "*mnazi*" (fermented coconut palm) have also been studied. Only *Lactobacilli* were isolated from "*Muratina*" and only *Streptococci* from "*Mnazi*" [123]. Currently, LAB are used to create new fermented products such as orange or grape juices, peach and apple juice, papaya and litchi juice. It was found that fruit juices contain bacteriostatic substances (mainly polyphenols), which restrain the multiplication of lactic acid bacteria [124, 125]. So these substances should be removed to enable the development of LAB to prepare the fermented juices. *Lactobacillus* strains in peach, apple and grape juice, while adjusting the concentration of 5 g/l by adding non-fermented fruit juice have been cultured [70]. Good tasting beverage from papaya and litchi using *Lactobacillus plantarum* strains selected from their ability to mainly produce L-(+)- lactic acid were developed [126].

A method for extension of postharvest life of mushroom by inducing lactic acid fermentation was developed, and sauce from fermented mushroom was prepared [77]. The sauce has a typical mushroom flavour, though lactic, acetic acid and other ingredients used in sauce preparation also imparted their respective taste and aroma to the product. The original texture of unfermented mushroom was spongy, and the use of lactic acid fermentation brought the consistency of mushroom sauce to the desirable level besides improved flavour.

Temperature influences the quality of sauerkraut. The optimal temperature for sauerkraut fermentation is around 21°C, and a variation of just a few degrees alters the activity of the microbial process and affects the quality of the final product. Its control is regarded as one of the most important factors in sauerkraut fermentation process [114, 127, 128]. At an optimal temperature of 18°C, sauerkraut fermentation takes about 2 months to complete. At 7.5°C, fermentation takes 6 months, while at 30°C, homofermentative bacteria will thrive, thus excessively increasing the acidity of the final product [63]. The *Leuconostoc* species, which initiate fermentation, have optimum temperature range of 18-22°C. Temperatures above 22°C favour the *Lactobacillus* species [15]. It has been revealed that the fermentation temperature below 10°C strongly influenced the quality of sauerkraut [129]. The fermentation

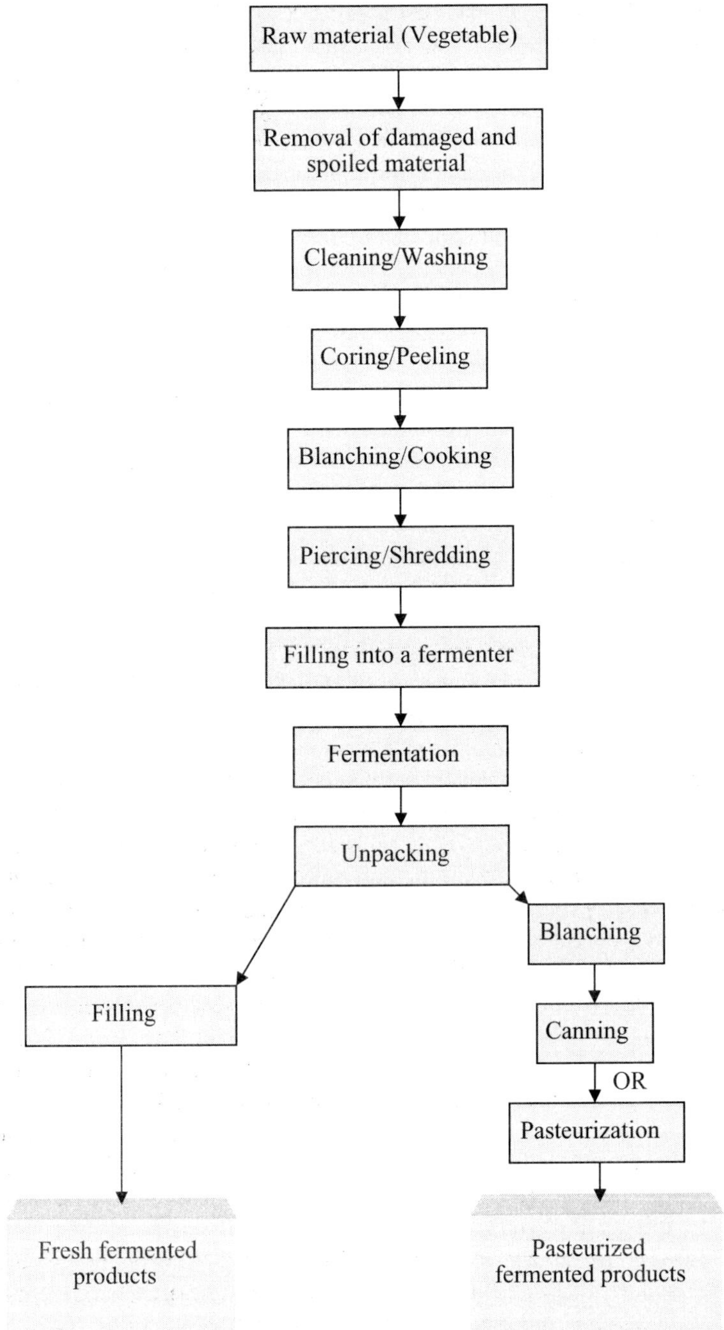


**Fig. 1:** A generalized flow sheet for fermented vegetable production.

is slow and only *Leuconostoc mesenteroides* occurs therefore the acid content remains low, whereas at high temperature, the kraut may be tough and show pink discoloration. The rate of fermentation in sauerkraut fermentation was found to be essentially temperature dependent [130]. The rate of acid production was considerably faster at 34°C than at 12°C. At 34°C, maximum acidity of 1.30% was obtained in red cabbage and only 0.81 and 0.61% acidities were obtained at 18-25°C and 12°C after 6 days of fermentation. It was further confirmed when the lots previously kept at 12°C were transferred to room temperature. There was a quick rise in acidity in these lots, although the maximum acidity in them remained much below than in the lots kept throughout the room temperature. The major share of the acid produced has been attributed to the growth of homofermentative bacteria, *Lactobacillus plantarum* and *Pediococcus cerevisiae,* which may preponderate at higher temperature [114] such as at 34°C, where the acidities in kraut started falling after reaching their maximum at 1.30% level. But the flavour of these krauts was not good and sensory grading was also poor [130]. When it is below 15.6°C, the fermentation will be slow and incomplete; if it is above 26-29°C, abnormal fermentation may result. An excessively high temperature may also inhibit the growth of *Leuconostoc,* and consequently the flavour production by that organism, and may permit the growth of *Pediococcus cerevisiae* and the development of undesirable flavours. An excessively low temperature may prevent adequate activity of the desired succession of lactic acid bacteria and encourage the growth of contaminants from the soil. Too long fermentation may favour the growth of gas-forming *Lactobacillus brevis*, which yields a sharply acid flavour. High temperature during fermentation can cause defects in sauerkraut such as softening, darkening and pink kraut [56]. Graham [128] recommended the use of lower temperature for processing and for cooling rapidly to reduce the darkening from the breakdown of proteins and liberation of sulphur compounds in sauerkraut canning.

Under commercial conditions, existence of definite correlation between the atmospheric temperature of day proceeding, filling a tank and other factors particularly temperature within the building during filling of vat is known to influence the results [131]. In carrot fermentation, the acidity development was faster at 37°C, but the product had a cooked flavour, while the acidity development was slower at 22°C than that at room temperature i.e., at 30°C [102]. So the fermentation of carrots at 30 ± 2°C for 6 days yielded an acceptable product. Joshi *et al.* [53] reported that amongst the temperatures tried, 26°C for LAB fermentation of carrot was the best, as at this temperature higher acidity, low pH and low reducing sugars were recorded. Further, the products had higher score for colour, flavour, taste, texture and over-all acceptability. Among the salt concentrations and the type of fermentation, 2.5% along with sequential culture fermentation was the best in carrot fermentation (Plate 2). The fermented products have a better flavour, texture and taste than the products fermented naturally or with other lactic cultures such as *Pediococcus cerevisiae, Lactobacillus plantarum* or *Streptococcus lactis* individually. Fig. 2 shows the flow sheet for production of lactic acid fermented carrot.

In dill pickle also, considerable attention is given to the effect of temperature on fermentation and storage of finished product. The fermentation at lower temperature of 18.5°C or even lower is initiated by *Leuconostoc mesenteroides,* while those at higher temperature may be initiated by *Streptococcus faecalis*, followed by the homofermenters [96]. While the rate of fermentation is slow at lower temperature, the quality of product is superior especially with respect to storage of pickles in brine for a longer period of time before softening occurs. Storage of dill pickles at lower temperature of even 2-4°C has been found to keep the pickles firm and crisp for an extended period of time.

In controlled cucumber fermentation, various factors influencing bloater formation have been studied [132]. Conditions were provided whereby fermentation of brined cucumber was directed to the predominant growth by *Lactobacillus plantarum*, the added starter culture, which produces little carbon dioxide. Bloater damage was greater at 32°C than at 27°C and depth damage was also enhanced at 32°C. Later, effect of temperature (17°C and 26°C) and purging with nitrogen or air on the development of fermentation and on the incidence of bloater damage in cucumber was studied on pilot scale [133]. The incubation temperature of cucumber affected the rate of fermentation, and both temperature and purging strongly affected incidence of bloater damage. At 17°C, purging was not

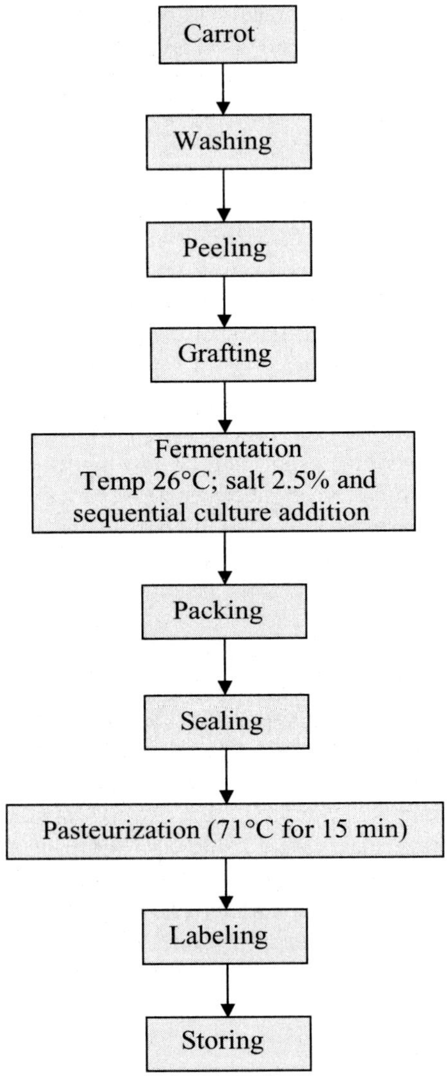


**Fig. 2:** Unit operations for preparation of fermented carrot.

necessary. Sharma [19] reported that for cucumber fermentation, the temperature of 32°C, salt concentration of 3% and addition of sequential starter culture were the best conditions to get the product with better quality (Plate 3).

## 7. FERMENTED CEREAL PRODUCTS

Cereals meet a considerable part of protein and carbohydrate needs of the world population. The microbial activity in these foods makes them attractive and adds variety to the raw cereals. Some of the lactic acid bacteria fermented products made from cereals are: *Warries*, *Dosa*, *Idli*, *Bhallae*,

*Bhatura*, *Jalebie*, *Kulcha* and *Nan* which are mainly consumed in India. Several bacteria and yeasts constitute the natural microflora of cereals associated with fermentation. The major bacteria involved in the fermentation and production of the above products are given Table 7.

## 7.1 *Warries*

These are in the form of spicy, hollow, brittle balls 5-8 cm in diameter, very popular in north India and are used as condiment in cooking with vegetables, legumes, or rice. The dehulled black gram grains after being washed and soaked overnight in water are ground to paste. The paste is then spiced with various species such as asafoetida, caraway, cardamom, cloves, fenugreek, ginger and red pepper and moulded into small balls [134, 135], which are allowed to undergo fermentation by various microorganisms (*Leuconostoc mesenteroides* along with *Lactobacillus fermentum, Streptococcus faecalis, Bacillus subtilis* and other yeasts) and dried in open air for 4-8 days. During fermentation, both yeast and bacteria, whenever present, increase significantly with the progress in fermentation followed by a decrease in pH and rise in volume of warri dough. The predominant bacteria present initially go on reducing with the fermentation time. The successive changes in microflora during *warri* fermentation are presented in Table 8.

## 7.2 *Idli*

These are prepared by steaming the fermented batter consisting of rice and black gram grains and are served with butter and honey or jams and sauces. The proportion of rice to black gram grains used in idli fermentation ranges from 1:4 to 4:1. The amount of water required to prepare the batter of desirable and uniform consistency varies from 1:5 to 2:2 times the dry weight of the ingredients.The paste prepared is then allowed to ferment for 18-30h by means of natural microflora or by inoculating with fermented batter of previous batch. Both bacteria and yeast have been reported to participate in idli fermentation. Bacteria identified include *Leuconostoc mesenteroides, Lactobacillus fermentum, Streptococcus faecalis, Lactobacillus coryneformis, L. delbrueckii, L. lacis* and *Pediococcus cerevisiae* [134, 136-140, 143]. The effect of yeast domination in idli batter fermentations revealed that adding *Saccharomyces cerevisiae* along with natural bacterial flora of the ingredients is the best method for standardizing idli fermentation in terms of improved organoleptic characteristics, leavening and nutritional constituents.

**Table 7:** Major bacteria involved in the fermentation of cereal products.

| Product | Microorganisms | Reference(s) |
|---|---|---|
| *Warries* | *Leuconostoc mesenteroides* along with *Lactobacillus fermentum, Streptococcus faecalis, Bacillus subtilis* and other yeasts | 135, 136 |
| *Idli* | *Leuconostoc mesenteroides, Lactobacillus fermentum, Streptococcus faecalis, Lactobacillus coryneformis, L. delbrueckii, L. lacis, Pediococcus cerevisiae* and other yeast species | 134, 136-140 |
| *Dosa* | *Leuconostoc mesenteroides, Lactobacillus fermentum, Streptococcus faecalis, L. delbrueckii, L. lactis, Pediococcus cerevisiae, Bacillus subtilis, Bacillus polymyxa, Enterobacter* sp. and other yeast species | 134, 141 |
| *Bhallae* | *Bacillus subtilis, Lactobacillus fermentum, Leuconostoc mesenteroides, Streptococcus faecalis* along with other yeast and mold species | 142 |

**Table 8:** Changes in microflora during *warri* fermentation.

| Incubation period (days) | pH | Bacteria | | Yeast | |
|---|---|---|---|---|---|
| | | Count (CFU/g) | Predominant types | Count (CFU/g) | Predominant types |
| 0 | 5.65 | $1.3 \times 10^{10}$ | *Leuconostoc mesenteroides, Lactobacillus fermentum, L. delbruekii, Bacillus subtilis, Flavobacter* sp. | 0-$8.0 \times 10^{4}$ | *Trichosporon beigelii, Saccharomyces cerevisiae, Candida krusei, Pichia membranaefaciens, Hansenula anomala* |
| 1 | 4.70 | $2.1 \times 10^{12}$ | *L. mesenteroides, L. fermentum, Bacillus subtilis, Streptococcus faecalis* | 0-$1.7 \times 10^{6}$ | *Trichosporon beigelii, Saccharomyces cerevisiae, Candida krusei, Pichia membranaefaciens* |
| 2 | 3.90 | $3.0 \times 10^{12}$ | *L. mesenteroides, L. fermentum, Streptococcus faecalis* | 0-$2.1 \times 10^{6}$ | *Trichosporon beigelii, Saccharomyces cerevisiae, Candida krusei* |
| 3 | 3.50 | $4.1 \times 10^{12}$ | *L. mesenteroides, L. fermentum, Streptococcus faecalis* | 0-$9.1 \times 10^{6}$ | *Trichosporon beigelii, Saccharomyces cerevisiae, Candida krusei* |
| 4 | 3.25 | $4.7 \times 10^{12}$ | *L. mesenteroides, L. fermentum* | 0-$6.9 \times 10^{7}$ | *Trichosporon beigelii, Saccharomyces cerevisiae* |
| 5 | 3.20 | $6.5 \times 10^{12}$ | *L. mesenteroides, L. fermentum* | 0-$9.6 \times 10^{6}$ | *Trichosporon beigelii, Saccharomyces cerevisiae* |

*Source:* 135

## 7.3 *Dosa*

*Dosa* is a thin, crisp, fried pancake like staple food of South India. It is prepared by soaking overnight, equal quantities of rice and dehulled black gram separately in water, grinding to a fine paste by adding 2.0-2.5 parts (w/w) of water and mixing the two together. The batter mixture is then allowed to ferment for another 24 h by means of the natural microflora, or by inoculating with the fermented batter of the previous batch which serve as starter. The batter is then spread on the hot greasy griddle where it assumes the form of a crisp pancake.

## 8. FERMENTED DAIRY PRODUCTS

The International Dairy Federation (IDF) defines fermented milks as products prepared from milks (whole, partly or fully skimmed milk, concentrated milk, or milk reconstituted from partly or fully skimmed milk powder) homogenized or not, pasteurized or sterilized and fermented by specific microorganism. The most important fermented milk will be discussed here. Some of the common bacterial cultures used in the manufacture of a variety of fermented milk products throughout the world are listed in Table 9.

## 8.1 Yoghurt

Fermented dairy products such as yoghurt, which is often made with fruit pulp, are known to be inhibitory to both pathogenic and spoilage causing microorganisms. The starter culture for most yoghurt production in North America is a symbiotic blend of *Streptococcus salivarius* subsp. *ther-*

**Table 9:** Common starter cultures used for preparation of fermented dairy products and their role.

| Starter culture | Role function | Fermented diary product |
|---|---|---|
| *Lactococcus lactis* subsp. *lactis* *Lactococcus lactis* subsp. *cremoris* | Acid | Cultured butter milk, sour cream, cottage cheese, all type of cheeses, starter culture |
| *Lactococcus lactis* subsp. *diacetylactis* | Acid and flavour | Sour cream, ripened cream, butter, cheese, butter milk, starter culture, etc. |
| *Streptococcus thermophilus* | Acid | Emmental, cheddar and Italian cheese and yoghurt |
| *S. durans, S. faecalis* | Acid and flavour | Soft Italian, cheddar and some Swiss cheese varieties |
| *Leuconostoc citrovorum, Leuco. dextranicum* | Flavour | Cultured butter milk, sour cream, cottage cheese, ripened cream, butter and starter cultures |
| *Lactobacillus delbrueckii* subsp. *bulgaricus, L. lactis, L. helveticus* | Acid and flavour | Bulgarian butter milk, yoghurt, kefir, kumiss, Swiss, emmental and Italian cheese |
| *Lactobacillus acidophilus* | Acid | Acidophilus butter milk |
| *Propionibacterium shermanii* | Flavour and eye formation | Emmental and Swiss cheese |

*Source:* 144

*mophilus* and *Lactobacillus delbrueckii* subsp. *bulgaricus*. Although they can grow independently, the rate of acid production is much higher when used together than either of the two organisms grown individually. *Streptococcus salivarius* subsp. *thermophilus* grows faster and produces both acid and carbon dioxide. The formate and carbon dioxide produced stimulates the growth of *Lactobacillus delbrueckii* subsp. *bulgaricus*. On the other hand, the proteolytic activity of *Lactobacillus delbrueckii* subsp. *bulgaricus* produces stimulatory peptides and amino acids for use by *Streptococcus salivarius* subsp. *thermophilus*. These microorganisms are ultimately responsible for the formation of typical yoghurt flavour and texture. The yohurt mixture coagulates during fermentation due to the drop in pH. The streptococci are responsible for the initial pH drop of the yoghurt mix to approximately 5.0. The *Lactobacilli* are responsible for a further decrease to pH 4.0. The various fermentation products that contribute to flavour are lactic acid, acetaldehyde, acetic acid and diacetyl. The common flow diagram for the manufacture of different types of yoghurt is presented in Fig. 3.

## 8.2 Cultured Butter Milk

Butter milk was originally the fermented by-product of butter manufacture, but today it is more common to produce cultured buttermilks from skimmed or whole milk. The common flow diagram for the manufacture of cultured butter milk is presented in Fig. 4. The culture most frequently used is *S. lactis*, perhaps also *S. cremoris*. Milk is usually heated to 95°C and cooled to 20-25°C before the addition of the starter culture. Starter is added at 1-2% and the fermentation is allowed to proceed for 16-20 hours, to an acidity of 0.9% lactic acid. This product is frequently used as an ingredient in the baking industry, in addition to being packaged for sale in the retail trade.

## 8.3 Acidophilus Milk

It is the traditional milk fermented with *Lactobacillus acidophilus*, which has been thought to have therapeutic benefits in the gastrointestinal tract; skimmed or whole milk may be used. The common

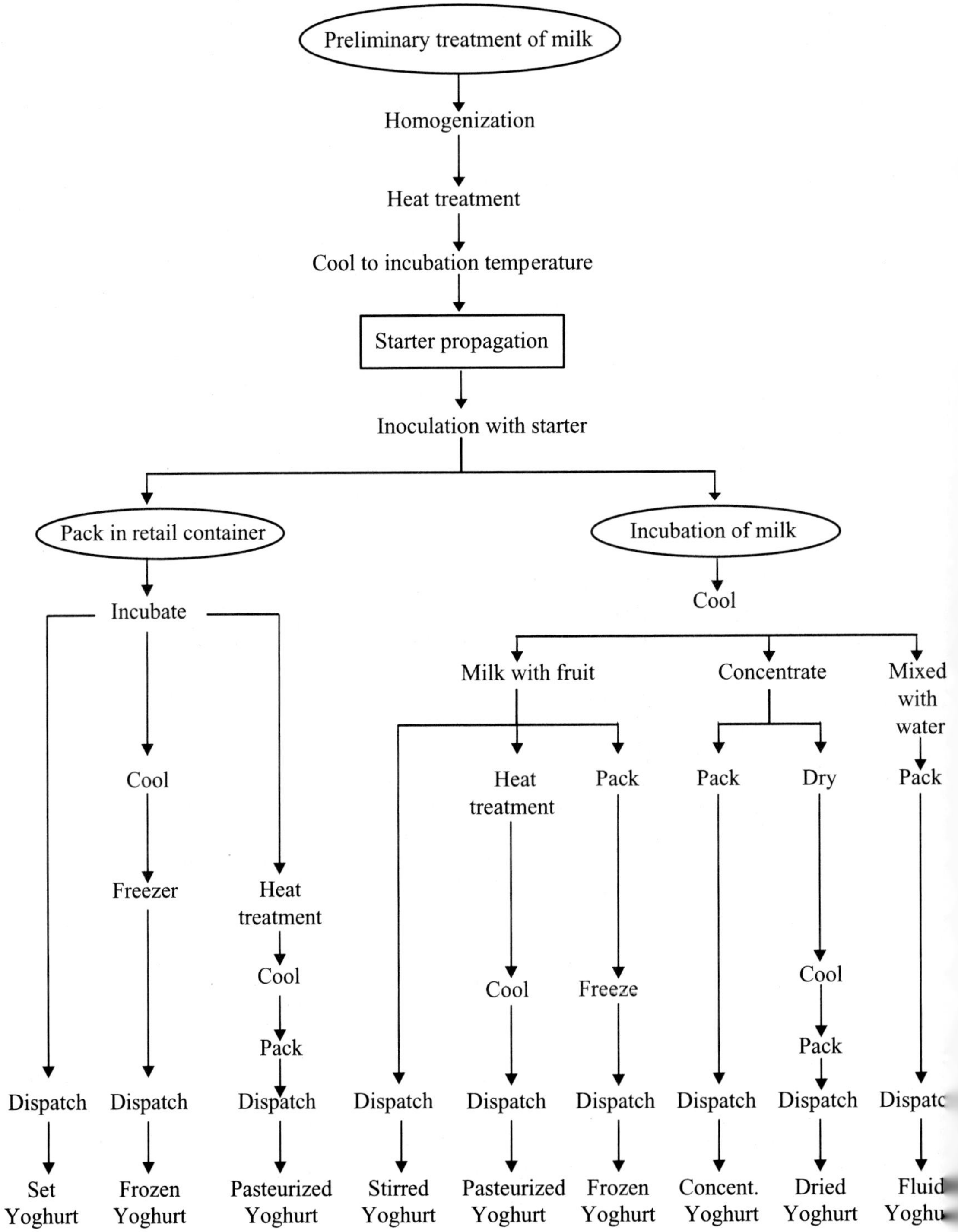


**Fig. 3:** Unit operations for preparation of different yoghurts.

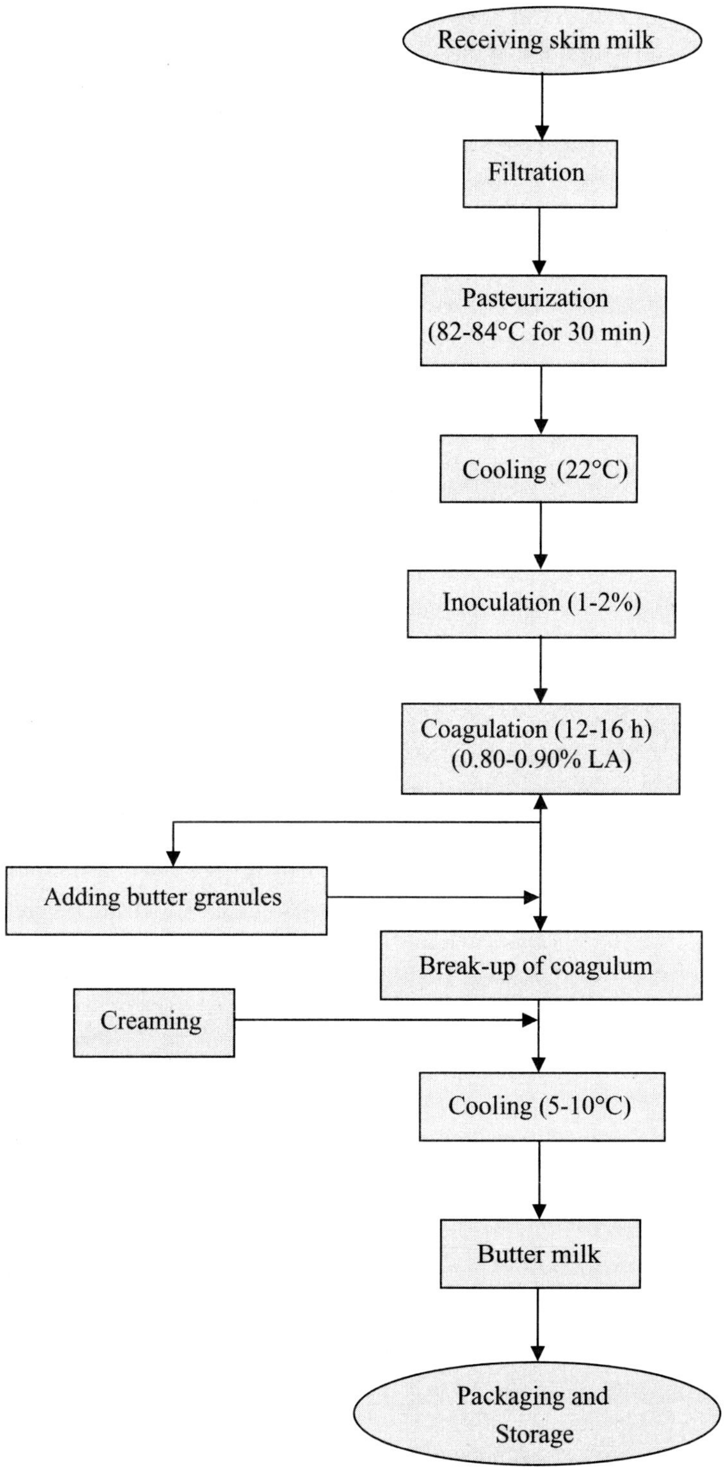


**Fig. 4:** Flow diagram for the manufacture of cultured butter milk.

flow diagram for the manufacture of acidophilus milk is presented in Fig. 5. The milk is heated to high temperature e.g., 95°C for 1 hour to reduce the microbial load and favour the slow growing *Lactobacillus acidophilus* culture. Milk is inoculated at a level of 2-5% and incubated at 37°C, until coagulated. Some acidophilus milk have acidity as high as 1% lactic acid, but for therapeutic purposes 0.6-0.7% is more common. Another variation has been the introduction of a sweet acidophilus milk, one in which the *Lactobacillus acidophilus* culture has been added, but there has been no incubation. It is thought that the culture will reach the gastrointestinal tract where its therapeutic effects will be realized, but the milk has no fermented qualities, thus delivering the benefits without the high acidity and flavour, considered undesirable by some people.

Cultured cream usually has a fat content between 12-30%, depending on the required properties. The starter is similar to that used for cultured buttermilk. The cream after standardization is usually heated upto 75-80°C and is homogenized at >13 MPa to improve the texture. Inoculation and fermentation conditions are also similar to those for cultured buttermilk, but the fermentation is stopped at an acidity of 0.6%. There are a great many other fermented dairy products, including *kefir*, *koumiss*, sour cream, beverages based on *bulgaricus* or *bifidus* strains, *labneh*, and a host of others. Many of these have developed in regional areas and, depending on the starter organisms used, have various flavours, textures, and components from the fermentation process, such as gas or ethanol.

### 8.4 Bulgarian Milk

It is manufactured from pasteurized standardized or skimmed milk. After the temperature has been brought to 40-45°C, the milk is inoculated with 2 to 3% of a culture of *Lactobacillus bulgaricus*. The milk is then incubated at about 40°C until it has reached the desired acidity, which will take about 12 hours. At that moment, it is stirred and cooled to a temperature below 10°C to prevent further acid production. After cooling, the milk is bottled and put in cold store.

## 9. FERMENTED MEAT

The most common lactic acid microorganisms found in fermented meats are various strains of *Lactobacilli, Leuconostoc, Pediococci* and *Streptococci.* Commercial meat starter cultures are selected microorganisms that have been isolated from meat, purified, grown to large numbers under controlled conditions, concentrated, and then preserved by freezing or drying prior to use. These starter cultures are not all the same. They differ in form, strain, purity, activity and consistency, depending on the manufacturer. The functions of commercial meat starter cultures are to allow fermentation with acid development and reduced pH, besides development of flavour (proteolytic, lipolytic), colour and provide stability. The commercial cultures most common in the US, and the temperature of fermentation used are given in Table 10.

In Europe, naturally fermented sausages have a long tradition originating from Mediterranean countries during Roman times [145]. Production then spread to Germany, Hungary and other countries including the United States, Argentina and Australia [146]. Europe is currently the major producer and consumer of dry fermented sausages [147]. There is a wide variety of dry fermented products on the European market as a consequence of variations in the raw materials, formulations and manufacturing processes, which come from the habits and customs of different countries and regions. However, Northern products have pH below 5, while Mediterranean products have a pH of 5.3-6.2 and are highly desiccated. In both categories, industrial development has led to the use of starter cultures to standardize and control sausage manufacturing. However, artisanal slightly fermented sausages form a group of traditional Mediterranean products with a great regional diversity, both between and within the countries. These traditional sausages can be defined as a meat product made of a mixture of meat (often pork), pork fat in variable ratio including salt, sugar, nitrate and/or nitrite [148, 149]. The fermentation and ripening/drying do not always constitute two separated steps and they can be carried out in a natural dryer depending on local climatic conditions [149, 150]. Traditional dry sausages rely on natural contamination by environmental microflora. This contamination occurs during slaughtering

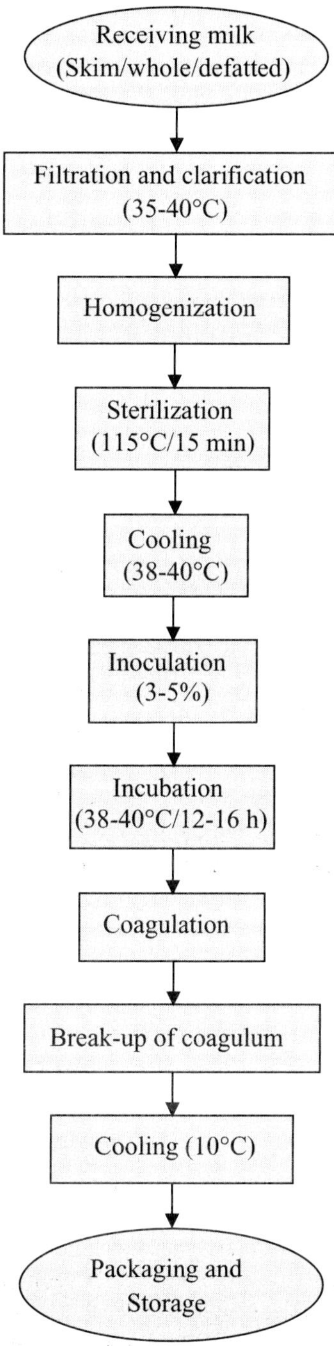


**Fig. 5:** Flow diagram for manufacture of acidophilus buttermilk.

**Table 10:** Cultures most commonly used in the US and the temperature of fermentation.

| Microorganisms | Temperature (°C) |
|---|---|
| *Pediococcus acidilactici* | 32-46 |
| *Pediococcus pentosaceus* | 21-38 |
| *Lactobacillus plantarum* | 21-38 |
| *Staphylococcus carnosus/Mirococcus varians* | 21-38 |
| *Lactobacillus curvatus, L. sake, S. xylosus* | 21-38 |

and increases during manufacturing. A wide range of "waste" products are fermented to produce edible food products in Sudan. These include bones, hides and locusts. Fresh bones are fermented into a variety of products. "*Dodery*" is produced by chopping bones into small pieces and placing them into fermenting vats. They are subsequently covered in water, left for three days, removed, crushed into a paste and mixed with the ash from burnt sorghum stalks. The mixture is returned to the fermenting vat for a period of two to five days. The final product is rolled into balls and has a shelf-life of upto two months. Another product "*Kaidu digla*" is made from the vertebrae of the backbone. These are chopped into smaller pieces and then sun-dried. After drying they are pounded with stones, mixed with water and salt, moulded into balls and allowed to ferment.

## 10. HEALTHFUL EFFECTS OF LACTIC ACID BACTERIAL FERMENTED PRODUCTS

### 10.1 Probiotic Effect

One of the reasons for the increasing interest in fermented foods is their ability to promote the functions of the human digestive system in a number of positive ways. Lactic acid bacteria have strong inhibitory effects on the growth and toxin production of other bacteria. This antagonistic activity can be a result of different properties outlined in Table 11. Each of these properties and especially the combination of some of them can be used to extend the shelf-life and safety of food products [151]. As early as 1900, Metchnikoff pointed out the use of fermented milks in the diet for prevention of certain diseases of the gastrointestinal tract and promotion of healthy day-to-day life. Since then, a number of studies have shown that the fermented food products do have a positive effect on health status in many ways. The human intestinal microflora is estimated to weigh about 1000 grams and may contain $10^{16}$ to $10^{17}$ colony forming units representing more than 500 strains. For physiological purposes, it is considered to be a specialized organ of the body with a wide variety of functions in nutrition, immunology and metabolism [152]. Studies on mice have shown that the indigenous microorganisms in the stomach are *Lactobacillus, Streptococcus* and *Torulopsis*, while in the small intestine, ceacum and colon, several different species (*Bacteroides, Fusobacterium, Eubacterium, Clostridium,* etc.) coexist [153]. The gastrointestinal microflora in humans is also known to contain hundreds of species. Even though there is a wide variation among individuals, the number of species and size of the population are usually kept stable in normal healthy subjects. But, there is a constant struggle in maintaining the desirable balance and a dynamic equilibrium between microbial populations within the intestinal flora [154]. The anaerobic organisms, which outnumber the Gram negative enteric bacteria by about 10000:1, are associated with the intestinal epithelium limiting adherence of potential pathogens by effective colonization [155-157]. The stability of the intestinal microflora is affected by many factors including dietary habits. Decrease in the number of anaerobic bacteria is

**Table 11:** Various properties responsible for antagonistic activity of lactic acid bacteria.

- Competition for available nutrients
- Decrease in redox potential
- Production of lactic acid, acetic acid and the resulting decrease in pH
- Production of other inhibitory primary metabolites, such as hydrogen peroxide, carbon dioxide or diacetyl
- Production of special antimicrobial compounds such as bacteriocins and antibiotics

Source: 151

associated with increase in the number of Gram negative pathogens in the intestinal tract and their translocation to extraintestinal tissues. Under normal conditions, the intestinal wall prevents translocation of organisms, both dead and living as well as microbial products like toxins from the gut to the blood. However, in patients with systemic abuse like starvation, shock, injury and infection or specific insult of the gastrointestinal canal through inflammation, chemotherapy or radiation, the gut mucosal permeability was increased leading to translocation of microbes [158-161]. A fermented food product or live microbial food supplement, which has beneficial effects on the host by improving intestinal microbial balance is generally understood to have probiotic effect [162].

## 10.2 Flatulence Reducing Activity

The lactic acid fermented products do have flatulence reducing effect. During fermentation of the beans for preparation of *tempeh*, the trypsin inhibitor is inactivated and the amount of several oligosaccharides which usually cause flatulence are significantly reduced [163]. Bean flour inoculated with *Lactobacillus* and fermented with 20% moisture content showed a reduction of the stachyose content [164].

## 10.3 Hypercholesteremic Effect

Hepner *et al.* [165] reported hypercholesteremic effect of yoghurt in human subjects receiving a one-week dietary supplement. Studies on supplementation of infant formulation with *Lactobacillus acidophilus* showed that the serum cholesterol in infants was reduced from 147 mg/ml to 119 mg/100 ml [166]. In an *in vitro* study, the ability of 23 strains of lactic acid bacteria isolated from various fermented milk products to bind bacterial cells to cholesterol showed that no cholesterol was found inside the cells [167]. The ability of yoghurt to lower the cholesterol in serum by controlled human trials has also been reported [168]. The possible role of LAB in lowering cholesterol concentration and various mechanisms by which it may be possible has been discussed earlier [169]. A gene encoding cholesterol oxidase from *Streptomyces lividans* has been cloned into *Bacillus, Lactobacillus* and *E. coli* [170].

## 10.4 Anticarcinogenic Effect

Fruits and vegetables are known to be effective in the prevention of cancer. They contain a multitude of compounds that have been shown to possess potentially cancer preventive activity [171]. Numerous investigations demonstrate that frequent intake of fruits and vegetables, ranging from daily to three times or more per week, may lower the risk of cancer [172, 173]. One of the most consistent finding has been an inverse association between cancer and vegetable intake, in particular the intake of cruciferous vegetables [174-176]. Apart from this, there are interesting data on anticarcinogenic effect of fermented foods showing a potential role of *Lactobacilli* in reducing or eliminating procarcinogens and carcinogens in the alimentary canal [177-179]. A statistically significant inverse association with oral and pharyngeal cancer was found for the consumption of total green vegetables (OR 0.37) and total fruit (OR 0.34) [180], while high consumption of cruciferous vegetables is associated with a reduced risk of prostate cancer [181]. Research indicates that the anti-tumour effect of a compound with multiple vegetables and fruits is stronger than that of a single component, and this may be related to the synergetic effect of multiple ingredients of vegetables and fruits [172]. The enzymes, glucuronidase, azoreductase and nitroreductase, which are present in the intestinal canal, are known to convert procarcinogens to carcinogens [182]. Oral administration of *Lactobacillus rhamnosus* GG was shown to lower the faecal concentration of glucuronidase in humans [183], implying a decrease in the conversion of procarcinogens to carcinogens. Fermented milk containing *Lactobacillus acidophilus* given together with fried meat patties significantly lowered the excretion of mutagenic substances compared to ordinary fermented milk with *Lactococcus* fed together with fried meat patties [184]. Further, the vegetable fruit ferment liquid has an anti-tumour effect and an extended duration of survival in human hepatoma-22 (H22)-bearing mice, which may associate with the inhibition of tumour cell growth and inducing apoptosis. The tumour weight of the three VFFL groups was index in VFFL 16.7 ml/kg group and the spleen index in 33.3 ml/kg group were significantly larger when compared with that of the control group (Table 12) [185].

## 10.5 Mutagenicity

The fermentation of foods is also reported to reduce the mutagenicity of foods by degrading the mutagenic substances during the process. Lactic acid bacteria isolated from *dadih*, traditional Indonesian fermented milk, were found to be able to bind mutagens and inhibit mutagenic nitrosamines. Milk fermented with *Lactobacillus acidophilus* LA-2 was demonstrated to suppress faecal mutagenicity in the human intestine. Studies on antimutagenic activity of milk fermented with mixed cultures of various lactic acid bacteria and yeast showed that the fermented milks produced with mixed cultures of lactic acid bacteria had a wider range of activity against mutagens than those produced with a single strain of lactic acid bacteria [186]. However, a review by McIntosh [187] concludes that there is only limited data to support the hypothesis that probiotic bacteria are effective in cancer prevention. The probiotic bacteria have the ability to bind carcinogenic substances, as *E. faecalis* was able to bind aflatoxin $B_1$, $B_2$, $G_1$ and $G_2$ as well as some pyrolytic products of tryptophan.

## 10.6 Immune System

Some LAB present in fermented milk products are found to play an important role in the immune system of the host after colonisation in the gut [188]. Oral administration of *Lactobacillus casei*

**Table 12:** Effects of VFFL on tumour weight, thymus and spleen index in $H_{22}$–bearing mice (Mean ± SD).

| Group | Dosage (ml/kg) | N | Thymus index (mg/g) | Spleen index (mg/g) | Tumour weight (g) | IR (%) |
|---|---|---|---|---|---|---|
| Control | - | 7 | 1.40 ± 0.75 | 3.97 ± 1.21 | 1.71 ± 0.51 | - |
| VFFL | 16.7 | 9 | 2.03 ± 1.11$^b$ | 4.31 ± 1.21 | 1.27 ± 0.21$^b$ | 25.7 |
| | 33.3 | 9 | 1.51 ± 0.32 | 5.18 ± 1.10$^b$ | 1.11 ± 0.41$^b$ | 35.0 |
| | 66.6 | 9 | 1.97 ± 0.44 | 4.77 ± 1.15 | 0.87 ± 0.34$^a$ | 49.1 |

$^a$ p<0.01; $^b$ <0.05 compared with control group

*Source:* 185

caused an improvement in the function of the peritoneal macrophages and increased the production of IgA [189]. The mechanism of this effect is not clearly known, but it is speculated that the *Lactobacilli*, their enzymes or the metabolic products present in the fermented food product may act as antigens, activating the production of antibodies. Marin *et al.* [190] have studied the influence of *Lactobacilli* used in fermented dairy products on the production of cytokines by macrophages. The results indicated that for most strains, direct interaction with macrophages caused a concentration dependent increase in tumour necrosis factor and interleukin. It was concluded that there is a potential of using lactic acid bacteria for therapy and immunomodulation in mucosal diseases, especially in the gastrointestinal tract. Isolauri [191] has presented a study suggesting that *Lactobacillus* sp. strain G could be used in the prevention of food allergy. It is suggested that dietary antigens induce immunoinflammatory response that impairs the intestine's barrier function, and probiotic organisms could be a means of introducing a tool to reinforce the barrier effect of the gut.

## 11. FUTURE TRENDS

Lactic acid fermentation is one of the oldest forms of food preservation technology in the world. Indigenous fermented foods such as sauerkraut, cheese and *gundruk* have been prepared and consumed for thousands of years and are strongly linked to culture and tradition, especially in rural households and village communities. Lactic acid fermented foods are popular throughout the world and in some regions make a significant contribution to the diet of millions of individuals. Lactic acid fermentation is a relatively efficient, low energy preservation process, which increases the shelf-life and decreases the need for refrigeration or other forms of food preservation technology. It is therefore, a highly appropriate technique for use in developing countries and remote areas where access to sophisticated equipment is limited. There is a tremendous scope and potential for the use of microorganisms towards meeting the growing world demand for food, through efficient utilization of available natural food and feed stocks and the transformation of waste materials. There is a danger that the introduction of 'western foods' with their glamorous image will displace these fermented foods. Although fermentation of foods has been in use for thousands of years for preservation and improvement of a range of foods, the microbial and enzymatic processes responsible for the transformations were, and still are largely unknown. Lactic acid bacteria fermented products play an important role in food preservation and due to their potential to contribute to the growing food needs of the world, it is essential that the knowledge of their production is not lost. Moreover, it is essential to increase the knowledge and understanding of the methods of preparation, in order to improve the efficiency of fermentation, especially the traditional processes as the yields of traditional fermentation processes are often low and sometimes the products are unsafe. The potential areas for improvement of lactic acid fermented products are better understanding of fermented products, refining the process and creating a supportive environment for production of fermented food products. With many of the fermented products, knowledge of the processes involved is poor. It is likely that the basic principles apply across the board, but production conditions vary enormously from region-to-region, giving rise to numerous variations of the basic fermented product. It is not the intention or the desire to standardize the process and thereby lose this huge diversity, rather it is to harness the tremendous potential these methods have to contribute to increasing not only the quantity, but also the quality of food available to the world population. If the processes are to be refined, with a view to production on a larger scale, it is essential to have a scientific understanding of the fermentation processes. This can be developed by the isolation and characterization of the essential microorganisms involved; determination of the role of external factors in fermentations and the effects of these on the metabolism of microorganisms; investigation of the effects of pre-treatments of raw materials on the fermentation process; identification of the options for further processing and how these affect the taste and texture of the product; biotechnological need to be developed which are affordable by the poor, who are likely to be benefited most by improvements in the traditional processes; art of traditional processes needs to be refined to incorporate objective methods of process control and to standardize quality of the final product without losing their desirable attributes such as improved keeping quality, taste and nutritional qualities. Science based fermentation research has often focused mainly on improving the metabolic properties of the microorganisms used as a starter culture, using the techniques of selection, mutation and genetic modification. To achieve this, a full understanding of the fermentation process is needed. The properties of the starter culture must be known and culture conditions must be controllable. This means that science based fermentation research currently has little to offer traditional food processing. Developing countries need to build their own resources of trained, knowledgeable individuals, who are able to apply the basic microbiological principles to the production of fermented foods. Extra support should be made available to train professionals in this discipline. Fermented foods should be recognised as a part of each countries' heritage and culture, and efforts made to preserve the methods of production. A recognised body (government or non-government) should take the responsibility for the collection of details and the promotion of fermented food products. More research is needed to ascertain the factor or combination of factors responsible for the positive consequence of fermented diets. Besides, consumers need to be made aware of the numerous benefits of fermented foods, and their prejudices against fermented foods, especially those traditionally produced at the home scale, are dispelled.

## REFERENCES


1. Pederson CS (1971) *Microbiology of Food Fermentation*, AVI Publishing Co Inc, Westport, USA, pp 108-152.
2. Anderson RE, Eriksson CE, Salmonsson AC & Theander D (1990) Lactic acid fermentation of fresh and stored carrot—chemical, microbial and sensory evluation of products. *Lebensm-Wiss-Technol* 23: 34-40.
3. Fleming HP & Mc Feeters RF (1981) Use of microbial cultures: Vegetable products. *Food Technol* 35: 84-88.
4. Daeschel MA & Fleming HP (1984) Selection of lactic acid bacteria for use in vegetable fermentations. *Food Microbiol* 1: 303-313.
5. Hajjar IM, Grim CE, George V & Kotchen TA (2001) Impact of diet on blood pressure and age-related changes in blood pressure in the US population. *Analysis of NHANES III* 161: 589-593.
6. Wattenberg LW (1992) Inhibition of carcinogenesis by minor dietary constituents. *Cancer Res* 52: 2085-2091.
7. Buckenhuskes HJ (2001) Fermented vegetables. In: *Food Microbiology: Fundamentals and Frontiers*, Doyle MP, Beuchat LR & Montiville TJ (eds), ASM Press, Washington, USA, pp 665-679.
8. Fleming HP, McFeeters RF, Thompson RL & Senders DC (1983) Storage stability of vegetable fermented with pH control. *J Food Sci* 48: 975-980.
9. Fleming HP (1982) Fermented vegetables. In: *Economic Microbiology: Fermented Foods,* Rose AH (ed), Academic Press, New York, USA, pp 227-258.
10. Kim JM, Kim IS & Yang HC (1987) Storage of salted Chinese cabbages for Kimchi. I. Physico-chemical and microbial changes during salting of Chinese cabbages. *J Korean Soc Food Nutri* 16: 75-82.

11. Adams MR & Moss MO (1996) Fermented and microbial foods. In: *Food Microbiology*, New Age International (P) Ltd Publishers, New Delhi, India, pp 252-302.
12. Ha J-Ho, Hawer WS, Kim YJ & Nam YJ (1989) Changes of free sugars in kimchi during fermentation. *Korean J Food Sci Technol* 21: 633-638.
13. Hawer WS, Ha JH, Seog HM, Nam YJ & Shin DH (1988) Changes in taste compounds of kimchi during fermentation. *Korean J Food Sci Technol* 20: 511-517.
14. Kim MR & Rhee HS (1993) Decrease of pungency in 'Radish Kimchi' during fermentation. *J Food Sci* 58: 128-137.
15. Battcock M & Azmi-Ali S (1998) Fermented fruits and vegetables. In: *Fermented Fruits and Vegetables. A Global Perspective,* FAO Agricultural Services Bulletin No 134, Food and Agriculture Organization of the United Nations, Italy, Rome, pp 135-150.
16. Eklund T (1983) The antimicrobial effect of dissociated and undissociated sorbic acid. *J Appl Bacteriol* 54: 383-389.
17. Kodama R (1952) Studies on lactic acid bacteria. 2. Lactolin a new antibiotic substance produced by lactic acid bacteria. *J Antibiot* 5: 72-74.
18. Tagg JR, Dajani AS & Wannamaker LW (1976) Bacteriocins of Gram positive bacteria. *Bacteriol Rev* 36: 722-756.
19. Sharma S (2004) Lactic acid fermentation of some vegetables for product development and bacteriocin production, Ph.D. thesis, Dr. Y.S. Parmar University of Horticulture and Forestry, Nauni, Solan, (HP), India.
20. Joshi VK, Sharma S & Rana N (2006) Production, purification, stability and efficiency of bacteriocin from the isolate of natural lactic acid fermentation of vegetables. *Food Technol Biotechnol* 44: 435-439.
21. Gross E & Morell JL (1967) The presence of dehydroalanine in the antibiotic nisin and its relationship to activity. *J Am Chem Soc* 89: 2791-2792.
22. Gross E & Morell JL (1971) The structure of nisin. *J Am Chem Soc* 93: 4634-4635.
23. Hirsch A (1951) Growth and nisin production of a strain of *Streptococcus lactis. J Gen Microbiol* 5: 208-221.
24. Hirsch A, Grimsted E, Chapman HR & Mattick ATR (1951) A note on the inhibition of an anaerobic sporeformer in a Swiss type cheese by a nisin producing *Streptococcus. J Dairy Res* 18: 205-206.
25. Kozak W, Bardowski J & Dobrzanski WT (1978) Lactostrepcins-acid bacteriocins produced by lactic streptococci. *J Dairy Res* 45: 247-257.
26. Dahiya RS & Speck ML (1968) Hydrogen peroxide formation by *Lactobacilli* and its effect on *Staphylococcus aureus. J Dairy Sci* 51: 1568-1572.
27. Price RJ & Lee JS (1970) Inhibition of *Pseudomonas* species by hydrogen peroxide producing Lactobacilli. *J Milk Food Technol* 33: 13-18.
28. Kandler O (1983) Carbohydrate metabolism in lactic acid bacteria. *Antonie van Leeuwenhoek* 49: 209-224.
29. Jay JM (1982) Antimicrobial properties of diacetyl. *Appl Environ Microbiol* 44: 525-532.
30. Axelsson L, Chung TC, Dobrogosz WJ & Lindgren S (1988) Discovery of a broad spectrum antimicrobial substance provided by *Lactobacillus reuterii*, Unpublished manuscript, North Carolina State University, Raleigh, USA.
31. Joshi VK, Bhutani VP & Thakur NK (1999) Nutrition and composition of fermented products. In: *Biotechnology: Food Fermentation*, Vol 1, Joshi VK & Pandey A (eds), Educational Publishers and Distributors, New Delhi, India, pp 259-319.

32. Adams MR (1990) Topical aspects of fermented foods. *Trends Food Sci Technol* 1: 141-144.
33. Obizoba IC & Atii JV (1991) Effect of soaking, sprouting, fermentation and cooking on nutrient composition and some anti-nutritional factors of sorghum (Guinesia) seeds. *Plant Foods Human Nutr* 41: 203-212.
34. Hesseltine CW & Wang HL (1980) The importance of traditional fermented foods. *Bioscience* 30: 402-404.
35. Reed G (1981) Use of microbial cultures: yeast products. *Food Technol* 35: 89-94.
36. Alm L (1982) Effect of fermentation on B-vitamin content of milk in Sweden. *J Dairy Sci* 65: 353-359.
37. Murdock FA & Fields ML (1984) B-vitamin content of natural lactic acid fermented cornmeal. *J Food Sci* 49: 373-375.
38. Speek AJ, Saichua S & Schreurs WHP (1988) Total carotenoid and beta-carotene contents of Thai vegetable and the effect of processing. *Food Chem* 27: 245-257.
39. Sahlin P (1999) Fermentation as a method of food processing: Production of organic acids, pH-development and microbial growth in fermenting cereals, Licentiate thesis, Division of the Applied Nutrition and Food Chemistry, Lund University, Sweden.
40. Joshi VK, Chauhan SK & Sharma R (1993) Preservation of fruits and vegetables by lactic acid fermentation. *Bev Food World* 19: 9-12.
41. Gibbs PA (1987) Novel uses for lactic acid fermentation in food preservation. *J Appl Bacteriol Symp Suppl* 63: 51S-58S.
42. Daeschal MA, Andersson RE & Fleming HP (1987) Microbial ecology of fermenting plant materials. *FEMS Microbiol Rev* 46: 357-367.
43. Kandler O, Hammes WP, Schneider M & Stetter KO (1987) Microbial interaction in sauerkraut fermentation. In: *Proceedings of the 4th International Symposium on Microbial Ecology*, Kyoto, Japan, pp 302-308.
44. Nettles CG & Barefoot SF (1993) Biochemical and genetic characteristics of bacteriocins of food associated lactic acid bacteria. *J Food Prot* 56: 338-356.
45. Eijsink VGH, Skeeie M, Middelhoven PH, Brurberg MB & Nes IF (1998) Comparative studies of class IIa bacteriocins of lactic acid bacteria. *Appl Environ Microbiol* 64: 3275-3281.
46. Joshi VK & Thakur S (2000) Lactic acid fermented beverages. In: *Postharvest Technology of Fruits and Vegetables*, Vol 2, Verma LR & Joshi VK (eds), Indus Publishing Company, New Delhi, India, pp 1102-1127.
47. Silva J, Carvalho AS, Teixeira P & Gibbs PA (2002) Bacteriocin production by spray-dried lactic acid bacteria. *Lett Appl Microbiol* 34: 77-81.
48. Scheider M (1988) Microbiology of sauerkraut fermented in small ready-to-sell containers, Dissertation (In German), Hohenheim University, Stuttgart, Germany.
49. Sagarika E & Pradeepa J (2003) Fermented foods of Sri Lanka and Maldives. In: *Proceedings of International Seminar and Workshop on Fermented Foods, Health Status and Social Well Being,* Anand, India, pp 25-27.
50. Anne S & Maike L (2003) Probiotics in fermented milk. In: *International Seminar and Workshop on Fermented Foods, Health Status and Social Well-Being*, Anand, India, pp 20-22.
51. Sankaran R (1998) Fermented foods of the Indian subcontinent. In: *Microbiology of Fermented Foods*, Wood BJB (ed), Blackie Academic and Professional, London, UK, pp 753-789.

52. Sharma S & Joshi VK (2007) Influence of temperature and salt concentration on lactic acid fermentation of radish (*Raphnus sativus*). *J Food Sci Technol* 44: 611-614.
53. Joshi VK, Sharma S & Thakur NS (2008) Effect of temperature, salt concentration and type of microorganism on lactic acid fermentation of carrot. *Acta Alimentaria* 37: 205-219.
54. Buckenhuskes H & Gierschner K (1989) Vegetable juice production-state of the art. *Fliis Obst* 56: 751-764.
55. Buckenhuskes HJ (1993) Selection criteria for lactic acid bacteria to be used as starter cultures for various food commodities. *FEMS Microbiol Rev* 12: 253-272.
56. Frazier WC & Westhoff DC (1998) *Food Microbiology*, Tata Mcgraw Hill Publishing Company, New Delhi, India.
57. Joshi VK & Bhat A (2000) Pickles: Technology of Preparation. In: *Postharvest Technology of Fruits and Vegetables*, Verma LR & Joshi VK (eds), Indus Publishing Company, New Delhi, pp 777-798.
58. Banwart GJ (1981) *Basic Food Microbiology*, AVI, Westport, Connecticut, USA.
59. Vaughn RH (1975) Lactic acid fermentation of olives with special reference to California conditions. In: *Lactic Acid Bacteria in Beverages and Food*, Care JG, Cutting CV & Whitang GC (eds), Academic Press, New York, USA, pp 307-326.
60. Cruess WV (1948) Spoilage and discolouration of sauerkraut. In: *Commercial Fruit and Vegetables Products*, Cruess WV (ed), McGraw-Hill Book Company, New York, USA, pp 724-732.
61. Vaughn RH, Doughlas HC & Gilliland JR (1943) Production of Spanish type green olives. Bull Univ of California, Agr Exp Sta, Berkeley, pp 678-684.
62. Buckenhuskes HJ & Hammes WP (1990) Starter cultures in fruit and vegetable processing. *Bioengineering* 6: 34-42.
63. Montet D, Loiseau G, Zakhia N & Mouquet C (1999) Fermented fruits and vegetables. In: *Biotechnology: Food Fermentation*, Vol 2, Joshi VK & Pandey A (eds), Educational Publishers and Distributors, New Delhi, India, pp 951-969.
64. Fulde RC & Fabian FW (1953) Influence of Gram negative bacteria on the sauerkraut fermentation. *Food Technol* 7: 486-488.
65. Leon Rde, Cabrera SS, de Rolz C (1975) Controlled fermentation of cucumbers by bacterial cultures. *Ravista Latinoamericans de Microbiologia* 17: 33-37.
66. Rodrigo M, Alvarruiz A, Villa R & Feria A (1985) Lactic acid controlled fermentation of fresh cucumbers and product quality. *Revista de Agroquimica Y. Technologia de Alimentos* 25: 104-116.
67. Chavasit V, Hudson JM, Torres JA & Daeschel MA (1991) Evaluation of fermentative bacteria in a model low salt cucumber juice brine. *J Food Sci* 56: 462-465.
68. Cho NC, Jhon DY, Shin MS, Hong YH & Lim HS (1987) Effect of garlic concentration on growth of microorganisms during kimchi fermentation. *Korean J Food Sci Technol* 20: 231-235.
69. Yoon JS & Rhee HS (1977) A study on the volatile flavour compounds in kimchis. *Korean J Food Sci Technol* 9: 116-120.
70. Wiesenberger A, Kolb E, Schildmann JA & Dechent HM (1986) *Production of lactic fermented fruit juice products*. German Federal Republic Patent Application DE 3503742 Al.
71. Sethi V (1990) Lactic fermentation of black carrot juice for spiced beverage. *Indian Food Packer* 44: 7-12.

72. Boonlong N (1986) Traditional technologies of Thailand: Traditional Fermented Food products. In: *Traditional Foods: Some Products and Technologies*, Central Food Technological Research Institute, Mysore, India.
73. Joshi VK, Sharma S & Thakur NS (2003) Technology of fermented fruits and vegetables. In: *International Seminar and Workshop on Fermented Foods, Health Status and Social Well-Being,* Anand, India.
74. Karkri TB (1986) Fermented vegetables. In: *Concise Handbook of Indigenous Fermented Foods in the ASCA Countries,* Canberra, Government of Australia.
75. Makayama T (1957) *Nippon Nogeikagakukaishi*, Japan, 23: 497-499.
76. Buckner H, Salmen A, Amar A & Buckenhuskes HJ (1993) Fermented dairy like products from advanced soy proteins. In: *Proceedings of Euro Food Chem*, VII, Valencia, Spain, pp 215-222.
77. Joshi VK, Kaur M & Thakur NS (1996) Lactic acid fermentation of mushroom (*Agaricus bisporus*) for preservation and preparation of sauce. *Acta Alimentaria* 25: 1-11.
78. Etchells JL, Costilow RN, Anderson TE & Bell TA (1964) Pure culture fermentation of brined cucumbers. *Appl Microbiol* 12: 523-535.
79. Etchells JL, Costilow RN, Anderson TE, Bell TA & Fleming HP (1966) Pure culture fermentation of green olives. *Appl Microbiol* 14: 1027-1041.
80. Lal G, Siddappa GS & Tondon GL (1986) Chutneys, sauces and pickles. In: *Preservation of Fruits and Vegetables*, ICAR, New Delhi, India, pp 235-269.
81. Fleming HP, Etchells JL, Thompson RL & Bell TA (1975) Purging of $CO_2$ from cucumber brines to reduce bloater damage. *J Food Sci* 40: 1304-1310.
82. Buckenhuskes H, Gessler A & Gietrschner K (1988) Analytical characterization of canned and pasteurized sauerkraut. *Ind Obst Gemuserverwert* 73: 434-463.
83. Fleming HP, Thompson RL, Bell TA & Hontz LH (1978) Controlled fermentation of sliced cucumbers. *J Food Sci* 43: 888-891.
84. Thompson RL, Fleming HP & Monroe RJ (1979) Effects of storage conditions on firmness of brined cucumbers. *J Food Sci* 44: 843-846.
85. Hudson JM & Buescher RW (1985) Pectic substances and firmness of cucumber pickles as influenced by $CaCl_2$, NaCl and brine storage. *J Food Biochem* 9: 211-215.
86. Fleming HP, Mcfeeters RF & Thompson RL (1987) Effects of sodium chloride concentration on firmness retention of cucumbers fermented and stored with calcium chloride. *J Food Sci* 52: 653-657.
87. Hubert JC & Dupuy P (1994) Conservation des fruits et legumes. In: *Bacteries Lactiques: Aspects Fondamentaux et Technologiques*, Vol 11, Roissart Hde & Luquet FM (eds), Lorica, Uriage, pp 233-240.
88. Harris LJ (1998) The microbiology of vegetable fermentation. In: *Microbiology of Fermented Food*, Vol 1, Wood BJB (ed), Elsevier, UK, pp 45-62.
89. Pederson CS & Albury MN (1953) The influence of salt and temperature on the microflora of sauerkraut fermentation. *Food Technol* 8: 1-5.
90. Pederson CS & Kelly CD (1938) Development of pink colour in sauerkraut. *Food Technol* 6: 46-49.
91. Kumari A, Kalia M & Attri S (1993) Studies on chemical and organoleptic evaluation of cabbage and its product 'Sauerkraut'. *Indian Food Packer* 47: 11-14.
92. Steinkraus KH (1983) Lactic acid fermentation in the production of foods from vegetables, cereals and legumes. *Antonie van Leeuwenhoek* 49: 337-348.
93. Niketic-Aleksic GK, Bourne MC & Stamer JR (1973) Preservation of carrots by lactic acid fermentation. *J Food Sci* 38: 84-86.

94. Joshi VK, Sharma S, Bhushan S, Sharma R & Singh RP (2005) Preparation and evaluation of olive pickles (Indian Style): Selection of suitable cultivars and recipe. *Processed Food Ind* 8: 14-16.
95. Pederson CS & Ward L (1949) The effect of salt upon the bacteriological and chemical in fermenting cucumbers, New York State Agri Sta Bull, pp 273-278.
96. Pederson CS & Albury MN (1950) The effect of temperature upon bacteriological and chemical changes in fermenting cucumbers, New York State Agri Expt Sta Bull, pp 744-750.
97. Suresh ER, Yakashekhara E & Ethiraj S (1997) Preparation of brine stock pickling cucumbers by fermentation. *Indian Food Packer* 51: 13-17.
98. Sewon PK (1991) Reusable brine for conserving cucumbers. US Patent 5039: 545.
99. Suresh ER, Rajashekhara E, Naik LB, Pitchaimuthu M & Prabhakar BS (2001) Evaluation of pickling cucumber varieties for brine-stock preservation. *Indian Food Packer* 55: 49-53.
100. Mcfeeters RF, Senter MM & Fleming HP (1989) Softening effect of monovalent cations in acidified cucumber mesocarp tissue. *J Food Sci* 54: 366-370.
101. Anders C (1977) Controlled fermentation reduced salt usage and improved pickle quality. *Food Proc* 38: 50-52.
102. Ramdas AR & Kulkarni PR (1987) Fermentative preservation of carrots. *Indian Food Packer* 4: 40-48.
103. Anand JC & Das L (1971) Effects of condiments on lactic fermentation in sweet turnip pickle. *J Food Sci Technol* 8: 143-148.
104. Aubert C (1985) Les aliments fermentes traditionnels: Une richesse meconnue. *Collection Les Vrais Aliments d'Aujourd'hui et de Demain.* Terre Vivante, Paris, France, pp 263-268.
105. Schoneck A (1988) Des crudites toute l'annee: les legumes Lacto- fermentes. *Collection Les Vrais Aliments d'Aujourd'hui et de Demain.* Terre Vivante, Paris, France, pp 96-102.
106. Laencina J, Guzman G, Guevara LA & Flores J (1985) Accion de los aceites esenciales sobre levaduras de la fermentation de aceitunas (*Olea europaea* L.). *Essenze Derivati Agrumari* 27: 35-40.
107. Sethi V & Anand JC (1984) Effect of mustard and its components on the fermentation of cauliflower. *Indian Food Packer* 38: 41-46.
108. Kesker O, Esselen EBJR & Fellers CR (1951) Effect of allylisothiocynate and related substances on thermal resistance of *Aspergillus niger, Saccharomyces ellipsoidius* and *Bacillus thermoacidurans*. *Food Res* 16: 510-513.
109. Anand JC & Johar DS (1957) Effect of condiments on control of *Aspergillus niger* in mango pickle. *J Sci Ind Res* 16A: 370-375.
110. Mikki MS (1971) Studies on Sauerkraut, MSc Thesis, Division of Horticulture and Fruit Technology, IARI, New Delhi.
111. Sethi V & Anand JC (1972) Effect of mustard on curing of longmelons for pickles. *Pros Hort* 3: 11-15.
112. Hang YD & Jackson H (1967) Preparation of soyabean cheese using lactic starter organisms- General characteristics of the finished cheese. *Food Technol* 21: 1033-1038.
113. Pederson GS (1930) Floral changes in the fermentation of sauerkraut, *New York State Agri Expt Sta Bull*, pp 168-172.
114. Pederson CS & Albury MN (1969) The sauerkraut fermentation, *New York State Agri Sta Bull*, pp 824-830.
115. Wallace DH (1977) Controlled fermentation of cucumbers reduces defects from 40 per cent down to five per cent. *Food Process* 38: 70-71.

116. Duran MC, Garcia P, Brenes M & Garrido A (1994) Induced lactic acid fermentation during the preservation stage of ripe olives from Hojiblanca cultivar. *J Appl Bacteriol* 76: 377-382.
117. Takagi K, Toyoda M, Saito Y, Niwa M & Morimoto H (1990) Composition of fermented grape juice continuously produced by immobilized *Lactobacillus casei. J Food Sci* 55: 455-457.
118. Bucker ER, Mitchell JH & Johnson MG (1979) Lactic fermentation of peanut milk. *J Food Sci* 44: 1534-1538.
119. Cheng YJ, Thompson LD & Brittin HC (1990) Sogurt - a yoghurt like soyabean product - Development and properties. *J Food Sci* 55: 1178-1182.
120. Naewbanij JO, Stone MB & Chambers E (1990) *Lactobacillus plantarum* and *E. cloacae* growth in cucumber extracts containing various salts. *J Food Sci* 55: 1634-1640.
121. Gobbetti M & Rossi J (1989) Wheat and maize flours as media for lactic acid beverages. *Annals Fac Agri Univ Perugia* 3: 377-383.
122. Thapa TB, Jagat Bhadur KC, Gyamtsho P, Karki DB, Rai BK & Limbu DK (2003) Fermented foods of Nepal and Bhutan. In: *Proceedings of International Seminar and Workshop on Fermented Foods, Health status and Social Well-being*, Anand, India.
123. Miyamoto T, Girchuru SMS, Akimoto T & Nakae T (1986) Identification of lactic acid bacteria isolated from traditional fermented beverages in East Africa. *Japanese J Zootechnical Sci* (Nihin Chikusan Gakkai-ho) 57: 265-270.
124. Niwa M, Matsauoka M, Nakabayashi A, Shinagawa K, Tsuchida F, Saitoh Y & Katayama H (1987a) Lactic acid fermentation of fruit juices by lactic acid bacteria. *Chemie Mikrobiologie Technologie der Lebensmittel* 11: 81-85.
125. Niwa M, Matsuoka M, Nakabayashi A, Shinagawa K, Tsuchida F, Saitoh Y & Katayama H (1987b) Milchsaure vergarung von Fruchtsaften mit Lactobacillen. *Monatsschrift Fur Brauwissenschaft* 40: 373-377.
126. Runge H, Dera Y, Mattern PC, Idler F, Schade W & Annemueller G (1996) Screening of lactic acid bacteria for biological acidification of tropical fruit juices. *Fruit Process* 6: 279-284.
127. McConkie JE (1927) Types of spoilage in canned sauerkraut. *Canning Age, Convention Digest,* pp 240-241.
128. Graham WE (1941) The canning of sauerkraut. *Phys Rev* 30: 513-548.
129. Neuwhoff M (1969) Variation in quality. In: *Cole Crops,* Leonard- Hill, London, UK, pp 267-272.
130. Mikki MS & Anand JC (1974) The effect of temperature on fermentation and quality of sauerkraut. *Indian Food Packer* 28: 9-11.
131. Pederson CS (1956) The influence of temperature on sauerkraut fermentation. *Food Packer* 37: 26-31.
132. Etchells JL, Fleming HP, Hortz LH, Bell TA & Monroe RS (1975) Factors influencing bloater formation in brined cucumbers during controlled fermentation. *J Food Sci* 40: 569-575.
133. Rodrigo M, Lazaro MJ, Garcia G, Conesa F & Safon J (1992) Study of cucumber fermentation: diffusing gases and bloater damage. *J Food Sci* 57: 155-160.
134. Soni SK (1987) Studies on some Indian fermented foods: Microbiology and biochemical aspects, PhD Thesis, Guru Nanak Dev University, Amritsar, India.
135. Sandhu DK & Soni SK (1989) Microflora associated with Indian Punjabi warri fermentation. *J Food Sci Technol* 26: 21-25.

136. Soni SK & Sandhu DK (1999) Fermented cereal products. In: *Biotechnology: Food Fermentation*, Vol 2, Joshi VK & Pandey A (eds), Educational Publishers and Distributors, New Delhi, pp 895-949.
137. Lewis YS, Johar DS & Subrahmanyan V (1955) Studies on process simplification in the preparation of a fermented type of foodstuff, Idli. *Bull Central Food Technol Res Inst, Mysore* 4: 257-261.
138. Mukherjee SK, Albury MN, Perderson CS, van Veen AG & Steinkraus KH (1965) Role of *Leuconostoc mesenteroides* in leavening the batter of idli: A fermented food of India. *Appl Microbiol* 13: 227-231.
139. Batra LR (1981) Fermented creals and gram legumes of India and vicinity. In: *Advances in Biotechnology*, Moo-Young M & Robinson CW (eds), Pergamon press, Toranto, Canada, pp 547-558.
140. Ramakrishnan CV (1979) Studies on Indian fermented foods. *Baroda J Nutr* 6: 1-7.
141. Soni SK, Sandhu DK, Vilkhu KS & Kamra N (1986) Microbiological studies on dosa fermentation. *Food Microbiol* 3: 45-53.
142. Soni SK & Sandhu DK (1990) Indian fermented foods: Microbiological and biochemical aspects. *Indian J Microbiol* 30: 227-234.
143. Rajalakshmi R & Vanaja K (1967) Chemical and Biological evaluation of the effects of fermentation on the nutritive value of food prepared from rice and gram. *Brit J Nutr* 21: 467-473.
144. Batish VK, Grover S, Pattnaik & Anmed N (1999) Fermented Milk Products. In: *Biotechnology: Food Fermentation*, Vol 2, Joshi VK & Pandey A (eds), Educational Publishers and Distributors, New Delhi, India, pp 781-863.
145. Comi G, Urso R, Iacum in L, Rantsiou K, Cattaneo P & Cantoni C (2005) Characterisation of naturally fermented sausages produced in the North East of Italy. *Meat Sci* 69: 381-392.
146. Demeyer D (2004) Meat fermentation: principles and applications. In: *Handbook of Food and Beverage Fermentation Technology,* Hui YH, Meunier-Goddik L, Hansen AS, Josephsen J, Nip W-K & Stanfield PS (eds), Marcel Decker Inc, New York, USA, pp 353-368.
147. Talon R, Leroy-Se′trin S & Fadda S (2004) Dry fermented sausages. In: *Handbook of Food and Beverage Fermentation Technology*, Hui YH, Meunier-Goddik L, Hansen AS, Josephsen J, Nip W-K & Stanfield PS (eds), Marcel Dekker Inc, New York, USA, pp 397-416.
148. Fontana C, Cocconcelli PS & Vignolo G (2005) Monitoring the bacterial population dynamics during fermentation of artisanal Argentinean sausages. *Int J Food Microbiol* 103: 131-142.
149. Lebert I, Leroy S & Talon R (2007) Microorganisms in traditional fermented meats. In: *Handbook of Fermented Meat and Poultry,* Toldra F, Hui YH, Astiasaran I, Nip WK, Sebranek JG, Silveira ETF, Stahnke LH & Talon R (eds), Blackwell Publishing Professional, Chichester, UK, pp 1113-1124.
150. Corbie`re Morot-Bizot S, Leroy S & Talon R (2006) *Staphylococcal* community of a small unit manufacturing traditional dry fermented sausages. *Int J Food Microbiol* 108: 210-217.
151. Kalantzopoulos G (1997) Fermented products with probiotic qualities. *Anaerobe* 3: 185-190.
152. Gustafsson BE (1983) Introduction to the ecology of the intestinal microflora and its general characteristics. In: *Nutrition and the Intestinal Flora,* Hallgren B (ed), Almqvist and Wiksell International, Stockholm, Sweden, pp 110-122.

153. Savage DC (1983) Microbial ecology of the gastrointestinal tract. *Nutrition and the Intestinal Flora,* Hallgren B (ed), Almqvist and Wiksell International, Stockholm, Sweden, pp 145-157.
154. Robinson RK & Samona A (1992) Health aspects of `bifidus' products: A review. *Int J Food Sci Nutr* 43: 175-180.
155. Van der Waaij D, Berghuis-deVries JM & Wees LV (1972) Colonization resistance of the digestive tract and the spread of bacteria to the lymphatic organs in mice. *J Hyg* 70: 335-342.
156. Nord CE & Kager L (1984) The normal flora of the intestinal tract. *Neth J Med* 27: 249-252.
157. Swank GM & Dietch EA (1996) Role of the gut in multiple organ failure. Bacterial translocation and permeability changes. *World J Surg* 20: 411-417.
158. Carrico J & Meakin JI (1986) Multiple organ failure syndrome. *Arch Surg* 121: 196-208.
159. Alexander JW, Boyce ST & Babcock GF (1990) The process of microbial translocation. *Ann Surg* 212: 496-512.
160. Wells CL (1990) Relationships between intestinal microecology and the translocation of intestinal bacteria. *Antonie van Leeuwenhoek* 58: 87-93.
161. Kasravi FB, Adawi D, Molin G, Bengmark S & Jeppson B (1997) Effect of oral supplementation of lactobicilli on bacteria translocation in acute liver injury induced by D-galactosamine. *J Hepatology* 26: 417-424.
162. Fuller R (1989) Probiotics in man and animals. *J Appl Bacteriol* 66: 365-378.
163. Hesseltine CW (1983) The future of fermented foods. *Nutr Rev* 41: 293-301.
164. Duszkiewicz-Reinhard W, Gujska E & Khan K (1994) Reduction of stachyose in legume flours by lactic acid bacteria. *J Food Sci* 59: 115-117.
165. Hepner G Fried RR, Jeor SS, Fuseti L & Morin R (1979) Hypercholesteremic effect of youghurt and milk. *Am J Clin Nutr* 32: 19-24.
166. Harrison VC & Peat G (1975) Serum cholesterol and bowel flora in the newborn. *Am J Clin Nutr* 28: 1351-1355.
167. Taranto MP, Sesma F, Pesce de Ruiz Holgado A & Valdez GFD (1997) Bile salts hydrolase plays a key role on cholesterol removal by *Lactobacillus reuteri. Biotechnol Lett* 19: 845-847.
168. Poppel G & Schaafsma G (1996) Cholesterol lowering by a functional yoghurt. In: *Proceedings of the Conference of Food Ingredients*, Europe, pp 31-32.
169. Haberer P, Holzapfel WH & Wagner H (1997) Moegliche Rolle von Milchsaeurebakterien bei der Cholesterinsenkung im Blutserum. *Mitteilungsblatt der Bundesanstalt fuer Fleischforschung, Kulmbach* 36: 202-207.
170. Brigidi P, Bolognani F, Rossi M, Cerre C & Matteuzzi D (1993) Cloning of the gene for cholesterol oxidase in *Bacillus* spp., *Lactobacillus reuteri* and its expression in *Escherichia coli. Lett Appl Microbiol* 17: 61-64.
171. Steinmetz KA & Potter JD (1991) Vegetables, fruit, and cancer. I. Epidemiology. *Cancer Causes Control* 2: 325-357.
172. Anonymous (1997) World Cancer Research Fund and American Institute for Cancer Research. Food, Nutrition and the Prevention of Cancer: A Global Perspective. American Institute for Cancer Research, Washington DC, USA, pp 436-446.
173. Jiang H & Chess L (2004) An integrated view of suppressor T cell subsets in immunoregulation. *J Clin Invest* 114: 1198-1208.

174. Simth-Warner SA & Giovannucci E (1999) Fruit and vegetable intake and cancer. In: *Nutritional Oncology*, Heber D, Blackburn GL & Go VLW (eds), Academic Press, Boston, USA, pp 153-183.
175. Steinmetz KA & Potter JD (1991) Vegetables, fruit, and cancer. II. Mechanisms. *Cancer Causes Control* 2: 427-442.
176. Block G, Patterson B & Subar A (1992) Fruit, vegetables, and cancer prevention: a review of the epidemiological evidence. *Nutr Cancer* 18: 1-29.
177. Reddy BS, Ekelund G, Bohe M, Engle A & Domellof L (1983) Metabolic epidemiology of colon cancer: Dietary pattern and fecal sterol concentrations of three populations. *Nutr Cancer* 5: 34-40.
178. Shahani KM (1983) Nutritional impact of *Lactobacillus* fermented foods. In: *Nutrition and the Intestinal Flora*, Hallgren B (ed), Almqvist and Wiksell International, Stockholm, Sweden, pp 178-188.
179. Mital BK & Garg SK (1995) Anticarcinogenic, hypocholesterolemic and antagonistic activities of *Lactobacillus acidophilus. Crit Rev Microbiol* 21: 175-214.
180. Tavani A, Gallus S, La Vecchia C, Talamini R, Barbone F, Herrero R & Franceschi S (2001). Diet and risk of oral and pharyngeal cancer. An Italian case-control study. *Eur J Cancer Prev* 10: 191-195.
181. Cohen JH, Kristal AR & Stanford JL (2000) Fruit and vegetable intakes and prostate cancer risk. *J Nat Cancer Inst* 92: 61-68.
182. Goldin BR & Gorbach SL (1984) The effect of oral administration on *Lactobacillus* and antibiotics on intestinal bacterial activity and chemical induction of large bowel tumors. *Dev Indus Microbiol* 25: 139-150.
183. Salminen S, Deighton M & Gorbach S (1993) Lactic acid bacteria in health and disease. In: Lactic acid *Bacteria.* Salminen S & von Wright A (ed), Marcel Dekker Inc, New York, USA, pp 237-294.
184. Lidbeck A, Overvik E, Rafter J, Nord CE & Gustafsson JA (1992) Effect of *Lactobacillus acidophilus* supplements on mutagen excretion in faeces and urine in humans. *Microbial Ecol Health Dis* 5: 59-67.
185. Lu M, Toshima Y, Wu X-li, Zhang X & Cai Y (2007) Inhibitory effects of vegetable and fruit ferment liquid on tumor growth in Hepatoma-22 inoculation model. *Asia Pac J Clin Nutr* 16: 443-446.
186. Tamai Y, Oishi H, Nakagawa I, Watanabe Y, Shinmoto H, Kuwabara Y, Yamato K & Nagai S (1995) Antimutagenic activity of the milk fermented by mixed-cultured with various lactic acid bacteria and a yeast. *J Japanese Soc Food Sci Technol [Nippon Shokuhin Kogyo Gakkaishi]* 42: 383-387.
187. McIntosh GH (1996) Probiotics and colon cancer prevention. *Asia Pacific J Clin Nutr* 5: 48-52.
188. De Simone C (1986) Microflora, yoghurt and the immune system. *Int J Immunol* (Suppl II): 19-23.
189. Sato K, Saito H & Tomioka H (1988) Enhancement of host resistance against *Listeria* infection by *Lactobacillus casei*: Activation of liver macrophage and peritoneal macrophages by *Lactobacillus casei. Microbiol Immunol* 32: 689-698.
190. Marin ML, Tejada SMV, Murtha J, Ustunol Z & Pestka JJ (1997) Effects of *Lactobacillus* spp. on cytokine production by RAW 264.7 macrophage and EL-4 thymoma cell lines. *J Food Prot* 60: 1364-1370.
191. Isolauri E (1996) Studies on *Lactobacillus* GG in food hypersensitivity disorders. *Nutr Today* 31: 28S-31S.

# Postharvest Biotechnology of Food 13

**V.K. Joshi and Dev Raj**

*Department of Food Science and Technology, Dr. Y.S. Parmar University of Horticulture and Forestry, Nauni, Solan 173 230 (H.P.)*

## 1. INTRODUCTION

The large amount of food grown in India, which includes 84 million tonnes of rice, 72 million tonnes of wheat and 132 million tonnes of fruits and vegetables needs to be effectively utilized. So for achieving this goal, there is a need of a networking system for minimizing the spoilage as well as for value addition. Biotechnology has a major role to play in this operation. Biotechnological approaches are also required for augmenting yields of neutraceutically improved crops and livestock, processing the harvests, value addition, quality assurance, enhancement of the shelf-life as well as management of wastes of agro/food processing industry. Postharvest biotechnology involves all the bio-techniques for enhancing the shelf-life of the commodity after harvest and preventing their spoilage as well. Food is the basic necessity of all the living entities in the world.

Despite of the improvement in the methods of production, the postharvest biotechnology still remains the most important factor. This aspect assumes importance from toxin production, spoilage of fresh and processed products and quality control and as sanitation indicators in a processing unit. Microbiology is also on the top of different hazards, which are associated with the safety of food for consumption by human beings [1]. In striving to achieve the zero risk, microbial quality is of immense significance. At the same time, we know that a safe food has to pass through various steps and there are plenty of chances for missteps along the way. In this respect, Hazard Analysis Critical Control Point (HACCP) system is quite significant [2].

Various fruits and vegetables and their products may be spoiled by one or more factors like unsuitable packaging, chemical changes or the action of microorganisms, tissue enzymes, insects, rodents or improper methods of processing, under processing, etc. [3]. Different spoilage agents for various food products, their prevention and health hazards associated with spoilage are also discussed in detail in this chapter.

## 2. CAUSES OF SPOILAGE

A food is said to be spoiled if it has been damaged or injured so as to make it undesirable for human use. "A product is fit as a food if a discriminating consumer, knowing the story of its production and

*Food Biotechnology: Principles and Practices*
*V.K. Joshi & R.S. Singh (Eds.)*
*I.K. International Publishing House Pvt. Ltd., India*
*pp 417–470*

seeing the material itself, will eat it, and conversely, the same product is spoiled when such an examiner refuses it as a food" [4]. So, the fitness of the food will depend upon the judgment of person as per the above definition. In other terms, a food is said to be spoiled if it is not harvested at proper maturity, is contaminated with dirt, handled by dirty or diseased person, is fertilized with sewage and has objectionable changes from the activity of microorganisms or action of enzymes of the food [5]. So, the food is considered as `spoiled' when it is decayed or decomposed. The food considered unfit for sanitary reasons is not included in the spoiled category [5]. The causes of food spoilage [6] may be broadly grouped as follows:

- Growth and activity of microorganisms in food
- Insects, rodents or other animals
- Action of enzyme of the plant, microbes or animal food
- Purely chemical reactions that are not catalyzed by enzyme
- Physical changes such as those caused by burning, drying, pressure, freezing, etc.

Depending upon the easiness of spoilage, the foods have been categorized into non-perishable (do not spoil easily if handled properly i.e., sugar, flour, dry beans), semi-perishable (do not spoil if handled and stored properly i.e., potatoes, apple, nut meal, etc.) and perishable foods (spoil readily unless special methods of preservation are used i.e., meats, milk and most of the fruits and vegetables). Since most of the fruits and vegetables, meat and milk fall into the perishable group, application of suitable method of preservation becomes almost indispensable for them.

## 2.1 Spoilage Causing Agents

The spoilage of fruits and vegetables occurs due to various physical factors, action of their own hydrolytic enzymes or microbial contaminants, etc. Foods including fruits and vegetables are living commodity even after picking for certain times, which means that they are sensitive to their environment. The rate of metabolism of commodities is temperature dependent and they may be damaged by heat or cold. They are affected by levels of different gases in the atmosphere. Oxygen is taken in during respiration and as a result $CO_2$, heat and water vapours are given-off. Moisture is also lost by evaporation. After detachment of fruits from the mother plant, the continuity of the flow of sap is totally disrupted but the respiration and water loss continues leading to exhaustion of food reserve and moisture they contain. Because of such irreparable losses, the deterioration begins and eventually they perish. Spoilage is mainly of two types:

### *2.1.1 Biotic spoilage*

This type of spoilage is due to the microbial action viz., bacteria, yeasts and moulds on fruits and vegetables and the normal processes of aging.

### *2.1.2 Abiotic spoilage*

This type of spoilage is caused by different physical and chemical changes in the product viz., hydrolytic action of enzymes, oxidation of fats, purification of proteins, browning reaction between proteins and sugars, and the physical changes of wilting, caking and melting, etc. Temperature control is the major factor to provide longevity to the fruits and vegetables. Mechanical fruit grader also result in abiotic spoilage during grading (Plate 1).

## 2.2 Factors Responsible for Biotic Spoilage

Microorganisms are the main agents which cause spoilage of the food. Most of the spoilage in food occurs due to microbes. Bacteria cause most of the spoilage problems in food commodity as these are not killed at ordinary temperatures i.e., resistant to high temperature. The yeasts and moulds have less resistance to heat (processing temperatures). The main factors responsible for spoilage of food are discussed here.

**Plate 1:** Mechanical fruit grader causing abiotic spoilage during grading.

### *2.2.1 Number and kind of microorganisms*

The initial number of microorganisms present in the processed food has direct relationship with its spoilage. More the number of microorganisms present in food, rapid will be the spoilage. The effect of initial number of spores on time required to kill them is shown in Table 1. The heat resistance of bacteria involved in food poisoning is of major concern from public health view point (Table 2).

**Table 1:** Effect of initial number of spores on time required to kill them.

| Initial concentration of spores, number/ml | Thermal death time, or time required to kill all spores min. at 120°C |
|---|---|
| 50,000 | 14 |
| 5,000 | 10 |
| 500 | 9 |
| 50 | 8 |

*Source:* 5, 7

**Table 2:** Heat resistance of microorganism of major concern in food poisoning.

| Growth temperature | Heat resistance | |
|---|---|---|
| | Low | High |
| Low (8-10°C) | *Listeria monocytogenes* (Inf) | *Clostridium botulinum* E (Tox) |
| | *Yersinia enterocolitica* (Inf) | *Bacillus cereus* (Tox) |
| | *Vibrio parahaemolytica* (Inf) | *Bacillus subtilis* (Tox) |
| | *Salmonella* species (Inf) | *Bacillus licheniformis* (Tox) |
| Medium (25-40°C) | *Escherichia coli* (Inf) | *Clostridium perfringens* Inf) |
| | *Staphylococcus aureus* (Tox) | *Clostridium botulinum* A (Tox) |
| High (50-55°C) | *Campylobacter jejuni* (Inf) | |
| | *E. coli* (Inf) | |

Inf = Microorganisms causing food infection; Tox = Microorganisms causing food poisoning

*Source:* 8

**Table 3:** Thermal death times of bacterial spores.

| Spores | Time to kill (100°C), min |
|---|---|
| *Bacillus anthracis* | 107 |
| *Bacillus subtilis* | 15-20 |
| *Clostridium botulinum* | 520 |
| *Bacillus stearothermophilus* | over 1030 |

*Source:* 5

### 2.2.2 Temperature

Different microorganisms grow at different temperatures. Psychrophiles grow best at low temperatures (8-10°C), mesophiles grow best at medium temperatures (25-40°C), while thermophiles grow at higher temperatures (50-55°C) (Table 2). Due to this reason, processed foods are immediately cooled to avoid the growth of thermophiles. Therefore, storage temperature is very important in relation to microbial growth. Different time is required to kill the spores of different microorganisms (Table 3). The time required to kill all the spores of flat sour bacteria (*Bacillus stearothermophilus*) in relation to temperature is also shown in Table 3.

### 2.2.3 Food type

Different microorganisms prefer different kind of food. Some grow best on pertinacious foods, and some on starchy or fatty foods. The physical state of the food, whether heated, frozen, moistened or dried, also has an important influence on the type of spoilage. The moisture content also influences the type of microorganisms in the foods, since the requirements of moisture for their growth are different. Bacteria, yeast and moulds have different moisture requirements in terms of water activity (Table 4).

### 2.2.4 pH

The pH of the food influences the kind and growth of microorganisms. Among the microorganisms, spoilage can be caused by bacteria, moulds/yeasts, etc. depending upon the pH/acids present in the food. Foods having pH<4.5 (acid foods) do not require heat processing (particularly cooking under pressure). Foods having pH>4.5 (low acid foods) always require processing under pressure. This is because the thermophilic bacteria may not be killed at normal temperatures as most bacteria thrive best at pH of 4-7.5. The yeasts and moulds (however, have low resistance to heat) require a pH of 2.5-8.0 and 1.5-8.5 for their growth. The commercially canned foods are divided into 4 groups viz., low acid foods (pH above 5.3), medium acid foods (pH 4.5-5.3), acid foods (pH 3.7-4.5) and high acid foods (pH below 3.7).

### 2.2.5 Presence of air

Aerobic or anaerobic microorganisms can be found in the spoiled fruit and vegetable products, depending upon the presence or absence of air (oxygen) in the container or package.

**Table 4:** Minimum limit of water activity for growth of microbes.

| Microorganisms | Water activity, $a_w$ |
|---|---|
| Bacteria | 0.92 |
| Yeast | 0.88 |
| Mould | 0.80 |
| Halophilic bacteria | 0.75 |
| Osmophilic yeast | 0.60 |
| Xerophilic mould | 0.66 |

## 3. GROWTH OF MICROORGANISMS

When all the parameters for the growth of microorganisms are favourable, especially when a particular microorganism is present in the food, these organisms start growing and multiplying. They will pass through various phases of growth. When organisms are counted at regular interval and the results are plotted, a growth curve is obtained. This curve is divided into seven phases (Fig. 1).

In addition to destruction or removal of microorganisms, the delay in the initiation of growth also prevents microbial spoilage. This is done by keeping the microorganisms in lag phase as long as possible. Once the microorganisms enter the log phase, it is very difficult to control them.

### 3.1 Lag Phase (A to B)

During this phase, there is no growth or even there is decline in the number of organisms. This is the phase during which we can check the growth of microorganisms by following the suitable preservation methods. So, the decomposition of the food can be delayed very easily at this very phase.

### 3.2 Positive Acceleration Phase (B to C)

During this phase, growth rate of microorganisms is continuously increasing with the passage of time.

### 3.3 Logarithmic Phase (C to D)

This is also known as exponential phase. During this phase, the multiplication rate is constant and most rapid i.e., the generation time of microorganism is the shortest, and varies with the environmental conditions during growth as given below:

- Type of food
- pH of food
- Temperature
- Oxidation-reduction potential
- Available moisture
- Presence of inhibitors

When the environmental conditions are more favourable, short time will be required for their generation.

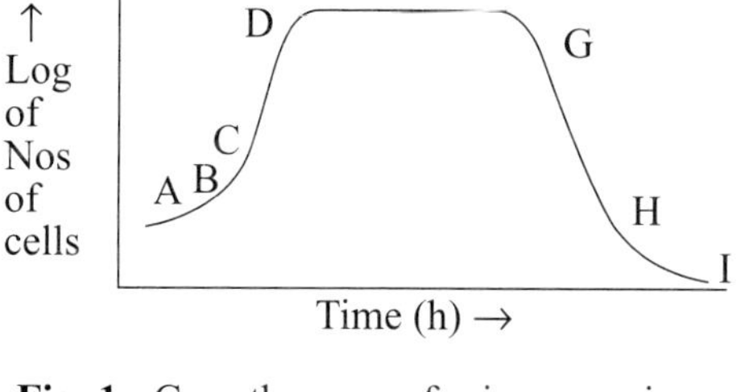


**Fig. 1:** Growth curve of microorganisms.
A to B lag phase, B to C positive acceleration phase, C to D logarithmic phase, D to E negative acceleration phase, E to F stationary phase, F to G accelerated death phase, G to H death phase, H to I survival phase
*Source:* 5

### 3.4 Negative Acceleration Phase (D to E)

During this phase, the rate of multiplication is decreasing. This phase is also useful from preservation point of view.

### 3.5 Stationary Phase (E to F)

During this phase, the number of microorganisms is constant. If we provide an environment unfavourable for the growth of a microorganism, then its number may even decrease with the passage of time. So, the spoilage of food can be delayed very easily by altering the environmental conditions.

### 3.6 Accelerated Death Phase (F to G)

During this phase, with the passage of time the number of microorganisms goes on decreasing.

### 3.7 Death Phase (G to H)

During this phase, the number of organisms decreases up to zero. With many bacteria, the number does not decrease at a fixed rate to zero, as indicated by the unbroken line in Fig. 1, but is tapered off gradually as low numbers are approached, as indicated by the broken line, and a few viable cells remain for some time.

## 4. BIO-DETERIORATION OF FOODS

### 4.1 Foods of Plant Origin

Fruits and vegetables are bulky, easily damaged by mechanical means and contain large quantity of water. They are sensitive to their environment viz., temperature, level of oxygen, carbon dioxide and ethylene, etc. Spoilage of fresh fruits and vegetables usually occurs during transport and storage while waiting to be processed. Fruits and vegetables get contaminated with spoilage organisms either from each other or from the containers in which they are packed i.e., baskets, lugs, boxes, etc. during harvesting. Mechanical injuries during transportation (Plate 2), further aggravate the deterioration process. Vegetables and fruits are available as fresh, dried, frozen, fermented, pasteurized or canned products and the spoilage can occur almost in any of these products.

**Plate 2:** Bio-deterioration of fruit/vegetables at field level and transportation level.

## 5. SPOILAGE MICROBIOLOGY

The spoilage microbiology of food largely depends upon different factors like kind and variety of fruits/vegetables, environmental condition viz., storage, temperature, relative humidity of the atmosphere and various gases of the atmosphere.

Microbial spoilage is of two types:

- Spoilage caused by plant pathogens, which attack various parts of the plant used as foods
- Spoilage caused by saprophytes

## 6. SPOILAGE OF FRUITS AND VEGETABLES

The prominent spoilage in fruits and vegetables is mainly mildew, brown rot, soft rot, black rot, green rot and mold rot, etc. Dry rots often lead to darkening and discolouring, and hardening of the surface of vegetables and fruits. In microbial spoilage, the vegetables often develop water soaked musky areas, while the fruits generally have brown or white coloured patches. The market diseases of fruits and vegetables are listed in Table 5.

**Table 5:** Market diseases of some fruits and vegetables.

| **Fruits** | **Market diseases** |
|---|---|
| Lemons, limes, oranges, grapefruit | Alternaria rot, anthracnose, blue mould rots, stem-end rots |
| Peaches | Alternaria (or green mould rot), gray mould rot, black mould rot |
| Apricots, plums, cherries | Blue mould rot, brown rot, cladosporium rot, rhizopus rot |
| Avocados | Anthracnose, rhizopus rot |
| Bananas | Anthracnose, Fusarium, Pestalozia |
| Figs | Alternaria rot, blue mould rot, Cladosporium |
| Dates | Yeasts, various moulds |
| Grapes | Black mould rot, gray mould rot, rhizopus rot, blue mould rot |
| Strawberries | Gray mould rot, leather rot (*Phytophthora cactorum*), rhizopus rot |
| Apples | Numerous moulds |
| Pears | Black rot, blue mould rot, brown rot, gray mould, rhizopus rot |
| **Vegetables** | **Market diseases** |
| Potatoes | Fusarium tuber rot, bacterial ring rot, bacterial soft rot |
| Cucumber | Rhizopus soft rot, bacterial soft rot, blue mould rot, gray mould rot |
| Cabbage | Bacterial soft rot, gray mould rot, black rot, watery soft rot |
| Cauliflower | Bacterial soft rot, gray mould rot, black rot, watery soft rot |
| Broccoli | Bacterial soft rot |
| Radishes | Bacterial soft rot, club root rot, rhizoctonia rot (*Rhizoctonia carotee*) |
| Turnips | Bacterial soft rot, club root rot, rhizoctonia root |
| Pumpkin | Rhizopus soft rot, diplodia rot, phytophthora rot, gray mould rot |
| Squash | Rhizopus soft rot, gray mould rot |
| Tomatoes | Alternaria rot, bacterial canker, bacterial spot, gray mould rot, green mould rot, rhizopus rot |
| Egg plant | Fruit rot (Phomopsis rot) |
| Asparagus | Bacterial soft rot, fusarium rot, gray mould rot, phytophthora rot |
| Onions | Bacterial soft rot, black mould rot, gray mould rot |
| Garlic | Bacterial soft rot, black mould rot |
| Green beans | Bacterial soft rot, mould rot, rhizopus rot |
| Lima beans | Soft rot |
| Carrots | Bacterial soft rot, black rot, fusarium rot, gray mould rot, watery soft rot |
| Beets | Bacterial soft rot, black rot, blue mould rot, fusarium rot |
| Lettuce | Bacterial soft rot |

*Source:* 5

## 6.1 Spoilage Microbiology of Bacteria, Yeasts and Moulds

### *6.1.1 Bacteria*

Bacteria have different phases of growth while growing. A typical growth curve (Fig. 1) is traced out by plotting the number of bacteria, periodically. The curve has lag phase representing nil growth of organisms, exponential phase or logarithmic showing a continuous increase in the number of microorganisms, and stationary phase where the number of microorganism gets constant, and decline phase which shows a decrease in the number of cells. The food can be preserved for longer time by prolonging the lag phase. This can be obtained by avoiding the contamination of the food and turning the environmental conditions e.g., temperature, moisture and pH, etc. unfavourable for the growth of contaminants (microorganism) as discussed earlier. Thus, by lowering the storage temperature of the fruits/vegetables, filling up the storage chamber with the inert gases will definitely lead to longer shelf life of the vegetables and fruits. Lowering of the storage temperature prolongs the lag phase of the bacterial growth and also reduces the slope of the growth phase (Fig. 1). The pH of food is another important factor governing the bacterial growth. Bacteria mostly grow between pH 4-8. Growth rate is lowered by decrease in pH. Fig. 2 shows the effect of changing pH on heat resistance of spores of *Bacillus subtilis* [10].

### *6.1.2 Yeasts*

Yeasts are widely found in the environment. These can be isolated from the surface of the fruits, leaves and flowers, etc. The yeast growth largely depends upon the nature of fruit product. The yeasts (Plate 3 a & b) are generally fermentative in nature and produce large quantities of carbon dioxide in addition to alcohol, glycerol, acetaldehyde, pyruvic acid and α-ketoglutaric acid. Some degrade starch, pectin and sugars, etc. thus, changing the composition of a particular food product and finally, leading to its degradation.

### *6.1.3 Moulds*

Moulds are frequently associated with food products. Many of the moulds secrete toxic compounds (mycotoxins) like aflatoxins, patulin, etc. Aflatoxins have been detected in dried figs and fig paste [11]. Another mycotoxin, patulin is the most common mycotoxin detected in processed

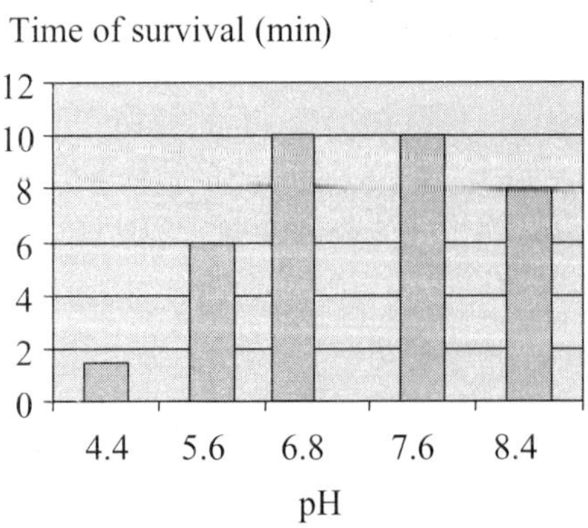


**Fig. 2:** Effect of pH on heat resistance of spores of *Bacillus subtilis.*

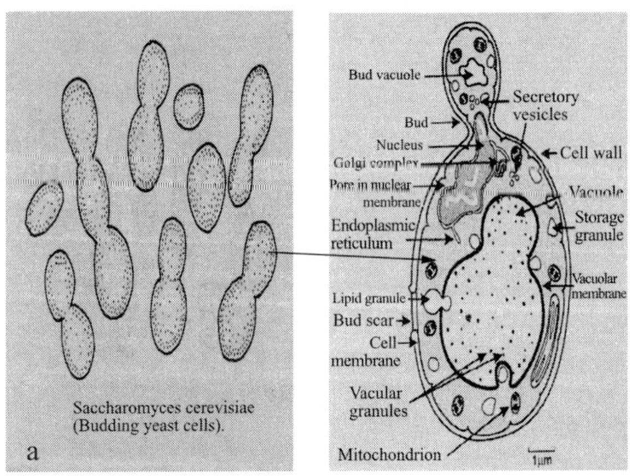


**Plate 3 (a) & (b):** Structure of typical brewing yeast cell.

**Table 6:** Mycotoxins associated with fruit and vegetable products.

| Fruit | Fungus | Mycotoxin |
|---|---|---|
| Apple juice | *Penicillium expansum* | Patulin |
| Banana | *Fusarium moniliforme* | Trichothecenes |
| Canned juices | *Byssochlamys* | Patulin, byssochlamic acid, byssotoxin, malformins |
| Canned juices | *Neosartorya, Fumitremorgins* | Mevinolins, terrein, terypacidin, tryptoquivalin, Verruculogen |
| Canned juices | *Talaromyces flavus* | Dehydroaltenusin, mitorubrins, vermicelline, vermitatine, worthmannin, worthmannolone |
| Preserves | *Eurotium* | Physicon, emodin |
| Figs | *Aspergillus flavus* | Aflatoxins |
| White beans | *Aspergillus, Penicillium* | Oxfiration A |

*Source:* 11, 14, 16-18

fruits (Table 6). Besides the production of aflatoxins, the moulds like *A. parasiticus* also produces compositional changes in the fruits [12]. Juices made from scabbed fruits (*Venturia inaequalis*), apple fruits, were found to be free from patulin, aflatoxin and microflora [13]. The mould *Penicillium expansum,* causes apple rot and some other moulds produce patulin. Their traces have been seen in the fruit juices. The mycotoxins are found deleterious to various animals, and presumably the human beings also [14]. Oils from eucalyptus, ginger, thymol, cinnamon anise, rosemary, orange gitter and nutmeg have shown inhibition of growth and aflatoxin producing capacity of *Aspergillus parasiticus*. Anise and cinnamon oils had pronounced effect on the growth and toxin production at 200 and 300 ppm concentration [15].

## 7. SPOILAGE OF PROCESSED FRUITS AND VEGETABLE PRODUCTS

The microbial deterioration of processed fruit and vegetable products is usually manifested by alterations in the appearance, texture, colour, odour, flavour or slime formation. The appearance includes colour changes, visible growth of microorganisms, formation of pockets of gas or swelling of cans and microbial growth, especially that of moulds on the surface of food. As some food products deteriorate, they tend to become soft or mushy. Degradation of foods results in the formation of compounds, which havc odours and flavours different from those of the fresh food.Usually bacteria, yeast and moulds are involved in the spoilage of processed fruit and vegetable products. The important genera of bacteria include *Bacillus, Clostridium, Acetobacter, Aeromonas, Brevibacterium, E. coli, Halobacterium, Lactobacillus, Leuconostoc, Microbacterium, Micrococcus, Pedicoccus, Propionibacterium, Proteus, Psuedomonas, Salmonella, Serratia, Shigella, Staphylococcus, Streptococcus* and *Streptomyces*. The important yeasts are *Pichia, Hansenula, Candida, Torulopsis, Zygosaccharomyces* and *Saccharomyces,* while moulds genera include *Mucor, Rhizopus, Aspergillus, Penicillium, Trichothecium, Neurospora, Botrytis, Trichoderma, Cladosporium, Helminthosporium, Alternaria, Fusarium, Endomyces,* etc. [5].

### 7.1 Spoilage of Fruits and Fruit Products

Most of the fruits and fruit products have higher acidity (pH <4.5), which influences the composition of their microflora. Generally, the fruit trees grown in cooler regions are more acidic than those grown in warmer areas [2]. Fungi are often the most predominant microorganisms in fruit products because yeasts and moulds grow better than bacteria under acidic conditions. Only a few aciduric bacteria are found in fruit-based products.

About 215 yeast (unicellular fungi) species are associated with fruits and fruit products [19] but only a few are pathogenic to human beings [20]. The growth of yeast depends upon the nature of the product and the strain of yeast. Growth of some yeasts in a clear fruit juice may produce only little haze and sediment. But, strong fermentative strains of *Saccharomyces cerevisiae* may produce enough $CO_2$, which may even burst the container. Major yeast species found in apple products are presented in Table 7. About 200 moulds species produce more than 100 kinds of toxic compounds [22]. Table 6 shows some important mycotoxins associated with fruit products. Patulin produced by a number of mould species including *Penicillium expansum*, the most usual causal organism of apple rots [23], appears to be the most common mycotoxin in processed fruit products. Enzymes present in fungi may result in breakdown of starch and pectins. However, more growth can be manifested in the form of colonies in the headspace or as strands in the beverages or other similar products.

Moulds (filamentous fungi) play an important role in the spoilage of fruit products during their storage [24]. Infected fruit may become soft after processing, since pectinases may not be inactivated by thermal application. Ascospores of some species may survive commercial pasteurization treatments. Some species of mould (xerophilic) spoil dried fruits and fruit juice concentrates. Some species may produce off-flavours in juices, wines and other fruit products, as these may come up on processing equipment (e.g., wooden tanks), and metabolites of many moulds are also toxic to human beings. Some important moulds found in apple products are shown in Table 8, while characteristics of some of the bacteria causing spoilage of the fruit products are shown in Table 9. Among bacteria, the major types responsible for fruit products spoilage are shown in Table 10.

**Table 7:** Major yeasts isolated from apple and apple products.

| Source | Organism |
|---|---|
| Fresh apples | *Candida pulcherrima, Candida* sp., *Torulopsis* sp., *Rhodotorula* sp. |
| Stored apples | *Kloeckera* sp., *Candida* sp., *Torulopsis* sp., *Rhodotorula* sp. |
| Apple juice | *Kloeckera* sp., *Saccharomyces* sp., *Torulopsis* sp. |
| Spoilage yeasts | *Saccharomyces* sp., *Torulopsis* sp., *Rhodotorula* sp. |

*Source:* 21

**Table 8:** Major moulds isolated from apples and apple products.

| Source | Organisms |
|---|---|
| Apples | *Penicillium expansum, Monilla* sp., *Mucor* sp., *Cladosporium* sp. |
| Apple juice | *Penicillium* sp., *Aspergillus* sp., *Paecilomyces* sp. |
| Heat resistant moulds | *Byssochlamus* sp., *Aspergillus* sp., *Penicillium* sp. |

*Source:* 21

**Table 9:** Characteristics of some of the bacteria causing spoilage of fruit products.

| Bacteria | Characteristics/Incidences |
|---|---|
| Lactic acid bacteria | All Gram positive |
| *Lactobacillus plantarum* | Rod shaped, homofermentative |
| *Lactobacillus brevis* | Rod shaped, heterofermentative |
| *Lactobacillus fermentum* | Rod shaped, heterofermentative |
| *Lactobacillus fructivorans* | Rod shaped, heterofermentative |
| *Lactobacillus hilgardi* | Rod shaped, heterofermentative |
| *Leuconostoc mesenteroides* | Cocci, heterofermentative |
| *Leuconostoc paramesenteroides* | Cocci, heterofermentative |
| *Leuconostoc oenos* | Cocci, heterofermentative |
| *Pediococcus pentosaceus* | Cocci-tetrad, homofermentative (All the above spoil fruit juices, wines and ciders) |
| Acetic acid bacteria | |
| *Acetobacter aceti* and *A. pasteurianus*<br>*Gluconobacter oxydans* | These also spoil juices, wines |
| Coliforms | Short rods, gram negative |
| *Escherichia coli, Enterobacter aerogenes* | Contaminate water, spoil milk, cheese juices |
| Spore forming bacteria | |
| *Clostridum pasteurianum* | Pear and apricot juices |
| *Bacillus macerans, B. polymyxa* | Acidified vegetables, canned peaches and pears having pH 3.8-4.0 |
| *B. coagulans, B. subtilis, B. pumilus* | Can spoil dessert wines that contain as much as 20% ethanol by volume |
| Pathogenic bacteria | Cause food-borne illnesses |
| *Salmonella oranieberg*<br>*Shigella sonnei, Shigella flexneri*<br>*Shigella dysenteriae*<br>*Shigella derby*<br>*Shigella typhi* | Present on wrapped watermelon, papaya, watermelon, etc. |

*Source:* 5, 25-37

**Table 10:** Some of the important fungi commonly isolated and found in vegetables.

| Microorganism | Name of vegetable |
|---|---|
| *Alternaria* | Cabbage, carrots, cauliflower, peppers, tomato |
| *Aspergillus* | Onions, tomatoes |
| *Colletotrichum* | Beans, onions, tomatoes |
| *Fusarium* | Asparagus, carrots, potatoes |
| *Rhizopus* | Carrots, sweet potatoes, tomatoes |
| *Sererotinia* | Carrots, celery |

*Source:* 39, 43

## 7.2 Spoilage of Vegetables and Vegetable Products

Spoilage of vegetables and vegetable products may be due to the reasons mentioned below [38-42]:

- Fresh vegetables owing to their higher water content ($a_w$), neutral pH and nutrients, are capable of supporting the growth of any type of microorganisms.

- The outer surfaces of vegetables are normally contaminated with bacteria and fungi.
- Microorganisms contaminating vegetables are *Pseudomonas, Erwinia, Xanthomonas, Enterobacter, Bacillus, Clostridium, Corynebacterium* and *Cytophaga* (Table 10).
- To reduce the spoilage of vegetable products, the measures that can be applied include careful harvesting (least handling), proper culling of damaged produce, equipment cleaning, post harvest treatments, packaging and proper storage.

### *7.2.1 Pathogenic organisms associated with vegetables*

Vegetables can be infested/infected with a variety of pathogenic organisms including bacteria, viruses and parasites through contaminated soil, irrigation water, unhygienic handling, contaminated machinery and containers resulting in human diseases [44]. Common disease causing bacteria associated with vegetables are *Salmonella, Shigella, E. coli, Listeria monocytogenes, Aeromonas* and *Clostridium botulinum.* Parasites found in vegetables are *Entamoeba histolytica, Giardia lamblia* and *Ascaris lumbricoides*. Hepatitis virus is also associated with vegetables.

### *7.2.2 Indicator organisms*

Indicator organisms are used to avoid the laborious process and expensive equipment needed to isolate pathogenic bacteria and viruses from food, and to analyze all the pathogens would be a tremendous task. Therefore, the testing for indicator organisms can be of value in assessing the microbial condition of a food. However, the absence of an indicator in a food sample cannot be regarded as absolute proof that all intestinal pathogens are also absent.

The indicator organisms have been used since 1892 when Schardinger tested water for *Escherichia coli*, instead of testing for *Salmonella typhi*, as it was difficult to test the latter and the presence of the former could be taken as the presence of both the types. The *E. coli* count in water should not exceed 10 per ml for safety [45]. The main organisms used as indicators of spoilage include *E. coli, Staphylococci, Pseudomonads, Clostridia, Enterobacteriaceae, Streptococci* sp. Machinary mould, *Geotrichum candidum* is used as an indicator of sanitation of food processing operations.

## 7.3 Spoilage of Fruit and Vegetable Juices

### *7.3.1 Fruit juice/squashes*

The fruit juices are more prone to spoilage by yeasts and moulds and less to bacteria since they have lower pH, while vegetable juices are more prone to bacteria than yeasts and moulds because of very low acidity. If the fruit and vegetable juices are not processed after extraction, they are spoiled because of enzymatic changes and microbial actions.

Moulds can grow on the surface of fruit juices if exposed to air. Most fruit juices have sufficient sugar to favour the growth of yeasts. Concentrates of fruit and vegetable juices favour the growth of yeasts species, which are tolerant to high concentration of acid and sugar. These include *Leuconostoc* and *Lactobacillus* bacteria. Heat treatment during the canning of these concentrates usually kills these microorganisms, and freezing prevents the growth of such organisms. In case of fruit squashes, if preservatives are not in proper concentration, some yeast (e.g., *Zygosaccharomyces*) can spoil such products.

### *7.3.2 Spoilage of canned fruit juices*

The acidity of the juice is the single most important factor affecting microbial spoilage [46]. Most bacteria have an optimum pH near 6.8, but may grow at pH values ranging from 4-8. Yeasts and moulds can grow at pH <2. Table 11 shows pH usually found in different fruit juices and the principal acids. As can be seen, the pH values are lower than 4.5 in nearly all the cases. In canned foods, a pH of 4.5 is used as a borderline between acid and low acid foods i.e., foods not requiring and those requiring respectively, the minimum botulinum cook (12 D). The typical spoilage flora of fruit juices is represented by some *Clostridium, Bacillus*, members of *Enterobacteriaceae*, lactic acid bacteria, *Acetobacteriaceae*, yeasts and moulds. The most commonly involved strains of butyric anaerobes in fruit juice spoilage belong to *Clostridium butyricum* and *Clostridium pasteurianum*, and the spoilage is characterized by lowering of pH (0.2-0.4 units), development of very high volumes of hydrogen and $CO_2$ and strong cheesy (butyric) odour. The spoilage due to vegetative growth of these bacteria may occur at pH >3.6, while spore germination may occur at pH <3.9 and >0.97 [49, 50]. The vegetative cells of butyric anaerobes are unable to grow at pH 4.7 if $a_w$ is <0.97, although *Clostridium acetobutylicum* may grow at $a_w = 0.95$ and pH 4.7. Three types of *Bacillus* are usually involved in spoilage of fruit juices i.e., *B. coagulans* (flat sour, between 30-65°C at pH >4, $a_w$ >0.97), *B. macerans* and *B. polymyxa* (both by storage at temperature <46°C) [46]. *E. coli* and *E. aerogenes* may grow at pH <4.5 and several strains of *Salmonella* at pH <3.9 in a wide range of chilling temperatures to 46°C. The spoilage by *Enterobacteriaceae* is characterized by the production of lactic acid, acetic acid, formic acid, $H_2$, $CO_2$ with some fruit flavour enhancement [45, 51]. Lactic acid bacteria causing the spoilage of fruit juices include *Lactobacillus* (*L. plantarum, L. fermenti, L. brevis* and *L. buchneri*), *Leuconostoc mesenteroides* and *Streptococcus viscosum*. Most of these species grow at pH >3.0. However, *Leuconostoc* can grow at pH 2-7.5 and at temperatures ranging from 10-40°C. *Acetobacter* also grow in juices with pH >3.6 and at temperature from 5-43°C. *Pediococci* grow at pH 2.9-8.5 upto a temperature of 50°C.

**Table 11:** Major acids and pH of some fruits.

| Fruit | pH | Acid(s) |
|---|---|---|
| Apple | 3.3-4.1 | Malic, citric, lactic |
| Apricot | 3.4-3.6 | Malic, citric |
| Banana | 4.5-5.2 | Citric, malic, tartaric |
| Blackberry | 3.0-4.2 | Malic, citric, isocitric |
| Cherry | 3.2-4.7 | Malic, citric, isocitric |
| Grapefruit | 2.9-3.4 | Citric |
| Grape | 3.0-4.5 | Tartaric, malic |
| Guava | 3.0-3.2 | Citric, malic, lactic |
| Lemon | 2.2-2.4 | Citric |
| Lime | 2.2-2.6 | Citric, malic |
| Mango | 3.3-3.7 | Citric, tartaric |
| Orange | 3.0-4.0 | Citric, malic |
| Papaya | 4.5-6.0 | Citric, malic, ketoglutarate |
| Passion fruit | 2.6-3.3 | Citric, malic |
| Peach | 3.1-4.2 | Malic, citric |
| Pear | 3.4-4.7 | Malic, citric |
| Pineapple | 3.2-4.0 | Citric, malic |
| Plum | 2.8-4.6 | Malic, quinic |
| Raspberry | 2.7-3.3 | - |
| Strawberry | 3.1-3.9 | - |
| Watermelon | 5.8-6.0 | Citric, malic |

*Source:* 47-48

Among the yeasts, the most often involved species in juice fermentation belong to *Saccharomyces, Torulopsis, Candida, Pichia, Hansenula* and *Hanseniaspora*. Fruit juice concentrates are fermented almost exclusively by *Saccharomyces rouxii, S. mellis, Torulopsis* and *Hansenula*. Typical fermentation products are ethanol and $CO_2$. The yeast growth is favoured more in fruit juices because of: (i) increase in oxidation reduction potential by removal of solids, (ii) temperature of incubation is generally 15-30°C, and (iii) deficiency of vitamin B in juices favours yeast than bacteria [52].

Among the moulds, the species belonging to genera found on surfaces of fresh fruit include *Mucor, Rhizopus, Penicillium, Aspergillus, Alternaria, Cladosporium, Acremonium* and *Botrytis*. These can develop at pH 1.5-11, and temperatures −11°C to 55°C. Moulds require $O_2$ for their development, so they usually grow at the surface of processed fruit products. However, *Byssochlamys fulva* and *Penicillium expansum* have been found in canned foods and the latter can grow under vacuum also. Recently, several fruit juices (tomato, orange, etc.) packed in brik containers were found to be spoiled by *Mucor spinescens* and *Acremonium roseogriseum* [53]. Many of these species produce highly toxic substances, and some of the species e.g., *A. flavus, A. versicolor, A. parasiticus*, etc. produce aflatoxins and sterigmatocystin, which are most potent carcinogens, thus causing a serious concern for health of the consumers [46].

## 7.4 Bottled Beverages

The spoilage of non-alcoholic bottled beverages depends upon their composition. Yeasts like *Torulopsis* and *Candida* are the most likely causes of spoilage of soft drinks because of acid and sugar contents. Cloudiness and ropiness are main types of spoilage of soft drinks. Cloudiness may result from marked growth of various yeasts and bacteria, and ropiness are the consequence of the development of capsulated bacteria e.g., *Bacillus*. Bacteria entering through ingredients, bottles or closures may be *Gluconobacter, Lactobacillus* or *Leuconostoc*. *Achromobacter* sp. was also found responsible for musty odour and taste in root beer. Fruit concentrates may also be spoiled by yeasts.

Carbonated beverages are usually not spoiled because of inhibitory effect of $CO_2$ on microorganisms. The acidity (resulting from carbonation and addition of acids) also inhibits microorganisms. Non-acid drinks (like root beer) are better culture media for spoilage organisms than acid drinks such as cola drinks, ginger ale, and fruit flavoured drinks. Since moulds require air, they do not grow on carbonated beverages, but may develop at the surface of uncarbonated soft drinks, which contain air above liquid surface.

## 7.5 Spoilage of Canned Food Products

Like other foods, canned fruit and vegetable products are also liable for spoilage by different causes. A number of microorganisms surviving the heat process, storage temperature, suitability of the canned food to support the growth of the microorganisms, pH of the food, oxygen tension, etc. may affect the spoilage of canned foods. Broadly, there are four causes for the spoilage of canned fruit and vegetable products viz., microbiological, physical, chemical and miscellaneous type of spoilage.

### *7.5.1 Microbiological spoilage*

The spoilage caused by the growth of different kinds of microorganisms may be affected by under-processing, inadequate cooling, infection resulting from leakage and pre-processing spoilage.

**7.5.1.1 Under processing:** Insufficient or improper heat treatment may result in survival of certain microorganisms causing spoilage of food product during subsequent storage. Some lot or a part of the lot may remain under-processed or not processed at all (gross under-processing) by mistake. As a general rule, only one type of microorganism (bacteria, yeast or mould) may be involved in such spoilage. However under severe under-processing, other kinds of microorganisms may also be present. Most of these are facultative anaerobes, usually heat resistant spore formers. Faulty operation during processing may also cause under-processing. For example, retort may not be showing true pressure (and hence temperature inside) if air/steam is not expelled initially, or heat transfer within the cans in the retort (due to over-stacking) may not be proper. Diagnosis of under processing or leakage can be made based on certain characteristics listed in Table 12. This type of spoilage can be avoided by ensuring proper heat transfer, retorting and stacking of cans in the retort.

**7.5.1.2 Inadequate cooling:** Immediately after processing, the cans are to be cooled using cold water. It gives a sort of shock to the surviving microorganisms and kills them. It also checks overcooking and hence saves the food from textural disintegration. If cooling is not proper, spoilage can occur. In addition, the cans are cooled down upto 35-38°C only to allow the water to evaporate readily from the can surface. Fans can also be used for this purpose. It will check the rusting of cans during storage, which may otherwise lead to some spoilage. We can avoid such types of spoilages by proper cooling after processing.

**7.5.1.3 Infection due to leakage:** Post-processing contamination may take place if there is leakage in the cans due to faulty seam, faulty lock seam or pinholes as a result of corrosion from inside of the can or rusting of can from outside [55, 56]. In this type of spoilage, all types of microorganisms can be present in the food. A lot of oxygen can enter the can. Such type of cans should be rejected from consumption point of view. To avoid such kinds of spoilage, seam tests should be carried out regularly while in operation, hygiene should be maintained and handling should be proper. The cooling water should be chlorinated. Usually, 5-7 ppm of chlorine is employed, particularly against slime bacteria [5].

**7.5.1.4 Pre-process spoilage:** If the raw material is already highly infected with microorganisms, many microorganisms may survive heat treatment and also the finished product may not be of desirable flavour or quality. It may also be due to faulty procedure during washing (of raw materials, cans and

**Table 12:** Comparison of characteristics of leaker spoilage vs. those of insufficient heat treatment.

| | Characteristics | |
|---|---|---|
| **Criterion** | **Leakage** | **Insufficient heat treatment** |
| Container condition | Usually swelled container quality excessive, can abuse may be involved | Flat or swelled, container quality rarely involved |
| Spoilage level | Low, usually not more than 1 or 2% even with significant container defects | Usually quite highly i.e., 10-20% sometimes almost 100% |
| Spoilage pattern | Randomly scattered through various production codes | Often sporadic and limited to certain production codes |
| Product odour | Sour or faecal | Normal, sour, putrid |
| Product pH | Usually varies widely | Relatively small variations |
| Direct smears | Mixed flora of non-spore forming rods and cocci | Only rods, usually 0.5-1.0 wide and 2.4-4.5 long, possibly clostridial cells with spores |

*Source:* 54

equipment), blanching and filling of cans. Many respiratory gases can develop causing swelling of cans during storage. To avoid such type of spoilage, proper testing and washing of raw material as well as equipment, chlorination of water and proper sterilization of cans should be practiced.

7.5.1.5 Thermophilic spore forming anaerobic spoilage: The microbiological spoilage of canned fruits and vegetables can be of different types:

7.5.1.5.1 Flat sour: This type of spoilage of canned products derives its name from the fact that the ends of the can of food remain flat but the contents become sour. This is mostly found in non-acid foods like canned vegetables by the action of microorganisms (flat sour bacteria). So it cannot be detected without opening of can and culturing the microorganisms [18, 56]. *Bacillus stearothermophilus* and *Bacillus coagulans* are the thermophilic spore-forming bacteria responsible for flat sour type of spoilage. The latter are found in tomato juices [5]. The immediate source of the flat sour bacteria is usually the plant equipment besides sugar, starch or soil.

7.5.1.5.2 TA spoilage: It is the short name for spoilage caused by thermophilic anaerobes which do not produce $H_2S$ e.g., *Clostridium thermosaccharolyticum*. It is a sugar-splitting, obligate thermophilic, spore forming anaerobe which produces $CO_2$ and $H_2$, causing the swell and even bursting of cans (Plate 4). The spoiled food usually have sour odour. The source of contamination may be the plant equipment, sugar, starch and soil like that in case of flat sours [5].

7.5.1.5.3 Sulphide or sulphur stinker spoilage: Such spoilage is caused by *Clostridium nigrificans*, mostly found in low acid foods like peas and corn. The spores of this bacterium are considerably less heat resistant, and hence their appearance in canned foods is indicative of gross under-processing. $H_2S$ can be detected by its characteristic odour on opening the can and the organism can also be detected in the form of black (FeS) colonies formed on iron sulphite agar at 55°C. In case of peas, no marked discoloration may be detected. The source of this organism includes sugar, starch, soil, manure and the plant equipment [5].

**Plate 4:** TA spoilage of canned pineapple.

7.5.1.5.4 Mesophilic spoilages: Some species of *Bacillus* and *Clostridium*, non-spore forming bacteria and even yeasts or moulds may spoil the under-processed canned foods. The spore forming mesophilic thermoduric species of bacteria include *C. pasteurianum, C. butyricum, C. botulinum, C. sporogenes, C. putrefaciens, B. sublitis, B. mesentericus, B. polymyxa, B. macerans,* etc.

The non-spore forming mesophilic bacteria involved in the spoilage of tomato products, pears and some other fruits were found to be *Lactobacillus* and *Leuconostoc*. Other such genera may be *Pseudomonas, Micrococcus, Flavobacterium,* etc., which may come from water and leaks in the cans [5].

7.5.1.5.5 Spoilage by yeasts: The yeasts have been found to spoil canned fruits, jams, jellies, fruit juices, etc. under leakage or under-processing conditions. Film yeasts like *Candida, Pichia, Hansenula* can grow on acid products like sauerkraut and pickles; osmophilic yeasts like *Saccharomyces rouxii, S. mellis* can spoil dry fruits, concentrated fruit juices and honey, etc; salt tolerant yeasts like *Torulopsis* and *Brettanomyces* can grow in brine solution.

7.5.1.5.6 Spoilage by moulds: The moulds are the common spoilage organisms of home canned foods like jams, jellies, marmalades. The common ones are *Aspergillus, Penicillium, S. fulva,* etc.

7.5.1.5.7 Physical and chemical spoilage: Some physical deformation of the container can lead to spoilage. Faulty technique in operation, under exhausting, over-filling and panelling or buckling of the cans can cause spoilage of processed food products. Chemical spoilage includes reactions among ingredients, reactions between can and ingredients, hydrogen swell, etc. Rusting and perforation of can body and metallic contamination can also cause spoilage of processed fruit and vegetable products. After cooling the processed cans in water, if some water remains on the surface of cans, rusting can take place resulting in danger for spoilage of foods contained. Sometimes, the hygroscopic nature of the labels can also add to the rusting. To avoid rusting, the cans should not be cooled below 35°C in water, or fans should be used to evaporate water from the can body. The labels used may not be hygroscopic in nature. Cause of physical and chemical spoilage of the canned food is detailed earlier [56].

## 7.6 Spoilage of Jams, Jellies, Marmalades and Preserves

Jams and jellies have usually low water activity (0.75-0.86). In addition, due to added acid, the pH is lowered. Sometimes, preservatives (benzoic acid, sorbic acid) are also added. All these factors lower down the risk of spoilage. Also due to heating, spoilage organisms (yeasts, moulds) are eliminated until the package is opened whereupon recontamination could be expected [57]. Usually, *Torulopsis, Xeromyces* and many other osmophilic yeasts have been reported to spoil jams, jellies and preserves [58] (Plate 5). Among moulds, *Aspergillus, Penicillium* and *Citromyces* can grow in sugar concentrations up to 67.5% [5]. Lowering down $a_w$ from 0.94 to 0.90 combined with reduction in storage temperature from 25°C to 15°C was found effective in extending the shelf-life of such products [59]. Market samples

**Plate 5:** Spoilage of jam and jelly by osmophilic yeast.

of carrot and amla preserves were found to contain *Saccharomyces rouxii* var. *polymorphus* and *Bacillus cereus* [60].

## 7.7 Spoilage of Frozen Fruits

Many fruits and fruit products preserved by freezing include cherries, fruit juice concentrates, purees (°B>45) and some sliced fruits. Usually, dry sugar or syrup is added to fruits prior to freezing. The predominant microorganisms are usually yeasts and moulds besides lactic acid bacteria (e.g., in orange juice concentrates). However, the microbial counts in frozen products indicate poor sanitation maintenance of belts and other processing equipment, while most of the microorganisms are killed by heat treatment (in case of concentrates) or freezing process. In addition, different types of microorganisms may be picked up by food in transportation during various unit operations [47].

## 7.8 Spoilage of Dried Fruits and Vegetables

In dried fruits (apples, apricots, dates, figs, peaches, prunes, resins, etc.), a number of microorganisms can be present. In whole dried fruits, these may vary from a few hundred per gram of fruit to thousands. Due to decreased water activity (<0.65 in case of sun dried products), heat treatment during dehydration and fumigation, the microorganisms may be killed or are unable to cause spoilage. But, spores of bacteria and moulds are likely to be the most numerous. Dried fruits may be spoiled due to the development of rancidity, as concentrated flavonoids may undergo oxidation.

Dried or partially dried fruits such as dates, figs and prunes are susceptible to yeast spoilage i.e., *Zygosaccharomyces*. In dates, spoilage may occur in moisture level exceeding 23-25%. In prunes (18-20% moisture), *Monascus bisporus* has been found to be the most frequently isolated spoilage causing microorganism. Attempts to provide very tender and more palatable products have resulted in 26-28% moisture containing prunes, which are highly susceptible to spoilage and require chemical preservative. The shelf-life of such products has been extended by dipping high-moisture prunes in 2% potassium sorbate solution or 0.1% sodium benzoate [61, 62].

In case of dried vegetables, microbial counts can be few millions per gram, since microbial loads are usually higher than fruits. During drying, if trays are not loaded properly, higher microbial contamination can occur. During blanching also, if the water is not properly chlorinated, infection can occur. For chlorination, 5.7 ppm of chlorine is recommended in water [5]. In case of onion, risks are higher as these are not blanched. The main genera responsible for spoilage are *Escherichia, Bacillus, Clostridium, Micrococcus, Pseudomonas, Streptococcus, Lactobacillus, Leuconostoc,* etc.

Both dried fruits and vegetables (including low moisture foods like potato chips) are liable for damage by insects and rodents. Before packing the dried foods, if fumigation is not done properly or package itself is containing heavy microbial loads, insect-pest eggs, the microorganisms can grow (spores can grow, eggs can be hatched) during storage. Hence, fumigation after packaging may be of no use.

Another reason for the spoilage of low moisture foods (dried/powdered fruits and vegetables, chips, etc.) is the presence of oxygen, light, relative humidity and higher temperature. At very low water activity (<0.1), no microbial or non-enzymatic spoilage is there, but lipid oxidation can occur. At $a_w$ between 0.2-0.6, only non-enzymatic browning can be there. Enzymatic deterioration starts if water activity exceeds 0.3, and mould and yeast activity usually starts at $a_w$ >0.7, but bacterial growth is still not apparent, which takes place at $a_w$ >0.8. The rate of reaction also depends upon the temperature (in general, for every 10°C increase in temperature, the rate of reaction is doubled).

Dried apricots treated with $SO_2$ were stored after canning at different temperature. The samples stored at 46.1°C darkened in 3 weeks, but those stored at room temperature (21.1°C) did not darken for 3 months, while those stored at 0°C showed no darkening after 6 months [63]. Light can also cause deterioration in colour, vitamins, etc. The fragile products e.g., chips may be crumbled due to compression within the package. So, the selection of package is also important, which should be a good barrier to moisture, oxygen, and robust enough to withstand compression damages during handling, storage or transportation. However, gas packaging and aseptic packing is the alternative to avoid such damages. In selecting a package, care should be taken that it itself should not impart any odour or undesirable effects to the product.

## 7.9 Spoilage of Fermented Vegetables

### *7.9.1 Sauerkraut*

It is the fermented product (chiefly lactic) of cabbage in the presence of 2-3% salt. It may be canned at 74°C, followed by exhausting, sealing and cooling. The sauerkraut resulting from abnormal fermentation is of low quality. Soft kraut is the result of faulty fermentation and exposure to air, while black kraut formation takes place due to oxidation and combined action of plant enzymes and microorganisms. It is encouraged by uneven salting (salt burn) and high temperature. Pink kraut is caused by red yeasts, while ropy kraut may be caused by *Lactobacillus plantarum*. In fact, sauerkraut is subject to spoilage at its surface, where it is exposed to air. The film yeasts and moulds destroy the acidity and permit other microorganisms to grow and cause softening, darkening and bad flavours.

### *7.9.2 Pickles*

Cucumber and dill pickles are subjected to spoilage. Shrivelling results from too strong salt, sugar, or vinegar solution. Hollow pickles are consequences of keeping cucumber for a longer time before fermentation, loose packing in vats, too rapid fermentation or too strong or week brine. Floaters (or bloaters) may result from hollow cucumbers or from gases produced by yeasts, and hetero-fermentative lactic acid bacteria. Slippery pickles may result from broken scums of film yeasts. Soft pickles result due to pectolytic enzymes from moulds like *Penicillium, Fusarium, Cladosporium,* and *Alternaria*. Black pickles are caused due to $H_2S$ production by bacteria (Plate 6).

**Plate 6:** Spoilage of pickle by bacteria.

## 7.10 Spoilage of Wine

The wine can be spoiled due to microbial as well as non-microbial causes. The non-biological defects include those due to metals (Zn, Cu), their salts, enzymes and materials employed in the wine clarification. Tin and copper (and their salts) may cause cloudiness in wine. Gelatin used in clarification may also be responsible for cloudiness. Iron may cause precipitation (grey black, blue, white). The microbial spoilage of wine is summarized in Tables 13 and 14.

**Table 13:** Faults caused due to yeasts in wine.

| Defect | Causal organism | Prevention |
|---|---|---|
| Ester taint | *Hanseniaspora uvarum*<br>*Metschnikowia pulcherrima*<br>*Hansenula anomala* | Minimize the growth of indigenous yeasts (avoid damage to grapes)<br>Monitor clarification, inoculate juice with *S. cerevisiae* slightly higher |
| Film formation | *Candida krusei*<br>*Pichia farinosa*<br>*P. membranaefaciens*<br>*Hansenula anomala* | Good hygiene and sanitation<br>Avoid air contact with wine |
| Refermentation turbidity, sediment | *Zygosaccharomyces bailii*<br>*Z. rouxii* | Avoid pre-contamination |
| Haze, turbidity, volatile acidity, mousiness, taint | *Brettanomyces intermedius* | Avoid pre-contamination |

*Source:* 64-66

**Table 14:** Faults caused due to bacteria in wine.

| Defect | Causal organisms | Prevention |
|---|---|---|
| Vinegary taint | *Gluconobacter oxydans*<br>*Acetobacter pasteurianus* | Avoid contamination of grapes<br>Avoid exposure of wine to *Acetobacter aceti* |
| Acidification (Mannitol taint) of wine | *Lactobacillus brevis* | Lower the pH (by adding acid) |
| Ropiness | *Streptococcus, Mucilaginosus, Pediococcus cerevisiae* | -do- |
| Buttery/whey flavour | *Leuconostoc oenos* | Avoid contamination<br>Referment with *Saccharomyces diacetyl* if there is not severe spoilage |
| Bitterness (acrolein formation) | *Pediococcus parvulus, Lactobacillus cellobiose, Clostridium* | Avoid long term storage in acid solution |
| Tartaric acid degradation | *Lactobacillus brevis, L. plantarum* | Avoid contamination |
| Geranium off-odour | *Lactobacillus, Leuconostoc, Pediococcus* | Do not use base wines containing sorbic acid for sparkling wine production |
| Butyric acid taint | *Clostridium butyricum* | Do not deacidify the juices heavily |
| Volatile acidity | *Bacillus coagulans, B. subtilis, B. circulans, B. latics* | -do- |
| Mousiness | *Lactobacillus hilgardii, L. brevis, L. cellobiose* | Increase acidity of wine |

*Source:* 64, 66

## 7.11 Spoilage of Vinegar

Like wines, metals and their salts can cause cloudiness and discoloration of vinegar. Iron may react with tannins and proteins to produce a haze or darkening of vinegar. Among the microorganisms, *Leuconostoc* and *Lactobacillus* already present in fruit juices may cause vinegar defects. Vinegar bacteria like *Acetobacter aceti* subsp. *xylinum* can cause sliminess. Many of these species

**Table 15:** Aciduric and heat resistant microorganisms isolated from apple juice and cider process.

| Group | Number of isolates |
|---|---|
| Bacteria | |
| *Bacillus brevis* | 1 |
| *B. coagulans* | 2 |
| *Gluconobacter lik* | 3 |
| Moulds | |
| *Aspergillus* spp. | 6 |
| *Byssochlamys* | 7 |
| *Penicillium* spp. | 5 |
| Unidentified | 3 |
| Yeasts | |
| *Aureobasidium* | 4 |
| *Candida krusei* | 1 |
| *Cryptococcus albidus* | 1 |
| *Pichia vini* | 1 |
| *Rhodotorula rubra* | 1 |
| *Saccharomyces cerevisiae* | 2 |
| *S. chevalieri* | 2 |
| *S. rosei* | 1 |
| *Torulopsis glabrata* | 1 |
| Unidentified | 2 |

*Source:* 52

produce highly toxic substances, and some of the species e.g., *A. flavus, A. versicolor* and *A. parasiticus,* etc. produce aflatoxins and sterigmatocystin, which are most potent carcinogens, thus causing a serious concern for health of the consumers [46].

The presence of acidiuric and heat resistant microorganisms in samples of apple juice and cider has been shown in Table 15. The yeast growth is favoured more in fruit juices because: (i) increase of oxidation-reduction potential by removal of solids, (ii) temperature of incubation is generally 15-30°C, and (iii) deficiency of vitamin B in juices favours yeast than bacteria. Vegetable juices have more sugar, but acidic contents are low, their pH range from 5-5.8, and they have high contents of accessory growth stimulatory substances. Therefore, bacterial spoilage is more common in them as compared to yeasts and moulds.

## 7.12 Spoilage of Animal Originated Foods

Since fresh animal products are perishable, they are chilled and stored in ice or a refrigerator (0 to 4°C). This means that psychrotrophs become dominant and are the primary cause of spoilage. If these fresh products are mishandled and allowed to remain at room temperature (20°C), a more diverse flora may be present; not only many psychrotrophs will grow, but there will also be some mesophillic strains.

The organisms most often involved in spoilage of refrigerated fresh meat, poultry, fish and eggs are species of the genus *Pseudomonas. Pseudomonas putrefaciens* has been designated as being important in spoilage of meat and poultry. This organism does not confirm to the given definition of *Pseudomonas,* so it was assigned the name *Alteromonas putrefaciens.* The DNA base composition was somewhat higher than the range for *Alteromonas*, so the exact nomenclature was delayed [67].

### 7.12.1 Meat and meat products

In general, the muscle tissue of live, healthy animals is sterile. The organisms that are present in the carcass are concentrated in the lymph nodes. Researchers have isolated organisms from 58% of the 360 nodes tested [68]. Contaminated nodes are thought to be a source of bacteria for deep spoilage of post rigor meat. After slaughter, the defense mechanisms are slowed down or halted. This enables bacteria to multiply and spread throughout the tissues. The most common indications of spoilage of meat and meat products are: (1) off-odour and slime, usually due to the growth of aerobic bacteria on the cut surfaces of meat; (2) fungal growth, which is favoured at too low water activity for the growth of bacteria; (3) bone taint, or deep spoilage, due to anaerobic or facultative microorganisms; (4) discolouration, primarily due to alterations of myoglobin, the muscles pigment. There are some differences in the spoilage patterns of fresh and cured meat, but there is only limited information on cooked meat spoilage. Generally, cooked meat is spoiled by few organisms that can survive the cooking or those that gain access to the cooked product. Fresh, chilled meat spoilage is evidenced by off-odour and slime due to *Pseudomonas, Acinetobacter* and *Alcaligenes.* Cured meats become sour due to the activity of *Micrococcus, Lactobacillus* and *Microbacterium* species. Microbial defects of meat and meat products are listed in Table 16.

**Table 16:** Microbial defects associated with red meat and their products.

| Product | Defect | Organisms |
|---|---|---|
| Fresh, refrigerated (0-5°C) meat | Off-odour, slime, discoloration, lipolysis, pungent odour | *Pseudomonas, Aeromonas, Alcaligenes, Acinetobacter, Microbacterium, Moraxella, Proteus, Flavobacterium, Alteromonas, Saccharomyces* yeasts |
| | Mouldy | *Penicillium* |
| | Whiskers | *Thamnidium* |
| | Black spot | *Cladosporium* |
| | White spot | *Sporotrichum* |
| Vacuum packaged meat | Acid, sweet, rancid | *Lactobacillus, Microbacterium, Enterobacter, Hafnia* |
| Cured meat | | |
| Bacon | Cheesy, sour, rancid | *Micrococcus* |
| | Discolouration | Moulds |
| | Slight souring | *Lactobacillus, Micrococcus, Vibrio, Alcaligenes, Corynebacterium* |
| | Putrefaction | *Clostridium sporogenes* |
| Vacuum packaged meat | Cabbage odour | *Proteus inconstans* |
| | Tainted | *Vibrio* |
| Ham | Surface slime | *Micrococcus, Microbacterium,* Yeasts |
| | Gassy or puffy | *Clostridium* |
| | Green discolouration | *Lactobacillus, Streptococcus, Leuconostoc* |
| | Bone and meat "sour" | *Clostridium* |
| | Surface growth | Moulds |
| Sausages | Slime on surface | *Micrococcus,* Yeasts |
| | Gas production (vacuum packaged) | *Lactobacillus* |
| | Greenish discolouration | *Lactobacillus viridescens, Leuconostoc* |
| Fermented sausage | Slime | Yeasts |
| Canned meat | | |
| Commercially sterile meat | Gas putrefaction | *Bacillus, Clostridium* |
| Semi-preserved meat | Souring, discolouration | *Streptococcus* |
| | Putrefaction, gas | *Bacillus, Clostridium* |

**7.12.1.1 Fresh meat:** The rate of spoilage depends upon the number and types of organisms initially present, the conditions of storage (temperature and gaseous atmosphere), and characteristics (pH, $a_w$) of the meat. In general, the whole carcass presents a surface of fat and connective tissue affording little opportunity for bacterial growth. The surfaces of meat cuts can support the growth of large number of bacteria, and ground meat offers not only ample and desirable surfaces, but also a thorough inoculation of the meat grinding.

To retard microbial growth and biochemical changes, fresh meat is stored at temperatures near 0°C. However at times, the meat is or may be subjected to higher temperatures. The temperature of the animal is approximately 37°C. After slaughter, the temperature tends to increase and then change to that of the room in which the meat is held. During subsequent processing of the carcass, mishandling may occur and the temperature may rise. Furthermore, if a consumer purchases the meat and simply places it in a trunk on a hot sunny day, the temperature of meat may be raised to a dangerously high level, thus causing spoilage. At room temperature (20°C), the dominant organisms causing the spoilage of sheep meat are *E. coli* and *Acinetobacter*-like organisms [69]. Rao and Sreenivasamurthy [69] stated that *Enterobacter, Pseudomonas,* and *Staphylococcus* could also form a major part of this flora. To reduce the biochemical and microbial changes, fresh meat is stored at temperatures near 0°C. At low temperature, spoilage is evidenced on the surface. In storage conditions in which a moist meat surface is maintained, spoilage is due to Gram negative bacteria, primarily species of *Pseudomonas.* Species of *Aeromonas, Alcaligenes, Acinetobacter, Moraxella, Flavobacterium, Enterobacter, Microbacterium, Proteus* and *Streptococcus* occasionally are found on the surface flora of the spoiled meat.

The defects of the meat are off-odor, slime formation and discolouration. The flavour would be unacceptable to most people. In one study, proteolysis could not be detected prior to the appearance of spoilage odour and slime [70]. The important biochemical changes occur in the meat juices that contain free amino acids, nucleotides and peptides. These nutrients are sufficient for microbial growth, and the metabolism of these compounds leads to the formation of $H_2S$, $NH_3$, indole and various amines and diamines. The pigment stability in meat involves factors such as oxygen penetration, bacterial growth, lipolysis and the presence of flavin compounds. Exposure to light can enhance the destruction of oxymyoglobin on the surface of beef. The $H_2S$ produced during breakdown of the sulphur containing amino acids reacts with myoglobin at low oxygen levels, forming the green sulphmyoglobin.

Water activity can play a role in spoilage. Fresh meat will lose moisture if stored below 99% RH. Even at a RH of 99%, the growth of *Pseudomonas* is reduced. At an $a_w$ of 0.96 or less, most of the usual microorganisms causing spoilage of fresh meat are inhibited. When the surface of the meat contains $a_w$ of lower than 0.96, the slower growing fungi become evident. *Thamnidium chaetocladioides* and *T. elegans* (cause of whiskers), *Cladosporium herbarum* (black spot) and *Sporotrichum carnis* (black spot) growth produce mouldy meat.

Storage atmosphere can also influence meat spoilage. When the meat is vacuum packaged, the dominant flora tends to be *Lactobacillus* instead of *Pseudomonas.* Other microorganisms include *Leuconostoc, Bronchothrix, Enterobacter, Hafnia* and *Alteromonas.* Odours of stored, vacuum packaged meat have been described as acidic, sour, sweet, cheesy, stale or rancid.

**7.12.1.2 Cured meat:** Meat that is treated with salts such as NaCl, $NaNO_2$ and $NaNO_3$ differs from untreated fresh meat. The most apparent differences are in the colour and flavour of fresh and cured meat. The $a_w$ of cured meat (0.88-0.95) is lower than that of fresh meat (0.98-1.00), and the specific inhibitory effects of the added salts may account for the differences in the spoilage patterns of fresh and cured meats. Changes in colour during processing and handling of the fresh and cured meats are illustrated in Fig. 3.

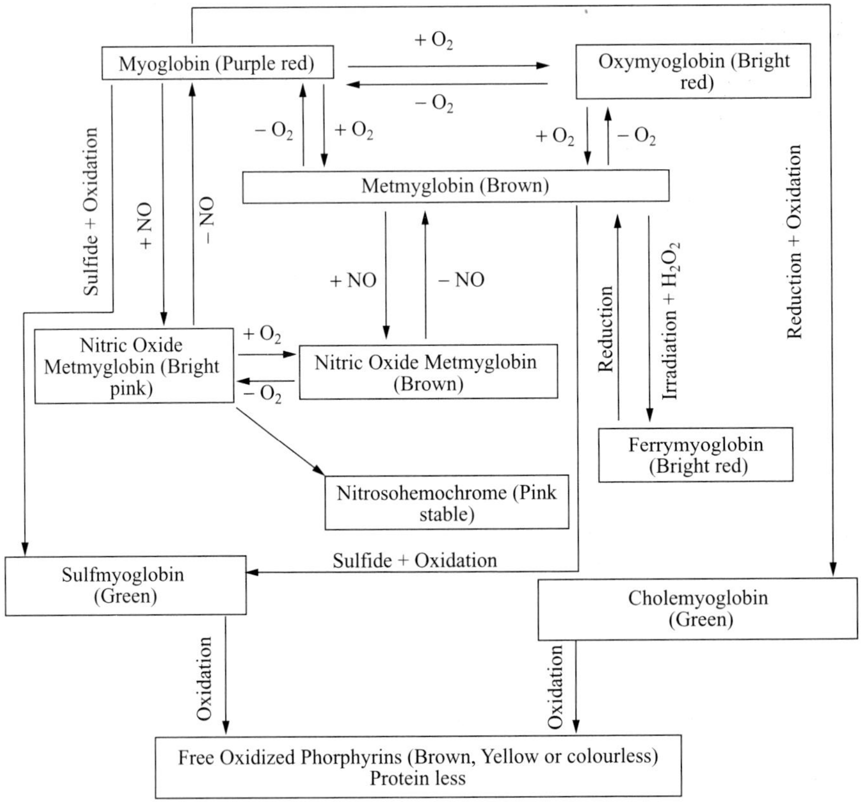


**Fig. 3:** Changes in pigments during processing and handling of fresh and cured meats.

## 7.13 Spoilage of Milk and Its Products

Milk obtained from the animal might not be sterile. During and after milking, the milk is subjected to organisms from various sources. The contaminated equipment used to handle, transport, store and process the milk seems to be the main source of microorganisms. Cold stored milk is usually spoiled by *Pseudomonas, Acinetobacter, Aerobacter, Alcaligenes* and *Flavobacterium.* Milk products during storage at refrigerated temperature are spoiled by the psychrophilic or psychrotrophic organisms [71]. The defects that can occur in milk due to microbial growth are:

- Off-flavour
- Lipolysis with development of rancidity
- Gas production
- Fermentation to lactic acid with souring
- Coagulation of milk proteins
- Viscous or ropy texture
- Discolouration

Some of the defects of the milk and milk products are listed in Table 17. Species of *Pseudomonas* are prominent in producing defects in the milk. They are associated with a bitter flavour and rancidity as well as a fruity odour [71]. The fruity odour is due to a mixture of ethyl butyrate and ethyl hexanoate.

**Table 17:** Microbial defects associated with milk and milk products.

| Product | Defect | Microorganisms |
|---|---|---|
| Pasteurized, refrigerated milk | Rancid | *Pseudomonas, Alcaligenes, Staphylococcus* |
| | Sour (acid) | Lactic acid bacteria |
| | Discolouration | *Chromobacterium* |
| | Bitter, fruity | *Pseudomonas, Alcaligenes, Flavobacterium, Proteus, Acinetobacter* |
| Canned milk | Swelling, gas | *Clostridium sporogenes* |
| Cream | Foamy | *Candida, Torulopsis* |
| Butter | Surface taint | *Pseudomonas, Alteromonas* |
| Cheese | Gassy, butyric acid, floating or split curd | *Clostridium tyrobutyricum, Leuconostoc, S. lactis* subsp. *diacetylactis* |
| Soft cheese | Black mould (cat's fur) | *Mucor* |
| | Surface growth | *Torulopsis, Debaryomyces* |
| Cottage cheese | Slimy curd, putrid odour | *Pseudomonas* |
| | Discolouration | *Flavobacterium,* yeasts, moulds |
| | Slimy, gelatinous | *Pseudomonas, Alcaligenes, Flavobacterium,* coliforms |
| | Fruity | Yeasts |
| Cheddar cheese | Sweet, yeasty, fruity | Yeasts |
| Swiss cheese | Gassy, sweet | Yeasts (*Torulopsis*) |
| | Off-odour | *C. sporogenes* |
| Yoghurt | Yeasty | Yeasts *(Torulopsis)* |

## 7.14 Spoilage of Poultry and Poultry Products

The spoilage of poultry is somewhat different from the spoilage of canned meat. Immediately after processing, any of the several hundred species of microorganisms might be found. However, as the poultry is chilled and held in cold storage, psychrotrophic microorganisms predominate and cause deterioration [72]. The main defects are off-odour and slime formation. Principal off-flavour causing microorganisms in the poultry meat are species of *Pseudomonas* (Table 18). Other microorganisms causing spoilage are *Aeromonas, Moraxella, Alcaligenes, Flavobacterium* and *Micrococcus.* When antibiotics are used in poultry, yeasts become the important spoilage microorganisms.

However, yeasts are not important on normally processed and eviscerated poultry. In general, it is accepted that when an egg (Fig. 4) is laid, its contents are free from bacteria though there may be exceptions because of ovarian infections. The eggs are protected by the shell and associated membranes and chemical inhibitors in the egg albumen. This means that microbes must penetrate these barriers and then be able to grow inside the egg to cause spoilage. The penetration inside the egg is aided by moisture on the shell. So if the eggs are not properly stored or washed, penetration may be quite rapid and spoilage can occur. The main defects of the eggs are black rot, white rot; sour, green white, musty, mouldy, red rot, custard rot; yellow and green rot, etc. and the eggs become inedible as reported earlier [73]. The defects along with the prevalent microbes causing their spoilage are summarized in Table 18.

**Table 18:** Microbial defects of poultry and poultry products.

| Product | Defect | Microorganisms |
|---|---|---|
| Poultry meat | Off-odour, slime | *Pseudomonas, Aeromonas, Alcaligenes* |
| Shell eggs | Black rot | *Proteus, Aeromonas* |
| | White rot (colourless) | *Citrobacter, Alcaligenes* |
| | Sour | *Pseudomonas* |
| | Green white | *P. fluorescens* |
| | Musty | *Pseudomonas* |
| | Mouldy | Many types of moulds |
| | Red rot | *Serratia marcescens* |
| | Yellow and green rot | *Alcaligenes, Flavobacterium* |
| Liquid whole egg | Fishy | *Pseudomonas, Flavobacterium, Chromobacterium* |
| | Off-odour, sour | *Proteus, Alcaligenes, Escherichia, Bacillus* |
| Liquid albumen | Off-odour | *Pseudomonas, Acinetobacter, Enterobacter* |

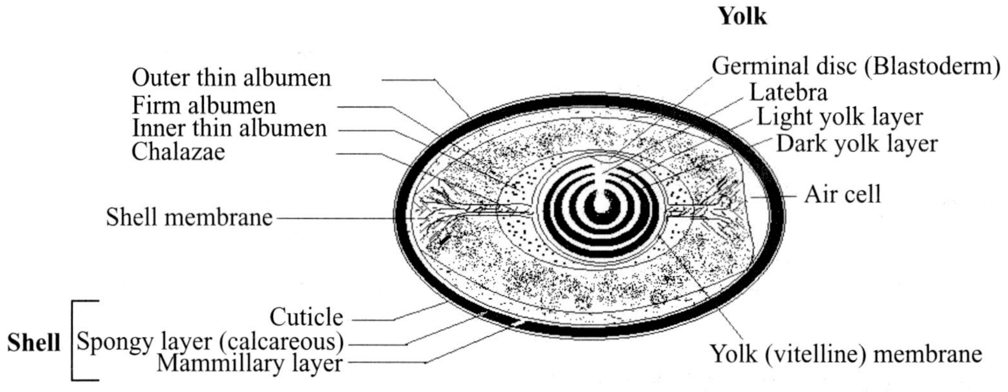


**Fig. 4:** Structure of the hen's egg.

## 7.15 Spoilage of Sea Foods

Fish and other sea foods are subject to contamination by microorganisms in their marine environment, as well as those acquired during catching, handling and processing. Certain types of organisms become dominant in the seafood products during their storage. The dominant types of microorganisms are those that can survive and multiply on or in the food at the prevailing temperatures. The main defects of the seafoods and the defect causing microorganisms are listed in Table 19. The main defects of sea foods are development of off-odour, fruity, ammoniacal and $H_2S$ odour, pink, yellow and red discolouration.

## 7.16 Spoilage of Miscellaneous Foods

Different products ranging from mushrooms to soya-protein extender have been found to be spoiled by microorganisms for which several reports have appeared. These are summarized in Table 20.

**Table 19:** Main microbial defects of sea foods.

| Product | Defect | Microorganisms |
|---|---|---|
| *Fish* | | |
| Fresh | Off-odour | *Pseudomonas, Alteromonas, Acinetobacter* |
| | Fruity odour | *Pseudomonas* |
| | Ammoniacal odour | *Pseudomonas, Alteromonas* |
| | $H_2S$ odour | *Pseudomonas, Alteromonas* |
| Salted | Pink | *Halobacterium, Halococcus* |
| | Dun (red growth) | *Hemispora stellata, Sporendonema epizoum* |
| | Cheesy, putrefactive | Red halophilic bacteria |
| *Oysters* | Pink | *Rhodotorula* |
| *Shrimp* | Off-odour | *Pseudomonas* |
| *Squid* | Yellow discolouration | *P. putida* |
| | Red discolouration | *Serratia marcescens* |

**Table 20:** Microbiological quality reports of miscellaneous foods.

| Commodity | Comments |
|---|---|
| Mushroom fresh | Growth of *Clostridium* |
| Breaded onion rings | Microbiological quality |
| Vegetable salads | Gram negative isolates |
| Salads (commercial) | Microbiological quality |
| Coconut products | Microbial profile |
| Dry soup and sauces | *Clostridium perfringens* |
| Soy protein extender | Microbiological quality |

*Source:* 74-81

### *7.16.1 Salad dressings*

Salad dressings being acidic (pH 3-4) are more prone to spoilage by yeasts or moulds than bacteria. These also contain oils which may be oxidized or hydrolysed. Pickles, sugars, starch, gums, gelatin, spices and other ingredients may add appreciable number of microorganisms. Species of *Zygosaccharomyces* and *Saccharomyces* can spoil salad dressings. *Bacillus* and *Lactobacillus* can also cause spoilage of such products.

### *7.16.2 Fats and oils*

The term 'rancidity' is used to denote spoilage of fatty parts of the foods. It refers to spoilage due to oxidation, chemical or microbial changes, bringing in undesirable flavours. Hydrolytic rancidity is caused by the presence of moisture and is brought about by lipases resulting in the formation of off-flavour ketones and aldehydes. Ketonic rancidity is due to the action of microorganisms (moulds) on the lipid surface causing β-oxidation of lipids. Oxidative rancidity occurs in unsaturated fats (at unsaturation point) as oxidation results in the formation of ketones and aldehydes which give off-flavour [63]. Major bacterial genera causing such spoilages include *Pseudomonas, Micrococcus, Bacillus, Serratia, Achromobacter* and *Proteus,* while the yeasts include film forming yeasts like lipolytic and the moulds include *Geotrichum, Penicillium, Aspergillus, Cladosporium* and *Monilla.*

### *7.16.3 Spices and other condiments*

Usually, the dry spices are not subjected to spoilage. However, heavy loads of mould spores during drying can lead to spoilage during storage. Chip dips flavoured with vegetables or spices usually have higher total coliform and mould counts than those flavoured with cheese. Prepared mustard may be spoiled by yeasts and species of *Proteus* and *Bacillus*. Horseradish is seldom spoiled due to its bactericidal/bacteriostatic properties.

### *7.16.4 Fruit syrups*

Fruit and chocolate syrups may also be spoiled by yeast growth inspite of chemical preservatives (sorbic or benzoic acid) used. Spoilage of canned fruit syrup by *S. bailli,* was reported, which was containing >300 ppm benzoic acid, $a_w$ of 0.88 with a pH of 2.8 [82].

### *7.16.5 Spoilage of honey*

Since honey contains 70-80% sugars (glucose, levulose, etc.) and has high acidity (pH 3.2-4.2), the chief cause of its spoilage is osmophilic yeasts. The standard plate counts were found to be highest in unheated fresh honey and lowest in honey subjected to heating and stored at 5°C. Both temperature and period of storage affected the standard plate counts of honey [83], mostly by species of *Zygosaccharomyces* (*Z. mellis, Z. nussbaumeri, Z. richteri*). Most of the moulds are not able to grow on honey except *Penicillium* and *Mucor* [5]. It is a well known fact that sugar at higher concentrations (>65%) acts as a preservative. So, special theories have been advanced to explain the initiation of yeast growth in honey. Being hygroscopic, honey becomes diluted at the surface where yeasts begin to multiply and become adapted to the high sugar concentrations. Glucose hydrate on crystallization results in lowered sugar content in the solution. Yeasts become adapted to the higher sugar concentration gradually on keeping the honey for longer periods. The critical moisture content for yeast growth initiation has been found to be 21%. The fermentation process lasting for months results in the production of $CO_2$, alcohol and non-volatile acids, which give off-flavour to the honey. Crystallization as well as darkening may also take place with the fermentation process.

## 8. PREVENTIVE MEASURES/CONTROL OF BIO-DETERIORATION

Fresh fruits, vegetables, meat, milk, eggs and sea foods are spoiled by biological (microbiological) and non-biological causes. So the spoilage of these food products is checked by adopting suitable preservative techniques. But the processed products do not normally spoil unless the preservative technique applied is not proper or is applied in faulty manner or product is not stored properly. As there are various causes of postharvest losses of food, nevertheless, it is necessary to identify the spoilage vector or vectors on a case-by-case basis, in order to select the preservation technology appropriate for each situation. There are few generalized preventive measures which can control biodeterioration of foods [84]. Table 21 summarizes the major preventive measures for reduction of postharvest losses in different foods to reduce spoilage. Bio-deterioration of foods can be prevented by different food preservation method as discussed in latter section of the chapter.

**Table 21:** Major preventive measure for reduction of postharvest losses in different foods.

| Commodity | Preventive measures |
|---|---|
| Grains, legume seeds | Adequate drying, prompt drying and storage at dry place |
| | Good sanitation, use of contact insecticides or fumigants, air tight storage |
| | Good house keeping, rodent proof stores, baits, traps, fumigation |
| Fruits and vegetables | Gentle harvesting and handling, protective packaging |
| | Good sanitation, cool storage, use of fungicides, maintain intact skin |
| | Cool or CA storage, prompt marketing, processing |
| | Maintenance of high humidity surroundings |
| Roots and tubers | Maintenance of intact skin, promotion of suberization and wound cork formation, good sanitation, cold storage, application of fungicides |
| | Cold storage, anti-sprout agents |
| | Gentle harvesting and handling, protective packaging |
| | Cool or CA storage, prompt marketing, processing into stable forms |
| | Good sanitation, use of insecticides |
| Fresh fish, meat, poultry, milk, juice | Refrigeration, good sanitation, pasteurization, processing into stable forms |
| Dry products | Good sanitation, use of insecticides |
| | Adequate drying, prompt drying, storage at dry places |

## 9. FOOD PRESERVATION

In accomplishing the preservation of food by various methods [5, 6], three main principles are involved: (a) delay or prevention of microbial decomposition of food, (b) delay or prevention of self decomposition of food, (c) prevention of damage to food caused by insects, rodents, birds and mechanical causes. The principles and methods involved in preservation of the food are discussed as under:

### 9.1 Prevention of Microbial Decomposition

Most of the methods of food preservation depend not only on the destruction or removal of microorganisms but also on the delay in the initiation of their growth, and hindrance to growth once it has begun. Knowledge of growth curve of microorganisms (as discussed earlier) is very helpful for developing the appropriate preservation method to delay the microbial decomposition of the food. Preservation methods for delay or prevention of microbial decomposition of food are discussed here in detail.

#### *9.1.1 Keeping out microorganisms (Asepsis)*

There are so many examples of asepsis in the nature as a preservative factor. The inner tissues of healthy plant parts usually are free from microorganisms. The microbial decomposition is delayed or prevented if there is a protective covering about the food. Such coverings are the skins, membranes and shells of nuts, etc. When the protective covering is damaged and decomposition has spread from the outer surface, then the inner tissues are subjected to decomposition by microorganisms. Much attention is being given to the prevention of contamination of foods from the raw material to the finished product in the food industry. The food technologist is concerned with the kind of organism and their load on or in a food. The kinds of microorganisms are equally important, may be spoilage causing, pathogenic organisms or may be desirable organisms in food fermentation. The numbers of microorganisms also play an important role. More the spoilage organisms, more likely the food spoilage will be, and more difficult would be to preserve the food. The microbial load in the food may be due to the multiplication of organisms, by their contamination or both. In many kinds of food, the quality is judged partly by the numbers of microorganisms present. There are various examples of aseptic method of preservation in food industries. Packaging of foods is a widely used application of asepsis. Packaging may be like a loose carton or wrapping, which prevents microbial contamination during handling, whereas hermetically sealed cans of canned foods protect the processed products from contamination by microorganisms.

The load of microorganisms in the canning industry determines the heat process necessary for the preservation of a food, especially if the contamination introduces heat resistant spoilage microorganisms, such as spore forming thermophiles that may come from equipment, and the sealed can prevents recontamination after the heat treatment. Now-a-days, the tetra packs are being used for packing various brand of mango and other fruit drinks in the market. This is aseptic packaging where the pre-sterilized ready-to-serve drink is aseptically filled in the aseptic packaging i.e., microorganisms are kept out. In case of large packaging like bag-in-box, pre-sterilized tomato paste is packed in presterilized aseptic bag in aseptic conditions.

### *9.1.2 Removal of microorganisms*

Under special conditions, removal of microorganisms in food preservation may be helpful, but for the most part it is not very effective. Removal of microorganisms may be carried out by washing, trimming, peeling, filtration and centrifugation (sedimentation).

**9.1.2.1 Washing:** Washing of raw foods with potable water can be helpful in their preservation, but may be harmful under some conditions. Washing of cucumbers or cabbage heads removes most of the soil microorganisms from the surface before their fermentation into pickles and sauerkraut, respectively. Washing of fresh fruits and vegetables removes those soil organisms, which might be resistant to the heat processing during the canning of these foods. Germicidal treatment of the apparatus and equipment is an essential and effective procedure during the handling of all kinds of foods. It is an effective means of removal of microorganisms. Washing of foods may be dangerous if the water adds spoilage organisms to the raw food.

**9.1.2.2 Trimming:** Discarding spoiled samples or trimming away the spoiled portion of a food is essential from the effective food preservation point of view. Although a large numbers of spoilage causing organisms are removed in this way, even then contamination of the remaining food may take place if the food is not handled properly. Trimming the outer leaves of cabbage heads has been considered essential for the manufacture of sauerkraut.

**9.1.2.3 Peeling:** Peeling operation involved in many of the fruit and vegetable products leads to the removal of most of the microorganisms from the surface of the raw fruits and vegetables. Microorganisms remained after washing are removed during peeling. Lye peeling even kills most of the microorganisms.

**9.1.2.4 Filtration:** In case of clear liquids, filtration is the only successful method for the complete removal of organisms. The liquid is filtered through a pre-sterilized filter made of asbestos pads, diatomaceous earth, sintered glass, unglazed porcelain, membrane pads, and the liquid is forced through by positive or negative pressure. This method has been used successfully in soft drinks, beer, wine and fruit juices. Filtration of juice through diatomaceous earth may reduce microbial population by two folds [52, 85], while ultrafiltration removes all yeasts, moulds and bacteria [86]. If the process is properly controlled after filtration, ultrafiltration serves as a mean of cold sterilization of apple juice.

**9.1.2.5 Centrifugation or sedimentation:** Centrifugation or sedimentation is not very effective means of removal of microorganisms. However, some of the microorganisms are removed but not all. Sedimentation is used in the treatment of drinking water but is insufficient by itself. Centrifugation of chilled and heated apple juice may reduce microbial populations by more than 99.8% [87]. Enzymatic clarification of juice may reduce microbial populations [52] and is reported to improve the keeping quality of the product [88].

### *9.1.3 Hindering the growth and activity of microorganisms*

Microbial decomposition can be delayed by hindering the growth and activity of microorganisms which can be done by changing the environmental conditions unfavourable to them. Various methods of preservation are based on this principle, which is given below:

**9.1.3.1 Low temperature preservation:** Low temperature is used to slow down or stop the growth and activity of microorganisms in the food. However, it also retards the chemical reactions and action of food enzymes. The microbial growth will also be the slowed down as the temperature is lowered. Any raw material is considered to be containing a variety of mould, yeast and bacteria, which need only suitable conditions for their growth to bring about undesirable changes in the food. Each microorganism has an optimum or best temperature for growth and a minimum temperature below which it cannot multiply. As the temperature drops from optimum towards the minimum, the growth rate of the organism decreases and is slowest at the minimum temperature. But its slow metabolic activity may continue. So, the cooling down of food from ordinary temperature has a varying effect on the various microorganisms present. A drop of 10ºC might stop the growth of some organisms and slow down the growth of others but to an extent that would vary with the kind of organism. A further decrease of 10ºC temperature would stop the growth of more organisms and still reduce the growth of others. The low temperature storage, can thus, act as a significant environmental factor influencing the type of spoilage organism. After washing of chicken and meat, they are rapidly chilled from 32 to 2ºC to prevent bacterial spoilage and to preserve quality [89, 90].

The growth and metabolic reactions of microorganisms are directly affected by temperature, which depends upon their enzymes and the rate of enzyme reactions. So when the temperature is lower, there is a decrease in the rate of growth of a microorganism besides reduction of rate of reactions. Freezing prevents the growth of most food-borne microorganisms, and refrigeration temperature slows down the growth rate. Temperatures of 5 to 6ºC or less effectively retard the growth of all food poisoning organisms except *Clostridium botulinum* type E.

**9.1.3.2 Drying and dehydration:** Microbial decomposition of food can be delayed by hindering the growth and activity of microorganisms following the drying and dehydration method of food preservation, as different microorganisms require specific moisture content expressed as water activity. Fruits and vegetables contain a lot of moisture to permit action by microorganisms and by their own enzymes. Moisture removal or binding (by solutes) is necessary for their successful preservation. Moisture from the fruits and vegetables is removed by drying, but any other method that reduces the amount of available moisture (water activity) in the food is also a form of drying. One of the examples is osmotic dehydration. In sun drying, moisture from the food is removed by exposure to the sun's rays without any artificially produced heat and without controlled temperatures, relative humidity or air velocities. A dehydrated or desiccated food has been dried by artificially produced heat under controlled conditions of temperature, relative humidity and air flow [3].

The moisture from foods is primarily removed by application of heat, but new methods employing other sources of energy are also being followed like osmotic dehydration, etc. Preparation of food prior to drying greatly affects the quality of the finished products. Foods are dehydrated to protect against spoilage by microorganisms or the enzymes from plant or microbes and to reduce the costs of packaging, handling, storage and transportation.

The whites, yolks or whole eggs after pasteurization may be dried by any of the several methods including spray drying, tray drying or freeze drying (Plate 7). Egg white contains traces of glucose, which during dehydration process combine with protein resulting in Maillard browning in the egg white. This can be prevented by removing glucose through fermentation by yeasts, or with commercial enzymes [89].

### 9.1.3.3 Maintenance of anaerobic conditions/fermentation

9.1.3.3.1 Anaerobic conditions: Anaerobic conditions in the container can be created if the food is sealed and packaged properly. These conditions are brought about by the evacuation of the unfilled space (headspace in a can), or replacement of the air by $CO_2$ or an inert gas like nitrogen in the container. Spores of some of the aerobic spore formers may survive in canned foods, but they cannot germinate or grow in the absence of oxygen.

Some of the microorganisms require oxygen for their growth, like majority of moulds and salt tolerant yeasts, but most of the bacteria and yeasts can grow even in the absence of oxygen. So, simple exclusion of oxygen cannot be considered a satisfactory method of food preservation. However, this method can be used in conjunction with other methods to prevent the growth of oxygen requiring microorganisms. For instance, jams and preserves which can support the growth on the surface, are packed under vacuum to prevent growth of these spoilage organisms. Similarly, the fermented or pickled vegetables, such as cucumbers and olives are packaged in oxygen free atmosphere to avoid fungal growth on their surface.

9.1.3.3.2 Fermentation: Food fermentation like drying has been an ancient method of preservation. Fermentation using natural or starter culture has emerged as a cheap method of preservation to increase the shelf-life of the products, besides preparation of new products with diversified taste and flavour [91-93]. Fermented foods are more nutritious than unfermented ones [94]. Fermented foods prepared by using lactic acid bacteria (LAB) have a better acceptability [3]. Fermented food products have a positive effect on health of the consumers [95]. In contrast to the other techniques discussed where efforts are made to avoid, remove or kill the microorganism in preservation by fermentation, the microorganisms are encouraged to multiply and continue their metabolic activities in foods. But the organisms used are selective and their end products are known and desirable. To cite an example, alcoholic fermentation where ethanol and carbon dioxide is produced [96]. $CO_2$ production during fermentation and its accumulation at the surface will make the conditions anaerobic, thereby preventing the growth of aerobes in the food. Most vegetables can be LAB fermented, but so far only

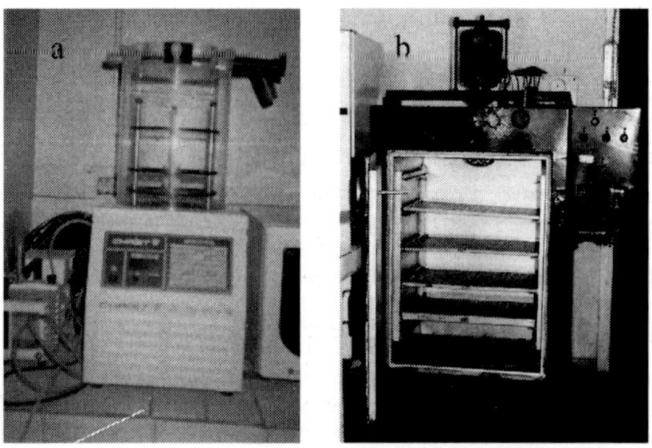


**Plate 7:** Freeze drier (a) and cabinet drier (b) used for drying of food.

beet, turnip, cauliflower, cabbage, olive and cucumber are fermented on commercial scale for human consumption [97]. The radish fermented with sequential culture of LAB, 2.5% salt concentration at 26°C had a higher sensory score.

### 9.1.3.4 Chemical preservation added or developed

9.1.3.4.1 Added preservatives: Microbial decomposition of food can also be delayed by hindering the growth and activity of microorganisms by following the chemical methods of preservation. The use of preservatives to extend the shelf-life of foods has been known for many years. They retard the growth of most of the spoilage microorganisms, but may not stop the growth completely. So, preservatives are most effective when added to the products that initially have low microbial populations, and are less effective when added to poor quality products with high microbial loads. Addition of preservative to apple products may also reduce heat resistance of yeasts [98]. One of the main purposes of the use of chemicals is to inhibit the growth and activity of microorganisms. Preservatives may inhibit microorganisms by interfering with their cell membranes, enzyme activity or genetic mechanisms. Their maximum permissible limits are given in Table 22, whereas their detailed mechanism of action is tabulated in Table 23. Preservatives may also be used to keep out microorganisms or hinder undesirable microbial spoilage by using a coating or wrappers. Besides, the growth of microorganisms can be checked by changing the surrounding environment i.e., the pH of the medium. A low pH attained with the addition of acid is widely used to preserve the foods, as in pickled or fermented foods. Most of the microorganisms grow rapidly when the pH is neutral, but the growth rate decreases as the pH decreases, until a point is reached where growth can no longer be initiated. Sodium nitrate and/or sodium nitrite, which is used to preserve meat helps to develop unique flavour, fix red colour of the cured meat and has anti-botulinum activity [89]. Factors that influence the effectiveness of chemical preservatives in killing microorganisms or inhibiting their growth and activity are as under:

- Chemical concentration
- Kind, number, age and previous history of organism
- Time
- Temperature
- The chemical and physical characteristics of the substrate in which the organism is found (pH, moisture, kinds of solutes, colloids, surface tension and other protective substances)

**Table 22:** Concentration of additives permitted to be used in specified food products under PFA Act, 1954 and Rule, 1955.

| Name | Max. limit (ppm) | ADI (mg/kg body weight) |
|---|---|---|
| Sulphur dioxide | 40-3000 | 0.0.7 |
| Benzoic acid | 50-750 | 0-5.0 |
| $NaNO_2/KNO_2$ | 200 | 0-0.2 |
| $NaNO_2/KNO_3$ | 500 | 0-5.0 |
| Sorbic acid | 1000-1500 | 0-25-0 |
| Propionic acid | 5000 | no limit |
| p-hydroxy benzoic acid (Netty ethyl cites) | not permitted | 0-10.0 |
| Nisin | 1000 | 0-33000 units/kg b.w. |

*Source:* 99

**Table 23:** Mechanism of action of food preservative and related chemicals.

| Compound | Probable action |
|---|---|
| Cationic and anionic active agents | Destruction of cell membrane denaturation of enzyme proteins |
| Phenols, chlorophenols, naphthol sulphonates, cinnamic acid | Destruction of cell membrane, reaction with protein in genetic mechanism |
| Fatty acids, alcohols and long chain aldehydes | Destruction of cell membrane, competitive inhibition of enzymes by short chain acids |
| Chloracetic acids | Membrane action, competitive inhibition |
| Benzoates, chlorobenzoate, hydroxy benzoate and its esters | Competition with co-enzymes |
| Salicylate | Membrane action, competition with coenzyme, competitive interference with enzymic utilization of amino acids |
| Borates | Reaction with enzyme involved in phosphate metabolism |
| Sulphur dioxide, sodium sulphite or pyrosulphite | Reaction with aldehyde formed in carbohydrate dissimilation, reduction of S-S links in enzyme protein |
| Chlorine, chloramines, nitrogen trichlorides, peroxides, nitrates and other oxidizing agents | Destruction of sulphydryl groups in enzyme protein |
| Ethylene oxide and other epoxides | Reaction with carboxyl and other active groups on enzyme protein, oxidizing action |
| Fluorides, fluosilicates and fluoborates | Interference with prosthetic groups, reaction with active groups on enzyme protein |
| Formaldehyde | Reaction with active groups on protein of enzyme (or gene) |
| Salt | Precipitation of enzyme protein |

*Source:* 100

A chemical agent may kill microorganisms at a certain concentration, but are only inhibitory at a lower level, and ineffective at still greater dilutions. Preservatives added to inhibit or kill microorganisms may be classified depending upon their chemical composition, effectiveness, specificity, mode of action and legality. For example, sugars are effective preservative because of their physical action and sodium benzoate because of its chemical action. Some of the preservatives are incorporated into foods and are antiseptic in action rather than germicidal, while others are used only to treat the outer surfaces and may kill or inhibit the organisms. Some chemicals are employed to treat wrappers or containers used for foods, whereas other are used as gases or vapours to treat the food. Preservatives may be specific against certain microorganism i.e., they may be more effective against moulds or yeast and less against bacteria, or vice versa. They may act against definite groups or species of organisms. Sodium benzoate is more effective at acidic pH and less effective or rather relatively ineffective at neutral pH (Table 24). Sorbic acid and its salts are known to inhibit yeasts and moulds but are less effective against bacteria. They are most effective at low pH. These compounds are most effective than sodium benzoate above pH 4.0. Acetic acid is more effective against yeasts and bacteria than against moulds. In aqueous solution, various sulphites like sodium sulphite, potassium sulphite, sodium bisulphite, potassium bisulphite, sodium metabisulphite and potassium metabisulphite form sulphurous acid which is an active antimicrobial compound used widely by the food industry. Sulphurous acid is very effective at low pH. Burning sulphur fumes are used to treat most of the light coloured dehydrated fruits. $SO_2$ has also been used in fruit juices, wines, syrups and other beverages (Table 25). The amount of $SO_2$ required at different pH values is given in Fig. 5.

**Table 24:** Effect of varying concentration of sodium benzoate and pH on sugar utilization by *Saccharomyces rouxii* var. *polymorphus*.

| Concentration of sodium benzoate (%) | Sugar utilization (%)* | | | | |
|---|---|---|---|---|---|
| | pH 6.0 | pH 5.5 | pH 5.0 | pH 4.5 | pH 4.0 |
| 0.025 | 21 | 29 | 26 | 25 | 12 |
| 0.050 | 23 | 18 | 19 | 13 | 13 |
| 0.075 | 15 | 14 | 7 | 0 | 0 |
| 0.10 | 13 | 9 | 5 | 0 | 0 |

* lower sugar utilization means high efficacy
*Source:* 6

**Table 25:** Concentration of $SO_2$ used for preservation of different fruit and vegetable products.

| Fruit product | $SO_2$ (ppm) |
|---|---|
| Squashes, syrups, cordials, juices | 350 |
| Wine | 250 |
| Sweetened RTS | 70 |
| Sharbats | 350 |

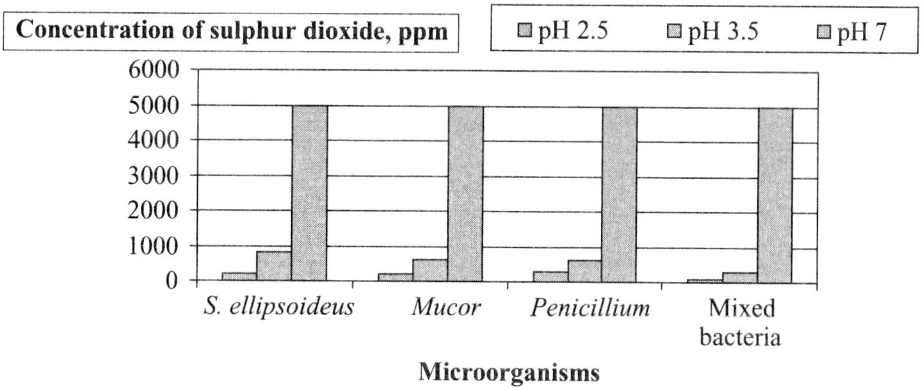


**Fig. 5:** Sulphur dioxide concentration for the prevention of microbial growth at different pH values.

Salt and sugar tend to tie up moisture and thus have an adverse effect on microorganisms. Ions of sodium chloride are harmful to the microorganisms, whereas sugars are considered effective preservative because of their ability to make water unavailable to microorganisms. Halogens, which are used mostly for washing of foods and equipment, kill organisms by injury to cell membranes, or by direct combination with their cell proteins. Hypochlorite of calcium or sodium, which yields hypochlorous acid, is a powerful oxidizing and germicidal agent.

9.1.3.4.2 Developed preservative (fermentation): During fermentation, a number of substances having preservative action are produced, and longer shelf life of fermented products is attributed to them. These can be called developed preservatives and include lactic acid as in pickles, acetic acid, ethyl alcohol as in alcoholic beverages, acetic acid in the vinegar, bacteriocin in the lactic fermented products, $H_2O_2$ produced by lactic acid bacteria, etc. Production of metabolites like acid lowers the pH in the fermented food to less than 4.6 where pathogens like *Clostridium botulinum* cannot produce toxin [96]. Besides, alcohols and acids are inhibitory to common pathogens. Thus, these preservatives developed during fermentation play a significant role in preservation.

9.1.3.5 Killing the microorganisms: Microbial decomposition of food can be prevented by killing the microorganisms, which are already present in the food. The food can be preserved by following methods of preservation.

9.1.3.5.1 Heat preservation: Application of heat treatment kills the microorganisms because of denaturation of the proteins and especially, the inactivation of enzymes required for metabolism. Different microorganisms have different heat resistance as shown by their thermal death time at a specified temperature (Table 26). The requirement of heat treatment to kill organisms varies with the kind of organism, its state, the effect of heat on food and environmental conditions during heating. Canning is one of the major methods of food preservation based on destruction of microorganisms by heat treatment and prevention of recontamination. The lethal temperature for microorganisms begins at about 46-49°C, with the exception of certain heat tolerant bacteria. The single cell microorganisms, bacterial cells and spores are destroyed at a constant uniform rate by heat, which is characteristic for a given microorganism and for a given medium in which it is heated. Certain factors that affect the heat resistance of cells or spores are given below:

A. Initial concentration of spores
B. The temperature-time relationship
C. Previous history of the vegetative cells or spores like:
   - Culture medium
   - Temperature of incubation
   - Phase of growth or age
   - Desiccation

D. Composition of the substrate in which cells or spores are heated like:
   - Moisture
   - pH
   - Other constituents of the substrate

When the temperature is increased, the time required for killing the cells or spores under the given conditions decreases (Fig. 6). The greater amount of heat is required to kill more number of cells or spores of microorganisms (Table 26). The resistance of microorganism to heat will depend upon the conditions in which the cells have been grown and spores have been produced and their treatment thereafter. In general, better the medium for growth, more resistant the cells or spores are. The heat resistance of microorganism increases when there is a small amount of glucose present in the medium, but it may decrease when sugar concentration is high. Lowering the water activity by increasing sugar content generally increases the heat resistance of yeasts, moulds and bacteria [101, 102]. Resistance also increases as the incubation temperature is raised towards optimum for a particular organism, and for many organisms increases further as the temperature approaches the maximum for growth. For example, during the late lag phase, bacterial cells show their greatest resistance, but almost as great resistance during their maximum stationary phase, followed by a decline in resistance. However, cells show least resistance during log phase of growth. It has been seen that it is very difficult to kill the dried spores of some bacteria by heat. Moist heat is a very effective killing agent than dry. Heat

**Table 26:** Thermal death times of bacterial cells having significance in foods.

| Bacterium | Time (min) | Temperature (°C) |
|---|---|---|
| *Salmonella typhi* | 4.3 | 50 |
| *Staphylococcus aureus* | 18.8 | 60 |
| *Escherichia coli* | 20-30 | 57.3 |
| *Streptococcus thermophilus* | 15 | 70-75 |
| *Lactobacillus bulgaricus* | 30 | 71 |

*Source:* 5

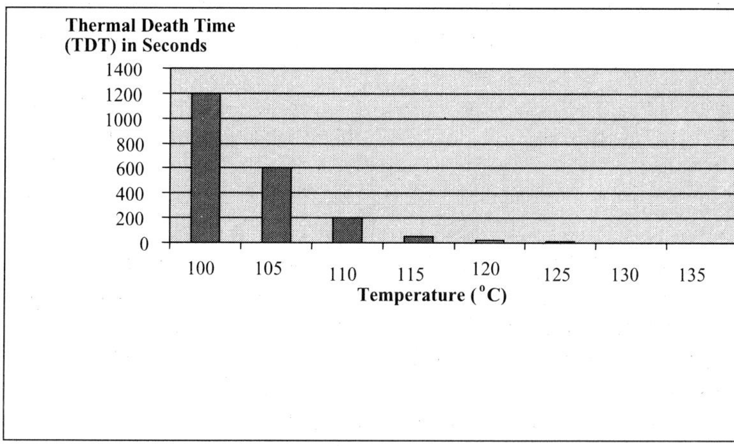


**Fig. 6:** Effect of heating temperature on time to kill spores of flat sour bacteria.

resistance of microorganism is influenced by pH alone. Generally as the pH declines, thermal tolerance of microorganism also declines (Fig. 2). The extent of reduction in heat resistance varies with different species [103]. The heat resistance of microorganisms is also influenced by the presence of organic acids. Malic, citric and tartaric acids protected ascospores of *Byssochlamys fulva* heated in water at pH 3.0, whereas fumaric, lactic and acetic acids reduced the heat resistance of ascospores [85]. The citric, malic and tartaric acids had no effect on thermal tolerance of yeasts in fruit juices [104]. Although the bacterial spores are most heat resistant (Table 3), they are generally incapable of growth at the normal pH of apple products. So, moulds are considered the most significant heat resistant organisms in apple product spoilage. Yeasts which produce ascospores also demonstrate considerable thermal resistance. Yeast ascospores may be 30-350 times more resistant than vegetative forms of the same species.

*9.1.3.5.1.1 Application of heat treatments in processing foods:* Greater the heat treatment, the more organisms will be killed. If all the organisms are not killed, heating must destroy either all potential spoilage organisms or the food must be handled properly thereafter, so as to delay or prevent the growth of surviving spoilage organisms. An attempt is made to kill all organisms which may spoil the food during later handling in canning, whereas in pasteurization most of the spoilage organisms are killed but some survive. So, if the spoilage is to be prevented, they must be inhibited by other suitable preservation methods. The various degrees of heating used in food are discussed as under:

**(a) Pasteurization:** Pasteurization is a heat treatment that kills some but not all of the microorganisms present and usually, the temperature is below 100°C. The heating may be in the form of hot water, dry heat, steam or electric currents and the food products are cooled immediately after the heat treatment. Application of electric current [105], though not a new concept, is getting increased attention as a method of preservation due to many innovations in science, especially engineering. When an electric current is passed through food, heat is generated (ohmic heating). The product is heated upto 140°C in less than 90 sec, and then cooled to ambient temperature. The particular interest in this technology stems from ongoing food industry interest in aseptic processing of particulate food [105]. Pasteurization inactivates the vegetative cells of pathogenic or spoilage organisms like yeast and mould in fruit juices and yeast in beer. Pasteurization is used in combination with other preservation methods, like fermentation (pickles), refrigeration (milk) and maintenance of anaerobic conditions (beer) as the food is not sterile. Pasteurization must also be used in the products like fruit juices, where the environment is not particularly suited for growth of health hazards and spoilage causing organisms. Pasteurization of surface of the intact eggs by applying a thermo ultrasonic treatment yields a product with a safe level of *Salmonella entritidis* and causes negligible damage to the thermolabile compounds [106].

**(b) Blanching or heating at about 100°C:** Pressure cookers are used for less acid foods by most of the home canners. But most of the acid foods like sauerkraut can successfully be processed at 100°C or less without the use of pressure cooker. Blanching is frequently applied to food prior to freezing, drying or canning, usually done in water or steam at a temperature of less than 100°C. The objectives of blanching depend upon the subsequent treatment of food e.g., blanching prior to freezing or drying is used primarily to inactivate enzymes besides killing microorganisms to reduce the microbial load in the food.

**(c) Sterilization or heating above 100°C:** Steam under pressure in steam pressure sterilizers or boilers provide the temperature above 100°C. The temperature in the retorts increases with rising steam pressure. The temperature at sea level is 100°C at atmospheric pressure, but with 15 lbs pressure, temperature of 121.5°C can be obtained. When liquid foods are sterilized before their filling in the sterile cans, a high steam pressure is used to apply a high temperature for a few seconds to the food before it is introduced into the sterilized cans. Sterilization is done to inactivate the vegetative cells and spores of microorganisms that would otherwise grow and cause spoilage during storage, which may also lead to the health hazards. In the strictest sense of the word, food products are not sterile because viable spores of microorganisms are present, but they cannot germinate under the conditions of storage. Most of the commercially sterilized foods are packed in such a manner to ensure anaerobic conditions, because spores of anaerobic organisms are generally less heat resistant than spores of aerobic organisms, resulting in less severe thermal processes.

*9.1.3.5.1.2 Irradiation preservation:* Irradiation kills the total microbial flora or eliminates the pathogenic microorganisms in the food. However, the organisms killed by radiation treatment are not necessarily the same as those killed by heat sterilization (Table 27). The storage life of irradiated foods is long only if they are stored under commercial cold storage conditions [107]. Infusion of chicken breasts with green tea (GT) extract at 3000 ppm is effective in preventing and maximizing major sensory changes of meat during irradiation. It improves the colour of the meat [108]. 8 kGy dose of gamma irradiation was reported better to improve appearance, flavour, texture, etc. in chicken breast [109]. There are several factors responsible for the destruction of microorganisms by irradiation treatment of food [5, 110]. They are as given below:

**Table 27:** Killing doses of ionizing radiations (kilo Grays) of different organisms.

| Organism | Lethal dose (kGy) |
|---|---|
| Insects | 0.22-0.13 |
| *Saccharomyces cerevisiae* | 5 |
| *Hansenula* | 4.7 |
| *Candida krusei* | 11.6 |
| *Penicillium* sp. | 1.4-2.5 |
| *Aspergillus* sp. | 1.4-3.7 |
| *Rhizopus* sp. | 1.4-3.7 |
| *Mycobacterium tuberculosis* | 1.4 |
| *Salmonella* sp. | 3.7-4.8 |
| *Staphylococcus aureus* | 1.4-7.0 |
| *Escherichia coli* | 14.0-2.3 |
| *Bacillus subtilis* | 12-18 |
| *Clostridium botulinum* (A) | 19-37 |
| *Bacillus stearothermophilus* | 10-17 |
| Viruses | 10-40 |

*Source:* 5

**(a) Kind and species of organisms:** Gram positive bacteria have been found to be more resistant to radiation than Gram negatives. Generally, the sporeformers are more resistant than non-sporeformers, but with the exception of *Micrococcus radiodurans*, which is one of the most radio-resistant bacteria. Among the sporeformers, *Bacillus* larvae have been reported to possess a higher degree of resistance than most aerobic sporeformers. Spores of *Clostridium botulinum* type A appear to be the most resistant of all clostridial spores. One of the most resistant vegetative bacteria is *Streptococcus faecalis*, *Micrococci* and the homofermentative *Lactobacilli* besides *M. radiodurans*. The *Pseudomonas* and *Flavobacteria* groups are most sensitive to radiation and other Gram negative bacteria being intermediate in radio-resistance between these genera and the *Micrococci*. Yeasts are considered more resistant to irradiations than moulds. However, both groups in general are less sensitive than Gram positive bacteria [5].

**(b) Number of organisms:** The larger the number of microbial cells, the less effective is the given dose of radiation.

**(c) Composition of food:** Generally, microorganisms are more sensitive to radiation when they are in buffer solutions than in protein-containing media. It has been observed that the presence of nitrites tends to make bacterial endospores more sensitive to radiation.

**(d) Presence or absence of oxygen:** The resistance of microorganisms to radiation is lesser in the presence of oxygen than its absence. Three-fold radiation resistance of *E. coli* has been reported when oxygen from the cell suspension was removed.

**(e) Physical state of food:** The dried cells are more resistant to radiation than the moist. This is due to the radiolysis of water by ionizing radiations. The frozen cells have been reported to be more resistant to radiation than non-frozen.

**(f) Age of organism:** Bacteria in the lag phase are most resistant to radiation. The cells become more sensitive to radiation as they progress through the log phase. The inactivation of microorganisms by various types of irradiation in apple juice has been reported. While evaluating the use of ultraviolet radiation for pasteurization of apple cider Harrington and Hills [111] reported that clarification of the cider was required for maximal inactivation of microorganisms. Although sterility of the product was not achieved but the microbial population was reduced by 99% and shelf-life was prolonged without altering the flavour. Although the ultraviolet sterilization of apple juice was achieved, but the flavour of the product was slightly changed, and there was a reduction in ascorbic acid, thiamin, and riboflavin levels [112].

*9.1.3.5.1.3 High pressure technology:* Application of high pressure to food processing is a recently developed tool. Though the technology was discovered in 1899, the food technologists accepted this technique only in late 1980's and today it is acknowledged worldwide. The pressure treatment may vary between 1000 and 10000 kg f/cm [104]. The pressure is transmitted rapidly and uniformly throughout the food that is being processed in an ecostatic hydraulic medium. Microbial death occurs due to permeability of cell membrane applying higher pressure, whereas changes that are induced in cell morphology of the microorganisms are reversible at low pressure [113]. The application of high pressure technology resulted in injury and death of microorganisms, death of bacterial spores when used in combination with moderate heating (50-70ºC) and denaturation of proteins and gelatinization of starch [113]. The possible applications of high pressure technology as detailed earlier [113] are given below:

- Preservation of fruit juices by inhibiting the growth of pressure resistant bacterial spores
- Freezing depression at high pressure gives a benefit in storing foods without formation of ice and avoiding damage due to freezing
- Pressure induced gels from egg white and egg yolk retain its original colour

The main advantages of high pressure technology are given below:

- Elimination or reduction of heat damage to food
- Eliminating the need of chemical additives
- Uniform penetration of pressure in the product

However, the commercialization of the technology would take its own time.

## 9.2 Preservation of Food by Delay or Prevention of Self Decomposition

This principle is explained clearly in the following sub-principles:

### *9.2.1 Destruction or inactivation of food enzymes*

Besides microbial decomposition, food gets self decomposed because of various enzymes and other chemical reactions. Fruits and vegetables are living entities. Various catabolic and anabolic reactions take place in them right from their inception. These reactions are catalyzed by various enzymes present in them. Various enzyme reactions are involved in fruits and vegetables right from their formation to senescence stage. In case of apple, enzymes present in the fruit continue their activity even when the fruit is ripe and ready for consumption. If it is not used or converted into a suitable product, its (enzyme) activity will lead to the decomposition of the fruit, which becomes unfit for consumption. That is, the activity of enzymes involved in the respiration reaction of the fruit will continue, thus leading to the over ripening of the fruit. There are certain enzymes which do not show their activity when the fruit is intact, but their activity becomes very fast when the fruit is peeled, cut or ruptured. For example, polyphenol oxidase enzyme present in apple, pear, peach, potato and mushrooms, etc. leads to the formation of brown pigments once these foods are cut, peeled and exposed to the air, ultimately affecting the quality of the product by altering the colour and flavour, etc.

Textural changes in fruits and vegetables are brought about by the enzyme e.g., when the fruit is immature, it is harder, but when it develops into a fully mature stage and finally to ripe stage, it become soft. It is because of the textural enzymes, which are responsible for certain reactions that result in the softening of fruits during ripening. Some enzymes are involved in the reactions, which are responsible for the production of aromatic compounds in the fruits and give rise to the pleasant flavour during ripening of the fruit. These enzymes are not of much concern here. Whereas, some enzymes are also involved in certain reactions leading to the production of off-flavours. From the processing point of view, we are more concerned with the enzymes which lead to the spoilage of food during the preparation of a particular product and during storage. These enzymes are responsible for the formation of brown pigments. When fruit and vegetable tissues are injured because of bruising, cutting or peeling during processing and exposed to air, they undergo a darkening of tissues, which is called enzymatic browning.

9.2.1.1 Browning reactions: These enzymatic reactions occur only in fresh living tissues or at least in tissues that still contain active enzymes. Thus when enzymes are inactivated by the application of heat or by any other agent, this reaction no longer occurs. For example, fresh peach will turn brown after peeling, whereas, a canned peach will not. Oxidation of phenols or polyphenols by the enzyme polyphenol oxidase is the principal reaction in enzymatic browning.

## 9.3 Effect of Preservation Methods on Enzymes

### *9.3.1 Influence of freezing on enzymes*

Activity of enzyme depends upon the temperature. The activity of enzyme or a system of enzymes can be destroyed at the temperature near 93°C, whereas they retain some activity at temperature as low as −73°C, although reaction rates are extremely low at this temperature. The velocity of enzymatic reactions is greater in super-cooled water than in crystallized water at the same temperature. Enzyme activity is also retarded easily by blanching prior to freezing and storage. For frozen fruits, colour is an important quality factor, and to maintain a bright colour in the final product, blanching by heat is replaced by chemical treatments or additives to inactivate the enzymes. For instance, ascorbic acid is added to the syrup in the freezing preservation of peach slices. A combination of chemical treatment plus a mild heat treatment can be used in some fruits. In some cases, enzymes are only partially inactivated, especially on the surface of the fruit, and then it is consumed quickly after thawing. Changes affecting the colour, flavour and texture of fruit may occur at any stage of handling.

### *9.3.2 Influence of heat on enzyme*

Nearly, all the enzymes are irreversibly destroyed by heating at 93°C. The effect of heat on the rate of coagulation of protein and the rate of inactivation of enzymes are two phenomena, which have high energies of inactivation, and may be due to similar chemical reactions. The heat inactivation of enzymes is associated with alteration of the surface of the molecules, breaking bonds and opening rings in the protein molecules, with dissociation and loss of structure. In some cases, the inactivation of enzymes may be used as an index of degree of heating of foods. For example, the pasteurization of milk can be evaluated by its phosphatase enzyme activity. The destruction of phosphatase in milk coincides with the heat treatment designed to kill tuberculosis and other human pathogenic organisms. In evaluating the relative efficiency of acid food canning processes, the peroxidase enzyme system in fruits may be useful. Unless the enzymes are destroyed, they will continue to function in the container, causing deterioration.

### 9.3.3 Influence of delayed processing on enzymes

As long as the processing of raw fruit is delayed, there is defect in the final product either from microbial growth or enzymatic action. Bruising in fruit will cause discolouration due to polyphenol oxidase activity. However, if the fruit is processed rapidly after harvest, the bruises may be hardly noticeable in the final product. If there is a period enough longer for enzymatic oxidation to proceed before the fruit is processed, the bruises may become quite obvious.

### 9.3.4 Influence of drying on enzymes

Generally, the enzymes are sensitive to moist heat especially where temperature reaches above the maximum for enzyme activity. Moist heat temperatures near the boiling point of water nearly inactivate the enzymes. When enzymes are exposed to dry heat such as that used in drying, they are notably insensitive to the effect of heat. Short exposures to temperature near 204°C have little effect on enzymes if the heating medium and the enzyme preparation is dry. So, it is important to control enzyme activity either by inactivating enzymes chemically or by subjecting the food material to moist heat conditions. Catalase and peroxidase are used as indicator enzymes. Catalase is less resistant to heat treatment than peroxidase. Enzyme activity is reduced with decreasing moisture in the dried products, but a concentration of enzyme and substrate simultaneously increases. When the moisture level is below 1%, there is no enzyme activity.

### 9.3.5 Influence of radiation on enzymes

Applications of ionizing radiation to the food inactivate the enzyme directly or indirectly. The decrease in inactivation dose in the presence of water is taken as a measure of destruction by indirect effects. In natural substrates, enzymes are more resistant to the effects of ionizing radiation than in pure solutions. This is taken as a measure of natural protective substances present in living tissues. A radiation dose which does not destroy the enzyme in a suitable substrate finds the reaction proceeding. Enzymes have low sensitivity to the radiation effects, which is of practical interest in sterilization because it is necessary to prevent enzymatic as well as microbial spoilage in preserving food. A complete inactivation of enzymes requires approximately five times the dose required for the destruction of microorganisms.

## 9.4 Approaches of Controlling or Inhibiting Enzymatic Browning Reactions

For the control of enzymatic browning reactions, various methods have been evolved which are given below:

### 9.4.1 Selection of raw material

The enzymatic browning is controlled by the selection of varieties of fruits least susceptible to discolouration either because of the absence of phenolic substrates or the presence of substrates or enzyme at low concentration and selection of fruit and vegetable at the stage of maturity at which discolouration is at a minimum. In the home preparations, lemon juice and pineapple juices have long been used to prevent browning. Pineapple juice contains relatively high percentage of sulphahydryl compounds which are considered active antioxidants, whereas lemon juice contains both ascorbic acid and citric acid.

### 9.4.2 Application of heat

Blanching denatures enzymes and also reduces the air content of the food and rinses away unwanted surface fluids. The inactivation of polyphenol oxidase as well as other spoilage enzymes can be achieved by subjecting the food to high temperature for an adequate time period to denature the protein. The pretreatment of vegetable for canning, freezing preservation, dehydration or in the manufacture of fruit juices and purees by blanching and HSTS pasteurization essentially achieves this goal.

Caution must be taken when using heat treatment to avoid cooking the fruit and vegetable, as this could result in unfavourable textural changes and development of off-flavour. This could occur particularly in the pre-processing of potatoes, apple, pears and peaches. The heat inactivation of enzymes in foods is not only dependent on time, but is also affected by pH.

### 9.4.3 Lowering of pH

Acids are used extensively to control enzymatic browning reactions. The acids found naturally in plant tissues include citric, malic, phosphoric and ascorbic acids. This method is based on the fact that lowering the pH of tissues will reduce or retard the development of enzymatic browning reactions. Citric acid has been used in combination with ascorbic acid or sodium sulphite as a chemical inhibitor of enzymatic browning reactions. Cut fruits like peaches are immersed in dilute solutions of these acids just prior to processing. This is particularly important in the case of lye peeled peaches, where the acid dip counteracts the effect that any residual lye might have on enzymatic browning. Citric acid also inhibits polyphenol oxidase by chelating the copper moiety of enzyme. Compared to citric acid, malic acid in apple juice is a much more effective inhibitor of enzymatic browning. But effective inhibitor of polyphenol oxidase is ascorbic acid.

### 9.4.4 Exclusion of oxygen

Immersion of peeled products like potato and apple in water prior to cooking or further processing is the simplest method of controlling enzymatic browning reactions. This operation limits the access of oxygen to the cut potato or apple tissues. This method is limited as the fruits and vegetables will brown on re-exposure to air or via oxygen occurring naturally in the plant tissues. The removal of $O_2$ from fruit or vegetable tissues could lead to anaerobiosis if they are stored for extended periods, which in turn could lead to abnormal metabolites and tissue breakdown. In case of frozen sliced peaches, the surfaces are treated with excess ascorbic acid to use up the surface oxygen.

### 9.4.5 Application of sulphur dioxide and sulphites

Sulphur dioxide and sulphites are powerful inhibitors of polyphenol oxidase enzyme. Sodium bisulphite by itself or in combination with citric acid is commonly used commercially to inhibit enzymatic blackening of pre-peeled potatoes or in the processing of apples and peaches. The use of $SO_2$ and sulphites is particularly advantageous when heating would result in unfavourable texture and off-flavour in a product. $SO_2$ may directly inhibit the enzyme or it may convert quinones to quinols.

### 9.4.6 Use of salts and sugars

Solutions of salt and sugar provide a high osmotic potential. They have the effect of dehydrating the cells and also lowering the concentration of dissolved oxygen. About 1% salt has been found to be inhibitory to the enzymatic browning.

## 9.5 Delay or Prevention of Purely Chemical Reactions

There are certain reactions that are purely chemical, which lead to the self-decomposition of the food. Among these are non-enzymatic reactions which lead to the browning of food. Non-enzymatic browning reactions have been the subject of numerous investigations, but these reactions are also least understood. There have been many attempts to control the browning caused by these reactions, which occur on storage of citrus fruit juices, in grapes, raspberry and strawberry juices, and in dried peaches, apricots, apples and pears. Various studies have been carried out on the possible course of browning reactions in fruits. But the mixture of compounds in a fruit is so complex and the factors which must be controlled are so large that the knowledge is in fragmentary state. There are three pathways, which have been suggested to explain non-enzymatic browning reactions.

### 9.5.1 Maillard reaction

The formation of brown pigments and melanoidins was first observed by French chemist Louis Maillard following the heating of a solution of glucose and lysine. This reaction was subsequently referred to as Maillard reaction and essentially covers all those reactions involving compounds with amino groups and carbonyl groups present in a food. These include amines, amino acids and proteins interacting with sugars, aldehydes and ketones, as well as with products of lipid oxidation. Maillard reaction can take place in acidic and basic conditions, but is favoured under basic conditions e.g., pickles, which are highly acidic, show little browning. Besides this, buffers usually promote browning reactions. Increase in temperature leads to the promotion of browning reactions, and it has been seen that at 0 to 80°C, the relationship between browning and temperature is linear. Browning reaction is faster in aqueous solution but during dehydration, it is slowed down or inhibited. Optimum moisture in Maillard reaction browning is found to be 13%, and beyond it browning slows down.

In Maillard reaction, reducing sugars favour the browning. Aldoses and pentoses are most reactive followed by aldo-hexoses, but disaccharides are very less active. The formation of metal complexes with amino acids can influence the Maillard reaction. This reaction is catalysed by copper and iron, while manganese and tin inhibit this reaction. In addition to the catalytic effect of iron on the browning reaction, it has been seen that iron also participates as a chromophore of pigment. The possible interaction of iron with hydroxypyridone and hydroxypyranone, both capable of chelating iron in the melanoidin polymer, may be responsible for colour formation.

### 9.5.2 Caramelization

Caramelization is another pathway of non-enzymatic browning involving the degradation of sugars. When sugars are heated above their melting points, they darken to a brown colouration under alkaline or acidic conditions. If this reaction is not carefully controlled, it could lead to the production of unpleasant, burnt and bitter products. So, it is important to control this reaction during food processing, while still retaining the pleasant qualities of caramel. Chemical composition of caramel is extremely complex and still poorly understood, although caramel products from different sugars showed similarity in composition. Continued heating results in the dehydration of sugars, leading to the formation of hydroxymethyl furfural, levulinic acid and humin.

### 9.5.3 Ascorbic acid oxidation

Ascorbic acid plays a central role in the browning of citrus juices and concentrates e.g., orange, lemon and grape fruit. The reaction of ascorbic acid in fruit juices and concentrates is very much dependent on pH, as the browning process is inversely proportional to pH over a range of 2 to 3.5. Juices with a higher pH are much less susceptible to browning e.g., orange juice at a pH of 3.4. Below pH 4.0, browning is primarily due to the decomposition of ascorbic acid to furfural. The degradation of ascorbic acid was investigated when 17 decomposition products, including dehydroascorbic acid and 2, 3-diketo gulonic acid and oxalic acids were identified [114].

Ascorbic acid degradation can occur under both aerobic and anaerobic conditions. While the level of air in juice is kept as low as possible by the use of vacuum deaeration and live steam injection, there is still some dissolved oxygen in the juice [115]. Ascorbic acid degradation also depends on head-space. Oxidative degradation of ascorbic acid in canned, pasteurized juice occurred during the first few days until the free oxygen was utilized. Following this, anaerobic breakdown of ascorbic acid proceeds, but at one-tenth of the original rate.

## 9.6 Inhibition of Non-enzymatic Browning Reactions

The major concern of postharvest technologist is to control or minimize the non-enzymatic browning reactions in food processing. Various means have been proposed for controlling these reactions which are described here.

### *9.6.1 Temperature*

An increase in temperature or the time of heat treatment accelerates the rate of these reactions. So, lowering the temperature during processing and storage can lengthen the period needed for the formation of brown coloured products.

### *9.6.2 Moisture*

The browning reactions which are dependent on moisture content provide a convenient method for control. A reduction of moisture content in dried products reduces the mobility of reactive components [116]. Whereas in case of solutions, an increase in water activity will diminish the reaction velocity, which will dilute the effect of reactions.

### *9.6.3 pH*

Alkaline conditions generally favour the Maillard reaction, so lowering the pH provides a useful method for the control of these reactions [116].

### *9.6.4 Gas packing*

Packing of the products under an inert gas exclude the oxygen. This reduces the formation of oxidation products of lipids, which are capable of interacting with amino acids. The exclusion of oxygen is thought to affect these reactions involved in the browning process and not the initial carbonyl amino reaction step.

### *9.6.5 Biochemical agents*

Removal or conversion of one of the reactants of the sugar-amino acid interactions forms the basis of the biochemical method of controlling the non-enzymatic browning reactions. For instance, in commercial production of egg white, glucose is removed by yeast fermentation prior to drying. The direct application of enzymes such as glucose oxidase and catalase, mediate the conversion of glucose to gluconic acid, which is no longer capable of combining with amino acids. Glucose oxidase has additional advantage of removing any residual oxygen and is used to reduce headspace oxygen during the production of bottled products.

### *9.6.6 Chemical inhibitors*

A variety of chemical inhibitors have been used to limit browning during the production and storage of a number of different food products. The most widely used chemicals are $SO_2$, sulphites, etc.

9.6.6.1 $SO_2$/Sulphites: $SO_2$ is unique in its ability to inhibit the Maillard reaction, and can be applied as a gas or in solution as sulphite/bisulphite. It is not only capable of partially bleaching of chromophores already formed, but also inhibits colour formation at the beginning of reaction [117]. The mechanism involved is binding of $SO_2$/sulphite with glucose to form hydroxy sulphonate and other compounds from which $SO_2$/sulphite can be reversibly released. This results in the blocking of the carbonyl group of sugar, rendering it unavailable for interaction in the typical Maillard reaction. As the reaction proceeds, $SO_2$/sulphite becomes irreversibly bound.

9.6.6.2 Aspartic and glutamic acids: Aspartic and glutamic acid decrease the rate of browning [118]. Dipping of potato chips into either acid solution prior to frying was accompanied by less darkening.

## 10. CONTROL OF BIODETERIORATION BY BIOTECHNOLOGICAL APPROACHES

Biotechnology has been defined broadly as the utilization of biologically derived molecules, structures, cells, or organisms to carry out a specific process. Many conventional food processing techniques utilize organisms, especially in brewing and fermentation industries. Soybean sauce, fish sauce and *kimchi* have been made for over a thousand years in Asia by using naturally occurring microorganisms, which grow under specific environmental conditions, similar to cheese production. Now-a-days by-products obtained from the microorganisms such as nisin, subtilin, lactic acid, etc. are used for preservation of the foods. Biotechnological approaches like genetic engineering make the food more nutritious and more self stable and help in the control of biodeterioration, as the food prepared by these approaches are resistant to insects, pests and diseases. Modern biotechnology is represented by the production of genetically modified organisms and their use in bio-industry [119, 120]. The enzymes produced by microorganisms and their applications in food industry are presented in Table 28.

**Table 28:** Application of enzymes produced by genetically modified microorganisms in food industry.

| Enzyme | Microorganism | Food application |
|---|---|---|
| Aminopeptidase | *Trichoderma reesei or T. longibrachiatum* | Cheese, dairy |
| Amylase | *Bacillus amyloliquefaciens* or *B. subtilis* | Baking, beverages |
| Catalase | *Aspergillus niger* | Egg-based products |
| Chymosin | *Aspergillus niger* | Cheese |
| β-glucanase | *Bacillus amyloliquefaciens* or *B. subtilis* | Beverages |
| Glucoamylase | *Aspergillus niger* | Baking, beverages |
| Glucose isomerase | *Streptomyces lividans* | Starch |
| Glucose oxidase | *Aspergillus niger* | Baking |
| Hemicellulase | *Bacillus amyloliquefaciens* or *B. subtilis* | Baking, starch |
| Lipase, triacylglycerol | *Aspergillus oryzae* | Fats |
| Maltogenic amylase | *Bacillus amyloliquefaciens* or *B. subtilis* | Baking, starch |
| Pectin lyase | *Aspergillus niger* | Beverages |
| Pectinesterase | *Trichoderma reesei* or *T. longibrachiatum* | Beverages |
| Phospholipase A | *Trichoderma reesei* or *T. longibrachiatum* | Baking, fats |
| Polygalacturonase | *Trichoderma reesei* or *T. longibrachiatum* | Beverages |
| Protease | *Aspergillus oryzae* | Cheese |
| Pullulanase | *Bacillus licheniformis* | Starch |
| Xylanase | *Aspergillus niger* | Baking, beverages |

## 11. PREVENTION OF DAMAGE CAUSED BY INSECTS, RODENTS, BIRDS AND MECHANICAL CAUSES

In the food industry, main attention is paid only to the microbial control. Although the control of microorganisms may be of uppermost concern, controlling other pests such as insects, rodents, and birds is also essential. These pests threaten to contaminate food products directly or with their excreta or by serving as vectors for a wide variety of microorganisms. The insect damage also leads to spoilage of such fruits and vegetables by bacteria, yeasts and moulds. To cite an example, a single hole in the melon is not bad itself, but can result in the total decay from bacterial invasion [96]. Insects like moths, house fly, drosophila, wasps and ants are attracted more towards the various food products in the factory. If their entry in the production section of the factory is not prevented, they can contaminate the products, which is considered to be highly objectionable from quality point of view. Rodents consume the raw material, packaging material as well as finished products wrapped in the paper and packed in the fibre board cartons. Besides spoilers, rodent urine and droppings harbour several kinds of disease producing bacteria, and rats spread such human diseases as *Salmonella leptospirosis*, typhus fever and plague [96]. So, pest proof buildings should be constructed to avoid insect and rodent manifestation in the plant. Nevertheless, it is virtually impossible to keep every pest out of plant. Even if all conceivable precautions are taken to keep free wandering insects and rodents out of a plant, some will always enter through incoming raw products or supplies. So, it is necessary to ensure that all habitats and breeding areas within a plant are eliminated. Birds are attracted towards the factory where the waste disposal is not done properly. They also take shelter in the factory buildings and infest the raw material as well as finished products at various stages. So, proper disposal of waste should be done and shelter places of birds should be blocked properly. For some kinds of raw materials, fumigation prior to entry into the processing plant may be beneficial in reducing pest entry; however, for most vegetable products to be processed, such a pretreatment is not practical. A thorough system for controlling free-flying insects within plants should be used. There are commercially available systems, which incorporate an attractant in the form of specific wavelength of light and an electrostatic field that kills the pests like house-flies and drosophila. In such cases, there is no use of chemicals. But for rodents and some insects as well, chemical control methods must still be used. Since the machinery involved in the manufacture of various products may become source of contamination of the product, its proper cleaning needs to be carried out regularly. The trickling of grease and seal oils of various motors in food should be avoided to prevent the contamination of food products.

## 12. FUTURE OUTLOOK

In future, the principles of preservation are likely to remain the same but changes in preservation methods would be an exciting possibility. Going by the present status, future research need to be focused on development of safe preservatives. Applications of enzymes in preservation or developments in phonic pressure preservation seem to be the future techniques of preservation of perishables. In light of consumers response to get a product in as natural form as possible, techniques which result in minimally processed product should be preferred in future.

## REFERENCES


1. Gnardia EJ & Hopper PA (1990) Food safety—An international concern. Industry leadership in producing safe food. *Food Microbiol* 7: 69-74.
2. Splittstoesser DF (1996) Microbiology of fruit products. In: *Processing Fruit Science and Technology, Biology, Principles and Applications,* Vol 1, Somogyi LP, Ramaswamy HS & Hui YH (eds), Technomic Publishing Co Inc, Lancaster, USA, pp 261-292.
3. Sudheer KP & Indira V (2007) *Post Harvest Technology of Horticultural Crops, Horticulture Science Series,* Vol 7, New India Publishing Agency, New Delhi, India.
4. Thom C & Hunter AC (1924) *Hygienic Fundamentals of Food Handling*, Williams and Wilkins, Baltimore, USA.

5. Frazier WC & Westhoff DC (1996) *Food Microbiology*, Tata McGraw Hill Publishing Co Ltd, New Delhi, India.
6. Desrosier NW & Desrosier JN (1987) *The Technology of Food Preservation,* CBS Publishers and Distributors, New Delhi, India.
7. Bigelow WD & Esty JR (1920) Thermal death point in relation to time of typical thermophilic organisms. *J Infect Dis* 27: 602-617.
8. Gould GW & Russell NJ (1991) Major, new and emerging food poisoning and food spoilage microorganisms. In: *Food Preservatives,* Russell NJ & Gould GW (eds), Plenum Publishers, New York, USA, pp 1-13.
9. Cameron EJ (1940) Report on canned vegetables. *JAOAC* 23: 607-608.
10. Williams OB (1929) The heat resistance of bacterial spores. *J Infect Dis* 44: 421-465.
11. Sharman M, Patey AL, Bloomfield DA & Gilbert J (1991) Surveillance and control of aflatoxin contamination of dried figs and fig paste imported into the United Kingdom. *Food Addit Contam* 8: 299-304.
12. Sinha KK & Singh A (1987) Chemical changes in apples due to aflatoxin producing Aspergilli. *J Food Sci Technol* 24: 44-49.
13. Azad KC, Vyas KK, Joshi VK & Sharma RP (1987) Some observations on the properties of juice and les. *Indian Food Packer* 41: 56-61.
14. Liewen MB & Bullerman LB (1992) Toxigenic fungi and fungal toxins. In: *Compendium of Methods for the Microbiological Examination of Foods,* Vanderzant C & Splittstoesser DF (eds), American Public Health Association, Washington, USA, pp 811-819.
15. Tiwari R, Dixit RP, Chandan NC, Saxena A, Gupta KG & Vadhera DE (1983) Inhibition of growth and aflatoxin B1 production of *Aspergillus parasiticus* by spice oils. *J Food Sci Technol* 20: 131-132.
16. Chakrabarti DK & Ghosal S (1986) Occurrence of free and conjugated 12, 13-epoxytrichothecenes and zearalenone in banana fruits infected with *Fusarium moniliforme. Appl Environ Microbiol* 51: 217-219.
17. Davis ND & Diener UL (1987) Mycotoxins. In: *Food and Beverage Mycology,* Beuchat LR (ed), Van Nostrand, Reinhold, New York, USA, pp 517-570.
18. Frisvad JC & Samson RA (1991) Filamentous fungi in foods and feeds: Ecology, spoilage and mycotoxin production. In: *Handbook of Applied Mycology, Foods and Feeds,* Vol 3 Arora DK, Mukerji KG & Marth EH (eds), Marcel Dekker Inc, New York, pp 31-68.
19. Deak T & Beuchat LR (1987) Identification of food-borne yeasts. *J Food Protec* 50: 243-264.
20. Pfaff JI, Miller MW & Mrak EM (1978) *The Life of Yeasts,* Harvard University Press, Cambridge, UK.
21. Swanson KMJ (1989) Microbiology and preservation: In: *Processed Apple Products,* Downing DL (ed), AVI Van Nostrand Reinhold, New York, USA, pp 343-363.
22. Krogh P (1989) The role of mycotoxins in disease of animals and man. *J Appl Bacteriol* (Symposium supplement) 67: 99S-104S.
23. Harwig J, Chen YK, Kennedy PC & Scott PM (1973) Occurrence of patuni and patulin producing strains of *Penicillium expansum* in natural rots of apple in Canada. *Can Inst Food Sci Technol J* 6: 22-25.
24. Ogawa JM, Rumsey J, Manji BT, Tate G, Toyoda J, Bose E & Dugger L (1974) Implications and chemical testing of two rhizopus fungi in softening of canned apricots. *Calif Agric* 28: 6-7.

25. Anon (1979) *Salmonella oranienburg* gastroenteritis associated with consumption of precut watermelons. *Morbidity Mortality Weekly Report* 28: 522-523.
26. De Ley J, Swings J & Golsete F (1984) Genus I. *Acetobacter Beijerinck* 1898. In: *Bergey's Manual of Systematic Bacteriology,* Vol 1, Holt JG (ed), Williams and Wilkins, Baltimore, USA, pp 276-278.
27. Drysdale GS & Fleet GH (1988) Acetic acid bacteria in wine making: A review. *Am J Enol Vitic* 39: 143-154.
28. Escartin EF, Ayala AC & Lozano JS (1989). Survival and growth of *Salmonella* and *Shigella* on sliced fresh fruit. *J Food Prot* 52: 471-472.
29. Fuentes FA, Hazen TC, Lopez-Torres AJ & Rechani P (1985) *Klebsiella pneumoniae* in orange juice concentrate. *Appl Environ Microbiol* 49: 1527-1529.
30. Gini B & Vaughn RH (1962) Characteristics of some bacteria associated with the spoilage of California dessert wines. *Am J Enol Vitic* 13: 20-31.
31. Hatcher WS, Weihe JL, Splittstoesser DF, Hill EC & Larish ME (1992) Fruit beverages. In: *Compendium of Methods for Microbiological Examination of Foods,* Vanderzant C & Splittstoesse DF (eds), American Public Health Association, Washington, USA, pp 565-568.
32. Juven BJ & Shomer I (1985) Spoilage of soft drinks caused by bacterial flocculation. *J Food Prot* 48: 52-53.
33. Luthi H (1959) Microorganisms in non citrus juices. *Advan Food Res* 9: 221-284.
34. Townsend CT (1939) Spore-forming anaerobes causing spoilage in acid canned foods. *Food Res* 4: 231-237.
35. Vaughn RH & Stadtman TC (1946) A note on pH tolerance of *Aerobacter aerogenes* and *Aerobacillus macerans* as related to natural ecology and decomposition of acid food products. *J Bacteriol* 51: 263-263.
36. Vaughn RH, Kreulevitch IM & Mercer WA (1952) Spoilage of canned foods caused by the *Bacillus maceranspolymyxa* group of bacteria. *Food Res* 17: 560-570.
37. Wolford ER (1956) Certain aspects of the microbiology of frozen concentrated orange juice. *Am J Public Health* 46: 708-715.
38. Adams MR, Hartley AD & Cox LJ (1989) Factors affecting the efficacy of washing procedures used in the production of prepared salads. *Food Microbiol* 6: 69-77.
39. Brackett RE (1992) Shelf stability and safety of fresh produce as influenced by sanitation and disinfection. *J Food Prot* 55: 808-814.
40. Deak T, Heaton EK, Hung YC & Beuchat LR (1987) Extending the shelf life of fresh and irradiated sweet corn by shrink-wrapping and refrigeration. *J Food Sci* 52: 1625-1631.
41. Hurst WC & Schuler GA (1992) Fresh produce processing—an industry perspective. *J Food Prot* 55: 824-827.
42. Sumner SS & Peters DL (1997) Microbiology of vegetables. In: *Processing Vegetables, Science and Technology*, Smith DS, Cash JN, Nip WK & Hui YH (eds), Technomic Publishing Co Inc, New York, USA, pp 87-116.
43. Jay JM (1992) Spoilage of fruits and vegetables. In: *Modern Food Microbiology*, Chapman & Hall, New York, USA, pp 187-198.
44. Madden JM (1992) Microbial pathogens in fresh produce—the regulatory perspective. *J Food Prot* 55: 821-823.
45. Pandey A, Joshi VK, Poonam N & Soccol CR (1998) Enterobacteriaceae, coliform and *E. coli.* In: *Encyclopedia of Food Microbiology,* Richard R, Carl B & Patel P (eds), Academic Press, London, UK, pp 604-610.

46. Casolari A (1989) Microbial spoilage of canned fruit juices. In: *Biotechnology Applications in Beverage Production*, Canterelli C & Lanzarini G (eds), Elsevier Applied Science, Essex, UK, pp 66-75.
47. Somogyi LP, Ramaswamy HS & Hui YH (1996) *Processing of Fruits: Science and Technology, Biology, Principles and Applications*, Vol 1, Technomic Publishing Co Inc, Westport, USA.
48. Hulme AC (1971) *The Biochemistry of Fruits and Products,* Vol 2, Academic Press, London, UK.
49. Blocher JC & Busta FF (1983) Bacterial spore resistance to acid. *Food Technol* 37: 87-99.
50. Jakobsen M & Jensen HC (1975) Combined effect of water activity and pH on the growth of butyric anaerobes in canned pears. *Lebensmittel Wiss Technol* 8: 158-160.
51. Joshi VK, Pandey A & Nigam P (1998) Enterobacteriaceae, coliform and *E. coli.* In: *Encyclopedia of Food Microbiology,* Robinson RC & Batt P (eds), Academic Press, London, UK, pp 125-136.
52. Swanson KM, Leasor SB & Downing DL (1985) Aciduric and heat resistant microorganisms in apple juice and cider processing operations. *J Food Sci* 50: 336-339.
53. Vicini E, Barbuti S, Spotti E, Campanini M, Castelvetri F, Gola S, Manganelli E, Cassara A & Casolari A (1992) Alterazione di succhi di frutta causata da muffe gasogene. *Industria Conserve* 59: 4-7.
54. Segner WP (1979) Mesophilic aerobic spore forming bacteria in the spoilage of low acid canned foods. *Food Technol* 33: 55-59.
55. Joshi VK, Kaushal NK & Sharma N (2000) Spoilage of fruits, vegetables and their processed products. In: *Postharvest Technology of Fruits and Vegetables: Handling, Processing, Fermentation and Waste Management,* Vol 1, Verma LR & Joshi VK (eds), Indus Publishing Company, New Delhi, India, pp 235-385.
56. Lal G, Siddappa GS & Tandon GL (1986) Spoilage in canned foods. In: *Preservation of Fruits and Vegetables,* ICAR Publications, New Delhi, India, pp 82-102.
57. Walker HW & Ayres JC (1970) Yeasts as spoilage organisms. In: *The Yeasts,* Vol 3, Rose AH & Harrison JH (eds), Academic Press, New York, USA, pp 463-528.
58. Tilbury RH (1976) The microbial stability of intermediate moisture foods with respect to yeasts. In: *Intermediate Moisture Foods*, Davies R, Birch GG & Parker KG (eds), Applied Science Publishers, Essex, UK, pp 138-165.
59. Horner KJ & Anagnostopoulos GD (1973) Combined effect of water activity, pH and temperature on the growth and spoilage potential of fungi. *J Appl Bacteriol* 36: 427-436.
60. Sethi V & Anand JC (1982) Physico-chemical and microbiological quality of carrot and amla preserves. *Indian Food Packer* 36: 38-43.
61. Stekly J & McBean DM (1976) Sorbate treatment of most prunes. *CSIRO Food Res Quarterly* 36: 31-37.
62. Schade JE, Stafford AE & King AD (1973) Preparation of high moisture dried prunes with sodium benzoate instead of potassium sorbate. *J Sci Food Agric* 24: 905-911.
63. Meyer LH (1978) *Food Chemistry*, AVI Publishing Co Inc, Westport, USA.
64. Amerine MA, Berg HW, Kunkee RE, Ough CS, Singleton VL & Webb AD (1980) *The Technology of Wine Making*, AVI Publishing Co Inc, Westport, USA.
65. Joshi VK (1998) *Fruit Wines,* Directorate of Extension Education, Dr YS Parmar University of Horticulture & Forestry, Nauni, Solan, India.
66. Sponholz WR (1994) Wine spoilage by microorganisms. In: *Wine Microbiology and Biotechnology,* Fleet GH (ed), Harwood Academic Publishers, Switzerland, pp 394-420.
67. Krieg NR & Holt JG (1984) *Bergey's Mannual of Systematic Bacteriology*, Vol 1, Williams and Wilkins, Baltimore, USA.

68. Moo D, O'Boyle D, Mathers W & Frost AJ (1980) The isolation of *Salmonella* from jejunal and caecal lymph nodes of slaughtered animals. *Aust Vet J* 56: 181-183.
69. Rao DN & Sreenivasamurthy V (1985) A note on microbial spoilage of sheep meat at ambient temperature. *J Appl Bacteriol* 58: 457-460.
70. Dainty RH Shaw BG, DeBoer KA & Scheps ESJ (1975) Protein changes caused by bacterial growth on beef. *J Appl Bacteriol* 39: 73-81.
71. Banwart GJ (1989) *Basic Food Microbiology*, CBS Publishers and Distributors, New Delhi, India.
72. Barnes EM & Shrimpton DH (1957) Causes of greening of uneviscerated poultry carcases during storage. *J Appl Bacteriol* 20: 273-285.
73. Hauver WE & Hamann JA (1955) *Egg Grading Manual*, Agricultural Handbook No. 75, US Department of Agriculture, Washington, USA.
74. Kajs TM, Hagenmaier R, Vanderzant C & Mathil KF (1976) Microbiological evaluation of coconut and coconut products. *J Food Sci* 41: 352-356.
75. Kinderler Z (1984) Spoilage of desiccated coconut resulting from growth of xerophilic fungi. *Food Microbiol* 1: 23-28.
76. Nakamura M & Kelly KD (1968) *Clostridium perfringens* in dehydrated soups and sauces. *J Food Sci* 33: 424-425.
77. Sugiyama H & Yang KH (1975) Growth potential of *Clostridium botulinum* in fresh mushrooms packaged in semi permeable plastic film. *Appl Microbiol* 30: 964-969.
78. Swartzentruber AH, Schwab AH, Wentz BA, Duran AP & Read RB (1984) Microbiological quality of biscuit dough, snack cakes, and soy protein meat extender. *J Food Prot* 47: 467-470.
79. Terry RC & Overcast WW (1976) A microbiological profile of commercially prepared salads. *J Food Sci* 41: 211-213.
80. Wentz BA, Duran AP, Swartzentruber A, Schwab AS & Read RB (1984) Microbiological quality of frozen breaded onion rings and tuna pot pies. *J Food Prot* 47: 58-60.
81. Wright C, Kominos SD & Yee RB (1976) *Enterobacteriaceae* and *Pseudomonas aeruginosa* recovered from vegetable salads. *Appl Environ Microbiol* 31: 453-454.
82. Sand FEMJ (1977) Spoilage of fountain syrup by *Saccharomyces bailii. Brauwelt* 117: 238-243.
83. Kaushik R, Joshi VK & Gupta JK (1993) Total soluble solids, acidity, pH and standard plate counts in the Indian honey as affected by different treatments and storage conditions. *J Food Sci Technol* 30: 442-443.
84. Bourne MC (2004) Selection and use of post harvest technologies as a component of the food chain. *J Food Sci* 69: 43-46.
85. Splittstoesser DF, Einset A, Wilkison M & Preziose J (1974) Effect of food ingredients on the heat resistance of *Byssochlamys fulva* ascospores. In: *Proceedings of $IV^{th}$ International Congress on Food Science and Technology,* Madrid, Spain, pp 79-85.
86. Heatherbell DA, Short JL & Strueb P (1977) Apple juice clarification by ultrafiltration. *Confructa* 22: 157-169.
87. Kosisowski FV & Moreno V (1970) Removing microorganisms in fresh apple juice by bacterial centrifugation. *J Food Sci* 35-36.
88. Grampp E (1978) Studies on hot clarification (in German). *Fluss Obst* 45: 336-340.
89. Potter NN & Hotchkiss JH (1996) *Food Science,* CBS Publisher and Distributors, New Delhi, India.
90. DeEll JR, Prange RK & Peppelenbos HW (2003) Postharvest physiology of fresh fruits and vegetables. In: *Handbook of Postharvest Technology: Cereals, Fruits, Vegetables, Tea*

*and Spices,* Chakraverty A, Majumdar AS, Raghavan GSV & Ramaswamy HS (eds), Marcel Dekker Inc, New York, USA, pp 455-484.

91. Joshi VK, Kaur M & Thakur NS (1996) Lactic acid fermentation of mushroom (*Agaricus bisporus)* for preservation and preparation of sauce. *Acta Alimentaria* 25:1-13.
92. Yoon KY, Woodams EE & Hang YD (2006) Production of probiotic cabbage juice by lactic acid bacteria. *Biores Technol* 97: 1427-1430.
93. Panda SH, Parmanick M, Ray RC (2007) Lactic acid fermentation of sweet potato (*Ipomoea batatas* L.) into pickle. *J Food Proc Preserv* 31: 83-101.
94. Joshi VK, Bhutani VP and Thakur NK (1999) Composition and nutrition of fermented products. In: *Biotechnology: Food Fermentation. Microbiology, Biochemistry and Technology,* Vol 1, Joshi VK & Pandey A (eds), Educational Publishers and Distributors, New Delhi, India, pp 259-320.
95. Soomoro AH, Masud T & Anwaar K (2002) Role of lactic acid bacteria (LAB) in food preservation and human health–A review. *Pakistan J Nutr* 1: 20-24.
96. Potter NN (1987) *Food Science*, CBS Publisher and Distributors, New Delhi, India.
97. Joshi VK & Thakur S (1999) Lactic acid fermented beverages. In: *Post Harvest Technology of Fruits and Vegetables,* Vol 2, Verma LR & Joshi VK (eds), Indus Publishing Co, New Delhi, India, pp 1102-1127.
98. Beuchat LR (1981) Synergistic effects of potassium sorbate and sodium benzoate on thermal inactivation of yeasts. *J Food Sci* 46: 771-777.
99. Kapur OP & Nagaraja KV (1984) Current trends in the use of food additives. In: *Proceedings of the 4th Indian Convention of Food Scientists and Technologists on Processed Foods Additives*, AFST (1) and CFTRI, Mysore, India, pp 30-33.
100. Ranganna S (1984) Modern concept and practice of chemical preservatives. In: *Proceedings of the 4th Indian Convention of Food Scientists and Technologists on Processed Food Additives*, AFST (1) and CFTRI, Mysore, India, pp 60-62.
101. Gibson B (1973) The effect of high sugar concentrations on the heat resistance of vegetative microorganisms. *J Appl Bacteriol* 36: 365-376.
102. Corry JEL (1976) The effect of sugars and polyols on heat resistance and morphology of osmophilic yeasts. *J Appl Bacteriol* 40: 269-276.
103. Cerny G (1980) Dependence of thermal inactivation of microorganisms on the pH value of media. I. yeasts and moulds. *Z Lebensm Unters Forsch* 170: 173-179.
104. Beuchat LR (1983) Conbined effects of food preservatives and organic acids on thermal inactivation of yeasts in fruit juices. *Lebensmittel Wiss Technol* 16: 51-54.
105. Bhat A & Joshi VK (1998) Ohmic processing of foods. The concept, application, present status and future outlook. *Alimentaria* 289: 83-88.
106. Cabeza MC, Garcia ML, Hoz LDL, Cambero I & Ordonez JA (2005) Thermoultrasonication eliminates Salmonellae from intact eggshells without changing to functional properties of their compounds. *J Food Sci* 70: M292-M295.
107. Lacroix M, Marcotte M & Ramaswamy HS (2003) Irradiation of fruits, vegetables, nuts and spices. In: *Handbook of Postharvest Technology: Cereals, Fruits, Vegetables, Tea and Spices,* Chakraverty A, Majumdar AS, Raghavan GSV & Ramaswamy HS (eds), Marcel Dekker Inc, New York, USA, pp 623-652.
108. Rababah T, Hettiarachchy NS, Eswaranandam S, Meullenet JE & Davis B (2005) Sensory evaluation of irradiated and non-irradiated poultry breads meat infused with plant extracts. *J Food Sci* 70: S228-S235.

109. Toledo TCFD, Cannialti-Brazaca SC, Spoto MHF & Arthur V (2005) Sensory evaluation of chicken breast under gamma irradiation at commercial doses. *J Food Sci* 70: S8-S12.
110. Thakur NS, Joshi VK & Thakur NK (2000) Principles and Methods of Preservation of Fruits and Vegetables. In: *Postharvest Technology of Fruits and Vegetables–Handling, Processing, Fermentation and Waste Management,* Vol 1, Verma LR & Joshi VK (eds), Indus Publishing company, New Delhi, India, pp 148-181.
111. Harrington WO & Hills CH (1968) Reduction of the microbial population of apple cider by ultraviolet irradiation. *Food Technol* 22: 117-120.
112. Mack SD, Albrecht JJ, Litchfield JA & Paker ME (1959) Studies in the cold sterilization of liquid foods using mercury resonance radiation. II. Apple juice. *Food Res* 24: 383-391.
113. Rastogi NK, Subramanian R & Raghavarao KSMS (1994) Application of high pressure technology in food industry. *Indian Food Packer* 13: 30-34.
114. Herrmann J & Andrae W (1963) Oxidative Abbauprodukre der L-Ascorbinsaure. I. Paper chromatographischer Nachweis. *Nahrung* 7: 243-248.
115. Nagy S (1980) Vitamin C contents of citrus fruit and their products: A review. *J Agric Food Chem* 28: 8-18.
116. Fox M, Loncin M & Weiss M (1983) Investigations into the influence of water activity, pH and heat treatment on the velocity of the Maillard reactions in foods. *J Food Qual* 6: 103-118.
117. Mc Weeny DJ (1984) Sulfur dioxide and the Maillard reaction in food. *Proc Food Natr Sci* 5: 395-399.
118. Nafisi K & Markakis P (1983) Inhibition of sugar-amine browning by aspartic and glutamic acids. *J Agric Food Chem* 31: 1115-1123.
119. Robinson C (2001) *Genetic Modification Technology and Food—Consumer Health and Safety*, ILSI Press, Brussels, Belgium.
120. Lee CH (2004) The role of biotechnology in modern food production. *J Food Sci* 96: 92-95.

# Baker's Yeast Production

# 14

**S. Bhushan[1], R.C. Kasana[1] and V.K. Joshi[2]**

[1]*Division of Biotechnology, Institute of Himalayan Bioresource Technology (CSIR), Palampur 176 061 (H.P.)*
[2]*Department of Food Science and Technology, Dr. Y.S. Parmar, University of Horticulture and Forestry, Nauni, Solan 173 230 (H.P.)*

## 1. INTRODUCTION

Historically speaking, existence of bread dates back to around 2600 BC in Babylonian when bakery scenes were discovered in the carving among the ruins of ancient Memphis. By 12$^{th}$ century, baking developed into a specialized craft. Earlier 'yeast' (known as *lees*) generally related with the residues of wine, beer and other alcoholic fermentation and were not associated with bread making [1]. The significance of bread making was realized when it moved out of home and developed into one of the most significant food processing sector i.e., "Baking Industry". 'Leavened bread' from ancient Egyptians, developed into sour-dough process and spread to other areas of the Mediterranean world [2, 3]. The accumulated evidence clearly revealed that Egyptians were familiar with the use of yeast (known as *barm*) in brewing and baking, but reported to have poor quality and performance [4]. Moreover, the practice of using beer yeast in sour-dough fermentation continued since 19$^{th}$ century [5]. But due to its bitter taste and poor fermentation activity, it was replaced by distiller's yeast and finally by baker's yeast [6].

The earliest report on baker's yeast production dates back to 1781 in Holland by a Dutch-process, with a yield of 4-6 per cent to that of the total amount of raw material used. Later on, Mautner developed a *Vienna process* in 1846. In 1919, a fed-batch culture process was developed (*Zulaufverfahren*) by Sako in Denmark and Hayduck in Germany. The sugar was incrementally fed to an aerated suspension of yeast. Simultaneously, the grain mash was replaced by molasses and the yield increased to a theoretical

**Table 1:** Major breakthrough in the production of brewer's yeast.

- Sour-dough fermentation conducted with brewer's yeast
- Replacement of brewer's yeast with distiller's yeast
- Changes in breweries yeast strain from top fermenting (*Saccharomyces cerevisiae*) to bottom fermenting (*Saccharomyces uvarum*)
- Pasteur's effect: Aeration led to higher biomass production
- Development of fed-batch culture (*Zulaufverfahren* process) by Sako (in Denmark) and Hayduck (in Germany)
- Introduction of pure culture yeast strains, with superior quality
- Effect of fermentation under sugar limiting medium to avoid '*Crabtree*' effect
- Replacement of malt and grain by molasses and addition of growth factors

*Source:* 1, 4-8, 10-14

*Food Biotechnology: Principles and Practices*
*V.K. Joshi & R.S. Singh (Eds.)*
*I.K. International Publishing House Pvt. Ltd., India*
*pp 471–494*

maximum of 50 per cent by weight of raw material used, without concomitant production of spirit [7-9]. The molasses, however, as a substrate has become an unattractive proposition because of improved sugar recovery from cane juice resulting in high BOD fermenter effluents [1, 10]. The major breakthrough made in the development of baking industry or baker's yeast production is summarized in Table 1. Recently, baker's yeast is coaxed to form biofilm, which adheres avidly to a number of plastic surfaces. Several strains have been tested alongwith nutrients manipulation of growth medium to form a robust biofilm [15]. Baker's yeast *Saccharomyces cerevisiae* initiated biofilm when grown on low glucose medium and formed 'mats' on semi-solid medium [16]. Presently, baker's yeast production is done by making full use of its respiratory system and incremental feeding of sugar solution and forms the subject matter of this chapter.

## 2. PRODUCTION STATISTICS

A tremendous rise in world wheat production simultaneously increased the amount of flour used for baking. The total wheat production in India was 71.01 MT during 1998-99 [17]. In baking, each 100 kg of flour would require around 0.5-2.0 kg of compressed yeast for the manufacture of leavened bread [1]. It emphasizes the total requirement of compressed yeast needed for the production of leavened bread, which has been an essential part of present fast growing world. In developing countries, the growth of bread consumption has been reported to increase at a rate of 10-20 per cent, annually. Whereas, India exhibited an increase in bread consumption at an annual compound rate of 13.2 per cent [1, 18]. In comparison to world production (1.8 MT), Indian production is merely 0.011 MT during 1992. Thus, it is obvious that bakery industry requires some extra attention.

## 3. TYPES OF BAKER'S YEAST

Basically, three types of baker's yeast are produced viz., wet compressed yeast (WCY), active dry yeast (ADY) and instant dry yeast (IDY) [10, 18]. Wet compressed yeast (WCY) has been reported to contain 70-72 per cent moisture content with a limited shelf-life (4-5 weeks). Compressed yeast is a perishable commodity and must be refrigerated or frozen at all times. Refrigerated compressed yeast remains useful for several weeks before moulds begin to grow. Frozen compressed yeast can be stored and used for up to a month, but some softening of the yeast cake occurs. The ADY (moisture content <8%) has a quite longer shelf-life (about 1 years without refrigeration), but lower baking activity than compressed yeast. The instant dry yeast was found to have a longer shelf-life (>1 year without refrigeration) than ADY even under ambient conditions.

## 4. YEAST HANDLING CHARACTERISTICS

Different types of baker's yeasts need different handling techniques, as summarized in Table 2.

**Table 2:** Handling characteristics of baker's yeast.

| Aspect/Criteria | Fresh compressed | Fresh cream (liquid) | Active dry | Instant |
|---|---|---|---|---|
| Receiving temperature | 33° to 45°F | 33° to 45°F | Room temperature | Room temperature |
| Storage temperature | 33° to 42°F | 33° to 42°F | Room temperature | Room temperature |
| Storage life | 2 to 4 weeks | 2 to 4 weeks | 2 to 12 months | 1 year plus |
| Water content | 66 to 73% | 80 to 85% | 6 to 8% | 4 to 6% |
| Conversion factor (compared with compressed) | 1 | 1.5 to 1.8 | 0.4 to 0.5 plus water | 0.33 to 0.4 plus water |

Contd...

Contd...

| | | | | |
|---|---|---|---|---|
| Handling requirements | Weigh and add other ingredients or disperse in water before mixing. | Meter and add other ingredients. | Must rehydrate in 105° to 110°F water for 10 to 15 minutes before use. | Dry blend with other dry ingredients or delay addition. |

## 5. BAKER'S YEAST CHARACTERISTICS

Various characteristics of baker's yeast are summarized in Table 3.

**Table 3:** Characteristics of baker's yeast of different types.

| Aspect/Criteria | Fresh compressed | Fresh cream | Active dry | Instant |
|---|---|---|---|---|
| Composition (total solids %) | | | | |
| Protein (N x 6.25) | 50 to 56 | 50 to 56 | 41 to 47 | 44 to 53 |
| Phosphorus (as $P_2O_5$) | 2.8 to 3.2 | 2.8 to 3.2 | 2.2 to 2.5 | 2.3 to 3.2 |
| Ash | 6 to 8 | 6 to 8 | 6 to 8 | 6 to 8 |
| Fat | 4 to 6 | 4 to 6 | 4 to 6 | 4 to 6 |
| Moisture | 66 to 73 | 80 to 85 | 6 to 8 | 3.5 to 6 |
| Gassing power (g glucose fermented/g yeast solids/h) | | | | |
| Straight dough | 2.2 to 2.3 | 2.2 to 2.3 | 1.4 to 1.5 | 1.6 to 1.8 |
| Lean dough | 2.4 to 2.6 | 2.4 to 2.6 | 1.2 to 1.3 | 1.8 to 2.0 |
| Sweet dough | 1.0 to 1.1 | 1.0 to 1.1 | 0.4 to 0.5 | 0.6 to 0.8 |
| Equivalent amounts (to produce equal gassing power) | | | | |
| % based on flour | 2.5 | 1.0 to 1.25 | 0.8 to 1.0 | 3.75 to 4.5 |

1 lb compressed = 1.5 to 1.8 lb cream or 6 to 8 oz active dry or 5 to 6.5 oz instant.

## 6. YEAST STYLES

### 6.1 Fresh Cream/Liquid Yeast

In yeast production, after batch growth is completed, the yeast cells are washed, cooled and concentrated by centrifugation. It can be pumped and metered with ordinary liquid-handling equipment. All equipment should be sanitary stainless steel, arranged for clean-in-place (CIP) sanitation weekly. Storage tanks should be cleaned immediately after emptying.

### 6.2 Fresh Compressed Yeast

Cream yeast is dewatered, raising the solids content to approximately 30% to make the traditional form of yeast: compressed yeast. It is homogenized and extruded to form 1-lb, 2-lb or 5-lb blocks and wrapped in waxed paper or cellophane before packaging in various case sizes. The yeast are scaled only as needed for each dough. The exposure to room temperature should be minimized. Partial blocks at the end of the shift are discarded. It is also available in a crumbled form, packed in 50 lb or 25 kg multi-wall paper bags with polyethylene liners. Do not expose to air for extended times without refrigeration.

### 6.3 Active Dry Yeast

Active dry yeast is a compressed yeast dried to 92-94% solids. It must be rehydrated by slurrying in water.

### 6.4 Instant Yeast

New drying methodology gives a form of dry yeast that does not have to be rehydrated before use. It can be added directly to dough before mixing. It is packed under vacuum in 1 lb, 10 lb or 25 lb bricks. Opened packages should be refrigerated in a closed container and used within three to five days.

## 7. MORPHOLOGY OF BAKER'S YEAST

Yeast is the first and most significant microorganism exploited by man to understand the role of different microorganisms and their significance in human life. The word 'yeast' indeed defines a wide range of fungi, whose usual growth form is unicellular [19]. The *Saccharomyces* yeast comes under Ascomycetes group of fungi. As it forms ascospores, it is also called ascosporogenous yeast. The yeasts comprising about 350 species in 39 genera are classified according to a number of morphological (Growth form, size and shape, number of ascospores and shape, and formation of a pellicle or skin) and physiological (fermentation in synthetic media, carbohydrate assimilation, pigment formation, enzyme activity, production of esters or acid etc.) characteristics [20]. A wide range of shape (round or oval) has been reported in *Saccharomyces*. The division of cells is typically by budding. Strains of *Saccharomyces cerevisiae* and *S. carlsbergensis* used for brewing have been classified on the basis of their performance during fermentation. Yeasts are reported to depend strictly on the presence of organic carbon as an energy source, being non-photosynthetic organisms [21-23].

The research has been directed towards the effects of environmental stresses on the morphology of baker's yeast [24-28]. Effect of aeration conditions, sugar limitation, etc. on cytoplasmic membrane [29, 30] and mitochondria [25] were found more pronounced and reported to affect the activity and cryotolerance of baker's yeast [31]. Baker's yeast has been differentiated from laboratory and other industrial strains by principal component analysis, and later on genetic analysis of haploids obtained from industrial strains of baker's yeast were reported and their genotypes were also described [32, 33]. In recent years, improvement in *S. cerevisiae* strains has been made to increase the fermentation efficiency by using protoplast fusion techniques [34].

## 8. PHYSIOLOGY OF BAKER'S YEAST

### 8.1 Strain of Baker's Yeast

Initially, brewer's yeast (*Saccharomyces uvarum* or *S. carlsbergensis*) was used for baking purpose, which later on was replaced with distiller's yeast, and finally, by baker's yeast [4, 5]. *Saccharomyces cerevisiae* was evaluated from about 75 cultures as having superior baking activity [35].

### 8.2 Preservation of Yeast Cultures

Conventional agar slant storage at refrigerated temperature (4-5°C) has been reported to be a simple method of preservation, which reduced the metabolic activity of culture for a period of 3-6 months. A number of growth media have been formulated and advocated for the preservation of yeast [20]. Besides this practice, other preservation methods such as storage under oil, in water (4°C), in soil, by dehydration or freezing (–55°C, using protective agents such as glycerol, dimethyl sulphoxide, etc.) and lyophilization or freeze drying (a few drops of propagating yeast suspension in a protective agent of 1:1 skim milk and water into an ampoule, placing a piece of sterile cotton plug and freezing to –35°C and removal of moisture from ampoule by sublimation under vacuum of 200 μm Hg and storing under refrigeration) etc. have been reported for better and longer storage life [36-40].

## 8.3 Carbon Source and Nutritional Requirements

From the elemental composition (Table 4) of baker's yeast reported by various authors, a general idea about its nutritional and energy requirement can be made [4, 8, 41-43]. A material balance equation for production of baker's yeast was formulated on the basis of its elemental composition which indicated the overall mass flow during fermentation [44].

| Sucrose + | $NH_3$ + | $O_2$ + | Ash → | Yeast biomass + | $CO_2$ + | Water + 387 kCal |
|---|---|---|---|---|---|---|
| (200g) | (10.4g) | (100.5g) | (7.5g) | (dry weight basis) (100 g) | (140.14 g) | (78.12 g) |

**Table 4:** Elemental composition of baker's yeast.

| **Constituents** | **Concentration (% dry weight)** |
|---|---|
| Carbon | 46.0 |
| Oxygen | 32.0 |
| Nitrogen | 8.5 |
| Hydrogen | 6.0 |
| Ash | 4.7-11.5 |
| *Minerals* | |
| $P_2O_5$ | 1.9-5.5 |
| $K_2O$ | 1.4-4.3 |
| MgO | 0.1-0.7 |
| CaO | < 0.2 |
| $SiO_2$ | < 0.2 |
| Others | < 0.1 |

*Source:* 4, 8, 41, 42, 56

An empirical formula i.e., $C_6H_{10.9}N_{1.03}O_{3.06}$ for baker's yeast was developed and from this equation, the nutritional requirement of baker's yeast was calculated [13]. It was reported that the growth medium must supply these constituents to achieve higher biomass. Besides these basic requirements, supplementation of medium with P, K, Mg, Na, S, Fe, Cu, and Zn for proper growth was also suggested [45]. The vitamin content in yeast cells has also been determined and six compounds in the vitamin B complex (i.e., biotin, pantothenic acid, inositol, thiamin, pyridoxine and niacin) were reported to be essential for the growth of certain species and strains of yeast including baker's yeast [46-55].

Baker's yeast propagation usually depends upon glucose, fructose, mannose, galactose, sucrose, maltose, hydrolyzed lactose and fructose moiety of raffinose as well as ethanol [57-63]. Fermentation and/or respiration of these carbon sources by yeasts is only possible when transported into the yeast cells. Monosaccharides are reportedly transported by carrier mediated or facilitated diffusion, while disaccharides such as sucrose are firstly hydrolyzed in the periplasmic space by invertase and then transported into the cells [57, 64]. Glucose as a substrate during aerobic conditions is catabolized by *Saccharomyces cerevisiae* through glycolytic-TCA cycle pathway (about 94%) or pentose phosphate (6%) pathway [57]. In trisaccharide such as raffinose (only the fructose moiety) present in molasses is utilized by the yeast after its hydrolysis by invertase [60].

## 8.4 Reserve Carbohydrate

Trehalose, a non-reducing disaccharide, is well documented in fungi [65]. In *Saccharomyces cerevisiae,* accumulation of this disaccharide as a reserve carbohydrate had been generally associated with the complete exhaustion of glucose from the medium and/or period of reduced growth, especially starvation of cells for nitrogen, phosphate or sulphur [66-70]. The biosynthesis of trehalose reportedly took place when the glucose levels in the medium fell below a critical level [69, 72]. Addition of glucose during stationary phase resulted into a transient increase in trehalase activity, however, its role in the degradation of trehalose is still unknown [72, 73]. *Saccharomyces cerevisiae* has been known to synthesize and degrade glycogen as well as trehalose [68, 74-75]. Both of these carbohydrates are reported to function as important carbon and energy source in starving cells [76, 77], in cells undergoing respiratory adaptation [69, 78], and in sporulating or vegetative cells during emergence from stationary phase [79]. Cytoplasm of yeast cell has been reported to contain trehalose [80], though it is known to be present in cytosol [81]. The survival of yeast cells during prolonged starvation mainly depends on the trehalose content [68]. It also had effect on cell viability and activity of pressed yeast [82, 83]. Intracellular trehalose has been reported to exert a protective effect on yeasts under extreme environmental conditions such as desiccation, freezing, osmotic stress and heat shock besides providing thermal stability to the cells [84-87]. Trehalose has been reported to act by replacing water molecules required for maintenance of the protein structure through multiple external hydrogen bonds [88-90]. Besides its main role as storage carbohydrate or general stress compound, it is also reported to function as carbon source, compatible osmolite or even as a structural component [91].

## 8.5 Growth Kinetics

Baker's yeast production has developed from anaerobic batch fermentation to an aerobic fed-batch or continuous fermentation process for higher cell biomass [10]. In a batch process, the log or exponential phase has been described [10, 92] by the equation given below:

$$dx/dt = \mu x \text{ or } dN/dt = \mu_n N \tag{1}$$

where,

$x$ = is the concentration of cell mass (g $l^{-1}$),

$N$ = Number of cells ($l^{-1}$)

$t$ = time (h)

$\mu$ = specific growth rate of cell mass ($h^{-1}$)

$\mu_n$ = specific growth rate for cell number ($h^{-1}$)

In a fed-batch process, nutrients or substrate(s) are supplied under limiting conditions by incremental feeding. Various types of fed batch culture system have been described [93]. The term fed-batch culture means incremental feeding of medium continuously without the removal of culture fluid. Broadly, the fed-batch culture is divided into: (i) Variable volume fed-batch culture (same medium added, resulting in an increase in volume and or solution of limiting substrate at the same concentration as that in initial medium added, resulting in an increase in volume), and (ii) fixed volume fed-batch culture (a concentrated solution of limiting substrate is added at a lower rate resulting in an insignificant increase in volume). A number of scientists have developed the kinetics of fed-batch culture [94-96]. The biomass at any time of interval considering a batch culture having substrate limiting growth [95] is given by:

$$x_t = x_0 + y\,(S - S_R) \tag{2}$$

where,

$x_t$ = is the biomass concentration,

$x_o$ = is the inoculum concentration,

$y$ = is the yield factor,

$S$ = is the initial sugar concentration,

$S_R$ = is the residual sugar concentration.

$X_{max} = Y.\ S_R$ (3)

If $x = x_{max}$, a medium feed started, where dilution rate is less than $\mu_{max}$ ($\mu = \mu\ S/K_s + S$) and expectedly all the substrate is consumed as it enters the culture, then

$F.S_R = \mu\ (X/Y)$ (4)

where,

$F$ = is the flow rate of medium,

$X$ = total biomass in the culture i.e., $xV$.

Equation (16.4) indicated that input of substrate is equal to its consumption by cells, therefore, ds/dt = 0. It is stated that total biomass increased with time, cell concentration remained constant (dx/dt = 0) and therefore $\mu = D$. This situation is termed as quasi-steady state. With the progress in time, dilution rate will decrease and is given by;

$D = F/C\ (Vo + Ft)$

where,

$Vo$ = Original volume,

$Ft$ = Flow rate at time '$t$'.

Therefore, according to Monod's kinetics, residual substrate should decrease as $D$ decreases, resulting in an increase in cell concentration. A major difference between the steady state of a chemostat and the quasi-steady state of a fed-batch culture is that 'μ' is constant in a chemostat, but decreases in the fed-batch [94, 96].

The doubling time for cell mass is given as:

$$\text{Doubling time} = \frac{0.693}{\mu}$$

where μ = specific growth rate ($h^{-1}$).

A decreased specific growth rate has been observed under fed-batch culture because of the continuous increase in cell mass concentration and decreased substrate concentration in the fermented broth [13]. As the cell mass increased and the yeast productivity approached the limit value of $O_2$ transfer capacity, the growth efficiency of biomass has been reported to decrease, since aerobic growth would tend to shift gradually toward anaerobic growth [12]. The relationship between growth rate and substrate concentration has been found to be similar to the saturation kinetics exhibited by monomolecular absorption [97]. *S. cerevisiae* reported to have 25 mg glucose $l^{-1}$ $K_s$, thus, μ became a strong function of substrate concentration below S = 10 Ks or 250 mg glucose $l^{-1}$ [14].

In a fermentation time of 10-20 h, yeast has been reported to multiply by 6 to 7 folds with yeast solids of 3-8 per cent, depending upon the aeration capacity of the fermenter and the concentration of substrate used [10, 98, 99]. Recently, some modification was introduced in Cybernectic models for the interpretation of the results during fed-batch cultures, especially when the oxygen concentration is quite low as a consequence of the mass transfer limitation [100]. This is really important when a different strain or a different substrate is employed for the said purpose. Therefore, it is a good effort in relation to scale up studies or for production of baker's yeast at industrial level.

## 9. MANUFACTURE OF BAKER'S YEAST

### 9.1 Basic Raw Material

Presently, the commercial baker's yeast production depends mainly upon the beet and cane sugar molasses (containing upto 50% invert sugar). However, various other cheaper substrates like sulphite waste, starch hydrolysate, hydrolyzed cheese whey [1, 18] corn starch and flour etc., have also been advocated for this purpose [101]. These substrates were used as such or in combination after diluting to 30-40°B. There are a number of reports suggesting the use of such basic raw materials for the production of baker's yeast [102-104]. Originally, the basic raw material used was the filtrate recovered from mashed cereal grains [1]. Later on, grains became prohibitively expensive and the processes for utilizing discarded molasses were developed, which were inexpensive and readily available. The proximate composition (Table 5) of molasses has been studied in detail. Molasses, besides being cheaper, have been found to contain a large number of growth supporting substances like nucleic acid bases, organic acids, amino acids, and sugars other than sucrose and alcohols. They have also been reported to contain some growth inhibitory compounds such as furfuraldehyde, hydroxyl methyl furfural, sulphuric acid, imido-sulphonate and sodium sulphite [18, 105].

Beet molasses (deficient in vitamins like biotin) were mixed with cane molasses (rich sources of vitamins) for the growth of baker's yeast during fermentation. However, the clarification and heavy supplementation with ammonia and phosphate was found necessary to overcome the deficiency in the media [1, 18]. The growth of baker's yeast on molasses medium was studied and the utilization of molasses from the sugar industry as a carbohydrate source to fulfill the ever increasing demand of yeast for the bakeries that have been set-up in recent years has also been reported [102]. Production

**Table 5:** Proximate composition of molasses.

| Constituents | Concentration |
|---|---|
| Moisture (%) | 17.0-28.0 |
| Total sugar (%) | 42.5-63.0 |
| Glucose (%) | 4-9 |
| Sucrose (%) | 30-40 |
| Total solids (%) | 75-85 |
| Volatile acidity (% acetic acid) | 0.18 |
| Total nitrogen (%) | 0.1-4.5 |
| Total ash (%) | 3.5-15.0 |
| Protein (%) | 0.5-1.5 |
| Total $SO_2$ (%) | 0.05 |
| Amino acids (%) | 0.3-0.5 |
| pH | 5.4 |
| *Minerals (wt %)* | |
| Phosphorus ($P_2O_5$) | 0.01-2.0 |
| Calcium (CaO) | 0.15-1.7 |
| Magnesium (MgO) | 0.25-0.8 |
| Potassium ($K_2O$) | 0.80-2.3 |
| *Vitamins (μg/g)* | |
| Biotin | 0.9-1.8 |
| Calcium pantothenate | 16.0 |
| Inositol | 2,500 |
| Thiamin* | 1.4-8.3 |
| Pyridoxin* | 6-7 |
| Riboflavin* | 2.5 |
| Nicotinamide* | 20-25 |
| Folic acid | 0.04 |
| Non-nitrogenous organic acid (%) | 1.5-6.0 |
| Aconitic acid (%) | 0.1-1.0 |

*Black strap cane molasses

*Source:* 10, 98, 109, 110, 111, 112

of yeast biomass from saccharified Tapioca processing waste (about 20% of the tapioca tuber/chip) was reported to contain 50-65 per cent starch content [106]. But it required saccharification treatment, thus, adding to the cost of final product. However, sugar cane juice with added biotin has been used successfully for production of baker's yeast [107]. Other substrates employed include the waste date products [108]. It was reported that the modified medium enriched with date-seed ash and date-seed lipid as carbon and nitrogen source was found to be satisfactory.

Scanty information on the utilization of apple pomace for the production of baker's yeast is available. However, attempt to utilize it for the production of yeast biomass has been made [113-115]. Different yeasts viz., *Saccharomyces cerevisiae, Candida utilis* and *Torula utilis* were tried and it was found that addition of 1.8 per cent ammonium sulphate along with mineral nutrients to the diluted apple pomace (1:6) resulted in higher biomass yield. Since, it was a simple batch process, the yield was quite low. The composition of dried apple pomace documented in the literature revealed that the total carbohydrates ranged between 48.0-62.0 per cent with a fermentable sugar content of 35-48 per cent [112, 116, 117]. In earlier studies, different substrates for the production of baker's yeast viz.,

**Table 6:** Proximate composition of different substrates used for baker's yeast production.

| **Constituents** | **Composition (Mean+SE) (on dry weight basis)** | | |
|---|---|---|---|
| | **Apple pomace powder** | **Molasses** | **Jaggery** |
| Moisture content (%) | 4.60 ± 0.521 | 22.40 ± 0.334 | 11.50 ± 0.469 |
| Total carbohydrate (%) | 56.20 ± 1.272 | 64.47 ± 1.69 | 71.96 ± 1.465 |
| Reducing sugar (% glucose) | 41.60 ± 0.941 | 47.52 ± 1.43 | 54.28 ± 1.340 |
| Total soluble solids (TSS, $^{o}$B) | 61.44 ± 1.136 | 74.60 ± 1.181 | 81.44 ± 1.273 |
| Titratable acidity (% malic acid) | 2.71 ± 0.244 | 1.37 ± 0.221 | 1.23 ± 0.208 |
| pH (1:5 dilution) | 3.86 ± 0.111 | 5.76 ± 0.105 | 6.55 ± 0.110 |
| Crude protein (%) | 6.91 ± 0.286 | 1.92 ± 0.08 | 2.16 ± 0.115 |

*Source:* 118

apple pomace, molasses and jaggery were used. The approximate composition of different substrates (Table 6) revealed that molasses had higher moisture content (22.4%) than both the jaggery (11.5%) and apple pomace powder (4.6%). It is expected, since both the jaggery and apple pomace were in solid forms, while molasses was a thick viscous liquid [118]. Although, fresh apple pomace has more moisture, it was dried for latter use in the off-season. There were higher total carbohydrate contents in jaggery than molasses and apple pomace powder. It is apparently due to the more concentration of sugar cane juice during its preparation. Interestingly, compared to molasses and jaggery, the crude protein content (6.91%) was higher in apple pomace powder (APP). Crude protein contents can serve as a nitrogen source for the production of baker's yeast. Perhaps, due to this, the supplementation of molasses medium with additional growth stimulators has been advocated time and again [4, 10, 11, 13, 41, 42]. The average cellular yield coefficient (CYC) was recorded significantly higher in glucose medium i.e., 0.28 g/g sugar consumed. The rest of the media were, however, statistically at par with each other. Figure 1 showed that maximum sugar consumed during the growth by yeast was in glucose based medium, which produced more ethanol and biomass than all other media. The reason can possibly be traced to the presence of glucose, which is readily metabolized by the yeast into ethanol and carbon dioxide besides for its own growth. Molasses medium produced the lowest CYC and this decrease might be due to the presence of some growth inhibitory substances in the medium. Among the different substrates, glucose medium significantly enhanced the biomass yield than other carbon source employed. Moreover, amongst the other substrates, apple pomace extract (APE) had supported higher biomass yield though at par with jaggery, clearly showing its suitability for the production of baker's yeast [114]. Cheese whey has also been used for the total or partial replacement of molasses. Molasses, no doubt is far cheaper, but have to be imported from its place of origin adding transportation and custom duties, thus increasing its price and consequently enhanced the cost of baker's yeast. Changes in the basic raw material have necessitated the search for an alternative substrate or carbohydrate source for the production of baker's yeast. Need of alternate substrate which can reduce the cost of production compared to molasses requiring additional unit operations such as clarification, separation, addition of minerals and other growth supplements, has been felt earlier too. Unfortunately, such efforts are still lacking and require immediate attention. However, our finding will lay down the foundation for the utilization of apple pomace as an alternative substrate for baker's yeast production, but it still requires industrial scale up studies.

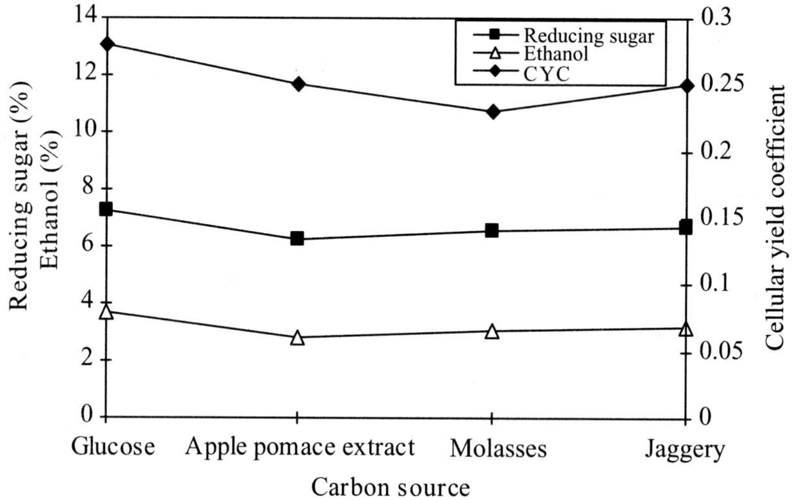


**Fig. 1:** Effect of different carbon sources on the growth characteristics of baker's yeast (irrespective of sugar concentration).
*Source:* 114

## 9.2 Fermentation Process

Various breakthroughs made in the baker's yeast production have led to the development of standard process. Exploitation of respiratory enzyme system of the yeast and the incremental feeding upto the metabolic ability of the yeast formed the basis for baker's yeast fermentation process. It has been reported that when glucose concentration (> 70 mg/L) in an aerobic culture medium increased, the ability of yeast to metabolize glucose tended to be partially converted into ethanol and $CO_2$, and below the specific level of glucose concentration fermentation ceased [12].

Propagation of baker's yeast starting from a slant culture to a commercial fermenter is conducted under limited aeration [18]. To minimize the glucose effect or the so-called 'crab tree' effect (to produce high biomass yield) contributed to the development of fed-batch culture. For proper control of these fermentation parameters and to enhance the productivity in baker's yeast cultivation, a computer control system had been proposed in fed-batch fermentation process [12, 14]. Fed-batch culture fermentation in a bench scale fermenter was carried out with an inoculum of 4.0 g $L^{-1}$ prior to start with medium having an initial concentration of 1.0 g glucose $L^{-1}$. The feeding of fresh medium into fermenter was started after half an hour at an adequate rate using two peristaltic pumps. Computer aided baker's yeast fermentations have also been carried out [13]. It was reported that constant maintenance of high cellular yield and volumetric productivity of the process for each batch was difficult and thus, advocated the use of Zulauf or fed-batch process. An online computer monitoring, as an indirect method of measuring cell density and growth rate, had also been suggested as a viable alternative. The control of sugar feed rate to the fermenter along with an online computer coupled with several process sensors was used; employing material balance to assess indirectly the cellular yield and growth rate during the fermentation. The information obtained from this indirect monitoring was further used to develop an anticipatory online computer strategy to add molasses during fed-batch culture fermentation with the objective of maintaining high volumetric productivity along with maximum cell yield [14]. Vijyalakshmi and Basappa [107] also carried out a fed-batch fermentation process with 10 L fermenter using molasses medium at a constant rate of 4.17 ml $min^{-1}$ over a period of 12 h.

A fuzzy logic-based control system for the production of compressed baker's yeast has been employed in which fuzzy set theory provided an effective tool for translating vague and imprecise linguistic expression in a numerical form that a computer system could easily assimilate [119]. The developed system was based on feeding of molasses and urea along with aeration to achieve the desired productivity. Recently, a fault detecting intelligent system has also been developed for baker's yeast fermentation.

## 9.3 Fermentation Parameters

Optimum conditions during fermentation led to the high productivity of yeast using any process. During baker's yeast production, fermentation parameters such as fermentable substrate concentration i.e., carbon source, temperature, pH, oxygen concentration and agitation, etc., significantly effected the cellular yield [10]. Although adequate sensors are available, but so far no satisfactory method for direct on-line measurement of microbial biomass in the fermenter could be developed [18]. The information available on various fermentation parameters has been reviewed under the following heads.

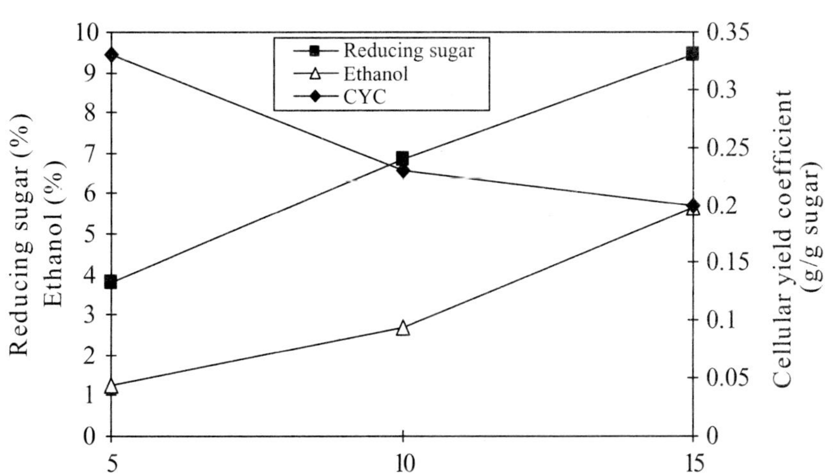


**Fig. 2:** Effect of different initial sugar concentrations on the growth characteristics of baker's yeast (irrespective of carbon source).
*Source:* 114

### *9.3.1 Substrate concentration*

Baker's yeast production under nutrient limiting conditions, especially sugar has resulted in the development of a fed-batch process as early as the onset of $19^{th}$ century (known as Zulaufverfahren the process). Under aerobic conditions, 'glucose effect' or 'Crabfree effect' could not be prevented during the growth of *Saccharomyces cerevisiae.* In an aerobic culture medium, if glucose concentration reached 70 mg $l^{-1}$, it tended to be partly metabolized, fermentatively. A maximum cellular yield of 0.55 g cell/g glucose with a specific growth rate of 0.24 $h^{-1}$ was found using semi-synthetic and molasses media having 30 per cent sugar concentration for 10 l fermenter [12]. A sugar concentration of 0.13 and 0.2 g $l^{-1}$ in the fermenter medium was found to achieve an acceptable cellular yield under fed-batch culture and continuous fermentation with a specific growth rate ($\mu$) of 0.14-0.18 $h^{-1}$ [13]. The initial sugar concentration (ISC) of 5 per cent recorded significantly higher cellular yield coefficient (0.33 g/g sugar consumed) than at 10 or 15 per cent. The higher initial sugar concentration (10 or 15%) failed to produce higher biomass. The trend (Fig. 2) clearly indicated that as the ISC increased from 5 to 15 per cent, the quantity of sugar utilized increased along with ethanol production [114]. But opposite trend was observed in cellular yield coefficient. It clearly indicates that increase in sugar concentration increased the amount of ethanol content, but reduced the cellular yield coefficient. Above these critical limits of glucose concentration, fermentation has been reported to result in the production of more ethanol and intermediate organic acids, which tend to lower the pH of fermentation medium.

A medium having lower sugar concentration has been reported to give higher yield, in general. Earlier it was mentioned that above 50-100 mg glucose concentration, the yeast metabolism tended to be fermentative despite intense aeration [11]. A glucose concentration of around 5 per cent in medium resulted in a residual sugar of 5-10 mg $l^{-1}$ than 45 mg $l^{-1}$ at 15 per cent. A lower concentration of glucose has been suggested by a number of researchers throughout the world [12-14, 107]. At high substrate concentration of the medium, a higher specific growth rate was observed, which gave lower cell yield [10]. It has been due to inhibition of yeast respiration and enhancement of alcoholic fermentation (even in the presence of oxygen), a phenomenon called 'Crabtree effect' [120, 121]. As the glucose concentration increased from 0.6 to 20 per cent in the fermentation medium, the activities of glycolytic enzymes (aldolase, triose phosphate isomerase, pyruvate kinase and pyruvate decarboxylase) increased from 25 to 100 folds [122].

### *9.3.2 Specific growth rate (μ)*

Specific growth rate for biomass production, is considered as a main controlling parameter. A specific growth rate of 0.117 $h^{-1}$ was reported in molasses media for the production of baker's yeast [31], while most of the workers found it to be 0.20-0.30 $h^{-1}$ [12-14] under aerobic conditions of fed-batch culture. As specific growth rate (μ) exceeded 0.18 $h^{-1}$ in a continuous fermentation, yield decreased sharply due to increased substrate concentration at a higher dilution rate, resulting into a higher quantity of ethanol production [79]. A lower cellular yield was also reported as specific growth rate exceeds 0.23 $h^{-1}$. However, the experimental values of specific growth rate ranged from 0.11-0.32 $h^{-1}$ with critical values between 0.2-0.3 $h^{-1}$ on different media [10, 13, 18, 118].

### *9.3.3 Oxygen supply and agitation*

Baker's yeast production, being an aerobic fermentation, largely depends upon oxygen supply and its concentration as well as retention in the medium. Agitation is directly correlated with the supply of oxygen to some extent. Oxygen demand of an industrial fermentation process is normally satisfied by aerating and agitating the fermentation broth [92]. Both the parameters work in cascade during the cultivation of the organisms in a bioreactor. The effect of oxygen concentration in the medium, on the production of biomass had been realized since long, as evident from Pasteur's discoveries known as

*'Pasteur's effect'* i.e., for higher biomass yield, the yeast is cultivated aerobically. Full use of respiratory enzyme system of yeast gave a new direction to the production of baker's yeast [4, 11].

Different studies conducted on the effect of oxygen on the growth of baker's yeast revealed that when yeast was cultivated under different intensities of aeration, its capacity to perform, varied considerably [11, 123, 124]. During an experiment on computer aided baker's yeast production, 29.3 per cent average dissolved oxygen (DO) was used as a controlling parameter along with temperature and pH [13]. Our studies revealed that the dissolved oxygen concentration of 20 per cent (agitation speed ranging 50-450 rpm) is enough in getting the optimum cellular yield under specified conditions of rest of the fermentation parameters [118]. Morphology of baker's yeast under two aeration conditions [28, 31] during fed-batch growth showed that strongly aerobic conditions (16-17 l $min^{-1}$ or 3.1 vvm) gave 35-90% dissolved oxygen, while partially aerobic (about 1 l $min^{-1}$ equivalent to about 0.25 vvm) gave 0 per cent oxygen saturation. It was also found that partial aerobic conditions caused a slight reduction in the volume of outgrowths on the surface of protoplast with fewer invagination on cytoplasmic faces and reduction in the volume of the single giant mitochondrion observed per cell. The presence of giant mitochondrion was generally associated with logarithmic growth of *S. cerevisiae*. Further, slight oxygen deprivation did not have any serious effect on the overall metabolism of the cells, but morphological characters such as invagination on cytoplasmic faces, reduces the mitochondrion volume (which indicates logarithmic growth of cells) [31].

Generally, 1 g $O_2$ is required for production of 1 g yeast solids, according to the material balance and empirical formula [4, 8, 13, 41], which is equal to the experimental value. To get a volumetric productivity of 5 g $l^{-1}$ of yeast solids, about 5 g/156 m mol of $O_2$ $l^{-1}$ $h^{-1}$ would be needed. Aeration rate of 1.5 vvm with a 300 rpm agitation rate had been considered as optimum for a fed-batch fermentation process of baker's yeast production during the studies on effect of carbon compounds added during ripening on activity of baker's yeast and biomass yield [107]. The aeration flow rate at 2.45-3.27 l $min^{-1}$ and 600 rpm impeller rotation speed was found optimum for cellular yield [12], while some workers during their work on baker's yeast growth on molasses media used aeration of 1 vvm and found a satisfactory growth [102]. The process of baker's yeast production and a minimum oxygen transfer rate of 5 g $l^{-1}$ $h^{-1}$ was considered essential to obtain 5 g yeast solid $l^{-1}$ $h^{-1}$ with a sugar feed rate of 10 g $l^{-1}h^{-1}$ during fed-batch fermentation [10, 18].

### *9.3.4 pH*

Variations in pH are reported to have a significant effect on the biomass yield and various workers optimized its value for baker's yeast production. Baker's yeast is known to grow at wide pH range of 3.6-6.0 [6]. However, low pH levels have been suggested to restrict the growth of unwanted microorganisms [18]. But at low pH values, yeasts have been reported to absorb pigments from molasses [4]. Reviewing the literature, a little variation in optimum pH could be easily seen (Table 7).

The author recorded pronounced effect of pH on specific growth rate as well as cellular yield coefficient during the production of baker's yeast in a 5 l bioreactor [118]. The highest biomass production was observed in the medium having pH 4.5.

**Table 7:** Optimum pH values for the growth of baker's yeast.

| pH | Reference(s) |
|---|---|
| 3.0-6.5 | 28, 31 |
| 4.5 | 11, 102, 107, 114 |
| 4.5-5.0 | 6, 12 |
| 5.0 | 13, 14 |
| 4.8-5.0 | 98 |

### 9.3.5 Temperature

Every microorganism has its own environmental conditions for optimum growth, and baker's yeast is no exception. An optimum temperature of 30°C or below was recommended for baker's yeast growth as revealed from the research work conducted from time-to-time (Table 8). The optimum temperature for yeast growth generally varied between 28-32°C. However, a lower generation time was obtained during the yeast growth at 30-36°C. Specific growth rate ($\mu$) was found to be maximum at 32.5°C, while the maximum growth efficiency has been reported at 20°C [4]. Optimal temperature of 28.5°C at pH 4.1 has also been recommended by other investigators [125]. However, a lower cell yield at higher temperature was recorded which indicated the greater maintenance energy requirement. Over 35°C, a rapid decrease in biomass yield was observed [4]. As the temperature increases above its growth optima, the studies revealed that the yeast generally produces more quantity of ethanol, which can be correlated to shift in biosynthetic pathway [118].

**Table 8:** Optimum temperature for baker's yeast growth.

| Temperature (°C) | Reference(s) |
|---|---|
| 28 | 11 |
| 28.5 | 125 |
| 30 | 12-14, 28, 31 |
| 30±1 | 102, 107 |

### 9.3.6 Antifoaming substances

Foam is generated during aerobic fermentation by the protein of liquid-gas interface which has to be controlled. Various synthetic antifoaming agents such as silicones, fatty acid derivatives, glyceride oils etc. have been injected into the fermenting media to control the foaming [126]. Antifoams with higher solubility and little or no effect on oxygen-transfer rate are preferred. The proteins present in medium were found to denature at air-broth interface, developed into a skin which is not easily ruptured. It caused a reduction in the working volume due to outgoing oxygen gas bubbles circulating in the fermentation vessel besides deposition of yeast cells in the upper part of fermenter, resulting into blockage of air filter exits [92].

## 9.4 Cellular Yield

Research and developments in the production of baker's yeast had resulted in a significant increase in the biomass from mere 11.5 per cent during 1947 to 40 per cent with the efforts made for optimization of the medium, using different substrates and supplementation with various growth factors [11]. The theoretical yield has been reported to be 50 per cent to that of substrate (glucose) used, which could only be made possible when the medium was supplemented with amino acids or their amides, yeast extract or minerals [45, 79]. A maximum yield of 51.8 per cent per gram glucose used has been reported and was considered as an adequacy parameter for the composition of the medium [11]. A 26 to 40 per cent yield of *Saccharomyces cerevisiae* (bakery strain) was obtained, when grown on saccharified tapioca processing waste with a specific growth rate of 0.22 $h^{-1}$, and generation time of 3.04 h was obtained [106]. The cell yield had been reduced by the presence of metabolic by-products and other substances such as $CO_2$, if exceeded $1.6 \times 10^{-2}$ g $l^{-1}$ [127]. Generally, a theoretical yield of 0.5 g/g sugar consumed has been reported for baker's yeast [11-14].

## 10. HANDLING AND HARVESTING OF BAKER'S YEAST

### 10.1 Separation, Washing and Cream Storage

After the completion of fed-batch fermentation phase, the yeast slurry is separated by centrifugation from the fermented broth followed by one or more washings [10]. Washing reduces the non-yeast solids, which otherwise hinder filtration, drying, besides darkening the colour of finally produced yeast. The separation and washing process yielded a viscous light coloured yeast cream having 22 per cent yeast solids [10, 18]. Yeast cream is reportedly stored at 2-4°C for a few weeks without appreciable deterioration. The pH of yeast cream has been found to vary between 2.5 to 3.5.

### 10.2 Filtration

Cream yeast obtained by filtration through filter press or rotary vacuum filters to a solid content of 28.33 per cent solids is then extruded into yeast cakes or packaged in large multi-walled paper bags to a size of 10.5-2.25 kg. The compressed yeast is refrigerated at 2-4°C [10].

### 10.3 Production of Active Dry Yeast (ADY)

The wet compressed yeast (WCY) being perishable in nature requires refrigerated storage conditions and has a very short life span [128]. It also needs careful handling, packaging and transportation. To overcome all these problems, production of ADY came into existence as it is known to retain good stability and fermentation activity for quite a longer period [10]. The first successful attempt was made during World War-II to prepare ADY [129]. A number of attempts had been reported earlier for its production and the process was patented in 1920s. In the present bakeries, WCY has been replaced to a large extent by ADY, considering its high stability and quality [18]. Besides its higher cost, it still enjoys the major share of present market. In comparison to WCY, it has many advantages, of course, with a few drawbacks also i.e., drying process cost, rehydration conditions and low activity on dry weight basis [10]. The dehydration of yeast to lower moisture content (4-6%) on one hand improved its thermostability, while on the other hand, it made ADY more sensitive for oxidative deterioration [130].

#### *10.3.1 Drying procedure*

For the production of ADY, drying temperature and time are the major points of concern, which are reported to be directly related with its quality. ADY production process has been described and reviewed by various workers [10, 18, 71, 128, 129] from time-to-time. The yeast cream is pressed to about 68 to 70 percent moisture content and then extruded into strands of various sizes and shapes. The extruded strands are dried in an air stream under controlled conditions of time and temperature as well as moisture content. The living vegetative cells of microorganisms like yeast are reported to be killed above 50°C, and hence, should not be dried at such high temperature [6].

#### *10.3.2 Methods of drying*

A number of drying methods have been employed to produce ADY. In commercial drying, the yeast strands are dried by a flow of warm air (35-60°C) through the bed to a final moisture content of about 8 per cent. The drying time has been reported to vary between 3 to 6 h depending on the original moisture content [128]. A number of methods employed for the production of ADY along with their main features are presented in Table 9. With the improvement in the method, drying time has been greatly reduced with an improvement in quality of the yeast produced. Addition of certain emulsifiers and swelling agents has also been suggested [131].

**Table 9:** Various methods of drying and their important features.

| Drying equipment | Drying time | Drying temperature | Feature |
|---|---|---|---|
| Roto-Louvre drier | For a batch operation time vary (10-20 h) | Air (50-60°C) and yeast particles (<45°C) | Batch and continuous operation, slow drying |
| Through-circulatory drier | 2-4 h | 35-50°C | Batch and continuous operation, slow drying |
| Air-lift fluidized-bed drier | 10 min-4 h | 24-40°C | Rapid drying with a final moisture content of about 7 per cent |
| Spray drier | Few minutes | 60-120°C | Moisture content 5-6 per cent |

### *10.3.3 Factors affecting the quality of ADY*

The quality of ADY is the function of its propagating conditions and drying procedure. The main factors summarized in Table 10 are known to influence the overall performance of baker's yeast. A greater amount of cell constituents has been found to leach out from low moisture (5.3%) ADY than from a high moisture (8.1%). Addition of emulsifier reduced the amount of leached substances from a low-moisture ADY [132].

**Table 10:** Factors affecting the quality of ADY.

| Properties | Effect on quality |
|---|---|
| High protein content | Loss in yeast activity and poor stability due to hydrolysis of phospholipid by lecithinase |
| Degree of catabolite repression | High degree leads to the reduction in glycolytic enzyme activity, which as a result affects fermentation efficiency |
| High trehalose content | Good gassing activity |
| Addition of antioxidant | Reduces the oxidative deterioration |
| *Physical appearance* | |
| Pellet | High storage stability due to coating developed by the tumbling yeast particles during drying process |
| Granular | Inferior storage stability, as yeast particles are not dense |
| Fine yeast particle | Poor storage stability due to large surface area |
| Rehydration temperature | Below optimum rehydration temperature (45°C), rate of leaching is high and above 45°C yeast begins to be inactivated by heat |

*Source:* 71, 74, 82, 128-130, 133, 134

## 11. QUALITY OF BAKER'S YEAST

The production of baker's yeast biomass at high efficiency and productivity did not necessarily have the most desirable properties for baking. Primarily, baker's yeast is found to impart appetizing flavour and aroma to the baked product besides its leavening.

### 11.1 Flavour and Aroma Development

Many of the flavour compounds, volatile and non-volatile compounds such as acids, alcohols, aldehydes, esters, ether derivatives, furan derivatives, hydrocarbons, ketones, lactone derivatives, pyrazines, pyrajole derivatives and sulphur compounds have been found in bread. However, little information is available to relate the formation of these compounds to the origin and propagating conditions of baker's yeast [135].

## 11.2 Leavening of Baked Products

The leavening in baked products is generally said to have occurred by the production of $CO_2$ during fermentation and its retention in dough systems. The yeast leavened baked products are produced from dough system containing flour, water, yeast, salt, sugar, shortening, emulsifier and other optimal ingredients. The rate of $CO_2$ production is dependent upon the quality and type of yeast [99, 123, 124, 130]. Composition and concentration of dough ingredients as well as environmental factors such as temperature, pH, etc. were reported to influence the quality of final product [6, 136-138]. Decrease in fermentation activity of baker's yeast was observed when transferred from an anaerobic to aerobic growth conditions, which is attributed to the decrease in the activities of several glycolytic enzymes in the aerobic yeasts such as pyruvate decarboxylase [99, 139, 140], hexokinase and alcohol dehydrogenase [11, 123]. During aerobicity of yeast propagation, the activities of glycolytic enzyme are affected by the DO concentration and growth medium composition. The highest fermentation activity and pyruvate decarboxylase activity were obtained from baker's yeast grown with glucose substrate, followed by that grown with ethanol and pyruvate substrates [123, 124]. The baker's yeast grown at high temperature, high $CO_2$ tension or high aconitic acid content in the molasses substrate usually had lower fermentation activity [127, 141].

## 11.3 Specifications

Besides the flavour/aroma development and leavening activity, baker's yeast must fulfill the general requirements. Wet compressed yeast (WCY) is reported to contain 70-75 per cent moisture content and 7.5-8.0 per cent of nitrogen, which are inversely related to its stability. According to the specifications, WCY should not have more than 75.0 per cent moisture content [142]. Nitrogen, moisture content and trehalose have been considered as the most important factors affecting the activity and stability of active dry yeast [10]. There is a direct relationship between baking activity of ADY, nitrogen and moisture contents. ADY is reported to have good stability at 6.5 to 7.0 per cent of nitrogen with a moisture content of about 8 per cent [71, 128]. The established standards revealed that ADY should contain less than 8.0 per cent moisture content. The maximum contamination (i.e., microflora other than yeast) allowed in WCY and ADY is 1.5 and 8.0 millions cells per g (dry weight basis), respectively [142]. The other quality characteristics included high dough fermentation capacity and good dispersability in water to satisfy the test. WCY should have creamy/white colour with fine even texture, while ADY, either in the powdered form or small granular, pellets or flakes with a good characteristic odour and should not have any sign of deterioration and contamination.

# 12. FUTURE PROSPECTS

It is apparent from the review of research work carried out on various aspects of baker's yeast production that molasses are still the only substrate used commercially with significant industrial output. Other substrates such as waste date products, apple pomace etc. have not so far been utilized to produce baker's yeast at commercial levels. The utilization of these substrates would not only reduce the extra burden from sugar refineries, but would also lead to reduction in the environmental/hazardous pollution occurring from such processing industries and economical gains from these bio-wastes. Therefore, efforts are needed to develop a method, which could be put to commercial usages, as the value of a technology without commercial realization is worthless. The exploitation of baker's yeast application in other areas of life sciences is also gaining momentum. The use of baker's yeast in biotransformation and biosorption of heavy metals by waste baker's yeast biomass present in the environment, mainly in the aquatic systems are a few examples to highlight [143, 144]. Thus we can say that efforts are still required in the exploration of eco-friendly and commercially suitable substrates for the production of Baker's yeast besides increasing the application horizon of this low volume and high value product.

## REFERENCES

1. Ellis B (1980) The role of baker's yeast in developing countries. *Indian Food Packer* 3: 16-17.
2. Jacob HE (1944) *Six Thousand Years of Bread*, Double day, Donam and Co, Garden City, New York, USA.
3. Pyler EJ (1958) *Our Daily Bread*, Siebel Publishing Co, Chicago, USA.
4. White J (1954) *Yeast Technology*, Chapman and Hall, London, UK.
5. Frey CN (1930) History and development of the modern yeast industry. *Ind Engg Chem* 22: 1154-1162.
6. Reed G & Peppler HJ (1973) *Yeast Technology*, VVI Publishing Co, Westport, UK.
7. Pyke M (1957) Baker's yeast. In: *Yeasts*, Roman W (ed), W. Junk Publishers, the Hague, Netherlands, pp 60-75.
8. Harrison JS (1971) Yeast production. *Prog Ind Microbiol* 10: 129-177.
9. Rosen K (1977) Production of baker's yeast. *Process Biochem* 12: 10-12.
10. Reed G (1982) Production of bakers yeast. In: *Prescott and Dunn's Industrial Microbiology,* Reed G (ed), AVI Publishing Co, Westport, USA, pp 593-633.
11. Oura E (1974) Effect of aeration intensity on the biochemical composition of baker's yeast I. Factors affecting the type of metabolism. *Biotechnol Bioengg* 16: 1197-1206.
12. Aiba S, Nagai S & Nishizawa Y (1976) Fed-batch culture of *Saccharomyces cerevisiae.* A perspective of computer control to enhance the productivity in baker's yeast cultivation. *Biotechnol Bioengg* 18: 1001-1016.
13. Wang HY, Cooney CL & Wang DLC (1977) Computer aided baker's yeast fermentation. *Biotechnol Bioengg* 19: 69-86.
14. Wang HY, Cooney CL & Wang DLC (1979) Computer control of baker's yeast production. *Biotechnol Bioengg* 21: 975-996.
15. Helmuth L (2001) Baker's yeast blooms into biofilms. *Science* 291: 806-807.
16. Reynold TB & Fink GR (2001) Baker's yeast, a model for fungal biofilm formation. *Science* 291: 878-881.
17. Venkataramani G (2000) Overview-scaling new heights. In: *The Hindu Survey of Indian Agriculture*, National Press, Chennai, India, pp 5-8.
18. Basappa SC (1999) Baker's yeast: production, quality and utilization. In: *Biotechnology: Food Fermentation - Microbiology, Biochemistry and Technology,* Joshi VK & Pandey A (eds), Educational Publishers and Distributors, New Delhi, pp 1113-1144.
19. Henrici AT (1947) *Molds, Yeasts and Actinomycetes*, Wiley, New York, USA.
20. Wickerham LJ (1951) *Taxonomy of Yeasts, US Department of Agriculture Tech Bull* 1029: 1-56.
21. Carmo-sousa L (1969) *The Yeasts,* Rose AH & Harrison JS (eds), Academic Press, London, UK.
22. Gentles JC & Touche C (1969) *The Yeasts,* Rose AH & Harrison JS (eds), Academic Press, London, UK.

23. Last FT & Price D (1969) *The Yeasts,* Rose AH & Harrison JS (eds), Academic Press, London, UK.
24. Douce R (1985) Mitochondria in higher plants-structure, function and biogenesis. In: *American Society of Plant Physiologists Monograph Series,* Nooden LD & Leopold AC (eds), Academic Press, Orlando, USA, pp 332-345.
25. Harrison DEF (1976) *The Oxygen Metabolism of Microorganisms*, Meadowfield Press Ltd, Shildon, UK.
26. Luzikov VM & Galkin AV (1985) *Mitochondrial Biogenesis and Breakdown*, Plenum Press, New York, USA.
27. Kappeli O, Gschwend-Betrik M & Fiechter A (1985) Transient response of *Saccharomyces uvarum* to a change of the growth limiting nurient in continuous culture. *J Gen Microbiol* 131: 47-52.
28. Gelinas P & Goulet J (1991) Morphology of baker's yeast and dissolved oxygen saturation during fed-batch growth. *Lett Appl Microbiol* 12: 164-170.
29. Moor H & Muhlethaler K (1963) Fine structure in frozen etched yeast cell. *J Cell Biology* 17: 609-628.
30. Willison JHM & Johnson GC (1985) Ultrastructure of *Saccharomyces cerevisiae* strain AGI 7 and its responses to changes in environment. *Can J Microbiol* 31: 109-118.
31. Gelinas P, Fiset G, Leduy A & Goulet J (1989) Effect of growth conditions and trehalose content on cryotolerance of baker's yeast in frozen doughs. *Appl Envi Microbiol* 55: 2453-2459.
32. Oda Y & Ouchi K (1989) Principle component analysis of the characteristics desirable in baker's yeast. *Appl Envi Microbiol* 55: 1495-1499.
33. Oda Y & Ouchi K (1989). Genetics analysis of haploids from industrial strains of baker's yeast. *Appl Envi Microbiol* 55: 1742-1747.
34. Sathe S, Shivaraman H & Gokhale DV (1992) Protoplast fusion in yeast: strain improvement in *Saccharomyces. Indian J Microbiol* 32: 15-27.
35. Mitchell J, Jr H & Enright JJ (1957) Effect of low moisture levels on thermostability of active dry yeast. *Food Technol* 11: 359-362.
36. Martin SM (1964) Conservation of microorganisms. *Ann Rev Microbiol* 18: 1-16.
37. Dalby DK (1982) Pure culture methods. In: *Prescott and Dunn's Industrial Microbiology,* Reed G (ed), AVI Publishing Co, West Port, USA, pp 44-62.
38. Hayner WC, Wickerham LJ & Hesseltine CW (1955) Maintenance of cultures of industrially important microorganisms. *Appl Microbiol* 3: 361-368.
39. Beech FW & Davenport RP (1971) Isolation, purification and maintenance of yeasts. In: *Methods in Microbiology,* Booth B (ed), Vol 4, Academic Press, New York, USA, pp 153-182.
40. Collins CH & Lyne PM (1976) *Microbiological Methods,* Butterworths, London, UK.
41. Frey CN, Kirby GW & Schultz A (1936) Yeast - physiology, manufacture and uses. *Ind Eng Chem* 28: 879-884.
42. Harrison JS (1967) Aspects of commercial yeast production. *Process Biochem* 2: 41-45.
43. Chen SL (1964) Energy requirement for microbial growth. *Nature* (London) 202: 1135-1136.
44. Cooney CL, Wang HY & Wang DIC (1977) Computer-aided material balancing for production of fermentation parameters. *Biotechnol Bioengg* 19: 55-67.
45. Olson BH & Johnson MJ (1949) Factors producing high yeast yields in synthetic media. *J Bacteriol* 57: 235-246.

46. Eddy AA (1958) Aspects of chemical composition of yeast. In: *The Chemistry and Biology of Yeast,* Cook AH (ed), Academic Press, New York, USA.
47. Burkholder PR, McVeigh I & Moyer D (1944) Studies on some growth factors of yeasts. *J Bacteriol* 48: 383-392.
48. Schultz AS & Atkin I (1947) The utilization of Bios response in yeast classification and nomenclature. *Arch Biochem* 14: 369-380.
49. Oura E & Suomalainen H (1978) Biotin and the metabolism of baker's yeast. *J Inst Brew* 84: 283-287.
50. Oura E (1978) The effect of culture conditions on the requirement of biotin by baker's yeast. Research paper presented at 6$^{th}$ International Specialist Symposium on Yeast. Mompellier, France, Abst Sl, pp 1-2.
51. Challinor SW & Power DM (1964) Effect of inositol deficiency on yeast in particular reference to chemical composition of the cell and of the cell wall. *Nature* 203: 250-251.
52. Power DM & Challinon SW (1969) The effect of inositol deficiency on the chemcial composition of the yeast cell wall. *J Gen Microbiol* 55:169-176.
53. Schultz AS, Atkin I & Frey CN (1937) A fermentation test for vitamin B. *J Am Chem Soc* 59: 948-949.
54. Schultz AS, Atkin I & Frey CN (1939) Vitamin B and growth promoting factor for yeast. *J Am Chem Soc* 61: 1931-1932.
55. Melnick DM, Hockberg HM & Oser BL (1945) The multiple nature of vitamin B critique of methods for the determination of the complex and its components. *J Biol Chem* 160: 1-4.
56. Richard OW & Troutman MC (1940) Spectroscopic analysis of the mineral content of yeast grown on synthetic and natural media. *J Bacteriol* 39: 739-746.
57. Chen SL (1959) Carbohydrate assimilation in actively growing yeast, *S. cerevisiae.* I: Metabolic pathways for $C^{14}$ glucose utilization by yeast during aerobic fermentation. *Biochem Biphys Acta* 32: 470-479.
58. Chen SL (1959) Carbohydrate assimilation in actively growing yeast, *S. cerevisiae.* II: Synthesis of polysaccharides from $C^{14}$ glucose. *Biochem Biophys Acta* 32: 480-483.
59. Stineman TL, Edwards JD & Grosskopf JC (1980) Production of baker's yeast from acid whey. US Patent 4: 192918.
60. Barnett JA (1976) The utilization of sugars by yeasts. *Adv Carbohydr Chem Biochem* 32: 125-234.
61. Haarasilta S & Qura E (1975) On the activity and regulation of anaplerotic and gluconeogenetic enzymes during the growth process of baker's yeast. *Eur J Biochem* 52: 1-7.
62. Mor JR & Fiechter A (1968) Continuous cultivation of *S. cerevisiae* I: Growth on ethanol under steady state conditions. *Biotechnol Bioengg* 10: 159-176.
63. Suomalainen H & Oura E (1978) Ethanol as substrate for baker's yeast. In: *Proceedings of XII International Congress of Microbiology,* Munich, Germany, pp 60-65.
64. Goyal RK (1999) Biochemistry of fermentation. In: *Biotechnology: Food Fermentation Microbiology, Biochemistry and Technology,* Vol 1, Joshi VK & Pandey A (eds), Educational Publisher and Distributor, New Delhi, India, pp 87-171.
65. Thevelein JM (1984) Regulation of trehalose mobilization in fungi. *Microbiol Rev* 48: 42-54.
66. Trevelyan WE & Harrison JS (1956) Studies on yeast metabolism: The trehalose content of baker's yeast during anaerobic fermentation. *Biochem J* 62: 177-183.

67. Manners DJ (1971) The structure and biosynthesis of storage carbohydrates in yeast. In: *The Yeast,* Vol 2, Rose SH & Harrison JS (eds), Academic Press, New York, USA, pp 419-437.
68. Lillie S & Pringle J (1980) Reserve carbohydrate metabolism in *Saccharomyces cerevisiae*: Response to nutrient limitation. *J Bacteriol* 143: 1384-1384.
69. Panek AD & Mattoon JR (1977) Regulation of energy metabolism in *Saccharomyces cerevisiae*: Relationship between catabolite repression, trehalose synthesis and mitocondrial development. *Arch Biochem Biophys* 183: 306-316.
70. Francois J, Eraso P & Gancedo C (1987) Changes in the concentration of cAMP, fructose 2,6-bisphosphate and related metabilites and enzymes in *Saccharomyces cerevisiae. Eur J Biochem* 164: 369-373.
71. Mitchell J, Jr. H & Enright JJ (1957) Effect of low moisture levels on thermostability of active dry yeast. *Food Technol* 11: 359-362.
72. Francois J, Neves M & Hers H (1991) The control of trehalose biosynthesis in *Saccharomyces cerevisiae*: Evidence for a catabolite inactivation and repression of trehalose-6-phosphate synthase and trechalose-6-phosphate phosphatase. *Yeast* 7: 575-587.
73. Thevelein JM & Beullens M (1985) Cyclic AMP and the stimultion of trehalase activity in the yeast *S. cerevisiae* by carbon sources, nitrogen sources and inhibitors of protein synthesis. *J Gen Microbiol* 131: 3199-3209.
74. Grba S, Oura E & Suomalainen H (1975) On the formation of glycogen and trehalose in baker's yeast. *Eur J Appl Microbiol* 2: 29-38.
75. Grba S, Oura E & Suomalainen H (1979) On the formation of glycogen and trehalose in baker's yeast. *Eur J Appl Microbiol* 2: 29-38.
76. Chester VE (1963) The dissimilation of the carbohydrate reserves of a strain of *Saccharomyces cerevisiae. Biochem J* 86: 153-160.
77. Panek AD (1969) Adenosine triosphosphate inhibition of yeast trehalase. *J Bacteriol* 99: 904-905.
78. Quain DE & Haslam JM (1979) Changes in glucose beta-phosphate and storage carbohydrates during catabolite depression in *Saccharomyces cerevisiae. J Gen Microbiol* 113: 195-198.
79. Von Meyenburg HK (1969) Energetic of the budding cell of *Saccharomyces cerevisiae* during glucose limited aerobic growth. *Arch Microbiol* 84: 254-260.
80. Sougza NO (1968) Locations of trehalose and trehalose in yeast cell. *Arch Biochem Biophysics* 125: 122-147.
81. Keller F, Schellenery M & Wiomken S (1982) Localization of trehalose in vacuoles and trehalose in the cytosol of yeast *Saccharomyces cerevisiae. Arch Microbiol* 131: 298-301.
82. Suomalainen H & Pfaffli S (1961) Changes in carbohydrate reserves of baker's yeast during growth and on standing. *J Inst Brew* 67: 249-254.
83. Ya Ling Z, Morimura S & Kida K (1995) Effect of fermentation temperature on relationship between cell viability and trehalose content of *Saccharomyces cerevisiae* KF-7 in repeated-batch fermentation. *J Ferment Bioengg* 80: 204-207.
84. Hottiger T, Boller T & Wiemken A (1989) Correlation of trehalose content and heat resistance in yeast mutants altered in the RAS/adenylate cyclase pathway: is trehalose a thermoprotectant? *FEBS Lett* 255: 431-434.
85. Van Laere A (1989) Trehalose: reserve and/or stress metabolite. *FEMS Microbiol Rev* 63: 201-210.
86. Wiemken A (1990) Trehalose in yeast, stress protectant rather than reserve carbohydrate. *Antonie Van Leeuwenhock* 58: 209-217.

87. Hounsa CG, Brandit EV, Thevelein J, Hohmann S & Priar B (1998) Role of trehalose in survival of *Saccharomyces cerevisiae* under osmotic stress. *Microbiology* 144: 671-680.
88. Arakawa T & Timasheff SN (1982) Stablization of protein structure by sugars. *Biochemistry* 21: 6536-6544.
89. Fernandez J, Soto T, Vicente-Soler J Cansado J & Gacto M (1995) Increased thermal stability of the enzyme content in permeabilized whole cells from the fission yeast *Schizosaccharomyces pombe* by exogenous trehalose and other compounds. *Can J Microbiol* 41: 636-641.
90. Cerrutii P, Segovia de Huergo M, Galvagno M, Schebor C & del Pilar Buera M (2000) Commercial baker's yeast stability as affected by intracellular content of trehalose, dehydration procedure and the physical properties of external matrices. *Appl Microbiol Biotechnol* 54: 575-580.
91. Arguelles JC (2000) Physiological roles of trehalose in bacteria and yeasts: a comparative analysis. *Arch Microbiol* 174: 217-224.
92. Stanbury PF, Whitaker A & Hall SJ (1997) *Principles of Fermentation Technology*, Aditya Books Pvt Ltd, New Delhi, India.
93. Yoshida F, Yamane T & Nakamoto K (1973) Fed-batch hydrocarbon fermentation with colloidal emulsion feed. *Biotechnol Bioengg* 15: 257-270.
94. Dunn IJ & Mor JR (1975) Variable-volume continuous cultivation. *Biotechnol Bioengg* 17: 1805-1822.
95. Pirt SJ (1975) *Principles of Microbe and Cell Cultivation*, Black Well Publishing, Oxford, UK.
96. Pirt SJ (1979) Fed-batch culture of microbes. *Ann NY Acad Sci* 326: 119-125.
97. Monod J (1949) The growth of bacterial culture. *Ann Rev Microbiol* 3: 371-374.
98. Burrows S (1970) Baker's yeast. In: *The Yeast*, Rose AH & Harrison JS (eds), Vol 3, Academic Press, New York, pp 349-419.
99. Suomalainen H (1963) Changes in the cell constitutions of baker's yeast in changing growth conditions. *Pure Appl Chem* 7: 634-654.
100. Di Serio M, Tesser R & Santacesaria E (2001) A kinetic and mass transfer model to stimulate the growth of baker's yeast in industrial bioreactor. *Chem Engg J* 82: 347-354.
101. Parisi F (1981) An evaluation of two simple methods for the measurement of fermenting power of baker's yeast. *Pure Applied Chem* 61: 603-610.
102. Srinivasa Rao PN, Basappa SC, Lewis YS & Ramanathan PK (1979) Studies on the growth of baker's yeast on molasses media. *J Food Sci Technol* 16: 70-73.
103. Ejiofor AO, Chisti Y & Moo-Young M (1996) Fed-batch production of baker's yeast using millet (*Pennisetum typhoides*) flour hydrolyzate as the carbon source. *J Ind Microbiol* 16: 102-109.
104. Infanzon A, Degiula M, Anon A & Varela H (1996) Baker's yeast production on whey permeate with *in situ* lactose hydrolysis. In: *Proceeding of XI SINAFERM*, Carlos, Brazil.
105. Kazuhiko T & Kozo O (1995) Factors affecting the ethanol productivity of yeast in molasses. *J Ferment Bioengg* 79: 449-452.
106. Kunhi AAM, Suseela T, Ghildyal NP, Lonsane BK & Ahmed SY (1982) Yeast biomass from saccharified tapioca processing waste. *Indian Food Packer* 36: 79-85.
107. Vijayalakshmi G & Basappa SC (1996) Enhancement of yield activity baker's yeast by carbon compounds added during ripening. *Indian J Microbiol* 36: 189-191.
108. Nancib N, Nancib A & Boudrant J (1997) Use of waste Date products in the fermentative formation of baker's yeast biomass by *Saccharomyces cerevisiae. Biores Technol* 60: 67-71.

109. Tauro P (1981) Production of ethanol from agricultural by products. In: *Recycling Residues of Agriculture and Industry*, Karla MS (ed), RRAI, Symposium, Punjab Aricultural University, Ludhiana, India, pp 199-210.
110. Bardiya MC, Chandna S & Hashia R (1987) *Yeast Biotechnology-II*, Basappa SS & Tauro P (eds), CFTRI, Mysore, India.
111. Joshi VK (1998) Apple pomace utilization - Present status and future strategies. In: *Advances in Biotechnology,* Pandey A (ed), Educational Publishers & Distributors, New Delhi, India, pp 141-155.
112. Devarajan A (1997) Solid state fermentation of apple pomace: Methods of ethanol recovery, physico-chemical evaluation and acceptability of animal feed produced, M.Sc. Thesis, Dr. Y.S. Parmar University of Horticulture and Forestry, Nauni, Solan, India.
113. Joshi VK, Jaswal S & Lal BB (1998) Apple pomace: Effect of sulphur dioxide and temperature on its preservation and medium optimization for yeast biomass production. *J Sci Ind Res* 57: 692-697.
114. Joshi VK & Bhushan S (2003) Apple pomace utilization for the production of baker's yeast: Effect of substrate concentrations and growth stimulator. *Ind J Biotechnol* 2: 220-226.
115. Bhushan S & Joshi VK (2006) Baker's yeast production under fed batch culture from apple pomace. *J Sci Ind Res* 65: 72-76.
116. Smock RM & Neubert AM (1950) *Apples and Apple Products,* Interscience Publishers, New York, USA.
117. Kaushal NK (1992) Apple pomace utilization for edible products, M.Sc. Thesis, Dr. Y.S. Parmar University of Horticulture and Forestry, Nauni, Solan, India.
118. Bhushan S (2002) Apple Pomace Utilization for the Production of Baker's Yeast: Process Optimization, Evaluation and Performance, Ph.D. Thesis, Dr. YS Parmar University of Horticulture and Forestry, Nauni, Solan (HP) India.
119. Mukherjee S, Basu S & Chattopadhya P (1998) A fuzzy logic-based control system for compressed baker's yeast fermentation. *J Food Sci Technol* 35: 164-168.
120. DeDeken RJ (1966) The crabtree effect: a regulatory system in yeast. *J Gen Microbiol* 44: 149-156.
121. Rickard PAD & Hogan BJ (1978) Effect of glucose on the activity and synthesis of fermentative and respiration pathways of *Saccharomyces* species. *Biotechnol Bioengg* 20: 1105-1110.
122. Hommes FA (1966) Effect of glucose on the level of glycolytic enzymes in yeast. *Arch Biochem Biophys* 114: 221-234.
123. Oura E (1972) The production of energy during aerobic growth of baker's yeast on glucose and ethanol. In: *Proceedings of the 2nd Congress on Yugoslav Microbtology,* Opatija, Croatia, pp 237-246.
124. Oura E (1972) The effect of aeration on the growth energetic and biochemical composition of baker's yeast. Ph.D. Thesis, Research Lab of the State Alcohol Manopoly (Alko), Helsinki, Finland.
125. Eroshin VK, Utkin IS, Ladynichev SV, Samoylov VV, Kuvshinnikov VD & Skoyabin GK (1976) Influence of pH and temperature on substrate yield coefficient of yeast growth in chemostat. *Biotechnol Bioengg* 18: 289-295.
126. Solomons GL (1967). Antifoams. *Process Biochem* 2: 47-48.
127. Chen SL & Gutmanis F (1976) Carbon dioxide inhibition of yeast growth in biomass production. *Biotechnol Bioengg* 18: 1455-1462.

128. Thorn JA & Reed G (1959) Production and baking techniques for active dry yeast. *Cereal Sci Today* 4: 198-201.
129. Frey CN (1957) History of the development of active dry yeast. In: *Yeast - Its Characteristics, Growth and Function in Baked Products,* McWilliams CS & Peterson MS (eds), Quartermaster Food and Container Institute for the Armed Force, Chicago, USA, pp 7-33.
130. Chen SL, Cooper EJ & Gutmanis F (1966) Active dry yeast: Protection against oxidative deterioration during storage. *Food Technol* 20: 79-83.
131. Clement P & Rossi JP (1982) Active dried baker's yeast. US Patent 4: 328250.
132. Mitchell J, Jr. H & Enright JJ (1959) Thermostable active dry yeast composition. US Patent 2: 894842.
133. Pollock GE & Holmstrom CD (1951) The trehalose content and the quality of active dry yeast. *Cereal Chem* 28: 498-505.
134. Harrison JS & Trevelyan WE (1963) Phospholipid breakdown in baker's yeast during drying. *Nature* 200: 1189-1190.
135. Suomalainen H & Keranen AJA (1978) Yeast as producer of aroma compounds. In: *Proceedings of the 1st European Congress on Biotechnology*, Interlaken, Switzerland, Monograph, pp 207-220.
136. Sato T, Tsumura N, Tanaka T, Onada T & Koyanag Y (1961) Comparison of practical characters of baker's yeast of the world. *Nippon Jozo Kvokai* 5: 74-80.
137. Watson TG (1970) Effect of sodium chloride on steady state growth and metabolism of *S. cerevisiae. J Gen Microbiol* 64: 91-99.
138. Atkin L, Schultz AS & Frey CN (1946) Yeast fermentation. In: *Enzymes and Their Role in Wheat Technology,* Anderson JA (ed), Wiley Interscience, New York, USA, pp 327-329.
139. Suomalainen H (1975) Some enzymological factors influencing the leavening capacity and keeping quality of baker's yeast. *Eur J Appl Microbiol* 1: 1-12.
140. Polakis ES & Bartley W (1965) Changes in the enzyme activities of *S. cerevisiae* during aerobic growth on different carbon sources. *Biochem J* 97: 284-297.
141. Chen L & Peppler HJ (1956) Destruction of cocarboxylase in active dry yeast. *Arch Biochem Biophys* 62: 299-304.
142. Anonymous (1981) Indian standard specification for baker's yeast. In: *Bakery and Confectionery Industry Sectional Committee,* III Revision, *AFDC* 3, IS: 1320.
143. Fadnavis NW, Vadivel S, Kumara & Bhalerao UT (1997) Biotransformations with baker's yeast: pH effects on diastereo selectivity during c-hydroxy-[5-ketoester reductions and carbon-carbon bond cleavages. *Tetrahedr Asym* 8: 2355-2359.
144. Goksungur Y, Uren S & Guvenc U (2004) Biosorption of cadmium and lead ions by ethanol treated waste baker's yeast biomass. *Biores Technol* 96: 103-109.

# Cereal Based Alcoholic Beverages

# 15

**T.C. Bhalla, Navdeep Thakur, Amit Seth and Amit Pratush**

*Department of Biotechnology, Himachal Pradesh University, Summer Hill, Shimla 171 005 (H.P.)*

## 1. INTRODUCTION

Alcoholic beverages have played an important role in the spiritual and cultural life of people, both in eastern and western societies. In Europe and the Middle East, the indigenous alcoholic beverages are primarily produced from fruits, while these are made from cereals in Asia-Pacific region and serve as important source of nutrients. The cereal based alcoholic beverages include a large variety of beers, beer-like beverages and a number of indigenous and traditional beverages in various parts of the world. European beers mainly use barley malt as the primary raw material. The cereal based alcoholic beverages vary from crystal clear products to turbid thick granules and pastes. Cereal crops are energy dense, containing 10000–15000 kJ/kg, about 10–20 times more energy than most succulent fruits and vegetables. Nutritionally, they are important sources of dietary protein, carbohydrates, vitamins B & E, iron, trace minerals and fibre.

The origin of alcoholic beverages is as old as the human history and civilization. The artefacts taken from the history of ancient cities such as Mesopotamia indicate that the formal practice of brewing existed well before 5000 years ago. The ancient writing and drawings recovered from Egypt show the details of commercial brewing and distillation of beers [1, 2]. In 2000 BC, Indian Aryans produced a beer-like beverage called 'Sura', and as per historical records, the Chinese used to make a kind of beer called 'Skiu'. The major ingredients of these fermented beverages were barley, wheat and rice.

Beer making is one of the oldest methods of fermentation, dating back to thousand of years and probably originating in the Nile valley [3]. In ancient times these beverages were like geographic indicators, because a particular fermentation product was produced in a particular locality, and the fermentation process was more of an art as no specific inoculum was used and it was entirely driven by natural microflora [4]. In earlier times, some characteristics of beer-like flavour could not be controlled due to limited knowledge of flavour imparting or enhancing agents.

It is one of the earliest processes undertaken on a commercial scale that involved transformation of art of beer making at household scale into a technology for its production in industry. The microbiological and biochemical studies on fermentation were initiated by scientists like Louis Pasteur, Liebig and Wohler. In 1876, Pasteur published a paper "Etudes sur la biere" explaining how fermentation was carried out by yeast, and he also reported that souring of beer and wine was caused by undesirable microorganisms [3].

*Food Biotechnology: Principles and Practices*
*V.K. Joshi & R.S. Singh (Eds.)*
*I.K. International Publishing House Pvt. Ltd., India*
*pp 495–524*

The fermentation starters prepared by growing yeast on raw or cooked cereals are more commonly used for the production of traditional fermented beverages for initiating fermentation. These fermentation starters are referred to as *chu* in Chinese, *nuruk* in Korean, and *koji* in Japanese and *ragi* in South East Asian countries. In India the starters used in the preparation of traditional fermented beverages are mostly confined to North Eastern and North Western Himalayan regions and they are commonly known as *bakhar, ranu, marcha* (*marchaar*) or *phab*.

Pure culture fermentations have been adopted by alcoholic beverage industries for the production of wine, beer, cider, *sake* and distilled alcoholic beverages [5]. The raw material for the preparation of alcoholic beverages varies according to its regional availability. It generally comprises either sugary materials (fruit juices, plant saps, honey) or starches (grain or root) which are saccharified by amylases. In this chapter, an overview of general characteristics and production of cereal based beverages is presented.

## 2. CLASSIFICATION OF CEREAL BASED ALCOHOLIC BEVERAGES

The cereal based alcoholic beverages can be classified on the basis of distillation, major raw material and texture or appearance.

### 2.1 Classification on the Basis of Distillation

The cereal based alcoholic beverages are mainly classified into two groups (Fig. 1) depending on the distillation criterion as below:

#### *2.1.1 Non-distilled beverages*

These are produced by fermentation of starch containing materials such as grains. After fermentation, the brew is either filtered or centrifuged or consumed as such. As a result alcohol is produced. The process for the production of beer is called brewing, e.g., beer, *sake, sour.*

#### *2.1.2 Distilled alcoholic beverages*

For the preparation of such beverages, the fermented broth/brew is subjected to distillation that results in an increase in the concentration of alcohol and flavouring component in the distillate. The resultant distillate is the beverage, e.g., whisky and arrack.

- Beer → Lager, Ales
- Takju
- Sour
- Shchi
- Boza
- Shaosinghjiu
- Chongju
- Sake
- Kvas

**(a) Non-Distilled Beverage**

- *(I) Whisky*
- Smoky scotch whisky → Malt whisky, Grain whisky
- Irish whisky
- Rye whisky
- Corn whisky
- Light whisky
- Blended whisky
- Canadian whisky
- Bourbon whisky
- *(II) Arrack (Rice beer)*

**(b) Distilled Beverages**

**Fig. 1:** Classification of cereal based alcoholic beverages based on distillation.

## 2.2 Classification on the Basis of Raw Cereal Ingredients

The cereal based alcoholic beverages can be classified on the basis of major raw material i.e., cereal grains used for their preparation as below:

| Raw material (cereal grains) | Beverages |
|---|---|
| Wheat based beverage | *Takju, Sourich Shchi, Boza* |
| Rice based beverage | *Shaosinghjiu, Chongju, Sake, Kvas, Takju* |
| Maize based beverage | *Chicha, Sora, Sekete, Boza* |
| Millet based beverage | *Thumba, Jaanr, Burukutu.* |
| Sorghum based beverage | *Gowe, Merissa.* |
| Barley based beverage | Beer, *kvas, Sourich Schi* |

## 2.3 Classification on the Basis of Texture or Appearance

The cereal based alcoholic beverages can be grouped into two/three categories on the basis of texture or appearance i.e., some beverages are clear liquids, others are turbid and some traditional beverages are like thin slurry or in paste form as mentioned below:

| Texture/Appearance | Beverage |
|---|---|
| Clear liquid beverage | Whiskey, Beer, *Shaosinghjiu, Chongju, Sale* |
| Turbid liquid beverage | *Takju* |
| Thin slurry/paste like beverage | *Tapuy, Jaanr* |

We will restrict our discussions to beer, whisky and some traditional beverages in the subsequent sections.

## 3. NON-DISTILLED BEVERAGES

These are prepared using a number of cereal grains like barley, rice, maize, rye and millet. These beverages are quite diverse owing to the type of grains used for their preparation. Rice based beverages are common in Eastern countries like China and Japan, whereas barley based alcoholic beverages are more prevalent in the African continent. Beer is the type beverage in this category of beverages, and its various aspects related to production are discussed below.

### 3.1 Beer

It is a fermented extract of malted barley grains, and the process of beer production is called brewing. It is a multi-step process and not all the steps are microbially controlled. The process of brewing differs from other industrial fermentations as it is not just transformation of sugars to alcohol, but also involves imparting flavour, aroma, clarity, colour, foam formation, foam stability, desired concentration of alcohol and storage stability. The ethanol content of a beer generally varies from 2 to 6%. It has a typical flavour that originates from constituents of malt, extract of hops and products of yeast metabolism. Its colour ranges from golden to black and the final product is usually clear. One of the unique features of beer is formation of white foamy layer at the surface (head) when it is poured into glass. A number of other malt beverages like ale, porter, stout and malt tonics are available in the market, which are made using processes similar to beer manufacturing.

The major ingredients for making beer comprise malted barley, hops and water. Malt is ground with water to form mash. The saccharifying enzymes present in the mash convert starch into an easily fermentable sugar, maltose. This sugar rich mash is called 'wort'. The hops are added to this wort and then boiled. The hops impart typical flavour to beer. This mixture is allowed to cool down and the brewers yeast (*Saccharomyces* sp.) is added to it that ferment maltose to alcohol and carbon dioxide [6]. It is followed by maturation of beer in storage tanks at 0°C for several weeks and subsequently the finishing of beer is done.

### *3.1.1 Raw materials for beer*

**3.1.1.1 Barley:** Barley belongs to the family Poaceae. The barleys used for beer brewing have a thick covering i.e., husk which is very hard and strong. The whole process of brewing involves malting of barley during which barley grains/seeds are allowed to initiate germination and subsequently stopped by addition of water. Two types of barley are used in beer making i.e., two row barley and six row barley, and both of these types possess distinct properties. The beer made from two row barley has lower protein content as compared to the beer produced from six row barley. In United States, generally six row barley is used which has higher enzymatic activity and comprises 80% of the total barley used in its brewing industry. The two row barley varieties preferably used in beer brewing are Piroline, Betzes, Klages, Firlbecks III, Vanguard, Hannchen and Shabet, while six row barley varieties commonly used in beer making include Larker, Dickson and Conquest [7].

**3.1.1.2 Adjuncts:** Adjuncts are "*...any carbohydrate source other than malted barley which contributes sugar to the wort*" such as corn, rice, rye, oats, barley, wheat and sugar. Adjuncts are mainly used because they are readily available and provide fermentable carbohydrates at low cost as compared to sugars available from malted barley. The most common adjunct materials used are corn (46% of total adjunct), rice (31%), barley (1%), and sugars and syrups (22%) and in the United States, non-malt adjuncts have been estimated to be 38% of total brewing materials employed [8].

The addition of adjunct in the wort results in beers with greater physical stability, better chill–proof qualities and higher brilliancy. Adjuncts like rice and corn contribute little or no soluble protein to the wort, while wheat and barley adjuncts add more soluble proteins and thus enhance protein level of wort. Most of the adjuncts, except barley, add little or no polyphenolic substances to the beer. Adjuncts beside carbohydrates also contribute to the flavour of the beer. The addition of rice adjuncts to wort may impart neutral aroma and taste, corn provides full flavour to beer. Wheat tends to impart dryness to beer, whereas semi-refined sugars as adjunct add flavour to ales. Adjuncts will also change the carbohydrate and nitrogen ratio of the wort, which in turn affects the formation of byproducts especially esters and higher alcohols [9].

The use of dark sugars as adjuncts serves dual purpose i.e., these adjuncts increase the level of fermentable sugars on one hand and impart dark colour to the final product on the other hand. However, adjuncts such as rice and pure starches and sugars are used to dilute malt colours to produce light coloured beers [7].

In general, the use of adjuncts results in increased brewing capacity, reduced labour costs, improved hot and cold breaks and shorter brewing cycles. However, if the level of adjuncts used is too high, the brewer has the risk of producing wort with insufficient insoluble nitrogen for yeast growth.

The proportion of adjuncts used by brewers varies from 10 to 30% in Europe, 40 to 50% in United States, to as high as 50 to 75% in some African continent, but not generally practiced by the brewers. This will need the addition of exogenous starch hydrolyzing enzymes. However, in some countries like Germany, only malt is the permitted source of fermentable extract, which is used as an adjunct in brewery [9].

**3.1.1.3 Hop plant:** The female flowers of hop plant (*Humulus lupulus*) are popularly known as hops, which are added as a flavouring source and stabilizer during beer brewing. The hop plant belongs to family Cannabaceae and is a perennial herbaceous plant. Three species are known *Humulus japonicus* (Asian hop), *Humulus lupulus* (Common hop) and *Humulus yunnanensis* (Yunnan Hop; endemic in Yunnan, China). China, USA and Germany are the major hop producers in the world. Flowers or cones of female hop plants are picked and dried in the autumn season. The cone comprises of bracts and seed–bearing bracteoles. At the base of bracteole lupulin glands develop, which contain resins and essential oils. The principal components of hop oil are hydrocarbons like myrecene, farnesene, humulene and caryophyllene. The resins and essential oils in the hop cones get dissolved on their addition to the wort. The unique bitterness of hops is because of the isomerisation of α-acids [7]. More than 900 compounds have been isolated from hops [9].

The bitter components of hop comprise α-acids (humulones), β-acids (lupulones), soft resins and hard resins. The α-acids are slightly soluble in wort but not at the pH of beer. In boiling water these acids form valuable bitter substances of beer. These aroma materials derived from hop inflorescence include many chemical compounds such as aliphatic terpenes, sesquiterpenes, aliphatic and terpenes esters, aldehydes, ketones, alcohols, epoxides, lactones, ethers and sulphur containing compounds [9–14]. Hops can be stored for only a limited period of time because many of its chemical constituents are sensitive to oxidation.

**3.1.1.4 Water:** Water is extensively used in the brewing process. Therefore, the quality of water used in breweries is a very important component as it significantly affects the quality of beer. The composition of water suitable for brewing is described in Table 1. Water is primarily used for steeping/soaking malted barley and for diluting wort to desired sugar level prior to fermentation. Breweries have traditionally been established near rivers, but brewers generally use river water only for running and cooling machinery. For brewing, ground water from well or spring is used in most of the breweries. The ground water varies in pH and concentration of some important ions from one geographical location to the other, and thus it immensely affects both styles and tastes of beer [15]. However, the use of ionically depleted i.e., soft water of Plzen in brewing in the Czech Republic resulted in the development of the light and clean-tasting lager beer known as Pilsner that became the standard for hoppy, pale and dry lagers. Lager has been derived from the German word lagern meaning "to store". It is a beer that is stored for some period in cold storage as part of the brewing process. Earlier to modern refrigeration systems, brewers stored beers in caves for lagering [7].

The porter brand beer was first developed in $18^{th}$ century in London from water containing higher concentration of calcium and carbonate and low level of sulphate and chloride. This beer was exported to Ireland and was modified in 1759 at Dublin to the brand Guinness. The name of this beer

**Table 1:** Chemical characteristics of water used for brewing.

| Chemical | Accepted range (ppm) |
|---|---|
| Sodium carbonate | None |
| Calcium and magnesium carbonate | Up to 50 |
| Sodium and magnesium sulphate | Up to 50 |
| Calcium sulphate | 240 to 400 |
| Sodium chloride | Up to 100 |
| Calcium and magnesium chloride | Not over 75 |
| Iron, as $Fe^{2+}$ | Less than 0.2 |
| Silica, as $SiO_2$ | Less than 10 |

was after the name of its brewer, Arthur Guinness [9]. The brewing water composition varies from place to place. The Irish brewing waters are Lower Carboniferous, while those of the London brew are Cretaceous Chalk, which results in slightly different levels of magnesium, chlorine, sodium and potassium and accounts for the sweeter taste of the London variety. For example, Dublin has hard water well-suited to making stout, such as Guinness, and Pilzen has soft water well-suited to making pale lager, such as Pilsner Urquell [16].

The production of any type or style of beer with a soft water supply is easy, because we can easily add the desired components to water to simulate water of a particular geographical location. However, to brew a Czech-style pilsner with a hard water supply is a bit more difficult since adding ions is easier than removing them.

The quality and taste of beer is determined by the composition of water used for brewing [9]. Many of the popular brands of beers throughout the world such as Pilsner, Munchner, Dortmunder and Weiner beers owed their nature primarily to the water used in brewing. The concentration of bicarbonate is mainly considered by the brewer. In the preparation of water for brewing, the bicarbonates that are harmful for brewing are removed. This is generally carried out by the precipitation process with milk of lime or ion exchange, and sometimes acid treatment is also done [7].

3.1.1.5 Yeast and fermentation: The most well-known and commercially important yeasts used in brewing are the strains of *Saccharomyces cerevisiae* and its related species. The *S. cerevisiae* is a unicellular and heterotrophic fungi. This yeast has long been utilized to ferment the sugars of rice, wheat, barley and corn to produce alcoholic beverages and in the baking industry to raise the dough.

In brewing, *S. carlsbergensis* (named after the Carlsberg Brewery in Copenhagen, where it was first isolated in pure culture by Emil Christian Hansen in 1883) is used in the production of several types of beers including lagers. This is a bottom yeast and carries bottom fermentation. *S. cerevisiae* used for the production of ales is a top yeast, which rises to the surface of the brewing vessel and conducts top fermentation. Many of the original top fermentation strains modified to be bottom fermenters are now available in the market for brewing. The *S. carlsbergensis* designation is no more in use now, and only the *S. cerevisiae* nomenclature is followed in modern text.

Ethanol production from cereal grains involves milling and hydrolysis of starch grains to release fermentable sugars followed by inoculation with yeasts. Chemically, starch is a polymer of glucose, and yeast cannot use starch directly for ethanol production. Therefore, grain starch needs to be broken down to glucose by a combination of two enzymes viz., amylase and amyloglucosidase before it is fermented by yeast to ethanol. The biochemical reactions and the process involved in starch hydrolysis and fermentation are shown in Figs. 2 and 3.

$$\underset{\text{Starch}}{(C_6H_{10}O_5)_n} + \underset{\text{Water}}{n\,H_2O} \xrightarrow[\text{Amyloglucosidase}]{\text{Amylase}} \underset{\text{Glucose}}{n\,C_6H_{12}O_6}$$

**Fig. 2:** Enzymatic hydrolysis of starch to glucose.

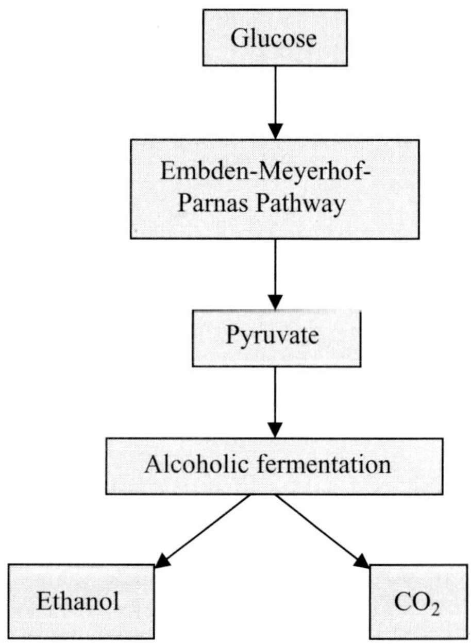


**Fig. 3:** Alcoholic fermentation by yeast.

### *3.1.2 Biochemistry of beer fermentation*

Alcoholic fermentation is the production of alcohol from sugars through a series of intermediates in the absence of oxygen. The pathway followed for alcohol production from glucose is referred to as glycolytic pathway or Embden-Meyerhof-Parnas (EMP) pathway (Fig. 4).

Up to conversion of glucose to pyruvate (Fig. 4), the steps of glycolytic pathway are the same for utilization of glucose in oxidative or fermentative mode. Under anaerobic conditions, pyruvate is

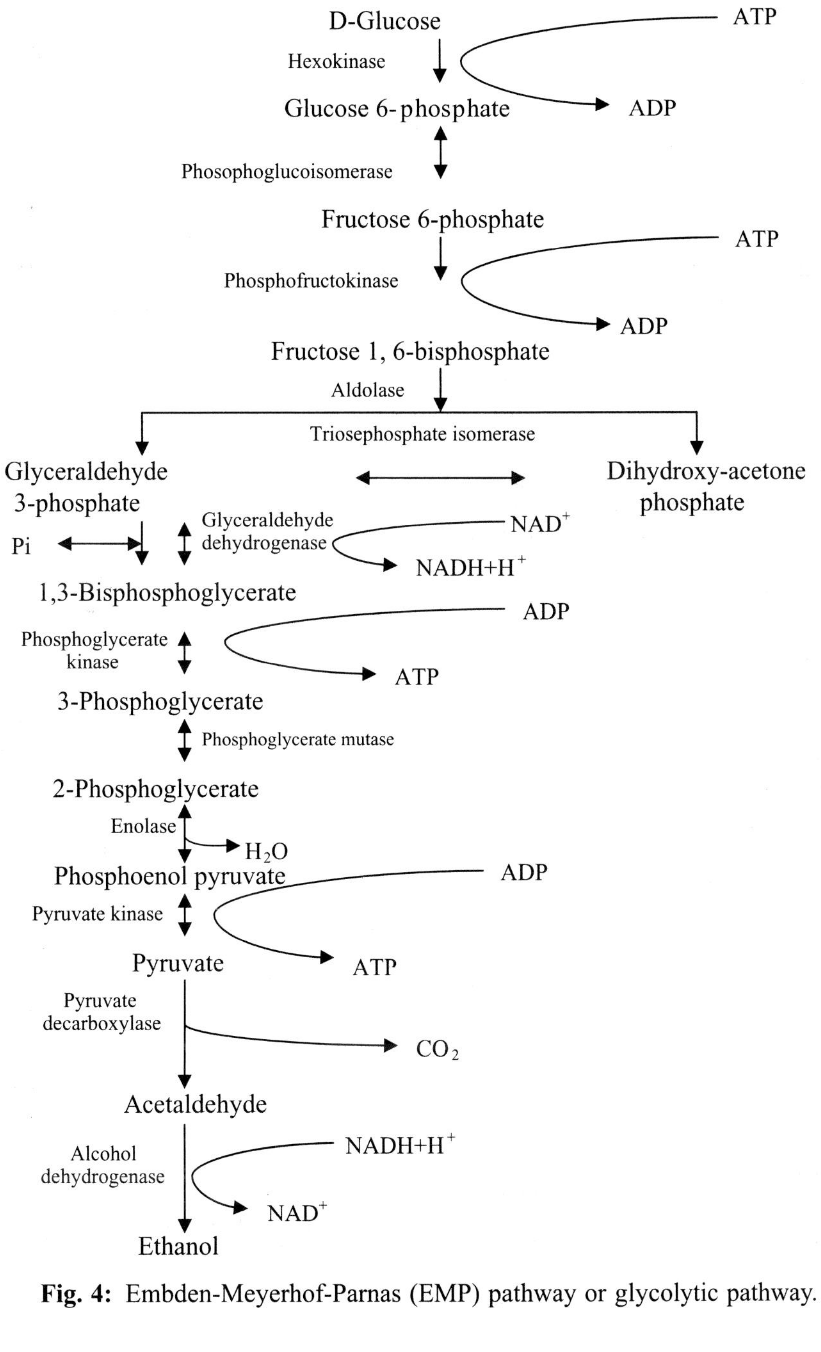


**Fig. 4:** Embden-Meyerhof-Parnas (EMP) pathway or glycolytic pathway.

first converted to acetaldehyde and finally ethanol and carbon dioxide in the yeast. The conversion of pyruvate to ethanol is a two step reaction. In the first step, pyruvate is transformed to acetaldehyde in the presence of enzyme, pyruvate decarboxylase. This reaction also involves $Mg^{2+}$ ions and the coenzyme thiamin pyrophosphate (TPP). In the subsequent step, acetaldehyde is converted into ethanol. It is a reversible reaction during which NADH is converted to $NAD^+$. This conversion is mediated by the enzyme, alcohol dehydrogenase.

### *3.1.3 Production of beer*

An overview of beer production is given in Fig. 5. The various steps involved in beer production [17] are described below:

3.1.3.1 Malting: In malting, barley is steeped to increase its water content from 12% to about 45%. It takes about 48 hours and consists of 2 or 3 rounds when grain is immersed in water with air rests in between. Moist grain is allowed to germinate for 3–4 days. During germination, gibberellins diffuse from embryo into the endosperm and this stimulates aleurone cells to produce hydrolytic enzymes viz., amylases, proteases, cellulases and hemicellulases enzymes. Malting process is represented in Fig. 6. During malting, the amylolytic enzymes in malt hydrolyse starch into fermentable sugars [3]. Mash is prepared

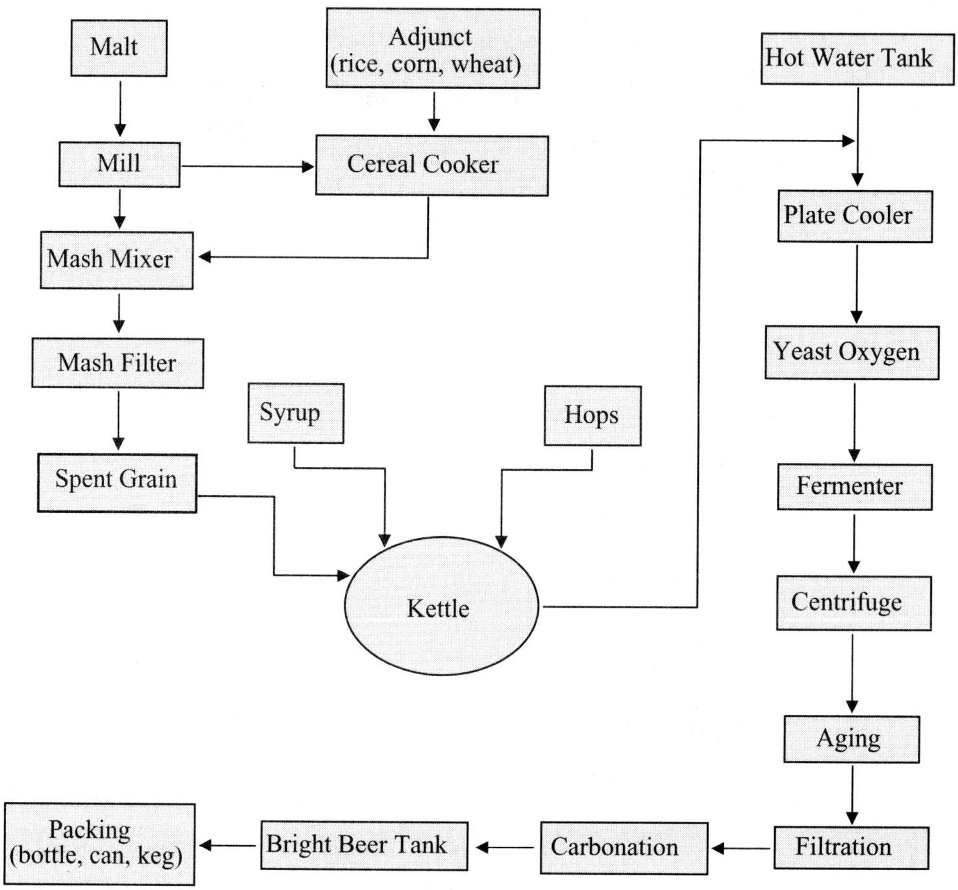


**Fig. 5:** Overview of beer production process.

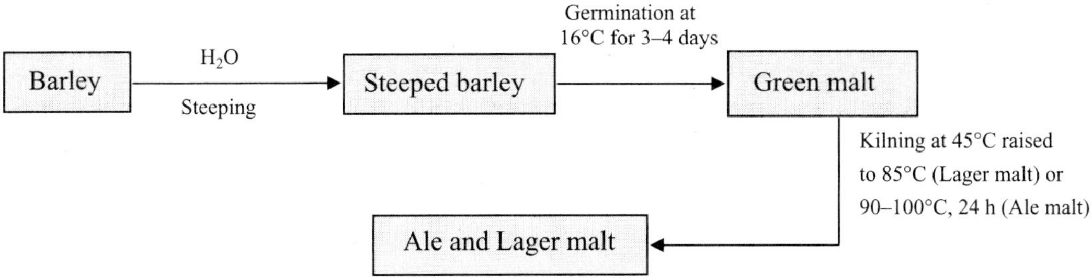


**Fig. 6:** Flow diagram of malting process.

by mixing grains and liquid in a ratio of 1 kg to about 3.2 litres of liquid at 65 to 70°C. The liquid comprises last two washings of previous mash. Mash is thoroughly mixed during the milling operation and is kept for about one hour at 64°C for amylolytic enzymes in the malt to convert starch to maltose. The mash is drained and the resulting wort is cooled to 21°C and sent to wash back and is inoculated with brewer's yeast. The residue on plate is washed with 76°C liquid, drained, cooled and pumped to wash back. This procedure is repeated twice with hot water and washings are sent to the hot water tank to be used as liquid for next mash. Spent grain (left out residues after washing) is removed from tank, and is sold as cattle feed [18]. The malting yields a sweet syrupy liquid called 'wort' which mainly contains fermentable sugars like maltose and glucose and significant quantities of small branched dextrins which are non fermentable. The 'wort' also contains soluble proteins, polypeptides and amino acids [12].

3.1.3.2 Mashing: The main objective of mashing process is the maximum solubilization of malt and malt adjuncts. Further breakdown of starches, proteins and other polysaccharides takes place during mashing [19]. The main malt mash is first prepared by mixing the malt with water at 38 to 50°C. To this mash, steam cooked starchy malt adjunct having a temperature around 100°C is mixed. The temperature of cereal–malt mash comes to 65 to 70°C, and at this temperature saccharification of starch to sugar takes place. The saccharification increases the temperature up to 75°C and this leads to inactivation of starch hydrolyzing enzymes. After some time the heavy and insoluble materials settle down at the bottom of the vessel and liquid 'wort' becomes clear. The clarified 'wort' is filtered; hops are added to it and processed further [20].

3.1.3.3 Boiling the wort with hops: The main objectives of wort boiling with hops are as follows: (i) enzyme inactivation, (ii) wort sterilization, (iii) protein coagulation (Hot break fermentation), (iv) hop extraction (isomerization), (v) flavour compound formation, and (vi) undesired volatile evaporation. In this process, the liquid having wort and hops is boiled for about two and a half hours. It is filtered to remove the hop residues, hops solids and precipitated proteins. The precipitates are washed with hot water to remove most of the soluble material and the washing is added to the original filtrate. The resulting mash or wort is now ready for fermentation.

3.1.3.4 Fermentation: The yeast is added to wort to drive fermentation. Traditionally yeast has been characterized according to morphological and physiological characteristics. The yeast taxonomists have grouped brewing (both ale and lager), wine, baking and numerous wild yeasts under the name *S. cerevisiae* [21]. Special beer yeast of bottom type (a strain of *S. cerevisiae*) is used for inoculation or 'pitching' of wort. The pitching yeast used as inoculum is generally recovered from the previous fermentation. However, brewer's yeast is also cultivated by the brewer's in house or it can be procured commercially. A heavy inoculum, about 1 lb per barrel (31.5 gal) of beer is added. The temperature of the wort during fermentation differs from one brewery to other, and it ranges from 3.3 to 14°C. The fermentation process is usually completed in a period of 8–10 days [20]. During fermentation, yeast primarily converts sugar in wort to alcohol and carbon dioxide. In the absence of oxygen, small amounts of glycerol and acetic acid are also produced. The sequence of enzymatic reactions that converts glucose to pyruvate is referred to as Embden–Meyerhof–Parnas (EMP) pathway Fig. 4 [22].

Not all glucose is metabolized via EMP pathway; some goes through a different pathway called Hexose Monophosphate pathway (HMP). This releases not only energy but also forms pentose sugars needed for the synthesis of nucleotides and nucleic acids.

The fermentation of sugars is carried out according to Gay–Lussac's equation:

$$C_6H_{12}O_6 \rightarrow 2C_2H_5OH + 2CO_2 \text{ and } G = -230 \text{ kJ}$$

It is an exothermic (energy releasing) process and therefore cooling is essential to maintain constant temperature of fermentation [20]. The beer at this stage is called green or young beer.

**3.1.3.5 Aging or maturing:** The young or green beer is stored in vats at about 0°C/32°F for few weeks to several months, which leads to precipitation of proteins, yeast, resin and other unwanted substances and the beer becomes clear and mellowed. This process is called aging or maturing. Esters and other compounds are produced, which add taste and aroma to the beer [22]. During the maturation, chill proofing is generally done to prevent development of turbidity on later exposure of the finished beer to cold. The turbidity or haze formation is attributed to an unstable protein in the beer. Antioxidants like sulphur dioxide and ascorbic acid are added during maturation to curtail or avoid the oxidative changes in the beer, which may affect its taste and flavour [3, 23–25].

**3.1.3.6 Finishing:** After aging, the lager beer is carbonated to enhance its carbon dioxide content to about 0.45 to 0.52%. This is commonly done either by injecting clean carbon dioxide recovered from the fermentation gas or by the activity of fermenting yeast. The process is also called "natural carbonation." This carbon dioxide replaces dissolved oxygen, which otherwise affects the shelf stability of the beer, and it also helps in the formation and retention of foam needed for the preservation of beer [3, 20].

### *3.1.4 Beer spoilage*

A few microbes can tolerate the relatively hostile environmental conditions during brewing (Hops, alcohol, low pH 4.0–4.5) and may spoil the beer. The spoilage problem is mostly encountered during the maturation and packaging of beer. The bottled or canned beers, which are properly processed by pasteurization or filtration, do not normally have spoilage problem. Growth of any bacteria, fungi or even species of *S. cerevisiae* other than the desired brewer's yeast may cause the spoilage of beer [26–28]. Some microbes involved in the spoilage of beer are enlisted in Table 2.

### *3.1.5 Microbial content of barley and malted barley*

It is very important to maintain the moisture level of barley, which is generally not allowed to exceed 12.5% during storage. High moisture content favours microbial growth and causes contamination [35]. A number of microorganisms have been isolated from barley and malt. The types of microorganisms that are found on barley depend on the growth and environmental conditions of barley. Even post-harvest and storage conditions have an immense bearing on the natural microbial flora of barley [36]. The water content of the kernels increases during the malting process, which enhances bacterial ($10^7$–$10^8$ cfu/kernel) and yeasts ($10^6$–$10^7$ cfu/kernel) counts of the barley. The principal bacterial species at the end of malting are *Flavobacterium esteroaromaticum, Erwinia herbicola, Pseudomonas fluorescens, Pseudomonas putida, Serratia plymuthica, Lactobacillus alimentarius* and *Lactobacillus plantarum*. The main yeasts present at the end of malting include *Debaryomyces hansenii* and *Candida catenulata* [35]. Occurrence of a large number of microorganisms at various

**Table 2:** List of some microbes involved in the spoilage of beer.

| Microorganism | Type of microorganism | Effect on beer |
|---|---|---|
| *Lactobacillus brevis* and *Lactobacillus lindneri* | Bacteria | Silky turbidity |
| *Pediococcus damnosus* | Bacteria | Buttery aroma due to diacetyl formation |
| *Zymomonas mobilis* | Bacteria | Rotten apple–fruity off–flavours |
| *Pectinatus* sp. | Bacteria | Turbidity, $H_2S$ production |
| *Selenomonas* sp. and *Zymophilus* sp. | Bacteria | Spoilage of packaged beer |
| *Acetobacter* sp. and *Gluconobacter oxydans* | Bacteria | Oxidation of ethanol to acetic acid and turbidity and ropiness |
| *Dekkera* and *Brettanomyces* | Yeast | Produce high levels of acetic acids and esters |
| *Hansenula* sp. | Yeast | Excessive ester production |

*Source:* 29–34

stages of beer fermentation can adversely affect the fermentation process vis–à–vis quality of beer because these microbes produce undesired metabolites. Some microbes produce plant hormones like indole–3–acetic acid, gibberellic acid and abscissic acid that may also affect the normal course of fermentation [37].

A number of microbial contaminants including *Bacillus coagulans, Bacillus stearothermophilus* and *Clostridium* have been isolated from wort. If proper handling and hygiene precautions are not observed, these bacteria grow and produce lactic acid and butyric acids in the fermenting wort [38–39].

### *3.1.6 Enzymes in brewing*

A number of enzymes like amylases, glucanases, proteases and cellulases are involved in brewing. Some enzymes e.g., β-amylase, exo-peptidase and carboxy-peptidase of barley endosperm also get activated during brewing. During malting, the aleurone layer of barley grain also synthesizes certain enzymes like β-glucanase, endo-proteases, α-amylase and pentosanases, which migrate into the starchy endosperm. Cell wall is degraded by hemicellulases and β-glucanases. During mashing, the insoluble starch molecules are cleaved into soluble fermentable sugars and unfermentable dextrins by the action of α- and β-amylase. In order to improve the pace of saccharification of starch, some exogenous bacterial or fungal α- and β-amylases are added during mashing. The mash produced at higher temperature is of very fine quality and can be easily filtered. Fungal α-amylase is only added if there is slow fermentation and saccharification is incomplete. The enzyme β-glucanasc (EC 3.2.1.4) is not able to tolerate high brewing temperatures, and thus it is not able to act efficiently. However, certain β-glucanases of fungal origin are known to withstand high temperatures and therefore find their application in the mashing process. Soluble proteolytic enzymes such as papain (EC 3.4.22.2), bromelain (EC 3.4.22.32) and ficin (EC 3.4.22.3) are used mostly in conjunction with other stabilization agents such as polyvinyl pyrrolidone (PVP) or silica gels (lucilite) before final filtration to enhance the colloidal stability of the beer. This also controls chill haze and improves the shelf-life of beer. Papain, extracted from the fruit of *Carica papaya* is effective within the temperature range of 30–45°C and the pH range of 4.0–5.5. It assists in cleavage of high molecular weight proteins that otherwise react with polyphenols and add to haze formation. The use of immobilized proteases has been studied, but has not been extensively adopted in brewing so far [40].

Some endo- and exo-peptidases breakdown high molecular weight proteins into simple peptides by splitting peptide bonds. These proteolytic enzymes are responsible for lysis of albumins and globulins. This helps in decreasing the haziness of the final product. The proteases from *Bacillus amyloliquefaciens* are widely used during malting.

### *3.1.7 Shelf-life improvement*

During fermentation, all the dissolved oxygen in the wort is consumed by yeast. The oxygen concentration in beer is generally below 200 ppm. After packaging, the concentration of oxygen in beer may range from less than 500 to 1000 ppm. Traces of entrapped oxygen result in the formation of volatile aldehydes imparting stale flavour to the beer. Addition of antioxidants such as sulphite, ascorbic acid and catechin after fermentation protects beer from becoming stale in the presence of oxygen [41].

The antioxidants are added to green beer either to remove oxygen or to negate its effects. Ascorbic acid (vitamin C) at 1.5 g/hl reduces oxidation of haze and its effect is similar to lowering the concentration of dissolved oxygen. Other antioxidants e.g., sodium hyposulphite and sodium metabisulphite are also used for the same purpose. Glucose oxidase plus catalase are also used to deplete the oxygen level of green beer. The glucose oxidase uses oxygen of beer to oxidise glucose to gluconic acid and hydrogen peroxide and the latter is transformed by catalase into water and oxygen [42].

## 3.2 Types of Beer

The major types of beers on the basis of type of yeast fermentation are as follows:

Beer

- Lager beer (Bottom yeast fermented)
- Ales (Top yeast fermented)

### *3.2.1 Lager beer (Bottom yeast fermented beer)*

The term lager is derived from a German verb, 'lagern' which means to lay down or store. It is a clear amber-coloured beer, which is fermented with a particular strain of *S. cerevisiae* (formerly called *S. carlsbergensis* and *S. uvarum*). The fermentation is carried out at low temperature i.e. 5–12°C. Traditionally these were known as 'bottom yeasts' because they settle (flocculate) to the base of tank during fermentation [22]. The two types of Lager beers are light and dark. These terms refer to the body of final beer product.

### *3.2.2 Ales (Top yeast fermented beer)*

This type of beer is formed by the top yeast which is a strain of *S. cerevisiae*. The top yeast forms clumps, which collects carbon dioxide and are carried to the top of the wort. There are several varieties of ales which are distinguished on the basis of body and colour [25]. The wort for ales beer production is fermented at 18–25°C and the fermentation is more rapid and takes place in 5 to 7 days. At intervals, the yeast clumps (scum) is skimmed off. The quantity of hops used is more in ales than the lager beer and its alcohol content is higher as compared to lager beers. Ales are generally darker in colour and tart in taste [22].

### *3.2.3 Other types of beers*

3.2.3.1 Porter/Stout beer: The Porter beer has its origin in London, while the strong Porter beer produced in Dublin is called Stout. Both of these beers are made from roasted malts and roasted barley and are dark in colour. The Stout beer possesses strong coffee or burnt caramel flavour. There are four main types of Stout beers:

3.2.3.1.1 Dry stout: The Dry Stout is prepared from the roasted and unmalted barley. The alcoholic content of Dry Stout is more than Sweet or Oatmeal Stouts. Another modification of Dry Stout beer is 'Milk Stout' since lactose or milk sugar is added to the wort during the brewing process (www.beeradvocate.com/beer/style).

3.2.3.1.2 Imperial stout: This beer is considered as the king of stouts having high alcoholic content varying from 8–12% and plenty of malt character. It has low to moderate levels of carbonation with huge roasted, chocolate and burnt malt flavours. The flavours of higher alcohols are quite evident in the imperial stout. Hop character ranges from none to aggressive.

3.2.3.1.3 Oatmeal stout: It is brewed using London ale yeast, resulting in smooth dark malty ale with a touch of finesse. It is prepared by adding oats maximum up to 5%, during the brewing process. If the content of oats is above 5%, it leads to a bitter or astringent taste [43]. Oatmeal Stouts are now made in several countries, including Australia producing 5% Oatmeal Stout and Denmark brewing 5.9% Oatmeal Stout using roasted barley and chocolate malt. Oatmeal stouts are usually bitter in taste due to the taste of oats, and its smoothness is due to the higher content of proteins and lipids (includes fats, waxes and gums) from oats.

3.2.3.1.4 Sweet stout: The Sweet Stouts are basically stouts that have a larger amount of residual dextrins and unfermented sugars that give the brew more sweetness that counters the roasted character. These beers were first documented in the 1900s in England. Sweet Stouts are very similar to Milk Stouts, but brewers add unfermentable sugars, usually lactose, to the brew kettle to add some sweetness. The average alcohol concentration varies between 4.0–7.0% (volume by volume).

**3.2.3.2 Weiss beer:** The Weiss beer is prepared with a significant proportion of wheat in the mash that also contains malted barley. The addition of wheat into the fermenting wort imparts a light flavour and pale colour to the beer. In Germany, law enforces the fermentation of wort for wheat beers by top yeast (*S. cerevisiae*) [44]. Wheat beer has a creamy texture and sweet flavour, and some styles have flavour of banana and clove. It has alcoholic content around 2.8%. The two most common varieties of wheat beer are hefeweizen and sour varieties.

3.2.3.2.1 Hefeweizen: The Hefeweizen is a German wheat beer and its main feature is that the fermenting yeast is not filtered out. Higher alcohol content varieties of this type of beer are *Kristallweizen* (clear), *Dunkelweizen* (dark) and *Weizenstarkbier* or *Weizenbock.* These are considered as true Hefeweizen only if the final product is left unfiltered. The filtration which takes the yeast out of *Kristallweizen* also separates the wheat proteins that are used to make Hefeweizen cloudy. There are many varieties of Hefeweizen in the market, but Rothaus is still known to be the best [45].

3.2.3.2.2 Sour varieties: A minor variety of wheat beer is represented by *Berliner Weisse* (Berlin White), which has low alcohol (2.5% to 3.0%) and quite tart. Often sometime sweetened syrups of lemon, raspberry or woodruff herb are added to the beer to change the taste and flavour of beer. Leipziger Gose is similar to Berliner Weisse, but it is slightly stronger as it has 4% alcohol. In addition to yeast, lactic acid bacteria are added to wort for fermentation [46].

## 3.3 Traditional Non-Distilled Alcoholic Beverages from Cereals

Indigenous fermented products form an important part of human diet in Indian Sub-continent, the Middle East Africa and Latin America. People of Asia are pioneers in the development of indigenous fermentations. There are several varieties of fermented cereals commonly employed in the preparation of traditional alcoholic beverages. Alcoholic and acidic beverages are the important fermentation products in Asian region.

The agricultural, climatic, ethnic, geographical/topological and religious diversities might have contributed to the development of a number of traditional fermented foods and beverages in various regions of the world. In some tribal areas of India including Himachal Pradesh, the preparation of customary alcoholic beverages and their consumption are permitted at home scale [47]. In preparation of such beverages, natural source of microflora is employed as fermenting agent, using locally available raw materials. Fermentation starters used in different countries are given in Table 3. Fermented products have not only emerged as a source of nutrition, but also as functional and probiotic foods, which besides nutritional value have health promoting effects or provide protection against various diseases.

The traditional fermentation processes are increasingly attracting the attention of scientists and policy makers as a vital part of food security strategies [48]. Indigenous food/beverage fermentation is one of the ancient 'food processes' dependent on the biological activity of microorganisms [49]. Millions of people in third world supplement their diet by indigenous foods and beverages, many of which have not yet been studied and characterized in detail [50]. Some of the common cereal based alcoholic beverages of world are given in Table 4. The alcoholic fermentation is also used for the preparation of paste-type food products and are used as desserts in South East Asian countries.

### 3.3.1 Sake

It is a traditional alcoholic beverage of Japan and is also produced in China and South-East Asia. Steamed rice is the raw material for its fermentation, which serves as a substrate for the fungal culture *Aspergillus oryzae* and yeast *S. sake. S. sake* has several characteristics which differ from other strains of *S. cerevisiae* [51]. It is heterothallic in its mating response, grows in vitamin-free medium, exhibits resistance to yeastcidin, tolerates up to 20–30% ethanol, and is specially found in the sake mash environment [52, 53]. Sake is normally consumed within one year of its production, otherwise its taste starts deteriorating. The production of sake from steamed rice is one of the traditional examples of simultaneous saccharification of starch by *A. oryzae* and fermentation by yeast. The combination of saccharification and fermentation processes at low temperature (below 10°C) is called parallel fermentation. It has a high alcohol content of about 15–20% [54, 55].

**Table 3:** Fermentation starters used in different countries.

| Country | Name | Ingredients | Shape | Microorganisms |
|---|---|---|---|---|
| China | *Chu* | Wheat, barley, millet, rice | Cake | *Rhizopus* sp., *Amylomyces* sp. |
| Korea | *Nuruk* | Wheat, rice, barley | Large cake | *Aspergillus* sp., *Rhizopus* sp., Yeasts |
| Japan | *Koji* | Wheat, rice grains | Large ball | *Aspergillus* sp. |
| Indonesia | *Ragi* | Rice flour | Small cake | *Amylomyces* sp., *Endomyces* sp. |
| Philippines | *Bubod* | Rice, glutinous rice flour | Small cake | *Mucor* sp., *Rhizopus* sp., *Saccharomyces* sp. |
| Thailand | *Loog–pang* | Bran | Powder | *Amylomyces* sp., |
| India | *Marcha/Phab* | Rice | Flat cake | *Hansenula anomala, Mucor fragilis* and *Rhizopus arrhizus* |

**Table 4:** Some of the traditional cereal based alcoholic beverages prepared in various countries.

| Cereal based beverages | Substrates | Country |
|---|---|---|
| *Sake* | Rice | Japan |
| *Kvass* | Rice, barley | Russia |
| *Raleshi* | Rice | Peru |
| *Sonti* | Rice | India |
| *Thumba* | Millet | India |
| *Binuburan* | Rice | Philippines |
| *Chhang* | Rice | India |
| *Tapuy* | Rice | Philippines |
| *Takju* | Rice, wheat | Korea |
| *Sora* | Maize | Peru |
| *Tuborg and San Miguel* | Barley, millet, rice | Nepal and Tibet |
| *Shaosinghjiu* | Rice | China |
| *Chongju* | Rice | Korea |
| *Jaanr* | Finger millet | Indian Himalayas |
| *Boza* | Maize, wheat rice | Turkey |
| *Sourish Shchi* | Barley, rye, wheat | Russia |
| *Sobia* | Mixture of cereals | Saudi Arabia |
| *Merissa* | Sorghum | Sudan |
| *Burukutu* | Sorghum | Ethiopia |
| *Chicha* | Maize | Latin America |
| *Sekete* | Maize | Nigeria |
| *Gowe* | Sprouted sorghum | West Africa |
| *Brembali* | Glutinous rice | Indonesia |

### 3.3.2 Kvass

It is a non-distilled alcoholic beverage and is the national drink of Russia. It is also consumed throughout Eastern Europe. The raw material for making *Kvass* is grains - wheat, rye, barley, etc. or dark rye bread, and often sugars, birch sap, berries, and fruits are added to it. It is also known as bread beer. This beer has been a popular beverage due to its longer shelf-life up to 2–3 years. It has a low alcohol content that may vary from 0.5 to 2.5%. It is often flavoured with herbs or fruits to suppress its bitter tastes out.

### 3.3.3 Rakshi

It is an alcoholic drink made out of millet by putting fermented millet seeds in a plastic or wooden mug filled with hot water. It is clear in colour and can cause blindness when not prepared properly. It is not only an alcoholic drink, but also used as an antiseptic. The alcohol concentration ranges from 10–15% [56–59].

### 3.3.4 Tuborg and San Miguel

These traditional home-brewed beers, made from barley, rice or millet are called *jaad* in Nepali and *chhang* in Tibetan. *Tongba* is another beverage prepared by pouring hot water onto fermented millet mash in a cylindrical and lidded wooden container and sipped through a bamboo straw. All home made liquors vary in strength and have alcoholic content in between 5–10%.

### 3.3.5 Sonti

It is an Indian rice-based alcoholic beverage. Its preparation is similar to sake, and as far as its alcoholic content is concerned, it is similar to wine. Sonti is made from steaming of rice followed by its mixing with a fungus, *Rhizopus sonti* for saccharification of starches to simple sugars, and subsequently yeast is added to derive fermentation.

### 3.3.6 Thumba

It is a very popular beverage produced in India, mainly in Sikkim and Darjeeling. It is believed to be a good tonic, especially for post-natal women [60]. Thumba has 5.8% alcoholic content, which is much lower than other traditional beverages. Its actual flavour is a bit fruity, but it can be blended with lager malt and spicy hops, which gives an impression of fresh bread fragrance [57].

### 3.3.7 Binuburan (Philippines rice beer)

It is a local alcoholic and indigenous fermented beverage of Philippines. It is also called Philippines rice beer. Other alcoholic beverages of this region are *tuba, tapuy* and *andbasi. Binuburan* is characterized by its sweet taste and juiciness. This is the result of pre-fermentation of rice for 2–3 days. The preparation of *Binuburan* is carried out by mixing of the starter called *bubod* to the steamed rice. The whole process is shown in Fig. 7.

### 3.3.8 Chhang

It is a traditional beverage of tribal regions (Lahaul, Spiti and Kinnaur) of Himachal Pradesh. Barley, millet (finger-millet) or rice is used to brew the drink. It is an indispensable hospitality beverage among the tribals of Lahaul valley and is considered to provide protection against cold during winter months. The preparation of *Chhang* involves solid state fermentation of cooked rice and *phab* (thc traditional inoculum) for 4–5 days. After 4–5 days, it is filtered and the filtrate is called *Chhang* (Fig. 8). The distilled form of *Chhang* is called *sra* [61]. Many of the fermented foods and beverages, which

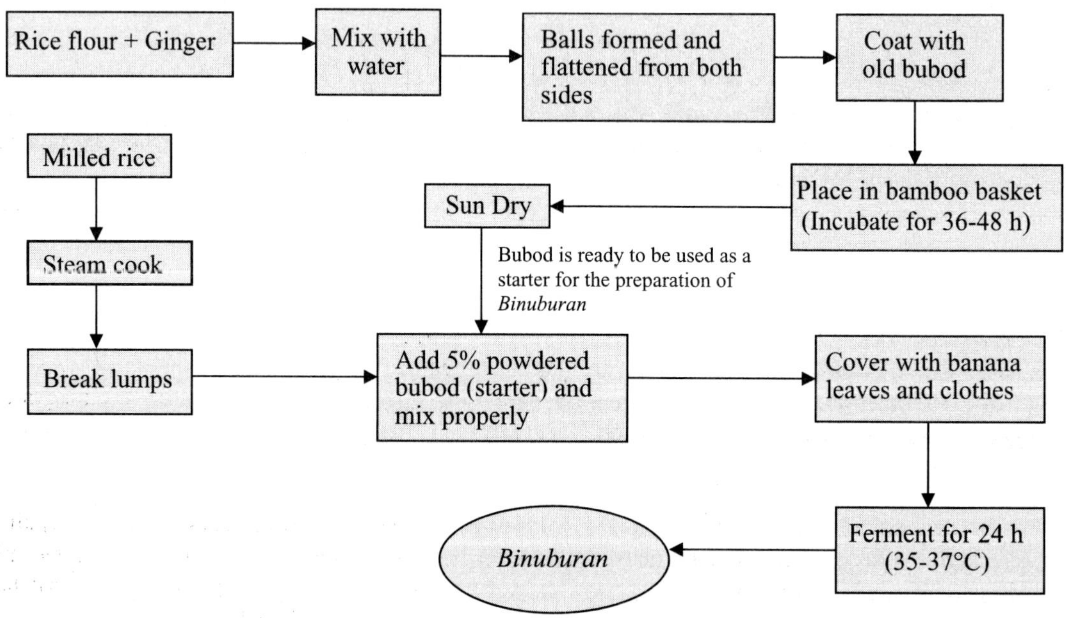


**Fig. 7:** Flow diagram of preparation of Philippines *Binuburan*.

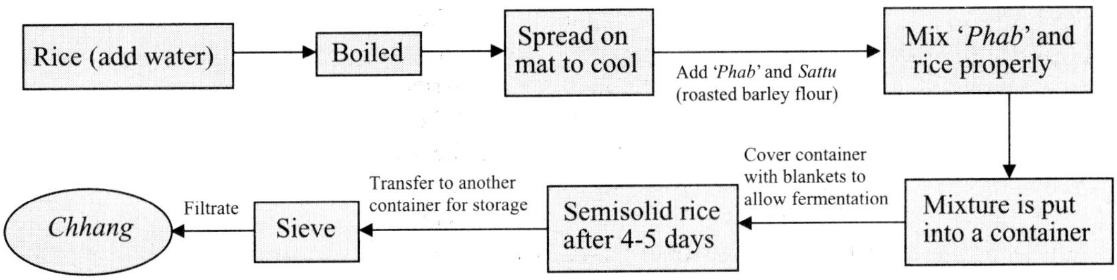


**Fig. 8:** Traditional method of *Chhang* preparation as practiced by tribal people of Lahaul valley of Himachal Pradesh.

are prepared by the action of microorganisms on plant materials are little known outside their native countries [62].

### *3.3.9 Sura*

*Sura* is a millet (*Eleucine coracana*) based fermented beverage mostly prepared in Lug valley of Kullu district of Himachal Pradesh (India). It is prepared by natural fermentation of finger millet flour. Flour is knead in the form of dough and is kept for 10 days in a container. After 10 days, half-baked *roties* are made, put into container and water is added. After two days, *dehli* made from traditional herbs is added and it is kept for 8–10 days for fermentation. Sura is consumed during local festivals like *shoeri saja* and marriages in rural area of Kullu especially, lug valley [61].

### *3.3.10 Pito*

It is a traditional cereal based alcoholic beverage from Africa. It is particularly popular in Nigeria and Ghana. It involves fermentation of maize, sorghum or combination of both. It is dark brown in colour and has a sweet to bitter taste. It has an alcoholic content of about 3%, and the microorganisms involved in the preparation of *Pito* are *Geotrichum candidum, Lactobacillus* sp. and *Candida* sp. The *G. candidum* and *Lactobacillus* sp. cause souring, whereas *Candida* sp. carries out alcoholic fermentation [63].

### *3.3.11 Merissa*

It is a cereal based alcoholic beverage of Sudan. It is largely prepared from sorghum and millet. The production of *Merissa* takes place in three successive steps: (i) Ajeen fermentation, lactic souring of sorghum; (ii) Debosa fermentation, starter activating phase; (iii) Merissa fermentation, an alcoholic fermentation. The Merissa fermentation is exactly similar to the fermentation procedure of other cereal based alcoholic beverages of Africa. At the start of fermentation, the alcoholic content of Merissa is 1%, but towards the end, the alcoholic content rises up to 6% [64].

### *3.3.12 Bouza*

It is an Egyptian fermented alcoholic beverage produced from wheat. The Egyptians have known it since the days of the Pharaohs [65]. It is a thick, pasty yellow beverage and has an agreeable taste. It is prepared by placing coarsely ground wheat grains in a wooden basin and kneading them with water into dough. The dough is cut into thick pieces and very lightly baked. Meanwhile, some of the wheat grains (¼ of the total grains) are moistened with water, germinated for 3–5 days, sun-dried, ground and mixed with the pieces of bread which are soaked in water in a wooden vessel. *Bouza* from a previous brew is added to serve as an inoculum. The mixture is allowed to ferment at room temperature for a 24 hour period, following which the product is sieved to remove large particles and diluted with water to a desired consistency. *Bouza* has a very short shelf-life and thus should be consumed within a day of its preparation [66].

### 3.3.13 Chicha

It is produced in the Andean regions and sometimes in the lower altitude regions of countries such as Argentina, Bolivia, Brazil, Colombia, Ecuador and Peru. It has a flavour similar to that of cider. Its colour varies from red to purple depending on the variety of the maize used for its preparation. The alcoholic content of *chicha* varies between 2 and 12% (*v/v*). Traditional *chicha* production is a unique fermentation process in which saliva serves as the source of amylase for hydrolyzing starch to fermentable sugars. Presently malting (germination) of maize kernels is widely done to produce the amylase required for starch conversion. Frequently, salivation is combined with malting to yield *chicha*. Yeasts, particularly *S. cerevisiae*, and bacteria of the genus *Lactobacillus*, are the primary fermenting organisms in *chicha* [67].

### 3.3.14 Tesguino

It is a slurry-like, alcoholic beverage prepared by fermentation of germinated maize or maize stalk juice. The term *tesguino* or *tejuino* comes from the *Aztec tecuin*, meaning heartbeat. *S. cerevisiae* is an important microorganism in the alcoholic fermentation of *tesgüino.* Other yeast species consistently identified at various stages of the fermentation are *Candida guilliermondii, Hansenula anomala, S. cerevisiae* and *S. kluyveri. Tesguino* is consumed by several ethnic groups of northern and northwestern Mexico. It is the preferred beverage at any celebration, religious ritual, funeral and sporting games in the everyday life of indigenous groups in Mexico [68]. The preparation of *tesguino* is schematized in Fig. 9.

## 4. DISTILLED BEVERAGES

The whisky is one of the popular cereal based distilled beverage. The term whisky is derived from the Gaelic word for 'Water' (Uisce) and is called *uisge–beatha* in Scotland that means "water of life" (www.wickipedia.com). The exact origin of whisky is not known, but the method of whisky production was known to the people of Scotland, Irish, Coruish and Welsh since 800 BC. The distilled beverages are believed to have been produced from rice around 800 BC in China. The Arabic chemists developed the method of distillation and made the first real distilled beverage known as *alembic.* The distillation was practiced by alchemists in Europe and this art of distillation became known to many by $13^{th}$ and $14^{th}$ centuries [9]. The first written record of whisky comes from 1405 BC in Ireland (www.theanswerbank.co.uk), where it was distilled by the monks [69]. Some of the researchers are of the view that distilled spirits were first produced in Middle East between $8^{th}$ – $9^{th}$ centuries AD [70].

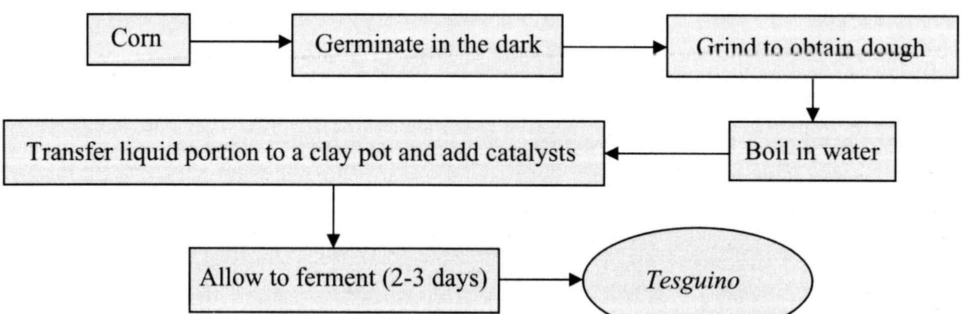


**Fig. 9:** Flow diagram for the preparation of *Tesguino.*

The whisky is generally defined as a group of alcoholic beverages that are distilled from fermented grain mash. The grain mash is subjected to aging in wooden tubs known as casks usually made of oak barrels before bottling [71, 72]. Whisky is normally made by distillation of a fermented aqueous infusion of malted barley and other cereals like rye, wheat and maize.

In whisky production, the grains are a major source of starch which is hydrolyzed in a similar way as malting/mashing of grains for beer production. The whisky fermentation is carried out with a known strain of *S. cerevisiae*. This yeast is available in dry and wet form from commercial producers of yeast.

## 4.1 Distillation

The separation of components of a solution on the basis of their volatilities is known as distillation [73]. Distillation thus separates, selects, and concentrates the ethanol and congeners from the fermented grain mash (drop beer).

Two type of distillation systems are used for the production of whisky. The first type of distillation system involves batch still for the production of highly flavoured spirit. The second type of distillation system uses continuous column distillation still, and is mainly used to produce light flavoured spirits [74, 75].

### *4.1.1 Continuous column system*

It is used for the production of grain neutral spirits (light flavoured spirits) that are used as a base for blending. The continuous column distillation system is used for producing Grain whisky. The various steps in continuous column distillation are given in Fig.10.

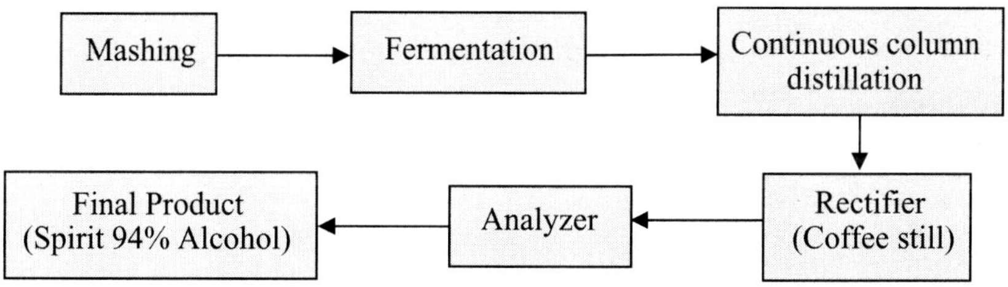


**Fig. 10:** Flow diagram of continuous column distillation.

### *4.1.2 Batch or pot still system*

Batch or pot still system is normally a double distillation and is used to produce a highly flavoured spirit (Fig. 11). In a batch system, a large kettle with a capacity of 50,000 gallons is used with vapour pipe fitted to it that leads to a concentrating column having as many as 55 bubble cap plates. The large capacity of kettle is important to maintain product uniformity. This process is also known as time-cycle distillation and involves the removal of heads, product and residual distillate [73].

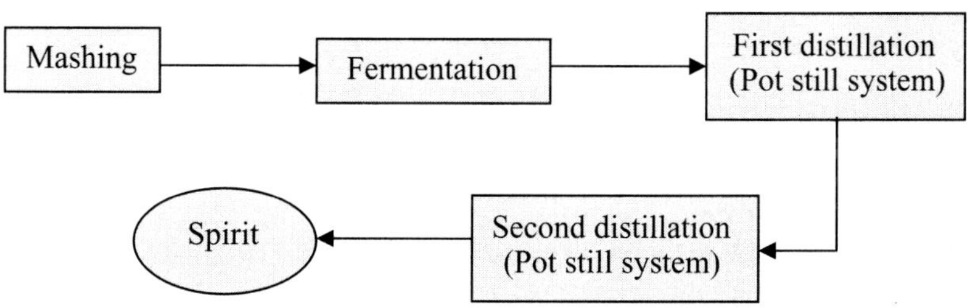


**Fig. 11:** Flow diagram of pot still, double stage distillation.

## 4.2 Neutral Spirit Production

Corn grains are mostly used for the preparation of neutral spirit. Slurry is first prepared from the ground grains and is subjected to a temperature of 100°C for atmospheric cooking or 160°C for pressure cooking. The resulting slurry is generally cooled under vacuum conditions [76]. Fungal α–amylase is added to mash and the temperature is set at 32°C or 38°C. It is inoculated with yeast and left for fermentation. Yeast strains used in this process are efficient enough to produce about 14% (v/v) alcohol in about 60 h of fermentation.

The finished beer is distilled in a batch still and further processed to remove most of its congeners present. Additional distilling columns are also employed depending on the amount of head and tail fractions in the product [77].

Normally batch rectifying column and kettle unit are used in the production of grain neutral spirits. The neutral spirit is left for maturation till the desired flavour and aroma appear. The products at 192 to 192.5° proof are collected from condenser, or the distillate is also removed as a side draw off from 5 to 7 plates below the top. This step makes it possible to take a 3 to 5% heads cut, yielding a more neutral product. Fusel oil is collected from the side of column at 130° proof. In this fraction, phenolic esters with high and intermediate boiling temperatures and traces of acids with high boiling points are present.

The most common distilling system for producing grain neutral spirits used in distilleries consist of continuous four column system comprising the following parts:

1. Aldehyde column
2. Rectifying column
3. Fusel oil concentrating column
4. Head concentrating column

The columns 1 and 2 are used to purify the distillate, while columns 3 and 4 serve to concentrate and remove the congeners.

### *4.2.1 Aldehyde column*

This column mainly refines the beer still distillates and recycles streams from other columns. The product is collected from the base at 35 to 40°C. Temperature is controlled by diluting feed with water. The dilution water comes from the base of rectifying column. The product is fed to the rectifying column. The heads containing low boiling congeners at 193° proof are collected from the condenser, and transferred to the heads concentrating column. A portion of head condensate is sent back to the top plate of aldehyde and rectifying column as reflux.

### *4.2.2 Rectifying column*

The rectifying column concentrates the alcohol in the feed coming from the base of aldehyde column to the required product proof. It also removes some volatiles comprising fusel oil, intermediate and high boiling esters, acids and phenolics. Besides this, some of the low boiling components present in the feed are also separated from the product. The product at 192 to 192.5° proof is obtained from condenser or may be collected as a side draw off from 5 to 7 plates below the top, and this makes it feasible to take a 3 to 5% heads cut, yielding a more neutral product. Fusel oil can be collected from the side of column at 130° proof. This side draw-off mainly carries phenolics; high and intermediate boiling esters and traces of high boiling acids are present.

### *4.2.3 Fusel oil concentrating column*

This column removes fusel oil, high and intermediate boiling esters and phenolics in feed stream. Alcohol present in the feed can be reclaimed from the condenser and transferred to the aldehyde column. The removed alcohol should invariably contain only traces of fusel oils. At 130°C, oil is removed from the column, which is optimum for the concentration of these congeners. This results in separation of an oil layer at the top and an aqueous layer at the bottom. The oil layer is removed as concentrated fusel oil and the aqueous phase is transferred to the feed plate of fusel oil concentrating column. The fusel oil is ultimately collected in a storage tank [78].

#### 4.2.4 Heads concentrating column

The function of this column is mainly to separate and remove low boiling esters, aldehydes, acetals, diacetyl and methanol from the feed and to reclaim the ethanol in the feed for eventual recovery as product. The feed streams for this column come from aldehyde and rectifying columns. Heads are taken out from the condenser and transferred to a storage tank for eventual disposal as "heads for destruction". A part of the condensate is brought back to the top plate as reflux [79].

#### 4.2.5 Batch rectifying column

This type of column normally comprises 50 to 60 plates and is used batch-wise for the production of neutral spirits. The use of such a large number of plates reduces the percentage of heads and tails. The kettle size and column diameter is selected such that the amount of ethanol retained on plates of the column is around 8% of the kettle. Condensing assembly comprises a dephlegmator, condenser, a vent condenser and a product cooler [80].

### 4.3 Grain Spirits

Neutral spirits obtained by the distillation of grains are called 'grain spirits'. Grain spirits are stored in oak containers and have a fine taste. The period of aging in oak barrels may be from months to years. The process for production of grain spirit is described in Fig. 12.

#### 4.3.1 Flavour

The flavour of whisky is partly dependent on the presence of congeners (toxic chemicals formed during fermentation) and fusel oils (a mixture of unpleasant, volatile, oily liquids formed during fermentation). Fusel oils have a higher alcoholic content and a strong undesired smell and taste. The presence of excess fusel oil changes the flavour of the whisky. The level of fusel oils can be brought down by distillation [81].

The American distillers pass the whisky through charcoal, gravel, sand or linen to subtract unpleasant distillates. The Canadian distillers use the column stills to produce an almost pure ethanol. The ethanol produced by these column stills is called neutral grain spirit or grain neutral spirit (GNS) (http://books.google.com/books). Typical whisky flavour is restored by blending the neutral grain spirits with flavouring whiskies (http://www.beerbrewer.co.uk/2007/06/). Acetals and diketones diacetyl are major components of flavour in whisky [82, 83] which impart fruitiness and buttery aroma to this beverage (http://www.wine–pages.com/guests/tom/taste5.htm).

#### 4.3.2 Chemistry of whisky

Whiskies and other distilled beverages like cognac and rum are complex beverages containing a vast range of flavouring compounds. Around 200 to 300 flavouring compounds can be easily detected by chemical analysis. The flavouring compounds comprise carbonyl compounds, alcohols, carboxylic acids and their esters, nitrogen- and sulphur-containing compounds, phenolic compounds, terpenes, and oxygen-containing heterocyclic compounds and esters of fatty acids [84]. The nitrogen compounds include pyridines, picolines and pyrazines [13, 76, 85].

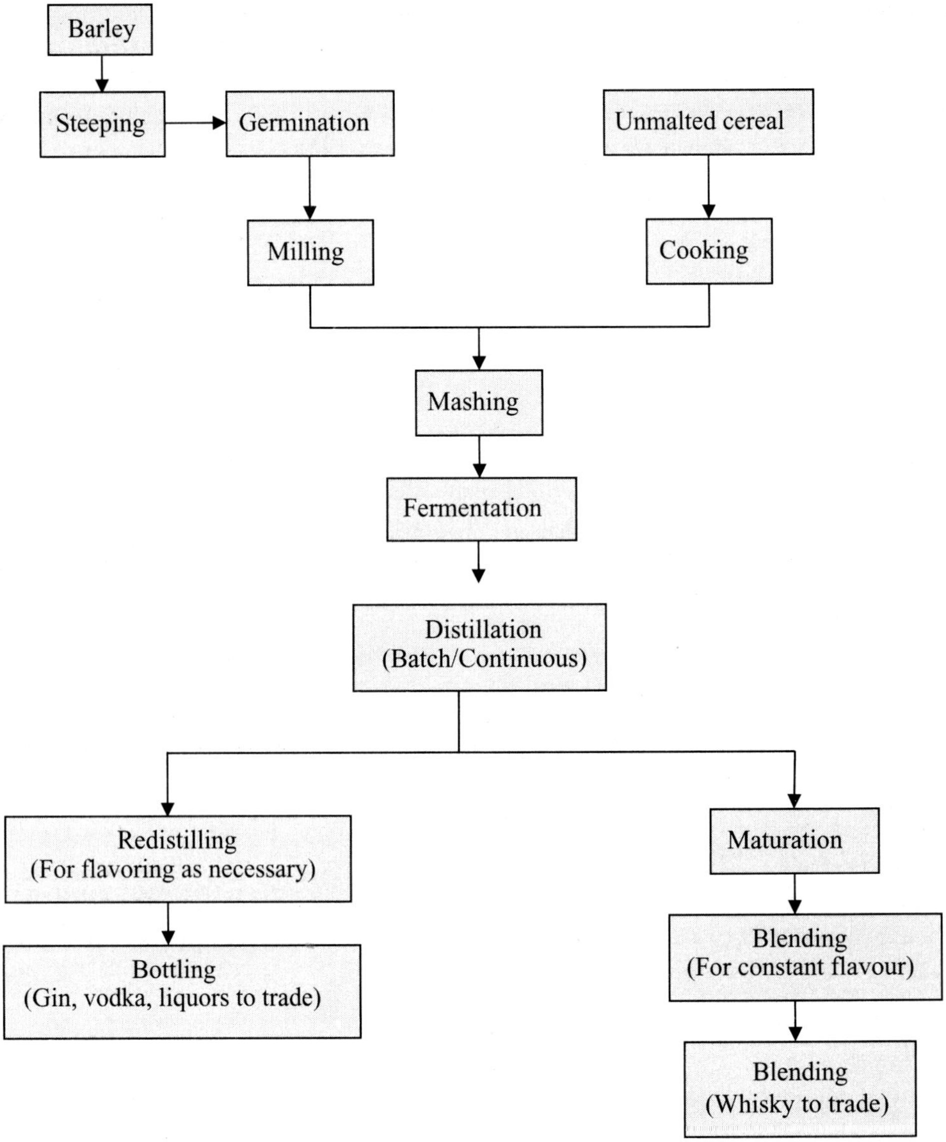


**Fig. 12:** Flow diagram of grain spirit production.

### *4.3.3 Types of whisky*

Depending on the variation in grains, production conditions and locations, a number of types or brands of whiskies exists in the alcoholic beverage market. Some popular brands of whisky are briefly described below:

4.3.3.1 Smoky scotch whisky: Scotland has internationally protected the brand name "Scotch" for its whisky. For a whisky to be labeled Scotch, it has to be made in Scotland. Excellent whiskies are produced by similar processes in other countries, notably Japan, but these cannot be labeled Scotches. Scotch whisky is a full malt whisky and specially kilned malted barleys are used in its preparation. The mash of milled grain and water is not cooked to gelatinize the starch as it is well modified during malting. Mashing is conducted at temperature between 60–65°C, after which the wort is separated from the mash and subjected to fermentation [13]. It is followed by distillation and the distillate is aged at least for three years [76]. The majority of Scotch whiskies are sold as a blend of malt and grain whisky. The blending recipies of these whiskies are the trade secrets of the blender.

4.3.3.1.1 Malt whisky: It is made from barley malt, and after fermentation by yeast, it is distilled using pot still. In the production of malt whiskies, peated barley malt is the sole source of starch. During mashing, the enzymes present in malt convert starch into sugar. As the fermentation is carried out in batches, the flavour of each batch product mainly depends upon the factors like type of yeast, temperature and location of fermentation.

4.3.3.1.2 Grain whisky: The grain whiskies are mainly produced in Scotland and USA. These are made from malted and unmalted barley along with other grains usually in continuous distillation stills. The unique difference between whiskies of USA and Scotland is that there is no microbial saccharification of starch in the Scotch whisky.

**4.3.3.2 Irish whisky:** The whiskies that are prepared from native grains of Ireland and distilled only in Ireland are called Irish whiskies. Irish whisky is considered as one of the oldest distilled alcoholic beverages of Europe, dating back to the mid 12$^{th}$ century. It is produced in the world's oldest licensed distillery 'The Old Bushmills Distillery' since 1608. Irish and Scotch whiskies differ at the malting stage. The pure form of dried malt is used in Irish whisky. This whisky is distilled thrice and the resulting final product is smoother in its taste. The flavour of the whisky is affected by the utensils in which they are stored during aging. Irish whisky is usually stored for three years in oak wood casks. Distillers avoid using new oak casks because the new oak wood imparts a rough taste to the whisky. Many distillers select used barrels that once held the contents of another type of beverages like bourbon or rum [9].

Irish whisky has several commercial forms. There is single malt whisky produced from 100% malted barley distilled in a pot still and grain whisky made from grains distilled in a column still. Grain whisky is comparatively lighter and more natural in flavour than single malt. It is blend with single malt to make a lighter blended whisky. "Pot-still" whisky differs from "single malt" in that the former uses both malted and unmalted barley, while the latter uses just malted barley [9]. This produces a heavy flavoured whisky with high ethanol content. The ratio of malted to unmalted barley may vary, but there is more unmalted barley in a pure pot-still whisky blend.

Under modern Irish law, any whisky distilled in a pot-still can be named as "pot-still whisky". Therefore, it is not a legal compulsion for a distiller to add unmalted barley to the grain mix. Some whiskies produced by the Cooley Distillery deviate from the tradition and do not use unmalted grain. Thus, they are actually single malt whiskies, but can be legally labeled "pot-still whisky". However, having being distilled in pot-still, most of the "pot-still whiskies" confirm to the traditional definition [9].

The "green" unmalted barley imparts the traditional pure pot-still whisky a spicy, uniquely Irish quality. Like single malt, pure pot-still is marketed as such or blended with grain whisky. Nevertheless, there is no real distinction between whether a blended whisky was made from single malt or pure pot-still [86].

**4.3.3.3 Rye whisky:** It is a type of whisky that must be distilled from fermented mash of rye grain. In this mash there are 51% of rye grains and remaining ingredients are usually malted barley grains and corn. The higher concentration of rye grains makes the distillation heavier. The taste of modern rye whisky mainly depends on the quality and type of grains used. Rye whisky was the common whisky of the northeastern states of USA, especially Pennsylvania and Maryland. Its production is now being revived in Kentucky by some special producers [87]. Canadian whisky is also called "Rye whisky" in Canada as its major content is rye. The major brands of Rye whisky are Canadian Club and Crown Royal. In Canada, most of the whiskies are labeled as Rye whisky with the exception of those that contain no rye. There is no legal requirement in Canada for rye to be included in the blend used to make "Rye Whisky". The only pure 100% Rye whisky made in Canada is Alberta Premium [9].

4.3.3.4 Corn whisky: It is an American whisky made from corn. The main constituent (up to 80%) of mash is corn (maize). The alcohol content in whisky is not more than 80% by volume [76]. Usually this whisky is not aged, but to change the taste and blend it is aged. The length of aging period is small as compared to other types of whiskies. Aging usually is for a short period of about six months. During aging the whisky, its colour, flavour and harshness is decreased. Corn whisky differs from Bourbon by the type of wood casks used for its aging. Bourbon is aged in new and charred barrels, whereas corn whisky should not have any contact with such barrels. This whisky is also popularly known as "Corn Liquor" or "Corn Squeezin's" [87].

4.3.3.5 Bourbon whisky: It is the most common category of distilled beverage made from grains in the family of American whisky. Major focus in the production of Bourbon whisky is to make a consistent product with a desired flavour. For each brand of Bourbon whisky, its flavour is defined by the individual product manager. A range of flavours can be produced primarily by varying the types and percentage of grains used and selection of a yeast strain that produces desired volatile flavour compounds (congeners). Bourbon mash contains 51% corn [73]. An air-dried corn with 14% moisture and 72% starch (dry) is generally used for the production of this whisky. Only grains having such characteristic produce a product of a desired volatile flavour. In addition to corn, other grains (rye, barley, and wheat) also give a specific flavour to the whisky. The most commonly used other grain in Bourbon production is rye that imparts unique spicy and ester flavour to the beverage.

Bourbon mashes are fermented with yeast strains of *S. cerevisiae* that are chosen for their capability to produce unique congeners (toxic chemicals formed during fermentation) besides ethanol. These strains produce at least 6% (v/v) of ethanol in a medium containing 11% (w/v) starch. About 2% (v/v) yeast inoculum is used. Stepwise propagation of yeast inoculum is done successively in the vessels of larger volumes till an adequate inoculum is produced to inoculate the main fermenter [88].

4.3.3.6 Canadian whisky: It is produced under the regulations of Canadian laws and must be aged for at least three years. It is made from mash of cereal grains and is produced solely in Canada. The type of grain and distillation methods may vary, but the processes used are basically the same as those in USA. The Canadian whisky is generally sold as a blend of approximately 90% neutral grain whisky and 10% bourbon or rye type whisky.

4.3.3.7 Light whisky: The process for manufacturing light whisky is basically the same as for neutral spirits, but only difference is that light whisky is distilled between 160°C to 190°C and is stored in used uncharred oak casks. The light whisky is made to meet the drift in consumer preference towards the light flavoured products. When light whisky is blended with less than 20% straight whisky, the mixture is designated as blended light whisky.

4.3.3.8 Blended whisky: The blended whisky is prepared from bourbon, rye, corn and light whiskies. Blending is the process of mixing different whiskies according to the predetermined proportion so as to get a product with desired flavour. The blended whisky contains lower level of ethyl acetate, total higher alcohols (fusel oil) and aldehydes than Bourbon whisky. Depending on the material being mixed, whisky may be classified as Blended Rye whisky or Blend of Straight Rye whisky [89].

## 5. ARRACK (RICE BEER)

Arrack is primarily produced and consumed in South and South East Asia. In most South-East Asian countries, arrack is made by fermenting fruits, grain, sugarcane and even coconut. However, in Indonesia, local red rice is fermented to produce the beverage named as 'Batavia Arrack' or 'Rum of Indonesia'. Batavia arrack is used in confectionary and flavour industries in Indonesia. Arrack is prepared by pot still distillation and is distilled to 70% alcohol volume. It is formed by blending of different components (http://en.wikipedia.org/wiki/Arrack#). Some of the popular brands of Arrack in Sri Lanka and Philippines are: V.S.O.A (Very Special old Arrack), Old Reserve, Extra Special, Double Distilled and Lambanog.

## 6. INDIAN MARKET SCENARIO IN PRODUCTION OF CEREAL BASED ALCOHOLIC BEVERAGES

It has been estimated that global cereal consumption directly provides about 50% of protein and energy necessary for the human diet, with cereals providing an additional 25% of protein and energy via livestock intermediaries. Major cereal crops produced worldwide include wheat, rice, maize, barley, sorghum, oats, millet and rye [90]. The cereals are also used to produce alcoholic beverages. The alcoholic content of these beverages varies between 4–12%. Due to lack of basic knowledge of fermentation, the technology of producing many indigenous fermented foods/beverages from cereals remains a household art in India or other Asian countries [90]. In India there are 12 joint venture companies which have a license to produce 33919 kilo litres cereal based alcoholic beverages per annum. The major beer making companies are as follows: United Breweries Limited (50.3%), SAB Miller (34.2%), Mohan Meakin (10.1%) and others (5.3%) [91].

In India, beer market grew by 6.7% in 2006 and reached a value of $874.2 million. It is expected to reach a value of $1,213.1 million by 2011. The market cost, production and % growth is given in Table 5.

**Table 5:** Market cost, production volume and % growth of Indian beer market from 2002–2006.

| Year | INR (billion) | Volume (million litres) | Growth (%) |
|---|---|---|---|
| **2002** | 29.9 | 658.7 | — |
| **2003** | 31.9 | 700.5 | 6.60 |
| **2004** | 33.9 | 744.0 | 6.50 |
| **2005** | 36.1 | 790.6 | 6.50 |
| **2006** | 38.6 | 840.6 | 6.70 |

## 7. CONCLUSION

The cereal based alcoholic beverages are widely prepared at household and industrial scale. There exists a large diversity of these beverages due to regional variations in availability of raw materials and processes developed for their preparation. Beer, whiskey and *sake* are the beverages that have a major share in alcoholic beverage market. However, most of the cereal based beverages have still remained confined to some region or geographical locations. Many of such beverages need to be explored for their biochemical, microbiological and nutritional characteristics. Scientific, engineering and technological intervention is must for widening their popularity, acceptability and production to national and international level. The yeast diversity of traditional fermented beverages has potential scope for further exploration to identify and use these for quality addition and further improvement of cereal based alcoholic beverages. The coming years will witness scientific community to address some of these issues.

## REFERENCES


1. Clerck JD (1957) *Textbook of Brewing* (Translated by Kathleen Barton–Wright), Vol 1, Chapman & Hall, London, UK.
2. Clerck JD (1958) *Textbook of Brewing* (Translated by Kathleen Barton–Wright), Vol 2, Chapman & Hall, London, UK.
3. Casida LE (1989) *Industrial Microbiology*, Wiley Eastern Limited, New Delhi, India.
4. Bamforth CW (1998) *Beer: Tap into the Art and Science of Brewing*, Plenum Press, London, UK.
5. Rose AH (1977) History and scientific basis of alcoholic beverage production. In: *Economic Microbiology*, Rose AH (ed), Academic Press, New York, USA, pp 1–41.
6. Moll M, Flayyeux R, Lipus G & Marc A (1981) Biochemistry of mashing. *Tech Q Master Brew Assoc Am* 18: 166–173.
7. Peppler HJ & Perlman D (1979) *Microbial Technology: Fermentation Technology,* Academic Press, New York, USA.
8. Marchbanks C (1987) A review of carbohydrate sources for brewing. *Brew Distilling Int* 17: 16–33.
9. Rusell I & Stewart R (1999) Cereal based alcoholic beverages. In: *Biotechnology: Food Fermentation: Microbiology, Biochemistry and Technology,* Joshi VK & Pandey A (eds), Educational Publishers & Distributors, New Delhi, India, pp 745–779.
10. Macdonald J, Reeve PTV, Ruddlesden JD & White FH (1984) Current approaches to brewery fermentations. In: *Progress in Industrial Microbiology Series 19*, Bushell ME (ed), Elsevier Publishing Co, UK, pp 47–198.
11. Walker MD (1992) Estimation of volatile sulphur compounds in beer. *J Inst Brewing* 98: 283–287.
12. Hammond JRM (1993) Brewer's Yeast. In: *Yeast Technology*, Rose AH & Harrison JS (eds), Academic Press, London, UK, pp 8–67.
13. Berry DR (1984) The physiology and microbiology of scotch whiskey production. In: *Progress in Industrial Microbiology Series 19*, Bushell ME (ed), Elsevier Publishing Co, UK, pp 199–224.
14. Cole VC & Noble AC (1985) Flavour chemistry and assessment. In: *Fermented Beverages Production,* Lea AGH & Piggott JR (eds), Blackie Academic and Professional, Glasgow, UK, pp 361–385.
15. Scholefield AJB (1956) *The Treatment of Brewing Water*, Published privately, Rockcliff, Liverpool, England.
16. Maltman A (2003) Wine, beer and whiskey: the role of geology. *Geol Today* 19: 22–29.
17. Lewis MJ & Young TW (1995) *Brewing,* Chapman and Hall, London, UK.
18. Lyons TP & Rose AH (1977) Whiskey. In: *Economic Microbiology, Alcoholic Beverages*, Rose AH (ed), Academic Press, London, UK, pp 44–137.

19. Singh BD (1998) *Biotechnology: Enzyme in Fruit Juice and Brewery Industry,* Kalyani Publishers, Ludhiana, India, pp 496–498.
20. Frazier WC & Westhoff DC (1978) *Food Microbiology,* Hill Publishing Co, New York, USA.
21. Barnett J (1992). The taxonomy of the genus *Saccharomyces* Meyen ex Rees: a short review for non–taxonomists. *Yeast* 8: 1–23.
22. Fleet GH (1998) *Microbiology of Fermented Alcoholic Beverages,* Wood JBB (ed), Blackie Academic & Professional, London, UK, pp 217–262.
23. Hammond JRM (1995) Genetically modified brewing yeasts for the 21st century. Progress to data. *Yeast* 11: 1613–1627.
24. Hammond JRM (1996) Yeast genetics. In: *Brewing Microbiology,* Priest FG & Campbell I (eds), Chapman and Hall, London, UK, pp 43–82.
25. Waites MJ, Morgan NL, Rockey JS & Higton G (2001) Food and beverage fermentation. In: *Industrial Microbiology: An introduction*, Blackwell Publishing Ltd Co, England, pp 179–209.
26. Hass GJ (1960) Microbial control methods in the brewery. *Adv Appl Microbiol* 2: 113–162.
27. Thomas SD (1993) Yeast as spoilage organisms in beverages. In: *The Yeasts, Yeast Technology*, Rose AH & Harrison JS (eds), Academic Press, London, UK, pp 517–562.
28. Fleet GH (1992) Spoilage yeast. *Crit Rev Biotechnol* 12: 1–44.
29. Priest FG (1996) Gram positive brewery bacteria. In: *Brewing Microbiology,* Priest FG & Campbell I (eds), Chapman and Hall, London, UK, pp 127–161.
30. Van Vuuren HJ (1996) Gram negative spoilage bacteria. In: *Brewing Microbiology,* Priest FG & Campbell I (eds), Chapman and Hall, London, UK, pp 163–191.
31. Chelack BJ & Ingledew WM (1987) Anaerobic Gram negative bacteria in brewing - a review. *J Am Soc Brew Chem* 45: 123–127
32. Waitier D, Leguerinel I, Hornez JP, Chowdhury I & Dubourguier HC (1995) Heat resistance of *Pectinatus* spp., a beer spoilage anaerobic bacterium. *J Appl Bacteriol* 78: 164–168.
33. Tholozan JL, Membre JM & Kubaczka M (1996) Effects of culture conditions on *Pectinatus cerevissiiphilus* and *Pectinatus frisingensis* metabolism: a physiological and statistical approach. *J Appl Bacteriol* 80: 418–424.
34. Campbell I (1996) Wild yeasts in brewing and distilling. In: *Brewing Microbiology,* Priest FG & Campbell I (eds), Chapman and Hall, London, UK, pp 193–208.
35. Petters HI, Flannigan B & Austin B (1988) Quantitative and qualitative studies of the microflora of barley malt production. *J Appl Bacteriol* 65: 279–297.
36. Flannigan B (1996) The microflora of barley and malt. In: *Brewing Microbiology,* Priest FG & Campbell I (eds), Chapman and Hall, London, UK, pp 83–126.
37. Toumi T, Laakso S & Rosenquist H (1995) Plant hormones in fungi and bacteria from malting barley. *J Inst Brew* 101: 351–357.
38. Hawthrone DB, Shaw RD, Davine DF, Kavanagh TE & Clarke BJ (1991) Butyric acid off–flavours in beer: origin and control. *J Am Soc Brew Chem* 49: 4–8.
39. Smith NA, Smith P & Woodruff CA (1992) The role of *Bacillus* sp. in N–nitrosamine formation during wort production. *J Inst Brew* 98: 409–414.
40. Wolfgang A (2004) *Enzymes in Industry: Production and Applications,* Wiley–VCH Verlag GmbH & Co, Weinheim, Germany.
41. Bamforth CW, Muller RE & Walker MD (1993) Oxygen and oxygen radicals in malting and brewing: a review. *J Am Soc Brew Chem* 51: 79–88.

42. Uchida M & Ono M (1999) Determination of hydrogen peroxide in beer and its role in beer oxidation. *J Am Soc Brew Chem* 57: 145–150.
43. Lewis MJ (1995) *Stout,* Brewers Publications, Gazelle Book Services Limited, Lancaster, UK.
44. Warner E (1992) *German Wheat Beer,* Brewers Publications, Gazelle Book Services Limited, Lancaster, UK.
45. Roger P (1998) *The Taste of Beer*, Weidenfeld and Nicholson Publishing, London, UK.
46. Eric W (1992) *German Wheat Beer,* Brewers Publications, Gazelle Book Services Limited, Lancaster, UK.
47. Joshi VK & Sandhu DK (2000) Quality evaluation of naturally fermented alcoholic beverages, microbiological examination of source of fermentation and ethanol productivity of isolates. *Acta Aliment* 29: 324–334.
48. Van de Sande T (1997) Socio–economic pitfalls of enhancing indigenous capabilities in household fermentation. *Food Cont* 8: 303–310.
49. Ross RP, Morgan S & Hill C (2002) Preservation and fermentation: past, present and future. *Int J Food Microbiol* 79: 3–16.
50. Steinkraus KH (1986) Fermented foods, feeds and beverages. *Biotech Adv* 4: 219–243.
51. Kodama K & Yashizawa K (1977) Sake. In: *Economic Microbiology, Alcoholic Beverages*, Rose AH (ed), Academic Press, London, UK, pp 423–477.
52. Onishi H (1990) Yeast in fermented foods. In: *Yeast Technology,* Spencer FT & Spencer DM (eds), Springer–Verlag, Berlin, Germany, pp 167–198.
53. Kodama K (1993) Sake–brewing yeast. In: *The Yeast Technology,* Vol 5, Rose AH & Harrison JS (eds), Academic Press, London, UK, pp 620–632.
54. Nunokawa Y & Hirotsune M (1993) Production of soft sake by an immobilized yeast Reactor system. In: *Industrial Application of Immobilized Biocatalysts,* Tanaka A, Tosa T & Kobayashi T (eds), Marcel Dekker, New York, USA, pp 235–244.
55. Ohba T & Sato M (1986) *Handbook of Food and Beverage Stability. Food Science and Technology–A Series of Monographs,* Charalambous G (ed), Academic Press, New York, USA.
56. Beuchat LR (1983) Indigenous fermented foods. In: *Biotechnology: Food and Feed Production with Microorganisms*, Reed G (ed), Verlag Chemie, Weinheim, Germany, pp 477–528.
57. Padmaja G & George M (1999) Oriental fermented foods: biotechnological approaches. In: *Food processing. Biotechnological Applications*, Marwaha SS & Arora JK (eds), Asiatech Publishers Inc, New Delhi, India, pp 143–189.
58. Beuchat LR (2001) Traditional fermented foods. In: *Food Microbiology: Fundamentals and Frontiers*, Doyle MP, Beuchat LR & Montville TJ (eds), ASM Press, Washington DC, USA, pp 701–719.
59. Dahal NR, Karki TB, Swamylingappa B, Li Q & Gu G (2005) Traditional foods and beverages of Nepal–A review. *Food Rev Int* 21: 1–25.
60. Tamang JP, Sarkar PK & Hesseltine CW (2006) Traditional fermented foods and beverages of Darjeeling and Sikkim–a review. *J Sci Food Agric* 44: 375–385.
61. Savitri & Bhalla TC (2007) Traditional foods and beverages of Himachal Pradesh. *Ind J Trad Knowl* 6: 17–24
62. Hesseltine CW & Wang HL (1980) The importance of traditional fermented foods. *Bioscience* 30: 402–404.
63. Orji MU, Mbata TI, Aniche GN & Ahonkhai I (2003) The use of starter cultures to produce 'Pito', a Nigerian fermented alcoholic beverage. *World J Microbiol Biotechnol* 19: 733–736.

64. Haard NF, Odunfa SA, Lee C–H, Quintero–Ramirez R, Lorence–Quinones A & Wacher–Raderte C (1999) Fermented cereals, a global perspective. In: *FAO Agricultural Services Bulletin No 138*, FAO, Rome, Italy.
65. Maksoud SA, Hadidi MN & Amer WM (1994) Beer from the early dynasties (3500–3400 cal B.C.) of Upper Egypt, detected by archaeochemical methods. *Veg His Archaeobot* 3: 219–224.
66. Blandino AME, Aseeri Al SS, Pandiella DC & Webb C (2003) Cereal–based fermented foods and beverages. *Food Res Int* 36: 527–543.
67. Steinkraus KH (1996) *Handbook of Indigenous Fermented Foods*, Marcel Dekker, New York, USA.
68. Ulloa M, Herrera T & Lappe P (1987) Fermentaciones tradicionales indígenas de México. *Serie de Investigaciones Sociales,* No. 16. Instituto Nacional Indigenista, Mexico City, pp 13–20.
69. Lloyd J & Mitchinson J (2006) *The Book of General Ignorance*, Faber & Faber Ltd, London, UK.
70. Hanson DJ (2007) History of Alcohol and Drinking around the World, PhD Thesis, California University, USA.
71. Lyons TP & Rose AH (1977) Whiskey. In: *Economic Microbiology*, Rose AH (ed), Academic Press, London, UK, pp 635–692.
72. Piggott JR & Conner JM (1995) *Whiskies in Fermented Beverage Production*, Lea AGH & Pigott JR (eds), Blackie Academic and Professional, Glasgow, UK, pp 247–274.
73. Johnson AH & Peterson MS (1974) Some Asian fermented foods and beverages and associated fungi. *Mycologia* 66: 942–950.
74. Suomalainen H & Nurminen T (1975) Some aspects of the structure and function of the yeast plasma membrane. In: *Aikon Keskuslaboratoria Report 8101,* Aiko, Helsinki, Finland.
75. Harrison JS & Graham JCJ (1970) *The Yeast*, Rose AH & Harrison JS (eds), Academic Press, New York, USA, pp 283–348.
76. Bluhm L (1983) Distilled beverages. In: *Biotechnology, Food and Feed Production with Microorganisms,* Reed G (ed), Verlag Chemie, Weinheim, Germany, pp 447–476.
77. Watson DC (1985) Current developments in the potable distilling industry. *Crit Rev Biotechnol* 2: 147–192.
78. Max G (1952) Relationship of fusel oil, aldehydes, and esters in distilling material, Hi–Proof Brandy and Dessert Wines. *Am Soc Enol Viticulture* 3: 167–171.
79. James FG (1960) The composition of plate samples from distilling columns with particular reference to the distribution of higher alcohols. *Am Soc Enol Viticulture* 11: 105–112.
80. Bernot CM, Doherty M & Malone MF (1990) Patterns of composition change in multicomponent batch distillation. *Chem Engg Sci* 45: 1207–1212.
81. Nykanen L (1986) Formation and occurence of flavour compounds in wine and distilled alcoholic beverages. *Am J Enol Viticulture* 37: 84–96.
82. Rice JF & Helbert JR (1975) *Proceedings of American Society of Brewing Chemists,* USA, pp 11–17.
83. Branen AL & Keenan TW (1971) Diacetyl and acetoin production by *Lactobacillus casei. Appl Environ Microbiol* 22: 517–521.
84. Varman AH & Sutherland JP (1994) *Beverages–Technology, Chemistry and Microbiology,* Chapman and Hall, London, UK.
85. Barbour EA & Priest FG (1988) Some effects of *Lactobacillus* contamination in Scotch whiskey fermentations. *J Inst Brewing* 94: 89–92.

86. Toumi T, Laakso S & Rosenquist H (1995) Plant hormones in fungi and bacteria from malting barley. *J Inst Brewing* 101: 351–357.
87. Prescott SC & Dunn CG (1959) *Industrial Microbiology*, McGraw–Hill, New York, USA.
88. James MJ, Loessner MJ & David AG (2005) *Modern Food Microbiology*, CBS Publishers & Distributors, New Delhi, India.
89. Simpson AC (1977) Gin and vodka. In: *Economic Microbiology, Alcoholic Beverages,* Rose AH (ed), Academic Press, London, UK, pp 537–94.
90. Chaven JK & Kadam SS (1989) Nutritional improvement of cereals by fermentation. *Crit Rev Food Sci Technol* 28: 349–357.
91. Beer in India (2006) Industry profile. Electronic Resource Reference Code No. 0102–0744.

# Wine and Brandy

# 16

**V.K. Joshi, Somesh Sharma and Vishal S. Rana**

*Department of Food Science and Technology, Dr. Y.S. Parmar University of Horticulture and Forestry, Nauni, Solan 173 230 (H.P.)*

## 1. INTRODUCTION

Wine is regarded as a gift from God and a divine fluid (*Soma*) in Indian mythology. It has been prepared and consumed by man since antiquity, espoused for centuries as a superior beverage [1] and has been used as a medicine [2]. Vedas have also mentioned the medicinal power of wine. Ancient scriptures suggest that the origin of wine might be accidental when the juice of some fruit must have transformed itself into such a beverage having exhilarating or stimulating properties. Clearly, the first wine master must have been a caveman who discovered the magic of fermentation. Consumption of wine in real sense is a form of returning to nature and escape from the social bonds [3]. Thus, grape fermentation is a suitable option for transformation of perishable seasonally available fruit into relatively non-perishable beverage with novel and potentially intoxicating properties [4]. Today, wine is an integral component of the cultures of many countries. Wine is a completely or partially fermented juice of grapes, but fruits other than grapes are also utilized for the production of wines [5-12]. Another grape product is brandy made by distillation of wine. The word 'brandy' is a Dutch word *'brandewein'* which means burnt wine, while *al-koh'l* is Arabic word [13]. The fruit is considered most suitable for wine production if it contains soluble sugars, available for fermentation by yeast into alcohol and the acids, which limit the growth of most spoilage causing bacteria and fungi [4].

The perception that wine is good for body is now scientifically supported and there is no dispute that moderate wine consumption is associated with lower mortality from coronary heart diseases [14]. The health benefits of wine are now largely attributed to the antimicrobial activities of ethanol, and antioxidant properties of phenolic components and flavonoids. Besides, these are nutritious, safer and provide psychotropic effects, and are preferable to distilled liquors [15, 16]. Wine is an important adjunct to the human diet, increases satisfaction, contributes to the relaxation necessary for proper digestion and absorption of food and has glucose tolerance factor (GTF), which is considered beneficial in the cure of diabetes [8]. Excessive alcohol consumption, both acute and chronic, however can have devastating effects on the physical and mental well-being of the consumers leading to cirrhosis of the liver, enhancement of chances of hypertension, strokes and development of breast and digestive tract cancers [17]. A total of 47 phenolic compounds including non-flavanoids (hydroxybenzoic and hydroxycinnamic acids) and their derivatives, stilbenes and phenolic alcohols were identified in wines from *Vitis vinifera* L cv Tempranillo, Graciano, Cabernet, Sauvignon, and Merlot by HPLC-DAD/ESI-MS [18]. The leading wine-producing countries include France, Italy, Spain, Argentina, Portugal, Germany, South Africa, and the United States. Cider and perry are important products in England and

*Food Biotechnology: Principles and Practices*
*V.K. Koshi & R.S. Singh (Eds.)*
*I.K. International Publishing House Pvt. Ltd., India*
*pp 525–561*

northern France; fortified cherry and black currant wines are produced in Denmark. Wine production is also picking up in India [19-21].

## 2. DEFINITION, CLASSIFICATION AND TYPE OF WINE

Generally, there is no accepted system of classification of wines, though, these may be classified broadly based on the geographic origin, grape variety used, fermentation, maturation process or may bear generic names [4, 22]. Based on colour, wines could be red, rosé (pink) or white and could be differentiated as table (red, rosé or white), sparkling or fortified wines. There are specific varieties for wine, especially for red and white wine making [6]. The alcohol content of table wine generally ranges from 10 to 11 per cent, while that of cider and perry is usually 2-8 per cent [23]. Fortified or dessert wines contain added brandy and could be red or white with alcoholic content ranging from about 16 to 23 per cent. Such wines with additional flavouring are called aromatic wines like vermouth. Sweet wines are high in sugar content while dry wines have negligible amount of sugar.

Unless otherwise specified, the term wine is applied here to the product made by the alcoholic fermentation of grape or grape juice by yeast and a subsequent aging process [12, 19]. Wines, however, could be produced by the fermentation of juices/extracts of many fruits such as cherries, berries, rhubarb, banana, pineapple, cashew nut, pomegranate, lemons, tangerines, oranges, dates and figs. Wines from grapes are usually of red or white types [6, 22]. Since the pigments of most of the red grapes are localized in the skin, therefore, in making red wine, the juice is fermented along with the skin to extract maximum colour. In white wine preparation, the juice is fermented free of skin in order to extract the minimum colour and tannin. Most wines are *still wines,* as they retain no carbon dioxide produced during the fermentation, in contrast to sparkling wines, which contain a considerable amount of carbon dioxide. *Champagne* is the sparkling wine made in champagne region of France. The old province of *Champagne* in Northern France has given its name to the most famous sparkling wine of the world. It is prepared either by bottle or tank fermentation for carbonation of base wine by secondary fermentation [22]. *Dry wines* contain little or no unfermented sugar, in contrast to sweet wines which have either residual sugar or added later on. Wines usually contain 11 to 16% of alcohol, but may have as low as 7 per cent. *Fortified wines*, to which distillate of wine called *Brandy* is added, may contain 19 to 21 per cent of alcohol [19, 24]. *Table wines* have a comparatively low alcoholic content and little or no sugar, while *dessert* wines are *fortified* sweet wines. *Sherry* is produced by a special processing technique, containing 18 to 21 per cent alcohol and could be sweet or dry [10]. The technique used is secondary fermentation by special type of *Flor* yeast or by *baking. Cider* (Cider or Cidre) in Great Britain and France means apple wine, hard cider or fermented apple juice, but in USA it may mean fermented or unfermented apple juice based on the definitions and available products [12, 19, 25]. *Perry* is the wine prepared from pear juice. It could be sweet or dry. Since the pear fruits are quite astringent, the same characteristic is imparted to the wine also. *Medicinal wines* are flavoured wines containing extract of several herbs of medicinal importance and may include even bitter compounds like quinine. These wines are usually sweet and have as high as 18-20 per cent alcohol. *Natural Wines* are the products obtained by the use of approved formulas with a natural wine base, herbs, spices, fruit juices, aromatics, essences and other flavouring materials, and are intended to attract the non-wine drinking public. Many of these wines contain less than 14 per cent alcohol. *Vermouth* is a wine with 15 to 21 per cent alcohol, flavoured with a characteristic mixture of herbs and spices, some of which impart an aromatic flavour and odour, and others, a bitter flavour. Vermouth may be sweet or dry [26].

*Brandy* is an alcoholic distillate from the fermented juice or wine. Fruit brandy derived from fruits other than grapes is designated after the name of such fruits like Peach brandy/Apricot brandy [10, 11].

## 3. SIGNIFICANCE AND ROLE OF WINE

Wines have played an intricate role in many aspects of world history as well as its culture. Its development has closely followed the pattern of western civilization. As civilizations became more advanced, so did the wine industry. Earlier wine was only available to a selected section but eventually it entered into almost every home throughout the world. Its consumption had influenced political and religious beliefs. Wine has food value and the major nutritional value comes from the rapidly utilizable caloric value of its alcoholic content. Alcohol does not need to be digested and can be absorbed directly through the intestine wall. Wine has some mineral elements, vitamins, sugars, acids, and phenols [15]. It has small quantities of the B-vitamins such as $B_1$ (thiamin), $B_2$ (riboflavin) and $B_{12}$ (cyanocobalamine), but is devoid of vitamins like A, D and K [27]. Indigestible polysaccharides are major constituent of dietary fibres. Significant amount of soluble dietary fibres were found in wines from Spanish region. The content was higher in red wine (0.94-1.37 g/l) than in white wine (0.19 to 0.39 g/l [28]. Volatile esters are a major by-product in the yeast fermentation of grape juice and can significantly impact wine flavour. The kinetics of ester formation for each nitrogen treatment (Diammonium phosphate or amino acids) was dependent upon the yeast strain. The DAP addition resulted in more significant increase in volatile esters than did the amino acid addition [29]. Consumption of wine with food has an advantage of slowing the rate of alcohol absorption into the blood, stimulates the production of gastric juices, enhances the release of gastrin hormone, fosters healthy appetite, and activates the release of bile into the intestine [30], etc.

The consumption of wine affects the *Heliocobacterium pylori*, the bacterium considered as primary cause of stomach ulceration. Moderate wine consumption appears to have prophylactic effect on the ulcer development and this action of wine against gastrointestinal infections was well established, even before the microbial nature of several diseases was discovered [31]. Recently, apple wine has been evaluated for antimicrobial activity against *E. coli, Staphylococcus, Bacillus, Aspergillus* and *Candida* [32]. Most of the antimicrobial effects of wine have been attributed to phenolic compounds such as *p*-coumaric acid, which are particularly active against Gram positive bacteria, while other phenols inhibit Gram negative bacteria, like *Escherichia coli, Proteus* and *Vibrio* which cause various forms of diahorrea [33]. Red wine was documented to be active against *H. pylori* [34]. The anti microbial activity of the red and white wine extract was related to total phenolic content and was more effective against *S. aureus* and less effective against *E. coli* and *C. albicans*. Further, some phenolic acids have the potential to inhibit the growth of certain pathogens such as *S. aureus, E. coli* and *C. albicans* [35]. Wine has also been found as a tranquillizer, diuretic, reduces muscle spasms and stiffness associated with arthritis, delaying the development of some forms of diabetes, cardiovascular diseases, have antioxidant effects and inhibition of platelet aggregation [22]. The moderate wine consumption is known to increase the plasma levels of high-density lipoprotein (HDL), which are helpful in lowering blood cholesterol level and the incidence of arteriosclerosis [36-39]. Wine is a particularly rich source of dietary flavonoid phenolics including resveratrol that could reduce CHD mortality by its ability to inhibit platelet aggregation, eicosanoid synthesis and oxidation of human low density lipoproteins (LDL). Trans-resveratrol concentration in the range of 0.148 to 4.547 mg/l in 15 Nova Scotia red wines (three produced from vinifera varieties and 12 from French hybrid) has been reported [40]. Since the wine production is anaerobic process, the phenolics are not degraded and are retained as antioxidants [1, 41]. The benefits of antioxidants are more pronounced in red wines than white wines due to their higher phenolic content. The antioxidants in wine have recently been linked to the prevention of cancer [4, 10]. The negative effects of excessive consumption of wine on the health including the effect on liver, blood pressure, etc. have also been well documented [10, 11, 16, 19, 43-49]. The most important contradiction relates to those with the past history of alcohol abuse or alcoholism.

The use of wine in religion was made in the Ancient Egypt, in funeral rites of hero, kings or such dignitaries. The wine was considered as a gift from the gods by the Ancient Persians who used to make toasts to praise their gods. The Greeks and Romans did not lag behind in their religious beliefs and they considered wine as a gift from the Greek god, Dionysus, and same was the case with the Romans who continued this practice with their god, Bacchus. It played a major role in the Jewish, Islamic, and Christian cultures too. In Jewish ritual, each Sabbath begins with a blessing chanted over a cup of wine, which the whole family then shares but the specific amounts of wine must also be drunk at weddings and several other functions. The Islamic Empire, however, halted the production of wine since its consumption was prohibited by the Islamic code of law and theology, especially in an area that stretched from Northern India to Spain [50, 51]. In contrast to prohibitions though, Muhammad's vision of the paradise promise to believers does include wine. The wine has been described as a divine fluid most clearly in the Indian mythology than in Mesopotamia, Egypt or Europe by the polymorphic deity *Soma* [52].

In the early civilizations of Mesopotamia and Ancient Egypt, wine played a significant role. Grape harvest depicted on many of the tomb walls of Egyptians reveals that the production of wine was an intricate part of their daily agriculture. The Greeks also celebrated the cultivation of grapes and held many religious festivals to ensure a good harvest [3]. Gods of wine were also found associated with agrarian fertility rituals in all the major religions where viticulture was practiced. Wine has inspired many poets and writers. Literary writings have also illustrated the wine such as the Odyssey by Homer. Wine drinking in Roman society is well attested by the numerous references in the works of Roman poet, Griffin.

## 4. MICROBIOLOGY OF WINE MAKING

The knowledge of microbiology in wine making is essential to understand the process of quality wine production. In the process of wine making, different microorganisms belonging to moulds, yeast and bacteria are involved. Diversity of various microorganisms associated with wine making has been reviewed earlier also [10, 53] and discussed here.

### 4.1 Moulds

Grapes like other fruits are also subjected to mould spoilage at different stages of harvesting and transportation. Besides damaging the fruits, the moulds can also affect the flavour of wine by their growth on the cooperage. However, the moulds cannot grow in the wine mainly due to inhibitory effect of ethanol and the anaerobic conditions as most of the moulds are aerobic in nature [10, 11].

### 4.2 Bacteria

The bacteria in wine making are more associated with spoilage than the wine production [6, 10]. The acetic acid and lactic acid bacteria are the important groups of bacteria, influencing the quality of wine either by spoilage or by improving the quality of wine by malolactic acid fermentation. However, all the lactic acid bacteria do not cause spoilage. A group of these, called 'Malolactic acid bacteria', can degrade the malic acid and reduce the acidity of high acid wines to make them palatable.

These bacteria belong to genus *Acetobacter* and *Gluconobacter. Acetobacter* genus includes three distinct species like *A. aceti, A. pasteurianus* and *A. peroxydans*. These bacteria are of great concern to the wine makers, as production of acetic acid increases the volatile acidity of wine, which affects its sensory quality. The volatile acid is a good indicator of growth and activity of acetic acid bacteria.

At lower concentration, only the sensory quality is affected which can be rectified by blending. But at higher concentration, the wine remains suitable only for vinegar production. The acetic acid bacteria can spoil the must, fermenting wine and the wine undergoing maturation. The wine stored in casks not filled completely and sealed tight would be converted into vinegar and spoiled. To prevent this, the contact of must with air during fermentation and storage of wine should be minimum possible, and the alcohol content of the wine should not be less than 15 per cent. Since the pomace undergoes acetification process rapidly, its disposal should be prompt. Addition of sulphur dioxide or pasteurizing the wine having low alcohol level can also prevent the spoilage of wine by the acetic acid bacteria [10, 19].

Lactic acid bacteria cause the spoilage of wine and also reduce the acidity of high acid wines to an acceptable level, and therefore, are of great significance to the winemakers [10, 11, 53]. The appearance and flavour of the wine is altered by the growth and activity of lactic acid bacteria. The appearance of wine in a bottle or a test tube becomes silky/cloudy due to their growth and it gives a disagreeable smell, called mousy with formation of flocculent or sediment [6]. The lactic acid bacteria belonging to genus *Lactobacillus, Leuconostoc* and *Pediococcus* are responsible for the wine spoilage by producing lactic acid. These bacteria are acid tolerant, rods or cocci, Gram positive and could be homo or heterofermentative. The former convert glucose chiefly to lactic acid without the formation of appreciable amount of carbon dioxide or acetic acid, while the latter form lactic acid, carbon dioxide, acetic acid, ethanol and glycerol. *Leuconostoc,* due to its ability to ferment sucrose into dextran can cause ropiness in wine.

## 4.3 Yeast

Yeasts are important microorganisms associated not only with the production of wine but its spoilage too. While the yeast strains of *Saccharomyces* predominantly carry out the alcoholic fermentation of fruit juices but the wild yeasts may be responsible for spoilage of wine. Growth of wild yeast during fermentation, especially apiculate yeast, is considered detrimental to the wine quality. Besides, the yeast might cause cloudiness and microbiological unstability. Yeasts are of immense importance to the wine makers as these conduct the alcoholic fermentation and affect the wine quality, but these yeasts can also spoil the wines. Out of 500 species of yeasts recognised by the current taxonomists, only 15-20 are relevant to the wine making [53, 54].

Morphologically, *Saccharomyces* is spherical to ellipsoidal in shape with an approximate dimension of 8 x 7$\mu$m, depending on the organism and the growth medium. Asexual reproduction takes place by multilateral budding. Most strains of *S. cerevisiae* are capable of producing alcohol levels upto 16 per cent (v/v). Although these yeasts are of fermentative type, but may also grow oxidatively as a part of the surface film community. *S. beticus*, for example, is commercially available as 'Flor Yeast' for the sherry preparation. Different related species, based on the fermentation pattern of various carbohydrates could be distinguished e.g., *Saccharomyces bayanus* and *Saccharomyces fermentati* with inability to ferment galactose, *Saccharomyces diastaticus* fermenting starch and *Saccharomyces uvarum* with ability to assimilate melibiose, etc. [6]. The yeast used mostly for sparkling wine production belongs to *Saccharomyces bayanus.* In the production of sherry, 'Flor Yeast', usually of species of *Saccharomyces bayanus, Saccharomyces capensis* and *Saccharomyces fermentatii* are used [55]. Yeast strain selection was demonstrated to be important for colour preservation in red wines during vinification. Differences in the loss of total anthocyanin compounds, as a consequence of their removal by yeast, were not significant for the different strain studied with RGJ medium. However, colour indices showed significant differences among strains. The production of these stable anthocyanin pigments may be a significant contribution by yeast to red wine colour, allowing longer ageing [56].

*Brettanomyces* resembles *Saccharomyces cerevisiae* microscopically, but is somewhat smaller in size. Vegetative cells are described as oval in shape. *Brettanomyces* and its sporulating counterpart, *Dekkera* are capable of producing 10-11 per cent alcohol. It is usually viewed as a problem in barrel-aging wines. The spoilage by the yeast imparts either metallic or a 'horsey' tone to the wine. Insects, particularly fruit flies, are the important vectors of the yeasts, besides the insanitary conditions of crushing equipment and resulting contamination [6]. The apiculate (lemon) shape cells characteristic of *Kloeckera/Hanseniaspora* arise from repeated budding at both the poles. Both *Kloeckera* and *Hanseniaspora* are capable of producing substantial and inhibitory levels of acetic acid and ethyl acetate.

*Zygosaccharomyces bailli,* an osmophilic yeast, may be troublesome in wines sweetened with concentrate due to its survival at a pH of less than 2.0 and in alcohol levels greater than 15 per cent [6]. Microscopically, *Candida* appears as long cylindrical cells of approximate dimensions 3-10 x 2-4 μm and reproduces asexually by budding. The asporogenous (imperfect) genus *Candida* includes numerous species. In wine, *Candida* typically grows as surface yeast, producing a chalky white film on the surface of low alcohol wines. In the presence of oxygen, it uses ethyl alcohol in addition to wine acids which act as carbon sources, it produces a mixture of oxidized end products. As a group, *Candida* are weak fermenters [10, 53, 58].

The important species of *Pichia,* an ascospore producing yeast include *P. membranaefaciens, P. vini* and *P. farinosa*. Most species are inhibited by alcohol levels of about 10 per cent. Growth on the surface of wine appears as a very heavy 'balbonlike' chalky film imparting an aldehydic character to the wine. Actively growing cells appear as short ellipsoidal to cylindrical shaped rods. Reproduction is usually by multilateral budding, leading to an extensive pseudomycellium development.

*Hansenula* appears as an oval to oblong shaped cell of approximate dimension of 2.5 x 5 $\mu$m. Members of this genus reproduce by asexual budding. *Hansenula anomala* has both oxidative and fermentative metabolism. Most species of *Hansenula* form film on liquid culture media and the wines of low alcohol content. *Hansenula* sp. forms large amounts of acetic acid ranging from 1-2 g/h as well as volatile esters, particularly ethyl acetate upto 2.15 mg/L, and also forms isoamyl acetate. The wines fermented by this yeast possess too much of aroma, which is improved with aging. This yeast might, however, be useful in increasing the ester content of the wine in the mixed culture [10, 11].

Fission yeast comprises of four species, out of which *Schizosaccharomyces pombe* is well studied for its ability to utilize malic acid fermentatively with the production of ethanol though at a slower rate. The reproduction takes place by fission similar to the bacteria. The deacidifying ability to utilize malic acid by the yeast has been used to reduce the acidity of grape wines. However, in grape wine fermentation, the yeast imparts off-flavour and cause difficulties due to its slow rate of fermentation. The capacity to resist high levels of sulphur dioxide has made use of it to carry out the two stage fermentation of grapes to produce quality wine. Possibility of reducing acidity in plum wine by the use of this yeast has also been explored, wherein the overall acidity in the wines made from fruits with high malic acid content such as plum was reduced [24, 59]. Deacidification of grape musts and wines is important for the production of well balanced wines. The bacterial malolactic fermentation (MLF) process is unreliable and stuck MLF often leads to spoilage of wines and the production of biogenic amines. The genetically engineered wine yeast, MLO1, is a *Prise de Mousse* strain that produced wine having lower volatile acidity, and improved colour properties than wines produced with parental yeast and bacterial MLF. The strain contains the malate transport gene (mael) from *Schizosaccharomyces pombe* and malolactic gene (mleA) from *Oenoccoccus oeni*, stably integrated into the genome at the URA3 locus [60].

## 5. GENERAL ASPECTS OF FRUIT WINES PRODUCTION

For production of wines, the raw material required is the fruits, and besides this, sweetening agent like concentrate, sugar, acid, nitrogen source, clarifying enzyme, filter-aid, etc. are also required. The sugar is the second most important raw material, which also dictates the final cost of production of these beverages. Almost all the fruits, their juices or extracts being low in sugar have to be ameliorated to the desired sugar level. Even when the sugar level is satisfactory, the high acidity demands dilution, consequently requiring more addition of sugar to the must [5, 61]. Spices and herbs or their extract constitute an essential raw material after fruit juices or their extract for the preparation of fortified wines like vermouth [5, 24]. Mostly, the enzyme pectin esterase is used for clarification of wines including those from stone fruits like plum and peach, which might later be used to produce wines of better quality having higher yield than the conventional wines. Basic technique for the production of fruit wines is essentially the same involving the routine alcoholic fermentation of juice or the pulp, but modifications with respect to the physico-chemical characteristics can be made depending upon the type of wine to be made and fruit used [61-63]. The major differences in the techniques of production of these wines arise from the difficulty to extract the sugar from the pulp of some of the fruits [5]. The juices obtained from most of the fruits other than grapes have poor fermentability. The higher acidity in some of the fruits makes it all the more difficult to prepare the wine of acceptable qualities. On the other hand, many tropical and subtropical fruits have low acidity and it is difficult to make attractive and stable wines from them [64, 65]. The fermentation of such fruits is very slow or may cease even before the completion due to the lack of certain nitrogenous compounds and other yeast growth factors [66]. In case of blackberries, harvest maturity is one of the important factors, because the fruits have higher pigment concentration towards the end of the season, leading to pigment deposits in the bottle. To avoid this, blackberries should better be picked before full maturity. Deficient irrigation appears to be an important tool for improving grape wine making quality. The Merlot berry quality was impaired by extreme water stress. The optimal level of stress is attained by maintaining the available water between 30-15%. Merlot grapes could also benefit from longer maceration as it increased the concentration of polymerized polyphenol, stabilized colour and improved the mouth-feel properties of the resulting wines [67]. In case of citrus fruits, bitterness of the juice is a major problem, besides darkening of wine. The juice extracted by rack and cloth press or screw expeller contains so much essential oil from the peel that the fermentation is reduced considerably.

The composition of fruit and its maturity are dependent upon the cultivation practices. A good wine yeast strain of *Saccharomyces cerevisiae* in the form of slants, tablet or compressed yeast is a prerequisite for making quality wine. It is activated in the juice/pulp appropriately intended for wine making at a rate of 2-5 per cent to the must. The must is prepared depending upon the fruits (pulp or juice) used and type of wine to be made. Addition of $SO_2$ to the must in the concentration of 100-150 ppm as potassium metabisulphite (KMS) is also required. In plum wine fermentation, addition of sodium benzoate gives better quality wine than the potassium metabisulphite [62]. In apple juice, ammonium sulphate with thiamin (4 mg/l) gives much greater effect on the fermentation rate [5, 58]. Fining with gelatin is found to be a good remedy for the juices like pear which have high tannin content. Heating of the juice for 35 minutes at 122°F can also increase the fermentation rate [68]. Six wine making varieties and techniques to increase phenolic concentration have also been reported [69]. These are fermentation temperature, thermovinification, must freezing, saignee, pectolytic enzyme treatments and carbonic maceration. In contrast, sulphur dioxide levels and cold soak treatments have frequently been shown to have no or little lasting effect, or lead to a decrease in phenolic level. The influence of selected factors viz., effect of solvent contact time, extraction temperature concentration of ethanol and hydrochloric acid on extraction of total phenols and anthocyanins has been studied [70]. Both extraction temperature and concentration of the solvents, ethanol and hydrochloric acid exerted highly significant effects on the extraction of both total phenols and anthocyanins.

The must is allowed to ferment at a temperature (20-25°C) after inoculation with yeast culture, and the container in which fermentation is carried out is equipped with an air bag. The sugar content is measured periodically to monitor the progress of fermentation. After the completion of fermentation, wine is siphoned/racked or filtered through a cheese/muslin cloth in case of pulpy must which is followed by siphoning. The process of maturation makes the wine mellow in taste and fruity in flavour besides the clarification. The period of maturation may range from 6 months to 2-3 years. The process of maturation is complex and the formation of esters takes place, which improves the flavour of such beverages [5]. Wine from fruits like apricot or plum is acidic and requires some amount of sweetening prior to final bottling. For apricot and plum wine, the extent of sweetening up to a TSS of 12°B has been found to be optimum [19]. The wine is filtered, if needed. Wines being low alcoholic beverages are pasteurized at 62°C for 15-20 minutes after keeping some headspace in the bottle followed by crown corking the same. The pasteurized wines once opened have to be kept at low temperature to prevent their spoilage. Alternately, the table wines can be preserved by addition of preservatives like sulphur dioxide, sodium benzoate, sorbic acid, etc. [5]. The fortified wines generally need no preservatives. A validated SPME-GC method for the analysis of essential trace components in pinot gris and chardonnay wines using the DVB/Carboxen PDMS mixed coating fibres has been presented [71]. Good results were obtained for esters, alcohols and terpenes. Organic acids were not affectively analyzed using this method, and limits of detection for most target aroma components were in the μg/l range.

## 6. WINERY EQUIPMENT

### 6.1 Crushing Unit

One of the best systems for extraction of juice combines a tank with a cylindrical perforated core for juice draining, an inclined screw dejuicer and coq press. An electric hydraulic press used for extraction of juice is shown in Plate 1. Pressing increases the malate content of the juice. Recovery of alcohol from fermented pomace is a major problem. Among the various procedures used is grinding the pomace in a hammer mill or disintegrator and distilling in a special still. However, with counter current remarkable recovery could be obtained where it is desired to recover as much wine as possible. The pomace may be discharged through the gate either into a conveyer or a pomace pump. When distillation material is needed, the pomace is usually sent into the pomace conveyor. Uncrushed grapes are also placed in tanks to allow them to ferment that results in the production of carbon dioxide in sufficient quantities to asphyxiate the cells giving greater colour and flavour. However, it also results in hydrolysis of pectins with liberation of methanol and increase in amino acids. Macerating enzymes and postfermentation enological tannin were used on Cynthiana (*Vitis aestivalis*) to examine the effect on colour extraction and retention. Out of the five enzyme preparations, Trenolin colour DF, Lallzyme Ex-V, and Vinozyme G, the last preparation had higher percentages of red colour and lower percentage of blue and yellow colour than control (with no enzyme). The addition of Grap Tan PC mimicked the effect of the enzyme preparations and yellow colour. Time of retention in bottle had a greater impact on colour than enzyme or tannin [72]. For apple fruit, a grater instead of juice extractor/crusher is needed, which grates the fruit and then, the same is pressed to get the juice. The pomace remains in the basket itself and the juice is used in the must preparation. Stone fruits like apricot and plum require steam operated kettles for cooking the fruits followed by pulping the cooked material in a suitable pulper. The seeds and skin get separated [10, 19].

**Plate 1:** Hydraulic press.

**Plate 2:** Fermenters.

**Plate 3:** Oak barrels and storage tanks.

## 6.2 Fermentation Room

It includes fermenters, juice separators, presses, conveyers, pumps, yeast propagator and cooling equipment. A good strain of *Saccharomyces cerevisiae* var. *ellipsoideus* is employed in fermentation. The yeast is available either in tablet form or a culture slant. For a successful fermentation, the yeast must be in a purified form and free from contamination. It can be propagated by growing the culture from a slant to 250 ml juice, to be shifted sequentially to 1 l, 5 l, 50 l and 500 l and so on as per the requirement of the specific fermentation. In the winery, usually, there are stainless steel tanks with increasing capacities, which are interconnected by piping. All the tanks have steam connections for proper sterilization. The sterilized juice is inoculated with the slant culture, and then the same is sequentially transferred to other tanks after 24 to 48 h incubation. The last tank is connected to the fermenter. All the tanks have the provision for the removal of $CO_2$, which is produced in the fermentation.

Fermenters of lined iron and stainless tanks predominate in California (Plate 2). Smaller wineries use fermenters made of oak or redwood. If metal tanks are used, they are normally closed for their later use in storage. Their size shall depend upon the number of varieties to be handled and total production capacity of the winery. Prior to pressing, the pomace is separated from the liquid by using the drag screens, horizontal rotating cylinders and flat rotating cylinder.

Systems in which the raw material is continuously introduced into the fermenters and removed continuously from the same or another fermenter are now widely used in fermentation industries. These systems save considerable labour and space, and process can be automated, but the problem of sterilizing the raw material and development of mutant yeast need attention. The continuous fermentation of sparkling wine has also been practiced, though the lack of yeast multiplication was recorded. A microscale fermentation technique was developed using 4 l fermenters that accommodated 3.5 kg fruit. During the course of fermentation, the microscale fermenters showed delayed extraction of phenolic components into the wine, but after eight days, concentrations of skin derived flavon-3-ols were the same as in commercial fermenters. Analysis of proanthocyanidins revealed that the total amount and relative proportion of seed proanthocyanidin extracted during microscale fermentations was lower than in commercial fermentations [73].

Alcoholic fermentation librates heat, which in a smaller container is lost to the environment but in larger vats increases the temperature of the must to critical levels, which if not controlled, can stop the fermentation (sticking). Consequently, cooling should be instituted as soon as the temperature approaches 29.4°C. The three main systems employed for cooling the musts are internal cooling coils in the fermenter, external shelf and multitude of coolers and jacketed tanks. Cooling can be in single operation or 2 to 3 times during the fermentation. With stainless steel tanks, water may be run over to cool them by exchange and evaporation. After cooling, the stuck wine can be refermented or added sequentially to vigorous fermentation at an installment of 10 per cent. However, if the fermentation is sticking due to low temperature, warming can serve the purpose [5, 10, 19].

## 6.3 Storage Method and Containers

Wines are stored in oak, concrete, red wood lined iron or steel, stainless steel and polyester containers (Plate 3). The storage area should be air conditioned, and if wood is employed, humidity control is also needed. In earlier times, caves provided the temperature and humidity control. If the tanks are kept out door, they should be insulated. It is important that all the equipment should be installed for the ease of operation. Storage rooms should be arranged with permanent pipes of stainless steel or fibre glass.

The type and size of containers used depends upon the quality of wines and the purpose of aging programme. Wineries with a focus on quality may profitably and necessarily use oak cooperage. Lined iron or stainless steel containers minimize aging, but they offer permanent storage with minimum loss or alteration. For wine of fine quality, oak is the preferred material for storage. Not only the wines clarify better in small containers, but they also acquire slight woody character and the necessary air for aging. The preferable size of oak barrel for wine storage is from 190 to 1900 l. Small sized containers are known as barrels, while large sized containers are known as ovals, butts or casks. When properly cared for, the thicker staves will last for centuries. Due to improvement in fabricating techniques, the cost of stainless steel tanks has been reduced considerably. But the steel has to be corrosion free. Use of polyester resin corks in wine storage has also been found.

The casks used for aging have to be cared for properly in order to get a wine of desirable quality. Alkaline solutions are more effective in removing tannins from new barrels. Soaking with 1 per cent sodium carbonate is the usual treatment followed by superheated steam of 2 to 3 atmospheric pressure. However, the prevalent practice is to wash the barrels with warm water in order to leave the tannin for extraction by the wine. New red wood tanks should be soaked with a warm dilute solution i.e., 1 per cent of soda ash for few days, followed by soaking with water for several days with two or more changes. Storage of clean distilling material before using the tank is also desirable.

In case of used cooperage, the interior is washed with water, and then spray of hot (49°C), 20 per cent solution of winery special i.e., 90 per cent soda ash +10 per cent caustic soda is made followed by washing with hot water, and then with chlorine water (44 mg/l of available chlorine). The cooperages should be rewashed and inspected for smell before use. For removal of cream of tarter deposit, most of the wineries use alkaline hot solution. The best method of cleaning and sanitization of cooperage is by soaking with warm 5 per cent soda ash solution for 24 h, again washing and filling with water at 85°C for a few hours, followed by draining. Burning a sulphur wick in the barrel is advisable if it is not to be used immediately. However, to clear stainless steel tanks, washing with 200 mg/l free chlorine having pH 8.9 is useful. To avoid the moulding of cooperage, washing with quaternary ammonium compounds is essential [5].

Recoopering is required when there is leakage of tanks or corks. There may be scrapping of staves or they may become leaky or mouldy. A single mouldy tank may contaminate and damage the flavour of a whole tank of wine. Mould could also be scrapped-off followed by washing with hypochlorite solution (500-1000 mg/l available chlorine). After undergoing these treatments, the corks or tanks should be washed properly, several times with hot water. In case of severe leakage, replacement of staves or head may be needed. Concrete fermentation tanks are left open and dry when not in use. Washing as such or in sprays with detergent cleans this type of tanks.

Storing empty barrels and tanks during hot, dry summer months is also a serious problem. Addition of sulphur dioxide or burning of sulphur wick is advisable. The inside of the barrels should be kept wet and these should be stored indoors or in the shade. Open wooden tanks should be thoroughly cleaned and then filled with saturated lime water, which is prepared by mixing about 1.8 g/l of lime. It will keep the tank sweet. If the storage is expected to be longer, the solution needs to be replaced periodically. It is advisable to inspect the interior of tanks or casks occasionally to prevent the drying-out of the top. Painting lime on the interior surfaces of open tanks is not recommended [19].

In order to prevent acetification, the storage tanks should be filled up completely and sealed. They could be provided with air bangs for this purpose. Small containers of wine should be kept aside to top up the tank already filled up to compensate the losses due to evaporation of wine during storage. In warmer months, some wines may need to be removed due to expansion in larger size containers.

## 6.4 Bottling, Labelling and Packing

The wines, after maturation, are clarified using filter press (Plate 4) or other filtration equipment, and clarified wines are bottled in appropriate bottles. The polysaccharides extracted from red and brown seaweeds can be effectively used for the clarification of white wines. Concentration of 0.005% of dried carrageenophyte purified carrageenan, algenaphytes or purified alginic acid can flocculate and precipitate proteins as effectively as bentonite [74]. Microbial analysis of older and recently bottled wines demonstrated the ability of certain wine microbial species to survive and grow in the bottle. Among them, the spoilage yeast *Brettanomyces bruxellensis* predominated, necessitating effective methods for removing these microorganisms before bottling. The work report on various filtering assays the smaller the pore size of filter, the more microbes were eliminated. Elimination of bacteria required a 0.3 μm filter, but a 1.0 μm filter was efficient for yeast elimination [75]. The wines or ciders can be carbonated using a carbonation machine. The bottling can be carried out manually or by the automatic machine (Plate 5). The bottles are then corked using crown corks or pith corks, as the case may be. In the latter case, however, care has to be taken to prevent any breakage. Care is

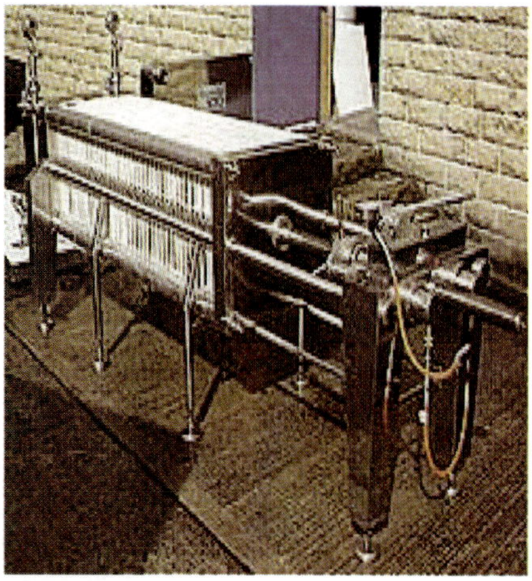

**Plate 4:** A filter press.

**Plate 5:** Bottling unit.

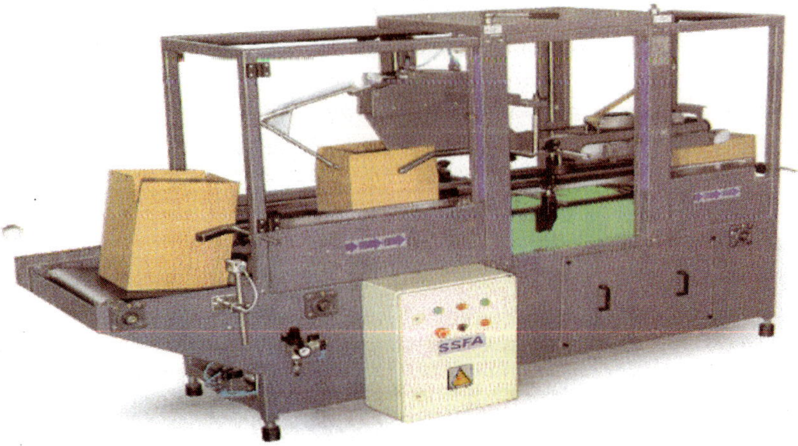


**Plate 6:** Automatic carton packing machine.

essential to completely scan the filled bottles for any extraneous matter which might come in the bottle during the bottling. The filled bottle should be labelled with suitable labels. These have to be got approved from the excise authorities before getting them printed. The bottles before sending to the market should be packed in a suitable package, either mono carton or multi cartons. Automatic carton machine (Plate 6) is also used. These cartons can also be printed to make them attractive.

## 7. TECHNOLOGY OF WINE PRODUCTION

### 7.1 Table Wines

#### *7.1.1 Red dry table wine*

Red wines are produced from the red or black skinned varieties of grapes [4]. The red colour owes its origin to the skin, and for red wine, the crushed grapes should be fermented along with their skin for some time i.e., until reasonable amount of colour gets extracted to impart the wine dark red colour [5, 10, 19]. The entire process of making a red wine is given in Fig. 1. Some of the varieties used for making red wine are *Cabernet, Sauvignon, Gamay noir, Nebbiolo, Pinot noir, Syrah, Beauty Seedless* and the *Bangalore Blue* or the *Bangalore Purple* [9]. In India, *Beauty Seedless* grown in Haryana and Punjab, and the *Bangalore Blue* or the *Bangalore Purple* grown in Mysore and Bangalore are the varieties found suitable for the red table wine production [19]. Other grape varieties like *Gulabi* and *Karachi Gulabi* can also be utilized for this purpose.

The production of grape wine involves following steps:

7.1.1.1 Crushing: After removal of the stem, the grape berries are crushed by hand, or when the large quantity is to be crushed, a roller crusher is used.

7.1.1.2 Amelioration: The juice should be tested for sugar in terms of Brix and total acidity. A reading of 22°B to 24°Brix is suitable for preparing a wine of about 12 per cent alcohol (v/v). The pH should be adjusted between 3.2 to 3.3, if the juice shows a higher pH i.e., 3.6-3.9 by adding tartaric acid which is readily available with laboratory chemicals dealers. Usually no acid addition is required in case of *Beauty Seedless* grape. However, *Bangalore Blue* grapes are too much sour (pH 2.8-3.0) and it is advisable to reduce the acidity by adding 1/4$^{th}$ teaspoonful of sodium bicarbonate (baking soda)/l grape juice prepared out of 2 kg of grapes approximately.

7.1.1.3 Addition of sulphur dioxide: Addition of potassium metabisulphite (KMS) at the rate of 200 ppm (100 ppm $SO_2$) to the juice or juice and skin is recommended. It should be weighed out very carefully, dissolved in a small volume of juice which is mixed thoroughly with the main bulk. Allow about 2-3 hours before inoculating the juice with yeast culture.

7.1.1.4 Yeast starter: The preparation of the yeast culture inoculum should be started at least 2-3 days prior to crushing of grapes. The fermentation will start within 24 hour and the juice will be rich in yeast cells and can be used as inoculum within 2-3 days. In other words, for every 10 l of must, 200 to 500 ml of inoculum is needed.

7.1.1.5 Fermentation: The must along with the skin is then allowed to ferment at a warm place (20 to 25°C). Temperatures higher than 26°C should be avoided because the fermentation proceeds rapidly at higher temperature thereby, decreasing, the alcohol yield. Low temperature can be maintained either by using air conditioners in the houses or by placing the container having the fermenting juice in a pan having water. The water in the pan should be changed periodically and can be cooled by addition of ice, if it is available. At the places where the temperature is lower than optimum, it should be maintained using heat convectors, blowers, etc. Direct heating should be avoided. The temperature of the room

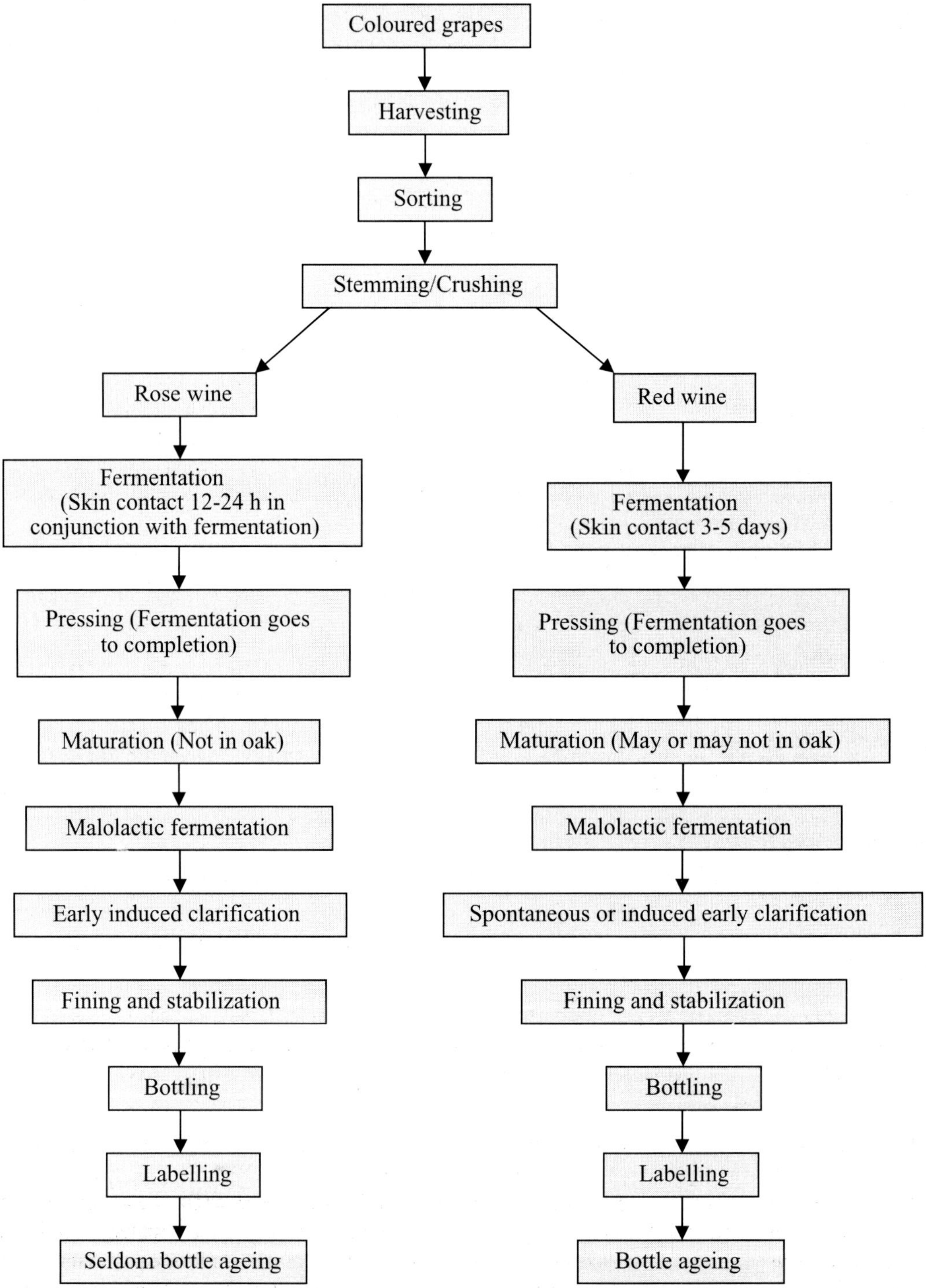


**Fig. 1** Flow chart for the preparation of rose and red wine.

of place should be kept near to this range. When the fermentation is vigorous, as evidenced from the frothing of the must, the sugar content of the juice should be measured periodically by Brix hydrometer or refractometer. The skin will rise to the top of fermenting liquid and foam, called 'Cap', due to brisk evolution of carbon dioxide. It should be punched down at least twice a day to extract the colour and to avoid its surface getting contaminated and temperature rising. The fermentation should be allowed to proceed till the Brix reading is about 8, which takes about 4 to 5 days at warm place and about 8-10 days at cooler environment. At this stage, separate the skin by first straining the juice with a suitable strainer (e.g., cheese cloth) and then by pressing the skin either with hand or with a basket press. The expressed liquid is then mixed with previously strained juice and transferred into another glass or plastic container, filling it approximately to 90 per cent of its capacity. It is once again allowed to ferment till whole of its sugar is converted into alcohol, which might take another week or so. After the wine becomes still i.e., no gas bubbles are seen rising, the yeast cells and other materials start settling at the bottom of the vessel as sediments.

7.1.1.6 Siphoning clarification or filling: Carefully rack (siphon) the wine, avoiding the sediment at the bottom into another plastic or glass container and fill-up completely. After about 6 weeks, rack the wine into another container to its full capacity. During the remaining period of the year, the wine is to be racked till it is fairly clear. During storage in the bottles or any other glass or plastic containers, care should be taken to avoid partial filling of containers as contact with air may result in spoilage. The containers should preferably be narrow necked or mouthed. Before the final bottling of wine, which can be done usually after about 4-6 months, as after the fermentation has been completed, the wine has to be clarified or fined with a suitable fining agent (e.g., Bentonite, Celite, etc.) to make it sparkling clear. The wine is then finally racked or filtered, if necessary, and bottled for consumption.

7.1.1.7 Bottling and corking: Perfectly clean bottles should be used. In case of red wine, green or brown coloured bottles are preferred. In white bottles, there are chances of red colour fading out. The bottles should be completely filled with wine and corked with bark cork or crown cork. The bark corks should be free from holes and crevices, which might serve as the breeding place for microorganisms. The bottles for home consumption may be sealed with wax after corking.

### *7.1.2 White dry table wine*

For the preparation of white wines, only the free run juice should be fermented. Production of white table wine is slightly difficult as compared to red table wine, as improper handling of grapes, juice or wine may bring about oxidation resulting in yellowish brown to brown colour of the wine. Varieties suitable for making white table wine are *Chardonnay, Muscat blanc, Pinot blanc, Riesling* [9], *Anab-e-shahi, Perlette* or *Thompson Seedless*. "Syrah" in red and 'Chenin Blanc' in white have been reported to be the two superior cultivars for wine making [76]. Care should be taken to select uniformly ripened bunches of grapes, and these should be free from moulds, have good acidity (0.6-0.8%) with reasonable amount of sugar (20-22°B). The complete process for white wine making is depicted in Fig. 2. Various steps involved in the preparation of white dry wines are discussed here under:

7.1.2.1 Crushing: The berries are separated from the stems and leaves. The grapes are then weighed and thoroughly mixed with desired amount of KMS (75-100 ppm $SO_2$) before crushing to control undesirable changes. It is possible to increase the yields from 50 to 60 per cent by hand pressing with the help of a small basket press. The juice is transferred to narrow necked glass or plastic containers after amelioration for sugar and acid to fill about 75 per cent of their capacity and their mouths are closed either with cotton plugs or by covering with clean muslin cloth.

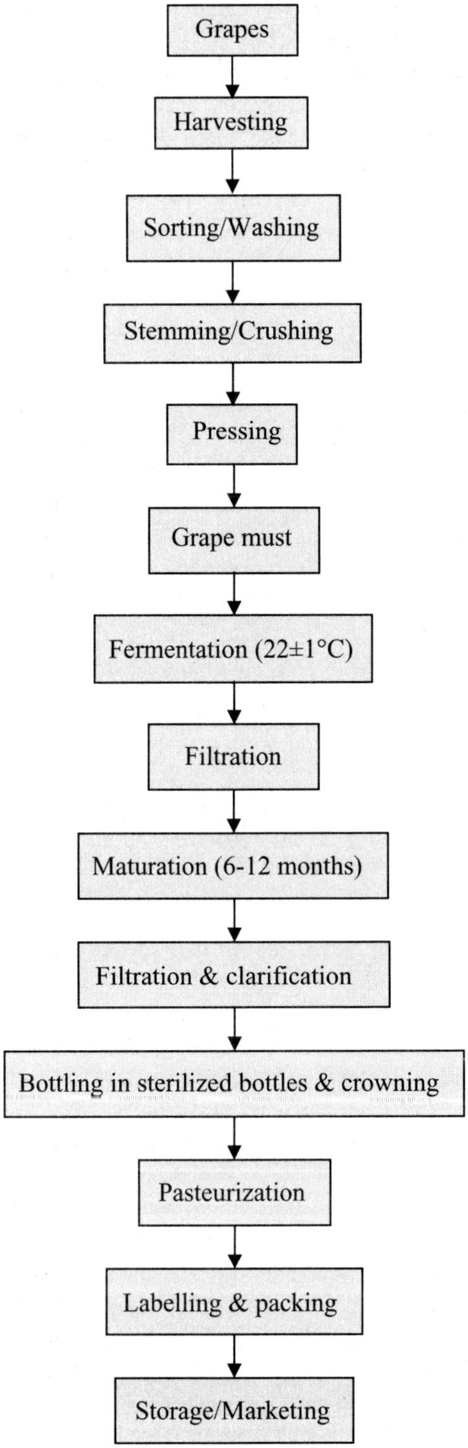


**Fig. 2:** Flow chart for the preparation of white table wine.

7.1.2.2 Yeast: Inoculum of yeast should be prepared. Its required quantity should be prepared and used, as described earlier.

7.1.2.3 Fermentation: Lower temperature (15-18°C) of fermentation is recommended for the production of white table wine. When the fermentation slows down i.e., frothing ceases, transfer the wine to another set of narrow mouth containers, filling upto approximately 90 per cent, allowing completing the fermentation which may take about a week.

7.1.2.4 Racking: The wine after fermentation is racked i.e., transferred to new containers made up of glass or plastic by siphoning carefully and filling them to the brim, leaving aside the deposit at the bottom (Lees).

7.1.2.5 Fining: Suitable fining agents like bentonite (quarter of teaspoonful/L), celite, etc., can be used, if required.

7.1.2.6 Bottling and corking: After 2-3 rackings, the wine would become fairly clear and ready for bottling, and final storage before consumption. Both types of corks i.e., bark or crown may be used.

## 7.2 Cider and Apple Wine

Cider is a low alcoholic drink produced by fermentation of apple juice and is known by different names around the world such as cider (France), *sidre* (Italy), *sidra* (Spain) and apple wein (Germany and Switzerland). It could be sweet or dry, soft cider (1-5%) or hard cider (6-7%), sparkling cider (with low sugar and $CO_2$) and still cider (without $CO_2$) [24, 25]. The dry cider is without sugar having an alcohol content of 6-7 per cent. The optimum temperature for cider fermentation ranges from 69 to 64°F (15-18°C). Various factors such as apple juice (variety, maturity, fresh or juice concentrate) [77], other ingredients (sugar, amelioration with acid, yeast nutrients), sulphur dioxide ($SO_2$) concentration, yeast strains (natural or inoculated strains), fermentation temperature, sugar, acid, fermenter design and operation, maturation and processing factors influence fermentation. $SO_2$ controls the microorganisms in the must and prevents the enzymatic browning of the juice. Mostly stainless steel tanks are used these days for fermentation of cider, though traditionally oak barrels were used for this purpose. A temperature of 4°C for bulk storage of cider is desirable. After fermentation, the cider is racked and filtered. The cider after aging and clarification is pasteurized at 60°C for 20-30 minutes or preserved with $SO_2$. Consumer acceptability of cider prepared from apple juice concentrate has been reported [78].

Wine is another product made from apple juice by alcoholic fermentation (having 11 to 14% alcohol content). Amelioration with sugar or juice concentrate is essential. Similar to cider, natural fermentation is also practised, but pure culture of yeast *Saccharomyces cerevisiae* is added to make wine of predictable quality. Addition of ammonium salts to fermenting solution reduces the higher alcohol production in wine due to non-degradation of amino acids of the must. Bitterness is related with fermentation conditions in apple wine, particularly yeast nutrients. To prepare table apple wine (11% alcohol), apples are washed, grated, pressed to extract juice, which is ameliorated to 24°B and fermented with *Saccharomyces cerevisiae* @ 5 per cent. To hasten clarification, 0.25 per cent pectinase enzyme is added. The sensory characteristics of raw, pasteurized and irradiated apple cider, with or without sorbate were studied. The processing treatments have no effect on characters such as sweetness, sourness, astringency and caramelized flavour. Irradiated cider had fewer apple flavour and lower ester content than raw cider and higher musty flavour; however addition of sorbate (0.05%) increases sweetness. Further, the addition of sorbate to irradiates cider increases musty flavour [79].

## 7.3 Pear Wine

To prepare perry, the fruits are grated and pressed in rack and cloth press. The juice with original TSS of about 14°B and 0.25% acid is ameliorated to 21°B and 0.5% citric acid along with 100 ppm $SO_2$. The fermentation is carried out with a pure wine yeast culture. The wine is allowed to settle after the completion of fermentation, racked and finally clarified using bentonite and preserved by pasteurization. It can be sweetened, fortified or blended with other fruit wine, if needed. Perry having 5% alcohol, 10°B and 0.5% acid is most acceptable [10, 11, 19,]. The procedure includes raising the TSS to 20°B, addition of DAHP @ 0.1% and carrying out the alcoholic fermentation at 22 ± 1°C by addition of yeast culture @ 5%. Addition of different sugar sources influences the fermentation behaviour and sensory quality of perry.

## 7.4 Coconut Beverage (Toddy)

The alcoholic beverage obtained by the natural fermentation of coconut palm inflorescence sap is Toddy [10, 11, 19,]. The latter is traditionally collected in clay pots, sterilized by inverting over a flame for 5 minutes and allowed to ferment in open pots for upto 2 days. During this period, microorganisms from the atmosphere enter the clay pots and multiply in the palm sap, which contains 15-18% sucrose, yielding about 7% ethanol. Such a fermentation of coconut sap is brought about by a succession of heterogeneous microorganisms consisting of yeast and bacteria [80].

## 7.5 Plum Wine

Plum wine is quite popular in many countries, particularly in Germany and Pacific coastal states. The fruit gives wine of appealing colour and acceptable quality. For wine preparation, to every pound of plum, 1 litre of water is added followed by the addition of starter culture. The mixture is allowed to ferment for 8-10 days before pressing. Addition of pectolytic enzyme before fermentation facilitates the pressing by increasing the yield of juice and hastening wine clarification. Additional sugar is added to the partially fermented juice, depending upon the type of wine (table or dessert) required. Aging, filtration, bottling and processing is similar to that for any other wine. Another method specifies that the plum fruits should be fully ripe, diluted with water in a 1:1 ratio (whole fruit basis), and added 0.3% pectinol, 150 ppm $SO_2$ and sugar added to raise TSS of the must to 24°Brix. It is fermented into wine. Different varieties of plum fruits have also been evaluated, and Santa Rosa was found to be the best [81]. The quality of the plum wine can be improved by raising the sugar content of the wine to 12°Brix [11, 19 24, 82]. The use of sodium benzoate instead of KMS gives a wine with better colour and sensory qualities [81, 83].

Plum fruit being highly acidic and pulpy in nature needs dilution of the pulp with water, but a yeast like *Schizosaccharomyces pombe* can be employed as a biological deacidifying agent besides the use of *Saccharomyces cerevisiae* var. *ellipsoideus* to produce alcohol [10, 84]. The deacidification of the plum must by this yeast was independent of sugar content of the must, but was adversely affected by higher concentrations of ethanol, pH 2.5 and $SO_2$ content of more than 150 ppm, while the addition of ammonium sulphate enhanced the malic acid degradation [26]. Biological deacidification of the plum must also have been made to produce wine of variable acidity [85].

## 7.6 Kinnow Wine

An effective method of kinnow wine making without bitterness includes the use of cyclodextrin or Amberlite XAD-16, which considerably reduces the extent of bitterness in wine by degrading limonin and naringin [11, 19, 24, 86].

## 7.7 Peach Wine

The peach fruit has lesser acid than plum or apricot, but being pulpy needs dilution with water [87]. The method for making peach wine consists of dilution of the pulp in the ratio of 1:1, raising the initial TSS to 24°Brix, adding pectinol and DAHP at the rate of 0.5 and 0.1%, respectively. To the must, 100 ppm of $SO_2$ is added to control the undesirable microflora [12]. The physico-chemical characteristics of peach wines prepared from various cultivars have been shown in Table 1. Treatment of wines with wood chips of *Quercus* gives the wine of best sensory quality [11]. The types of phenols present in peach wine are given in Table 2.

## 7.8 Strawberry Wine

Strawberry wine of good quality has the appealing colour of premium rose wine. Various varieties such as Chandler, Camarosa and Doughlas can be utilized for the preparation of wine [90]. In one method, the juice is ameliorated to 22°Brix by the addition of cane-sugar added in stages e.g., 25% of the total sugar is added to prepare the must and the rest is added during fermentation [5]. However, addition of sugar after fermentation dilutes the alcohol level. The must is mixed with 1% ammonium phosphate and the fermentation is initiated by addition of 5% of yeast culture at a temperature of 16°C. The juice is transferred to the jar and the fermentation is continued till 0.1-0.2% reducing sugar contents are obtained. After the completion of fermentation, it is racked, bottled and stored in the dark. Treatment with enzymes, mainly pectinases inhibits polymerization and increases colour extraction and colour density in strawberry wine. The use of enzymes and premash treatments for juice and colour extraction from strawberries has been recommended [68]. Juice yield, which is affected by the addition of pectic enzyme, is an important cost consideration for wine production

**Table 1:** Physico-chemical characteristics of wines of different peach cultivars.

| Cultivars | TSS (°B) | Titratable acidity (% MA) | pH | Total Sugars (%) | Ethanol (%) | Colour intensity | Higher Alcohol (mg/l) | Volatile Acidity (% aa) | Total esters | Total phenols | Total aldehyde |
|---|---|---|---|---|---|---|---|---|---|---|---|
| Richahaven | 8.33 | 0.66 | 3.60 | 1.33 | 10.73 | 0.53 | 113.6 | 0.023 | 98.23 | 241.8 | 48.23 |
| J.H.Hale | 7.73 | 0.71 | 3.80 | 1.24 | 10.67 | 0.33 | 126.9 | 0.029 | 90.97 | 259.0 | 43.30 |
| Red Haven | 9.00 | 0.72 | 3.71 | 1.26 | 11.10 | 0.86 | 121.4 | 0.023 | 98.40 | 301.7 | 45.53 |
| Flavorcrest | 8.20 | 0.80 | 3.61 | 1.19 | 11.33 | 0.33 | 134.4 | 0.029 | 94.33 | 306.1 | 49.32 |
| Sunhaven | 8.06 | 0.63 | 3.71 | 1.22 | 11.67 | 0.36 | 131.6 | 0.022 | 101.50 | 278.6 | 53.97 |
| Stark early giant | 8.20 | 0.74 | 3.81 | 1.32 | 10.77 | 0.19 | 144.2 | 0.026 | 97.80 | 206.0 | 65.73 |
| Kateroo | 7.60 | 0.61 | 3.84 | 1.22 | 11.83 | 0.23 | 2154.3 | 0.029 | 92.37 | 230.4 | 56.83 |
| July Elberta | 9.13 | 0.70 | 0.74 | 1.36 | 10.87 | 0.48 | 151.5 | 0.020 | 95.50 | 203.3 | 63.10 |
| CD ($p<0.05$) | 0.26 | 0.01 | 0.02 | 0.02 | 0.46 | 0.01 | 0.70 | 0.002 | 0.845 | 1.3 | 0.77 |

*Source:* 88

**Table 2:** Different type of phenols present in peach wine.

| Type of phenol | RF | Control wine | Oak treated wine | Albizia treated wine | Bombax treated wine |
|---|---|---|---|---|---|
| Epicatechin | 0.52 | +1 to 8* | +1 to 8* | +1 to 8* | +1 to 8* |
| Catechin | 0.46 | - | - | - | - |
| Leucoanthocyanidin | 0.42 | +1 to 8* | +1 to 8* | +1 to 8* | +1 to 8* |
| Gallic acid | 0.60 | +3,4,5,7* | +3,4,5,7* | +3,4,5,7* | +3,4,5,7* |
| Quercitin | 0.70 | - | +3,4,5,7* | +1 to 8* | - |
| Vanilline | 0.86 | - | +1,2,6,8* | - | - |
| Rutin | 0.46 | - | +3,4,5,7* | - | - |
| Ferulic acid | 0.86 | - | +3,4,5,7* | - | +1,2,6,8* |
| Caffeic acid | 0.80 | - | +3,4,5,7* | - | - |
| Syringic acid | 0.83 | - | - | - | - |

*1 Richahaven; *2 J.H. Hale; *3 Red Haven; *4 Flavorcrest; *5 Sunhaven; *6 Stark early giant; *7 Kateroo; *8 July Elberta; + Present; - Absent;

*Source:* 89

**Table 3:** Composition of strawberry wine.

| | |
|---|---|
| Total soluble solids (°B) | 8.1-9.7 |
| Total sugars (%) | 0.6-1.7 |
| Reducing sugars (%) | 0.124-0.135 |
| Titratable acidity(% citric acid) | 0.63-0.73 |
| pH | 3.18-3.26 |
| Ethanol (%) | 9.2-11.5 |
| Higher alcohols (mg/l) | 151-169 |
| Volatile acidity (% A.A) | 0.025-0.032 |
| Esters (mg/l) | 78.3-102.4 |
| Phenols (mg/l) | 126.8-144.7 |
| Anthocyanins (OD/ml of wine at 535 nm) | 0.10-0.15 |

*Source:* 90

as the price of strawberry juice is quite high. It has been found that thermovinification is the best method for the preparation of strawberry wine [91]. Descriptive analysis of the wines showed that out of 14 descriptors, 6 had very high intensity for strawberry like alcoholic, phenolic, higher alcoholic, astringency and bitterness [92]. Various constituents present in strawberry wines are presented in Table 3.

## 7.9 Apricot Wine

Wine is made both from the cultivated and wild apricots. The method consists of diluting the wild apricot pulp in the ratio of 1:2, addition of DAHP @ 0.1%, 0.5% pectinol and fermentation with *Saccharomyces cerevisiae* [93]. In the cultivated apricot wine preparation, the extraction of pulp either by hot method or addition of enzyme and water to the fruits could be adopted. Dilution of pulp in the ratio of 1:1 with addition of DAHP @ 0.1% and 0.5% (liquid pectinase), raising the initial TSS to 30°B, made the wine of superior quality [94].

### 7.10 Pineapple Wine

The pineapple juice has only 12-15°Brix and therefore sugar content is raised by the addition of sugar upto 22-23°Brix to produce a wine having 12-13% alcohol (v/v). The wine is preserved by pasteurization, but can be fortified and sweetened. Since the flavour of pineapple is not stable, oxidation can occur easily [10, 19].

## 8. SPARKLING WINE TYPES AND METHODS OF PRODUCTION

A wine is called sparkling if it is surcharged with carbon dioxide not less than 5g/l at 20°C. Four types of sparkling wines can be made: Type 1 includes those with excess carbon dioxide produced by fermentation of residual sugar from the primary fermentation (Australian, German, Loire and Italian and the Muscato, Ambila of California); second type includes those with excess $CO_2$ from a malolactic fermentation (Vinho verela wines of northern Portugal); third type of sparkling wines contain excess of $CO_2$ from fermentation of sugar added after the process of fermentation. Most of the sparkling wines of the world are of this type. The fourth type of sparkling wines includes those in which excess of $CO_2$ is added, e.g., carbonated wines [5].

Different steps involved in the production of sparkling wine are preparation of a base wine, sugaring, yeasting, bottling, proper blending, secondary fermentation in bottles or tanks during maturation, finishing and disgorging [5, 82]. Sparkling wine such as champagne from grape is well known all over the world. It is produced by secondary fermentation in closed containers such as bottles or tanks to retain the carbon dioxide produced. The secondary fermentation is influenced by a number of factors such as ethanol content of base wine, ability of yeast to ferment sugar in closed containers, ability to withstand higher carbon dioxide concentration under low temperature, amount of fermentable sugar and the availability of nitrogenous source in the base wine [5, 95, 96]. The maturation is a very important step in the production of sparkling wine and may extend upto 3 years. Bottle aging results in acquiring the physico-chemical and sensory characteristics usually associated with such wines. In the production of sparkling wine, preparation of base wine is the first and foremost step. The quality of sparkling wine depends largely on the characteristics of the base wine, which in turn are dictated by a number of factors such as cultivar, yeast culture, preservatives, temperatures, nitrogen source, etc. The composition of base e.g., titratable acidity, alcohol, pH, colour, mineral content, biochemical characteristics and sensory characteristics influence the quality of sparkling wine [82]. Out of the tank and bottle fermented wine, the latter is considered superior. However, yeast strain, nutrient status of base wine, ethanol content, carbon dioxide and extent of aeration influence the production of sparkling wine by bottle fermentation.

Description of the '*Methode Champanoise*' to the finest details of each unit operations can be found in many standard books. The tank method is called Charmat process used mostly in the Soviet Union and California [5]. It is described to be more economical than the bottle fermentation. The production of low quality wine by this method is attributed to the introduction of oxygen during finishing, increasing the aldehyde content and darkening the colour. The most important and desirable character of sparkling wine is the retention of $CO_2$ in the form of the foam head, which contains proteins and polysaccharides as one of the compounds.

### 8.1 Sparkling Cider

Sparkling cider is prepared either by the '*Methode Champanoise*' or artificial carbonation or a combination of both. To prepare sparkling cider, a secondary fermentation requiring 3-4 days at a temperature of 75°F (21.1°C) is utilized in '*Methode Champanoise*'. Raw materials for this secondary fermentation include dry cider, apple juice concentrate, apple juice, sugar, ammonium phosphate, tannic acid and the yeast starter [10].

## 8.2 Sparkling Apple Wine

Sparkling wines have been produced by secondary fermentation of apple wines (5 to 6% ethanol content) under normal pressure, resulting in alcohol level of 9-10% by volume. Ammonium dihydrogen phosphate as well as ammonia solutions are added as nitrogen source during the production of apple wines. Sparkling apple wines are prepared by either carbonation in tank or bottle by secondary fermentation of base wine. For preparation of this type of wine, a clear juice is obtained by pectinase enzyme treatment followed by rough filtration and amelioration to 16% solids. This is fermented at 25°C for 4 days and chilled to –1.7°C to stop fermentation. Alternatively, cold wine is polish filtered and carbonated artificially at 35 lbs pressure, bottled and pasteurized. In this case, it would however be called a carbonated wine.

Studies on the secondary fermentation of apple wine with low alcohol content (5-6%) have revealed that hydrolysed autolysate had a favourable effect on the secondary fermentation of apple wine and was found suitable as a nitrogen source for secondary fermentation of apple wine, as about 40 mg/l ensured the sparkling process [97]. Use of pectinase enzyme reduced the pectic substances to the optimum level, which is needed for product clarification and foaming characteristics.

## 8.3 Sparkling Plum Wine

After evaluating various physico-chemical and sensory characteristics of plum wines from must prepared with two preservatives i.e., potassium metabisulphite and sodium benzoate, it was concluded that both the types of wine were suitable for conversion into sparkling wine. However, the sodium benzoate treated wine was considered better. In preparation of sparkling wine from plum, effect of different concentrations of sugar (1, 1.5, 2%) and diammonium hydrogen phosphate (0.1 to 0.3%), two strains of *Saccharomyces* viz., UCD 595 and UCD 505 in two base wines showed the increase in ethanol content upto 6% in plum base wine, and decreasing the temperature of incubation from 24 to 8°C reduced the viable count of both the yeast strains, though it remained above 80%. Strain UCD 595 was found to give better acclimatization to ethanol and low temperature than UCD 505. A sugar concentration of 1.5% and DAHP of 0.2% in 200 ml juice bottles at 15 ± 2°C was found to be optimum for bottle fermentation. Bottle fermented sparkling wine was rated better both in physico-chemical and sensory quality [98]. There is potential of plum fruits for sparkling wine production by the *'Methode Champagnoise'*, though more work is needed to commercialize the same.

## 8.4 Sparkling Mead

For preparation of sparkling mead, the honey is dissolved in the warm water and dry mead is made as described. After storing for six months in champagne bottles, the wine is sweetened with honey and to each bottle is added a small quantity of champagne yeast. The bottles are allowed to stand in a warm place for one week only, followed by laying them on their sides in a cool place for a year. The sparkling wines are served cool [99].

## 8.5 Sparkling Gooseberry Wine

Sparkling gooseberry wine is called *English Champagne* and is known to be a very palatable wine. To make such a wine, 3 pounds of green round gooseberries are taken in a pan, 1 gallon of water is poured over them with addition of 4 pounds of sugar. It is allowed to ferment for three weeks by putting into the bottles, tying down the corks and laying the bottles on their sides for carrying out secondary fermentation.

## 9. FORTIFIED WINES

*Vermouth* is a fortified wine (alcohol 15 to 21%) flavoured with characteristic mixture of herbs and spices. Some of these impart an aromatic flavour while others give a bitter flavour [5]. The word '*Vermouth*' is derived from the German word '*Wermut*' ('Wer' means man 'Mut' means courage, spirit, manhood) or English '*worm wood*' which in Latin stands for *Artemesia absinthium* [100]. However, in USA it is mandatory that product should be free from thujone. Details on the history of vermouth preparation and names of different herbs and spices used for this purpose have also been described elsewhere [5, 100]. The vermouth could be sweet or Italian type and the dry or French type. In the Italian type, alcohol content vary from 15 to 17% with 12 to 15% sugar while the French have 18% alcohol with 4% reducing sugar [101]. The dry vermouth contains less herbs and spices than the sweet. Traditionally, vermouth or aperitif wines are compounded from grape wine by adding herb and spice mixture or their extracts. The essential requirements of a base wine for conversion into vermouth are that the wine should be sound, neutral and cheap [55]. The wines are popular in European countries and in USA. Aromatized wines are also produced commercially in USSR and Poland [102]. These type of beverage are served straight in European countries while these are mostly used in the cocktails in America.

*Sherry* is a fortified wine that is fractionally blended while being given prolonged oxidative maturation. It is produced in two basic styles and are sub-divided into fino and oloroso style. Finos are dry and consumed primarily as *apertifs* before meals. However, olorosos are often sweetened and consumed after or with dessert. The major factor affecting the ultimate direction of sherry evolution is the degree of fortification. Increasing the alcohol content to 15-15.5 per cent promotes the development of flor yeast essential for the development of fino sherry. But fortification to 18 per cent inhibits the flor development, but permits wine oxidation in the absence of microbial spoilage. The production of fino sherries requires frequent transfers of wine through many stages (criaderas) of fractional blending. Twenty per cent of wine in the oldest criadera (called the solera) is removed upto three tonnes per year. The volume of wine that is removed is in turn replaced from the next oldest criadera. The process continues until the youngest criadera is reached. The wine obtained is finally fined and prepared for bottling. Sherry is also produced outside Europe, but is seldom made by the slow, complex procedure of fractional blending. Often, they do not resemble Spanish sherries that are heated during their maturation. This gives them a baked flavour, which resembles madeira more than sherry.

### 9.1 Grape Vermouth

The vermouth is prepared by making the base wine, extracting the herbs and spices in wine and brandy mixture, blending the extract with base wine, fortifying the base wine to the desired level of alcohol and finally maturation of the prepared vermouth [5]. The base wine is prepared from grape juice or a concentrate as per the routine method. The grapes used should be well balanced in acidity and sugar content. Better American vermouth is produced if a wine of natural higher acidity is used. The wine should be fortified with neutral high proof brandy. The extract can be prepared by direct extraction method in which the calculated amount is placed in the base wine till the wine has absorbed the desired flavour and aromas. The wine may be heated during the extraction process upto 60°C, and to minimize the loss of aroma, the container should be covered [100]. Instead of directly flavouring a large quantity of wine, another approach would be to prepare a small volume of more concentrated extract, which is then used to flavour a relatively large volume of base wine. About 0.5 to 1.0 oz of mixed dry flavouring materials per gallon of base wine is sufficient to adequately flavour the sweet or Italian style vermouth [101]. Brandy or alcohol extracts of spices and herbs are available for flavouring the vermouth. Use of brandy in the extraction of flavour can dispense with the need of heating usually carried out in the process with wine. Dry vermouth should not be aged very long, but should be finished and bottled very young. Sweet or dry vermouth should be low in pH and preserved with sulphur dioxide so as to prevent the spoilage by *Lactobacillus trichodes*. The common steps involved in the preparation of vermouths are discussed here.

### *9.1.1 Preparation of base wine*

Wines are prepared from the respective fruits as described earlier under the section table wine. These are cheap and have neutral flavour, which meets the requirements of vermouth preparation.

### *9.1.2 Maturation of base wine*

The wine is matured normally for a minimum period of 3-4 months before conversion into vermouth. Racking in-between is done after an interval of 15-20 days, as usual.

### *9.1.3 Brandy preparation*

A part of the base wine is distilled to make brandy. The method of brandy making is described under the section of 'Brandy.' At home scale, distillation is not permitted, therefore, it cannot be carried out. But any commercially prepared and available brandy can be used for extraction process.

### *9.1.4 Preparation of spices/herbal extract*

The spices/herbal mixture is immersed in the wine containing brandy in 1:1 ratio. The volume is made keeping in view the total quantity of the extract required. The container is closed with a cork and heated gently at 60°C for 5-7 minutes for 10 days. For apple vermouth, the spices extract is prepared with quantities twice than that used for plum and sand pear. After extraction, the extract is filtered appropriately and allowed to sediment and then again filtered.

### *9.1.5 Addition of spices extract*

The clarified extract is added to the wine fortified with brandy to raise its alcohol content from 12 to 15-16% along with its sugar contents by 6 to 8% for plum; the quantity of 5-10% has been found to be optimum. Apple needs 2.5% extract to make an acceptable product. If needed, it can be further filtered.

### *9.1.6 Maturation and clarification*

Potassium metabisulphite @ 100 ppm is then added and the vermouth is matured further for atleast 3-4 months.

## 9.2 Mango Vermouth

Sweet and dry vermouths have been prepared successfully from the mango juice pulp [103]. Comparison of 4 formulae has been made to prepare the vermouth. The base wine was prepared from cv. *Banganpalli*, raising TSS to 22°Brix, adding 100 ppm $SO_2$, 0.5% pectinol enzyme and carrying out fermentation at 22 ± 1°C with Montrachet strain 522 of *Saccharomyces cerevisiae*. Herbs and spices mixture is used in the preparation of mango vermouth (Table 4). The formula A and C were found suitable for the preparation of dry and sweet vermouth, respectively.

**Table 4:** Herb mixture for dry and sweet mango vermouths.

| | Amount used (g/l base wine) | | | |
|---|---|---|---|---|
| | Dry wine | | Sweet wine | |
| **Herbs** | **Formula A** | **Formula B** | **Formula C** | **Formula D** |
| Black pepper | 0.75 | 1.25 | 2.5 | 5.0 |
| Coriander | 0.70 | 1.25 | 2.5 | 5.0 |
| Cumin | 1.25 | 2.50 | 3.0 | 4.0 |
| Bishop's weed | 0.50 | 1.00 | 1.50 | 2.0 |
| Clove | 0.25 | 0.50 | 0.75 | 1.0 |
| Large cardamom | 0.50 | 1.00 | 1.50 | 1.0 |
| Saffron | 0.10 | 0.10 | 0.10 | 0.10 |
| Fenugreek | 0.50 | 1.50 | 2.0 | 2.50 |
| Nutmeg | 0.25 | 0.50 | 0.50 | 0.75 |
| Cinnamon | 0.50 | 1.00 | 1.50 | 2.00 |
| Poppy seeds | 1.00 | 1.50 | 2.0 | 2.50 |
| Ginger | 1.00 | 1.50 | 2.0 | 2.50 |
| Flame of Forest | 0.25 | 0.50 | 0.75 | 0.75 |
| Lichen | 0.25 | 0.50 | 1.00 | 2.00 |

*Source:* 103

## 9.3 Plum Vermouth

Vermouth of commercial acceptability can be made from plum similar to grape vermouth [26]. The spices extract contributes to the increase in total phenols, esters, and aldehyde content. The best product contains about 15% alcohol and is sweet in nature. The method for plum vermouth has been standardized and is similar, as discussed for mango vermouth [24, 26]. Comparison of physico-chemical characteristics of dry and sweet plum vermouth is made in Table 5.

## 9.4 Sand Pear Vermouth

The fruit is found in abundance in many northern Indian states and is neither relished for table purpose nor processed into any product, commercially. The juice of sand pear is also blend, and as such does not command any market. It can, however, be converted into vermouth as per the technology standardized [11, 24,104,]. The product having 15% alcohol and sweet in nature was found to have high sensory acceptability [104]. Composition of sand pear base wine and its vermouth is compared in Table 6. Economically, the product has high feasibility for commercialization. The herbs and spices, and their quantity used were same as that used for plum [26].

## 9.5 Apple Vermouth

The apple fruit can be converted into *aperitif* type wine, known as '*Vermouth*'. Basically, the technique of vermouth preparation resembles that of grape vermouth, but due to differences in the type of fruit, the technique is modified. Efforts made to prepare the apple vermouth showed that the base wine had all the desirable characteristics needed to make vermouth. The process includes preparation

**Table 5:** Physico-chemical characteristics of dry and sweet plum vermouth.

| Characteristics | Type of vermouth | |
|---|---|---|
| | **Dry** | **Sweet** |
| Total sugar (%) | ND | 4.8 |
| Titratable acidity (% malic acid) | 0.81 | 0.79 |
| Ethanol (%, v/v) | 15.00 | 14.5 |
| Volatile acidity (% acetic acid) | 0.03 | 0.04 |
| pH | 3.38 | 3.34 |
| Vitamin C (mg/100 ml) | 3.5 | 3.2 |
| Total phenols (mg/l) | 417 | 390 |
| Aldehyde (mg/l) | 411 | 112 |
| Esters (mg/l) | 204 | 219 |

ND: Not detected

*Source:* 26

**Table 6:** Physico-chemical characteristics of sand pear base wine and sweet vermouth.

| Characteristics | Wine | Vermouth |
|---|---|---|
| Total soluble solids (°B) | 6.1 | 13.0 |
| Titratable acidity (% malic acid) | 0.81 | 0.79 |
| pH | 3.99 | 3.95 |
| Reducing sugars (%) | - | 4.17 |
| Total sugars (%) | - | 4.35 |
| Alcohol (%, v/v) | 10.80 | 14.95 |
| Volatile acidity (% acetic acid) | 0.04 | 0.04 |
| Ascorbic acid (mg/1000 ml) | 6.6 | 5.5 |
| Aldehyde (mg/l) | 103 | 133 |
| Total phenols (mg/l) | 226 | 264 |
| Esters (mg/l) | 197 | 268 |
| Optical density (at 440 nm) | 0.64 | 0.58 |

*Source:* 104

of base wine, herbs and spices extract, fortification and sweetening, blending with the extract keeping in view the available material, the taste perception and finally, maturation. Vermouth from apple juice concentrate has been prepared and product of best quality contained having 15% alcohol, 2.5% spices extract and 4% sugar [11, 19]. The composition of vermouth is influenced by different alcohol levels, spices extract or sweetness level, and differences in sensory quality were also observed [11]. The differences in the preparation of apple vermouth compared to plum and sand pear were due to an increase in concentration of herbs and spices to double, resulting in reducing the dilution levels with the availability of more minerals in apple vermouth than plum or sand pear vermouth [11, 61]. The sensory evaluation further indicated that by increasing the acid content, the acceptability of a product with 18% alcohol content can be increased easily.

## 9.6 Tamarind Vermouth

Tamarind (*Tamarindus indica* L.) is one of the important tropical fruit tree and is found wildly in India [105]. Tamarind wine is not preferred due to high level of acidity. The fruits have been converted into vermouth of acceptable quality by the method developed [105, 106]. To prepare the base wine, appropriate quantity of tamarind fruits (50g/L) are soaked in water for 24 hours, followed by boiling for 15 minutes and extraction of pulp by hand pressing. The pulp is diluted with water to give 0.9% acidity followed by amelioration with sugar to 23°Brix, 150 ppm of $SO_2$ and the fermentation is initiated with *Saccharomyces cerevisiae* var. *ellipsoideus* Montrachet at a temperature of 27 ± 1°C. The vermouth is prepared as per the methods described for preparation of grape and mango vermouth except for the quantities of the herbs and spices extract used [5, 103, 105]. Both dry and sweet vermouths are acceptable.

# 10. BRANDY TYPES AND PRODUCTION TECHNOLOGY

The term 'Brandy' is applied to the distillate obtained by distillation of wine or any other fermented fruit juice or residue thereof. Grapes are the most frequently employed fruits for preparation of brandy. Fruit brandy other than grapes, derived from more than one variety of fruit, shall be designated by the brandy qualified by the name of such fruit e.g., 'Plum brandy'. The brandy made by the distillation of apple wine is called apple brandy or applejack. According to the USA Standard of Identity, a brandy should be produced at less than 190° Proof and bottled at not less than 80°Proof. A fruit brandy is a product prepared from fermented juice of grape, citrus or any other fruit with addition of not more than 20% by weight of the pomace or the lees of wine not exceeding 30% by weight. Depending upon the source, it may be termed as 'Fruit Brandy' 'Cognac' or Cognac, Armognac brandy from grape. Fruit brandy from the dried fruits such as raisin called raisin brandy, residue brandy/pomace brandy. A neutral brandy is one made at more than 170° Proof and shall have to be qualified by the source used for fermentation. The ethanol is the major component of alcoholic beverages and its strength is expressed in different systems in different countries like % by volume in France, Belgium, Italy, Austria, USSR and US, proof in USA and UK, % by weight in Germany, specific gravity and Cartier in Switzerland [5].

## 10.1 Technology of Fruit Brandy Preparation

When a wine is distilled, it is called 'Brandy'. There are many brandies other than grape like plum, grape and apricot, which are very popular, but brandies being the distillate do not contain any nutrient in contrast to the fruit wines. A general method of brandy making is described here.

### *10.1.1 Wine preparation*

Wine, which is produced by the fermentation of the fruits, described earlier, is used.

### *10.1.2 Distillation*

There are different types of distillation apparatus, but basically these have a container linked to the condenser, water circulation pipes and systems of heating. Condenser and container should be of glass or other metals. Supporting stand and a temperature monitoring device should also be incorporated. The container is filled with the wine upto its 60% capacity. The condenser is then filled with water, pipes are connected and water is run. Connections are made properly. Temperature during distillation is kept 78 ± 2°C. After the heating starts, the water is run in the pipes.

### *10.1.3 Collection of distillate*

Wine is heated uniformly and the first distillate amounting to 1/10 of the expected distillate i.e., 1/4th of the total volume of the wine taken is called 'Head' and is discarded. The middle fraction of 'Heart' of the sample is collected. The 'Tail' is not collected, as it contains more acid and higher alcohols.

### *10.1.4 Analysis for alcohol*

The alcohol contents of the middle fraction should normally be about 35-40%. If the alcohol concentration of wine is 10-11% (v/v), it can be checked up with an alcoholometer. It can also be converted into proof for comparison with other products and to calculate the volume required for blending by using the values.

### *10.1.5 Addition of oak wood*

To the middle fraction, 2.5 gm/l of slightly roasted oak wood is added.

### *10.1.6 Maturation*

The container is closed with lid and brandy is matured for a period of atleast a year or so for development of proper flavour and mellowness.

### *10.1.7 Standardization*

The matured brandy is checked up for the alcohol concentration, filtered and is ready for use.

## 10.2 Production of Typical Brandies

### *10.2.1 Cognac*

The grapes are crushed and immediately pressed in vertical or horizontal process. In its production, various factors such as timely vineyard sulphuring, absence of botrytis on the fruit, high must fixed acidity, clean fermentation without using pure yeast cultures, little or no $SO_2$, little press wine, low alcohol, removal of seeds and dry stems, low volatile acidity, and storage of the wine in the absence of air should be considered [107]. The must is fermented in cement tanks at a relatively high temperature, 25°C without prior sulphiting or clarification. The fermentation of the must is spontaneous and has low concentrations of sugars (less than 180 g/l), tannins and high concentrations of malic acid. The wine, thus, produced is of high acidity but with relatively low ethanol concentration. The yeast species present in the must and during the fermentation are *S. uvarum, S. rasie, S. capensis, S. chevalieri, S. globosus* and quite often *Saccharomycodes ludwigii* besides the dominant *Saccharomyces cerevisiae*.

The malo-lactic fermentation, usually occurs after the alcoholic fermentation wherein the bacterial population of the must consists of 60% of *Leuconostoc mesenteroides*, 20% of *L. oenos* and 20% of *Lactobacillus plantarum*. Until the wines are distilled, they are kept on the lees. Distillation starts immediately after the vintage and continues until all the newly fermented wine is distilled. Normally, only direct-fired pot stills of small capacity, not exceeding 30 hl (792 gal) are employed. The new wine with its lees (not more than 8% of added lees) is placed in the still and brought to boiling. The distillation is allowed to continue until the vapours contain negligible amount of alcohol. The distillation takes 8 or more hours and the main distillate (brouillis or low wines) contains about 24-32% alcohol. A tail fraction may be separated. The still is then emptied and refilled with fresh wine and also often with the tails of the previous distillation. A second distillation is made and is followed by third distillation. These three distillates are finally combined and then redistilled. This last distillation takes longer time than the original distillation i.e., about 14 or more hours. Near about 1-2%

heads are separated. The main distillate (*coeur*) contains 58-60% ethanol and cannot exceed 72%. A tails fraction (*seconde*) is also separated. To control the amount of each fraction, continuous testing, using an alcoholic hydrometer and recycling of the heads and tails are practised. The distillation is carried out gently to avoid mechanical entrainment, and no more than three distillations are recommended in 24 h [108]. The fatty acids produced during the fermentation and extracted by heating the lees from insoluble salts with the copper of the pot of the stills result in undesirable taste, therefore, they are eliminated. Since the experiments with pots made of glass or stainless steel have given poor results, copper is considered to be indispensable for the production of a quality brandy [109].

The brandy with 70% alcohol by volume is first stored in new oak barrels from the Limousin or the forest of Troncais. The staves are prepared from 40-50 years old trees. The trees are split, not sawed and pieces of wood are used after prolonged exposure to the air. This permits aging of the wood, elimination of part of the highly astringent tannins, oxidation of tannins and phenols, development of moulds which darkens the wood and degrades the lignin. Later, these brandies are transferred to old casks to avoid any excessive enrichment in tannins. The very old brandies are kept in casks made from dense wood fibre. The dissolution of soluble oak components especially tannins, polyphenols, lignin, protein, pectins and minerals are reflected in the colour of the brandy. The aging process has been divided schematically into two phases. During the first phase, the acidity increases, acetal is formed, and extracted tannins are oxidized and intensify the colour of the brandy. The hemi-celluloses are hydrolyzed, the lignins and esters are alcoholised, and a vanilla-like or flowery odour appears in the second phase. The aroma is highly strengthened by evaporation of water and ethanol between 10 and 30 years of storage. The taste of the brandy becomes sweeter due to the lower alcohol content and due to the formation of sugars by hydrolysis of hemicellulose. The alcohol concentration is reduced by 6-8% in a period of 15 years and the pH is lowered from 5 to 3.5 in 50 years [110]. Cognac is sold at an alcoholic concentration of 40%, which can be obtained by very long aging in casks. All Cognac houses maintain inventories of old vintage. The cognac is filtered at 5°C before being bottled and once bottled; it does not "age" any more.

### *10.2.2 Apple brandy*

Apple brandy is called 'Applejack' locally in the United States. It has been a commercial product of colonial and 19$^{th}$ century America, but at present, its production is quite small. Applejack is thought by many to be the first spirit produced in the British colonies. This colonial tradition has continued on the East Coast with the Laird's Distillery in New Jersey which was established in 1780 and in the oldest distillery in America. Apple brandies that are more like *eau-de-vie* are produced in California and Oregon. In France, Normandy is one of the few regions which does not have a substantial grape wine industry but produces and distills cider into an apple brandy known as Calvados with 45% alcohol. France is by far the leading exporter of brandy to the United States of America. Greece is the second largest exporter followed by Spain, Germany and Italy [111]. The local cider apples, which tend to be small and tart, are closer in type to crab apples than modern table apples, are used. The method of apple brandy production has been described as early as 1939 and distillation technology is similar to the grape brandy. Mostly the cull fruits are used for apple brandy production. Since it is the wine, which is distilled, so its quality influences the brandy quality. To make quality apple brandy, the use of fresh, mould free and ripe apples for fermentation and distillation after clarification has been recommended. The first distillation gives a product of 60° proof, which is redistilled to 110-130° proof after cutting suitable head and tail. At present, continuous columns are used to produce apple brandy. The abnormal characteristics of naturally fermented beverages are correlated with raw material source and methods of distillation [10].

All varieties of Calvados are aged in oak casks for a minimum period of two years and has its own appellations, with the best brands coming from *Appellation Controlee Pays d'Auge* near the Atlantic Seaport of Deauville and the rest in 10 adjacent regions which are designated *Appellation Reglementee*. Most *Pays d'Auge* and some of the better *Appellation Reglementee* are produced in Pot stills. Pomace brandy has also been attempted, but it contains higher methanol content and efforts are being made to control its quality [112].

### *10.2.3 Pear brandy*

For preparing pear brandy, tree ripened Williams pear (Bartlett in America) from local North-west orchards are hand picked, carefully fermented and distilled. The distillation is done in copper pot stills and century old traditional methods. The resulting distillates have rich aroma and fruit flavours that are typical of Williams pear, considered one of the most precious throughout Europe. Brandy made from William pears like Poire Williams and Williamine are especially good. Some bottles have an entire pear in the bottle. Pear brandies (Poire) is now being produced in United States in California and Oregon. Application of $GA_3$ spray to frost affected trees at blossom induced parthenocarpic fruit development in Williams L Bon Chretien L pear without affecting the flavour of the flesh and that of brandy prepared from the fruits [113].

### *10.2.4 Peach brandy*

The brandy produced from the peach fruit is called peach brandy and is popular in some countries. Commercial demand for small quantities of fruit brandies had been noted in California and sporadic attempts to prepare brandies from fruits including peach in California have been made. Methods for preparing fruit brandy along with peach have been described [5, 114-116]. Peach brandy can be prepared in a similar way as the grape brandy, except for the method of wine preparation. Peach fruits are very pulpy, so for fermentation, these have to be diluted. Consequently, the sugar content is reduced to a very low level, unlike grape and apple. Production of brandy can soak up excessive production of peach, especially the cullage, but the use of poor quality waste fruit usually has little chance of making quality brandy. Brandy of commercial quality can be made from cv. July Elberta [115]. However, the dried peach has not been found suitable for wine making [5]. The process of distillation is similar to any brandy. A blend of tree ripened Sunny slope peaches is carefully distilled to preserve the delicate aroma. Peach brandy is distilled in pot stills by the centuries old traditional methods and has rich aroma and fruit flavour. Peach brandy was produced by using peach pulp of July Elberta variety ameliorated with sugar, jaggary and molasses as fermentable sugar sources in peach must and distilled to brandy [115]. The biochemical characteristics of peach brandy from must of different sources of sugar is given in Table 7. Sugar based peach must gave the highest ethanol content (Table 8). Brandy treated with different wood chips viz., *Quercus, Albizia* and *Bombax* recorded the higher esters, tannins, higher alcohols, ethanol and furfural contents than the untreated brandy. Peach wine and brandy should be matured with *Quercus* wood chips for improvement in all the quality characteristics [117]. Composition and properties of peach brandies are affected by peach variety [118]. For more details regarding the production of peach brandy, see references cited [87, 117-119]. A simple chromatography method to estimate ethyl carbomate in spirits including peach has also been developed [87].

### *10.2.5 Plum brandy*

The generic name for plum brandy is slivovitz, which is made from small blue Sljiva plum throughout the Eastern Europe and the Balkans. It is the national drink of the Middle European countries of Rumania, Hungary and particularly, Yugoslavia [120]. Plum brandy is usually colourless and quite potent. It is made in Serbia and Bosnia from blue plums and is highly regarded. Other varieties include

**Table 7:** Biochemical characteristics of peach brandy from must with different sources of sugar.

| Treatment | Ethanol (%, v/v) | Total esters (mg/l) | Fusel alcohol (mg/l) | Furfural (mg/l) | Aldehyde | Titratable acidity (mg/l acidity) | pH |
|---|---|---|---|---|---|---|---|
| Peach brandy (sugar) | 40.17 | 21.68 | 48.71 | 15.12 | 24.54 | 58.47 | 2.81 |
| Peach brandy (molasses) | 30.72 | 31.63 | 36.33 | 9.45 | 19.19 | 92.22 | 2.80 |
| Peach brandy (jaggery) | 37.50 | 27.60 | 41.43 | 12.90 | 21.43 | 60.76 | 3.04 |
| CD ($p<0.05$) | 0.19 | 0.44 | 0.35 | 0.16 | 0.48 | 0.50 | 0.10 |

*Source:* 115

**Table 8:** Recovery of brandy from peach wine Cv July Elberta.

| Method | Wine | | Brandy | |
|---|---|---|---|---|
| | Volume (l) | Alcohol (%, v/v) | Volume (l) | Alcohol (%, v/v) |
| Sugar | 18.6 | 10.67 | 3.50 | 40.86 |
| Molasses | 15.6 | 7.75 | 3.10 | 31.65 |
| Jaggery | 17.0 | 10.00 | 3.40 | 37.00 |

*Source:* 115

Mirabelle, which are made from yellow plums, Ouetsch from Alsatian plums and light green prunelic. Yugoslavia slivovitz is often aged in oak, mulberry or acacia cooperage [121]. It is mostly light yellow in colour without distinct evidence of wood-aged character. Plum brandies produced in the Alsace region of France are named after the variety of fruit used e.g., Quetsch and Mirabelle [122]. Besides the over-production, the growers are interested to turn cullage fruit, usually having defects of size or side-split, into profitable products acceptable to American consumers. Although for brandy production, plums must have high sugar content, but the pattern of sugar is also important. Some plums contain sorbitol (0-36 % of total sugar) which is not a fermentable sugar. If the sorbitol content is high, the sucrose content is low and *vice-versa.* The cultivars with high sorbitol content include German Prune, Italian Prune, Chrudimes, Tulver Gras and Buhler Fruhzwetsche. Cultivars with little or no sorbitol are Stanley, Ersinger, Bluefire and Grand Prize. There are differences between the clones of one cultivars. So the two cultivars having the same percentage of sugar may result in production of different quantities of alcohol depending upon the level of sorbitol. The content of sorbitol is also influenced by climatic conditions i.e., in warm and sunny weather, the yield is low and the sorbitol content is high that may cause bad fermentation [123]. At present, 50% of all plums grown in Yugoslavia belong to cv. Pozegaca and another 49.5% are local *cvs.* used for brandy production (Slivoviz) [124]. Comparison of five local varieties with *Grase romanesti* showed that *Gogosele de munte* gave the highest yields, was the hardiest and found suitable for plum brandy production [125].

### *10.2.6 Apricot brandy*

Apricots are greatly valued in Hungary as fresh and processed. Fruits are used for making renowned apricot brandy [126]. Brands include the French Abricotine and the Hungarian Barack Palinka. In 10 authentic laboratory brandies of stone fruit i.e., sour and sweet cherry, damson plum and apricot, 0 to 6.4 mg of benzaldehyde per 100 ml of absolute alcohol and traces to 8.6 mg of hydrocyanic acid have been revealed [127]. In 12 commercial samples, benzaldehyde ranged from traces to 13.4

and hydrocyanic acid from traces to about 10. The use of λ-undecalactone in certain apricot liqueurs has been made. Brandy can also be prepared from wild apricot [128]. The pulp is diluted in 1:1 ratio with water fortified with 0.1% DAHP and fermented to completion with TSS of 25°B. The wine is then distilled to make brandy. The brandy made from 1:1 diluted pulp was superior to that from 1:2 dilution. The brandy produced had intense flavour of apricot and high sensory acceptability. The product was treated with slightly roasted oak wood chips during maturation to impart colour and flavour. Comparison of the methanol content of brandy made experimentally with that prepared locally showed a considerable reduction in the quantity of this alcohol [129].

## REFERENCES

1. Waterhouse AL (1995) Wine and heart disease. *Chem Ind* 5: 338-341.
2. Vyas SR & Chakravorty SR (1971) Wine making at home. Haryana Agricultural University Bulletin, Hissar, India.
3. Unwin T (1991) *Wine and the Vine: An Historical Geography of Viticulture and the Wine Trading*, Routledge Press, London, UK.
4. Jackson RS (2000) *Wine Science – Principles, Practices, Perception*, Academic Press, San Diego, USA.
5. Amerine MA, Berg HW & Cruess WM (1967) *The Technology of Wine Making,* AVI Publishing, Westport, USA.
6. Amerine MA, Berg HW, Kunkee RE, Ough CS, Singleton VL & Webb AD (1980) *The Technology of Wine Making*, AVI Publishing, Westport, USA.
7. Jackson D & Schuster D (1981) *Grape Growing and Wine Making: A Handbook for Cool Climates,* Martinborough, Alister Tagcor, New Zealand.
8. Sandhu DK & Joshi VK (1995) Technology, quality and scope of fruit wines with special reference to apple. *Indian Food Ind* 14: 24-28.
9. Jackson RS (1999) Grape based fermentation products. In: *Biotechnology: Food Fermentation - Microbiology, Biochemistry and Technology*, Joshi VK & Pandey A (eds), Vol 2, Educational Publishers and Distributors, New Delhi, India, pp 583-646.
10. Joshi VK, Sandhu DK & Thakur NS (1999) Fruit based alcoholic beverages. In: *Biotechnology: Food Fermentation - Microbiology, Biochemistry and Technology,* Joshi VK & Pandey A (eds), Vol 2, Educational Publishers and Distributors, New Delhi, India, pp 647-744.
11. Joshi VK, Chauhan SK & Bhushan S (2000) Technology of fruit based alcoholic beverages. In: *Postharvest Technology of Fruits and Vegetables,* Verma LR & VK Joshi (eds), Indus Publishing Company, New Delhi, India, pp 1019-1101.
12. Joshi VK, Sharma S, Bhushan S & Attri D (2004) Fruit based alcoholic beverages. In: *Concise Encyclopedia of Bioresource Technology*, Pandey A (ed), Haworth Press, New York, USA, pp 335-345.
13. Braudel F (1981) *The Structure of Everyday Life: The Limits of the Possible Civilization and Capitalism 15th – 18th Century*, Vol 1, University of California Press, London, UK.
14. Delin CR & Lee TH (1991) The J shaped curve revisited: Wine and cardiovascular health update. *Aust New Zealand Wine Ind* 6: 15-16.
15. Gasteineau FC, Darby JW & Turner TB (1979) *Fermented Food Beverages in Nutrition.* Academic Press, New York, USA.
16. Joshi VK & Sharma S (2004) Importance, nutritive value and medicinal contribution of wines. *Bev Food World* 34: 36-41.
17. Seixas R (1975) Alcohol, a carcinogen. *Ca Cancer J Clin* 25: 62-65.

18. Monagas M, Suarez R, Cordoves CG & Bartolome B (2005) Simultaneous determination of non-anthocyanin phenolic compounds in red wine by HPLC-DAD/ESI-MS. *Am J Enol Vitic* 56: 139-147.
19. Joshi VK (1997) *Fruit Wines,* Dr. Y.S. Parmar University of Horticulture and Forestry, Nauni, Solan, India.
20. Verma LR & Joshi VK (2000) An overview of post-harvest technology. In: *Postharvest Technology of Fruits and Vegetables*, Vol 1, Indus Publishing Co, New Delhi, India, pp 1-76.
21. Joshi VK & Parmar M (2004) Present status, scope and future strategies of fruit wines production in India. *Indian Food Ind* 23: 48-52.
22. Jackson RS (2003) Wines: Types of table wine. In: *Encylopedia of Food Sciences and Nutrition*, Caballero B, Trugo L & Figlas PN (eds), Elsevier Science Ltd, London, UK.
23. Amerine MA & Toslyer MA (1970) *'Table Wines' The Technology of their Production,* University of California Press, Berkeley, USA.
24. Joshi VK, Sharma SK, Goyal RK & Thakur NS (1999) Sparkling plum wine: Effect of method of carbonation and the type of base wine on physico-chemical and sensory qualities. *Braz Arch Biol Technol* 42: 315-322.
25. Downing DL (1989) Apple cider. In: *Processed Apple Products,* AVI Van Nostrand, New York, USA, pp 169-188.
26. Joshi VK, Attri BL & Mahajan BVC (1991) Studies on the preparation and evaluation of vermouth from plum. *J Food Sci Technol* 28: 138-141.
27. Bhutani VP, Joshi VK & Chopra SK (1989) Mineral composition of experimental fruit wines. *J Food Sci Technol* 26: 332-333.
28. Rubio-Diaz ME & Calixto-Saura F (2006) Dietary fiber in wine. *Am J Enol Vitic* 57: 69-72.
29. Miller AC, Wolff SR, Bisson LF & Ebeler SE (2007) Yeast strain and nitrogen supplementation: Dynamics of volatile ester production in chardonnay juice fermentations. *Am J Enol Vitic* 58: 470-483.
30. McArthur K, Hogan D & Isenber JI (1982) Relative stimulating effects of commonly infested beverages on gastric acid secretion in humans. *Gastroenterology* 83: 199-203.
31. Brenner H, Rothenbacher D, Bode G & Adler G (1997) Relation of smoking and alcohol and coffee consumption to active *Helicobacerium pylori* infection: Cross-section study. *Br Med J* 315: 1489-1492.
32. Joshi VK & John S (2002) Antimicrobial activity of apple wine against some pathogenic and microbes of public health significance. *Alimentaria* 2: 67-72.
33. Masquelier J (1988) Effects physiologiques du Vin, sa port dans L'alcoolisme. *Bulletin de l'OIV* 689-690: 554-578.
34. Fugelsang KC & Muller IJ (1996) The *in vitro* effect of red wine on *Helicobacterium pylori*. In: *Wine in Context: Nutrition, Physiology, Policy, Proceedings of the Symposium on Wine and Health*, Waterhouse AL & Rantz JM (eds), American Society of Enology and Viticulture, Davis, CA, USA, pp 43-45.
35. Papadoboulou C, Soulti K & Roussis IG (2005) Potential antimicrobial activity of red and white wine phenolic extracts against strains of *S. aureus, E. coli* and *C. albicans. Food Technol Biotechnol* 43: 41-46.
36. Rimm EB, Giovannucci EI, Willett WC, Colditz GA, Ascerio A, Rosner B & Stampfer MJ (1991) Prospective study of alcohol consumption and risk of coronary heart disease in men. *Lancet* 338: 464-468.
37. Kinsella JE, Frankel EN, German JB & Kanner J (1993) Possible mechanisms for the protective role of antioxidants in wine and plant foods. *Food Technol* 47: 85-89.

38. Steinberg D (1993) Modified forms of low-density lipoprotein and atherosclerosis. *J Int Med* 233: 227-232.
39. Soleas GJ, Diamondis EP & Goldberg DM (1997) Wine as a biological fluid: History, production and role in disease prevention. *J Clin Lab Anal* 11: 287-313.
40. Naugler C, McCallum JL, Klassen G & Strommer J (2007) Concentration of trans-resveratrol and related stilbenes in Nova Scotia wines. *Am J Enol Vitic* 58: 117-119.
41. Watkins TR (1997) Wine: Nutritional and therapeutic benefits. In: *ACS Symposium Series 661, American Chemical Society*, Washington, USA, pp 1, 45, 152, 232.
42. Bisson LF, Butzke CE & Ebler SE (1995) The role of moderate ethanol consumption in health and nutrition. *Am J Enol Vitic* 46: 449-462.
43. Liber CS, Jones DP & Decarli LM (1965) Effect of prolonged ethanol intake: Production of fatty liver despite adequate diets. *J Clin Invest* 44: 1009-1021.
44. Cook GL & Hutt MSR (1967) The liver after Kwashiorkor. *Br Med J* 3: 454-457.
45. Hillman RS, McGuffin R & Campbell C (1977) Alcohol intereference with the folate eutronepatic cycle. *Clin Res* 25: 518A.
46. Klatsky AL, Friedman GD, Sieglaub AB & Gezard MJ (1977) Alcohol consumption and blood pressure. *N Engl J Med* 296: 1194-1200.
47. Marmot MG, Rose G, Shipley MJ & Thomas BJ (1981) Alcohol and mortality: A U-shaped curve. *Lancet* 1: 580-583.
48. Klein H & Pittman D (1990) Drinker prototypes in American society. *J Subst Abuse* 2: 299-316.
49. Smart RG & Walsh G (1999) Heavy drinking and problems among wine drinkers. *J Studies Alcohol* 60: 467-471.
50. Dawood MJ (1974) *The Koran,* Penguin, Harmond Sworth, England, UK.
51. Ali AY (1983) *The Holy Quran: Text, Translation and Commentry*, Bretwood, Amana Corp, Maryland, USA.
52. Masson-Oursel P & Morin L (1968) Indian mythology. In: *New Larousse Encyclopedia of Mythology*, Grimal P (ed), Paul Hamlyn, London, pp 325-378.
53. Fleet GH (1993) The microorganisms of wine making - isolation, enumeration and identification. In: *Wine Microbiology and Biotechnology,* Fleet GH (ed), Harwood Academic Publishers, GmbH, Switzerland, pp 1, 461.
54. Amerine MA & Kunkee RE (1968) Microbiology of wine making. *Ann Rev Microbiol* 22: 323-358.
55. Goswell RT & Kunkee RE (1977) Fortified wines. In: *Alcoholic Beverages,* Rose AH (ed), Academic Press, London, UK, pp 477-485.
56. Medina K, Boido E, Dellacassa E & Carrau E (2005) Yeast interaction with anthocyanins during red wine fermentation. *Am J Enol Vitic* 56: 104-109.
57. Thomas S & Davenport RR (1985) *Zygosaccharomyces baili* - a profile of characteristics and spoilage activities. *Food Microbiol* 2: 157-169.
58. Joshi VK (1995) *Fruit Wines*, Dr. Y.S. Parmar University of Horticuture and Foresty, Nauni, Solan, India.
59. Joshi VK, Sharma PC & Attri BL (1991) A note on deacidification activity of *Schizosaccharomyas pombe. J Appl Bacteriol* 70: 385-390.
60. Husnik JI, Delaguis PJ, Cliff MA & Vuuren Van HJJ (2007) Functional analysis of the malolactic wine yeast MLOI. *Am J Enol Vitic* 58: 42-52.
61. Joshi VK & Bhutani VP (1990) Evaluation of plum cultivars for wine preparation. In: *Proceedings of XXIII International Horticulture Congress,* Italy, Abst 3336.

62. Bhutani VP & Joshi VK (1995) Plums. In: *Handbook of Fruit Science and Technology,* Salunkhe DK & Kadam SS (eds), Marcel Dekker Inc, New York, UK, pp 203-232.
63. Salunkhe SS & Kadam SS (1995) *Handbook of Fruit Science and Technology,* Salunkhe DK & Kadam SS (eds), Marcel and Dekker Inc, New York, USA.
64. Faparusi SI (1973) Origin of initial microflora of palm wine from oil palm trees (*Elaeis guineensis*). *J Appl Bacteriol* 36: 559-565.
65. Moralesdeleon JC (1976) Tropical fruits: characteristics and physico-chemical properties. *Technol Aliment Mexico City* 11: 205-214.
66. Luthi H (1953) *Gnafu Nung and Behandlung der Obstwein in Kleinbrieh,* Verlag Huber and Co, Frauenfeld, Switzerland.
67. Sivilotti P, Bonetto C, Paladin M & Petertuger E (2005) Effect of soil moisture availability on merlot: From leaf water potential to grape composition. *Am J Enol Vitic* 56: 9-18.
68. Flores JH & Heatherbeth DA (1984) Optimizing enzymes and pre-mash treatments for juice and colour extraction from strawberries. *Fluessiges Obst* 7: 327-328.
69. Sacchi KL, Bisson LF & Adams DO (2005) A review of the effect of wine making techniques on phenolic extraction in red wines. *Am J Enol Vitic* 56: 197-206.
70. Jensen J S, Blachez B, Egebo M & Meyer AS (2007) Rapid extraction of polyphenols from red grapes. *Am J Enol Vitic* 58: 451-461.
71. Howard K L, Mike JH & Riesen R (2005) Validation of solid phase micro extraction method for head space analysis of wine aroma compounds. *Am J Enol Vitis* 56: 37-45.
72. Main GL & Morris JR (2007) Effect of macerating enzymes and postfermentation grape seed tannin on the colour of Cythiana wines. *Am J Enol Vitic* 58: 365-372.
73. Sampaio TL, Kennedy JA & Vasconcelos MC (2007) Use of microscale fermentations in grape and wine Research. *Am J Enol Vitic* 58: 534-539.
74. Pasini AC, Cota NV, Carranza VM, Garibay EH & Salazar RM (2005) Clarification of wines using polysaccharides extracted from seaweeds. *Am J Enol Vitic* 56: 52-59.
75. Renouf V, Perello MC, Revel deG & Funel-LA (2007) Survival of wine microorganism in the bottle during storage. *Am J Enol Vitic* 58: 379-386.
76. Patil NS, Kotecha PM & Chavan JK (2007) Wine making quality of different grape cultivars. *J Food Sci Technol* 44: 125-126.
77. Joshi VK & Sandhu DK (1997) Effect of different concentrations of initial sugar soluble solids on physico-chemical and sensory qualities of apple wine. *Indian J Hort* 54: 116-123.
78. Joshi VK, Sandhu DK, Attri BL & Walia RK (1991) Cider preparation from apple juice concentrate and its consumer acceptability. *Indian J Hort* 48: 321-327.
79. Yalianti F, Reitmeier CA, Glatz BA & Boylston TD (2005) Sensory flavour and microbial analysis of raw, pasteurized and iirradiated apple ciders. *J Food Sci* 70: 153-158.
80. Atputharajah JD, Samarajeewa U & Vidanpathirana S (1986) Efficacy of ethanol production by coconut toddy yeasts. *J Food Sci Technol* 23: 5-7.
81. Joshi VK & Bhutani VP (1990) Effect of preservatives on the fermentation behaviour and quality of plum wine. In: *Proceedings of XXIII International Horticulture Congress,* Italy, Abst 2585.
82. Joshi VK, Sharma SK & Thakur NS (1995) Technology and quality of sparkling wine with special reference to plum - An Overview. *Indian Food Packer* 49: 49-55.
83. Joshi VK & Sharma SK (1995) Comparative fermentation behaviour, physico-chemical and sensory characteristics of plum wine as effected by the type of preservatives. *Cheme Mikrobiol Technol Lebensmittel* 17: 45-53.

84. Joshi VK, Thakur NS & Thakur KS (1994) Effect of different yeast strains on the physico-chemical and sensory qualities of plum wine. In: *International Horticulture Congress*, Japan, Abst 148.
85. Joshi VK, Attri D & Bhushan S (2002) Biological deacidification of plum (*Prunus saliciana*) must and quality of wine produced with variable acidity. *Bev Food World* 11: 33-36.
86. Joshi VK, Thakur NK & Kaushal BBL (1997) Effect of debittering of kinnow juice on physico-chemcial and sensory quality of kinnow wine. *Indian Food Packer* 50: 5-10.
87. Joshi VK & Bhutani VP (1995) Peach. In: *Hand Book of Fruit Science and Technology*, Salunkhe DK & Kadam SS (eds), Marcel & Dekker, USA, pp 230-240.
88. Joshi VK, Shah PK & Kumar K (2005) Evaluation of peach cultivars for wine preparation. *J Food Sci Technol* 42: 200-208
89. Joshi VK & Shah PK (1999) Effect of wood treatment on chemical and sensory quality Peach wine during ageing. *Acta Alimentaria* 27: 307-318.
90. Joshi VK, Sharma S & Krishan K (2006) Technology for production and evaluation of strawberry wine. *Bev Food world* 33: 77-78.
91. Joshi VK, Sharma S & Bhushan S (2005) Effect of method of preparation and cultivar on the quality of strawberry wine. *Acta Alimentaria* 34: 339-353.
92. Sharma SK & Joshi VK (1996) Optimization of some parameters of secondary fermentation for production of sparkling wine. *Indian J Exp Biol* 34: 235-238.
93. Joshi VK, Bhutani VP & Sharma RC (1990) Effect of dilution and addition of nitrogen source on chemical, mineral and sensory qualities of wild apricot wine. *Am J Enol Vitic* 41: 229-231.
94. Joshi VK & Sharma SK (1994) Effect of method of must preparation and initial sugar levels on the quality of apricot wine. *Res Ind* 39: 250-252.
95. Nagadowithana TW & Stainkraus KH (1976) Influence of the rate of ethanol production and accumulation on the viability of *Saccharomyces cerevisiae* in "rapid fermentation". *Appl Environ Microbiol* 31: 158-162.
96. Juroszek JR, Feuillat ME & Charpentier C (1987) Effect of the champagne method of starter preparation on ethanol tolerance of yeast. *Am J Enol Vitic* 38: 194-198.
97. Tchorbanov B, Mitchev G, Lazarova G & Popov D (1993) Studies on the secondary fermentation of low alcohol sparkling apple wine. *Am J Enol Vitic* 44: 93-98.
98. Sharma S & Joshi VK (2004) Flavour profiling of strawberry wine by qualitative descriptive analysis technique. *J Food Sci Technol* 41: 22-26.
99. Foster C (1967) *Home Wine Making*, Ward Lock, London, UK.
100. Pilone FJ (1954) Production of vermouth. *Am J Enol Vitic* 5: 30-45.
101. Joslyn MA & Amerine MA (1964) *Dessert, Appetizer and related Flavoured wines*, University of California, Division of Agricultural Sciences, Berkeley, USA.
102. Jarczyk A & Wzorek W (1977) Fruit and honey wines. In: *Alcoholic Beverages*, Vol 5, Rose AH (ed), Academic Press, London, UK, pp 387-421.
103. Rose R (1992) Orange based sparkling alcoholic beverage. French Patent Application, FR 2657, 878 AL (FSTA 24(2): 2H, 88.)
104. Attri BL, Lal BB & Joshi VK (1994) Technology for the preparation of sand pear vermouth. *Indian Food Packer* 48: 39-44.
105. Lingappa K, Padshetty NS & Chowdary NB (1993) Tamarind vermouth - a new alcoholic beverage from Tamarind (*Tamarindus indica* L.). *Indian Food Packer* 47: 23-26.
106. Kalyananda MK, Jaya Raj GS & Jansz ER (1977) Factors affecting the yield of alcohol from coconut sap. In: *Proceedings of the Institute of Chemistry,* Colombo, Sri Lanka, pp 7-16.

107. Lafon J, Coulland P & Gaybellile F (1973) *Le Cognac et Sa Distillation*, JB Bailliere et Fils, Paris, France.
108. Lafon R, Lofan J & Couilland P (1973) *Le Cognac et Sa Distillation,* JB Bailliere et Fils, Paris, France.
109. Lafon R, Lofan J & Couilland P (1964) *Le Cognac et Sa Distillation*, JB Bailliere et Fils, Paris, France.
110. Lafon S (1983) Wine and brandy. In: *Biotechnology: Food and Feed Production with Microorganisms*, Vol 9, Rehm HJ & Reeds G (eds), VCH Weinheim, Germany, pp 629-666.
111. Sponholz WR & Dittrich HH (1979) Analytiche Vag Leiche Von Mosten and Weinen aus gesunden und ensigstichigen Tran benbervan. *Wein-Wissenchaft* 34: 279-292.
112. Tolu LA (1962) If mentanolo acqueviti di Vinacia. *Ann Fac Sci Agric Univ Torino* 1: 199-203.
113. Hurter J, Uoniker T & Tanner H (1983) Does the application of gibberellic acid reduce the quality of williams pear? *Scheizerische-Zeitsohrift-fur obst. Und-Weinban* 119: 517-521.
114. Patel JD, Venkataramu K & SubaRao MS (1977) Studies on the preparation of cider and brandy from some varieties of Indian apples. *Indian Food Packer* 31: 5-8.
115. Shah PK & Joshi VK (1999) Influence of different sugar sources and addition of wood chips on physico-chemical and sensory quality of peach brandy. *J Sci Ind Res* 58: 995-1004.
116. Stanciulescu G, Rusnac D & Bortes G (1975) *Technologia Distilator Alcoolial din Fruete si Vin*, Cares, Bucharest, Romania.
117. Kumar P (1995) Evaluation of peach varieties for preparation of peach wine and brandy, M.Sc. Thesis, Dr. YS Parmar University of Horticulture and Forestry, Nauni, Solan, India.
118. Paunovic R & Bulatovic S (1978) The effect of peach variety on the composition and properties of peach brandy. *Zbornik-Radova-Pol.ioprivrednog-Fakulteta-Beograd* 586: 15-20.
119. Bulatovic S, Paunovic R & Mijaoika M (1978) Pomological technological studies on fruits of various peach cultivars as raw material for brandy production. *Archiv-20-Pol Iop-brivredne-Nauke* 31: 113-118.
120. Anonymous (1970) Yugoslavia and its growing prune exports. *Pacific Fruit News* 148: 2-7.
121. Litchev V (1961) Aldehyde composition of Bulgarian plum brandy. *Lozarstvo L Vinarstvo* 7: 42-46.
122. Crowell EA & Guyman JF (1973) Aroma constituents of plum brandy. *Am J Enol Vitic* 24: 159-165.
123. Hartmann W (1982) The importance of sugar content and patterns of plum for brandy production. In: *XXIst International Horticultural Congress*, Vol 1, Abstr 1159 A.
124. Paunovic S, Gavrilovic M, Kapetanovic N, Adamic F, Stampar K, Krgovic L & Paunovio SA (1974) Plum Cultivars suggested for Yugoslavia. *Jugoslovensko-Vooarstvo* 8: 29-30.
125. Mihaeecu G & Prodan G (1977) Contribution to the study of some local plum varieties. *Luorari-Stiintifioe Institutul-Agronomic 'Nicolae Balcescu' B* 17: 50.
126. Nyujto F & Guranyi D (1981) Apricots. *Kajszibarack,* p 465.
127. Bandian F, Valenta M & Kain W (1976) Zur Beurteilung des Benzatdehydgehaltes in Steinobst-Edelbranntweinen und Steinobst-Fruchlikoren. Mitt. Rebe Wein-Obstban Friichteverw. *Klosterneuburg* 26: 131-135.
128. Joshi VK, Sharma PC & Chauhan SK (1989) Preparation of brandy from wild apricot fruit. In: *National Seminar on Recent Advances in Postharvest Management of Temperate Fruits, Vegetable and Ornamental Plants*, Nauni, Solan, India, p 47.
129. Joshi VK & Sandhu DK (2000) Quality evaluation of naturally fermented alcoholic beverages, microbiological examination of source of fermentation and ethanolic productivity of the isolates. *Acta Alimentaria* 29: 323-334.

# Organic Acids, Gums and Vitamins

# 17

**Parmjit S. Panesar[1], Sawinder Kaur[2], Gurpreet Kaur[1] and Reeba Panesar[1]**

*[1]Department of Food Engineering & Technology, Sant Longowal Institute of Engineering & Technology, Longowal 148 106 (Pb.)*
*[2]Department of Food Technology, K.M. Vidyalaya, Jalandhar 144 004 (Pb.)*

## 1. INTRODUCTION

Microorganisms have long played a major role in the production of diverse range of fermentation products, which have multifarious applications in food, pharmaceutical and beverage industries. Besides several commodity and specialty chemicals, microbes are also being directed towards the production of high-value products for the fine chemicals and pharmaceutical industries. The production of chemicals using industrial biotechnology can replace multi-step chemical synthesis with a single step involving low energy and less material input and even enables the synthesis of products that are not possible chemically [1, 2]. Several products of microbial fermentation are incorporated into food as additives and supplements (antioxidants, flavours, colourants, preservatives, sweeteners, etc.). There is a great demand in the development and use of natural food and additives derived from microorganisms, since they are more desirable than the synthetic ones produced by chemical processes [3]. Organic acids, gums and vitamins constitute the important components of microbial products. These products play a very important role and have a wide range of applications in food, pharmaceutical and beverage industries as additives, acidulants, preservative and pickling agents, flavour enhancers, packaging, emulsifiers, stabilizers, gelling agents, coagulants, lubricants, thickening and suspending agents. Besides this, vitamins are essential for growth and normal metabolic functioning of an organism. In this chapter, these important fermentation products (organic acids, gums and vitamins) are discussed in context of microorganisms involved, production, recovery and their applications.

## 2. PRODUCTION OF ORGANIC ACIDS

Organic acids widely used in the food, pharmaceutical and chemical industries are important chemicals. Fermentation technology for the production of organic acids in particular has been known for more than a century, and these acids have been produced through different microbial processes. Several microorganisms possess the ability to convert carbohydrates to organic acids with high yields. For example, *Lactobacillus* sp. produce lactic acid, *Acetobacter* produce acetic acid, gluconic acid etc., and *Aspergillus* sp. are known to produce citric acid (Table 1). All acids of tricarboxylic acid cycle can be produced microbiologically in high yields. Some acids derived indirectly from krebs

*Food Biotechnology: Principles and Practices*
*V.K. Joshi & R.S. Singh (Eds.)*
*I.K. International Publishing House Pvt. Ltd., India*
*pp 563–608*

**Table 1:** Common microorganisms and substrates used in the production of organic acids.

| Organic acid | Common Substrate | Microorganism (s) |
|---|---|---|
| Acetic acid | Glucose<br>Sucrose | *Acetobacter aceti*<br>*Gluconobacter oxydans*<br>*Acetomonas* sp. |
| Citric acid | Glucose<br>Sucrose<br>Molasses | *Aspergillus niger*<br>*A. awamori*<br>*Trichoderma viride* |
| Gluconic acid | Glucose | *Aspergillus niger*<br>*Acetobacter* sp. |
| Lactic acid | Lactose<br>Whey | *Penicillium* sp.<br>*L. bulgaricus*<br>*L. delbrueckii*<br>*L. acidophilus*<br>*L. helveticus*<br>*L. casei*<br>*S. thermophilus* |
| Propionic acid | Glucose<br>Lactose | *Propionibacterium shermanii*<br>*P. acidipropionici*<br>*Clostridium propionicum* |

cycle such as itaconic acid can likewise be produced. Other organic acids are produced directly from glucose, or they are formed as end products from pyruvate or ethanol.

Many of the commercial production processes for organic acids are excellent examples of fungal biotechnology. However, unlike penicillin, organic acids had a less visible impact on well being of humans. Indeed, organic acid fermentations are often not even identified as fungal bioprocesses, having been overshadowed by the successful deployment of the β-lactam processes. Yet, in terms of productivity, fungal organic acid processes may be the best examples of all. Perhaps a greater understanding of this extraordinary capacity of filamentous fungi to produce organic acids in high concentrations will allow greater exploitation of these organisms via application of new knowledge in this era of genomics-based biotechnology.

## 2.1 Lactic Acid

Lactic acid (2-hydroxy propionic acid) is an organic hydroxy acid whose occurrence in nature is widespread. It was first discovered in 1780 by Swedish chemist Scheele [4], but it was first produced commercially by Charles E. Avery at Littleton, Massachusetts, USA in 1881. Lactic acid can be manufactured by chemical synthesis or carbohydrate fermentation. Chemical synthesis produces a racemic mixture of lactic acid, but stereospecific acid can be made by carbohydrate fermentation.

Of the 80,000 tonnes of lactic acid produced worldwide every year, about 90% are made by lactic acid bacterial fermentation and the rest is produced synthetically by the hydrolysis of lactonitrile [5]. Microbial fermentation has the advantage that by choosing a strain of lactic acid bacteria (LAB) producing only one of the isomers, an optically pure product can be obtained, whereas synthetic production always results in a racemic mixture of lactic acid.

After fermentation, the broth containing calcium lactate is filtered to remove cells, carbon treated, evaporated and acidified with sulphuric acid to get lactic acid and calcium sulphate. The insoluble calcium sulphate is removed by filtration; lactic acid is obtained by hydrolysis, esterification, distillation and hydrolysis. The chemical and physical properties of lactic acid have been reviewed by Holten *et al.* [6]. Lactic acid exists in two optically active isomeric forms. The L(+) isomer is present in humans, although both L(+) and D(-) isomers are found in biological systems. Lactic acid is soluble in water and water miscible organic solvents but insoluble in other organic solvents.

### *2.1.1 Microorganisms*

Pasteur discovered in 1857 that the souring of milk was caused by microorganism. In 1860's and 1870's, the presence of lactic acid bacteria was noted and investigation was made regarding their optimum growth temperature. A wide range of bacteria are involved in the production of lactic acid [7]. Lactic acid bacteria can be classified (Table 2) as homofermentative or heterofermentative [8]. Out of these only homofermentative are of industrial importance. These are from the genera *Lactobacillus, Streptococcus* and *Pediococcus* [9]. The homofermentative lactic acid bacteria catabolize glucose via EM pathway. Organism may produce D(-), L(+) or DL lactic acid. The formation of racemic mixture could arise from action of two stereospecific lactate dehydrogenases [10].

The desirable characteristics of industrial microorganisms are their ability to rapidly and completely ferment cheap raw materials, requiring minimal amount of nitrogenous substances, providing high yields under conditions of low pH and high temperature, production of low amounts of cell mass and negligible amounts of other by-products. The choice of an organism primarily depends on the carbohydrate to be fermented. *Lactobacillus delbreuckii* are able to ferment sucrose. *Lactobacillus delbreuckii* subsp. *bulgaricus* is able to use lactose. *Lactobacillus helveticus* is able to use both lactose [10] and galactose. *Lactobacillus amylophilus* is able to ferment starch [11]. *Lactobacillus lactis* can ferment glucose, sucrose and galactose. Organisms such as *Rhizopus oryzae* have less limiting nutritional requirements and can utilize starch feed stocks. They are able to produce pure L(+) lactic acid [12].

### *2.1.2 Method of production*

2.1.2.1 Raw materials: A large number of carbohydrates and nitrogenous materials have been studied for the production of lactic acid by fermentation. These have been investigated on the basis of high lactic acid yields, optimum biomass production, negligible by-product formation, fast fermentation rate, less pre-treatment, easy downstream processing, low cost, ease of availability, etc. The choice of the raw material to be used depends on the microorganisms studied and also on the product desired.

Sucrose (from syrups, juices and molasses), lactose (from whey), maltose (produced by specific enzymatic starch conversion processes), glucose (from starch conversion processes), mannitol, etc. have been commercially used. Molasses are cheap but give low yields of lactic acid and require laborious purification procedures. Whey is also cheap and easily available, but like molasses has expensive purification processes, although different modern technologies like ultra filtration and electrodialysis have been employed [13]. Cordon *et al.* [14] employed fungal amylases to hydrolyze potato starch

**Table 2:** Common lactobacilli classified by fermentation group.

| | Heterofermentative | |
|---|---|---|
| **Homofermentative** | **Facultative** | **Obligate** |
| *L. acidophilus* | *L. plantarum* | *L. brevis* |
| *L. helveticus* | *L. rhamnosus* | *L. buchneri* |
| *L. delbrueckii* subsp. *bulgaricus* | *L. coryneformis* | *L. fermentum* |
| *L. delbrueckii* subsp. *lactis* | *L. curvatus* | *L. kefir* |
| *L. delbrueckii* | *L. casei* | *L. reuteri* |
| | *L. paracasei* | |

*Source:* 8

for lactic acid production. Cellulosic materials like corn, straw, cottonseed hulls, grapefruit, sulphite waste liquor, etc. have also been investigated.

Most of lactic acid bacteria have generally complex nutrient requirements, due to their limited ability to synthesize vitamin B and amino acids. Therefore, these are generally associated with habitats rich in nutrients, such as various food products (milk, meat, beverages, vegetables, etc.), but some are also normal flora of the mouth, intestine, and vagina of mammals. A complex natural environment renders a microorganism able to metabolize many different carbohydrates [15]. For normal growth they need, besides a carbon source, nitrogen, partly in the form of amino acids, several vitamins, growth substances and minerals. Since relatively small amounts of free lactic acid inhibit growth, the presence of a buffer is necessary to maintain the culture [16].

Nitrogenous materials such as whey permeate, yeast extract, malt sprouts, malt combing nuts, grass extract, peptones, beef extract, casein hydrolysate, corn steep liquor, N-Z-amine, soybean hydrolysate with supplementation of vitamins to supplement carbohydrate sources to give fast and heavy growth have been studied. However, yeast extract seems to be the most effective supplement. Additional minerals are occasionally required when the carbohydrate and nitrogenous sources lack sufficient quantities. Various amounts of vitamin B were studied to replace yeast extract [17].

2.1.2.2 Fermentation: The most common method used industrially is the batch fermentation. The simplified flow diagram has been given in Fig. 1 [18]. Fermentation conditions are different for each microorganism, but are generally in range of 45-60°C with a pH of 5.0-6.5 for *L. delbreuckii* [19] and 30-50°C and a pH below 6.0 for *Rhizopus* [20]. The fermenters are typically steamed, heated with water to boiling [19] or chemically sterilized [21] before filling with pasteurized medium. Inoculum size is usually 5-10% of the liquid volume of fermenter. The acid formed is neutralized either by calcium hydroxide or calcium carbonate. Fermentation time is 1-2 days for 5% whey and 2-6 days for 15% glucose or sucrose. The yield of lactic acid depends on initial sugar or starch concentration. L(+) lactic acid is also commercially produced in fermentation processes using lactic acid bacteria or fungi such as *Rhizopus oryzae* in submerged culture. *Rhizopus* sp. can produce L(+) lactic acid from starch, but the yield is very less in comparison with lactic acid bacteria.

Temperature is one of the most important factors influencing the growth of lactic acid bacteria. The optimal temperature for growth varies between the genera from 20 to 45°C [22]. Depending on the optimal temperatures for fermentation and growth, lactic acid bacteria are classified as thermophilic or mesophilic. Most of the lactobacilli come under the mesophilic category; however, *L. bulgaricus, L. thermophilus* and *L. delbruckii* can be grouped under the thermophilic category. The lactic acid fermentation can be carried out at comparatively higher temperatures using suitable bacteria. In fermentations using *L. delbrueckii* and *L. bulgaricus,* a temperature of 45°C or higher may be maintained [16]. *L. helveticus* and *L. acidophilus* can be used in a temperature range of 37-45°C. However, in case of other bacteria such as *L. casei*, a temperature of 28-35°C is preferred.

Another important parameter is pH, which has the strong effect on the lactic acid fermentation. Lactic acid produced during fermentation has to be continuously neutralized. Generally, calcium carbonate is added as a buffering agent during batch fermentations. For rapid and complete fermentation, the optimal pH range is between pH 5.5-6.0, and in some cases 6.0-6.5, depending upon the culture used. Fermentation is strongly inhibited at pH and ceases at pH values below 4.5 [23]. However, their acid tolerance gives them a competitive advantage over other bacteria.

### *2.1.3 Recovery*

Lactic acid is sold in three major grades: technical, food and pharmaceutical. The first step in almost all recovery processes is to raise the temperature of fermentation liquor to 80-100°C and pH to 10-11 in order to kill organism, coagulate protein and degrade residual sugars. It also solubilizes calcium lactate, and the liquid is then decanted.

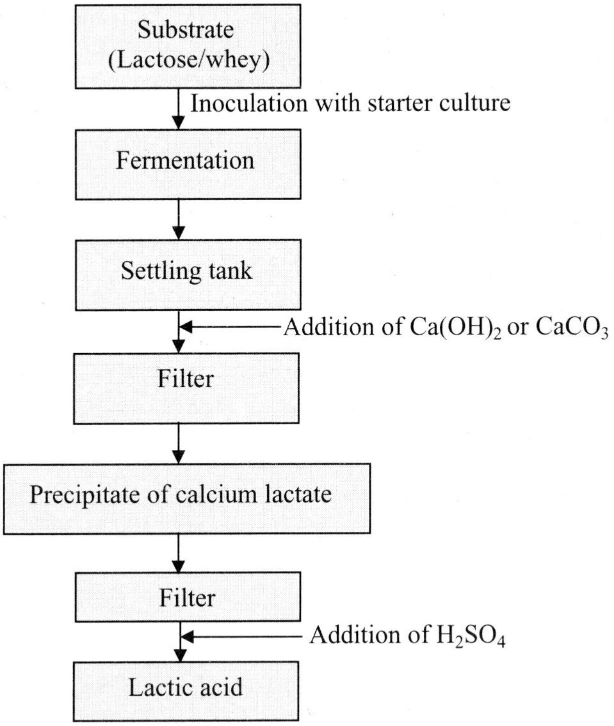


**Fig. 1:** Major steps in the production of lactic acid.
*Source:* 18

2.1.3.1 Filtration, carbon treatment and evaporation: The process is used to produce technical food grade acid. In this process, activated vegetable carbon is added to the filtered broth in order to bleach the calcium lactate for food grade acid production. Calcium lactate is then concentrated to 37% at 70°C and treated with sulphuric acid to precipitate calcium sulfate, which is removed by continuous filter. The lactic acid is then evaporated from 8 to 52% in evaporators.

2.1.3.2 Liquid-liquid extraction: In this process, the acid is extracted by solvent and then recovered from solvent by either distillation or back extraction. Lactic acid refined by liquid-liquid extraction is substantially free from ash, but contains other impurities from raw materials and needs additional treatment [24].

2.1.3.3 Other recovery processes: Lactic acid may be recovered by ion-exchange separation, separation by adsorption, separation by vacuum distillation, and membrane separation [25]. Each of these exhibits some advantages and disadvantages. The choice of the separation process should be based on the efficient and economical usage of these extractants [26].

### *2.1.4 Applications*

Lactic acid is widely used as acidulant/flavouring/pH buffering agent or inhibitor of bacterial spoilage in a variety of processed foods. In contrast to other food acids, it has a mild acidic taste. It is classified as GRAS (generally regarded as safe) by FDA because of its non-volatile and odourless nature. It is a very good preservative and pickling agent. Addition of lactic acid aqueous solution to the packaging of poultry and fish increases their shelf-life [27]. The esters of lactic acid are used as emulsifying agents in baking foods (stearoyl-2-lactylate, glyceryl lactostearate, glyceryl lactopalmitate). Only heat stable fermentation grades can be used for this application [28, 29]. Technical grade lactic acid is used as an acidulant in vegetable and leather tanning industries. Lactic acid is being used in many small scale applications like pH adjustment, hardening baths for cellophanes used in food packaging, terminating agent for phenol formaldehyde resins, lithographic and textile printing developers, adhesive formulations, electroplating and electropolishing baths and detergent builders. Lactic acid has many pharmaceutical and cosmetic applications and formulations in topical ointments, lotions, anti-acne solutions, humectants and dialysis applications. Calcium lactate can be used for calcium deficiency therapy and as anti-caries agent. Its biodegradable polymer has medical applications as sutures, orthopaedic implants, controlled drug release, etc. Polymers of lactic acids are biodegradable thermoplastics. These polymers are transparent and their degradation can be controlled by adjusting the composition and the molecular weight. Poly L-lactic acid with low degree of polymerization can help in controlled release or degradable mulch films for large-scale agricultural applications [28]. In addition, optically pure L(+)lactic acid is polymerized to a high crystal polymer suitable for fibre and oriented film production, and is expected to be useful in the production of liquid crystal as well [30].

## 2.2 Citric Acid

The production of citric acid is the oldest and most thoroughly studied filamentous fungal fermentation [31]. It was first isolated from lemon juice by Scheele in 1784. Its widespread presence in citrus fruits ensures its non-toxicity and has been used as an acidulent in various food industries. Among the industrially produced organic acids, citric acid is the most important in quantitative terms with an estimated annual production of about 1000,000 tonnes [32]. The annual growth of its demand/consumption rate is around 3.5-4.0% [33, 34]. About 70% of the marketed citric acid is used in diverse food and beverage products, with carbonated beverages accounting for nearly 50% of the total production in the 1990s. Almost entire quantity of citric acid is obtained through submerged fermentation of starch or sucrose based media using the filamentous fungus *Aspergillus niger* [35]. Solid state fermentation (SSF) has been an alternative method for citric acid production using agro-industrial residues [36]. Commercially citric acid is produced either as an anhydrous form or as the monohydrate. Anhydrous form is obtained by crystallizing from hot aqueous solutions, whereas monohydrate form is obtained by crystallization at temperatures below 36.6°C.

### *2.2.1 Microorganisms*

*Aspergillus niger* is a good and most popular source of citric acid production. Apart from *A. niger,* several other strains and yeasts [37] have been employed for the production of citric acid (Table 3). Examples are *Aspergillus wentii, A. clavatus, Penicillum luteum, P. citrinum, Mucor piriformis, Paecilomyces divaricatum, Trichoderma viride* and *Corynebacterium* sp. Tabuchi and Abe [38] filed a patent for manufacture of citric acid using eight genera of yeasts including *Candida.* Chas, Pfizer & Co. Inc. [39] claimed the use of strains of *Endomycopsis, Torulopsis, Candida, Hansenula* and *Pichia.*

### *2.2.2 Method of production*

2.2.2.1 Raw materials: Beet molasses is generally used as substrate for surface fermentation. In order to obtain good yields of citric acid, the levels of heavy metals such as iron and manganese are kept below certain critical levels [40]. For submerged fermentation, beet and cane molasses or glucose syrup

**Table 3:** Different microorganisms involved in citric acid production.

| Bacteria | Yeasts | Fungi |
|---|---|---|
| *Arthrobacter paraffineus* | *Candida tropicalis* | *Aspergillus niger* |
| *Bacillus licheniformis* | *C. oleophila* | *A. wentii* |
| *Corynebacterium* sp. | *C. guilliermondii* | *A. carbonarius* |
| | *C. citroformans* | *A. aculeatus* |
| | *Hansenula anomala* | *A. awamori* |
| | *Saccharomyces lipolytica* | *Pencillium luteum* |
| | | *Paecilomyces divaricatum* |
| | | *Trichoderma viride* |

*Source:* 37

may be used. Ion exchange technique is used to remove heavy metals in case of glucose syrup as substrate [41].

2.2.2.2 Fermentation: The production can be carried out through both surface as well as submerged modes. The major steps involved in the production of citric acid are shown in Fig. 2 [18].

2.2.2.2.1 Surface fermentation: In the surface process, beet molasses is diluted to a suitable sugar concentration and the pH is adjusted at 5 to 7. At low pH, *A. niger* does not germinate [42]. Additional nutrients and alkali ferrocyanide are added, if necessary, and media is sterilized. Spores of *A. niger* grown on sporulation medium are collected and distributed over surface of medium. Incubation temperature is kept at 30°C and sterile air is supplied to the incubation chamber. Fermentation lasts for 7-15 days. The liquor obtained is then pumped forward to the recovery section.

2.2.2.2.2 Submerged fermentation: In this process, a suspension of spores of *A. niger* grown on a solid media is introduced into sterilized medium in the inoculum fermenter. The medium is aerated and the mould is allowed to grow at a temperature of about 30°C for a period of 18 to 30 h [43]. The kinetics of citric acid fermentation with *A. niger* have been studied by Rohr *et al.* (1981). A growth phase is followed by a citric acid producing phase during which only small amount of growth occurs. When the rate of increase in citric acid concentration reaches the maximum point, the fermentation is discontinued and broth is pumped to a filter or centrifuge. After filtration, the liquor is forwarded to recovery section. Tabuchi & Abe [38] filed a patent for the manufacture of citric acid from glucose and molasses using

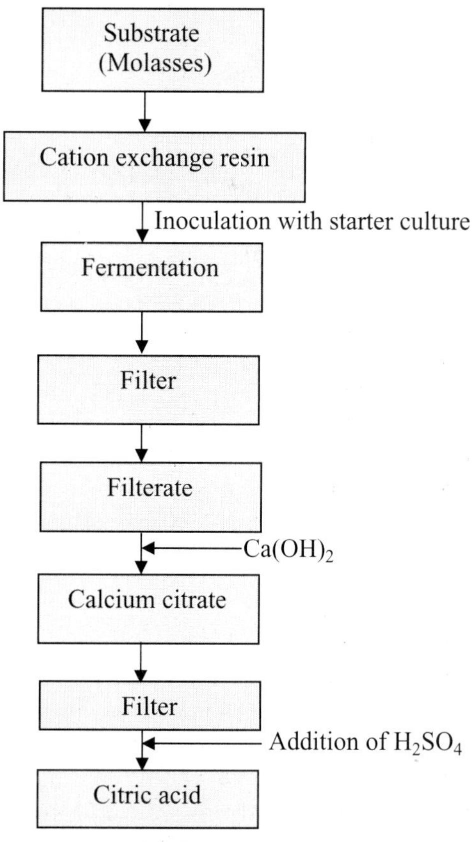


Fig. 2: Major steps in the production of citric acid.
*Source:* 18

eight genera of yeast including *Candida*. The advantages of using yeast rather than *A. niger* are the possibilities of using high initial sugar concentration with faster fermentation rate. Fermentation is conducted at temperature of 25-37°C depending on the organism. In continuous process, the pH is controlled at 3.5 [44].

#### *2.2.3 Recovery*

The liquor obtained after the fermentation and filtration contains aqueous solution of citric acid contaminated by various organics and inorganics depending upon the initial carbon source. There are generally two methods employed for the recovery of citric acid.

**2.2.3.1 The classical process:** This process is suitable for use with very impure liquors derived from molasses. To the hot fermented liquor, lime is added in order to precipitate calcium citrate tetrahydrate. These precipitates after washing are treated with $H_2SO_4$ yielding an aqueous solution of citric acid and a precipitate of by-product gypsum ($CaSO_4$).

**2.2.3.2 Solvent extraction:** This method is generally applied to the products from glucose or alkane-based substrates. The main advantage of this method is that it avoids the use of lime and $H_2SO_4$ and thus the problem of gypsum disposal. The use of butan-2-ol as an extractant for citric acid has been proposed [45]. Also, tributyl phosphate diluted with a minor amount of kerosene could be used [46]. In this process, the extraction can take place at a low temperature and after extraction, the solvent is stripped with hot water. In fermentations with n-alkanes and in particular where sodium hydroxide is used for pH control, monosodium or trisodium citrate can be directly crystallized from the clarified fermented liquor [47].

#### *2.2.4 Applications*

Citric acid is the most versatile and widely used food acidulant. The use of citric acid as a food acidulent depends in part on its strength as an acid. It has a pleasant taste and is a flavour enhancer. This acid is increasingly used as a stabilizer of oils and fats because of its ability to complex heavy metals such as iron and copper, thereby reducing oxidation catalyzed by these metals. Citric acid forms a wide range of metallic salts, which find use in various applications such as trisodium and tripotassium citrates. It is also employed as an aid to emulsification in the manufacture of processed foodstuff e.g., cheese. Mixtures of citric acid and its salts have good buffering capacity and are used in pharmaceutical, food industry and toiletry.

Citric acid esters of alcohols, in particular the triethyl, tributyl and acetyltributyl esters, are used as non-toxic plasticizers in plastic films used to protect food stuffs. Monostearyl citrate can be used instead of citric acid as an antioxidant in oils and fats. The properties and uses of citric acid have been reviewed extensively [48-50].

### 2.3 Acetic Acid

The production of acetic acid from alcoholic liquids has been known as long as the production of wine. Vinegar, which is the aqueous solution of acetic acid, is a well known household product. Vinegar was produced only for local consumption until middle ages. Earlier acetic acid was obtained from natural carbohydrates by biochemical oxidation of carbohydrate and destructive distillation of wood. Today acetic acid is one of the fastest growing commercial organic chemical amongst the important aliphatic intermediates. Microbial production of acetic acid was started to cope up with the increasing industrial demand [51]. The acetic acid was produced from alcohol which was produced mainly from biomass.

### 2.3.1 Microorganisms

The practical approach to the manufacture of vinegar has been known for a long time. Vinegar had been produced by natural fermentation from the alcoholic mash long before the discovery of acetic acid bacteria. Earlier, it was thought that the conversion was brought about by one species of bacteria '*Mycoderma aceti*'.

Now a days, many fermentative bacteria are known for the production of acetic acid, however, acetic acid bacteria are the popular bacteria for its commercial production. The acetic acid bacteria can be divided into two genera, *Gluconobacter* and *Acetobacter*. The first group oxidizes ethanol solely to acetic acid, and the second group is able to oxidize ethanol first to acetic acid and then further to $CO_2$ and $H_2O$. The acetic acid bacteria belong to the family *Pseudomonadaceae*. The cells are generally Gram negative, acid tolerant, rod-shaped, but elongated, swollen or branched forms may occur. They may be motile or non-motile; they do not form endospores. Examples of strains which are used commercially are *Acetobacter aceti, A. pasteurianus* or *A. peroxidans*. *Gluconobacter oxydans* and several subspecies of this species are also used commercially [52].

### 2.3.2 Method of production

**2.3.2.1 Raw materials:** Starting materials with low alcohol content such as wine, whey, malt, or cider do not require any additional components to constitute a complete nutrient solution. However, if potato or grain spirits or technical alcohol is used, nutrients must be added in many cases to obtain optimal growth and acetic acid production. The importance of the addition of yeast extract to the medium is its higher ethanol tolerance and higher productivity and also shortening of the lag phase in acetic acid fermentation [53]. Lactic acid, pyruvic acid, glycerol, α-glycerophosphate, alanine, succinic acid and fumaric acid are reported to exercise such effects. Decreased lag effect is accompanied by an increase in the ratio of acid produced to cell growth implying that the flow of metabolism in cells is changed in favour of acetic acid production rather than cell growth. Acceleration in acetic acid fermentation has been observed by supplying low-priced malt sprouts to the medium. *Tetracoccus soyae*, a homofermentative lactic acid bacterium, can utilize glycerol as an energy source only under aerobic conditions, and both growing and resting cells produce acetic acid as the end product from glycerol [54]. The acid yield is about 70%. Glycerol can be completely dissimilated to acetic acid; lactic acid is produced initially but it readily undergoes oxidative degradation to acetic acid. Glycerol dissimilation is catalyzed by inducible enzymes and cells possess high levels of catalase and cytochromes. $H_2O_2$ produced in the oxidation of L-glycerol 3-phosphate is decomposed by catalase.

**2.3.2.2 Fermentation:** The production of acetic acid (vinegar) actually requires two fermentation processes. The organisms *Saccharomyces cerevisiae* var. *ellipsoideus*, and *Acetobacter* sp. are involved in this process. The first step is carried by yeast, and is an anaerobic process, which involves the fermentation of sugar to ethyl alcohol. Sulphur dioxide is sometimes added to the fermentation to prevent the growth of bacteria during alcoholic fermentation. The second step is the oxidation of alcohol to acetic acid, which is an aerobic reaction carried out by acetic acid bacteria i.e., *Acetobacter* sp. Thus process of acetic acid production can be explained by the following equations:

$$\begin{array}{rcl} & \text{Yeast} & \\ C_6H_{12}O_6 & \rightarrow & 2C_2H_5OH + 2CO_2 \\ \text{(Glucose)} & \rightarrow & \text{(Ethyl alcohol)} \\ C_2H_5OH + O_2 & \rightarrow & CH_3COOH + H_2O \\ \text{(Ethyl alcohol)} & \rightarrow & \text{(Acetic acid)} \end{array}$$

Acetic acid (vinegar) production can be carried by either of the following methods:

2.3.2.2.1 The 'slow' process: In this process, vinegar is produced by keeping partially filled containers of wine in open [55]. Later it was found that addition of some vinegar to the partially filled casks increases its reaction. This concept was already incorporated into the process known as the 'Orleans process', the 'French method' or the 'slow process'. In this process, barrels are filled partially with good quality vinegar, which acts as inoculum and wine is added at regular intervals for four weeks. After five weeks, a portion of vinegar is withdrawn and the same amount of wine is introduced and the process is repeated number of times, thus resulting in a very slow continuous process. The acetic acid bacteria grow on the surface of the liquid and a gelatinous mat known as the 'mother of vinegar' is formed. This contains a large number of bacteria. Air is admitted through holes just above the surface of the liquid. The costly 'slow process' was used extensively for a long time and was eventually replaced by the 'quick process.'

2.3.2.2.2 The 'quick' process: The 'quick process', also known as the 'German process', is the basis for modern methods of manufacture of vinegar. The generator consists of a wooden or metal-coated tank packed with wood shavings on which cells are allowed to grow. The feed trickles from the top through the wood shavings. A large volume of air is sprayed into the tank through perforations in the bottom. Employing 12% (v/v) alcohol, 98% conversion into acetic acid is attained in 5 days by this process [56].

2.3.2.2.3 Frings process: Major improvements in the quick process took place in 1929 with the emergence of trickling generator, forced aeration and temperature control. The significant advantages of this process include the low cost, relatively simple process, and higher concentration of acetic acid. The process is easy to control, and evaporation losses are low.

2.3.2.2.4 Submerged fermentation: The application of submerged fermentation to the oxidation of ethanol to acetic acid was also the significant technical advancement in the commercial production of vinegar. It has been seen that *Acetobacter* species in submerged conditions were very sensitive to oxygen deficiency, and the level of oxygen in the gas phase was less than 5%, resulting in the stoppage of fermentation [57]. The success of this process depends largely on the efficiency of the aeration of broth. During submerged exponential fermentation, the acetic acid bacteria have an average oxygen uptake rate of 7.75 oxygen/gm cell/h. Submerged cultivation have several advantages over the trickling generator. It permits faster oxidation (80 times) of alcohol with greater efficiency, and yield is 5-8% higher, and more than 90% of the theoretical yield is obtained. Moreover, a smaller reactor volume is needed (about 16% of the trickling generator) to produce an equivalent amount of vinegar. This process can be economical with elimination of clogging by shavings, interruptions, etc. The ratio of productivity to capital investment is much higher in the case of submerged cultivation.

### 2.3.3 Recovery

The physical methods of separation of acetic acid from water include fractional distillation, azeotropic dehydration distillation, solvent extraction, extractive distillation, and carbon adsorption. In extractive distillation counter current, washing of mixed vapors in a distillation column takes place via descending stream of high boiling point liquid, which is a preferential solvent for one of the components. The most economic method for the recovery of acetic acid from a dilute aqueous stream is by extraction with a hydrocarbon followed by distillation [58]. To avoid product inhibition in the reactor, a concurrent extraction process is required which will remove acetic acid continuously to keep its level below the critical point in the fermenter. The extraction methods suggested are ion-exchange resins, solvent extraction and membrane separation. A membrane system has been developed having a hollow fibre membrane device consisting of silicone rubber tubing. This system is resistant to biological attack and does not suffer from mechanical fatigue [59].

### 2.3.4 Applications

Acetic acid has a wide range of applications [60]. The major use of acetic acid is in the production of vinyl acetate monomer. This application consumes approximately 40 to 45% of the world's production of acetic acid. Vinyl acetate can be polymerized to polyvinyl acetate or to other polymers, which are applied in paints and adhesives. The major esters of acetic acid are commonly used solvents for inks, paints and coatings. They are typically produced by catalyzed reaction from acetic acid and the corresponding alcohol. The condensation product of two molecules of acetic acid is acetic anhydride. Acetic anhydride is a strong acetylation agent. As such, its major application is for cellulose acetate, a synthetic textile and it is also used for photographic films. Acetic anhydride is also a reagent for the production of aspirin, heroin and other compounds. In the form of vinegar, acetic acid solutions are used directly as a condiment, and also in the pickling of vegetables and other foodstuffs.

## 2.4 Gluconic Acid

Gluconic acid (pentahydroxycaproic acid) is an oxidation product of D-glucose having a specific rotation $\alpha_D^{20}$ = -6.7° by simple dehydrogenation reaction catalyzed by glucose oxidase [61]. Oxidation of the aldehyde group on the C-1 of β-D-glucose to a carboxyl group results in the production of glucono-δ-lactone ($C_6H_{10}O_6$) and hydrogen peroxide. Glucono-δ-lactone is further hydrolyzed to gluconic acid either spontaneously or by lactone hydrolyzing enzyme, while hydrogen peroxide is decomposed to water and oxygen by peroxidase.

The individual products may be separated as crystalline solids from a gluconic acid solution by obtaining supersaturated solutions at defined temperatures and seeding them with crystals of the desired compound [62].

### 2.4.1 Microorganisms

Hlasiwetz and Habermann discovered gluconic acid in 1870 [63]. In 1922, Molliard showed that gluconic acid was produced by a strain of *Aspergillus niger.* Other microorganisms include bacteria such as *Pseudomonas ovalis* [64], *Acetobacter methanolicus* [65], *Zymomonas mobilis* [66], *Acetobacter diazotrophicus* [67], *Gluconobacter oxydans* [68], *Gluconobacter suboxydans* [69], *Azospirillum brasiliense* [70], fungi such as *Penicillium funiculosum* [71] and yeasts such as *Aureobasidium pullulans* [72].

### 2.4.2 Method of production

Different approaches are available for the production of gluconic acid, namely, chemical, electrochemical, biochemical and bioelectrochemical [73]. Although the conversion is a simple one-step process, the chemical method is not favoured. However, the fermentation method is fully competitive with chemical techniques, and is at present the method of choice for manufacturing gluconic acid. Among various microbial fermentation processes, the method utilizing the fungus *A. niger* is one of the most widely used ones.

*A. niger* produces all the enzymes required for the conversion of glucose into gluconic acid, which include glucose oxidase, catalase, lactonase and mutarotase. During the process of glucose conversion, glucose oxidase present in *A. niger* undergoes self-reduction by the removal of two hydrogens. The reduced form of the enzyme is further oxidized by the molecular oxygen, which results in the formation of hydrogen peroxide, a by-product in the reaction. *A. niger* produces catalase which acts on hydrogen peroxide releasing water and oxygen. Hydrolysis of glucono-δ-lactone to gluconic acid is facilitated by lactonase. Removal of lactone from the medium is recommended, as its accumulation in the media has a negative effect on the rate of glucose oxidation and the production of gluconic acid and its salt. There are reports stating that the enzyme gluconolactonase is also present in *A. niger* [74], which increases the rate of conversion of glucono-δ-lactone to gluconic acid. Production of gluconic acid is directly linked with the glucose oxidase activity.

**2.4.2.1 Raw materials:** The carbohydrate source for gluconate production is either glucose crystals [75] or dextrose syrup [76]. However, hydrolysates of various raw materials such as agro-industrial waste have also been used as substrate. Kundu and Das [71] obtained a high yield of gluconic acid in media containing glucose or starch hydrolysate as the sole carbon source. Vassilev *et al.* [77] used corn starch hydrolysate as the fermentable sugar to produce gluconic acid by immobilized *A. niger*. Rao and Panda [78] used Indian cane molasses as a source of glucose. Gluconic acid synthesis was influenced by various metal ions such as copper, zinc, magnesium, calcium, iron, etc. Singh *et al.* [79] observed that grape must and banana pulp resulted in significant levels of gluconic acid production i.e., 63 and 55 g/l respectively. The purification of grape and banana must leads to a 20-21% increase in gluconic acid yield.

**2.4.2.2 Fermentation:** The first fermentation stage is the growth of vegetative inoculum. To this a spore suspension of selected strain of *A. niger* is introduced into the inoculum fermenter. The initial pH of the inoculum medium is set at 6.5 with alkali. After the growth of spores at 30-33°C, the inoculum is transferred to the production fermenter. The correct time for this may be judged by the amount of mycelial growth [76]. During the fermentation, the broth is aerated and agitated. The progress of gluconate production may be monitored by observing the rate of addition of sodium hydroxide solution. The fermentation period can be as short as 19 h [75]. At the end of the fermentation, the broth is filtered to remove the cells and the clarified liquor pumped to the recovery stage. The mycelium removed can be used as a source of glucose oxidase [80].

Commercial production of sodium gluconate uses submerged fermentation with *A. niger* and is based on the modified process developed by Blom *et al.* [75]. It involves fed-batch cultivation with intermittent glucose feedings and the use of sodium hydroxide as a neutralizing agent. pH is held at 6.0-6.5 and the temperature at about 34°C. The productivity of this process is very high, since glucose is converted at a rate of 15 g/l.h. Gluconic acid yields up to 90% in 48-60 h employing submerged culture of *Penicillium luteum* and *A. niger* have also been reported [81].

There are two key parameters which influence the gluconic acid production. These are oxygen availability and pH of the culture medium. Oxygen is one of the key substrates in the oxidation of glucose as glucose oxidase uses molecular oxygen in the bioconversion of glucose. The concentration of oxygen gradient and the volumetric oxygen transfer coefficient are the critical factors, which monitor the availability of oxygen in the medium. These two factors highly influence the rate of transfer of oxygen from gaseous to aqueous phase. Gluconic acid production is an extremely oxygen-consuming process with a high oxygen demand for the bioconversion reaction, which is strongly influenced by the dissolved oxygen concentration. It has been reported that immobilized mycelium of *A. niger* grown using pure oxygen produced high titres of gluconic acid in comparison with mycelium grown in air [82]. Generally, during the course of fungal growth, the distribution of oxygen becomes uneven, as the size of gas bubbles increases, resulting in insufficient oxygen supply [83]. The oxygen absorption rate is also influenced by the viscosity of the culture. A rapid decrease is observed in the absorption rate of oxygen with an increase in mycelial concentration [84].

Studies have shown that *A. niger* produced high yields of gluconic acid when it was neutralized by calcium carbonate, and the yield was found to be highly pH dependent [85]. However, it was found that with *Penicillium* sp., the pH dependence is not as critical when compared to *A. niger*, indicating that there was some correlation between the amount and time-dependent appearance of organic acids, such as gluconic acid, citric acid, oxalic acid, which are formed under different conditions. Therefore, pH is another important parameter that influences the gluconic acid production. *A. niger* produces weak organic acids such as citric acid, gluconic acid and oxalic acid, and their accumulation depends on the pH of the nutritive medium [86]. pH below 3.5 triggers the TCA cycle, which facilitates the citric acid formation. The pH range of the fungi for the production of gluconic acid is around 4.5 to 7.0. The pH 5.5 is generally considered as optimum for *Aspergillus niger* [87].

### 2.4.3 Recovery

The recovery process depends on the method followed for broth neutralization and the nature of carbon sources used. Generally, the downstream process is similar for the fermentation processes using fungal and bacterial species. Gluconic acid, glucono-δ-lactone, calcium gluconate, and sodium gluconate are some of the important products and their extraction process is briefly mentioned below. For the recovery of free gluconic acid from calcium gluconate, the broth is clarified, decolourized, concentrated and exposed to -10°C in the presence or absence of alcohol. Thus the calcium salt of gluconic acid crystallizes, and then it is recovered and further purified. Gluconic acid can also be obtained by precipitating the calcium gluconate from hypersaturated solutions in the cold and released subsequently by adding sulphuric acid stoichiometrically, removing the calcium as calcium sulphate.

For obtaining calcium gluconate as a product, calcium hydroxide or calcium carbonate is used as the neutralizing agent. They are added to the nutritive broth accompanied by heating and vigorous stirring. The broth is concentrated to a hot supersaturated solution of calcium gluconate, followed by cooling at 20°C, and adding water miscible solvents, which crystallizes the compound. A treatment with activated carbon facilitates the crystallization process. Finally they are centrifuged, washed several times and dried at 80°C.

Sodium gluconate is prepared using ion exchange resins. In a process, sodium gluconate from the filtered fermented broth is concentrated to 45% (mass per volume), followed by the addition of sodium hydroxide solution, thus raising the pH to 7.5, and then drying by drum dryers [75]. Refined product can be obtained by carbon treatment of the hot solution before drying process.

### 2.4.4 Applications

Gluconic acid is a non-corrosive, non-volatile, non-toxic, mild organic acid which finds applications in the food industry. It imparts a refreshing sour taste in many food items such as wine, fruit juices, etc. Glucono-δ-lactone imparts an initially sweet taste, which later becomes slightly acidic and is extensively used as a latent acid in baking powders for use in dry cake mixes. It is used as a flavouring agent and also finds application in reducing fat absorption in doughnuts and cones. Foodstuffs containing D-glucono-δ-lactone include bean curd, yoghurt, cottage cheese, bread, confectionery and meat. Although many metallic salts of gluconic acid have been studied [88], commercially the most important are sodium, calcium and iron (II) gluconates.

The sodium salt finds a major use in the sequestering of calcium and iron. The sequestering action for calcium is rather poor, even up to pH 14, but in the presence of free sodium hydroxide, it becomes outstanding [61]. It is used in the bottle washing preparations, where it helps in the prevention of scale formation and its removal from glass. If sodium gluconate is added to the alkali at the rate of about one part gluconate to ten parts NaOH, the alkaline earth metals are sequestered and the problem does not arise [89, 90]. It is a good chelator at alkaline pH; its action is comparatively better than EDTA and other chelators. Its property of sequestering iron over a wide range of pH [61] is exploited in the textile industry, where it prevents the deposition of iron and for desizing polyester and polyamide fabrics.

Calcium gluconate is used in pharmaceutical industry as a source of calcium for treating calcium deficiency by oral or intravenous administration. It also finds a place in animal nutrition. Iron gluconate and iron phosphogluconate are used in iron therapy.

## 2.5 Propionic Acid

The propionic acid is a naturally occurring carboxylic acid with chemical formula $CH_3CH_2COOH$. In the pure state, it is a colourless, corrosive liquid with a sharp, somewhat unpleasant odour. Propionic acid is used in the manufacture of herbicides, chemical intermediates, artificial fruit flavours, pharmaceuticals, cellulose acetate propionate, and preservatives for food, animal feed, and grain. Its present US market is estimated at 235 million pounds per year [91] and is growing at 4% annually.

### *2.5.1 Microorganisms*

Propionic acid bacteria have long been used in the dairy industry. These bacteria play important roles in the development of characteristic flavour and eye production in Swiss type cheeses. *Propionibacterium* are Gram positive, non-spore forming, rod-shaped, facultative anaerobes, but are slow growing species. This can be accelerated by using mixed inoculum of a lactic acid producing organism such as *Lactobacillus casei* [92].

### *2.5.2 Method of production*

**2.5.2.1 Raw materials:** Several carbon sources have been used for this fermentation such as glucose [93], xylose [94], maltose [95], sucrose [96], and whey lactose [97].

**2.5.2.2 Fermentation:** Commercial production is mainly carried out by chemical synthesis from petroleum feedstocks [98], but fermentation is an attractive alternative route to produce this acid from renewable resources. Like most organic acid fermentations, the propionic acid fermentation is inhibited by acidic pH and the major fermentation product is propionic acid [99]. The conventional fermentation technology for propionate production is thus limited by low fermentation rate and low product concentration. Furthermore, the fermentation is heterogeneous i.e., propionate is produced along with other by-products. This not only results in a low product yield, but also renders product purification difficult and expensive. Consequently, the conventional fermentation route for propionic acid production is inefficient and it competes with difficulty with petrochemical routes. Presently, only small amounts of propionate are produced by fermentation of whey and are used as a natural product in foods for the labelling purpose. In order to make the fermentation route economically viable, it is necessary to develop novel fermentation processes that use highly efficient bioreactors and separations techniques. Integrated fermentation-separation systems have been successfully used to reduce end-product inhibition and thus, to improve the overall process efficiency [100].

An extractive fermentation process was developed to produce propionate from lactose. The bacterium *Propionibacterium acidipropionici* was immobilized in a spiral wound, fibrous matrix packed in the reactor. Propionic acid is the major product from lactose fermentation, with acetic acid and carbon dioxide as by-products. Propionic acid is a strong inhibitor to this fermentation [93]. Despite the great variety of substrates studied for propionic acid production, glycerol has rarely been considered [101]. However, a better efficiency of propionic acid production from glycerol could be expected because of its higher reduction level compared to conventional substrates. Effectively, a propionic acid yield of 0.84 mol/mol, a final concentration in propionic acid of 42 g/l and a low acetic acid production have been obtained from glycerol with *P. acidipropionici* [101]. Similar data have also been reported with *P. thoeni* [102]. The kinetics of glycerol and glucose fermentation and product formation for *P. acidipropionici*, and *P. freudenreichii* have been compared [103]. In all cases, fermentation end-products were the same and consisted of propionic acid as the major product, acetic acid as the main by-product and two minor metabolites, n-propanol and succinic acid. Facts were provided that greater production of propionic acid was obtained with glycerol as carbon and energy sources. *P. acidipropionici* showed higher efficiency in glycerol conversion to propionic acid with a faster substrate consumption. The almost exclusive production of propionic acid from glycerol by this bacterium suggested a homo propionic tendency of this fermentation.

### *2.5.3 Recovery*

Economics of the process depends on the development of an effective recovery method for the organic acids from the broth. The recovery technique should provide high selectivity and yield. Solvent extraction processes as a promising recovery technique have been proposed as an alternative to the conventional precipitation process, and many solvents have been tried to improve such recovery. The effect of the mixed solvents to separate propionic acid from aqueous solution was investigated [104]. Propionic acid extraction from the aqueous phase is important in view of its number of uses. A number of techniques like distillation, membrane dialysis, electrodialysis, and reactive extraction are available. Membrane-based extraction methods have effectively removed acids from fermentation broth and have maintained acids at subinhibitory levels in the fermenter, so that fermentation times can be extended. The profound success of reactive extraction has been noted in the recovery of carboxylic acids from dilute solutions. Recently, equilibrium parameters (distribution coefficients, loading ratio, degree of extraction, etc.) with propionic acid using tri-n-butyl phosphate, an organophosphorous compound, in eight different diluents have been studied [105]. The data obtained can be useful in understanding the equilibrium characteristics and efficient design of the recovery process of propionic acid by reactive extraction.

### *2.5.4 Applications*

Propionic acid inhibits the growth of moulds and some bacteria. As a result, most propionic acid produced is used as a preservative for both animal feed and food for human consumption, and can be used as a preservative for Ballistics Gel. For animal feed, it is used either directly or as its ammonium salt. In human foods, especially bread and other baked goods, it is used as its sodium or calcium salt. Similar usage occurs in some of the older anti-fungal foot powders. Propionic acid is also useful as a chemical intermediate. It can be used to modify synthetic cellulose fibres. It is also used to make pesticides and pharmaceuticals. The esters of propionic acid are sometimes used as solvents or artificial flavourings.

## 3. PRODUCTION OF GUMS

Microbial exopolysaccharides produced by a variety of microorganisms (Table 4) have attracted worldwide attention due to their novel and unique physical properties. These exopolysaccharides have diverse range of applications in industries, especially in food and pharmaceutical industries, as emulsifiers, stabilizers, binders, gelling agents, lubricants, film formers, and thickening and suspending agents (Table 5). These are rapidly emerging as new and industrially important source of polymeric materials, which are gradually becoming economically competitive. Some microbial polysaccharides are commercially accepted, while others are at various stages of development. Currently a small number of biopolymers are produced commercially on a large scale [106]. Among the biopolymers which have been the subject of extensive studies are xanthan, pullulan, alginates and gellan.

## 3.1 Classification of Microbial Polysaccharides

Microbial polysaccharides are formed from the condensation of monosaccharides by removal of water. These can be classified [107] into:

### *3.1.1 Homopolysaccharides*

These contain only one type of carbohydrate moiety, and their synthesis involves a single or simple enzyme system. For example pullulan, dextan.

### *3.1.2 Heteropolysaccharides*

These contain two or more than two carbohydrate moieties. A complex enzyme is involved in their synthesis. Xanthan is the most popular example of this polysaccharide.

Polysaccharides can also be classified on the basis of their properties. They can be either anionic like xanthan and alginate; neutral like pullulan, levan; or cationic. Polysaccharides also differ on the basis of structural arrangements of their monomeric units.

## 3.2 Xanthan

Xanthan is an extracellular heteropolysaccharide, which is produced by the aerobic fermentation of *Xanthomonas campestris* commonly from glucose or sucrose. It is the most important microbial polysaccharide from the commercial viewpoint, with a worldwide production of about 30,000 tonnes per year, corresponding to a market of $408 million [106, 108]. This polymer is employed in many industrial processes because of its unique rheological behaviour. Solutions of xanthan are highly pseudoplastic and show very good suspending properties.

### *3.2.1 Structure and biosynthesis*

The primary structure of xanthan is a cellulose-like main chain with trisaccharide side chains composed of D-mannose/D-glucuronic acid/D-mannose linked to alternate glucose residues (Fig. 3). Side chains of xanthan usually carry an O-acetyl group ester linked attached to the internal mannose, while terminal mannose may contain a ketal-linked pyruvate group [109, 110]. The composition of xanthan

**Table 4:** Different substrates and microorganisms used in the production of exopolysaccharides.

| Exopolysaccharide | Common Substrate | Microorganism (s) |
|---|---|---|
| Alginate | Sucrose | *Azotobacter vinelandii* |
| Curdlan | Glucose | *Alcaligenes faecalis* |
| Dextran | Sucrose | *Leuconostoc mesenteroides*<br>*Leuconostoc dextranicum* |
| Gellan | Glucose, fructose, maltose, sucrose | *Pseudomonas elodea*<br>*Sphingomonas paucimobilis* |
| Levan | Lactose<br>Sucrose | *Alcaligenes viscosus*<br>*Zymomonas mobilis* |
| Pullulan | Glucose, sucrose | *Aureobasidium pullulans*<br>*Cyttaria harioti*<br>*Cyttaria darwinii*<br>*Tremella mesenterica*<br>*Rhodotorula bacarum* |
| Scleroglucan | Glucose, sucrose | *Sclerotium glucanium* |
| Xanthan | Glucose, sucrose | *Xanthomonas campestris* |

**Table 5:** Applications of exopolysaccharides.

| Property | Applications |
|---|---|
| *Physical properties* | |
| Emulsion stabilization | Foods, thixotropic paints |
| Fibre strength | Acoustic membranes |
| Film formation | Food coatings |
| Flocculant | Water clarification, ore extraction |
| Foam stabilization | Beer, fire-fighting fluids |
| Gelling agents | Foods, cell and enzyme technology, oil recovery |
| Hydrating agent | Cosmetics, pharmaceuticals |
| Inhibitor of crystal formation | Frozen foods, sugar syrups |
| Shear thinning and viscosity control | Oil-drilling 'muds', jet printing |
| Suspending agent | Food, paper coatings, agrochemical pesticides and sprays |
| *Chemical properties* | Enzyme substrates |
| | Oligosaccharide preparation |
| *Biological properties* | Antitumour agents |
| | Eye and joint surgery |
| | Heparin analogues |
| | Wound dressings |

*Source:* 106

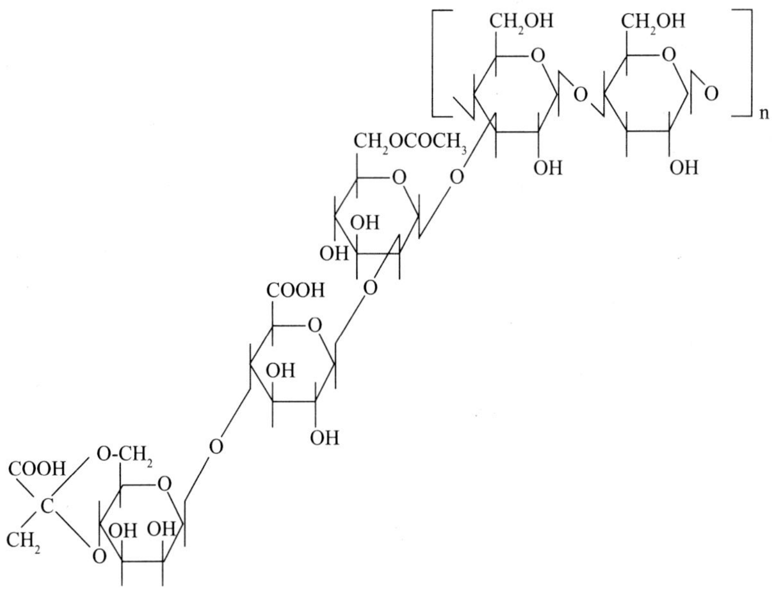


**Fig. 3:** Structural unit of xanthan gum.
*Source*: 110

is affected by several factors such as strain, batch, culture media and downstream processing. The changes in composition (mainly the extent of acetylation and pyruvylation) affect the properties of xanthan solutions [111], and a wide range of different behaviour may be obtained.

The biosynthesis of microbial heteropolysaccharides such as xanthan is a complicated process involving a multienzyme system. The initial step in the biosynthesis of xanthan is the uptake of carbohydrate, which may occur by active transport or facilitated diffusion. This is followed by phosphorylation of the substrate with a hexokinase enzyme that utilizes adenosine 5'-triphosphate. The biosynthesis involves conversion of the phosphorylated substrate to the various sugar nucleotides required for assembly of the polysaccharide-repeating unit via enzyme such as UDP-Glc pyrophosphorylase. UDP-glucose, GDP-mannose and UDP-glucuronic acid are necessary for the synthesis of xanthan with the appropriate repeating unit [112]. In *X. campestris*, the Entner-Doudoroff pathway (Fig. 4) in conjunction with the tricarboxylic acid cycle pathway is the predominant mechanism for glucose catabolism [113, 114]. A small portion of glucose is routed via the pentose phosphate pathway.

### 3.2.2 Production

Xanthan gum is a water-soluble polysaccharide produced industrially from sucrose or glucose by fermentation using the Gram negative bacterium *X. campestris*. The organism is usually having yellow, smooth and butyrous or viscid colonies [115]. Cells are aerobic, straight rods (usually 0.4-0.7 wide × 0.7-1.8 μm long) with a single polar flagellum. *Xanthomonas* can be cultured at different temperatures ranging from 25 to 35°C in neutral pH [116].

Different carbohydrates have been used as an alternative to glucose for xanthan production including molasses, hydrolyzed cereal grains, and hydrolyzed acid whey. The inorganic materials like ammonium or nitrate salts may be used as nitrogen source. Further, the addition of phosphorus is necessary and is usually added in the form of phosphate buffer.

In addition, magnesium as $MgSO_4$ and various trace elements are also needed in the medium. Different fermentation systems like batch [117] or continuous [118] using various types of bioreactors such as stirred tank [119], bubble column [120], airlift reactor [121] have been used for xanthan gum production. The production and properties of xanthan gum are influenced by different process conditions like bacterial strain, substrate, temperature, pH and agitation, etc. [122, 123].

Amongst the various process parameters, temperature affects xanthan production significantly, and the optimum temperature for obtaining maximum yield of xanthan is 30-33°C. However, commercial fermentations are usually carried out at about 28°C [124]. The pH of the medium is maintained at 7 using buffer or addition of base during the process. The pH of the fermentation medium decreases with time, therefore control of pH is also very important. Other parameters like aeration and agitation also play an important role during fermentation. The possibility of xanthan production using solid state fermentation (SSF) has also been explored [125, 126]. *X. campestris* strains were cultivated on a great variety of solid substrates or by products such as spent malt grains, apple pomace, grape pomace and citrus peels in order to evaluate their ability to produce the exopolysaccharide, xanthan. With most of the substrates, xanthan yields were comparable to those obtained from conventional submerged cultivation [3].

### 3.2.3 Recovery

The recovery step is very important part of microbial polysaccharide production. The cost of recovery including concentration, isolation and purification is a significant part of the total production cost, which needs to be minimized. Different methods have been employed for the recovery of polysaccharides (Table 6). The fermentation broth of xanthan gum is apparently very viscous. The main steps in most commercial recovery processes involve deactivation and removal of microbial cells, precipitation, dewatering, drying and milling. The separation of cells from viscous broth by centrifugation, filtration or flocculations is very expensive. In the laboratory method, the fermentation broth is diluted with ethanol (33%) to 100 cP (Centipoise) viscosity and centrifuged to remove the cells, and KCl is added to the broth. Then, chilled ethanol (95%, 3 volumes) is added to precipitate the gum [112]. Solvent precipitation, precipitation using divalent metal ion salts of the polymer at alkaline pH,

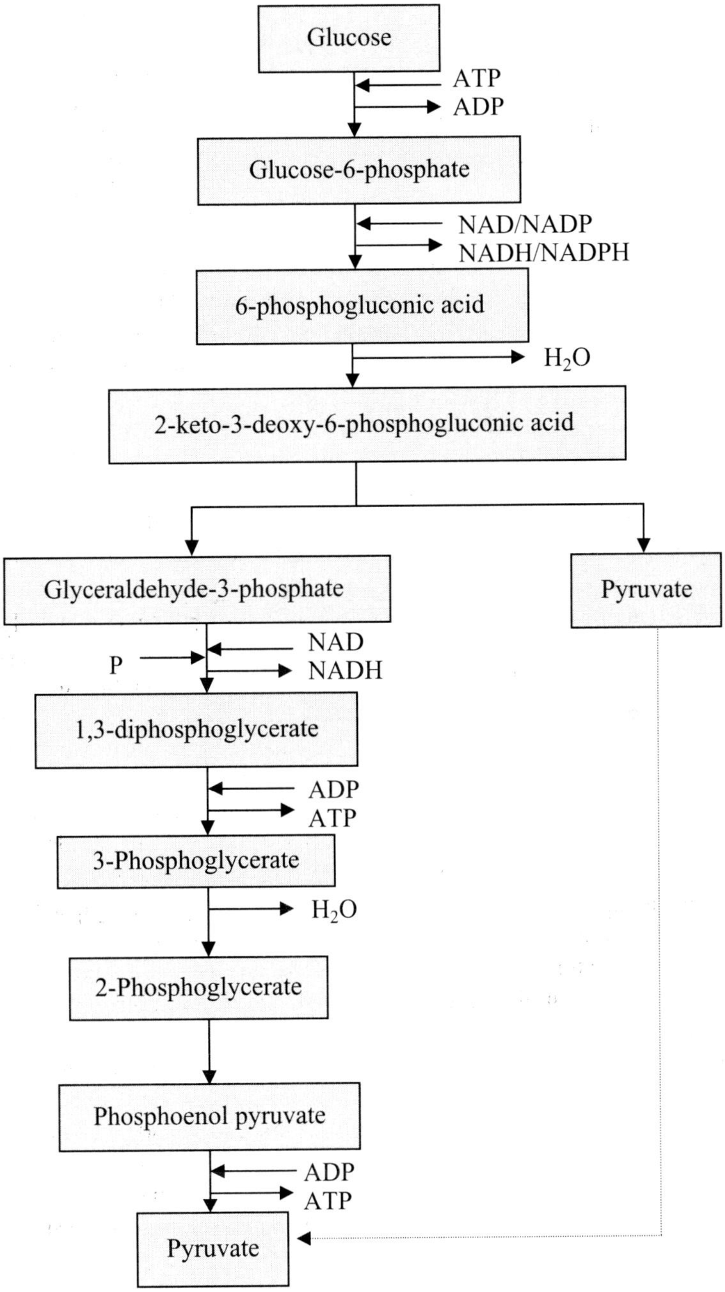


**Fig. 4:** Entner-Doudoroff pathway.
*Source*: **113**

**Table 6:** Different methods used for recovery of microbial polysaccharides.

| Method | Chemical used | Comments |
|---|---|---|
| Acid precipitation | Hydrochloric acid | Used in some commercial processes for extraction of algal alginate |
| Alcohol precipitation | Isopropanol | Precipitation conditions and washing can affect subsequent processing and final product |
| Salt precipitation | Calcium hydroxide | Insoluble salt of alginate or xanthan formed; required acid titration to form soluble salt of polymers |
| | Quaternary ammonium compounds | Precipitate washed with methanol to give soluble salt |

precipitation as a quaternary ammonium complex at acid pH, or precipitation as fatty amine complex at acid pH have also been successfully attempted. After precipitation, xanthan still contains large amount of liquid, and hence centrifugation or pressing may be used to remove the excess liquid. Final drying of xanthan precipitates may be obtained by using batch or continuous driers [127]. Whole broth containing xanthan can also be spray or drum dried.

### *3.2.4 Applications*

The excellent rheological properties of xanthan gum contribute to its wide-range of applications in the food industry as a suspending, stabilizing, and/or thickening agent and its use as an emulsifier, lubricant, thickening agent, and/or mobility control agent to enhance oil recovery [128, 129]. The approval for the use of xanthan in food industry was given by United States Food & Drug Administration (FDA) in 1969 [130]. This polysaccharide is also on FDA list of GRAS (Generally regarded as safe) substances [130]. The major applications of xanthan gum are in food industry as a suspending and thickening agent for fruit pulp and chocolates. Many of today's foods require the unique texturization, viscosity, flavour release, appearance and water-control properties. Xanthan gum improves all these properties and additionally controls the rheology of the final food product. Xanthan provides favourable emulsion, suspension, stability and viscosity, improves freeze-thaw stability of starch thickened products. In many foodstuffs, xanthan provides useful attributes, including rapid flavour release, good mouthfeel and compatibility with other food ingredients such as proteins, lipids and polysaccharides [106]. Xanthan has wide applications in the chemical industry. Xanthan acts as a stabilizer, emulsifier for thixotropic paints, compatible with other thickeners and water-based emulsion inks, prevents sagging and allows easy and even application. It is a good suspending agent for dyes, pigments, controlling agent for printing. A mixture of xanthan and locust bean gum can be used in deodorant gels. A special gel of xanthan can form in the presence of borate, and this combination has been used for the preparation of explosives [112].

The rheological properties of xanthan are particularly suitable for the 'Enhanced Oil Recovery' (EOR) application [131, 129]. At low concentration, the gum forms high viscosity solution that displays pseudoplasticity. The oil is held in the tiny pores of the small sand stone rocks. For the efficient displacement of oil, the pumping of xanthan gum solution in the rocks is necessary. As a result, oil held in the pores of the sand stone rocks is displaced. Further, as xanthan can withstand extreme conditions of reservoirs for many years together e.g., high temperature, salinity, pressure, etc., it has been found to be an ideal candidate for use in EOR.

## 3.3 Pullulan

Pullulan is a water-soluble glucan composed of linear α-(1→6) linked maltotriose units and a small number of α-(1→4) linked maltotriose units [132]. The regular alternation of α-(1→4) and α-(1→6) bonds results in two distinctive properties, structural flexibility and enhanced solubility [133]. Pullulan is one of the few neutral water-soluble microbial polysaccharides, which can be produced in large quantities by fermentation [134].

### 3.3.1 Production

Pullulan is produced by strains of *Aureobasidium pullulans*. Two varieties of this species i.e., var. *pullulans* and var. *aubasidani* exists which are distinguished by molecular characteristics, nutritional assimilation patterns and structure of the exopolysaccharide produced by them [135]. Exopolysaccharides produced by these two varieties are pullulan and aubasidan, respectively. Aubasidan is a highly branched polysaccharide comprising main core of β-1,3-linked D-glucopyronosyl residues, to which side chains of α-1,4-D-glucosyl residues are attached through α-1,6-glucosidic bonds [136]. Recently, melanin free pullulan production by *Rhodotorula bacarum* has also been reported [137]. This polymer is produced commercially by Hayashibara Corp in Japan with the use of *Azotobacter pullulans*. Apart from *A. pullulans*, a number of other microorganisms are also known for pullulan production. Pullulan has also been isolated from the saprophytic (sometimes mycoparasitic) fungus *Tremella mesenterica* [138], obligate tree parasitic fungi *Cyttaria harioti* and *C. darwinii* [139, 140] and the fungal agent of chestnut blight, *Cryphonectria parasitica* [141].

Different substrates such as glucose, sucrose, starch syrup, etc. have been used for the production of pullulan using different strains of *A. pullulans* [142, 143, 144]. The agro-industrial wastes like potato starch waste, peat hydrolyzate, whey, molasses, brewery wastes and olive oil waste effluents as substrates for pullulan production have also been explored by many researchers [145-147]. The use of these substrates would definitely be environmental friendly and economically advantageous. However, the pullulan produced from different substrates may vary in purity, which is more pronounced when agro-industrial wastes are used as a carbon source for fermentation. Depending upon the substrate used and fermentation conditions employed, the yield and molecular weight of pullulan may vary widely. Different process parameters such as temperature, initial pH of the growth medium, oxygen supply to the fungi, nitrogen concentration, source of carbon, nature of microorganism, as well as the content of certain trace elements influence the production of pullulan [148].

The bioreactor design and nitrogen sources affect pullulan production by *A. pullulans*, but its morphology has no affect on the yield [149]. Ammonium and complex nitrogen sources have been reported to be superior to nitrate for pullulan production from *A. pullulans* [150, 151]. Yeast extract has been found as a useful nitrogen source in increasing the production of pullulan using *A. pullulans*. The supplementation of soybean pomace, which is an agro-industrial by-product from the soy sauce industry, has also shown promising results as nitrogen source [152]. The soybean pomace was better than yeast extract in terms of productivity as well as economics. It can be a good substitute for yeast extract as a nitrogen source for the production of pullulan. Further, it can reduce the cost of production, increase productivity, and solve environmental problems that result from the disposal of soybean pomace. The other process parameters also effect the pullulan production. It has been reported that pH 4.5 is optimal for pullulan production, though yeast-like growth is favoured at pH 6.5 [153]. However, optimal yield of pullulan at an initial pH of 6.0 using *A. pullulans* has also been reported [154]. The optimal temperature for pullulan production varies slightly in the range of 24-30°C depending on the specific strain [155-157].

### 3.3.2 Recovery

The pure biopolymer from the fermentation broth is obtained through downstream processing. This process consists harvesting of cell from culture broth, removal of melanin pigments produced during fermentation, and precipitation of the polymer with a suitable solvent. Melanin is one of the major hindrances in pullulan production. Recovery and purification is achieved with one precipitation step in strains producing melanin free pullulan using a suitable organic solvent. The organic solvent usually used is from the group of alcohols, esters and ethers with three or more carbon atoms. The solvents of relatively higher molecular weight and slightly lower hydrophilicity are more suitable for such processes in comparison with lower molecular weight and higher hydrophilicity solvents. Further purification of pullulan can be completed by ultrafiltration and ion-exchange resins [158].

#### *3.3.3 Applications*

Pullulan and its derivatives have many potential food, pharmaceutical, clinical, and health care applications [157, 159]. The unique physical property of pullulan is its low oxygen permeability when cross linked with other materials to form thin polymeric films. Thus, films formed from pullulan are suitable for coating foods and pharmaceuticals, especially where exclusion of oxygen is desirable. Other typical industrial uses of pullulan are as an ingredient of low calorie foods and as a starch substitute, as an adhesive in the form of pastes with water, as a construction material (after esterification) with fibres similar in strength and elasticity to those in nylon, and as a bulking agent and stabilizer for tablets in the pharmaceutical industry [146]. It has been reported that dietary pullulan functions as prebiotic, promoting the growth of beneficial bifidobacteria [160, 161]. Pullulan may be incorporated in food to replace starch, imparting the characteristics to food such as consistency, dispersibility, moisture retention, etc. It can also be used as a partial replacement for starch in pastas or baked goods. Pullulan improves the shelf-life of the food as it is not a readily degradable carbon source for bacteria, moulds and fungi responsible for spoilage of food. Pullulan is also superior to starch in water retention, thus retarding the spoilage of food [162, 163]. Adhesive properties are also exhibited by pullulan and its derivatives [163]. It can be used as a binder and stabilizer in food pastes; it can also be used to adhere nuts to cookies. It can be employed as a binder for tobacco [164], seed coatings and plant fertilizers. Pullulan can be used as low-viscosity filler in beverages and sauces.

Pullulan can be used in pharmaceutical coatings, including sustained-release formulations [165, 166]. Novel preparations such as tablets, pills, granules, which contain pullulan in the sugar layer, serve the purpose of preventing brownish colour change of the composition. Oral care products based on pullulan films have been commercialized [167]. In addition, pullulan and its derivatives have photographic, lithographic, and electronic applications [168-170]. Pullulan substituted with cholesterol or fatty acids can be used to stabilize fatty emulsions [171]. Pullulan derivatives are promising as non-toxic conjugates for vaccines [172] and interferon [173], and can facilitate liposome delivery [174].

### 3.4 Dextran

Dextran is a D-glucose polymer composed of 95% (1→6) linkages in the linear chain. Commercially, dextran is usually produced using *Leuconostoc mesenteroides* and *Leuconostoc dextranicum* [175]. Other bacteria like *Aerobacter* sp., *Streptococcus bovis* and *S. viridens* also produce this polymer.

#### *3.4.1 Production*

Dextran fermentation by *Leuconostoc mesenteroides* has been studied under both anaerobic and aerobic conditions. The maximum dextran yield was obtained as a result of higher yield, stability, and activity of dextransucrase when the oxygen transfer rate in the bioreactor was equal to the maximum oxygen uptake rate of the organism, and the pH of the culture medium was 5.5 [176]. The rheological behaviour of the broth depends on the type of microorganism and the process conditions. Different results were obtained for different strains and process conditions during the rheological behaviour studies of cultures of *Leuconostoc mesenteroides* rich in dextran [177-179].

*Leuconostoc mesenteroides* grown in a sucrose-rich media releases an enzyme, dextransucrase, which converts excess sucrose to dextran and fructose [180]. High sucrose concentration in media can result in the increase in viscosity and process control becomes more difficult. Acceptor molecules (such as maltose present in the culture media) can influence dextran molecular weight by allowing the growing chain to be separated from the enzyme active site and transferred to the acceptor [181, 182]. A decrease in dextran molecular weight was seen together with less viscous and more Newtonian solutions at lower temperatures [183]. However, dextran produced at lower temperatures was less branched and as a result more viscous [177]. The polymer is further solvent precipitated and subsequently degraded with the help of enzymes, by acid hydrolysis or by heat to generate products with lower molecular weight.

#### *3.4.2 Recovery*

After completion of fermentation, dextran can be recovered using precipitation with solvents. In the first step, equal amount of ethanol is added, and the mixture is centrifuged. The supernatant is decanted. Then again, chilled ethanol is added with constant stirring and precipitates of dextran appear. This process can be repeated till the fine form of dextran appears. The precipitated dextran can be dried under vacuum over calcium chloride at 30°C.

#### *3.4.3 Applications*

Commercial applications for dextran are generally in the pharmaceutical industry, but new applications are being explored in the food and textile industries [152, 184]. The commercial applications of dextrans include their use as plasma substitutes. It may also be used as a hydrophilic layer on surface of burns to absorb fluid exudates. It is also used as polyelectrolyte after sulphonation. Cross-linked dextran known as sephadex is widely used for separation and purification of various products like protein in research and industry. It finds various other applications as adjuvant, emulsifier, carrier and stabilizer in food, pharmaceutical and chemical industries. In food industry, it is being used as a thickener for jam and ice cream. It prevents crystallization of sugar, improves moisture retention, maintains flavour and appearance of various foodstuffs.

### 3.5 Alginate

Alginate is a general term used to describe the salts of alginic acid, the most important of which is sodium alginate. It is a linear block copolymer composed of two hexuronic acid monosaccharide residues, D-mannuronic acid and L-guluronic acid, and the distribution of the two monosaccharides along the chain is non-random but involves relatively long sequences containing only one hexuronic acid [185, 186]. The presence of L-guluronic acid in alginates is important because the increase in its content improves the gelling characteristics of the alginate obtained. In the presence of divalent cations, such as calcium, alginic acid/alginate is able to form gels because of association between the calcium ion and two guluronic acid residues [186]. The cross-linking in alginate gels is therefore ionic, based on calcium bridges, with the degree of cross-linking depending upon the concentration of calcium ions and the number of guluronic acid sequences in the polysaccharide.

#### *3.5.1 Production*

This polymer can be extracted from marine algae, however there is the problem of variability associated with this product. Hence the production of alginate from microbes has been explored. Microbial alginates have physical properties quite similar to those of algal alginates, so they can be used for the same applications as algal alginates. The microbial alginates possess O-acetyl groups attached to D-mannuronic acid, which is absent in algal alginates [187]. In the bacterial world, alginates are produced by *Pseudomonas* and *Azotobacter* species. This biopolymer has a wide range of molecular weights and the extracellular enzyme alginate lyase may play an important role in the molecular weight of alginates. A considerable work has been reported on the alginate production using *Pseudomonas* species. In contrast to *P. aeruginosa, A. vinelandii* is a non-pathogenic soil bacterium, which can be used for the development of biotechnological process to produce alginate. *Azotobacter vinelandii* produces an alginate with a high molecular mass (and therefore high viscosity), when it is cultivated in unbaffled flasks [188]. The culture broth exhibits a high viscosity, which is determined mainly by the alginate molecular mass and its concentration. From a rheological point of view, the bacterial alginate solutions show a pseudoplastic behaviour and therefore the viscosity depends on the shear rate [189]. Specific power input is one of the most important operational parameters in fermentation of aerobic microorganisms, as it may be associated with a number of process parameters [190], such as oxygen transfer rate, carbon dioxide removal rate, mixing degree, dispersion of a second organic liquid phase, and hydromechanical stress. It has been observed that keeping equal the initial power input (0.27 $kWm^{-3}$) in shake-flasks and a stirred fermentor did not reproduce the behaviour of alginate production by *Azotobacter vinelandii* [191]. A lower mean molecular mass ($1.1\times10^6$ Da) of the polymer was obtained in stirred-fermentor as compared to the alginate obtained in shake-flasks ($1.9\times10^6$ Da), which may be due to the fact that the evolution of power input in shake-flasks can be considerably different to that observed in the stirred bioreactor.

#### 3.5.2 Recovery

The objectives of the recovery processing are cell removal, removal of other solid matter, concentration and purification of alginate from fermentation broth. The high viscosity of a polysaccharide fermentation broth results in conventional cell separation steps such as centrifugation, filtration and flocculation being expensive, and to give practical rates of processing, it is often necessary to dilute the broth. The isolation of a polysaccharide is achieved by lowering the solubility of polymer through solvent (chilled ethanol) precipitation [192]. The precipitates are recovered by centrifugation and dissolved in distilled water.

#### 3.5.3 Applications

Alginates are important family of biopolymers, of both technological and scientific interest and have a wide range of applications as stabilizing, thickening, emulsifying, gelling, and gel-or-film-forming agents in food, pharmaceutical, textile and paper industries [193]. These are mainly used in food industry, for example, in ice-creams, frozen custards, as well as cream, cake mixtures and in beer manufacture to enhance the foam and fruit drinks to assist the suspension of fruit pulp, which makes the product more appealing to the consumer [194]. Alginate (sodium alginate) is widely used for immobilization of microbes, plant and mammalian cells [195]. New applications for these polymers are also being explored. Alginates have historically been known to have a haemostatic function and are capable of absorbing specific solutes, therefore it is logical that they should be evaluated as a suitable components for modern wound management aids [196]. The calcium alginate gels have a large pore size and high water absorbency, making them potentially useful as wound management aids, of the 'moist' dressing type. The alginate gels are thermally stable, deodorize the wounds and absorb pain-stimulating compounds. Large quantities of alginate dressings are used each year to treat exuding wounds, such as leg ulcers, pressure sores and infected surgical wounds [197-199]. The degree of swelling is determined principally by the chemical composition of the alginate, which depends on its botanical source.

### 3.6 Gellan

Gellan is one of the industrially useful exopolysaccharides, which is a sphingan group of heteropolysaccharides secreted by members of the bacterial genus *Sphingomonas* [200]. This polysaccharide is composed of glucose, rhamnose and gluconic acid and has attached O-acetyl groups. Gellan gum is the generic name for extracellular polysaccharide produced by the bacterium *Pseudomonas elodea.* It is currently produced by C.P. Kelco in Japan and the USA.

### *3.6.1 Production*

Gellan gum can be produced by fermentation using *Sphingomonas paucimobilis,* and *Pseudomonas elodea.* The industrial production of gellan gum is carried using the bacterium *Sphingomonas paucimobilis* ATCC 31461 [201]. Different new strains such as *Sphingomonas paucimobilis* ATCC 31461, *Sphingomonas paucimobilis* E2 (DSM 6314), *Sphingomonas paucimobilis* NK2000 and *Sphingomonas paucimobilis* GS1 have been isolated for gellan gum production, but their use for commercial production has not been reported [201-204]. The media used for the production of this polysaccharide is simple media containing carbon source, nitrogen source and inorganic salts. Carbon source is the most important component of the media used for the production of exopolysaccharides, because it directly affects yield, composition, structure, and properties of polysaccharide [205]. Different carbohydrates (such as glucose, fructose, maltose, sucrose and mannitol) can be used either alone or in combination as carbon source [201]. The comparison of soluble starch, glucose, lactose, maltose and sucrose as carbon source for gellan gum production depicted that soluble starch is the best carbon source for gellan gum production [204, 202]. Other low cost substrates such as molasses and whey can also be used as raw material for its production. Studies have been reported to improve gellan production using mutant strains [206] and whey and corn starch as substrates [207, 208] as well as optimization of culture conditions. Nitrogen source is another important medium component for gellan production. Organic nitrogen sources like corn steep liquor [201] and inorganic nitrogen sources like ammonium nitrate and potassium nitrate [203] have been tested for gellan production. The soybean pomace (an agro-industrial by-product) has also been explored as a nitrogen source for gellan gum production from *Sphingomonas paucimobilis* [209].

Different process parameters also affect the gellan production. Most of the studies involving gellan gum production are performed at 30°C [209]. However, it is reported that gellan yield reaches its peak at 20°C, remains quite stable at 25°C, and significantly decreases above 30°C [210]. The pH plays a very important role in the production of gellan by *Sphingomonas paucimobilis*, and the optimal pH value for bacterial exopolysaccharide production is somewhat higher than that of the fungal glucan production [211]. The pH value usually recommended for gellan production ranges from 6.5-7 [201, 202].

### *3.6.2 Recovery*

The recovery of the gellan gum after fermentation is a very crucial process. The fermentation broth is first heated (90-95°C) for 10-15 min [201]. The heating step gently reduces the viscosity of the broth, which facilitates mixing during precipitation. The polysaccharide is separated from the broth by filtration or centrifugation. Then ice-cold isopropyl alcohol is added to cell-free supernatant and the mixture is kept at 4°C for complete precipitation of gellan, and the precipitate formed is recovered by centrifugation. After gellan recovery, the product is dried. Lyophilization of gellan can also be carried out for formulating dry gellan powder [209]. Further purification can be carried using dialysis technique.

### *3.6.3 Applications*

Gellan has various applications in food and pharmaceutical industry [212]. It forms thermo-reversible gels, which are useful in food industry as thickening and binding agents and to maintain water capacity during cooking and storage [213,106]. The major function of gellan in confectionery products is to provide structure and texture to reduce the set time of starch jellies. Gellan gum offers dessert gels with mouth-feel characteristics similar to those of gelatin. The application of high clarity gellan gum is favoured in this application. Gellan can also successfully replace pectin in jams and is effective at lower concentrations. It can be employed to partly replace starches in pie fillings and pudding. Gellan also has applications in dairy industry. The interaction of gellan with milk proteins, especially casein and whey lactoglobulins, increases the total yield of cheese and reduces the loss of solids in whey. GELRITE, a commercial form of gellan is used for solidifying microbial media, especially those used for the isolation of thermophiles [215]. The potential role of gellan in pharmaceutical applications such as controlled drug delivery and ophthalmic solutions has also been reported [216, 217].

## 3.7 Scleroglucan

Scleroglucan is a neutral exopolysaccharide composed of glucose units. It is highly branched and has a backbone of β-(1→3) linked glucopyranosyl residues, and every third or fourth backbone residue is linked to a single D-glucopyranosyl side groups by β-(1→3) glucosidic bond.

### *3.7.1 Production*

Scleroglucan can be produced using fungus *Sclerotium glucanium, Sclerotium rolfsii* and *Sclerotium delphinii. Sclerotium glucanicum* [218] and *Sclerotium rolfsii* [219] are the two main species for its production. *Corticium rolfsii* and *Schizophyllum commune* produce other polysaccharides, which are structurally very similar to scleroglucan [220]. The polymer chain length, and hence the molecular mass varies according to the type of microbial culture. Scleroglucan has a structure, which is identical to schizophyllan (produced by *Schizophyllum commune).* Schizophyllan has a slightly higher molecular mass than scleroglucan [221]. The properties of scleroglucan may be influenced by the molecular mass and recovery methods.

Different carbohydrate materials can be used as substrates for the production of scleroglucan. Glucose and sucrose are commonly used carbon sources for biopolymer production, although other carbohydrates can also be utilized. Different inorganic (inorganic salts) and organic compounds such as yeast extract, casein hydrolyzate, soya hydrolyzate and corn steep liquor can be used as nitrogen sources [153]. High concentrations of nitrogen in the form of ammonium sulphate have been reported to reduce the scleroglucan production. Moreover, phosphorus is an important element in the medium, and an increase in total phosphorus has been reported to improve scleroglucan production [222]. Scleroglucan displays pseudoplastic behaviour over a wide range of pH and temperature. Temperature is a crucial parameter, which affects both culture growth and polysaccharide production. The temperature of 20-37°C has been reported optimal for exopolysaccharide production [223]. The pH of medium can affect both the rate of production and the synthesis of polysaccharides. As with temperature, the appropriate pH for maximum production of the polysaccharide can differ from that for optimum growth. In polysaccharide production, the fermentation medium becomes very viscous and exhibits non-Newtonian (pseudoplastic) behaviour. The increase in agitation and aeration is generally helpful for polysaccharide production, although conflicting reports are also available.

### *3.7.2 Recovery*

Recovery of polysaccharide from fermentation broth is an important step. The method used for recovery depends on the type and characteristics of the producing organisms, the type of polysaccharide and desired grade of purity. There are three different methods reported in the literature for the recovery of scleroglucan. Pretreatment of fermentation broth in all the methods is a common step, which involves neutralization, heating, homogenization and centrifugation. The supernatant obtained is used for the recovery of scleroglucan. In the first method, the supernatant is cooled and precipitated by adding an equivalent volume of ethanol (96%) or iso-propanol. The crude exopolysaccharide can be purified twice by ethanol re-precipitation. Finally, the precipitated polymer is either dried or freeze-dried and milled to whitish glucan powder [219].

The second method involves the use of divalent cations such as calcium, magnesium, manganese, iron, etc. along with the organic solvent. Among these, calcium chloride is the preferred divalent cation. Addition of calcium chloride results in insoluble precipitate of calcium oxalate, which is removed by centrifugation or filtration. Other steps are similar to the first method. In the third method, recovery of glucan is carried by using calcium chloride, and then adjusting the solution to an alkaline pH by addition of sodium hydroxide or potassium hydroxide. The precipitated water-soluble polysaccharide is obtained by centrifugation or filtration, and further purity can be increased by repeated precipitation [224].

#### *3.7.3 Applications*

The preliminary application of scleroglucan was in the oil recovery, where it displayed better stability than xanthan over a wide range of temperature and pH [225]. It is employed in stabilization of bentonite drilling muds and enhanced oil recovery. Besides this, it also finds applications in ceramic glazes, printing inks, latex paints, etc. This biopolymer can be used in food as a thickener, gelling or stabilizing agent. However, its cost is the major constraint, and xanthan has similar properties, which at present dominates the market. If the problem of high cost could be overcome, then it could replace xanthan in many foods. The antitumor, antiviral and antimicrobial properties of scleroglucan have also been studied [226]. It has also shown immune stimulatory effects, and its potential contribution in the treatment of many diseases. Recently, the polysaccharide has shown applications in controlled drug release and especially in immunopharmaceutical market [227].

## 4. PRODUCTION OF VITAMINS

The term "vitamin" is used to describe certain organic compounds, which are required by the body, but cannot be synthesized by the body. They mainly serve as catalysts for certain reactions in the body. Most vitamins are essential for the metabolism of all living organisms, and they are synthesized by microorganisms and plants.

Most vitamins and related compounds are now industrially produced and widely used as food or feed additives, medical or therapeutic agents, health aids, cosmetic and technical aids, and so on. Thus, vitamins and related compounds are important products for which many biotechnological production processes (i.e., fermentation and microbial/enzymatic transformation) as well as organic chemical synthetic ones have been reported. Some of them are now applied for large-scale production. Biotechnological processes allow a fully biological, single step fermentation method to replace the chemical, multi-step process, thereby saving time, money and natural resources. Different vitamins have been produced through biotechnological means (Table 7). The most important vitamins which can be produced through biotechnological processes are riboflavin, and vitamin $B_{12}$.

### 4.1 Riboflavin

Riboflavin (Vitamin $B_2$) is widely distributed in plants and animals, and plays an important role in living organisms. It is a precursor of the coenzymes flavin adenine dinucleotide (FAD) and flavin mononucleotide (FMN), which are required in reactions such as the enzymatic oxidation of carbohydrates. It is used for human nutrition and therapy and as an animal feed additive. Riboflavin is often used as a colourant because of its bright yellow colour. The crude concentrated form is also used for feed. Riboflavin deficiency in humans is associated with loss of hair, inflammation of the skin, vision deterioration, and growth failure. This vitamin has also been helpful in the treatment of migraine [228] and malaria [229].

**Table 7:** Different substrates and microorganisms used in the production of vitamins.

| Vitamin | Common Substrate | Microorganism (s) |
|---|---|---|
| Riboflavin | Glucose, molasses | *Eremothecium ashbyii, Ashbya gossypii, Bacillus* sp. |
| Vitamin $B_{12}$ | Glucose, fructose, molasses, whey | *Propionibacterium shermanii, P. freudenreichii, Pseudomonas denitrificans* |
| Vitamin C | Glucose, sorbitol | *Gluconobacter oxydans, Erwinia* sp., *Corynebacterium* sp. |
| β-Carotene | Corn starch, vegetable oil | *Blakeslea trispora, Phycomyces blakesleeanus, Choanephora cucurbitarum* |

### *4.1.1 Microorganisms*

Annually, thousands of tonnes of riboflavin are produced commercially for fortification of both food and feed. The first microorganism reported to be capable of producing riboflavin was *Clostridium acetobutylicum* and subsequently two other microorganisms, ascomycetes, namely, *Eremothecium ashbyii* and *Ashbya gossypii* have been reported for its production [230]. More popular organisms used for riboflavin production are bacterium *Bacillus subtilis*, the fungi *Eremothecium ashybyii* and *Ashbya gossypii*, and the yeasts *Candida flareri, Candida famata, Candida tropicalis* and *Saccharomyces cerevisiae* [230-233].

### *4.1.2 Raw materials*

A large number of cheap natural materials and industrial wastes have been used for riboflavin fermentation. Vegetable oils like palm oil, glucose or molasses may act as the source of carbon [235, 236]. Natural substances such as yeast extract, corn steep liquor have been known to stimulate both growth and flavinogenesis [237]. Emulsifying agents exerted variable effects on yeasts producing riboflavin from hydrocarbons. Similarly, metal ions, oils, fatty acids, amino acids and vitamins influenced the riboflavin productivity in different manners [238, 239].

Fermentation residues such as from the ethyl alcohol and butyl alcohol-acetone fermentations sugar containing substances such as molasses and corn syrup; fish meal, dried yeast; skim milk, cotton-seed meal, soybean meal, peas, wheat bran, animal protein and malt extract are some of the raw material employed in industrial processes for riboflavin production by *Eremothecium ashbyii* [240]. The riboflavin production was reported to be appreciably enhanced with the use of corn oil by *Candida guilliermondii* (about 13% increase), but the use of cotton seed oil, olive oil and some fatty acid showed no stimulatory effect [241].

Beet molasses were also utilized as the sole carbon source for riboflavin production using *Mycobacterium phlei.* The highest vitamin productivity was achieved by incubating the inoculated medium containing 9% molasses [242]. Yields much higher than 10 g of riboflavin per litre of culture broth are obtained in a sterile aerobic submerged fermentation with a nutrient medium containing molasses or plant oil as a major carbon source [243]. Recently, the use of agro-industrial wastes such as orange rind has also been used in the production of riboflavin using *Ashbya gossypii* mutants [244].

### *4.1.3 Method of production*

Commercially riboflavin has been produced by chemical synthesis, fermentation and a combination of fermentation and chemical synthesis. But in recent years, fermentation has been widely used as it produces the vitamin in a single step, resulting in substantial cost savings, while on the other hand chemical processes are multistage, and incur a lot of cost [245]. More than 90% of riboflavin is currently produced with a biotechnology fermentation process that replaced a conventional chemical process, which employed several highly toxic chemicals. Various biotechnological processes have been developed for industrial scale riboflavin biosynthesis using different microorganisms. The fermentation process is conducted in four phases:

- During the first phase, which is the growth phase, glucose is utilized. Pyruvic acid is formed due to which the pH level decreases. At the end of this phase, the growth almost ceases, as the carbon source of the medium get exhausted.
- In the second phase, the sporulation of cells occurs. The level of the pyruvate is also reduced and due to deaminase activity, ammonia is formed in the medium, which results in a rise in pH level.
- Third phase involves the synthesis of cell bound riboflavin in the form of FAD and FMN.
- The last and the fourth phase involves the autolysis of the cells, which causes the release of free riboflavin into the medium. Vitamin $B_2$ is yielded as yellow crystals directly in the fermentation broth.

#### *4.1.4 Recovery*

Riboflavin is recovered from fermentation broth by diluting the broth with a predetermined quantity of water, and heating the broth. Pasteurization of the broth ensures that no viable cells of the production organism are present in the final product. After heating, the broth is centrifuged to form a sludge and cell mass is separated. Differential centrifugation leads to separation of cells and riboflavin crystals because of differences in size and sedimentation behaviour. Riboflavin is then recovered from cell-free broth by using evaporation and vacuum drying. Another method of purifying ferment produced riboflavin has been reported, wherein the impure riboflavin is suspended in water or dilute aqueous acid, and the suspension is heated along with stirring, during which the crystal transformation takes place, after which the mixture is cooled and the purified product is isolated [246].

### 4.2 Vitamin $B_{12}$

Vitamin $B_{12}$ (cobalamin) is an important vitamin for humans and animals. This vitamin is important for the normal functioning of the brain and nervous system, and for the formation of blood, besides an important animal feed supplement as growth enhancer. It is widely used to describe compounds of the cobalamin group. Vitamin $B_{12}$ is synthesized in nature exclusively by microorganisms (bacteria) and cannot be produced by plants or animals. These bacteria are thought to live in water, soil, and the digestive tracts of animals. In animals, $B_{12}$ is normally attached to a protein either for transport or storage. Today, cobalamin is exclusively produced by biosynthetic fermentation processes, using selected and genetically optimized microorganisms. A genetically engineered highly effective cobalamin-producing strain of *Pseudomonas denitrificans* has a productivity that reaches 300 mg/l and accounts for 80% of the cobalamin production in the world [247].

#### *4.2.1 Microorganisms*

Industrial production of Vitamin $B_{12}$ is through fermentation of selected microorganisms [247]. Species from the following genera *Aerobacter, Agrobacterium, Alcaligenes, Arthrobacter, Azotobacter, Bacillus, Butyribacterium, Citrobacter, Clostridium, Corynebacterium, Escherichia, Flavobacterium, Klebsiella, Lactobacillus, Micromonospora, Mycobacterium, Nocardia, Propionibacterium, Protaminobacter, Proteus, Pseudomonas, Rhizobium, Rhodopseudomonas, Salmonella, Serratia, Streptomyces, Streptococcus* and *Xanthomonas* are known to synthesize vitamin $B_{12}$ [247, 248]. However, mainly two genera, *Propionibacterium (P. shermanii* and *P. freudenreichii) and Pseudomonas* (*Ps.*

*denitrificans*), are used for industrial production of vitamin $B_{12}$ because of their rapid growth and high productivity of selected mutants [249-251].

### 4.2.2 Raw materials

Glucose is commonly used as a carbon source, but other substrates i.e., fructose, mannose, galactose, glycerol and milk can be used for growth. Dairy waste like whey has also been used for the production of vitamin $B_{12}$. The vitamin $B_{12}$-enriched fermented whey was found to replace the animal protein factor (APF) as a vitamin $B_{12}$ source in poultry rations. Therefore, the technology developed is a significant attempt to utilize whey for vitamin $B_{12}$ production and in turn, reduce the water pollution problem [252].

Propionic organisms are grown on carbohydrate-based media in anaerobic conditions. Cobalt supplement is necessary for cobalamin production. It also depends on the internal formation or external supply of 5,6-dimethylbenzimidazole (5,6-DBI). Since cobalt is an essential component of vitamin $B_{12}$, it was at first added to the medium with the purpose of enhancing the vitamin $B_{12}$ activity produced by these organisms. Another preferred supplement for the culture medium is 5,6-dimethylbenzimidazole (DBI). Preferably, DBI is added to the culture medium at the beginning of the aerobic phase or during that phase.

The role of amino acids, betaine and choline in vitamin $B_{12}$ biosynthesis by the three strains of *Propionibacterium viz., P. shermanii* 566, *P. shermanii* and *Propionibacterium arl* AKU 1251 was also investigated [252]. They supplemented whey permeate medium with eleven amino acids (0.05%, by mass per volume). Betaine was found to be a better stimulator than choline. Sugar beet molasses containing 5.0% betaine (trimethyl-glycine) can also be used as carbohydrate source, as it stimulates the production of vitamin $B_{12}$.

### 4.2.3 Method of production

Traditionally, cobalamin was isolated from residues of antibiotic fermentations. But today, it is exclusively produced by fermentation. *Propionibacterium* and *Pseudomonas* are frequently used for the production of cobalamin. All *Propionibacterium* strains employed for vitamin $B_{12}$ production are microaerophilic and produce vitamin $B_{12}$ in high yields only under very low oxygen concentrations. The organism grows anaerobically and produces propionic acid which accumulates in the medium. Most industrial processes for the production of vitamin $B_{12}$ by *P. shermanii* require neutralization of the accumulated propionic acid to keep the culture growing logarithmically. However, the biosynthesis of dimethylbenzimidazole (DMB) requires oxygen. Hence, the bioprocess of vitamin $B_{12}$ production using *Propionibacterium* strains is divided into two stages. During first 3 days of fermentation, the bacteria are grown anaerobically to produce vitamin $B_{12}$ precursor cobamide, an intermediate vitamin $B_{12}$ in the absence of DMB moiety. Practically all of the cobamides formed in the fermentation are retained in the cells, and the first step is the separation of the cells from the fermentation medium. Large high speed centrifuges are used to concentrate the bacteria to a cream, while filters are used to remove streptomycete cells. Later, vitamin $B_{12}$ formation is completed by gentle aeration of the whole culture for 1-3 days, allowing the bacteria to undertake the oxygen-dependent synthesis of the DMB and to bond it to cobamide [253].

In contrast to the *Propionibacterium* fermentation process, *Ps. denitrificans* shows oxygen dependent growth and gives high vitamin $B_{12}$ production rates. The culture is aerated during the whole fermentation process and maintained at 30°C, pH 6-7 for 3-4 days [253]. This is the single stage process in which vitamin $B_{12}$ is produced during the whole fermentation process. During the fermentation, cobalt and DMB must be added as supplements. Generally, the whole broth or an aqueous suspension of harvested cells is heated with cyanide or thiocyanate at 80-120°C at pH 6.5-8.5, and the conversion to cyanocobalamin (vitamin $B_{12}$) is obtained [254]. The addition of the precursor dimethylbenzimidazole and aerobic incubation in the latter phase of fermentation led to improvement in the production of vitamin $B_{12}$ [96, 255, 250, 256].

### 4.2.4 Recovery

The extraction process depends on the release of unstable cobalamin from the lysed cells and subsequent treatment with cyanide to convert into more stable cyanocobalamin. Cells are separated from the culture broth by centrifugation. Then cells are lysed by heating the centrifuged cell mass in aqueous solution at 80 to 120°C for 10-30 minutes at pH 6.5-8.5. The cells on lysis release various cobalamin. Then, these are solubilized with potassium cyanide in the presence of sodium nitrite and heat. The obtained cobalamin gets converted into cyanocobalamin. The vitamin solution is clarified afterward by filtration, treatment with zinc chloride, and then precipitated out by the addition of tannic acid or cresol to give the product of 80% purity, which is suitable for use as an animal food additive.

Further purification via different extraction steps, using organic solvents such as cresol, carbon tetrachloride and water/butanol, is often supplemented by adsorption to ion exchangers or activated carbon. Finally, vitamin $B_{12}$ is crystallized by the addition of organic solvents, leading to a product of recommended quality for food and pharmaceutical applications [253]. Since some *Propionibacterium* species do not produce either endo- or exotoxins [257], *Propionibacterium* species are the preferred species for the production of food additives or medicines. Thus, processes of vitamin $B_{12}$ production using *Propionibacterium* species have the advantage; *Propionibacterium* species allow the production of vitamin $B_{12}$ together with the biomass, in which vitamin $B_{12}$ is produced [258].

## 4.3 Vitamin C

Vitamin C (Ascorbic acid) has various applications in the food and pharmaceutical industries. It is also being used in cosmetics due to its antioxidant properties and its potential to stimulate collagen production [259]. L-Ascorbic acid has been industrially produced for around 70 years. Over the past two decades, several innovative bioconversion systems have been proposed in order to simplify the long time market-dominating Reichstein method, a largely chemical synthesis by which still a considerable part of L-ascorbic acid is produced.

At present, vitamin C is manufactured commercially via the Reichstein process, which is highly energy-consuming, involves considerable quantities of organic solvents and gives an overall yield of about 50%. Therefore, during the past decades, a lot of research was done to extend biotechnological alternatives for the production of Reichstein intermediates by enzymatic or fermentative means, which show some advantages regarding costs and environmental friendliness.

### 4.3.1 Microorganisms

Preferred microorganisms for the production of ascorbic acid are bacteria and fungi. Microorganisms suitable for the production of ascorbic acid can be of the following genera: *Pantoea, Escherichia, Bacillus, Erwinia, Lactobacillus, Bifidobacterium, Brevibacterium, Xanthomonas, Streptococcus, Leuconostoc,* and *Saccharomyces* [260]. *Gluconobacter oxydans* and *Corynebacterium* sp. are the popular organisms used for the production of vitamin C.

### 4.3.2 Raw materials

Typically, the medium is aqueous comprising, for instance salts, substrate(s), and a certain pH. Carbon source such as, D-glucose, D-sorbitol, L-sorbose, L-sorbosone, L-gulose or L-gulono-gamma-lactone can be used as substrates. Furthermore, the culture medium usually contains inorganic salts for example, magnesium sulphate, potassium phosphate and calcium carbonate. The production medium may also contain digestible nitrogen sources such as peptone, yeast extract, baker's yeast, urea, amino acids, corn steep liquor and inorganic substances e.g., ammonia, ammonium sulphate, and sodium nitrate at such concentrations that the cells kept in a resting cell mode can also be used. Various inorganic substances may also be used as nitrogen sources for example, nitrates and ammonium salts.

#### *4.3.3 Method of production*

At commercial scale, vitamin C is produced using semi-synthetic method from D-glucose. All the processes are operated via 2-keto-L-gulonic acid [260]. The lactonization and isomerization of 2-keto-L-gulonic acid yields the production of L-ascorbic acid. The 2-keto-L-gulonic acid is also produced using different oxidative or reductive reactions. It is obtained by oxidation of D-glucose without inversion of carbon sequence or by reduction of D-glucose involving the inversion of the carbon sequence. The 2-keto-L-gulonic acid is effectively produced in double fermentation through 2,5-diketo-D-gluconic acid pathways [261], which eliminates all the chemical stages, and involves only two stages.

2,5-Diketo-D-gluconic acid pathway is the fermentation of Ca-2, 5-diketo-D-gluconate. *Erwinia* sp. ATCC 31626 is employed for this fermentation. The first phase is followed by second phase of fermentation of Ca-2,5-diketo-D-gluconate using *Corynebacterium* sp. ATCC 31090, which produces Ca-2-keto-L-gulonate (Fig. 5) The fermentation of *Erwinia* sp. is carried at 28°C for 26 h with agitation (160 rpm) and aeration. It should be noted that after 3 h of the fermentation, culture media should be supplied intermittently with glucose. The addition of glucose solution should be stopped after 15 h. The fermentation with *Corynebacterium* sp. ATCC 31090 is carried out at 28°C with agitation and aeration. Sodium nitrate is also supplemented at 16 h of incubation.

#### *4.3.4 Recovery:*

The vitamin C thus produced and accumulated in the reaction mixture may be separated and purified as a free acid or as a salt of sodium, potassium, calcium, and ammonium. Specifically, the separation may be performed by any suitable combination of the following steps: by the formation of a salt, by using differences in properties between the product and the surrounding impurities, such as solubility, absorbability and distribution coefficient between the solvents, by absorption, for example, on ion-exchange resin. Any of these procedures alone or in combination constitute a convenient means for isolating the product. The product thus obtained may be further purified in a conventional manner, for example, by re-crystallization or chromatography [262].

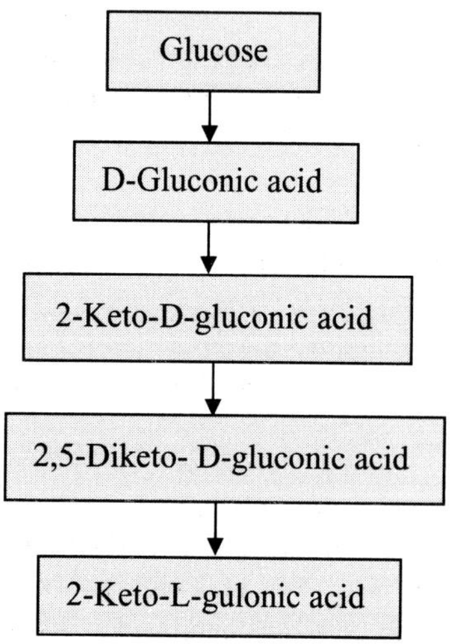


**Fig. 5:** Production of 2-keto-L-gulonic acid.

### 4.4 β-Carotene

Carotenoids are highly unsaturated isoprene derivatives. Naturally occurring carotenoids are tetraterpenoids consisting of 8 isoprene residues [263]. Carotenoids occur in many plant and animal tissues, but originate exclusively from plant or microbes. Carotene (provitamin A) is converted into vitamin A in the intestinal mucous membrane and is stored as the plamitate ester in the liver. The deficiency of vitamin A in human beings causes night blindness, xerophthalmia (dry eyes), increased susceptibility to infections, rough and dry skin.

#### *4.4.1 Microorganisms*

β-Carotene is produced by many organisms, but particularly by members of Choanephoraceae family of the Phycomycetes.

*Phycomyces blakesleeanus, Choanephora cucurbitarum* and *Blakeslea trispora* have been extensively studied for β-carotene production [264-266].

### 4.4.2 Method of production

Carotenoids have been produced through microbial fermentation, however, this method does not seem to be economical in the present market. The main reason is that the yields of currently available processes do not compete with cost-effective chemical synthetic processes. However, depending on the market, it appears that a competitive process might be successful, using the halophilic alga *Dunaliella salina* [267]. Among the different microorganisms, the highest yields have been obtained with the *Blakeslea trispora* process [263]. The observation that carotene production occurs during the process of zygospore formation in this organism has had an impact on process development. When cultures of both sexual forms, (+) and (−) strains, are mixed, a significant increase in carotene production in the (−) strain is achieved.

Production is induced by trisporic acids, a mixture of closely related substances, which are derived biosynthetically from β-carotene. The submerged fermentation is employed for β-carotene production using a mixed culture of (+) and (−) strains (Fig. 6). The (−) strain, which produces the β-carotene can be present in great excess, so it is not necessary that proportions of the two strains should be equal. Isoniazid is another activator of β-carotene synthesis, particularly in combination with β-ionone. Other substances, such as terpenes or cyclohexane and cyclohexanone and their trimethyl derivatives can also be used as replacement of β-ionones. The addition of purified kerosene to the medium can increase the yield through increased solubility of the hydrophobic substrate. Due to the low stability of β-carotene within the cells, the addition of an antioxidant is necessary for the fermentation process.

### 4.4.3 Recovery

The mycelium is first separated or removed, dehydrated (for example, with methanol), and then extracted with methylene chloride (75-90% yield) to get pure β-carotene. The crude product can also be further purified. The carotenoid-rich mycelium can also be used directly as a feed additive.

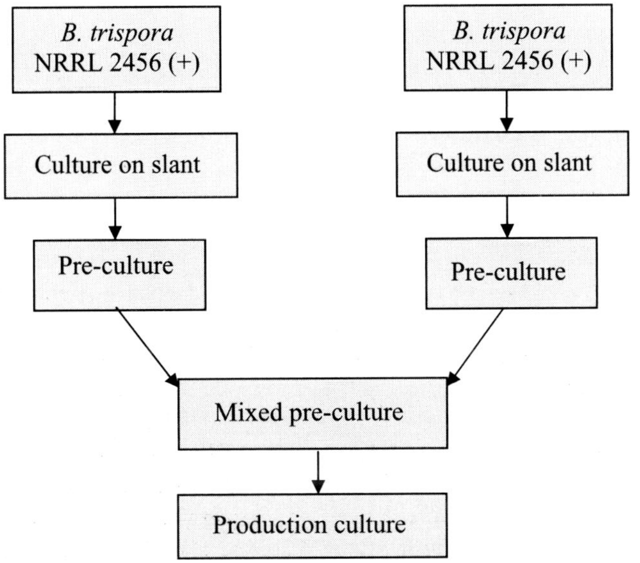


**Fig. 6:** Simplified process for β-carotene production using *B. trispora.*
*Source:* 263

## 5. FUTURE TRENDS

Biotechnology has played a big role in the production of different useful products. The chemical industry in Europe, which contributes ~28% of the demand for chemicals in the world, has recognized industrial biotechnology as a key emerging technology area [268]. The production of chemicals using industrial biotechnology offers several advantages over chemical processes, such as less waste generation, biodegradable products, besides reduced energy consumption. However, major constraint in the microbial products is final cost of the product, which needs to be economized. The success of microbial processes relies on the optimization of all fermentation parameters in order to achieve high stability, along with high productivity, and low operating and capital costs. The selection of suitable high yielding microbial strains, use of cheaper raw materials like agricultural/industrial waste along with efficient product recovery method can help to reduce the cost of production. Engineering of metabolic pathways along with design of bioreactors can also play a pivotal role, especially in exopolysaccharide production. Therefore, development of innovative and sustainable process technologies will be needed, which could improve the quality, increase the productivity and also reduce the cost for realizing the shift to bio-based production of chemicals.

## REFERENCES


1. Sijbesma F (2003) *White Biotechnology: Gateway to a More Sustainable Future*, European Association for Bioindustries, Lyon, France.
2. Hatti-Kaul R, Törnvall U, Gustafsson L & Börjesson P (2007) Industrial biotechnology for the production of bio-based chemicals-a cradle-to-grave perspective. *Trends Biotechnol* 25: 119-124.
3. Couto SR & Sanroma, MA (2006) Application of solid-state fermentation to food industry-A review. *J Food Engg* 76: 291-302.
4. Lockwood LB, Yoder DE & Zienty M (1965) Lactic acid. *Ann N Y Acad Sci* 119: 854-867.
5. Hofvendahl, K & Hahn-Hagerdal B (2000) Factors affecting the fermentative lactic acid production from renewable resources. *Enzyme Microbial Technol* 26: 87-107.
6. Holten CM, Muller A & Rehbinder D (1971) *Lactic Acid*, Verlag Chemie, Weinheim, Germany.
7. Prescott SC & Dunn CG (1959) *Industrial Microbiology*, McGraw Hill, New York, USA.
8. Curry B & Crow V (2003) *Lactobacillus* sp.: General characteristics. In: *Encylcopedia of Dairy Sciences*, Vol 3, Roginski H, Fuquay JW & Fox PF (eds), Academic press, London, UK, pp 1479-1484.
9. Stanier RY, Adleburg EA & Ingraham JL (1976) *The Microbial World*, Prentice Hall, Englewood Cliffs, USA.
10. Gasser F (1970) Electrophoretic characterization of lactate dehydrogenases in the genus *Lactobacillus*. *J Gen Micrbiol* 14: 233-252.
11. Nakamura LK & Crowell CD (1979) *Lactobacillus amylophilus*, a new starch hydrolyzing species from spices waste-corn fermentation. *Dev Ind Microbial* 20: 531-540.
12. Skory CD, Freer SN & Bothast V (1998) Production of L-lactic acid by *Rhizopus oryzae* under oxygen limiting conditions. *Biotechnol Lett* 20: 191-194.
13. Kulozik U & Wilde J (1999) Rapid lactic acid production at high cell concentration in whey ultrafiltrate by *Lactobacillus helveticus*. *Enzyme Microbial Technol* 24: 297-302.
14. Cordon TC, Treadway RH, Walsh MD & Osborne MF (1950) Lactic acid from potatoes. *Ind Eng Chem* 42: 1833-1836.

15. Viniegra-Gonzalez G & Gomez J (1984) Lactic acid production by pure and mixed bacterial cultures. In: *Bioconversion Systems*, Wise DL (ed), CRC Press, Boca Raton, USA, pp 17-39.
16. Buchta K (1983) Lactic acid. In: *Biotechnology*, Vol 3, Rehm HJ & Reed G (eds), VCH Verlag GmBH, Weinheim, Germany, pp 409-417.
17. Hujanen M & Linko YY (1996) Effect of temperature and various nitrogen sources on L (+) lactic acid production by *Lactobacillus casei*. *Appl Microbiol Biotechnol* 45: 307-313.
18. Sethi V & Maini SB (1999) Production of organic acids. In: *Biotechnology: Food Fermentation,* Joshi VK & Pandey A (eds), Educational Publishers & Distributors, New Delhi, India, pp 1259-1289.
19. Inskeep GC, Taylor GG & Breitzke WC (1952) Lactic acid from corn sugar. *Ind Eng Chem* 44: 1955-1966.
20. Snell RL & Lowry (1964) Calcium (L+) lactic acid production. US Patent 3 125 494.
21. Burton LV (1937) By products of milk. *Food Ind* 9: 571-575.
22. Wood BJB & Holzapfel WH (1995) *The Genera of Lactic Acid Bacteria*, Blackie Academic and Professional, Glasgow, UK.
23. Buchta K (1974) Die biotechnologische Gewinnung von organischen Sauren. *Chem Ztg* 98: 532-538.
24. Schopmeyer HH (1954) *Industrial Fermentations,* Chemical Publishing Co, New York, USA.
25. Eyal AM, Starr JN, Fisher R, Hazen B, Canari R, Witzke DR, Gruber PR & Kolstad JJ (2001) Lactic acid processing; methods, arrangements, and product. US Patent 6 320 077.
26. Roychoudhary PK, Srivastava A & Sahai V (1995) Extractive bioconversion of lactic acid. In: *Advances in Biochemical Engineering/Biotechnology,* Vol 53, Fiechter A (ed), Springer-Verlag, Weinheim, Germany, pp 61-87.
27. Anon (1992) Ecochem (Dupont and Conagra Joint Venture, Wilmington, DE) completes lactic acid plant. *Chemical Engineering News* August, pp 20-25.
28. Datta R (1995) Technological and economical potential of polylactic acid and lactic acid derivatives. *FEMS Microbiol Rev* 16: 221-231.
29. Sodegard A (1998) Preparation of poly (L-lactide-graftacrylic acid) by pre-irradiation grafting. *Polymer Prep* 39: 215-216.
30. Amass W, Amass A & Tighe B (1998) A review of biodegradable polymers: Uses, current developments in the synthesis and characterization of biodegradable polymers, blends of biodegradable polymers and recent advances in biodegradation studies. *Polym Int* 47: 89-114.
31. Currie JN (1917) Citric acid fermentation. *J Biol Chem* 31: 15-37.
32. Soccol CR, Prado FC, Vandenberghe LPS & Pandey A (2003) *Concise Encyclopedia of Bioresource Technology*, The Haworth Press, New York, USA.
33. Pandey A, Soccol CR, Rodriguez-León JA & Nigam P (2001) Production of organic acids by solid-state fermentation. In: *Solid-state Fermentation in Biotechnology-Fundamentals and Applications,* Asiatech Publishers, New Delhi, India, pp 113-126.
34. Vandenberghe LPS, Soccol CR, Pandey A & Lebeault JM (2000) Cassava bagasse, an alternative substrate for citric acid production in solid-state fermentation. In: *Proceedings of International Biotechnology Symposium and Exhibition*, Frankfurt, Germany, pp 153-155.
35. Vandenberghe LPS, Soccol CR. Pandey A & Lebeault JM (2000) On-line monitoring of citric acid production using respirometry in solid state fermentation with cassava bagasse.

In: *Proceedings of International Symposium on the Bioconversion of Renewable Raw Materials*, Hannover, Germany, pp 210-212.
36. Prado FC (2002) Desenvolvimento de Bioprocesso em escala semipiloto para produção de ácido cítrico por fermentação no estado sólido a partir do bagaço de mandioca, *MSc Thesis*, Universidade Federal do Paraná, Brazil.
37. Grewal HS & Kalra KL (1996) Fungal production of citric acid. *Biotechnol Adv* 13: 209-234.
38. Tabuchi T & Abe M (1968) Process for the production of citric acid. Japanese Patent 68 20 707.
39. Pfizer, Chas. & Co. Inc. (1970) Fermentation process for the production of citric acid. Br. Pat 1 182 983.
40. Clark DS, Ito K & Tymchuk P (1965) Effect of potassium ferrocyanide addition on the chemical composition of molasses mash used in the citric acid fermentation. *Biotechnol Bioeng* 7: 269-278.
41. Swarthout EJ (1966) Citric acid production by fungal fermentation of sugar. US Patent 3 285 831.
42. Fenci Z & Leopold J (1959) Inhibition of spore germination of the mould *A. niger* by acetic acid. *Folia Microbiol* 4: 7-15.
43. Hustede H & Rudy H (1976) Manufacture of citric acid by submerged fermentation. US Patent 3 940 315.
44. Miall LM & Parker GF (1975) Citric acid production. Br Patent 1 418 561.
45. Les Usines de Melle (1961) Process for extracting citric acid from aqueous solutions thereof. Br Patent 874 030.
46. Les Usines de Melle (1963) Process for extracting carboxylic acids produced by fermentation. Br Patent 936 339.
47. Ferrara L, Cesari LD & Salvini E (1977) N- Paraffine. *Chim Ind (Milan)* 59: 202-206.
48. Grayson M & Eckroth D (1979) *Kirk-Othmer Encyclopedia of Chemical Technology*, Vol 6, Wiley, New York, USA, pp 150-178.
49. Schulz G & Rauch J (1975) *Ullmans Encyklopaedie der technichen Chemie,* Vol 9 Verlag Chemie, Weinheim, Germany, pp 624-636.
50. Fricke H & Jensen SB (1975) Citric acid- a versatile food additive. *Food Process Ind* 44: 37-44.
51. Filho AP & Sattamin LC (1981) *Proceedings of the 2nd World Congress of Chemical Engineers,* Montreal, Canada, p 39.
52. Ghose TK & Bhadra A (1985) Acetic acid. In: *Comprehensive Biotechnology*, Vol 3, Moo-Young M (ed), Pergamon Press, New York, USA, pp 701-729.
53. Hijikata Y, Okumura H & Terui G (1972) Acceleration of acetic acid fermentation by malt sprouts. *J Ferment Technol* 49: 577-579.
54. Kawasaki H, Okano S & Omata S (1975) Oxidative dissimilation of glycerol in *Tetracoccus soyae. J Ferment Technol* 53: 207-217.
55. Allgeier RJ & Hildebrandt FM (1960) Developments in vinegar manufacture. *Adv Appl Microbiol* 8: 163-182.
56. Conner HA & Allgeier RJ (1976) Vinegar-its history and development. *Adv Appl Microbiol* 20: 81-133.
57. Hromatka O & Ebner H (1950) Acetic fermentation. Lack of oxygen in generator fermentation. *Enzymologia* 14: 96-106.

58. Helsel RW (1977) Removing carboxylic acids from aqueous wastes. *Chem Eng Prog* 73: 55-59.
59. Sanderson JE, Wise DL & Augenstein DC (1979) Organic chemicals and liquid fuels from algal biomass. *Biotechnol Bioeng Syrup* 8: 131-151.
60. Lancaster M (2002) *Green Chemistry, an Introductory Text*, Royal Society of Chemistry, Cambridge, UK, pp 262 - 266.
61. Prescott FJ, Shaw K, Bilello JP & Cragwall GO (1953) Gluconic acid and its derivatives. *Ind Eng Chem* 45: 338-342.
62. Yamauchi T & Shimizu K (1975) Glucono-δ-lactone crystallization. Japanese Patent 75 123 674.
63. Röhr M, Kubicek CP & Kominek J (1983) *Biotechnology,* Verlag Chemie, Weinheim, Germany.
64. Bull DN & Kempe LL (1970). Kinetics of the conversion of glucose to gluconic acid by *Pseudomonas ovalis. Biotechnol Bioeng* 12: 273-290.
65. Loffhagen N & Babel W (1993) The influence of energy deficiency-imposing conditions on the capacities of *Acetobacter methanolicus* to oxidize glucose and to produce gluconic acid. *Acta Biotechnol* 13: 251-256.
66. Kim DM & Kim HS (1992) Continuous production of gluconic acid and sorbitol from *Jerusalem artichoke* and glucose using an oxidoreductase of *Zymomonas mobilis* and inulinase. *Biotechnol Bioeng* 39: 336-342.
67. Attwood M, Dijken JP & Pronk JT (1991). Glucose metabolism and gluconic acid production by *Acetobacter diazotrophicus. J Ferment Bioeng* 72: 101-105.
68. Velizarov S & Beschkov V (1994). Production of free gluconic acid by cells of *Gluconobacter oxydans. Biotechnol Lett* 16: 715-720.
69. Shiraishi F, Kawakami K, Kono S, Tamura A, Tsuruta S, Kusunoki K (1989) Characterization of production of free gluconic acid by *Gluconobacter suboxydans* adsorbed on ceramic honeycomb monolith. *Biotechnol Bioeng* 33: 1413-1418.
70. Rodriguez H, Gonzalez T, Goire I & Bashan Y (2004) Gluconic acid production and phosphate solubilisation by the plant growth promoting bacterium *Azospirillum* spp. *Naturwissenschaften* 91: 552-555.
71. Kundu P & Das A (1984) Utilization of cheap carbohydrate sources for calcium gluconate production by *Penicillium funiculosum* mutant MN 238. *Ind J Exp Biol* 22: 279-281.
72. Anastassiadis S, Aivasidis A, Wandrey C & Rehm HJ (2005) Process optimization of continuous gluconic acid fermentation by isolated yeast-like strains of *Aureobasidium pullulans. Biotechnol Bioeng* 91: 494-501.
73. Wilt HGJ (1972) Oxidation of glucose to gluconic acid. *Ind Eng Prod Res Dev* 11: 370-378.
74. Cho YK & Bailey JE (1977) Glucoamylase and glucose oxidase preparations and their combined application for the conversion of maltose to gluconic acid. *Biotechnol Bioeng* 19: 185-198.
75. Blom RH, Pfeifer VF, Moyer AJ, Traufler DH & Conway HF (1952) Sodium gluconate production – Fermentation with *Aspergillus niger. Ind Eng Chem* 44: 435-440.
76. Hatcher JH (1972) Gluconic acid production. US Patent 3 669 840.
77. Vassilev NB, Vassileva MC & Spassova DI (1993) Production of gluconic acid by *Aspergillus niger* immobilized on a polyurethane foam. *Appl Microbiol Biotechnol* 39: 285-288.
78. Rao S & Panda T (1994) Critical analysis of the metal ions on gluconic acid production by *Aspergillus niger* using a treated Indian cane molasses. *Bioprocess Eng* 10: 99-107.

79. Singh OV, Kapur N & Singh RP (2005) Evaluation of agro-food by-products for gluconic acid production by *Aspergillus niger* ORS-4410. *World J Microbiol Biotechnol* 21: 519-524.
80. Lockwood LB (1975) Organic acid production. In: *Filamentous Fungi,* Vol 1, Smith JE & Berry DR (eds), Wiley, New York, USA, pp 140-157.
81. Curie JN, Kane JH & Finlay A (1931) Process for producing gluconic acid by fungi. US patent 1 893 819.
82. Sakurai H, Lee HW, Sato S, Mukataka S & Takahashi J (1989) Gluconic acid production at high concentrations by *Aspergillus niger* immobilized on a nonwoven fabric. *J Ferment Bioeng* 67: 404-408.
83. Philips DH & Johnson MJ (1959) Oxygen transfer in agitated vessels. *Ind Eng Chem* 51: 83-88.
84. Steel R & Maxon WD (1962) Some effects of turbine size on novobiocin fermentations. *Biotechnol Bioeng* 4: 231-240.
85. Bernhauer K (1940) *Handbook of Enzymology*, Akademische Verlagsgesellschaft, Leipzig, Germany.
86. Lockwood LB (1979) Production of organic acids by fermentation. In: *Microbial Technology*, Vol 1, Peppler HJ & Perlman D (eds), Academic Press, New York, USA, pp 355-387.
87. Znad H, Markos J & Bales V (2004) Production of gluconic acid from glucose by *Aspergillus niger*: Growth and non growth conditions. *Process Biochem* 39: 1341-1345.
88. Sawyer DT (1964) Metal gluconate complexes. *Chem Rev* 64: 633-643.
89. Dvorkovitz V & Hawley TG (1952) Washing composition. US Patent 2 584 017.
90. Dvorkovitz V and Hawley TG (1952) Washing composition. US Patent 2 615 846.
91. CMR (1997) Propionic acid (chemical profile), *Chemical Marketing Reporter*, February 24, 1997.
92. Whittier EO & Sherman JM (1923) Propionic acid and ketones from whey. *Ind Eng Chem* 15: 729-731.
93. Lewis PV & Yang ST (1992) Propionic acid fermentation by *Propionibacterium acidipropionici*: Effect of growth substrate. *Appl Microbiol Biotechnol* 37: 437-442.
94. Carrondo MJT, Crespo JPSG & Moura MJ (1988) Production of propionic acid using a xylose utilizing *Propionibacterium. Appl Biochem Biotechnol* 17: 295-312.
95. Babuchowski A, Hammond EG & Glatz BA (1993) Survey of propionibacteria for ability to produce propionic and acetic acids. *J Food Prot* 56: 493-496.
96. Quesada-Chanto A, Afschar AS & Wagner F (1994) Microbial production of propionic acid and vitamin $B_{12}$ using molasses or sugar. *Appl Microbiol Biotechnol* 41: 378-383.
97. Colomban A, Roger L & Boyaval P (1993) Production of propionic acid from whey permeate by sequential fermentation, ultrafiltration, and cell recycling. *Biotechnol Bioeng* 42: 1091-1098.
98. Playne MJ (1985) Propionic and butyric acids. In: *Comprehensive Biotechnology*, Moo-Young M (ed), Pergamon, New York, USA, pp 731-759.
99. Blanc P & Goma G (1987) Propionic acid fermentation: Improvement of performances by coupling continuous fermentation and ultrafiltration. *Bioprocess Engg* 2: 137-139.
100. Daugulis AJ (1988) Integrated reaction and product recovery in bioreactor systems. *Biotechnol Prog* 4: 113-122.
101. Barbirato F, Chedaille D & Bories A (1997) Propionic acid fermentation from glycerol: comparison with conventional substrates. *Appl Microbiol Biotechnol* 47: 441-446.
102. Boyaval P, Corre C & Madec MN (1994) Propionic acid production in a membrane bioreactor. *Enzyme Microbial Technol* 16: 883-886.

103. Himmi EH, Bories A, Boussaid A & Hassani L (2000) Propionic acid fermentation of glycerol and glucose by *Propionibacterium acidipropionici* and *Propionibacterium freudenreichii* subsp. *shermanii. Appl Microbiol Biotechnol* 53: 435-440.
104. Kim JK, Han SH & Park DW (2004) A study on the recovery of propionic acid by solvent extraction. *Theories Appl Chem Eng* 10: 1398-1401.
105. Keshav A, Chand S & Wasewar KL (2008) Equilibrium studies for extraction of propionic acid using tri-n-butyl phosphate in different solvents. *J Chem Eng Data* 53: 1424-1430.
106. Sutherland IW (1998) Novel and established applications of microbial polysaccharides. *Trends Biotechnol* 16: 41-46.
107. Marwaha SS & Arora JK (1999) Production of gums, amino acids and vitamins. In: *Biotechnology: Food Fermentation-Microbiology, Biochemistry & Technology,* Joshi VK & Pandey A (eds), Educational Publishers & Distributors, New Delhi, India, pp 1231-1255.
108. Demain AL (2000) Small bugs, big business: The economic power of the microbe. *Biotechnol Adv* 18: 499-514.
109. Jansson PE, Kenne L & Lindberg B (1975) Structure of the extracellular polysaccharide from *Xanthomonas campestris. Carbohyd Res* 45: 275-282.
110. Sutherland IW (1977) Microbial exopolysaccharide synthesis. In: *Extracellular Microbial Polysaccharides,* Sandford PA & Laskin A (eds), American Chemical Society, USA, pp 40-57.
111. Sutherland IW (1994) Structure – function relationships in microbial exopolysaccharides. *Biotechnol Adv* 12: 393-448.
112. Lachke A (2004) Xanthan – A versatile gum. *Resonance* 9: 25-33.
113. Flickinger FC & Drew SW (1999) *Encyclopedia of Bioprocess Technology*, Vol 5, Wiley Interscience, New York, USA, pp 2706-2707.
114. Rosalam S & England R (2006) Review of xanthan gum production from unmodified starches by *Xanthomonas comprestris* sp. *Enzyme Microbial Technol* 39: 197-207.
115. Bradbury JF (1984) Genus II: *Xanthomonas*. In: *Manual of Systematic Bacteriology*, Krieg NR & Holt CG (eds),Williams & Wilkins, Baltimore, USA, pp 199-210.
116. Garcia FO, Santos VE, Casas JA & Gomez E (2000) Xanthan gum: Production, recovery, and properties. *Biotechnol Adv* 18: 549-579.
117. Peters HU, Suh IS, Schumpe A & Deckwer WD (1992) Modeling of batchwise xanthan production. *Can J Chem Eng* 70: 742-750.
118. Jana AK & Ghosh P (1995) Xanthan biosynthesis in continuous culture: citric acid an energy source. *J Ferment Bioeng* 80: 485-491.
119. Serrano LC, Corona RM, Sanchez A & Galindo E (1997) Prediction of xanthan fermentation development by a model linking kinetics, power drawn and mixing. *Process Biochem* 33: 133-146.
120. Suh SI, Schumpe A & Deckwer WD (1992) Xanthan production in bubble column and air-lift reactors. *Biotechnol Bioeng* 39: 85-94.
121. Kawase Y & Hashimoto N (1996) Gas hold-up and oxygen transfer in three-phase external-loop airlift bioreactors: non-Newtonian fermentation broths. *J Chem Technol Biotechnol* 65: 325-334.
122. Chaitali M, Kapadi M, Suraishkumar GK & Gudi RD (2003) Productivity improvement in xanthan gum fermentation using multiple substrate optimization. *Biotechnol Progr* 19: 1190-1198.
123. Psomas SK, Kyriakides LM & Kyriakidis DA (2007) Optimization study of xanthan gum production using response surface methodology. *J Biochem Eng* 35: 273-280.

124. Pace GW & Coote SDJ (1981) Xanthan gum, and method of producing it by fermentation. European Patent 32,293 A.
125. Stredansky M & Conti E (1999) Xanthan production by solid state fermentation. *Process Biochem* 34: 581-587.
126. Stredansky M, Conti E, Navarini L & Bertocchi C (1999) Production of bacterial exopolysaccharides by solid substrate fermentation. *Process Biochem* 34: 11-16.
127. Pace GW & Righelato RC (1980) Production of extracellular microbial polysaccharides. *Adv Biochem Engg* 15: 41-70.
128. Margaritis A & Pace GW (1985) Microbial polysaccharides. In: *Comprehensive Biotechnology,* Young M, Blanch M, Drew WH & Wang S (eds), Pergamon Press, Toronto, Canada, pp 1005-1043.
129. Katzbauer B (1998) Properties and applications of xanthan gum. *Poly Degrad Stabil* 59: 81-84.
130. FASEB Committee (1980) Evaluation of GRAS monographs (scientific literature reviews) PW 80-203789, NTIS, US Department of Commerce, Washington, USA, Federal Register, 34 (1969) 5379 (March 19).
131. Kulicke WM, Oertel R, Otto M, Kleinitz W & Littmann W (1990) Characterization of xanthan solutions for application in EOR. *Erdol und Kohle, Erdgas, Petrochemie vereinigt mit Brennstoff-Chemie* 43: 471-476.
132. Roukas T (1998) Pretreatment of beet molasses to increase pullulan production. *Process Biochem* 33: 805-810.
133. Leathers TD (1993) Substrate regulation and specificity of amylases from *Aureobasidium* strain NRRLY-12,974. *FEMS Microbiol Lett* 110: 217-222.
134. Pollock TJ, Thorne I & Armentrout RW (1992) Isolation of new *Aureobasidium* strains that produce high-molecular-weight pullulan with reduced pigmentation. *Appl Environ Microbiol* 58: 877-883.
135. Yurlova NA & De Hoog GS (1997) A new variety of *Aureobasidium pullulans* characterized by exopolysaccharides structure, nutritional physiology and molecular features. *Antonie van Leeuwenhoek* 72: 141-147.
136. Elinov NP, Marikhin VA, Dranishnikov AN, Myasnikova LP & Maryukhta YB (1975) Peculiarities of the glucan produced by a culture of *Aureobasidium (Pullularia) pullulans. Dokl Akad Nauk SSSR* 221: 213-216.
137. Chi Z & Zhao S (2003) Optimization of medium and new cultivation conditions for pullulan production by a new pullulan-producing yeast strain. *Enzyme Microbial Technol* 33: 206-211.
138. Fraser CG & Jennings HJ (1971) A glucan from *Tremella mesenterica* NRRL-Y6158. *Can J Chem* 49: 1804-1807.
139. Waksman N, De Lederkremer RM & Cerezo AS (1977) The structure of an α-D-glucan from *Cyttaria harioti* Fischer. *Carbohyd Res* 59: 505-515.
140. Oliva EM, Cirelli AF & De Lederkremer RM (1986) Characterization of a pullulan in *Cyttaria darwinii. Carbohyd Res* 158: 262-267.
141. Corsaro MM, Castro DC, Evidente A, Lanzetta R, Molinaro A, Parilli M & Sparapano L (1998) Phytotoxic extracellular polysaccharide fractions from *Cryphonectria parasitica* (Murr) Barr strains. *Carbohyd Pol* 37: 167-172.
142. Heald PJ & Kristiansen B (1985) Synthesis of polysaccharide by yeast-like forms of *Aureobasidium pullulans. Biotechnol Bioeng* 17: 1516-1519.
143. Moscovici M, Ionescu C & Oniscu C (1993) Exopolysaccharide biosynthesis by a fast-producing strain of *Aureobasidium pullulans. Biotechnol Lett* 15: 1167-1172.

144. Youssef F, Roukas T & Biliaderis CG (1999) Pullulan production by a nonpigmented strain of *Aureobasidium pullulans* using batch and fedbatch culture. *Process Biochem* 34: 355-366.
145. Le Duy A & Boa JM (1983) Pullulan production from peat hydrolyzate. *Can J Microbiol* 29: 143-146.
146. Israilides C, Scanlon B & Smith A (1994) Characterization of pullulans produced from agro-industrial wastes. *Carbohyd Pol* 25: 203-209.
147. Roukas T (1999) Pullulan production from brewery wastes by *Aureobasidium pullulans. World J Microbiol Biotechnol* 15: 447-450.
148. Shingel KI (2004) Current knowledge on biosynthesis, biological activity, and chemical modification of the exopolysaccharide, pullulan. *Carbohyd Res* 339: 447-460.
149. Gibbs PA & Seviour RJ (1992) Influence of bioreactor design on exopolysaccharide production by *Aureobasidium pullulans. Biotechnol Lett* 14: 491-494.
150. Reed-Hamer B & West TP (1994) Effects of complex nitrogen sources on pullulan production relative to carbon source. *Microbios* 80: 83-90.
151. West TP & Reed-Hamer B (1991) Ability of *Aureobasidium pullulans* to synthesize pullulan upon selected sources of carbon and nitrogen. *Microbios* 67: 117-124.
152. Seo HP, Son CW, Chung CH, Jung DI, Kim SK, Gross RA, Kaplan DL & Lee JW (2004) Production of high molecular weight pullulan by *Aureobasidium pullulans* HP-2001 with soybean pomace as a nitrogen source. *Biores Technol* 95: 293-299.
153. Wang Y & McNeil B (1996) Scleroglucan. *Crit Rev Biotechnol 16:* 185-215.
154. Lee KY & Yoo YJ (1993). Optimization of pH for high molecular weight pullulan. *Biotechnol Lett* 15: 1021-1024.
155. McNeil B & Kristiansen B (1990) Temperature effects on polysaccharide formation by *Aureobasidium pullulans* in stirred tanks. *Enzyme Microbial Technol* 12: 521-526.
156. West TP & Reed-Hamer B (1993) Effect of temperature on pullulan production in relation to carbon source. *Microbios* 75: 261-268.
157. Singh RS, Saini GK & Kennedy JF (2008) Pullulan: Microbial sources, production and applications. *Carbohydr Pol* 73: 515-531.
158. Kachhawa DK, Bhattacharjee P & Singhal RS (2003) Studies on downstream processing of pullulan. *Carbohyd Pol* 52: 25-28.
159. Leathers TD (2003) Biotechnological production and applications of pullulan. *Appl Microbiol Biotechnol* 62: 468-473.
160. Mitsuhashi M, Yoneyama M & Sakai S (1990) Growth promoting agent for bacteria containing pullulan with or without dextran. Canadian Patent 2,007,270.
161. Sugawa KY, Kondou F, Mandai T & Yoneyama M (1994) Effects of pullulan, polydextrose and pectin on cecal microflora. *Oyo Toshitsu Kagaku* 41: 413-418.
162. Hijiya H & Shiosaka M (1975b) Process for the preparation of food containing pullulan and amylose. US Patent 3, 872, 228.
163. Kato K & Shiosaka M (1975a) Food compositions containing pullulan. US Patent 3, 875, 308.
164. Miyaka T (1979) Shaped matters of tobaccos and process for preparing the same. Canadian Patent 1 049 245.
165. Miyamoto Y, Goto H, Sato H, Okano H & Iijima M (1986) Process for sugar-coating solid preparation. US Patent 4, 610, 891.
166. Izutsu Y, Sogo K, Okamoto S & Tanaka T (1987) Pullulan and sugar coated pharmaceutical composition. US Patent 4, 650, 666.

167. Leung S-HS, Leone RS, Kumar LD & Kulkarni N (2000) Fast dissolving orally consumable films. PCT International Application WO 00/18365.
168. Tsukada N, Hagihara K, Tsuji K, Fujimoto M & Nagase, T (1978) Protective coating material for lithographic printing plate. US Patent 4 095 525.
169. Sasago M, Endo M, Takeyama K & Nomura N (1988) Water-soluble photopolymer and method of forming pattern by use of the same. US Patent 4, 745, 042.
170. Vermeersch JT, Coppens PJ, Hauquier GI & Schacht EH (1995) Lithographic base with a modified dextran or pullulan hydrophobic layer. US Patent 5, 402, 725.
171. Yamaguchi S & Sunamoto J (1991) Fatty emulsion stabilized by a polysaccharide derivative. US Patent 4 997 819.
172. Mitsuhashi M & Koyama S (1987) Process for the production of virus vaccine. US Patent 4, 659, 569.
173. Suginoshita Y, Tabata Y, Matsumura T, Toda Y, Nabeshima M, Moriyasu F, Ikada Y & Chiba T (2002) Liver targeting of human interferon-with pullulan based on metal coordination. *J Control Release* 83: 75-88.
174. Sunamoto J, Sato T, Hirota M, Fukushima K, Hiratani K & Hara K (1987) A newly developed immunoliposome-an egg phosphatidylcholine liposome coated with pullulan bearing both a cholesterol moiety and an IgMs fragment. *Biochim Biophys Acta* 898: 323-330.
175. Barker PE & Ajongwen NJ (1991) The production of the enzyme dextransucrase using nonaerated fermentation techniques. *Biotechnol Bioeng* 37: 703-707.
176. Lazic ML, Veljkovic VB, Vucetic JI & Vrvic MM (1993) Effect of pH and aeration on dextran production by *Leuconostoc mesenteroides. Enzyme Microb Technol* 15: 334-338.
177. Choplin L & Sabatie J (1986) The effect of synthesis temperature on the rheological properties of native dextran. *Biotechnol Lett* 6: 425-430.
178. Veljkovic VB, Lazic ML, Rutic DJ, Jovanovic SM & Skala DU (1992) Effects of aeration on extracellular dextransucrase production by *Leuconostoc mesenteroides. Enzyme Microbial Technol* 14: 665-668.
179. Landon RS, Law RCS & Webb C (1993) Fermentation broth rheology during dextran production by *Leuconostoc mesenteroides* B512. *Appl Microbiol Biotechnol* 40: 251-257.
180. Koepsell HJ & Tsuchiya HM (1952) Enzymatic synthesis of dextran. *J Bacteriol* 63: 293-295.
181. Robyt JF & Walseth TF (1978) The mechanism of acceptor reactions of *Leuconostoc mesenteroides* NRRL B-512F dextransucrase. *Carbohyd Res* 61: 433-445.
182. Robyt JF & Eklund SH (1983) Relative, quantitative effects of acceptors in the reaction of *Leuconostoc mesenteroides* NRRL B-512F dextransucrase. *Carbohyd Res* 121: 279-286.
183. Tecante A, Lopez MA & Garcia RA (1986) Rheological characterisation of dextran-enzymatic synthesis media. *J Appl Polym Sci* 31: 2337-2350.
184. Shamala TR & Prasad MS (1995) Preliminary studies on the production of high and low viscosity dextran by *Leuconostoc* spp. *Process Biochem* 30: 237-241.
185. Collins PM (1998) *Dictionary of Carbohydrates*, Chapman and Hall, London, UK.
186. Miraftab M, Qiao Q, Kennedy JF, Groocock MR & Anand SC (2002) Advanced wound care materials: developing an alginate fibre containing branan ferulate. *J Wound Care* 11: 353-356.
187. Pena C, Hernandez L & Galindo E (2006) Manipulation of the acetylation degree of *Azotobacter vinelandii* alginate by supplementing the culture medium with 3-(N-morpholino)-propane-sulfonic acid. *Lett Appl Microbiol* 43: 200-204.

188. Pena C, Campos N & Galindo E (1997) Changes in alginate molecular mass distribution, broth viscosity and morphology of *Azotobacter vinelandii* cultured in shake flasks. *Appl Microbiol Biotechnol* 48: 510-515.
189. Clementi F, Mancini F & Moresi M (1998) Rheology of alginate from *Azotobacter vinelandii* in aqueous dispersions. *J Food Eng* 36: 51-62.
190. Buchs J, Lotter S & Milbradt C (2001) Out of phase operating conditions, a hitherto unknown phenomenon in shaking bioreactors. *J Biochem Eng* 7: 135-141.
191. Reyes C, Pena C & Galindo E (2003) Reproducing shake flasks performance in stirred fermentors: production of alginates by *Azotobacter vinelandii. J Biotechnol* 105: 189-198.
192. Khanafari A & Sepahei AA (2007) Alginate biopolymer production by *Azotobacter chroococcum* from whey degradation. *Int J Env Sci Tech* 4: 427-432.
193. Gacesa P (1998) Bacterial alginate biosynthesis-recent progress and future prospects. *Microbiology* 114: 1133-1143.
194. Neidleman SL (1991) Microbial production of biochemical. The genetic engineer and biotechnologist. *Biopaper J* (May-June): 20-22.
195. Hartmeier W (1986) *Immobilized Biocatalysts: An Introduction*, Springer-Verlag, Heidelberg, Germany.
196. Schmidt RJ (1986) Xerogel dressings - an overview. In: *Advances in Wound Management*, Turner TD, Schmidt RJ & Harding KG (eds), Wiley, Chichester, UK, pp 65-71.
197. Thomas S (2000) Alginate dressings in surgery and wound management. Part 1. *J Wound Care* 9: 56-60.
198. Thomas S (2000) Alginate dressings in surgery and wound management. Part 2. *J Wound Care* 9: 115-119.
199. Thomas S (2000) Alginate dressings in surgery and wound management. Part 3. *J Wound Care* 9: 163-166.
200. Pollock JT (2002) Sphingan group of exopolysaccharides (EPS). In: *Biopolymer,* Vol 5, Vandamme EJ, Beats DS & Stainbuchel A (eds), Wiley-VCH Verlag, Weinheim, Germany, pp 239-257.
201. Kang KS, Colegrove GT & Veeder GT (1982) Deacetylated polysaccharide S-60. US patent 4385123.
202. Bajaj IB, Saudagar PS, Singhal RS & Pandey A (2006) Statistical approach to optimization of fermentative production of gellan gum from *Sphingomonas paucimobilis* ATCC 31461. *J Biosci Bioeng* 102: 150-156.
203. Ashtaputre AA & Shah AK (1995) Studies on the exopolysaccharide from *Sphingomonas paucimobilis* GS 1: Nutritional requirements and precursors forming enzymes. *Curr Microbiol* 31: 234-238.
204. Nampoothiri KM, Singhania RR, Sabarinath C & Pandey A (2003) Fermentative production of gellan using *Sphingomonas paucimobilis. Process Biochem* 38: 1513-1519.
205. Fialho A, Martins L, Donval M, Leitao J, Ridout M, Jay A, Morris V & Sá-Correia I (1999) Structures and properties of gellan polymers produced by *Sphingomonas paucimobilis* ATCC 31461 from lactose compared with those produced from glucose and from cheese whey. *Appl Environ Microbiol* 65: 2485-2491.
206. West TP (2002) Isolation of mutant strain of *Pseudomonas* sp. ATCC 31461 exhibiting elevated polysaccharide production. *J Ind Microbiol Biotechnol* 29: 185-188.
207. Dlamini AM & Peiris PS (1997) Production of exopolysaccharide by *Pseudomonas* sp. ATCC 31461 (*Pseudomonas elodea*) using whey as fermentation substrate. *Appl Microbiol Biotechnol* 47: 52-57.

208. Madhavan NK, Reeta RS, Sabarinath C & Pandey A (2003) Fermentative production of gellan using *Sphingomonas paucimobilis. Process Biochem* 38: 1513-1519.
209. Hyuck J, Nam-Kyu L, Myung-Kyo S, Sung-Koo K, David LK & Jin-Woo L (2003) Production of gellan gum by *Sphingomonas paucimobilis* NK2000 with soybean pomace. *J Biochem Eng* 16: 357-360.
210. Martins LO & Sacorreia I (1993) Temperature profiles of gellan gum synthesis and activities of biosynthetic enzymes. *Biotechnol Appl Biochem* 20: 385-395.
211. Giavasis I, Harvey LM & McNeil B (2000) Gellan gum. *Crit Rev Biotechnol* 20: 177-211.
212. Banik RM, Kanari B & Upadhyay SN (2000) Exopolysaccharides of gellan family: prospects and potential. *World J Microbiol Biotechnol* 16: 407-414.
213. Sanderson GR & Clark RC (1983) Laboratory-produced microbial polysaccharide has many potential food applications as a gelling, stabilizing and texturing agent. *Food Technol* 37: 63-70.
214. Shangu D, Valiant M, Tutlane V, Weinberg E, Weissberger B, Koupal L, Gadebusch H & Stapley E (1983) GELRITE as an agar substitute in bacteriological media. *Appl Environ Microbiol* 46: 840-845.
215. Lin CC & Casida LE (1984) GELRITE as a gelling agent in media for the growth of thermophilic microorganisms. *Appl Environ Microbiol* 47: 427-429.
216. Sanzgiri YD, Maschi S, Crescenzi V, Callegaro L, Topp EM & Stella VJ (1993) Gellan-based systems for ophthalmic sustained delivery of methylprednisolone. *J Control Release* 26: 195-201.
217. Wataru K, Shozo M & Attwood D (2003) Sustained delivery of paracetamol from *in situ*-gelling gellan and sodium alginate formulations. *Int J Pharm* 258: 55-64.
218. Wang Y & McNeil B (1995) Effect of temperature on scleroglucan synthesis and organic acid production by *Sclerotium glucanicum. Enzyme Microbial Technol* 17: 893-899.
219. Farina JI, Sineriz F, Molina OE & Perotti NI (2001) Isolation and physicochemical characterization of soluble scleroglucan from *Sclerotium rolfsii* – Rheological properties, molecular weight and conformational characteristics. *Carbohydr Polym* 44: 41-50.
220. Ferguson RB, Westover JD & Paul S (1969) Fungal polysaccharide composition and method for making the same. US Patent 3, 436, 311.
221. Bluhm TL, Deslands Y, Marchessult RH, Perz S & Rinaudo M (1982) Solid state and solution conformations of scleroglucan. *Carbohydr Res* 100: 117-130.
222. Farina JI, Sineriz F, Molina OE & Perotti NI (1998) High scleroglucan production by *Sclerotium rolfsii*: Influence of media composition. *Biotechnol Lett* 20: 825-831.
223. Halleck FE (1967) Polysaccharides and methods for the production thereof. US Patent 3, 302, 848.
224. Johal SS & Coleman GM (1990) Recovery of glucan by employing a divalent cation at an alkaline pH. US Patent 4, 950, 749.
225. Brigand G (1993) Scleroglucan. In: *Industrial Gums,* Academic Press, New York, USA, pp 461-472.
226. Marchetti M, Pisani S, Petropaolo V, Seganti L, Nicoletti R, Degener A & Orsi N (1996) Antiviral effect of polysaccharide from *Sclerotium glucanicum* towards herpes simplex virus type I infection. *Planta Med* 62: 303-307.
227. Coviello T, Palleschi A, Matricardi P, Bocchinfuso G, Grassi M & Alhaique F (2005) Scleroglucan: A versatile polysaccharide for modified drug delivery. *Molecules* 10: 6-33.
228. Krymchantowski A & Barbosa J (2002) Rizatriptan combined with rofecoxib vs. rizatriptan for the acute treatment of migraine: an open label pilot study. *Cephalalgia* 22: 309-312.

229. Akompong T, Ghori N & Haldar K (2000) *In vitro* activity of riboflavin against the human malaria parasite *Plasmodium falciparum. Antimicrob Agents Chemother* 44: 88-96.
230. Prabhakar G, Raju DC & Satyanarayana M (1993) Bioproduction of riboflavin from molasses and lentils. *Bioprocess Eng* 8: 247-249.
231. Crueger W & Crueger A (1989) *Biotechnology,* Sinauer Associates Inc, Sunderland, UK, pp 219-228.
232. Bretzel W, Schurter W, Ludwig B, Kupfer E, Doswald S, Pfister M & van Loon APGM (1999) Commercial riboflavin production by recombinant *Bacillus subtilis*: Down-stream processing and comparison of the composition of riboflavin produced by fermentation or chemical synthesis. *J Ind Microbiol Biotechnol* 22: 19-26.
233. Buzzinil P & Rossil J (1998) Semi-continuous and continuous riboflavin production by calcium-alginate-immobilized *Candida tropicalis* in concentrated rectified grape must. *World J Microbiol Biotechnol* 14: 377-381.
234. Stahmann KP, Revuelta JL & Seulberger H (2000) Three biotechnical processes using *Ashbya gossypii, Candida famata*, or *Bacillus subtilis* compete with chemical riboflavin production. *Appl Microbiol Biotechnol* 53: 509-516.
235. Park EY, Kato A & Ming H (2004) Utilization of waste activated bleaching earth containing palm oil in riboflavin production by *Ashbya gossypii. J Am Oil Chem Soc* 81: 57-62.
236. Pujari V & Chandra TS (2000) Statistical optimization of medium components for enhanced riboflavin production by a UV- mutant of *Eremothecium ashbyii. Process Biochem* 36: 31-37.
237. Ertrk E, Erkmen O & Oner MD (1998) Effect of various supplements on riboflavin production by *Ashbya gossypii* in whey. *Turk J Eng Environ Sci* 22: 371-376.
238. Olczyk C (1978) Investigations of the dynamics of the flavinogenesis in chosen yeasts of the genus *Candida. Pol J Pharmacol Pharm* 30: 83-88.
239. Keller PT, Quang LV, Bacher A, Kozlowski JE & Floss HG (1983) Biosynthesis of riboflavin analysis of biosynthetically carbon 13-labelled riboflavin by double-quantum and two dimentional NMR. *J Am Chem Soc* 105: 2505-2507.
240. Gutcho SJ (1973) *Chemicals by Fermentation, Chemical Technology Review No. 19*, Noyes Data Corporation, Park Ridge, USA, pp 323-325.
241. Ghanem KM, Sabry SA & Gamati Y (1992) Physiological study on riboflavin production by hydrocarbon-utilizing *Candida guilliermondii* Wickerham. *Zentrabl. Microbiol.* 147: 282-287.
242. Ghozlan HA (1992) Utilization of beet molasses for riboflavin production by *Mycobacterium phlei. J Basic Microbiol* 34: 157-162.
243. Perkins JB, Sloma A, Hermann T, Theriault K, Zachgo E, Erdenberger T, Hannett N, Chatterjee NP, Williams V, Rufo GA Jr, Hatch R & Pero J (1999) Genetic engineering of *Bacillus subtilis* for the commercial production of riboflavin. *J Ind Microbiol Biotechnol* 22: 8-18.
244. Lizama G, Moguel-Salazar F, Peraza-Luna F & Oritz-Vazquez E (2006) Riboflavin production from mutants of *Ashbya gossypii* utilising orange rind as a substrate. *Annals Microbiol* 57: 157-161.
245. Vandamme EJ (1992) Production of vitamins, coenzymes and related biochemicals by biotechnological processes. *J Chem Technol Biotechnol* 53: 313-327.
246. Johannes G, Hans K & Christoph M (1991) Method of purifying ferment-produced riboflavin. US Patent 5210023.
247. Martin, JH, Barg H, Warren MJ & Jahn D (2002) Microbial production of vitamin $B_{12}$. *Appl Microbiol Biotechnol* 58: 275-285.

248. Keuth S & Bisping B (1994) Vitamin $B_{12}$ production by *Citrobacter freundii* or *Klebsiella pneumoniae* during tempeh fermentation and proof of enterotoxin absence by PCR. *Appl Environ Microbiol* 60: 1495-1499.
249. Hollriegl V, Lamm L, Rowold J, Horig J & Renz P (1982) Biosynthesis of vitamin $B_{12}$, Different pathways in some aerobic and anaerobic microorganisms. *Arch Microbiol* 132: 155-158.
250. Roth JR, Lawrence JG & Bobik TA (1996) Cobalamin (coenzyme $B_{12}$): Synthesis and biological significance. *Ann Rev Microbiol* 50: 137-181.
251. Hunik JH (2000) Improved process for the production of vitamin $B_{12}$, Van H. Margaretha. World Patent WO0037669.
252. Marwaha SS & Sethi RP (1984) Utilization of dairy waste for vitamin $B_{12}$ fermentation. *Agric Wastes* 9: 111-130.
253. Eggersdorfer M (1996) Vitamins. In: *Ullmann's Encyclopedia of Industrial Chemistry*, Vol A27, Elvers B & Hawkinds S (eds), Wiley VCH, Weinheim, Germany, pp 443- 613.
254. Spalla C, Grein A, Garofano L & Ferni G (1989) Microbial production of vitamin $B_{12}$. In: *Biotechnology of Vitamins, Pigments and Growth Factors*, Vandamme EJ (ed), Elsevier Publishing Co, London, UK, pp 257-284.
255. Quesada-Chanto A, Silveira MM, Schmid-Meyer AC, Schroeder AG, DaCosta JPCL, Lopez J, Carvalho-Jonas MF, Artolozaga MJ & Jonas R (1998) Effect of the oxygen supply on pattern of growth and corrinoid and organic acid production of *Propionibacterium-shermanii*. *Appl Microbiol Biotechnol* 49: 732-736.
256. Ye K, Shijo M, Miyano K & Shimizu K (1999) Metabolic pathway of *Propionibacterium* growing with oxygen: enzymes, 13C NMR analysis, and its application for vitamin $B_{12}$ production with periodic fermentation. *Biotechnol Progr* 15: 201-207.
257. Salminen S, von Wright A, Morelli L, Marteau P, Brassart D, de Vos WM, Fonden R, Saxelin M, Collins K & Mogensen G (1998) Demonstration of safety of probiotics - a review. *Int J Food Microbiol* 44: 93-106.
258. Bijl HL (1998) Microbial production of vitamin $B_{12}$ and production and use of composition comprising high concentrations of vitamin $B_{12}$ activity. European Patent 0824152 A2.
259. Chauhan AS, Ramteke RS & Eipeson WE (1998) Properties of ascorbic acid and its applications in food processing: A critical appraisal. *J Food Sci Technol* 35: 381-392.
260. Boudrant J (1990) Microbial processes for ascorbic acid biosynthesis: A review. *Enzyme Microbial Technol* 12: 322-329.
261. Hancock RD & Viola R (2002) Biotechnological approaches for L-ascorbic acid production. *Trends Biotechnol* 20: 299-305.
262. Hoshino T & Sugisawa T (2008) Microbial production of vitamin C. US Patent 7341854.
263. Ninet L & Renaut J (1979) Catrotenoids. In: *Microbial Technology*, Vol 1, Peppler HJ and Perlman D (eds), Academic Press, New York, USA, pp 529-544.
264. Barnett, HV, Lilly, VG & Krause RF (1956) Increased production of carotene by mixed + and – culture of *Choanephora cucurbitarum*. *Science* 123: 141-142.
265. Hesseltine, CW & Anderson RF (1957) Microbiological production of carotene carotenoids. I. Zygospores and carotene produced by intra-specific and inter-specific crosses of Chonephoraceae in liquid medium. *Mycologia* 49: 449-452.
266. Lilly VG, Barnett HL & Krause RF (1960) The production of carotene by *Phycomyces blakesleeanus*. *West Virginia University Agricultural Exp Stn Bull* 441 T, Morgantown.
267. Borowitzka LJ, Moulton TP & Borowitzka MA (1986) Salinity and the compound production of beta-carotene from *Dunaliella salina*. *Nova Hedwigia* 83: 224-229.
268. Lorenz P & Zinke H (2005) White biotechnology: Differences in US and EU approaches? *Trends Biotechnol* 23: 570-574.

# Microbial Production of Lactic Acid and its Applications

# 18

**Aravindan Rajendran and Viruthagiri Thangavelu**

*Department of Chemical Engineering, Annamalai University, Annamalai Nagar 608 002 (T.N.)*

## 1. INTRODUCTION

Lactic acid (2-hydroxypropanoic acid) is a naturally occurring multifunctional organic acid, a valuable industrial chemical used as an acidulant and preservative in the food industry, pharmaceutical, leather and textile industries, as well as a chemical feedstock, and has received increased attention for use as a monomer for the production of biodegradable polymer (lactic acid). It can be produced either by biotechnological fermentation or chemical synthesis, but the former route has received considerable interest recently due to environmental concerns and the limited nature of petrochemical feedstocks. There have been various attempts to produce lactic acid efficiently from inexpensive raw materials. A viable process based on a low cost production media is desired to enhance the economics of fermentative production of L-lactic acid. The interest in the fermentative production of lactic acid has increased due to the prospects of environmental friendliness and using renewable resources and various agroindustrial wastes instead of petrochemicals. The economics of the production of lactic acid and its derivatives is dependent on many factors of which the cost of raw material is very significant. Biological conversion has an important role in waste utilization, and it is likely that various food-processing wastes, agricultural wastes and different industrial wastes may contain useful substrates, which can be used for lactic acid production.

Lactic acid has a long history of application in fermentation and preservation of human foodstuffs. Lactic acid was first isolated in 1780 by a Swedish chemist, Carl Wilhelm Scheele. In 1789, Lavoisier named this milk component "*acide lactique*", which became the possible origin of the current terminology for lactic acid. In 1857, however, Pasteur discovered that it was not a milk component, but a fermentation metabolite generated by certain microorganisms. In 1839, Fremy, produced lactic acid by fermentation of carbohydrates such as sucrose, lactose, mannitol, starch and dextrin. The first commercial production of lactic acid started in the United States by a microbial process in 1881.

Lactic acid has a hydroxyl group adjacent to the carboxyl group, making it an alpha hydroxy acid. In solution, it can lose a proton from the acidic group, producing the lactate ion $CH_3CH(OH)COO^-$. It is miscible with water or ethanol, and is hygroscopic. Lactic acid is chiral and has two optical isomers, L(+)-lactic acid or (S)-lactic acid and the other, its mirror image, is D(-)-lactic acid or (R)-lactic acid but L(+)-lactic acid is the biologically important isomer. The L(+) form of lactic acid (Fig. 1) is used in food and drug industry, because the human body is adapted to assimilate only this form.

*Food Biotechnology: Principles and Practices*
*V.K. Joshi & R.S. Singh (Eds.)*
*I.K. International Publishing House Pvt. Ltd., India*
*pp 609–640*

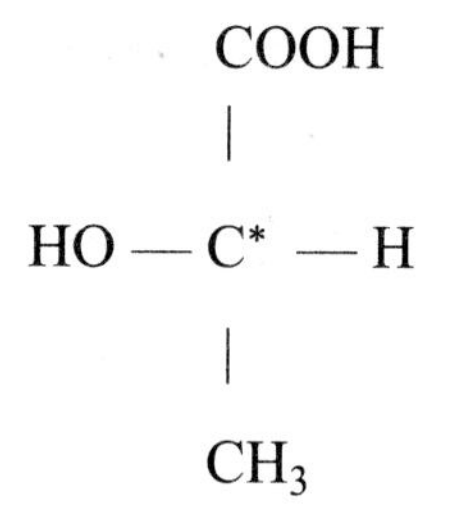

C* - asymmetric carbon atom

**Fig. 1:** Structure of L- (+) lactic acid.

The production of lactic acid by biotechnological fermentation process has received considerable interest recently due to environmental concerns and the limited nature of petrochemical feedstocks, low cost of substrates, low production temperature, and low energy consumption. At present, 90% of the world production of lactic acid is by bacterial fermentation and the rest is produced synthetically. The world wide market growth is increasing every year and the production in 2006 was about 68 million Kg per year. The worldwide market growth is expected to be between 10% and 15% per year.

The most commonly used synthetic method for chemical production of lactic acid is based on the hydrolysis of lactonitrile derived from acetaldehyde and hydrogen cyanide. Other possible chemical synthesis routes for lactic acid include base-catalyzed degradation of sugars, oxidation of propylene glycol, reaction of acetaldehyde, carbon monoxide and water at elevated temperatures and pressures, hydrolysis of chloropropionic acid, and nitric acid oxidation of propylene. None of these routes have led to technically and economically viable processes [1]. The biotechnological process can yield either form (D(+) or L(+)) alone, or a mixture in different proportions of two isomers, depending upon the microorganism, substrate and growth conditions used, whereas the chemical production only results in a mixture of the two isomers [2]. The significant advantage over the chemical synthesis is that biological production can use cheap raw materials, such as whey, molasses, starch waste, beet and cane-sugar and other carbohydrate-rich materials [3-6].

Lactic acid is used in a wide variety of specialized industrial applications where the functional speciality of the molecule is desirable. In 1999, Hepner rated the demand for lactic acid to grow continually at 5-8% annual clip, estimating its worldwide consumption at 130,000-150,000 metric tonnes per year [7]. Its use as a raw material for biodegradable plastics was identified in late 1940s and early 1950s [8]. Demand for lactic acid is expected to increase as rated by different surveys due to its use in biodegradable plastics and other large-scale industrial products. Yet the market is limited by cost in competition with polystyrene as prices for heat stable L(+) lactic acid and higher grades of lactic acid are more [7].

Lactic acid has got one of the prime positions in the family of carboxylic acids because of its widespread applications in food industries as a preservative and acidulant. Calcium lactate is a good dough conditioner, while sodium lactate is both a conditioner and an emulsifier. Lactic acid is being used as the pH controller in wine making. In dairy industry, lactic acid is used for the production of cheese. Its salts with calcium, sodium, potassium, zinc, iron, etc. find many applications in pharmaceutical, cosmetic and food industries. It is used as the raw material for different organic acids such as propionic acid, acetic acid, acrylic acid, etc. Technical grade lactic acid is extensively used in leather tanning industries for deliming. As a precursor for the biodegradable plastic, poly lactic acid (PLA), lactic acid has got much consideration [9]. There have been numerous investigations on the development of biotechnological processes for lactic acid production, with the ultimate objective to enable the process to be more effective and economical.

## 2. SIMULTANEOUS SACCHARIFICATION AND FERMENTATION

The bioconversion of carbohydrate materials to lactic acid can be made much more effective by coupling the enzymatic hydrolysis of carbohydrate substrates and microbial fermentation of the desired sugar into a single step, known as simultaneous saccharification and fermentation (SSF). SSF eliminates the bacterial contamination in the system. When raw starch is used as substrate, it has to be saccharified first, which is carried out by hydrolysis using enzymes such as α-amylase and glucoamylase following preliminary physio-chemical treatment. The starch is then used as a carbon source for lactic acid production by lactic acid bacteria (LAB). The cost reduction related to the saccharification is possible by direct lactic acid production from raw starch materials.

A direct benefit of the SSF is the decrease in inhibition caused by the mono or disaccharide accumulation, leading to an increase in the saccharification rate, that consequently increases the productivity, which in hand reduces the reactor volume and capital cost. Many factors such as pH, temperature, substrate and product concentration of glucose and lactic acid can affect the efficiency of SSF. SSF has been successfully employed for lactic acid production from raw starch materials using *Lactobacillus* or *Lactococcus*.

Another two major important functions are that they can take a role to carry out the SSF as both enzyme and lactic acid producer, and may secrete L-lactate as a sole isomer.

$$\underset{\text{Raw carbon (Starch/Cellulose)}}{C_{24}H_{46}O_{22}} + 2H_2O \rightarrow \underset{\text{Disaccharide}}{C_{12}H_{22}O_{11}} + 2H_2O \rightarrow \underset{\text{Glucose}}{C_6H_{12}O_6} \rightarrow \underset{\text{Lactic acid}}{CH_3 - CH\,(OH) - COOH} + CO_2 + \text{Biomass}$$

In SSF, starch is converted to disaccharide or oligosaccharide and hydrolysed to glucose by the action of glucoamylase. For liquefaction, α-amylase is used for the conversion of amylase into simple sugar by the breakdown of α-1,4 glycosidic bond. Amylopectins are not converted by this α-amylase enzyme because it contains α-1,6-glcosidic bond, that is cleaved or brokendown by the glucoamylase. Glucoamylase is added to the liquefied starch and fermentation is carried out simultaneously.

There are many *Lactobacillus* species, which can be used for the one-step lactic acid production. The starch utilizing *Lactobacillus* species are *L. amylophilus, L. amylovorus,* amylolytic strains of *L. plantarum* and *L. amylolyticus*. Semi-solid state fermentation is adopted in the case of certain amylolytic bacteria because they prefer to grow at higher moisture level. The filamentous fungus *R. oryzae* is very much utilized in the solid state fermentation for the production of L-lactic acid. *Rhizopus oryzae* and *R. arrhizus* can aerobically convert starch to lactic acid. Fungal species of *R. arrhizus* 36017 and *R. oryzae* 2062 produced lactic acid in a single stage simultaneous saccharification and fermentation process using potato, corn, wheat and pineapple waste streams as production media. *Rhizopus arrhizus* gave a high lactic acid yield upto 0.94-0.97 g/g of starch or sugars, whereas lactic acid yield of 0.65-0.76 g/g was produced by *R. oryzae* in 36 to 48 h of fermentation. Supplementation of 2 g/L of ammonium sulphate, yeast extract and peptone stimulated an increase in 8 to 15% lactic acid yield. Research on lactic acid production by *Rhizopus* species has continued basically because of the ease of product separation, product purification and the ability of the fungus to utilize both complex carbohydrate and pentose sugar [10]. Temperature of 30°C and pH 6.0 was found favourable for both saccharification and lactic acid fermentation, resulting in lactic acid yield 0.85-0.918 g/g associated with 1.5 to 3.5 g/l fungal biomass produced in 36 to 48 h fermentation [11]. The high concentration of starch used may result in a high concentrated mycelial biomass, leading to increased viscosity and consequent reduction of mass transfer efficiency. A disadvantage of SSF is the difference in cultivation conditions such as pH and temperature requirements. In many cases, conditions required for saccharification and fermentation have been found to be the low pH, temperature higher than 40°C for the enzymatic hydrolysis, whereas, the low pH inhibits the LA production and the high temperature may adversely affect the fungal cell growth [12]. So the SSF of lactic acid was enhanced by the optimal process condition, identification of microbial and biochemical kinetics.

## 2.1 Lactic Acid Producing Microorganisms

Microorganisms capable of producing lactic acid can be divided into two groups: bacteria and fungi [13]. The major advantage of using the fungi over the bacteria is the low cost. Due to the use of raw carbon source or waste materials of agrowaste, fermentation pH is very low as most fungi can tolerate low pH environment and easy downstream processing of biomass separation from fermentation broth.

Lactic acid bacteria can be divided into homo-fermentative and hetero-fermentative based upon the products produced from the fermentation of glucose. Homo-fermentative organisms ferment glucose to two moles of lactic acid, generating net 2 ATP per mole of glucose metabolized. Lactic acid is the major product of this fermentation. Hetero-fermentative lactic acid bacteria ferment 1 mole of glucose to 1 mole of lactic acid, 1 mole of ethanol, and 1 mole of $CO_2$. One mole of ATP is generated per mole of glucose, resulting in less growth per mole of glucose metabolized. Because of the low energy yields, lactic acid bacteria often grow more slowly than microbes capable of respiration, and produce smaller colonies of 2-3 mm. Table 1 shows the list of homo and hetero-fermentative lactic acid bacteria for lactic acid production. It is easy to determine whether a lactic acid bacterium has a homo or heterofermentative metabolism by the hot-loop test. A major endproduct of

**Table 1:** Homo and hetero-fermentative lactic acid bacteria used for lactic acid production.

| Genera and species | Homo-fermentative | Hetero-fermentative |
|---|---|---|
| ***Lactobacillus*** | | |
| *L. delbrueckii* | + | – |
| *L. lactis* | + | – |
| *L. bulgaricus* | + | – |
| *L. casei* | + | – |
| *L. plantarum* | + | – |
| *L. curvatus* | + | – |
| *L. brevis* | – | + |
| *L. fermentum* | – | + |
| ***Sporolactobacillus*** | | |
| *S. inulinus* | + | – |
| ***Streptococcus*** | | |
| *S. faecalis* | + | – |
| *S. cremoris* | + | – |
| *S. lactis* | + | – |
| ***Leuconostoc*** | | |
| *L. mesenteroides* | – | + |
| *L. dextranicum* | – | + |
| ***Pediococcus*** | | |
| *P. damnosus* | + | – |
| ***Bifidobacterium*** | | |
| *B. bifidum* | – | + |

heterofermentation is $CO_2$. In a medium containing glucose, this gas is highly soluble at high pH and will stay in solution. If the temperature of the solution is increased, $CO_2$ will become insoluble and will be released in the gaseous form. The hot-loop test consists of growing test isolates to saturation in a medium containing glucose. After incubation, a wire loop (inoculating loop) is heated to redness and plunged into the broth culture. This causes the liquid around the loop to heat up. If a test microorganism is heterofermentative, $CO_2$ bubbles will evolve close to the loop. The homofermentative lactic acid bacteria usually metabolize glucose via the Embden-Meyerhof pathway (i.e., glycolysis). The two major pathways for better assimilation of glucose and xylose in lactic acid are the Embden–Mayerhof-Parnas (EMP) pathway and the pentose phosphoketolase (PK) pathway. The lactic acid bacteria have limited biosynthetic ability, requiring amino acids, B vitamins, purines, pyrimidines and typically a sugar as carbon and energy source.

Various other classifications based on the nature of fermentation are:

(i) **Obligately hetero-fermentative rods:** These kinds of organisms produce DL-Lactate and they form gas from the glucose during the fermentation. This category includes *Lactobacillus* and *Weissella,* which lack the key enzyme (fructose-1,6 phosphate aldolase) of EMP pathway.

(ii) **Obligately homo-fermentative rods:** These organisms produce D-Lactate, L-Lactate or DL-Lactate. There is no gas production from glucose and they do not ferment any of the three pentose sugars (arabinose, ribose and xylose) tested.

(iii) **Facultatively hetero-fermentative rods:** Bacteria of this group produce DL-Lactate and they have mDAP on their cell wall, which do not produce gas from glucose. This group includes *Lactobacillus plantarum, L. pentosus* and *L. paraplantarum.*

(iv) **Hetero-fermentative cocci:** These organisms produce gas from glucose metabolism and D-Lactate, indicating that they belong to the genera *Leuconostoc* or *Weisella.*

(v) **Homo-fermentative cocci:** These groups do not produce any gas from glucose and include *Enterococcus, Lactococcus* or *Streptococcus.*

The microorganisms used in recent investigations for the production of lactic acid from various agro-industrial wastes are listed in Table 2. The major homofermentative LAB used in the lactic acid production from different carbon sources are *Lactococcus lactis* [14], *Lactobacillus delbrueckii* [15, 16], *L. helveticus* [17] and *L. casei* [18, 19]. Some of the homofermentative bacteria like *L. amylophilus, L. manihotivorans,* etc. can directly consume complex carbohydrates like starch [20, 21]. Amylolytic bacteria *Lactobacillus amylovorus* ATCC33622 had the efficiency of full conversion of liquefied corn starch to lactic acid with a productivity of 20 g/l h [22]. Huang *et al.* [11] produced lactic acid from potato starch waste water by using *Rhizopus oryzae* and *Rhizopus arrhizus*. A growth condition with pH 5.0-6.0 at 30°C was found to be favourable for both starch saccharification and lactic acid fermentation, resulting in lactic acid yield of 0.85-0.92 g/g associated with 1.5-3.5 g/l fungal biomass production at 36-48 h fermentation. John *et al.* [15] produced lactic acid from various agro residues by using *Lactobacillus delbrueckii,* and under the optimized conditions, the strain produced up to 249.1 mg lactic acid/gram dry substrate with a conversion efficiency of more than 99% of the total sugar available to lactic acid. Narita *et al.* [23] discussed the efficient production of L(+) lactic acid from raw starch by *Streptococcus bovis* 148. The productivity was highest at 37°C, with 14.7 g/l lactic acid produced from 20 g/l raw starch. The yield and optical purity of L-lactic acid were 0.88 and 95.6%, respectively.

Kourkoutas *et al.* [24] immobilized *Lactobacillus casei* cells on fruit (apple) pieces for the production of lactic acid. Wee *et al.* [25] obtained lactic acid from sugar molasses by using *Enterococcus faecalis* RKY1 by batch fermentation. The maximum lactic acid productivity was 5.3 g/l.h. Rivas *et al.* [26] produced lactic acid from corn cobs using *Lactobacillus rhamnosus* by simultaneous

**Table 2:** Microorganisms used in recent investigations on lactic acid production from agro-industrial wastes.

| Organism | Substrate | Lactic acid concentration | Yield | Productivity | Reference |
|---|---|---|---|---|---|
| *L. delbrueckii* | Wood | 108 $g/dm^3$ | - | 0.94 $g/dm^3$ h | 39 |
| *L. coryniformis* ssp. *torquens* | Wood card board | 23.4 $g/dm^3$ | 514 g/kg of glucose | 0.48 $g/dm^3$ h | 34 |
| *R. arrhizus* | Starch waste | 19.5-44.3 g/l | 0.85-0.96 g/g | - | 35 |
| *L. amylovorus* ATCC 33620 | Cassava starch | - | - | 7.7 g/l | 36 |
| *L.delbrueckii* | Alfalfa fibre | 34.5 g/l | 67.9% | - | 38 |
| *L. plantarum* | Alfalfa fibre | 42.8 g/l | 84.3% | - | 38 |
| *L. plantarum* NRRL B-787 | Solid waste | 17 g/l d | 0.42 g/g | - | 40 |
| *L. plantarum* NRRL B-787 | Solid waste | 19 g/l | 0.46 g/g | - | 40 |
| *R. oryzae* | Waste paper | 17.4 g/l | 0.59 g/g | - | 27 |
| *E. faecalis* | Molasses | 95.7[a] | 94.9% | 4.0 g/l h | 25 |
| *L. delbrueckii* CECT 286 | Molasses | 45.1 g/l | 87.8% | 2.7 g/l h | 41 |
| *L. delbrueckii* IFO 3202 | Molasses | 60.3 g/l | 95.4% | 3.4 g/l h | 42 |
| *S. bovis* | Raw starch | 14.7 g lactic acid/ 20 g raw starch | 0.88 g/g | - | 23 |
| *L. delbrueckii* | Agro wastes | 249 mg/gds | - | - | 15 |
| *E. faecalis* RKY1 | Barley | - | 0.94g/g | 0.88 g/l h | 29 |
| *B. licheniformis* TY7 | Kitchen refuse | 40.0 g/l | - | 2.5 g/l h | 43 |
| *L. delbrueckii* | Pineapple waste | 29.02 g/l | 92.7% | - | 44 |
| *L. delbrueckii* NRRL-B445 | Wet oxidized wheat straw | 31 $g/dm^3$ | 84% | 0.31 $g/dm^3$ h | 45 |
| *L. delbrium* | Corn cob | 34 $g/dm^3$ | 79% | 0.47 $g/dm^3$ h | 46 |

[a] 95.7 g/l of lactic acid produced from 200 g/l of molasses

saccharification and fermentation method. Park *et al.* [27] studied the conversion of waste office paper to L(+) lactic acid by the filamentous *Rhizopus oryzae*. The lactic acid yield and production rate were 0.59 g/g and 16.3 g/l.h, respectively. Garde *et al.* [28] used enzymatic and acidic treated hemi-cellulose hydrolysate from wet oxidized wheat straw as substrate for lactic acid production by *Lactobacillus brevis* and *Lactobacillus pentosus*. A lactic acid yield of 95% and complete substrate utilization were achieved with a mixed culture of *Lactobacillus brevis* and *Lactobacillus pentosus* on enzyme treated hemi-cellulose hydrolysate, which was a little higher than the yield of 88% on enzymatic treated hemi-cellulose hydrolysate, by the pure *L. pentosus* culture. Oh *et al.* [29] discussed the production of lactic acid from various agricultural resources such as barley, wheat, and corn, which were hydrolyzed by commercial amylolytic enzymes by using *Enterococcus faecalis* RKY1. In barley flour, 38 g/l of lactic acid could be produced with maximal dry cell weight of 2.3 g/l and lactic acid productivity of 0.88 g/l/h. When whole wheat flour was pre-treated by 0.3% (v/v) sulphuric acid treatment and sterilized by filtration, lactic acid productivity and maximal dry cell weight reached were 2.6 g/l/h and 5.90 g/l/h, respectively. By adaptation of *Enterococcus faecalis* RKY1 to corn steep liquor based medium, lactic acid productivity and maximal dry cell weight increased to 31% and 12% higher values than those of non-adapted *Enterococcus faecalis* RKY1. Ohkouchi *et al.* [21] studied the direct production of lactic acid from starch and food wastes using *Lactobacillus manihotivorans* LMG18011. Under these conditions, 19.5 g l(+) lactic acid was produced from 200 g food wastes by *L. manihotivorans* LMG18011. Gao *et al.* [30] investigated the production of lactic acid from fish wastes by using *Lactobacillus rhamnosus* NBRC3863. The unhydrolyzed fish wastes were having less lactic acid production, whereas the hydrolyzed fish wastes gave higher lactic acid production. Fermentation of soybean stalk hydrolysate by *L. sake* and *L. casei* yielded the lactic acid conversion of 48% and 56%, respectively, however, lactic acid conversion increased to 71% by co-inoculation of both strains [31].

Ohkouchi *et al.* [32] produced lactic acid from various organics by using *Lactobacillus manihotivorans* LMG18011. Naveena *et al.* [33] discussed the production of lactic acid by *Lactobacillus amylophilus* GV6 in semi-solid state fermentation using wheat bran. Maximum starch conversion to lactic acid was observed at 37°C and pH 6.5. Yanez *et al.* [34] obtained D-lactic acid from waste card board using *L. coryniformis*. SSF fed batch experiments led to D-lactic acid concentration up to 23.4 g/dm at a product yield of 514 g of lactic acid/kg of potential glucose and a volumetric productivity of 0.48 g/dm.h. Jin *et al.* [35] produced L(+) lactic acid from starch waste materials by simultaneous saccharification and fermentation by using *Rhizopus arrhizus*. Lactic acid yield of 0.95 g/g starch was achieved using soluble starch medium. The volumetric concentrations of lactic acid achieved varied from 25-40 g/l with increasing starch concentration from 20-60 g/l in 40 h fermentation. Xiaodong *et al.* [36] discussed the direct fermentative production of lactic acid from cassava and other starch substrates by using *Lactobacillus amylovorus. Lactobacillus amylovorus* utilized raw corn, rice and wheat starch medium to produce lactic acid with a productivity of 10.1, 7.9 and 7.8 g/l, but had lower productivities of 4.8 g/l and 4.2 g/l on cassava and potato starch in basal medium, respectively. When peptone is added to basal medium with cassava starch as substrate, conversion rate increased from 43% to 70% (7.7 g/l). Chatterjee *et al.* [37] produced lactic acid from waste potato mash by using *Lactobacillus cellobiosus*. Sreenath *et al.* [38] explained the production of lactic acid from various agricultural residues (alfalfa fibre, soy fibre) by using *Lactobacillus delbrueckii* and *Lactobacillus plantarum*. Lactic acid production was higher in alfalfa fibre and soy fibre compared to corn cobs and wheat straw. In *Lactobacillus plantarum,* the amount of lactic acid obtained from alfalfa fibre and soy fibre was 46 and 44 g/100 g of fibre, respectively. However, in *Lactobacillus delbrueckii*, the lactic acid production in soy fibre was 44 g/100 g fibre and that of alfalfa was 32 g/100 g fibre.

## 2.2 Lactic Acid Production from various Industrial Wastes

One of the main obstructions in the large-scale production of lactic acid is the cost of the raw material. The economics of production of lactic acid and its derivatives is dependent on many factors, of which the cost of raw material is very significant [14]. Application of agro-industrial waste residues in bioprocesses provides an alternative way to replace the refined and costly raw materials and the bulk use of such materials will help to solve many environmental hazards [4]. It is very expensive when purified sugars like glucose, sucrose, etc. are given as the feedstock for lactic acid production. Different food/agro-industrial products or residues form the cheaper alternatives to refined sugars for lactic acid production. Table 3 shows the different cheaper carbon sources used for the lactic acid production.

Sucrose-containing materials such as molasses are commonly exploited raw materials for lactic acid production. Some agricultural by-products that are the potential substrates for lactic acid production are corn starch [47], cassava starch [48], lignocellulose/hemicellulose hydrolyzates [49], cottonseed hulls, Jerusalem artichokes, corn cob, corn stalks [8], beet molasses [42, 50], wheat bran [20], rye flour [51], sweet sorghum [52], sugarcane press mud [53], cassava [36, 19, 15], barley starch [54], cellulose [55], molasses spent wash [58], corn fibre hydrolysate [57]. Like different agriculture resources, some waste materials of animal origin are also used as the carbon sources for the lactic acid fermentation. Mussel processing wastes are an under-exploited liquid by-product generated in large volume in the industrial steam treatment of mussels that contain glycogen as the main component.

**Table 3:** Effect of various carbon sources on the production of lactic acid.

| Organism | Substrate | Lactic acid concentration (g/l) | Yield (g/g) | Productivity (g/l/h) | Reference |
|---|---|---|---|---|---|
| *L. amylovorus* ATCC 33620 | Cassava starch, | 4.8 | 0.48 | 0.69 | 36 |
| | Corn starch, | 10.0 | 1.0 | 1.2 | |
| | Potato starch, | 4.2 | 0.42 | 0.14 | |
| | Rice starch, | 7.9 | 0.79 | 0.86 | |
| | wheat starch | 7.8 | 0.78 | 1.2 | |
| *L. delbrueckii* IFO 3534 | Hydrolyzed news paper, | 24 | 0.85 | 0.48 | 96 |
| | Hydrolyzed pure cellulose | 53 | 0.60 | 0.53 | |
| *L. paracasie* No. 8 | Glucose, | 95 | 5.6 | 0.95 | 55 |
| | Sweet sorghum | 91 | 10.0 | 0.91 | |
| *L. pentosus* | Hydrolyzed wood | 40 | 1.3 | 0.70 | 97 |
| *L. plantarum* | Hydrolyzed soluble starch, | 15 | - | 0.30 | 98 |
| | Hydrolyzed tapioca, | 15 | - | 0.35 | |
| | Hydrolyzed tapioca flour | 17 | - | 0.34 | |
| *L. acidophilus* R | Whey | 20 | - | - | 99 |
| *L. rhamnosus* ATCC 10863 | Hydrolyzed molasses | 16 | - | 0.81 | 100 |

Some industrial waste products, such as whey and molasses, are of interest as common substrates for lactic acid production. Whey is a major by-product of the dairy industry, and it contains lactose, proteins, fat, and mineral salts. For complete utilization of whey lactose, it is necessary to supplement whey with an additional nitrogen source [58]. Whey supplemented with yeast extract has been found suitable for rapid production of lactic acid using *L. helveticus* [59, 60, 61]. According to Fitzpatrick and O'Keeffe [62], the addition of whey protein hydrolysate to whey medium would make the fermentation more economically viable and would also reduce the amount of unused nutrients left during fermentation. Also, there have been several attempts to produce lactic acid from whey by batch culture of *L. casei* [63]. The strain most used for lactic acid production from whey was *L. delbrueckii* subsp. *bulgaricus* [64, 65], but in many studies *L. helveticus* [66, 67, 68] or *L. casei* [69, 70] was utilized. Molasses is a waste product from the sugar manufacturing process, and it usually contains a large amount of sucrose [58], but it is currently used almost as an animal feed. Although several studies on lactic acid production from molasses were previously reported, the lactic acid concentration (17-90 g/l) and volumetric productivity (2-4 g/l/h) in batch fermentation were relatively of low values [41, 42, 50, 71]. Moreover, the additional pretreatments of molasses, such as sulphuric acid, tricalcium phosphate, potassium ferrocyanide, and EDTA treatments, might be necessary to enhance the fermentation efficiency, because molasses contains heavy metals to some degree (e.g. iron, zinc, copper, manganese, magnesium, calcium, etc.), which might inhibit the cell growth, affect the medium pH, and inactivate the enzymes associated with product formation [72]. However, these additional steps for pretreatment are economically unfavourable because they may complicate the fermentation processes. *L. delbrueckii* and *E. faecalis* have recently been used for lactic acid production from molasses [25, 42, 50, 73]. Shukla *et al.* [74] also reported D(–)-lactic acid production from molasses with recombinant *E. coli* strain.

Sugarcane bagasse is one of the largest cellulosic agro-industrial byproducts, which was used as the inert support for the production of various metabolites [75, 76, 77]. Soccol *et al.* [76] studied the potential of sugarcane bagasse as inert support, which was impregnated with glucose and calcium carbonate medium for the production of lactic acid from a strain of *Rhizopus oryzae*. Cassava ranks as the world's sixth important food crop. A large part of cassava is used in industries to obtain starch, and the industrial application of cassava generates solid residue in the form of bagasse [78]. Cassava bagasse traps a large part of starch in its fibrous matter. The utilization of starchy materials in the place of expensive refined sugar is economical [79].

Corn cobs are important by-products of the sweet corn processing industry [80]. They are currently either used as animal feed or returned to the harvested field. Corn cobs are rich in cellulose and hemi-cellulose that have been used as a cheap source of raw material for production of soluble sugars and other value-added products by enzymatic and microbial fermentation processes [81]. Ram horns consist of k-keratin, which is relatively rich in cysteine (up to 22%). In addition, they contain most of the other common amino acids [82, 83]. Kurbanoglu *et al.* [84] produced lactic acid from ram horns hydrolysate by using *Lactobacillus casei* ATCC10863.

Agricultural products such as corn, potato and wheat containing large amounts of starchy substrate have been preferred as raw materials for the production of lactic acid [58]. Food wastes are also important raw material for the production of lactic acid. The food wastes contain polysaccharides such as starch, as well as various oligosaccharides. Pintado *et al.* [85] used amylolytic strains of lactic acid bacteria to direct utilization of food processing waste material. Sakai *et al.* [86] reported L(+) lactic acid production from food wastes pretreated by adding commercial glucoamylase.

In recent years, for the sake of decreasing environmental pollution and expense of lactic acid production, various wastes like kitchen waste [87, 88, 89], waste water sludge [90, 91], waste office paper [27], crop residue including corn cobs, wheat stalk and bran, have been used for lactic acid production [20, 26, 28]. High quality poly-l-lactic acid can effectively be produced from municipal kitchen refuse [92]. The thermophilic bacterium *Bacillus coagulans* is quite useful for producing optically active l-lactic acid from kitchen refuse under non-sterile condition [93]. Sakai *et al.* [94] isolated a thermotolerant bacterium and produced optically pure l-lactic acid from kitchen refuse by the bacterium under open condition. Idris *et al.* [44] produced lactic acid from pineapple wastes. Ueno [95] used grape invertase to improve the production of lactic acid using *Lactococcus lactis* utilizing canned pineapple syrup as substrate. The efficiency and economics of lactic acid production has always been a concern from various aspects, and the ability to use fruit waste such as liquid pineapple waste, that is rich in nutrients such as glucose, could prove beneficial.

## 2.3 Additional Nitrogen Sources for Enhancing Lactic Acid Production

Traditionally, the most common nutrients for the preparation of fermentative media are yeast extract and peptone, which turn out to be very expensive being able to account for almost 30% of the total cost of the process. Because of this, the search for alternative, financially competitive nutrient sources is particularly interesting. When *L. casei* was cultivated in a medium containing only the inorganic nitrogen source and vitamins, no lactic acid was produced. With 100 g/l glucose and malt sprouts as nitrogen source, a satisfactory lactic acid concentration of 88 g/l was obtained in 66 h as compared to 100 g/l in 48 h with yeast extract. Thus, malt sprouts were regarded as the best alternative to yeast extract as the source of nitrogen and growth factors. With 1.5% peptone (corresponding to 2.2% yeast extract), lactic acid production was lower (79 g/l in 66 h) than with medium containing 5.4% malt sprouts (88 g/l in 66 h). Grass extract, a waste material from silage production, also gave satisfactory results, with 74 g/l lactic acid in 73 h.

Jin *et al.* [35] studied the effect of various nitrogen sources in starch waste materials by using *R. arrhizus* 36017. *R. arrhizus* 36017 utilized $(NH_4)_2SO_4$, urea, yeast extract and peptone for lactic acid production in the soluble starch medium, but urea was the least favoured nitrogen source. The concentration of lactic acid, fungal biomass and other by-products was increased as the concentration of these nitrogen sources was enhanced, but there was only a minor impact on ethanol production. The lactic acid concentration did not vary significantly as the yeast extract and peptone concentrations varied from 5 to 10 g/l. An increase from 2 to 5 g $(NH_4)_2SO_4$/l decreased the lactic acid concentration. Xiaodong *et al.* [36] explained that *Lactobacillus amylovorus* had poor growth rates with 1% cassava starch as carbon source when sufficient yeast extract was used as the nitrogen source. When peptone (1%) was added to the basal medium, the strain gave a better growth on cassava starch. This is indicated by the faster rate of glucose consumption and higher final lactic acid concentration and conversion rate of 7.7 g/l and 70%, respectively as compared to 4.8 g/l and 43% conversion with only yeast extract. Ohkouchi *et al.* [32] reported that lactic acid production was improved by the supplementation of nitrogen sources with tryptone, beef extract, and MRS-complex. In particular, by supplementation with tryptone or MRS-complex, the total nitrogen contents were improved to 0.58% and 0.72%, respectively, and both the acceleration of the bioconversion rate and a doubling of lactic acid production, from 35 g/l up to 70 g/l, were observed. There was a little improvement on the bioconversion when supplemented with beef extract (the total nitrogen content; 0.48%). Nancib *et al.* [101] reported the effects of supplementing date juice with different nitrogen sources such as yeast extract, ammonium sulphate, tryptic soy, urea, peptone and casein hydrolyzate on the lactic acid production by *Lactobacillus casei* subsp. *rhamnosus*. None of the non-yeast extract nitrogen sources gave lactic acid concentrations as high as that of yeast extract. On the other hand, ammonium sulphate seems to be a good alternative to yeast extract.

## 2.4 Parameters Affecting Lactic Acid Production

### *2.4.1 Effect of temperature*

The effect of temperature on the production of lactic acid using various microorganisms has been shown in Table 4. The temperature giving the highest productivity was in some cases lower than the temperature resulting in highest lactic acid concentration and yield, whereas in others the same temperature gave the best results in all categories. John *et al.* [15] studied the production of lactic acid at various temperatures viz., 25, 30, 37, 44 and 51°C. The maximum production of lactic acid was 238.8 mg/g dry substrate at 37°C. However, the culture showed equally good production (234.35 mg/g dry substrate) at 30°C. Huang *et al.* [11] studied the effect of the cultivation temperature on the simultaneous saccharification and fermentation of potato starch waste water to lactic acid by *Rhizopus oryzae* and *Rhizopus arrhizus* by controlling the growth temperature at 22, 30, 35 and 40°C. The maximum lactic acid concentration was found at 30°C, while both cultures produced the lowest biomass concentration at 35°C. The fermentation temperature of 30°C was the optimum cultivation temperature for both saccharification and fermentation by the *Rhizopus* species in the solid state fermentation (SSF). Narita *et al.* [23] investigated the effect of temperature on the production of lactic acid from 20 g/l raw starch by *S. bovis* 148. The maximum lactic acid concentrations at 30, 37 and 45°C were 10.60, 14.73 and 10.77 g/l, respectively. Kourkoutas *et al.* [24] immobilized *Lactobacillus casei* cells on apple and quince pieces separately, and then 15 successive whey fermentation batches for each immobilized biocatalyst were carried out at 30, 37 and 45°C. At 45°C, the values obtained for yield and lactic acid productivity were higher as compared to process performance at lower temperature. Wee *et al.* [25] obtained the highest lactic acid yield (94.9%) and productivity (4.0 g/l/h) at 38°C from sugar molasses by using *Enterococcus faecalis* RKY1. Sakai *et al.* [43] isolated a thermotolerant L (+)-lactic acid producing bacterium identified as *Bacillus licheniformis* TY7. This organism shows optimum growth at 30°C and normal growth up to 65°C. Using non-sterile kitchen refuse at 50°C, the strain produced 40 g/l lactic acid with 97% optical purity and 2.5 g/l/h productivity.

Sakai *et al.* [94] discussed open L-lactic acid fermentation of food refuse using thermophilic *Bacillus coagulans* at various temperatures. Below 40°C, lactic acid accumulated is nearly racemic

**Table 4:** Effect of temperature on the production of lactic acid from various agro-industrial wastes.

| Organism | Substrate | Temperature (°C) | Lactic acid concentration | Yield | Productivity | Reference |
|---|---|---|---|---|---|---|
| *E. faecalis* RKY1 | Sugar molasses | 38 | 95.7a | 94.9% | 4.0 g/l h | 25 |
| *S. bovis* 148 | Raw starch | 30 | 10.6b | - | - | 23 |
| | | 37 | 14.7c | 0.88 g/g | - | |
| | | 45 | 10.7d | - | - | |
| *R. oryzae* | Agricultural commodities | 30 | - | 56% | - | 102 |
| *B. licheniformis* TY7 | Kitchen refuse | 50 | 40 g/l | - | 2.5 g/g | 43 |
| *B. coagulans* | Food refuse | 37 | 12.3 g/l | - | - | 94 |
| | | 45 | 16.7 g/l | - | - | |
| | | 55 | 18.0 g/l | - | - | |
| *L. delbrueckii* | Pineapple waste | 37 | 28.73 g/l | 91.7% | - | 44 |
| *L. paracasie* No.8 | Sweet sorghum | 30 | - | - | 1.5 g/l h | 52 |
| | | 36 | - | - | 1.9 g/l h | |
| | | 44 | - | - | 2.2 g/l h | |
| *L. casie* NRRL B-441 | Hydrolyzed barley starch | 37 | 140 g/l | 0.98 g/g | - | 54 |
| | | 41 | 117 g/l | 0.82 g/g | - | |
| *L. amylophilus* ATCC 49845 | Starch | 25 | 26 g/l | 0.52 g/g | 0.54 g/l h | 48 |
| | | 28 | 29 g/l | 0.58 g/g | 0.44 g/l h | |
| | | 35 | 30 g/l | 0.60 g/g | 0.33 g/l h | |
| *L. casie* subsp. *rhamnosus* | Malt sprout | 45 | 70 g/l | - | 3.5 g/l h | 103 |

[a] 95.7g/l of lactic acid produced from 200g/l of molasses; [b]10.6 g lactic acid/20 g raw starch; [c] 14.7 g lactic acid/20 g raw starch; [d] 10.7 g lactic acid/20 g raw starch.

mixtures and optical purity was very low. At temperature more than 50°C, a very small amount of D-lactic acid accumulated, and as a result, optical purity was more than 95%. The maximum accumulation of total lactic acid was observed at 37°C and 55°C, whereas optical purity was the highest at 55°C. Idris *et al.* [44] discussed the effect of temperature on lactic acid production from pineapple waste by using *Lactobacillus delbrueckii* at various temperatures (27, 30, 37, 40, 45 and 50°C). The highest lactic acid production and yield obtained were 28.73 g/l and 91.7%, respectively at 37°C. *L. delbrueckii* seems to grow well at 37°C, promoting maximum cell concentration and this contributes to maximum lactic acid concentration. Increasing the temperature beyond 40°C and below 37°C did not promote cell growth, and thus, lactic acid production is decreased.

Hujanen *et al.* [103] investigated the effect of temperature on the production of lactic acid by *L. casei* NRRL B-441 and *L. casei* subsp. *rhamnosus* NRRL B-445. The best production temperature for *L. casei* was 37°C, at which the highest maximum volumetric productivity of lactic acid, 4.1 g/l/h was obtained. The highest lactic acid concentration of 80 g/l was observed at 24 h. The best production temperature of *L. casei* subsp. *rhamnosus* was 37°C, at which the highest lactic acid productivity (3.5 g/l) and lactic acid concentration (70 g/l) were reached. Yu and Hang [102] studied the kinetics of direct fermentation of agricultural commodities to l (+)-lactic acid by *R. oryzae*. The rate of carbohydrate consumption and l (+)-lactic acid production from rice was much faster at 30°C than at 23°C or 37°C. The highest amount of l (+)-lactic acid was attained in 72 h at 30°C. The yield was 56% based on the amount of carbohydrate as glucose that was consumed. Under similar conditions, the yields from barley, cassava, corn and oats were 41%, 74%, 55% and 23%, respectively.

### 2.4.2 Effect of pH

The fermentation pH is either set at the beginning and then left to decrease due to acid production or is controlled by base titration, or by extraction, adsorption, or electro-dialysis of lactic acid. Table 5 shows the effect of pH on the production of lactic acid from various waste materials using different microorganisms. Wee *et al.* [25] discussed the production of lactic acid from molasses at various culture pH values. The optimum pH for cell growth of *E. faecalis* RKY1 was found to be 8.0. The lactic acid fermentation at pH 7.0 was completed faster than at pH 8.0, and the cell growth at pH 5.0 was almost ceased. The highest productivity was obtained at pH 7.0. Since the optimal pH for starch hydrolysis by amylase from *Streptococcus bovis* 148 was found to be between 5.6 and 6.5 [104], the pH was adjusted to 6.0 during the fermentation [23]. Huang *et al.* [11] observed the production of lactic acid from potato starch waste water by *Rhizopus oryzae* and *Rhizopus arrhizus*. A growth condition with pH 5.0-6.0 was found to be favourable for both the starch saccharification and lactic acid fermentation. Associated with 1.5-3.5 g/l fungal biomass produced, the lactic acid yield varying from 0.87 to 0.92 g/g was achieved in 36-48 h. John *et al.* [15] obtained lactic acid from agro-waste using *L. delbrueckii* at various pH values. The pH 6.5 was proved to be the optimum pH for lactic acid production (237.5 mg/gram dry substrate). Ohkouchi *et al.* [21] studied the effect of pH on the production of lactic acid from starch and food wastes by using *L. manihotivorans* LMG18011. The optimum pH was found to be in

**Table 5:** Effect of pH on the production of lactic acid from various agro-industrial wastes.

| Organism | Substrate | pH | Lactic acid concentration | Yield | Productivity | Reference |
|---|---|---|---|---|---|---|
| *E. faecalis* RKY1 | Sugar molasses | 7.0 | 95.7g | 94.9% | 4.0 | 25 |
| *S. bovis* 148 | Raw starch | 6.0 | 14.7 g lactic acid/ 20 g raw starch | 0.88 g/g | - | 23 |
| *R. oryzae* | Potato starch waste water | 5-6 | - | 0.85-0.92 g/g | - | 11 |
| *L. delbrueckii* | Agro wastes | 6.5 | 237.5 mg/gds | - | - | 15 |
| *L. manihotivorans* LMG 18011 | Food wastes | 4.5 | 41.4 g/l | 1.02 g/g | - | 21 |
| | | 5.0 | 48.7 g/l | 1.11 g/g | - | |
| | | 5.5 | 31.2 g/l | 0.93 g/g | - | |
| *B. licheniformis* TY7 | Kitchen refuse | 6.5 | 40.0 g/l | - | 2.5 g/l/h | 43 |
| *B. coagulans* | Food refuse | 6.5 | 86.0 g/l | 53% | - | 94 |
| *L. delbrueckii* | Pineapple waste | 6.5 | 29.02 g/l | 92.7% | - | 44 |
| *L. amylophilus* GV6 | Wheat bran | 6.5 | - | 98% | - | 33 |
| *L. delbrueckii* IFO 3534 | Newspaper | 4.7 | 24.0 g/dm$^3$ | 81% | 0.48 g/dm$^3$/h | 96 |
| *L. delbrueckii* NRRL-B445 | Switch grass | 5.0 | 15.0 g/dm$^3$ | 100% | 0.46 g/dm$^3$/h | 105 |
| *E. faecium* | Alfalfa | 6.2 | 8.3 g/l | 0.27 g/g | - | 106 |
| | | 5.8 | 27.0 g/l | 0.9 g/g | - | |
| *L. plantarum* | Sorghum | 5.5 | - | - | 2.1 g/l h | 107 |
| | | 6.0 | - | - | 2.0 g/l h | |
| | | 6.5 | - | - | 2.8 g/l h | |
| | | 7.0 | - | - | 5.1 g/l h | |
| | | 7.5 | - | - | 1.9 g/l h | |
| *L. salivarius* subsp. *salivarius* | Soy molasses | 5.6 | 5.5 g/l | 0.85 g/g | - | 108 |
| | | 6.0 | 5.4 g/l | 0.79 g/g | - | |
| | | 6.4 | 4.9 g/l | 0.82 g/g | - | |

between 5.0 and 5.5. Above pH 6.0 or below pH 4.5, this strain could not convert whole of the starch to lactic acid. Sakai *et al.* [43] isolated a thermotolerant L-lactic acid producing bacterium and identified it as *B. licheniformis* TY7. This strain showed optimum growth at pH 6.5 and temperature 30ºC. Using non-sterile kitchen refuse at 50ºC, the strain produced 40 g/l lactic acid and 2.5 g/l/h productivity. Sakai *et al.* [94] discussed open L-lactic acid fermentation of food refuse using thermophilic *Bacillus coagulans.* At pH 6.5, 86 g/l lactic acid with 97% optical purity was produced under non-sterile conditions. The yields of total lactic acid from total carbon and total sugar were 53% and 98%, respectively.

Idris *et al.* [44] studied the effect of various initial pH on the lactic acid production of the immobilized *L. delbrueckii* during the batch fermentation of liquid pineapple waste. The maximum lactic acid concentration is attained at initial pH of 6.5 with a yield of 29.02 g/l or 92.7%. Further increase in initial pH beyond 6.5 did not improve the lactic acid production. Naveena *et al.* [33] observed the maximum starch conversion to lactic acid by using *L. amylophilus* GV6 at optimum pH value of 6.5. Xiaodong *et al.* [36] observed highest lactic acid production from cassava and other substrates at optimum pH 5.5 by using *L. amylovorus.*

## 3. PURIFICATION OF LACTIC ACID

Lactic acid can be separated by several recovery processes, such as calcium precipitation, solvent extraction, and electro-dialysis [1, 13]. Among them, electro-dialysis has been found to be potentially useful in the recovery of lactic acid from fermentation broth [1, 109-111]. However, the treatment of wastewater, generated after the electro-dialysis operation, may result in an increase in lactic acid production costs, because this wastewater generally has a high biological oxygen demand (BOD >1000 ppm). Therefore, if the electro-dialysis wastewater can be used for subsequent fermentations, costs for lactic acid production and wastewater disposal can be reduced simultaneously. Many studies of electro-dialysis for lactic acid recovery have been conducted [66, 109-113]. The evaluation of lactic acid fermentation using electro-dialysis wastewater has rarely been investigated [25].

Lactic acid is sold in various commercial grades, and the better grades require that well-purified substrates be utilized in the fermentation medium in order to reduce the levels of impurities present during recovery which, without great difficulty, cannot be separated from the lactic acid. Also in this regard, the sugar should be depleted from the medium by harvest of the fermentation. One of the commercial grades of lactic acid, "crude" or "technical" grade is a coloured product prepared for commercial usage at mass fraction in water of 22, 44, 50, 66 and 80%. It is prepared by employing sulphuric acid to remove the calcium from the calcium lactate derived from the heated and filtered fermentation broth, followed by filtration, concentration, and refiltration to remove additional calcium sulphate. Thus, this grade of lactic acid contains many of the impurities from the fermentation medium, and it finds many industrial uses where purity of the product is not essential as, for example, in the deliming of hides in the leather industry. The 'edible' grade of lactic acid is snow coloured and is marketed at 50-80% strengths. Thus, it receives additional refining over that of technical lactic acid. Colourless, high purity lactic acids are the plastic grade, marketed at 50-80% strength, and "U.S.P" lactic acid is marketed at 50-80% strengths. Other commercial preparations of lactic acid are calcium lactate, sodium lactate, and copper lactate (a salt used in electroplating). The final recovered yields of technical and edible-grade lactic acids, based on the original carbohydrate of the medium, are approximately 85-90% and 80%, respectively. Plastic and U.S.P grades are prepared by further refining of technical grade lactic acid and therefore, slight to moderate yield losses are incurred during this refining. Lactic acid is commercially available in different grades (qualities). They are technical grade lactic acid (20-80%), food grade lactic acid (80%), pharmacopoeia grade lactic acid (90%), and plastic grade lactic acid. Pharmaceutical and food grade lactic acids are considered to be most important. For the recovery of lactic acid, additional calcium carbonate is added to the medium, the pH is adjusted to approximately 10, and the fermentation broth is heated and then filtered. This procedure converts whole of the lactic acid to calcium lactate, kills bacteria, coagulates protein of the medium, removes excess calcium carbonate and helps to decompose any residual sugar in the medium. Various processes are employed for the recovery and purification of the lactic acid:

(i) The heated and filtered fermentation broth is concentrated to allow crystallization of calcium lactate, followed by addition of sulphuric acid to remove the calcium as calcium sulphate. The lactic acid is then re-crystallized as calcium lactate, and activated carbon is used to remove coloured impurities. As an alternative to the latter step, the zinc salts of lactic acid are sometimes prepared because of the relatively lower solubility of zinc lactate.

(ii) The free lactic acid is solvent extracted with isopropyl ether directly from the heated and filtered fermentation broth. This is a counter current continuous extraction, and the lactic acid is recovered from the isopropyl ether by further counter-current washing of the solvent with water.

(iii) The methyl ester of the free lactic acid is prepared, which is separated from the fermentation broth by distillation followed by hydrolysis of the ester by boiling in dilute water solution (the methyl ester decomposes in water). The lactic acid is then obtained from the aqueous solution by evaporation of the water, and the methanol is recovered by distillation.

(iv) Secondary or tertiary alkyl amine salts of lactic acid are formed and then extracted from aqueous solution with organic solvents. The solvent is removed by evaporation, and then, the salt is decomposed to yield the free acid. An older procedure, not utilized commercially to any extent at present, involves direct high-vacuum steam distillation of the lactic acid from the fermentation broth, but decomposition of some of the lactic acid occurs.

The fermentation broth is generally heated to 70°C to kill the bacteria and then, acidified with sulphuric acid to pH 1.8. The clarified lactic liquor is then ion exchanged and concentrated to 80%. Smell and taste can be improved further by oxidative treatment with hydrogen peroxide. The lactic acid obtained at this stage is suitable for some food industries. The lactic acid produced from biological fermentation requires extensive purification operations. It is of particular importance that the recovery processing equipment be resistant to the corrosive action of the high concentrations of lactic acid that accumulates. Therefore, special stainless steel equipment is most often employed for this purpose.

Sun *et al.* [114] used two reactors with a rectifying column to carry out the recovery of lactic acid from the fermentation broth. Ammonium lactate obtained by fermentation was used directly to produce butyl lactate by reacting with butanol for 6 h, and the esterification yield of ammonium lactate was 87.7%. In this procedure, a cation exchange resin, which was modified by $SnCl_2$, replaced sulphuric acid as a catalyst, and neutral ammonium lactate replaced former lactic acid as a starting material, which not only eliminated corrosion of a reactor, but also avoided generating calcium salts as a by-product. Then, butyl lactate was rectified, and the purified butyl lactate was sequentially hydrolyzed into lactic acid in the presence of cation exchange resin in the $H^+$ form as a catalyst for 4 h. The hydrolysis yield was 89.7% and the purity of recovered lactic acid was 90%. Bouchoux *et al.* [115] investigated nanofiltration for usability in a specific lactic acid production process based on conventional and bipolar electrodialysis operations. Industrial fluids, corresponding to two potential integration levels and coming from an existing installation, were investigated. Nanofiltration was able to efficiently remove magnesium and calcium ions from a sodium lactate fermentation broth before its concentration and conversion by electro-dialysis (first potential integration level). Maximum impurities rejections and lactate recovery were obtained at maximum transmembrane pressures. $Mg^{2+}$ and $Ca^{2+}$ rejections were 64±7 and 72±7%, respectively, and lactate recovery flux reached 25±2 mol/m$^2$/h for pressure

$p = 20$ bar. Sulphate and phosphate ions were also partially removed from the broth (40 % rejection). At the invert, chloride ions were negatively retained by the membrane and were consequently more concentrated in the permeate. Nanofiltration also led to a nearly total decolouration of the fermentation broth. On the other hand, sulphate and phosphate rejections obtained from the filtration of a converted broth containing the lactic acid under its neutral form (second potential integration level) were also satisfactory, i.e. 47±5 and 51±5%, respectively. High recovery rates were observed in that case.

Tong *et al.* [89] reported the purification results of lactic acid from the fermentation broth with paper sludge as a cellulosic feedstock using weak anion exchanger, Amberlite IRA-92. Some factors such as flow rate, sample volume loaded, pH, and column were systematically examined to improve the purity, yield and productivity in lactic acid purification. Adsorption isotherm of standard lactic acid and lactic acid in the fermentation broth by anion exchanger IRA-92 were also investigated. Results indicate that in purification process the increase of pH of the fermentation broth ranging from 5.0 to 6.0 can significantly enhance the recovery yield, purity and productivity. The decrease of flow rate and sample volume loaded can also improve the recovery yield and purity, but apparently reduce the productivity. In addition, the scale-up of purification process in laboratory size had little influence on the recovery yield and purity. After optimization, the yield, purity and productivity were found to be about 82.6%, 96.2% and 1.16 g LA/(g resin/day), respectively.

Madzingaidzo *et al.* [113] studied the purification of cell free sodium lactate solutions by mono- and bi-polar electro-dialysis. Lactate was concentrated by mono-polar electro dialysis to a maximum of 150 g/l. At high feed mass lactate flux reached 300 $g/m^2/h$ with correspondingly high current efficiency in the 90% range. Relatively low water transport rates were observed during processing with mono-polar electro-dialysis. A low incidence of impurities was observed in the concentrate solutions with less than 2 g/l, glucose and 1.5 g/l acetate detected, respectively. Subsequent purification with bi-polar electro-dialysis yielded good performance parameters with water transport rates as low as 70 g $H_2O$ per mol per l lactate and lactate flux reaching a high of 300 $g/m^2/h$. Current efficiency for bi-polar electro-dialysis was in excess of 90%. Free lactic acid mass concentration reached a moderate of 160 g/l, while colour and other chemical impurities were significantly reduced. Additional bleaching and de-ionisation process steps should, however, be integrated to polish the free lactic acid for high-grade applications in the biodegradable thermoplastic and pharmaceutical industries. Acetic acid impurity remained at around 1 g/l. Significant reduction in colour and minerals in the product streams was observed during electro-dialysis purification.

The physico-chemical and operating effects of lactic acid, sodium lactate and ammonium lactate on the RO process have been investigated using a polyamide composite membrane by Liew *et al.* [116]. This particular type of membrane was found to swell at pH 2.2, but had no detectable solute-membrane affinity. The flux and permeate concentration were found to be an inverse function, whereas the solute reduction factor was a direct function of the pH value of feed. This was attributed to the greater concentration of ions dissociated into the solution as the pH was increased, and the fact that hydrated ions possess larger sized molecules than by merit of their charge densities, is believed to help enhance rejection by their lower rates of diffusion. At this stage, ammonium hydroxide is considered to be the optimum pH-controlling agent for lactic acid production by virtue of its reasonably high flux and solute rejection as well as its capability to augment cell growth by acting as a nitrogen source. In terms of operating conditions, an increase in pressure from 1 MPa to 7 MPa has contributed to a higher flux, which outweighs the effect of the increasing total solute loss and consequently, leads to an increase in solute rejection. On the contrary, an increase in feed mass fraction, especially to ≥3.80%, has incurred pronounced effects on concentration polarization, leading to a reduced solute rejection and concentration efficiency.

## 4. KINETICS OF LACTIC ACID FERMENTATION

The study of fermentation rates can be very useful for the design and control of both continuous and batch systems. The classic study of Luedeking and Piret [117] on the lactic acid fermentation by *Lactobacillus delbrueckii* indicated that the product formation kinetics combined growth associated and non growth-associated contributions. Luedeking and Piret states that the product formation rate depends upon both the instantaneous biomass concentration (X) and growth rate (dX/dt) in a linear fashion.

$$\frac{dP}{dt} = \alpha \frac{dX}{dt} + \beta X$$

where α and β represent the growth and non-growth associated product formation constants, with units of mole lactic acid/g cell and mole lactic acid/g cell/h, respectively. The production kinetics can be analyzed, according to Weiss and Ollis [118] by integrating the above equation using Malthus' law equation for X(t), which gives an equation with two initial conditions, ($X_0$, $P_0$), a final condition, $X_{max}$ and three parameters $\mu_0$, α and β,

$$P_t = P_0 + \alpha A(t) + \beta B(t)$$

where $A(t) = X_0 \dfrac{e^{\mu_0 t}}{1 - \left(\dfrac{X_0}{X_{max}}\right)\left(1.0 - e^{\mu_0 t}\right)} - 1.0$ and $B(t) = \dfrac{X_{max}}{\mu_0} \ln\left[1.0 - \dfrac{X_0}{X_{max}}\left(1.0 - e^{\mu_0 t}\right)\right]$

This two-parameter kinetic expression has proved extremely useful and versatile in fitting product formation data for many different fermentations.

Friedman & Gaden [119] demonstrated the inhibitory effect of lactate, by removal from the culture medium using dialysis, and this model may be improved by the addition of a term indicating dependence of the rate of acid production on inhibitor concentration, the lactic acid itself [73].

$$\frac{dP}{dt} = \left[\alpha \frac{dX}{dt} + \beta X\right]\left[1 - \frac{P}{P_{max}}\right]$$

According to this equation, dP/dt will become zero when P approaches $P_{max}$, the concentration greater than $P_{max}$ above which bacteria do not produce lactic acid.

The fermentation kinetics for the conversion of beet molasses, a valuable and economical fermentation substrate, to lactic acid by the homo-fermentative organism *Lactobacillus delbrueckii* CECT 286 was studied at controlled pH and temperature under anaerobic batch conditions by Monteagudo *et al.* [73]. An inhibitory effect of lactic acid on fermentation of beet molasses was observed, and the bacterium was able to produce lactic acid even after the growth ceased. A kinetic model for the fermentation is proposed. From this model, the maximum allowable lactic acid concentration above which growth stops and the lactic acid level above which bacteria stop producing lactic acid were found to be 45 $g/dm^3$ and 57 $g/dm^3$, respectively.

## 5. VARIOUS MODES OF FERMENTATION

The selection of various fermentation processes depends on the process condition and the microorganism and the volume required. The batch process is better in high lactic acid concentration than the continuous mode, but the volumetric productivity is low. If the raw material cost is expensive, the product yield should be high, as in batch or semi-continuous operation. When the volumetric productivity is maximized by continuous operation, the investment cost is low. The productivity is increased by recycling the cells, resulting in a high cell mass without reducing the yield. Repeated or semi-continuous batch modes increase the yield further. The free cell culture yield better productivity than the immobilized cell culture. The productivity and yield was increased by the mixed culture of free and co-immobilized *L. casei* and *L. lactis* cells in batch and fed-batch culture [69]. The maximum lactic acid production in fed-batch culture was obtained with both co-immobilized cells and free cell mixture at a substrate concentration of 100 g/l and a feeding rate of 250 ml/h. The accumulation of lactic acid leads to product inhibition, so the lactic acid produced was removed in fed-batch manner for higher productivity. In repeated fed-batch culture, co-immobilized cells gave a higher acid production compared with the free cell culture. The co-immobilized cells in Ca-alginate beads retained their ability to produce lactic acid for 20 days. In the continuous culture mode, the productivity and yield were high due to high dilution rate. However, the effect of dilution rate on the yield and productivities were inconclusive, and it has influence over the substrate and nutrient concentrations.

Table 6 shows the various modes of fermentation for microbial production of lactic acid. The reducing sugar concentration in the hydrolysate affects fermentations, as bacterial cells reduce lactic acid production in higher sugar concentration. The conversion of starch or cellulose to sugar consumes energy during liquefaction or saccharification and increase the cost of production. In simultaneous saccharification and fermentation mode, the hydrolyzing enzymes were added along with the inoculum. Many studies have carried out on SSF of starchy wastes using amylase and cellulosic waste using cellulose. This approach can reduce the cost on energy consumption, and lower the negative influence of higher glucose concentration in the media where the refined sugar and hydrolysate were used. In some instances, protease are employed to degrade the protein part of residues to reduce the supplementation of extra nutrients.

Fed batch culture proved to be a better fermentation system for the production of lactic acid. Maximum lactic acid concentration of 46 g/l in fed batch culture was obtained with both free cell mixture and co-immobilized cells at a substrate concentration of 100 g/l and a feeding rate of 250 ml/h. For fungal species, two types of pneumatic reactors (bubble column reactors and air lift loop reactor) were used. Compared to stirred tank reactors, pneumatic reactors have the advantages of simple construction and low energy transfer rate. Pneumatic reactors have demonstrated to be suitable for the cultivation of filamentous fungi with high oxygen demand and sensitivity to shear and for processes such as microbial biomass protein production [142]. In repeated fed batch culture, co-immobilized cells gave a higher overall lactic acid concentration compared with the free cell mixtures. The co-immobilized cells in calcium alginate beads retained their ability to produce lactic acid for 20 days. From whey there was a productivity of 29 g/l/h lactic acid in a continuous mode of operation using immobilized *L. helveticus.* Continuous recirculation of cells gave higher lactic acid concentration and higher yield. Continuous removal of the lactic acid with extraction or electro-dialysis results in even higher concentrations and yields. In SSF, recirculation of the cells is hampered by solid substrate residues fouling the equipment. That was the disadvantage of the cell recirculation in the SSF. In some studies, the recirculation of cell gave a higher lactic acid concentration and higher or equal yield [58]. Air lift loop reactor of *R. oryzae* NRRL 395 was used for the lactic acid production and produced concentration of 104.6 g/l and yield of 0.87 g/g and productivity of 1.8 g/l/h. The cell recycle system, together with repeated batch and continuous process, enables the achievement of a higher cell concentration and product productivity in the process. In cell-recycle repeated batch fermentation, the lactic acid production was about 6.4 g/l/h. In addition to that, 26% of yeast extract dosage, compared with conventional batch fermentation, was used. A successful approach to continuous production of lactic acid with cell retention has been reported to produce lactic acid by a two stage cell recycle culture of *L. rhamnosus.* The membrane cell-recycle bioreactors were connected in a series and obtained 92 g/l

**Table 6:** Various modes of fermentation for microbial production of lactic acid.

| Organism | Type of Fermentation | Lactic acid (g/l) | Yla/Tot (g/g) | Qv (g/l/h) | Reference |
|---|---|---|---|---|---|
| *R. arrhizus* | Batch | | 0.94-0.97 | | 120 |
| *R. oryzae* | Batch | | 0.65-0.76 | | 121 |
| *Lactobacillus delbrueckii* CECT 286 | Batch | 45.1 | | 2.7 | 41 |
| *Lactobacillus delbrueckii* NCIMB 8130 | Batch | 90.0 | | 3.8 | 50 |
| *Enterococcus faecalis* RKY1 | Batch | 95.7 | | 4.0 | 25 |
| *Lactobacillus casei* | Batch, immobilized on apple | 32.9 | 0.658 | | 24 |
| *Lactobacillus casei* | Batch, immobilized on quince | 36.1 | 0.722 | | 24 |
| *Lactobacillus casei* | Batch, free cells | 33.6 | 0.672 | | 24 |
| *Streptococcus bovis* | Batch | 14.7 | 0.88 | | 23 |
| *R. arrhizus* | Batch | 19.74 | 0.91 | | 11 |
| *R. oryzae* | Batch | 15.69 | 0.79 | | 11 |
| *Lactobacillus delbrueckii* NCIM 2365 | Batch | 120 | | 3.3 | 122 |
| *Rhizopus oryzae* NRRL 395 | Air lift loop | 104.6 | 0.87 | 1.8 | 123 |
| *Enterococcus faecalis* RKY1 | Batch | 144.0 | 0.96 | 5.1 | 124 |
| *Lactobacillus rhamnosus* ATCC 10863 | Batch | 67.0 | 0.84 | 2.5 | 125 |
| *Lactobacillus helveticus* ATCC 15009 | pH controlled batch | 65.5 | 0.66 | 2.7 | 61 |
| *Lactococcus lactis* ssp. *lactis* IFO 12007 | Circulating loop bioreactor | 90.0 | 0.76 | 1.6 | 126 |
| *Lactobacillus casei* SU No 22 + *Lactobacillus lactis* WS 1042 | Fed-batch, co-immobilization | 47.0 | | 2.0 | 72 |
| *Enterococcus faecalis* RKY1 | Repeated batch, cell-recycle *via* membrane | 93.2 | | 6.4 | 127 |
| *Enterococcus faecalis* RKY1 | Batch | 95.7 | | 4.0 | 127 |
| *Lactobacillus rhamnosus* ATCC 10863 | Batch | ~ 120.0 | | 2.1 | 128 |
| *Lactobacillus rhamnosus* ATCC 10863 | Continuous, cell-recycle *via* membrane | 92.0 | | 57.0 | 128 |
| *Lactobacillus casei* ssp. *rhamnosus* ATCC 11443 | Continuous, cell-recycle *via* immobilization | 22.4 | | 9.0 | 129 |
| *Lactobacillus delbrueckii* NRRL B445 | Fed-batch, *in situ* removal *via* solvent extraction | ~ 23. | | 10.2 | 130 |

Contd...

Contd...

| | | | | | |
|---|---|---|---|---|---|
| *Lactococcus lactis* IO-1 JCM 7638 | Batch, *in situ* removal *via* electrodialysis | ~ 39.0 | | 0.9 | 131 |
| *Lactobacillus rhamnosus* IFO 3863 | Batch | 98.0 | | 1.9 | 132 |
| *Lactobacillus rhamnosus* IFO 3863 | Continuous, *in situ* removal *via* electrodialysis | ~ 20.0 | | 8.2 | 132 |
| *Lactobacillus helveticus* CNRZ 303 | Continuous, cell-recycle *via* membrane | 55.0 | | 7.1 | 133 |
| *Lactobacillus delbrueckii* CECT 286 | Continuous, *in situ* removal *via* ion-exchange resin | 26.1 | | 10.4 | 73 |
| *Lactobacillus* sp. RKY2 KCTC 10353BP | Batch | 120.8±1.1 | 0.96±0.01 | 1.3 | 134 |
| *Lactobacillus casei* LA-04-1 | Batch culture | 112.5 | | 1.34 | 135 |
| *Lactobacillus casei* LA-04-1 | Pulse fed-batch | 130 | | 1.55 | 135 |
| *Lactobacillus casei* LA-04-1 | Constant feed rate fed-batch | 135 | | 1.61 | 135 |
| *Lactobacillus casei* LA-04-1 | Constant glucose concentration fed-batch | 152.5 | | 1.82 | 135 |
| *Lactobacillus casei* LA-04-1 | Exponential fed-batch | 180 | | 2.14 | 135 |
| *L. delbrueckii* NCIM 2365 | Fed-batch | 150 | 0.817 | 1.875 | 136 |
| *Lactobacillus plantarum* A6 | Batch | 14.25 | | | 137 |
| *Lactobacillus manihotivorans* OND 30 | Batch | 11.59 | | | 137 |
| *Lactobacillus fermentum Ogi* E1 | Batch | 6.66 | | | 137 |
| *Leuconostoc* St3-2 | Batch | 14.38 | | | 137 |
| *R. oryzae* | Airlift loop batch | 104.6 | | 1.8 | 123 |
| *R. oryzae* | Stirred tank reactor | 103.6 | | 1.7 | 138 |
| *R. oryzae* | Airlift loop repeat-batch | 85.7 | | 1.07 | 139 |
| *R. oryzae* | Stirred tank repeat-batch | 60 | | 2.9-6.2 | 140 |
| *R. oryzae* | Semi-continuous | | | 2.91 | 141 |
| *Lmanihotivorans LMG18011* | | 38.3 | 0.828 | | 21 |

$Y_{la}$/Tot (g/g): Yield; $Q_v$ (g/l/h): Productivity

of lactic acid with a productivity of 57 g/l. In solid state fermentation, the *Lactobacillus delbrueckii* produced at the maximum of 249 mg/gds L-Lactic acid after 5 days of fermentation using agro wastes like cassava, baggase and sugar cane baggase.

An electrokinetic bioreactor was developed combining conventional electrodialysis and bipolar membrane electrodialysis for product removal and pH control in lactic acid fermentation. In another work, they developed continuous electrodialysis fermentation system which gave the maximum lactic acid productivity of 8.18 g/l/h and yield of 0.685 g/g for ammonium lactate production. In other study, pH controlled fed-batch fermentation was carried out with *L. lactis* BMES-18M, which gave the final concentration and productivity of 161.2 g/l and 2.02 g/l/h, respectively. The lactic acid was produced by the various fed batch modes to increase the productivity, and widely used modes were pulse fed-batch, constant feed rate fed-batch and continuous glucose concentration fed-batch and exponential fed-batch. Another efficient bioreactor termed as "synchronized fresh cell bioreactor" was developed and consisted of a pH dependent substrate feed system coupled with cross flow filtration and turbidity control. At a cell concentration of 15 g/l and a feed glucose concentration of 53 g/l, volumetric lactic acid productivities of 8.2, 19.3 and 33.1 g/l/h were obtained at dilution rates of 0.21, 0.50 and 1.1 per h, respectively.

## 6. APPLICATIONS OF LACTIC ACID

Lactic acid is widely used in almost every segment of the food industry, where it serves in a wide range of functions, such as flavouring, pH regulation, improved microbial quality, and mineral fortification. The major use of lactic acid is in food and food-related applications, which in the US accounts for approximately 85% of the demand. The rest (15%) of the uses are in non-food industrial applications. Lactic acid occurs naturally in many food products. There are many properties of lactic acid which make it a very versatile ingredient in the food industry. Due to the mild acidic taste of lactic acid, it is also used as an acidulant/flavouring/pH buffering agent, inhibitor of bacterial spoilage in a wide variety of processed foods, such as candy, breads and bakery products, soft drinks, soups, sherbets, dairy products, beer, jams and jellies and processed eggs. Another potential application of lactic acid in the food industry is the mineral fortification of food products. Lactic acid or its salts are now used in the disinfection and packaging of carcasses, particularly those of poultry and fish, where the addition of their aqueous solutions during processing increased shelf-life and reduced microbial spoilage by *Clostridium botulinum* [1]. The esters of calcium and sodium salts of lactate with longer chain fatty acids have been used as very good dough conditioners and emulsifiers in bakery products. Fig. 2 shows the applications of lactic acid in various sectors.

Currently, lactic acid is considered the most potential feedstock monomer for chemical conversions, because it contains two reactive functional groups, a carboxylic group and a hydroxyl group. Lactic acid can undergo a variety of chemical conversions into potentially useful chemicals, such as propylene oxide (via hydrogenation), acetaldehyde (via decarboxylation), acrylic acid (via dehydration), propanoic acid (via reduction), 2,3-pentanedione (via condensation), and dilactide (via self-esterification). In the chemical industries, lactic acid is used in the dyeing of silks and other textile goods, as a mordant in the printing of woolens, in the bating and plumping of leathers, in the deliming of hides, in vegetable tanning, and as a flux for soft solders. The water-white grade is used in plastic industry. Lactic acid functions as a descaling agent, pH regulator, neutralizer, chiral intermediate, solvent, cleaning agent, slow acid-release agent, metal complexing agent, and humectant. Natural lactic acid has an emerging use as an excellent and safe solvent, which is alternative in many fine mechanical cleaning applications. Due to the high solvency power and solubility of lactic acid, it is an excellent remover of polymer and resins.

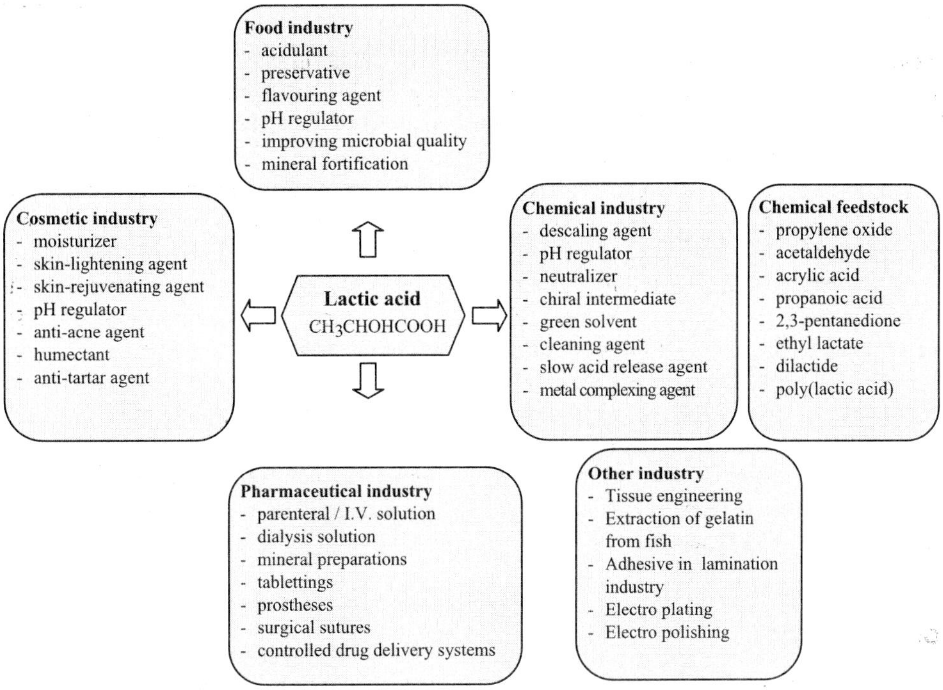


**Fig. 2:** Various industrial applications of lactic acid.

The water-retaining capacity of lactic acid makes it suitable for use as moisturizer in cosmetic formulations. The ability of lactic acid to suppress the formation of tyrosinase is responsible for its effects like skin lightening and rejuvenation. As humectants, the lactates are often superior to natural products and more effective than polyols [1]. Ethyl lactate is the active ingredient in many anti-acne preparations. The natural occurrence of lactic acid in human body makes it very useful as an active ingredient in cosmetics. Lactic acid is also used in the pharmaceutical industry as an electrolyte in many parenteral/I.V. (intravenous) solutions that are intended to replenish the bodily fluids or electrolytes. Examples include Lactated Ringer's or Hartmann's solutions, continuous ambulatory peritoneal dialysis solution, and dialysis solution for conventional artificial kidney machines. Technical grade lactic acid is used as an acidulant in vegetable and leather tanning industries. Lactic acid is being used in many small scale applications like pH adjustment hardening baths for cellophanes used in food packaging, terminating agent for phenol, formaldehyde resins, alkyl resin modifier, solder flux, lithographic and textile printing developers, adhesive formulations, electroplating and electro-polishing baths, detergent builders.

Environmentally friendly, 'green' solvents are another potential area for lactic acid derivatives, particularly lactate esters of low-molecular-weight alcohols such as ethyl, propyl and butyl lactate. Several specialized applications of lactate esters in electronics and precision cleaning were developed and commercialized by 'Purac.' Blends of these esters with other biologically active solvents with a wide range of solvating and cleaning properties have been developed and commercialized very recently by Vertec Biosolvents. The US Environmental Protection Agency (EPA) has approved the lactate ester solvents as inert ingredients with negligible toxicity and an excellent environmental profile.

## 6.1 Food Industry

Lactic acid occurs naturally in many food products. It has been in use as an acidulant, preservative and pH regulator for quite some time. There are many properties of lactic acid which make it a very versatile ingredient in the food industry. It has a pronounced preservative action, and it regulates the microflora. It has been found to very effective against certain type of microorganisms. Some times a combination of lactic acid and acetic acid is used as it has a greater bactericidal activity. The salts are very soluble, which gives the possibility of partially replacing the acid in buffering systems. The calcium salt of lactic acid, calcium lactate, has greater solubility than the corresponding salt of citric acid. In such products where turbidity caused by calcium salts is a problem, the use of lactic acid gives products which are clear. L(+) lactic acid is the natural lactic acid found in biological systems, and hence, its use as an acidulant does not introduce a foreign element into the body.

It is widely used in almost every segment of the food industry, where it serves in a wide range of functions, such as flavouring, pH regulation, improved microbial quality, and mineral fortification. Moreover, lactic acid is used commercially in the processed meat and poultry industries to provide products with an increased shelf-life, enhanced flavor, and better control of food-born pathogens. Due to the mild acidic taste of lactic acid, it is also used as an acidulant in salads and dressings, baked goods, pickled vegetables, and beverages.

## 6.2 Confectionery

Lactic acid finds use as an acidulant in the confectionery industry. It is a better acidulant than citric acid since the sugar inversion is less when used for hard boiled candies. It does not have the initial burst of flavor and tanginess of citric acid. Lactic acid imparts a mellower and lasting sourness. It enhances the flavor much more than others. The use of buffered lactic acid in continuous production lines for high boiled sweets is a more recent application. Liquid buffered lactic acid can be easily measured into the molten syrups, even at the high temperatures used in depositing lines. Sugar inversion will be minimal. In sugar confectionery, it is used in continuous production lines for high boiled sweets (like bonbons) to make perfectly clear sweets, with minimum sugar inversion and with no air trapped. Lactic acid is used in confectionery, not only for flavour, but also to bring the pH of the cooked mix to the correct point for setting. The advantages of adding lactic acid in confectionery include its low inversion rate, ease of handling, and ability to produce clear candies.

## 6.3 Beer and Wine

Lactic acid is a natural beer acid and hence it is used for pH adjustments during the mashing process and in wort cooking. Lactic acid improves the microbial stability and also enhances the flavor. Almost all the breweries in India use the L (+) lactic acid manufactured by Lactochem Limited.

## 6.4 Beverages

Lactic acid is the acidulant in delicately flavoured soft drinks and fruit juices. It does not mask or over power the natural flavour. Its flavour enhancing property makes the beverage more palatable and leaves a lingering taste. Lactic acid is preferred over citric acid for these reasons. Use of buffered lactic acid improves the taste and flavour of many beverages, such as soft drinks, mineral water, carbonated fruit juices, etc.

### 6.5 Olives, Pickles, Cabbage, Gherkins

Green olives, gherkins and others are often packed in a solution of salt, lactic acid and water. The lactic acid acts as a preservative and improves the clarity of the brine and flavour. A mixture of acetic acid and lactic acid in pickled products such as gherkins, silver skin onion, etc. imparts a milder taste and flavour, and improves microbial stability. Large quantity of gherkins in brine with Lactochems's L(+) lactic acid is exported from India to many countries worldwide. Calcium lactate was reported to be used as a firming salt, which have been used for canned fruits and vegetables.

### 6.6 Dairy Products

Direct acidification with lactic acid, in dairy products such as cottage cheese, is preferred to fermentation as the risks of failure and contamination can be avoided. The processing time can also be saved. Lactic acid and calcium lactate are used extensively in the production of *Channa* and *Panneer* by direct acidification. Lactic acid is also used as an acidulant in dairy products like *cheese* and *yoghurt* powder. In dairy products such as cottage cheese, addition of lactic acid is preferred to fermentation. For direct acidification of certain breads, lactic acid is the natural sour dough acid. The general appearance of a loaf of bread is greatly improved by the use of bacterial lactic acid, a larger loaf results per weight of bread with improved bloom, and colour of crust. Lactic acid is directly added to certain types of fermented dough crispy biscuits.

Lactic acid added to dough increases the shelf-life due to its retarding action on molds and bacteria. The sodium and calcium stearoyl lactyllates find use as emulsifiers in the baking industry as they provide substantial quality improvement of baked products besides reducing shortening levels. In bakery products, it is used for direct acidification of rye or rye-wheat breads. It increases butter stability and volume. Part of the egg albumen can be replaced by less expensive calcium lactate. A large fraction (> 50%) of the lactic acid for food-related uses goes to produce emulsifying agents used in foods, particularly for bakery goods. These emulsifying agents are esters of lactate salts with longer chain fatty acids, and the four important products are calcium and sodium, stearoyl-2-1actylate, glyceryl lactostearate, and glyceryl lactopalmirate. Out of the stearoyl lactylates, the calcium salt is a very good dough conditioner, and the sodium salt is both a conditioner and an emulsifier for yeast leavened bakery products. The glycerates and palmitates are used in prepared cake mixes and other bakery products and in liquid shortenings.

### 6.7 Meat and Meat Products

Lactic acid is widely used in meat products as an antimicrobial agent. Decontamination of beef, poultry and pork carcasses in slaughterhouse operations is practiced to reduce *Salmonella* infection. In sausages, sodium lactate is used to reduce water activity and achieve higher shelf-life. Recent research publication indicates the use of hot lactic acid spray on carcasses where reduction of over 99 % of *E. coli* has been observed. Lactic acid is also used in the improvement of shelf-life of buffalo meat. An emerging new use of lactic acid or its salts is in the disinfection and packaging of carcasses, particularly those of poultry and fish, where the addition of aqueous solutions of lactic acid and its salts during the processing increased shelf-life and reduced the growth of anaerobic spoilage organisms such as *Clostridium botulinum*.

## 7. POLY LACTIC ACID (PLA)

PLA is currently used in a number of biomedical applications, such as sutures, dialysis media and drug delivery devices, but it is also evaluated as a material for tissue engineering. Being biodegradable, it can also be employed in the preparation of bioplastic useful for producing loose-fill packaging, compost bags, food packaging and disposable tableware. In form of fibres and non-woven textiles, PLA also has many potential uses, for example, as upholstery, disposable garments, awnings, feminine hygiene products and nappies. It is important that PLA compares well with other popular plastics already used for packaging. It is clear and naturally glossy like the polystyrene used in "blister packs" for products such as batteries, toys, and many others [1]. PLA is a non-volatile, odourless polymer and is classified as GRAS (generally recognized as safe) by the Food and Drug Administration in the United States [1]. Polylactic acid is a versatile polymer that has many potential uses, including many applications in the textile and medical industries as well as the packaging industry. Four types of polylactic acid are available for packaging purposes: PLA polymers 4041D, 4031D, 1100D, and 2000D. These polymers are expected to be offered in common pellet form, which should allow for rapid adoption with conventional extruders. Polymer 1100D is a thermoplastic resin designed for extrusion coating on paper. Potential applications of PLA1100D include lawn and leaf bags, hot and cold drinking cups, picnic plates, bowls, straws, fried food boxes, frozen vegetable packaging, and liquid food packaging. Polymer 2000D is a thermoplastic resin designed for extrusion and thermoforming applications. Potential uses of 2000D include dairy containers, food service ware, transparent food containers, blister packs, and cold drink cups. Polylactic acid also has many potential uses in fibres and non-wovens. It is easily converted into a variety of fibre forms using conventional melt-spinning processes. Polylactic acid polymers are more hydrophilic than PET, have a lower density, and have excellent crimp and crimp retention. They also offer low flammability and smoke generation characteristics. Major applications of PLA fibres and non-wovens include clothing and furnishings such as drapes, upholstery and covers. Some exciting potential applications include household and industrial wipes, diapers, feminine hygiene products, disposable garments, and UV resistant fabrics for exterior use (awnings, ground cover, etc.) among others. Since PLA can be assimilated by the body, it has found important applications in sustained release drug delivery systems. Its mechanical properties and absorbability make PLA polymer an ideal candidate for implants in bone or soft tissue (facial traumatology, orthopedic surgery, ophthalmology, orthodontics, local implants for controlled release of anti-cancer drugs), and for resorbable sutures (eye surgery, conjunctional surgery, surgery of the chest and abdomen). The mechanical, pharmaceutical and bio-absorption characteristics are dependent on controllable parameters such as chemical composition and molecular weight of the polymer. Poly-L-lactic acid is currently being used in Europe and is approved by the Food and Drug Administration (FDA) in the United States for soft tissue augmentation in HIV-associated facial lipoatrophy. Poly-L-lactic acid is also used in the management of soft-tissue augmentation.

Lactic acid is the simplest hydroxyl carboxylic acid that can be derived from renewable carbohydrates and used to make a wide variety of useful products. There have been numerous investigations on the development of biotechnological processes for lactic acid production from various wastes. However, there are still several issues that need to be addressed in order to produce lactic acid biotechnologically within the targeted cost, such as the development of high-performance lactic acid-producing microorganisms and the lowering of the costs of raw materials and fermentation processes. The biotechnological processes for the production of lactic acid from cheap raw materials should be improved further to make them competitive with the chemically-derived one. The use of nutrient-rich waste materials such as various food wastes, agro-wastes and industrial wastes opens an avenue in a dual working manner for value addition through an eco-friendly green technology.

## REFERENCES

1. Datta RS, Sai PT, Patric B, Moon SH & Frank JR (1995) Technological and economic potential of polylactic acid and lactic acid derivatives. *FEMS Microbiol Rev* 16: 221-231.

2. Akerberg C, Hofvendahl K, Zacchi G & Hahn-Hagerdal B (1998) Modeling the influence of pH, temperature, glucose and lactic acid concentrations on the kinetic of lactic acid production by *Lactococcus lactis* ssp. *lactis* ATCC 19435 in whole-wheat flour. *Appl Microbiol Biotechnol* 49: 682-690.
3. Tsao GT, Cao NJ, Du J & Gong CS (1999) Production of multifunctional organic acids from renewable resources. *Adv Biochem Eng/Biotechnol* 65: 245-277.
4. Anuradha R, Suresh AK & Venkatesh KV (1999) Simultaneous saccharification and fermentation of starch to lactic acid. *Process Biochem* 35: 367-375.
5. Aristidous A & Penttila M (2000) Metabolic engineering applications to renewable resource utilization. *Curr Opin Biotechnol* 11: 187-198.
6. Huang LP, Jin B, Lant P & Zhou JT (2003) Biotechnological production of lactic acid integrated with potato wastewater treatment by *Rhizopus arrhizus. J Chem Technol Biotechnol* 781: 889-906.
7. Mirasol F (1999) Lactic acid prices falter as competition toughens. *Chemical Market Reporter* 255: 16-20.
8. Vickroy TB (1985) Lactic acid. In: *Comprehensive Biotechnology - The principles, Applications and Regulation of Biotechnology in Industry, Agriculture and Medicine*, Vol 3, Blanch HW & Drew S (eds), Dic Pergamon Press, Toronto, Canada.
9. Demirci A, Pometto III AL & Johnson KE (1993) Lactic acid production in a mixed culture biofilm reactor. *Appl Environ Microbiol* 59: 203-207.
10. Jin B, Yin P, Y Ma & Zhao L (2005) Production of lactic acid and fungal biomass by *Rhizopus* fungi from food processing waste streams. *J Ind Microbiol Biotechnol* 32: 678-686.
11. Huang LP, Jin B, Lant P & Zhou J (2005) Simultaneous saccharification and fermentation of potato starch wastewater to lactic acid by *Rhizopus oryzae* and *Rhizopus arrhizus. Biochem Eng J* 23: 265-276.
12. Stenberg K, Galbe M & Zacchi G (2000) The influence of lactic acid formation on the simultaneous saccharification and fermentation of softwood to ethanol. *Enzyme Microb Technol* 26: 71-79.
13. Litchfield JH (1996) Microbiological production of lactic acid. *Adv Appl Microbiol* 42: 45-95.
14. Nolasco-Hipolito C, Matsunaka T, Kobayashi G, Sonomoto K & Ishizaki A. (2002) Synchronised fresh cell bioreactor system for continuous L(+)-lactic acid production using *Lactococcus lactis* IO-1 in hydrolysed sago starch. *J Biosci Bioeng* 93: 281-287.
15. John RP, Nampoothiri KM & Pandey A (2006) Solid-state fermentation for L-lactic acid production from agro wastes using *Lactobacillus delbrueckii. Process Biochem* 41: 759-763.
16. Kadam SR, Patil SS, Bastawde KB, Khire JM & Gokhale DV (2006) Strain improvement of *Lactobacillus delbrueckii* NCIM 2365 for lactic acid production. *Process Biochem* 41: 120-126.
17. Tango MS & Ghaly AE (2002) A continuous lactic acid production system using an immobilized packed bed of *Lactobacillus helveticus. Appl Microbiol Biotechnol* 58: 712-720.
18. Hujanen M, Linko S, Linko YY & Leisola M (2001) Optimization of media and cultivation conditions for L(+)(S)-lactic acid production by *Lactobacillus casei* NRRL B-441. *Appl Microbiol Biotechnol* 56: 126-130.
19. John RP, Nampoothiri KM, Nair AR & Pandey A (2005) L(+)-Lactic acid production using *Lactobacillus casei* in solid-state fermentation. *Biotechnol Lett* 27: 1685-1688.
20. Naveena BJ, Altaf M, Bhadriah K & Reddy G (2005) Selection of medium components by Placket Burman design for the production of L(+)-lactic acid by *Lactobacillus amylophilus* GV6 in SSF using wheat bran. *Biores Technol* 96: 485-490.

21. Ohkouchi Y & Inoue Y (2006) Direct production of L(+)-lactic acid from starch and food wastes using *Lactobacillus manihotivorans* LMG18011. *Biores Technol* 97: 1554-1562.
22. Zhang DX & Cheryan M (1994) Starch to lactic acid in a continuous membrane bioreactor. *Process Biochem* 29: 145-150.
23. Narita J, Nakahara S, Fukuda H & Kondo A (2004) Efficient production of L(+)-lactic acid from raw starch by S*treptococcus bovis* 148. *J Biosci Bioeng* 97: 423-425.
24. Kourkoutas Y, Xolias V, Kallis M, Bezirtzoglou E & Kanellaki M (2005) *Lactobacillus casei* cell immobilization on fruit pieces for probiotic additive, fermented milk and lactic acid production. *Process Biochem* 40: 411-416.
25. Wee YJ, Kim JN, Yun JS & Ryu HW (2004) Utilization of sugar molasses for economical L (+)-lactic acid production by batch fermentation of *Enterococcus faecalis. Enzyme Microb Technol* 35: 568-573.
26. Rivas B, Moldes AB, Dominguez JM & Parajo JC (2004) Lactic acid production from corn cob by simultaneous saccharification and fermentation: mathematical interpretation. *Enzyme Microb Technol* 34: 627-634.
27. Park EY, Anh PN & Okuda N (2004) Bioconversion of waste office paper to lactic acid by the filamentous fungus *Rhizopus oryzae. Biores Technol* 93: 77-83.
28. Garde A, Jonsson G, Schmidt AS & Ahring BK (2002) Lactic acid production by *Lactobacillus brevis. Biores Technol* 81: 217-223.
29. Oh H, Wee YJ, Yun JS, Han SH, Jung S & Ryu HW (2005) Lactic acid production from agricultural resources as cheap raw materials. *Biores Technol* 96: 1492-1498.
30. Gao MT, Hirata M, Toorisaka E & Hano T (2006) Acid-hydrolysis of fish wastes for lactic acid fermentation. *Biores Technol* 97: 2414-2420.
31. Xu Z, Wang O, Wang P, Cheng G, Ji Y & Jiang Z (2007) Production of lactic acid from soybean stalk hydrolysate with *Lactobacillus sake* and *Lactobacillus casei. Process Biochem* 42: 89-92.
32. Ohkouchi Y & Inoue Y (2007) Impact of chemical components of organic wastes on L(+) lactic acid production. *Biores Technol* 98: 546-553.
33. Naveena BJ, Altaf M, Bhadrayya K & Reddy G (2004) Production of L(+)lactic acid by *Lactobacillus amylophilus* GV6 in semi-solid state fermentation using wheat bran. *Food Technol Biotechnol* 42: 147-152.
34. Yanez R, Alonso JL & Parajo JC (2005) D-Lactic acid production from waste cardboard. *J Chem Technol Biotechnol* 80: 76-84.
35. Jin B, Huang LP & Lant P (2003) *Rhizopus arrhizus* – a producer for simultaneous saccharification and fermentation of starch waste materials to L(+) lactic acid. *Biotechnol Lett* 25: 1983-1987.
36. Xiaodong W, Xuan G & Rakshit SK (1997) Direct fermentation of lactic acid from cassava or other starch substrates. *Biotechnol Lett* 9: 841-843.
37. Chatterjee M, Chakrabarty SL, Chattopadhyay BD & Mandal RK (1997) Production of lactic acid by direct fermentation of starchy wastes by an amylase-producing *Lactobacillus. Biotechnol Lett* 19: 873-874.
38. Sreenath HK, Moldes AB, Koegel RG & Straub RJ (2001) Lactic acid production from agriculture residues. *Biotechnol Lett* 23: 179-184.
39. Moldes AB, Alonso JL & Parajo JC (2001) Strategies to improve the bioconversion of processed wood into lactic acid by simultaneous saccharification and fermentation. *J Chem Technol Biotechnol* 76: 279-284.
40. McCaskey TA, Zhou SD, Britt SN & Strickland R (1994) Bioconversion of municipal solid waste to lactic acid by *Lactobacillus species. Appl Biochem Biotechnol* 45/46: 555-563.

41. Monteagudo JM, Rodriguez L, Rincon J & Fuertes J (1994) Optimization of conditions of fermentation of beet molasses to lactic acid by *Lactobacillus delbrueckii*. *Acta Biotechnol* 14: 251-260.
42. Goksungur Y & Guvenç U (1999) Production of lactic acid from beet molasses by calcium alginate immobilized *Lactobacillus delbrueckii* IFO 3202. *J Chem Technol Biotechnol* 74: 131-136.
43. Sakai K & Yamanami T (2006) Thermotolerant *Bacillus licheniformis* TY7 produces optically active L-lactic acid from kitchen refuse under open condition. *J Biosci Bioeng* 102: 132-134.
44. Idris A & Suzana W (2006) Effect of sodium alginate concentration, bead diameter, initial pH and temperature on lactic acid production from pineapple waste using immobilized *Lactobacillus delbrueckii*. *Process Biochem* 41: 1117-1123.
45. Schmidt AS & Thomsen AB (1998) Optimization of wet oxidation pretreatment of wheat straw. *Biores Technol* 64: 139-151.
46. Luo J, Xia L, Lin J & Cen P (1997) Kinetics of simultaneous saccharification and lactic acid fermentation process. *Biotechnol Progr* 13: 762-767.
47. Cheng P, Muller RE, Jaeger S, Bajpai R & Jannotti EL (1991) Lactic acid production from enzyme thinned cornstarch using *Lactobacillus amylovorus*. *J Ind Microbiol* 7: 27-34.
48. Yumoto I & Ikeda K (1995) Direct fermentation of starch to L(+)lactic acid using *Lactobacillus amylophilus*. *Biotechnol Lett* 17: 543-546.
49. Karel M, Jaroslav V, Vera H & Mojmir R (1997) Lactic acid production in a cell retention continuous culture using lignocellulosic hydrolysate as a substrate. *J Biotechnol* 56: 25-31.
50. Kotzanmanidis C, Roukas T & Skaracis G (2002) Optimization of lactic acid production from beet molasses *Lactobacillus delbrueckii* NCIMB 8130. *World J Microbiol Biotechnol* 18: 442-448.
51. Raccach M & Mamiro T (1997) The effect of temperature on the lactic acid fermentation of rye flour. *Food Microbiol* 4: 213-220.
52. Richter K & Trager A (1994) L(+)lactic acid from sweet sorghum by submerged and solid state fermentations. *Acta Biotechnol* 14: 367-378.
53. Xavier S & Lonsane BK (1994) Sugarcane press mud as a novel and inexpensive substrate for production of lactic acid in a solid state fermentation system. *Appl Microbiol Biotechnol* 41: 291-295.
54. Linko YY & Javanainen P (1996) Simultaneous liquefaction, saccharification, and lactic acid fermentation on barley starch. *Enzyme Microb Technol* 19: 118-123.
55. Venkatesh KV (1997) Simultaneous saccharification and fermentation of cellulose to lactic acid. *Biores Technol* 62: 91-98.
56. Sharma N, Wati L & Singh D (2003) Production of lactic acid during bioremediation of anaerobically digested molasses spent wash. *Ind J Microbiol* 43: 119-121.
57. Saha BC & Nakamura LK (2003) Production of mannitol and lactic acid by fermentation with *Lactobacillus intermedius* NRRL B-3693. *Biotechnol Bioeng* 82: 865-871.
58. Hofvendahl K & Hahn-Hagerdal B (2000) Factors affecting the fermentative lactic acid production from renewable resources. *Enzyme Microb Technol* 26: 87-107.
59. Amrane A & Prigent Y (1998) Lactic acid production rates during the different growth phases of *Lactobacillus helveticus* cultivated on whey supplemented with yeast extract. *Biotechnol Lett* 20: 379-383.
60. Kulozik U & Wilde J (1999) Rapid lactic acid production at high cell concentrations in whey ultrafiltrate by *Lactobacillus helveticus*. *Enzyme Microb Technol* 24: 297-302.

61. Schepers AW, Thibault J & Lacroix C (2002) *Lactobacillus helveticus* growth and lactic acid production during pH-controlled batch cultures in whey permeate/yeast extract medium. Part II: Kinetic modeling and model validation. *Enzyme Microb Technol* 30: 187-194.
62. Fitzpatrick JJ & O'Keeffe U (2001) Influence of whey protein hydrolyzate addition to whey permeate batch fermentations for producing lactic acid. *Process Biochem* 37: 183-186.
63. Buyukkilci AO & Harsa S (2004) Batch production of L(+) lactic acid from whey by *Lactobacillus casei* (NRRL B-441). *J Chem Technol Biotechnol* 79: 1036-1040.
64. Stieber RW & Gerhardt P (1981) Dialysis continuous process for ammonium lactate fermentation: simulated and experimental dialysate-feed, immobilized-cell systems. *Biotechnol Bioeng* 23: 535-550.
65. Podlech PAS, Luna MF, Jerke PF, De Souza Neto CAC, Dos Passos RF, Souza O & Borzani W (1990) Semicontinuous lactic fermentation of whey by *Lactobacillus bulgaricus*. I. Experimental results. *Biotechnol Lett* 12: 531-534.
66. Boyaval P, Corre C & Terre S (1987) Continuous lactic acid fermentation with concentrated product recovery by ultra filtration and electro dialysis. *Biotechnol Lett* 9: 207-212.
67. Aeschlimann A & Von SU (1989) The production of lactic acid from whey permeate by *Lactobacillus helveticus. Biotechnol Lett* 11: 195-200.
68. Aeschlimann A and Von Su (1991) Continuous production of lactic acid from whey permeate by *Lactobacillus helveticus* in a cell recycle reactor. *Enzyme Microb Technol* 13: 811-816.
69. Roukas T & Kotzekidou P (1991) Production of lactic acid from deproteinized whey by coimmobilized *Lactobacillus casei* and *Lactococcus lactis* cells. *Enzyme Microb Technol* 13: 33-38.
70. Tuli A, Sethi RP, Khanna PK & Marwaha SS (1985) Lactic acid production from whey permeate by immobilized *Lactobacillus casei. Enzyme Microb Technol* 7: 164-168.
71. Goksungur Y & Guvenc U (1997) Batch and continuous production of lactic acid from beet molasses by *Lactobacillus delbrueckii* IFO 3202. *J Chem Technol Biotechnol* 69: 399-404.
72. Roukas T (1998) Pretreatment of beet molasses to increase pullulan production. *Process Biochem* 33: 805-810.
73. Monteagudo JM, Rodríguez L, Rincón J & Fuertes J (1997) Kinetics of lactic acid fermentation by *Lactobacillus delbrueckii* grown on beet molasses. *J Chem Technol Biotechnol* 68: 271-276.
74. Shukla VB, Zhou S, Yomano LP, Shanmugam KT, Preston JF & Ingram LO (2004) Production of D(–) lactic acid from sucrose and molasses. *Biotechnol Lett* 26: 689-693.
75. Pandey A, Soccol CR, Nigam P & Soccol VT (2000) Biotechnological potential of agro industrial residues. I. Sugarcane bagasse. *Biores Technol* 74. 69-80.
76. Soccol CR, Marin B, Raimbault M & Lebeault JM (1994) Potential of solid state fermentation for production of L(+) lactic acid by *Rhizopus oryzae. Appl Biochem Biotechnol* 41: 286-290.
77. Nampoothiri KM & Pandey A (1996) Solid state fermentation for L-glutamic acid production using *Brevibacterium* sp. *Biotechnol Lett* 16: 199-204.
78. Pandey A, Soccol CR, Nigam P, Soccol VT, Vandenberghe LPS & Mohan R (2000) Biotechnological potential of agro-industrial residues. II. Cassava bagasse. *Biores Technol* 74: 81-87.
79. Vishu C, Seenayya G & Reddy G (2002) Direct fermentation of various pure and crude starchy substrates to L(+)-lactic acid using *Lactobacillus amylophilus* GV6. *World J Microbiol Biotechnol* 18: 429-433.

80. Inglett GE (1970) *Corn: Culture Processing and Products,* AVI Publishing Co, Westport, USA.
81. Barl B, Bilideris CG, Murray ED & Macgregor AM (1990) Combined chemical and enzymatic treatments of corn husk lignocellulosics. *J Sci Food Agric* 56: 195-214.
82. Baden HP & Kubilus J (1983) Fibrous proteins of bovine hoof. *J Invest Dermatol* 81: 220-224.
83. Dalev P (1990) An enzyme-alkaline hydrolysis of feather keratin for obtaining a protein concentrate for fodder. *Biotechnol Lett* 12: 71-72.
84. Kurbanoglu EB & Kurbanoglu NI (2003) Utilization for lactic acid production with a new acid hydrolysis of ram horn waste. *FEMS Microbiol Lett* 225: 29-34.
85. Pintado J, Guyot JP & Raimbault M (1999) Lactic acid production from mussel processing wastes with an amylolytic bacterial strain. *Enzyme Microb Technol* 24: 590-598.
86. Sakai K, Taniguchi M, Miura S, Ohara H, Matsumoto T & Shirai Y (2004) Making plastics from garbage. *J Ind Ecol* 7: 63-74.
87. Wang QH, Narita JY, Xie WM, Ohsumi Y, Kusano K, Shirai Y & Ogawa HI (2002) Effects of anaerobic/aerobic incubation and storage temperature on preservation and deodorization of kitchen garbage. *Biores Technol* 84: 213-220.
88. Wang QH, Narita JY, Ren NQ, Ohsumi Y, Kusano K, Shirai Y & Ogawa HI (2003) Effect of pH adjustment on preservation of kitchen garbage used for producing lactic acid. *Water Air Soil Pollut* 144: 405-418.
89. Tong WY, Fu XY, Lee SM, Yu J, Liu JY, Wei DZ & Koo YM (2004) Purification of L-lactic acid from fermentation broth with paper sludge as a cellulosic feedstock using weak anion exchanger Amberlite IRA-92. *Biochem Eng J* 18: 89-96.
90. Zakasaki K, Akakura N, Adachi T & Akiyama T (1999) Use of wastewater sludge as a raw material for production of L-lactic acid. *Environ Sci Technol* 33: 198-200.
91. Oda YJ, Park BS & Moon KH (1997) Recycling of bakery wastes using an amylolytic lactic acid bacterium. *Biores Technology* 60: 101-105.
92. Sakai K, Taniguchi M, Miura S, Ohara H, Matsumoto T & Shirai Y (2003) Making plastics from garbage: a novel process for poly-l-lactate production from municipal food waste. *J Ind Ecol* 7: 63-73.
93. Heriban V, Sturdik E, Zalibara L & Matus P (1993) Process and metabolic characteristics of *Bacillus coagulans* as a lactic acid producer. *Lett Appl Microbiol* 16: 243-246.
94. Sakai K & Y Ezaki (2006a) Open L-lactic acid fermentation of food refuse using thermophilic *Bacillus coagulans* and fluorescence *in situ* hybridization analysis of microflora. *J Biosci Bioeng* 101: 457-463.
95. Ueno T (2003) Lactic acid production using two food processing wastes, canned pineapple syrup and grape invertase as substrate and enzyme. *Biotechnol Lett* 25: 573-577.
96. Abe S & Tagaki M (1991) Simultaneous saccharification and fermentation of cellulose to lactic acid. *Biotechnol Bioeng* 37: 93-96.
97. Linko P, Stenroos SL, Linko YY, Koistinen T, Harju M & Heikonen M (1984) Applications of immobilized lactic acid bacteria. *Ann N Y Acad Sci* 434: 406-417.
98. Shamala TR & Sreekantiah KR (1988) Fermentation of starch hydrolysates by *Lactobacillus plantarum. J Ind Microbiol* 3: 175-178.
99. Gupta R & Gandhi DN (1995) Effect of supplementation of some nutrients in whey on the production of lactic acid. *Ind J Dairy Sci* 48: 636-641.
100. Aksu Z & Kutsal T (1986) Lactic acid production from molasses utilizing *Lactobacillus delbrueckii* and invertase together. *Biotechnol Lett* 8: 157-160.

101. Nancib A, Nancib N, Meziane-Cherif D, Boubendir A, Fick M & Boudrant J (2005) Joint effect of nitrogen sources and B vitamin supplementation of date juice on lactic acid production by *Lactobacillus casei* subsp. *rhamnosus. Biores Technol* 96: 63-67.
102. Yu RC & Hang YD (1989) Kinetics of direct fermentation of agricultural commodities to L(+)lactic acid by *Rhizopus oryzae. Biotechnol Lett* 2: 597-600.
103. Hujanen M & Linko YY (1996) Effect of temperature and various nitrogen sources on L(+) lactic acid production by *Lactobacillus casei. Appl Microbiol Biotechnol* 45: 307-313.
104. Satoh E, Nimura Y, Uchimura T, Kozaki M & Komagata M (1993) Molecular cloning and expression of two α-amylase genes from *Streptococcus bovis* 148 in *Escherichia coli. Appl Environ Microbiol* 59: 3669-3673.
105. Chen R & Lee YY (1997) Membrane mediated extractive fermentation for lactic acid production from cellulosic biomass. *Appl Biochem Biotechnol* 63/65: 435-438.
106. Cavazzoni V, Manzoni M & Craveri R (1988) Ammonium lactate from deproteinized alfalfa juice by *Streptococcus faecium. J Ind Microbiol* 3: 373-376.
107. Samuel WA, Lee YY & Anthony WB (1980) Lactic acid fermentation of crude sorghum extract. *Biotechnol Bioeng* 22: 757-778.
108. Montelongo JL, Chassy BM & McCord JD (1993) *Lactobacillus salivarius* for conversion of soy molasses into lactic acid. *J Food Sci* 58: 863-866.
109. Boniardi N, Rota R, Nano G & Mazza B (1997) Lactic acid production by electrodialysis part I: experimental tests. *J Appl Electrochem* 27: 125-133.
110. Ishizaki A, Nomura Y & Iwahara M (1990) Built-in electro dialysis batch culture, a new approach to release of end product inhibition. *J Ferment Bioeng* 70: 108-113.
111. Hongo M, Nomura Y & Iwahara M (1986) Novel method of lactic acid production by electrodialysis fermentation. *Appl Environ Microbiol* 52: 314-319.
112. Nomura Y, Yamamoto K & Ishizaki A (1991) Factors affecting lactic acid production rate in the built-in electro dialysis fermentation, an approach to high speed batch culture. *J Ferment Bioeng* 71: 450-452.
113. Madzingaidzo L, Danner H & Braun R (2002) Process development and optimization of lactic acid purification using electro dialysis. *J Biotechnol* 96: 223-239.
114. Sun X, Wang Q, Zhao W, Ma H & Sakata K (2006) Extraction and purification of lactic acid from fermentation broth by esterification and hydrolysis method. *Sep Purif Technol* 49: 43-48.
115. Bouchoux A, Balmann HR & Lutin F (2006) Investigation of nanofiltration as a purification step for lactic acid production processes based on conventional and bipolar electrodialysis operations. *Sep Purif Technol* 52: 266-273.
116. Liew MKH, Tanaka S & Morita M (1995) Separation and purification of lactic acid: Fundamental studies on the reverse osmosis down-stream process. *Desalination* 101: 269-277.
117. Luedeking R & Piret EL (1959) A kinetic study of the lactic acid fermentation. *J Biochem Microbiol Technol Eng* 1: 393-412.
118. Weiss RM & Ollis DF (1980) Extracellular microbial polysaccharides I. Substrate, biomass and product kinetic equations for batch xanthan gum fermentation. *Biotechnol Bioeng* 22: 859-873.
119. Friedman MR & Gaden ELJR (1970) Growth and acid production by *Lactobacillus delbrueckii* in a dialysis culture system. *Biotechnol Bioeng* 12: 961-974.
120. Kristoficova L, Rosenberg M, Vlnova A, Sajbidor J & Cetrik M (1991) Selection of Rhizopus strains for L (+) lactic acid and gammalinolenic acid production. *Folia Microbiol* 36: 451-455.

121. Hang YD (1989) Direct fermentation of corn starch to L (+) lactic acid by *Rhizopus oryzae*. *Biotechnol Lett* 11: 299-300.
122. Patil SS, Kadam SR, Patil SS, Bastawde KB, Khire JM & Gokhale DV (2006) Production of lactic acid and fructose from media with cane sugar using mutant of *Lactobacillus delbrueckii* NCIM 2365. *Lett Appl Microbiol* 43: 53-57.
123. Park EY, Kosakai Y & Okabe M (1998) Efficient production of L(+) lactic acid using mycelial cotton-like flocs of *Rhizopus oryzae* in an air-lift bioreactor. *Biotechnol Prog* 14: 699-704.
124. Yun JS, Wee YJ & Ryu HW (2003) Production of optically pure L(+) lactic acid from various carbohydrates by batch fermentation of *Enterococcus faecalis* RKY1. *Enzyme Microb Technol* 33: 416-423.
125. Berry AR, Franco, Zhang W & Middelberg APJ (1999) Growth and lactic acid production in batch culture of *Lactobacillus rhamnosus* in a defined medium. *Biotechnol Lett* 21: 163-167.
126. Roble ND, Ogbonna JC & Tanaka H (2003) L-lactic acid production from raw cassava starch in a circulating loop bioreactor with cell immobilized in loofa (*Luffa cylindrica*). *Biotechnol Lett* 25: 1093-1098.
127. Oh H, Wee YJ, Yun JS & Ryu HW (2003) Lactic acid production through cell-recycle repeated-batch bioreactor. *Appl Biochem Biotechnol* 107: 603-613.
128. Kwon S, Yoo IK, Lee WG, Chang HN & Chang YK (2001) High-rate continuous production of lactic acid by *Lactobacillus rhamnosus* in a two-stage membrane cell-recycle bioreactor. *Biotechnol Bioeng* 73: 25-34.
129. Cotton JC, Pometto III AL & Gvozdenovic-Jeremic J (2001) Continuous lactic acid fermentation using plastic composite support biofilm reactor. *Appl Microbiol Biotechnol* 57: 626-630.
130. Iyer PY & Lee YY (1999) Simultaneous saccharification and extractive fermentation of lignocellulosic materials into lactic acid in a two-zone fermentor-extractor system. *Appl Biochem Biotechnol* 78: 409-419.
131. Nomura Y, Iwahara M, Hallsworth JE, Tanaka T & Ishizaki A (1998) High-speed conversion of xylose to L-lactate by electrodialysis bioprocess. *J Biotechnol* 60: 131-135.
132. Min-Tian G, Koide M, Gotou R, Takanashi H, Hirata M & Hano T (2005) Development of a continuous fermentation system for production of lactic acid by *Lactobacillus rhamnosus*. *Process Biochem* 40: 1033-1036.
133. Jeantet R, Maubois JL & Boyaval P (1996) Semicontinuous production of lactic acid in a bioreactor coupled with nanofiltration membranes. *Enzyme Microb Technol* 19: 614-619.
134. Gonçalves LMD, Ramos A, Almeida JS, Xavier AMRB & Carrondo MJT (1997) Elucidation of the mechanism of lactic acid growth inhibition and production in batch cultures of *Lactobacillus rhamnosus*. *Appl Microbiol Biotechnol* 48: 346-350.
135. Ding S & Tan T (2006) L-Lactic acid production by *Lactobacillus casei* fermentation using different fed-batch feeding strategies. *Process Biochem* 41: 1451-1454.
136. Roy S, Ravindra D Gudi, Venkatesh KV & Shah SS (2001) Optimal control strategies for simultaneous saccharification and fermentation of starch. *Process Biochem* 36: 713-722.
137. Thomson MH, Guyot JP & Kiel P (2007) Batch Fermentation on synthetic mixed sugar and starch medium with amylolytic lactic acid bacteria. *Appl Microbiol Biotechnol* 74: 540-546.
138. Kosakai Y, Park YS & Okabe M (1997) Enhancement of L(+) lactic acid production using mycelial flocs of *Rhizopus oryzae*. *Biotechnol Bioeng* 55: 461-470.

139. Yin PM, Yahiro K, Ishigaki T, Park Y & Okabe M (1998) L(+) lactic acid production by repeated batch culture of *Rhizopus oryzae* in air-lift bioreactor. *J Ferment Bioeng* 85: 96-100.
140. Yang WC, Zhong JL & Tsao GL (1995) Lactic acid production by pellet-form *Rhizopus oryzae* in a submerged system. *Appl Biochem Biotechnol* 51-52: 57-71.
141. Martak J, Schlosser S, Sabolova E, Kristofkova L & Rosenberg M (2003) Fermentation for lactic acid with *Rhizopus arrhizus* in a stirred tank reactor with a periodical bleed and feed operation. *Process Biochem* 38: 1573-1583.
142. Zhang ZY, Jin B & Kelly JM (2007) Production of lactic acid from renewable materials by *Rhizopus* fungi. *Biochem Eng J* 35: 251-263.

# Biocolours: Chemistry, Production, Safety and Market Potential

# 19

**V.K. Joshi, Mutum Preema Devi, Devender Attri and Rakesh Sharma**

*Department of Food Science and Technology, Dr. Y.S. Parmar University of Horticulture & Forestry, Nauni, Solan 173 230 (H.P.)*

## 1. INTRODUCTION

Colour is one of the first characteristics perceived by senses. It is one of the most important constituents which help in determining acceptability, judging the quality and basic aesthetic value of commodities like food. Addition of colour to food ensures the uniformity and intensity of the naturally occurring colour in the food. The practice of colouring food dates back to ancient time and today it has become a vital constituent of food. Colour, besides preserving the identity of food, also helps to protect flavour and light sensitive vitamins during storage by a sun screen effect. Colour preferences are influenced by the geographical, sociological and cultural aspects of the populations. It aids the recognition of food and also has psychological appeal on the enjoyment of food [1]. Saffron, turmeric and vegetable dyes were used by our ancestors to make their food attractive [2].

### 1.1 Synthetic Dyes

Towards the end of $19^{th}$ century, the attention was diverted to develop synthetic colours called 'coaltar dyes' [1]. These synthetic colours have a widespread acceptability due to their good stability and colouring ability. But, these synthetic colours are not totally safe for human beings. High levels of erythrosine intake is known to cause thyroid tumours. Ponceau 4R, tetrazine and sunset yellow have proved allergic in several individuals even at low levels of intake. The incidence of tetrazine sensitivity appears to be higher in asthmatic persons [3]. From the sake of consumer safety, the use of synthetic colours in foods should be slowly phased out. It is not an unrealistic dream as Norway and Sweden have already banned the use of synthetic colours at all in food. India is also taking initiative towards this goal.

### 1.2 Natural Colours/ Biocolours

At end of $19^{th}$ century, the action in delisting a number of certified dyes and proposals for the specific labelling of foods containing dyes created a colour crisis in food industry [4], which created a strong demand of colourants from natural sources. Being natural, they are considered safe [5]. Moreover, in

*Food Biotechnology: Principles and Practices*
*V.K. Koshi & R.S. Singh (Eds.)*
*I.K. International Publishing House Pvt. Ltd., India*
*pp 641–689*

recent years, increasing consumer awareness for natural products with almost no chemical additive and certified dyes, which were earlier considered safe, has necessitated the need to exploit food colourants of natural origin. Natural colours are generally extracted from fruits, vegetables, seeds, roots and microorganisms and are often called 'Biocolours' due to their biological origin [2]. Anthocyanins, betalaines, carotenoids, chlorophyll and riboflavin are classified as the natural colourants. Among these, carotenoids occur widely in nature and their annual production is the highest among all other natural colours. They are responsible for many of the brilliant red, orange, and yellow colours of edible fruits, vegetables and mushroom [6]. The vegetables like agathi, amaranth, colocasia leaves, mayalu and carrot, and among fruits, mango ripe, orange and papaya are the potential sources of β-carotene [7]. In foods, carotene has been used in butter, margarine, salad oil, popcorn, baked goods, confectionery and candy, eggnog, coffee, whiteners, fruit juices and beverages, cake, mixes, and soup, etc. There are more than 600 identified carotenoids and only about 50 of these have vitamin A activity. "A colourant is natural if it exists in nature, be it agricultural, biological or mineral in form". The FDA, however, does not recognize any colourant as natural unless it is derived from the characterizing food. Title 21 of the Code of Federal Regulations (CFR), part 73, reads: Food ingredients, such as cherries, green or red peppers, chocolate, and orange juice, which contribute their own natural colour when mixed with other foods, are not regarded as colour additives, but where a food substance such as beet juice is deliberately used as a colour, as in pink lemonade, it is a colour additive. When strawberry juice is added to enhance the colour of strawberry ice cream, it is considered natural. But if beet juice extract, also a natural colourant, is used to enhance the colour of strawberry ice cream, the colourant is considered artificial as it is not derived from strawberries. European Union manufacturers define these terms on the basis of colourant preparation. FDA links the classification of colourants to whether a colourant requires or does not require certification. Certification applies to the regulatory need to do laboratory analysis to assure the identity and purity of each batch of a colourant produced. In 21 CFR, pt. 73, FDA lists the colour additives exempt from certification, while 21 CFR, pt. 74, lists the colour additives subject to certification.

## 1.3 Creating the Colour

Though natural products are the hottest topics these days in the marketplace, but colouring foods with naturally derived colourants is a challengeable task. Product developers normally find that hue, and colourant compatibility with the food system pose two primary challenges to successful application of naturally derived colourants. Hue is the first hurdle for a variety of factors, the most important of which is the absence of one of the primary colours. The approved colourants, regardless of form, can provide yellows, yellow-oranges, reds, red–violets and browns. True blue and green colourants though missing in the natural palette are necessary to complete the rainbow of colours. But, blending these naturally derived colourants is like blending on an artist's palette; it often requires hands-on experience to achieve the particular hue sought.

The second hurdle in the way is the attributes of the system to be coloured so as to achieve compatibility. It is a task that should be done early in product development. There are several questions that need answers: What physical form of a colourant will the product need? Should it be oil or water soluble? Or if its final pH affects the colourant selected? Will the product require heat processing, and can the colourant survive the level of processing? Will the product be distributed in a clear or coloured package and could achieve the targeted shelf-life? What regulations or approval kosher apply to the colourant selected? An understanding of the chemistry and diversity of the various natural colourants is necessary for finding solutions to these problems. This chapter will attempt to review all these aspects of naturally derived colourants for use in foods. Some features of naturally derived colours are shown in Table 1.

**Table 1:** Some features of naturally derived colours

| Pigment | Source | Colour shade | Comments |
|---|---|---|---|
| Cucumin | Turmeric rhizome (roots) India | Bright lemon yellow | • A non-soluble pigment<br>• Light sensitive once solubilised and in the presence of water<br>• Used widely in desserts and confectionary |
| Lutein | *Tagetes erecta* (marigold) and alfalfa | Golden yellow | • Oil-soluble carotenoid pigment naturally present in many foods<br>• The development of easy to use water-soluble form with increased oxidative stability has extended its use to applications such as desserts, soft drinks and bakery products<br>• Only permitted in the USA in chicken feed |
| Natural mixed carotenes | Palm oil, *D. salina* (algae) | Golden yellow to orange | • Another oil soluble mixture of carotenoids (principally β-carotene) which is naturally sensitive to oxidation<br>• The USA permits only the algal source; has provitamin A activity |
| Bixin/nor-bixin | *Bixa orelana* (Bush seeds South America) | Orange | • Bixin is oil-soluble and norbixin water dispersible<br>• Initially widely used as natural colour<br>• Use is now restricted |
| Capsanthin/ casorubin | Paprika/capsi-cum annum | Reddy orange | • Naturally oil-soluble carotenoids from the red pepper. The flavour level must be lower than 250 ppm in the EU<br>• Used in products and sauces and coating as well as desserts, confectionary and bakery products |
| Lycopene | Tomatoes | Orange red | • Naturally, oil-soluble carotenoid pigment<br>• Has found little commercialization as a colour owing to high cost, poor stability and orange rather than red shade<br>• Also not listed in USA as colour additive |
| Carminic acid | Cochineal insect (female) Peru | Orange to red | • Water dispersible pigment with colour shade becoming redder with increased pH<br>• Price sensitive owing to cultivation method<br>• Approximately 70,000 insects are needed to produce 500 g of 50% strength pigment |
| Carmine | Cochineal insect (female) Peru | Pink to red | • Carmine is the aluminium lake of carminic acid<br>• Produces very stable pink/red shades<br>• In acidic products delaking can occur resulting in colour change and precipitation<br>• Widely used in confectionary and savoury products<br>• As with carminic acid, limited to non-vegetarian food |

Contd...

Contd...

| Pigment | Source | Colour shade | Comments |
|---|---|---|---|
| Betanin | Red table beet root | Pink to red | • Water-soluble pigments that are degraded by prolonged heating<br>• Widely used in ice cream and desserts |
| Anthocyanin | Black grape skin, elderberries, black carrots, red cabbage | Pink/red to mauve/ depending on pH | • Natural pigments of many red fruits, flowers and vegetables<br>• Largest commercial source is grape skin (*V. vinifera*) as a by-product of the wine industry<br>• Widely used in soft drinks, to depict black currants flavoured products |
| Chlorophyll (in) | Grass, lucern and nettle | Olive green | • Chlorophylls are oil dispersible and chlorophyllins water-soluble<br>• Limited use owing to poor stability and dull colour shade<br>• In USA, permitted source is as a vegetable (e.g., Spinach) extract |
| Copper (in) | Grass, lucern and nettle | Bluish green | • Exchange of $Mg^{+2}$ at the porphorin centre of $Cu^{+2}$ leads to bright stable shade<br>• Not listed in USA |
| Carbon black | Vegetable material | Grey to black | • Insoluble black pigment with excellent stability<br>• Widely used in confectionary. Not listed in USA |
| Crocin | Saffron/Gardenia fruit | Yellow | • Water-soluble carotenoids pigment<br>• No longer listed in Eu<br>• Further safety testing is required. Saffron is listed in USA but is limited by its high cost |
| Titanium dioxide | Anatase | White | • Insoluble white colour derived from nature by various processes<br>• Used to add whiteness to confectionary and bakery decoration<br>• Also used for low fat products to add creaminess |

## 2. HISTORICAL DEVELOPMENTS IN BIOCOLOURS

**1856:** Perkin's mauve pigment was discovered and coaltar dyes were synthesized.

**1884:** *Monascus* sp. was traditionally cultivated and utilized in the orient for making red rice wine, red shaohsing wine and red Chinese rice.

**1954:** The first carotenoid pigment from *Cryptococcus* was marketed.

**1963:** Production of carotenoid pigments from *Rhodotorula* species started.

**Early 1970s:** Astaxanthin was isolated from *Phaffia rhodozyma* (in honour of Prof. Herman Jan Phaff) grown on exudates of deciduous trees in Japan and Alaska.

| | |
|---|---|
| **Late 1970's and early 80's:** | Production of β-carotene from *Dunaliella salina* took place. |
| **1985:** | Betatene Limited Corporation was established for cultivation of *D. salina* on large scale for producing natural carotene products. |

## 3. SOURCES OF BIOCOLOURS

Fruits, vegetables, seeds, roots, barks, leaves, flowers, microorganisms, algae, insects, e.g., butterfly, etc. are the major sources of natural or biocolours.

### 3.1 Classification of Natural Colours

Natural colourant is the colour that occurs naturally in foods and is either produced by extraction from natural sources or reproduced by chemical synthesis [8]. Natural colours are generally extracted from vegetables, fruits, seeds, roots, insects and microorganisms. Origin and examples of some of the natural colours are given in Table 2.

**Table 2:** Origin and examples of some of the natural colours.

| **Origin** | **Example** |
|---|---|
| Natural colours obtained by direct use of approved pigmented vegetables and fruit products | • Dehydrated beet powder<br>• Dried paprika powder<br>• Dried turmeric powder<br>• Dried saffron powder<br>• Vegetable juice and fruit juice |
| Natural colours obtained by extraction and concentration of colours derived from approved plant materials | • Anthocyanin, extract of berries, grapes, vegetables<br>• Betalaines, extract of vegetables<br>• Carotenoids: Annatto extract |
| Natural colours obtained by extraction and concentration of colours derived from approved animal materials | • Laccania acid pigments from *Laccifera lacca*<br>• Cocchineal extract and caramine from *Coccus cacti* |
| Natural colours obtained by extraction and concentration of colours derived from approved microorganisms | • Monascus pigment from mould *Monascus purpureus*<br>• Phycocyanin pigment from *Spirulina* sp. |
| Natural colours derived by heat processing of food grade carbohydrates | • Caramel colours |
| Natural pigment found in nature but chemically synthesized | • β-Carotene, β-apo-8 carotenal and canthaxanthin |
| Miscellaneous natural colourants | • Lycopene pigment<br>• Haemoglobin pigment<br>• Kokam pigment<br>• Phycocyanin, Riboflavin |

*Source:* 6

### 3.2 Major Sources of Biocolours

Following are the major sources of biocolours.

**Plants/fruits/vegetables**
- Annatto
- Anthocyanins
- Astaxanthin
- Beet juice (Betanin)
- β-apo-carotenal
- β-carotene
- Canthaxanthin
- Capsanthin [paprika extract]
- Caramel
- Lutein [tagetes extract]
- Lycopene

**Animals/insects**
- Cochineal
- Carmine

**Microorganisms**
- Bacteria
- Yeasts
- Moulds
- Algae

Chemistry, production technology, various properties of natural pigments, safety, applications, etc. have been discussed in the subsequent sections.

## 4. CHEMISTRY AND STRUCTURE

### 4.1 Chlorophyll

Chlorophyll is the most important pigment and has use in nature, in photosynthesis. It is capable of channelling the energy of sunlight into chemical energy through the process of photosynthesis. In this process, the energy absorbed by chlorophyll transforms carbon dioxide and water into carbohydrates and oxygen:

$$CO_2 + H_2O \longrightarrow (CH_2O)_n + O_2$$

The chemical energy stored by photosynthesis in carbohydrates drives biochemical reactions in nearly all living organisms. Chlorophyll is vital for photosynthesis, which allows plants to obtain energy from light. Chlorophyll molecules are specifically arranged in and around pigment protein complexes called photosystems, which are embedded in the thylakoid membranes of chloroplasts. Once extracted from the protein into a solvent (such as acetone or methanol), these chlorophyll pigments can be separated in a simple paper chromatography experiment, and based on the number of polar groups between chlorophyll-*a* and chlorophyll-*b*, will chemically separate out on the paper.

#### *4.1.1 Chemical structure*

Chlorophyll is a chlorin pigment, which is structurally similar to and produced through the same metabolic pathway as other porphyrin pigments such as haeme. At the centre of the chlorin ring is a magnesium ion. The chlorin ring can have several different side chains, usually including a long phytol chain. There are a few different forms that occur naturally, but the most widely distributed form in terrestrial plants is chlorophyll-*a*. Treatment of cholorophyll-*a* with acid removes the magnesium ion replacing it with two hydrogen atoms giving an olive-brown solid, phaeophytin-*a*. Hydrolysis of this (reverse of esterification) splits off phytol and gives phaeophorbide-*a* (Fig. 1). Similar compounds are obtained if chlorophyll-*b* is used.

Chlorophyll when reacts with a base, which yields phyllins, which are magnesium porphyrin compounds (Fig. 2). Treatment of phyllins with acid gives porphyrins.

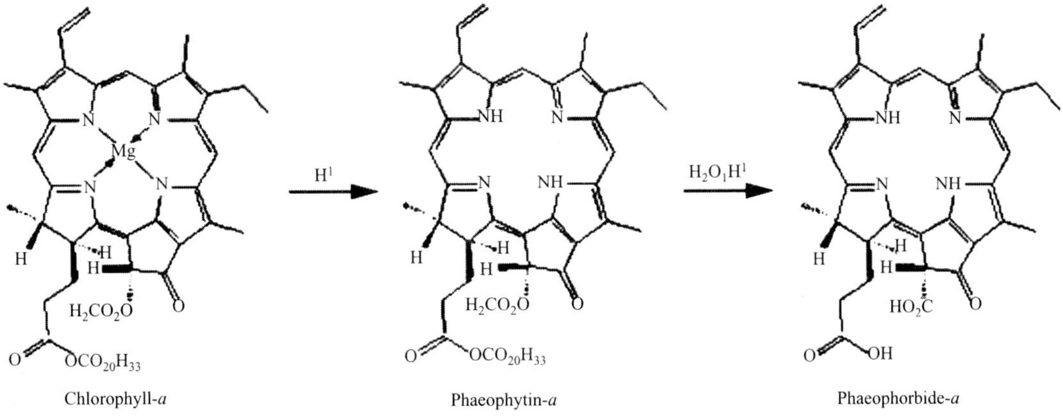


**Fig. 1:** Overall reaction scheme for the hydrolysis of chlorophyll.

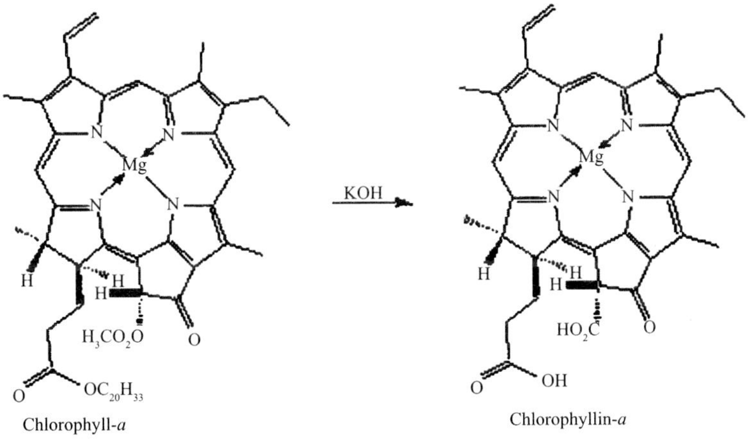


**Fig. 2:** Overall scheme for the reaction of alkali with chlorophyll.

### *4.1.2 Extraction of chlorophyll from plants*

In plants, chlorophyll is associated with specific proteins, for example, chlorophyll-*a* binding proteins are referred to as CP I, CP 47 and CP 43. With improved biochemical techniques for use on the membrane systems, there has been an ever increasing success in the isolation and characterization of these proteins.

Initially, detergents are used to break down the membrane into fragments, and these fragments are further broken down by the use of different detergents that work by replacing the membrane lipids, which surround integral membrane proteins. The resulting particles are separated by polyacrylamide gel electrophoresis (a standard biochemical method) in the presence of sufficient detergent to keep them 'solubilised'. The activity and polypeptide composition can then be assayed as the particle is purified. The detergents work by substituting lipids at different spots in the membrane; this is also affected by the concentration of the detergent. One such detergent that is very commonly used is SDS (sodium dodecyl sulphate). It has several advantages over other detergents: the separation can be carried out fairly rapidly and it also gives a good overall picture of the distribution of chlorophyll.

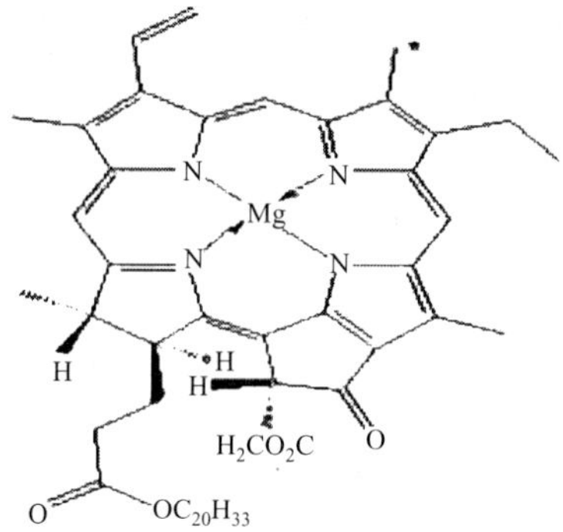


**Fig. 3:** Chlorophyll-*a* ($C_{55}H_{72}MgN_4O_5$, mol. wt.: 893.49). Methyl group marked with an asterisk is replaced by an aldehyde in chlorophyll-*b* ($C_{55}H_{70}MgN_4O_6$, mol. wt.: 906.51).

**Table 3:** Different structures of chlorophyll.

| Molecular formula | Chlorophyll-*a* | Chlorophyll-*b* | Chlorophyll-*c1* | Chlorophyll-*c2* | Chlorophyll-*d* |
|---|---|---|---|---|---|
| $C_3$ group | $C_{55}H_{72}O_5N_4Mg$ | $C_{55}H_{70}O_6N_4$ Mg | $C_{35}H_{30}O_5N_4Mg$ | $C_{35}H_{28}O_5N_4Mg$ | $C_{54}H_{70}O_6N_4Mg$ |
| $C_7$ group | $-CH=CH_2$ | $-CH=CH_2$ | $-CH=CH_2$ | $-CH=CH_2$ | -CHO |
| $C_8$ group | $-CH_3$ | -CHO | $-CH_3$ | $-CH_3$ | $-CH_3$ |
| $C_{17}$ group | $-CH_2CH_3$ | $-CH_2CH_3$ | $-CH_2CH_3$ | $-CH=CH_2$ | $-CH_2CH_3$ |
| $C_{17}$-$C_{18}$ group | $-CH_2CH_2COO$- Phytyl | $-CH_2CH_2COO$- Phytyl | -CH=CHCOOH | -CH=CHCOOH | $-CH_2CH_2COO$- Phytyl |
| Bond | Single | Single | Double | Double | Single |
| Occurrence | Universal | Mostly plants | Various algae | Various algae | Cyanobacteria |

The general structure of chlorophyll *a* was elucidated by Hans Fischer in 1940, and by 1960, when most of the stereochemistry of chlorophyll *a* was known, Robert Burns Woodward published a total synthesis of the molecule as then known. In 1967, the last remaining stereochemical elucidation was completed by Ian Fleming and in 1990, Woodward and co-authors published an updated synthesis. The structure of chlorophyll-*a* is shown in Fig. 3. The different structures of chlorophyll are summarized in Table 3. The absorption spectrum of chlorophyll-*a* is presented in Fig. 4.

## 4.2 Anthocyanins

Anythocyanins belong to a group of yellow to red plant compounds called flavonoids—a sub-class of plant polyphenols—and are the naturally occurring, water-soluble compounds that impart many of the orange, red, magenta, violet, purple and blue colours to a variety of fruits, vegetables and plants. The term "anthocyanin" is derived from the two Greek words *anthos*, meaning "flower," and *kyanos*, "dark blue". About 300 anthocyanins have been discovered in nature. Major food sources include apples, blackberries, black carrots, blackcurrants, black raspberries, blueberries, cherries, chokeberries, cranberries, eggplants, elderberries, grapes, radishes, raspberries, red cabbage, red currants, red-fleshed potatoes, red raspberries, red onions and strawberries and their content is shown in Table 4. Anthocyanins are commercially prepared from grape-skin extract (or the lees from the wine-making process), fruit juices (elderberry, black raspberry, blackberry and aronia) and vegetable juices (red cabbage and black carrot). Extraction is usually performed with a slightly acidic solution, and the extractive is a mixture of sugars, acids, salts and a pool of anthocyanins characteristic of the source material. All extracts act as pH indicators, changing from red to blue-red, purple, blue, and blue-green as pH increases. Usually, colour intensity is greatest at pH 4 or below. Absorption spectrum of anthocyanins at pH 4.5 and 1.0 is illustrated in Fig. 5.

Stability to heat and light improves with the use of extracts having naturally acylated anthocyanins, like those found in black-carrot and red-cabbage extracts. High sugar concentrations appear to provide a protective effect, while the presence of ascorbates will accelerate degradation. Non-carbonated and carbonated beverages, especially those retailed in cans that protect the colourant from light, have successfully employed anthocyanins. Anthocyanins have also been used in ice cream, yoghurt, soft candy,

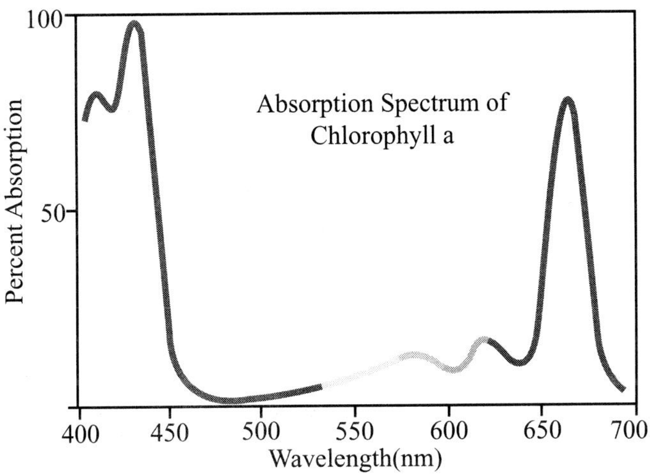


**Fig. 4:** Absorption spectrum of chlorophyll-*a*.

**Table 4:** Anthocyanin content of some common fruits and vegetables.

| Fruit/vegetable | Total anthocyanin (mg/100 g fresh tissue) |
|---|---|
| Apples | 10 |
| Bilberries | 300-320 |
| Blackberries | 83-326 |
| Blueberries | 25-495 |
| Grapes | 6-600 |
| Plum | 44-231 |
| Peach | 7.64-50 |
| Sour cherries | 27.8-80.4 |
| Strawberries | 15-35 |
| Kiwi | 100 |
| Red raspberries | 20-60 |
| Red onions | 7-21 |

*Source:* 9-12

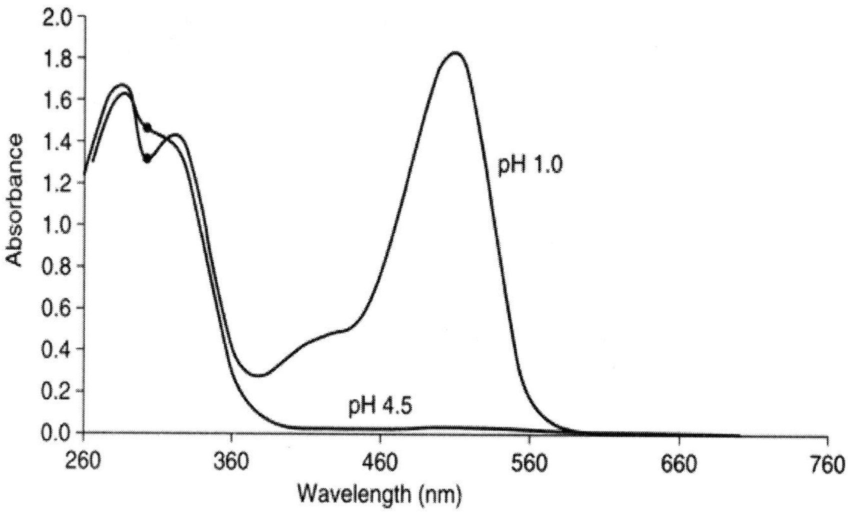


**Fig. 5:** Absorbance of anthocyanin at pH 4.5 and pH 1.0.

alcoholic beverages, juices, chewing gum, beverage powders, icings, and as a fruit-colour supplement in jams and jellies.

### *4.2.1 Stability of anthocyanins*

Anthocyanins were intensively investigated for beverage uses in model system which contained about 13-15% sucrose and/or glucose [13-18]. There is a continuing effort to make anthocyanin colourant having more stability. The formation of brown colour in plant juice is the result of two step process. In the first instant, plant phenolic from the juice is oxidized into colourless compounds by reacting with oxygen. This reaction occurs spontaneously and can be accelerated by a plant enzyme, the polyphenol oxidase. Secondly, the uncoloured oxidized phenolics undergo condensation and form brown pigments. The molecule of anthocyanin exists in a state of equilibrium between four primary structures. At low pH, the red form of the anthocyanin is favoured. As the pH increased, the red form converts into blue form and a non-coloured form. Therefore, the hue of an anthocyanin solution becomes bluer and less intense as pH increases. While the blue form reverts back to the red form, the non-coloured form of the anthocyanin can either revert back to the red form or can undergo an irreversible hydrolysis of the flavilium portion of the anthocyanin that prevents reversion to the red form. Prolonged heat exposure causes browning in some colours, particularly anthocyanins, grape skin extract and fruits and vegetable juices and destroy other. Palamidis and Markakis [14] have studied the effect of temperature on the stability of anthocyanin in soft drinks and have shown that increase in the storage temperature greatly accelerated the destruction of pigments in soft drinks. Maccarone *et al.* [19] studied the stability of anthocyanin in red orange juice in 15°C, 25°C and 35°C during a 15 day period and found that the increase in temperature accelerates the destruction of anthocyanins. Palamides and Markakis [14] have also studied the role of light on the stability of anthocyanin in grape juice, and showed that the exposure of the pigments to light accelerates their destruction. It was shown that after placing the juice samples containing anthocyanin in dark for 135 days at 20°C, almost 30% of the pigments were destroyed, but placing the same samples in the same temperature and same period of time in the presence of light destroyed more than 50% of total pigments. Sistrunk and Cash [20] attempted to stabilize the colour of strawberry puree and reported that the presence of ascorbic acid improves the colour retention during the first 6 hours of storage at 50°C.

#### 4.2.2 *Anthocyanins in fruits and effect of processing*

Anthocyanins are responsible for the attractive colour of black currant nectar and they have been hypothesized as important antioxidants [21]. Anthocyanins are present in most of higher plants and provide a wide range of colours from orange to blue [22, 23]. Concord grapes, grape skin extract, red cabbage, and fruit and vegetable juices (e.g., elderberries, blackcurrant, chokeberries, etc.) are used as commercial sources of natural colourants. The anthocyanin content of peaches ranges from 7.64 to 50.1 mg as cyanidin 3-glucoside per 100 g of fresh fruit tissues [5, 23-27]. Chiang *et al.* [12] studied the anthocyanin pigment composition of blackberries and found that the total anthocyanin content ranged from 70.3 to 201 mg/100 g with a mean of 137 mg/100 g.

Anthocyanins are very unstable [26]. Loss of colour in strawberries can occur after thawing of frozen fruits and during processing and storage of strawberry products i.e., juice, concentrates, wine, preserves and purees [28-30]. Anthocyanin disappears as monomeric compounds and is transformed into polymeric form [31-34]. This transformation results in a colour change to a more brownish shade [24]. The major anthocyanins of sunflower hulls have been identified as malonated glycosides of cyanidin [35], and colour stability experiments have shown that pigments are relatively stable. Blueberries are often processed into juice or juice concentrates for subsequent use in beverages, syrups and other food products. Knowledge of the changes that the anthocyanin pigments undergo with processing is important with respect to their role in colour quality. The stability of anthocyanin pigments is affected by oxygen, pH, temperature, light, metal ions, enzymes and ascorbic acid and may also be affected by the actual anthocyanin concentration [9].

#### 4.2.3 *Spray drying of anthocyanins*

Glicksman *et al.* [36] reported that powdered product has a long shelf-life when properly packaged. A concentrated powdered form of anthocyanin pigment is desirable for many reasons i.e., ease of handling, transportation, storage and especially for use in powdered form products. It has also been demonstrated that anthocyanin can be freeze-dried to a stable powder form [13]. Spray drying would be a practical and economical method of obtaining a large quantity of anthocyanin powdered food colourant, as the cost of freeze-drying is considerably higher. Glicksman *et al.* [36] reported that at present, the most practical and economical method for producing large quantity of a variety of powdered heat sensitive flavours is spray drying. Francis [37] studied that since anthocyanins in general do not show good stability under most storage conditions, they can be concentrated by using a vacuum pan at relatively low temperature in order to minimize the pigment degradation due to heat. Main *et al.* [38] reported that spray drying of anthocyanin with use of low D.E. dextrin as a carrier acts as an encapsulating agent much in the same way as when drying heat sensitive flavours. Glicksman *et al.* [36] recommended gum arabic as a carrier food coating agent for spray drying of flavours, but dextrose, starch, gum tragacanth, methylcellulose and mixture of these has also been used. It has also been shown that some carbohydrate carrier agent has the desired encapsulating properties when used in the spray drying of heat sensitive flavours [39].

#### 4.2.4 *Structure*

An anthocyanin pigment is composed of an aglycone (an anthocyanidin) esterified to one or more sugar. A generalized structure of anthocyanins is illustrated in Fig. 6. The molecular weights of different anthocyanins, anthocyanidins and acylating groups commonly found in nature are summarized in Table 5. Anthocyanins have four different structures, which are in equilibrium, and the relative amount of these structures in equilibrium are varied and dependent on the pH and anthocyanin structure, which is the reason why some anthocyanins are more stable than others depending on their molecular structures (Fig. 7).

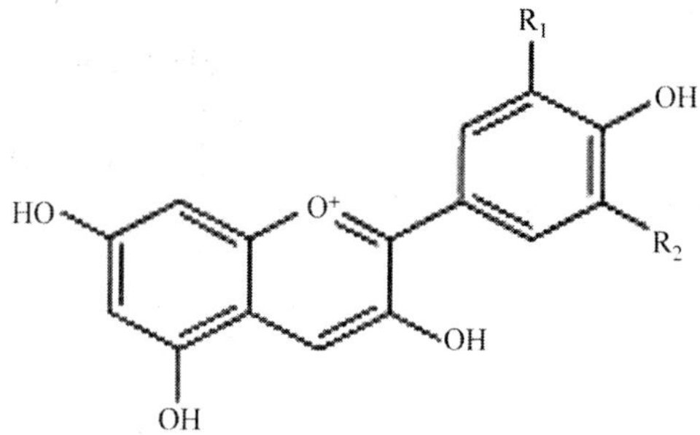


**Fig. 6:** Structural formula of anthocyanidins. Cyanidin, $R_1$ = OH, $R_2$ = H; delphinidin, $R_1$ & $R_2$ = OH; peonidin, $R_1$ = $OCH_3$, $R_2$ = H; petunidin, $R_1$ = $OCH_3$, $R_2$ = OH; malvidin, $R_1$ & $R_2$ = $OCH_3$.

**Table 5:** Molecular weights of anthocyanidins, anthocyanins and acylating groups commonly found in nature.

| **Anthocyanidins** | **Pelargonidin** | **Cyanidin** | **Peonidin** | **Delphinidin** | **Petunidin** | **Malvidin** |
|---|---|---|---|---|---|---|
| | 271 | 287 | 301 | 303 | 317 | 331 |
| Hex | 180.2 | 180.2 | 180.2 | 180.2 | 180.2 | 180.2 |
| Hex $-H_2O^a$ | 162.2 | 162.2 | 162.2 | 162.2 | 162.2 | 162.2 |
| Acd + 1 hex | 433.2 | 449.2 | 463.2 | 465.2 | 479.2 | 493.2 |
| Acd + 2 hex | 595.4 | 611.4 | 625.4 | 627.4 | 641.4 | 655.4 |
| Acd + 3 hex | 757.6 | 773.6 | 787.6 | 789.6 | 803.6 | 817.6 |
| Pent | 150.0 | 150.0 | 150.0 | 150.0 | 150.0 | 150.0 |
| Pent $-H_2O^a$ | 132.0 | 132.0 | 132.0 | 132.0 | 132.0 | 132.0 |
| Acd + 1 pent | 403.0 | 419.0 | 433.0 | 435.0 | 449.0 | 463.0 |
| Acd + 1 hex + 1 pent | 565.2 | 581.2 | 595.2 | 597.2 | 611.2 | 625.2 |
| Rhamnose | 164.2 | 164.2 | 164.2 | 164.2 | 164.2 | 164.2 |
| Rutinose | 326.2 | 326.2 | 326.2 | 326.2 | 326.2 | 326.2 |
| Ruiinose $-H_2O^a$ | 308.2 | 308.2 | 308.2 | 308.2 | 308.2 | 308.2 |
| Acd + rutinose | 579.2 | 595.2 | 609.2 | 611.2 | 625.2 | 639.2 |
| Acd + rutinose + 1 hex | 741.4 | 757.4 | 771.4 | 773.4 | 787.4 | 801.4 |
| Acd + rutinose + 1 pent | 711.2 | 727.2 | 741.2 | 743.2 | 757.2 | 771.2 |

*Common acylating groups*

| | | **$-H_2O^b$** |
|---|---|---|
| *p*-Coumaric acid | 164.2 | 146.2 |
| Caffeic acid | 180.2 | 162.2 |
| Ferulic acid | 194.2 | 176.2 |
| Sinapic acid | 224 | 206 |
| Acetic acid | 82 | 64 |
| Propionic acid | 96.1 | 78.1 |
| Malonic acid | 104.1 | 86.1 |
| Succinic acid | 118.1 | 100.1 |

Abbreviations: hex: hexose; pent: pentose; acd: anthocyanidin.

$^a$-$H_2O$ indicates a dehydrated sugar (water is lost upon forming a glycosidic bond)

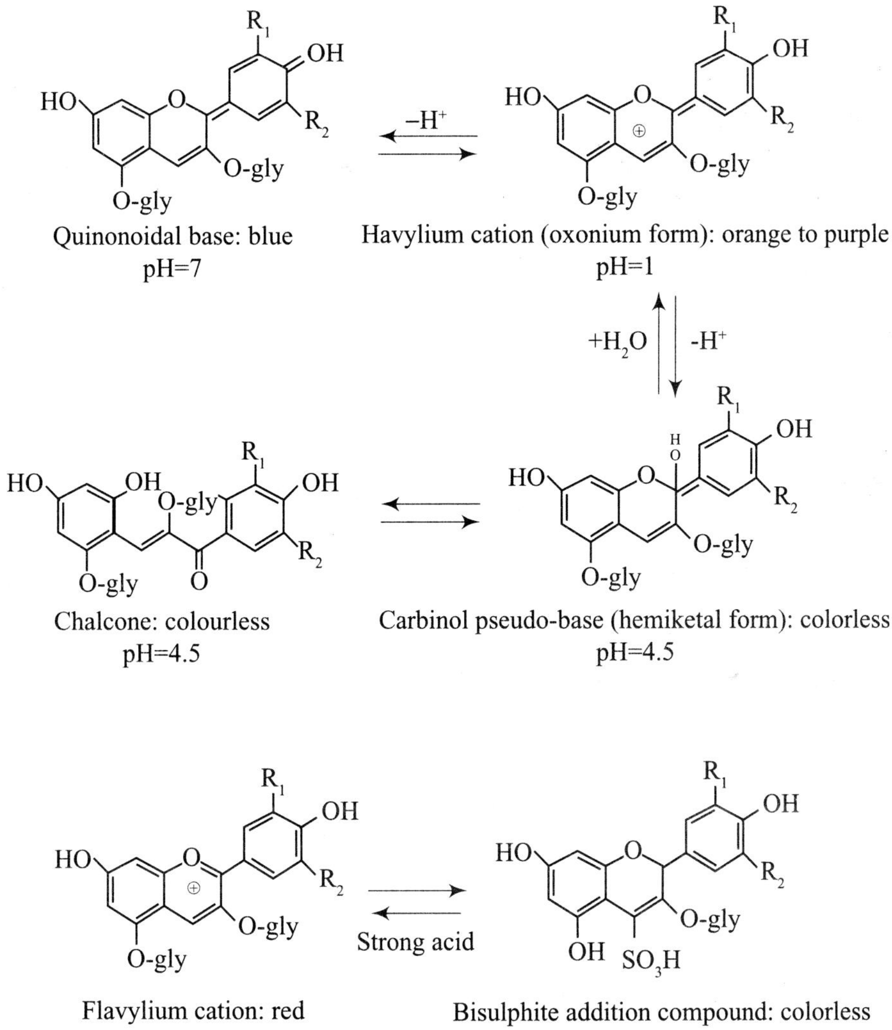


**Fig. 7:** Molecular structure of anthocyanins at different pH values.
*Source:* 35

The HPLC chromatogram of berry and mass chromatogram of bilberry are depicted in Fig. 8 (a-e) and Fig. 9 (a-e).

## 4.3 Carotenoids

Carotenoids are a class of natural fat-soluble pigments found principally in plants, algae, and photosynthetic bacteria, where they play a critical role in the photosynthetic process. They also occur in some non-photosynthetic bacteria, yeasts, and moulds, where they may carry out a protective function against damage by light and oxygen. Although animals appear to be incapable of synthesizing carotenoids, many animals incorporate carotenoids from their diet. Within animals, carotenoids provide bright colouration, serve as antioxidants, and can be a source for vitamin A activity [41, 42]. Carotenoids are responsible for many of the red, orange, and yellow hues of plant leaves, fruits, and flowers, as well as the colours of some birds, insects, fish, and crustaceans. Some familiar examples of carotenoid

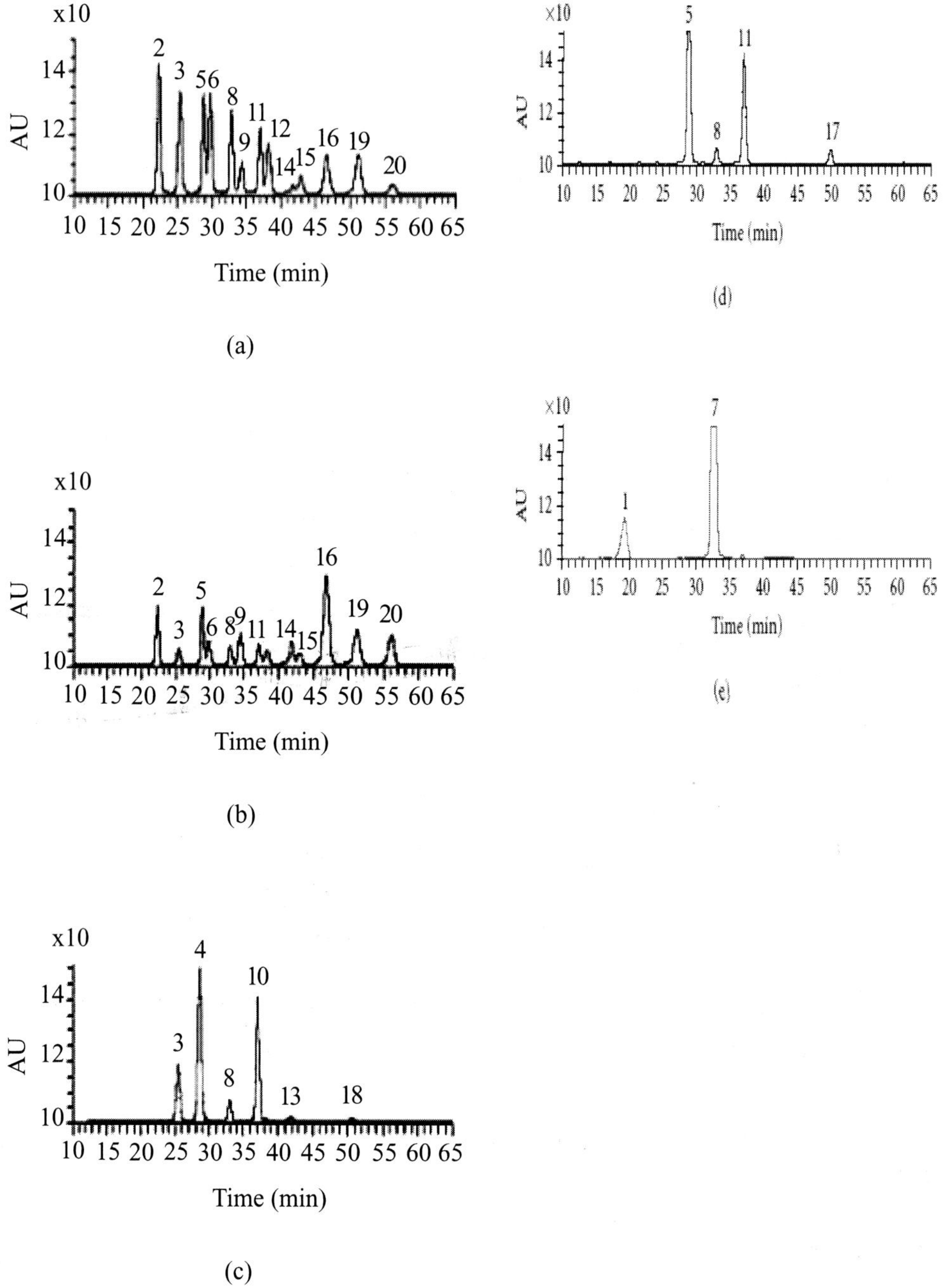


**Fig. 8:** HPLC chromatogram of berry extracts on PDA. The chromatograms were obtained on PDA of selective wavelength (500 to 550 nm). Numbers on the chromatograms indicate detected peak. (a) Bilberry, (b) Rabbiteye, (c) Black currant, (d) Chokeberry, (e) Elderberry

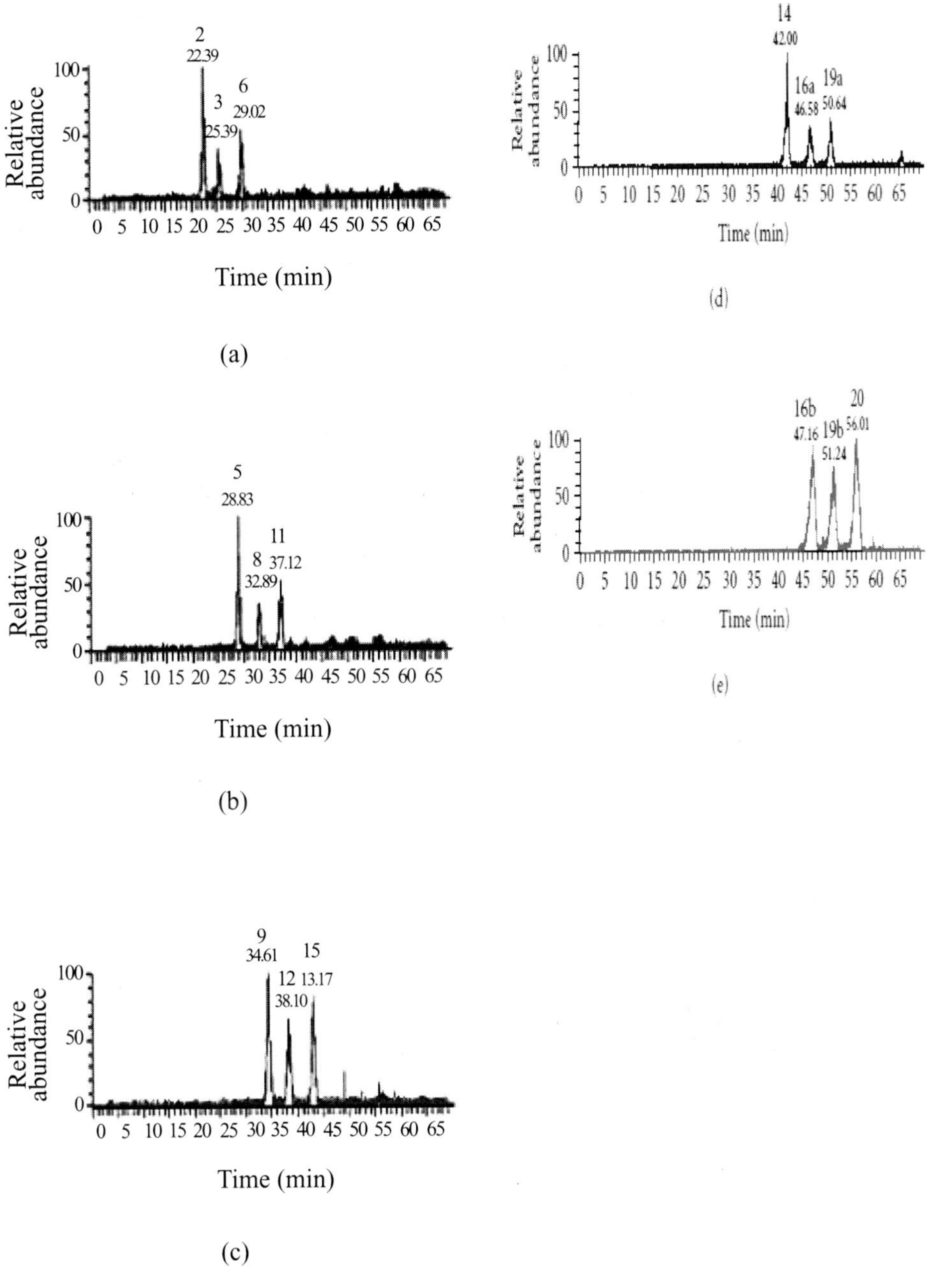


**Fig. 9:** Mass chromatogram of bilberry extract selected by m/z of each aglycon. Mass chromatogram revealed that bilberry extract contains 15 anthocyanins; nevertheless, the extracts showed only 13 peaks on PDA. The peaks of 16 and 19 overlap with 16a and 16b and 19a and 19b, respectively. (a) Delphinidin (m/z = 303); (b) Cyanidin (m/z = 287) (c) Petunidin (m/z = 317) (d) Peonidin (m/z = 301) (e) Malvidin (m/z = 331)
*Source:* 40

colouration are the oranges of carrots and citrus fruits, the reds of peppers and tomatoes, and the pinks of flamingos and salmon [43]. Some 600 different carotenoids are known to occur naturally [41], and new carotenoids continue to be identified [44].

### *4.3.1 Structure*

Carotenoids are defined by their chemical structure. The majority of carotenoids are derived from a 40-carbon polyene chain, which could be considered the backbone of the molecule. This chain may be terminated by cyclic end-groups (rings) and may be complemented with oxygen-containing functional groups. The hydrocarbon carotenoids are known as carotenes, while oxygenated derivatives of these hydrocarbons are known as xanthophylls. Beta-carotene, the principal carotenoid in carrots, is a familiar carotene, while lutein, the major yellow pigment of marigold petals, is a common xanthophyll (Fig. 10). The structure of a carotenoid ultimately determines what potential biological function(s) that pigment may have. The distinctive pattern of alternating single and double bonds in the polyene backbone of carotenoids is what allows them to absorb excess energy from other molecules, while the nature of the specific end groups on carotenoids may influence their polarity. Lutein and zeaxanthin, xanthophylls found in corn and in leafy greens such as kale and spinach, are believed to function as protective anti-oxidants in the macular region of the human retina [45]. Lutein is an oxycarotenoid, or xanthophyll, containing 2 cyclic end groups (one beta- and one alpha-ionone ring) and the basic C-40 isoprenoid structure common to all carotenoids. Lutein has been assigned Chemical Abstract Service (CAS) number 127-40-2 with a chemical formula of $C_{40}H_{56}O_2$ and a molecular weight of 568.88.

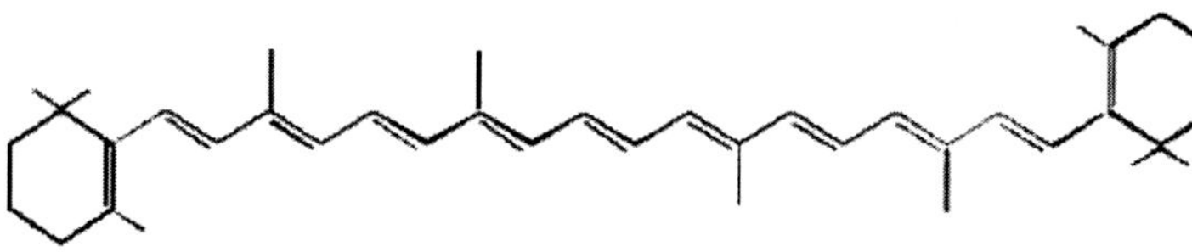

**Beta-carotene**

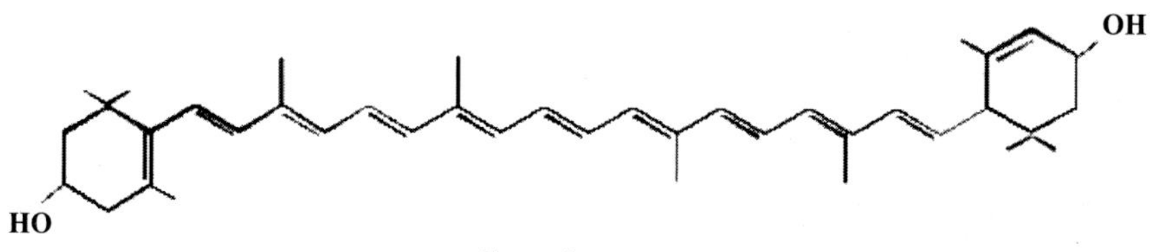


**Lutein**

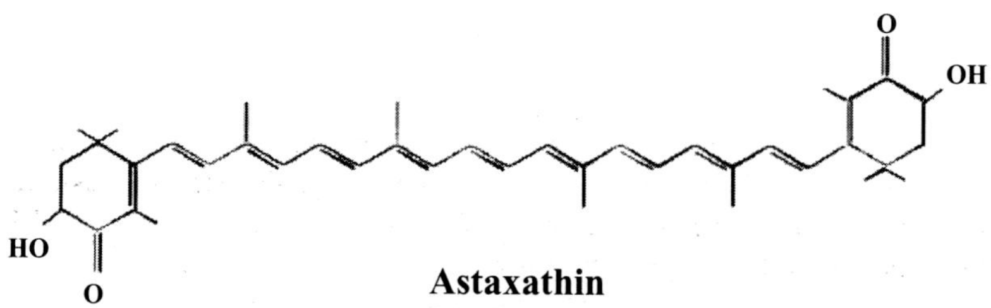


**Astaxathin**
(3,3'-dihydroxy 4,4'-dihetro -carotene)

**Fig. 10:** Structures of different carotenoids.

# 5. TECHNOLOGY OF BIOCOLOUR PRODUCTION

## 5.1 Methods of Extraction

One way of colouring a food would be to add a strongly coloured food (e.g., elderberry or black currant) to the food item (e.g., raspberry jam) that is to be coloured. This is the approach used in home-cooking where spices, besides flavour, may impart colour: turmeric and paprika are examples of this. However, for industrial food production, this approach presents a series of problems such as low concentration of pigments, unwanted flavour and insoluble matter (e.g., peel and seeds), which may not be a problem in some applications (jam) but is unacceptable in others (e.g., beverages). Therefore, pigments are extracted to overcome the problems of low concentration, flavour, and insoluble material. Lipid-soluble pigments like chlorophyll and carotenoids are usually extracted with organic solvents, which are subsequently removed, yielding an oleoresin rich in pigments, but also containing other material such as triglycerides, sterols, wax, and other lipid-soluble compounds. Water-soluble pigments such as carminic acid and anthocyanins are in general extracted with water or lower alcohols. The traditional methods for extraction of the colourings usually involve disruption of the plant material followed by acidified aqueous extraction for chlorophyll, carotenoids and annatto. However, efforts are being made to improve yield, quality and stability of these biocolours by using following improved methods of extraction [2, 5, 6, 46].

### *5.1.1 Enzyme aided extraction*

Macerating enzymes have been successfully applied to improve the extraction yield of valued compounds from natural products. Various enzymes like pectinase, cellulase, invertase, etc. are being used to facilitate the extraction, increase pigment yield (up to 50%) and improve the quality and stability of the resultant colour preparations [47, 48].

### *5.1.2 Microorganisms aided extraction*

Natural colouring extracts, especially those obtained through aqueous extraction or direct pressing of plant materials, contain pigments along with high proportion of unwanted dissolved solids like salts, organic acids, phenolics and sugars. Such high solid extracts on drying produce hydroscopic material, which affects the quality of colourants. Debicki-Pospisil *et al.* [49] found that sugars produce furfural and 5-hydroxymethyl furfural during concentration and drying of the extract, which accelerate pigment degradation. On the other hand, microorganism aided extraction using suitable yeasts and moulds reduces the solids substantially, giving a 5-7 fold increase in pigment content on dry weight basis [46].

### *5.1.3 Supercritical carbon dioxide extraction*

Current processes used for extracting colour such as solvent extraction, steam distillation include low recovery of pigment, risk of thermal degradation of the extract and risk of oxidation. These risks can be avoided by using a solvent that is in a supercritical condition with regard to temperature and pressure [50]. Chao *et al.* [51] used supercritical carbon dioxide as an extracting solvent as one of the modern approaches to avoid the detrimental affect of heat and light and also to reduce the interference of co-extracted material.

### *5.1.4 Membrane extraction*

Plant origin colourants generally lack sufficient colour strength, uniformity and stability. To overcome these deficiencies, membrane processing techniques have been developed for purification and concentration of natural colours like anthocyanin, batalaine, etc., which are heat sensitive. Ultrafiltration of anthocyanin and betanin containing extracts using cellulose acetate membranes with molecular weight cut off (MWCO) of 1000 Da and 500 Da, respectively removes 75-90% of water and

50-60% of sugars in single pass and gives two-fold increase in colour content [52]. For commercial production of natural colours, the application of membrane processing offers some advantages which are summarized in Table 6.

## 5.2 Production Technology of Some Biocolours

### *5.2.1 Grape-skin extract*

Different varieties of grapes, as well as different extraction techniques, are used for the production of grape-skin extract. Since it lacks uniformity among the producing countries, product-application results can differ with extracts produced by different manufacturers. Grape-skin extract is available as a liquid or powder (spray-dried onto maltodextrin), and both versions are water-soluble. The hue of the extract is pH and concentration dependent. This colourant will work in products with a low pH, minimal or no heat processing, limited exposure to light and a shelf-life of less than a year, but in the United States, the FDA limits its use to beverages.

### *5.2.2 Grape juice*

It consists of the extract obtained from the lees derived from grape juice production and is available in liquid or powdered form. Formulators can also obtain a kosher ingredient form. Normally, it tends to be made red than the extract derived from grape skin. In the United States, it can be used in all food products where it can be added (those without a standard of identity or whose standard of identity contains grape colour extract) except beverages.

### *5.2.3 Red cabbage*

It is the most expensive of the anthocyanin concentrates, but its high cost is offset by a high tinctorial strength. It is available in a liquid or spray-dried form. The extracts are water-soluble, and because of the presence of acylated anthocyanins, have improved stability to heat and light. Slight changes in pH have a large effect on the resulting hue and so, pH control of the final product becomes critical. For the correct hue, these extracts are best-suited for products that have low pH, require heat processing and/or will have minimal exposure to light.

### *5.2.4 Elderberry*

Using membrane technology, this water-soluble colourant is extracted and concentrated from elderberry fruit, and is available as a liquid or a spray-dried powder (on a maltodextrin carrier). As with all the anthocyanins, pH adjustment can move the colourant's hue from strawberry-red (pH 2.4 to 2.8) to raspberry-red (pH 2.9 to 3.3) and ultimately, cherry-red (pH 3.5). Above pH 3.5, the hue moves through a violet-blue to a deep, but unstable blue. Elderberry may survive HTST (High

**Table 6:** Advantages of membrane extraction of biocolours.

- Purification and concentration of pigment from dilute extract.
- Removal of microorganisms (contaminated or added for fermentation) from the extract.
- Removal of cells debris in plant extracts.
- Removal of objectionable flavour.
- Elimination of risk of transformation of pigment during processing of extract.
- Enhanced shelf-life of the colour preparation.
- Saving of energy.

*Source:* 46

temperature short time) processing and a temperature up to 110°C for a few minutes, but it is susceptible to light degradation and oxidation. Products needing some heat processing may use this colourant.

### *5.2.5 Aronia*

Extracted from the chokeberry, aronia is a violet black fruit berry native to North America. It is available as a liquid concentrated or a water-soluble powder. Its hue is concentration and pH-dependent. The tinctorial strength available is the weakest of the anthocyanin-based colours.

### *5.2.6 Black carrot*

The biocolour is water-extracted and concentrated from selected varieties of black carrots. The extract has good colour intensity and, due to the presence of acylated anthocyanins, is as stable to light and heat similar to that form of red cabbage. Like all anthocyanin extracts, the hue is pH-sensitive and tends to be more red at an acidic pH than red cabbage. Less expensive than red cabbage, black-carrot extract can often replace it.

### *5.2.7 Beet juice*

Red beet juice extract is prepared by mechanically pressing blanched ground beets, followed by filtering and vacuum-concentrating processes, to produce a solid-extract content of about 60-65%. Spray-drying beet juice extract onto maltodextrin produces an easily stored powder. The colour of the liquid extract, and the powder made from it, varies depending upon beet variety, root quality, age at harvest, handling practices and method of extraction. The hue in solution remains unchanged over the pH range of 3 to 7. Below pH 3, the colour becomes violet, while above pH 7, the colour takes on a blue tone. Beet juice extract becomes brown on heating, limiting its use to products that do not require cooking. Products distributed in a dry state have successfully used the powder form of the extract, which is hydroscopic. All beet juice extracts have found use in ice cream, no-cook fruit preparations as well as powdered beverages.

Besides chlorophyll, anthocyanins are probably the most important group of visible plant pigments. As colourants, anthocyanins date back to antiquity where Romans used richly coloured berries to enhance the aesthetic appeal of existing plant products or formulated substitutes for purposes such as to improve the colour of wine. In 1927, anthocyanin pigments were isolated and identified from three species of alpine *Primula*. The flowers of *P. viscosa* and *P. integrifolia* both were found to contain a pigment identical to malvin, a diglucoside of delphinidin-3': 5'-dimethyl ether, while from the flowers of *P. hirsuta,* a diglucoside of a new O-trimethyldelphinidin has been isolated, which has been named *'hirsutidin'*. Primulin, the anthocyanin pigment found in magenta and dark red flowers of *P. polyanthus*, is an interesting addition to this series. Like oenin, it consists chiefly of a 3-monoglucoside of malvidin, contaminated by traces of a lower methylated compound. Thus, although primulin furnishes the same aglucone as that found in two of Karrer's pigments, it differs from them in the nature and position of its glucosidal residue. *P polyanthus* is a horticultural hybrid probably derived from the cowslip and the common primrose. It is found in a large range of colours including various shades of yellow, magenta, crimson, orange and brick-red. Only magenta and crimson flowers were used for this investigation, as it is possible that a second anthocyanin is present in the orange and flame-coloured flowers. The method of extraction and isolation was adapted from Anderson's modification of Picrate method for the preparation of oenin chloride from purple grapes, in which 1% aqueous hydrochloric acid instead of glacial acetic acid is used for the extraction of the pigment from the flesh flowers. Despite the ubiquity of anthocyanins in nature, only grapes and red cabbage had commercial success in the extraction of colour with elderberries, aronia berries, and black carrots gaining strong ground. Grapes are the world's largest fruit crop from which an annual consumption of about 10,000 tonnes of anthocyanin-based colourants has been estimated. In the United States, grape skin extract, a by-product of the wine industry, is permitted in beverages only (regulated by FDA in 21CFR 73.170), whereas applications of grape colour extract derived from the fruit are limited to non-beverage food uses (21CFR 73.169). Different varieties of grapes are used for extraction and techniques vary. This, in turn, produces extracts with unique properties affecting colour shade and stability. The final extract is a particle-free, water-soluble liquid of high tinctorial strength, which may also be oven- or spray-dried to yield a water-soluble, dry powder.

## 5.3 Sources of Biocolours

### *5.3.1 Biocolours from plants*

5.3.1.1 Caramel: Available as a dark-brown to black liquid or powder, caramel is prepared by the controlled heat treatment of selected carbohydrates. The addition of acid, alkalis and salts can enhance this caramelization process (Table 7). Some do not consider it a natural colourant, but a reaction product. Its acceptance depends on the manufacturer's definition of natural. The industry has created four groups into which caramels are classified based on its application. These colours are not affected by heat, light or changes in pH, and are used in soft drinks, beers, distilled products, sauces, gravies and other products where a deeper, richer hue is beneficial to appearance.

5.3.1.2 Beet (E162): Beet, termed *Beta vulgaris*, is an annual root vegetable of the goosefoot family *Chenopodiaceae*. The name "beet" comes from the fact that when the seed pods swell, they look like the Greek letter "beta". The beet, originally grown only for its leaves, is native to Europe and North Africa and dates back to prehistoric times. Historic tales say that when the Greeks paid homage to Apollo, they served him beetroots on a silver platter. But the first recorded reference to beets is not found until the third century in Rome. Ancient civilizations along the shores of the Mediterranean sea grew beets for the medicinal quality of their tops. During the 16$^{th}$ century, beets were transported from France to Germany. Today, beets and beet dishes are popular throughout Europe and also used as a colourant for various applications. The pigments in beets are collectively known as betalains, which consist of two sub-classes: betacyanins (red pigments) and betaxanthins (yellow pigments). Both are water soluble and produced in abundance for use as an established colourant in foods and pharmaceuticals (Table 7). About 10% is converted into juice and colour. The use of red beets is restricted to vegetable juice, juice concentrate or powder. Beetroots are processed, using either pressing or diffusion techniques. The juice is then centrifuged, pasteurized and concentrated to yield a viscous liquid juice concentrate which is widely used as a food ingredient (regulated by FDA in 21CFR 73.260). Commonly spray dried with maltodextrin as a carrier, this vegetable juice is referred to as dehydrated beet or beet powder (exempt from certification according to 21CFR 73.40).

5.3.1.3 Turmeric (E100): The name "turmeric" finds its origin in the Latin term *Merita*, meritorious earth, referring to the colour of ground turmeric, which resembles a mineral pigment. It is a tall herb with large root stocks and consists of rhizomes. Turmeric is native to the humid tropical climate of southern Asia. It is an ancient spice and one of the cheapest, derived from the dried rhizomes - underground stems of several species of *Curcuma longa*, a perennial herb of the ginger-like *Zingiberaceae* family. It is grown in many parts of India as a commercial crop. It is a household spice for diverse cuisines throughout in India. The use of turmeric dates back nearly 4,000 years to the Vedic culture in India where the plant had religious significance and was used for its unique yellow colour, aroma, flavour, and health benefits as a body dye, cosmetic, culinary spice or medication. In Biblical times, turmeric was often used for its rather exotic fragrance to make perfume. Most of the production is concentrated in India

**Table 7:** Properties of some of the natural pigments used in foods.

| Pigment | Source | Solubility | Effect of Acid | Alkali |
|---|---|---|---|---|
| Anthocyanins | Fruits | Water soluble | Redness | Gray |
| Betataines | Beet root | Water soluble | - | - |
| Bixin | Seed coat of *Bixa orellana* | Fat soluble | - | - |
| Canthaxanthin | - | Fat soluble | - | - |
| Caramel | Heated sugar | Water soluble | - | - |
| Carotene or lycopene | Leaves | Fat soluble | - | - |
| Chlorophylls | Leaves | Water soluble | Darkness | Yellow |
| Curcumin | Turmeric | Water soluble | | |
| Flavone | Vegetables | Water soluble | Brightens | Browns |
| Norbixin | Seed coat of *Bixa orellana* | Water soluble | - | - |
| Polyphenols | Tea leaf | Water soluble | - | |
| Quinones | Roots, bark | Water soluble | - | - |
| Xanthophylls | Fruits | Water soluble | - | - |

*Source:* 53

where the country produces about 94% of the world's crop. Its powder is considered auspicious in many religious ceremonies in Hindu families. Besides, it is being often used in beauty creams for its antibiotic properties. It is also very commonly used as household remedy for minor cuts and wounds. The yellow pigmented fraction "Curcumin" is isolated from the rhizomes and is also used for diverse medicinal/ food purposes. Turmeric is usually marketed as the dried rhizome, which is subsequently milled to a fine powder also known as curcumin, curcuma, Indian saffron and yellow ginger. It imparts both flavour and colour to a food product.

The three principal types of turmeric extract are: essential oil of turmeric, turmeric oleoresin and curcumin. Essential oil of turmeric is obtained by the steam distillation of dried rhizome powder. Turmeric oleoresin, obtained by solvent extraction of the ground turmeric, is the most commonly produced extract of turmeric and contains both the flavour compounds and colour in the same relative proportion as that present in the spice. Curcumin, produced by crystallization from the oleoresin, is the pure colouring principal and a predominant colour compound of turmeric. The curcumin and the volatile oil are both extracted by the volatile solvent. Until the late 1970s, turmeric oleoresin was prepared only in a few of the consuming countries, mainly the United States and the United Kingdom, and their methods of processing were closely guarded secret. India has also developed a processing technique at the Central Food Technological Research Institute in Mysore. The extracted pigment, curcumin, is available in a number of functional forms: as an oleoresin of such as a powder (some powders contain as much as 95% curcumin), and as a suspension in a variety of carriers, such as vegetable oil (Table 8).

The addition of emulsifiers, such as polysorbates, converts the curcumin into water-soluble form widely used today. In aqueous solution, turmeric produces hues ranging from yellow to greenish-yellow, depending upon pH and curcumin concentration. It exhibits poor to moderate stability to light and variable stability to heat. Turmeric has been used successfully in products that feature opaque packaging. It is found in ice creams, yoghurts, dairy desserts, candies, pickles, snacks, soups and cakes. Turmeric is primarily cultivated in India, China, South America, and the East Indies. It is asexually propagated from its rhizomes. The yellowing of leaves and stems finally indicates that the rhizomes are mature and ready for harvesting which is done by manually lifting the clusters of rhizomes

**Table 8:** Various characteristics of turmeric oleoresin (curcumin).

| **Product** | **Turmeric oleoresin (95% curcumin)** |
|---|---|
| Latin name | *Curcuma longa* |
| The country of origin of the botanical | India |
| The country where the botanical extracted | India |
| Plant part used | Turmeric fingers |
| Colour | Yellow colour powder |
| Odour | Negligible |
| Loss on drying | Not more than 0.5% |
| Residual solvent | Less than 10 ppm |
| Method of analysis of curcumin | Spectrophotometric |
| Curcumin content | above 95% |
| Colour value | NA |
| Type of extraction | Solvent extraction |
| Carrier/excipients | Nil |
| Microbiology | |
| Total plant count | < 1000/G |
| *E.coli* | Absent |
| Yeast & mould | < 100/G |
| *Salmonella* | Nil |
| Maximum limits for the following | |
| Heavy metals | < 10 ppm |
| Arsenic | < 1 ppm |
| Lead | < 2 ppm |

from the soil. Soil, roots and leaf scales are removed from the rhizomes, and a portion is selected and stored for planting the next crop. The bulbs and fingers are segregated and transferred to curing and drying facilities.

5.3.1.4 Carotenoids: Carotenoids are one of the largest groups of pigments produced in nature. While many are known for the health promoting properties, such as β-carotene from algae or palm oil, others are used for colouring purposes such as annatto, lutein, lycopene, paprika, and saffron. All are approved for use in most of the countries except lutein and lycopene which are not approved as colourants in the United States. The β-carotene colourant available today is either an extract taken from a number of natural sources, such as palm oil, algae and carrots, or is made synthetically.

Typical of carotenoids, beta-carotene is susceptible to degradation by light, heat and oxidation. The natural form is insoluble in water and slightly soluble in oil. The development of a number of manufacturing approaches for the synthetic version made for easier suspension in oil, and with appropriate emulsification, make it water dispersible. Other carotenoids commonly found in food include Apo-8 carotenal, and canthaxanthin. Most carotenoids are produced by chemical synthesis [related to vitamin A process route], although microbial fermentation processes have been developed for astaxanthin and β-carotene, whilst lycopene can also be obtained by extraction from natural sources. Carotenoids play a very significant role in the well being of humans and some of these are summarized in Table 9.

Carotenoids have extensive applications as antioxidants in dietary supplements, and as colours in foods and beverages as well as pigments in poultry and fish. Six categories of carotenoids include:

- Astaxanthin
- β-Apo-carotenal

- Canthaxanthin
- β-carotene
- Lutein/zeaxanthin
- Lycopene

**Table 9:** Role of carotenoids in human health.

| Nutraceutical property | Type of careteniods |
|---|---|
| Antioxidant | Grape anthocyanins, canthaxanthin, crocin |
| Anti-cancer | β-carotene, astaxanthin, elderberry extract |
| Anti-free radical | Blackberry anthocyanin |
| Lowering the risk of degenerative diseases | Lycopene, β-carotene, canthaxanthin |
| Anti-atherosclerotic and anti-chronic heart diseases | Grape anthocyanins, crocin |
| Anti-inflammatory | Cyanidins |
| Blood vessel protection | Grape anthocyanins |
| Prevention of night blindness | β-carotene |
| Anti-viral action | Elderberry extract |
| Enhancement of immune system | Lycopene, β-carotene, canthaxanthin |

*Source:* 2, 6, 9, 54, 55

5.3.1.5 Paprika (E160c): Paprika is a deep red, pungent, powdered spice, obtained from the dried, ground pods of the sweet chile peppers (*Capsicum annum*), an annual shrub that belongs to the nightshade family *Solanaceae*. The colour of paprika is a result of a mixture of several carotenoids, principally β-carotene, capsanthin and capsorubin, both of which are unique to the *Capsicum* species. Oleoresin of paprika is the preferred form for colouring foods, though it also can contribute flavour. To make this colouring agent, conventionally dried, ground red peppers of the *Capsicum annum* variety are extracted with a volatile solvent, such as hexane. The oleoresin contains a mixture of capsanthin (yellow fraction), capsorubin (red fraction), and about 20 other carotenoids, in addition to any other material that solubilizes in the solvent used for extraction. Colour strength is usually specified by an ASTA (Washington, D.C.-based American Spice Trade Association) value; higher the number, more concentrated the extract. The oleoresin can be emulsified and made water-dispersible. It can be plated onto various carriers and used as a dry powder in everything from dry gravies to spice blends. But its oil-soluble nature often limits its use to oil-based formulations. The hue in oil is concentration dependent, varying from a light yellow to a bright red-orange. Oleoresin of paprika has been used in margarine spreads, butter, vegetable oils, soups, gravies, ice cream, dairy desserts and sauces.

Capsicum peppers are of American origin and one of the oldest cultivated crops, having been grown by Native Americans and used as a condiment with their food 5,000 to 7,000 years ago in southern Mexico. Christopher Columbus first mentioned the small hot peppers in a letter to the Chapter of Seville on his second voyage in 1494. Today, paprika is one of the most consumed spices worldwide and one of the most economically significant sources of natural food colour produced. Hues range from yellow-to-brown, purple and red. *Capsicum annum*, which includes most of the sweet peppers and many of the pungent, strong-flavoured types, is cultivated throughout the world with main growing areas in Hungary, Spain, South America, and California. In the 1950s, technology was developed that allowed the extraction and concentration of the paprika pigments responsible for the red-orange colour. Harvesting has traditionally been a labour-intensive operation as the pods have been hand-picked for decades in Spain and the United States but non-mechanical harvesters are used. Having been sorted, washed, de-stoned, and dehydrated in belt or tunnel driers and ground, the harvested paprika pods are destined for one of three markets: fresh produce, dehydrated-ground or oleoresin extraction. Paprika powders are manufactured by dry-blending or spray-drying with carriers and gums. To obtain paprika oleoresin (an orange-red oil-soluble extract), the dried and ground peppers are extracted with a volatile solvent, followed by removal of the solvent. Paprika (regulated by FDA in 21CFR 73.340 and 73.345) is subject to colour degradation that could occur during grinding and storage due to pigment oxidation, and when the extract is diluted for use in foods.

5.3.1.6 Annatto (E160b): Annatto is a natural carotenoid and extract of the red, oily outer layer of the seeds of *Bixa orellana*. It is extracted from a tropical shrub named after the Spanish conquistador Francisco Orellana, who discovered the Amazon River in 1541. It has been used in the United States and Europe for more than 100 years to colour dairy products, primarily cheeses. Annatto is protein-fast, does not bleed and is versatile. It is prepared by leaching the outer oily surface of seeds produced by a tropical shrub (*Bixa orellana*) with solvents, vegetable oil, fats or alkalized water. The extract is a mixture of the carotenoids bixin and norbixin, and other soluble materials specific to the solvent used in the extraction. Bixin is moderately oil-soluble, while norbixin is completely water-soluble. It is available in an oil-soluble form, a water-soluble form, a crystal (bixin) form, and as a water-soluble, sprayed-dried form. The oil-soluble form oxidizes over time, and degradation will accelerate with exposure to light or trace metals, resulting in a faded or uncoloured system. It gives a yellow to deep-orange hue to foods and beverages (depending on concentration), can mimic the colour of egg yolk in cake batters and enhance the golden hue of salad oils, citrus drinks, tropical beverages, health-bar centres, jelly candies, hard candies and extruded cereals. It can also provide a rich vanilla hue to soy- or milk-based protein drinks, toppings and desserts. To the product developer, annatto's many forms make it a very versatile colourant.

The addition of antioxidants, such as ascorbic acid or polyphenols, can slow degradation of annatto. It is reasonably stable to heat, making it suitable for baking, extrusion, spray-drying, pasteurization and aseptic processes, but heating the oil-soluble form can cause the formation of a more-yellow product. The norbixin form appears to precipitate at different pH levels in different food systems such as slow precipitation pH 6.5 in some foods, while in others precipitation will not appear until a pH of 5.

Indigenous to the Caribbean and Central America, *Bixa orellana* is also widely cultivated in Asia and Africa. The majority of the world's annatto seeds come from wild trees or trees planted on family farms, especially in northern South Africa. Today, annatto seeds and their extracts rank second in economic importance worldwide among all natural colours. Peru and Brazil are the leading suppliers and the United States and Western Europe are the main markets.

5.3.1.7 Lycopene: It is an open-chain unsaturated carotenoid belonging to a group of more than 600 lipid-soluble phytochemicals found in nature, commonly known for their brilliant yellow, orange and red colours. Lycopene is much more widely distributed in fruits and vegetables than other pigments. It is predominantly found in the chromoplast of plant tissues and occurs as the major pigment in red, fresh tomatoes, as well as in canned, condensed or processed tomato preparations. Even though more than 80% of lycopene consumed in the United States is derived from tomato products, watermelon, papaya, guava, pink grapefruit, chilli, apricots, rosehip and other fruits contribute to dietary intake of lycopene. The name "lycopene" is derived from the tomato's species classification *Lycopersicon lycopersicum*. Colorants containing lycopene were not commercially available because the carotenoid was believed to be susceptible to degradation by oxidation and light. Wide range of health benefits from lycopene in tomatoes has been demonstrated. Since then, the interest in lycopene has increased dramatically and has also kept its major food sources — tomatoes, tomato extracts and tomato phytonutrients — in the public eye [54].

**5.3.1.8 Saffron:** It is derived from the stigmas of the purple saffron crocus flower *Crocus sativus*, a bulbous perennial of the iris family *Iridaceae*. It is treasured both for its bright orange-yellow colour and its strong, intense flavour and aroma. The word "Saffron" is derived from the Arabic word *zafran* meaning "be yellow." The ancient Greeks knew the spice as *krokos*. It is not exactly known in which country saffron finds its origin and when its cultivation began. It is believed that saffron came from the Orient because its cultivation was widely spread in Minor Asia far before the birth of Christ. It is supposed to be the oldest of all cultivated spices, as excavations in Crete brought to light several frescoes dating back to at least 1700 BC, where yellow saffron flowers are depicted on the ruins of the walls of the palace of Knossos. In antiquity, saffron was considered to be the essence of youth and life and one of the most desired and expensive spices of ancient Greeks, Egyptians and Romans. In several cultures, saffron was highly appreciated for its aromatic, colouring, aphrodisiac and healing properties. The dye had even been used for royal garments. Today, the greatest saffron producing countries are Greece, Spain, Turkey, Iran, India and Morocco. The largest saffron importers are Germany, Italy, the United States, Switzerland, England and France. It contains crocetin, a discarboxylic carotenoid along with its digentiobiose ester crocin, a yellow and one of the few carotenoids found in nature that is freely water-soluble. It is the source of saffron's strong, yellow colouring property that finds application in the food and pharmaceutical industries.

**5.3.1.9 Tagetes (Lutein):** Tagetes is the botanical genus name of a stout branching annual with large yellow to orange flower heads, growing 3-4 feet tall in temperate climates. Marigolds, native to the Americas from Argentina north to New Mexico and Arizona, are recorded to have been used by the Aztec people to treat hiccups, and for other ailments such as being struck by lightning. The marigold species *Tagetes erecta*, which is synonymous with African, American, or Aztec marigold, contains large amounts of carotenoids and up to 80% lutein, and is therefore, most commonly used for its colouring properties. The principal producers of Tagetes and its derivatives are Mexico, Peru, Ecuador, the United States, Spain and India. Due to the seasonality of the crop, the mature marigold flowers are collected and the petals separated prior to storage in large bins containing up to 80 tonnes. To minimize pigment degradation, the petals are sprayed with an antioxidant, compacted and covered with a tarpaulin. The petals are removed as needed, pressed, then dried to less than 10% moisture, and can finally be ground and sold as Tagetes meal. A brown oleorosin, which can be incorporated directly to poultry feed, is produced by extracting the ground petals with a number of solvents and by later removal of these solvents. After further purification and washing procedures, the oleorosin can be absorbed by several substances to produce a dry powder that is suitable for use as a food colourant.

### *5.3.2 Biocolours from insects*

**5.3.2.1 Cochineal (E120):** Cochineal is purplish-red coloured extracted from the shells of the wingless female cochineal beetle termed *Dactylopius coccus costa*. This parasitic insect attaches itself to the green leaves of specific varieties of cacti growing in the semi-arid regions of Bolivia, Chile, the Canary Islands, Mexico and Peru. The insects contain a high concentration of carminic acid — a red hydroxy-anthraquinone dye, which is up to 24% of their dried body weight. Having been used for centuries as a source of red colour and dating back at least to the Aztec empire, carmine was the homage of kings and more valuable than gold in the ancient world of Aztecs. It is believed that in the year 1518 BC and with the Spanish conquest of Central and South America, Cortez discovered the Aztecs using cochineal as a paint and dye. The Spanish government turned it into a lucrative export. In 1970s and 1980s users of synthetic food colours were required, the FDA need to get tested these products to prove zero risk to the consumer. So a resurgence of interest in cochineal-based colourants began to grow. Today, with an annual production exceeding 500 tonnes, Peru is the principal supplier of dried cochineal, followed by Mexico, Chile and the Canary Islands. The cochineal is collected by hand. The female cochineal is brushed from the leaves, air-dried in the sun, and cleaned by screening. It is estimated that about 80,000 to 130,000 insects are required for each kilogram of raw, dried cochineal as exported from Central and South America. The average human body only consumes one to two drops of it in a given year. Cochineal extracted with water gives a solution of carminic acid, proteins, and minor quantities of several salts, referred to as cochineal extract. It is one of the few water-soluble colourants that resist degradation with time. Carminic acid is the colouring pigment of all cochineal-based colourants and the root of all carmine colours. It is one of the most light and heat stable of all the natural colourants and is even more stable than many synthetic food colours. Water-soluble carminic acid precipitated on a base of aluminum or calcium salts, forms a water- and oil-insoluble pigment called carmine — a lake that is traditionally dried to contain about 50 to 53% carminic acid. Depending upon the metal used, different shades of the carmine pigment can be obtained, ranging from bright strawberry magenta red to black currant purple.

5.3.2.2 Carmine: A deep-red powder, carmine is made by precipitating the carminic acid obtained from the cochineal insect onto an aluminum hydrate. Carmine powder is insoluble in water and so not suitable for beverages. It can however, be solubilized in alkalized water, producing a deep, rich, red-violet colour, but acidification of the carmine solution will cause it to precipitate. Of all the colourants, carmine/carminic acid is the most stable to light, heat and long-term storage. These compounds tend to hydrolyze in high-acid food systems, but an acid-proof form of the colourant is available. It is used in products requiring heat processing, including retorting and extrusion, and where more than a year of shelf-life is required.

### *5.3.3 Biocolours from microorganisms*

Currently, the pigments produced by microorganisms and used commercially are riboflavin (Vitamin B2, a yellow pigment permitted in most countries) by *Eremothecium ashbyii* and *Ashbya gossypii*, the pigment from *Monascus* produced by *M. purpureus* and *M. rubber*, carotenoids (yellow pigments produced by several microorganisms, but currently produced commercially only from microalgae) such as β-carotene (by *Dunaliella salina* and *D. bardawil*) and astaxanthin (by *Haematococcus pulvialis*), and ficobiliproteins such as phycocyanin (a blue pigment used in food and cosmetics) produced by *Spirulina sp.* Indigoids, anthraquinones and napthoquinones are pigments, which hold potential use in the near future [56]. In recent years, active search for microorganisms producing non-toxic metabolites has been performed by several researchers [57-59].

#### 5.3.3.1 Microorganisms and their pigments

5.3.3.1.1 <u>Bacteria:</u> Bacteria produce different colours/shades. Some of them alongwith their shades are *Chromobacter* sp. (red), *Bacillus* sp. (brown), *Brevibacterium* sp. (orange, yellow), corynebacterium *michiganense* (grayish to creamish), *Pseudomonas* sp. (yellow), *Rhodococcus maris* (bluish red), *Streptomyces* sp. (yellow, red, blue), *Micrococcus* sp. (light yellow) and *Sarcina* sp. (dark yellow) [60-62].

5.3.3.1.2 <u>Moulds:</u> The moulds which produce colour are *Aspergillus* sp. (orange red), *Blakslea trispora* (cream), *Monascus purpureus* (yellow, orange, red), *Helminthosporium catenarium, H. gramineum* (red), *H. cynodontis* (bronze colour), *P. cyclopium* (orange colour), *H. avenae* (bronze colour) *Aspergillus glaucus* (dark red), *H. caternarin* (dark maroon) and *P. nalgrovensis* (yellow) [63].

5.3.3.1.3 <u>Yeasts:</u> A few yeasts which are coloured are *Rhodotorula* (pink/red), *Yarrowia lipolytica* (brown), *Cryptococcus* sp. (red) and *Phaffia rhodozyma* (red) [64, 65].

5.3.3.1.4 Algae: A number of pigments are produced by algae (Table 10). Recently, natural β-carotene has been produced from algae *Dunaliella salina* occurring singly in marine environment, and belongs to class chlorophyta. Another microalgae known as *Rhodophyta* is characterized by the presence of necessary pigments belonging to phycocyanin and phycoerythrins apart from the main pigment chlorophyll [66].

Recently, there has been a considerable interest world wide in production of pigment from microbial sources, as their production is advantageous being independent of environmental conditions and can be grown in the cheap medium [67]. A large number of different species of bacteria, yeast, moulds and algae produce pigments, but only a few of them are considered suitable for this purpose (Table 11). Most of the microbial pigments are carotenoids in nature. The best known function of carotenoids such as α-carotene, β-carotein and β-cryptoxanthin is that of provitamin A [68]. Humans and animals convert these derivatives to retinol (vitamin A). Effect of mineral on pigment production is shown in Table 12. Carotenoids are effective antioxidants and quencher of singlet oxygen. In humans, these properties have been linked to a personal protection against certain forms of cancer.

5.3.3.2 Type of fermentation: The growth and pigmentation of the microorganism is greatly affected by the type of fermentation i.e., solid state fermentation or submerged fermentation besides other parameters. Some of these are described here:

**Table 10:** Uses of microbial colours produced by algae.

| Pigment | Algae | Use |
|---|---|---|
| β-carotene | *Dunaliella salina* | Food colourant |
| Lutein | *Chlorococcum*<br>*Chlamydomonas*<br>*Spongiococcum*<br>*Chlorella pyrenoidosa* | Poultry feed |
| Canthaxanthin | *Hematococcus*<br>*Chlorella*<br>*Chlamydomonas*<br>*Scenedesmus*<br>*Ankistrodesmus*<br>*Dictyococcus cinnanarinus* | Poultry feed and fish feed |
| Astaxanthin | *Hemacoccus pluvialis* | Fish feed |

*Source:* 65

**Table 11:** Carotenoids produced by *Phaffia rhodozyma.*

| Carotenoid | Percent of total carotenoids |
|---|---|
| Carotene | 2-2.5 |
| Carotene | 0.01 |
| Neurosporene | 0.01 |
| Lycopene | 0.01 |
| Echimenone | 2-4 |
| 3-Hydroxyechinone | 3-4 |
| Phoenicoxanthin | 5-7 |
| Astaxanthin | 83-87 |

*Source:* 69

5.3.3.2.1 Solid state fermentation: Solid culture of *M. purpureus* on steamed rice, bread, beans or cereal meals is the traditional method of pigment production [70], but the effect of cultural parameters on yield has not been reported earlier. John and Stuart [70] studied the production of *Monascus purpureus* in solid culture and observed that solid state fermentation yielded 3 times more pigment production than in the submerged fermentation. Pigment produced by *Sarcina* and *Rhodotorula* is demonstrated in Plate 1.

5.3.3.2.2 Submerged fermentation: In contrast, extensive submerged cultural studies of red pigment showed that the yields are affected markedly by many factors including medium composition, pH and agitation [71, 72]. The effect of pH and amino acid on conidiation and pigment production on *Monascus rubiginosus* in submerged shaken culture was studied by Carels and Shephered [71]. The liquid culture was prepared by homogenizing the mycelium on yeast agar slants with about 30 ml of yeast extract medium in a 50 ml Sorvall Ommimixer homogeniser for about 45 seconds. This homogenate was used to inoculate a shake flask containing 150 ml of yeast extract liquid medium. The cultures were incubated at 25°C for 48 hrs at 150 rpm on a New Brunswick G 53 rotatory shaker blade. These cultures were grown for 96 hrs. The general flow sheet for the microbial colour production from apple pomace based medium is shown in Fig. 11.

**5.3.3.3 Medium for pigment producing microorganisms:** Extensive studies on the production of pigment from different microorganisms have been conducted. Some of these have been summarized here:

5.3.3.3.1 *Monascus purpureus*: Carbon source affect the growth and production of the microbial cells. D-glucose has been proved better for the utilization in the presence of zinc, and sucrose was regarded equally good without adding zinc. But when zinc was supplemented to the media, D-glucose showed

**Table 12:** Effect of minerals on pigment production by *A. oryzae* var. *effuses*.

| Minerals | Pigment |
|---|---|
| KCl | Deep red |
| $CaCl_2$ | Deep red |
| $NH_4H_2PO_4$ | No growth |
| $ZnSO_4$ | Deep red |
| $MgSO_4$ | Pink |
| $FeCl_3$ | Pink |
| Sodium citrate | Light pink |

*Source:* 69

(a) *Sarcina* sp.

(b) *Rodotorula* sp.

**Plate 1** Production of microbial pigments.

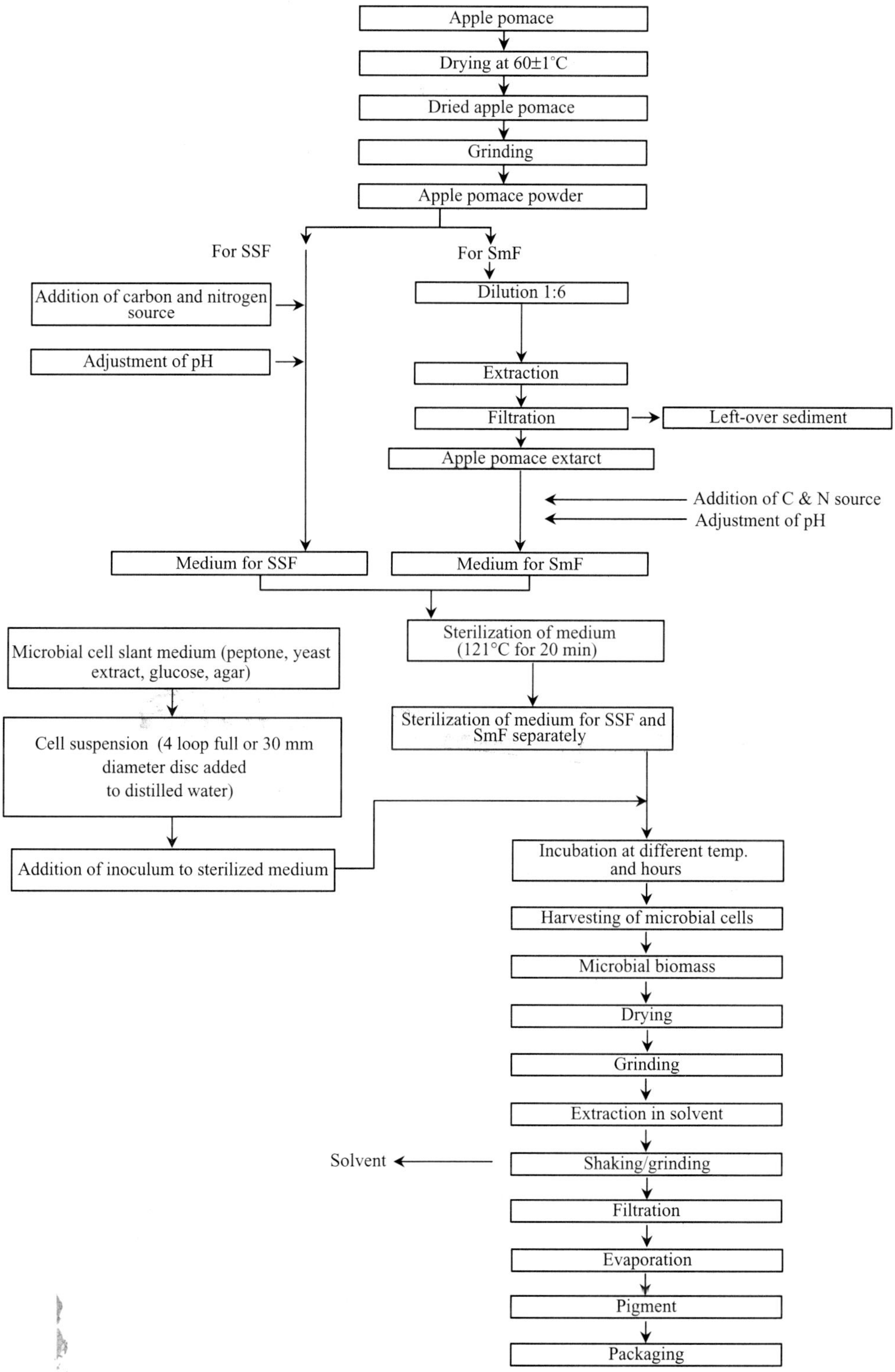


**Fig. 11:** General flow sheet for the microbial colour production from apple pomace-based medium.

the maximum growth on dry weight basis among the different carbon sources tried [73]. In another experiment, the different levels of glucose ranging from 40-200 g/l were studied and 40 g/l was found the best for the maximum pigmentation in *Monascus purpureus*. [74]. Maltose has been proved to be the best carbon source among the glucose, fructose, lactose and galactose for the growth and volumetric production of biomass [75]. The use of glucose at a concentration of 50 g/l led to poor growth in *Monascus purpureus* and less than 20 g/l achieved better growth and pigmentation [76], while glucose was found to be the most effective carbon source for stimulating pigment production. The pigment extracts were predominantly reddish orange. At 15% glucose level, *Monascus* was distinctly red in colour [60]. Carels and Shephered [71] utilized 4% glucose for the production of *Monascus purpureus* in addition to 0.1% $KHPO_4$, 0.05% $MgSO_4{\cdot}7H_2O$, 0.05% NaCl and 0.01% $FeSO_4{\cdot}7H_2O$. Both glucose and fructose were found as the best carbon source for *M. purpureus* for yield and maltose as the second best carbon source. So it can be concluded that for the growth and pigmentation of *Monascus purpureus,* 4-20% glucose is suitable [60, 71, 74, 75, 77]. *Monascus* production in trays can be done as shown in Fig. 12.

The pigment production is also affected by the type of nitrogen sources. The pigment produced is affected by different glucose and $NH_4NO_3$ concentrations. At 40 g/l glucose, pigment production was maximum with 0.5 g/l $NH_4NO_3$ concentration. However, higher glucose concentrations gave maximum pigment production. But, more than 10 g/l nitrogen concentration inhibited pigmentation [74].

In another study, ammonium chloride was observed as the best nitrogen source for the growth followed by ammonium nitrate and then glutamate. Potassium nitrate was the poorest nitrogen source among the four. Further, with respect to pigment production, glutamate was outstanding [75]. It was observed that peptone produced superior yields of monoscorubin compared to ammonium when glucose was the carbon source [76]. The production of pigment from *Monascus* species was enhanced by the use of monosodium glutamate as a sole nitrogen source [75], $NaNO_3$ (0.3%) and $NH_4Cl$ (0.5%) as nitrogen source were added into the basic media [71], and the presence of yeast extract in MP11 ($NO_3$) and ($NH_4$) medium reduced the uptake of nitrate and ammonium, whereas $NO_3$ based medium after adjustment of pH enhanced nitrate uptake. The presence of amino acid reduced the uptake.

The growth and pigmentation of the *Monascus purpureus* is also affected by the minerals or nutrients in the media. Zinc at concentration higher than 0.5 mg/ml greatly inhibited the germination of both the conidia and ascospores [78]. Growth of *Monascus purpureus* wild type was inhibited by zinc at $2 \times 10^{-3}$ M and $3 \times 10^{-3}$ M concentration. No significant inhibition was however, observed at lower concentration [79]. On the other hand, the uptake of glucose increased in the presence of

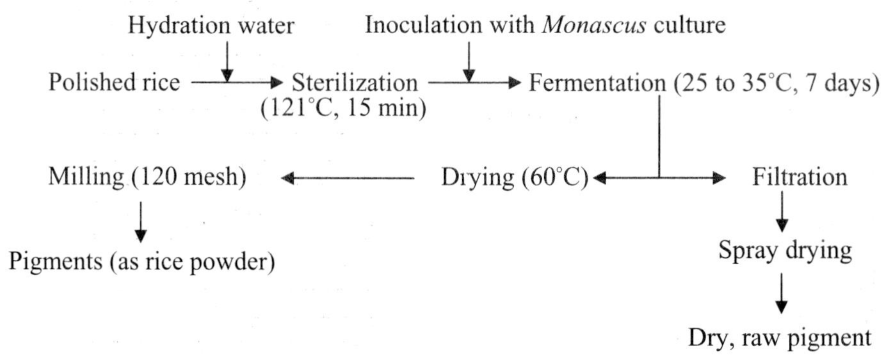


**Fig. 12:** *Monascus* production in trays.

zinc. This also enhanced the production of pigment and antibiotic. In another experiment, Tseng *et al.* [77], while studying the effect of salt, sodium nitrate and polyphosphate on growth and pigmentation, observed that at 25 g/l salt addition, protease activity was higher than the control. The pH of the culture decreased significantly compared with the control when 25 g/l salt was added to the culture and only decreased slightly when 50 g/l salt was added to the culture. The mycelial growth and pigmentation production were stimulated by the addition of 0.1-0.2 g/l sodium nitrate and more than 0.2% sodium nitrate decreased protease activity. The growth, pigment production and protease activity were inhibited by the addition of polyphosphate (1.0 g/l). The mould growth was almost stopped by the addition of 5.0 g/l polyphosphate.

Johnson and McHan [73] studied the effect of zinc on the utilization of carbohydrate and pigment production in *Monascus purpureus*. It was found that zinc stimulated the utilization of 8 out of 10 carbohydrates tested as a sole source of carbon. Among the carbohydrates tested, only D-galactose and lactose failed to support zinc retarding growth response. Ethyl alcohol was the only non-carbohydrate substrate which increased growth with added zinc. The inorganic salts have also been reported to affect the growth and pigmentation in *Manascus purpureus*. It was found that KCl concentration did not affect the cell growth or pigment production above the usual concentration of 6.3 mM, however, below this level, pigment production decreased. The level of $MgSO_4$ proved to be extremely important. A level of 2 mM limited the growth which was virtually a linear function of $MgSO_4$ concentration in the range of 0.5 to 16 mM. $MnSO_4$ at a concentration of 0.018 mM was found to be optimum for pigment production. The optimum concentration of ferrous sulphate was 0.036 mM for pigment production [75].

5.3.3.3.2 *Rhodotorula* species: When *Rhodotorula* was grown on apple pomace based media supplemented with different nitrogen sources like ferrous ammonium sulphate, sodium nitrate, diammonium hydrogen phosphate, ammonium sulphate, potassium nitrate and urea, it was observed that ferrous ammonium sulphate, yeast extract and peptone were comparable and gave the highest pigmentation, while ammonium sulphate and potassium nitrate resulted in lower pigmentation. The enhancement of pigmentation increased up to 0.3%, after which it declined [64]. The growth and pigmentation in *Rhodotorula* are affected by the minerals or trace elements. Sandhu and Joshi [64] have reported the use of different nitrogen sources/trace elements grown on apple pomace extract. Ferrous ammonium sulphate, yeast extract and peptone were observed to be comparable, but gave the highest pigmentation, while ammonium sulphate and potassium nitrate resulted in lower pigmentation. Joshi and Attri [80] reported that addition of various concentrations ranging from 0.1 to 0.5% of maltose to the medium caused significant difference in yield as well as in carotenoid production by *Rhodotorula* sp. (Table 13).

**Table 13:** Effect of nitrogen source on the yield and carotenoid production by *Rhodotorula* sp.

| Nitrogen source (0.1%) | Yield* (g/l) | Carotenoid ** (mg/100 ml) |
|---|---|---|
| Ferrous ammonium sulphate | 13.10 | 23.13 |
| Ammonium sulphate | 8.10 | 17.62 |
| Sodium nitrate | 12.67 | 19.22 |
| Peptone | 11.20 | 15.95 |
| Potassium nitrate | 8.40 | 15.61 |
| Control | 12.02 | 18.72 |
| $CD_{0.05}$ | 0.25 | 0.42 |

* Biomass; ** Dry weight basis

5.3.3.2.3 Micrococcus species: *Micrococcus* species were tried on different carbon sources like glycerol, mannitol, sucrose, lactose, starch, succinic acid, maltose, galactose, inulin, fructose, mannose arabinose, sorbose, cellobiose and ribose. Among all these, fructose was more effective than glucose. With glucose as the principal carbon source, culture attained an optical density of 240 klett units and cell contained 0.33 mg of pigment per mg dry weight, whereas fructose as a carbon source gave 700 klett units and 0.86 mg per mg, respectively. The concentration of fructose was 0.5 mg per 100 mg [67]. Devender and Joshi 2005 [81] studied the effect of carbon and nitrogen source on yield, and carotenoid production by *Micrococcus* sp. was also studied in apple pomace-based medium. The addition of fructose (0.2%) to apple pomace-based medium gave the maximum yield of biomass and carotenoid. *Micrococcus roseus* has been observed better in growth under the medium containing glutamic acid as a sole source of carbon, nitrogen and energy. It was found that medium containing adenine as the sole nitrogen base yielded good growth through supplementing adenine, but all the combinations of bases did not increase growth [67].

The growth of *Micrococcus roseus,* in addition to carbon and nitrogen source in the basal medium is also affected by the presence of trace elements or minerals. Adenine, arginine, glutamic acid, isoleucin, proline, serine and inorganic salts contributed to growth and pigment content when glycine was present. The lag phase was shorter and pigment synthesis was enhanced slightly [57]. When the defined medium was supplemented with carotenoid inhibitor like mevalonic acid at 1.25, 2.5 and 5.0 mg/100 ml, the mg pigment per mg dry weight decreased from 0.70 to 0.53.

5.3.3.3.4 *Phaffia rhodozyma*: The astaxanthin containing yeast *Phaffia rhodozyma* was isolated during the early 1970's from the exudates of deciduous trees in mountaineous region of Japan and Alaska, and was known as *Rhodozyma montanae*, but due to its two important properties like producing red colour and fermenting yeast, it was named after the name of Herman Jan Phaff *P. rhodozyma.* The pigmentation and growth of *P. rhodozyma* was supported more in cellobiose as carbon source (652 mg/g yeast) than any of the carbon sources tested. The other disaccharides, maltose and sucrose also promoted high pigmentation. Sucrose and glucose promoted more rapid growth of *P. rhodozyma* [82]. In another experiment, *P. rhodozyma* along with antimycin mutant were grown on four carbon sources viz., glucose, cellobiose, ethanol and succinate. It was observed that the two mutants had comparatively less cell growth than the parent. The mutants were highly pigmented on each of the energy sources [65].

*Phaffia* is a red colour producing yeast. It was observed that the ammonium sulphate concentration in the range of 0.25-5 mg/ml had little effect on the yeast growth rate, final yeast biomass or on the carotenoid production in the yeast nitrogen based medium (YNB supplemented with 1% D-glucose). The substitution of $(NH_4)_2HPO_4$ or peptone for $(NH_4)_2SO_4$ at various concentrations also did not affect these values. However, when the concentration of yeast extract was increased (0.1 to 10 mg/ml) to vitamin free medium, there was an increase in pigmentation from 156 to 5.24 mg/g [82], though there was an increase in the biomass, pigmentation and final yeast biomass when vitamin free yeast medium was used. In the same way, when the *Phaffia rhodozyma* was grown in a fermenter culture on a complex medium i.e., brewer's malt wort diluted to a specific gravity of 1.020 or addition of 1.01 of colourless tomato pressing to 2.01 of the standard medium, the carotenoid yields were 712 and 814 mg/g after 60 hour of growth respectively, from the 1.56 mg initially.

5.3.3.3.5 Chromobacter sp.: In the literature, there is no report on the pigment producing medium for *Chromobacter*. The nutrient agar medium which is usually used for the growth of bacteria contains peptone, yeast extract, dextrose and agar-agar, while apple pomace medium has carbohydrates sugar and nitrogen, and the yield of pigment as well as growth of biomass was also more in the medium with apple pomace than that without it. However, with increase in concen-

**Table 14:** Effect of nitrogen source on the yield and carotenoid production by *Chromobacter* sp.

| Nitrogen source (0.1%) | Yield* (g/l) | Carotenoid** (mg/100 g) |
|---|---|---|
| Ferrous ammonium sulphate | 4.40 | 41.85 |
| Ammonium sulphate | 4.03 | 41.23 |
| Sodium nitrate | 4.07 | 40.63 |
| Peptone | 4.70 | 41.17 |
| Potassium nitrate | 6.23 | 46.25 |
| Control | 4.07 | 41.25 |
| CD (P<0.05) | 0.20 | 0.63 |

* Biomass; ** Dry weight basis

*Source:* 62

tration of apple pomace beyond 20 g/l, the yield as well as pigment production was decreased. It shows that there is substrate level inhibition of biomass and pigment production (Table 14). Higher concentration of nitrogen source might have been toxic to the microorganism, as is known for other microorganisms [62].

5.3.3.4 Nature of pigment: Despite the availability of a variety of natural and synthetic carotenoids, there is currently a renewed interest in microbial sources of pigments. Processes for the large scale production of β-carotene and lutein from fungi and algae, respectively were developed in 1960's, but were later abandoned because they could not compete with synthetic colour. But β-carotene is now again commercially produced from microorganism (microalgae). The nature of different pigment producing organism is discussed in the subsequent sections.

5.3.3.4.1 Bacteria

*5.3.3.4.1.1 Chromobacter sp.:* These are violating red, creamy colour producing microorganisms, rods, which occur singly or occasionally in pairs. These are Gram negative, motile with single flagella occurring mostly in soil and water. Growth occurs on ordinary medium at 25°C. These have mainly fermentative metabolism, produce acid but no gas.

*5.3.3.4.1.2 Micrococcus sp.:* These are spherical cells which produce yellow or red colour shades, occurring in pairs or clusters. These produce little or no acid from carbohydrates and usually grow on simple media. The optimum temperature for growth is 25-37°C. They occurs in mammalian skin, but have been commonly isolated from food products and air.

The principal coloured carotenoid of this organism has been identified as canthaxanthin 4, 4-diketo β-carotene [83]. Ungers and Cooney [84] gave the method to separate pigment extracted from *Micrococcus roseus*. The pigment was dissolved in petroleum ether. The isolation of 4-hydroxy echinar has also been done [85]. The TLC of the pigment developed with petroleum ether showed two coloured bands - one yellow band with Rf of 0.65, and the second coloured band was observed as a faint band with Rf of 0.5. The absorption was found maximum in the 400-500 nm range [85].

*5.3.3.4.1.3 Sarcina sp.:* These are spherical and occur in cuboidal packets of eight or more. It divides in three perpendicular planes. These cells are Gram positive and non-motile, and produce yellow coloured pigment which is carotenoid in nature. The carotenoids of *Sarcina flava* have been found to have four fractions viz., hydrocarbons, mono and dihydroxylated compounds and polar fraction. Chloroform and methanol (95+5%) as solvents were used for fractionation [86].

5.3.3.4.2 Yeast: Yeasts are well known for pigment productions, which are carotenoids in nature. The different yeasts, their nature and pigmentation are described here.

*5.3.3.4.2.1 Phaffia rhodozyma:* The genus *Phaffia* possesses only one species, *P. rhodozyma*. The yeasts are of basidiomycetous origin, but have no perfect stage [87]. It can be identified by its particular characteristics such as ability to synthesize carotenoid and its possession of certain metabolic properties such as the ability to utilize urea (less common in ascomycetous yeast), and its cell wall structure and mode of bud formation. *Phaffia rhodozyma* is unusual among the pigmented yeasts because of two special properties: it ferments glucose and other sugars and synthesizes astaxanthin as its principal carotenoid. Recently, there has been considerable commercial interest in using *P. rhodozyma* as a pigment source of aqua cultured animals. Astaxanthin is an important constituent of aqua culture feeds because it provides necessary pigmentation, and it may also contribute to distinctive flavour of baked salmon.

Analysis of carotenoid by TLC using 20% acetone-80% petroleum ether was done. For the identification of pigment, Rf and chromatography was used [82]. The carotene consisted mainly of β-carotene (bb-carotene), g carotene, (b,g,carotene), torulene (3', 4' didehydro-bg' carotene) and torularhodin (3',4'didihydro-bg cartotene-16'-oic acid) as their major pigments [88, 89].

*5.3.3.4.2.2 Rhodotorula sp.:* The carotenoid composition of *Rhodotorula* is very similar to the genera *Cryptococcus, Rhodosporidium sporidiobolus* and *Sporolomyces* is pink to red colour producing yeast. They mainly contain β-carotene (bb carotene), g-carotene (b,g carotene) torulene (3',4'-dihydro-bg-carotene) and torularhodin (3', 4'-didehydro–b g carotene–16–oic acid) as the major pigment [88]. Recently, plectanixanthin (3', 4'-didehydro–1', 2'–b, g-carotene – 1,2,diole) has been found in *Cryptococcus laurentii* and 2 hydroxy plectanaxanthin in *Rhodotorula aurantiaca* [90], which has added some structural diversity to the carotenoid found in this group.

5.3.3.4.3 Moulds/fungi

*5.3.3.4.3.1 Monascus purpureus:* The fungus, *Monascus purpureus* produces six closely related polyketide pigments ranging in colour from bright yellow to deep red. The organism has attracted continuous attention as a microbial source of natural red pigment to replace synthetic dyes, particularly erythrosine [72, 91]. The pigments and the fungus have been used to colour food in Asia for several centuries, but have yet to gain regularly approval in the west. The pigment was isolated from fractions and thin layer chromatography on silica gel using chloroform/methanol/water (65/25/4, v/v) as solvent system were performed. The pigment consisted of one molecule of glutamic acid substituted respectively to monoscorubrine or rubropunctatine and were differentiated by the side chain [92].

**5.3.3.5 Conditions for growth:** Different organisms have different conditions for their growth and development. Different workers used different conditions of temperature, pH and time for same or different organisms.

5.3.3.5.1 Incubation temperature, pH and time: Temperature plays an important role in the growth of microbial pigment producers. The optimum temperature for the production of *Rhodotorula* was reported as 30+1°C temperature, pH 4.5 and 72 hours of incubation period [64], whereas when *Rhodotorula* was grown in co-culture with *Debaromyces castellii*, the optimum pH was 5.8 with incubation temperature of 30°C for 120 hours [93]. An incubation period of 10 days was reported for *Rhodotorula glutinis* 22P when cultivated with *Lactobacillus helveticus* 12A [94].

Broder and Koehlr [60] incubated the *Monascus purpureus* for 14 days at 28°C. The pH was set to 6.3 and drop in pH from 6.3 to 2.0 had little or no effect on the colour. An optimum pH of 5.5 was the best for the growth of *Rhodotorula rubra,* and it failed to grow at pH 3.0. Incubation period of 6 days at temperature 37°C and at pH 6.3 has been used for the production of *Monascus purpureus* as red pigment [75]. While working with *Monascus* red pigment formation by biological and semi-synthetic process, Lin *et al.* [95] incubated the organisms at 30°C for 4 days at pH 5.5. Later on Chen and John [76], while studying the effect of nitrogen and carbon source on ethanol and pigment production by *Monascus purpureus*, cultivated them at temperature 30°C with pH 6.5 and found that maltose (50 g/l) improved the production of the pigment. In the same year, Blanc *et al.* [92], while working with *Monascus purpureus*, grew the pigment at temperature 28°C at pH 6.5 for 10 days and found MSG as the best source of nitrogen, and glucose or ethanol as best carbon source.

Carriera *et al.* [61], while working with brown pigment from yeast *Yarrowia lipolytica*, found that growth of the yeast requires 25°C temperature at pH 5.5 for three days. In the same year, Buzzini, while working with the batch and fed-batch carotenoid production by *Rhodotorula glutinis* and *Debaromyces castellei* as co-culture, found that there was maximum production rate of pigment under batch carotenoid production with 24-72 hours at pH 5.8. The different conditions for growth used by different workers are tabulated (Table 15).

#### 5.3.3.6 Steps involved in pigment production

- Selection of suitable species
- Productivity and yield optimization
- Development of cheap industrial media
- To support optimal microbial yield
- Identification and optimization of pigment accumulation
- Bioreactor design development and scale-up of the process
- Downstream processing and storage of pigment
- Demonstration of the use of pigment

#### 5.3.3.7 Chemical modification of microbial pigments

5.3.3.7.1 Techniques of solubilization: As most of the microbial pigments are carotenoids in nature, they are also soluble in organic and inorganic solvents. Their solubility in different solvents is discussed here under:

The pigment from *Rhodotorula* cells was found to be soluble in all the organic solvents except ethanol, and the colour profile of the extracted pigment showed hexane to have the highest red and

**Table 15:** Incubation temperature, time and pH for different yeasts reported by different authors.

| Organism | Temperature (°C) | Time | pH | Reference |
|---|---|---|---|---|
| *Rhodotorula* | 30±1 | 72 h | 4.5 | 64 |
| *Rhodotorula* & *D. castellis* | 30 | 120 h | 5.8 | 94 |
| *Monascus purpureus* | 28 | 14 days | 6.3 | 60 |
| *Monascus purpureus* | 35 | 17 days | 5.5 | 74 |
| *Monascus purpureus* | 30 | 6 days | 6.3 | 75 |
| *Monascus* sp. | 30 | 4 days | 5.5 | 90 |
| *Monascus purpureus* | 30 | - | 6.5 | 76 |
| *Monascus purpureus* | 28 | 10 days | 6.5 | 93 |
| *Yarrowia lipolytica* | 25 | 3 days | 5.5 | 61 |
| *Phaffia rhodozyme* | 20-22 | 3 days | 5.0 | 88 |

yellow colour. The lowest colour was given by water, hydrochloric acid or ethanol when used separately [64]. For *Monascus purpureus*, four solvents namely chloroform, ethanol, methanol and acetone were added to 1 g sample ground in microwill grinder. Among them, methanol was found as preferred solvent [60].

*Micrococcus roseus*, a light yellow colour producing organism, is extracted with methanol [84, 96]. Similar trend was observed in *Sarcina flava*, which produces dark red colour. *Phaffia rhodozyma,* a red colour producing yeast, was extracted with petroleum ether and acetone [88]. The microbial pigments from *Rhodotorula* sp., *Chromobacter* sp., *Sarcina* sp. and *Micrococcus* sp. were modified by chemicals like amino acetic acid, benzoic acid and gelatin [97]. In case of *Rhodotorula* sp., maximum conversion of pigment in gelatin (51 %) at pH 9.2 was found. Maximum absorption also took place at pH 9.2 followed by pH 7. Similarly, maximum conversion (60%) of pigment obtained from *Chromobacter* and *Sarcina* was observed with amino acetic acid at pH 9.2 followed by pH 7 and 3. But the solubility of pigment (55, 56%) under gelatin of pH 7, 9.2 was lower. However, amino benzoic acid was equally good for conversion of pigment into water soluble having 54, 58 and 59% conversion at pH 3, 7 and 9.2, respectively (Table 16).

5.3.3.7.2 Storage stability of the microbial pigments in model solution: Colourants from *Monascus* fungi are well known in the Orient. Traditionally grown on rice or bread, *Monascus* species produce red colourants through solid state fermentation and have been used for hundred of years as general food colourants and medicinal agents [98]. Stability has been a problem with the use of natural colourants, which are often sensitive to heat, light, acidity, air or water activity [99-101]. Recently, *Monascus* pigments have been tested for their ability to colour different foods, including meat products [102]. They appeared to be a viable substitute for nitrate salts and nitrate, traditionally added to enhance meat colour. Fabre *et al.* [103] studied the application of *Monascus ruber* in sausage and pate (meat products) and reported that it remained stable for 3 months at 4°C. The stability was 92-98% and it was concluded that *Monascus* pigments can replace some traditional food additives such as nitrate salts or cochineal. Further, they reported that these colours are affected by light and temperature. Contrary to this, Francis reported that *Monascus* pigments were stable at high temperature [104]. The pH of the medium also has a major influence on colouration of the pigment in solution. The colour change in microbial pigments during the course of storage is shown in Plate 2.

In another study, heat and light stability was studied in 0.01 M citrate phosphate buffer at pH 3.0, 7.0 and 9.2, and it was reported that the stability is more at pH 7.0 and 9.2 than 3.0 [92]. When *Gul-*

**Table 16:** Chemical modification of microbial pigments for water solubility.

| | | *Rhodotorula* | | *Chromobacter* | | *Sarcina* | | *Micrococcus* | |
|---|---|---|---|---|---|---|---|---|---|
| **Chemical** | **pH** | **OD** | **% conversion** | **OD** | **% conversion** | **OD** | **% conversion** | **OD** | **% conversion** |
| | 3 | 0.48 | 46 | 0.62 | 56 | 0.42 | 40 | 0.36 | 38 |
| Amino | 7 | 0.56 | 49 | 0.67 | 59 | 0.46 | 43 | 0.39 | 39 |
| acetic acid | 9.2 | 0.58 | 50 | 0.68 | 60 | 0.48 | 45 | 0.39 | 40 |
| | 3 | 0.47 | 45 | 0.61 | 55 | 0.38 | 36 | 0.37 | 36 |
| Gelatin | 7 | 0.58 | 50 | 0.63 | 55 | 0.39 | 36 | 0.38 | 38 |
| | 9.2 | 0.59 | 51 | 0.63 | 56 | 0.37 | 35 | 0.38 | 39 |
| | 3 | 0.46 | 44 | 0.63 | 54 | 0.39 | 35 | 0.36 | 35 |
| Amino | 7 | 0.55 | 48 | 0.68 | 58 | 0.41 | 37 | 0.38 | 37 |
| benzoic acid | 9.2 | 0.53 | 47 | 0.66 | 59 | 0.42 | 37 | 0.38 | 38 |

Weight of sample: 1 g; % Solubility: Soluble pigment x 100/ Total pigment

*Source:* 97

**Plate 2:** Different shades of microbial pigment during storage.

*mohar* having β-carotene was stored in a model solution having sugar, acid and emulsifier, there was a decrease in carotenoid content as the storage period progressed [105].

5.3.3.8 Hurdles in the use of microbial pigments: Certain hurdles are faced during pigment production from microorganisms, which include the following:

- Toxicology of the pigment producer
- Methods/technology to produce pigment at affordable price
- Water solubility of the pigment
- Storage stability and behaviour in the food system

5.3.3.9 Utilization of waste in pigment production: The utilization of waste for the production of pigment can be done by soaking the soya waste water where cultivation of *Monascus purpureus* can take place. 1-2 percent starch containing syrup, which is available from food industry waste, is employed as a carbon source for the Hypomyces fungi. The apple pomace can also be utilized for the production of pigments by *Rhodotorula, Chromobacter, Sarcina* and *Micrococcus* [62, 64, 65, 80, 81].

## 6. MARKET DYNAMICS OF BIOCOLOURS

At 5 to 8% growth per annum, the market of biocolour is growing much faster than that of synthetic colours, which is currently growing at the rate of 2 to 3% per annum (Fig. 13). However, over the next six years, the growth rate for natural colours is not expected to rise dramatically, but will be sustained at the current levels. The natural food colour with the largest share of revenues in the Australian market is β-carotene. The consumption of foods with natural colours is more on account of the synthetic colour stigma than a result of consumer recognition of the benefits of natural colours.

The market for natural pigments produced by bioprocesses is hard to estimate, either due to the lack of statistics of regional, low technology products, or the fact that the production is pulverized over many small companies worldwide. In the specific case of *Monascus* pigments, the consumption of these pigments in Japan rose from 100 tonnes in 1981 to 600 tonnes in 1992, and was estimated as USD 12 million [106, 107]. A large global market exists for food colourants of about USD 5 billion annually of which synthetic colours enjoy a share of USD 4 billion, while only about USD 1 billion is expended for natural colour. Annual growth rate for natural colour is 4% as compared to 2-3% for synthetic colours.

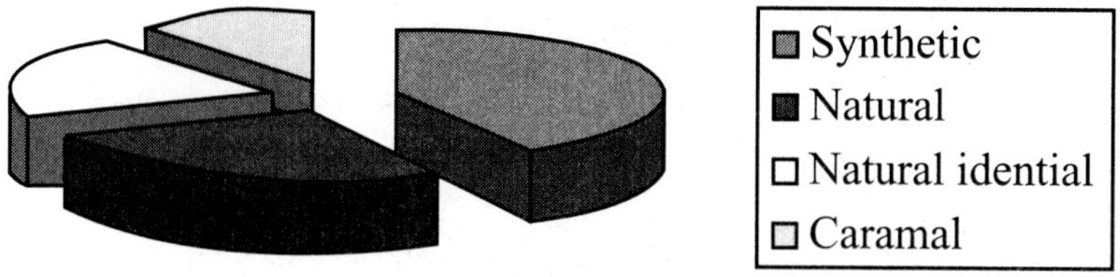


**Fig. 13:** Percentage market share of food colours.

The potential market for astaxanthin pigment is approximately USD 200 million per year and the natural material cost 20-30% less cost than the synthetic material. The current popularity of natural colour is due to the two factors: one is public distrust of food additives in general, and second is due to the perceived problem of safety with food colourants. In nutshell, it seems that the consumers make a close link between 'health' and 'natural' and believe that 'natural is better'. Due to this, it is expected that the demand for natural pigments will increase. Natural pigments frequently cost 5 to 10 times more than their synthetic equal. The best case is that of β-carotene produced by microalgae, which cost approximately USD 1000/kg against USD 500/kg produced by synthesis. Despite the higher price, β-carotene produced by fermentation may compete in markets where it is important that all the ingredients be "natural". Besides, the microbial pigment is the mixture of cis- and trans-isomers, with anti-cancer effects that synthetic β-carotene, mostly cis form, does not show [108]. The leading international natural colourant of natural pigments are:

1. ChrHansen's Lab (Denmark)
2. Warner-Jenkinson Europe Ltd. (UK)
3. Kalsec Inc. (USA)
4. Quest International (The Netherland)

Analysis of Table 17 shows that while several natural pigments are permitted in as much amount as needed in foods, none of the artificial pigments have derestricted use: they must be used at a controlled level. This indicates that the natural pigment market is rather promising, especially because the pigment now permitted does not span over the whole visible spectrum [109].

### 6.1 Commercially Produced Microbial Pigments

1. *Monascus* as Lactopurpurian
2. Astaxanthin as AstaXin by Igene Biotechnology Inc., Columbia, Maryland.
3. CAROPHYLL Pink of Hoffmann-La Roche Ltd., Cambridge, Ontario.
4. Astaxanthin of Mera Pharmaceuticals Inc.
5. Beta Carotene of ParryNeutraceuticals, Parry Agro Industries Ltd.
6. Spirulina of New Ambadi Estates Pvt Ltd
7. Chlorella of Taiwan, Chlorella Manufacturing Co. Ltd.

## 7. TOXICOLOGY OF COLOURED FOOD

Concerns have been expressed that some foods commonly consumed by children and adolescents may contain excessively high levels of synthetic food colours to increase their consumer appeal. The Joint Australia New Zealand Food Standards Code sets maximum permitted levels for synthetic colours. Under this code, most synthetic colours are permitted in beverages to a maximum concentration of

**Table 17:** Food colours permitted in the European Union.

| Used at Quantum statis in any food | Use limited to a maximum level, in some specific foods | Restricted use (permitted for some foods only) |
|---|---|---|
| Riboflavin (E101) | Curcumin (E100) | Amaranth (Bordeaux Red) |
| Chlorophylls and chlorophyllins (E 140) | Tartrazine yellow (E 102) | Eritrosin (E 127) |
| Copper complexes of chlorophylls and chlorophyllins (E 141) | Quinoline yellow (E 104) | R ed 2 G (E 128) |
| Caramel (E 150a) | Sunset yellow FCF, sunset orange S (E 110) | Brown FK (E 154) |
| Caramel sulphite (E 150b) | Cochonile, carminic acids, carmines (E 1120) | Canthaxantin (E 161g) |
| Caramel ammonia (E 150c) | Azorubin, carmoisin (E 122) | Aluminum (E 173) |
| Caramel sulphite ammonia (E 150d) | Ponceau Red 4R, Carmine Red A (E 124) | Silver (E 174) |
| Carbo medicinalis (E 153) | Allura red AC (E 129) | Gold (E 175) |
| Carotenoids (E 160a) | Sea Blue V (E 131) | Litorubin BK (E 180) |
| Paprika extracts, capsanthin, capsorubin (E 160c) | Indigo tin, indigo carmine (E 132) | Annato, bixin, norbixin (E 160b) |
| Beetroot red, betanins (E 162) | Brilliant blue FCF (E 133) | |
| Anthocyanins (E 163) | Green S (E 142) | |
| Calcium carbonate (E 170) | Brilliant black BN, black PN (E 151) | |
| Titanium dioxide (E 171) | Brown HT (E 155) | |
| Iron oxides and hydroxides (E172) | Lycopene (E 160d) | |
| | Beta 8'-carotenal (C30) (E 160c) | |
| | Ethyl Ester of beta-apo-8'-carotenoic acid (C30) (E 160f) | |
| | Lutein (E 161b) | |

*Source:* 108, 109

70 mg/l, and in other foods to a maximum concentration of 290 mg/kg. Exceptions are erythrosine, which is only permitted in preserved cherries (maximum concentration 200 mg/kg), and amaranth, which has a range of specific permissions for various foods, including 30 mg/l in beverages and 300 mg/kg in confectionery.

## 7.1 Tests to Measure Synthetic Colour Levels in Food

During the 1999/2000 year, a study was carried out to determine the levels of synthetic food colourings in food available in New Zealand to know the foods that:

- contain elevated levels of synthetic food colours (they were visually intensely coloured), and
- are targeted for consumption by children and adolescents (confectionery, beverages).

Six percent of samples analyzed (15/253) were found to contain amounts of food colouring in excess of the maximum levels specified in the Joint Food Standards Code. Most of the non-compliant samples (11 out of 15) were confectionery. While these results were non-compliant with the requirements of the Joint Food Standards Code, the health implications of these results are uncertain and so more informations were collected:

- to summarize the currently available toxicological information on the food colouring compounds, tartrazine, sunset yellow FCF, allura red AC, ponceau 4R, brilliant blue FCF, amaranth, green S, azorubine (carmoisine), and erythrosine;
- to estimate exposure to these food colouring compounds for children and adolescents aged 5-15 years; and
- to assess the public health risk by comparison with Acceptable Daily Intakes agreed by the WHO/FAO Joint Expert Committee on Food Additives (JECFA).

A huge number and range of animal toxicology studies have been carried out on synthetic food colours. For most colours no adverse effects were observed, even when colours were fed at the rate of 5% of the total diet. Typically, human consumption of food colours would be less than 0.01% of the diet. The exceptions were high dose effects related to amaranth (calcium deposits in the kidneys) and erythrosine (impaired thyroid function). Acceptable Daily Intakes (ADIs) for these colours have been set to be well below dose levels where effects were observed and include large safety factors.

## 7.2 Adverse Reactions of Synthetic Food Colours

Synthetic food colours, particularly tartrazine, have been implicated in cases of adverse reactions ascribed to food. Three reactive states (asthma, urticaria/hives, and hyperactivity) have been particularly associated with synthetic food colours. Cases of food colour induced asthma or other allergy-like reactions have been documented, but these cases are extremely rare and usually occur only in individuals who are sensitive to a wide range of environmental and dietary components (atopic). Studies examining links between synthetic food colours and hyperactivity are beset by difficulties, such as how the hyperactivity is assessed, and an Australasian Expert Committee has concluded that "despite popular belief, well designed studies have failed to show a consistent relationship between a wide range of behavioural problems in childhood and the consumption of food additives".

## 7.3 Dietary Exposure and Risk

Estimates of exposure have been derived by combining mean levels of food colours in confectionery and beverages from the 1999/2000 study with food consumption information for children and adolescents obtained from the 2002 National Children's Nutrition Survey (2002 CNS). The 2002 CNS collected information on food eaten in the previous 24 hours for more than 3000 New Zealand children and adolescents. Average dietary exposures ranged from as low as 0.02 mg/day (erythrosine) to 2.6 mg/day (tartrazine). Dietary exposure estimates were adjusted for differences in body weight and compared to the internationally agreed Acceptable Daily Intakes (ADIs). Average exposure estimates for all food colours considered were less than 5% of the relevant ADI. Dietary exposure estimates were also generated for two age-related subgroups (5-12 years and 13-15 years). Dietary exposure estimates for food colours, expressed on a per body weight basis, were generally higher for the younger age group, however, average estimates for all food colours were still less than 5% of the ADI. The 2002 CNS data is a valuable resource for risk assessment as the dietary records for 3000+ individuals allows estimation of not only the average population dietary exposure, but also dietary exposure levels for 'high' consumers of artificially coloured foods. High consumers are often deemed to be those at or above the $95^{th}$ percentile level. That is, if all dietary exposure estimates are placed in ascending order, the top 5% are taken to represent high consumers. High consumer exposure estimates were all equal to or less than 15% of the relevant ADIs. Given that there is a large safety margin built into the ADI, dietary exposure to food colours by New Zealand children and adolescents is highly unlikely to present an unacceptable health risk.

Tartrazine is a widely used yellow colouring, which has been associated with allergic reactions in some rare instances. Reactions have ranged from rashes and swelling to asthma and possibly even behavioural changes. If it is medically advised, diets that eliminate specific additives can be compiled. These diets should be maintained only under medical supervision, as prolonged exclusion of certain foods from the diet can lead to development of nutritional deficiencies. In Australia, tartrazine and other added food colouring can be identified by their code numbers. For example, 'colours (102, 110, 133)' in the ingredient list means that the food contains tartrazine, sunset yellow FCF and brilliant blue FCE. In the USA, foods containing tartrazine (where it is known as FD&C Yellow No. 5) are required to list this colouring by name. Many drugs also contain colouring additives, and this fact is sometimes overlooked when trying to avoid particular colourings.

## 7.4 Toxicity of Natural Pigments

There are several studies on the toxicity of *Monascus* pigments, which are apparently safe in the quantities tested. The many years for which these pigments have been used suggested low or non-existent toxicity [87]. Since the strains from the genus *Monascus* started to be studied systematically, it has been believed that the pigments produced also presented antibiotic properties; later, it was verified that this activity is mainly due to other substances, named monascidin-A [110]. Further studies showed that this substance was in fact, citrinin, a mycotoxin with nephrotoxic action produced by several fungi [111], but that not all *Monascus* strains produced citrinin. Unfortunately, it seems that all good pigment producers also produced citrinin. Finally, studies over the toxicity of purified fractions of pigments showed that there was indeed an antibiotic activity for *Monascus* pigments, especially the orange ones and in lower degree, the red ones [112].

Although there is not yet a definitive conclusion about pigment toxicity and citrinin production in industrial processes, several actions may be taken to prevent or reduce this problem: the use of strains which produced less citrinin; the control of the nitrogen source (organic nitrogen sources favour red pigment production and decrease citrinin production); the control of cultivation conditions (aeration, pH, solid-substrate vs. liquid fermentation); the transformation of orange pigments into red, non-toxic complexes, using amino acids [111, 113, 114] and the extraction in low citrinin-solubility conditions, which may be achieved by controlling pH, since citrinin is strongly acidic. There is much space for strain development, as may be noted by the examination of Table 18 [115, 116]. It is clear from the table that there is a high diversity among different *Monascus* strains, with a large span in production,

**Table 18:** Pigment productivity and citrinin production in rice for several *Monascus* strains.

| Strain | Relative absorbance (500 nm) | Productivity ($day^{-1}$) | Citrinin (μg/g product) |
|---|---|---|---|
| CCT 3802 | 0.15 | 0.021 | 20 |
| ATCC 6405 | 0.47 | 0.047 | 23 |
| ATCC 16365 | 0.47 | 0.067 | 83 |
| UFPE 3196 | 0.14 | 0.01 | 29 |
| Commercial product | 1.00 | 0.07 | 31 |
| NRRL 1991 | 0.16 | 0.023 | 25 |
| NRRL 2897 | 0.14 | 0.020 | 22 |
| CCT 3802 | 0.15 | 0.021 | 20 |
| LPB 31 | 0.87 | 0.124 | 18 |

productivity and citrinin production. It is very likely that future production will be made with pigment overproducers which do not produced citrinin.

## 8. APPLICATIONS OF BIOCOLOURS

Carotenoids have extensive applications as antioxidants in dietary supplements, and as colours in foods and beverages as well as pigments in poultry and fish. It is reported that carotenoids are associated with an array of biological functions, all requiring intestinal uptake [46]. These functional roles include antioxidant activity, modulation of detoxifying enzyme, enhancing immune apoptosis and regulating gene expression [90]. Therefore, in addition to colouring properties, these also have physiological importance. The primary requirements for the successful commercial use of appealing red-shaded, anthocyanins-based colourants are a low pH and the absence of a cloud. Anthocyanins gradually change from red to blue-red, purple, blue and green to yellow as the pH increases. An average anthocyanin-based colourant is red in acid, violet in neutral and blue in alkaline solution. In addition to pH sensitivity, anthocyanins are susceptible to degradation by light, heat, oxygen, iron, copper, tin, ascorbic acid and sulphur dioxide. Such colourants containing anthocyanins have been suggested for beverages - in particular fruit drinks which represent the biggest market of commercial applications in the United States, fruit preparations, jellies, jams, preserves, ice cream, yoghurt, gelatin desserts, canned fruits, etc. (Table 19). Annatto is considered essentially non-toxic. In the United States, it is a natural colourant exempt from certification (regulated by FDA in 21CFR 73.30). It has good heat, water and oil stability, and thus provides flexibility for use in a variety of applications. Due to its technological properties such as changing colour tone upon heating solutions of bixin, annatto allows food technologists great versatility in adjusting the hue of food products from red to orange to yellow as desired. Moreover, annatto acts as an antioxidant, sometimes improving the stability of the food product. Annatto is especially suited for colouring dairy products such as cheese, ice cream and yoghurt besides popcorn (oil/butter), confectionery products, snack foods, dry mixes and soft drinks. The susceptibility of the betanin colour pigments in beetroot to heat, oxygen, light and high water activity sometimes restricts its use as a food colourant. It typically cannot survive the hot-fill process for beverages or candies, standing to lose its colour and change shade. Considering these limitations, beetroot juices and the further processed beetroot colours are ideal for use in products that have either a short shelf-life or low water activity, and are packaged for minimal exposure to light, oxygen and high humidity. Ice cream and powdered beverages are the most important application of betanin. Most of the world's strawberry flavoured ice creams have betanin either from beetroot juice or beetroot colour. Betanin is also used in the colouration of some sugar confectionery products such as fondants, sugar strands, sugar coatings and fruit or cream fillings, dry mixes, sauces, salad dressings, seasoning bases, soup mixes, instant desserts and snack foods. Beet juice can be used successfully in hot, processed candies if added at the end of the process (Table 19).

Beet powder has good to excellent stability in the applications of colouring dry-mix beverages - even in the presence of oxygen. Most carotenoids are produced by chemical synthesis [related to vitamin A process route], although microbial fermentation processes have been developed for astaxanthin and β-carotene, whilst lycopene can also be obtained by extraction from natural sources. In almost every country of the world, cochineal extract and carmine are listed as exempt from certification (US: regulated by FDA in 21CFR 73.100) and permitted as safe colour additive mixtures for colouring foods. Cochineal and its derivatives are staging a comeback as food colourants because of the influence of the "natural" trend and their superior technological properties such as excellent heat and light stability, resistance to oxidation, clarity, and desirable, appealing hue. Its non-kosher status is an important limitation, although some

**Table 19:** Commercial applications of certain plant origin colours in various food products.

| Food | Shade | Natural colours | Addition level (%) |
|---|---|---|---|
| Hard candies | Pink rose | Beet root juice powder | 0.1 |
| | Red | Anthocyanin powder | 0.2 |
| | Yellow | Riboflavin | 0.01 |
| Jams | Lemon | Riboflavin | 0.01 |
| | Strawberry | Anthocyanin + Canthaxanthin 10% WS | 0.006 + 0.02 |
| Jelly | Black current | Anthocyanin powder | 0.6 |
| | | Beet root juice powder | 1.0 |
| Pectin | Violet | Anthocyanin powder | 0.25 |
| | Red | Canthaxanthin 10% WS | 0.02 |
| | Yellowish green | Beta-carotene 1% CWS | 0.025 |
| Gelatine | Raspberry red | Beetroot juice | 0.2 |
| Ice cream | Brown | Caramel | 2.0-3.0 |
| | Bluish green | Chlorophyll | 0.1-0.5 |
| | Pink | Caramine + Annatto | 1.0 + 0.006 |
| | Light cream yellow | Annatto | 0.01-0.05 |
| | Orange yellow | Beta-carotene | 0.1-0.2 |
| Pastry | Pale yellowish | Chlorophyll powder | 0.0002 |
| | White | Beta-carotene | 0.05 |
| Cake | Chrome orange | Chlorophyll powder + Dry Canthaxanthin 10% WS | 0.0055 + 0.055 |
| Biscuit | Mauve | Anthocyanin powder | 0.01 |
| Cream | Brown | Beet root juice powder | 1.6 |

*Source:* 2, 6, 117

kosher carmine exists commercially today. Cochineal extract and carmine are ideally suited and extensively utilized around the world for a variety of food products, including meat, sausages, processed poultry products (meat products cannot be coloured in the United States) surimi, and red marinades. Cochineal and its derivatives find further important application in fruit preparations, jams, preserves, gelatin desserts, juice beverages, baked goods, confectionary, icings, toppings and dairy products. Some pulp-containing, non-carbonated soft drinks, as well as certain alcoholic aperitifs such as Campari, are commonly coloured with cochineal extract varying from bright, stable orange shades to clear; magenta-red shades. Alpha-carotene, beta-carotene, lutein, cryptoxanthin and lycopene are most important carotenoids from nutritional point of view for human beings [78].

Turmeric and saffron find use by developers as natural yellow colourants. Both oil- and water-soluble forms exist, providing versatility in their application. Saffron colourant has not found wide application because of its high cost. The availability of the synthetic form and its standardized uniformity, as well as the consumer perceived value of β-carotene as a pro-vitamin, has led to its use in a large number of products that are coloured in the yellow to orange range. Applications for β-carotene include margarines, beverages, cakes, soups, milk-based products, desserts, candies, salad dressings, pastas and others. The unusual stability of caramel in most food applications favours broad use. But its cost, variable availability, non-kosher status and in some markets, its non-vegetarian source have limited its appearance in food products. Applications include aperitifs, candies, gelatin desserts, jams, jellies, pet foods, baked goods, icings and value-added products that can sustain the relatively high cost. Natural colourants are not like a Swiss army knife, they do not work under all sorts of conditions and with all sorts of items. Each naturally derived colorant is unique, its application limited to product systems having compatible physical and/or chemical properties. Since each colourant is available in numerous forms, choosing the correct form requires careful consideration. Knowledge of the factors governing colourant stability is essential to assess the compatibility of the food system to the colourant. Meeting a specific colour target is perhaps the most challenging aspect of this work, as not all of the colours found in nature have been exploited and the palette of available colourants will not accommodate the entire spectrum of potential hues. Available colourants may also have restrictions placed upon them for religious and regulatory reasons. Even with all these hurdles, successfully colouring foods with natural colourants is achievable. It is expected that with further technological advances, and the development of new pigment sources, the job will get easier in the future. The potential for a wider application of naturally derived colourants would seem to be unlimited.

## 9. RESEARCH GAPS

Although various works have been conducted on biocolours, there are certain research gaps to be overcome. This includes the following:

- Conversion of research outputs into technologies and their up scaling for commercial exploitations.
- Development of equipment for the extraction of colours.
- Development of techniques for preservation of natural colours during storage.
- Natural colours especially β-carotenes have problems of water solubility.
- Little information on toxic profile of colour extracted from natural sources.
- Maximizing the yield of biocolours obtained from plant and microbial sources.
- Colours which are used for beverages, jellies, and confectionary should give brightness, while vegetable oil based products give haziness upon emulsification.
- A majority, almost 70% of plant sources have not been exploited; only 0.5% plant sources could be fully investigated.
- Many of the waste from processing industry has not been evaluated for the production of microbial colours.
- Gaps exist with respect to optimum extraction medium, optimum stable emulsion, encapsulation technique, proper carrier of biocolours, milling of pigments, etc.
- Standardization of process for utilization, extraction, drying and storage is totally lacking.

## REFERENCES

1. Sankaranarayan R (1981) Food colours. *Indian Food Packer* 35: 25-29.
2. Pattnaik P, Roy U & Jain P (1999) New generation additives for food. *Beverage Food World* 3: 36-40.
3. Bhat RV & Mathur P (1998) Changing scenario of food colours in India. *Curr Sci* 74: 198-202.
4. Prasad PK, Bhat GS & Rao DS (1993) The colour crisis in food industry and search for safe natural colourants. *Indian Dairyman* 45: 352-356.
5. Murai K & Wilkins D (1990) Natural red colour derived from red cabbage. *Food Technol* 44: 131-132.

6. Mahajan R, Shukla FC & Shoree M (2000) Commercial use of natural food colours — A review. *Beverage Food World* 27: 44-51.
7. Srilakshmi B (1999) Vegetable and fruits. In: *Food Science,* New Age International Publishers, New Delhi, India, pp 225-276.
8. Bauernfeind JC (1981) *Carotenoid as Colourant and Vitamin A Precursor*, Academic Press Inc, New York, USA, pp 1-37.
9. Mazza G & Miniati E (1993) Introduction. In: *Anthocyanins in Fruits, Vegetables and Grains*, Mazza G (ed), CRC Press, Boca Raton, USA, pp 1-28.
10. Casal BAC, Byrne DH, Zevallos LC & Okie WR (2002) Total phenolic and anthocyanin content in red fleshed peaches and plums. *Acta Hort* 592: 589-595.
11. Federica B, Carmela G & Isabella N (2004) Sour cherry (*Prunus cerasus* L.) anthocyanins as ingredients for functional foods. *J Biomed Biotech* 5: 253-258.
12. Chiang HJF & Wrolstat RE (2005) Anthocyanin pigment composition of blackberries. *J Food Sci* 70: 198-202.
13. Chiriboga CD & Francis FJ (1973) Ion exchange purified anthocyanin pigments as a colourant for cranberry juice cocktail. *J Food Sci* 38: 464-465.
14. Palamidis N & Markakis P (1975) Stability of grape anthocyanin in carbonated beverages. *J Food Sci* 40: 1047-1049.
15. Calvi JP & Francis FJ (1978) Stability of concord grape (*V. labrusca*) anthocyanin in model system. *J Food Sci* 43: 1447-1456.
16. Clydesdale FM, Main JH & Francis FJ (1978) Concord grape pigments as colourants for beverages and gelatin desserts. *J Food Sci* 43: 1687-1692.
17. Bassa IA & Francis FJ (1987) Stability of anthocyanins from sweet potatoes in a model beverage. *J Food Sci* 52: 1753-1754.
18. Teh LS & Francis FJ (1988) Stability of anthocyanins from *Zebrina pendula* and *Ipomoea tricolor* in a model beverage. *J Food Sci* 53: 1580-1581.
19. Maccarone EA, Maccarone S & Rapisared P (1985) Stabilization of anthocyanin of blood orange fruit juice. *J Food Sci* 50: 901-904.
20. Sistrunk WA & Cash JN (1968) Stabilizing the colour of frozen strawberries. *Arkansas Farm Research* pp 17-25.
21. Rice-Evan AC, Miller NJ & Paganga G (1996) Structure-antioxidant activity relationships of flavonoids and phenolic acids. *Free Radical Biol Med* 20: 933-956.
22. Newsome RL (1986) Food colours. *Food Technol* 40: 49-56.
23. Francis FJ (1989) Food colourants: Anthocyanins. *Crit Rev Food Sci Nutr* 28: 273-314.
24. Shrede G, Wrolstsd RE, Lea P & Enersen G (1992) Color stability of strawberry and blackcurrent syrups. *J Food Sci* 57: 172-177.
25. LaBell (1993) Grapes provide brilliant red colour. *Food Proc* 54: 88-89.
26. Jackman RL & Smith JL (1996) Anthocyanin and betalanins. In: *Natural Food Colourant,* Hendry GAF & Houghton JD (eds), Blackie Academic Press, Great Britian, UK, pp 244-280.
27. Shrikhande AJ (1976) Anthocyanin in foods. *Crit Rev Food Technol* 24: 169-174.
28. Wrolstad RE, Putnam TP & Varseveld GW (1970) Colour quality for frozen strawberry: Effect of anthocyanin, pH, total acidity and ascorbic acid variability. *J Food Sci* 35: 448-450.
29. Abers JE & Wrolstad RE (1979) Causative factors of colour deterioration in strawberry preserve during processing and storage. *J Food Sci* 44: 75-78.
30. Sistrunk WA & Cash J (1970) The effect of certain chemicals on the color and polysaccharides of strawberry puree. *Food Technol* 24: 169-171.

31. Markakis P (1982) Stability of anthocyanin in foods. In: *Anthocyanins as Food Colours*, Markakis P (ed), Academic Press, New York, USA, pp 1-38.
32. Baranac JM, Petranovic NA & Dimitric-Markovic JM (1996) Spectrophotometric study of anthocyanin co-pigmentation reactions. *J Agric Food Chem* 44: 1333-1336.
33. Francia-Aricha EM,Guerra MT, Rivas-Gonzola JC & Santos-Buelga C (1997) New anthocyanin pigments formed after condensation of flavonoides. *J Agric Food Chem* 45: 2262-2266.
34. Johnston TV & Morris JR (1997) HPLC Analysis of cabernet savignon and nobel wine pigment fractions. *J Food Sci* 62: 684-687.
35. Mazza G & Gao L (1994) Malonylated anthocyanin in purple sunflower seeds. *Phytochemistry* 35: 237-238.
36. Glicksman M (1969) Gum technology in the food industry. In: *Food Science and Technology,* A series of monographs, Academic Press Inc, New York, USA, pp 126-130.
37. Francis FJ (1977) Anthocyanin. In: *Current Aspect of Food Colourants,* Furia E (ed), CRC Press, New Orleans, USA, pp 19-28.
38. Main JH, Clydesdale FM & Francis FJ (1978) Spray drying anthocyanin concentrates for use as food colourants. *J Food Sci* 43: 1693-1694.
39. Anon (1973) Exotic flavors for cocktail mixes captured with versatile spray drying aids. *Food Prod Dev* 6: 70-73.
40. Nakajima J, Tanaka I, Seo S, Yamazaki M & Saito K (2004) LC/PDA/ESI-MS Profiling and radical scavenging activity of anthocyanins in various berries. *J Biomed Biotechnol* 5: 241-247.
41. Ong ASH & Tee ES (1992) Natural sources of carotenoids from plants and oils. *Meth Enzymol* 213: 142-167.
42. Britton G, Liaaen-Jensen S & Pfander H (1995) Carotenoids, today and challenges for the future. In: *Carotenoids: Isolation and Analysis,* Vol 1A, Britton G, Liaaen-Jensen S & Pfander H (eds), Birkhäuser, Basel, Switzerland, pp 206-216.
43. Pfander H (1992) Carotenoids: an overview. *Meth Enzymol* 213: 3-13.
44. Mercadante A (1999) New carotenoids: Recent progress. *Pure Appl Chem* 77: 2263-2272.
45. Snodderly DM (1995) Evidence for protection against age-related macular degeneration by carotenoids and antioxidant vitamins. *Am J Clin Nutr* 62 (Suppl): 1448S-1461S.
46. Aparnathi KD & Borkhatriya VN (1999) Improved extraction and stabilization of natural food colorants. *Indian Food Ind* 18: 164-168.
47. Delgado-Vargas F & Paredes-Lopez O (1997) Effects of enzymatic treatments of marigold flowers on lutein isomeric profiles. *J Agric Food Chem* 45: 1097-1102.
48. Barzana E, Rubio D, Santamaria RI, Garcia-Correa O, Garcia F & Ridaura Sanz VE (2002) Enzyme-mediated solvent extraction of carotenoids from marigold flower (*Tagetes erecta*). *J Agric Food Chem* 50: 4491-4496.
49. Debicki-Pospisil J, Lovric T, Trinajestic N & Sabljic A (1983) Anthocyanin degradation in the presence of furfural and 5-hydromethylfurfural. *J Food Sci* 48: 411-416.
50. Degnan AJ, Elbe JH & Hartel RW (1991) Extraction of annatto seed pigment by supercritical carbondioxide. *J Food Sci*. 56: 1655-1659.
51. Chao RR, Mulvaney SJ, Sanson DR & Tempesta MS (1991) Supercritical $CO_2$ extraction of annatto (*Bixa orellana*) pigments and some characteristics of the colour extracts. *J Food Sci* 56: 80-83.
52. Philip T (1984) Purification and concentration of natural colourants by membranes. *Food Technol* 38: 107-108.

53. Fellows PJ & Worgan JT (1988) Growth of *Saccharomyces fibuliger* and *Candida utilis* in mixed culture on apple processing wastes. *Enzyme Microb Technol* 9: 434-437.
54. Di Mascio P, Murphy ME, & Sies H (1991) Antioxidant defense systems: The role of carotenoids, tocopherols, and thiols. *Am J Clin Nutr* 53: 194S-200S.
55. Mangels AR, Holden JM, Beecher GR, Forman MR & Lanza E (1993) Carotenoid content of fruits and vegetables: an evaluation of analytic data. *J Am Diet Assoc* 93: 284-296.
56. Jacobson G & Wasileski J (1994) *Bioprocess Production of Flavours, Fragrance and Colour Ingredients,* John Wiley & Sons, New York, USA.
57. Babitha S, Sandhya C & Pandey A (2004) Natural food colourants. *Appl Bot Abstr* 23: 258-266.
58. Pandey A & Babitha S (2005) Microbial pigments. *Adv Biotech* 3: 17-19.
59. Downham A & Collins P (2000) Colouring our foods in the last and next millennium. *Int J Food Sci Technol* 35: 5-22.
60. Broder CV & Koehler PE (1980) Pigment produced by *Monascus purpureus* with regard to quality and quantity. *J Food Sci* 45: 587-588.
61. Carreilla A, Feereira LM & Loureis V (2001) Production of brown tyrosine pigments by the yeast *Yarrowia lipolytica. J Appl Microbiol* 90: 372-379.
62. Devender A & VK Joshi (2006) Optimization of apple pomace based medium and fermentation conditions for pigment production by *Chromabacter* species. *J Food Sci Technol* 43: 515-520.
63. Manonimani HK & Sreektaniah KR (1984) Pigment production by a strain of *Aspergillus* sp. *J Food Sci Technol* 21: 195-197.
64. Sandhu DK & Joshi VK (1997) Development of apple pomace based medium, optimizing pigment production by *Rhodotorula* and its characterization. *Adv Food Sci (CMTC)* 19: 31-34.
65. Joshi VK, Attri D, Anju B & Shashi B (2003) Microbial pigments. *Ind J Biotechnol* 2: 362-369.
66. Nelis HJ & Leenheer AP (1991) Microbial sources of carotenoids used in food and feed. *J Appl Bacteriol* 70: 181-191.
67. Joshi VK (2006) Production and applications of microbial colours in food: Present status and future strategy. In: *Proceedings of National Symposium on New Horizons in Fermentation and Food Biotechnology,* March 21-22, Department of Biotechnology, Punjabi University, Patiala, India, pp 14-29.
68. Olson JH (1989) Provitamin A function of carotenoids: The conversion of beta-carotene into vitamin A. *J Nutr* 119: 105-108.
69. Anderwes AG, Phaff HJ & Starr MP (1976) Carotenoids of *Phaffia rhodozyma,* A red pigmented fermenting yeast. *Phytochemistry* 15: 1003-1007.
70. John MR & Stuart DM (1991) Production of pigments by *Monascus purpureus* in solid culture. *J Ind Microbiol* 8: 23-28.
71. Carels M & Shephered D (1978) The effect of pH and amino acids on conidiation and pigment production of *Monascus major* and *Monascus rubiginosus* in submerged shaken culture. *Can J Microbiol* 24: 1346-1357.
72. Lin CF (1973) Isolation and cultural conditions of *Monascus* sp. for the production of pigment in submerged culture. *J Ferment Technol* 6: 407-457.
73. Johnson GT & McHan F (1973) Some effects of zinc on the utilization of carbon sources by *Monascus purpureus. Mycologia* 67: 806-816.
74. Wong HC, Lin YC & Koehler PE (1981) Regulation of growth and pigmentation of *Monascus purpureus* by carbon and nitrogen concentration. *Mycologia* 73: 649-654.

75. Lin TF & Demain AL (1991) Effect of nutrition of *Monascus* species on formation of red pigment. *Appl Microbiol Biotechnol* 36: 70-75.
76. Chen MH & John MR (1994) Effect of carbon source on ethanol and pigment production by *Monascus purpureus. Enzyme Microb Technol* 16: 584-590.
77. Tseng YY, Chen MT & Lin CF (2000) Growth, pigment production and protease activity of *Monascus purpurens* as affected by salt, sodium nitrate, polyphosphate and various sugars. *J Appl Microbiol* 88: 31-37.
78. McHan F & Johnson GT (1970) Zinc and amino acids: Important components of a medium promoting growth of *Monascus purpurens. Mycologia* 62: 1018-1031.
79. Ban YS & Wong HC (1979) Zinc effect on growth, pigmentation and antibacterial activity of *Monascus purpureus. Physiol Plant* 46: 63-67.
80. Joshi VK & Devender A (2006) Optimization of apple pomace based medium and fermentation conditions for pigment production by *Rhodotorula* species. *Proc Nat Acad Sci India* 76: 171-177.
81. Devender A & Joshi VK (2005) Optimization of apple pomace based medium and fermentation conditions for pigment production by *Micrococcus* species. *J Sci Ind Res* 64: 598-601.
82. Johnson EA & Lewis MJ (1979) Astaxanthin formation by the yeast *Phaffia rhodozyma, J Gen Microbiol* 115: 173-183.
83. Cooney JJ, Marks HW & Surith AM (1966) Isolation and identification of canthaxanthin from *Micrococcus roseus. J Bacteriol* 92: 342-345.
84. Ungers GE & Cooney JJ (1968) Isolation and characterization of carotenoid pigments of *Micrococcus roseus. J Bacteriol* 96: 234-241.
85. Schwartzel EM & Cooney JJ (1972) Isolation of 4'-hydroxy echinenone from *Micrococcus roseus. J Bacteriol* 112: 1422-1424.
86. Thirkell D, Strang RHC & Chapman JR (1967) The pigment of *Sarcina flava,* A new series of $C_{50}$ carotenoids. *J Gen Microbiol* 49: 157-164.
87. Miller MW, Yoneyama M & Soneda M (1976) *Phaffia* a new yeast genus in the *Deuteromycotina* (Blastomycetes). *Int J Syst Bacteriol* 26: 286-291.
88. Simpson KL, Chichester CO & Phaff HG (1971) Carotenoid pigment of yeasts. In: *The Yeasts*, Vol 2, Rose AH & Harison JS (eds), Academic Press, New York, USA, pp 493-515.
89. Simpson KL, Nakamaya TOM & Chichester CD (1971) Biosynthesis of yeast carotenoids. *J Bacteriol* 11: 1688-1694.
90. Bae M, Lee TH, Yokayama H, Boettger HG & Chichester CO (1971) Different compounds in *Rhodotorula aurantiaca. Phytochemistry* 10: 625 627.
91. Wong HC & Koechler PE (1983) Production of red water soluble *Monascus* pigments. *J Food Sci* 48: 1200-1203.
92. Blanc PJ, Coret MO, Santerre AC, Pareilleux A, Prome D, Prom JC, Lausset JP & Goma G (1994) Pigments of *Monascus. J Food Sci* 59: 862-865.
93. Buzzini P (2001) Batch and fed batch carotenoid production by *Rhodolorula glutinis — Debaromyces castellii* co-cultures in corn syrup. *J Appl Microbiol* 90: 843-847.
94. Frengova G (1994) Formation of carotenoids by *R. glutinis* in whey ultrafiltrate. *Biotechnol Bioeng* 44: 888-894.
95. Lin TF, Yakushijin K, Buchi GH & Demain AL (1992) Formation of water soluble *Monascus* red pigments by biological and semisynthetic processes. *J Ind Microbiol* 9: 173-179.
96. Cooney JJ & Thierry OC (1966) A defined medium for growth and pigment synthesis of *Micrococcus roseus. Can J Microbiol* 12: 83-89.

97. Attri D & Joshi VK (2006) Chemical modification of microbial carotenoids into water soluble pigments for use in food products. In: *Proceedings of National Symposium on New Horizons in Fermentation and Food Biotechnology*, Punjabi university, Patiala, India, p 46.
98. Went FAFC (1895) *Monascus prepureus* le champignon de lang-quac unenouvelle. *Ann Sc Nat Bot* 8: 1-17.
99. Attoe V & Elbe JH (1971) Photochemical degradation of betanine and soluted anthocyanins. *J. Food Sci* 6: 1934-1936.
100. Carnevale J, Cole ER & Crank G (1980) Photocatalysed oxidation of paprika pigments. *J. Agri Food Chem* 28: 953-954.
101. Chang RQ (1977) Studies on the stability of *Monascus* pigment, Master Thesis, National Taiwan University, Taipai, Taiwan.
102. Fink-Gremmels J & Lustner L (1989) Biologiche wirkungen von *Monascus purpureus. Fleischjwirtsch* 69: 115-122.
103. Fabre CE, Santerce AL, Lorte MO, Barberian R, Pareilleux A, Goma G & Blanc PJ (1993) Production and food application of the red pigments of *Monascus ruber. J Food Sci* 58: 1099-1102.
104. Francis FJ (1987) Lesser-known food colourants. *Food Technol* 41: 62-68.
105. Thakur BR & Arya SS (1989) Stability of Gul Mohar (*Delonix regia*) carotenoids in isolated model system. *J Food Sci Technol* 26: 76-78.
106. Lee YK, Chen DC & Toshiomi Y (1995) Production of *Monascus* pigments by a solid-liquid state culture method. *J Ferment Bioeng* 79: 516-518.
107. Hajjaj H, Klaebe A, Loret MO, Tzedakis T, Goma B & Blanc P (1997) Production and identification of N-glucosylrubropunctamine and N-glucosylmonascorubramine from *Monascus ruber* and occurrence of electron donor-acceptor complexes in these red pigments. *Appl Env Microbiol* 63: 2671-2678.
108. Carvalho CJ, Soccol CR, Babitha S, Pandey A & Woiciechowski AL (2007) Production of pigments. In: *Current Developments in Solid State Fermentation*, Pandey A, Soccol CR & Larroche C (Eds), Asiatech Publishers Inc, New Delhi, India, pp 337-355.
109. Vargas FD, Jimenez AR & Lopez OP (2000) Natural pigments: Carotenoids, anthocyanins and betalains-Characteristics, Biosynthesis, Processing and Stability. *Crit Rev Food Sci Nutrit* 40: 173-189.
110. Wong HC & Koehler PE (1983) Production and isolation of red water soluble pigment from *Monascus purpureus. J Food Sci* 48: 1200-1203.
111. Blanc PJ, Loret O & Goma G (1995) Characterisation of monascidin A from *Monascus* as citrinin. *Int J Food Microbiol* 27: 201-213.
112. Martinkova L, Juzlova P & Veseley D (1995) Biological activity of polyketide pigments produced by the fungus *Monascus. J Appl Bacteriol* 79: 609-616.
113. Blanc PJ, Loret O & Goma G (1995) Production of citrinin by various species of *Monascus. Biotechnol Lett* 17: 291-294.
114. Juzlova P, Martinkova L & Shin CS (2002) Morphology control of *Monuscus* cells and scale up of pigment fermentation. *Process Biochem* 38: 649-655.
115. Carvalho CJ, Pandey A, Babitha S & Soccol CR (2003) Production of *Monascus* biopigments: an overview. *Agro Food Ind Hi-tech* 14: 37-42.
116. Carvalho JC, Oishi BO, Pandey A & Soccol CR (2005) Biopigment from *Monascus*: strain selection, citrinin production and colour stability. *Braz Arch Biol Technol* 48: 6-8.
117. Sharma R & Guleria SPS (2005) Plant origin colours—colours of the next millenium. *Young Indian* 25: 3-5.

# Microbial Flavours: Current Status and Future Prospects

# 20

**R.S. Singh and Ranjeeta Bhari**

*Department of Biotechnology, Punjabi University, Patiala 147 002 (Pb.)*

## 1. INTRODUCTION

Flavour is a distinctive property of a substance affecting the gustatory sense. Flavour may be defined as "the sum of those characteristics of any material taken in the mouth, perceived principally by the senses of taste and smell and interpreted by the brain" [1]. So flavour is the blend of taste and smell sensation evoked by a substance in the mouth. It can have three components—taste, odour and perception. It is perceived principally by aroma receptors in nose and taste receptors in mouth. Flavour perception is a complex process which requires the particles to be mixed with saliva and salivary enzymes to promote the release of flavour [2].

Flavour is usually the result of presence, within the complex matrices, of many volatile components exhibiting various chemical and physico-chemical properties. A high quality flavour is due to combined effect of a large number of different molecules, each with a different taste character but blended together to produce a desirable flavour. Flavour molecules are very diverse in nature. More than 5000 compounds have been identified as flavour components of food. In coffee, more than 700 compounds have been identified, however only a few have been identified to contribute significantly to the flavour profile. These are referred to as the 'character impact compounds' or 'key compounds.'

Flavour molecules can be detected at very low concentration, and are thus used in products at very low levels. Hexenols and hexenals, characteristic of green taste of fresh fruits and vegetables, can be perceived at a concentration of about 1 parts per million [3]. Naturally occurring flavour compounds are present at a very low concentration from parts per trillions to parts per millions level. Majority of these compounds are volatile, with boiling points in the range of 20-300°C. Most of them are lipophilic and possess a great diversity of chemical group and structure with molecular weight ranging from 50-250 [1].

The processing of raw food modifies the equilibrium between the different components, and thus the original flavour becomes weaker and requires subsequent supplementation [4]. These flavour supplements can be natural or chemically synthesized. Of all the available flavour compounds, around 80% are produced by chemical synthesis [5], which has certain drawbacks such as the formation of racemic mixtures and lack of substrate specificity. The growing aversion of consumers towards chemicals added to his food has led to the development of flavour compounds of biological origin, so-called natural flavours or bioflavours. Some of the important flavour compounds are listed in Table 1.

*Food Biotechnology: Principles and Practices*
*V.K. Joshi & R.S. Singh (Eds.)*
*I.K. International Publishing House Pvt. Ltd., India*
*pp 691–738*

**Table 1:** Some important flavour compounds.

| | | |
|---|---|---|
| 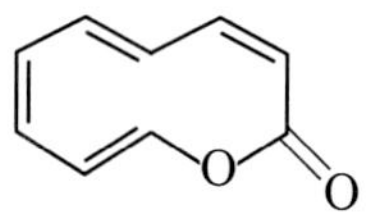 **Coumarin** | 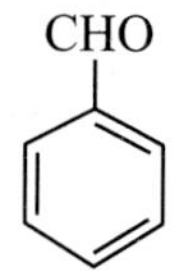 **Benzaldehyde** | 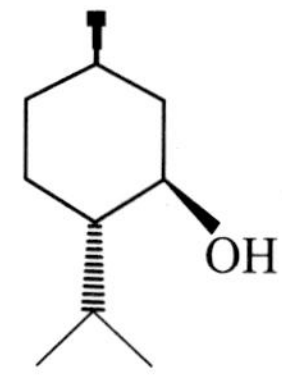 **L-Menthol** |
| 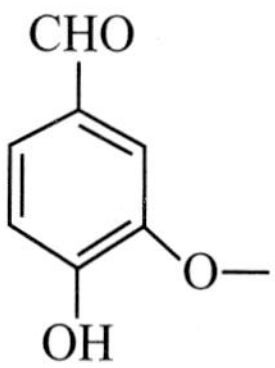 **Vanillin** | 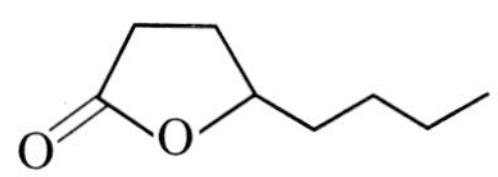 **γ-Octalactone** | 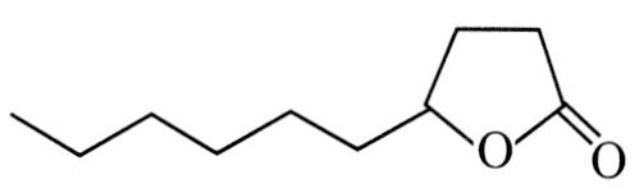 **γ-Decalactone** |
| 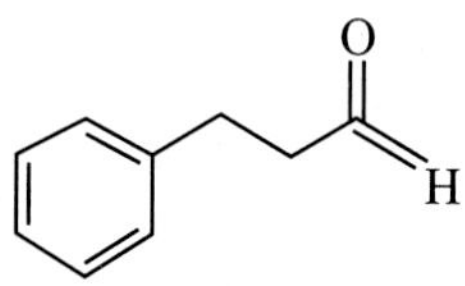 **Cinnamaldehyde** | 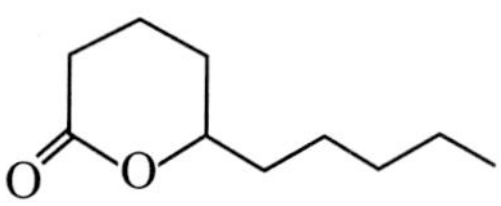 **δ-Decalactone** | 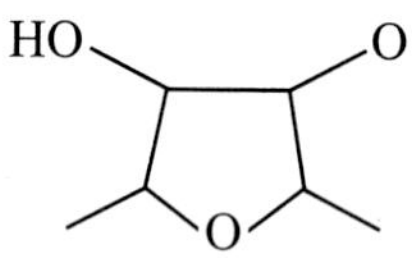 **Furaneol** |
| 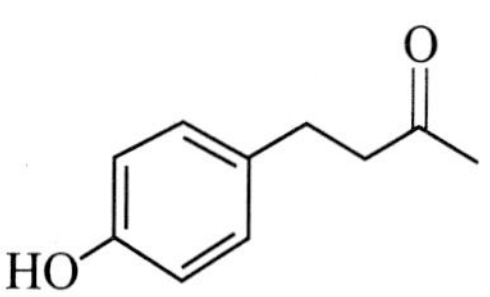 **Raspberry ketone** | 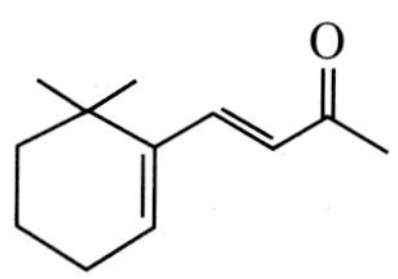 **β-Ionone** | 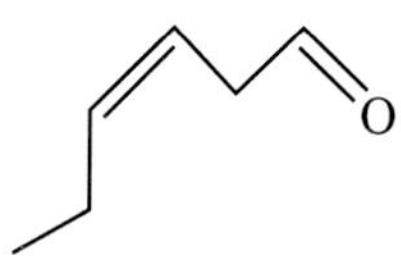 **Cis-3-hexenal** |

The code of federal regulations, USA, defines natural flavour as "essential oil, oleoresin, essence or extractive, protein hydrolysate, distillate of any product of roasting, heating or enzymolysis, which contains the flavouring constituents from a spice, fruit juice, vegetable or vegetable juice, edible yeast, herb, bud, bark, root, leaf or similar plant material, meat, seafood, poultry, eggs, dairy products or fermentation products, whose significant function in food is imparting flavouring rather than nutrition" [6]. According to the EC flavour directive (88/388/ECC), "Natural flavours are flavouring substances or preparations which are obtained by appropriate physical processes or enzymatic/microbiological processes from material of vegetal or animal origin." Thus, flavours prepared by biocatalytic routes from natural precursors can be considered 'natural' if they have been found in nature or traditional foods [7].

Amoore [8] has given a stereochemical theory based on lock and key hypothesis to explain characteristic flavours exhibited by different substances (See Chapter 25 for more details). Classes of compounds responsible for particular aroma notes are listed in Table 2. While thousands of molecules are known to have interesting aromas or tastes, only a few hundred are used regularly in flavours and still fewer are manufactured at large scale. The key compounds contributing to flavour in some of the foods are given in Table 3.

**Table 2:** Aroma notes and possible class of compounds.

| Aroma notes | Possible class of compounds |
|---|---|
| Fruity | Aliphatic short chain aldehydes and esters |
| Floral | Aromatic alcohol and esters |
| Green | Alcoholics, aldehydes, ketones, terpenes |
| Soily | Terpenes |
| Almond | Aromatic aldehydes, nitriles |
| Woody, musty | Aliphatic alcohols |
| Sugary | Furan derivatives |
| Sweet | Alcohols, thiols |
| Spicy | Terpenes |
| Camphoraceous | Camphor |
| Sulphurous | Sulphur compounds |

*Source:* 73

**Table 3:** Key compounds contributing to flavour in some foods.

| Food | Flavour component | Formula |
|---|---|---|
| Apple | Ethyl-2-methylbutyrate | 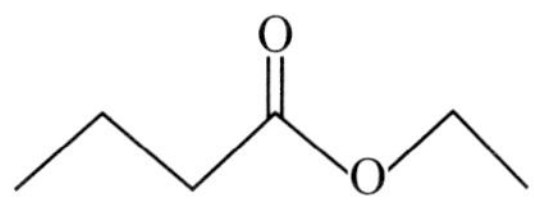 |
| Basmati rice | 2-Acetyl-1-pyrroline | 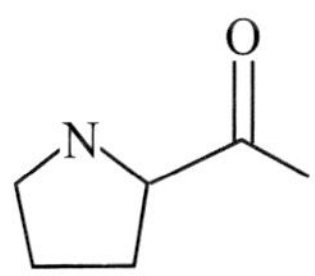 |
| Butter | Diacetyl | 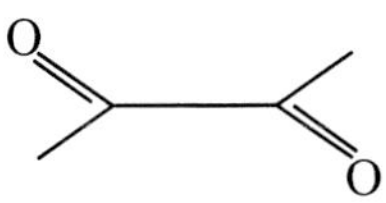 |
| Coffee | Furfuryl mercaptan | 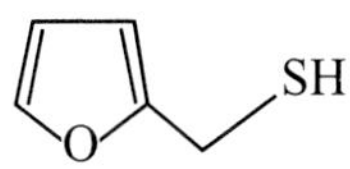 |
| Cucumber | Trans-2-cis-6 nonadinal | 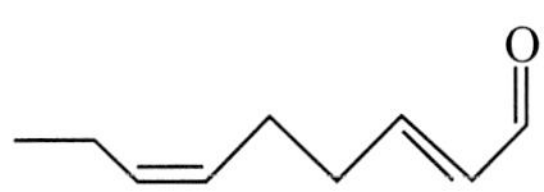 |
| Meat | 2,4,5-trimethyl-3-oxazoline | 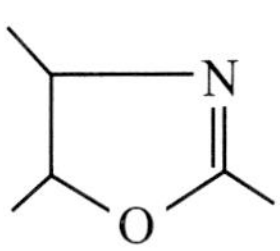 |
| Mushroom | 1-octane-3-ol | 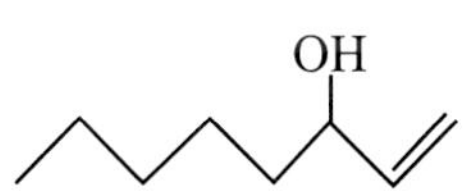 |

Contd...

Contd...

| Food | Flavour component | Formula |
|---|---|---|
| Popcorn | Methyl 2-pyridyl ketone | 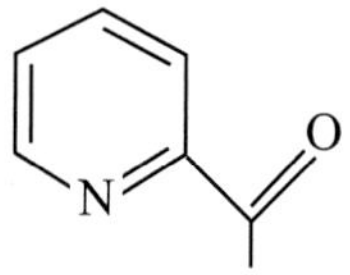 |
| Potato chips | 2,5-Dimethylpyrazine | 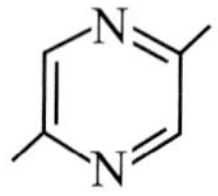 |
| Strawberry | Furaneol | 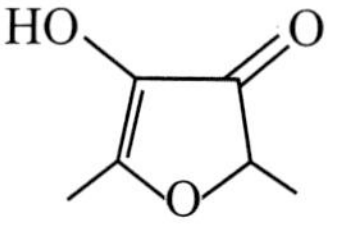 |
| Almond | Benzaldehyde | 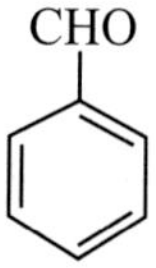 |
| Vanilla | Vanillin | 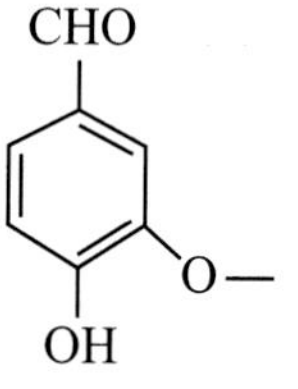 |
| Raspberry | 4-Phenyl-2-butanone | 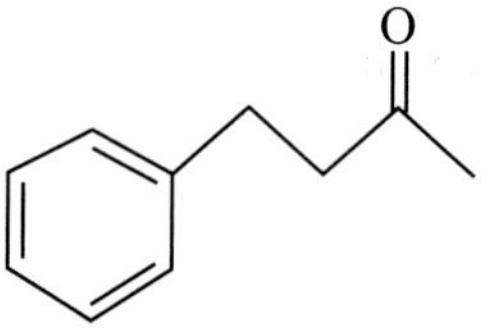 |
| Pineapple | 2,5-Dimethyl-4-hydroxy-3(2H)-furanone | 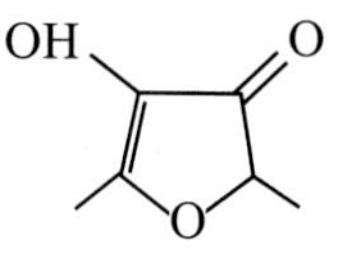 |
| | Nootkatone | 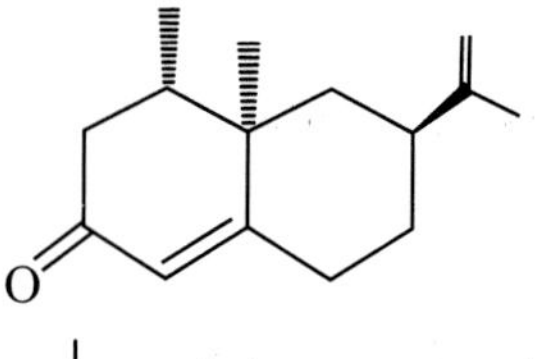 |
| Grapes | 1-p-Menthene-8-thiol | 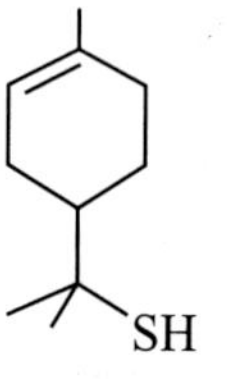 |

Contd...

Contd...

| Food | Flavour component | Formula |
|---|---|---|
| Orange | Limonene | 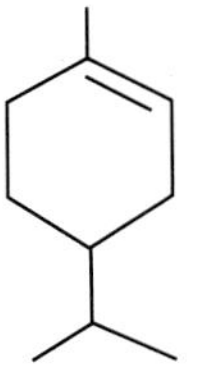 |
| Banana | Isoamylacetate | 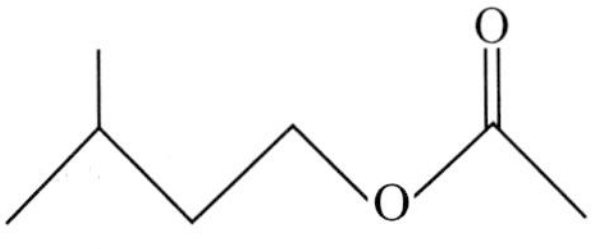 |
| Cinnamon | Cinnamic aldehyde | 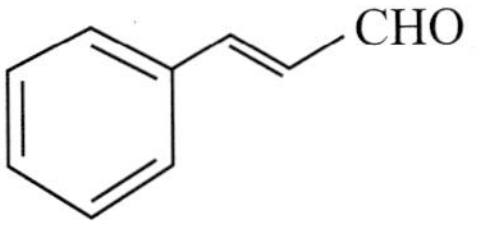 |
| Pear | Ethyl-(E,Z)-2,4-decadienoate | 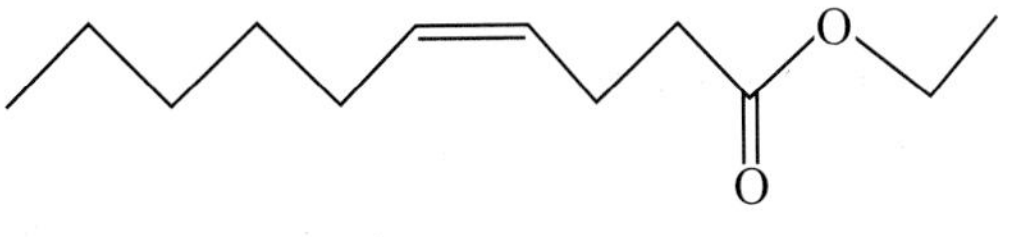 |

## 2. ORIGIN OF NATURAL FLAVOURANTS

Natural flavours are produced by biochemical processes in plants during growth, development, maturity and reproduction, like ripening of fruits and their storage or during the food processing operations, during which the physical, chemical or biological conditions imposed result in development of flavours, as in roasted food like coffee, cocoa, peanuts and popcorns (Fig. 1). Biogenesis of flavour in fruit and vegetables is genetically controlled, while flavour development during the processing operations depends upon the precursor and the condition applied. The aroma compounds formed as a result of catabolic reactions during the ripening phase of fruits are described in Table 4. Typical flavour of fruit develops during the ripening phase when there is an increase in the rate of respiration. Simple sugars such as glucose and fructose are catabolized to pyruvic acid, and the intermediates of protein hydrolysis form volatile compounds like keto acids, aldehydes and alcohols. Likewise, lipids are acted upon by lipases and lipo-oxygenases to form free fatty acids, which are further hydrolyzed to aldehydes, ketones and lactones. The specific interactions between the intermediates and end products of catabolism control the characteristic flavour of each fruit [9].

The mode of development of flavours in vegetables is slightly different than in fruits as the former do not have any ripening phase. Here, the flavours develop by enzymatic activation during the cellular disruption, which includes cutting and cooking operations. Thioglucosinolates and cysteine sulphoxides are the chief precursors to volatile sulphur flavours in vegetables.

The possible alternatives for the production of natural flavouring compounds are biotechnological methods and the use of biocatalysis (plant cell and tissue cultures, microorganisms and enzymes) for their synthesis. At the moment, the microbial processes seem to be most promising for the production of pure flavour compounds or complex mixtures. The bioflavours produced by microorganisms are referred to as microbial flavours.

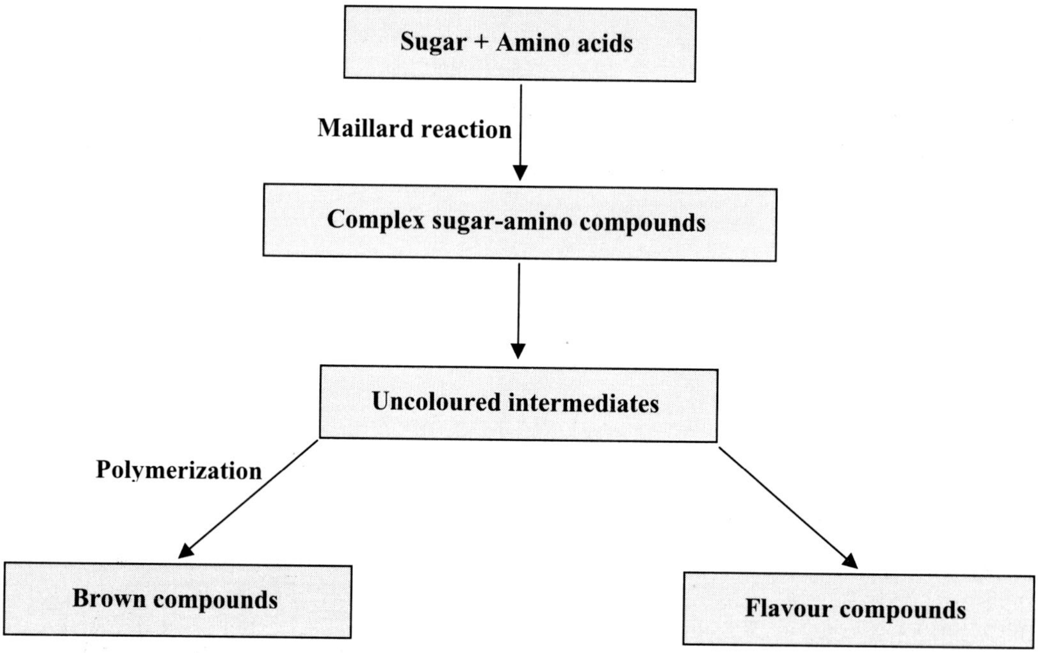


**Fig. 1:** Flavour formation during food processing operations.

**Table 4:** Aroma compounds produced by degradation of precursors in food.

| Nutrient | Aroma component |
|---|---|
| **Carbohydrates** | |
| Glucose | Organic acids |
| Fructose | Pyruvic acid, acetic acid, propionic acid, acetoacetic acid, butyric acid, hexanoic acid, |
| Sucrose | octanoic acid |
| | Esters |
| | Pyruvate, acetate, propionate, butyrate, acetoacetate, hexanoate, octanoate |
| | Alcohols |
| | Ethanol, propanol, butanol, hexanol, octanol |
| | Aldehydes |
| | Acetaldehyde, propanal, butanal, hexanal, octanal |
| | Terpenes |
| | Monoterpenes, linalool, limonene, α-pinene, citronella, citral, geranial |
| **Amino acids** | |
| Alanine | Pyruvic acid, acetaldehyde, alcohol |
| Valine | Isopropanal, isopropanol, α-ketoisobutyric acid |
| Leucine | 3-methyl butanal, 3-methyl butanol, α-ketoisocaproic acid |
| Isoleucine | 2-methyl butanal, 2-methyl butanol |
| Phenylalanine | Benzaldehyde, phenylacetaldehyde, cinnamaldehyde, hydrocinnamaldehyde, p-hydroxyphenyl acetaldehyde, p-hydroxyphenyl benzaldehyde, p-hydroxy cinnamaldehyde |
| Serine | Pyruvic acid |
| Threonine | Thiazoles |
| Glycine | Glyoxal |
| Cysteine/cystine | |

Contd...

Contd...

| Nutrient | Aroma component |
|---|---|
| **Fatty acids** | |
| Linoleic acid | Trans-2-trans-4-decadienal, hexanal, trans-2-octenal, trans-2-pentanal, trans-2-hexenol, cis-3-hexenal, cis-3-hexenol, trans-2-trans-4-heptadienal, propanal |
| **Organic acids** | |
| Citric acid | Glyoxylic acid |
| Oxaloacetic acid | Glyoxal |
| Malic acid | Pyruvic acid |
| Lactic acid | Acetaldehyde, Ethanol |
| **Vitamins** | |
| Carotenoids, β-carotene | β-ionone |

*Source:* 9

## 3. HISTORY OF MICROBIAL FLAVOURS

The microbial processes have played a significant role in the development of complex flavours since the advent of beer, wine, cheese, soy sauce, dry sausages, and related fermented products [10]. More than 150 years ago, benzaldehyde was the first flavour compound identified [11]. Isolation, identification and synthesis of vanillin marked the beginning of modern flavour industry [12, 13]. Replacement of classical methods of organic analysis with new chromatographic techniques facilitated the separation and structural elucidation of volatile compounds, and hence marked the beginning of increasing interest in use of microbes for the production of flavour compounds. Flavours are bioactive compounds and their chirality influences the odour perception, thus strengthening the use of biocatalysts (Fig. 2). Contemporary microbiological techniques, including genetic engineering, are now being increasingly realized to enhance the efficiency of biocatalyst. Advantages of biotechnological approach for production of flavour compounds include independence from climatic conditions, disease, application of fertilizers or pesticides and shortage due to local condition of production (climate, disease, pesticides, etc.), easy scale-up and easy product recovery.

Biotechnological processes involve conditions, which are less damaging to the environment than chemical processes and yield desirable enantiomeric flavour compounds [14]. Raspberry ketone (threshold: 1-10 parts per billion) is a characteristic flavour compound of raspberries. It could be produced chemically via condensation of p-benzaldehyde with acetone (Fig. 3a) or could be extracted from raspberries. Berries (1 kg) yield 3.7 mg of the ketone, thus posing economic restraints on commercialization due to high cost involved in the extraction procedure. However, the hydrolysis

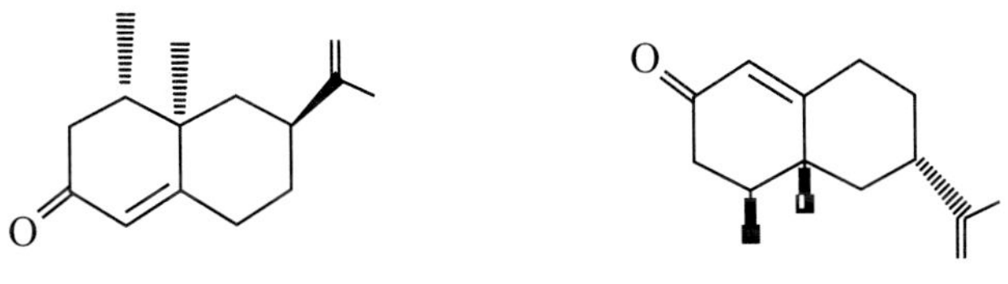


(–): Odour threshold 400-800 ppm (+): Odour threshold 0.6-1.0 ppm

**Fig. 2:** Effect of chirality on odour perception of nootkatone.

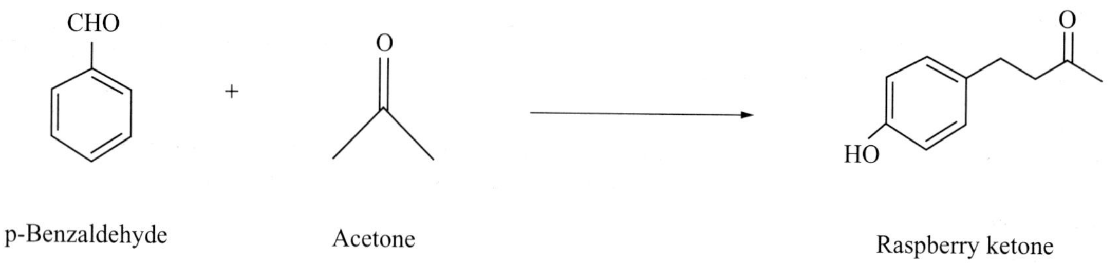


**(a) Chemical synthesis**

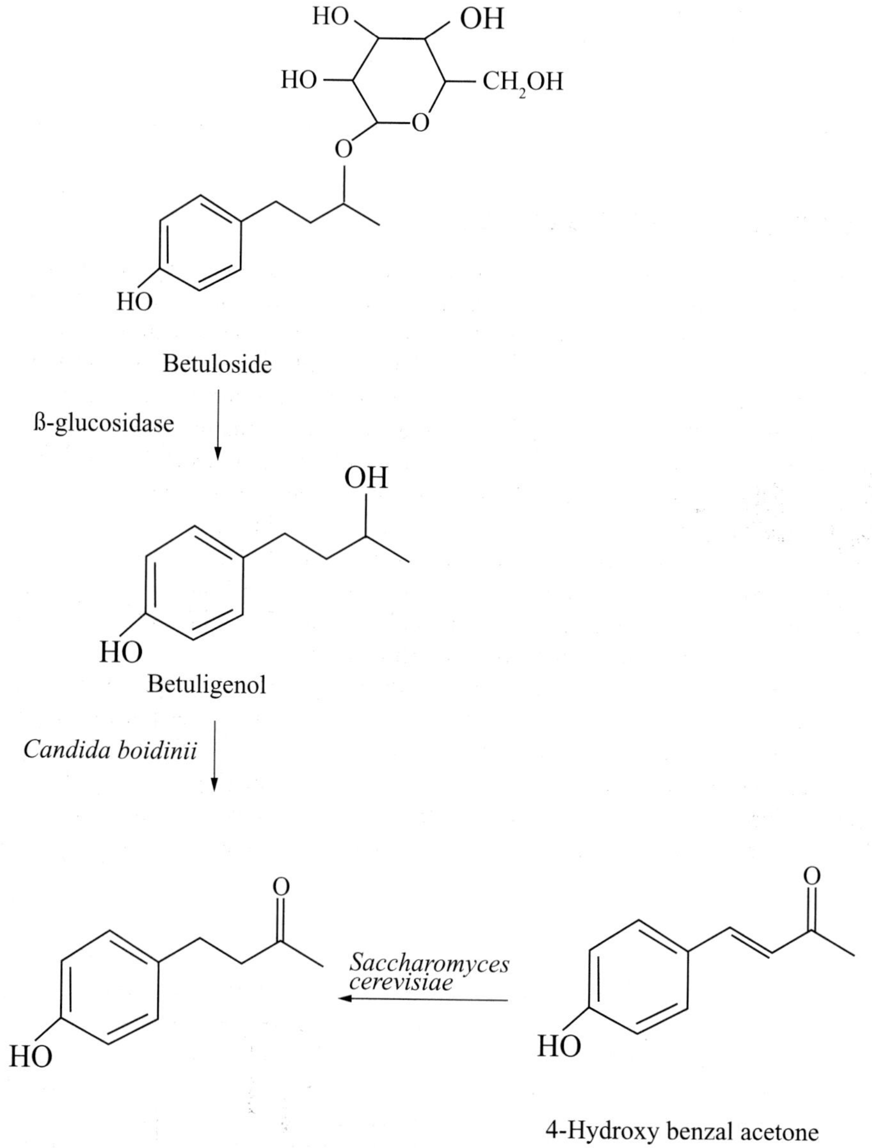


**(b) Biotransformation approach**

**Fig. 3:** Production of raspberry ketone.

of betuloside from *Rhododendron* using commercial β-glucosidase from *A. niger* yields betuligenol, which could be further converted into the ketone via secondary dehydrogenase of *Candida boidinii* (Fig. 3b).

## 4. TYPES OF FLAVOURS

The flavours can be classified variously based on the taste they provide, their source, extent of usage and chemical structure (Fig. 4).

### 4.1 Based on Taste

Flavours are the substances which provide any of the five basic attributes of taste i.e., sweet, salty, bitter, sour or savoury, and the flavours could be divided accordingly.

#### *4.1.1 Sour flavours*

The sour flavours in food are generated by hydronium ions, provided by organic acids such as citric, tartaric and acetic acids, while the phosphoric acid is the only inorganic acid used as sour flavourant in soft drink industry.

#### *4.1.2 Salty flavours*

The salty flavour in food is provided by low molecular weight inorganic salts like sodium chloride, potassium chloride, sodium bromide and sodium iodide. Taste is attributed by the ionic part of the salt, while the anionic character determines what taste would arise in addition to salty or otherwise mask the salty taste. Potassium chloride and potassium iodide are predominantly salty, while lead acetate and beryllium chloride salts are sweet in addition to being salty, potassium iodide and caesium chloride are bitter however, potassium bromide is salty and bitter.

#### *4.1.3 Sweet flavours*

The sweet flavour in food is imparted by carbohydrates, amino acids, polyalcohols (sorbitol, mannitol, xylitol) and synthetic sweeteners (aspartame, saccharin, cyclamate, etc.).

#### *4.1.4 Bitter flavours*

These flavours are usually perceived as disagreeable by the consumer. Bitter flavours are generally attributed by the presence of terpenes in food. Common bitter food and beverages include coffee, unsweetened chocolate, bitter melon, beer, uncured olive, citrus peel and many plants.

In a molecule, third position of a compound is responsible for its sweet or bitter taste i.e., in molecule AH/B-X, A & B are electronegative atoms, H is hydrogen atom and X, the remainder non-polar portion of the molecule is responsible for the sweet or bitter taste [15]. The sweet receptor on the tongue has an arrangement such that hydrogen bonds are formed between A and B, while X fits in a pocket in the receptor. If the distance between A and B is 2.5-4.0 Å, the compound tastes sweet, while for a bitter tasting compound, the distance between A and B lies between 1.0-1.5 Å [16].

#### *4.1.5 Savoury or umami flavours*

The word 'umami' in Japanese means 'delicious.' These flavourants are based on amino acids and nucleotides. The savoury flavourants, as recognized by European Union include glutamic acid, glycine, guanylic acid, inosinic acid and 5′ ribonucleotide salts. These savoury flavours, usually in conjunction with glutamic acid, are added in instant noodles, potato chips, snacks, savoury rice, tinned vegetable, cured meat and packet noodles.

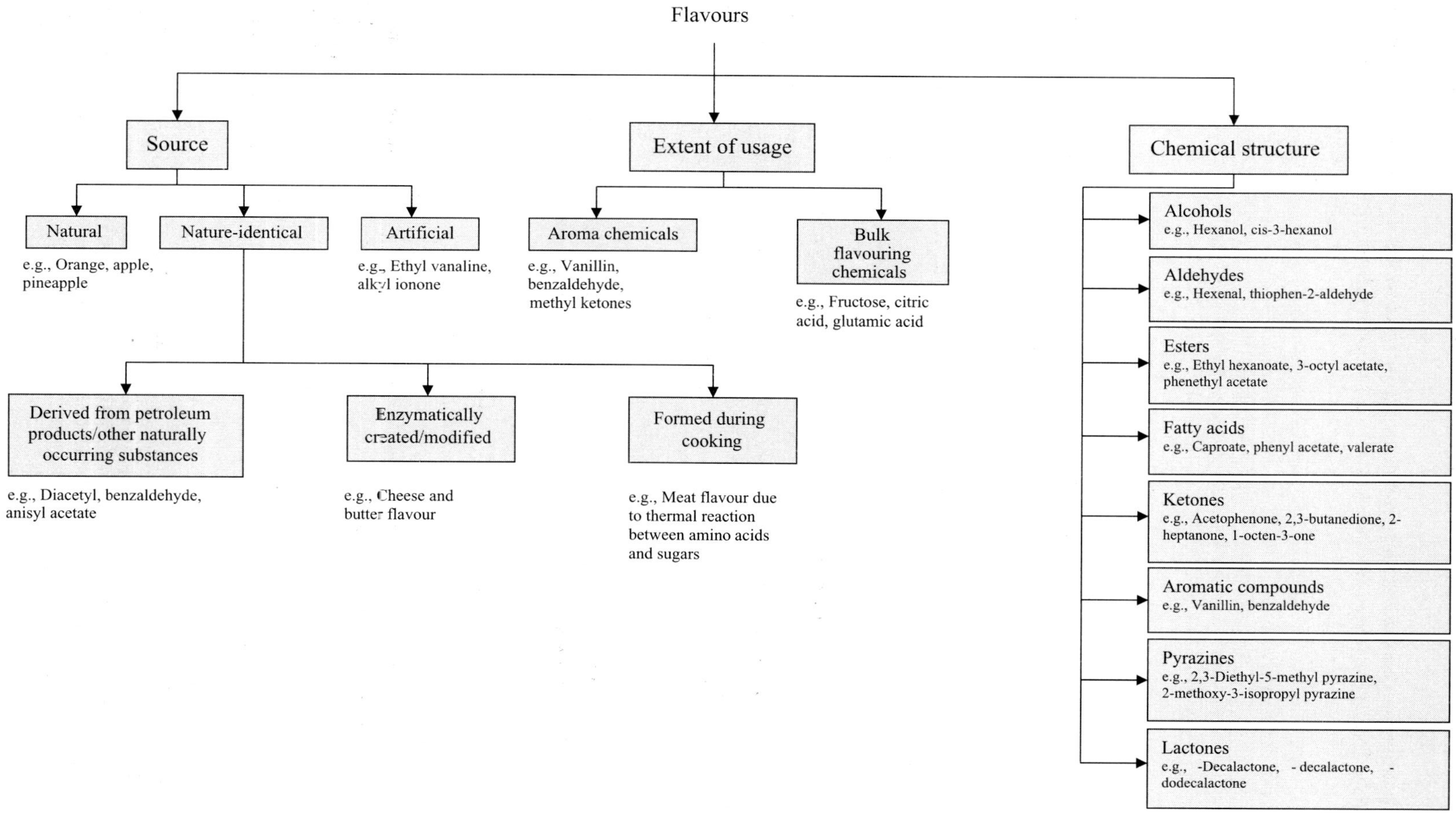


**Fig. 4:** Classification of flavour compounds.

## 4.2 Based on Source

The Ministry of Agriculture, Fisheries and Food, U.K., International Organisation of Flavour Industry, Switzerland and Codex Committee on Food Additives (FAO/WHO) has classified flavour compounds into natural, nature-identical and artificial.

### *4.2.1 Natural flavours*

These are substances obtained by appropriate physical processes (distillation and solvent extraction) or biotechnological processes (microbiological or enzymatic) from vegetables and fruits, either in natural state or processed for human consumption, e.g., orange, pineapple, apple flavours, etc.

### *4.2.2 Nature-identical flavours*

These are substances isolated from aromatic raw material or obtained synthetically. They are chemically identical to substances present in natural products, e.g., diacetyl and benzaldehyde.

### *4.2.3 Artificial flavours*

These are chemically synthesized compounds used to flavour food, but not yet identified in natural products. These are formulated with the same chemical compounds as present in natural flavours. Most of the artificial flavours are specific and complex mixtures of naturally occurring flavour compounds combined together to imitate or enhance a natural flavour, e.g., alkyl ionone, ethyl maltol and ethyl vanillin. Ethyl vanillin is a synthetic flavour, not present in vanilla pods, but it is 3.5 times more strong in flavour and more stable than natural vanillin.

## 4.3 Based on the Extent of Usage

Depending upon the extent of usage, flavour compounds can be grouped into bulk flavouring chemicals and specialty aroma chemicals.

### *4.3.1 Bulk flavouring chemicals*

These are used to provide basic taste attributes such as sweet, sour, bitter, salty or savoury. It includes sweeteners such as fructose, acidulants such as citric acid, malic acid and savoury compounds such as glutamic acid, which are produced by means of microbial biocatalysis, such as:

4.3.1.1 Fructose: It is a simple monosaccharide found in many foods and is amongst the three important blood sugars. It is the sweetest naturally occurring sugar, twice as sweet as sucrose and is present in honey, tree fruits, berries, melons and underground tubers as beet, sweet potatoes and onions. The free fructose present in honey and fruits accounts for their greater sweetness. It can be derived by degradation of sucrose, a disaccharide consisting of fructose and glucose, using hydrolase enzymes. It can also be synthesized from glucose by enzyme xylose isomerase (Fig. 5). The enzyme although prefers pentoses like xylose, but has activity high enough towards glucose making the conversion process economical. Owing to its low glycemic index, it is recommended to people with diabetes mellitus. Fructose as high fructose corn syrup is the most commonly used sweetener in soft drinks.

4.3.1.2 Citric acid: It is a tricarboxylic acid found in citrus fruits like lime and lemon, where it comprises 8% of the dry weight of fruit. It is the most widely used organic acid in the food industry as a sour flavourant. It gives tart, refreshing taste to many food and beverages. It enhances the taste of fruit flavours like lime, orange and berries. It was first commercially produced in 1923 by *Aspergillus niger*

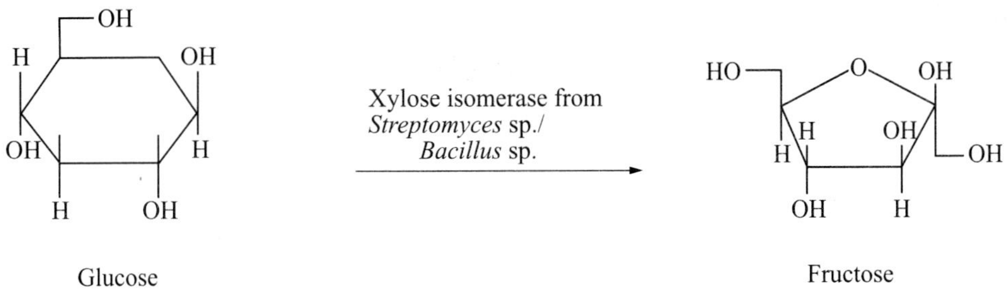


**Fig. 5:** Biotransformation of glucose to fructose.

fermenting glucose present in the medium to citric acid. It can be recovered from the fermented broth by precipitation as calcium salt followed by acidification and crystallization as free acid.

4.3.1.3 Malic acid: It is an α-hydroxy fruit acid having a mellow, smooth, persistent sourness and is often blended with multiple acids, sugars, high intensity sweeteners, flavours and seasonings to create a unique taste in food, beverages and confectioneries. It is the principal acid present in apples and many other fruits and vegetables. It is known for its flavour enhancing effect. It intensifies the impact of many flavours, thus reducing the amount of flavour needed and improves the after-flavour, thus extending the impact of some flavours. It boosts savoury flavours like cheese and also broadens the flavour profile of many products. It is used in beverages containing intense sweeteners to achieve desired sourness and flavour at higher pH. It is a preferred acidulant for still beverages like sports drinks, iced-tea, fortified juices, fruit nectars, etc. It is added to ciders to maintain consistent sharp taste. It gives tartness to candies and chewing gums and is also added to bakery products with fruit fillings.

4.3.1.4 Glutamic acid: It is a non-essential amino acid, present in a wide variety of food. It is one of the flavouring ingredients providing savoury character to the food and is also known to enhance the flavour of food like meat, fish and vegetables. It has the ability to induce starvation, which may be attributed to its role in flavour enhancement. Monosodium glutamate enhances the flavour and consistency of food. It is discovered as a major flavour constituent of Japanese seaweed. L-glutamic acid has been demonstrated as a non-volatile flavourant, along with guanosine 5' monophosphate in commercial mushroom concentrates from European countries [17]. It is mainly produced through fermentation.

### *4.3.2 Aroma chemicals*

These are the volatile organic compounds found in food and impart particular organoleptic characteristics to the food like fruity, floral and beefy. Some of the aroma compounds are mentioned below:

4.3.2.1 Vanillin: 4-hydroxy-3-methoxy benzaldehyde, obtained from a climbing orchid, *Vanilla fragrans*, a native tree of tropical American forests is the most widely used aromatic flavour compound in food, beverages, perfumes and pharmaceuticals. It develops in pods by the breakdown of glucosides in the curing process and contributes to about 2% (w/w) of the dry weight. Natural vanillin is priced at US $ 30/kg to US $ 120/kg, mainly due to limited availability of vanilla pods and labour intensive cultivation, pollination, harvesting and curing of vanilla pods [18]. Vanillin is also a metabolic intermediate in the biodegradation of natural products such as phenolic stilbenes, eugenol, ferulic acid and lignin. This provides an alternative approach for its commercial production. It can also be produced microbially by a fed batch fermentation employing strain *Amycolatopsis* sp. HR167, and 12g of vanillin has been reported per litre of the medium [19].

**4.3.2.2 Benzaldehyde:** It is the second most important flavour compound, next to vanillin and is the simplest aromatic aldehyde with a pleasant almond-like odour. It is the primary component of bitter almond oil extract and is present in cherry and other natural fruit flavours. It can be extracted from apricot, laurel leaves and peach seeds. Alternatively, it can be produced by biotransformation of phenylalanine using bacterium *Pseudomonas putida* and white rot fungi *Trametes suaveolens, Polyporus tuberaster* and *Phanerochaete chrysosporium*. It can also be synthesized from glycoside, amygdalin present in fruit kernels and nuts.

**4.3.2.3 Methyl ketones:** These are the primary components of cooked flavour associated with baked food, containing butter. The precursors to methyl ketones exist in fresh butter as alkanoic acids, which do not have flavouring characteristics. However upon heating, they are converted to methyl ketones which are the chief flavourants in heated and cooked food containing butter. Amongst this class of flavouring compounds, 2-heptanone, 2-nonanone and 2-decanone contribute to flavour in blue cheese and fruits. These are the largest contributors to stale flavour of UHT milk [20]. Coconut oil, rich in medium chain fatty acids, is a commercial raw material for the production of methyl ketones. These are produced by *Aspergillus bisporus, A. niger, Trichoderma viride* and *Penicillium roquefortii* [21].

**4.3.2.4 γ-Decalactone:** γ-decalactone, also called decan-4-olide, is an important lactone for flavour industry and is used as a fruit and dairy flavour. It has an extraordinarily tenacious odour and very powerful, creamy taste at a concentration below 5 parts per million. It is found in peach and used in peach, apricot, pear, maple, coconut, tropical, butterscotch and date flavours. It can be produced commercially by transformation of ricinoleic acid present in castor oil, the reaction being catalyzed by *Yarrowia lipolytica* or lipase from *Candida guillermondii* (Fig. 6). *Aspergillus niger, Pichia etchelisii* and *Cladosporium suaveolens* have been reported to produce γ-decalactone from castor oil [22]. Relatively lower product yields have been reported using other organisms such as *Monilia fructicola, Sporobolomyces odorus* and *Rhodotorula glutinis* and *Sporidiobolus* sp. [23]. In early 1980s, γ-decalactone was extremely expensive, priced at more than US $ 10,000/kg. Optimization of microbial processes has resulted in a price drop to US $ 300/kg [24].

**4.3.2.5 2-Phenyl ethanol:** It is an aromatic alcohol with a rose-like odour and is present in many essential oils. It is used as a flavourant in soft drinks, candies and cookies. Rose petals contain about 60% of this alcohol, but the natural extraction from rose petals bears economic constraints. In 2002, it was marketed at around US $ 1,000/kg. Although, certain species of *Corynebacterium* have been found to produce 2-phenyl ethanol [25]. It is being commercially produced using yeast strains of *Hansenula anomala, Kluyveromyces marxianus* and *Saccharomyces cerevisiae* [26, 27]. Yeasts are known to produce this alcohol by bioconversion of L-phenyl alanine through Ehrlich pathway (Fig. 7).

**4.3.2.6 Diacetyl:** Diacetyl (2,3-butanedione) is a diketone flavourant, providing a rich note in butter flavour and is used as an imitant of butter in dairy industry. It is a volatile yellow liquid ketone with a cheese-like smell. It is a characteristic flavour of buttermilk and sour cream. It imparts yellow colour to dairy products and a buttery flavour to popcorns, chips, candies and pasteries. Chardonnay wines contain a high concentration of diacetyl (0.005-1.7 mg/L) [28]. These are produced by bacteria such as *Lactococcus lactis, Lactobacillus* sp., *Streptococcus thermophilus* and *Leuconostoc mesenteroides*. Enzyme acetoin dehydrogenase present in lactic acid bacteria catalyzes dehydrogenation reactions in dough products forming diacetyl [29]. Genetic manipulation of gene encoding for diacetyl-acetoin reductase, involved in diacetyl production in lactic acid bacteria has been found to increase the diacetyl production in *Lactococcus lactis* [30].

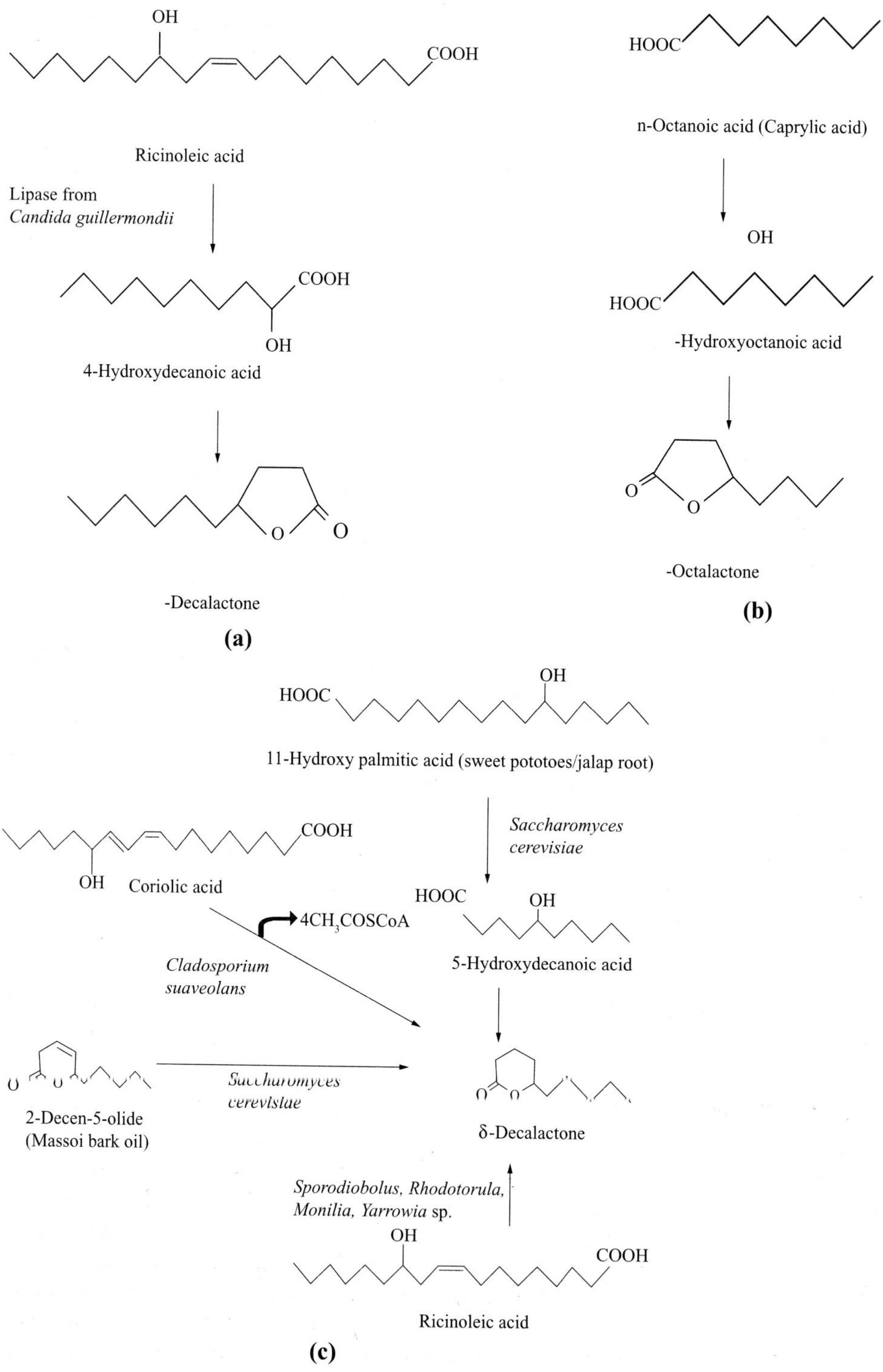


**Fig. 6:** Production of flavour lactones by bioconversion of fatty acids.

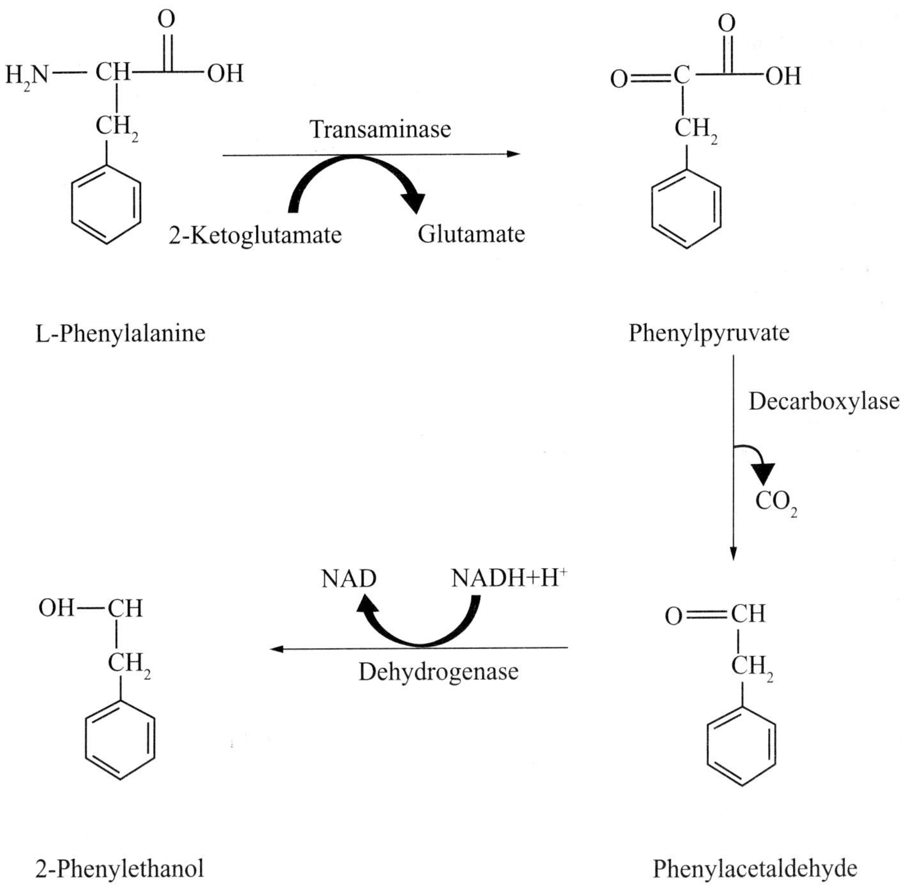


**Fig. 7:** Ehrlich pathway of yeasts converting amino acids to corresponding higher alcohols. *Source:* 24

4.3.2.7 Menthol: Menthol, a monoterpene, is a major constituent of peppermint or other mint oils and is one of the chief flavourants used extensively as a food additive in pharmaceuticals, cosmetics, toothpastes, chewing gums and candies [31]. Out of the seven possible isomers of menthol, (-)-(1R,3R,4S) isomer exhibits mint taste and gives a cooling sensation. (-)-menthol or L-menthol can be obtained by crystallization of *Mentha arvensis* oil upon freezing. It can also be produced using enantioselective enzymes [32].

## 4.4 Based on Chemical Structure

Based on their chemical structures, flavouring compounds can be classified as alcohols, aldehydes, esters, fatty acids, ketones, aromatic compounds, pyrazines and lactones.

### *4.4.1 Lactones*

These are cyclic esters of γ- and δ-hydroxy acids formed internally and are in equilibrium with the corresponding alcohols. These have a high flavour profile and contribute chiefly to fruity, coconut, buttery, creamy, sweet or nutty flavours, desirable in dairy products. Free lactones and their precursors are present in fresh butter, where they exist in the lipid phase. δ-lactones are also present in animal fats and auto-oxidized vegetable oils [33]. Upon heating, lactone precursors are converted to lactones and their total concentration increases, thus providing rich flavour notes associated with heated food containing butter fat. These impart perceptible fruity, sweet flavour to butter. These are major flavourants in candies and confectionaries, where they provide pleasant flavours. γ and δ-lactones are important flavours in many natural products. Although both chiral enantiomeric forms occur in nature, the "R" chiral forms tend to be predominant, especially as the alkyl chain length increases. δ-hexalactone or 5-methylpentanolide is a weak flavour, present in products such as butter, coconut, raspberry, strawberry and tea. Some of the other important flavour δ-lactones and their properties are described in

**Table 5:** δ-Lactones and their properties.

| Lactone | Structure | Sources | Taste |
|---|---|---|---|
| δ-Hexalactone (5-Methylpentanolide) | 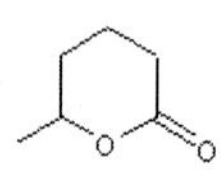 | Butter, Coconut, Raspberry, Strawberry, Tea | Coconut, cream, chocolate |
| δ-Octalactone (5-Propylpentanolide) | 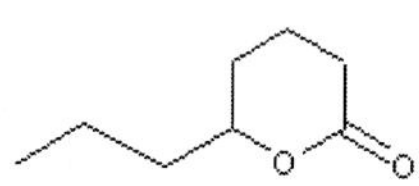 | Apricot, Blue Cheese, Burley Tobacco, Butter, Cheddar Cheese, Coconut, Cream, Pineapple, Raspberry, Strawberry, Yogurt | Sweet, creamy, fatty |
| δ-Nonalactone (5-Butylpentanolide) | 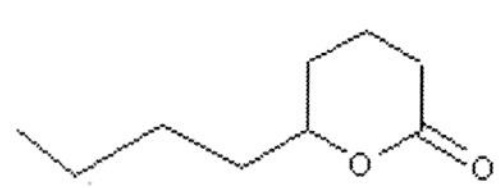 | Butter, Cooked Beef, Cooked Pork, Milk, Whiskey, Wine | Sweet, oily-fatty, coconut, fruity, dairy-creamy |
| δ-Decalactone (5-Pentylpentanolide) | 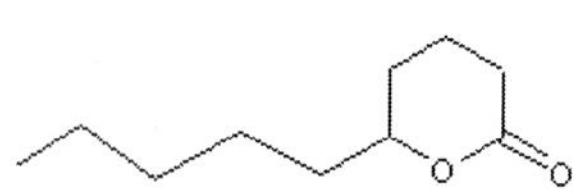 | Apricot, Blue Cheese, Burley Tobacco, Butter, Coconut, Mango, Peach, Raspberry, Rum, Strawberry, Tea | Creamy, coconut, lactonic, sweet and dairy-like with milky nuances |
| δ-Undecalactone (5-Hexylpentanolide) | 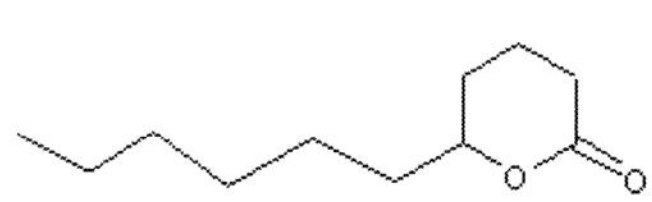 | Butter, Coconut, Cream, Milk | |
| δ-Dodecalactone (5-Heptylpentanolide) |  | Butter, Coconut, Cream, Milk | Sweet, creamy, lactonic, waxy, milky, coconut |
| Massoia lactone (5-Pentylpent-2-en-5-olide) | 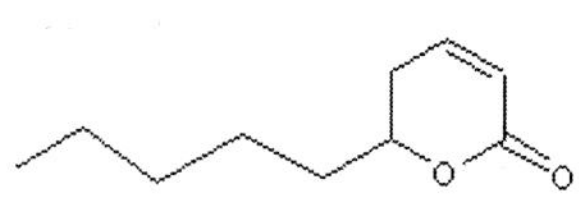 | Tobacco, Massoia oil, Osmanthus, Sugar Molasses, Tuberose, Wine | Creamy, coconut, milky, green and slightly fruity |
| Jasmine lactone (Z-Dec-7-en-5-olide) | 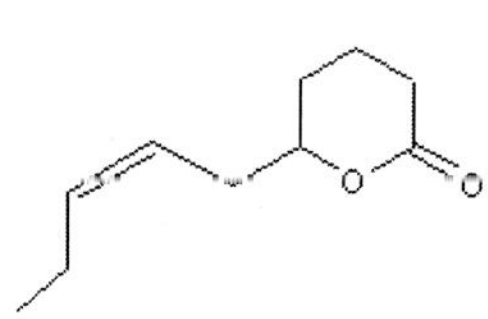 | Gardenia flowers, Jasmine, Osmanthus, Tea, Tuberose | Powerful, oily-fatty, fruity, Peach-Apricot |
| 6-Pentyl-alpha-pyrone (Deca-2,4-dien-5-olide) | 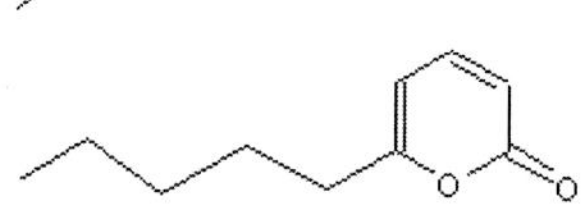 | Cooked Beef, Mushroom, Osmanthus, Peach | Coconut, creamy, sweet, brown with a fatty waxy |

Table 5. 6-pentyl-2-pyrone (6-PP), found as a major volatile constituent in *Trichoderma viride* cultures, exhibits a coconut aroma. Ricinoleic acid and linoleic acid can be transformed to coconut flavouring γ-decalactone (Fig. 6a). Octalactones have been formed by fermentation of n-octanoic acid (caprylic acid) present in coconut oil by *Mortinella* species [34]. The reaction is outlined in Fig. 6b. δ-decalactone can be produced from 11-hydroxy palmitic acid present in sweet potatoes by yeast species [35]. The various routes to δ-decalactone formation are illustrated in Fig. 6c.

### *4.4.2 Pyrazines*

These are heterocyclic, nitrogen-containing compounds contributing to nutty and roasted flavours. Methoxy pyrazines are grape derived flavours, which give characteristic herbaceous, bell pepper or earthy aroma to grape wines. 3-isobutyl-2-methoxy pyrazine and 3-isopropyl-2-methoxy pyrazine are important determinants of green flavours in sauvignon blanc wines [36]. 2,3,5-trimethyl pyrazine is used as a chocolate flavour enhancer, while 2-methyl-3-methoxy pyrazine enhances the flavour of nuts, 2-methyl-3-isobutyl pyrazine enhances the flavour of bell pepper and 2-acetyl-3-methoxy pyrazine is used as toasted corn flavour enhancer. These are formed through maillard reaction during the roasting of food. These are also synthesized by some microbes, such as bacteria *Corynebacterium glutamicum*, which produces tetramethyl pyrazine from amino acids. Some natural pyrazines produced by *Corynebacterium glutamicum* are described in Table 6.

**Table 6:** Natural pyrazines produced by *Corynebacterium glutamicum*.

| Pyrazine | Structure | Sensorial description |
|---|---|---|
| 2-Methoxy-3-isopropyl- | 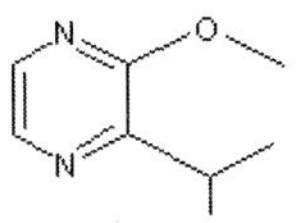 | Soil, pea, potato |
| Tetramethyl- | 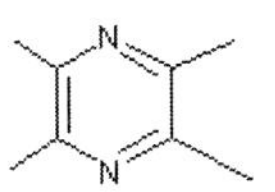 | Sharp |
| 2,5-Dimethyl- | 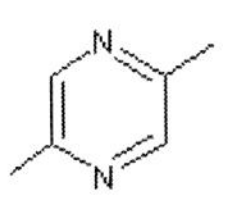 | Nutty, fatty |
| 2-Methoxy-3-isobutyl- | 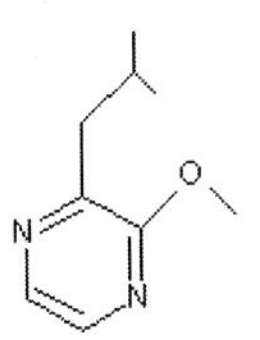 | Peppery |
| 2-Methoxy-3-methyl- | 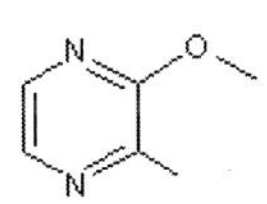 | Vegetable, popcorn, potato |
| 2-Methyl-6-ethoxy- | 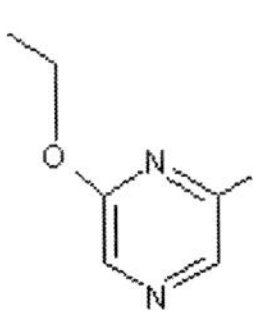 | Pineapple |

*Source:* 14

### *4.4.3 Esters*

Esters are class of organic compounds occurring naturally in plants and animals. These are common flavouring agents known for their fruity aroma. Ethyl butyrate is known for its pineapple flavour. These are used in beverages, jams, jellies, candies, baked products, wine and dairy products like yoghurt, sour cream and cheese. Ethyl, methyl, propyl, butyl, isobutyl, amyl and isoamyl esters are commonly used esters in food industry. Flavour of golden apples is attributed by hexyl-2-methylbutyrate. Methyl and ethyl cinnamates provide sweet, honey notes and are associated with strawberry flavour [37]. Ethyl esters are formed by esterification or interesterification reactions catalyzed by lipases (Fig. 8). In plants, esters are formed from amino acids or fatty acids by the action of lipases and esterases [38]. Acetate esters like isoamyl acetate and 2-phenyl ethyl acetate produced by yeasts *Hanseniaspora guillermondii* and *Pichia anomala*, respectively, are chief flavourants in grape-derived alcoholic beverages like wine [39].

### *4.4.4 Terpenes*

Terpenes have got their names from terpentine, fluid obtained by distillation of resins from certain trees. These are large and varied class of hydrocarbons, having general formula $(C_5H_8)_n$. Terpenes are referred to as monoterpenes if n = 2, sequiterpenes if n = 3, diterpenes if n = 4, triterpenes if n = 6 and tetraterpenes if n = 8. These are produced by a wide variety of plants, particularly conifers, though also produced by some insects like swallowtail butterflies, which secrete terpenes from their osmeterium. These are the primary constituents of essential oils in plants and are widely used as natural flavour additives for food in traditional and alternative medicines. These can be cyclic, open-chained, saturated or unsaturated. Wide range of natural derivatives of terpenes expands the variety of flavours as food additives, particularly the oxygenated terpenoids. Nerol, geraniol, citronellol and linalool are the most active bioflavourants. These terpenes ensure proper taste and aroma of wines, especially Muscat and Cabernet Sauvignon. Terpenes are commonly produced by fungi belonging to genus *Ceratocystis*. Some important terpenes produced by fungal species are enlisted in Table 7.

### *4.4.5 Ketones*

Ketones are the molecules containing the carbonyl group (=C=O) and are responsible for many natural flavours and odours. Methyl ketones, as already discussed, contribute an important class of ketone flavourants. Though a large number of saturated and unsaturated aliphatic, aromatic and cyclic ketones have been isolated from various cheese, odd numbered ($C_5$-$C_{11}$) 2-alkanones, along with free fatty acids and secondary alcohols give *Penicillium*-ripened cheese their distinctive flavour. A strong blue cheese flavour to dairy products is provided by non-8-en-2-one. Oct-1-en-3-one is responsible for metallic flavour in oxidized butter. Pineapple ketone (2,5-dimethyl-4-hydroxy-3(2H)-furanone) is

$$R_1\text{-COOH}+R_2\text{-OH} \xrightleftharpoons{\text{Esterification}} R_1\text{-CO-O-}R_2 + H_2O$$

$$R_1\text{-CO-O-}R_2+R_3\text{-OH} \xrightleftharpoons{\text{Transesterification}} R_1\text{-CO-O-}R_3 + R_2\text{-OH}$$

**Fig. 8:** Lipase-mediated esterification-transesterification reactions.

**Table 7:** Terpenes produced by the fungal species.

| Terpene | Structure | Sensoral description | Fungal species |
|---|---|---|---|
| Citronellol | 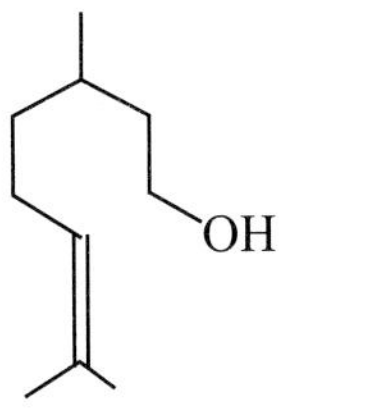 | Fresh, rose-like, bitter taste | *Ceratocystis variospora, C. moniliformis, C. virescens, Tramates odorata* |
| Geraniol | 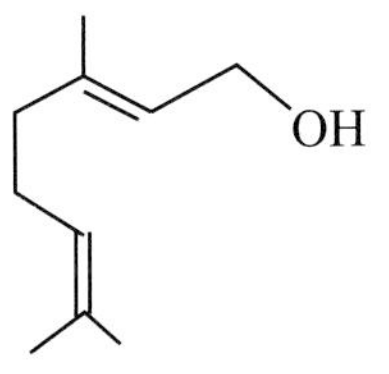 | Sweet, rose-like, fruity | *Ceratocystis variospora, C. moniliformis, C. virescens, Tramates odorata* |
| Linalool | 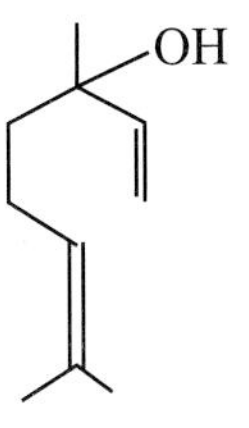 | Sweet, fresh, citrus, floral | *Ceratocystis variospora, C. moniliformis, C. virescens* |
| Nerol | 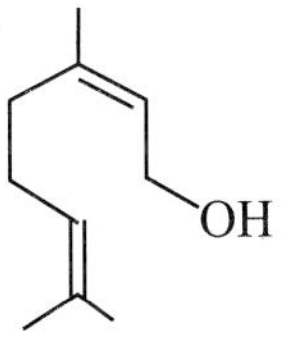 | Sweet, floral, rose-like, slightly bitter | *Ceratocystis variospora, C. moniliformis, Tramates odorata* |
| α-Terpineol | 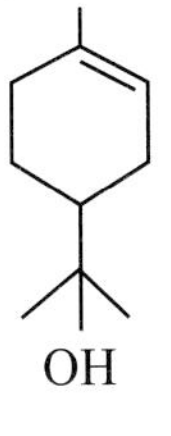 | Sweet, floral, fruity when diluted | *Ceratocystis moniliformis* |

*Source:* 14

found in pineapples, strawberries, raspberries and other food. It is found in cooked, roasted and fermented food such as coffee, roasted almond and soy sauce. It is formed by non-enzymatic browning during roasting and cooking operations. It has cotton-candy, caramelized-sugar flavour and is extensively used in food industry, particularly chewing gums.

### *4.4.6 Aldehydes and alcohols*

Aldehydes are chemical compounds containing –CHO group and constitute an important component of natural flavours. They have a low flavour threshold and produce desirable creamy, buttery flavours at low concentration, while at high concentrations, they lead to oxidized off-flavours. Vanillin, benzaldehyde, acetaldehyde, anisaldehyde, phenylacetaldehyde and heliotropic are some of the examples. Odour of citrus is due to aliphatic aldehydes like decanal and oxygenated terpenes such as terpineol and citral. *Cis* and *trans* deca-2,4-dienals impart a deep fat flavour to a variety of food such as cottonseed oil, soybean oil, milk fat, beef fat, cooked chicken and rye crisp bread. Benzaldehyde, major component of cherry and almond flavour, has been identified as secondary metabolite of microorganisms. Alcohol oxidase from *Pichia pastoris* oxidizes benzyl alcohol to benzaldehyde. *Streptococcus thermophilus* and *Lactobacillus bulgaricus*, responsible for yoghurt production possess enzyme threonine aldolase, which catalyses the conversion of threonine to acetaldehyde and glycine (Fig. 9). The most relevant compounds to obtain the "green" organoleptic characteristic are $C_6$ aldehydes and their corresponding alcohols [40]. Hexen-1-ol, also called leaf alcohol, has a strong odour of freshly cut grass and is an important flavourant for natural green notes. Bioproduction of green notes from linoleic acid and linolenic acid by enzymes lipooxygenases, hydroperoxide lyase and alcohol dehydrogenase is outlined in Fig. 10. In this bioconversion, the lyase reaction is rate-limiting. A novel strategy has been designed

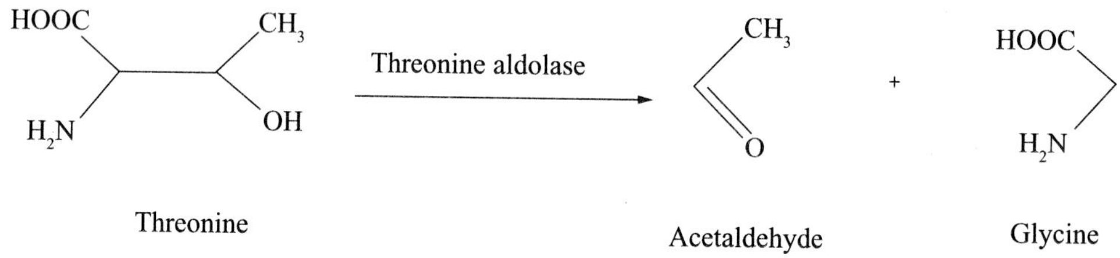


**Fig. 9:** Bioproduction of acetaldehyde from threonine using threonine aldolase from *S. thermophilus* and *L. bulgaricus*.

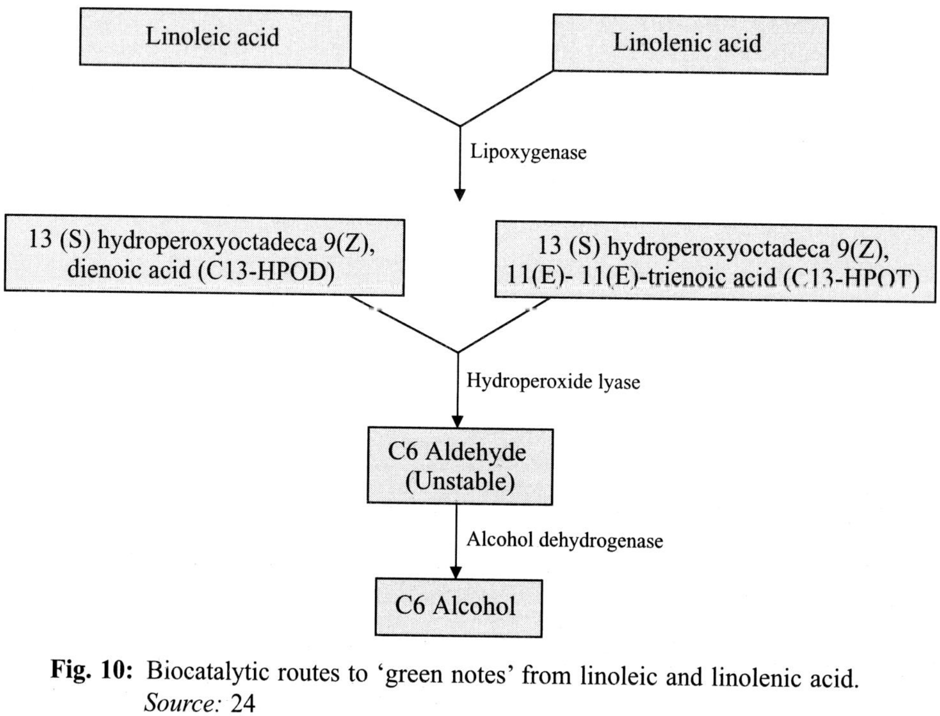


**Fig. 10:** Biocatalytic routes to 'green notes' from linoleic and linolenic acid. *Source:* 24

by recombinant DNA technology, in which *S. cerevisiae* has been genetically modified to co-express the hydroperoxide lyase gene from banana and the lipoxygenase gene [41, 42].

### *4.4.7 Fatty acids*

Fatty acids are carboxylic acids with long unbranched aliphatic tail, which is either saturated or unsaturated. Due to their sour taste and intense smell, they are of great importance in the flavour industry. These are produced by oxidation of corresponding alcohols and are inexpensive natural flavours. Upon heating of animal and vegetable oils in the presence of air, alkenoic acids are formed which impart flavour to the foodstuff. Of the various fatty acids, $C_1$-$C_{10}$ are important food flavourants. Short chain fatty acids have strongly pungent and irritating odours. As the molecular weight increases, flavour becomes more rancid, buttery and cheesy and then becomes more fatty [43]. Common acids include acetic, propionic, butyric, isobutyric, valeric, isovaleric and lactic acids. 2-methyl butyric acid is an important aroma constituent of cranberry. Methylpentenoic acid imparts a sweet, green, sharp, strawberry character to food. Propionic acid is an important flavour component of Swiss blue cheese. *Propionibacterium* has been reported to synthesize the propionic acid *de novo* [44]. A number of biotechnological processes have been developed for their production. Hex-2-enoic acid and hex-3-enoic acid are important constituents of raspberry flavour. Some of the acids produced by acetic acid bacteria and the characteristic flavours they impart are enlisted in Table 8. *Pseudomonas shermanii*, when co-cultured with *Lactobacillus casei* in whey, has been reported to produce 30 g/L propionic acid, 10 g/L acetic acid and 8 g/L lactic acid after 90 h of fermentation [45]. Besides, *Clostridium* sp. and *Peptostreptococcus anaerobius* convert leucine to isovaleric and isocaproic acids [46].

**Table 8:** Carboxylic acids produced by acetic acid bacteria.

| Carboxylic acid | Structure | Taste | Flavouring type | Bacteria |
|---|---|---|---|---|
| Propionic acid | 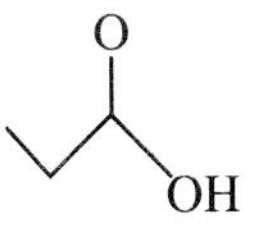 | Pungent, sour | Raspberry, strawberry, cognac, butter | *Gluconobacter oxydans*, *Acetobacter pasteurianus*, *Propionibacterium* |
| Butyric acid | 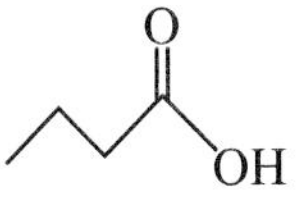 | Penetrating, diffusive sour, reminiscent of rancid butter | Butter, cheese, nuts, fruits | *Gluconobacter oxydans*, *Acetobacter pasteurianus* |
| Isobutyric acid | 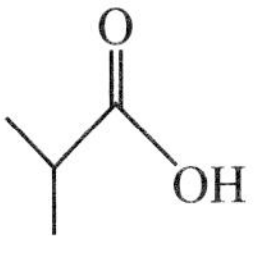 | Powerful, diffusive sour, in dilution pleasant and fruity | Cheese, fruit | *Gluconobacter oxydans* |
| 2-Methylbutyric acid | 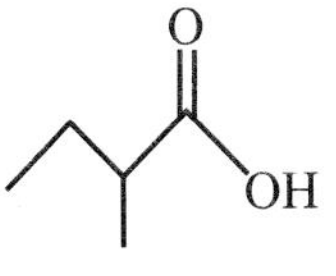 | Pungent, acrid, in dilution pleasant and fruity-sour | Cheese, butter, cram, chocolate | *Gluconobacter oxydans*, *Acetobacter pasteurianus* |
| Isovaleric acid | 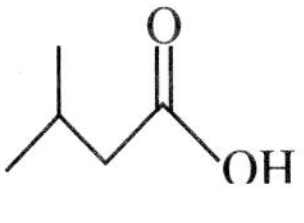 | Diffusive, acid-acrid, in dilution cheesy, unpleasant, at < 20 mg/L herbaceous, dry | Nut, coffee | *Gluconobacter oxydans*, *Acetobacter pasteurianus* |

*Source:* 24

## 5. APPROACHES FOR BIOPRODUCTION OF FLAVOURS

The natural flavours can either be directly extracted from plant material or can be produced biotechnologically. The extraction of natural flavours directly from plants is not much favoured due to low recovery of the desired compound, dependence on weather conditions and limitations imposed by spread of disease that makes the whole process expensive. Thus, this approach is not very promising. Biotechnological approaches outlined in Fig. 11 include production of flavour compounds by plant cell cultures, microbiological methods and enzymatic techniques.

### 5.1 Plant Cell Cultures

The production of bioflavours by plant cells *in vitro* exploits the idea of totipotency of plant cells i.e., ability of a somatic cell to develop into a whole plant. Plant tissue culture could be exploited in three ways for the production of useful products i.e., direct extraction of compounds from the cells/medium, biocatalysis using plant cells or enzymatic synthesis.

Valencene can be transformed by *Citrus* cells to nootkatone, a constituent of grape flavour [47]. In this reaction, 70% of the substrate could be transformed after about 6 h of incubation. Cocoa flavour can be produced from cultures of *Theobroma cacao* [48]. Vanillin flavour could be produced from ferulic acid using aerial root cultures of *Vanilla* [49]. Vanillin so formed could be recovered by adsorption onto activated charcoal and 400 mg/kg of vanillin could be obtained. Besides, ferulic acid could be transformed to vanillin by root cultures of *Capsicum frutescens* [50]. Other flavours produced by plant cell cultures are summarized in Table 9.

The plant cell cultures offer various advantages over the conventional extraction methods as independent of biological, seasonal and other environmental variations, rapid production, efficient recovery, high productivity and low cost. However, the method has certain limitations as large scale development of plant suspension cultures, sensitivity of plant tissue to shear stress, long growth cycles, progressive loss of biosynthetic activity and rare product secretion [21]. Optimization of cul-

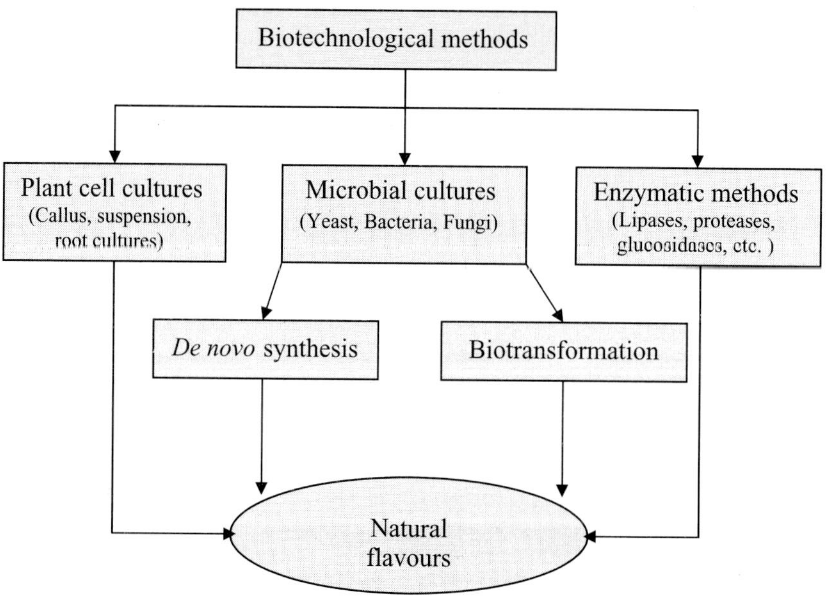


**Fig. 11:** Biotechnological processes for the production of flavour compounds.

**Table 9:** Flavour compounds produced by plant cell cultures.

| Flavour compound | Structure | Plant species | Culture type |
|---|---|---|---|
| 2,3-Butanedione | 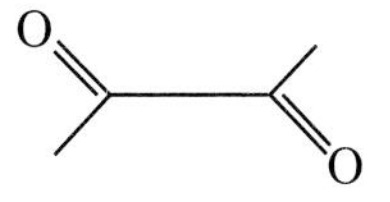 | *Agastache rugosa* | Suspension |
| Vanillin | 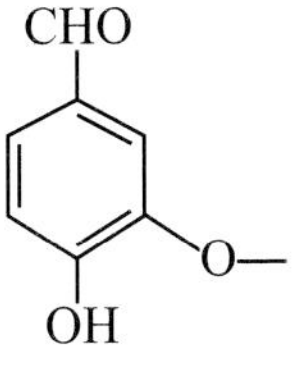 | *Vanilla planifolia* | Suspension |
| Apple aroma (Ethyl-2-methyl butyrate) | 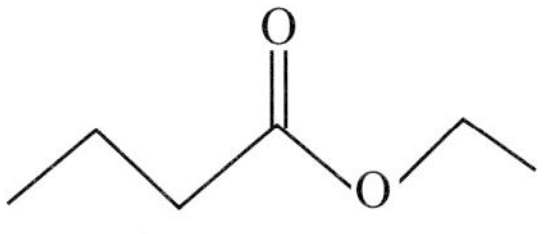 | *Malus sylvestris* | Suspension |
| Basmati flavour (2-acetyl-1-pyrroline) | 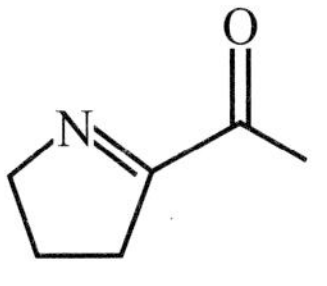 | *Oryza sativa* | Callus |
| Flavonol | 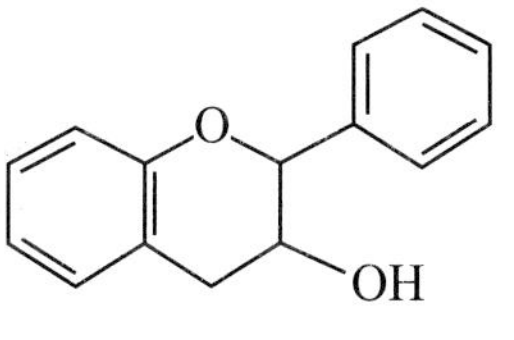 | *Polygonum hydropiper* | Suspension |
| Diallyl disulphide | 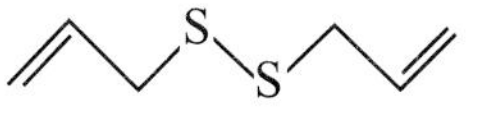 | *Allium cepa* | Callus |
| Limonene | 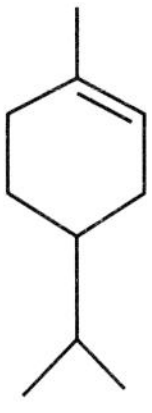 | *Perilla frutescens* | Callus |
| Glycyrrhizin | 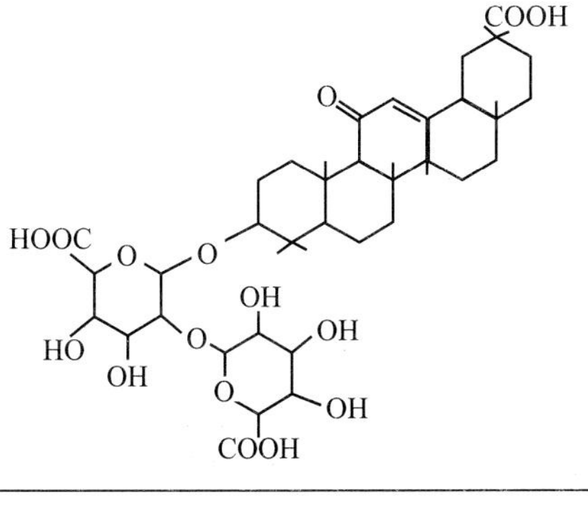 | *Glycyrrhiza glabra* | Suspension |

Contd...

Contd...

| Flavour compound | Structure | Plant species | Culture type |
|---|---|---|---|
| Cinnamic acid | 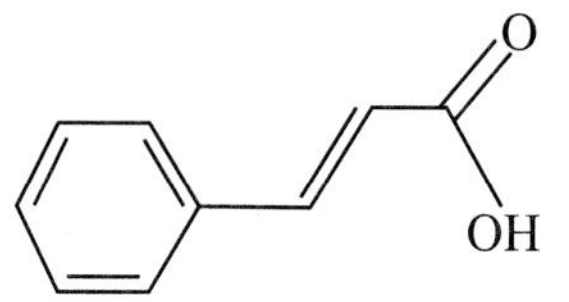 | *Nicotiana tobacum* | Callus |
| 4-Hydroxy-3-meth-oxybenzoic acid | 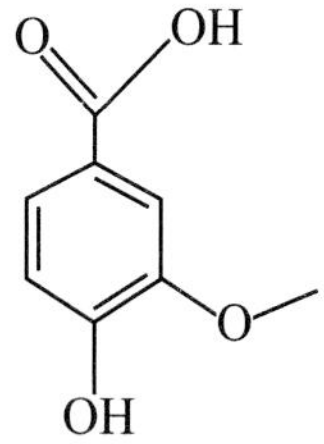 | *Linum usitatissimum* | Callus |
| 2-Undecanone |  | *Ruta graveolens* | Callus |
| Anethol | 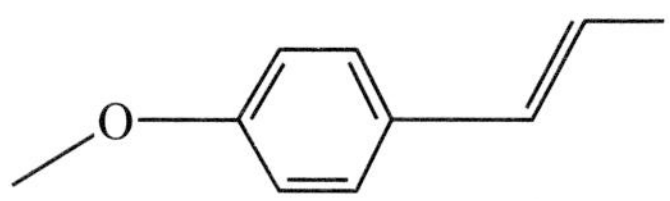 | *Foeniculum vulgare* | Callus |
| Farnesol | 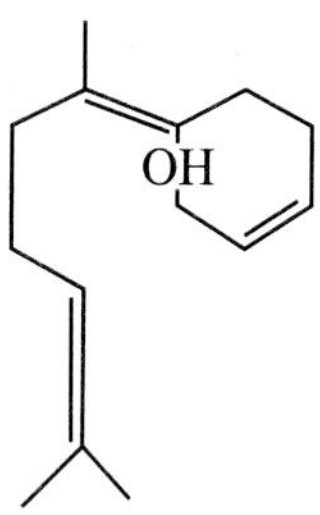 | *Andrographis paniculata* | Suspension |
| L-Glutamine | 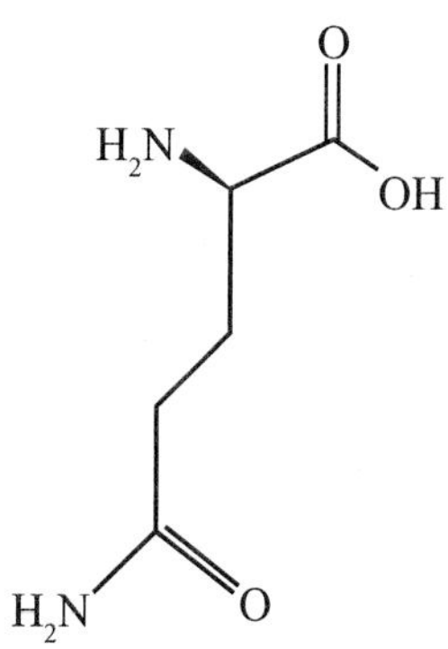 | *Symphytum officinale* | Suspension |
| 2-Phenyl ethanol | 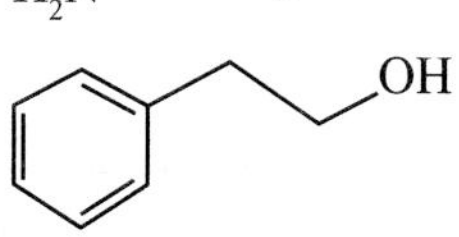 | *Symphytum officinale* | Suspension |

tural conditions, selection of high producing strains, precursor feeding and immobilization techniques have been employed to prolong the viability period of cells and reduce shear stress in cultures [51].

## 5.2 Microbial Cultures

The microorganisms can be used either for *de novo* synthesis of flavours i.e., production of flavour compounds by metabolizing cells or by biocatalysis i.e., the use of microbial cells to carry out specific chemical reactions.

### *5.2.1 De novo synthesis of flavour compounds by microorganisms*

Microorganisms produce a wide range of secondary metabolites of which flavour compounds constitute an important class. The typical flavour characteristics of most of the fermented food and beverages are attributed to a large number of microorganisms as in a special type of beer, Belgian gueze, where flavour characteristics depend upon mixed cultures [52]. Microbial pure cultures can also generate specific and complex aroma profile as in beers, where *Saccharomyces cerevisiae* produces the final complex aroma. Various microorganisms are capable of producing specific flavours *de novo*, when grown in a defined medium. These consist of representatives of various biological families.

5.2.1.1 Fungi: Several fungal species have been found to be able to produce complex flavours, including the floral flavours. Some of the fungal flavours are enlisted in Table 10. Several species of *Penicillium, Mucor* and *Oidium* produce fruity odours [53]. *Ceratocystis moniliformis* produces ethyl, propyl, isobutyl and isoamyl acetates and terpenes- citronellol and geraniol. The strain also produces τ-δ-decalactones having fruity banana, peach and citrus aroma [54]. Fungi of genus *Phellinus* produce ethyl and methyl benzoates and methyl salicylate [55]. Mucorales (*M. racemosus, M. piriformis*) impart a bitter almond flavour to butter. *Ceratocystis* species produce a variety of terpenes with floral/fruity odours [56]. Benzyl alcohol and 2-phenethyl alcohol metabolites are produced in the fermentation medium of *Trichothecium roseum, Phellinus ignarius, P. laevigatus* and *P. tremulae* [57]. Mushroom flavours have been found in cultures of *Aspergillus* and *Penicillium* species. *Penicillium caseiocolum* and *P. camberti* produce flavour alcohols like 1-octen-3-ol, oct-1,5-dien-3-ol and octan-3-ol [58]. *Tyromyces sambuceus* and *Cladosporium suaveolens* generate coconut-flavoured γ-decalactone and δ-dodecalactone from ricinoleic acid and linoleic acid, respectively [59, 60]. Methyl ketones can be produced by *Agaricus bisporus, Aspergillus niger, Penicillium roquefortii* and *Trichoderma viride* using coconut fat as substrate [61]. γ-decalactones have been produced from ethyl decanoate by *Mucor circillenoides* [62]. *Trichoderma viride* and *T. harzianum* produce coconut-flavoured lactone, 6-pentyl-α-pyrone [42]. *Marasmius alliaceus* has been reported to produce sulphur-containing volatile compounds imparting the sulphurous flavours [63]. Acetaldehyde and 3-methylbutanol can be produced by *Rhizopus oryzae* when grown on agro-industrial waste [64]. Pineapple aroma can be produced by growing *Ceratocystis fimbriata* on solid substrate such as coffee husk [65]. However, if leucine is added to the culture medium, a strong banana odour is produced. *Nidula niveotomentosa* produces raspberry ketone in a culture medium supplemented with phenylalanine [66].

5.2.1.2 Yeast: Pellicle forming yeast have been found to play an important role in the generation of aroma in various products [53]. Food flavourants produced by yeast strains are enlisted in Table 11. *Sporobolomyces odorus* has been reported to produce peach-smelling γ-lactones in cultures [67]. *Kluyveromyces lactis* produces terpenes as citronellol, linalool and geraniol [68]. γ-decalactones are formed by *Candida* sp., *Yarrowia lipolytica, Monilia fructicola, Sporobolomyces odorus* and *Rhodotorula glutinis* using castor oil as substrate [23]. *Geotrichum klebahnii* produces pleasant fruity flavoured ethyl esters of branched carboxylic acids [61]. Another species *Geotrichum fragrans* forms ethylisovalerates in cultures

**Table 10:** Flavour compounds generated by various fungal species.

| Flavour compound | Structures | Microbial species |
|---|---|---|
| Vanillin | 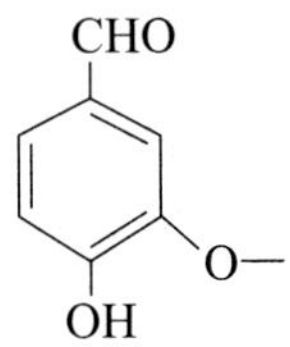 | *Pycnoporus cinnabarinus* |
| Benzaldehyde | 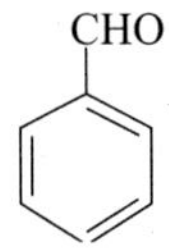 | *Ischnoderma benzoinum* |
| Methyl anthranilate | 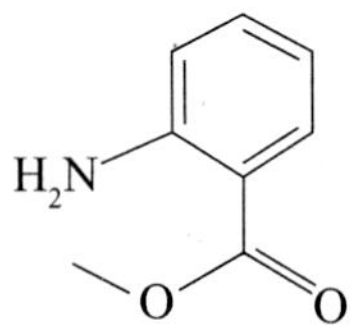 | *Pycnoporus cinnabarinus, Trametes* sp. |
| 4-(4-Hydroxyphenyl)-2-butanone | 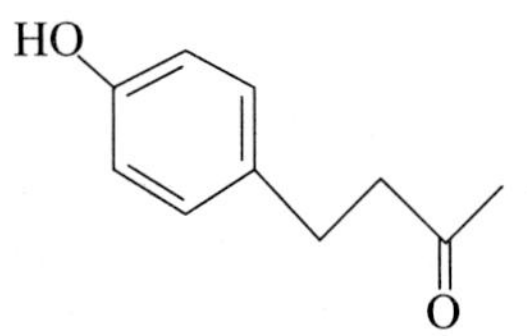 | *Nidula niveo-tomentosa* |
| Methyl salicylate | 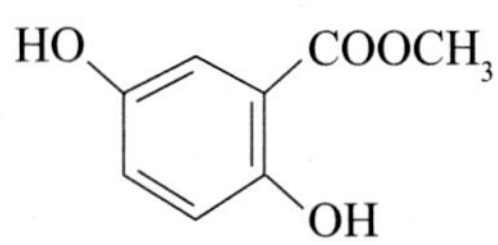 | *Phellinus* sp. |
| Methyl benzoate | 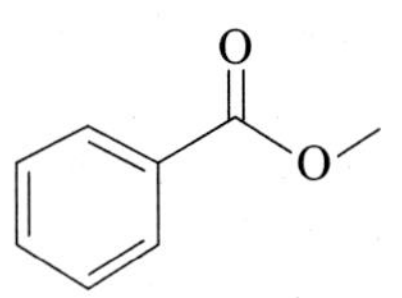 | *Polyporus tuberaster, Phellinus* sp. |
| Ethyl benzoate | 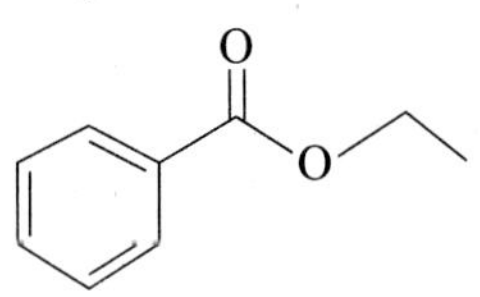 | *Polyporus tuberaster, Phellinus* sp. |
| 2-Phenylethanol | 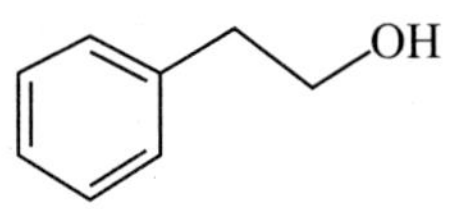 | *Ascoidea hylecoeti* |
| Coumarins | 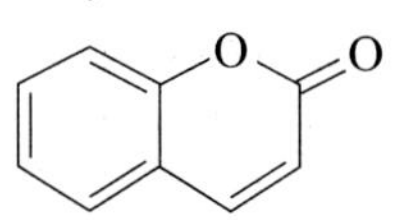 | *Pleurotus euosmus* |
| Methyl ketones | 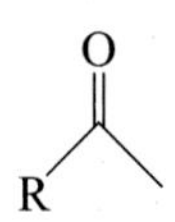 | *Aspergillus niger, Penicillium* sp., *Aureobasidium pullulans* |

Contd...

Contd...

| **Flavour compound** | **Structures** | **Microbial species** |
|---|---|---|
| Pyrazines | 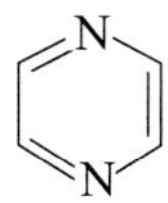 | *Aspergillus* sp. |
| Citronellol | 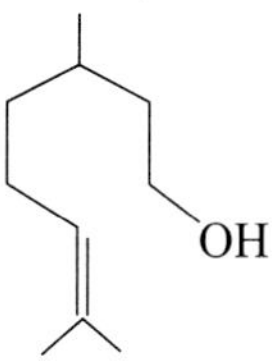 | *Mycena pura* |
| Linalool | 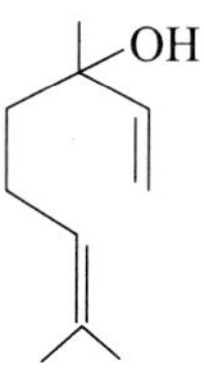 | *Wolfiporia cocos* |
| 1-Octen-3-ol |  | *Lentinus edodes, Grifola frondosa, Pleurotus pulmonarius* |
| 1-Octen-3-one |  | |
| 6-Pentyl-α-pyrone (6-PP) | 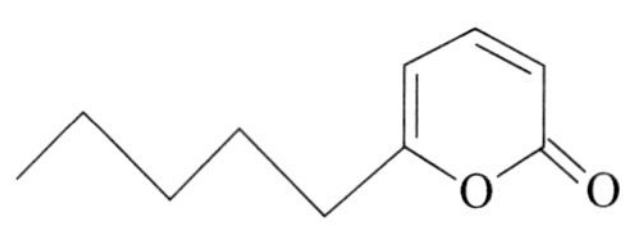 | *Trichoderma viride, Trichoderma harzianium* |
| 4-Methoxy benzaldehyde | 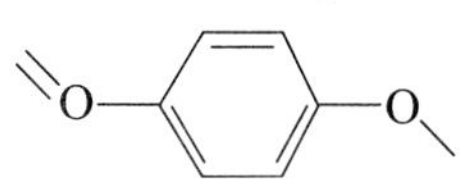 | *Ischnoderma benzoinum* |
| β-Ionone | 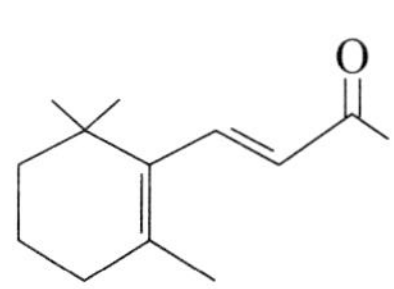 | *Gyromitra esculenta, Clitocybe geotrapa, Fibroporia vaillanti, Ischnoderma benzoinum* |
| Esters | 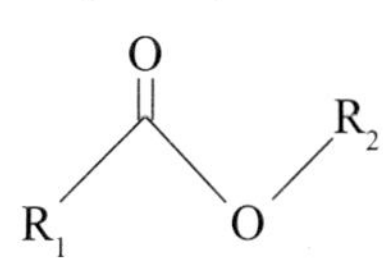 | *Rhizopus arrhizus* |
| γ-Decalactone | 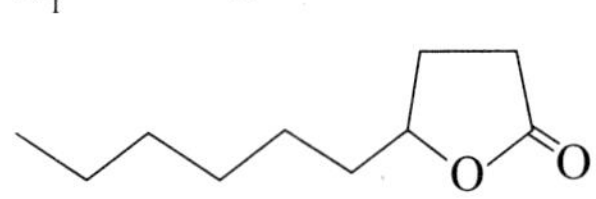 | *Penicillium* sp., *Polyporus odorus, Fusarium* sp., *Sporobolomyces* sp., *Aspergillus oryzae, Geotrichum klebahnii,* etc. |
| Raspberry ketone | 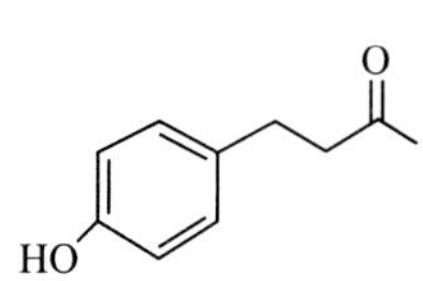 | *Nidula niveo-tomentosa* |

**Table 11:** Flavour compounds produced by yeast cells.

| **Flavour Compound** | **Structure** | **Yeast species** |
|---|---|---|
| Furaneol | 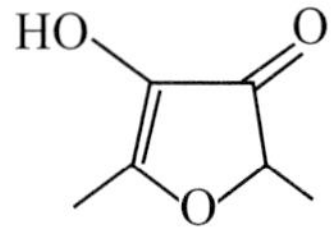 | *Zygosaccharomyces rouxii* |
| γ-Decalactone | 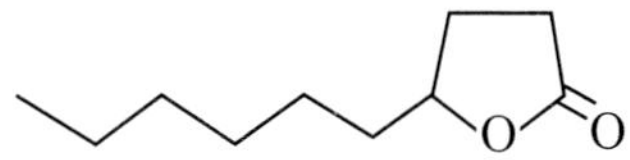 | *Candida, Yarrowia lipolytica, Monilia fructicola, Rhodotorula glutinis, Sporobolomyces odorus* |
| 2-Phenyl ethanol | 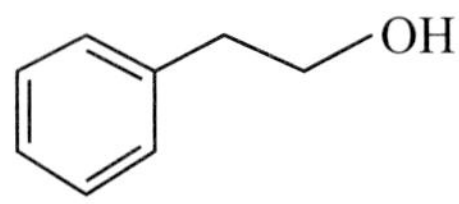 | *Kluyveromyces* sp. |
| Pyrazines | 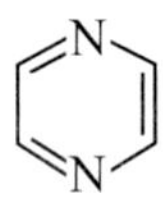 | *Kluyveromyces lactis* |
| Ethyl isovalerate | 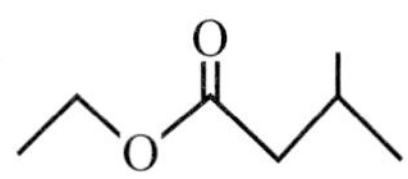 | *Geotrichum fragrans* |
| Ethyl esters | 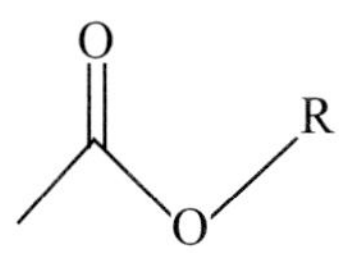 | *Geotrichum klebahnii* |
| Acetate esters | 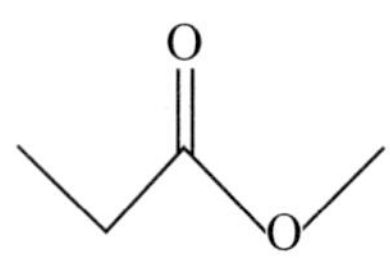 | *Hanseniaspora guillermondii, Pichia anomala* |
| Acetaldehyde | 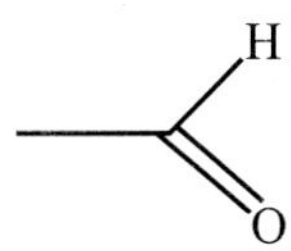 | *Candida utilis* |
| 3-Methyl-butyl acetate | 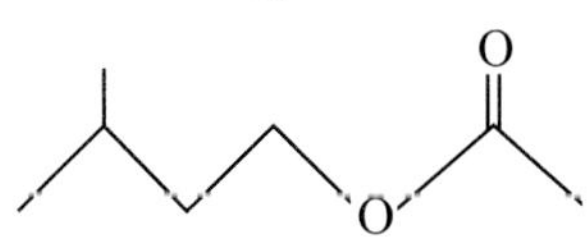 | *Williopsis saturnus* var. *mrakii* |
| Ethyl acetate | 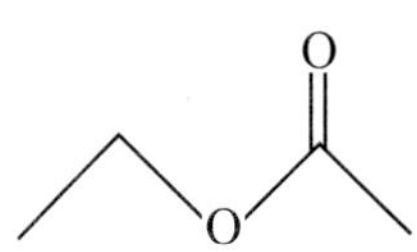 | *Saccharomycs cerevisiae* |

when supplemented with ethanol (Fig. 12). *Kluyveromyces marxianus* produces alcohol esters and aldehydes when grown on cassava bagasse [69]. *Hanseniaspora guillermondii* and *Pichia anomala* are potent producers of acetate esters [39]. *K. marxianus* also produces 2-phenylethanol, *Saccharomyces cerevisiae* synthesizes ethyl acetate and *Candida utilis* produces acetaldehyde when grown on ethanol [25].

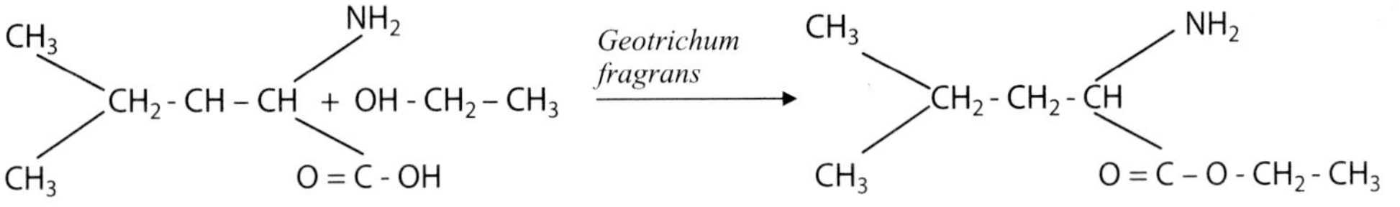

**Fig. 12:** Bioconversion of L-leucine into ethylisovalerate.

5.2.1.3 Bacteria: The major food flavourants produced by bacteria are given in Table 12. *Acetobacter* species produce methyl-butyric acid, which is a precursor to common flavour esters. Acetic acid bacteria form esters of ethyl acetate, while *Lactobacillus lactis-acidi* and *Streptococcus hollandicus* impart aromatic properties to milk and milk products. Butyric acid bacteria, *Clostridium pasteurianum* when grown in presence of oxygen, produce a complex ester aroma resembling the apple odour [54]. *Corynebacterium glutamicum* produces pyrazines in cultures [25]. High levels of propionic acids are produced by *Propionibacterium* sp. Butyric acid is primarily produced by obligate anaerobes of genus *Clostridium, Butyrivibrio, Eubacterium* and *Fusarium. Clostridium* and *Peptostreptococcus anaerobius* when co-cultured produce high levels of isovaleric acid [46]. High levels of butanol and isopropanol have been reported from batch fermentation of *Clostridium beijerinckii* [70]. Mutants of *Zymomonas mobilis* have been reported to produce acetaldehyde [71]. *Bacillus subtilis* produces pyrazine when grown on soybean substrate [72]. *Streptococcus* species produce various diketone flavours [73]. *Pseudomonas* sp. HR 199 carrying the genes encoding for enzymes responsible for bioconversion of eugenol to vanillin has been genetically engineered [74]. Benzaldehyde can be produced from benzoyl formate by whole cells of *Pseudomonas putida* encapsulated in calcium alginate [75]. *Lactobacillus acidophilus* and *Pediococcus pentosaceus* produce dairy flavours when grown on maize-based semi-solid medium [76].

### *5.2.2 Biotransformation approaches to flavour production*

The specific substrates when added to the fermentation medium can be transformed into flavour compounds by microbial cells. The commercialization of this approach is limited due to high cost of the process and low yields [77]. Lactones, vanillin and various esters have been produced successfully by the bioconversion approach.

5.2.2.1 Production of vanillin by biotransformation: Priefert and co-workers [18] have reviewed the various biotransformation approaches to vanillin synthesis. Biotransformation routes to vanillin synthesis are given in Fig. 13. Vanillin can be synthesized from phenolic stilbenes like isorhapontin, commonly found in spruce bark and isoeugenol by enzyme dioxygenase present in *Pseudomonas* species. Another approach for vanillin synthesis is deamination of m-methoxytyrosine using *Proteus vulgaris* and then aldehyde group formation by treatment with mild base. Clove oil contains eugenol, which can be converted to ferulic acid and then to vanillin by *Corynebacterium, Pseudomonas* and *Arthrobacter* strains. Biotransformation of eugenol to vanillin can also be achieved by *Fusarium solani* [78]. A much cost effective method for vanillin synthesis employs maize and wheat-based agricultural wastes (beet pulp and cereal bran), which can be converted to ferulic acid using polysaccharide-degrading enzymes and feruloyl esterases [79, 80]. Ferulic acid can then be converted to vanillin by the action of *Pycnoporus cinnabarinus* or via a two step fungal process in which *A. niger* transforms ferulic acid into vanillic acid, which is further converted to vanillin by basidiomycete, *Pycnoporus cinnabarinus* or *Phanerochaete chrysosporium* [14]. Biotransformation of vanillin can also be achieved by incubating aerial roots of vanilla plant tissue with ferulic acid, as already discussed in Section 5.1.

**Table 12:** Flavour compounds produced by bacteria.

| Flavour compound | Structure | Bacterial species |
|---|---|---|
| Diacetyl | 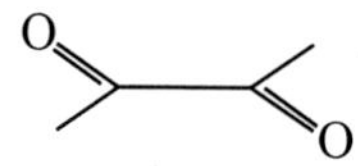 | *Lactobacillus lactis, Streptococcus diacetylactis* |
| Short chain fatty acids | 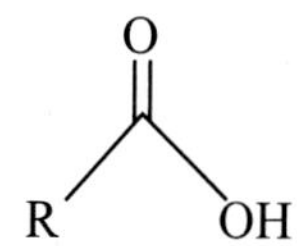 | *Acetobacter aceti, Gluconobacter oxydans, Propionibacterium* sp., *Clostridium* sp., *Fusarium* sp. |
| Methyl ketones | 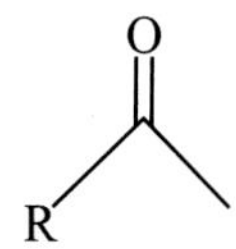 | *Pseudomonas oleovorans* |
| Geosmin | 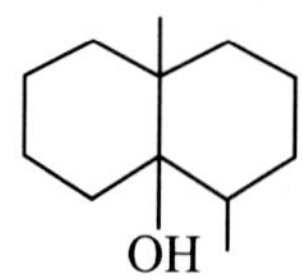 | *Streptomyces citreus* |
| Pyrazines | 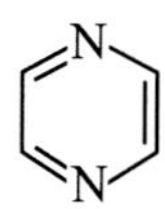 | *Bacillus* sp., *Penicillium* sp., *Pseudomonas* sp., *Corynebacterium* sp. |
| 2-Acetyl-1-pyrroline | 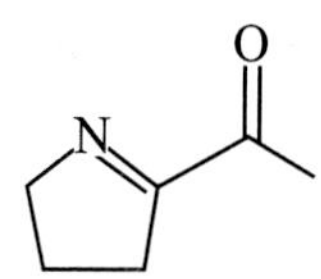 | *Bacillus cereus* |
| Nootkatone | 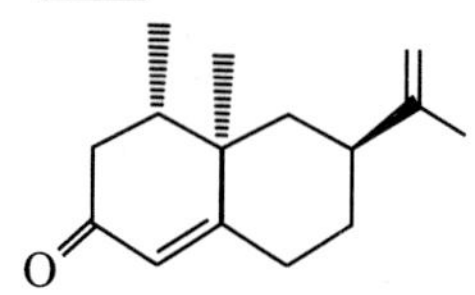 | Soil bacteria, *Enterobacteriaceae* |
| Borneol | 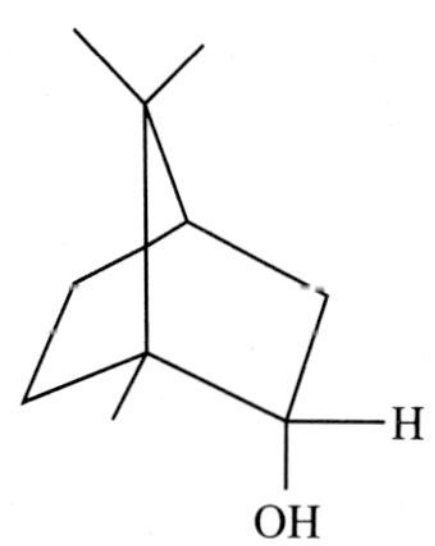 | *Pseudomonas pseudomallei* |
| Isoborneol | 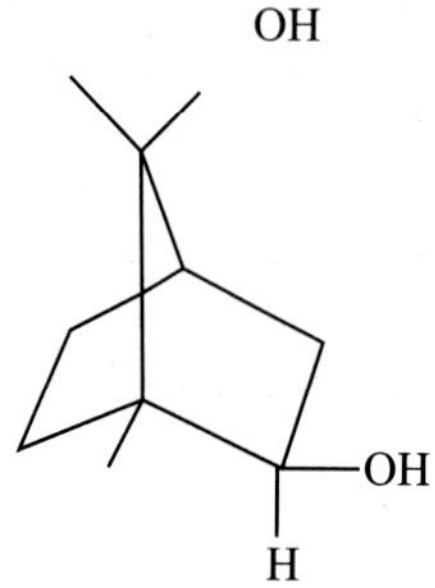 | *Pseudomonas pseudomallei* |

Contd...

Contd...

| Flavour compound | Structure | Bacterial species |
|---|---|---|
| Limonin | 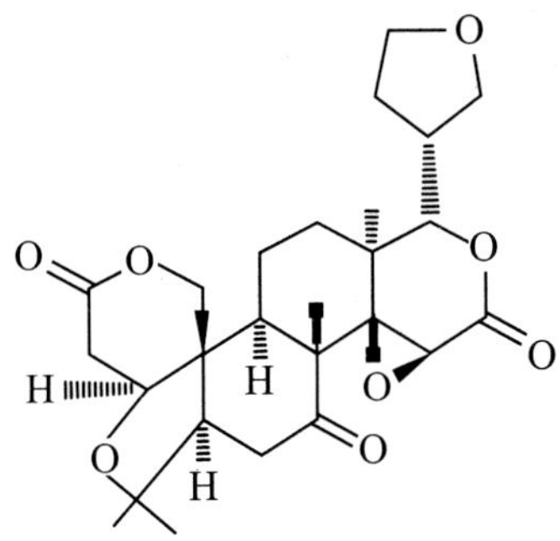 | *Arthrobacter globiformis* |
| Monosodium glutamate | 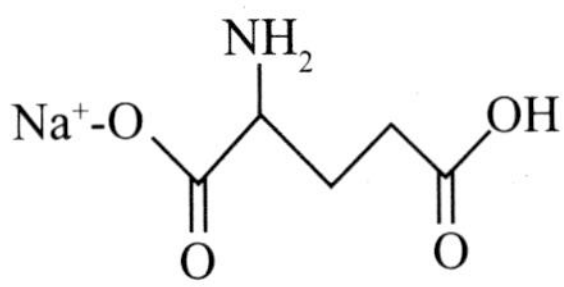 | *Corynebacterium glutamicum* |
| Butyric acid | 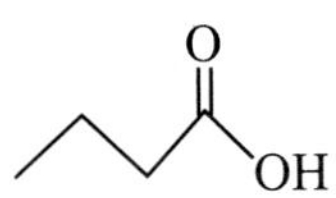 | *Clostridium* sp., *Butyrivibrio* sp., *Eubacterium* sp., *Fusarium* sp. |
| Propionic acid | 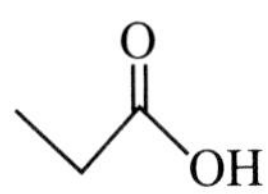 | *Propionibacterium* sp. |
| Isovaleric acid | 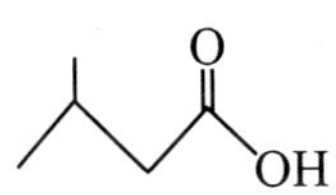 | *Clostridium* sp., *Peptostreptococcus anaerobius* |
| Diketones | 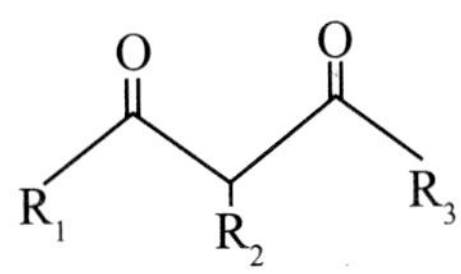 | *Streptococcus* sp. |
| Indole | 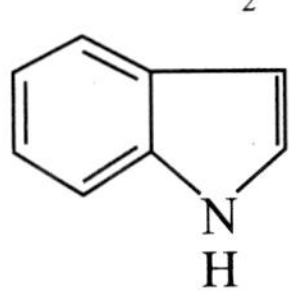 | *Pseudomonas* sp. |
| Benzaldehyde | 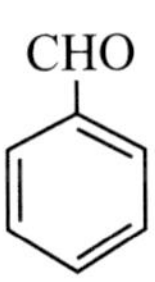 | *Pseudomonas putida* |
| 2-Phenyl ethanol | 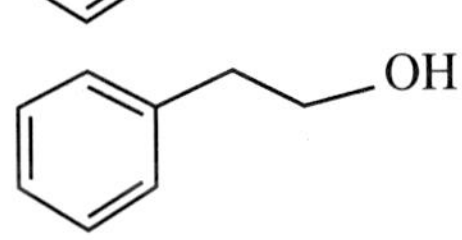 | *Corynebacterium* sp. |

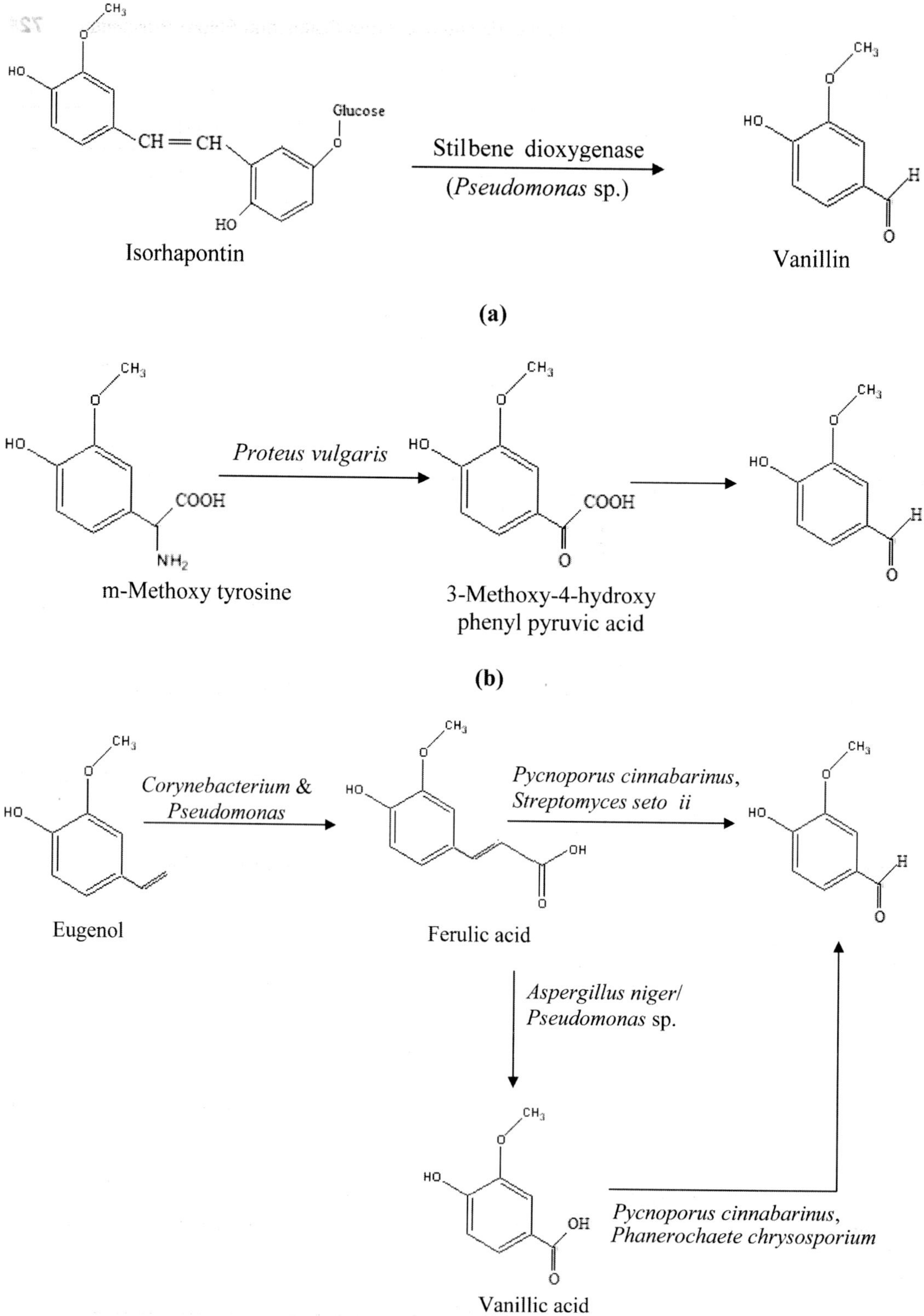


(c)

**Fig. 13(a-c):** Biotransformation approaches to vanillin synthesis.

Contd...

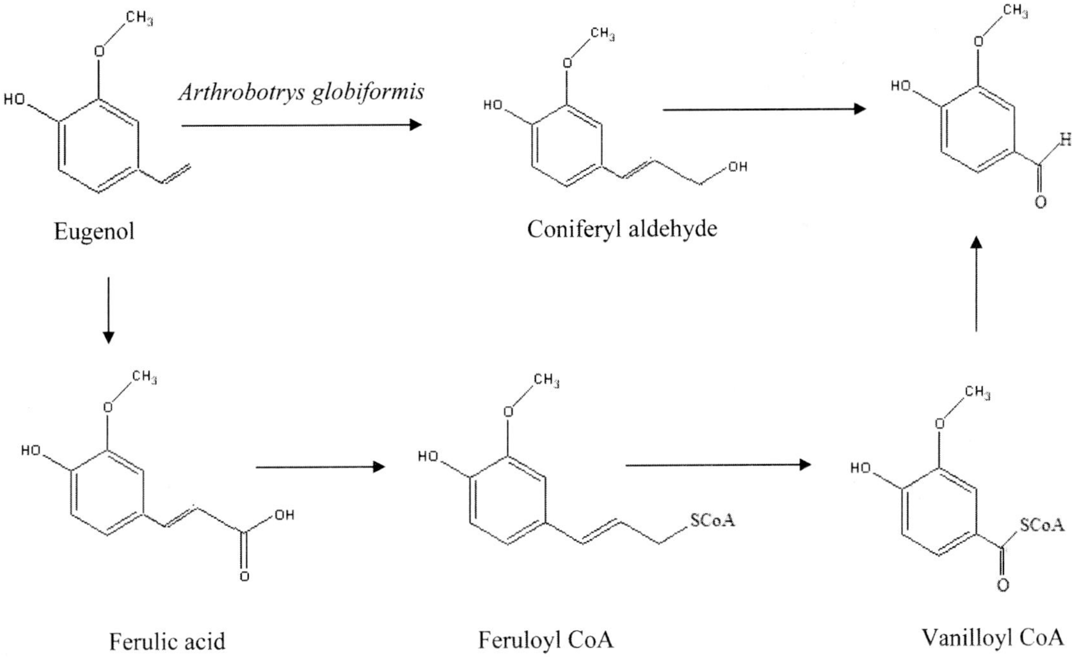


(d)

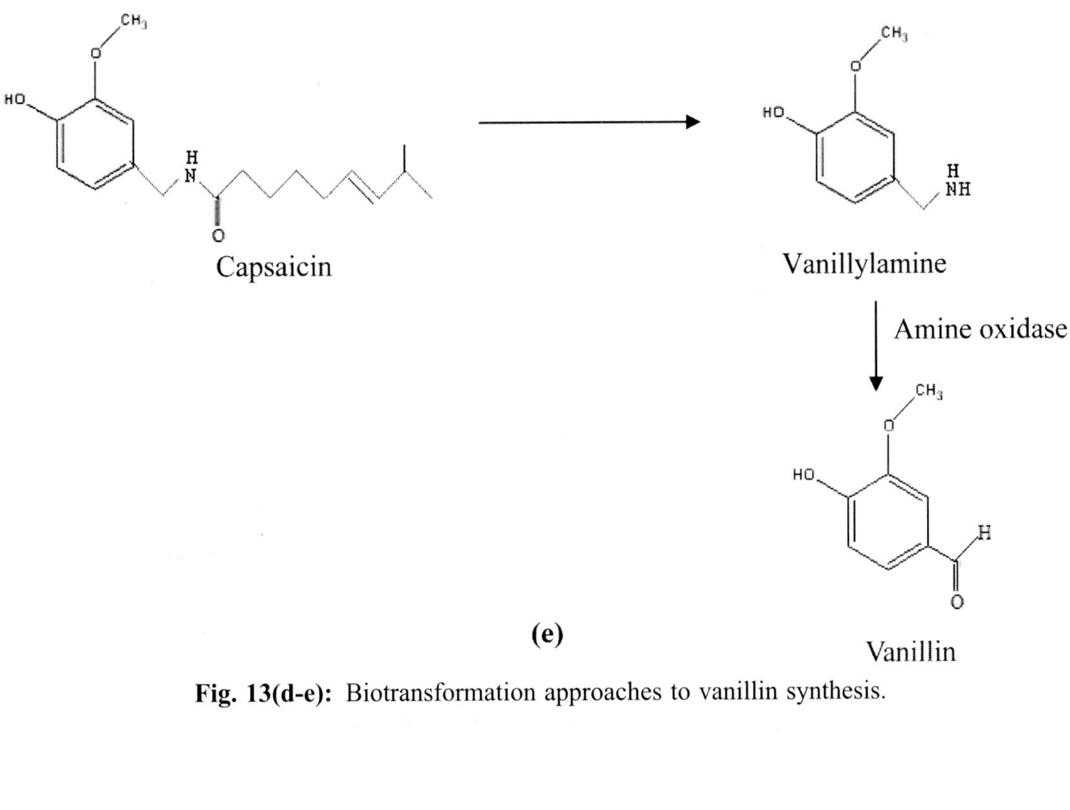


(e)

**Fig. 13(d-e):** Biotransformation approaches to vanillin synthesis.

**5.2.2.2 Production of higher terpenes and terpenoids from carotenoids:** Carotenoids, when acted upon by free radicals, undergo oxidation reactions and form volatile compounds important for the flavour industry. Ionones present in blackberry, peach and apricot are one of them. Fungi such as *Gyromitra esculenta, Clitocybe geotrapa* and *Fibroporia vaillantes* possess enzymes such as polyphenol oxidases and lipoxygenases, which carry out transformation of β-carotenoids into ionones by cleavage at $C_{13}$. In the culture supernatant of *Ischnoderma benzoinum, Marasmius scorodonius* and *Trametes versicolor*, β-ionone is produced as the main metabolite [4].

**5.2.2.3 Production of flavour lactones from fatty acids and unsaturated lactones:** Ricinoleic acid present in the castor oil can be converted into δ-decalactone (Fig. 6) by partial β-oxidation catalyzed by yeast strains of *Sporidiobolus salmonicolor, Monilia fructicola, Rhodotorula glutinis, S. ruineii, S. johnsonii, S. odorus, S. pararoseus* and *Yarrowia lipolytica* [81]. δ-decalactone can also be synthesized by bioconversion of unsaturated fatty acids such as 11-hydroxypalmitic acid, coriolic acid and 2-decen-5-olide using various yeast and fungal strains such as *Saccharomyces cerevisiae* and *Cladosporium suaveolens* [14]. Dodecalactone with whisky flavour is formed by lactic acid bacteria and yeast from unsaturated fatty acids [82]. Okui and co-workers [83] have demonstrated γ-decalactone in cultures

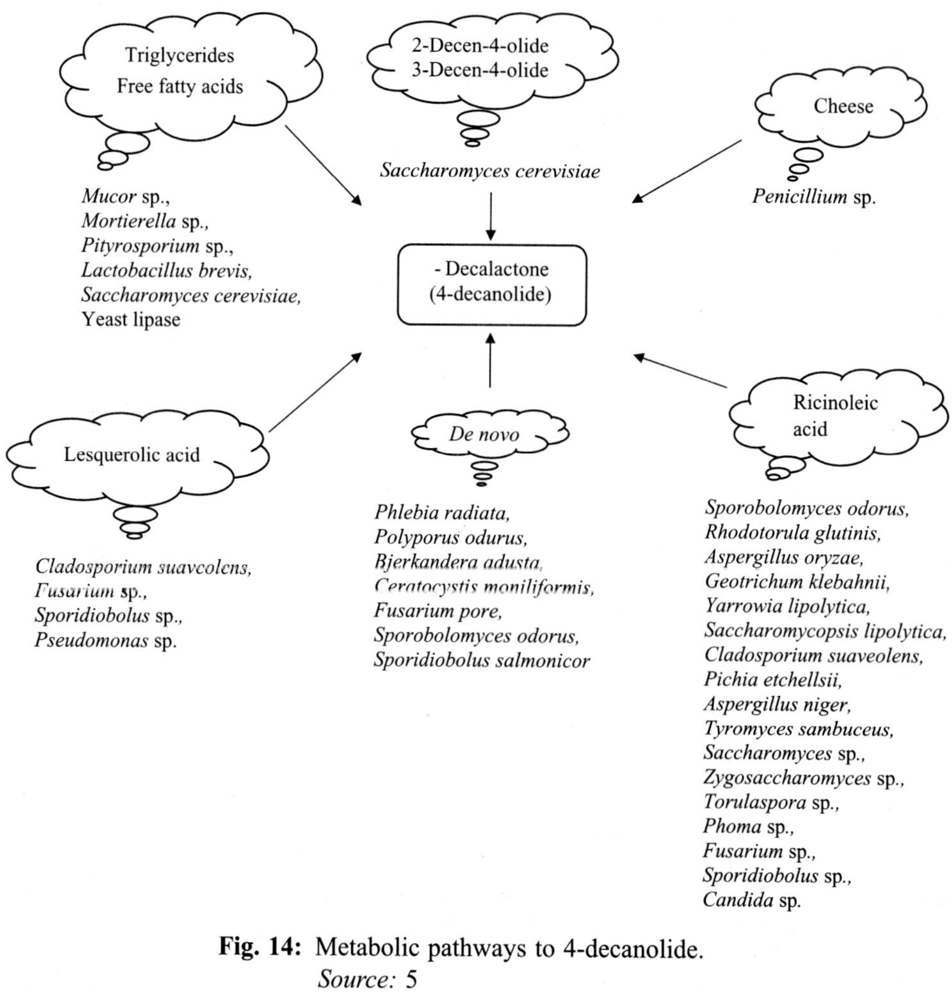


**Fig. 14:** Metabolic pathways to 4-decanolide.
*Source:* 5

of *C. guillermondii* growing in a medium rich in castor oil. Other organisms known to be involved in formation of lactones involve *P. chrysosporium, P. etchellsii, A. niger, Mucor* species and *Candida petrohillium* [84]. The various pathways to γ-decalactone synthesis are given in Fig. 14.

5.2.2.4 Biotransformation of fatty acids to methyl ketones: 2-heptanone, 2-nonanone and 2-undecanone can be produced by biotransformation of coconut oil, which is rich in medium chain fatty acids. Biotransformation of methyl ketones from fatty acids by *Penicillium roquefortii* is given in Fig. 15. Octanoic acid can be converted into 2-heptanone by *Penicillium roquefortii, Aspergillus niger* and *Trichoderma koningii* (Fig. 16). *Aureobasidium pullulans* produces 2-pentanone, 2-heptanone and 2-nonanone from octanoic and decanoic acids [84].

5.2.2.5 Biotransformation approaches to other flavour compounds: *Gluconobacter suboxydans* transforms glucose/gluconic acid into 5-ketogluconic acid, which is a good precursor for the chemical synthesis of savoury flavour, monomethyl furanone [14]. Benzaldehyde, an important flavour after vanillin, is produced biocatalytically by microbial degradation of L-phenylalanine by a basidiomycete, *Ischnoderma benzoinum* (Fig. 17). *Zygosaccharomyces rouxii* has been shown to form 2,5-dimethyl-4-hydroxy-3-dihydroxyfuranone (DMHF) when supplemented with fructose-1,6-bisphosphate and glucose [85]. Phenylalanine is transformed into raspberry ketone by cultures of fungus *Nidula niveo-tomentosa* [66]. Besides, raspberry ketone can also be produced from 4-hydroxy benzalacetone by cultures of *Saccharomyces cerevisiae* [86].

## 5.3 Use of Isolated Enzymes for Flavour Production

Enzymes isolated from microbes or other natural sources have been widely used in food industry for cheese manufacturing, baking and brewing industries. Lipases, proteases, esterases and nucleases aid in flavour extraction from the precursors. Flavour compounds produced by isolated enzymes are enlisted in Table 13.

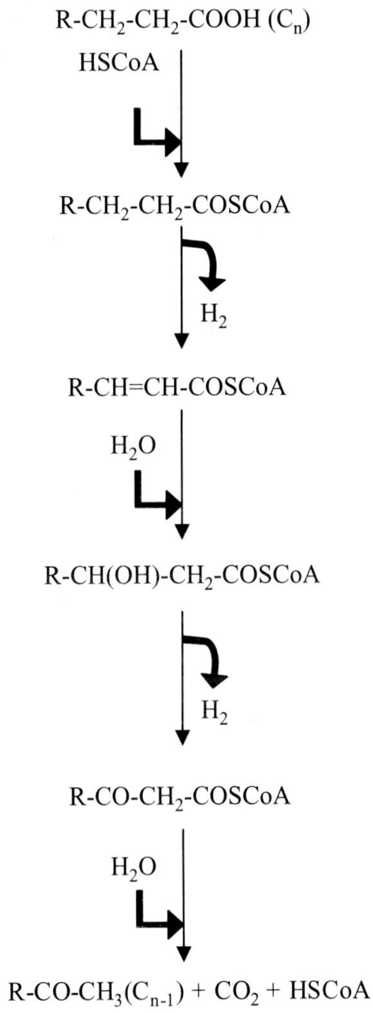


**Fig. 15:** Methylketone synthesis by *Penicillium roquefortii*. *Source:* 3

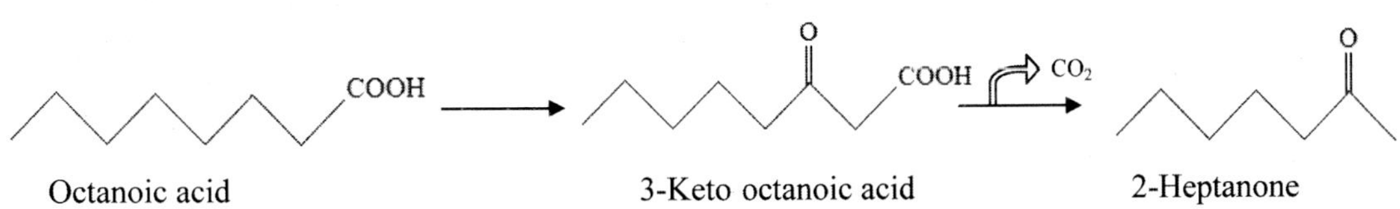


**Fig. 16:** Biotransformation of fatty acids to methyl ketones. *Source:* 84

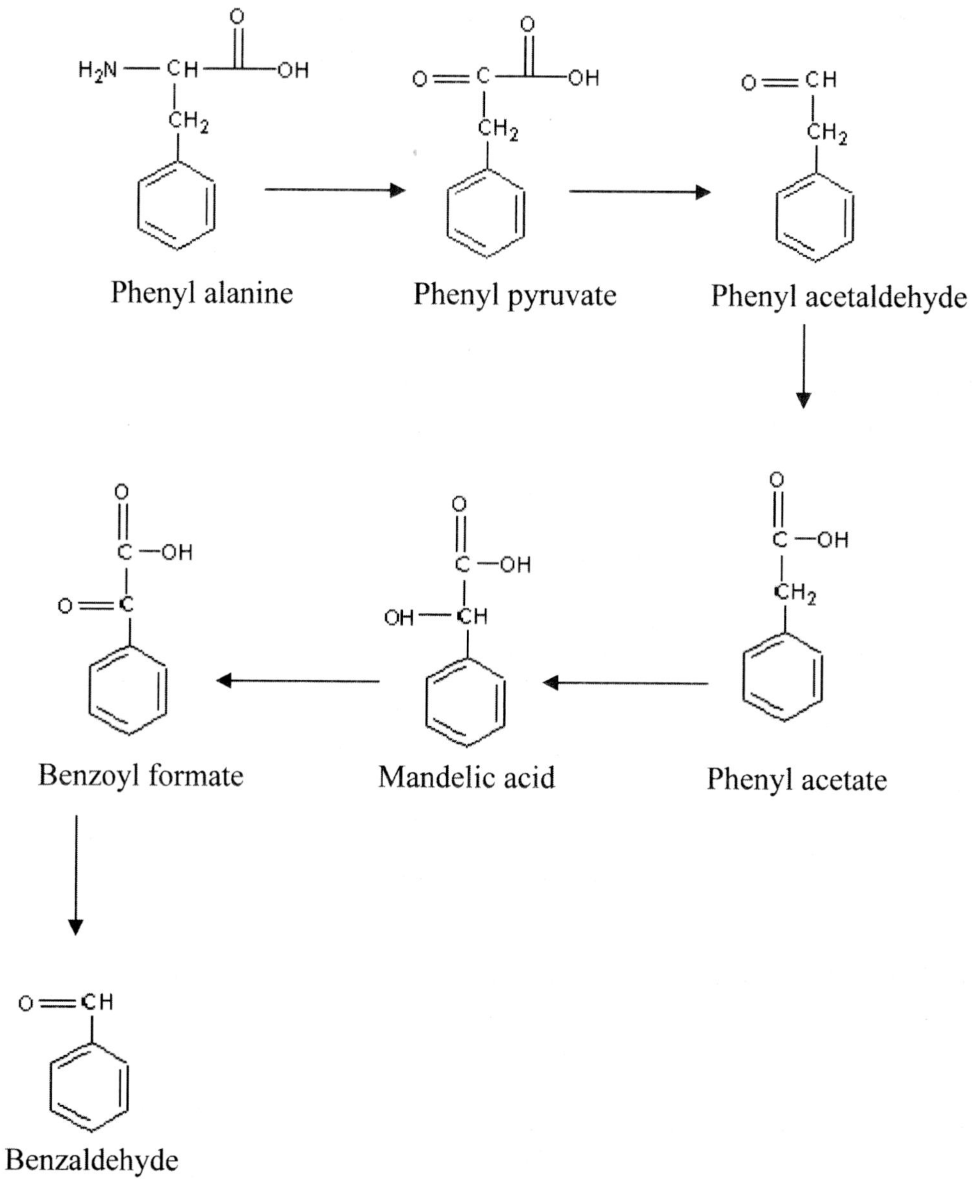


**Fig. 17:** Biotransformation of phenyl alanine to benzaldehyde by *Ischnoderma benzoinum*.

Limitations of enzyme catalyzed reactions include substrate specificity, regioselectivity, enantioselectivity, use under mild reaction conditions and high cost of the biocatalyst. However, enzyme catalysis offers several advantages such as flavour development in a single step, easy recovery and high process efficiency. Proteases, peptidases, lipases and esterases have been reviewed to produce enzyme modified flavour enhanced cheese [21]. Some of the enzymes that play major roles in flavour production are mentioned below.

**Table 13:** Flavourants produced by isolated enzymes.

| Enzyme | Source | Substrate | Flavour compound | Structure |
|---|---|---|---|---|
| Xanthine oxidase, lipooxygenase | Bacteria | β-carotene | β-ionone | 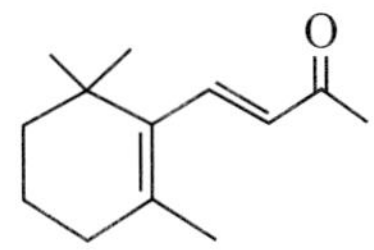 |
| Rhamnosidase | *Aspergillus niger* | Naringin/hesperidin | Furaneol | 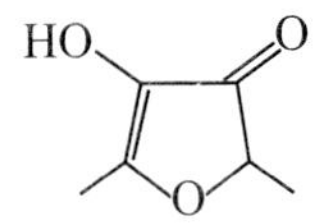 |
| β-Glucosidase, Mendelonitrile lyase | Sour cherry | Amygdalin | Benzaldehyde | 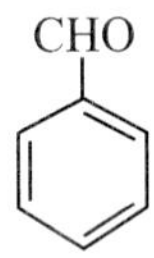 |
| Lipase | *Aspergillus niger*, etc. | L-Menthyl esters | L-Menthol | 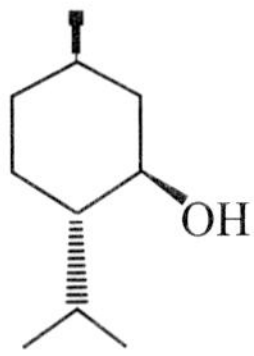 |
| Lipase | *Staphylococcus simulans* | Valeric acid and ethanol | Ethyl valerate | 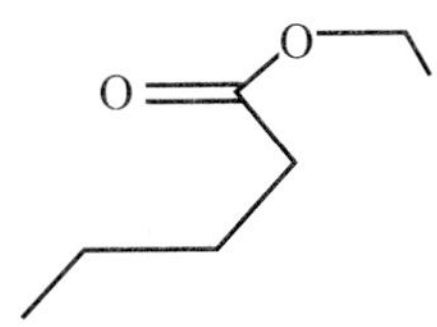 |
| Lipase | *Staphylococcus simulans* | Acetic acid and hexanol | Hexyl acetate |  |
| β-Glucosi-dase, Alcohol dehydrogenase | *Aspergillus niger, Candida boidinii* | Betuloside | Raspberry ketone | 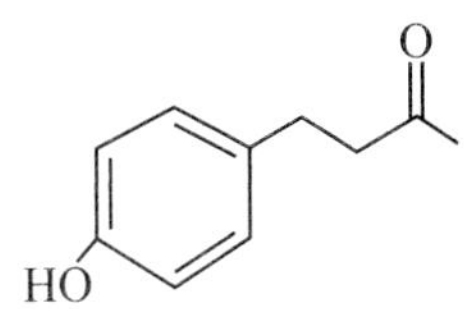 |
| Alcohol dehydrogenase | *Lactobacillus kefir* | Acetophenone | (R)-phenylethanol | 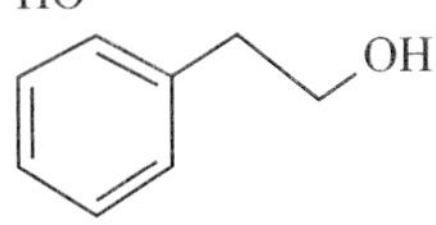 |
| (+)-Germacrene A hydroxylase | *Lactobacillus kefir* | Sequiterpenes | Nootkatone | 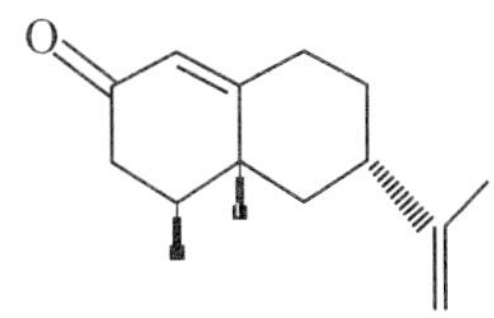 |
| Amine oxidase | *Aspergillus niger* | Vanillylamine | Vanillin | 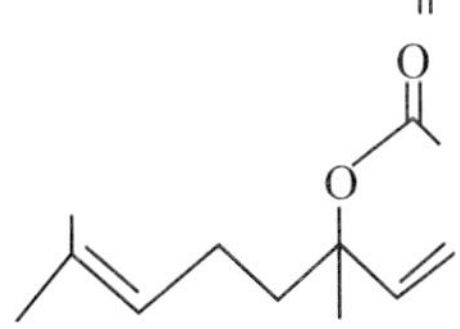 |

Contd...

Contd...

| Enzyme | Source | Substrate | Flavour compound | Structure |
|---|---|---|---|---|
| Ester synthase | Coriander | Linalool | Linalyl acetate | 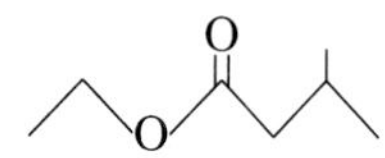 |
| Esterase | Hop | Isovaleric acid | Ethyl isovalerate | 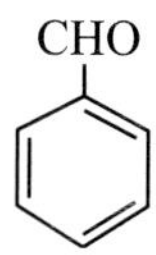 |
| Alcohol oxidase | *Pichia pastoris* | Benzyl alcohol | Benzaldehyde | 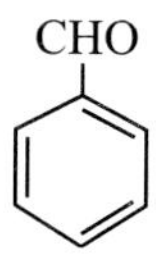 |

### *5.3.1 Lipases*

Use of lipases for preparation of chiral stereoisomers in enantiomeric mixtures is a convenient method, which requires only the identification of suitable alcohol or ester substrates for the transesterification reaction. Lipase from *Candida rugosa* has been found to catalyse L-methanol esterification with long chain unsaturated fatty acids to moderate its strong flavour. Immobilized lipase of *Mucor meihei* [87] and lipase of *Candida cylindracea* [88] catalyze the synthesis of geranyl acetate by transesterification reaction in organic solvents. Bacterial lipases are known to catalyze the enantioselective conversion of L-methyl esters in the racemic mixture to L-menthol, a major mint flavourant (Fig. 18). Lipases from *Rhizomucor meihei, Aspergillus niger* and *Aspergillus javanicus* catalyse the enantioselective synthesis of (S)-2-methylbutanoic acid methyl ester, a chief flavour component of apple and strawberry in presence of isooctane [89]. *Staphylococcus epidermidis* lipase catalyzes the synthesis of flavour esters in aqueous media [90]. Ethyl valerate (green apple flavour) and hexyl acetate (pear flavour) have been produced by crude lipases immobilized on different supports [91].

### *5.3.2 Proteases*

Free amino acids and relatively short oligopeptides play a significant role in flavour industry [92, 93]. Different sequences of amino acids can be used to recreate four basic taste sensations i.e., sweet, bitter, sour and salty. Proteases catalyze the hydrolysis of vegetable proteins, thereby forming a savoury flavour, mainly due to formation of pyrazines and alcohols, earlier formed by heating the protein source at acidic pH [94]. Serine and cysteine endopeptidases have been found to be particularly

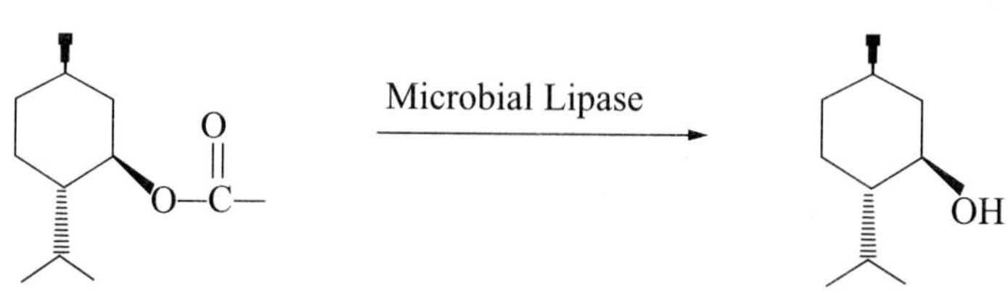


L-menthyl acetate L-menthol

**Fig. 18:** Stereoselective formation of L-menthol.

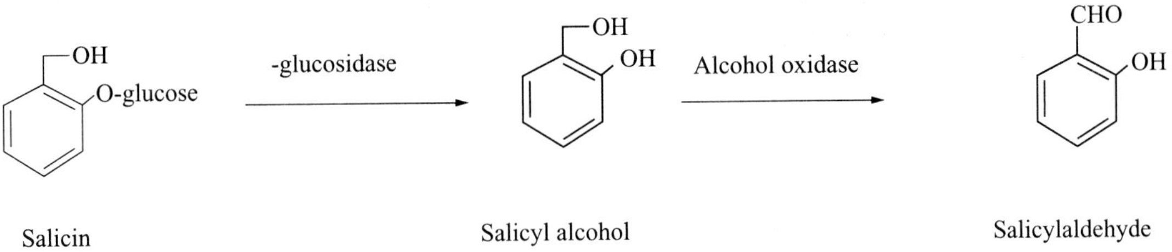


**Fig. 19:** Biotransformation of salicin to salicylaldehyde.

important for synthesis of peptides used as food additives [95]. Proteases act on cray fish processing by-products and form benzaldehyde and pyrazines [96]. Tetrapeptide Asp-Asp-Asp-Asp, isolated from beer yeast seasoning, is known to mask the bitter taste of beer. It can be synthesized by condensation of L-aspartic acid α,β-dipropyl ester hydrochloride, the reaction being catalyzed by α-chymotrypsin [97].

### *5.3.3 Glucosidases*

β-glucosidases from *Vitis, Saccharomyces, Oenococcus, Aspergillus* and *Candida* have been found to enhance wine aroma by hydrolysis of glycosidic-bound precursors like terpenols resulting in release of terpenes and other flavours [98]. Commercial β-glucosidase from *Aspergillus niger* converts betuloside (found in birch, maple and rhododendron) to betuligenol, which can then be transformed to raspberry ketone (Fig. 3) by alcohol dehydrogenase from *Candida boidinii* [14]. β-glucosidases and alcohol oxidases of *Candida methanolevescens* have been found to transform glycoside salicin to almond flavoured salicylaldehyde [99]. The biotransformation reaction is illustrated in Fig. 19.

### *5.3.4 Other enzymes*

Benzaldehyde can be produced from cyanogenic glycoside, amygdalin present in cherry stones, by the successive action of enzymes β-glucosidase and mendelonitrile lyase (Fig. 20) present in crude almond meal [3]. Rhamnosidases present in commercial *Aspergillus* pectinase preparations act on naringin and hesperidin, forming 6-deoxysugar rhamnose, which is a precursor to 2,5-dimethyl-4-hydroxy-2,3-dihydrofuran-3-one (furaneol®), exhibiting strawberry flavour in dilute solutions and caramel-like flavour in concentrated form (Fig. 21). Alcoholic dehydrogenases from *Lactobacillus kefir* immobilized on polyvinyl alcoholic gel beads synthesize (R)-phenylethanol from acetophenone in presence of hexane [100]. (+)-germacrene A hydroxylase produces nootkatone from sequiterpenes [101]. Vanillylamine from capsaicin (natural ingredient of pepper and capsicum) can be converted to vanillin by enzyme amine oxidase from *Aspergillus niger* [102]. L-glutamic acid, chief flavour ingredient of fermented soy sauce can be produced by action of microbial glutaminases [103]. Agricultural waste like beet pulp and cereal bran when treated with polysaccharide-degrading enzymes and esterases form ferulic acid, which is a precursor to vanillin [104].

## 6. FUTURE PROSPECTS

Biotechnological advancements have opened up new ways for the production of natural flavours. Due to consumer preference for natural products and better economic benefits, biotechnological production of flavours has gained more importance. Low cost natural precursors can be transformed to expensive flavour compounds by microorganisms and their enzymes, and thus these offer advantage over the traditional methodologies. Now, there is a need for better understanding of mechanisms so as to improve the processes significantly and to decrease the cost of production furthermore. Identification

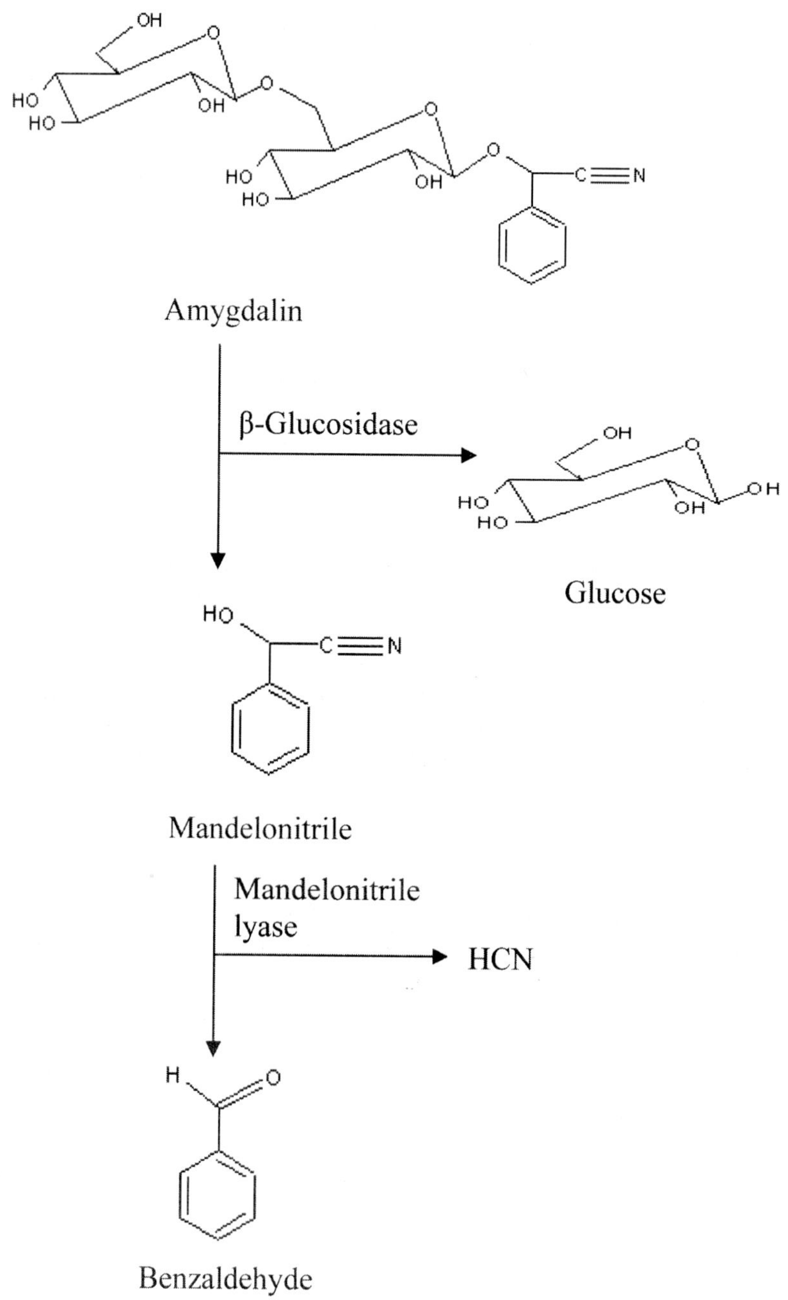


**Fig. 20:** Synthesis of benzaldehyde from amygdalin using isolated enzymes.

and characterization of novel strains producing flavour compounds needs to be optimized. Identification of alternative genes better suited for genetic engineering of existing strains could be pioneered. Enzymes that could be exploited in future are cyclases, which are involved in terpene synthesis in plants [105].

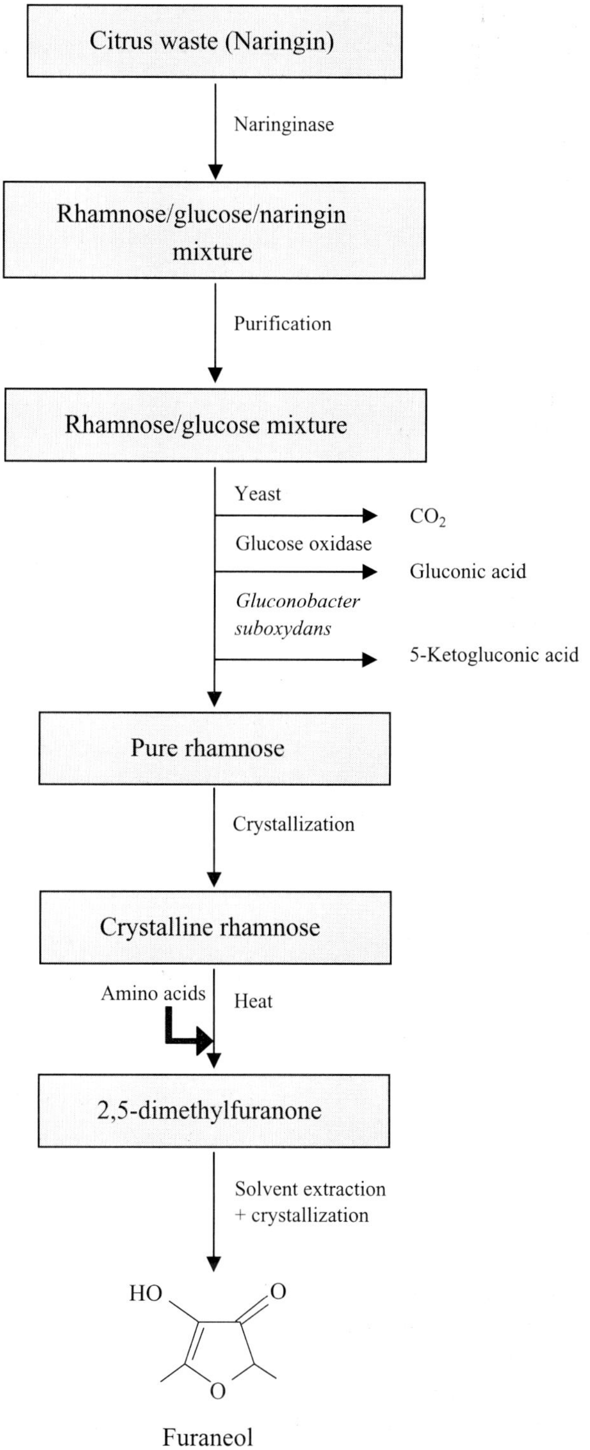


**Fig. 21:** Production of furaneol from citrus waste.

Genetic engineering methods are being looked upon for the development of strains with ability to produce flavours or incorporating genes in microorganisms to curtail the multistep reactions to a single step or development of significantly improved biocatalysts. *Leuconostoc* strain has been developed by protein engineering techniques, which is capable of lactose fermentation as well as diacetyl production [106]. In beer production, yeast supplemented with gene encoding α-acetolactate decarboxylase would result in removal of diacetyl, and thus, the buttery off-flavour and hence, eliminating the need for lagering [107]. Supplementing wine yeast with malolactic enzyme has resulted in improved composition of volatiles [108]. A multifunctional o-methyl transferase catalyzing the formation of a compound with strawberry-like flavour, 2,5-dimethyl-4-methoxy-3(2H)furanone (DMMF) has been cloned and expressed in *E. coli* [109]. Expression of flavour generating plant enzymes in microbes offer an opportunity for development of high level expression system. A pathway in the total synthesis of terpenoids involving 8 genes from *S. cerevisiae* has been engineered in *E. coli* [110].

Fed batch fermentation, immobilization and *in situ* flavour extraction technique can circumvent the conditions where flavour compounds and their precursors are inhibitory or toxic to producer strains. Accumulation of undesirable by-products and cost of raw materials pose economic constraints to the bioflavour production. For this, the undesirable intermediates could be metabolized to desirable products by other microbes e.g., when ricinoleic acid is converted to γ-decalactone, 3-hydroxy-γ-decalactone is formed as a by-product, which can be converted to 3,4-unsaturated γ-decalactone and then reduced to γ-decalactone by *S. cerevisiae* [14].

Biotechnological methods need to overcome several drawbacks like low water solubility, toxicity of precursors and products, and unwanted by-product accumulation. The industrial application of novel biocatalysis depends on scientific and technological feasibility and also the public perception on the use of genetic engineering techniques for improved food quality and more environmental friendly production processes. A better understanding of the mode of action of enzymes as well as the mechanism of development of flavour compounds will further enhance the use of biotechnological methods to develop specific and desired flavours in food.

## REFERENCES


1. Joshi KC (1996) The significance of food flavours. *Invention Intelligence* 4: 183-185.
2. Cometto-Muniz JE (1981) Odor, taste, flavour perception of some flavouring agents. *Chem Senses* 3: 215-223.
3. Cheetham PSJ (1993) The use of biotransformation for the production of flavours and fragrances. *Trends Biotechnol* 11: 478-488.
4. Aguedo M, Ly MH, Belo I, Teixeira JA, Belin JM & Waché Y (2004) The use of enzymes and microorganisms for the production of aroma compounds from lipids. *Food Technol Biotechnol* 42: 327-336.
5. Krings U & Berger RG (1998) Biotechnological production of flavours and fragrances. *Appl Microbiol Biotechnol* 49: 1-8.
6. Shaikh Y (2002) *Speciality Aroma Chemicals in Flavours and Fragrances*, Culinary and Hospitality Industry Publications Services, Texas, USA.
7. Brenna E (2002) Flavours and fragrances by biocatalytic routes. *Suppl Chem Today* 23: 8-11.
8. Amoore J (1971) Stereochemical and vibrational theories of odour. *Nature* 233: 270-271.
9. Salunkhe DK & Do JY (1976) Biogenesis of aroma constituents of fruits and vegetables. *Crit Rev Food Sci Nutr* 8: 161-189.
10. Vanderhaegen R, Neven H, Coghe S, Verstrepen KJ, Derdelinckx G & Verachtert H (2003) Bioflavouring and beer refermentation. *Appl Microbiol Biotechnol* 62: 140-150.

11. Liebig J & Wohler F (1837) Formation of the oil of bitter almonds. *Ann Chim Phys* 64: 185-209.
12. Tiemann F & Haarmann W (1874) Ueber das Coniferin und seine Unwandling in das aromatische Princip der Vanille. *Ber Dtsch Chem Ges* 7: 608-623.
13. Reimer K & Tiemann F (1876) Ueber die einwirkung von chloroform auf alkalische phenolate. *Ber Deutsch Chem Ges* 9: 824-828.
14. Vandamme EJ & Soetaert W (2002) Bioflavours and fragrances via fermentation and biocatalysis. *J Chem Technol Biotechnol* 77: 1323-1332.
15. Holtje HD & Kier LB (1974) Sweet taste receptor studies using model interaction energy calculations. *J Pharm Sci* 63: 1722-1725.
16. Kuboto T & Kubo I (1969) Bitterness and chemical structure. *Lett Nature* 223: 97-99.
17. Hanssen H-P & Klingenberg A (1983) Determination of some important flavour compounds in commercial mushroom concentrates. *Zeitschrift für Lebensmitteluntersuchung und- Forschung* A177: 333-335.
18. Priefert H, Rabenhorst J & Steinbüchel A (2001) Biotechnological production of vanillin. *Appl Microbiol Biotechnol* 56: 296-314.
19. Rabenhorst J (2000) Biotechnological production of aroma chemicals by fermentation processes. In: *Biotechnology*, Vol 8b, Rehm H-J & Reed G (eds), Wiley-VCH, Weinheim, Germany, pp 333-350.
20. Badings HT, Van der Pol JJG & Neeter R (1981) Aroma compounds which contribute to the difference in flavour between pasteurized milk and UHT milk. In: *Flavour*, Vol 81, Schreier P (ed), Walter de Gruyter, New York, USA, pp 683-692.
21. Longo MA & Sanromán MA (2006) Production of food aroma compounds: Microbial and enzymatic methodologies. *Food Technol Biotechnol* 44: 335-353.
22. Cardillo R, Fuganti C, Sacerdote C, Barbeni M, Cabella P & Squarcia F (1989) Process for the microbiological production of gamma-decalactone (R) and gamma-octalactone (R). EP Patent 356291.
23. Cheetham PSJ, Maume KA & de Rooji JF (1988) Production of lactones. EP Patent 0258 993.
24. Schrader J, Etschmann MMW, Sell D, Hilmer J-M & Rabenhorst J (2004) Applied biocatalysis for the synthesis of natural flavour compounds-Current industrial processes and future prospects. *Biotechnol Lett* 26: 463-472.
25. Jollivet N, Bèzenger M-C, Vayssier Y & Belin J-M (1992) Production of volatile compounds in liquid cultures by six strains of *Coryneform* bacteria. *Appl Microbiol Biotechnol* 36: 790-794.
26. Fabre CE, Blanc PJ & Goma G (1998) Production of 2-phenylethyl alcohol by *Kluyveromyces marxianus. Biotechnol Progr* 18: 514-523.
27. Stark D, Münch T, Sonnleitner B, Marison IW & Von Stockar U (2002) Extractive bioconversion of 2-phenylethanol from L-phenylalanine by *Saccharomyces cerevisiae. Enzyme Microb Technol* 32: 212-223.
28. Nielsen JC & Richelieu M (1999) Control of flavour development in wine during and after malolactic fermentation by *Oenococcus oeni. Appl Environ Microbiol* 65: 740-745.
29. Bratovanova P (2001) To the issue of formation of acetoin and diacetyl in dough part manufactured products. *Biotechnol Biotec Eq* 15: 124-127.
30. Aungpraphapornchai P, Griffin HG & Gasson MJ (1999) Cloning, DNA sequence analysis, and deletion of a gene encoding diacetyl-acetoin reductase from *Lactococcus lactis. DNA Sequence* 10: 163-172.
31. Serra S, Fuganti C & Brenna E (2006) Biocatalytic preparation of natural flavours and fragrances. *Trends Biotechnol* 23: 193-198.

32. Schreier P (1997) Enzymes and flavour biotechnology. *Adv Biochem Eng Biotechnol* 55: 51-72.
33. Dimick PS, Patton S, Kinsella JE & Walker NJ (1966) Prevalence of aliphatic delta-lactones or their precursors in animal fats. *Lipids* 1: 387-391.
34. Gregory P & Eilerman RG (1989) A process for the production of γ and δ-octalactone. PCT Int Patent 8912-134A.
35. Boog LGM, Peters ALJ & Roos R (1991) Process for producing delta-lactones. EP Patent CA2021270 BO.36025.
36. Marais J, Hunter JJ & Haasbroek PD (1999) Effect of microclimate, season and region on Sauvignon blanc grape composition and wine quality. *South Af J Enol Vitic* 20: 19-30.
37. Fisher C & Scott TR (1997) *Food Flavours: Biology and Chemistry,* Royal Society of Chemistry, Cambridge, USA.
38. Belitz H-D & Grosch W (1999) *Food Chemistry*, 2nd Engl edn, Springer, Berlin, Germany.
39. Rojas V, Gill JV, Pinaga F & Manzanares P (2001) Studies on acetate ester production by non-*Saccharomyces* wine yeasts. *Int J Food Microbiol* 70: 283-289.
40. Bauer K, Garbe D & Surburg H (2001) *Common Fragrance and Flavour Materials*, 4th edn, Wiley-VCH, Weinheim, Germany.
41. Muheim A, Häusler A, Schilling B & Lerch K (1997) The impact of recombinant DNA-technology on the flavour and fragrance industry. In: *Flavours and Fragrances,* Proceedings of a Conference, Warwick. Royal Society of Chemistry, Cambridge, pp 11-20.
42. Häusler A & Münch T (1997) Microbial production of natural flavours. *ASM News* 63: 551-559.
43. Patton S (1964) Flavour threshold of volatile fatty acids. *J Food Sci* 29: 679-680.
44. Jin Z & Yang S-T (1998) Extractive fermentation for enhanced propionic acid production from lactose by *Propionibacterium acidipropionici. Biotechnol Prog* 14: 457-465.
45. Bodie EA, Goodman N & Schwartz RD (1987) Production of propionic acid by mixed cultures of *Propionibacterium shermanii* and *Lactobacillus casei* in autoclave-sterilized whey. *J Ind Microbiol Biotechnol* 6: 349-353.
46. Britz ML & Wilkinson RG (1982) Leucine dissimilation to isovaleric and isocaproic acids by cell suspensions of amino acid fermenting anaerobes: The Stickland reaction revisited. *Can J Microbiol* 28: 291-300.
47. Drawert F, Berger RG & Goldenmann R (1984) Regioselective biotransformation of valencene in cell suspension cultures of *Citrus* spp. *Plant Cell Rep* 3: 37-40.
48. Townsley PM (1974) Chocolate aroma from plant cells. *J Inst Can Sci Technol Aliment* 7: 76-78.
49. Westcott RJ, Cheetham PSJ & Barraclough AJ (1994) Use of organized viable vanilla plant aerial roots for the production of natural vanillin. *Phytochemistry* 35: 135-138.
50. Suresh B, Ritu T & Ravishankar GA (2003) Vanilla flavour production through biotransformation using *Capsicum frutescens* root cultures. *Biocat Biotransform* 21: 333-340.
51. Discosmo F & Misawa M (1995) Alternatives for metabolite production. *Biotechnol Adv* 13: 425-453.
52. Verachtert H, Kumara HMC & Dawoud E (1989) Yeast in mixed cultures with emphasis on lambic beer brewing. In: *Yeast-Biotechnology and Biocatalysis*, Verachtert H & De Mot R (eds), Marcel-Dekker, New York, USA, pp 429-478.
53. Omeliansky VL (1923) Aroma-producing microorganisms. *J Bacteriology* 8: 393-419.
54. Lanza E, Ko KH & Palmer JK (1976) Aroma production by cultures of *Ceratocystis moniliformis. J Agric Food Chem* 20: 1247-1250.

55. Collins RP & Halim F (1971) An analysis of the odorous constituents produced by various species of *Phellinus. Can J Microbiol* 18: 65-66.
56. Collins RP (1976) Terpenes and odoriferous materials from microorganisms. *Lloydia* 39: 20-24.
57. Vanhaelen M, Vanhaelen-Faster R & Geeraerts J (1978) Volatile constituents of *Trichothecium roseum. Sabouraudia* 16: 141-150.
58. Karahadian C, Josephson DB & Lindsey RC (1985) Contribution of *Penicillium* sp. to the flavours of Brie and Camembert cheese. *J Dairy Sci* 68: 1865-1869.
59. Kapfer GF, Berger RG & Draweti F (1989) Production of 4-decanolide by semicontinuous fermentation of *Tyromyces sambuceus. Biotechnol Lett* 11: 561-566.
60. Allegrone G, Barbeni M, Cardillo R, Fuganti C, Grasselli P, Miele A & Pisciotta A (1991) On the steric course of the microbial generation of (Z6)-gamma-dodecenolactone from (10R,S) 10-hydroxyoctadeca-(E8,Z12)-dienoic acid. *Biotechnol Lett* 13: 765-768.
61. Janssens L, de Pooter HL, Vandamme EJ & Schamp NM (1992) Production of flavour and aroma compounds by biotechnology. *Process Biochem* 27: 195-122.
62. Kümin B & Münch T (1997) Process for the preparation of a lactone. EP Patent 0795607.
63. Rapior S, Breheret S, Talou T & Bessière J-M (1997) Volatile flavour constituents of fresh *Marasmius alliaceus* (Garlic *Marasmius*). *J Agric Chem* 45: 820-825.
64. Bramorski A, Christen P, Ramirez M, Soccol CR & Revah S (1998) Production of volatile compounds by the edible fungus *Rhizopus oryzae* during solid-state cultivation on tropical agro-industrial substrates. *Biotechnol Lett* 20: 359-362.
65. Soares M, Christen P, Pandey A & Soccol CR (2000) Fruity flavour production by *Ceratocystis fimbriata* grown on coffee husk in solid state fermentation. *Process Biochem* 35: 857-861.
66. Zorn H, Fischer-Zorn M & Berger RG (2003) A labeling study to elucidate the biosynthesis of 4-(4-hydroxyphenyl)-butan-2-one (raspberry ketone) by *Nidula niveo-tomentosa. Appl Environ Microbiol* 69: 367-372.
67. Tahara S (1978) Gamma-decalactone-one of constituents of volatiles in cultured broth of *Sporobolomyces odorus. Agric Biol Chem* 42: 879-883.
68. Drawert F & Barton H (1978) Biosynthesis of flavour compounds by microorganisms. 3. Production of monoterpenes by yeast *Kluyveromyces lactis. J Agric Food Chem* 26: 765-766.
69. Medeiros ABP, Pandey A, Freitas RJS, Christen P & Soccol CR (2000) Optimization of the production of aroma compounds by *Kluyveromyces marxianus* in solid-state fermentation using factorial design and response surface methodology. *Biochem Eng J* 6: 33-39.
70. Groot WJ & Lyben KChAM (1986) In situ product recovery by adsorption in the butanol/isopropanol batch fermentation. *Appl Microbiol Biotechnol* 25: 29-31.
71. Wecker MSA & Zall RR (1987) Production of acetaldehyde by *Zymomonas mobilis. Appl Environ Microbiol* 53: 2815-2819.
72. Besson I, Ceruly C, Gros JB & Larroche C (1997) Pyrazine production by *Bacillus subtilis* in solid state fermentation on soybeans. *Appl Microbiol Biotechnol* 47: 489-495.
73. Karmakar B, Sharma P, Gupta KG & Sobti RC (1999) Biotechnological generation of food flavours. *JPAS* 1: 35-46.
74. Achterholt S, Priefert H & Steinbüchel A (2000) Identification of *Amycolatopsis* sp. strain HR167 genes, involved in the bioconversion of ferulic acid to vanillin. *Appl Microbiol Biotechnol* 54: 799-807.
75. Park JK & Jung JY (2002) Production of benzaldehyde by encapsulated whole-cell benzoylformate decarboxylase. *Enzyme Microb Technol* 30: 726-733.

76. Escamilla-Hurtado ML, Valdes Martinez SE, Soriano-Santos J, Gomez-Pliego R, Verde-Calvo JR, Reyes-Dorantes A & Tomasini-Campocosio A (2005) Effect of culture conditions on production of butter flavour compounds by *Pediococcus pentosaceus* and *Lactobacillus acidophilus* in semisolid maize-based cultures. *Int J Food Microbiol* 105: 305-316.
77. Feron G, Bonnarme P & Durand A (1996) Prospects in the microbial production of food flavours. *Trends Food Sci Technol* 7: 285-293.
78. Nazareth S & Mavinkurve S (1986) Degradation of ferulic acid via 4-vinylguaiacol by *Fusarium solani* (Mart) Sacc. *Can J Microbiol* 32: 494-497.
79. Bonnin E, Brunel M, Gouy Y, Lesage-Meesen M, Asther M & Thiobault JF (2001) *Aspergillus niger* I-1472 and *Pycnosporus cinnabarius* MUCL39533, selected for the biotransformation of ferulic acid to vanillin, are also able to produce cell wall polysaccharide-degrading enzymes and feruloyl esterases. *Enzyme Microb Technol* 28: 70-80.
80. Mathew S & Abraham TE (2004) Ferulic acid: an antioxidant found naturally in plant cell walls and feruloyl esterases involved in its release and their applications. *Crit Rev Biotechnol* 24: 59-83
81. Lee SL, Cheng HY, Chen WC & Chou CC (1998) Production of gamma-decalactone from ricinoleic acid by immobilized cells of *Sporidiobolus salmonicolor. Process Biochem* 33: 453-459.
82. Wanikawa A, Hosoi K & Kato T (2000) Conversion of unsaturated fatty acids to precursors of gamma-lactones by lactic acid bacteria during production of malt whisky. *J Am Soc Brew Chem* 58: 51-56.
83. Okui S, Uchiyama M & Mizugaki M (1963) Metabolism of hydroxyl fatty acids. II. Intermediates of the oxidative breakdown of ricinoleic acid by genus *Candida. J Biochem* (Tokyo) 54: 536-540.
84. Hagedorn S & Kaphamer B (1994) Microbial biocatalysis in the generation of flavour and fragrance chemicals. *Ann Rev Microbiol* 48: 773-800.
85. Dahlen T, Hauck T, Wein M & Schwab W (2001) 2,5-Dimethyl-4-hydroxy-3(2H)-furanone as a secondary metabolite from D-fructose-1,6-diphosphate metabolism by *Zygosaccharomyces rouxii. J Biosci Bioeng* 91: 352-358.
86. Fuganti C & Zucchi G (1998) Product distribution in the microbial biogeneration of raspberry ketone from 4-hydroxybenzalacetone. *J Mol Catal B: Enz* 4: 289-293.
87. Chulalaksananukul W, Condoret JS & Combes D (1992) Kinetics of geranyl acetate synthesis by lipase-catalyzed trans-esterification in n-hexane. *Enzyme Microb Technol* 14: 293-298.
88. Huang SY, Chang HL & Goto M (1998) Preparation of surfactant-coated lipase for the esterification of geraniol and acetic acid in organic solvents. *Enzyme Microb Technol* 22: 552-557
89. Kwon DJ, Hong YJ & Yoon SH (2000) Enantiomeric synthesis of (S)-2-methylbutanoic acid methyl ester, apple flavour, using lipases in organic solvent. *J Agric Food Chem* 48: 524-530.
90. Chang RC, Chou SJ & Shaw JF (2001) Synthesis of fatty acid esters in recombinant *Staphylococcus epidermidis* lipases in aqueous environment. *J Agric Food Chem* 49: 2619-2622.
91. Karra-Châabouni M, Ghamgui H, Bezzine S, Rekik A & Gargouri Y (2006) Production of flavour esters by immobilized *Staphylococcus simulans* lipase in a solvent-free system. *Process Biochem* 41: 1692-1698.
92. Nishimura T & Kato H (1988) Taste of free amino acids and peptides. *Food Rev Int* 4: 175-194.
93. Maehashi K, Matsuzaki M, Yamamoto Y & Udaka S (1999) Isolation of peptides from an enzymatic hydrolysate of food proteins and characterization of their taste properties. *Biosci Biotechnol Biochem* 63: 555-559.

94. Aaslyng MD, Elmore JS & Mottram DS (1998) Comparison of the aroma characteristics of acid-hydrolyzed and enzyme-hydrolyzed vegetable protein produced from soy. *J Agric Food Chem* 46: 5225-5231.
95. Gill I, Lopez-Fandiño R, Jorba X & Vulfson EN (1996) Biologically active peptides and enzymatic approaches to their production. *Enzyme Microb Technol* 18: 162-183.
96. Baek HH & Cadwallader KR (1996) Volatile compounds in flavour concentrates produced from crayfish-processing by-products with and without protease treatment. *J Agric Food Chem* 44: 3262-3267.
97. Matsushita A & Ozaki S (1995) Purification and sequence determination of tasty tetrapeptide (Asp-Asp-Asp-Asp) from beer yeast and its enzymic synthesis. *Pept Chem* 32: 249-252.
98. Aryan AP, Wilson B, Strauss CR & Williams PJ (1987) The properties of glycosidases of *Vitis vinifera* and a comparison of their β-glucosidase activity with that of exogenous enzymes. An assessment of possible applications in enology. *Am J Enol Vitic* 38: 182-188.
99. Van den Bremt K, Delvaux FR, Verachtert H & Derdelinckx G (2001) Biogeneration of flavours: performance of *Candida methanolovescens* strains in non-alcoholic beer. *J Am Soc Brew Chem* 59: 80-83.
100. de Temiño DM, Hartmeier W & Ansorge-Schumacher MB (2005) Entrapment of the alcohol dehydrogenase from *Lactobacillus kefir* in polyvinyl alcohol for the synthesis of chiral hydrophobic alcohols in organic solvents. *Enzyme Microb Technol* 36: 3-9.
101. de Kraker JW, Schurink M, Franssen MCR, Konig WA, de Groot A & Bouwmeester HJ (2003) Hydroxylation of sequiterpenes by enzymes from chicory (*Cinchorium intybus* L.) roots. *Tetrahedron* 59: 409-418.
102. Yoshida A, Takenaka Y, Tamaki H, Frebort I, Adachi O & Kumagai H (1997) Vanillin formation by microbial amine oxidases from vanillylamine. *J Ferment Bioeng* 84: 603-605.
103. Nandakumar R, Yoshimune K, Wakayama M & Moriguchi M (2003) Microbial glutaminase: Biochemistry, molecular approaches and applications in the food industry. *J Mol Catal B: Enz* 23: 87-100.
104. Mathew S & Abraham TE (2005) Studies on the production of feruloyl esterase from cereal brans and sugarcane bagasse by microbial fermentation. *Enzyme Microb Technol* 36: 565-570.
105. Miyazaki JH & Croteau R (1990) Immobilization of cyclase enzymes for the production of monoterpenes and sequiterpenes. *Enzyme Microb Technol* 12: 841-845.
106. Tsai H-J & Sardine WE (1987) Conjugal transfer of lactose-fermenting ability from *Streptococcus lactis* C2 to *Leuconostoc cremoris* CAF7 yields *Leuconostoc* that ferment lactose and produce diacetyl. *J Ind Microbiol Biotechnol* 2: 25-33.
107. Gabelman A (1994) *Bioprocess Production of Flavour, Fragrance, and Color Ingredients*, Wiley, New York, USA.
108. Berger RG (1995) Application of genetic methods to the generation of volatile flavours. In: *Food Biotechnology,* Hui HY & Katchatourians G (eds), VCH, Weinheim, Germany, pp 281-295.
109. Wein M, Lavid N, Lunkenbein S, Lewinsohn E, Schwab W & Kaldenhoff R (2002) Isolation, cloning and expression of a multifunctional o-methyltransferase capable of forming 2,5-dimethyl-4-methoxy-3(2H)-furanone, one of the key compounds in strawberry fruits. *Plant J* 31: 755-765.
110. Martin VJJ, Pitera DJ, Withers ST, Newman JD & Keasling JD (2003) Engineering a mevalonate pathway in *Escherichia coli* for production of terpenoids. *Nat Biotechnol* 21: 796-802.

# Biopreservatives

# 21

**Neerja Rana,[1] V.K. Joshi[2] and L. Sophia[2]**

[1]*Department of Vegetable Crops,* [2]*Department of Food Science and Technology, Dr. Y.S. Parmar University of Horticulture and Forestry, Nauni, Solan 173 230 (H.P.)*

## 1. INTRODUCTION

The single most important development permitting the formation of civilization was the ability to produce and store large quantity of food. Hunter gatherer societies lived from day to day by either starving or gorging themselves based upon the amount of food they could find in a day. When it became possible for one person to produce more food, it is beneficial to be able to store as much food as possible in order to minimize the amount of time spent gathering food. Food storage has always been at odds with food spoilage. Some of the earlier evidences of food preservation come from the post glacial era, from 15000 to 10,000 BC. Over the years mankind has developed several processes to enhance the safety and prolong shelf life of perishable food. In developing societies food preservation is viewed as a convenience of an efficient food system. During the past few years, food safety has become an increasingly important international concern. Recent outbreaks of *Listeria monocytogenes* have promoted the food industry, public and government to question the adequacy of current methods of food preservation and preservatives.

A preservative means any substance which when added to food is capable of inhibiting, retarding or arresting the process of fermentation, acidification or decomposition of food. The preservative acts by interfering with the genetic mechanism of cell division, either by denaturation of vital genes [1] or by oxidation in autocatalytic synthesis of the polypeptide [2]; or by interfering with the cellular enzymes of the microorganism [3]; or by altering the permeability of the cellular membrane of organisms [4]. The preservatives may be chemical or biological. The chemical preservatives act with the interference of cellular machinery of food in such a way that may lead to loss of vital characteristics of food and it may even cause addition of some harmful products into the food because of chemical used. Therefore due to the adverse effect of chemical preservatives on health and also of increasing consumer awareness about more natural products without chemical preservative, it has necessitated the need to exploit the biologically derived antimicrobial substances as preservative commonly known as biopreservative [5]. Biopreservation means the application of biopreservatives at low concentration having generally recognized as safe (GRAS) status and not imparting any deleterious effect to the food in which it is used. The nutraceuticals benefits and the healing power of certain bioactive food ingredients that had been noticed for centuries came under close scientific scrutiny. A multifaceted application from food science, microbiology and medicine has established 'biopreservative' as an emerging area of study.

Many food-borne pathogens are known to cause spoilage in food. In recent years, *Listeria monocytogenes* has emerged as a food borne pathogen of major concern and several studies have been con-

*Food Biotechnology: Principles and Practices*
*V.K. Joshi & R.S. Singh (Eds.)*
*I.K. International Publishing House Pvt. Ltd., India*
*pp 739–780*

ducted dealing with the occurrence of *Listeria monocytogenes* in different food environments. *Listeria* multiplies at low temperature of 3°C and hence grow frequently in refrigerated and caned foods. *L. monocytogenes* is the cause of listerosis food-borne disease in human beings. In an American survey, this disease affected more than 1600 Americans and caused 415 deaths nation wide per year. One study has shown that bacteriocin of *Leuconostoc carnosum* could be the key for the eradication of this deadly disease from packaged food. *L. carnosum* also grow at temperature 1-5°C [6] and its bacteriocin kills 99.9 per cent of *Listeria* [7]. Some other Gram-positive bacteria such as *S. aureus* and *B. cereus* are known to cause food intoxification especially in heated and processed vegetables [8]. *Clostridium botulinum* types A and B were isolated from food of plant origin such as mushrooms, green pepper and cabbage. A few of food spoilage and pathogenic bacteria such as *A. hydrophila, Micrococcus, Enterococcus, S. typhimurium, Leuconostoc, Streptococcus, E. coli*, etc. are known to be present in various foods such as vegetables, milk, meat, cheese, etc. [9, 10]. Thus, in spite of the introduction of modern technologies and preservatives, the reported number of food-borne illness and intoxication are on the rise. Many of the ready to eat and novel products represent new food systems with respect to health and spoilage risks. In the light of above as well as improved understanding and knowledge of the complex microbial interactions, there is an increased tendency to use biopreservatives in the form of protective culture or their metabolites. Biological preservation implies a novel approach to improve the microbial safety of foods. This concept refers to the use of antagonistic microorganisms or their metabolic products to inhibit or destroy undesirable microorganisms in foods. Any biopreservative to be used commercially should fulfil the following requirements:

- The biopreservative to be used should not be toxic and must be acceptable by recognized authorities.
- It should not have deleterious effect on the organoleptic properties of product.
- Its use should be economical to industries.
- The biopreservative should be sufficiently stable if being stored.
- It should not have any medicinal use.

In this chapter an endeavour is made to highlight the features of natural biopreservatives and their possible applications in preservation of food.

## 2. BIOPRESERVATIVES VIS A VIS ANTIMICROBIALS

Biopreservatives are the antimicrobials, which are used to extend the lag phase or kill microorganisms. These have long been investigated for their effectiveness to kill or inhibit the growth of microorganisms in and on foods. This has been done as an effort to increase the shelf-life of food products. As part of process control activities in food manufacturing plant, antimicrobial agents have been successfully applied both in the product formulation stage as direct food additives designed to reduce or eliminate pathogens or spoilage organisms and as processing aids or secondary food additives during the food production process. Table 1 shows the naturally occurring and traditional food preservatives approved by food and drug administration. Naturally occurring antimicrobials include compounds that originate from microbial, plant and animal sources. Many additional antimicrobials, especially those derived from microorganisms, hold the potential for regulatory approval in future (Table 2). The ability of antimicrobials to be used is to prevent decontamination of food as described by US Department of Agriculture's Food Safety and Inspection Service Directives (USDA FSISD) and defined by the US Food and Drug Administration (FDA).

Antimicrobial agents can be classified into three categories:

(1) *Processing aids*: As by FDA processing aid, antimicrobial agent should be added to the food during processing and is either removed or converted into normal food constituents or functional additives that leave insignificant non-functional residuals.

(2) *Secondary direct food additives*: These are added during the manufacturing or processing of food as functional additives but are removed from the final food product. Residual of such additives cannot have any technical effect.
(3) *Direct food additives*: These agents provide technical effect to the final food product and should be listed on the food label by their common name.

**Table 1:** Traditional and naturally occurring food antimicrobials approved by the Food and Drug Administration.

| Compound | Microbial agent | Primary food application |
|---|---|---|
| Acetic acids, acetates, diacetates, dehydro acids, fats/oils, benzoic | Yeast, bacteria | Baked goods, condiments, confections, dairy products, meat products |
| acids, benzoate | Yeast, moulds | Beverages, fruit products, margarine |
| Dimethyl dicarbonate | Yeasts | Beverages |
| Lactic acids, lactates | Bacteria | Meats, fermented foods |
| Lactoferin | Bacteria | Meat |
| Lysozyme | *Clostridium botulinum* | Cheese, Frankfurters, cooked meat and poultry products |
| Natamycin | Moulds | Cheese |
| Nisin | *Clostridium botulinum* | Cheese |
| Nitrite | *Clostridium botulinum* | Cured meat |
| Parabens alkyl esters | Yeasts, moulds | Beverages, baked goods |
| Propionic acids and propionates | Moulds | Bakery products, dairy products |
| Sorbic acids, sorbates | Yeasts, moulds, bacteria | Most foods, beverages, wines |
| Sulphites | Yeasts, moulds | Fruits, fruit products |

*Source:* 11, 12

**Table 2:** Sources of some naturally occurring compounds having potential for food preservatives.

| Sources | Antimicrobial |
|---|---|
| **Animals** | |
| Milk | LP system |
| Milk, eggs | Lysozyme |
| Milk | Lactoferin |
| Honey | Glucose oxidase |
| **Plants** | |
| Spices, herbs | Essential oils, phenolics, isoprenoids |
| Onion, garlic, horse radish | Sulphur compound |
| *Brassica* | Isothiocyanates |
| Grape fruit seeds | Grape fruit seed extract |
| **Microorganisms** | |
| *Lactococcus* | Nisin, Lactin |
| *Pediococcus* | Pediocin |
| *Lactobacillus* | Lactocin, helviticin, sakacin, bavaracin, curvacin |
| *Leuconostoc* | Leucocin, Mesentricin |
| *Carnobacterium* | Carnocin |
| *Streptomyces natalensis* | Natamycin |

The use of biopreservatives depends on several factors such as desired effect, legal limits of use and effect on the food.

## 3. CLASSIFICATION OF BIOPRESERVATIVES

The different types of biopreservatives can be grouped into five classes: (1) Acid-biopreservatives, (2) Ovo-biopreservatives, (3) Lacto-biopreservatives, (4) Bacto-biopreservatives, and (5) Phyto-biopreservatives.

### 3.1 Acid-Biopreservatives

Some of the most widely studied biopreservatives include organic acids such as lactic acid, acetic acid and citric acid. These acids can also enhance or contribute to the flavour of acidified or fermentive foods such as sausages, cheese, pickles or sauerkrauts. The effectiveness of organic acids as antimicrobials differs widely based on concentration, pH, molarity and concentration of non dissociated form. These acids when used in combination with other compounds have exhibited synergism. Organic acids are amply found in a variety of fruits including apples, bananas, grapes and peaches, etc. Acids inhibit microbial growth by lowering the pH, affecting the proton gradient across biological membrane, acidifying cytoplasm and interfering with chemical transport across cell membrane. Acid environment not only limits microbial growth, but also enhances the destruction of microorganism with other antimicrobial system.

#### *3.1.1 Lactic acid*

Lactic acid, as the term implies, is a milk associated acid also known as 2-hydroxyl propionic acid present in almost all life forms. It is present as an intermediary of the carbohydrate and amino acid metabolism. Currently the market for lactic acid and its derivative amount to 100000 MT annually with a steady growth rate of 12 to 15 per cent per year.

3.1.1.1 Molecular characteristics: Lactic acid is a weak organic acid and one of the smallest molecules with an asymmetrical carbon in location α of the carboxylic function (Fig. 1). The L and D symbol indicate the spatial configuration of the molecule with reference to glyceraldehyde isomer. In addition to optical activity, this molecule has simultaneous carboxylic and hydroxyl group, allowing it to react either as an acid or an alcohol. This dual functionality results in spontaneous formation of intermolecular esters with discharge of water, leading to thermodynamic equilibrium of concentration and age of solution. The kinetics of this condensation reaction is dependent on the temperature, presence of catalyst and difference in equilibrium of solutions. The antimicrobial effect of acid is larger when it is protonized i.e., not dissociated.

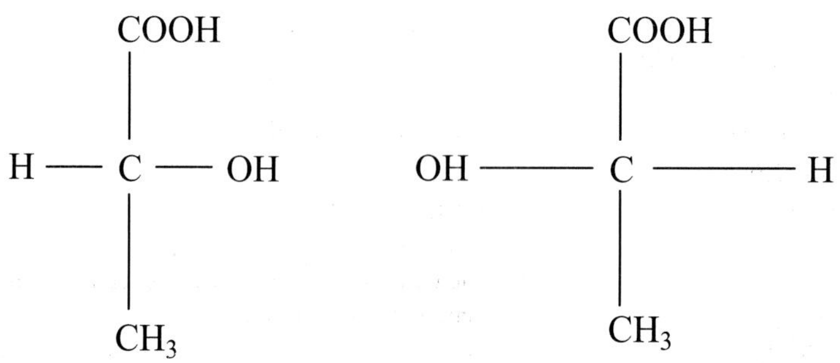

**Fig. 1:** Structure of D and L optical isomers of lactic acid.

The salts of lactic acid are numerous. More than fifty metallic salts and close to twenty double salts containing lactate ion have been mentioned [13]. But potassium and sodium lactate are widely used in food industry. Generally lactic acid is used by food industry due to the following properties: (1) The acidification potential, (2) pH regulation property of sodium and potassium lactate, (3) reduction of water activity, (4) synergism with common antioxidants, and (5) antimicrobial activity.

3.1.1.2 Mechanism of action as biopreservative: The purpose of most antimicrobial agents is to reduce or eliminate microbial activity by stasis or cidal mechanism. In the food industry, antimicrobial effect of lactic acid is considered at two stages: (1) decontamination of food, and (2) shelf-life enhancement of fresh food.

Although different mechanisms may be considered [14], the two main ones are: (1) capacity to re duce pH, which facilitates its penetration through the cell wall [15, 16, 17]; and (2) the effect of lactate on water activity in foods with intermediate humidity [18, 19].

Lactic acid elicits antimicrobial effects on a variety of microorganisms (Table 3). This antimicrobial activity is due to its ability to reduce pH. Thus, in fermented food, lactic acid with other antimicrobial factors inhibits competing microorganisms.

3.1.1.3 Uses: Lactic acid is a hygroscopic, syrupy liquid with a moderately strong acid taste. It is used in manufacture of jams, jellies, sherbets, confectionery products and beverages. It is used to adjust acidity in brines for pickles and olives. Salts of lactate are also used as firming agents for apple slices to prevent discoloration in fruit and baking powders. Its spray is effective in limiting the growth of microbes on meat carcasses [20, 21, 22]. Lactic acid spray and dip of prechilled birds have also been used successfully in poultry industry [23, 24].

Lactic acid bacteria (LAB) including the genera *Lactobacillus, Leuconostoc*, and *Pediococcus* have long been used in fermentations to preserve the nutritive qualities of various foods. The major metabolite of lactic acid bacteria is lactic acid. The production of this organic acid is responsible for the association of drop of pH, which may be sufficient to antagonize many microorganisms. In addition to direct effect on the pH, the undissociated form of the molecule can cause the collapse of the electrochemical proton gradient of susceptible bacteria, leading to bacteriostasis and eventual death. Antagonistic activities of lactic acid bacteria are summarized in Table 4. Outside its use in food fermentations, the main application of lactic acids in the food industry is in the decontamination of meat and poultry carcasses. Acetic acid and propionic acids are also produced in small amount by lactic acid bacteria. They act in a similar manner to lactic acid and are widely used as food additives,

**Table 3:** Antimicrobial spectrum of lactic acid and lactate on various microorganisms.

| Susceptible microorganism | Antimicrobial effect |
|---|---|
| *Mycobacterium tuberculosis* | Inhibit proliferation at low pH |
| *Bacillus coagulans* | Inhibit growth in tomato juice |
| *Yersinia enterocolitica* | Inhibit anaerobic growth (100 mM) |
| *Aeromonas hydrophila* | Inhibit aerobic growth at pH 5.5 |
| *Aspergillus* sp. | Inhibit aflatoxins BI & GI production |
| *Enterobacteriaceae* | 2% acid spray of beef caracasses |
| *Lactobacillaceae* | 0.2% acid dip of porcine livers |
| *Helicobactor pylori* | pH dependent growth inhibition |
| *Clostridium botulinum* | Inhibit growth in raw turkey |
| *Listeria monocytogenes* | Delay the growth |

however, they are not usually derived from LAB fermentations for this purpose. Acetic acid of fermented food has a synergistic antimicrobial effect as that of lactic acid. Diacetyl and acetyldehyde, in addition to imparting aroma and flavour to cultured dairy products, also have an antimicrobial effect. Acetyldehyde can inhibit the cell division in *E. coli,* and diacetyl inhibits yeasts and Gram-negative and Gram-positive bacteria. However, the use of diacetyl as a food preservative is precluded owing to its intense aroma and large amounts required for preservation. Hydrogen peroxide is also produced by a large number of lactic acid bacteria lacking the enzyme catalase, in particular by *Lactobacillus* sp., and inhibits other microorganisms such as *Staphylococcus aureus* and *Pseudomonas* sp. The lactoperoxidase system is a naturally occurring antimicrobial system in milk in which the antagonistic hypothiocyanite radical is generated by the action of lactoperoxidase on thiocyanate and peroxide. This system has been successfully used to extend the shelf life of raw milk and cottage cheese and to inhibit pathogens. Food preservation may be limited by the oxidizing nature of the molecule, and free radicals produced may have profound effect on the sensory quality, causing rancidity of fats and oils and discoloration reactions.

The other acids which can be used as biopreservatives are sorbic acid, acetic acid and citric acid which are briefly discussed here.

### *3.1.2 Sorbic acid*

Sorbic acid and its more water-soluble salts, especially potassium sorbate, are well established safe microbial agents and are used throughout the world as preservative for various foods, animal feeds pharmaceutical, cosmetic and various industrial applications [25]. Sorbic acid was first isolated from the oil of rowan berries (Sorbapple or mountain ash tree) by A.W. Von Hofman in 1859, while its antimicrobial properties were discovered in the late 1930 and 1940 in Germany and United States by E. Miller and C.M. Goodings, respectively [25-27]. The evaluation of sorbic acid as preservative increased in late 1940 and 1950 when it became commercially available.

**3.1.2.1 Molecular properties:** Sorbic acid is a naturally occurring trans unsaturated fatty acid (2-4 hexadienoic $CH_3CH=CH-CH=CHCOOH$) with its highly reactive carboxylic group forming salts and esters, while its conjugated double bonds are also reactive. The reactivity of sorbate in food system may influence its activity as an antimicrobial agent as well as the safety of products. The solubility of sorbic acids is higher in lipid than water (0.15°%, w/v at 20°C), whereas, its salts, especially potassium sorbate are more soluble in water. Solubility is high with increase in temperature and increasing concentration of ethanol, acetic acid and propylene glycol, whereas it decreases with increasing concentration of sucrose and sodium chloride. In the pure and dry form, sorbic acid and its derivatives are remarkably resistant to oxidation but in aqueous solution, they are relatively unstable and degrade by first order reaction kinetics [25]. The olefinic bonds of sorbate are attacked by oxidizing agents, resulting in the formation of peroxides followed by their degradation or polymerization. Products of sorbic acid oxidation include carbonyl compounds such as crotonyldehyde, malonyldehyde, acrolein, formic acid and malonic acid

**Table 4:** Antagonistic activities caused by lactic acid bacteria.

| Metabolic product | Mode of antagonistic action |
|---|---|
| Carbon dioxide | Inhibits decarboxylation, reduces membrane permeability |
| Oxidizes basic protein | Interact with arginine binding proteins |
| Diacetyl | Hydrogen peroxide / lactoperoxidase |
| Lactic acid | Undissociated lactic acid penetrates the membranes, lowering the intracellular pH. It also interferes with metabolic processes such as oxidative phosphorylation |
| Bacteriocins | Affect membranes, membrane associated replication, DNA and protein synthesis. |

[28]. Factors affecting the rate of oxidation include pH, sugar, salt, amino acids, antioxidants, metal ions, light, temperature, time, types of food material present and properties of packaging material. Low or no oxygen permeability is desirable in packaging to avoid oxidation. Diffusion of sorbic acid into food is desirable when preservative of the core of the product is necessary instead of only the surface layer. When there is no need for microbial inhibition in the interior of the product, then it is desirable that sorbate remain on the surface where its activity is needed. Uptake and diffusion of sorbic acid into model system or foods is affected by moisture content and water activity of substrate, its structure and hydration properties. Diffusivity appears to decrease with an increase in the concentration of gelling agents, decreasing concentration of sorbic acid, increasing concentration and molecular weight of solutes and decreasing temperature [25, 28, 29].

Sorbate may be applied by direct mixing into the food formulation by treating food wrappers and packaging material that come in contact with food by immersing, dipping or spraying of food with solutions or by dusting food with dry form of compound.

3.1.2.2 Mechanism of action: Sorbic acid has been described as a competitive and reversible inhibitor of amino acid producing bacterial spore germination by acting on the connecting reactions that follow the initiation of germination process. The exact mechanism of spore germination is unknown, thus exact mechanism of inhibition cannot be given. Indirect evidence suggests that inhibition could involve:

(1) Alteration of permeability of spore membrane or inhibition of spore lytic enzymes that may be involved in germination by causing cortex hydration and loss of refractivity [25, 28, 29, 30].
(2) Generally, the amount of sorbic acid employed in food preservation ($< 0.03\%$) inhibits or delays the growth of vegetative microorganisms without causing permanent or lethal damage to their cells, while higher concentration may cause inactivation and inhibition of microbial cell metabolic function. Different mode of inhibition may be operational in specific microbial species and different substrate and environmental as well as processing parameters [25, 28, 29, 31].
(3) Inhibition of microbial cell metabolism in the presence of sorbic acid has been detected as a decrease in assimilation of carbon from substrates such as glucose, acetate, lactate, ethanol, succinate, pyruvate and oxaloacetate. Sorbate is known to inhibit the *in vitro* activity of many enzymes, especially sulphydral containing enzymes [32, 33].
(4) Sorbic acid may also inhibit microorganisms by interfering with the mechanism of electron and substrate transport in the cell, leading to growth inhibition by starvation.
(5) Sorbic acid has been found as an inhibitor of mould growth through the depletion of ATP level in conidia [25, 29].

3.1.2.3 Applications: Sorbic acid and its potassium salt are widely used as antimicrobials throughout the world in the preservation of human food, animal feed, pharmaceutical, cosmetic products, packaging material and technical preparations that come in contact with food or human body, different foods like dairy products, bakery items, fruits and vegetable products, edible fat emulsion products, certain meat and fish products, sugar and confectionery items [25, 28].

### *3.1.3 Acetic acid*

Acetic acid is a naturally occurring and relatively inexpensive organic acid, which has been traditionally used as a food preservative. Whether naturally produced during fermentation or intentionally added, these acids retard microbial growth and contribute desirable sensory properties to a number of foods.

3.1.3.1 Production: Acetic acid historically diluted in the form of vinegar has been utilized perhaps longer than any other preservative for its antimicrobial effect that influences food keeping quality, wholesomeness and safety. Early uses of vinegar were for medicinal purposes such as wet compresses or consumption as a drink elixir [34]. Earlier processes of vinegar making are: (1) The let alone process, (2) The field process, (3) The Orleans process, (4) The quick process, and (5) The submerged culture process.

Of these, the quick process and submerged culture process were developed and used for commercial vinegar production [34]. Industrial development and advances in vinegar production has been reviewed by a number of co-workers [35, 36]. Acetic acid is formed in four step reaction involving conversion of starch to sugar by amylases, anaerobic conversion of sugar to ethanol by yeast fermentation, conversion of ethanol to hydrated acetaldehyde and dehydrogenation to acetic acid by aldehyde dehydrogenase [34, 37]. The last two steps are performed aerobically with the aid of acetic acid forming bacteria. Acetic acid yield from fermented sugar is approximately 40%, and yield improvement can be achieved with high aeration rates during continuous production [38].

Cheryan *et al.* [39] reviewed that anaerobic process of acetic acid production by *Clostridium thermoaceticum* is more economical and is typically used for the manufacture of industrial acetic acid. The microbial action of acetic acid bacteria to produce vinegar has been described difficult when the pure culture is used [34].

3.1.3.2 Structure and properties: Acetic acid (ethanoic acid $CH_3COOH$) is a colorless liquid having a pungent vinegar-like odour and sour taste. It is an widely ionized in aqueous solution. It is an excellent solvent for organic compounds and is miscible with water, alcohol, glycerol and lipids. Advantages of using acetic acid and acetate include relatively low cost, GRAS status and low toxicity. The major disadvantage of the use of acetic acid in food is its acidic taste and strong sourness intensity. Another problem is its potential to cause delaminating of packaging materials.

3.1.3.3 Mechanism of action: Acetic acid has been evaluated for *in vitro* antimicrobial efficacy against a specific pathogen using microbial media/food [40]. The various mechanisms of action may be:

(1) When acetic acid dissolves in solution, it releases free proton which decreases the pH. This may increase the number of protons on the outer surface of microorganisms, which disrupt the membrane of microorganisms by denaturing enzymes [41-44].
(2) Undissociated acetic acid can also transverse the lipid bilayer of bacteria and yeasts and release protons into the cytoplasm [45, 46].

Physiological response of certain microorganisms to varying concentration of acetic acid has been reported [47, 48].

3.1.3.4 Applications: Acetic acid, used and label listed as vinegar, has been employed as a flavouring ingredient, but has also functioned primarily as an acidulant or preservative in foods [49]. Excluding vinegar, the foremost food types important when considering acetic acid are mayonnaise, salad dressing, ketchups/catsups, sauces, condiments and pickles are relished [50-52]. Use of acetic acid as a sanitizing agent of food contact surface has shown promise.

(1) Decontamination of meats - use of acetic acid for decontamination of meat has been reviewed by Dickson and Anderson, [53], Frederick [54], Siragura [55], Bolder [56], Smulders and Greer [57].

(2) Decontamination of poultry - It increase the shelf-life of poultry by inhibiting spoilage bacteria [58], aid in controlling *Salmonella* [59, 60].
(3) Decontamination of sea food [61]
(4) Decontamination of produce [62, 63]
(5) Natural preservation system [64]

#### *3.1.4 Citric acid*

Citric acid was first isolated from lemon juice by a Swedish chemist in 1784. Citric acid plays an important role in the food and beverage industry, as well as in the manufacture of pharmaceuticals and cosmetics. Although citric acid was initially produced from citrus fruits, but due to increasing demand, alternate methods were also used. Wehmer [65] produced citric acid by fungal fermentation process using *Penicillium* species. In 1917, Currie used *Aspergillus niger* to produce citric acid in a high yielding fermentation process. Among various substrates, beet molasses was found economical for large scale production. In 1949, Perlman reported submerged fermentation process using *Aspergillus* and molasses. Since 1965, different methods were developed and only small amount is produced from citrus fruits.

**3.1.4.1 Production and properties:** Citric acid is produced as a bioactive molecule by plants and animals during metabolism. Acetyl CoA from various oxidation processes enter citric acid cycle and with oxaloacetate from citric acid.

$$\text{Acetyl CoA} + \text{Oxaloacetic acid} \rightarrow \text{Citroyl CoA} \rightarrow \text{Citric acid} + \text{CoA}$$

Citric acid does not accumulate in normal physiological process, however *Aspergillus,* yeast and certain bacteria accumulate citric acid in large quantities. A rapid flux of carbon through glycolysis to pyruvate and restriction TCA prevents the breakdown of citric acid and result in its rapid accumulation. Its chemical formula is $C_6H_8O_7$ and nomenclature is 2-hydroxyl-1,2,3-propane tricarboxylic acid. Its molecular weight is 192.13. It has a pleasant odour and sour taste. The physical and chemical properties of citric acid are listed in Table 5.

**3.1.4.2 Mechanism of action:** Citric acid, like many other organic acids, possesses antimicrobial activity. This activity of citrate has been attributed to its ability to chelate metal ions [66]. *Lactobacillus casei* is inhibited by sodium citrate induced metal chelation at 12-18 μM concentration. Chelation effect also appears to be an influencing factor in the growth inhibition by *Staphylococcus aureus* [67] and *Arthrobactor pascens* [68]. The antimicrobial outcome is influenced by various factors including pH, concentration, temperature, chain length, etc. Fabian and Graham [69] found that citric acid was more inhibitory to thermophilic bacteria than lactic acid and acetic acid. Concentration of 12-12.5 per cent sodium citrate was inhibitory to *Salmonella anatum* and *Salmonella oraniennburg. Staphylococcus*

**Table 5:** Properties of citric acid.

| | |
|---|---|
| Molecular weight | 192.1 |
| Melting point | 153°C |
| Solubility | 146 g dissolves in 1l water at 20°C |
| Dissociation constant | $pk_1$ 3.14 |
| | $pk_2$ 4.77 |
| | $pk_3$ 6.40 |

*aureus* is inhibited by increasing concentration of citric acid and decreasing pH [70]. Sodium citrate could elicit bacteriostatic activity. Ahmad and Marth [71] examined the ability of acetic acid, citric acid and lactic acid to prevent the growth of *Listeria monocytogenes* in TB during extended incubation of 7-35°C. Some studies have demonstrated synergism between citric acid, polyphosphoric acid and EDTA with antimicrobial such as monolaurin, monocarpin, sucrose dicaprylate and sucrose monolaurate. This exhibits a high antibacterial activity against *E. coli* and other Gram-negative bacteria. It was concluded that transport of monolaurin into the cell membrane of *E. coli* was enhanced by the action of citric acid, polyphosphoric acid and EDTA [72, 73].

Citric acid acts as buffering agent to maintain the pH 4.5-5.5, thus increasing the storage life of refrigerated food. El Shenawy and Marth [74, 75, 76] investigated the anti-listerial activity of sodium propionate in combination with common organic acid at 13 and 35°C.

3.1.4.3 Industrial production: Currently there are two organisms that are widely used in the production of citric acid i.e, yeast and *Aspergillus niger*. Many other fungi are also able to produce citric acid such as *A. clavatus, A. fumigatus, A. citrinum* and *Candida* species. Yeast has certain advantages over *Aspergillus* due to its independence from trace metals and higher fermentation rates. It also has the potential for higher sugar concentrations using osmophilic variants [77]. Citric acid is also produced by submerged liquid fermentation, liquid surface and Koji methods [78, 79].

## 3.2 Ovo-Biopreservatives

In contrast to the immune system of animals, which produce antimicrobial polypeptides when needed, the egg efficiently resists microorganism over a long period in absence of such innate and systemic host defence. The wide diversity in the nature of microorganism implies that ovoantimicrobial possess multiple functions and many different types of protein are found in egg albumen, most of which appear to have biopreservative properties. The main biopreservatives under this class are lysozymes.

### *3.2.1 Lysozyme*

Lysozyme (1,4 β-N acetylmuramidase) is a lytic enzyme found in many natural systems. It is a small and stable enzyme whose dimensional structure and sequence were completely analysed [80]. Lysozyme has a high potential in food preservation because of its stability over a wide range of pH and temperature [81]. It is attractive as a natural food preservative because it is endogenous to many foods, specific to bacterial cell walls and harmless to humans. There are microbial, viral, phage, insect, plant, and animal tissue lysozymes. Body fluids such as tears, saliva, urine and human milk contain 2.6, 0.13, trace and 0.2-0.4 mg/ml of lysozyme, respectively [82].

3.2.1.1 Sources: Egg-white lysozyme is the representative of the lysozyme family and the related enzymes are called C type (chicken or conventional type) lysozyme. One egg contains about 0.3-0.4 g of lysozyme. The eggs have been the easiest and most economical source for recovering lysozyme. The radically different lysozyme called g type (Embedengoose) was found only in birds [83]. Further, a v type (viral) family has also been identified and defined as phage lysozyme. The lambda is different from other lysozymes as it is not a hydrolase but a transglycosidase. Several other phage lysozymes have been identified and named after the first member of the family investigated. In plants, h type (hevamine type) and b type (barley type) families have been identified [84].

Since we restrict the application of lysozyme as biopreservative in food, only c and g family lysozyme will be discussed herein.

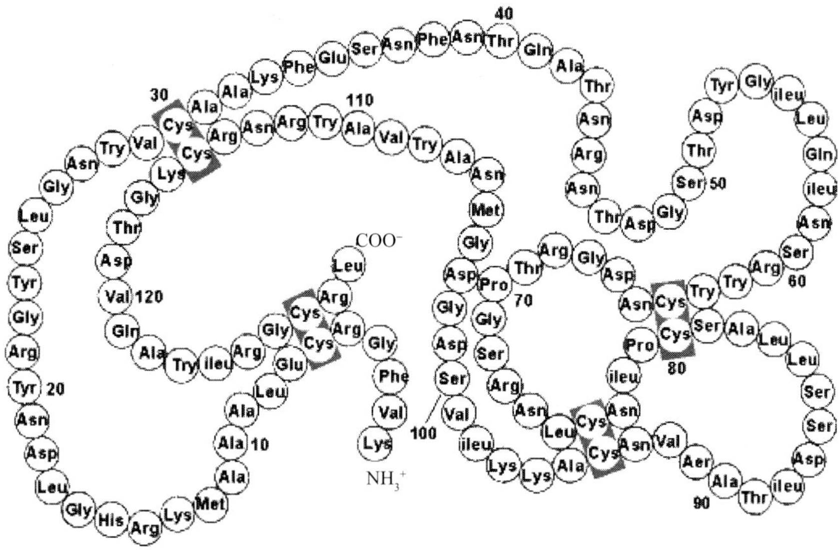


**Fig. 2:** The primary structure of lysozyme indicating the four disulphide bonds.

3.2.1.2 Physico-chemistry: The molecular weight of chicken lysozyme computed from the amino acid sequence of 129 residues is 14,307, and isoelectric point is 10.7. Chicken lysozyme has a higher extinction coefficient EI % at 280 nm than most proteins because of higher content of aromatic amino acids in the lysozyme molecule. Canfield and Mc Murry [85] reported that the g type lysozyme (185 residues) is three times more active than c type (129 amino acid residue). Lysozyme have four disulphide bonds, thereby making the molecule unusually compact. It is considerably heat stable. In dried powder form, lysozyme can be stored for a long time (>6 month) at temperature upto 30°C without losing lytic activity. Chicken lysozyme is stable at 100°C for 2 min at pH 4.5 and at 100°C for 30 min at pH 5.29. But the enzyme is inactivated quickly in alkaline solution. Proctor and Cunningham [81] reviewed the effect of heat, pH and ionic strength on lysozyme activity. Lysozyme is rapidly inactivated in the presence of thiol compounds that split the disulphide linkage, which are essential for maintaining the structure. Proteolytic enzymes such as trypsin, chymotrypsin and papain do not hydrolyse lysozyme, but pepsin does. Chicken lysozyme contains 4 disulphide bonds at 6-12, 30-115, 64-80 and 76-94 with no free sulphydryl (SH) group (Fig. 2). The lysozyme molecule consists of two domains or lobes i.e., alpha and beta domain linked by a long alpha helix between which lies the active site.

3.2.1.3 Mechanism of action: Lysozyme promotes catalysis by inducing stearic stress in the substrate. The two amino acids Asp52 and Glu35 are the precise amino acids that participate in the catalysis. Glu35, which lies in hydrophobic environment, participates in catalysis in the protonated form, and Asp52 which lies in hydrophilic environment, does so in the dissociated form. In case of g type lysozyme with 185 amino acids, it was suggested that Glu73 and Asp86 correspond to Glu35 and Asp52 of c type, respectively. However, Weaver *et al.* [86] found no counterpart of Asp52 in c and suggested that only one acidic residue is essential for the catalytic activity of lysozyme g and further that Asp52 might not be required for catalysis by lysozyme c.

The rate of lysozyme catalysis depends upon the pH of the medium. The pH profile is bell-shaped with a maximum at pH 5.0 and inflections at pH 3.8 and 6.7. The osmotic uptake of water leads to the expansion and eventual rupture of the cytoplasmic membrane, which causes cell death. Water then

**Table 6:** Antimicrobial spectrum of lysozyme against Gram-negative bacteria.

| Family | Species | Sensitivity |
|---|---|---|
| *Enterobacteriaceae* | *E. coli* | –/ ndr /+ |
| | *Salmonella typhimurium* | + / – |
| | *Shigella* | + |
| | *Proteus vulgarius* | – |
| | *Serratia marcescens* | – |
| | *Erwinia* sp. | + |
| | *Yersinia enterocolitica* | $m^+$ |
| *Pseudomonadiaceae* | *P. aeruginosa* | + |
| | *P. fluorescens* | ndr |
| | *Achromobacter* sp. | – |
| *Pasteurella* | *Pasteurella* sp. | |
| *Neisseriaceae* | *Neisseria* sp. | + |
| *Vibrionaceae* | *Vibrio cholerae* | – |
| *Campylobacleriaceae* | *Campylobacter jejuni* | $m^+$ |
| *Klebsielleae* | *Klebsiella pneumoniae* | – |

*Sources:* 88, 90

reacts with the acylium ion and the enzyme is ready for another round of catalysis. Gram-positive bacteria are very susceptible to lysozyme because the cell wall of Gram-positive bacteria consists of about 90% peptidoglycan. In some Gram-positive bacteria such as *Bacillus cereus,* the absence of NAG residue confers resistance to lysozyme [87]. The antimicrobial spectrum of lysozyme is given in Table 6.

In Gram-negative bacteria, the peptidoglycan that makes only 5-10% of the cell wall lies beneath the outer membrane of the cell envelope. The lipopolysaccharide layer of the outer membrane acts as a barrier against macromolecules and hydrophobic compounds. The lipid component of the inner core of the lipopolysaccharides molecules through their content of phosphate and carboxyl groups and their electrostatic interactions with divalent cations reinforce the stability of Gram-negative bacterial cell walls towards microbicidal agent such as lysozyme. The antiviral activity of lysozyme is not associated with its lytic activity and was reported to be associated with the +ve charge of lysozyme [88].

During *et al.* [89] discovered that heat treatment denatured $T_4$ lysozyme as well as hen egg white lysozyme in which the enzymatic activity was abolished, preserved its antimicrobial activity. The membrane perturbing activity of denatured lysozyme was demonstrated on bacterial, fungal and plant cells. The amphipathic c terminal domains of lysozyme seem to have mediated the bactericidal and fungicidal activities of denatured lysozyme. Therefore, the antimicrobial activity of lysozyme may not be limited by heat treatment encountered during food processing operations.

3.2.1.4 Uses: Lysozyme is a protein endogenous to most foods, is safe and is stable under pH conditions mostly encountered in foods. It is available in ready to use form such as liquid or spray dried powder. Lysozyme is not known to undergo any adverse chemical or organoleptic changes that would be of concern under normal food processing conditions. Lysozyme has proven to be an economically viable alternative to chemical preservatives in a number of food applications, most notably in cheese making. Lysozyme is selective at industrial level for certain Gram-positive spoilage bacteria and may not interfere with a number of organisms that are considered to be beneficial for human health. Lysozyme added to foods is destroyed in the human stomach, making it harmless to humans. Lysozymes of both human and avian origin kill Gram-positive

**Table 7:** Uses of lysozymes.

| Industry | Use | Reference(s) |
|---|---|---|
| Cheese industry | (1) To prevent butyric late blowing by killing butyric acid bacteria (*Clostridium tyrobutyricum*) present in milk.<br>(2) Prevent the development of undesirable microorganisms in butter and cheese for more than 2 months<br>(3) Delays the growth of 4 strains of *Listeria monocytogenes* | 94, 95, 106 |
| Alcoholic beverages | (1) To inhibit malolactic fermentation by using lysozymes<br>(2) Microbial stabilization and to prevent acetic acid and biogenic amines lysozyme<br>(3) As an antimicrobial in unpasteurized beer<br>(4) Lysozyme and less $SO_2$ may allow good control of bacterial flora and production of quality wine with less $SO_2$ | 94-97 |
| Fish and sea food products | (1) Lysozyme and NaCl have been used to preserve low quality salmon or trout roe<br>(2) Lysozyme and glycine in combination are effective preservative of fish paste | 98-100 |
| Fruits and vegetables | (1) Vegetable and fruit products decontaminated by using hen egg white lysozyme<br>(2) Chelators in combination with lysozyme delayed toxin production in potato suspension<br>(3) Fresh vegetables, *tofu, kimchi,* Japanese potato salad, *sushi,* Chinese noodles and creamed custard have been preserved by using lysozyme or combination of lysozyme with amino acid<br>(4) Lysozyme in combination with glycine preserves fruit juice, uncooked noodles, bean jam, and custard cream<br>(5) Fruit and vegetable were preserved by lysozyme chloride, lower fatty acid, monoglyceride phytic acid and sodium acetate | 97-103 |
| Chewing gum and tooth paste | (1) Lysozyme in chewing gum formulations was effective against periodontilis causing bacteria and gingivitis associated with inflammation and bleeding<br>(2) Lysozyme used in toothpaste in combination with chloride, flouride, thiocyanate and bicarbonate to lyse *Streptococcus* mutants<br>(3) Lysozyme with EDTA effective for gum infection | 104-106 |
| Meat products | (1) Nisin in combination with lysozyme was more effective than lysozyme alone for preserving meat products<br>(2) Vienna sausages were best preserved by a combination of dipping the casings in 0.05%<br>(3) Lysozyme plus phosphate buffer at pH 6.5 to cure meat | 103 |
| Animal feed | Lysozyme (5-20mg/Kg feed) administered simultaneously with 100,000 to 5,000,000 viable cells of *Bacillus cereus* CIP 5832 in feed has improved the growth of animals and chicks | 104 |
| Plants | Hen egg, white lysozyme in combination with compounds such as EDTA help permeate the cell wall of Gram–ve bacteria and provide protection against bacterial spoilage. Ruminant lysozyme has been genetically engineered and was shown to provide protection to plants against fungal pathogens | 105 |

bacteria. Lysozyme orally taken has proven to be an effective immuno-stimulant, antiviral and anti inflammatory in cancer patients and individuals with herpetic lesions [91, 92, 93].

Hen egg white lysozyme has several desirable properties as food preservative. The FAO/WHO and several countries have acknowledged the non-toxicity of lysozyme, and have therefore, approved its use in some foods. Several applications have been developed and patented including the treatment of fresh fruits, vegetables, seafoods, meat, *tofu,* and sape food. However, the most important food application of lysozyme is in the cheese industry. Various uses of lysozyme are enlisted in Table 7.

3.2.1.5 Limitations: The activity of lysozyme against bacterial spores may be limited by the ability of the enzyme to penetrate the cell wall of Gram-negative bacteria and the intact bacterial sporecoat. The effect is sometimes overcome when lysozyme is used in combination with metal chelators which disrupt the integrity of cell wall thus allowing lysozyme to lyse the remaining cell wall. The use of lysozyme including a genetically modified with increased heat resistance and other lytic enzyme preservatives increased the risk of growth of non proteolytic *C. botulinum* [102]. Lysozyme is probably not used extensively in processed cheese making because the processing temperature was thought to be typically high enough to cause significant reduction in enzyme activity. Cost has been another factor that has limited the use of lysozyme. However, consumer demands for more natural and healthy food products has increased over time. This hurdle may be overcome as the nutraceutical attributes and lysozyme become more evident.

## 3.3 Lacto-Biopreservatives

Milk is the first complete functional food devised by nature for the protection and development of new-born mammal. The antimicrobial and nutraceutical potential of many components of milk has been unravelled in recent years. These are lactoferrin, lactoperoxidase, lactoglobulins and lactolipids. Lactoferrin has been discussed in this chapter.

### *3.3.1 Lactoferrin*

Lactoferrin is a naturally occurring iron binding protein present in milk. Lactoferrin has a multifunctional role in a variety of physiological pathways and is considered a major component of the pre immune innate defence in mammals. The ability of LF to bind two $Fe^{+3}$ ions with high affinity in cooperation with two $HCO^{-}_{3}$ ions is an essential characteristic that contributes to its major structure functional properties including antimicrobial activities. Lactoferrin is a component of the natural protection system of humans and animals with small amount of lactoferrin found in external secretions such as in saliva, tears, and mucous secretions [107]. The range of lactoferrin present in various biological fluids is given in Table 8.

**Table 8:** Range of lactoferrin in various biological fluids.

| Biological fluids | Range (mg/ml) |
|---|---|
| Colostral breast milk | 7 |
| Mature breast milk | 1-2 |
| Tear fluid | 2.2 |
| Seminal plasma | 0.4-1.9 |
| Synovial fluid | 10-80 |
| Saliva | 7-10 |
| Cow's milk | 20-200 |

*Source:* 113

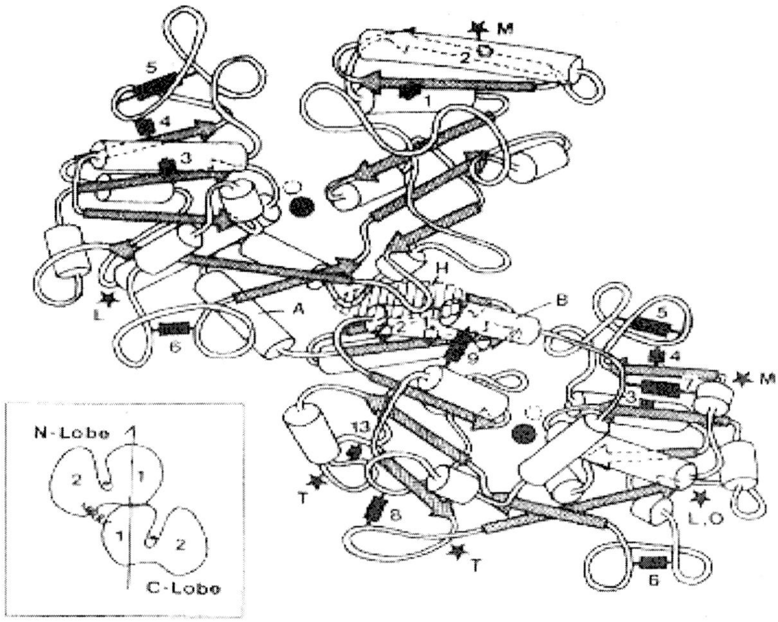


**Fig. 3:** Structure of lactoferrin molecule.

3.3.1.1 Chemical nature and structure: Lactoferrin is a glycoprotein that belongs to iron transporter in a transferrin family. It was first identified in 1939 in milk whey [108]. It is a monomeric metal binding glycoprotein synthesized as a 708-amino acid protein with a 19 amino acid signal peptide [109, 110]. It consists of two globular lobes, each of which contains one iron binding site. These lobes designated as N lobe and C lobes represent as N and C terminal halves of polypeptide [109]. The two metal binding sites are highly similar to each other [111]. Iron binding occurs concomitally with the binding of two bicarbonate anions that appears to play a prominent structural role [112]. In natural state, it is only partially saturated with iron (15-20% for bovine lactoferrin, 4-5% for human lactoferrin) and has a pink colouration due to bound iron. Three-dimensional ribbon structure is shown in Fig. 3.

3.3.1.2 Functions: It is a natural iron binding protein, therefore initial research on lactoferrin was concentrated on its role in iron absorption. Several functions like antimicrobial and antibacterial activity have also been reported. The physiological role of mechanism involved is often unclear and in some cases controversial. Many of its functions are shown in Fig. 4.

3.3.1.3 Consumption studies and safety: The *in vivo* efficacy as well as safety of lactoferrin has been documented in human clinical trials and in several consumption studies in experimental animal models. Infant food formulations have been supplemented with bLF in four clinical trials to human infants with no adverse effects. bLF has also been safely administered orally to cancer patients at 1.6 g/day for six months. The literature describes no report on toxicity of LF.

3.3.1.4 Future perspective as biopreservative: Lactoferrin is an antimicrobial glycoprotein present in milk and various exocrine secretions that bathe the mucosal surfaces. This metal binding protein is a

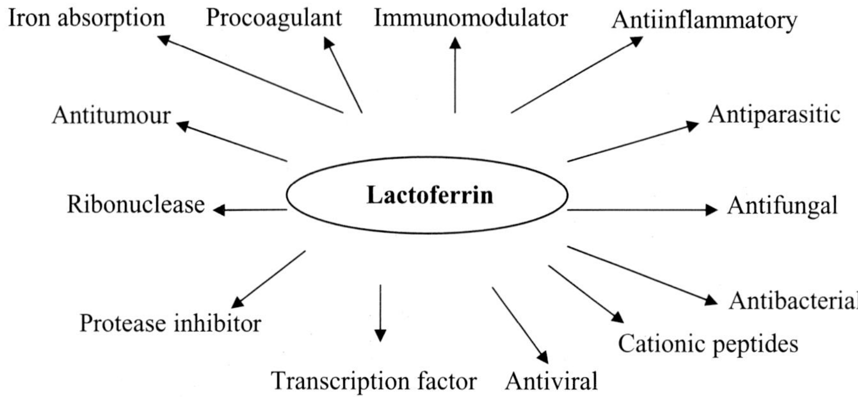


**Fig. 4:** Functions of lactoferrin.

multifunctional bioactive molecule with a critical role in many important physiological pathways. LF could elicit a variety of inhibitory effects against microorganisms comprising stasis, cidal, adhesion, blockade, cationic, synergistic and opsonic mechanisms. This protein is finding its application as an active ingredient in infant formulae and health food in South East Asian countries. Recent advances in LF research technologies for large scale protein isolation and biotechnology is opening unlimited opportunities for this natural antimicrobial molecule in the development of new products and formulations. A number of efficacy studies and trials are ongoing in various laboratories with over 100 patents filed on this molecule in the past 10 years. Undoubtedly, LF is emerging as one of the leading natural microbial blocking agents in food safety and preservation.

### 3.4 Bacto-Biopreservatives

Probiotic strains of lactic acid bacteria are generally enteric flora, believed to play a beneficial role in the ecosystem of the human gastrointestinal tract. LAB are capable of preventing the adherence, establishment, replication and/or pathogenic action of specific enteropathogens. Probiotic LAB have been used to treat disturbed intestinal microflora and increased gut permeability, which are characteristic of many intestinal disorders such as acute rota viral diarrhoea, food allergy, colon disorders, metabolic changes during pelvic radiotherapy and changes associated with colon cancer development. These antagonist properties may be manifested by: (a) decreasing the luminal pH through the production of volatile short chain fatty acids, (b) rendering specific nutrients unavailable to pathogen, (c) decreasing the redox potential of luminal environment, (d) producing hydrogen peroxide, and (e) producing specific inhibitory substances such as bacteriocins.

Bacteriocins are ribosomally synthesized extracellularly released bioactive peptides, which have bactericidal or bacteriostatic effect on others. Bacteriocin producing strains can be readily identified in a deferred antagonism assay in which colonies of the putative producer are overlaid with bacterial lawn of sensitive strain. A number of schemes for differentiating between different bacteriocin types have been proposed, but it is generally agreed that at least three definable classes exist:

*Class I* encompasses the small translationally modified broad host range, lantibiotics, of which nisin is the best known example and described in this chapter.

*Class II* includes the small heat stable unmodified peptides

*Class III* contains the larger heat labile molecules

**Table 9:** Bacteriocins characterized from lactic acid bacteria.

| Bacteriocin | Producer | Inhibitory spectrum | Food source |
|---|---|---|---|
| | *Class I Lantibiotics* | | |
| Nisin (A and Z) | *Lactococcus lactis* | Broad | Milk |
| Lacticin 481 | *Lactococcus lactis* | Broad | |
| Lactocin S | *Lactococcus sake* | Broad | Fermented sausages |
| Carnocin U 149 | *Carnobacterium pisicola* | Broad | Fish |
| Variacin | *Micrococcus varians* | Broad | Meat fermentations |
| | *Class II: Non-lantibiotic, small, heat stable* | | |
| Lactococcin A | *Lactococcus lactis* | Narrow | Cheese |
| Lactococcin B | *Lactococcus lactis* | Narrow | Cheese |
| Lactococcin M | *Lactococcus lactis* | Narrow | Cheese |
| Lactocin F | *Lactobacillus johnsonii* | Narrow | |
| Mesenterocin52 B | *Leuconostoc mesenteroides* | Narrow | Raw milk |
| Curvaticin F S 47 | *Lactobacillus curvatus* | Medium | Ground beaf |
| | *Pediocin-like bacteriocins* | | |
| Sakacin A | *Lactobacillus sake* | Medium | Meat |
| Sakacin P | *Lactobacillus sake* | Medium | Fermented sausage |
| Carnobacteriocin A, B | *Carnobacterium piscicola* | Medium | Meat |
| Pediocin AcH/PA-1 | *Pediococcus acidilactici* | Medium | Meat |
| Leucocin A-UAL | *Leuconostoc gelidum* | Medium | Meat |
| Enterocin 1146 /A | *Enterococcus faecium* | Medium | Fermented milk/milk |
| Piscicolin 126 | *Carnobacterium piscicola* | Medium | Ham |
| Mesenterocin Y 105 | *Leuconostoc mesenteroides* | Medium | Goat's milk |
| | *Class III Large heat labile* | | |
| Helveticin J | *Lactobacillus helveticus* | Narrow | - |

Table 9 shows the non-exhaustive list of bacteriocin characterized to date. Narrow spectrum indicates bacteriocins that only affect the producer genus, and medium spectrum indicates bacteriocin that affects the producer genus and members of one or two other genera. Bacteriocin fulfils the requirement of GRAS. To date, nisin is the only bacteriocin to be commercially exploited on a large scale having gained FDA approval in the USA in 1988. World health organization approved the use of nisin in 1961.

### *3.4.1 Nisin*

Nisin is an antimicrobial substance that has found widespread use as a food preservative over many years. It is a naturally occurring heat stable antimicrobial polypeptide produced by a common dairy organism *Lactococcus lactics*. Nisin demonstrates a selective antimicrobial activity against a range of Gram-positive food spoilage and food poisoning bacteria, including the heat resistant spore forming *Bacillus* sp. and *Clostridium* sp. It has no activity against yeasts and moulds.

The first successful use of nisin as a food preservative was demonstrated in the 1950's by preventing the spoilage of pasteurized processed cheese products. Progress over the last forty years has led to many new food applications being discovered and nisin is now used in a variety of food products in approximately fifty countries of the world. Toxic studies carried out using levels far in excess of those required for food preservation have shown that nisin is non-toxic and non-carcinogenic. Nisin was approved as safe for food use in 1969 by the FAO/WHO joint Expert Committee on Food Additive. It was also granted GRAS status by USFDA in 1988 for use in cheese products.

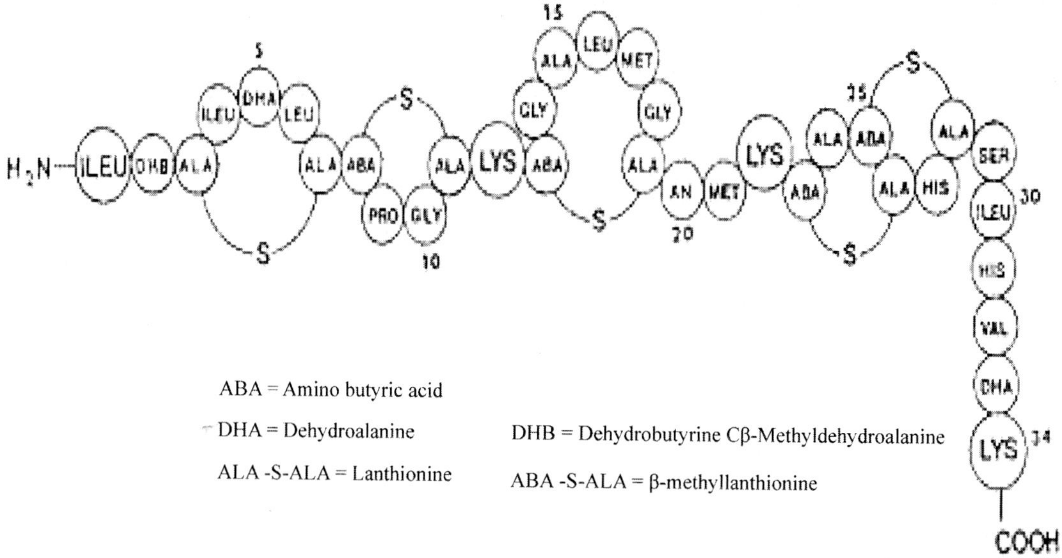


**Fig. 5:** Structure of nisin.
*Source:* 110

3.4.1.1 Structure and stability: The complete structure of the nisin molecule (Fig. 5) was elucidated by Gross and Morell [114]. At that time it was a novel oligopeptide, but since then a number of similar bacteriocins have been characterized [115]. Nisin is a small, heat stable antimicrobial peptide described as a class 1 bacteriocin. Nisin consists of 34 amino acid residues [116, 117]. The molecule possesses amino and carboxyl end groups and five thioether bonds form internal rings. The molecular weight of the structure is 3353 Dalton [118]. It occurs more stable as dimer form (7 kDa), and method for purification and concentration of nisin has been described [119, 120]. Nisin has now been produced synthetically [121].

Nisin molecule is acidic in nature and exhibits greatest stability under acid conditions. It is also more stable at lower pH [122]. Several workers have studied the conformation of nisin in aqueous solution using mostly two-dimensional NMR [123-127]. In solution, nisin is flexible with no overall characteristic conformation. In a solvent that mimics membranes, nisin adopts a rod-shaped amphiphilic structure with clusters of charged and hydrophobic regions over the surface [120, 128] and also occurs as a 14 kDa tetramer [115, 129, 130].

Nisin was reported to be stable to autoclaving at 115.6°C at pH 2, but 40% of its activity was lost at pH 5, and more than 90% was lost at pH 6.8. Large molecules such as milk or broth have a protective effect, so the degree of inactivation is less drastic in foods than in buffers. Loss of nisin activity also occurs during the storage of foods. Losses are more pronounced at high pH and temperature.

3.4.1.2 Antimicrobial properties: Nisin, like other bacteriocins, possesses antimicrobial activity against a limited range of microorganisms. It does not inhibit Gram-negative bacteria, yeasts or fungi. It inhibits a wide range of Gram-positive bacteria, particularly those that produce spores. In general, spores are more sensitive than cells and sensitivity increases in acidic substrates. Less is understood about the activity of nisin against spores. The effect is normally sporostatic, but sporocidal effects have been reported depending upon both the strain and severity of any heat treatment [113]. Nisin inhibits certain strains of species of *Staphylococcus, Streptococcus, Micrococcus* and *Lactobacillus* [131, 132]. Majority of spore forming species of *Clostridium* and *Bacillus* and *Listeria monocytogenes* are sensitive to nisin. There can be a wide variation in nisin sensitivity even among the strains of the same species [127-135]. Meghrous *et al.* [135] compared the activities of nisin A, nisin Z and pediocin and found that lactic acid bacteria, *Clostridium* and *Bacillus* (spores and cells) are more effectively inhibited by nisin variants compared to pediocin. In other studies, *Listeria* species including *Listeria monocytogenes* have shown a wide variation in sensitivity [136, 137]. Scott and Taylor [138] demonstrated that nisin is effective in preventing the outgrowth of *Clostridium botulinum* Type A, B and E spores in bacterial media. Type E spores are the most sensitive with order of sensitivity to nisin being Type E greater than Type B greater than Type A.

Certain Gram-positive bacteria are nisin resistant due to their ability to synthesize enzymes that could inactivate nisin, although this has never emerged as a practical problem in food systems. Spontaneous resistance has been observed in strains of *Clostridium butyricum, Lactobacillus lactis* sub sp. *cremoris, Lactobacillus monocytogenes* and *Streptococcus agalactiae* [139-142]. Resistance was linked with the production of anionic phosphate polysaccharide with subunits of rhamnose and galactose.

3.4.1.3 Mode of action: Numerous studies have been done on the mode of action of both susceptible vegetative cells and spores [140, 143, 144]. Research on vegetative cells indicates that the point of action is the cytoplasmic membranes, and nisin causes disruption resulting either in leakage of essential cellular material such as adenosine triphosphate from the cell or in more severe cases lysis [145-147]. The flexible nature and amphiphilic properties of nisin molecule are important for this character. The discovery that nisin inhibits N-acetyl glucosamine synthesis in *Bacillus subtilis* and DNA, RNA and protein synthesis in *M. luteus* [148] has prompted speculation that nisin interacts with the phospholipid component of membranes. More precise studies on the bimolecular mode of action studied by various workers are summarised in Table 10.

Action of nisin against spores is sporocidal rather than sporostatic. Unlike other preservatives, nisin is believed to inhibit the germination process at the stage of pre-emergent swelling [149, 150]. Germination is not affected by nisin. Nisin activity in food matrix may depend on several factors. These include: (1) changes in nisin solubility and charge, (2) binding of nisin to food components, (3) inactivation of nisin by food ingredients, and (4) environmental factors [151].

**Table 10:** Mode of action of nisin on vegetative cells.

| **Mode of action** | **Reference(s)** |
|---|---|
| Biomolecular mode of action indicates disruption by nisin due to inactivation of sulphydryl groups in the cytoplasmic membrane. | 152 |
| Nisin is a cationic molecule, which is believed to interact with the charged head groups of membrane phospholipid via ionic forces. The C-terminal is responsible for interaction. | 153 |
| The association of nisin with membrane may be charge dependent. | 154, 155 |
| The nisin may form pores in cytoplasmic membranes with the help of energy supplied by transmembrane potential of microorganism. | 156–158 |
| The mechanism of pore formation is still unclear. | 156 |
| The barrel stare mechanism of pore formation. | 159, 160 |
| Nisin is also reported to inhibit cell wall biosynthesis. | 148 |
| Nisin may absorb the teichoic acid of the cell wall and activate autolytic enzyme | 161 |

### 3.4.1.4 Uses of nisin in food products

3.4.1.4.1 Processed cheese: Nisin has proved to be a most effective preservative in pasteurized processed cheese. Anaerobic spores formers in processed cheese are *Clostridium butyricum, Clostridium lyrobutyricum* and *Clostridium sporogenes* [162, 163]. *Clostridium botulinum* in processed cheese spreads with the formation of toxins. Somer and Tayor [164] indicated that nisin at level of 500 to 10,000 IU/g is effective in pasteurized processed cheese spread by delaying the growth of *Clostridium botulinum* strain Type A and B. The level of nisin used to control non-botulinum spoilage varies from 250 to 500 IU/g. Degradation profile of nisin in pasteurized processed cheese spreads is dependent on temperature.

3.4.1.4.2 Dairy dessert: Other applications of nisin in dairy products include the extension of shelf-life of dairy products (Fig. 6). Pasteurised products have a limited shelf-life, but the addition of nisin can significantly increase the shelf-life [165].

3.4.1.4.3 Milk: Addition of nisin to milk is not permitted in the UK and other countries with temperate climates. It is permitted in some Middle East countries [166] where special shelf-life problems are experienced. The shelf-life of milk is relatively short and nisin has been demonstrated to have a beneficial effect, more than doubling the shelf at chilled, ambient and elevated temperature at nisin levels of 30 to 50 IU/ml [103]. Milk contains low thermophilic heat resistant spores, which can be prevented from outgrowth by low levels of nisin [167]. Canned evaporated milk is another product in which nisin efficiency increasing shelf-life has been demonstrated [168].

3.4.1.4.4 Canned foods: UK regulations [169] on the use of nisin in low acid foods (pH above 4.5) recommend that it should always receive a heat treatment sufficient to ensure the destruction of *Clostridium botulinum*. The heat resistant spores of thermophiles *Bacillus stearothermophillus* and *Clostridium thermosaccharolylicum* can survive and cause spoilage of canned foods, particularly under warm storage conditions [170]. By using nisin, it is possible to control thermophilic spoilage of canned foods. Additional levels of nisin used in canned foods are generally between 100 and 200 IU/g.

3.4.1.4.5 Cured meat: Nisin is a good alternative to nitrite for curing meat [138, 171, 172]. Only the high level of nisin, which is uneconomical, achieves good control of *Clostridium botulinum*. Thus nisin can be used as a partial replacement of nitrite. The poor preservative effects of nisin include the binding of nisin onto meat particles and surface, uneven distribution, poor solubility in meat systems and possible interference with the mode of action of phospholipids [138, 144, 172].

3.4.1.4.6 Alcoholic beverages: Research in the United Kingdom and Germany has reported that nisin has potential in controlling spoilage lactic acid bacteria in beer [131,173, 174] and wine [132, 175]. The studies indicate that although the spoilage lactic acid bacteria are sensitive to nisin, the yeasts are completely unaffected. This factor allows the nisin to be introduced during fermentation.

3.4.1.4.7 Other applications: Nisin has also been shown to have a potential use in other segments of the fermentation industry such as production of fruit brandies [148]. These are basically mixed fermentations. The use of nisin to prevent Gram-positive bacteria containing fermentations carried out by nisin insensitive microorganisms such as Gram-negative bacteria, yeasts or fungi could be applicable in other industrial fermentation such as production of single cell proteins, organic acids, polysaccharides, amino acids and vitamins [176].

### *3.4.2 Pediocin*

These are ribosomally synthesized antibacterial peptides or bacteriocin by the genus *Pediococcus* [177]. Strains of two species in this genus, *Pediococcus pentosaceus* and *Pediococcus acedilactici* naturally found in many plant sources have been used for a long time, both in natural and pure culture

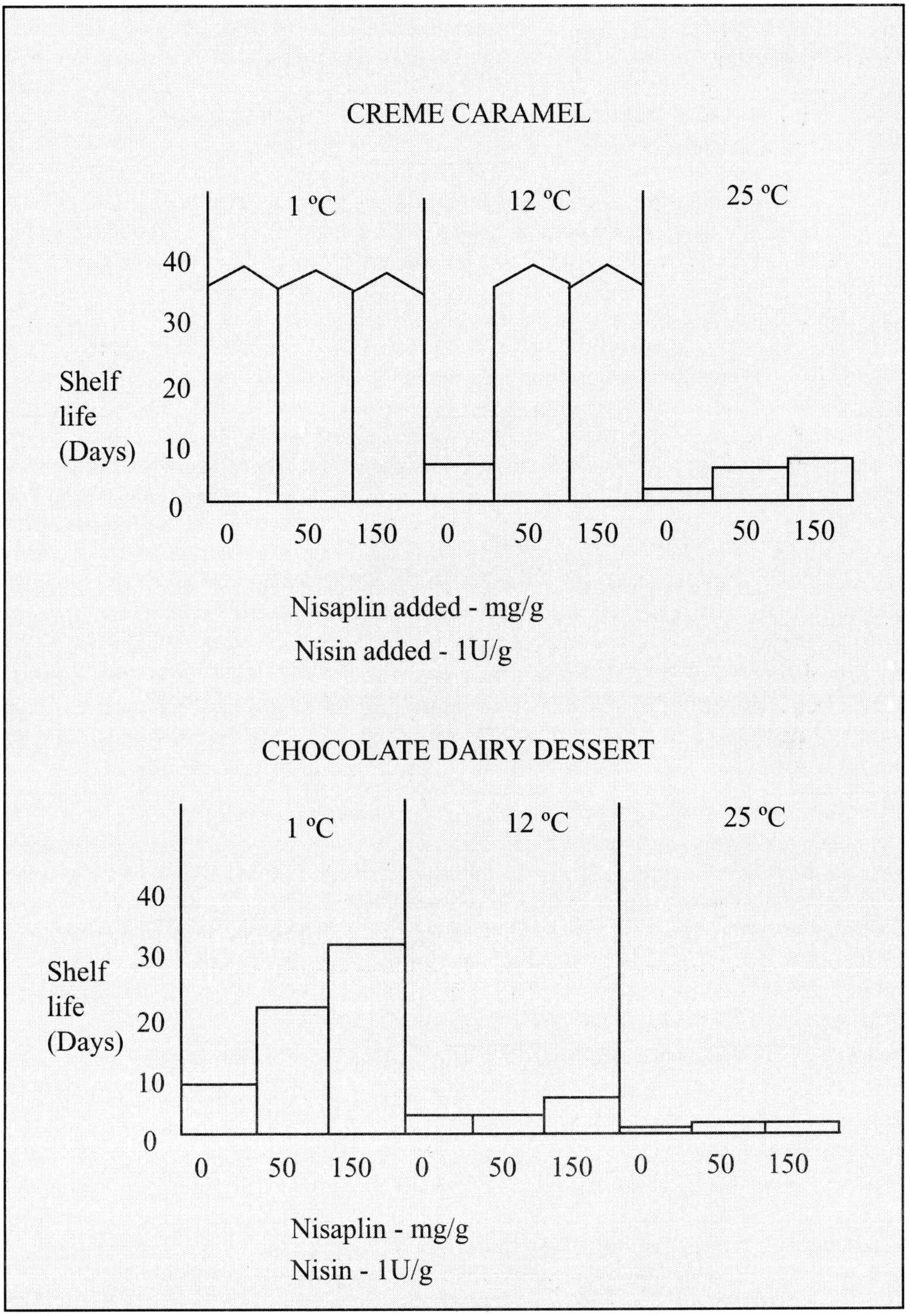


**Fig. 6:** Preservation of chilled dairy dessert using nisin.
*Source:* 147

fermentation to produce fermented food of plant, meat, fish, cereal and dairy origin [178]. Their cells are spherical, and appear as tetrad and pair, but never single cells or chains. They are Gram-positive, non-motile, facultative anaerobes and homolactic fermentative. Several strains of *Pediococcus acidilactici* and *Pediococcus pentosaceus* have been reported to produce pediocin. In recent years, several chapters and reviews have been published that narrate the early studies on pediocin [178-184]. In this chapter, only brief account of pediocin is given.

3.4.2.1 Structure: The primary structure i.e., the amino acid sequence of pediocin from at least nine *Pediococcus acidilactici* strains have been elucidated. All of them have same amino acid sequence as pediocin PA-1/ACH. The cytoplasmic form of molecule has 62 amino acids and is designated as pre-pediocin. During passage through the membrane, the 18 amino acid leader peptide at the N-terminus is removed and 44 amino acid pediocin PA-1/ACH excreted to the environment [106]. Once outside, the molecule forms secondary and tertiary structure that are essential for its bactericidal property. The four cysteine residues form two disulphide bonds and may cause the molecule to fold. The hydropathy plot shows the pediocin molecule to be moderately amphipathic. The N-terminal half (upto amino acid 21 to 22) is relatively polar, while the C-terminal half is relatively non-polar. The N-terminal half is predicted to form a three strand antiparallel β-sheet in which the second and third strands are linked by a disulphide bond between Cys 9 and Cys 14, and the first and second strands are also connected by disulphide bond between Cys 9 and Cys 14 to form a β-turn [185, 186]. A helical wheel plot of the C-terminal region of the molecule indicates that it could form amphipathic alpha-helix [187]. Analysis of different mutants of pediocin indicate that the bactericidal property of pediocin PA-1/ACH is not confined to any specific segment on the primary structure. The five amino acid sequence, -Tyr3-Gly4-Asn5-Gly-6-Val-7 are essential for the bactericidal action, and so are the two S-S bridges between Cys 9 and Cys14 and Cys 24, Cys 44. It is evident that the entire 44 amino acids sequence is necessary to confer the specific secondary and tertiary structure to the molecule, which is essential for the high level of bactericidal activity [188].

### 3.4.2.2 Mechanism of action

3.4.2.2.1 Antibacterial: Early studies [194] indicated that the antibacterial effect of pediocin PA-1/Ach towards sensitive Gram-positive bacterial cell is caused by its bactericidal action. Pediocin induced cell death occurred due to impairment of the cytoplasmic membrane of sensitive bacteria resulting in the leakage of cellular ions and small molecules and UV absorbing material [194]. Cells of some sensitive strains are also lysed, probably due to a secondary effect. In resistant Gram-positive cells, adsorption of pediocin at lower level may not allow the molecules to act on the cell membrane. Bacterial spores of *Bacillus* and *Clostridium* strains are resistant to pediocin PA-1/ACH. However, after the spores germinate, they are killed by pediocin [196].

3.4.2.2.2 Bactericidal: Christensen and Hutkins [198, 199] reported that the treatment of *Listeria* strains with pediocin caused rapid collapse of the membrane potential, resulting in the death of cell. It has been suggested that the pediocin molecules form small pores in membrane and cause cell death. The mechanism of bactericidal action is not clearly understood. The membrane lytic hypothesis was given with the available information [187, 188]. The effect of factors on bactericidal activity of pediocin is given in Table 11.

3.4.2.2.3 Lysis of sensitive cells: At 200 AU/ml of pediocin, the cells lysed at a rate of 0.19 OD unit/h for *Listeria monocytogenes,* and 0.29 OD unit/h for *Pediococcus pentasaceus.* It was suspected that while cell death results from the direct action of pediocin, lysis occurs by an indirect mechanism. Cell death probably triggers the activation of autolytic enzymes in some strains, which causes the cells to lyse.

**Table 11:** The influence of several factors on bactericidal properties of pediocin PA 1/ACH.

| | |
|---|---|
| Enzymes | • Being peptide, it is sensitive to proteolytic enzymes and loses bactericidal activity. It is sensitive to chymotrypsin, trypsin, ficin, papain and proteases [189-193].<br>• Not sensitive to lipase amylases, DNase, RNases. |
| pH | Pediocin from *Pediococcus acidilactici* strains retain their bactericidal activity at a pH range between 3 to 9. At higher pH values, pediocin lost activity within 24 h at pH 10.0 and within 2 h at pH 12.0 [194]. |
| Heat | The bactericidal effect remains stable after heating upto 60 min at 80°C. Partial inactivation occurs following heating at 121°C for 15 min. The level of inactivation is higher in culture broth than in purified preparation [189, 192-194]. |
| Organic solvents and chemicals | • Pediocin retains the activity with several organic chemicals like acetone (100%), chloroform (40%), isopropanol (80%), ethyl alcohol (80%), hexane (80%), aceto nitrite (70%), EDTA (10Mm), SPS (1%) and Tris (0.6M) [195].<br>• Formaldehyde (10%) reduces the activity [196, 197] |
| Ions | Cations such as $Ca^{2+}$, $Mg^{2+}$, $Mn^{2+}$, $K^+$ and $Na^+$ did not reduce the bactericidal activity of pediocin PA-1/ACH, whereas anions, especially $Cl^-$ and $PO_4^{3-}$, reduce the bactericidal activity to a greater extent [195]. |
| Hydrostatic pressure | Pediocin PA-1/ACH, both purified and in culture broth, has been subjected to high hydrostatic pressure of 345 Mpa for 30 min at 25°C. Pediocin retains full activity. During subsequent storage for 5 months at 2°C, the preparation retains full activity [197]. |
| Storage time and temperature | Pediocin retains full activity at 20°C or below for 6 months. Vacuum packed products are stable at room temperature [179, 181]. |

3.4.2.2.4 Immunity and resistance against pediocin: Compared to most bacteriocins produced by the pediocin family of LAB, both pediocin PA-1/ACH and pediocin-A have relatively wide range against Gram-positive bacteria. Yet there are many species and strains of Gram-positive bacteria that are not sensitive to them. A producer strain is not sensitive to its own pediocin. Gram-negative bacteria in general are resistant to pediocin and other bacteriocins of Gram-positive bacteria. Lipopolysaccharide molecule or the outer membrane itself may simply prevent pediocin from coming in contact with cytoplasmic membrane.

3.4.2.3 Applications: As a food biopreservative, pediocin has been reported to prevent the spoilage of food systems namely ground beaf, sausage mix, milk and ice cream [200]. Gonzalez [188] have also reported the role of pediocin in controlling spoilage of salad dressing and salad caused by *L.fermentum* and extend the shelf life of coleslaw, macroni salad and potato salad. The possible use of pediocin producing *P. parvulus* strain to control the growth of spoilage and pathogenic bacteria in fresh, ready to eat vegetables has been suggested by Bennik *et al.* [201]. The effectiveness of pediocin from *P. acidilactici* strains in controlling *L. monocytogenes* in processed meat products has been studied by several researchers [202, 203].

### 3.5 Phyto-Antimicrobial

The potential benefits of edible plants as well as their phytochemicals in food preservation and improvement of organoleptic qualities of certain traditional foods have been known for centuries. Plant products with natural antimicrobial properties notably have obtained emphasis for a possible application in food production in order to prevent bacterial and fungal growth. Reports on the early use of spices or herbs as antimicrobial preservative can be traced to 1550 BC, when the ancient Egyptians used cinnamon, cumin and thyme both for food preservation and mummification. Scientific evidence on the preservation potential of spices emerged in early 19$^{th}$ century. Spices and herbs are used for adding desirable sensory properties to food products. Although ancient civilization acknowledged the antiseptic and antimicrobial potential of many plant extracts, but recently the various phytochemicals are characterized with the advanced molecular separation techniques. The increased demand for minimally processed extended shelf life food has further revived interest in exploitation of these natural antimicrobial agents.

The various phytochemicals of plants, which can be used in biopreservation and act as antimicrobials, are divided into six subgroups viz., phytophenols, saponins, flavonoids, thiosulphinates, catechins and agar, out of which only four major groups have been discussed briefly in this chapter.

### 3.5.1 Phytophenols

Among various phytochemiclas, phytophenols are used commercially as biopreservatives in many food products. The compounds responsible for the antimicrobial activity of spices and plants are primarily phenolic compounds associated with various plant sources and are listed in Table 12. The antimicrobial activity of cinnamon, allspice, and cloves is attributed to eugenol (2 methoxy 4 allyl phenol) and cinnamic aldehyde, which are major constituents of the volatile oils of these spices. Cinnamon contains 0.5-1.0% volatile oil, which contains 65-75% cinnamic aldehyde and 8% eugenol. Allspice contains upto 4.5% volatile oil of which 80% is eugenol. Clove buds have an average essential oil content of 17% that is 93-95% eugenol [204]. Oregano, savory and thyme demonstrated antifungal activity. Terpenes carvacrol, pcyme and thymol are the volatile components of oregano, thyme and savory, which account for their antimicrobial activity. Recently, a number of potential phytophenolic compounds such as phytophenols and glycosides have been isolated from olive and virgin olive oil. The chemical structures of commonly occurring phytophenols of essential oils are shown in Fig. 7. The compounds identified are tyrosol, p-hydroxy phenyl lactic acid, p-coumaric acid and feralic acid [206]. Table 13 shows phytophenolic compounds from various plant sources and their antimicrobial spectrum.

#### 3.5.1.1 Mechanism of action:

The mode of action of phenolic compounds is generally thought to involve interference with functions of the cytoplasmic membrane including proton motive force and active transport [207]. In fact, Juven *et al.* [208] hypothesized that the inhibition by thymol of *Salmonella typhimurium* was due to a reaction of the compound with protein in the cytoplasmic membrane of microorganism. This caused changes in permeability of the membrane resulting in possible leakage and affected the proton motive force. Conner and Beuchat [209] reported that spice essential oils may inhibit enzyme systems in yeasts, including those involved in energy production and synthesis of structural

**Table 12:** Phenolic compounds associated with various plant sources.

| Phenolic compound | Plant source |
|---|---|
| Apigenin-7-glucoside | *Oleaceae* sp. |
| Benzoic acid | Spices |
| Berbamine | Beriberis |
| Caffeine | Tea |
| Caffeic acid | *Oleaceae* sp./avocado/artichoke/apple |
| O-Caffeylquinic acid | Artichoke/plum/carrots |
| Caryophelene | Hops |
| Catechin | Skin and seeds of wine grapes/tea |
| Cinnamic acid | *Brassica* oil seeds |
| Chlorogenic acid | *Brassica* oil seeds/artichoke/apple |
| Chelldonic acid | Berberis |

Contd...

Contd...

| Phenolic compound | Plant source |
|---|---|
| Chicorin | Chicory |
| Columbamine | Berberis |
| Coumarin | Spices |
| p-Coumaric acid | *Oleaceae* sp./avocado/*Brassica* oil seeds/apple |
| Cyanarine | Artichoke |
| Dihydrocafeic acid | *Oleaceae* sp. |
| Dimethyloleuropein | *Oleaceae* sp. |
| Esculin | Chicory |
| Ferulic acid | *Oleaceae* sp./avocado/*brassica* oil seeds |
| Gallic acid | *Oleaceae* sp./grapes/ avocado/*Brassica* oil seeds |
| Gingerols | Spices |
| Humulon | Hops |
| Hydroxytyrosol | *Oleaceae* sp. |
| 4-hydroxybenjoic acid | Avocado/*Brassica* oil seeds/vanilla/carrots |
| 4-hydroxycinnamic acid | *Oleaceae* sp./carrots |
| Isovanillin | Avocado |
| Linalool | Boldo |
| Lupulon | Hops |
| Luteoline-5-glucoside | *Oleaceae* sp. |
| Ligustroside | *Oleaceae* sp. |
| Myricetin | Tea |
| 3-methoxybenzoic acid | *Oleaceae* sp. |
| Oleoside | *Oleaceae* sp. |
| Oleuropein | *Oleaceae* sp. |
| Paradols | Spices |
| Procatechuic acid | *Oleaceae* sp./avocado |
| O-pyrocatechuic | Avocado |
| Querecetin-3-rutinoside | Tea |
| Quercetin | *Oleaceae* sp./tea |
| Resocrylic acid Avocado | |
| Sesamol | Sesame oil |
| Shogoals | Spices |
| Syringic acid | *Brassica* oil seeds |
| Sinapic acid | Avocado/*Brassica* oil seeds |
| Tannins | Wine grapes/spices/*Brassica* oil seeds/boldo/herbs |
| Tannic acid | Skin and seeds of wine grapes |
| Thymol | Thyme |
| Trimethoxybenzoate | *Oleaceae* sp. |
| Trihydroxyphyenlacetate | *Oleaceae* sp. |
| Tryrosol | *Oleaceae* sp. |
| Verbascoside | *Oleaceae* sp. |
| Vanillin | Spices/vanilla |
| Vanillic acid | *Oleaceae* sp./avocado/*Brassica* oil seeds/vanilla |

*Source:* 205

components. Conner *et al.* [210] demonstrated energy depletion in yeast caused by essential oil of several spices including clove, thyme, oregano and cinnamon. Effective phytophenolic compounds preserve food by various mechanisms including stasis (growth inhibition of microorganism) and cidal (direct

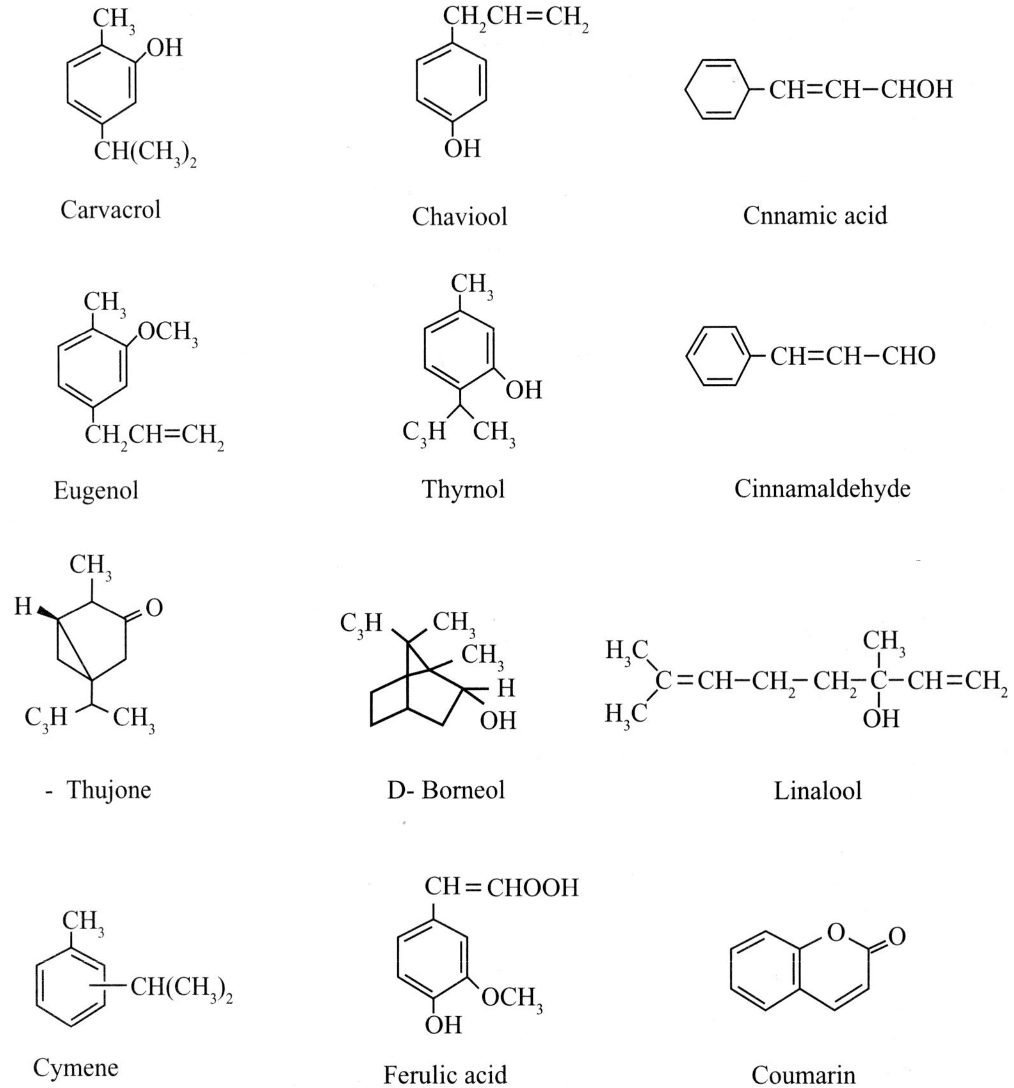


**Fig. 7:** Chemical structure of some phytophenols.

**Table 13:** Different types of phytoantimicrobials.

| | | |
|---|---|---|
| Garlic | Allicin | *Salmonella typhimurium, E. coli, Staphylococcus aureus, Bacillus cereus, Bacillus subtilis,* Mycotoxigenic *Aspergillus, Candida albicans* |
| Mustard | Allyl isothiocynate | Mycotoxigenic *Aspergillus* |
| Cinnamon | Eugenol | Mycotoxigenic *Aspergillus, Aspergillus parasiticus* |
| Cloves | Eugenol | Mycotoxigenic *Aspergillus, Aspergillus parasiticus* |
| Sage | Thymol, Eugenol | *Bacillus cereus, Staphylococcus aureus, Vibrio parahaemolyticus* |
| Oregano | Thymol, Carracol | Mycotoxigenic *Aspergillus, Salmonella* sp., *Vibrio parahaemolyticus* |

destruction of microorganism) effects. Certain phytophenolic agent such as olive oil delivers multifunctional physiological benefits to the consumer, and therefore is attractive to the health and food industry. Essential oils and herbs active against microbes are listed in Table 14.

### *3.5.2 Saponins*

These are naturally occurring glycosides which are predominantly found in the plant kingdom; however certain lower marine animals also contain these compounds. They are found in different plant parts i.e., roots, shoots, flowers and seeds. The important characteristic of saponins is the formation of soapy lather when shaken in water solution. Depending on the structure, saponins may show haemolytic, antimicrobial, membrane depolarizing, cholesterol binding and allelopathic activities. Saponins as natural phytoantimicrobial agents elicit antibacterial, antifungal and antiviral effects. They constitute defence mechanism for the host. These activities can be utilized in natural food preservation against susceptible strains of bacteria and fungi, and are also therapeutic for certain chronic diseases.

**Table 14:** Minimum inhibitory concentrations (%, v/v) of selected essential oils of herbs and spices against various bacteria and yeasts as determined by agar dilution.

| Spice | *Enterococcus faecalis* | *E. coli* | *Pseudomonas aeruginosa* | *Salmonella typhimurium* | *S. aureus* | *Candida albicans* |
|---|---|---|---|---|---|---|
| Basil | >2.0 | 0.5 | 2.0 | 2.0 | 2.0 | 0.5 |
| Bay | 0.5 | 0.12 | 1.0 | 0.25 | 0.25 | 0.12 |
| Black pepper | 1.0 | >2.0 | >2.0 | >2.0 | >2.0 | >2.0 |
| Clove | 0.5 | 0.25 | >2.0 | >2.0 | 0.25 | 0.12 |
| Coriander | 0.25 | 0.25 | >2.0 | 1.0 | 0.25 | 0.25 |
| Fennel | >2.0 | 0.5 | >2.0 | 1.0 | 0.25 | 0.5 |
| Ginger | >2.0 | >2.0 | >2.0 | >2.0 | >2.0 | >2.0 |
| Lemongrass | 0.12 | 0.06 | 1.0 | 0.25 | 0.06 | 0.06 |
| Marjoram | 2.0 | 0.25 | >2.0 | 0.5 | 0.5 | 0.25 |
| Oregano | 0.25 | 0.12 | 2.0 | 0.12 | 0.12 | 0.12 |
| Peppermint | 2.0 | 0.5 | >2.0 | 1.0 | 1.0 | 0.5 |
| Rosemary | >2.0 | 1.0 | >2.0 | >2.0 | 1.0 | 1.0 |
| Sage | 2.0 | 0.5 | >2.0 | 2.0 | 1.0 | 0.5 |
| Spearmint | 2.0 | 0.25 | >2.0 | 0.5 | 0.25 | 0.12 |
| Tea tree | 2.0 | 0.25 | >2.0 | 0.5 | 0.5 | 0.5 |
| Thyme | 0.5 | 0.12 | >2.0 | >2.0 | 0.25 | 0.12 |
| Wintergreen | >2.0 | 0.5 | >2.0 | 0.5 | 2.0 | 0.25 |

3.5.2.1 Sources: Saponins are widely distributed in plant species, from 100 families. Gubanov [211] reported that 76% of Asian plant families contain saponins. They are found in edible plants such as soya, beans, oat, *Solanum* and *Allium* species, tomato, asparagus, tea, peanut, spinach, sugarbeet, yam and blackberry in plants. Their concentration in plants ranges from traces upto 10% of dry matter. Depending upon the origin of the variety in saponin mixture, soya sapogenal glycosides, medicagenic acid glycosides or zanhic acid tri desmoside may dominate. Figure 8 shows the skeleton of some saponins.

3.5.2.2 Molecular properties: Saponins consist of non-sugar aglycone coupled to sugar chain units. These sugars could be attached to the aglycone either as one, two or three side chains, and termed monodesmoside, bidesmoside, and so on. According to the nature of aglycone, they can be divided into the groups of saponins containing steroidal or triterpene aglycones sapogenins.

STEROIDAL GROUP

TRITERPENE GROUP

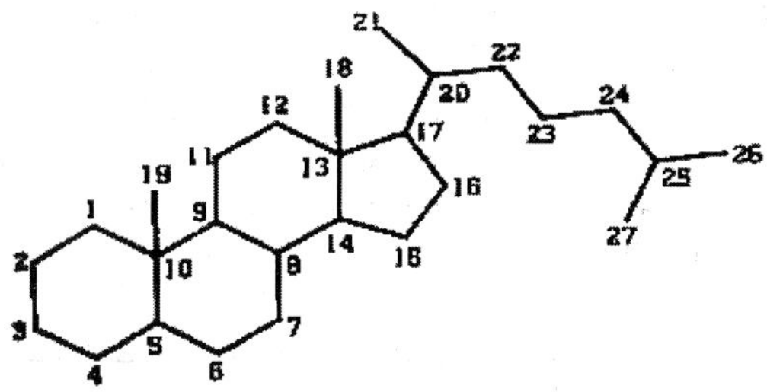


Cholestan

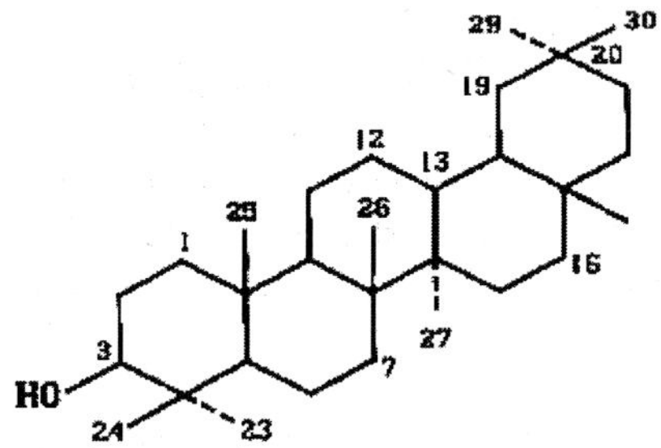


β-Amyrin type
(Oleanane)

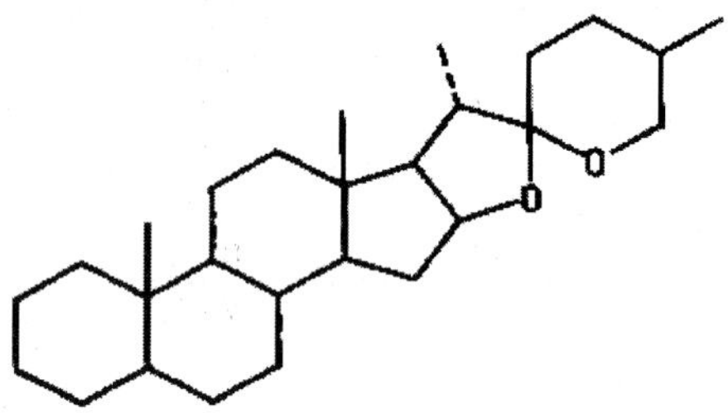


Spirostan

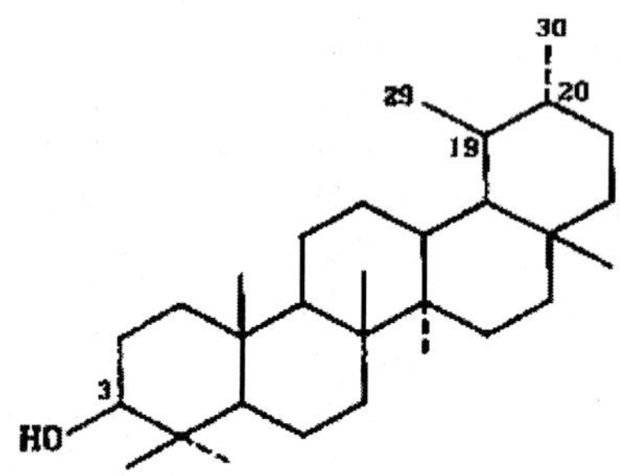


α-Amyrin type
(Ursane)

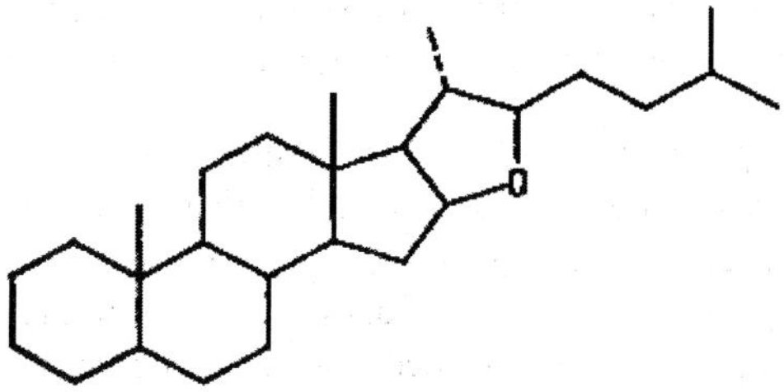


Furostan

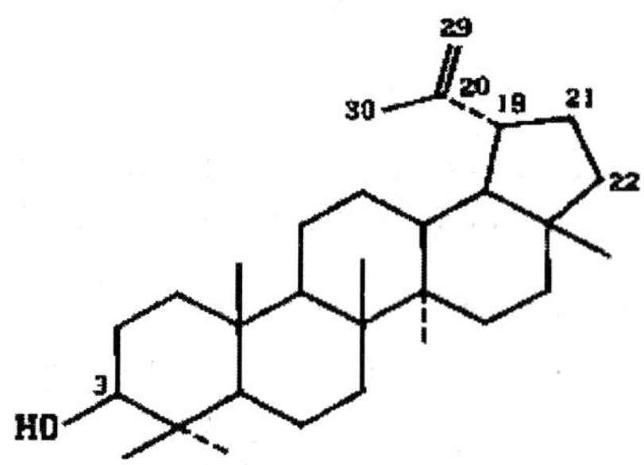


Lupeol type
(Lupanes)

**Fig. 8:** Skeleton of some saponins.

3.5.2.3 Mechanism of action: There are three different types of antimicrobial activities reported by the saponins viz., antibacterial, antiviral and antifungal. The general principle of the action of saponin on microorganism is their interaction with membrane sterols. Thus, the bacteria should not be sensitive to saponin as their membranes are low in cholesterol. However, reports on saponin indicate that fatty acid composition of membrane may be equally important in this respect. Thus a number of saponins, including both triterpenes and steroidal, show antibacterial activities. The antibacterial activity of saponin has been extensively reviewed by Takeya [212] and Tamura [213]. Among the saponins, those with triterpene and spirostanol skeleton generally demonstrate antifungal activities. The fungicidal activity of saponins is determined by the aglycone structure and also by the number, quality and branching patterns of sugar substitution.

Most of the work has been performed on antiviral activity of plant saponins. The most studied compound was glycyrrhizin, which was shown to inhibit some DNA and RNA viruses *in vitro* and to have therapeutic and prophylactic effects on some chronic viral hepatitis [214]. This compound was shown to inhibit several enveloped viruses giving 100% inhibition of cytopathic effect at concentration 5.33 μg/ml, HIV and VZV. Okubo *et al.* [215] have shown that soyasapogenol and an aglycone, the common saponin of legumes have partially inhibitory effect on HIV.

3.5.2.4 Applications: Several plant sources have been commercially used, but only two of them are actually permitted as food additives. These are *Quillaya saponaria* and *Yucca schidigera* extracts, containing triterpene and steroidal saponins, respectively. They are regarded as GRAS products. In the UK, the use of quillaya extracts in food was permitted as early as 1962 in beverages. The use of such extracts as emulsifier and foaming agents was restricted to not more than 20 ppm by weight. These two extracts are also permitted in the USA by the FDA for use in food and beverages. They are used as flavour enhancers in some categories of beverages including root beer, flavour teas and juices and addition to non-fat dairy products such as chocolate milk and vanilla pudding. As natural emulsifiers, they have been used to emulsify oil based flavours for candy, to prevent precipitation in a protein containing liquid composition, used as a leavening agent in bakery industry, and to disperse oil in water type oil emulsion composition. They can also play the role of natural preservative against certain strains of bacteria (*E. coli, Staphylococcus aureus, Salmonella, Pseudomonas aeruginosa*) and fungi (*Aspergillus niger, Penicillium chrysogenum*).

### *3.5.3 Flavonoids*

Flavonoids are group of C15 aromatic compounds with numerous derivatives of basic forms (Fig. 9). They are biosynthesized by a confluence of the acetate/malonate and shikimate pathways. They are found in all land based green plants, and although not produced by animals may be accumulated from food source. A single plant may contain different types of flavonoids and play different roles. They include chalcones, dihydroflavones, flavones, biflavonoids, dihydroflavonols, flavonols, anthocyanins and proanthocyanidin tannin. Predominant among them is glycosidic form located in cell vacuoles of plants.

Lipophilic flavonoids display antimicrobial activity, a property associated with their ability to penetrate biological membranes. Methylation of 5 OH has been considered a structural feature essential for its antimicrobial activity [216]. A reactive mechanism of action, pronounced and broad spectrum antimicrobial activity (antibacterial, antiviral and antifungal) and the lack of resistance induction promoted many laboratories to explore flavonoids as food antimicrobials. Isoflavonoids as natural products and as potent fungal growth inhibitors could be extremely useful in controlling plant diseases. Due to this highly specific antimicrobial activity, a possible application of isoflavonoids as alternative to conventional fungicides in the control of storage fungi to prevent loss of grains in

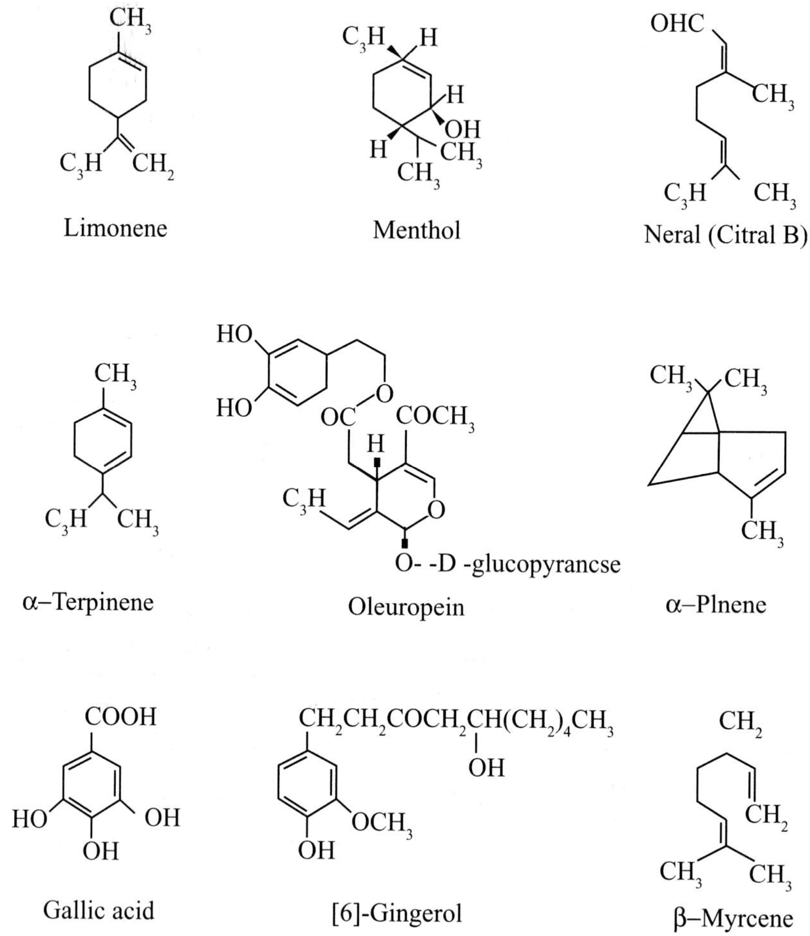


**Fig. 9:** Basic structural classes of flavonoids.

developing countries has been suggested. Flavonoids are also of great interest in food technology, as they contribute sensory and nutritional quality to food products.

### *3.5.4 Thiosulphinates*

It has been demonstrated that garlic and its derivatives may be multifunctional as well as flavourful food additives. Garlic has been used in the past and is presently being used as a medicine, curative, preservative, and flavouring agent. The antimicrobial properties of garlic and its derivatives have been documented. It has been shown that garlic and its derivatives exhibit both cidal and static effects. A variety of thiosulphinate compounds have been isolated and identified from garlic (Fig. 10 A & B). These include allicin, alliin, deoxyalliin, alliin, thio-2-propene-1-1 sulphinic acid S allyl ester + S allyl L cysteine sulphoxide, ajoene, organic trisulphur, diallyl thiosulphate, etc., The chemical formula and molecular mass of alliin ($C_6\ H_{10}\ O_3$ SN), Allicin, ($C_6\ H_{10}\ O_3$ S2 = 162), ajoene ($C_9\ H_{14}\ O_3$ S3 = 234), and diallyl disulphinate ($C_6\ H_{10}\ O_3$ S2 = 146) are well documented [218]. Table 15 shows the potential use of garlic compounds. If garlic sulphinates are to be used as preservatives or

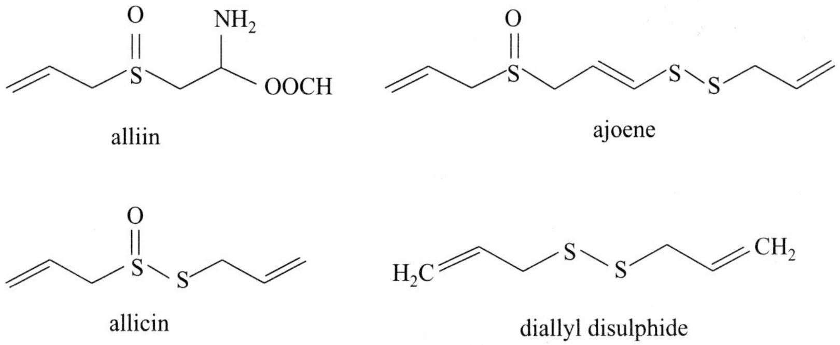


**(A)**

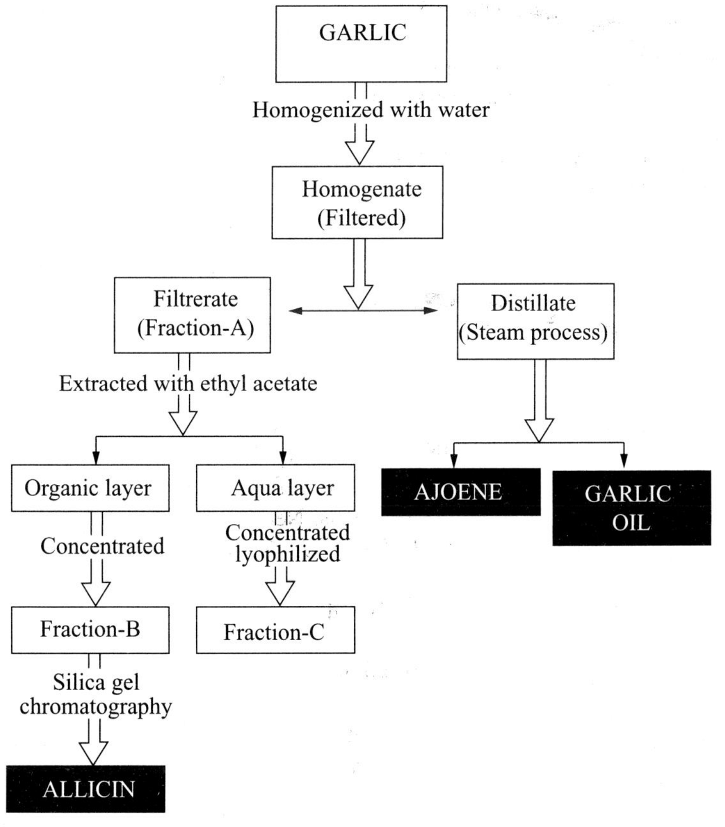


**(B)**

**Fig. 10:** (A) Chemical structure of major thiosulphinate from garlic; (B) Fractionation of thiosulphinate from raw garlic.
*Source:* 217

as antimicrobial agents, the purification techniques and quality control methods should be standardized. Currently many companies are marketing garlic extract in various forms. During large scale production, many of the thiosulphinates seem to loose their antimicrobial activity depending upon the processing conditions [219].

**Table 15:** Uses of garlic compounds.

| Activity | Uses |
|---|---|
| Food preservative | 100 μg/ml ajone was found to be inhibitory to *B. cereus;* Garlic extract, lactoperoxidase and ethanol was useful in preserving raw buffalo's milk; Garlic juice 2-10% effective in eliminating coliform bacteria of fermented cabbage. |
| Antibacterial | Thiosulphinates from garlic are effective antibacterial agents against pathogens *Salmonella, Listeria, E. coli, B. cereus, Streptococcus, S. aureus, Pseudomonas* and *H. pylori,* which cause food-borne illness; Garlic has also been shown to inhibit various strains of food spoilage organisms. |
| Antifungal | Garlic was found fungistatic at 50, 150, 300 μg/ml and fungicidal at 400 μg/ml against *C. albicans.* |
| Antiviral | *In vitro* viricidal effects of fresh garlic extract and its associated diallyl thiosulphinate, Allyl methyl thiosulphinate, methyl allyl thiosulphinate, ajoene, alliin, deoxyalliin. |
| Antiprotozoal | Garlic and garlic derivatives have been shown to elicit antiprotozoal activity against *Entamoeba histolytica* and *Trypanosoma cruzi, T. brucei, T. congolense and T. vivax.* |

*Source:* 106

## REFERENCES


1. Zink DL (1997) The impact of consumer demands and trends on food processing. In: *Ways to Prevent Food Spoilage,* http://is6.pacific.net.hk/pp laung/chem./preservation.html.
2. Corlett DA & Brown MH (1980) Factors affecting the life and death of microorganisms. In: *Microbial Ecology of Food,* Vol 1, Academic Press, New York, USA, pp 92-111.
3. Gould GW (1989) *Mechanisms of Action of Food Preservation Procedures,* Elsevier Applied Science, London, UK.
4. Smith BFM (1992) Bacteriocins: Applications in food preservation. *Trends Food Sci Technol* 3: 297-304.
5. Singhal RS & Kulkarni PR (1999) Production of additives by fermentation. In: *Biotechnology Food Fermentation: Microbiology, Biochemistry and Technology,* Vol 2, Joshi VK & Pandey A (eds), Educational Publishers and Distributors, New Delhi, India, pp 1145-1193.
6. Balow A (1992) *Prokaryotes: Handbook on the Biology of Bacteria,* Springer-Verlag, New York, USA.
7. Bjorkroth K (1998) Identification and characterization of *Leuconostoc carnosum. Appl Environ Microbiol* 64: 3313-3319.
8. Zollinger W (1990) Mikrobiologie-spezielle Aspekte bei Fruchten und Gemuse. *Lebensmittel-Technologie* 23: 262-264.
9. Lewus CB, Kaiser A & Montville TJ (1991) Inhibition of food-borne pathogens by bacteriocins from lactic acid bacteria isolated from meat. *Appl Environ Microbiol* 57: 1683-1688.

10. Yildirim Z & Johnson MG (1998) Detection and characterization of a bacteriocin produced by *Lactococcus lactis* subsp. *cremoris* R. isolated from radish. *Lett Appl Microbiol* 26: 297-304.
11. Davidson PM & Branen AL (1993) *Antimicrobials in Foods*, Marcell Dekker, Inc, New York, USA.
12. CFR (2001) *Code of Federal Regulations,* National Archives and Records Administration, US Government Printing Office, www.acces.gpo.gov/cgi-bin/cfrassemble.cgi title=200121.
13. Holten CH (1971) *Lactic Acid,* Verlag Chemie, Weinheim, Germany.
14. Shelef AL (1994) Antimicrobial effects of lactates: A Review. *J Food Prot* 57: 445-450.
15. Freese E, Sheu CW & Galliers E (1973) Function of lipophilic acids as antimicrobial food additives. *Nature* 241: 321-325.
16. Hunter DR & Segal IH (1973) Effect of weak acids on amino acid transport by *Penicillium chrysogenum*: Evidence for proton or charge gradient as the driving force. *J Bacteriol* 113: 1184-1192.
17. Salmond CV, Kroll RG & Booth IR (1984) The effects of food preservatives on pH homeostasis in *Escherichia coli. J Gen Microbiol* 130: 2845-2850.
18. Debevere JM (1989) The effect of sodium lactate on the shelf life of vacuum-packed coarse liver pate. *Fleischwirtsch* 69: 223-224.
19. Loncin M (1975) Basic principles of moisture equilibria. In: *Freeze Drying and Advanced Food Technology,* Goldblith SA, Rey L & Rothmayr WW (eds), Academic Press, New York, USA, pp 599-617.
20. Smulders FJM & Woolthus CHJ (1983) Influence of two levels of hygiene in the microbiological condition of veal as a product of two slaughtering/processing sequences. *J Food Prot* 46: 1032-1035.
21. Woolthus CHJ & Smulders FJM (1985) Microbial decontamination of calf carcasses by lactic acid sprays. *J Food Prot* 48: 832-837.
22. Dixon ZR, Vanderzant C, Acuff GR, Savell JW & Jones DK (1987) Effect of acid treatment of beef trip loin steaks on microbiological and sensory characteristics. *Int J Food Microbiol* 5: 181-186.
23. Van der Marel GM, van Logtestijn JG & Mossel DAA (1988) Bacteriological quality of broiler carcasses as affected by in-plant lactic acid decontamination. *Int J Food Microbiol* 6: 31-34.
24. Zeitoun AAM & Debevere JM (1990) The effect of treatment with buffered lactic acid in microbial decontamination and on shelf life of poultry. *Int J Food Microbiol* 11: 305-309.
25. Sofos JN (1989) *Sorbate Food Preservatives,* CRC Press, Boca Raton, USA.
26. Luck E (1980) *Antimicrobial Food Additives, Characteristics, Uses, Effects*, Springer-Verlag, Berlin, Germany.
27. Sofos JN & Busta FF (1981) Antimicrobial activity of sorbate. *J Food Prot* 44: 614-620.
28. Sofos JN & Busta FF (1993) Sorbic acid and sorbates. In: *Antimicrobials in Food,* Davidson PM & Branen AL (eds), Marcel Dekker, Inc, New York, USA, pp 49-94.
29. Sofos JN (1992) Sorbic acid, mode of action. In: *Encyclopedia of Microbiology*, Vol 4, Lederberg J (ed), Academic Press, San Deigo, USA, pp 43-52.
30. Blocher JC & Busta FF (1985) Multiple modes of inhibition of spore germination and outgrowth by reduced pH and sorbate. *J Appl Bacteriol* 59: 469-478.
31. Sofos JN, Pierson MD, Blocher JC & Busta FF (1986) Mode of action of sorbic acid on bacterial cells and spores. *J Food Microbiol* 3: 1-17.
32. Kouassi Y & Shelef LA (1995) Listeriolysin O secretion by *Listeria monocytogenes* in the presence of cysteine and sorbate. *Lett Appl Microbiol* 20: 295-299.

33. Kouassi Y & Shelef LA (1995) Listeriolysin O secretion by *Listeria monocytogenes* in broth containing salts of organic acids. *J Food Prot* 58: 1314-1319.
34. Nickol GB (1979) Vinegar. In: *Microbial Technology,* Peppler HJ & Perlman D (eds), Academic Press, New York, USA, pp 155-172.
35. Allgeier RJ & Hildebrandt FM (1960) Newer developments in vinegar manufacture. *Adv Appl Microbiol* 11: 163-181.
36. Conner HA & Allgeier RJ (1976) Vinegar: Its history and development. *Adv Appl Microbiol* 20: 81-133.
37. Canning A (1985) Vinegar brewing. *Nutr Food Sci* 9-10: 20-21.
38. Ghommidh C, Cutayar JM & Navvvaro JM (1986) Continuous production of vinegar I. Research strategy. *Biotechnol Lett* 8: 13-18.
39. Cheryan M, Parekh S, Shah M & Witjitra K (1997) Production of acetic acid by *Clostridium thermoaceticum. Adv Appl Microbiol* 43: 1-33.
40. Doores S (1993) Organic acids. In: *Antimicrobials in Food,* Davidson PM & Branen AL (eds), Marcell Dekker, Inc, New York, USA, pp 95-136.
41. Booth IR (1985) Regulation of cytoplasmic pH in bacteria. *Microbiol Rev* 49: 359-378.
42. Cassio F, Leao C & van Uden N (1987) Transport of lactate and other short chain monocarboxylates in the yeast *Saccharomyces cerevisiae. Appl Environ Microbiol* 53: 503-513.
43. Ray B & Sandine WE (1992) Acetic, propionic, and lactic acids of starter culture bacteria as biopreservatives. In: *Food Biopreservatives of Microbial Origin*, Ray B (ed), CRC Press, Boca Raton, USA, pp 103-136.
44. Casal M, Cardoso H & Leao C (1998) Effects of ethanol and other alkanols on transport of acetic acid in *Saccharomyces cerevisiae. Appl Environ Microbiol* 64: 665-668.
45. Young KM & Foegeding PM (1993) Acetic, lactic and citric acids and pH inhibition of *Listeria monocytogenes* Scott A and the effect on intracellular pH. *J Appl Bacteriol* 74: 515-520.
46. Guldfeldt LU & Arneborg N (1998) Measurement of the effects of acetic acid and extracellular pH on intracellular pH of non-fermenting, individual *Saccharomyces cerevisiae* cells by fluorescence microscopy. *Appl Environ Microbiol* 64: 530-534.
47. Huang L, Forsberg CW & Gibbins LN (1986) Influence of external pH and fermentation products of *Clostridium acetobutylicum* intracellular pH and cellular distribution of fermentation products. *Appl Environ Microbiol* 51: 1230-1234.
48. Deiz-Gongalez F & Russell JB (1997) The ability of *Escherichia coli* O157:H7 to decrease its intracellular pH and resist the toxicity of acetic acid. *Microbiology* 143: 1175-1180.
49. Stroup WH, Dickersons RW, Jr & Johnston MR (1985) Acid equilibrium development in mushrooms, pearl onions and cherry peppers. *J Food Prot* 48: 590-594.
50. Fabian FW & Wadsworth CK (1939) Experimental work on lactic acid in preserving pickles and pickle products. II. Preserving value of acetic acid and lactic acid in the presence of sucrose. *Food Res* 4: 511-519.
51. Gardner WH (1968) Acidulants in food processing. In: *Handbook of Food Additives*, Furia TE (ed), The Chemical Rubber Co, Cleveland, USA, pp 247-287.
52. Kabara JJ & Eklund T (1991) Organic acid and esters. In: *Food Preservatives*, Russell NJ & Gould GW (eds), Blackie and Son Ltd, Glasgow, Scotland, UK, pp 44-71.
53. Dickson JS & Anderson ME (1992) Microbiological decontamination of food animal carcasses by washing and sanitizing systems: a review. *J Food Prot* 55: 133-140.
54. Frederick TL (1994) The effects of organic acid and temperature on the microbiological properties of pork. *J Food Sci* 59: 300-302.

55. Siragusa GR (1995) The effectiveness of carcass decontamination systems for controlling the presence of pathogens on the surfaces of meat animal carcasses. *J Food Safety* 15: 229-238.
56. Bolder NM (1997) Decontamination of meat and poultry carcasses. *Trends Food Sci Technol* 8: 221-227.
57. Smulders FJM & Greer GG (1998) Integrating microbial decontamination with organic acids in HACCP programs for muscle foods: prospects and controversies. *Int J Food Microbiol* 44: 149-169.
58. Mountney GJ & O'Malley J (1965) Acids as poultry meat preservatives. *Poultry Sci* 44: 582-586.
59. Lillard HS, Blankenship LC, Dickens JA, Craven SE & Shackeland AD (1987) Effect of acetic acid on the microbiological quality of scalded picked and unpicked broiler carcasses. *J Food Prot* 50: 112-114.
60. Dickens JA & Whittemore AD (1994) The effect of acetic acid and air injection on appearance, moisture, pick-up, microbiological quality, and *Salmonella* incidence on processed poultry carcasses. *Poultry Sci* 73: 582-586.
61. Kim CR, Hearnsberger JO, Vickery AP, White CH & Marshall DL (1995) Sodium acetate and bifidobacteria increase shelf life of refrigerated catfish fillets. *J Food Sci* 60: 25-27.
62. Sholberg PL & Guance AP (1995) Fumigation of fruit with acetic acid to prevent postharvest decay. *Hort Sci* 30: 1271-1275.
63. Entani E, Asai M, Tsujihata S, Tsukamoto T & Ohta M (1997) Antibacterial action of vinegar against food-borne pathogenic bacteria including *Escherichia coli* O157:H7 (part 2). Effect of sodium chloride and temperature on bactericidal activity. *Kansenshogaku Zasshi* 71: 451-458.
64. Daeschel M (1989) Antimicrobial substances from lactic acid bacteria for use as food preservatives. *Food Technol* 43: 164-167.
65. Wehmer C (1919) Verlust des oxalsaure-bildungs-vermogens bei einem degenerieten *Aspergillus niger. Zentrabl Bacteriol* 49: 145-148.
66. Branen AL & Keenan TW (1970) Growth stimulation of *Lactobacillus casei* by sodium citrate. *J Dairy Sci* 53: 593-597.
67. Rammel CG (1962) Inhibition by citrate of the growth of coagulase positive staphylococci. *J Bacteriol* 84: 1123-1127.
68. Imai K, Banno I & Ujima T (1970) Inhibition of bacterial growth by citrate. *J Gen Appl Microbiol* 16: 479-487.
69. Fabian FW & Graham HT (1953) Viability of thermophilic bacteria in the presence of varying concentrations of acids, sodium chloride and sugars. *Food Technol* 7: 212-217.
70. Minor TE & Marth EH (1970) Growth of *Staphylococcus aureus* in acidified pasteurized milk. *J Milk Food Technol* 33: 516-519.
71. Ahamad N & Marth EH (1987) Behaviour of *Listeria monocytogenes* at 7, 13, 21, and 35°C in tryptic soy broth acidified with acetic, citric or lactic acid. *J Food Prot* 52: 688-695.
72. Shibasaki I & Kato N (1978) In: *The Pharmacological Effect of Lipids*, Kabara JJ (ed), The American Oil Chemist Society, Champaign, USA, pp 15.
73. Blaszyk M & Holley RMA (1998) Interaction of monolaurin, eugenol and sodium citrate on growth of common meat spoilage and pathogenic organisms. *Int J Food Microbiol* 39: 175-183.
74. El-Shenawy MA & Marth EH (1989) Inhibition and inactivation of *Listeria monocytogenes* by sodium benzoate together with some organic acids. *J Food Prot* 52: 771-776.

75. El-Shenawy MA & Marth EH (1991) Organic acids enhance the antilisterial activity of potassium sorbate. *J Food Prot* 54: 593-597.
76. El-Shenawy MA & Marth EH (1992) Behaviour of *Listeria monocytogenes* in the presence of sodium propionate together with food acids. *J Food Prot* 55: 241-245.
77. Pfizer Inc (1974) Fermentation process for the production of citric acid. Br Pat 1,364,094.
78. Kubicek CP & Rohr M (1986) Citric acid fermentation. *Crit Rev Biotechnol* 3: 331-339.
79. Kristiansen B & Sinclair CG (1979) Production of citric acid in continuous culture. *Biotechnol Bioeng* 21: 297-315.
80. Blake CCF, Koeing DF, Mair GA, North ACT, Phillips DC & Sarma VR (1965) Structure of hen egg-white lysozyme. *Nature* 206: 757-761.
81. Proctor VA & Cunningham FE (1988) The chemistry of lysozyme and its use as a food preservative and a pharmaceutical. *Crit Rev Food Sci Nutr* 26: 359-395.
82. Grossowicz N & Ariel M (1983) Methods for determination of lysozyme activity. *Meth Biochem Anal* 29: 435-446.
83. Prager EM & Jolles (1996) Animal lysozymes c and g: An overview. In: *Lysozymes: Model Enzymes in Biochemistry and Biology*, Jolles P (ed), Birkhauser Verlag, Basel, Switzerland, pp 9-31.
84. Beintema JJ & van Scheltinga ACT (1996) Plant lysozyme. In: *Lysozymes: Model Enzymes in Biochemistry and Biology,* Jolles P (ed), Birkhauser Verlag, Basel, Switzerland, pp 75-86.
85. Canfield RE & McMurry S (1967) Purification and characterization of lysozyme from goose egg white. *Biochem Biophys Res Commun* 26: 38-42.
86. Weaver LH, Grutter MG & Mattews BW (1995) The refined structures of goose lysozyme and its complex with a bound trisachharide show that the goose type lysozyme lack a catalytic aspartic residue. *J Mol Biol* 245: 54-68.
87. Board RG (1995) New methods of food protection preservation, Gould GW (ed), Blackie Academic and Professional, London, UK, pp 41-57.
88. Cisani G, Varaldo PE, Ingiani A, Pompei R & Satta G (1984) Inhibition of herpes simplex virus-induced cytopathic effect by modified hen egg white lysozymes. *Curr Microbiol* 10: 35-40.
89. During K, Porsch P, Mahn A, Brinkmann O & Gieffers W (1999) The non-enzymatic microbicidal activity of lysozymes. *FEBS Lett* 449: 93-100.
90. Carini S, Muchetti G & Neviani E (1985) Lysozyme: Activity against *Clostridia* and use in cheese production - A Review. *Microbiologie Aliments Nutrition* 3: 299-320.
91. Inou K (1987) The effect of lysozyme chloride on the immune response in patients with head and neck cancer. *Gan No Rinsho* 33: 627-632.
92. De Douder C & Marias J (1974) On lysozyme therapy. I. Lysozyme tablets in treatments of some localized and generalized viral skin diseases. *Medikon* 3: 19-20.
93. Satoh J, Okano T, Kikuchi T, Suzuki N & Goto Y (1980) *In vitro* and *in vivo* effect of lysozyme chloride on mitogenic responses of lymphocytes of mice. *Tokohu Igaku Zasshi* 93: 32-37.
94. Makki F & Durance TD (1996) Thermal inactivation of lysozyme as influenced by pH, sucrose and sodium chloride and inactivation and preservative effect in beer. *Food Res Int* 29: 635-645.
95. Johnson E (1994) Egg white lysozyme as a preservative for use in foods. In: *Egg Uses and Processing Technologies. New Developments*, Sim JS & Nakai S (eds), CAB International, Wallingford, UK, pp 177-191.

96. Gebraux CV, Villa A, Monamy C & Bertrand A (1997) Use of lysozyme to inhibit malolactic fermentation and to stabilize wine after malolactic fermentation. *Am J Enol Vit* 48: 49-54.
97. Villa A (1996) Controlling malolactic fermentation (MLF) with lysozyme: Application and Results. In: *Wine Spoilage Microbiology Conference*, California State University, Fresno, USA, March 8, 1996.
98. Eisai Co Ltd (1971) Ikura and suzuko by lysozyme treatment of inferior roe. Japanese Patent JP7103732.
99. Eisai Co Ltd (1972) Preserving raw fishery product by saline lysozyme solution. Japanese Patent JP7203732.
100. Eisai Co Ltd (1973) Preservation of foods by adding a bacteriolytic enzyme and glycine. Japanese Patent JP73016613
101. Dell'Aqua E, Bruzzes T & Van Den Heuvel HH (1989) Process for producing foods free of *Listeria* bacteria. US Patent 4,810,508.
102. Johnson EA, Dell'Aqua E & Ferrari L (1991) Process for bacterial decontamination of vegetable foods. US Patent 5,019,411.
103. Ueno R, Fujita Y, Nnagamura U, Kamino Y & Tabata A (1996) Preservation of foods by the combined action of a natural antimicrobial agent and separately packaged deoxidizing agent. US Patent 5,549,919.
104. Cunningham FE, Proctor VA & Gretsch SJ (1991) Egg-white lysozyme as a food preservative: A review. *World Poultry Sci J* 47: 141-163.
105. Sava G (1996) Pharmacological aspects and theraupeutic applications of lysozymes. In: *Lysozymes: Model Enzymes in Biochemistry and Biology*, Jolles P (ed), Birkhauser Verlag, Basel, Switzerland.
106. Mirkov TE, Tex H & Fitzmaurice LC (1998) Protection of plants against plant pathogens. US Patent 5,850,025.
107. Kutila T (2004) Role of lactoferrin in treatment of bovine mastitis, Ph.D. Thesis, Department of Clinical Veterinary Sciences, University of Helsinki, Finland.
108. Soerensen M & Soerensen SPL (1939) The protein in whey. *CR Trav Lab Carlsberg* 23: 55-99.
109. Mead PE & Tweedie JW (1990) cDNA and protein sequence of bovine lactoferrin. *Nucleic Acid Res* 18: 7167-7167.
110. Pierce A, Colavizza D, Benaissa M, Maes P, Tartar A, Montreuil J & Spik G (1991) Molecular cloning and sequence analysis of bovine lactoferrin. *Eur J Biochem* 196: 177-184.
111. Rabussay DP (1982) Method of dental treatment. US Patent 4355022.
112. Moore SA, Anderson BF, Groom CR, Haridas M & Baker EN (1997) Three-dimensional structure of diferric bovine lactoferrin at 2.8 Å resolution. *J Mol Biol* 274: 222-236.
113. Anderson BF, Baker HM, Norris GE, Rice DW & Baker EN (1989) Structure of human lactoferrin: Crystallographic structure analysis and refinement at 2.8 Å resolution. *J Mol Biol* 209: 711-734.
114. Gross E & Morell JL (1971) The structure of nisin. *J Am Chem Soc* 93: 4634-4635.
115. Anonymous (1988) Preservation of milk and milk products with nisaplin. *Tech Info Leaflet No 11/88*, Aplin and Barret Ltd, Dorset, UK.
116. Klaenhammer TR (1993) Genetics of bacteriocins produced by lactic acid bacteria. *FEMS Microbiol Rev* 12: 39-86.
117. Shiba T, Wakamiya T, Fukase K, Ueki Y, Teshima T & Nishikawa M (1991) Structure of the lanthionine peptides nisin, ancovenin and lanthiopeptin. In: *Nisin and Novel*

*Lantibiotics,* Jung G & Sahl HG (eds), Escom Science Publishers, Leiden, Netherlands, pp 113-122.

118. De Vuyst L & Vandamme EJ (1994) Nisin, a lantibiotic produced by *Lactococcus lactis* subsp. *lactis*: Properties, biosynthesis, fermentation and applications. In: *Bacteriocins of Lactic Acid Bacteria: Microbiology, Genetics and Applications*, de Vuyst L & Vandamme EJ (eds), Blackie Academic and Professional, London, UK, pp 151-221.
119. Jung G (1991) Lantibiotics - ribosomally synthesized biologically active polypeptides containing sulfide bridges and didehydro amino acids. *Angew Chem Int Ed Engl* 30: 1051-1068.
120. Cheeseman GC & Berridge NJ (1975) An improved method of preparing nisin. *Biochem J* 65: 603-608.
121. Bailey FJ & Hurst A (1971) Preparation of highly active form of nisin from *Streptococcus lactis. Can J Microbiol* 17: 61-67.
122. Fukase K, Kitazawa M, Sano A, Shimbo K, Fujita H, Horimoto S, Wakamiya T & Shiba T (1988) Total synthesis of peptide antibiotic nisin. *Tetra Lett* 29: 795-798.
123. Hurst A (1981) Nisin. *Adv Appl Microbiol* 27: 85-123.
124. Kitazawa M, Fujita H, Kubo A & Shiba T (1991) Conformations of nisin and its fragments using synthesis, NMR and computer simulations. In: *Nisin and Novel Lantibiotics,* Jung G & Sahl HG (eds), Escom Science Publishers, Leiden, Netherlands, pp 59-75.
125. Van de Van FJM, van de Hooven HW, Konings RNH & Hilbers CW (1991) NMR studies of lantibiotics. The struture of nisin in aqueous solution. *Eur J Biochem* 202: 1181-1188.
126. Van de Van FJM, van de Hooven HW, Konings RNH & Hilbers CW (1991) The spatial structure of nisin in aqueous solution. In: *Nisin and Novel Lantibiotics*, Jung G & Sahl HG (ed), Escom Science Publishers, Leiden, Netherlands, pp 35-42.
127. Lian LY, Chan WC, Morley SD, Roberts GC, Bycroft BW & Jackson D (1991) NMR studies of the solution structure of nisin A and related peptides. In: *Nisin and Novel Lantibiotics,* Jung G & Sahl HG (eds), Escom Science Publishers, Leiden, Netherlands, pp 43-58.
128. Lian LY, Chan WC, Morley SD, Roberts GC, Bycroft BW & Jackson D (1992) NMR studies of the solution structure of nisin A. *Biochem J* 283: 413-420.
129. Van de Hooven HW, Doeland CCM, van de Kamp M, Konings RNH, Hilbers CW & van de Van FJM (1996) Three-dimensional structure of the lantibiotics micelles of dodecylphocholine and sodium dodecylsulphate. *Eur J Biochem* 235: 382-393.
130. Ingram L, Tombs MP & Hurst A (1967) Mobility-molecular weight relationships of small proteins and peptides in acrylamide-gel electrophoresis. *Anal Biochem* 20: 24-29.
131. Javis B, Jeffcoat J & Cheeseman GC (1968) Molecular weight distribution of nisin. *Biochim Biophys Acta* 168: 153-155
132. Ogden K & Tubb RS (1985) Inhibition of bear spoilage lactic acid bacteria by nisin. *J Inst Brew* 91: 390-392.
133. Radler F (1990) Possible use of nisin in wine making. I. Action of nisin against lactic acid bacteria and wine yeasts in solid and liquid media. *Am J Enol Vitic* 41: 1-6.
134. Gupta RK & Prasad DN (1989) Incorporation of nisin in stirred yoghurt II. Effect on biochemical activities during storage. *Cult Dairy Prod J* 24: 9-10.
135. Meghrous J, Lacroix C & Simard RE (1999) The effects on vegetative cells and spores of three bacteriocins from lactic acid bacteria. *Food Microbiol* 16: 105-114.
136. Ferreira MASS & Lund BM (1996) The effect of nisin on *Listeria monocytogenes* in culture medium and long life cottage cheese. *Lett Appl Microbiol* 22: 433-438.
137. Ukuku DO & Shelef LA (1997) Sensitivity of six strains of *Listeria monocytogenes* to nisin. *J Food Prot* 60: 867-869.

138. Scott VN & Taylor SL (1981) Temperature, pH and spore load effects on the ability of nisin to prevent the outgrowth of *Clostridium botulinum* spores. *J Food Sci* 46: 121-126.
139. Hirsch A (1950) The assay of the antibiotic nisin. *J Gen Microbiol* 4: 70-83.
140. Ramseier HR (1960) The action of nisin on *Clostridium butyricum. Arch Microbiol* 37: 57-94.
141. Hirsch A & Grinsted A (1951) The differentiation of the lactic streptococci and their antibiotics. *J Dairy Res* 18: 198-204.
142. Harris LJ, Fleming HP & Klaenhammer TR (1991) Sensitivity and resistance of *Listeria monocytogenes* ATCC 19115, Scott A, and UAL500 to nisin. *J Food Prot* 52: 384-387.
143. Thorpe RH (1960) The action of nisin on spoilage bacteria. I. The effect of nisin on the heat resistance of *Bacillus stearothermophilus* spores. *J Appl Bacteriol* 23: 136-143.
144. Henning S, Metz R & Hammes WP (1986) New aspects for the application of nisin to foods based on its mode of action. *Int J Food Microbiol* 3: 135-141.
145. Ruhr E & Sahl HG (1985) Mode of action of the peptide antibiotic nisin and influence on the membrane potential of whole cells and on cytoplasmic and artificial membrane vesicles. *Antimicrob Agents Chemother* 27: 841-845.
146. Abee T, Rombouts FM, Hugenholtz J, Guibard G & Letellier G (1994) Mode of action of nisin Z against *Listeria monocytogenes* Scott A grown at high and low temperatures. *Appl Environ Microbiol* 60: 1962-1968.
147. Abee T (1995) Pore-forming bacteriocins of Gram-positive bacteria and self-protection mechanisms of producer organisms. *FEMS Microbiol Lett* 129: 1-10.
148. Henning S, Metz R & Hammes WP (1986) Studies on the mode of action of nisin. *Int J Food Microbiol* 3: 121-134.
149. Hitchens AD, Gould GW & Hurst A (1963) The swelling of bacterial spores during germination and outgrowth. *J Gen Microbiol* 30: 445-453.
150. Lipinska E (1997) Nisin and its applications. In: *Antibiotics and Antibiosis in Agriculture,* Woodbine M (ed), Butterworth, London, UK, pp 103-130.
151. Morris SL, Walsch RC & Hansen JN (1984) Identification and characterization of some bacterial membrane sulphydral group which are targets of bacteriostatic and antibiotic action. *J Biol Chem* 259: 135190-135195.
152. Ganzle MG, Weber S & Hammes WP (1999) Effect of ecological factors on the inhibitory spectrum and activity of bacteriocins. *Int J Food Microbiol* 46: 207-217.
153. Breukink E, van Kraaij C, Demel RA, Seizen RJ, Kuipers OP & de Kruijff B (1997) The C-terminal region of nisin is responsible for the initial interaction of nisin with the target membrane. *Biochemistry* 36: 6968-6976.
154. Garcia-Garcera MJ, Elferink MJL, Driessen AJM & Konings WN (1993) *In vitro* pore-forming activity of the lantibiotic nisin. Role of proton motive force and lipid composition. *Eur J Biochem* 212: 417-422.
155. Demel RA, Peelen T, Seizen RJ & de Kruijif OP (1996) Nisin Z, mutant nisin Z and lacticin 481 interactions with anionic lipids correlate with antimicrobial activity. A monolayer study. *Eur J Biochem* 235: 267-274.
156. Jack RW, Bierbaum G & Sahl HG (1998) Biological activities. In: *Lantibiotics and Related Peptides*, Springer-Verlag, USA, pp157-190.
157. Sahl HG (1991) Pore formation in bacterial membranes by cationic lantibiotics. In: *Nisin and Novel Lantibiotics,* Jung G & Sahl HG (eds), Escom Science Publishers, Leiden, Netherlands, pp 347-358.
158. Maisnier-Patin S, Deschamps N, Tatini SR & Richard J (1992) Inhibition of *Listeria monocytogenes* in Camembert cheese made with a nisin-producing starter. *Lait* 72: 249-263.

159. Ojcius DM & Young JDE (1991) Cytolytic pore-forming proteins and peptides: is there a common structural motif. *Trends Biochem Sci* 16: 225-229.
160. Nissen-Meyer J, Holo H, Haversteuin LS, Sletten K & Nes IF (1992) A novel bacteriocin whose activity depends on the complementary action of two peptides. *J Bacteriol* 174: 5686-5692.
161. Bierbaum G & Sahl HG (1991) Induction of autolysis of *Staphylococcus simulans* 22 by Pep5 and nisin and influence of the cationic peptides on the activity of the autolytic enzymes. In: *Nisin and Novel Lantibiotics,* Jung G & Sahl HG (eds), Escom Science Publishers, Leiden, Netherlands, pp 386-396.
162. Meyer AM (1973) *Processed Cheese Manufacture*, Food Trade Press Ltd, London, UK.
163. Thomas MA (1977) *The Processed Cheese Industry*, Department of Agriculture, New South Wales, Australia.
164. Somers EB & Taylor SL (1987) Antibotulinal effectiveness of nisin in pasteurized process cheese spreads. *J Food Prot* 50: 842-848.
165. Heinemann B, Voris L & Stumbo CR (1965) Use of nisin in processing food products. *Food Technol* 19: 592-596.
166. Anonymous (1989) International acceptance of nisin as a food preservative. *Tech Info Leaflet No.*4/89/11, Aplin and Barret Ltd, Dorset, UK.
167. Wajid HRA & Kalra MS (1976) Nisin as an aid for extending the shelf life of sterilized milk. *J Food Sci Technol* 13: 6-8.
168. Williams AH (1973) Preservatives in food. A commentary on the report of the Governmental food additives and contaminants committee on the review of the preservatives in food regulations, 1962. *J Royal Soc Prom Health* 93: 92-96.
169. Gregory ME, Henry KM & Kon SK (1964) Nutritive properties of freshly prepared and stored evaporated milks manufactured by a normal commercial procedure or by reduced thermal processes in the presence of nisin. *J Dairy Res* 31: 113-119.
170. Eyles MJ & Richardson KC (1988) Thermophilic bacteria and food spoilage. *CSIRO Food Res Quart* 48: 19-24.
171. Calderon C, Collins-Thompson DL & Usborne WR (1985) Shelf-life studies of vacuum packaged bacon treated with nisin. *J Food Prot* 48: 330-333.
172. Bell RG & de Lacy KM (1987) The efficacy of nisin, sorbic acid and monlaurin as preservatives in pasteurized cured meat products. *Food Microbiol* 4: 277-283.
173. Ogden K (1986) Nisin: A bacteriocin with a potential use in brewing. *J Inst Brew* 92: 379-383.
174. Ogden K, Waites MJ & Hammond JRM (1988) Nisin and brewing. *J Inst Brew* 94: 233-238.
175. Radler F (1990) Possible use of nisin in wine making. II. Experiments to control lactic acid bacteria in the production of wine. *Am J Enol Vit* 41: 7-11.
176. Delves-Broughton J (1990) Nisin and its uses as a food preservative. *Food Technol* 44: 100-117.
177. Garvie EI (1986) Genus *Pediococcus* Claussen 1903. In: *Bergey's Manual of Systematic Bacteriology*, Vol 2, Sneath PHA (ed), Willams and Wilkins, Baltimore, USA, pp 1075-1079.
178. Ray B (1995) *Pediococcus* in fermented foods. In: *Food Biotechnology Microorganisms,* Hui YH & Khachatourians GG (eds), VCH Publishers Inc., New York, USA, pp 745-796.
179. Ray B (1992) Pediocin(s) of *Pediococcus acidilactici* as a food biopreservative. In: *Food Biopreservatives of Microbial Origin*, Ray B & Daeschel MA (eds), CRC Press, Boca Raton, USA, pp 265-322.
180. Ray B (1994) Pediocins of *Pediococcus* species. In: *Bacteriocins of Lactic Acid Bacteria*, de Vuyst L & Vandamme EJ (eds), Blackie Academic and Professional Inc, New York, USA, pp 465-496.

181. Ray B (1996) Characteristics and applications of pediocin(s) of *Pediococcus acidilactici*: Pediocin PA-1/AcH. In: *Lactic Acid Bacteria*, NATO ASI Series, Vol 98, Bozoglu TF & Ray B (eds), Springer-Verlag, New York, USA, pp 155-204.
182. Ray B & Hoover DG (1993) Pediocins. In: *Bacteriocins of Lactic Acid Bacteria*, Hoover DG & Steenson LR (eds), Academic Press, San Deigo, USA, pp 108-210.
183. De Vuyst L (1994) Bacteriocins of *Pediococcus*. In: *Bacteriocins of Lactic Acid Bacteria*, De Vuyst L & Vandamme EJ (eds), Blackie Academic and Professional Inc, New York, USA, pp 461-464.
184. Jack WR, Tagg GR & Ray B (1995) Bacteriocins of Gram-positive bacteria. *Microbiol Rev* 59: 171-200.
185. Chen Y, Ludescher RD & Montville TJ (1998) Influence of lipid composition on pediocin PA-1 binding to phospholipid vesicles. *Appl Environ Microbiol* 64: 3530-3532.
186. Montville TJ & Chen Y (1998) Mechanistic action of pediocin and nisin: Recent progress and unresolved questions. *Appl Microbiol Biotechnol* 50: 511-519.
187. Naidu AS (2000) *Natural Food Antimicrobial Systems,* CRC Press, Boca Raton, New York, USA, pp 803.
188. Gonzalez CF & Kunka BS (1987) Plasmid-associated bacteriocin production and sucrose fermentation in *Pediococcus acidilactici. Appl Environ Microbiol* 53: 2534-2538.
189. Bhunia AK, Johnson MC & Ray B (1987) Partial purification and characterization of antimicrobial substances from *Pediococcus acidilactici* and *Lactobacillus plantarum.* In: *Inst Food Technol Ann Meeting*, June 16-19, Las Vegas, Abstract #143.
190. Bhunia AK, Kim WJ, Johnson MC & Ray B (1987) Direct detection of an antimicrobial peptide of *Pediococcus acidilactici* in sodium dodecyl sulfate-polyacrylamide gel electrophoresis. *J Ind Microbiol* 2: 319-322.
191. Coventry MJ, Muirhead K & Hickey MW (1995) Partial characteristics of pediocin $PO_2$ and comparision with nisin for biopreservation of meat products. *Int J Food Microbiol* 26: 133-145.
192. Schved F, Lalazar A, Henis Y & Juven BJ (1993) Purification, partial characterization and plasmid linkage of pediocin SJ-1, a bacteriocin produced by *Pediococcus acidilactici. J Appl Bacteriol* 74: 67-77.
193. Bhunia AK, Johnson MC & Ray B (1998) Purification, characterization and antimicrobial spectrum of a bacteriocin produced by *Pediococcus acidilactici. J Appl Bacteriol* 65: 261-268.
194. Bhunia AK, Johnson MC, Ray B & Kalchayanand N (1991) Mode of action of pediocin AcH from H on sensitive bacterial strain. *J Appl Bacteriol* 70: 25-33.
195. Conventry MJ, Muirhead K & Hickey MW (1994) Partial characteristics of pediocin $PO_2$ and comparison with nisin for biopreservation of meat products. *Int J Food Microbiol* 26: 130-140.
196. Kalchayanand N, Ray B, Sikes T & Dunne CP (1995) Bacteriocin based biopreservatives add an extra dimension in food preservation by hydrostatic pressure. In: *Proceedings of the Spring Meeting*, April 24-26, Research and Development Associates for Military Food and Packaging System, Inc, San Antonio, Texas, USA, pp 282-286.
197. Ray B & Miller KW (2000) Pediocin. In: *Natural Food Antimicrobial Systems*, CRS Press, Boca Raton, New York, USA, pp 525-566.
198. Christensen DP & Hutkins RW (1992) Collapse of the proton motive force in *Listeria monocytogenes* caused by bacteriocin produced by *Pediococcus acidilactici. Appl Environ Microbiol* 58: 3312-3315.
199. Christensen DP & Hutkins RW (1994) Glucose uptake by *Listeria monocytogenes* Scott A and inhibition by pediocin. *J Appl Environ Microbiol* 60: 3870-3873.

200. Motlagh AM, Holla S, Johnson MC, Ray B & Field RA (1992) Inhibition of *Listeria* spp. in sterile food system by pediocin AcH a bacteriocin produced by *Pediococcus acidilactici* H. *J Food Prot* 55: 337-343.
201. Bennik MHJ, Smid EJ & Gorris LGM (1997) Vegetable associated *Pediococcus parvulus* produces pediocin PA I. *Appl Environ Microbiol* 63: 2074-2076.
202. Berry ED, Liewen MB, Mandiog RW & Hutkins RW (1990) Inhibition of *Listeria monocytogenes* by bacteriocin producing *Pediococcus* during the manufacture of fermented semidry sausage. *J Food Prot* 53: 194-197.
203. Pucci MJ, Vedamuthu ER, Kunka BS & Vandenbergh PA (1988) Inhibition of *Listeria monocytogenes* by using bacteriocin PAI produced by *Pediococcus acidilactici* PAC 10. *Appl Environ Microbiol* 54: 2349-2353.
204. Farell KT (1985) *Spices, Condiments and Seasonings,* AVI Publishing Co, Westport, USA.
205. Nychas GJ (1994) Natural antimicrobial from plants. In: *Natural Antimirobial System and Food Preservation*, Dillon VM & Board RG (eds), CAB International, Wallingford, UK, pp 58-83.
206. Keceli T, Robinson RK & Gordon MH (1998) The antimicrobial activity of some selected phytophenols from virgin olive oil. IFT Annual Meeting, Atlanta, Abstract 46-B-20.
207. Davidson PM (1993) Parabens and phenolic compounds. In: *Antimicrobials in Foods*, Davidson PM & Branen AL (eds), Marcel Dekker Inc, New York, USA, pp 263-306.
208. Juven BJ, Kanner J, Schved F & Weisslowicz H (1994) Factors that interact with the antibacterial action of thyme essential oil and its active constituents. *J Appl Bacteriol* 76: 626-631.
209. Conner DE & Beuchat LR (1984) Effects of essential oils from plants on growth of food spoilage yeasts. *J Food Sci* 49: 429-434.
210. Conner DE, Beuchat LR, Worthington RE & Hitchcock HL (1984) Effects of essential oils and oleoresins of plants on ethanol production, respiration and sporuralation of yeasts. *Int J Food Microbiol* 1: 63-74.
211. Gubanov IA, Libizov NI & Gladkikh AS (1970) Search for saponin containing plants among flora of central Asia and Southern Kazakhistan. *Farmatsiya (Moscow)* 19: 23-31.
212. Takeya K (1997) Antimicrobial activities of natural medicines. *Kagaku Kogyo* 48: 888-897.
213. Tamura Y (1996) Antimicrobial activities of extract from Yucca Kurin. *Tekunoroji* 6: 67-71.
214. Pompei R, Flore O, Marciallis MA, Pani A & Loddo B (1979) Glycyrrhizic acid inhibits virus growth and inactivates virus particles. *Nature* 281: 689-690.
215. Okubo K, Kudau S, Uchida T, Yoshiki Y, Yoshikoshi M & Tonomura M (1994) Soybean saponin and isoflavonoids structure and antiviral activity against human immunodeficiency virus *in vitro*. *ACS Symp Ser* 546: 330-339.
216. Tomas-Barberan FA, Maillard M & Hostettmann K (1988) Antifungal flavonoids from the leaf surface of *Helichrysum nitens* and from the stem bark of *Erythrina berteroana.* In: *Plant Flavonoids in Biology and Medicine,* Cody V, Middleton Jr E, Harbrne JB & Beretz A (eds), Alan R Liss, New York, USA, pp 61-65.
217. Yoshida S, Kasuga S, Hayashi N, Ushiroguchi T, Matsura H & Nakagawa S (1987) Antifungal activity of ajone derived from garlic. *Appl Environ Microbiol* 53: 615-617.
218. Feldberg RS, Chang SC, Kolik AM, Nadler M, Newtrth Z, Sundstrom DC & Thomson NH (1988) *In vitro* mechanism of inhibition of bacterial cell growth by allicin. *Antimicrob Agents Chemother* 32: 1763-1768.
219. Tayler VE (1993) Garlic and other alliums. *The Honest–Herbal*, Pharmaceuticals Products Press, New York, USA, pp 139-143.

# Probiotics: Concepts and Applications in Food

# 22

**S.K. Sharma[1], V.K. Joshi[2] and Somesh Sharma[2]**

[1]*Department of Horticulture, G.B. Pant University of Agriculture & Technology, Hill Campus Ranichauri, Tehri-Garhwal 249 199, Uttarakhand*
[2]*Department of Food Science and Technology, Dr. Y.S. Parmar University of Horticulture and Forestry, Nauni, Solan 173 230 (H.P.)*

## 1. INTRODUCTION

Eating balanced diet and nutritious food containing carbohydrates, proteins, fats, vitamins, minerals and sufficient roughages keeps a person healthy and is known to most, but whether our body possesses the important digestive aids needed for the breakdown of these specific foods and to absorb the nutrients from it or not, seldom scratches a common man's mind. But now it has been shown by various studies on humans and animals that for the absorption of these nutrients there are two important substances/components or digestive aids i.e. friendly microorganisms (mainly bacteria) and specific enzymes. Whatever we eat and how much nutritious it may be, unless and until it is absorbed by our body, it is all useless. With the changing lifestyles, increasing urbanization, changing food habits and modern food preparations, more and more people are going for the ready-to-serve and ready-to-cook type of foods. This has some-how led to the depletion of the two digestive aids in our bodies. Poor eating habits, dietary deficiencies, chronic stress (emotional or physical), lack of exercise, insufficient rest and indiscriminate or excessive use of antibiotics further intensifies this problem. Therefore, under the present circumstances, restoration of such digestive aids has become necessary for healthy and problem free living [1].

There is a famous saying that "Preventing or reducing the risk of a disease is preferable to treating the disease". Probiotics have emerged as a major factor influencing gastro-intestinal physiology and functions. This development introduces many challenges, but also creates new opportunities for food and nutrition scientists to improve food quality and develop new products with specific health benefits for different sub-populations [2].

The ready availability of antibiotics in the 1950's resulted in their widespread use as therapeutic agents and growth stimulants for farm animals. Since that time, there has been growing concern that the use of antibiotics as growth promoters was resulting in the development of resistant populations of bacteria, which made subsequent use of antibiotics for therapy difficult. The possibility of antibiotics ceasing to be used as growth stimulants for farm animals and the concern about the side effects of their use as therapeutic agents has produced a climate in which both consumer and manufacturer are looking for alternatives. Probiotics are being considered to fill this role, and some people are already using them in preference to antibiotics [3].

*Food Biotechnology: Principles and Practices*
*V.K. Joshi & R.S. Singh (Eds.)*
*I.K. International Publishing House Pvt. Ltd., India*
*pp 781–798*

## 2. HISTORY AND DEFINITION

### 2.1 Probiotics

The concept of probiotics evolved at the turn of the 20th century from a hypothesis first proposed by Nobel Prize winner Russian scientist 'Elie Metchnikoff' who suggested that long healthy life of Bulgarian peasants resulted from their consumption of fermented milk products [4]. He believed that when consumed, the fermenting *Lactobacillus* positively influenced the microflora of the colon, decreasing the toxic microbial activity. The historical association of probiotics with the fermented dairy products stems from these early observations. However, the term *probiotic*, meaning "for life," is derived from the Greek language. It was first used by Lilly and Stillwell in 1965 to describe "substances secreted by one microorganism which stimulates the growth of another" and thus, was contrasted with the term *antibiotic* [5]. Sperti [6] applied the term to tissue extracts that stimulate microbial growth. Parker [7] defined probiotics as "organisms and substances that contribute to intestinal microbial balance" and was first to use this term in the sense that it is used today. Fuller [3] attempted to improve Parker's definition of probiotic with a distinction, "A live microbial feed supplement which beneficially affects the host animal by improving its intestinal microbial balance." However, Havenaar and co-workers [8] defined probiotics as "a viable mono- or mixed culture of microorganisms, which when applied to animal or man, beneficially affects the host by improving the properties of the indigenous microflora". According to Salminen [9], a probiotic is "a live microbial culture or cultured dairy product which beneficially influences the health and nutrition of the host." Schaafsma [10] defined probiotic as "oral probiotics are living microorganisms, which upon ingestion in certain numbers, exert health effects beyond inherent basic nutrition." Thus, Salminen and Schaafsma broadened the definition by including the health effects. Schrezenmeir and Vrese [11] defined it as "a preparation of or a product containing viable, defined microorganisms in sufficient numbers, which alter the microflora (by implantation or colonization) in a compartment of the host, and by that event exert beneficial health effects to this host."

### 2.2 Prebiotics

In 1995, one more term was introduced by exchanging pro for pre, which means "before" or "for". Prebiotics were defined as "a non-digestible food ingredient that beneficially affects the host by selectively stimulating the growth and/or activity of one or a limited number of bacteria in the colon." This definition more or less overlaps with the definition of dietary fibre, with the exception of its selectivity for certain species. This selectivity was shown for bifidobacteria, which may be promoted by the ingestion of substances such as fructo-oligosaccharides and inulin, transgalactosylated oligosaccharides, and soybean oligosaccharides [12, 13, 14, 15, 16, 17, 18].

### 2.3 Synbiotics

The term *synbiotic* is used when a product contains both probiotics and prebiotics, and the term should be restricted for products in which the probiotic compound selectively favours the prebiotic compound [11]. In this strict sense, a product containing oligofructose and probiotic bifidobacteria would fulfill the definition, whereas a product containing oligofructose and a probiotic *Lactobacillus casei* strain would not.

## 3. HYPOTHESIS FOR HEALTH EFFECTS

Various friendly bacteria benefit the human body by improving the absorption of nutrients, supporting the immune system functions of the body, encouraging healthy digestion, protecting the body system from yeast infections and retarding the growth and development of other harmful bacteria, etc. The numerous health effects of probiotics can hardly be explained by a single unifying hypothesis based on a single quality or mechanism and remains valid for all microorganisms exerting one or the other effect as given below:

- Reduction of *Helicobacter pylori*
- Relief from constipation
- Reduction of allergic symptoms
- Cancer prevention
- Beneficial effects on mineral metabolism, particularly bone density and stability
- Relief from irritable bowel syndrome
- Reduction of cholesterol and triacylglycerol plasma concentrations
- Lowering frequency and duration of diarrhoea
- Decrease in unfavourable metabolites e.g., ammonium and procancerogenic enzymes in the colon
- Stimulation of humoral and cellular immunity.

Probiotics containing products are common in Japan and Europe [19, 20]. In the United States of America, probiotics are just now receiving attention by the food industry as healthful food ingredients for an increasingly health conscious consumer. India, however, is far behind all these nations as far as the use of probiotics is concerned. Only a few reports from India regarding probiotics fermentations and use in animal feed are available [21, 22]. In the list of probiotic microorganisms (Table 1), all are intestinal strains except *Lactobacillus bulgaricus* and *Streptococcus thermophilus* which are yoghurt starter organisms and are included on the basis of claims made for yoghurt [3, 23].

**Table 1:** Microorganisms used in probiotic preparations.

| Microorganism | Strain |
|---|---|
| *Lactobacillus acidophilus* | NCFM |
| | DDS-1 |
| | SBT-2062[a] |
| | LA-1 (same as strain LA-5 sold in Europe) |
| *Lactobacillus bulgaricus* | Shirota |
| *Lactobacillus casei* | Immunitas |
| *Lactobacillus fermentum* | RC-14 |
| *Lactobacillus helveticus* | |
| *Lactobacillus johnsonii* | La 1 (same as Lj 1) |
| *Lactobacillus lactis* | L1A |
| *Lactobacillus paracasei* | CRL 431 |
| *Lactobacillus plantarum* | 299 V |
| *Lactobacillus renteri* | SD1221 (same as MM2) |
| *Lactobacillus rhamnosus* | GG [a] |
| | GR-1 |
| | 271 |
| | LB21 |
| *Lactobacillus salivarius* | UCC118 |
| *Bifidobacterium lactis* | Bb-12 |
| *Bifidobacterium longum* | BB536 [a] |
| | SBT-2928 [a] |
| *Bifidobacterium breve* | Yakult |
| *Enterococcus faecium* | – |
| *Enterococcus faecalis* | – |
| *Escherichia coli* | – |
| *Streptococcus thermophilus* | – |

[a] Strains have been awarded *Foods for Specific Health Use Status* in Japan

*Source:* 23

## 4. SCIENTIFIC BASIS OF FUNCTIONALITY

Hundreds of microbial species live in association with humans i.e., on skin and in oral, intestinal and vaginal tracts. Studies with germfree (genobiotic) animals prove that microbial colonization is not required for survival. In fact, microbial colonization can have negative effect as a result of toxic, genotoxic, mutagenic or carcinogenic potential of microbial metabolites [24]. Germ free animals, however, are more susceptible to infections than conventional counterparts [25]. According to Vollaard and Clasener [26], the increased susceptibility to infection is attributed to poor immune function and absence of 'colonization resistance' (competition of normal microflora with invading microorganisms).

Probiotic hypothesis states that addition of certain exogenous microorganisms to the intestinal ecosystem will have a positive effect. The intestinal tract is a fairly stable microbial ecosystem in the adult [27]. Acute perturbations might result from antibiotic use, disease or certain dietary changes seem to be self correcting [28]. Probiotic bacteria consumed even in high numbers do not become permanent colonizers and are rarely detectable in faecal or intestinal samples beyond a couple of weeks after ingestion. Whether or not a specific probiotics bacterium will have beneficial, detrimental or no effect on the health cannot be presumed strictly through determinations of its genus or species. Probiotic effects are strain specific. In addition, physiological conditions of the host are likely to be as important to probiotics efficacy as the microbial strain.

The beneficial effects of probiotics may be mediated by a direct antagonistic effect against a specific groups of organisms, resulting in a decrease in numbers or by an effect on their metabolism or stimulation of immunity (Table 2). For all these mechanisms, there is some support from experimental data. The suppression of bacterial numbers could result from the production of antibacterial substances. Several high molecular weight antibacterial substances have been described as being produced by lactic acid bacteria, but some authors think that the observed inhibitory effects can be, in many cases, attributed to low pH and primary metabolites [29]. Another mechanism for preventing colonization by pathogens is competition for adhesion sites on the gut epithelial surfaces.

Lactic acid bacteria are known to be associated with the gut wall of chickens and pigs [30, 31]. This sort of evidence suggests that it is desirable to use adhering strains when designing probiotics supplements. Feeding of *Lactobacillus acidophilus* to humans suppressed the activity of β-glucuronidase, nitroreductase and azoreductase [32]. Similarly, in rats associated with human gut-flora, there was a reduction in β-glucuronidase and β-glucosidase when they were dosed with same strains of *Lactobacillus acidophilus.* Probiotics may also have their influence by increasing the activity of useful enzymes e.g., β-glactosidase in the elevation of lactose intolerance. Some health benefits of probiotics and prebiotics are enlisted in Tables 3 and 4.

**Table 2:** Mode of action of probiotics.

| Effect | Mode of action |
|---|---|
| • Suppression of viable count | (a) by production of antibacterial compounds |
| | (b) by competition for nutrients |
| | (c) by competition for adhesion sites |
| • Alteration of microbial metabolism | (a) increases enzyme activity |
| | (b) decreases enzyme activity |
| • Stimulation of immunity | (a) increases antibody levels |
| | (b) increases macrophage activity |

*Source:* 83

**Table 3:** Strength of evidence for improvement of body functions and disease risk reduction by probiotics and prebiotics.

| Functional effects | Strength of evidence | | Disease risk reduction | Strength of evidence | |
|---|---|---|---|---|---|
| | Probiotics | Prebiotics | | Probiotics | Prebiotics |
| • Lactose intolerance | Strong | Unknown | Diarrhoea | Promising | Unknown |
| • Immunostimulation | Preliminary | Unknown | Constipation | Unknown | Promising |
| • Faecal mutagenesis | Preliminary | Unknown | Colon cancer | Preliminary | Preliminary |
| • Hypocholesterolemia | No effect | Preliminary | Osteoporosis | Unknown | Unknown |
| • Hypolipidemia | Unknown | Promising | Lipid associated chronic disease | Probably no reduction | Unknown |
| • Colonic flora | Preliminary | Strong | | | |
| • Calcium bioavailability | Unknown | Promising | | | |

*Sources:* 15, 28, 30-32

**Table 4:** Improvement in various metabolic functions by probiotics and prebiotics.

| Probiotics | Prebiotics |
|---|---|
| • Alleviation of lactose intolerance | • Fate in gastrointestinal tract |
| • Immune enhancement | • Effect on mineral absorption |
| • Decrease in faecal enzyme and mutagenicity | • Effect on metabolism of lipids |
| • Hypocholesterolemic effect | • Reducing disease risk |
| • Reduction of risk of disease | |

Probiotics also increase the immunity and phagocytic activity of the body. A strain of *Enterococcus faecium* (a species frequently used in probiotics preparations) established a monoassociate in germfree mice and was able to reduce the amount of *Salmonella typhimurium* in the spleen implying a systemic effect [33]. Yoghurt has been shown to increase the antibody levels when fed to germ free mice [34]. *Lactobacilli* are also involved in the phagocytic activity.

## 5. COMPOSITION OF PROBIOTICS

Probiotics can be presented to animals in various ways. The type of preparation will depend on the intended use. They can be either included in the pelleted feed or produced in the form of capsules, paste, powders or granules which can be used for dosing animals directly or through their food. Target species are cattle, sheep, goats, pigs, poultry, horses and domestic pets [3]. Nearly all probiotics currently in the market contain *Lactobacilii* or *Streptococi;* a few contain *Bifidobacteria.* Probiotic preparations may consist of single strain or may contain any number up to eight strains. The attraction of multiple strain preparations is that they are active against a wider range of conditions and in a wider range of animal species.

Besides their use in the growth promotion and for treatment of various ailments of farm and pet animals, probiotics are also being used for the treatment of many human diseases like constipation, cancer, tumour, intestinal and vaginal infections, allergy, stomach health, hypertension, cholesterol and other health effects [35].

## 5.1 Strain Characteristics and Habitat Specificities

Different strains of probiotic bacteria may exert different effects based on specific capabilities and enzymatic activities, even within one species [36, 37]. Different microorganisms express habitat preferences that may differ in various host species [38]. Lactobacilli are among the indigenous microflora colonizing the chicken's crop, the stomach of mice and rats, and the lower ileum of man. Bacteria colonizing such high-transit-rate sites must adhere firmly to the mucosal epithelium and must adapt to the milieu of this adhesion site [39, 40, 41]. The competition for adhesion receptors between probiotic and pathogenic microorganisms, therefore, is dependent on such habitat specificities. On the other hand, bacteria are found in much higher numbers in the colon, particularly in the faeces, than are lactobacilli. It is self-evident that effects bound to this luminal site of action may be exerted even more efficiently by such microorganisms, which do not necessarily need to adhere to the mucosa. Moreover, preferences for microhabitats have to be considered. Four microhabitats in the gastrointestinal tract were outlined earlier [38] as follows: (1) the surface of epithelial cells; (2) the crypts of the ileum, caecum, and colon; (3) the mucus gel that overlays the epithelium; and (4) the lumen of the intestine.

As mentioned earlier, several indigenous, pathogenic, or probiotic microorganisms target the surface of the epithelium by specific adhesion, often mediated by special organelles, e.g, fimbriae [41, 42]. The crypts are typically colonized by motile, spiral-shaped bacteria of the genera *Borellia, Treponema, Spirillium* and others, e.g., *H. pylori* [43, 44, 45]. The mucus layer can form a microbial habitat and can protect the host against colonization in some circumstances. As a result of its complex and varying composition, and for technical reasons, its function in this context is least clarified.

The luminal content of bacteria depends greatly on bowel transit. Therefore, the microbial density in the small bowel is low, whereas it is abundant in the lumen of the colon, which gives space to microorganisms without adhesion molecules. When the great variety of species, strain characteristics, and the habitat specifics are considered, it becomes clear that a proven probiotic effect on one strain or species cannot be transferred to other strains or species.

## 5.2 Effects of Probiotics on Human and Animal Health

Probiotics are known to have a very positive effect in curing various human and animal ailments by various mechanisms as detailed in Tables 5 and 6.

### *5.2.1 Cancer/tumours*

The claimed anti-tumour properties of (mainly) cultured dairy products are supposed to be due to:

- Inactivation or inhibition of carcinogenic compounds in the GI (gastrointestinal) tract.
- Stimulation or enhancement of immune response.
- Reduction of faecal bacterial enzyme activities i.e., β-glucuronidase, azoreductase and nitroreductase.

Faecal enzyme activities are important because they are associated with conversion of procarcinogens to proximal carcinogens. Suppression of β-glucuronidase might reduce 1,2-dimethyl hydrazine (a chemical carcinogen) activation and subsequently, tumour formation [32]. The effect of probiotics and prebiotics against cancer is given in Table 7. Studies also show a significant trend of decreasing breast cancer risk with increasing consumption of fermented milk [46]. Research has demonstrated that dietary components may increase or decrease cancer incidence [47]. One study demonstrated that a powder preparation of *L. casei* ($10^{10}$ cfu three times a day for one year) increased the recurrence free period among human subjects with superficial bladder cancer [448]. Results suggest that probiotics bacteria

**Table 5:** Effects of probiotics on animal and human health.

| Animals | Humans |
|---|---|
| | *Disturbance of indigenous flora* |
| (Non-) infectious disbacteriosis after (high) dosages of antibiotics | (Non-) infectious disbacteriosis after (high) dosages of antibiotics |
| Stress (e.g., feed change, transport) | After radiation |
| | *Growth promotion* |
| Higher feed conversion | Promotion of health |
| Destruction of anti-nutritional factors (ANFs) | Carcinogenesis inhibition |
| Synthesis of various vitamins | Anti cholesteremic compounds |
| Pre-digestion of proteins | Increased resorption of calcium |
| | Lactose intolerance |
| | Destruction of anti-nutritional factors (ANFs) |
| | Synthesis of various vitamins |
| | Pre-digestion of proteins |
| | *Deficient development of microflora* |
| Restricted access to mother | Neonate intensive care unit |

*Source:* 46

**Table 6:** Potential and established effects of probiotic bacteria.

| Target health benefit | Postulated mechanism |
|---|---|
| (1) Aid in digestion of lactose | (a) Bacterial lactase enzyme hydrolyses lactose |
| (2) Resistance to enteric pathogens | (a) Secretory immune effect<br>(b) Colonization resistance<br>(c) Alteration of intestinal conditions to be less favourable for pathogenicity (pH, short chain fatty acid, bacteriocins)<br>(d) Alteration of toxin binding sites<br>(e) Influence on gut flora populations<br>(f) Adherence to intestinal mucosa, interfering with pathogen adherence<br>(g) Up regulation of intestinal mucin production, interfering with pathogen attachment to intestinal epithelial cells |
| (3) Small bowel bacterial overgrowth | (a) Alteration of intestinal conditions to be less favourable to overgrowth of flora activities or populations<br>(b) Influence on activity of overgrowth of flora, decreasing toxic metabolite production |
| (4) Immune system modulation | (a) Strengthening of non-specific defence against infection and tumours<br>(b) Adjuvant effect in antigen specific immune responses<br>(c) Enhancement of secretary Ig A production |
| (5) Blood, lipid, heart diseases | (a) Assimilation of cholesterol within bacterial cell<br>(b) Increases excretion of bile salts due to deconjugation of bile salt hydrolase<br>(c) Antioxidant effect |
| (6) Antihypersensitive effect | (a) Peptidase action on milk proteins yields tripeptides which inhibit angiotension-1 converting enzyme |

Contd...

Contd...

| Target health benefit | Postulated mechanism |
|---|---|
| | (b) Cell wall components act as angiotension converting enzyme inhibitors |
| (7) Urogenital infections | (a) Adhesion to urinary and vaginal tract cells<br>(b) Colonization resistance<br>(c) Inhibitor production ($H_2O_2$, biosurfactants) |
| (8) Infection caused by *Helicobacter pylori* | (a) Production of inhibitors of *H. pylori* (lactic acid and others) |
| (9) Anti-colon cancer effect | (a) Mutagen binding<br>(b) Carcinogen deactivation<br>(c) Inhibition of carcinogen producing enzymes of colonic microbes<br>(d) Immune response<br>(e) Influence on secondary bile salt concentration |
| (10) Allergy | (a) Prevention of antigen translocation into blood stream |

*Source:* 20

**Table 7:** Activities of probiotics or probiotics containing products that may play a role in reducing the risk of cancer.

**Activity suppressed**

- Cell growth and differentiation of tissue culture cells
- Aberrant crypt foci in colonic tissue of animals
- Tumours in mice (colon, liver and mammary)
- Recurrence of superficial bladder cancer in humans
- Enzyme activity (nitroreductase, β-glucuronidase, azoreductase, 7-α-dehydroxylase, glycoholic acid hydrolase) involved in conversion of procarcinogens into carcinogens in faeces of laboratory animals and humans
- Mutagenic activity
- Genotoxicity

*Sources:* 48, 84-94

are able to counteract mutagenic and genotoxic effects in the colon and other organ sites. Additionally, mechanistic studies suggest that probiotics bacteria or their by-products influence epithelial cell kinetics in the colon, decreasing cancer cell proliferation [35]. Probiotic bacteria may be dietary constituent that reduce cancer risk; this conclusion has been derived from several lines of evidence.

Dietary administration of prebiotics such as oligofructose, inulin and lyophilized cultures of *B. longum* inhibits the formation of preneoplastic lesions in the colon [49]. In addition, dietary administration of lyophilized cultures of *B. longum* suppressed colon and mammary carcinogenesis in the laboratory animal models. Inhibition of colon carcinogenesis by lyophilized cultures of *B. longum* is associated with the modulation of colonic cell proliferation and colonic mucosal and tumour ODC and *ras* p-21. Further studies are required to investigate the efficacy of prebiotics in combination with probiotics on the inhibition of colon tumours. Although pre- and probiotics comprise a diverse group with different modes of action, their ability to inhibit colon carcinogenesis may be important for the development of potential nutritional and related food supplements against colon cancer.

### 5.2.2 Allergy

Preliminary research on probiotics mediated modulation of certain allergic reactions has been reported. A break down of intestinal mucosal barrier function, allowing extensive antigen challenge, may be a factor in some allergic reactions. Since probiotics bacteria have been shown to improve mucosal barrier functions, the hypothesis that they may play role in moderating allergic response was tested. In a study, change in receptors mediating phagocytosis was measured in eight milk-hypersensitive (not lactose intolerant) adults consuming milk and milk with *Lactobacillus* GG [50]. Consumption of probiotics containing milk did not result in a significant increase in receptor expression, whereas, consumption of regular milk did, suggesting that this strain may suppress induced immune inflammatory response. A significant decrease in allergy symptoms was observed in individuals consuming yoghurt containing live, active bacteria compared to individuals consuming pasteurized yoghurt and those who did not consume yoghurt [51].

### 5.2.3 Immune function

Improperly directed or balanced immunological activity can lead to serious health problems; the suffering brought on by allergic and inflammatory reactions is a testimony to these occurrences. The oro-gastro-intestinal system is a prime means of interface between foreign antigens (either pathogens, harmless bacteria or food antigens) and mucosal surfaces. The gut associated lymphoid tissue (GALT) transforms gastrointestinal tract into largest immune organ in the human body. Approximately $10^{10}$ immunoglobulin-producing cells per metre of small bowel are estimated to be present accounting for approximately 80 per cent of all immunoglobulin producing cells in the body [52]. Subsequently, the intestinal tract can evoke a variety of immunological responses from many different local immune cells [53]. It is through this mechanism that probiotics bacteria are postulated to influence the immune response although genera in addition to *Lactobacillus* and *Bifidobacterium*, have been evaluated as immune stimulators or biological response modifiers with varying degree of success.
Probiotic bacteria are able to:

- Enhance both non-specific and specific immune responses by activating macrophages.
- Increase the level of cytokinins.
- Increase natural killer cell activity.
- Increase the level of immunoglobulins, especially secretory Ig A.

Viable probiotic cells, dead cells or fermentation products have been shown to mediate immune activity. Important in this capability is the positive influence of probiotics on immune activity without eliciting a harmful inflammatory response [35]. Biological effects correlated with enhanced immune function such as protection against viral and bacterial pathogens, and tumour inhibition have also been measured [54].

### 5.2.4 Intestinal tract functions

Gastrointestinal disturbances can range from annoying to life threatening. Diarrhoeal illness have a host of microbiological, immunological and physiological causes, some related to the disruption of normal microecology. Disruption of intestinal permeability barriers can lead to translocation of bacteria into the blood stream [55]. Consumption of non-digestible foodstuffs (e.g., lactose for lactose intolerant people, soluble fibres) can provide fermentable substrates for growth of intestinal microbes, sometimes with deleterious effects. Probiotic bacteria have been shown to improve the clinical outcome in many intestinal diseases targets [56, 57]. Results in Table 8 show that probiotic fermentation of RSMT mixture with *L. acidophilus* reduces the contents of phytic acid and polyphenols, and effected considerable enhancements in the digestibilities of starch and proteins [21].

**Table 8:** Changes in phytic acid and polyphenol contents and *in vitro* digestibility of proteins (%) and starch (mg maltose released/g mixture).

| Processing treatments | Phytic acid (mg/100 g) | Polyphenols (mg/ 100g) | Protein digestibility | Starch digestibility |
|---|---|---|---|---|
| Raw blend (control) | 231.33 | 455.73 | 51.79 | 36.26 |
| Unfermented autoclaved mixture | 186.75 | 404.95 | 62.65 | 59.95 |
| Fermented autoclaved mixture | 86.67 | 367.19 | 78.33 | 75.01 |
| CD ($P < 0.05$) | 16.96 | 18.49 | 1.83 | 3.27 |

*Source:* 21

### 5.2.5 Lactose intolerance and diarrhoea

Fermented foods such as acidophiles yoghurts, acidophiles yeast milk, acidophilin, sweet acidophilus milk, acidophilus powders, etc. have been developed and are consumed as an acceptable means of ingesting probiotic cells. These food products containing a microorganism i.e., *Lactobacillus acidophilus* having probiotic property are known to have high nutrient contents and therapeutic value in checking diarrhoea [58].

### 5.2.6 Cholesterol and cardiovascular diseases

Elevated levels of certain blood lipids are a risk factor for cardiovascular diseases. The effects of probiotics bacteria on reduction of serum cholesterol and mechanisms of any effects are unknown. One hypothesis suggests that some strains of *Lactobacillus acidophilus* can assimilate cholesterol molecule [59]. This activity has been reported in *in vitro* assays. The ability of certain probiotics bacteria, *Lactobacillus* and *Bifidobacteria* to enzymatically deconjugate bile acids has also been suggested to have a role in regulating cholesterol levels in humans. Deconjugated bile acids are more efficiently excreted [60]. Since cholesterol is a precursor of bile acids, this could lead to reduction of serum cholesterol, as cholesterol molecules are converted to bile acids to replace those lost through excretion. The desirability of bile salt hydrolysis as probiotics attribute, however, is questionable. Deconjugated bile acids especially after their re-entry into hepatic circulation enter the colon where they can be converted into secondary bile acids by colonic microbes. Secondary bile acids are known as cancer promotors, hence this activity could result in health risk.

### 5.2.7 Urogenital health

Similar to the intestinal tract, the urogenital tract in women is highly colonized and susceptible to infection upon colonization disruption [61]. Both urinary and genital tract infections have been linked to bacteria originated from the colon. This has led to the hypothesis that modulation of gut microflora through consumption of bacteria can have effect on urogenital ecosystems [62]. Populating the colon with lactobacilli enables the colon to serve as a source of beneficial, not just harmful bacteria. Several studies have correlated vaginal health (absence of infections) with the presence of lactobacilli [63]. Most of the clinical substantiations of the ability of probiotics to decrease recurrence of urogenital infections in women were conducted with intravaginal installation of probiotics bacteria. Oral consumption of certain probiotics containing products, however, was found to mediate decreased recurrence of *Candida* infections and bacterial vaginosis [64, 65]. To confirm the dietary role of probiotics bacteria in the prevention of urogenital infections, placebo-controlled, blinded studies using oral routes of introduction with sufficiently large study groups are needed.

### 5.2.8 Hypertension

Dietary recommendations accompany more aggressive strategies to control hypertension, and some preliminary evidence suggests that food products derived from probiotic bacteria could possibly contribute to blood pressure control [66]. This anti-hypertensive effect has been documented with studies in spontaneous hypertensive rats and one human clinical study [67, 68]. Two tripeptides, valine-proline-proline and isoleucine-proline-proline, isolated from fermentation of milk based medium by *Saccharomyces cervisiae* and *Lactobacillus helveticus* have been identified as the active components. These tripeptides function as angiotensin-I-converting enzyme inhibitors and reducing blood pressure. Based on this technology, Calpis (Kangama, Japan) developed a pasteurized product (Ameal-S) having Foods for Specific Health Use (FOSHU) status, unlike many other probiotic-induced effects. It is important to note that this effect is mediated by a fermentation end-product, and not viable probiotic cells. Some of the studies also show that probiotic bacteria or the fermentation end-products may be effective in mediating a mild antihypertensive effect.

#### *5.2.9 Stomach health*

The ability of probiotic bacteria to influence colonization and activity of *Helicobacter pylori*, which is associated with chronic gastric, peptic valves, and risk of gastric cancer has been evaluated [69]. Conventionally, colonized (stomach microflora is dominated by Lactobacilli) mice resisted infection by *H. pylori,* whereas germ free mice did not [70]. Results from animals and humans studies suggest that probiotic bacteria or their end products inhibit *H. pylori* infection. Aiba and co-workers [71] found that *L. salivarius* but not *L. casei* or *L. acidophilus* initiate *H. pylori* colonization in a mouse model. Lactic acid production was thought to correlate with inhibition. Lactic acid was shown to be more inhibitory to *H. pylori in vitro* than acetic or hydrochloric acids [72]. These results suggest that probiotic bacteria and their metabolic end products can inhibit *H. pylori* and its physiological effects. Results regarding the influence of probiotics on *H. pylori* infection must be extended and confirmed.

#### *5.2.10 Other health effects*

The influence of probiotic on a variety of other clinical targets has been evaluated. Probiotic mediated reduction in the severity of reaction to exposure to radioactive isotopes has been shown in mice and humans [73, 74]. The effects of endotoxemia associated with alcoholic liver disease were reduced by probiotics [75]. Probiotics have been used in external feeding to provide nutritional support to surgery patients [76]. In addition to the proposed direct effects on humans, probiotics may also have implications for human health through their use in animal agriculture. Probiotics have been tried for prevention of food, animal colonization with pathogens. Probiotics may also benefit animals, agriculture through greater resistance to farm animals to infectious diseases, increased growth rate, improved feed conversion, and increased yield of milk and eggs [77]. Commercial probiotic products for use in animal agriculture are available also.

### 5.3 Safety Considerations of Probiotics

Several of the bacteria used in United States as probiotics are listed by the Food and Drug Administration (FDA) on its partial list of microorganisms and microbial-derived ingredients that are used in foods. The usual approach of safety assessment for marketing probiotic bacteria in the United States is presumption of safety, reasoned by long history of safe use in fermented dairy products. The safety of lactobaccili and bifidobacteria has been recently reviewed [78]. The general conclusion is that the pathogenic potential of lactobacilli and bifidobacteria is quite low. This conclusion is based upon the widespread consumption of these microorganisms in fermented foods, their presence as normal colonizers in the human body, failure to isolate these bacteria as primary pathogens and lack of negative side effects of the bacteria when fed in high levels to immuno-compromised humans (premature infants, elderly, AIDS patients, chronic disease patients and people with enteric infections).

Safety concerns were expressed regarding the use of enterococci as probiotic microbes [79]. The association of *Enterococcus faecium* and *E. faecalis* with the bacteremia and the increased incidence of antibiotic resistance in these strains provide rationale for excluding them from food fermentation. Enterococci, however, are associated with some traditional food fermentations and are currently used in some dietary supplement formulations [80].

## 6. COMMERCIAL PRODUCTION OF PROBIOTICS

The commercial production of probiotics was started during the last few years in different countries, and the products are being made available for use under various trade names. Rosell Institute is one such name that has got mixed strain formulation of probiotics for various uses. They have also started coating the capsules of their preparations with some material that protects them from stomach acids so that their active ingredients are released at the optimal rate, delivering the probiotic further down the digestive tract for added effectiveness.

## 7. FUTURE TRENDS

The increased worldwide interest in probiotics has set the stage for expanded marketing of these products, even though much research remains to be done. Some research goals are shown in Table 9. The benefit of probiotic as a healthful ingredient could be enhanced if used in combination with other health promoting dietary strategies. Whole foods, however, represent a very attractive vehicle for delivery of probiotics, although delivery through a traditional dietary supplement is convenient for some (especially clinical) applications. Probiotic bacteria have always had a natural association with dairy products. Dairy products containing probiotics provide not only viable bacteria, but also high quality macronutrients and unique micronutrients (such as calcium, fermentation end-products,

**Table 9:** Future research needs in the probiotic area.

- Determine the physiological role, mechanism of action, and extent of influence of probiotics in human health using human feeding studies. Studies on high-risk human populations for colon cancer or cancer recurrence would be a possible target area.
- Validation of biomarkers used for assessing probiotic function. Testing of predictions based on biomarker studies with actual results in human clinical evaluation is needed. Biomarker validation in the areas of immune system, cancer, and gut microecology is especially important. Once validated, biomarkers will be useful tools to asses dose-dependence and strain specific responses. Biomarkers commonly used to select strains for probiotic use (adherence in tissue cultures, cholesterol assimilations, competitive exclusion of pathogens in tissue culture, inhibitor (bacteriocins, organic acids production, lactose activity) have never been tested in controlled studies to determine if mutants without these characteristics perform differently in clinical evaluations.
- Assessing effects of probiotics on populations and activity of gut microbes. The application of gene based methods hold much promise in their field.
- Determine the role of probiotics as part of whole food compared to isolated component.
- Improve reliability and case of taxonomic classification of probiotic bacteria. Improve strain performance and activity.
- Marketing studies with consumers to understand how best to communicate the concept of probiotics/prebiotics and to determine favourable probiotic preparations.
- Improvement in product formats, consumer acceptance, stability and efficacy of probiotic containing products.
- Incorporation of probiotics/prebiotics in snack foods, confectioneries and other food items.

*Source:* 35

bioactive peptides, sphingolipids and conjugated linoleic acids) found in fermented milk products [81, 82]. At the same time, natural buffering of stomach acid by the food carrier would enhance the stability of probiotics after consumption. Formulations of food products with additional vitamins, non-digestive carbohydrates, soluble fibre, phytochemicals or other bioactive ingredients could further enhance the dietary value of the product. Besides this, there are some minimum requirements, which a good probiotic should possess (Table 10).

**Table 10:** Characteristics of a good probiotic.

- Should be a strain which is capable of exerting a beneficial effect on the host animal e.g., increased growth or resistance to disease
- Should be non-pathogenic and non-toxic to the consumers
- Should be easy to culture and have high viability and good conservation
- Should be capable of surviving passage through the intestinal tract
- Should be capable of establishing a (temporary) niche in the gastrointestinal tract by competing with the resident flora
- Should be capable of surviving and metabolizing in the gut environment e.g., resistance to low pH and organic acids
- Should be stable and capable of remaining viable for long periods under storage and field conditions
- Should be available in multi-strain preparations

*Sources:* 3, 83

Probiotics have a potential to be exciting ingredient for foods with a 'healthy' image. Probiotics could be combined with other healthful ingredients or simply used to complement the natural functional attributes of whole foods. The combination of probiotics bacteria with nutrient-dense foods, such as dairy products will have the added benefit of enhancing consumer nutrition. Consumer awareness of the probiotics in various nations including India is minimal. Therefore, success with probiotics will require consumer education, which could also contribute to changing the common perception that the bacteria are harmfull. The vast array of health benefits attributed to probiotics may seem implausible. However, when considering that many of the effects of probiotic bacteria have strong influence on the balance or activity of gut microflora, it is easier to comprehend the broad-reaching-postulated effects. The ability of probiotic bacteria to modulate these activities is being established. In a climate of trend towards reducing antibiotic use, awareness of disease resulting directly from microecosystem disruption, emergence of pathogens with enhanced virulence, clinical treatment and awareness that some infections lead to serious sequelae, probiotic bacteria may add a low-cost, low risk layer of protection from infection and disease.

## REFERENCES


1. Anonymous (2003) Healthy Sunshine Inc. Morrison, TN 37357. www.healthy-sunshine.com.
2. Diplock AT, Agget P, Ashwell M, Bornet F, Fern E & Roberfroid M (1999) Scientific concepts of functional food science in Europe: Consensus document. *Brit J Nutr Suppl* 1: 1-28.
3. Fuller R (1989) Probiotics in man and animal: A review. *J Appl Bacteriology* 66: 365-378.
4. Bibel DJ (1988) Elie Metchkinoff's *Bacillus* of long life. *ASM News* 4: 661-665.
5. Lilly DM & Stillwell RH (1965) Probiotics – growth promoting factors produced by microorganisms. *Science* 147: 747-748.
6. Sperti GS (1971) *Probiotics*, AVI Publishing Co, New York, USA.
7. Parker RB (1974) Probiotics, the other half of the antibiotic story. *Ani Nutr Health* 29: 4-8.

8. Havenaar R & Huis In't Veld MJH (1992) Probiotics: A general view. In: *Lactic Acid Bacteria in Health and Disease*, Vol 1, Wood BJB (ed), Elsevier Applied Science Publishers, Amsterdam, Netherlands, pp 271-281
9. Salminen S (1996) Uniqueness of probiotic strains. *IDF Nutr News Lett* 5: 16-18.
10. Schaafsma G (1996) State of art concerning probiotic strains in milk products. *IDF Nutr News Lett* 5: 23-24.
11. Schrezenmeir J & Vrese M De (2001) Probiotics, prebiotics and synbiotics – approaching a definition. *Am J Clin Nutr* 73: 361S-364S.
12. Gibson GR & Roberfroid MB (1995) Dietary modulation of the human colonic microbiota. Introducing the concept of prebiotics. *J Nutr* 125: 1401-1412.
13. Hidaka H, Eida T & Takizawa Teta I (1986) Effects of fructooligosaccharides on instestinal flora and human health. *Bifidobacteria Microflora* 5: 37-50.
14. Gibson GR, Beatty ER, Wang X & Cummings JH (1995) Selective stimulation of bifidobacteria in the human colon by oligofructose and inulin. *Gastroenterology* 108: 975-982.
15. Tanaka R, Takayama H & Morotomi M (1983) Effects of administration of TOS and *Bifidobacterium breve* 4006 on the human fecal flora. *Bifidobacteria Microflora* 2: 17-24.
16. Ito M, Kimura M, Deguchi Y, Miyamori-Watabe A, Yajima T & Kan T (1993) Effect of trans-galactosylated disaccharides on the human intestinal microflora and their metabolism. *J Nutr Sci Vitaminol (Tokyo)* 39: 279-288.
17. Rowland IR & Tanaka R (1993) The effects of transgalactosylated oligosaccharides on gut flora metabolism in rats associated with human faecal microflora. *J Appl Bacteriol* 74: 667-674.
18. Hayakawa K, Mizutani J & Wads K (1990) Effects of soybean oligosaccharides on human faecal microflora. *Microb Ecol Health Dis* 3: 293-303.
19. Lee YK, Nemoto K, Salminen S & Gorbach SL (1999) *Handbook of Probiotics*, John Wiley & Sons, Inc, New York, USA.
20. Sanders ME & Huis in't Veld J (1999) Bringing a probiotic containing functional food to the market: Microbiological product, regulatory and labelling issues. *Antonie van Leeuwenhoek* 76: 293-315.
21. Rani B & Khetrapaul N (1999) Probiotic fermentation of indigenously developed RSMT mixtures: Effect on antinutrients and digestibility of starch and proteins. *J Food Sci Technol* 36: 71-73.
22. Mahajan P, Sahoo J & Panda PC (2000) Effect of probiotic (Lacto-Sacc) feeding, packaging methods and reasons on the microbial and organoleptic qualities of chicket meat balls during refrigerated storage. *J Food Sci Technol* 37: 67-71.
23. Yeung PSM, Cano R, Tong PS & Sanders ME (1999) Comparison of API, 16S rDNA sequencing and fatty acid analysis as methods to speciate commercial probiotic bacteria. Annual Meeting of the American Dairy Science Association, Memphis, USA, p 82-86.
24. Hill MH, Draser BS, Aries V, Crowther JS, Hawksworth G & Williams RED (1971) Bacteria and etiology of cancer of large bowel. *Lancet* 1: 95-100.
25. Hentges DJ (1992) Gut flora and disease resistance. In: *Probiotics: The Scientific Basis,* Fuller R (ed), Chapman and Hall, London, pp 87-109.
26. Vollaard EJ & Clasener HAL (1994) Colonization resistance. Antimicrobial agents *Chemother* 38: 409-414.
27. Tannock GW (1990) The microecology of lactobacilli inhabiting the gastrointestinal tract. *Adv Microb Ecol* 11: 147-171.
28. Tannock GW (1983) Effects of dietary and environmental stress on the gastrointestinal microbiota. In: *Human Intestinal Microflora in Health and Disease*, Hentges Dj (ed), Academic Press Inc, London, UK, pp 517-539.

29. Ten Brink B, Minekus M, Bol J & Huis in 't Veld JHJ (1987) Production of antibacterial compounds by lactobacilli. *FEMS Microbiol Rev* 46: 64-74.
30. Fuller R & Turvey A (1971) Bacteria associated with the intestinal wall of the fowl (*Gallus domesticus*). *J Appl Bacteriol* 34: 617-622.
31. Barrow PA, Brooker BE, Fuller R & Newport MJ (1980) The attachment of bacteria to the gastric epithelium of the pig and its importance in the microecology of the intestine. *J Appl Bacteriol* 48: 147-154.
32. Goldin BR & Gorbach SL (1984) The effect of milk and lactobacillus feeding on human intestinal bacterial enzyme activity. *Am J Cli Nutr* 39: 756-761.
33. Roach S & Tannock GW (1980) Indigenous bacteria that influence the number of *Salmonella typhimurium* in the spleen of intravenously challenged mice. *Can J Microbiol* 26: 408-411.
34. Wade S, Corthier G, Moreau L & Besiner MO (1984) L' ingestion de yaout vivant modifie-t-elle la response immunitaire? *IDF Bulletein* 179: 1-10.
35. Sanders ME (1999) Probiotics. *Food Technol* 53: 67-71.
36. Ouwehand AC, Kirjavainen PV, Grönlund M-M, Isolauri E & Salminen SJ (1999) Adhesion of probiotic micro-organisms to intestinal mucus. *Int Dairy J* 9: 623-630.
37. Bernet MF, Brassart D, Neeser JR & Servin AL (1993) Adhesion of human bifidobacterial strains to cultured human intestinal epithelial cells and inhibition of enteropathogen-cell interactions. *Appl Environ Microbiol* 59: 4121-4141.
38. Freter R (1992) Factors affecting the microecology of the gut. In: *Probiotics: The Scientific Basis*, Fuller R (ed), Chapman & Hall, London, UK, pp 111-144.
39. Savage DC (1972) Associations and physiological interactions of indigenous microorganisms and gastrointestinal epithelia. *Am J Clin Nutr* 25: 1372-1379.
40. Fuller R (1973) Ecological studies on the lactobacillus flora associated with the crop epithelium of the fowl. *J Appl Bacteriol* 36: 131-139.
41. Beachey EH (1980) *Bacterial Adherence,* Chapman and Hall, London, UK.
42. Gibbons RJ & van Houle J (1975) Bacterial adherence in oral microbial ecology. *Ann Rev Microbiol* 29: 19-44.
43. Lee A (1980) Normal flora of animal intestinal surfaces. In: *Adsorption of Microorganisms to Surfaces*, Bitten G & Marshall KC (eds), Wiley & Sons, New York, USA, pp 145-173.
44. Lee A (1985) Neglected niches, the microbial ecology of the gastrointestinal tract. *Adv Microb Ecol* 8: 115-162.
45. Blaser MJ (1990) Epidemiology and pathophysiology of *Campylobacter pylori* infections. *Rev Infect Dis* 12: 99-106.
46. Huis in't Veld JHJ & Havenaar R (1991) Probiotics and health in man and animal. *J Chem Tech Biotechnol* 51: 563-567.
47. Williams GM & Wynder EL (1996) Diet and Cancer: A synopsis of causes and prevention strategies. In: *Nutrition and Cancer Prevention,* Watson RR & Muffi SI (eds), CRC Press Inc, Boca Raton, USA, pp 1-23.
48. Aso Y & Akazan H (1992) Prophylactic effect of a *Lactobacillus casei* preparations on the recurrence of superficial bladder cancer. *Urol Intl* 49: 125-129.
49. Reddy BS (1999) Possible mechanisms by which pro and pre biotics influence colon carcinogenesis and tumour growth. *J Nutr* 129: 1478S-1482S.
50. Pelto L, Salminen SJ & Isolauri E (1996) *Lactobacillus* GG modulates milk-induced immune inflammatory response in milk-hypersensitive adults. *Nutr Today Suppl* 31: 455-465.
51. Trapp CL, Chang CC, Halpen GM, Keen CL & Gershwin ME (1993) The influence of chronic yogurt consumption on populations of young and elderly adults. *Int J Immunother* 9: 53-64.

52. Tragon R & Shanahan F (1994) *Inflammatory Bowel Disease From Bench To Bedside*, Williams and Wilkins, Baltimore, USA.
53. Mc Cracken BJ & Gaskins HR (1999) Probiotics and the immune system. In: *Probiotics: A Critical Review,* GW Tannock GW (ed), Horizon Scientific Press, Norfolk, England, pp 85-111.
54. Marteau P & Ramband JC (1993) Potential of using lactic acid bacteria for therapy and immuno-modulation in man. *FEMS Microbiol Rev* 12: 207-220.
55. Welli CL, Maddaus MA & Simmons RL (1988) Proposed mechanisms for the translocation of intestinal bacteria. *Rev Infect Dis* 10: 958-979.
56. Elmer GW, Surawicz CM & Mc Farland IV (1996) Biotherapeutic agents: A neglected modality for the treatment and prevention of selected intestinal and vaginal infections. *J Amer Med Assn* 275: 870-876.
57. Salminen S, Bouley C, Bourton-Rault MC, Cummings JH, Franck A, Gibson GFR, Isolauri E, Morean MC, Roberfroid M & Rowland I (1998) Functional food science and gastrointestinal physiology and functions. *Brit J Nutr* 80: 5147-5171.
58. Khedekar CD, Dave JM & Sannabhadti SS (1990) Antibacterial activity of human strains of *Lactobacillus acidophilus* grown in milk against selected pathogenic and spoilage type bacteria. *Cult Dairy Prod J* 25: 29-31.
59. Gilliland SE, Nelson CR & Maxwell C (1985) Assimilation of cholestrol by *Lactobacillus acidophilus. Appl Env Microbiol* 49: 377-381.
60. De Smet I, Vanhoorde L, De Saeyer N, Van de Woestyne M & Verstraete W (1994) *In vitro* study of bile salt hydrolase (BSH) activity of BSH isogenic *Lactobacillus plantarum* 80 strains and estimation of cholesterol lowering through enhanced BSH activity. *Microbial Ecol Health Dis* 7: 315-329.
61. Reid G, Bruce AW & Smeianov V (1998) The role of lactobacilli in preventing urogenital and intestinal infections. *Intl Dairy J* 8: 555-562.
62. Sobel JD (1999) Biotherapeutic agents as therapy for vaginitis. In: *Biotherapeutic Agents and Infectious Disease,* Elmer GW, Mc Farland L & Surawicz C (eds), Humana Press Inc, Totowa, USA, pp 221-244.
63. Hiller SL, Krohn MA, Klebanoff SJ & Eschenbach DA (1992) The relationship of hydrogen peroxide producing lactobacillus to bacterial vaginosis and genital microflora in pregnant women. *Obstetrics Gynecol* 79: 369-373.
64. Hilton E, Isenberg HD, Alperstein P, France K & Borenstein MT (1992) Ingestion of yogurt containing *Lactobacillus acidophilus* as prophylaxis for candidal vaginitis. *Ann Intern Med* 116: 353-357.
65. Shalev E, Battino S, Weiner E, Cologner R & Keness Y (1996) Ingestion of yogurt containing *Lactobacillus acidophilus* compared with pasteurized yougurt as prophylaxis for recurrent candidal vaginitis and bacterial vaginosis. *Arch Fam Med* 5: 593-596.
66. Takano T (1998) Milk derived peptides and hypertension reduction. *Intl Dairy J* 8: 375-381.
67. Nakamura Y, Masuda O & Takano T (1996) Decrease of tissue angiotensin-I-converting enzyme activity upon feeding sour milk in spontaneously hypertensive rats. *Biosci Biotechnol Biochem* 60: 448-489.
68. Hata Y, Yamamoto M, Ohni M, Nakajima K, Nakamura Y & Takano T (1996) A placebo-controlled study of the effect of sour milk on blood pressure in hypertensive subjects. *Amer J Clin Nutr* 64: 767-771.
69. Marshall BJ (1994) *Helicobacter pylori. Amer J Gastroenterol* 89: 5116-5128.

70. Kabir AMA, Aiba Y, Takagi A, Kamiya S, Miwa T & Koga Y (1997) Prevention of *Helicobacter pylori* by the oral administration of *Lactobacillus salivarius* as a probiotic in a gnotobiotic murine model. *Am J Gasteroenterol* 93: 2097-2101.
71. Aiba Y, Suzuki N, Kabir AMA, Takogi A & Koga Y (1998) Lactic acid mediated suppression of *Helicobacter pylori* by the oral administration of *Lactobacillus salivarious* as a probiotic in a gnotobiotic murine model. *Amer J Gastroenterol* 93: 2097-2101.
72. Midolo PD, Lambert JR, Hull R, Luo F & Grayson ML (1995) *In vitro* inhibition of *Helicobacter pylori* NCTC 11637 by organic acids and lactic acid bacteria. *J Appl Bacteriol* 79: 475-479.
73. Dong MY, Chang TW & Gorbach SL (1987) Effects of feeding lactobacillus GG on lethal irradiation in mice. *Diagn Microbial Infect Dis* 7: 1-7.
74. Henriksson R, Franzen L, Sandstrom K, Nordin A, Arevarn M & Grahn E (1995) Effects of active addition of bacterial cultures in fermented milk to patients with chronic bowel discomfort following irradiation. *Support Care Cancer* 3: 81-83.
75. Nanji AA, Khettry U & Sadrzadeh SMH (1994) *Lactobacillus* feeding reduces endotoxemia and severity of experimental alcoholic liver (disease). *Proc Soc Exp Biol Med* 205: 243-247.
76. Bengmark S & Gianotti L (1996) Nutritional support to prevent and treat multiple organ failure. *World J Surg* 20: 474-481.
77. Fuller R (1998) Probiotics for farm animals. In: *Probiotics: A Critical Review,* Tannock GW (ed), Horizon Scientific Press, Wymondham, UK, pp15-22.
78. Salminen S & Wright A (1998) Current probiotics-safety assured? *Microbiol Ecol Health Dis* 10: 68-77.
79. Adams MR & Marteau P (1995) On the safety of lactic acid bacteria from food. *Intl J Food Microbiol* 27: 263-264.
80. Giraffa G, Carminati D & Neviani E (1997) Enterococci isolated from dairy products: A review of risks and potential technological use. *J Food Protect* 60: 732-738.
81. Mc Bean ID (1999) Emerging dietary benefits of dairy foods. *Nutr Today* 34: 47-53.
82. Vander Meer R, Bovee-Oudenhoven IMJ, Sesink ALA & Kleibeuker JH (1998) Milk products and intestinal health. *Intl Dairy J* 8: 163-170.
83. Hose H & Sozzi T (1991) Biotechnology group meeting: Probiotics *J Chem Tech Biotechnol* 51: 540-544.
84. Baricault L, Denariaz G, Houri J-J, Bouley C, Sapin C & Trugnan G (1995) Use of HT 29 a cultured human colon cancer cell line, to study the effect of fermented milks on colon cancer cell growth and differentiation. *Carcinogenesis* 16: 245-252.
85. Rowland IR, Rumney CJ, Couts JT & Lievense LC (1998) Effect of *Bifidobacterium longum* and inulin on gut bacterial metabolism and carcinogen induced aberrant crypt foci in rats. *Carcinogenesis* 19: 281-285.
86. Rao CV, Sanders ME, Indranie C, Simi B & Reddy BS (1999) Prevention of indices of colon carcinogenesis by probiotic *Lactobacillus acidophilus* NCFM in rats. *Int J Oncol* 14: 939-944.
87. Reddy BS & Rivenson A (1993) Inhibitory effect of *Bifidobacterium longum* on colon, mammary and liver carcinogenesis induced by 2-amino 3 methylimidazo (4,5) quinoline, a food mutagen. *Cancer Res* 53: 3914-3918.
88. Goldin BR, Gualtieri LJ & Moore RP (1996) The effect of *Lactobacullus* GG on the initiation and promotion of DMH induced intestinal tumours in the rat. *Nutr Cancer* 25: 197-204.

89. Adachi S (1992) Lactic acid bacteria and the control of tumours. In: *The Lactic Acid Bacteria in Health and Disease,* Vol 1, Wood BJB (ed), Elsevier Applied Science, London, UK, pp 233-261.
90. Ling WH, Korpela R, Mykkanen H, Salminen S & Hanninen O (1994) *Lactobacillus* strain GG supplementation decrease colonic hydrolytic and reductive activities in healthy female adults. *J Nutr* 124: 18-23.
91. McConnell MA & Tannock GW (1993) A note on lactobacilli and β glucuronidase activities in the intestine contents of mice. *J Appl Bacteriol* 74: 649-651.
92. Goldin BR, Swenson L, Dwyer J, Sexton M & Gorbach S (1980) Effect of diet and *Lactobacillus acidophilus* supplements on human fecal bacterial enzymes. *J Natl Cancer Inst* 64: 255-261.
93. Renner HW & Munzner R (1991) The possible role of probiotics in dietary antimutagens. *Mutation Res* 262: 239-245.
94. Venturi M, Hambly RJ, Glinghammer B, Rafter JJ & Rowland IR (1997) Genotoxic activity in human fecal water and the role of bile acids: A study using the alkaline comet assay. *Carcinogenesis* 18: 2353-2359.

# Biosensors in Food Industry

23

**Manisha Kaushal**

*Department of Food Science and Technology, Punjab Agricultural University, Ludhiana 141 001 (Pb.)*

## 1. INTRODUCTION

Measurement is fundamental to all scientific and technical disciplines. It quantifies an unknown variable by comparing it with a generally accepted standard. Rapid, accurate and reliable estimation of specific analytes is very important for efficient process, environment monitoring, food, medical applications, etc. Conventional analytical methods are labourious, expensive, non-specific, time consuming and often involve sample pre-treatment [1]. The consequent need for a large number of rapid and precise means of analysis has created a demand for techniques, which give immediate and accurate results and could force the processors to adopt alternate quality control means within the process system. With the advances made in the field of electronics, research in the area of measurement by incorporating biological components to electronic sensing device has been growing rapidly. Such devices are commonly called biosensors and the development of biosensors for different type of estimations constitute biosensor technology [2]. Here in this chapter, need for biosensors, history of biosensor, various types of biosensors and their applications in agriculture and food industry has been described.

## 2. NEED FOR BIOSENSORS

It is essential to measure various metabolites in the fermentation process with accuracy and reliability. Due to the lack of any such suitable technique for rapid and accurate measurement, the biosensors have now attracted a considerable attention as they are able to monitor the single analyte even in the extremely complex processes. Biosensors are thus, a new type of monitoring and measuring device whose speed, sensitivity, stability and ease of use exceeds the current methodologies and the results obtained by them are rapid, specific, accurate and reliable.

## 3. HISTORY OF BIOSENSORS

Biosensors have been under intense development since the middle of 1960's with prospects of commercial potential offered by biotechnology. Professor Leland Clark Jr. is regarded as the father of biosensor concept. In 1956, he published his paper on the oxygen electrode [3] and made a land

*Food Biotechnology: Principles and Practices*
*V.K. Joshi & R.S. Singh (Eds.)*
*I.K. International Publishing House Pvt. Ltd., India*
*pp 799–816*

mark address in 1962 [4] at New York Academy of Sciences Symposium in which he described how to make electrochemical sensors (pH, polarographic, potentiometric or conductometric transducers). Later, Clark and Lyons coined the term 'enzyme electrode.' Guilbault and Montalvo [5] were the first to detail a potentiometric enzyme electrode. The biosensor took a further fresh evolutionary route in 1975, when it was suggested that bacteria could be used as a biological element in microbial electrodes for measurement of alcohol [6]. Lubbers and Opitz [7] coined the term "optade" to describe a fibre optic sensor to measure $CO_2$ or $O_2$. An electrochemical glucose biosensor in a bedside artificial pancreas was incorporated which was later marketed by MILES (Elkhart) as 'biostator' [8]. The biostator was then replaced by cathetor-base blood glucose analyser introduced by VIA Medical, San Diego. The lactate analyser LA 640 was introduced in the same year. A major advancement in the *in vivo* application of glucose biosensor was reported by describing the first needle type enzyme electrode for sub-cutaneous implantation [9]. In 1984, a paper was published [10] in which the use of ferrocene and its derivatives as an immobilised mediator was shown with oxido-reductases in the construction of inexpensive enzyme electrodes. This formed the basis for the screen-printed enzyme electrodes launched by Medisense (Cambridge, USA) in 1987 with a pen sized meter for home blood glucose monitoring.

## 4. WHAT ARE BIOSENSORS?

Biosensors have been defined differently by various workers such as "a biosensor is a device that transduces or converts into another form a physical or a chemical parameter to a detectable signal" [11] or the biosensor might simply be described as "a device which makes use of biochemical specificity to detect and often to measure the concentration of a particular or a group of closely related chemicals" [12]. Thus it is a device which responds to the presence of a specific analyte (a chemical substance that needs to be measured) by producing an electric signal proportional to the concentration of analyte [2]. The term 'biosensor' also implies a device containing a reactive surface (i.e. a biochemical transducer) which generates a measured response upon binding with an analyte. The device and associated instrumentation (physical transducer component) convert analyte binding event into qualitative or quantitative units of analyte concentration. So, the conversion of analyte by the biochemical transducer into another chemical species and/or physical property is sensed and converted into a measurable signal by the physical transducer [13]. In simple words, 'biosensor' can be described as an analytical device that converts a biological response into an electrical signal by means of a transducer enabling the detection of a wide range of chemical and biochemical parameters.

### 4.1 Basic Components of Biosensor

There are 3 basic components of a biosensor:

#### *4.1.1 Bioelement (Receptor)*

It is usually composed of immobilised materials like enzymes, antibodies, proteins, DNA, organelles, microbial cells or whole sections of mammalian or plant tissues [11, 13, 14]. Depending upon their use, many different types of biochemical entities can be used alternatively [2, 13, 14]. Individual enzymes which specifically catalyze the conversion of one or more substrates into defined products, multi enzyme systems become useful for conversions not possible with single enzyme, where several enzymes work in an organised way. Whole cells become particularly useful when chemical entities are able to bind onto their surfaces, where chemical conversion is catalysed by virtue of a single enzyme or multi enzyme systems located in them. Even sub-fractions of cells

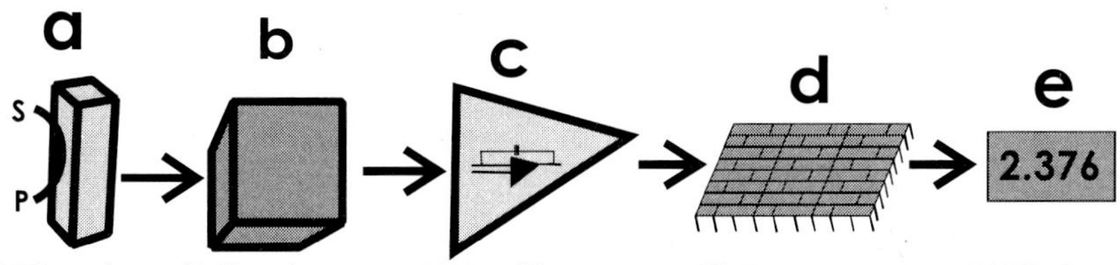


**Fig. 1:** Schematic diagram showing the working the of a biosensor.
*Source:* 76

can be used for the purpose. Antibodies and antigens are also very specific in recognition and can be used for the detection of particular components. To achieve this, the biological sensing element may be protected by a membrane, which has selective permeability to certain molecules in the environment.

#### *4.1.2 Transducer or detector system*

It is an electrical device that converts the effects of biocatalyst reaction into forms, which can be electronically amplified, stored and further display the changes such as in pH, ionic concentration, light emission, etc. It takes energy impulse from the receptor and translates it into a usable signal relative to the conditions of a measured analyte.

#### *4.1.3 Electronic device*

It amplifies the signal and transfers it into an output that can be interpreted by the user.

### 4.2 Principle of a Biosensor

The working of a biosensor (Fig. 1) can be viewed as convertion of chemical information into an electrical information flow through a series of events [2, 6-7]. Biocatalyst (a) converts the substrate (S) to product (P). This reaction is determined by the transducer (b), which converts it to an electrical signal that is then amplified (c), processed (d) and finally, displayed (e).

### 4.3 Properties of a Biosensor

The biological components used in the biosensors must be specific and stable in their functioning. The reaction should be independent of physical parameters such as pH, temperature, and stirring and the response should be accurate, precise and reproducible. The sensing element should be tiny and biocompatible. The complete unit should be cheap and portable.

### 4.4 Merits and Demerits of Biosensors

The biosensors have several advantages over the conventional methods of estimation. These are rapid, accurate and reliable and are simple to operate. The biosensors can be easily fabricated with minimal sample pre-treatment. These biosensors offer specificity and sensitivity of biological based assays, and can be mass produced at low cost. There is a reduced risk of contamination because sample is not returned to the bioreactor and there is an automatic recalibration of the sensor to counter any problem with drift.

The biosensors have certain drawbacks also, i.e. the reactions which involve consumption of energy are not suitable for biosensors because the operation of the sensing devices usually requires the release of chemical energy in order to activate the detector. As compared to some conventional methods, biosensors are costly and have limiting stability. Another problem is sterilization of the biosensor probe, which is usually solved by a complex engineering design. Sometimes microbial growth of the upstream side of the membrane causes plugging.

## 5. CLASSIFICATION OF BIOSENSORS

Theoretically, it is possible to device biosensors for any organic molecule having an ability to interact with a biological entity but, practically the most common and successful systems of sensors are based on enzyme catalysts. There are various analytes which are detected and measured by the biosensors (Table 1).

Analyte concentrations measurable with various types of biosensors range from $10^{-7}$ mol $l^{-1}$ to as little as $10^{-15}$ mol $l^{-1}$ [22-23]. The biosensors are classified into different types on the basis of transducer and receptor system (Fig. 2). The receptor system involves a biological system, while the transducer system is based on the physical component. Table 2 shows the basic principle involved in the functioning of different biosensors based on the transducer system.

**Table 1:** Analytes detected by biosensors.

| | | |
|---|---|---|
| • Carbohydrates | • Amines | • Microorganisms |
| • Organic acids | • Heterocyclic compounds | • Gases |
| • Alcohols | • Proteins | • Pesticides |
| • Phenols | • Enzyme cofactors | • Drugs |
| • Inorganic ions | • Hormones | • Xenobiotics |
| • Vitamins | • Peptides | • Cholestrol |

*Source:* 15-21

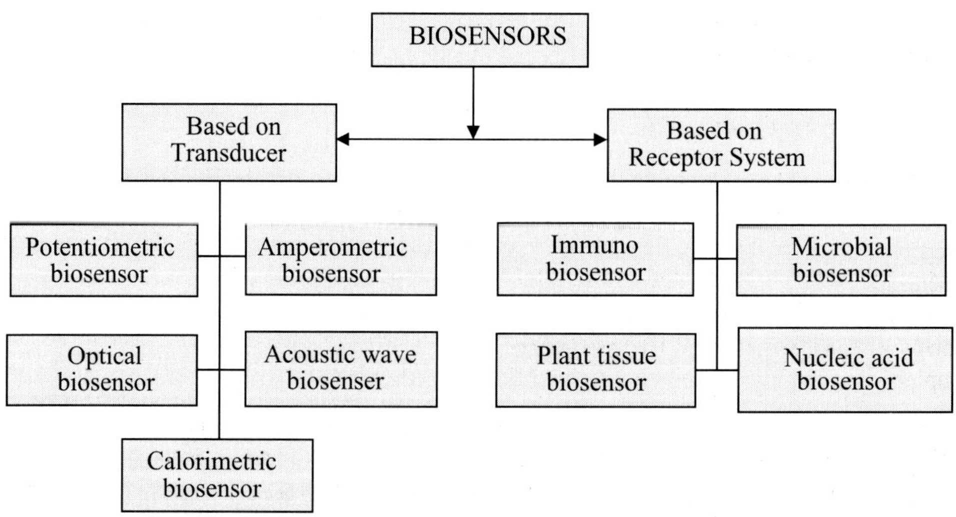


**Fig. 2:** Classification of biosensors.
*Source:* 2

**Table 2:** Transducer systems used in the design of a biosensor and the changes indicated and their causes.

| Transducer | Indicating change in | Caused by |
|---|---|---|
| Amperometric electrodes | Electron transport | Enzyme-catalyzed oxidation or reduction of an analyte |
| Potentiometric electrodes | Ion concentration | Enzyme catalyzed oxidation, hydrolysis or cleavage of C-C, C-O, C-N or other bonds |
| Calorimetric devices | Enthalpy | Enzyme catalyzed reactions |
| Piezoelectric crystals | Mass | Bioaffinity reactions |
| Fibre optic sensor | Fluorescing of enzyme substrate, products or immunotables | Enzyme and bioaffinity reactions |
| Devices based on surface plasmon resonance | Resonance | Bioaffinity reactions |

*Source:* 15

The physico-chemical changes produced by the biological material must be converted into an electrical output signal by an appropriate transducer. On one hand, the transducer may be unspecific and indicate general parameters such as reaction enthalpy (thermistor), mass change (piezoelectric crystal oscillator) or layer thickness (reflector). On the other hand, a specific indication may be achieved with ion-selective electrodes responding to species such as hydrogen peroxide or oxygen. Different types of biosensors based on both transducer and receptor system are explained in the subsequent sections.

## 5.1 Receptor System Based Biosensors

### *5.1.1 Microbial biosensor*

Microbial biosensors are composed of immobilised microorganism and an electrochemical device, which are suitable for the online determination of organic compounds [2, 24-28]. Microbial biosensors utilise microbial cells for the detection of various analytes in the process. These cells serve as an alternative biocatalyst replacing enzyme electrode, which are highly expensive and have only limited life time. A schematic diagramme of a microbial biosensor is shown in Fig. 3.

Every equipment in use has some merits and demerits and so does the biosensors. Microbial biosensors have several advantages. Since the enzymes are already contained in optical environment provided by the nature, the use of bacteria and other microorganisms may help in extending the life of biosensors. These include low cost with respect to biocatalyst involved because no enzyme purification steps are required in microbial biosensors. The microbial sensors have ability to affect complex reactions too, due to the presence of a number of enzymes in the microorganism. Due to the presence of complex enzyme system in the microbial cells, these biosensors possess the ability of regeneration of enzyme activity.

Apart from various advantages, there are few drawbacks also. The microbial cell contains a variety of enzymes, which may compromise the selection of these biosensors. However, this problem can be overcome by the use of enzymes and transport inhibitors such as benztropinamine analogs, carbonic anhydrase and cholinesterase [29-30]. Even the problem of contamination can be solved by the use of appropriate inhibitors.

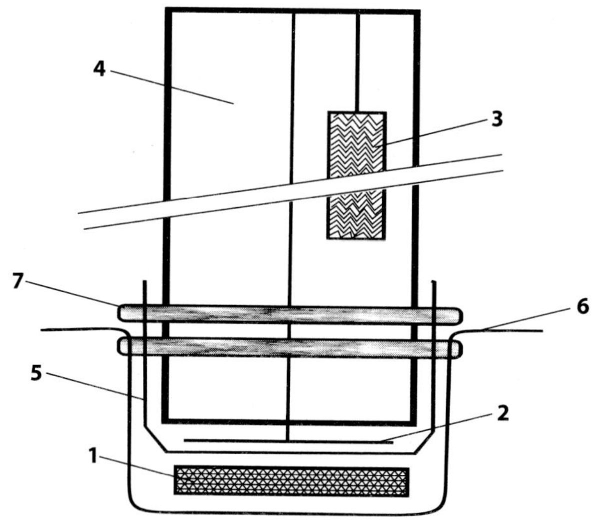


1. Membrane containing immobilised microorganism 2. Pb (cathode) 3. Pb (anode) 4. KOH 5. $O_2$ permeable teflon membrane 6. Dialysis membrane 7. Rubber ring

**Fig. 3:** Schematic diagram of a microbial biosensor.
*Source:* 28

**Table 3:** Examples of microbial electrodes.

| Substrate | Biocatalytic material | Internal sensing element | Life time (days) |
|---|---|---|---|
| L-arginine | *Streptococcus faeciun* | $NH_3$ | 20 |
| L-aspartate | *Bacterium cadaveris* | $NH_3$ | 10 |
| L-glutamate | *E. coli* | $CO_2$ | 21 |
| Nitrilo-triacetic acid | *Pseudomonas* sp. | $NH_3$ | 30 |
| Serine | *Clostridium acidurici* | $NH_3$ | 3 |
| Acetic acid | *Trichosporon cutaneum* | $NH_3$ | 9 |

*Source:* 2, 16, 30

The principal application of microbial biosensors is the measurement of BOD in waste water by utilising *Trichosporon* sp. or *Issatchenkia* (yeasts) and the bacterium *Rhodococcus erythropolis*. The measurement time is very short (20 sec to several minutes) compared to the classical BOD test, which requires five days. So the application of biosensor results in very rapid estimation. A few examples of microbial electrodes are given in Table 3.

### *5.1.2 Immuno biosensor*

Antibodies are known to have characteristics of specific biological and chemical components. The specific binding reaction between antibody and antigen has been known for a long time by immunising an animal with a foreign macromolecule. A protein reaction is initiated that leads to the formation of antibodies where the mechanism involves activation of lymphocytes to produce specific type of antibodies. These immobilized antibodies can then be used as sensitive reagents in immunosensors [31]. An antibody-based immunosensor is outlined in Fig. 4.

Typically, antibodies specifically directed against the substance of interest (desired analyte) are immobilised on the transducer of a biosensor. Then, the sensor is exposed to the intended media (blood or other biological fluid). If antigen is present in that medium, it will be bound by the antibody to form a larger antigen-antibody [Ag-Ab] complex. This will change some physico-chemical parameter of environment at transducer surface of the biosensor and that change is subsequently detected [16]. In contrast to enzymes, antibodies do not (usually) catalyse chemical transformations but rather undergo a physical transformation i.e. they bind tightly to the foreign material (antigen) that provoke the response and mark it for attack by other elements of the immune system. Antibodies are also very specific-they need to be in recognising and binding only to the foreign substances and not to material native to the organism. The analytes detected and measured by immunosensor include many medical diagnostic molecules such as hormones (e.g. pregnancy related steroids), clinical disease markers, drugs (therapeutic and abused), bacteria and some environmental pollutants like pesticides [15, 32].

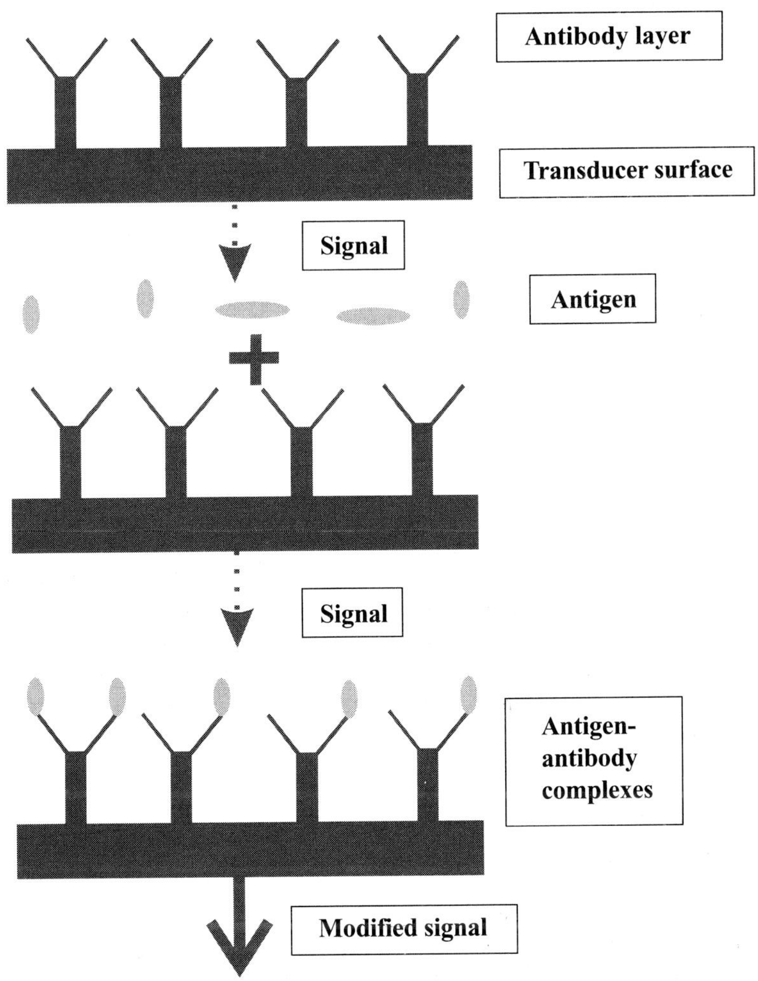


**Fig. 4:** Schematic diagram of an antibody-based immunosensor.
*Source:* 16

### 5.1.3 Nucleic acid biosensor

It is based on the detection of unique sequence of nucleic acid bases through hybridization. If a DNA probe hybridizes when added to an unknown sample because of pairing, complementary base recognising detection and identification are possible by visualizing hybridization through colour development. Commercial nucleic acid probes are available for detection of *Salmonella, Listeria, E. coli, S. aureus* and other fungi [33-34]. Highly sensitive detection of nucleic acids in biological and biotechnological samples is done with the help of nucleic acid biosensors. In these biosensors *Serratia marcescans* endonuclease is used, which hydrolyses DNA and RNA non-specifically but with high efficacy. In the biosensors, the variants of *Serratia* nuclease are immobilised which are further coupled to an electrode. These techniques are mostly used to detect minute nucleic acid contaminants especially in biotechnological drugs [35].

### 5.1.4 Plant tissue biosensor

Plant tissue biosensor can act as an alternative to enzyme and microbial sensors and can be partially selective, if the substrate to be detected/estimated is either a major nutrient or functional

**Table 4:** Plant tissue based biosensors.

| Substrate/ Analyte | Biocatalytic material | Sensing element |
|---|---|---|
| Ascorbic acid | Cucumber peel/cabbage/ yellow squash | $O_2$ probe |
| Tyrosine | Sugar-beet tissue | $O_2$ probe |
| Pyruvate | Corn kernel | $CO_2$ sensor |
| Cysteine | Cucumber leaf | $NH_3$ sensor |
| Glutamate | Yellow squash | $CO_2$ sensor |

*Source:* 38-39

metabolite of enzyme containing tissue. Table 4 shows various plant tissues being used as a biocatalyst material for the estimation of various analytes. Plant tissue biosensors have been constructed to analyse nutrients like amino acids, vitamins and intoxicants like pesticides and phenolics in food [36]. Whole plant tissue bioreactors are also used for enzymatic degradation of mono and polyphenols to orthohydroxyphenols with subsequent oxidation of latter to quinones in edible oils. These quinines are finally transformed into less toxic molecules easily degradable in environment. These biosensors are also used to detect phosphate in foods like wine, fresh and powered milk and tomato puree [37]. The plant tissue biosensors are inexpensive, simple to construct and require few cofactors commercially.

## 5.2 Transducer Based Bissensors

### 5.2.1 Potentiometric biosensor

It measures a potential between the sensing element and reference element when there is no current flowing between them [40]. It makes use of ion-selective electrodes in order to transduce the biological reaction into electrical signal. It consists of a membrane containing immobilised enzyme, surrounding the probe such as the pH meter, where a catalysed reaction generates or absorbs hydrogen ion. The electrodes can be simple metal wires whose surfaces are modified to make them selective for a particular ion. They are dipped directly into sample or separated from the sample by a membrane or a porous plug and are placed in an electrolyte solution of defined composition.

The most common potentiometric devices are pH and ion-selective electrodes. The ion sensitive (ISE) electrodes are commonly used for monitoring $H^+$, $Na^+$, $K^+$, $Ca^+$, $Li^+$, $NH_4^+$ and a variety of anions, drugs and other materials. The analytically range of these ISE sensors is generally from $10^{-1}$ to $10^{-5}$ Mol $l^{-1}$ [41]. These potentiometric biosensors are used in the monitoring of water quality and microbial contamination.

The potentiometric biosensors are simple and reliable to use, but require a stable electrode for their functioning, where the glass electrode could break causing contamination of food products. The potentiometric biosensors are slow in responding apart from being susceptible to electronic noise.

### *5.2.2 Amperometric biosensor*

Amperometric biosensors function by the production of current when a potential is applied between the two electrodes [42]. Amperometric detection of microorganisms involves the measurement of the current (flux of electrons) generated through oxidation or reduction catalyzed by their enzymes or by their involvement in a bioaffinity reactions when the voltage is applied between the working and reference electrode. The potential of the working electrode is maintained with respect to a reference electrode, usually Ag/AgCl. The most common working electrodes are noble metals, graphite and modified forms of carbon or conducting polymers [43, 44].

The well known amperometric biosensor employs glucose oxidase and produces $H_2O_2$ as given below:

$$\text{Glucose} + O_2 \xrightarrow{\text{Glucose oxidase}} H_2O_2 + \text{Gluconic acid}$$

Thus, these biosensors can measure either $O_2$ depletion or $H_2O_2$ concentration and can be used to quantify glucose in concentration range of $10^{-7}$ to $10^{-3}$ mol $l^{-1}$. These amperometric biosensors have been used to measure glucose, uric acid and cholesterol levels in blood.

The amperometric biosensors have different advantages like these are simple to use, have high selectivity and reduced instrumentation risks. These amperometric biosensors also have some drawbacks like low sensitivity and dependence on dissolved oxygen, which can be overcome by using mediators like electron transfer molecules that shuttle electrons from enzyme to the electrode,

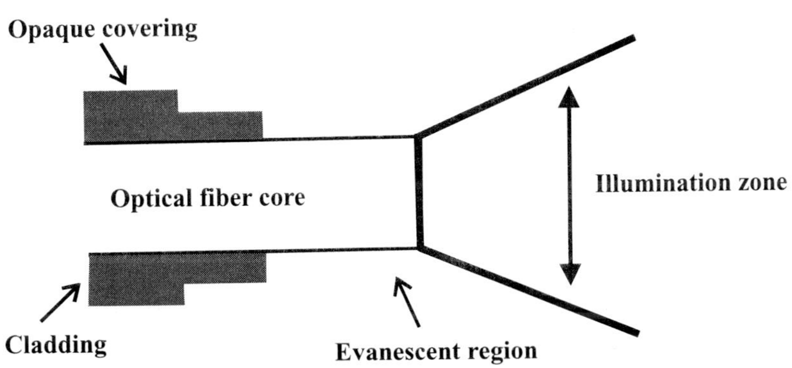


**Fig. 5:** Basic construction of an optical fibre sensor.
*Source:* 2

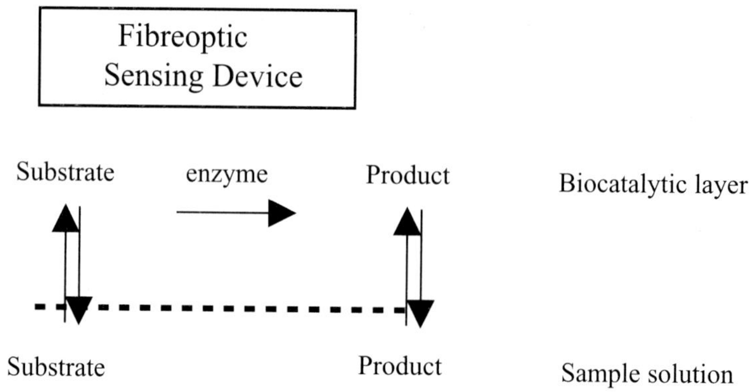


**Fig. 6:** Schematic representation of a fibreoptic biocatalytic biosensor.
*Source:* 49

such a ferrocene, N-methyl phenazinium cation ($NMP^+$) and tetracyanoquinodimethane radical anion ($TCNQ^-$).

Although amperometric biosensor has been the most successful category of biosensors, they have not made an impact on the market where continuous monitoring is required.

### *5.2.3 Optical biosensor*

Signal transduction by monitoring changes in the optical parameter at the biological level has been taken as a basis for the development of these biosensors. Fibreoptic biosensor is the device (Fig. 5) in which a biocatalyst is immobilised at the distal end of a fibreoptic sensing device [47-48]. The biocatalyst mediates between the sensing device and analyte of interest by either forming a detectable species from the analyte or by consuming a detectable co-substrate (Fig. 6).

Figure 6 shows the various processes that occur at the sensing tip when a detectable species is produced from the biological reaction. The analyte or substrate of the biocatalysed reaction diffuses from the bulk solution to the biocatalytic layer where the substrate is converted to detectable species. A steady state concentration of the acceptable species is established and a corresponding steady state signal is obtained [49]. The magnitude of this signal is related to the analyte concentration in a bulk solution through a calibration curve.

The fibreoptic biosensors have various merits such as they avoid direct electrical connections to the system being measured, as the system is optical and is not subject to electric interference, so these are safer to use. These neither require any reference electrode nor does the immobilised biocatalyst needs to be in direct contact with the optical fibre that makes fibreoptic biosensors smaller and cheaper [49-50].

### *5.2.4 Calorimetric biosensor or thermal sensors (Thermistors/Thermophiles)*

Calorimetric biosensors (Fig. 7) are used when the enzyme catalysed reactions are exothermic, generating heat in the range of 5-100 KJ $mol^{-1}$ [51]. Modern thermistors can be used to accurately measure temperature changes as low as $10^{-4}$ °C [52]. The biogenic heat can be detected by thermistors or temperature sensitive semiconductor devices. This can be performed either by immobilizing enzyme on particles in a column around the heat sensing device or by direct attachment of the immobilized enzyme to the temperature transducer [53-56]. The schematic diagram of a calorimetric biosensor is shown in

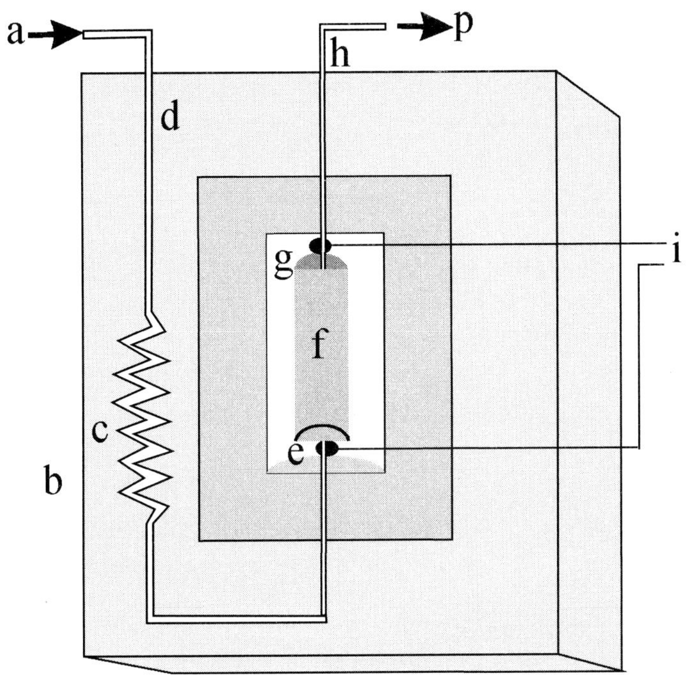


**Fig. 7:** Schematic diagram of calorimetric biosensor.
*Source:* 16

**Fig. 7.** The sample stream (a) passes through the outer insulated box (b) to the heat exchanger (c) within an aluminium block (d). From there, it flows through the reference thermistor (e) and into the packed bed bioreactor (f) containing the biocatalyst (1ml volume), where the reaction occurs. The change in temperature is determined by thermistor (g) and the solution passed to waste (h). External electronics (I) determines the difference in the resistance, and hence temperature, between the thermistors.

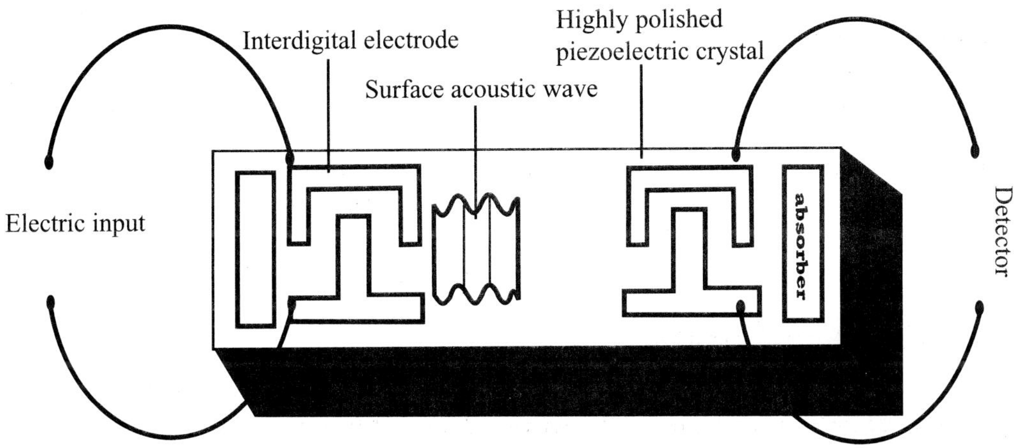


**Fig. 8:** Schematic diagram of surface acoustic wave sensor.

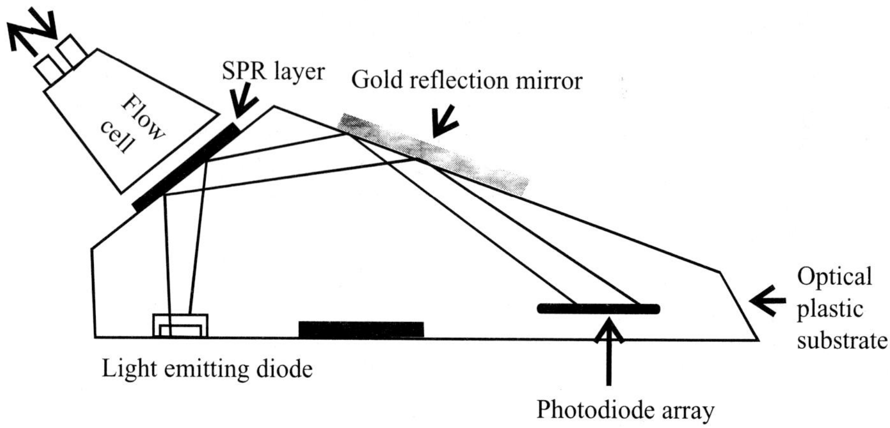


**Fig. 9:** Diagram of a commercial biosensor.

The calorimetric biosensors involve simple procedures, possibility of continuous flow operation and reversibility of a biocatalyst, while some drawbacks of calorimetric biosensors are that these require large amounts of enzyme thus incurring high costs and slow response. There are also non-specific thermal effects resulting in overestimation of the substrate concentration, or there may be base line drifts due to the heating of measurement apparatus. Calorimetric biosensors have been applied to analyse ascorbic acid, ATP, cellubiose, cephalosporin, cholesterol, creatine, ethanol, galactose, glucose, lactate, lactose, oxalic acid, penicillin, sucrose, triglycerides, urea and uric acid [53-57].

### *5.2.5 Acoustic wave-based biosensor*

The piezo-electrical effect of deformations of quartz under alternating current (at frequency in the order of 10 MHz) is used by coating the crystal with a selectively binding substance, e.g. an antibody. When exposed to the antigen, the antibody-antigen complex will be formed on the surface and shift the resonance frequency of the crystal in correlation with the mass increment, which is proportional to the antigen concentration [58-59]. A surface acoustic wave (SAW) immunosensor (Fig. 8) consists of piezoelectric crystals such as quartz or lithium niobate carrying thin film inter-digital electrode assays with surface waves and an electrical input.

These biosensors are small, robust and give rapid response, but their long term stability under working conditions is generally poor and there is difficulty in handling and sterilization. Currently, the technique is under investigation for gaseous substances such as HCl, iso-cyanates, CO, $NO_2$ and $SO_2$ [60].

## 6. COMMERCIAL BIOSENSORS

There are a number of commercial successes of biosensors, the most successful to-date being the home blood glucose monitor for use by people suffering from diabetes [16]. A commercial biosensor is schematized in Fig. 9.

## 7. APPLICATIONS OF BIOSENSORS IN FOOD INDUSTRY

The ability of biosensor to provide real-time information and their relatively simple design results in a wide variety of applications for monitoring the quality and safety of our food supply. Applications of biosensor right from fermentation to genetic manipulation is becoming increasingly popular such as in the development of food and beverage, chemical, pharmaceutical, waste management, medical diagnostics and several other areas of environmental and socio-economic importance. Various applications of biosensors in food industry are enlisted in Table 5. Problems relating to food safety, water quality, chemical and biochemical analysis, toxins produced by food bacteria can be easily detected and resolved within a short span of time using a biosensor. These instruments already have the ability to provide instant information related to the presence and concentration of certain chemical species. There exists at least three significant areas of applications namely, processing, quality control, and environmental monitoring [61-64].

**Table 5:** Applications of biosensors in food industry.

| Area/field of application | Type of biosensor | Applications |
|---|---|---|
| Food | Microbial biosensors | • On-line measurement of ethanol concentration in food by utilising *Trichosporon brassicae*, which do not utilise methanol.<br>• Rapid determination of vitamin $B_6$ in marine products by using *Saccharomyces uvarum* 4288 ATCC9080.<br>• Measurement of acetic acid by using *Trichosporon brassicae,* which can measure the concentration as low as 5mg per litre. |
| Organic pollution/environmental monitoring | Microbial biosensors | • Measurement of organic pollution by utilizing yeast strain of *Trichosporon cutaneum.*<br>• Determination of $CO_2$ required in some industrial processes by using chemoautotrophic bacterium, which can grow with only carbonate as the carbon source.<br>• Detection of cyanide by immobilising cells of *Saccharomyces cerevisiae* and an oxygen electrode. |
| Food | Fibreoptic biosensor | • Measurement of hydrogenation or degree of unsaturation in oils.<br>• The oil is passed through the optical cavity where its refractive index and effective path length changes, which changes the extent of spectral modulation.<br>• The refractive index correlates to the extent of hydrogenation.<br>• Detection of pH by using enzyme penicillase.<br>• Detection of ammonia. |
| Food | Calorimetric biosensor | • Successfully used for monitoring ascorbic acid, glucose, lactose, triglycerides, cholesterol, galactose, ethanol, sucrose, antibiotic and therapeutic agents and oxalic acid.<br>• These could also be applied to detect and monitor microbial contamination.<br>• Monitoring penicillin, glucose and ethanol in bioreactor vessel.<br>• Analysis of urea as pollutant by immobilising enzyme urease. |
| Food | Amperometric biosensor | • Determination of ascorbic acid in fresh grape juice.<br>• Analysis of glucose, D-galactose, glycolate, glutamine and L-amino acid.<br>• Determination of sulphite (preservative, antioxidant) in food and beverages by immobilising sulphite oxidase enzyme. |

Contd...

Contd...

| Area/field of application | Type of biosensor | Applications |
|---|---|---|
| | | • Analysis of ethanol in foods by co-immobilising alcohol dehydrogenase and bovine serum albumin (BSA) in carbon paste containing entrapped $NAD^+$.<br>• Detection of microbial biomass in liquid samples of milk, water and wine<br>• Measurement of rate of fermentation<br>• On-line measurement of monosaccharides, disaccharides, lactate, and amino acid during the production of alkaline protease and penicillin. |
| Food | Piezoelectric biosensor | • Detection of *Salmonella typhimurium* in foods. A crystal with pre coated glutaraldehyde is immobilised with antibody of *Salmonella* sp. The antibody will react specifically to the *Salmonella* cells and will bind to the crystal surface. With a specific amount of antibody immobilised, the cell mass binding to the crystal is directly related to the concentration of the solution.<br>• Detection of *Vibrio cholerae - Listeria monocytogenes.*<br>• Detection of toxic amines in sea foods. |
| Food | Acoustic biosensor | • Detection of food spoilage.<br>• Detection of herbicides (triazine and phenyl urea) in drinking water. |

*Source:* 58-84

## 8. FUTURE THRUST

Although the concept of biosensors holds considerable promise and potential for on-line quality and safety monitoring in food industry, the commercialisation of this technology is still in its infancy. Very few biosensors are presently available in the market that have been commercially successful. Their commercial realisation has certain basic constraints such as technical and manufacturing feasibility, market development and investment strategies inspite of extensive research being carried out.

The promising avenues of biosensors in food industry will arise only when they can fulfil the common needs for analysis of foods. Presently, there are some constraints in the long term stability, sterilizability, cost, maintenance and reusability of biosensors, which needs further research to make it commercially viable for on-line monitoring of food processes.

## REFERENCES


1. Thakur MS & Karanth NG (1998) Biosensors and bioelectronics and their applications. In: *Advances in Biotechnology*, Pandey A (ed), Education Publishers and Distributors, New Delhi, India, pp 257-266.
2. Tavva DP, Babu RP & Kumar GC (2000) Biosensors as on line monitors in food and dairy industries. *Processed Food Industry* 5: 12-17.
3. Clark LC Jr (1956) Monitor and control of blood and tissue oxygen measurements. *Trans American Soc Artif Intern Organs* 2: 41-48.
4. Clark LC Jr & Lyons C (1962) Electrode system for continuous monitoring in cardiovascular surgery. *Ann NY Acad Sci* 102: 29-45.
5. Guilbault GG & Montalvo J (1969) Urea-specific enzyme electrode. *JACS* 91: 2164-2569.

6. Divies C (1975) Remarques sur l'ethanol par une electrode microbinne *d'Acetobacter zylinum. Annals Microbiol* 126A: 175-186.
7. Lubbers DW & Opitz NZ (1975) The $pCO_2/pO_2$ optode. *Naturforsch C: Biosci* 30c: 532-533.
8. Clemens AH Chang PH & Myers RW (1976) Development of an automatic system of insulin infusion controlled by blood sugar, its system for the determination of glucose and control algorithms. *Proc Journes Ann de Diabtologic de l' Hotel-Dieu,* Paris, France, pp 269-278.
9. Shichiri M, Kawamori R, Yamaski R, Hakai Y & Abe H (1982) Wearable artificial endocrine pancreas with needle-type glucose sensor. *Lancet* 2: 1129-1131.
10. Cass AEG, Francis DG, Hill HAO, Aston WJ, Higgins IJ, Plotkin EV, Scott LDL & Turner APF (1984) Ferrocene-mediated enzyme electrode for amperometric determination of glucose. *Anal Chem* 56: 667-671.
11. Nylander C (1985) Chemical and biological sensors. *J Physics E: Scientific Instrum* 18: 736-749.
12. Kimura J, Kuriyama T & Kawana Y (1986) An integrated SOS/ISFET multibiosensor. *Sensors Actuators* 9: 373-378.
13. Deshpande SS & Rocco RM (1994) Biosensors and their potential use in food quality control. *Food Technol* 48: 146-150.
14. Debnath S, Kumar P & Verma SS (2000) Biosensor: a novel technique for measuring food contaminants. *Processed Food Industry* 3: 19-22.
15. Bose S, Khare RA & Fegade SL (2004) Biosensors: a great achievement in detecting pathogens in food and water. *Indian Food Ind* 23: 51-59.
16. Turner APF (2002) Biosensors: Past, present and future. http:// www. cranfield ac. uk/ biotech/ chinap. htm.
17. De Boer E & Buemer RR (1999) Methodology for detection and type of food borne microorganisms. *Int J Food Microbiol* 50: 119-130.
18. King RJ (1992) Microwave sensors for process control. Part II: Open resonator sensors. *Sensors* 9: 25-29.
19. Swatland HJ (1992) Fibre optics in the food industry. *Food Res Intl* 25: 227-235.
20. Babacan S & Pivamik P (2000) Evaluation of antibody immobilisation methods for piezoelectric sensor application. *Biosens Bioelectron* 15: 615-621.
21. Tor R & Freedman A (1986) A new enzyme membrane for enzyme electrodes. *Anal Chem* 58: 1042-1046.
22. Pennington SN (1976) A small volume micro-calorimeter for analytical determination. *Anal Biochem* 72: 230-237.
23. Schultz JS & Meyerhoff M (1987) Status of monitoring in biotechnology. *Enzyme Microb Technol* 9: 697-699.
24. Turner AP & Newman JD (1998) An introduction to biosensors. In: *Biosensors for Food Analysis*, Scott AO (ed), The Royal Society of Chemistry, London, UK, pp 28-36.
25. Endo H, Matsunaga C, Hoshi M, Hayashi T, Takai R & Watanabe E (1994) Determination of dehydroascorbic acid by an enzyme sensor system with a parallel flow cell. *Biosens Bioelectron* 9: 501-507.
26. Karube I, Wang Y, Tamiya E & Kawarai M (1987) Microbial electrode sensor for vitamin $B_{12}$. *Anal Chim Acta* 199: 93-97.
27. Rabinowotz J & Snell EE (1947) Extraction procedures for the microbial determination of vitamin $B_6$. *Anal Chem* 19: 277-280.
28. Endo H, Kamata A, Hoshi M, Hayashi T & Watanabe E (1995) Microbial biosensor system for rapid determination of vitamin $B_6$. *J Food Sci* 60: 554-557.

29. Kitagawa Y, Tamiya E & Karube I (1987) Microbial FET alcohol sensor. *Anal Lett* 20: 81-96.
30. http://www. Bentham.org/cmciema1-1/clandiu/clandiu.htm
31. Turner APF (2000) Biosensors: sense and sensitivity. *Science* 290: 1315-1317.
32. North JR (1985) Immunosensors: Antibody based immunosensors. *Trends Biotechnol* 3: 180-186.
33. Downs MEA, Kobayashi S & Karube I (1987) New DNA technology and the DNA bio-sensor. *Anal Lett* 20: 1897-1927.
34. Janata J & Blackburn GF (1984) Immunochemical potentiometric sensors. *Ann NY Acad Sci* 428: 286-292.
35. http://www.fuctionalgenomics.Org.uk/sections/activites/Reports/report_florence_2003.htm
36. http://w3.uniroma/.it/biobrobe/original%20 paper htm
37. http://eb.tiscali.it/biosensors/Plant%20Tissue%20electrodes%20partially%20disposal.htm
38. Sidewall JS & Rechniz GA (1986) Progress and challanges for biosensors using plant-tissue materials. *Biosensors* 2: 221-233.
39. Jarvis MT (1985) *Biosensors: Today's Technology, Tomorrow's Products: Technology Insights*, Inc Fort Lec NT and SEAT Technical Publications, Madison, USA.
40. Keating MY & Rechnitz GA (1985) Potentiometric enzyme immunoassay for digoxin using polystyrene beads. *Anal Lett* 18: 1-10.
41. Fonong G & Rechnitz GA (1984) Homogenous potentiometric enzyme immunoassay for human immunoglobin. *J Anal Chem* 56: 2586-2590.
42. Kress RE (1986) Instrumentation in food industry: Chemical biochemical and immunological determinations. *J Physics E: Scientific Instrum* 19: 13-21.
43. Luchner G, Sonnleitner B & Fiechter A (1992) On line measurement in biotechnology: Techniques. *J Biotech* 25: 23-53.
44. Invitski D, Abdel-Hamid I, Atanasov P & Wilkins E (1999) Biosensors for detection of pathogenic bacteria. *Biosens Bioelectron* 14: 599-624.
45. Rishpon J & Ivnitski D (1997) An amperometric enzyme channelling immunosensor. *Biosens Bioelectron* 12: 195-204.
46. Arnold MA (1984) Biocatalytic membrane electrodes. *Am Lab* 15: 34-40.
47. Arnold MA & Meyerhoff ME (1988) Recent advances in the development and analytical applications of biosensing probes. *Crit Rev Anal Chem* 20: 149-196.
48. Wangsa J & Arnold MA (1988) Fibreoptic biosensor based on the fluorometric detection of reduced nicotinamide adenine dinucleotide. *Anal Chem* 60: 1080-1082.
49. Arnold MA (1990) Fibreoptic biosensors. *J Biotechnol* 15: 219-228.
50. Carr PW & Bowers LD (1980) *Immobilized Enzymes in Analytical and Clinical Chemistry*, John Wiley & Sons, New York, USA.
51. Danielsson B & Mosbach K (1986) Theory and applications of calorimetric sensors. In: *Biosensors: Fundamentals and Applications*, Turner APF, Karube I & Wilson GS (eds), Oxford University Press, Oxford, UK, pp181-197.
52. Mattiasson B, Danielsson B, Mandenius CF & Winquist F (1981) Enzyme thermistors for process control. *Ann NY Acad Sci* 369: 295-305.
53. Mosbach K & Danielsson B (1974) An enzyme thermistor. *Anal Lett* 9: 867-889.
54. Weaver JC, Cooney CL, Futtons SP, Schuleer D & Tannenbaun SR (1976) Experiments and calculation concerning a thermal enzyme probe. *Biochim Biophys Acta* 452: 285-291.
55. Yerian TD, Christian GD & Ruzicka J (1988) Flow injection analysis as a diagnostic tool for development and testing of a penicillin sensor. *Anal Chem* 60: 1250-1256.
56. Danielsson B & Mosbach K (1988) Enzyme thermistors. In: *Methods in Enzymology*, Mosbach K (ed), Academic Press, London, UK, pp 181-197.

57. Danielsson B (1990) Calorimetric biosensors. *J Biotechnol* 15: 187-200.
58. Guilbault CA & Luong JHT (1991) A flow through analysis biosensor system for cyanide. *J Biotechnol* 21: 161-172.
59. Jianming Y, Stephon VL & Arthur GR (1997) Piezoelectric biosensor for detection of *Salmonella typhimurium*. *J Food Sci* 62: 1067-1071.
60. Lemeur R & Fredrick F (1996) Cultivation of ornamental plants and biosensors. *Agricontact* 281: 7-11.
61. Milka P, Krest I & Keusgen M (1999) Development of alliin biosensor. *Pharmazie* 54: 435-438.
62. Rose MA (1998) Performance of heat-balance sap flow gauge on rose. *Acta Horic* 421: 201-207.
63. Groom CA, Luong JHT & Masson C (1993) Development of a flow injection analysis mediated biosensor for sulphite. *J Biotechnol* 27: 117-127.
64. Boujtita M & Murr N El (1995) Biosensors for analysis of ethanol in foods. *J Food Sci* 60: 201-203.
65. Esti M, Volge G, Massingnan L, Campagnone D, Palleschi G & La Notte E (1998) Determination of amines in fresh and modified atmosphere packaged fruits using electrochemical biosensors. *J Agric Food Chem* 46: 4233-4237.
66. Esti M, Messia MC, Sinesio F, Nicotra A, Conte L & La-Notte E (1997) Quality evaluation of peaches and nectarines by electrochemical and multivariate analysis: Relationship between analytical measurements and sensory attributes. *Food Chem* 60: 659-666.
67. Mulchandani A, Groom CA & Luong JHT (1991) Determination of sulphite in food products by an enzyme electrode. *J Biotechnol* 18: 93-102.
68. Karube I, Matsunaga T, Mitsuda S & Suzuki S (1977) Microbial electrode BOD sensors. *Biotechnol Bioeng* 19: 1535-1547.
69. Hikuma M, Kubo T, Yasuda T, Karube I & Suzuki S (1979) Microbial electrode sensor for alcohols. *Biotechnol Bioeng* 21: 1845-1853.
70. Hikuma M, Kubo T, Yasuda T, Karube I & Suzuki S (1979) Amperometric determination of acetic acid with immobilised *Trichosporon brassicae*. *Anal Chim Acta* 109: 33-35.
71. Hikuma M, Kubo T, Yasuda T, Karube I & Suzuki S (1979) Amperometric estimation of BOD by using living immobilised yeasts. *Eur J Appl Microbiol Biotechnol* 8: 289-297.
72. Suzuki H, Tamiya E & Karube I (1987) An amperometric sensor for carbon dioxide. *Anal Chim Acta* 199: 85-91.
73. Suzuki H, Tamiya E, Karube I & Oshima T (1988) Carbon dioxide sensor using thermophilic bacteria. *Anal Lett* 21: 1323-1336.
74. Endo H, Kamata A, Hoshi M, Hayashi T & Watanabe E (1995) Microbial biosensor system for rapid determination of vitamin $B_6$. *J Food Sci* 60: 554-557.
75. Arnold MA & Ostler TJ (1986) Fibre-optic ammonia gas sensing probe. *Anal Chem* 58: 1137-1140.
76. Rhines TD & Arnold MA (1988) Simplex optimisation of a fibre-optic ammonia sensor based on multiple indicators. *Anal Chem* 60: 76-81.
77. Goldfinch MJ & Lowe CR (1984) Solid phase opto-electronic sensors for biochemical analysis. *Anal Biochem* 138: 430-436.
78. Kulp TJ, Camines I, Angel SM, Munkholm C & Walt DR (1987) Polymer immobilised enzyme optrodes for the detection of penicillin. *Anal Chem* 59: 2849-2853.
79. Fuh MS, Burgess LW & Christian GD (1988) Single fibre-optic fluorescence enzyme-based sensor. *Anal Chem* 60: 433-435.

80. Brooks SL, Ashby RE, Turner APF, Calder MR & Clarke DJ (1987) Development on-line glucose sensor for fermentation monitoring. *Biosensors* 3: 45-56.
81. Cattaneo MV & Luong JHT (1993) Monitoring glutamine in animal cell cultures using a chemiluminescene fiber optic biosensor. *Biotechnol Bioeng* 41: 659-665.
82. Huang YL, Li SY, Bremel BAA, Bilitewski U & Schmid RD (1991) On-line determination of glucose concentration in animal cell cultures based on chemiluminescent detection of hydrogen peroxide coupled with flow injection analysis. *J Biotechnol* 18: 161-166.
83. Kleman GL, Chalmers JJ, Luli GW & Strohl WR (1991) A predictive and feedback control algorithm maintains a constant glucose concentration in fed-batch fermentations. *Appl Environ Microbiol* 57: 910-917.
84. Rank M, Gram J & Dannielsson B (1993) Industrial on line monitoring of penicillin V, glucose and ethanol using a split flow modified thermal biosensor. *Anal Chim Acta* 281: 521-526.

# Instrumental Analysis of Foods

# 24

**Umesh K. Narta and Wamik Azmi**

*Department of Biotechnology, Himachal Pradesh University, Summer Hill, Shimla 171 005 (H.P.)*

## 1. INTRODUCTION

Food analysis is an important area to study the physical, chemical, biochemical and microbiological composition of food, required as a part of quality and safety assessment of food. Globally, food security and safety concerns are currently at an all time high due to world wide publicity and awareness about health, nutrition and food-borne illnesses. The rapid globalization of food production and trade has further increased the likelihood of international incidents involving contaminated, adulterated, nutritionally poor and unsafe food. Therefore, the analysis of chemical composition and other properties of food has become necessary in order to determine nutritive value, functional characteristics and safety of the food products. Food characteristics like nutritional value, wholesomeness and safety cannot be readily determined by consumers, but if this information is given to the consumer, it will influence the acceptability of the product.

Science of food analysis has shown a rapid development in recent years. New analytical and diagnostic techniques are being developed and existing techniques are being improvised so as to fulfil the requirement of various sectors such as food manufacturers, ingredient suppliers, analytical service laboratories, government laboratories and R&D organizations. There is a variety of methods for analysis of each component of the food; the choice of the analytical method is, however, dictated by the nature of food sample and specific reasons for analysis. The choice of a particular method should be able to answer some basic questions: Does it take less time for an assessment than the reference method? Is a faster but less sensitive method sufficient? Can operational sequences be better organized with other analysis systems e.g., by automation? Is it simple and user friendly? Does it have a potential to deliver online results of measurement? etc. These questions need to be prioritised depending upon the actual purpose of the food analysis.

In the present chapter, an overview of most commonly used methods of food analysis is given, for which it has been divided into four sections: (1) Sampling and sample preparation; (2) Physical characterization of food; (3) Nutritive analysis of food; (4) Analysis of the contaminants and residues. Each method has been supported by its application in food analysis as cited by various workers in their studies.

*Food Biotechnology: Principles and Practices*
*V.K. Joshi & R.S. Singh (Eds.)*
*I.K. International Publishing House Pvt. Ltd., India*
*pp 817–859*

## 2. SAMPLING AND SAMPLE PREPARATION

Food is a complex and heterogeneous composition of hundreds, if not more of different biochemical compounds, and hence pose a great challenge to analysis in the isolation and measurement of individual ingredients. Accurate analysis of the various attributes of food material depends on the successful completion of three steps prior to the analytical measurement: (1) sampling; (2) homogenization and (3) sample preparation. These three steps take care of the requirement that the test portion taken for analysis reflects the composition of the whole material.

### 2.1 Sampling

It is the practice of selecting a fraction of whole material (bulk) for analysis with the assumption that its properties are representative of the whole material. It requires the development of a proper sample plan to make sure that the properties tested from a sample are a good representative of the true value of the whole population (material).

The important considerations which should be taken into account while developing a sample plan include idea about the purpose of analysis (official, raw material, process control, R&D), nature of measured property (colour, weight, fat content, microbial count), nature of population from which sample is to be taken (homogeneous or heterogeneous, finite or infinite, continuous or compartmentalized) and features of the test procedure to be adopted (speed, precision, accuracy, cost, destructiveness). The commonly specified feature in official sampling plants is sample size, sample location and mode of collection.

### 2.2 Homogenisation

Homogenisation is required to get rid of the complexity and heterogeneity (texture, viscosity) of a sample before subjecting it to any sophisticated analysis. It is generally effected by using mechanical devices (grinder, mixer, slicer, blenders), enzymatic treatment (proteases, cellulose, lipases) and chemical methods (strong acids, bases, detergents). The $CO_2$ milling techniques and liquid $N_2$ milling techniques, which use cryogenic embrittlement are also being used now a days for the tests requiring stringent homogeneity. The two functions of homogenization, reduction of particle size and mixing are helpful in achieving the gain in accuracy and precision, which are brought about by improving sampling at the test portion and overcoming cohesive forces to avoid aggregate formation, thereby assuring the inclusion of entire sample.

### 2.3 Sample Preparation

It involves the operations performed on the test portion prior to the analytical measurement, which generally includes storage, preservation, weighing, dilution, cleanup, digestion, purification, separation and derivatization depending upon the type of analytical method to be used and requirements thereof.

## 3. PHYSICAL CHARACTERIZATION OF FOOD

It is not only the nutritive status and safety assurances that determine the quality of a food, but the physical properties also play an important role in quality determination. Physical properties of a food product play a decisive role in securing its marketability and acceptance by the consumers. Physical properties deciding the quality of a food product may include mechanical parameters, visual parameters and aromatic and flavour parameters [1]. Moisture content is another factor which is frequently recognized as an index of economic value, stability and quality of food product.

## 3.1 Measurement of Food Texture

The quality of a food product determines its ultimate acceptance by consumers or in other words its marketability. Therefore, it becomes important to comprehend quality in its wider context. Quality has been defined as the composite of those characteristics that differentiate individual units of a product having significance in determining the degree of acceptability of that unit by the buyer [2].

Texture is one of the most important food characteristics, which plays a critical role in determining the acceptability of that very food by consumers. It is defined as the sensory and functional manifestation of structural, mechanical and surface properties of food detected through the senses of vision, hearing, touch and kinesthetic [3]. The desired textural quality of food is economically important and plays a crucial role in deciding the prices of food products [4-6]. Food texture, being one of the most important quality-determining characteristics, makes food producers and wholesalers particularly interested in its measurement. There are generally two approaches by which the texture of foods is measured: one is the direct measurement through sensory analysis and the other is the instrumental measurement. The former is very complex and time consuming and has been discussed in relevant sections of the book. In the present section, the most commonly used instrumental methods of textural analysis of food are being elaborated.

### *3.1.1 Mechanical deformation methods*

These methods apply compression or penetration for measuring various texture-related attributes of foods. Mechanical deformation methods are destructive in nature as they impart some sort of injury or destruction to the sample being analysed and render it unfit for consumption and sale. These methods apply mechanical force to the sample and measure the response (force/deformation or stress/strain), which is ultimately correlated with the sensory evaluation to measure the textural attributes (e.g., firmness) of a particular food.

**3.1.1.1 Puncture test:** Puncture test is a penetrometry-based method, which involves the penetration of a food sample with a cylindrical probe or punch having a diameter at least three times smaller than the sample at a specific speed up to a constant depth. The various specifications, however, are subject to the type of food sample being evaluated. The force applied and the responses in terms of deformation are recorded through the sensors and plotted as a curve. The force/deformation curve is typical for a particular food and gives the value of different parameters that have a correlation with various sensory description of food texture. Mehinagic *et al.* [7] observed that parameters viz., $F_f$ (flesh rupture force) and $W_f$ (work associated with $F_f$) obtained by penetrometry were effective in measuring the firmness of apple in terms of crunchiness and touch resistance. Magness-Taylor (MT) firmness tester is routinely used for measuring the firmness of fruits in association with TA.$XT_2$ texture analyser.

**3.1.1.2 Texture profile analysis:** Texture profile analysis (TPA) involves a two-cycle compression of food sample with a flat-based probe. In fact, texture profile analysis imitates the action of jaw motion to bite-size piece of food. A bite-size food sample having standard size and shape is placed on the base plate and compressed and decompressed twice by the plate attached to the drive system with a high compression to imitate the chewing action of the teeth. The force applied on the sample at different time intervals during the course of double compression is plotted as force-time curve, which is sample (food) specific and gives invaluable information about the texture of a particular food sample (Fig. 1).

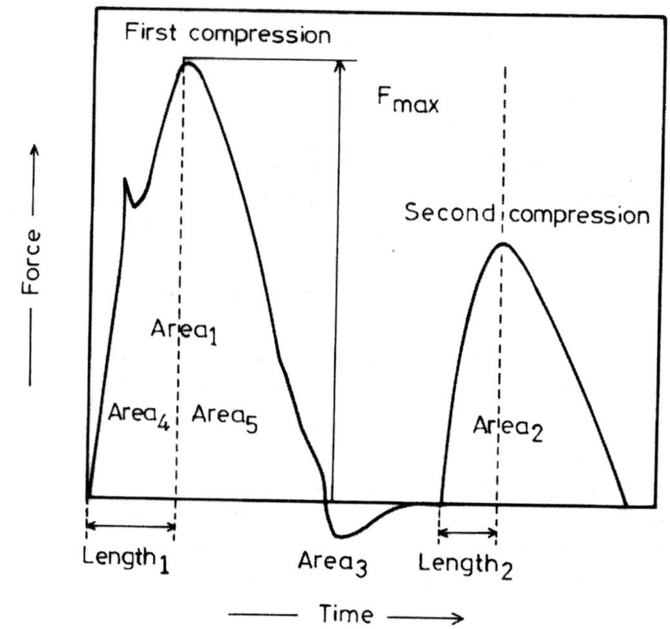


**Fig. 1:** A generalised texture profile analysis curve showing various primary texture measurements.

TPA has the perceived advantage of measuring a number of parameters from a single test with the help of automated instruments like TA.$XT_2$ texture analyzer (Fig. 2). TPA is useful in testing the hardness of flour made food items such as white floured noodles in which hardness is defined as the peak force on the first compression cycle in a typical TPA curve of that food [8]. TPA can also be used to determine springiness, cohesiveness and resilience of foods. These parameters were evaluated by Epstein *et al.* [8] using length, force and area of a TPA curve as Cohesiveness = $area_2/area_1$; Springiness = $length_2/length_1$; Resilience = $area_5/area_4$. The other parameters of food texture such as fracturability, chewiness and gumminess of different foods can also measured by length, area and force ratios from TPA curve as Gumminess = $area_2/area_1$ + hardness; Chewiness = gumminess + $length_2/length_1$ (Fig. 1). Stickiness of food is measured by applying a predetermined force to the food sample with a flat probe and then measuring the force required to pull the probe away from the surface of the sample. This is known as compression-decompression test [9]. Despite their wide application in the analysis of various textural characteristics of a variety of food materials, the mechanical force/deformation techniques are not very suitable for on-line sorting of food material (fruits and vegetables), as they are of low speed and destructive nature.

### *3.1.2 Non-destructive instrumental methods*

In recent years, non-destructive methods of quality evaluation have gained momentum and considerable attempts are being made to develop them. These methods have an advantage of on-line and rapid measurement of different physical properties of foods without causing damage to the food sample. These techniques are user friendly and have become popular in the measurement of different attributes of food texture, especially that of fruits and vegetables [10].

3.1.2.1 Spectroscopic methods: Spectroscopic methods of analysis use interaction between atoms and molecules of the food sample and electromagnetic radiation to furnish qualitative and quantitative

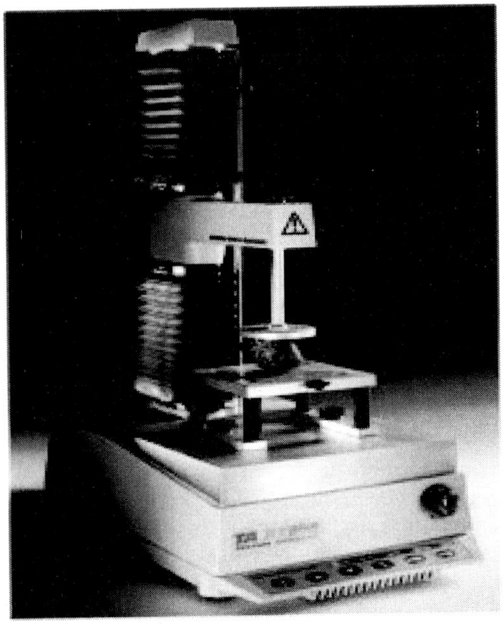


**Fig. 2:** TA.XT2 Texture Analyser (Courtesy: Texture Technologies Corp.).

information about the texture. These are the most preferred non-destructive methods of texture measurement.

3.1.2.1.1 Visible/Near infrared spectroscopy: Visible (VIS) and near-infrared (NIR) reflectance spectra of the surface of various intact fruits and vegetables provide important information about their surface properties (e.g., firmness). The fruits and vegetable are scanned at equatorial area with radiations of visible and near infrared (0.8-2.5 μm) wavelengths. The reflectance spectra is detected by a sensor and conveyed to visible/NIR spectrometer, which can be made to furnish useful information regarding the firmness of fruits and vegetables on comparison with the suitable standard. NIR reflectance spectroscopy has been used in determination of the firmness and internal quality of various fruits such as apple, pears, etc. [11, 12, 13]. VIS/NIR spectroscopy was also used in comparison of the firmness of fruit in association with other non-destructive methods [14]. Thybo *et al.* [13] used NIR spectroscopy in combination with uniaxial and NMR techniques to measure the sensory texture of the cooked potatoes.

3.1.2.1.2 Nuclear magnetic resonance: The nuclear magnetic resonance (NMR), also known as magnetic resonance imaging (MRI), measures resonant magnetic energy absorption by nuclei placed in an alternating magnetic field. The amount of energy absorbed by the nuclei is directly proportional to the number of a particular nucleus ($^{1}H$, $^{13}C$, $^{31}P$) in a sample, such as protons in water or oil. The theory of NMR is presented in detail elsewhere [16].

NMR finds applications in determining the moisture content and oiliness of food materials [17, 18]. It possesses a unique advantage of detecting the concentration and state of water and fats in fruits and vegetables, which is very useful for the assessment of ripeness and defect or decay in them. High-resolution images of these internal characteristics of fruits and vegetables can contribute to evaluation of maturity and quality parameters related to texture [19]. Mealiness in apples and pears can be assessed by using MRI techniques [20]. Three-dimensional MRI is also used to analyse external (textural) and internal structure of chocolates, which helps in selection of chocolate and other similar confectioneries [21]. The sensory texture of cooked foods such as potatoes has been predicted by using NMR techniques [15].

3.1.2.2 Impulse response and ultrasonic techniques: Recent research of texture analysis concentrates on impulse response and ultrasonic techniques. Mechanic or acoustic impulses applied to fruits produce acoustic responses such as resonance, which can be related to elasticity, internal friction or damping, shape, size and density. Hence, sonic vibration measurements represent the mechanical properties of the entire fruit [22]. For ultrasonic waves, their propagation velocity, attenuation and reflection are important parameters to evaluate tissue properties of fruits and vegetables [23]. Some other workers have reported the use of vibration to measure the firmness of tomatoes [24]. A system and method based on ultrasonic energy absorbance was suggested by Mizrach *et al.* [25] and Bechar *et al.* [26] for rapid non-destructive measurement of mealiness in apples. The method enables the fruit quality attributes to be tested by measuring the change in ultrasonic sound waves while passing through the fruit tissue over a short distance across the peel [27].

In addition to the above mentioned non-destructive techniques, there are a number of different non-invasive methods of texture analysis. Mehinagic *et al.* [7] used acoustic measurements (stiff-coefficient) to characterize the parameters of texture (crunchiness and touch resistance) in various apple cultivars. A laser air-puff firmness detector has been found to be useful in sensing the firmness of fruits and a variety of other food products. In this method, a small puff of air at a regulated pressure is made to impinge on the fruit surface, which depressurizes it by a measurable amount. The resulting deformation is measured on-line by a laser displacement sensor [28].

## 3.2 Optical Properties of Foods

Visual parameters play a pivotal role in defining the quality of food product and the optical properties occupies a special place among other parameters. Optical properties refer to those sight perceived attributes that depend on geometric or chromatic modification of the light striking the sample in question e.g., colour, translucency, gloss and their uniformity on the surface of the product. Needless to say that visual evaluation precedes the evaluation of other parameters (flavour, aroma, texture) while selecting a particular food product. It can be further understood by the fact that with the sense of sight we obtain more than 80% of the information about our environment, in comparison to a less than 20% obtained from the other senses [29].

### *3.2.1 Measurement of colour*

Colour is undoubtedly an indispensable and first attribute of food quality, which plays a decisive role in the customer acceptance of a particular food. The importance of colour in foods can be judged by the fact that the food legislations in almost all countries of the world have made it mandatory to describe it on every food material. Although the easiest way of evaluating the colour of food material is by visual observation, but here we shall focus only on the instrumental measurement. However, the visual evaluation of the colour will be discussed somewhere else in the sensory evaluation section.

3.2.1.1 Spectrophotometric method: Generally spectrophotometers are built to measure the transmission of light through liquid samples. In order to measure the light reflected by opaque solid samples, spectrophotometer is connected to a suitable accessory, capable of measuring the quantity of light reflected from the surface, over the whole visible spectrum, as a fraction of the light reflected by the white standard (white marble or hard white plastic) illuminated in the same way. This fraction is called reflectance. The measurements done with these instruments provide a detailed reflectance spectrum from which complete information about the colour of a sample measured can be obtained. The modern instruments display the spectral reflectance curve and the data for spectral sweep on the screen. It is however, difficult and often impossible to insert samples like an apple, or asparagus into a spectrophotometer for colour measurement for which other instruments like tristimulus colorimeters are effective.

3.2.1.2 Tristimulus colorimetric method: Tristimulus colorimeter is one of the most commonly employed instruments in colour measurements of food products. In this apparatus, the integration procedures that incorporate the standard observer curves (weighted ordinates and selected ordinates) are replaced by the use of various suitable filters that stimulate the response. A beam of white light from a source (Wolfram incandescent lamp or Xenon flask lamp) is made to strike the sample at an angle of 45°, which is reflected at an angle of 90° to the surface of sample. This beam of diffuse light then passes through three filters, X, Y, Z, and is measured by photocell.

Tristimulus colorimeter is calibrated with white plate and black plate. Depending on the lighting provided by lamp (light beam or circular light), it is possible to measure the colour of both smooth surfaced food articles (peanut butter, tomato puree, canned pimientos, etc.) as well as the uneven surfaced food articles (peas, rice, bean, etc.). Although the measurement of colour difference by tristimulus colorimeter is exact, but it does not give exact colour measurement, which is one of the basic disadvantages of this instrument.

### *3.2.2 Measurement of gloss*

Gloss may be defined as the fraction of light reflected in the specular direction or the degree to which it resembles the surface of a mirror [29]. It contributes to the variation in the apparent colour of an object when viewed from a different angle. Despite its prime importance in relation to the appearance of products, the measurement of gloss and its application in the field of food technology is not widespread. The instrument used for measuring gloss is called goniophotometer. It is designed to measure the quantity of light reflected when viewing angle changes. A goniophotometric curve produced by measuring the directional reflectance with changing angle of a surface of food sample is the basic determination of gloss in a similar manner as the spectrophotometeric curve determine colour. The sample with highest gloss is the one which produces highest peak, at a particular angle in goniophotometric curve. Now a days a simplified goniometer called gloss meters are available, which take measurement only at three angles (20°, 60°, and 85°).

Matsuoka *et al.* [30] measured the gloss of eggplant by a spectral radiometer system and found it to be a visible parameter for determining freshness. A unique gloss meter for measuring the gloss of curved surfaces was used in parallel with a conventional flat surface gloss meter to measure peel gloss of ripening banana. This has also been used to measure the gloss of other fruits and vegetables such as oranges, green bell pepper, tomato, eggplant and onions [31, 32].

### *3.2.3 Measurement of translucency*

There are a number of food products that are translucent. They are neither totally opaque nor completely transparent by virtue of which light is absorbed, reflected, transmitted and dispersed or scattered. For such food products objective measurement of colour presents serious difficulties.

A colorimeter cannot measure visual appearance of the translucent foodstuffs. It simply records the proportion of light striking the photodetector in relation to the standard. However, the instrumental difficulties posed by definition and the measurement of the colour of cloudy or translucent food material can be overcome by making use of Kubelka-Munk thin layer analysis [33]. A measurement for the reflection of a thin sample against a white background and a black background by relating it to coefficient absorption (k) and a coefficient of scattering (s):

$$\frac{k}{s} = \frac{(1 - R_\infty)^2}{2R_\infty}$$

$R_\infty$ denotes the reflectance of a layer, which is described as infinitely thick. $R_\infty$ can be calculated from the measurements made with the black and white backgrounds using the following equation:

$$R_\infty = a - b, \; a = 1/2 \; \frac{R + (R_0 - R + R_g)}{R_o R_g} \qquad b = (a^2 - 1)^{1/2}$$

$R$ = reflectance of the sample against a white background
$R_o$ = reflectance of the sample against a black background
$R_g$ = reflectance of the white background

## 3.3 Moisture Content of Foods

Moisture content is one of the most commonly measured characteristic of food material. The analysis of moisture content in a food material is important for a number of reasons such as legal and labelling requirements, economic reasons (as an inexpensive ingredient), microbial stability, food quality and processing operations. It becomes therefore important to measure the moisture content accurately. A number of analytical techniques exist for this purpose, but the choice of application depends on the accuracy, cost, speed, sensitivity, type of food and purpose of analysis.

Food, being a heterogeneous mixture, contains different proportion of chemically bound, physically bound, capillary and trapped or bulk water in different physical states (solid, liquid, gas). The suitability of the analytical procedure depends on the type of molecular environment in which the water exists and also the form of water which is intended to be measured. The methodologies of moisture determination have been reviewed by many authors [34, 35]. However, the most commonly used methods of moisture determination in different food stuffs are being discussed here.

### *3.3.1 Evaporation methods*

These methods rely on measuring the mass of water in a known sample by measuring the mass of the food before and after the water is removed by evaporation. Hence the instruments employed intend to dry the food sample so that the weight loss could be correlated to the amount of water present. The evaporation devices employed varies with the type of food sample being analysed. Air-oven, vacuum oven and microwave oven are the commonly used drying instruments. However, freeze drying is practiced in case of high value liquid foods such as coffee and juices; and solid foods like strawberry, whole shrimp, chick dries and mushroom slices. Sublimation characteristics of water vapours forms the basic principle of freeze drying in which water from frozen food sample is evaporated without melting it under low vapour pressure in a vacuum chamber (4.7 mm Hg). The evaporation methods are precise, cheap and user friendly. Destructive nature and time consuming operation procedures are the only constraints of evaporation method.

### *3.3.2 Distillation methods*

Distillation methods are based on direct measurement of amount of water removed from food sample by evaporation. In distillation methods, the weighed food sample is heated in the presence of immiscible organic solvent (xylene or toluene) having higher boiling point than water. The evaporated water of the food sample is collected in a graduated glass tube where its mass is determined. Dean and Stark method is the best illustration of distillation method in which moisture content is measured volumetrically. Distillation methods are best suitable to foods with low moisture content and the food containing volatile oils such as herbs or spices.

### 3.3.3 Chemical reaction methods

In chemical reaction methods, reaction between water and certain chemical reagents is used as a basis for determining the concentration of moisture in foods. The chemical reagent added to the food sample reacts specifically with water to produce a measurable change in the properties of the system, e.g., colour, mass, volume, pH, conductivity, etc. The measurable changes observed in the system are correlated to the moisture content using calibration curves. The specific affinity of reagent to water molecule without interacting with other components of the food is important for the accurate assessment of the moisture content. Two methods that are commonly used in food industry are Karl-Fisher titration and gas production methods.

**3.3.3.1 Karl-Fisher method:** Karl-Fisher titration has become a standard method for determining the moisture content of foods having low water contents e.g., dried fruits and vegetables, coffee, confectionary, oils and fats, wheat flour, cocoa powder, spices, etc. This titration method makes use of reduction of iodine by sulphur dioxide in the presence of water:

$$2H_2O + SO_2 + I_2 \text{ (red brown)} \longrightarrow H_2SO_4 + 2HI \text{ (colourless)}$$

The food to be analysed is placed in a beaker containing solvent ($C_5H_5N$) to keep $SO_4$ and $I_2$ in solution and is titrated against Karl-Fisher reagent (iodine solution). Presence of any moisture tends to decolourise $I_2$ solution by forming HI. The volume of iodine solution required to titrate the water in food is measured and can be related to moisture content using calibration curve. Now a days, the precision of the technique is improved by using electrical method to follow the end point. Relatively inexpensive commercial instruments have also been developed which are based on Karl-Fisher titration.

**3.3.3.2 Gas production method:** Automatic instruments are also available that measure the moisture content by detecting the production of a gas due to specific reactions between chemical reagents and water present in a food sample. One such example is the production of acetylene gas when powdered calcium carbide reacts with water:

$$CaC_2 + 2H_2O \longrightarrow C_2H_5 (\uparrow) + Ca(OH)$$

The gas thus produced is related to the moisture content of foods in a number of different ways such as by measuring the volume of gas produced, by measuring the decrease in the mass of the sample after the gas is released and by recording the increase in pressure of closed vessel containing the reactants.

### 3.3.4 Spectroscopic methods

Spectroscopic methods utilize the interaction of electromagnetic radiation with the water present in the food sample. The spectroscopic methods developed to measure the moisture content of food are based on the fact that water absorbs electromagnetic radiation at a characteristic wavelength that is different from the other components of the food matrix. A measurement of the absorption of radiation at this wavelength can then be used to determine the moisture content.

**3.3.4.1 Infrared absorption spectroscopy:** The infrared spectrum of a chemical compound is probably the most characteristic physical property of that compound. The infrared beam is passed through an optical filter that transmits energy at the wavelength of maximum absorption which is 0.7 to 2.4 μm and has been successfully used for the measurement of moisture content in cereal grains. Absorption bands for water have been shown to occur at 0.76, 0.97, 1.118, 1.45 and 1.94 μm [36]. Moisture contents in a food sample can be determined by comparison of the depth of the band of interest with that of the same band for standard concentrations of water.

**3.3.4.2 Near-infrared reflectance spectroscopy:** The high resolving power of reflectance spectra in the near-infrared (NIR) range (0.8-2.5 μm) makes NIR spectroscopy a powerful and rapid analytical technique not only in moisture measurement but also for the oil, and protein contents of various food samples. Non-requirement of wet chemical analysis is an additional advantage of this method. Ground samples are packed into a sample holder maintaining direct contact with a concentric IR transmitting quartz window. The reflected radiation signals of the diffused spectra from glass window are collected with four lead sulphide detectors equally spread around the incident beam. Signal from the detector is amplified, digitized and fed to digital computer. The infrared reflectance (R) curve is recorded as the second derivative of original log (1/R) absorption curve.

Chemical analysis is used to make a calibration curve and the data of sample is recorded in the computer, which analyses it with a stepwise multiple regression method to develop prediction equations by regression analysis of IR spectral data against chemical data. This method has been successfully used in measuring the water content of beef.

**3.3.4.3 Nuclear magnetic resonance spectroscopy:** It is a non-destructive method of moisture determination in which nuclear magnetism of hydrogen atom of water molecules is utilized. The instrument consists of four main units: a permanent magnet, a magnetic sweep system, a radiofrequency generator and a radiofrequency receiver.

The extra characteristic of spin possessed by most nuclei form the basis of NMR spectroscopy. The magnetized food sample absorbs energy when radiofrequency radiations are passed through it. The instrument exhibits a linear relationship between indicator readout in milli-volts and the moisture content. It is a rapid, accurate and non-destructive method of moisture measurement and control in raw material and finished food products. NMR spectroscopy has been applied for determination of moisture in virtually unlimited types of foods, including wheat flour, rice, corn, candy, sugar, starch cheese and many more food products.

### *3.3.5 Chromatographic methods*

These are the direct methods of moisture measurement, which involve the removal of moisture from solid food sample and then the removed moisture is estimated. Gas chromatography finds wide application in moisture analysis of the foods. The moisture content is quantified by determining the peak areas of water and solvent followed by comparison with the standard curve for known amount of water. A weighed amount of food such as meat is homogenized in a known amount of solvent (ethanol, anhydrous methanol or isopropanol). The extract is then subjected to analysis by a GC equipped with a column such as Poro-pack-Q and thermal conductivity detector.

Reineccius and Addis [37] compared the efficiency of gas chromatography with conventional method of moisture determination and found the former to be better in most of the cases. In addition to the methods discussed so far, there exist a number of other methods, which are used in moisture determination but to a lesser extent. The physical methods like density measurement, refractive index and electrical conductivity are also useful in moisture determination.

## 4. ANALYSIS OF FOOD NUTRIENTS

In this section, the analysis of nutritionally important components of human diet are being discussed which include, carbohydrate, protein, lipid, vitamin and minerals.

### 4.1 Analysis of Carbohydrates

All foods with few exceptions contain carbohydrates, which may vary in form from a simple monosaccharide to more complex polysaccharide. Carbohydrates are generally defined as poly hydroxyl aldehydes or poly hydroxyl ketones. They form one of the major classes of food components along with lipids and proteins. Besides being the major source of energy in human diet, they also impart physical properties to food. Carbohydrates (digestible) account for more than 70% of total daily caloric intake by people in many parts of the world [38].

Although the majority of carbohydrates are in the form of polysaccharides, the starch is however the only polysaccharide, which is digestible by human. A major portion of digestible carbohydrates is contributed by monosaccharides and disaccharides. The indigestible carbohydrates form part of a group of substances known as dietary fibres, which also include lignin. These dietary fibres constitute indispensable part of human nutrition by reducing risks of various deadly diseases including cancer, coronary heart diseases, diabetes and constipation. In addition to an important source of energy and dietary fibres, carbohydrates also contribute to the sweetness, appearance and textural characteristics of many foods.

There is thus a continuing need to monitor the types and levels of carbohydrates in foods in order to predict and control the properties of the food for nutritional labelling, detection of adulteration, maintenance of standards of identity and for economic and food processing reasons. The analysis of carbohydrate therefore, frequently involves not only the determination of total amount of sugar present in the sample, but often the identity, configuration and conformation of the carbohydrate components are also determined. A large number of analytical techniques have been developed over the years to identify and quantify the carbohydrates in the food. There are various traditional methods which include physical and chemical methods available for carbohydrate analysis, but given the nature of analysis, accuracy and convenience involved, chromatographic, enzymatic and other automatic methods are being employed now a days. A list of various methods along with the scope of their application is given in Table 1.

#### *4.1.1 Chromatographic methods*

The earlier methods, which included paper chromatography (PC), thin layer chromatography (TLC), and open column chromatography, have largely been replaced by instrumental methods such as high performance liquid chromatography (HPLC), and gas liquid chromatography (GLC), which in general provide rapid analyses with greater precision and specificity [39].

Both of these techniques need a careful sample treatment. In HPLC, extraction and clean up (removal of interfering compounds) is required, whereas GLC requires the samples to be volatile which is achieved by derivatization of the sugars. The most commonly used derivatizing agent is o-trimethylsilyl (TMS), which is used to convert aldonic acid form of carbohydrates into TMS ethers. Acetates and trifluoroacetates are also used as derivatizing agents. The reaction mixture is directly subjected to the chromatograph to achieve the separation and identification of individual component by programming the temperature. Flame ionization detector (FID) is generally used in carbohydrate analysis with GLC, which is a preferable choice for monosaccharide analysis in bread, honey, milk, bun mix, glucose syrup, soya bean and other syrups [40].

**Table 1:** Methods used for the carbohydrate analysis in foods.

| Type of method | Nature of analysis/application |
|---|---|
| *Traditional methods* | |
| *Physical* | |
| Polarimetry (specific rotation) | All soluble carbohydrates |
| Refractometry (refractive index) | Total carbohydrate (soluble) |
| Hydrometry (density) | Total carbohydrate (soluble) |
| *Chemical* | |
| Picric acid | Reducing sugars, sucrose |
| Anthrone | Free hexoses |
| Clegg-Anthrone | Hexoses |
| Nelson-Somogyi | Reducing sugars |
| Dinitrosalicylate | Reducing sugars |
| Cu-reduction | Sucrose, reducing sugars |
| Phenol-sulphuric acid | All soluble carbohydrates |
| Neocuproine | Reducing sugars |
| Ferricyanide | Reducing sugars |
| Iodometric | Lactose, glucose, maltose and galactose |
| Triphenyltetrazolium | Reducing sugars |
| *Recent methods* | |
| Chromatographic | Separation & analysis of soluble carbohydrates |
| Enzymatic | Most carbohydrates |
| Flow injection analysis | A wide range of carbohydrates |
| Electrophoretic | Separation of derivatised carbohydrates |
| Immunoassays | Antibody based test kits for the quantification of specific carbohydrates |

HPLC has been found to be more suitable for the detection and determination of medium and high molecular weight polysaccharides along with mono and oligosachharides, whereas the larger polysaccharides require a prior hydrolysis for chromatographic analysis. Refractive index detector (RID) is a detector of choice for HPLC analysis of carbohydrates. The ultraviolet spectrophotometer ($\lambda_{max}$ = 190 nm) has also been used in the analysis of glucose, fructose and galactose in juices, and mass detector (MD) in coffee. HPLC has been found applicable in carbohydrate analysis in cereal beverages, cream, fruit juices, fruits and vegetables, coffee, milk, melon and yoghurt. A variety of different columns can be used with bonded amino phases to separate carbohydrate with molecular weight up to about 2500Da, depending upon carbohydrate composition and solubility properties. Such columns have been successfully used in the analysis of carbohydrates in fruits, vegetable, confectionaries, beverages and breakfast cereals [38]. Survey of literature suggests gel filtration chromatography to be the most suitable technique for the analysis and characterization of larger polysaccharides.

#### *4.1.2 Enzymatic methods*

Analytical methods based on enzymes rely on their ability to catalyze specific reactions. These methods are rapid, highly specific and sensitive to low concentration, and are therefore ideal for determination of carbohydrates in foods. However, the specificity of enzymatic method depends on the purity of the enzyme used, which assumes a special importance during the analysis of food carbohydrates where a number of many other carbohydrates exist, and hence may behave as interfering compounds by acting as a substrate to the impure enzyme.

**Table 2:** Specific enzymatic assays for carbohydrate analysis.

| Analyte | Enzyme system (Immobilized) | Principle of detection |
|---|---|---|
| Glucose | GOD; membrane | ISFET, $O_2$ or $H_2O_2$ probe |
| Glucose | Concanavalin A+ fluorescein-labled dextran | Fluorometric (affinity sensor), optical fibres |
| Glucose | Luciferase; reactor | Luminometeric |
| Glucose | GOD; reactor | Enthalpimeteric |
| Glucose | GOD; membrane | Micro-$O_2$ sensor |
| Sucrose | Invertase + GOD; membrane | $H_2O_2$ probe |
| Sucrose | Invertase + mutarotase + GOD; membrane | $O_2$ probe |
| Maltose | Glucan 1, 4-glucosidase + GOD + POD; membrane | $H_2O_2$ probe |
| Lactose | Lactase + GOD; membrane | $O_2$ probe |

ISFET = ion-sensitive field effect transmitter; GOD = glucose oxidase; POD = peroxidase

Enzymatic determination requires very small amount of material. Sampling is therefore, important to ensure the holistic representation of the material. The immobilized enzymes have been able to reduce the cost incurred in enzymatic analysis vis-à-vis the other conventional techniques. Since the enzymes are substrate specific, the elaborate clean up procedures to remove interfering compound can be by passed. However, the presence of heavy metal ions (Hg, Pb) and oxidizing agents may inactivate enzyme, which may have negative bearings in reaction rate based method of analysis. The enzymatic methods of quantitative carbohydrates analysis can be carried out by making use of two options: (i) measuring the initial rate of enzyme catalysed reaction, which is proportional to the initial substrate concentration, and (ii) determining the concentration of the end product of enzyme-substrate reaction at the completion of the reaction [41]. Product may either be measured directly or combined with suitable indicator or dye to produce coloured complex, which is then monitored spectrophotometrically. This type of approach in which a biological material like enzyme is used is termed as a biosensing mode of detection. Suitable biosensors have been developed by assembling all required components into a probe in which oxidases and peroxidases are used.

$$\text{(i) Glucose} + O_2 \xleftrightarrow{\textit{Glucose oxidase}} \text{Gluconic acid} + H_2O_2$$

$$H_2O_2 + \text{indicator/dye} \longleftrightarrow \text{Colour} + H_2O_2 \qquad \textbf{(Indirect method)}$$

$$\text{(ii) Glucose} + \text{ATP} \xleftrightarrow{\textit{Hexokinase}} \text{Glucose-6-phosphate} \qquad \textbf{(Direct method)}$$

$$\text{Glucose-6-phosphate} + \text{NADP} \xleftrightarrow{\textit{G-6-P,dehydrogenase}} \text{Gluconate-6-Phosphate}$$

$$+ \text{NADPH } (\lambda \text{ 340 nm})$$

Glucose, fructose, galactose, sucrose, sorbitol, lactose, maltose, raffinose and starch have been analysed by making use of various enzymes by direct and indirect quantification methods (Table 2).

### *4.1.3 Immunological methods*

Immunoassays are finding increasing use in the food industry both in qualitative as well as quantitative analysis of food components. Immunoassays specific for low molecular weight carbohydrates are developed by attaching the carbohydrate of the interest to a protein, and then injecting it into a suitable animal. With the passage of the time the animal will develop antibodies specific for the carbohydrate molecule, which can be extracted from the animal and used as part of a test kit for detecting and

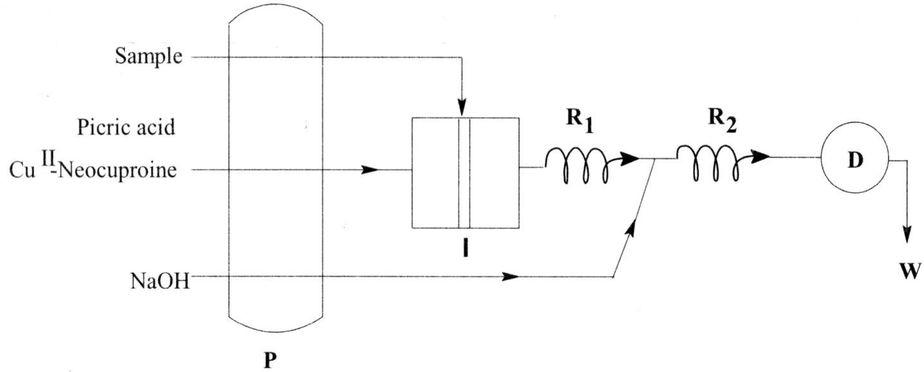


**Fig. 3:** Normal FIA configuration for picric acid and Cu(II)-neocuproine method. R1, R2 (reactors); D (detector); P (peristaltic pump); I (injection valve); W (waste).

determining the concentration of the specific carbohydrate in the food samples. Immunoassays are extremely sensitive, specific, rapid and user friendly methods of analysis.

#### *4.1.4 Flow injection analysis (FIA)*

Flow injection analysis (FIA) is an automated, continuous flow approach to perform chemical analysis based on injecting a small volume of sample into a continuously flowing carrier stream to which appropriate reagent can be added. In the analysis of carbohydrates especially reducing sugars, $Cu^{II}$-neocuproine and Nelson-Somogyi method based FIA are used. It comprises of an assembly wherein reactors are fitted for facilitating the reaction between reducing sugar and reagents. The resulting complex is detected by a detector fitted in it. The flow of sample and reagents is ensured by the peristaltic pump fitted. An injection valve in the peristaltic pump helps in injecting the sample or the reagents into the system (Fig. 3).

Undoubtedly the future of carbohydrate analysis belongs to flow injection analysis (FIA) along with enzymatic assays and HPLC. FIA require relatively inexpensive materials for its excellent performance and has been the basis for a large number of methods used in the analysis of carbohydrates. Some workers in 1987 proposed and described automatic flow injection methods for the determination of reducing sugars in wine. These methods were based on classical chemical methods of reducing sugar analysis viz., the picric acid method, the $Cu^{II}$-neocuproine and Nelson-Somogyi method. This work was further extended to the determination of reducing sugars in different food samples [42]. In addition to this, different workers are pursuing research on the automation of enzymatic determination of some carbohydrate by means of FIA techniques. Several biosensors have also been designed on this basis [43].

### 4.2 Analysis of Proteins

Proteins are the polymers of amino acids. Twenty different types of amino acids occur naturally in proteins. The variation in the molecular structure, physiochemical properties and nutritional attributes of proteins is due to the variation in number, type and sequence of these amino acids, which makes up the polypeptide backbone. Proteins are the most abundant cell components and play important biological functions within the organism. Nutritionally proteins are also the source of energy and essential amino acids in humans. They are major structural components of many natural foods, often determining their overall texture e.g., tenderness of meat or fish products. Proteins are also used as gelling agents, emulsifiers, foaming agents and thickeners. The analysis of proteins in foods should be aimed at determining the total concentration, type, molecular structure and functional properties.

In fact, nitrogen is the most distinguishing element in proteins. It contributes about 13 to 19% in the protein [44], and therefore for decades the protein analysis has been performed by determining nitrogen contents. However, in last few years, some new instrumental methods have also been developed to quantify and characterize the specific proteins from different types of foods.

### *4.2.1 Determination of total protein*

The quantitative determination of total protein in food products can be achieved directly (spectrophotometrically) or by determining the total organic nitrogen followed by conversion of total nitrogen to crude protein content using a suitable conversion factor. The indirect methods are based on two assumptions that dietary carbohydrates, fats and any other material present in food do not contain nitrogen, and that nearly all of the nitrogen in the diet is present as amino acids in protein.

4.2.1.1 Kjeldahl method: It has been the method of choice for total protein estimation for past several decades. The protein estimation is performed by determining the nitrogen content of food product after complete acidic hydrolysis and digestion, and subsequent conversion to protein content using various conversion factors [44, 45]. The schematic representation for determination of nitrogen is given in Fig. 4. Once the nitrogen content has been determined, it is converted to a protein content using appropriate conversion factor: % Protein = F x % N (nitrogen). Not withstanding the drawbacks such as the use of toxic and expensive catalysts, time consuming nature and interference of non-protein nitrogen, Kjeldahl method is still a widely used international standard method of protein estimation. It is often used as a standard method in comparison with all other methods.

4.2.1.2 Enhanced Dumas method: It is an automated instrumental technique, capable of rapidly measuring the protein concentration of food sample. This technique is based on a method first developed in 1831 by a scientist called Dumas. It involves the combustion of nitrogen containing sample at elevated temperature in the presence of oxygen, which leads to release of $CO_2$, $H_2O$ and $N_2$. Carbon dioxide and water are removed by a special column through adsorption, whereas nitrogen gas is measured by passing it through a column which has a thermal conductivity detector at the end. The signal from thermal conductivity detector is converted into nitrogen content, which is ultimately converted into the total protein content using suitable conversion factors.

4.2.1.3 Colorimetric methods: These methods are based on the colour development by proteins or parts thereof in the presence of reagents. The intensity of the colour development is proportionated to the protein content of the food sample. Biuret reaction is an example of colorimetric method of protein estimation in which a violet purplish colour complex is formed by the interaction of cupric ion complex with peptide bond at an alkaline pH. The absorbance of the coloured complex thus formed is measured at 540-560 nm, which has a direct relationship to the protein content [44]. The sensitivity of the method for determining the proteins in whole milk was increased by combining it with liquid scintillation counter [46].

The Lowry method is another colorimetric method that works on the basis of the reduction of Folin–Ciocalteau phenol reagent (Phosphomolybdic–phosphotungstic acid) by tryptophan and tyrosine residues in proteins. The characteristic blue colouration has absorption maximum at 745-750 nm, which is highly sensitive to low protein concentration in solutions. The continuous measurement of protein in milk using modified Lowry method has been described [47].

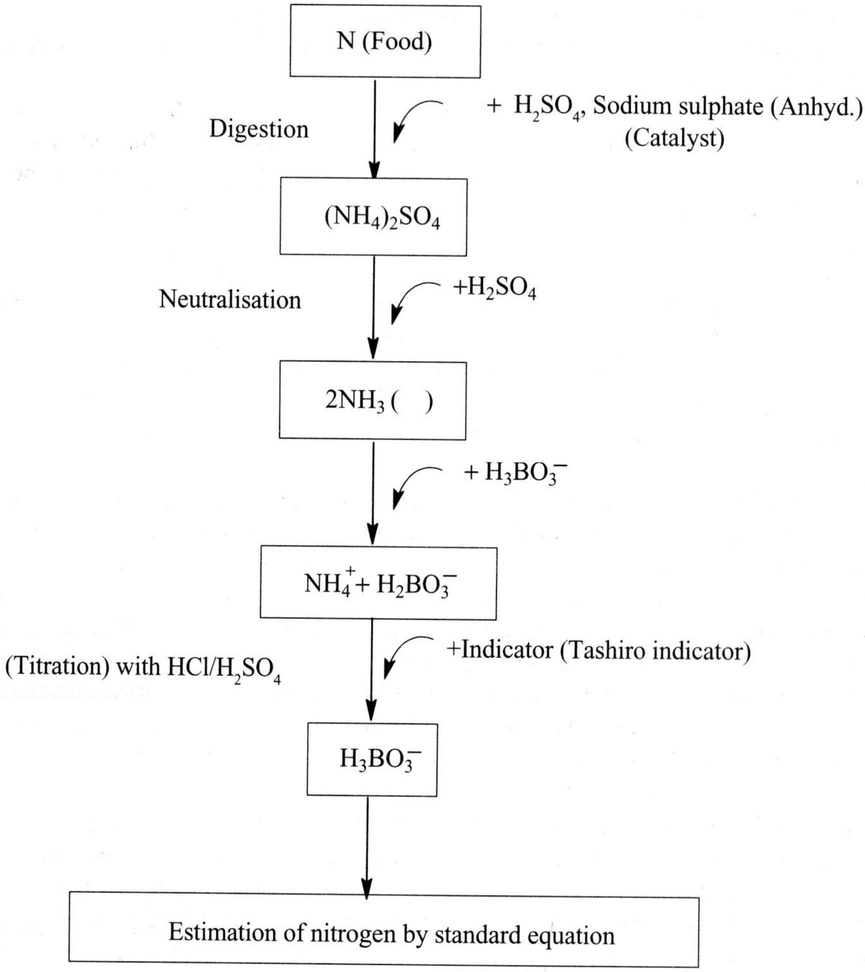


**Fig. 4:** Flow diagram showing nitrogen estimation in food by Kjeldahl method.

**4.2.1.4 Ultraviolet spectrophotometric method:** This is the simplest method of protein estimation, which exploits the tendency of proteins to absorb strongly in the ultraviolet spectrum (250 nm) due to the phenolic group of tyrosine residues and indolic group of tryptophan residues of proteins. It has the advantages of being non-destructive and reagent-free method of protein estimation. Despite the spectral interferences by the UV-absorbing compounds (buffer, salts nucleic acids, nucleotides), the method has gained significant importance in protein identification and quantification particularly in the eluants from chromatographic columns.

**4.2.1.5 Infrared spectrophotometric method:** The infrared spectrophotometeric method can be used to determine the concentration of proteins in food samples. Proteins absorb infrared radiations naturally due to characteristic vibrations (stretching and bending) of certain chemical groups along the polypeptide backbone. Measurement of absorbance of radiation at 2.5-16 μm (i.e., 4000-625 $cm^{-1}$) can thus be used to quantify the protein in a food sample. It is particularly useful for rapid on-line analysis of protein contents. IR absorption can be used to determine protein contents in whole milk by measuring the intensity of the absorption band at the well defined wavelength of 1545 $cm^{-1}$, which corresponds to the absorption maximum of the peptide bond.

Irradiation of food products with near infrared (NIR) light (0.8-2.5 μm) followed by measurement of reflected radiations from the sample at different wavelengths produce a reflectance spectrum, which contains quantitative information on different ingredients present in the sample. NIR spectroscopy has become a popular method for the determination of protein in cereal products because of its speed, simplicity, safety non-requirement of wet chemicals and low running costs. There is however, the interferences shown by water, starch and oil, which are taken care of either by using double beam spectrophotometer or by using an adapted calibration procedure in which specific absorption bands for protein, starch, oil and water, are taken into account.

### *4.2.2 Determination of specific proteins*

In the preceding sections, the methods used to determine total concentration of protein in a food sample were discussed. These techniques however fail to characterize and quantify the actual type of protein in a food. The most recently used methods of specific protein detection and quantification come from a varied area, which includes chromatography, immunology, electrophoresis and a combination of all these methods (Table 3).

4.2.2.1 Chromatographic methods: Among the chromatographic techniques, HPLC finds wide application in the specific protein analysis of food products. It separates the specific protein from a mixture of proteins on the basis of varied affinity for the stationary phase in the column when the mixture is introduced into the mobile phase. The proteins thus separated can be quantified and identified by various spectrophotometric methods depending on the nature of protein being analysed. There are various versions of the HPLC that are used in the separation of specific proteins. However, the most frequently used principles are ion exchange HPLC, size exclusion (gel permeation) HPLC, and reverse phase HPLC. The columns for each of these chromatographic techniques are now commercially available. Now a days fast protein liquid chromatography (FPLC), capable of working with low back pressure has become more popular. It avoids denaturation of proteins caused by shearing forces at high pressure.

4.2.2.1.1 Ion exchange HPLC: This technique makes use of the net charge on a particular protein for separation, which depends on the pH of the medium in which protein is suspended. At isoelectric point (pI), protein is neutral. It acquires positive charge in a solution having pH below its pI and becomes negatively charged at a pH above its isoelectric point. Therefore, depending on the type of charge, a protein is bound to show affinity towards cationic and anionic exchangers of the columns and can be removed from a particular exchanger through selective desorption achieved by the change in pH or increasing the ionic strength of the eluting buffer. The application of ion exchange HPLC in the protein analysis of food has been found to be limited to separation and quantification (280 mm) of albumin and globulins of wheat flour [49].

4.2.2.1.2 Size exclusion HPLC: This technique of separation also known as gel permeation separates proteins on the basis of their molecular size. A protein solution is poured into a column packed with porous beads made of a cross linked polymeric material (dextran or agarose). Protein molecules larger than the pores size in the beads are excluded, and move quickly through the column, whereas the movement of protein molecules, which are smaller in size is retarded as they pass through the pores. Thus the protein molecules are eluted off the column in the order of their decreasing size. Beads of different average pore size are available for separating proteins of different molecular weights.

**Table 3:** Techniques used for the characterization and separation of various food proteins.

| Technique | Determination method | Protein component(s) analyzed |
|---|---|---|
| Chromato-graphic | Ion exchange | 1 g BSA, α-La, β-Lg (whey/280 nm); α-, β-, whole casein, albumin, globulin (wheat/280 nm) |
| | Size exclusion | Total casein (whole milk/210 nm); 1 g BSA, α-La, β-Lg (skimmed milk powder/220 nm); zein & glutenin (maize/210 nm); glycomacropeptide (skimmed milk powder & buttermilk powder/210) |
| | Hydrophobic interaction | α-gliadin, total gliadin (rice, maize, oat, rye, wheat/280 nm) |
| | Reverse phase | BSA, α-La, β-Lg (whey/glycomacropeptide (milk powder/210 nm); albumin, globulin; prolamin, glutenin (rice/210 nm); zein glutenin (maize/210 nm); gliadins, glutenin (wheat/210 nm); avenin (oat/210 nm) |
| Electrophoretic | SDS-PAGE (silver, CBB) | BSA, α-La, β-Lg (whey); hordeins (barley); zein (maize); albumin; globulin; gliadin (wheat); casein, whey proteins (processed milk) |
| | Native PAGE (silver, CBB) | α-, β-, κ-casein (whole milk); casein and whey proteins (processed milk); avenins (oat); zein (maize); gliadin (wheat) |
| | Two-dimensional electro-phoresis (CBB) | Zeins (maize) |
| | Isoelectric focusing (CBB) | Casein, whey proteins (processed milk, whey protein concen-trate); BSA, α-La, β-Lg (whey); α-, β-, kappa casein |
| | Capillary electrophoresis (uv absorption) | BSA, α-La, β-Lg (whey/214 nm) |
| Immunological | *Precipitation reactions* | |
| | Double diffusion (Ouchterlony Kaminski) | α-, β-, whole casein (meat products, sheep cheese) |
| | Radial immunodiffusion | Prolamins, glutelins (wheat, barley & malt); α-La, β-Lg (whole milk) |
| | Immuno electrodiffusion | α–casein (skimmed milk, animal foods); α-La, β-Lg (milk); α-La (meat product) |
| | *Antigen antibody labeling* | |
| | Immunofluorescence | Prolamins, glutelins (wheat, barley & malt) |
| | Enzyme-linked immuno-sorbent assay (peroxidase) | α–gliadin (bread dough); α-La, β-Lg (milk serum); hydrolysed milk protein (meat products, sausages) |
| | Radioimmunoassay | Prolamins, glutelins (wheat, barley & malt) |
| Combined | Affinity chromatography | Purification of various proteins |

*Source:* 48

When proteins are usually separated by size exclusion HPLC on a Zorbax GF-250 column and are detected in the far (220 nm) or near UV (280 nm) region [50]. Size exclusion chromatography has also been used for the separation of unreduced flour proteins into its glutenin, gliadin, albumin and globulin components by which proteins were detected by UV absorption at 210 nm [51].

4.2.2.1.3 Reverse phase and hydrophobic interaction HPLC: In reverse phase HPLC, separation is based on the difference in the surface hydrophobicity between protein molecules. Amino acid residues having hydrophobic groups get bound through hydrophobic interaction to a stationary matrix (usually $C_2$—$C_{18}$) and silica-based core material. The retention time of protein is controlled by increasing the concentration of organic solvents (ethanol, methanol, etc.) in solution, which results in altered polarity of the aqueous mobile phase. Protein separation by hydrophobic interaction chromatography is based on the same principle, but the elution of bound proteins is achieved by decreasing the ionic strength of the mobile phase. The gliadin and glutenin fractions of cereals proteins have been analysed by reverse phase HPLC, detection of which can be done by absorption in far UV region (210 nm).

Reverse phase HPLC with UV absorption (210 nm) as detector has been used in determining the adulteration of skim milk powder and butter milk powder [52]. Reverse phase HPLC has also been found to be useful in varietal identification of food crops, the selection of which is an important aspect for a farmer as it influences the returns from the crop he has sown [45].

4.2.2.2 Electrophoretic methods: Electrophoresis forms one of the powerful and most frequently used techniques of separation and analysis of specific protein component in food products. It relies on the difference in migration of charged molecules in a solution when subjected to an electrical field. The separation may take place on the basis of electric charge, size and shape of the protein. The separation is generally effected in a suitable support material (polyacrylamide, starch or agarose), which nullifies the influence of convection and diffusion during their migration. The separated proteins are detected by staining with Coomassie brilliant blue R-250 (CBB) or silver stain. There are sophisticated apparatus available for electrophoresis, which address the various practical hurdles associated with the technique.

4.2.2.2.1 SDS-PAGE and Native-PAGE: Denaturing polyacrylamide gel electrophoresis or SDS-PAGE is a useful analytical method for low solubility food proteins. Proteins are denatured prior to the analysis by mixing them with mercaptoethanol (breaks disulfide bond by reduction) and sodium dodecyl sulphate (SDS), which brings about unfolding of the proteins. After the reduction, protein molecule binds approximately the same amount of SDS per unit length, which results in similar charge per unit length and similar molecular conformation for all proteins. Hence SDS-PAGE separates proteins on the basis of molecular size. Bands of separated proteins on the gel are detected by treating the protein bearing gel with protein dyes such as Coomassie brilliant blue (CBB) or silver stain. Molecular weight of separated proteins is determined by SDS-PAGE, whereas in native PAGE, the characterization of soluble and non-denatured protein takes place on the basis of charge and molecular weight. The PAGE techniques have proved to be very useful for the characterization of gliadin and glutenin in cereal proteins. Caseins and whey proteins are easily separated on PAGE and SDS-PAGE gels using a discontinuous buffer system [53].

4.2.2.2.2 Isoelectricfocusing: Isoelectricfocusing is a modification of electrophoresis in which separation of protein is achieved by means of a pH gradient generated across the gel matrix using ampholytes. Migration of charged protein molecules takes place in response to the electric field until they reach the location where pH corresponds to their isoelectric point (pI) and becomes electrically neutral. This technique is known to exhibit highest resolution of all other methods used in protein separation [54].

Isoelectricfocusing has been found to be effective in separation of protein both in one and two-dimensional formats [54, 55]. The technique has also been used to separate and identify various genetic variants of casein and whey protein. Sometimes for a better resolution in the separation of proteins of food products, the isoelectricfocusing is followed by application of SDS-PAGE in perpendicular direction. Such type of strategy exemplifies two-dimensional version of electrophoresis.

4.2.2.2.3 Capillary electrophoresis: The advent of capillary electrophoresis has revolutionised the science of protein separation by reducing the time of separation and by making on-line detection

(UV absorption and fluorescence) possible. Use of narrow capillary tube (fused silica 20-200 μm x 10-100 cm) in place of polyacrylamide gel distinguishes it from conventional electrophoresis. Capillary electrophoresis has emerged as a method of choice for the separation of whey proteins (β-Lg, α-La and BSA) with UV absorption (214 mm) as detector. It has also been found to be effective in detection of adulteration of cow milk with goat and sheep milk by separation and detection of genetic variants of milk proteins [56].

4.2.2.3 Immunological method: Immunology based methods are highly specific methods of protein analysis, which are based on antibody and antigen interactions. There exists a plethora of immunological methods of identification and quantification of food proteins and are being used quite regularly for a variety of foods.

4.2.2.3.1 Single/double immunodiffusion: The immunochemical reaction between antigens and antibodies is allowed to take place in a suitable support material (usually agarose gel). Separation and quantification of different proteins of a sample can be done simultaneously. In single immunodiffusion, which is also known as radial immunodiffusion (RID), the antibodies are equally distributed in agarose gel and the sample to be analysed is filled into the wells on the same agarose plate and allowed to diffuse into the gel. The precipitation rings formed as result of antibody-antigen interaction can be visualized by staining. Diameter of the ring gives an idea about the amount of the specific protein present in the food sample. In case of double immunodiffusion, both the protein sample and serum (antibodies) are placed in separate wells in the gel and are allowed to diffuse through it. The precipitation band formed due to their interaction on gel can be detected by staining with Coomassie brilliant blue (CBB). Double immunodiffusion technique is often used to detect the cow milk casein in cheese of sheep.

4.2.2.3.2 Immunoelectrodiffusion: Immunoelectrodiffusion is a combination of immunochemistry and electrophoresis, which exhibits better sensitivity and selectivity over conventional immunodiffusion method. In this method the compatible antibodies for the protein of interest are equally distributed within the agarose gel, while the food sample (antigen) is placed in the wells of the gel. Antigens (food protein) are forced to migrate into the gel by diffusion and electrophoretic mobility. The height and surface of the precipitation peaks are used to detect and quantify the protein of interest in the food sample. α-La and β-Lg in milk and whey have been quantified by immunoelectrodiffusion [57]. The whey protein concentrate (WPC) contents in a commercialized meat product (frankfurters) are estimated by using α-La antiserum.

4.2.2.3.3 Enzyme linked immunosorbent assay (ELISA): ELISA is one of the most sensitive immunochemical methods of protein analysis. Protein antigen or antibodies are labelled with a specific enzyme capable of converting a colourless substrate into a coloured product, and one of these enzyme labelled immunoreactants are immobilized onto a support material and the other is put into the support afterwards. Immediately after the completion of the immunochemical reaction, the unbound components of sample are washed out and the bound material, which essentially happens to be the protein of interest, can be quantified spectrophotometrically by measuring the activity of the enzyme (usually peroxidase, glucose oxidase, alkaline phosphatase) in terms of the coloured product formed. Owing to its high sensitivity and simplicity of operation, ELISA has become highly popular for routine analysis of a large number of food samples. Now a days the specific ELISA test kits are available in market for the analysis of various protein components in food samples. It has been used in estimation of α-La and β-Lg in bovine serum and skim milk by using peroxidase enzyme as a tag and tetramethylbenzidine as a substrate [58, 59]. ELISA has also been applied for the determination of milk protein in meat products [60].

## 4.3 Analysis of Lipids

Lipids are defined as the components that are soluble in various organic solvents but insoluble in water. This group of substances therefore includes triacylglycerols (triglycerides), di-, monoacylglycerols, free fatty acids, phospholipids, sterols and carotenoids. However, triglycerides are the major components of most foods, making up more than 95-99% of the total lipids in the form of the fats and oils.

Lipids are one of the major constituents of foods and form an indispensable constituent of human diet for a number of reasons such as an energy source and as a determinant of physical characteristic (flavour, texture, mouthfeel and appearance) of the food. Nevertheless, overconsumption of certain lipid components (cholesterol, saturated fats) can be detrimental to health. Analysis of lipids in food involves the determination of total lipid concentration and the determination of lipid composition.

### *4.3.1 Determination of total lipid content*

It is important to accurately estimate the total lipids present in foods for health, economic, legal, quality and processing point of view. The analytical techniques based on various physicochemical properties e.g., solubility in organic solvents, immiscibility with water and spectroscopic properties have been developed for the measurement of total lipids present in a particular food. The choice of the method employed however, varies with the type of food and the purpose of the analysis.

**4.3.1.1 Solvent extraction methods:** These methods take advantage of solubility of lipids in organic solvent and their insolubility in water, and hence lipids can be separated from water soluble food components like proteins, carbohydrates and minerals. Solvent extraction methods are in fact the most commonly used methods of lipid isolation and quantification.

The solvent extraction methods require a few steps for its accomplishment viz., sample drying, particle size reduction, acid hydrolysis (for complex, lipids viz., lipoproteins, glycolipids), and solvent selection. Soxhelt method, a semi-continuous solvent extraction method, is one of the most commonly used method in which the prepared sample is taken into the extraction chamber to be completely submerged in solvent for about fifteen minutes. The extracted lipid and solvent are transferred back into boiling flask reservoir. This process is repeated until the removal of whole fat is complete. At the end of the process, solvent is evaporated and mass of lipid remaining behind is measured from the weight lost by the sample.

Super critical fluid extraction is another method of estimation of total fat in which solvent extraction is carried out by special instrument that uses compressed gas (commonly $CO_2$) as a solvent. Pressurized $CO_2$ when heated above certain critical temperature becomes a super critical fluid, and shares some of the features of a gas and a liquid. The instruments used in this method heat the food sample in a pressurized chamber and mix super critical $CO_2$, which results in extraction of the fats. The dissolved fat is then separated from the compressed $CO_2$ by dropping the pressure of the chamber. The precipitated lipid can be quantified as percent lipid by measuring the weight [61].

**4.3.1.2 Non-solvent liquid extraction methods:** Those liquid extraction methods that do not make use of organic solvent and instead use other chemicals to separate the lipids from the rest of the food are called non-solvent liquid extraction methods. These methods are very popular in determination of lipid content of milk and some other dairy products. One of the most commonly used non-solvent liquid extraction method is Babcock method in which a specified amount of milk is put into a specially designed flask (the Babcock bottle) followed by mixing with sulphuric acid. This treatment with acid generates heat, digests protein, breaks down fat globule membrane surrounding the droplet, thereby releasing the fat. The sample is centrifuged while it is still hot (50-60°C), which results in the rise of liquid fat up to the neck of graduated Babcock bottle. The graduated Babcock bottle gives the measure of the milk fat in terms of the percent weight.

Gerber method is a slightly modified version of Babcock method in which a mixture of sulphuric acid and isoamyl alcohol is used in a slightly different shaped bottle. The modifications incorporated make the analysis faster and prevents the charring of sugars during heating, which often hinders the proper reading from the graduated bottle. Use of surfactants in place of corrosive acid to displace the fat globules for separation is another merit of the Gerber method.

**4.3.1.3 Instrumental methods:** A whole range of instrumental methods are available for the determination of total lipid content in foods and food products. These methods use different physicochemical principles for the analysis of lipids. Some of the instrumental methods measure the various bulk physical properties of the food for the determination of total lipid content. The electrical conductivity and the density of food sample decreases with the increasing lipid content of food and thus has been used to estimate the total fat contents of foods. The change in the speed of ultrasonic waves when allowed to pass through a sample containing fat is another feature which can give an idea about the concentration of lipids in a food sample. This technique is capable of rapid, non-destructive on-line measurement of lipid contents.

Infrared (IR) absorption is another instrumental method of total lipid estimation in food products. It is based on the absorbance of IR energy at a wavelength of 5.73 μm due to the molecular vibrations or rotations associated with lipid molecules [61]. The energy absorbed at 5.73 μm has a direct proportionality with the fat content of the food sample. It is therefore, a useful method for rapid and on-line estimation of food lipids. A suitable standard curve is however, required. Near infrared spectroscopy has been successfully applied in determining the lipid content of cereals, oilseeds and meat products without disrupting the sample.

Nuclear magnetic resonance (NMR) spectroscopy is a routinely used instrumental method for the determination of total lipid content of foods. The measurement of the area under a peak in an NMR chemical shift spectra corresponds to the lipid fraction in the food sample. X-ray absorbance finds application in determining the fat content in meat and meat products. Lean meat is reported to absorb X-rays strongly than fat containing meat. This relationship has therefore been used in determining the lipid content in meat and meat products.

### *4.3.2 Determination of triglycerides*

It is not only the amount of lipid in a food, which matters but also the type of lipid present in a particular food that is equally important. As discussed earlier, lipids are an extremely diverse and complex group of compounds. The majority of the food lipids comprise of triglycerides; the other components however comprise a small fraction. The characterization of the lipid in food is essential from food quality, adulteration processing and even legal point of view. There are a number of methods available for the analysis of triglycerides in foods. Some of these are instrumental methods while others are chemical methods. Chemical methods are in fact crude in nature as they give information about the average properties of lipid components e.g., determination of iodine value gives a measure of average degree of saturation, saponification number gives an idea about the molecular weight of the triglycerides and the acid value is a measure of free acids. On the other hand, the instrumental methods give specific information about the lipid composition. Chromatographic methods are considered to be the most useful analytical procedures for separating and analyzing the various lipid components of food. The identification of these separated lipid components is effected by combining the separation techniques with a suitable detector (UV-visible, flame ionization, fluorescence).

**4.3.2.1 Gas-liquid chromatography:** Lipids are derivatized prior to analysis to increase their volatility, which is the pre-requisite of GLC analysis. Therefore, the triglycerols are saponified to break them down to glycerol and free fatty acids, which are then methylated. The fatty acid methyl esters thus formed are dissolved in a suitable organic solvent and injected into GC chamber, from where they enter into the separation chamber and are heated there. As they pass through the column, they are separated into different components revealed by peaks in chromatogram. The separated components are identified and quantified by using a suitable detector (flame ionisation detector or light scattering detector). Hence GLC help in characterization of the food lipids in terms of their fatty acid composition [62].

4.3.2.2 Thin layer chromatography: Thin layer chromatography (TLC) is mainly used to separate and quantify different types of lipids in foods i.e., triglycerides, diglycerides, monoglycerides, cholesterol and phospholipids. A thin layer chromatographic plate is coated with a suitable adsorbing material (silica gel, silver nitrates) which is placed in an appropriate solvent. A small amount of food sample to be analysed is spotted on a TLC plate. The capillary action forces the solvent to move up the plate and different lipid fractions of the sample are separated on the basis of their differential affinity for the adsorbent material. The separated fraction can be visualized by treating the plate with suitable dye. The distance covered by various fractions from the spot when compared with the standard separation profile of lipid gives an idea about the type of lipid components present. The separated spots can be further analyzed by using other sophisticated techniques like NMR and GC. Various oxygenated triglycerols are also separated by TLC. Triglycerols containing long chain fatty acids and acetic or isovaleric acids have been separated by TLC [63].

4.3.2.3 High performance liquid chromatography: HPLC has gained the reputation of being the most efficient method of lipid analysis (65% analysis). The molecules of the analyte are made to distribute according to their solubility between two non-miscible solvents, one mobile and other stationary. The fractions of triglycerides emerging from the HPLC system can be detected and quantified by various detectors such as differential refractometer, UV or IR spectrophotometer, mass or light scattering detector, flame ionization detector and mass spectrophotometer. The solvent system comprising of acetonitrile as a weak solvent is best suited mobile phase solvents. Triacylglycerol analysis by HPLC has also been performed by using solvent gradients. This procedure is suitable for a complex mixture of triacylglycerols exhibiting wide range of polarities e.g., butter oil triacylglycerols comprise a mixture of saturated triacylglycerol of high molecular weight. The triacylgycerols get resolved into different peaks and can be detected by a suitable non-destructive detector. By making use of HPLC, it was established that under isocratic condition the logarithm of elution volume of triglycerol is directly proportional to total number of carbon atoms and inversely proportional to the total number of double bonds in fatty acyl chains [64]. The elution behaviour of triacylglycerol depends upon the number of carbon atom and double bonds present in the fatty acid chain. Triglyceride in vegetable oils can be successfully separated by reversed phase HPLC and detected by differential refractometery. The elution order is determined by calculating equivalent carbon numbers [64].

4.3.2.4 Mass spectrophotometery: Mass spectrophotometry (MS) is an additional method for the analysis of triacylglycerols. It gives qualitative information about the triglycerides e.g., molecular weight, empirical formula and the structure of the triglycerides. It can also furnish quantitative information on the composition of triglycerol mixture. It studies the molecular ions produced by ionization. The reaction is brought about in the ion source of the instrument from the sample in the gaseous form by making use of electron impact procedure or chemical ionization. Recently the field desorption and fast atom bombardment techniques are being used for the ionization of molecules, which is a solid phase ionization technique utilizing the production of charged particles from the surface of condensed sample. In the early years of its application in lipid analysis, mass spectrometery was made to analyse the triglycerides by directly introducing the samples into heated ionization chamber [65]. Later on, the mass spectrometery was coupled to gas liquid chromatography (GLC) and high performance liquid chromatography (HPLC) through different interfaces. By means of HPLC-MS in a chemical ionization mode, the major triacylglycerol molecular species from a number of fats and oils of commercial interest were identified and conveniently separated [66, 67].

#### *4.3.3 Determination of non-glyceride lipids*

Although the non-glyceride lipids are not important sources of energy in foods, still there is a considerable interest in the role they play in human metabolism. Sterols are the important non-glyceride components of the lipid present in food.

4.3.3.1 Analysis of phytosterols: Accurate determination of phytosterols in edible oils or in foods containing vegetable lipids provides useful information about the presence of unexpected lipids in foods. Indeed, several sterols are specific to certain oils and hence can be used as markers for checking the purity of oils. For example, brassica sterol is exclusively present in rapeseed oil and its presence in sunflower oil proves an act of adulteration with rapeseed oil. Sterol fractions are isolated from unsaponified matter using TLC and HPLC, which need further separation and quantification. Capillary gas chromatography (CGC) is considered to be most suitable method for the separation of sterols. The analysis of sunflower oil by capillary gas chromatography revealed the presence of twelve types of sterols and were identified by means of mass spectrometry supplemented by standard curves [68]. Reversed phase liquid chromatography is another useful instrumental method for the analysis of sterols. It offers the opportunity to separate sterol fraction for further analysis by GC alone or by GC-MS.

4.3.3.2 Analysis of cholesterol: Analysis of the presence of cholesterol in foods is important for health reasons. High incidences of heart diseases are generally due to high levels of plasma cholesterol in human. Therefore all those factor which may be responsible for raising blood cholesterol levels need to be checked in diet. Food cholesterol is one such factor which is required to be monitored and hence underlines the importance of determination of amount of cholesterol in human diet. There exist a plethora of procedures for rapid and accurate determination of cholesterol in foods. However, colorimetric and enzymatic methods are preferred in routine analysis as the commercial kits for these methods are readily available in the market.

The colorimetric method of Lieberman-Burchard is a widely used method of cholesterol determination. The simplicity and cost effective nature makes it quite popular among food analysts. A green coloured complex is developed by cholesterol in the presence of acetic anhydride and concentrated sulphuric acid in dark, which can be estimated with absorption spectrophotometery at 620 nm [69].

The enzymatic method is preferred by more than 99% of analysts. The method is based on cholesterol oxidase catalysed oxidation of free cholesterol to Δ4-cholestenon and hydrogen peroxide. The hydrogen peroxide thus formed is made to oxidize a leukodye (triarylimidazole or aminophenazone-phenol) to a coloured dye in the presence of peroxidase enzyme. The colour developed has a direct proportionality to the concentration of cholesterol and is measured by absorption spectrophotometery.

Various chromatographic methods are also used for the estimation of cholesterol and other phytosterol in different foods. Gas chromatography on capillary column is perhaps the most widely used chromatographic method of cholesterol analysis [70]. Supercritical fluid chromatography (SFC) has also become equally popular instrumental method of cholesterol analysis due to its application in thermally labile compounds. Just like supercritical fluid extraction (SFE), supercritical fluid chromatography also makes use of $CO_2$ as a mobile phase, which is inert in nature and separation and extraction is done at low temperature. In many cases, both these methods are used in the extraction and characterization of cholesterol in foods and food products e.g., Ong *et al.* [71] applied SFE and SFC in the extraction and analysis of cholesterol of egg yolk samples using SE-52 fused silica capillary column with supercritical $CO_2$ as mobile phase. Detection was done by flame ionization detector. Capillary SFC coupled to mass spectrometery (MS) has also been used in the analysis of cholesterol [72].

## 4.4 Analysis of Vitamins

Vitamins are the organic compounds of low molecular weight that are required to be obtained from external sources in the diet and play an important role in normal physiological and metabolic activities of human body [73]. Our food must contain adequate quantity of different vitamins. The inadequate levels of some of the vitamins may lead to deficiency diseases like scurvy and pellagra. Additionally, vitamins have been found to play crucial role in health promotion by acting as antioxidants. Therefore, the basic objective of the vitamin analysis in foods is for nutrient labelling, regulatory compliances and change in vitamin contents during various steps of processing and packaging [74]. On the basis of solubility, vitamins are classified into two groups viz., water-soluble vitamins (B and C) and fat-soluble vitamins (A, D, E and K).

### *4.4.1 Analysis of water-soluble vitamins*

Vitamin B and C get solubilised in water due to which they are classified as water-soluble vitamins. The extraction of vitamins from the complex food is the first step of vitamin analysis for which the food sample is subjected to different treatments depending upon their stability. The commonly used extraction techniques are the acidic or alkaline hydrolysis ($B_1$, $B_3$, $B_6$ and pantothenic acid), enzymatic hydrolysis with papain, clarases, proteases, takadiastases, phosphatases, conjugases ($B_1$, pantothenic acid, biotin, $B_{12}$ and folacin), treatment with elevated temperatures ($B_1$, $B_{12}$ and biotin) and a combination of these treatments. There exists a variety of different analytical methods for the detection and determination of water-soluble vitamins, but HPLC based methods are the most commonly used methods in food samples. Of late, flow injection analysis (FIA) has also become quite popular.

**4.4.1.1 HPLC based methods:** Survey of scientific literature reveals the abundance of HPLC based techniques for the analysis of water-soluble vitamins [73]. The popularity of HPLC based methods may be attributed to the high speed of analysis, increased precision, reasonable accuracy, and increased specificity including simultaneous separation of multiple vitamins, vitavers and metabolites [75]. Non-volatile and hydrophilic nature of water-soluble vitamins makes them extremely suitable analytes for reversed phase HPLC [73]. HPLC based analytical methods for water-soluble vitamins in different food products such as juices, milk, meats, fortified break fast cereals, cheese, etc. have been thoroughly reviewed. UV absorbance is the most widely used vitamin detection method (vitamin C) however, fluorescence and electrochemical methods are also used particularly in case of B vitamins [76].

**4.4.1.2 Flow injection analysis (FIA):** As described earlier, it is an automated continuous flow strategy of chemical analysis in which a small well defined volume of sample is injected into a continuously flowing carrier stream to which appropriate auxiliary reagent stream can be added. Flow injection analysis (FIA) methods have been developed for the analysis of vitamin C in fruit juices, based on different types of chemistry exhibited by vitamin C. Analysis of vitamin C in fruit juices makes use of the chemiluminescence and immobilized enzyme (ascorbate oxidase) based detection system [77, 78]. Total thiamine content in chicken, pork, rice, bread, meat pie and fortified breakfast cereal has been quantified by using flow injection analysis with florescence based detection system [79]. Flow injection analysis methods are also reported for riboflavin analysis in various meat products with fluorescence based chemistry of detection [80].

In addition to the above mentioned methods, there are a number of other methods, which have not been widely used in food analysis e.g., gas chromatography–fragmentographic method has been developed for the estimation of pantothenic acid in rice, green tea and dried yeast. The method uses helium as the carrier gas and the same is coupled to mass fragmentography with multiple ion detectors [81]. Capillary electrophoresis is becoming a viable alternative to conventional and micro column HPLC for the analysis of water-soluble vitamins. It has been used in the estimation of vitamin C content of fruit juices with UV-absorbance dependent detector. Low solvent volume, rapid analysis, high resolution and the requirement of a small sample volume gives it an advantage over the conventional methods.

### *4.4.2 Analysis of fat-soluble vitamins*

Fat-soluble group of vitamins comprises vitamin A, D, E and K. It also includes those carotenoids, which act as precursors of vitamin A synthesis. In spite of the number of different techniques of vitamin analysis, the universally recognized standard methods for determining the fat-soluble vitamins in foods are lacking [74]. The chromatographic methods especially the HPLC based are the extensively used methods of fat-soluble vitamin analysis. The generalized protocol adopted for the HPLC based analysis of various fat-soluble vitamins in foods is illustrated in Fig. 5. Depending on the type of vitamin to be analyzed and the method of separation adopted, there are different detection systems available for fat-soluble vitamins. Generally three types of detectors viz., absorbance dependent, fluorescence dependent and electrochemical based detector are utilized for the on-line vitamin determination in the HPLC column effluent of food samples. The photodiode array detection, which permits simultaneous absorbance based detection at several wavelengths and continuous memorizing of spectra during the evolution of a peak, has proved invaluable in the assessment of peak purity and identity. Ball [82] has thoroughly reviewed the HPLC methods used in the determination fat soluble vitamins in a variety of food products. The normal phase or reversed phase HPLC are generally employed for the assay of fat-soluble vitamins, but the selection depends on the extraction/cleanup procedures adopted, type of vitamins to be determined, and the mode of system used (semi preparative or analytical separation).

4.4.2.1 Normal phase HPLC: The normal phase HPLC separates the vitamins by making use of adsorption in which the stationary phase consists of the polar surface of micro-particulate silica like adsorbent (nitrile and amino bonded) and in the mobile phase, generally hexane containing small fractions of more polar solvent (isopropanol) is used, which acts as a modifier. The difference in the polar functional groups enables the separation of solutes (fat soluble vitamins) in a normal phase HPLC. It is also effective in separation of geometrically distinct (cis-trans) molecules. Slow equilibration towards water is an operational disadvantage of the normal phase HPLC, which to some extent has been addressed by using isohydric mobile phases. Procedures have been devised for the determination of vitamin A and E in which total lipid fraction of food sample is extracted with non-aqueous solvent. An aliquot of the non-polar lipid extract containing fat-soluble vitamins is then injected into chromatographic column without further purification. The normal phase HPLC is used in determination of vitamin A compounds (retinyl palmitate, 13-cis-trans retinols), carotenes in fortified fluid milk, margarine, infant formulas and cheese using UV (313-328 nm) and visible (436-453 nm) absorbance detection systems [83, 84, 85]. Vitamin D ($D_2$, $D_3$ and 25-OH $D_3$) in infant formulas, fortified fluid milk, meat and fats from live stocks has also been analysed and quantified by using normal phase HPLC [86, 87], coupled with UV-absorbance detection system.

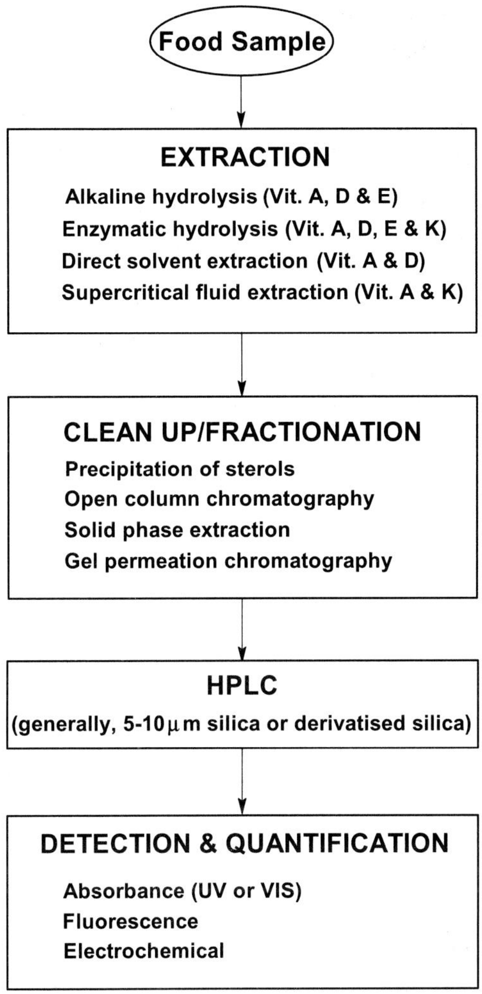


**Fig. 5:** Schematic HPLC protocol for analysis of fat-soluble vitamins.

Although, fluorescence-based detection is preferred with normal phase HPLC separation of vitamin E and its analogs from various food products [88-92], however UV-absorbance based detection is also reported in few cases [93,94]. Normal phase HPLC with UV-absorbance is also useful in the analysis of vitamin K (phylloquinone), particularly in infant formulas.

4.4.2.2 Reverse phase HPLC: In reverse phase HPLC systems, a bound octadecylsilane (ODS) stationary phase is employed in which octadecyl (C18) hydrocarbon acts as a functional group. Separation in reverse phase chromatography depends on the hydrophobicity of solutes, which shows a continuous increase with the number of carbon atoms in the molecule. Generally, reverse phase columns are operated with semi-aqueous mobile phases consisting of methanol or acetonitrile and water. Non-aqueous reverse phase (NARP) HPLC is usually preferred for the lipophilic vitamin K compounds and carotenes. A typical NARP eluant consists of a polar base (usually acetonitrile), a solvent of lower polarity (dichloromethane) to act as a solubiliser, which controls retention by adjusting the strength of solvent. Sometimes a small amount of third solvent with hydrogen bonding capacity (methanol) is used to optimize the selectivity. The non-aqueous mobile phase has an advantage over semi-aqueous mobile phase for increased solubility of less polar compounds, which results in their early elution and hence avoiding the risk of precipitation. It is essential to remove triglycerides from sample before injecting into the reverse phase HPLC due to their insolubility or sparing solubility in water, methanol and acetonitrile, which may allow the retention of some amount of triglycerides in columns. The presence of triglycerides in column impairs the efficiency and reproducibility of the method, and also distorts the peak shape. The reverse phase HPLC has a broad range of application in the determination of fat-soluble vitamins in a variety of foods and food products (Table 4).

4.4.2.3 Two-dimensional HPLC: Sometimes while determining the traces of naturally occurring vitamins D and K in foods, the food analysts fail to achieve the requisite separation on a single column. This usually happens due to the lack of the efficient clean up techniques adopted prior to chromatographic separation. It eventually results in the incomplete removal of lipoidal substances having a polarity similar to the vitamins to be separated. These lipoidal substances interfere in HPLC analysis. The solution to this problem lies in the performance of HPLC in two systems. In a two-dimensional HPLC, the use of normal phase and reverse phase columns in series provides better resolution and selectivity than obtained while using two similar columns. The first system (semi-preparative HPLC) is designed to isolate a fraction of sample extract that contains the analyte (vitamin) and the internal standard for the reference. The fraction can be collected by referring to the retention time of the internal standard.

The elution of vitamin K is detected by comparing it with UV-absorbing profile given by the standard [99]. The fraction containing the vitamin of interest and the internal standard is collected in a small tapered tube. A drop counter system of fraction collection is considered to be more effective than volume collector system. The solvent from the fraction is carefully evaporated using a stream of nitrogen, and the residue leaving behind is dissolved in a small volume of a suitable solvent. This solution is now subjected to the second (analytical) stage of HPLC, which gives better results than what are achieved on a single column. Analysis of vitamin D and K in various food samples by two-dimensional HPLC are given in Table 4.

4.4.2.4 Gas chromatography: Gas chromatography was widely used in the 1960s for the determination of fat-soluble vitamins (especially D and E). But thin layer chromatography and open column techniques were still required for preliminary separation of the vitamins, followed by derivatization to increase the volatility and thermal stability of these fat-soluble vitamins [74]. This perhaps curtailed the prospect of gas chromatography in routine analysis of fat-soluble vitamins in food. Recent development of fused-silica, open tubular capillary columns has once again revived the use of gas chromatography in the analysis of various fat-soluble vitamins specially, vitamin E [100, 101, 102]. Gas chromatography becomes an effective method of fat-soluble vitamins analysis when it is coupled with supercritical fluid extraction (SFE).

## 4.5 Analysis of Minerals

Minerals are naturally present in most of the food materials and they represent the inorganic components of the food. Minerals play an important role in structure and physiology of the body. Minerals like calcium and phosphorous are the chief constituents of the bones, which form the whole of our skeletal system [103]. Depending on their requirement in the body, minerals are classified as macro minerals (magnesium, sodium, potassium, phosphorous, calcium, chlorine and sulphur) and micro or trace minerals (selenium, fluorine, molybdenum, manganese, chromium, copper, zinc, iodine, iron, and silicon). Each one of them carries out a specific function in our body and deficiency of minerals

**Table 4:** Application of reversed-phase HPLC in the determination of fat-soluble vitamins in different foods.

| Vitamin(s) analysed | Food sample | Semipreparative HPLC | Quantitative HPLC | |
|---|---|---|---|---|
| | | | **Column** | **Detector** |
| Total vitamin A (retinol) | All foods | - | μ Bondapak $C_{18}$ 10 μm, 0.39×30 cm MeOH/$H_2O$ (90:10) | UV 325 nm [95] |
| Total vitamin A (as retinol), α- and β-carotene | Foods of animal origin, processed foods | - | μ Bondapak $C_{18}$ 10 μm, 0.39×30 cm MeCN/ MeOH/EtOAc (88:10:2) | UV 313 nm (r)* VIS 436 nm (c)* [96] |
| Vitamin $D_2$ or $D_3$ | Margarine, infant formulas | Supelcosil LC-Si 5 μm, 0.46×15 cm cyclohexane/hexane (50:50) + 2-PrOH (0.25%) | Radial-pak cartridge containing Resolve $C_{18}$, 5 μm Or Spherisorb ODS, 10 μm MeCN/MeOH (90:10) | UV 254 nm [97] |
| Vitamin E α-, (β+γ)-, δ-T | All food types | - | Zorbax ODS, 5 μm 0.8×25 cm MeCN/ $CH_2Cl_2$ containing 0.001% triethylamine/MeOH, 700+300+50 | Fluorescence ex. 290 nm em. 330 nm [92] |
| Phylloquinine | Milk | μ Porasil Silica 10 μm, 0.8×30 cm heptane/EtOAc (99:1) | Resolve $C_{18}$, 5 μm 0.39×15 cm 100% MeCN | UV 254 nm (sph) UV 270 nm (qh) [98] |

*: (c) = carotenoids, (r) = retinol; MeOH (methanol); EtOH (ethanol); EtOAc (ethyl acetate); MeCN (acetonitrile); $CH_2Cl_2$ (dichloromethane); PrOH (isopropanol); ex. (excitation); em. (emission); sph (semipreparative HPLC); qh (quatitative HPLC).

also leads to various deficiency diseases. Given the importance of mineral in our diet for the proper functioning of body, some foodstuffs like breakfast cereals are often supplemented with iron, calcium and zinc.

In South Asia and North America, table salt is fortified with iodine to control goitre, an iodine deficiency disease prevalent in these regions. Processing of food materials sometimes leads to the loss of invaluable minerals and the resulting product is rendered to be mineral deficient. Therefore the rapid, precise and simple methods of mineral analysis are always in demand for the nutritional evaluation, labelling, quality control, and for ensuring the proper processing of food material.

#### *4.5.1 Determination of total mineral content*

The total mineral content of a food material is the inorganic residue remaining after the water and the organic matter have been removed by heating in the presence of oxidizing agents [104]. It is also known as the ash content. Total mineral content of a food material is determined by various ashing (heating) processes, which are based on the fact that minerals are not destroyed by heating and have a low volatility as compared to other food components. Three main types of methods used to determine the total mineral content of food are dry ashing, wet ashing and low temperature plasma ashing.

Dry ashing method uses high temperature muffle furnace capable of maintaining temperatures between 500 and 600°C in which the food samples are incinerated in crucibles. Water and other volatile materials get vaporized; organic substances are burnt to $CO_2$, $H_2O$ and $N_2$ in the presence of oxygen. Minerals get converted into oxides, sulphates and phosphates. The food sample is weighed before and after ashing to determine the amount of mineral present, which is expressed on dry or wet weight basis. Some of the volatile minerals like Cu, Fe, Pb, Hg, Ni and Zn are partially lost during dry ashing. Microwave heating instruments are now used in place of furnaces in dry ashing of the foods. Wet ashing utilizes strong acids and oxidizing agents along with heating to remove (oxidize) organic material present as matrix in foods. Heating (350°C) continues until the organic matter is completely digested leaving only the mineral oxides in solution, which can be analyzed further for different minerals.

In low temperature or plasma ashing, a food sample is placed into a glass chamber evacuated with a vacuum pump. The food samples are oxidized by nascent oxygen formed by an electromagnetic radio frequency field. Relatively low temperature (150°C) used in plasma ashing causes less loss of volatile minerals than other ashing procedures. The weight of the residue left after getting rid of the organic matter gives the measure of the total mineral content of a food sample.

### *4.5.2 Analysis of specific minerals*

The methods discussed so far are unable to differentiate the actual mineral elements present in a food sample. It is, however, possible to determine the specific minerals of a food by evaluating various physico-chemical characteristics of minerals such as volatility, specific reactivities and unique electromagnetic spectra. Although there exist a number of classical methods of mineral analysis like gravimetric, colorimetric and titration-based (redox reactions) methods, the spectroscopic methods being sensitive, specific and exceptionally quick are considered to be the most effective methods of mineral analysis.

**4.5.2.1 Atomic absorption spectroscopy (AAS):** The basic principle of atomic absorption spectroscopy is the absorption of UV-visible radiations by free atoms (minerals) in gaseous state. A thoroughly ashed food sample is dissolved in an aqueous solution, which is atomized into flames by heating in a special instrument. The atomized sample is subjected to a beam of UV-visible radiation and the absorption of these radiations (of specific wavelength) by atomized minerals is related to the concentration of a particular element in the food sample by a photomultiplier tube. The atomic absorption spectroscopy (AAS) became popular in the 1960s with an ability to measure the trace amounts of mineral elements in biological samples [105].

**4.5.2.2 Atomic emission spectroscopy (AES):** Atomic emission spectroscopy differs from atomic absorption spectroscopy in a way that it measures the radiations emitted by a sample rather than the absorption of the radiation. Food sample is heated to higher temperatures so that the mineral atoms are excited but not ionised. The atomic emissions are produced when the electrons in an excited atom falls back to lower energy levels. Each mineral atom has a specific emission spectrum, which can differentiate the various minerals present in a food sample. Flame and inducively coupled plasma (ICP) are the commonly used atomisation-excitation sources in the analysis of food minerals.

**4.5.2.3 Ion selective electrodes (ISE):** The on-line analysis of the mineral composition of foods has become possible with ion-selective electrodes (ISE). These devices work on the same principle as the pH meters work, except their sensitivity to a specific type of ion rather than $H^+$ ion. Special glass electrodes are commercially available for the direct measurement of various anions and cations such as sodium, calcium, potassium, bromides, chlorides and sulphides [103]. The concentration of a specific mineral is determined from a calibration curve plotted between voltage and log of ion concentration. Ion selective electrodes have been used for determining the salt concentration of butter, cheese and meat. It also finds application in determining the concentration of calcium in milk and $CO_2$ in soft drinks.

**Table 5:** Some important food-borne microbial pathogens.

| Pathogen | Characteristics | Food commonly contaminated | Reference |
|---|---|---|---|
| Bacterial | | | |
| *E. coli* O157:$H_2$ | Gram (–) | Dairy products, ground beef, unpasteurised cider, lettuce, radish sprouts, salami. | 106 |
| *Campylobacter* | Gram (–) | Poultry, eggs | 107 |
| *Listeria monocytogenes* | Gram (+) | Sauerkraut, coleslaw, soft cheese, meats | 108 |
| *Salmonella* (non-typhoid) | Gram (–) | Poultry, raw meat, eggs, milk | 109 |
| *Clostridium perfringes* | Gram (+) | Pork and protein rich foods | 110 |
| *Staphylococcus* | Gram (+) | Sea foods, poultry, milk | 111 |
| *Yersinia enterocolitica* | Gram (–) | Raw milk, pork | 110 |
| Fungal | | | |
| *Aspergillus* | Ascomycetes | Groundnut, corn, fig | 112 |
| *Fusarium* | Deuteromycetes | Cereals, corn, wheat, barley | 113 |
| *Penicillium* | Ascomycetes | Apple juice, cereals, corns | 113 |
| Viral | | | |
| Hepatitis | A & E | Raw shellfish, raw oyster, fruits & vegetables, contaminated water | 113 |
| Norwalk | - | Variety of foods & water | 113 |
| Avian influenza | H5N1 | Raw & undercooked poultry products | 114 |
| Protozoans | | | |
| *Cryptosporidium* | Sporozoa | Raw sausage & milk, offal | 115 |
| *Cyclospora* | Sporozoa | Contaminated raspberries, mesculun lettuce | 116 |
| *Sarcocytis* | Coccidian parasite | Raw & undercooked meat | 117 |

## 5. ANALYSIS OF CONTAMINANTS

Detection and determination of harmful contaminants and residues in foods is another important aspect of food analysis. A healthy food cannot be defined merely by its nutritional status. It has to be free from any life threatening contaminants and thus should be safe for consumption. Three main categories of food safety concerns in food industry comprise microbiological, chemical and physical hazards. The microbiological contaminants are the food-borne pathogens; chemical contaminants include pesticides, herbicides, antibiotics, heavy metals, etc., and physical contaminants represents foreign object in foods likely to result in injury or illness when consumed with food. It is quite impossible to address the analysis of all types of food contaminants simultaneously in this section, however the methods used in the detection of microbial contaminants i.e., food-borne pathogens will be discussed in detail.

### 5.1 Detection of Microbial Contaminants

Microbial contaminants of food refer to food-borne pathogens that cause illness in human when the contaminated food is ingested. The microbial contaminants not only pose a serious threat to consumers, they are also equally troublesome to food industry as they may result in food recalls, which often proves expensive. Despite various efforts made to prevent food contamination, it is impossible to completely eliminate the microbes from food supply. The ubiquitous distribution of microbes (pathogens), lapses in hygiene or sanitation and the emergence of new pathogens are considered to be the prime causes of increasing reports of the outbreaks of food-borne diseases. Amongst the food-borne pathogens, bacteria dominate the scenario, although the viral and fungal contaminants have also created havoc in past. Some of the common food-borne pathogens are listed in Table 5. The access to efficient tools and techniques that allow rapid detection and identification of food-borne pathogens is essential for both food industry and the public health official so that any incidence of the outbreak of food-borne illness is avoided. In the present section, the most commonly used techniques of pathogen detection are being presented.

#### *5.1.1 Culture-based methods*

In cultural-based methods, a food sample is put into suspension by grinding it in a Stomacher™ or similar apparatus. A serially diluted suspension is then plated onto an agar medium incubated for a suitable period for growth. The microbial colonies thus appearing on the agar plate are subjected to enrichment and selection medium for enumeration and detection. Commercial selection media are available in market for the detection of various food-borne pathogens. Culture-dependent methods of microbial detection are sensitive, inexpensive and capable of providing qualitative and as well as quantitative information. But the time consuming and labour intensive nature of these methods restrict their routine use in food industry [118]. The use of chromogenic substrates in commercial selection media, which change colour in the presence of an particular enzyme to be found in a particular strain of the pathogen, has made culture-based methods more easy, rapid and specific for the identification of pathogenic strains like *Listeria* [119].

#### *5.1.2 Nucleic acid-based methods*

These methods detect DNA or RNA of the pathogen in question. These methods are broadly divided into amplification-based and direct probe-based methods. The basic difference between these two categories is that in the former, a particular target DNA or RNA sequence is enzymatically amplified (PCR-based), whereas in the latter, the target gene or nucleotide sequence is directly detected by hybridization (base pairing principle). However in some of the newer methods, both of these principles (i.e., amplification and hybridization) are integrated.

5.1.2.1 Conventional PCR methods: Polymerase chain reaction (PCR) has become an extremely successful amplification-based method for detection of pathogenic microbes in a variety of samples, including food. It involves extraction of DNA of the potential pathogen present in food sample and the binding of oligonucleotide primers to a specific target gene. The amplification of the specific gene of the pathogen is brought about by subjecting the mixture of target gene, primer, four deoxyribonucleotides and a DNA polymerase enzyme in suitable buffer to a thermocycler, which is capable of changing temperature from 94°C to 50-60°C and then to 72°C in a cyclic manner. This cyclic change in temperature is conducive for denaturation, annealing and extension of the target DNA in the presence of a thermostable DNA polymerase. In this way, the genetic fragment of pathogen gets amplified exponentially as the thermocycler runs through a series of these temperatures.

The presence of a particular pathogen in a food sample is detected by analyzing the band of the amplified gene of pathogen on an agarose gel, which is visualized after staining with ethidium bromide under UV light. PCR has been found to be a better diagnostic method than culture dependent method for the detection of food-borne pathogens like *Listeria monocytogenes* in terms of sensitivity and rapidity [120]. It is not easy to detect pathogen in food if the level of contamination is quite low (early stage of contamination), but PCR capable of amplifying the gene of the pathogen makes it possible. PCR-based diagnostic kits for the detection of *Salmonella* sp., *Listeria* sp., and *E. coli* O157:$H_7$ are now commercially available in the market. However, a few limitations such as inability to distinguish between live and dead microorganism (pathogen), inability to quantify the level of contamination and inability to detect PCR products during the course of reaction (real time detection), demand for some changes in conventional PCR method. These limitations have been addressed by incorporating various modifications in the conventional PCR methods (Table 6).

5.1.2.2 DNA chips and micro-arrays: DNA chips and micro-arrays are the intrinsically miniaturized formats of the conventional tests of nucleic acid hybridization. Micro-arrays comprise of a number of discretely located DNA probes immobilized on a solid surface such as glass. Each probe corresponds to an oligonucleotide specific to a target DNA sequence of a pathogen [122]. The DNA sample of pathogen is tagged with a radioactive or fluorescent oligonucleotide by using labelled deoxyribonucleotides during the amplification. The amplified fluorescently labelled DNA fragments of the food-borne pathogen when exposed to micro-arrays or DNA chips, bind only to those probes for which they have a complementary sequence. The washing is done to remove the unbound DNA, which helps to keep only the bound DNA of the pathogen on probes. Probes bearing the labelled bound DNA fragments fluoresce and the pattern of fluorescing spots is used to identify the pathogen. Micro-arrays are comparatively cheaper, and are capable of identifying a number of pathogens or serotypes at once, but require genetic amplification by PCR or culture enrichment.

### *5.1.3 Pathogen sub-typing and verification*

Once the food sample is suspected of harbouring a pathogen, various molecular methods such as amplified fragment length polymorphism (AFLP), restriction fragment length polymorphism (RFLP), randomly amplified polymorphic DNA (RAPD), ribotyping, pulsed field gel electrophoresis (PFGE) and fluorescence *in situ* hybridisation (FISH) are used for further verification of the pathogens. However, these methods are not used routinely in food analysis; they are employed only for the further characterization of the pathogen in order to know the sub-type and serotype of the microbial contaminant, which has epidemiological importance. Each isolate however, requires culturing before

**Table 6:** Modified versions of PCR-based methods of detection and their advantages.

| Method | Modifications incorporated | Detection | Advantages |
|---|---|---|---|
| Real-time PCR | Use of fluorescent oligonucleotide probe having reporter and quencher dye at 5′ and 3′ end. | Fluorescence on cleavage of 5′end | Real-time detection of PCR product i.e., pathogens |
| Competitive PCR | Addition of a small competitor fragment of DNA matching the target gene | Comparison of intensity of competitor & pathogenic DNA bands on gel | Quantification of contamination levels |
| FRET-PCR | Use of two types of DNA probe one with fluorescent label and other with quencher at the end of PCR | Fluorescence by free fluorogens | Estimation of the number of pathogens without running agarose gel |
| Multiplex PCR (microsphere sorting) | Use of multiple sets of primers together for different pathogens; primers are labeled with biotin and have pathogen specific oligonucleotides with different spectral labels. At the end, streptovirid-in-R-phyco-erythrin is added. | Fluorescence by spectrally sorted pathogenic DNA | Simultaneous quantification of a number of pathogens |
| RT-PCR | Use of reverse transcriptase to create complementary DNA (cDNA) | Same as the conventional PCR | Detection of live pathogens |

FRET= fluorescence resonance energy transfer

*Source:* 121

the analysis for sub-typing. These methods use difference in the genetic profile of each pathogen when the genetic material (DNA) is subjected to enzymatic fragmentation (restriction endonucleases). Difference in the electrophoretic pattern of the DNA fragments or the automatic analysis by laser fluorescence assay elucidates the different serotypes of the pathogen of the same group. The comparative account of these methods along with their application in food analysis is described in detail by Churchill *et al.* [121].

### *5.1.4 Immunology-based methods*

Immunology-based methods manipulate the immune response and its products (antibodies) in detection and analysis of various food contaminants and components. Antibodies are the proteins produced by B-cells of mammalian immune system in response to the presence of foreign proteins or other compound to which they have an ability of specific binding. There are five classes of antibodies: IgA, IgD, IgG, IgM and IgE. But IgG is the most commonly used in immunodiagnostics. It is a Y shaped protein that has a constant region (in stalk of Y) and a variable region (in forked top of Y). In fact, it is the variable region of the antibodies, which bestows the specificity on them towards the different foreign molecules. Antigens (usually foreign proteins) have numerous epitopes, representing different regions of antibody specificity. Immunology-based methods capitalize on the specific reaction between antibodies and antigens for the detection of various pathogens in foods. Antibodies therefore, act as molecular tools in immuno-diagnostics.

Apart from conventional heavy chain antibodies (IgG), there are a number of other antibodies such as polyclonal, monoclonal and recombinant antibodies, which can be successfully produced and applied in immuno-detections. Polyclonal antibodies are raised by immunizing an animal host with a specific antigen several times and then harvesting the serum from the animal. The serum will contain a mixture of antibodies some of which were already present prior to immunization. The raising of the monoclonal antibodies also begins with immunization of an animal with a specific antigen. Once immune response is on the peak, the B-cells producing antibodies are isolated from spleen of immunized host animal and fused with myeloma cancer cells. The resultant cell is called hybridoma cell. It possesses the immortality of a cancer cell and antibody producing capability of B-cell. Unlike the antibodies obtained from the serum, antibodies produced by hybridoma cells are monoclonal, which are highly specific to a particular antigen. The production of monoclonal antibodies is however, a costly proposition.

The bacterial expression systems are capable of producing a variety of antibodies in a short period of time than hybridoma cells. The production of recombinant antibodies by bacterial expression system is more economical. It also gets rid of the requirement of specialized growth apparatus needed by hybridoma cells [123]. A considerable reduction in the size of antibodies has been achieved by separate expression of different fragments of antibodies followed by recombination, and also by the removal of constant region of the antibodies. These recombinant antibodies are effective in detection of a wide range of pathogens such as *Streptococcus* [124], *Bacillus* [125], *Staphylococcus aureus* [126] and *Listeria monocytogenes*. The most commonly used immunological methods of detection of food-borne pathogens are being discussed in the sections that follow.

5.1.4.1 Latex agglutination assay: It represents the simplest format of the immunology-based assay that has been used for a long time in detection of food-borne pathogens. This method makes use of the antibodies bound to latex beads, which are specific to the antigen of interest (pathogen). An appropriate solution containing these beads is added to the food sample where agglutination takes place due to the network created by antigens while binding to more than one latex-bound antibody. The agglutination is readily visible as a suspension of particles [108]. *Listeria monocytogenes* has been detected by using listeriolysin (LLO) specific antibodies in latex beads in various foods and pure cultures [127]. This is a simplest method in which the expensive apparatus and reagents are not required.

5.1.4.2 Enzyme-linked immunosorbent assay: Enzyme-linked immunosorbent assay (ELISA) is a more sensitive (between $10^3$ and $10^5$ cfu/ml of sample) and quantitative immunoassay in which antibodies labelled with an enzyme capable of converting colourless substrate into coloured product are used. The colour developed as a result of enzyme activity helps in detection and estimation of the pathogen of interest. Just like the PCR, there are two common strategies possible with ELISA – one is called direct ELISA and another is known as competitive ELISA. In direct ELISA, the pretreated food sample is coated on the bottom of the wells in a multi-well plate. The plate is blocked and washed, and then antibodies specific to the antigen of pathogen are added into the wells. The suitable incubation period facilitates the binding of antibodies to the specific antigen immobilized on the wells. The unbound antibodies are removed by washing. The secondary antibodies labelled with an enzyme and having specificity to the antigen bound primary antibody when added, change a colourless substrate into a coloured product. The intensity of colour developed is proportional to level of contamination present in the food sample [128]. There exists a variation of direct ELISA called sandwich ELISA in which the antigen specific antibodies are coated onto a well and the food sample is added followed by enzyme labelled antibody and the rest of procedure remains unchanged.

In competitive ELISA, a known amount of antigen is coated on the well. The sample to be detected and analyzed is first incubated separately with primary antibody, and then added to the well. Enzyme labelled secondary antibody is added to the plate followed by washing. The detection procedure however, remains the same as for the direct ELISA. The presence of free antigen in the sample results in the reduction of signal produced by labelled antibodies in the wells, as a large number of the enzyme labelled antibodies, which would have contributed in the formation of coloured product, are washed away along with the antigen of the food. The inhibition in the colour produced is compared to the inhibition curve obtained with known amount of antigen, which ultimately gives a quantitative value of the pathogen present in a sample.

5.1.4.3 Fluorescently labelled immunoassay: The basic working principle of fluorescently labelled immunoassay is similar to ELISA except that the label of antibody is a fluorescent dye instead of an enzyme. It is more sensitive and rapid than ELISA as it bypasses the colorimetric detection at the end, which is necessary in ELISA. The detection of *Listeria monocytogenes* has been successfully done by using fluorescently labelled immunoassay with 95% accuracy as confirmed by selective media culturing [129]. Another version of fluorescently labelled immunoassay makes use of microsphere having different spectral labels and coated with antibodies exhibiting specificity each to *Salmonella typhimurium, C. jejuni, Listeria monocytogenes* and *E. coli* O157: H7, respectively. Enrichment and amplification is required in this procedure. Fluorescent bead sorter using spectral label is used for sorting the microsphere, and the addition of fluorophore labelled antibody helps in the detection of the food-borne pathogen in question [130].

5.1.4.4 Surface plasmon resonance: Surface plasmon resonance (SPR) is based on an optical phenomenon to measure the binding kinetics between the molecules. It is capable of detecting and measuring the real time kinetics of interaction between unmodified proteins [131]. The kinetics and the level of interaction is measured by an instrument called BIAcore (biomolecular interaction assay). The procedure of detection involves the passage of sample buffer over a disposable chip coated with pathogen specific antibodies. The sensor in the instrument measures the change in the angle of refraction of light over the antibody coated chip in response to the changed mass resulting due to binding of the antigen of interest from the food sample. The signal i.e., the change in angle of refraction is displayed in real time as optical response, which is measured in resonance units (RU). It is plotted against time and is proportional to the amount of antigen bound to the chip [132, 133]. Although SPR is a useful method for the detection of pathogens and their toxins in a food sample, the exorbitant instrumentation cost, tedious and time consuming optimization of the sample purification restricts its routine use in food analysis.

### *5.1.5 Bioluminescence-based methods*

Bioluminescence-based assays hold a great potential in the area of food safety assurance. The numbers of diagnostics that allow the measurement of adenosine triphosphate (ATP), as an indicator of microbial load are plenty. ATP approach of measuring microbial biomass is possible because all organisms including microbes use ATP as energy currency. Hence, the presence of ATP on inert materials like food stuffs, indicates the presence of microbe. Though it gives no information about their type, but definitely reflects the standard of hygiene. The presence of ATP is detected by an enzyme luciferase, which results in light emission through the following reaction:

$$\text{Luciferin} + \text{ATP} + \text{Mg}^{2+} \xrightarrow{\textit{luciferase}} \text{Oxyluciferin} + \text{AMP} + \text{CO}_2 + \text{light}$$

Use of recombinant phages having bacterial luciferase gene is another version of the bioluminescence-based assays. The specificity of phages to a particular class of bacteria provides strength to this approach. Presence of the phage specific bacterial pathogen (*E. coli, Salmonella*) in a food sample will result in the transfer of luciferase gene to the food-borne bacterial pathogen and subsequently its expression. This will result in the emission of detectable light as depicted in the above mentioned equation.

### *5.1.6 Biosensor-based assays*

A biosensor may be defined as an analytical device, which employs a biological material to specifically interact with analyte. This interaction produces some detectable physical changes, which are converted into electrical signal by a transducer. There are a number of formats of biosensor available, but one of the most commonly used biosensor for the detection of food-borne pathogens is the affinity-based biosensor. It relies on the specific recognition between the sensor and the target.

Antibodies are more frequently used than nucleic acids in affinity-based biosensors, which have been developed for major food-borne pathogens such as *Salmonella, E. coli* O157:H7 and *Listeria monocytogenes* [108]. One of the antibody-based biosensor makes use of antibodies immobilized on piezoelectric crystals, which detect the change in mass due to binding of the pathogen of interest to antibody-crystal complex. The change in the mass of the crystal results in oscillation of the piezoelectric crystal and produces electrical potential. The electrical signal thus produced confirms the presence of a particular food-borne pathogen in a sample.

Enzyme-linked antibodies can also be used in biosensors. Pathogen-specific antibodies are immobilized in an electrode and made to bind the pathogen in question. Enzyme-labelled secondary antibodies added get attached to the pathogen, already bound to primary antibodies on the electrode. Another enzyme (glucose oxidase) is also immobilized on the same electrode. Shifting the electrode bearing primary antibody, antigen (pathogen), enzyme-linked secondary antibody and glucose oxidase to a solution containing amino salicylic acid, produces hydrogen peroxide by facilitating reaction between glucose and oxygen. The horseradish peroxidase on secondary antibody catalyses the reaction between hydrogen peroxide and amino salicylic acid to form 5–amino salicylic acid quinoeimine, which is reduced by an electron supplied by the electrode. The flow of electron is detected by electronic instrument and confirms the presence of a particular pathogen like *Salmonella aureus* in a food sample. Biosensors capable of detecting the changes in optical properties like refractive index as a result of antibody-antigen interaction and nucleic acid hybridisation have also been successfully used in the detection of food-borne pathogens.

## 5.2 Detection of Non-Pathogenic Contaminants

Non-pathogenic contaminants of food include a whole range of physical and chemical contaminants such as pesticides, weedicides, heavy metals, growth regulators, adulterants, antibiotics and colorants, etc., which are either added deliberately (adulterants) or for the protection and preservation of food materials and persist as residues in food articles. Sometimes, these enter into food via water and during processing of foods. The analysis and detection of these non-pathogenic contaminants is discussed elsewhere [134].

## REFERENCES


1. Kramer A & Twigg BA (1970) *Fundamentals of Quality Control for the Food Industry,* AVI Publishing Co, Westport, USA.
2. Joshi VK & Bhushan S (2002) Sensory evaluation of fruits, vegetables and their products. In: *Postharvest Technology of Fruits & Vegetables,* Verma LR & Joshi VK (eds), Indus Publishing Co, New Delhi, India, pp 286-336.
3. Szezesniak AS (2002) Texture is a sensory property. *Food Qual Pref* 13: 215-225.
4. Daillant-Spinnler DHB, Macfie HJH & Beyts PK (1996) Relationships between perceived sensory properties and preference directions of 12 varieties of apple from the southern hemisphere. *Food Qual Pref* 7: 113-126.
5. Jaeger SR, Andani Z, Wekeling IN & Macfie HJH (1998) Consumer preferences for the fresh and aged apples: A cross cultural comparison. *Food Qual Pref* 9: 355-366.
6. Bourne MC (2002) *Food Texture and Viscosity: Concept and Measurement,* Academic Press, San Diego, USA.
7. Mehinagic E, Royer G, Symoneaux R & Jourjon F (2006) Relationship between apple sensory attribute and instrumental parameters of texture. *J Fruit Ornam Plant Res* 14: 25-37.
8. Epstein J, Morris CF & Huber KC (2002) Instrumental texture of white salted noodles prepared from recombinant inbred lines of wheat differing in the three granule bound starch synthetase (waxy) genes. *J Cereal Sci* 35: 51-63.
9. Dunneewind B, Janssen AM, van Vliet T & Weenen H (2004) Relative importance of cohesion and adhesion for sensory stickiness of semisolid foods. *J Texture Stud* 35: 603-620.
10. Jha SN & Matsuoka T (2000) Non-destructive techniques for quality evaluation of intact fruits and vegetables. *Food Sci Technol Res* 6: 248-251.
11. Lu R & Ariana D (2002) A near-infrared sensing technique for measuring internal quality of apple fruit. *Appl Eng Agri* 18: 585-590.
12. Lu R, Guyer DE & Beaudry RM (2000) Determination of firmness and sugar content of apples using near-infrared diffuse reflectance. *J Texture Stud* 31: 615-630.
13. McGlone VA, Abe H & Kawano S (1997) Kiwifruit firmness by near-infrared light scattering. *J NearInfrared Spectr* 5: 83-89.
14. Lu R & Peng Y (2005) Comparison of multi spectral scattering and visible/NIR spectroscopy for predicting apple fruit firmness. In: *7th Fruit, Nut, and Vegetable Production Engineering International Symposium,* September 12-16, Montpellier, France, Paper No. 62, pp 493-501.
15. Thybo AK, Bechmann IE, Martens M & Engelsen SB (2000) Prediction of sensory texture of cooked potatoes using uniaxial compression, near infrared spectroscopy and low field H NMR spectroscopy. *Food Sci Technol* 33: 103-111.

16. Farrar TC & Becker ED (1971) *Pulse and Fourier Transform NMR: Introduction to Theory and Methods*, Academic Press, New York, USA.
17. Mousseri J, Steinberg MP, Nelson AI & Wei LS (1974) Bound water capacity of corn strand and its derivatives by NMR. *J Food Sci* 39: 114-116.
18. Miller BS, Lee MS, Huges JW & Pomer UZ (1980) Measuring high moisture content of cereal grain by pulsed nuclear magnetic resonance. *Cereal Chem* 57: 126-129.
19. Faust M, Wang PC & Maas J (1997) The use of resonance imaging in plant sciences, In: *Horticultural Review*, Janicke J (ed), Wiley, Chichester, UK, pp 226-266.
20. Barreiro P, Ruiz-Cabello J, Fernandez-Valle ME, Ortiz C & Ruiz Altisent M (1999) Mealiness assessment in apples using MRI techniques. *Mag Reson Imag* 17: 275-281.
21. Miquel ME, Evans SD & Hall LD (1998) Three-dimensional imaging of chocolate confectionery by magnetic resonance methods. *Food Sci Technol* 31: 339-343.
22. Lu R & Abbott JA (1996) A transient method for determining dynamic viscoelastic properties of solid foods. *Trans ASAE* 39: 1461-1467.
23. Upchurch BL, Miles GE, Stroshine RL, Furgason ES & Emerson FH (1987) Ultrasonic measurement for detecting apple bruises. *Trans ASAE* 30: 803-809.
24. De Ketelaere B & De Baerdemaeker J (2001) Tomato firmness estimation using vibration measurements. *Math Comp Simul* 56: 385-394.
25. Mizrach A, Bechar A, Grinshpon Y, Hofman A, Egozi H & Rosenfeld L (2003) Ultrasonic mealiness classification of apples. *Trans ASAE* 46: 397-400.
26. Bechar A, Mizrach A, Barreiro P & Landahl S (2005) Determination of mealiness in apples using ultrasonic measurement. *Biosys Eng* 91: 329-333.
27. Mizrach A, Galili N & Rosenhouse G (1994) Methods and system for non-destructive determination of quality parameters in fresh produce. Israel patent 109406, issued Oct 10, 1998.
28. Hung YC & Prussia SE (1995) Firmness measurement using a non destructive laser-puff detector. In: *Proceedings of $4^{th}$ ASAE Food Processing and Automation Conference*, November 3-5, Chicago, USA, pp 145-154.
29. Calvo C (1996) Optical properties. In: *Handbook of Food Analysis,* Vol 1, Nollet LML (ed), Marcel Dekker Inc, New York, USA, pp 139-161.
30. Matsuoka T, Miyauchi K & Yano T (1996) Basic studies on the quality evaluation of agricultural products: Quantification of gloss of eggplants by spectral characteristics. *J Jpn Sci Agric Mach* 57: 33-40.
31. Ward G & Nussinovitch A (1996) Peel as a potential indicator of banana ripeness. *Food Sci Technol* 29: 184-186.
32. Nussinovitch A, Ward G & Mey-Tal E (1996) Gloss of fruits and vegetables. *Food Sci Technol* 29: 184-186.
33. Judd DB & Wyszeki G (1967) *Colour in Business Science and Industry*, John Wiley & Sons, New York, USA.
34. Park SF (2002) The physiology of *Campylabacter* species and its relevance to their role as food-borne pathogens. *Int J Food Microbiol* 74: 177-188.
35. Pande A (1974) *Handbook of Moisture Determination and Control*, Marcel Dekker Inc, New York, USA.
36. Karmas E (1980) Techniques for measurement of moisture content of foods. *Food Technol* 34: 52-59.
37. Reineccius GA & Addis PB (1973) Rapid analysis of moisture in meat by gas-liquid chromatography. *J Food Sci* 38: 354-354.
38. BeMiller JN & Low NH (1998) Carbohydrate analysis. In: *Food Analysis*, Nielsen SS (ed), Aspen Publication Inc, Maryland, USA, pp 167-187.

39. Peris-Tortajada M (2000) HPLC determination of carbohydrates in foods. In: *Food Analysis by HPLC,* Nollet LML (ed), Marcel Dekker Inc, New York, USA, pp 287-302.
40. Peris-Tortajada M (1996) Carbohydrates. In: *Handbook of Food Analysis,* Vol 1, Nollet LML (ed), Marcel Dekker Inc, New York, USA, pp 533-550.
41. Pomeranz Y & Meloan CE (1977) *Food Analysis: Theory and Practices,* AVI Publishing Co, Westport, USA.
42. Peris-Tortajada M, Puchades R & Maquieira A (1992) Determination of reducing sugars by neocuproine method using flow injection analysis. *Food Chem* 43: 65-69.
43. Alvarez-Icaza M & Bilitewski U (1993) Mass production of biosensors. *Anal Chem* 65: 525-533.
44. Chang SKC (1998) Protein Analysis. In: *Food Analysis,* Nielsen SS (ed), Aspen Publication Inc, Maryland, USA, pp 237-249.
45. Diercky S & Huyghebaert A (2000) HPLC of food proteins. In: *Food Analysis by HPLC,* Nollet LML (ed), Marcel Dekker Inc, New York, pp 127-167.
46. Noble RC, Shand JH & West IG (1981) Novel determination of protein, fats and lactose of milk by liquid scintillation counter. *J Dairy Sci* 64: 14-18.
47. Anderson ME & Marshal RT (1975) An automated continuous protein analysis: Modification of Lowry method. *J Food Sci* 40: 728-731.
48. Camp VJ & Huyghebaert A (1996) Protein. In: *Handbook of Food Analysis*, Vol 1, Nollet LML (ed), Marcel Dekker Inc, New York, USA, pp 277-309.
49. Bietz JA (1985) High performance liquid chromatography: How protein look in cereals. *Cereal Chem* 62: 201-212.
50. van den Bedem JW & Leenheeler J (1988) Heat treatment classification of low heat extra heat skim milk powder by HPLC. *Neth Milk Dairy J* 42: 345-350.
51. Singh NK, Donovan GR, Batey IL & MacRitchie F (1990) Use of sonication and high performance liquid chromatography in study of wheat flour protein II: Dissolution of total protein in the absence of reducing sugars. *Cereal Chem* 67: 150-161.
52. Olieman C & van Riel JAM (1989) Detection of rennet whey solids in skim milk powder and buttermilk powder with reversed-phase HPLC. *Neth Milk Dairy J* 43: 171-184.
53. Basch JJ, Douglas FW, Procino LG, Holsinger VH & Farrel Jr HM (1985) Quantitation of casein and whey proteins of processed milk and whey protein concentrates application of gel electrophoresis and comparison with Harland-Asworth procedure. *J Dairy Sci* 68: 23-31.
54. Paulis JW & Bietz JA (1988) Characterization of zein fractionated reversed-phase high performance liquid chromatography. *Cereal Chem* 65: 215-222.
55. Wall JS, Fey DA & Paulis JW (1984) Improved two-dimensional electrophoretic separation of zein proteins: Application to study zeins. *Cereal Chem* 61: 141-146.
56. Olieman C (1993) Capillary electrophoresis in dairy research. *Voedingsmiddelen technologie* 26: 10-11.
57. Dougherty TM (1977) Quantitative determination of α-lactalbumin and β-lactoglobulin in whey protein fractions and whey protein concentrates in frankfurters using electro-immunoassay. *J Food Sci* 42: 1611-1614.
58. Mao FC & Bremel RD (1991) Enzyme-linked immunosorbent assays for bovine α-lactalbumin and β-lactoglobulin in serum and tissue culture media. *J Dairy Sci* 74: 2946-2951.
59. Mao FC, Bremel RD & Dentine MR (1991) Serum concentration of milk proteins α-lactalbumin and β-lactoglobulin in pregnancy and lactation correlation with milk and fat yield in dairy-cattles. *J Dairy Sci* 74: 2952-2958.
60. Teufel P & Sacher V (1982) Zum Nachweis von Milchweiss in Fleischerzeugnissen mit ELISA. *Fleischwirtschaft* 62: 1474-1476.

61. Min DB & Steenson DF (1998) Crude fat analysis. In: *Food Analysis 2nd Edition*, Nielsen SS (ed), Aspen Publication Inc, Maryland, USA, pp 425-442.
62. Pike OA (1998) Fat characterization. In: *Food Analysis*, Nielsen SS (ed), Aspen Publication Inc, Maryland, USA, pp 217-235.
63. Gruger EH Jr, Malinns DC & Gauglitz EJ Jr (1960) Glycerolysis of marine oils and the preparation of acetylated monoglycerides. *J Am Oil Chem Soc* 37: 214-217.
64. Marini D (2000) HPLC of lipids. In: *Food analysis by HPLC,* Nollet LML (ed), Marcel Dekker Inc, New York, USA, pp 169-249.
65. Ryhage R & Stanhagen E (1960) Mass spectrometry in lipid research. *J Lipid Res* 1: 361-390.
66. Kuksis A, Marai L & Myher JJ (1983) Strategy of glycerolipid separation and quantitation by complementry analytical techniques. *J Chromatogr* 273: 43-66.
67. Marai L, Myher JJ & Kuksis A (1983) Analysis of triglycerol by reversed-phase high pressure liquid chromatography with direct liquid inlet mass spectrometry. *Can J Biochem Cell Biol* 61: 840-849.
68. Sempore G & Bezard JA (1996) Analysis of neutral lipids: Unsaponified matter. In: *Handbook of Food Analysis,* Vol 1, Nollet LML (ed), Marcel Dekker Inc, New York, USA, pp 491-506.
69. Fieser LF & Fieser M (1959) *Steroids,* Chapman & Hall Ltd, London, UK.
70. Derks HJ, van Heiningen A & Koedam HC (1985) Gas-chromatographic determination of cholesterol in serum: Candidate reference method. *Clin Chem* 31: 691-694.
71. Ong CP, Lee HK & Li SFY (1990). Supercritical fluid extraction and chromatography of cholesterol in food samples. *J Chromatogr* 515: 509-513.
72. Porter NA, Dussault P, Breyer RA, Kaplan J & Morelli J (1990) The resolution of racemic hydroperoxides: A chromatography-based separation of perketals derived from arachidonic, linoleic, and oleic acid hydroperoxides. *Chem Res Toxicol* 3: 236-243.
73. Russell LF (2000) Quantitative determination of water-soluble vitamins. In: *Food analysis by HPLC,* Nollet LML (ed), Marcel Dekker Inc, New York, USA, pp 463-476.
74. Ball GF (2000) Fat-soluble vitamins. In: *Food analysis by HPLC,* Nollet LML (ed), Marcel Dekker Inc, New York, USA, pp 321-402
75. Eitenmiller RR (1990) Strengths and weaknesses of assessing vitamin content of foods. *J Food Qual* 13: 7-20.
76. Russell LF (1996) Water-soluble vitamins. In: *Handbook of Food Analysis,* Nollet LML (ed), Marcel Dekker Inc, New York, USA, pp 649-713.
77. Alwarthan AA (1993) Determination of ascorbic acid by flow injection with chemiluminescence detection. *Analyst* 118: 639-642.
78. Daily S, Armfield SJ, Haggett BGD & Downs MEA (1991) Automated enzyme packed-bed system for the determination of vitamin C in foodstuffs. *Analyst* 116: 569-572.
79. Fox JB, Ackerman SA & Thayer DW (1992) Fluorometric determination of thiamine vitamers in chicken. *J Assoc Off Anal Chem* 75: 346-354.
80. Russell LF & Vanderslice JT (1992) Comments on the standard fluorometric determination of riboflavin in foods and biological tissues. *Food Chem* 43: 79-83.
81. Banno K, Matsuoka M, Horimoto S & Kato J (1990) Simultaneous determination of pantothenic acid and hopantenic acid in biological samples and natural products by gas chromatography-mass fragmentography. *J Chromatogr* 525: 255-264.
82. Ball GFM (1996) Determination of fat-soluble vitamins in foods by high performance liquid chromatography. In: *Handbook of Food Analysis,* Nollet LML (ed), Marcel Dekker Inc, New York, USA, pp 601-647.

83. Thompson JN, Hatina G & Maxwell WB (1980) HPLC determination of vitamin A in margarine, milk, partially skimmed milk and skimmed milk. *J Assoc Off Anal Chem* 63: 894-898.
84. Thompson JN & Duval S (1989) Determination of vitamin-A in milk and infant formula by HPLC. *J Micronutr Anal* 6: 147-159.
85. Stancher B & Zonta F (1982) High-performance liquid chromatographic determination of carotenes and vitamin-A and its geometric isomers in foods: Application to cheese analysis. *J Chromatogr* 238: 217-225.
86. Aggarwal VK (1989) Liquid cromatographic determination of vitamin D in infant formula. *J Assoc Off Anal Chem* 72: 1007-1009.
87. Thompson JN & Hatina G (1979) Determination of tocopherols and tocotrienols in foods and tissues by high performance liquid chromatography. *J Liquid Chromatogr* 2: 327-344.
88. Thompson JN & Plouffe L (1993) Determination of cholecalciferol in meat and fat from livestock fed normal and excessive quantities of vitamin D. *Food Chem* 46: 313-318.
89. Tuan S, Lee TF, Chou CC & Wei QK (1989) Determination of vitamin-E homologs in infant formulas by HPLC using fluorometric determination. *J Micronutr Anal* 6: 35-45.
90. Indyk HE (1988) Simplified saponification procedure for the routine determination of total vitamin E in dairy products, foods and tissues by high-performance liquid chromatography. *Analyst* 113: 1217-1221.
91. Piironen V, Syvaoja EL, Varo P, Salminen K & Koivistoinen P (1985) Tocopherols and tocotrienols in Finnish foods: Meat and meat products. *J Agric Food Chem* 33: 1215-1218.
92. Hogarty CJ, Ang C & Eitenmiller RR (1989) Tocopherol content of selected foods by HPLC/fluorescence quantitation. *J Food Comp Anal* 2: 200-209.
93. Woollard DC & Blott AD (1986) The routine determination of vitamin E acetate in milk-powder formulations using high-performance liquid chromatography. *J Micronutr Anal* 2: 97-115.
94. Zonta F & Stancher B (1983) High-performance liquid chromatography of tocopherol in oils and fats. *Riv Ital Sost Grassse* 60: 195-199.
95. Brubacher G Muller Mulot Souithgate DAT (1985) *Methods for Determination of Vitamins in Foods,* Elsevier, London, UK.
96. Tee ES & Lim CL (1992) Reanalysis of vitamin A values of selected Malaysian foods by the AOAC and HPLC methods. *Food Chem* 45: 289-296
97. Thompson JN, Hatina G, Maxwell WB & Duval S (1982) High performance liquid chromatographic determination of vitamin D in fortified milks, margarine and infant formulas. *J Assoc Off Anal Chem* 65: 624-631.
98. Fournier B, Sann L, Guillaumont M & Leclercq M (1987) Variations of phylloquinone concentration in human milk at various stages of lactation and in cow's milk at various seasons. *Am J Clin Nutr* 45: 551-558.
99. Shearer MJ (1983) High-performance liquid chromatography of K vitamins and their antagonists. *Adv Chromatogr* 21: 243-301.
100. Mark C (1988) Determination of free tocopherols in deodorizer by capillary gas chromatography. *J Am Oil Chem Soc* 65: 1936-1939.
101. Kmostak S & Kurtz DA (1993) Rapid determination of supplemental vitamin E acetate in feed premixes by capillary gas chromatography. *J Assoc Off Anal Chem Int* 76: 735-741.
102. Miriani C & Bellan G (1996) Content of tocopherols, deidrotocopherols, tocodienols, tocotrienols, in vegetable oils. *Riv Ital Sostanze Grasse* 73: 533-543.
103. Hendrick DG (1998) Mineral analysis. In: *Food Analysis*, Nielson SS (ed), Aspen Publication Inc, Maryland, USA, pp 151-165.

104. Harbers LH (1998) Ash analysis. In: *Food Analysis*, Nielsen SS (ed), Aspen Publication Inc, Maryland, USA, pp 141-149.
105. Miller DD (1998) Atomic absorption and emission spectroscopy. In: *Food Analysis*, Nielson SS (ed), Aspen Publication Inc, Maryland, USA, pp 425-442.
106. Altekruse SF, Cohen ML & Swerdlow DL (1997) Emerging food borne diseases. *Emerg Infect Dis* 3: 285-293.
107. Park YW (1996) Determination of moisture and ash contents of foods. In: *Handbook of Food Analysis,* Vol 1, Nollet LML (ed), Marcel Dekker Inc, New York, USA, pp 59-92.
108. Johnson-Green P (2002) *Introduction to Food Biotechnology*, CRC Press, Boca Raton, USA, pp 151-180.
109. Tauxe RV (1998) Public health perspective on immunoprophylactic strategies for *Escherichia coli* O157:$H_7$: who or what would we immunize? In: *Escherichia coli O157:$H_7$ and Other Shiga Toxin-Producing E. coli Strains*, Kaper JB & O'Brien AD (eds), ASM Press, Washington, USA, pp 445-452.
110. Granum PE & Brynstad S (1999) Bacterial toxins as food poisons. In: *The Comprehensive Source Book of Bacterial Protein Toxins,* Alouf JE & Freer JH (eds), Academic Press, Toronto, Canada, pp 669-690.
111. Balaban N & Rasooly A (2000) *Staphylococcal* enterotoxins. *Intl J Food Microbiol* 61: 1-10.
112. Krishnamachari KAVR, Bhat RV, Nagarajon V & Tilek TBG (1975) Hepatitis due to aflatoxicosis. *Lancet* 1: 1061-1063.
113. Odumera J (2006) Inside microbiology: Current microbial concerns in dairy industry. *Food Safety Magazine* 8: 20-23.
114. WHO Report (2007) *Avian Influenza: Food Safety Issues*, Department of Food safety, Zoonoses and Foodborne Diseases (FOS), World Health Organisation, Geneva, Switzerland.
115. Casemore DP (1990) Epidemiological aspects of human cryptosporidiosis. *Epidemiol Infect* 104: 1-28.
116. Herwaldt BL & Beach MJ (1997) *Cyclospora* working group. The return of *Cyclospora* in 1997: Another outbreak of cyclosporiasis in North America associated with imported raspberries. *Ann Intrl Med* 130: 210-220.
117. Tenter AM (1995) Current research on *Sarcocystis* species of domestic animals. *Intl J Parasitol* 25: 1311-1330.
118. de Boer E & Beumer RR (1999) Methodology for detection and typing of food-borne microorganisms. *Intl J Food Microbiol* 50: 119-130.
119. Jinneman KC, Hunt JM, Eklund CA, Wernberg JS, Sado P, Johnson JM, Richter RS, Torres ST, Ayottee E & Eliasberg SJ (2003) Evaluation and interlaboratory validation of a selective agar for phosphatidylinositol-specific phspholipase C activity using achromogenic substrate to detect *Listeria monocytogenes* from foods. *J Food Prot* 66: 441-445.
120. Wan J, King K, Forsyth S & Coventry MJ (2003) Detection of *Listeria monocytogenes* in Salmon using probelia polymerase chain reaction system. *J Food Prot* 66: 436-440.
121. Churchill RLT, Lee H & Hall A (2006) Emerging bacterial foodborne pathogens and methods of detection. In: *Food Biochemistry & Food Processing,* Hui YH (ed), Blackwell Publishing, Iowa, USA, pp 507-544.
122. Call DR, Borucki MK & Loge FJ (2003) Detection of bacterial pathogens in environmental samples using DNA-microarrays. *J Microbiol Meth* 53: 235-243.
123. Breitling F & Dűbel S (1999) *Recombinant Antibodies,* John Wiley and Sons Inc, New York, USA.

124. De Greeffe A, van Aiphen L & Smith HE (2000) Selection of recombinant antibodies specific for pathogenic *Streptococcus suis* by subtractive phage display. *Infect Immun* 68: 3949-3955.
125. Turnbough Jr CL (2003) Discovery of phage display peptide ligands for species-specific detection of *Bacillus* spores. *J Microbiol Meth* 53: 263-271.
126. Bjerketorp J, Nilsson M, Ljungh A, Flock J-I, Jacobsson K & Frykberg L (2002) A novel von willembrand factor binding protein expressed by *Staphylococcus aureus. Microbiology* 148: 2037-2044.
127. Matar GM, Hayes PS, Ribb WF & Swaminathan B (1997) Listeriolysin O-based latex agglutination test for the rapid detection of *Listeria monocytogenes* in foods. *J Food Prot* 60: 1038-1040.
128. Hess J, Miko D, Catric A, Lehmensiek V, Russell DG & Kaufmann SHE (1998) *Mycobacterium bovi* bacilli calmette- Gűerin strain secreting listeriolysin of *Listeria monocytogenes. Proc Natl Acad Sci USA* 95: 5299-5304.
129. Sewell AM, Warburton DW, Boville A, Daley EF & Mullen K (2003) The development of an efficient and rapid enzyme linked fluorescent assay method for the detection of *Listeria* spp. from foods. *Intl J Food Microbiol* 81: 123-129.
130. Dunbar SA, Vander Zee CA, Oliver KG, Karem KL & Jacobson JW (2003) Quantitative, multiplexed detection of bacterial pathogens: DNA and protein applications of the Luminex LabMAP system. *J Microbiol Meth* 53: 245-252.
131. Hashimoto S (2000) Principles of BIACORE. In: *Real-Time Analysis of Biomolecular Interactions: Applications of BIACORE,* Nagata K & Handa H (eds), Springer-Verlag, Tokyo, Japan, pp 23-32.
132. Bamdad C (1997) Surface plasmon resonance for measurements in biological interests. In: *Current Protocols in Molecular Biology,* Ausubel FM, Brent R, Kingston RE, Moore DD, Seidinan JA, Smith JA & Struhl K (eds), John Wiley and Sons, Inc, Canada, pp 2001-2042.
133. Nice EC & Catimel B (1999) Instrumental biosensors: New perspectives for analysis of biomolecular interactions. *Bio Essays* 21: 339-352.
134. Nollet LML (1996) *Handbook of Food Analysis,* Vol 2, Marcel Dekker Inc, New York, USA.

# Methods in Sensory Evaluation

# 25

**V.K. Joshi, N.S. Thakur and Shashi Bhushan**

*Department of Food Science and Technology, Dr. Y.S. Parmar University of Horticulture and Forestry, Nauni, Solan 173 230 (H.P.)*

## 1. INTRODUCTION

Quality is the combination of those characteristics that differentiate individual units of a product, which has significance in determining the degree of acceptability of that unit by the buyer. In the past, human sensory perception was the only means of assessment of food quality as illustrated by the classical examples of cooks who knew how to mask the off-notes long before microbial assessment of quality came into existence. Everybody relied on the sensory evaluation of foods as a criterion of quality assessment. Even the king use to depend upon "tasters" to assure that the food was suitable for his consumption [1]. But later on, with the development of instrumentation, sensory analysis was overlooked and the instrumental methods were preferred to as they were supposed to provide absolute measures of quality. The product quality was being traditionally determined by physical, chemical and microbiological criteria with the assumption that the products meeting the prescribed specifications in the above parameters would automatically meet the desired sensory quality. However, very soon it was realized that without sensory evaluation, the acceptability of a food product cannot be determined.

Integration of sensory analysis into a quality control programme makes it a powerful tool for product assessment, integrity and consistency [2]. Though many food processing plants are aware of this, few have really made attempts to make it a routine procedure to get objective and actionable data. It is perhaps due to the feeling that sensory analysis is 'Subjective' method of evaluation i.e., not repeatable or reproducible and is biased. This was so at the beginning for want of objective sensory methods. But now the case is different as sensory analysis has developed into a full fledged discipline [1]. It has been defined as "A scientific discipline used to evoke, measure, analyze and interpret reactions to those characteristics of foods and materials as they are perceived by the senses of sight, smell, taste, touch and hearing." This field has now developed into an exact science that can provide meaningful results if the guidelines are followed properly [3].

Sensory analysis could be integrated into a well-balanced quality control programme to serve as a tool to maintain the product quality and pattern matching with the consumer concepts. It provides a mean to optimize the product successfully. Not only this, the product is rendered more flexible for defined adjustments. The technique can identify and interpret quality control problems by effective evaluation and monitoring of incoming raw ingredients, packaging materials and consumer thinking

*Food Biotechnology: Principles and Practices*
*V.K. Joshi & R.S. Singh (Eds.)*
*I.K. International Publishing House Pvt. Ltd., India*
*pp 861–880*

besides assistance in distribution, sales and abuse prevention of products [4]. Sensory evaluation plays a very significant role in the metabolism of food in human beings. Sensory evaluation is involved in the development of attraction, salivation or stimulation, etc. in humans towards the food which gives them satisfaction, and can also affect the health either by over-dieting or under-nutrition. The instruments can also measure many quality parameters, besides showing that the ingredient levels are within their specifications or not. But the quality and acceptance of the product will be entirely different and can only be determined by sensory evaluation. However, lack of human perceptions makes instrumental analysis meaningless and invalid. Ingredients and processing conditions result in innumerable number of interactions, synergistic or antagonistic, which entirely change the product profile. For example, a product can be manufactured as per the specifications of salt and sugar or with other such ingredients, but if each were at their extreme end (though within the limits), their interaction may either boost the sweetness or saltiness or both. Besides application in food evaluation, sensory analysis is being employed in a number of fields including evaluation of new varieties, process, and products, genetically modified foods, flavours and fragrances, etc. Sensory science has got impetus from genetic engineering resulting in production of new metabolites with variable properties including taste/flavours. Molecular biology tools have also been employed to understand the mechanism of sensory perception. The recombinant DNA technology has opened new vistas for the field with exciting results. All these aspects have been discussed in one or the other way in this chapter.

## 2. SENSORY EVALUATION

The emergence of sensory evaluation as a legitimate scientific discipline has been made possible by adherence to well defined objectives, which led to selecting appropriate methods, panelists and appropriate statistical hypotheses and analysis of data (Fig. 1). For a successful sensory evaluation programme, there are some pre-requisite which must be met. Some of these are well responsive panel with not less than 10 members for laboratory testing or 200 members for consumers testing, well designed proforma for getting responses with defined objectives, training of panel for laboratory work using standard sensitivity tests for the specific sensory quality characteristics as mentioned in Fig. 2 for salt, sugar, sweetness, bitterness tests are normally carried out for screening of panel for taste evaluation, presentation of coded samples at suitable places/chambers without the possibility of discussion and development of consumers response is possible in the descriptive analysis of the products, objectivity of data collection, compilation and statistic analysis is the most important aspect in sensory analysis.

### 2.1 Components of Sensory Evaluation

The components of sensory evaluation programme have been depicted in Fig. 1 and explained under the following heads:

#### *2.1.1 Human sense organs*

In sensory evaluation, the human sense organs serve as an instrument. Human beings possess five senses - sight, smell, hearing, taste and touch. Accordingly, we have five sense organs viz., eye, ear, skin, nose and tongue. Sense organs consist of sensory neurons, which are tied into some suitable detector for getting the response.

2.1.1.1 Eye/sight: The sensory neurons are activated only when light rays are focused on the retina by the lens of the eye. Eye is a visual receptor, working similar to a camera. Light first passes through the pupil of variable size, depending primarily on light intensity. Iris around the pupil controls the size

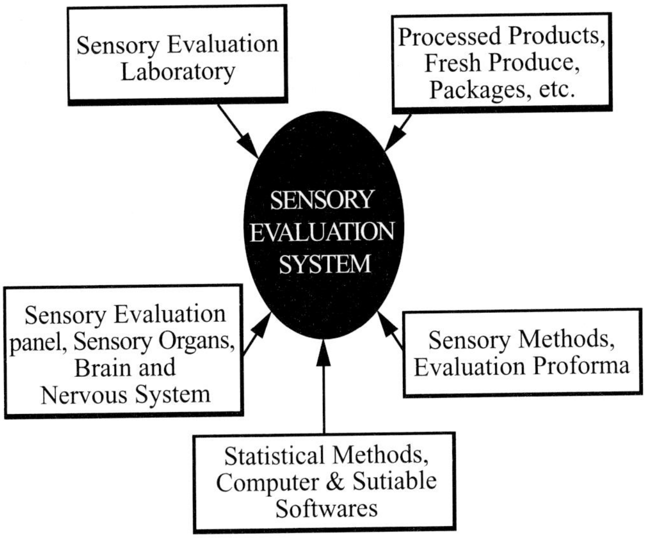


**Fig. 1:** An overview of sensory evaluation system.

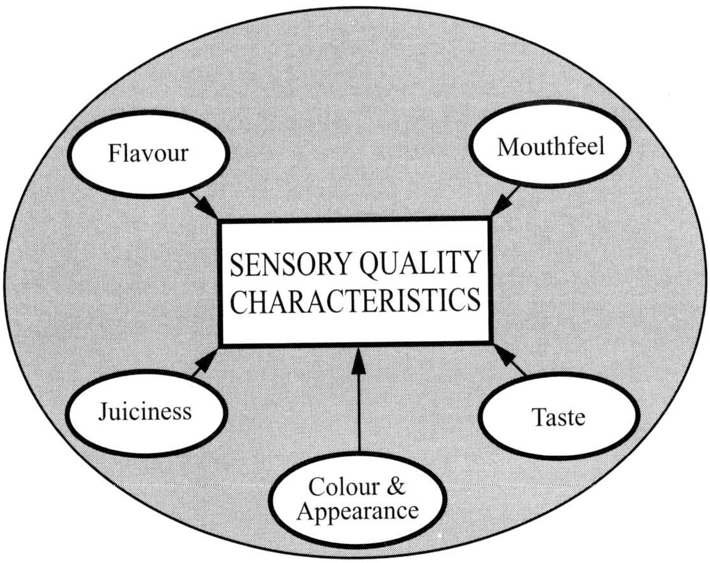


**Fig. 2:** A representative of sensory quality attributes of a food product.

or contraction of pupil and the amount of light that enters the eye. Behind the pupil iris, there is a lens, which with the help of outer cornea, focus the light rays in such a manner that the focal length exactly equals the distance between the lens and the light receptive areas on the back of the eye. Retina, having a layer of sensory cells covering the interior of eye ball, is the focal point of light. The deepest layer of

**Plate 1:** A panelist evaluating the aroma by sniffing.

retina (consists of a conglomerate of cells) is responsive to light directly. There are two types of neurons: (i) rods, elongated, having no colour sensing capacity, found mostly in peripheral regions of retina, and (ii) cones, shorter, colour sensitive and found in the centre of retina, directly behind the lens. Out of the four visual pigments, one is found in rods which is very sensitive to low levels of light but has no colour detecting capacity. Cones have three types each responding to red, blue and green colouring pigment, respectively. Stimulation of more than one type at different intensities allows the individual to distinguish a whole series of intermediate colours. The interpretation as vision begins, when light causes a physical straightening of some of the visual pigment molecules, generating an action potential to two successive types of neurons in the retina, and it is transmitted through the optic nerve to the thalamus of the brain and hence to the visual cortex of the back of head. However, the colour is a function of light source, optical properties of a food product and pigments in retina which act as photoreceptors and absorb energy, the neurological responses to the stimulus, and finally, the interpretation is made based on the signals transmitted to the brain.

**2.1.1.2 Ear/hearing:** It is a type of pressure receptor. It is known that any vibrating object causes bands of compression and rarefaction, and if the vibration is of sufficient intensity, it is detectable as sound. Sound waves are intercepted by the eardrum at the base of ear canal, causing it to vibrate. A set of three middle ear bones are attached to the eardrum, transmitting sound to inner ear and serve as to amplify the intensity of the vibration about 18 times over that received by the eardrum. Oval window (the point of contact between the middle ear bones and the inner ear) is coiled, filled with a fluid that received the vibration (at original frequency) with the membrane dividing the inner ear, down the middle, vibrating at the same frequency. But because of its structure, sound of different frequency causes maximum displacement at different points along with membrane. Membranes consist of two layers (not fused), as one layer slides over the other during displacement, deflecting tiny hairs (hair cells), which are connected to auditory nerve. The hair deflection generates an action potential, as a result, activating different fibres in the auditory nerve, which ultimately allow the brain to make discriminations according to the sound frequency. The detection of complex sounds by the simultaneous deflection of hair cells is responsible for the particular frequencies in the sound.

**2.1.1.3 Skin feeling/touch:** The sense of touch is perceived by tiny organs surrounding an individual neuron known as pacinian corpuscles. As the rate of action potential produced by the deflecting neuron/neurons, the magnitude of stimulus conveyed to the brain and the touch receptors fatigue quickly.

**2.1.1.4 Taste buds/taste:** Tongue is the seat of taste evaluation. The taste buds are on the tongue located at different places and are specific for various types of basic tastes. The receptors of taste in man and vertebrates are mainly situated on the surface of the tongue and to a small extent spread over oral cavity and down in oesophagus. The taste buds in man are located chiefly in three types of papilla, the foliate, fungi forms and circumvallated. The pathway of perception of taste is simpler than olfactory pathway, but the mechanism of decoding in the brain is still unknown. The tongue, however, is well established to have definite sensitive area to the basic taste. However, stimulating human papillae indicate that tip can perceive bitter, while posterior or basal part can taste sweet also. Taste buds are immersed in the flat epithelia of papillae besides other parts of the mouth, pharynx and oesophagus. The interior tip shows sensitivity to sweetness, the sides to salty and sour taste, while bitterness and pungency is perceived in posterior portion of the tongue and initial part of oesophagus. Taste buds are a type of chemical receptors. All are aware of four basic tastes i.e. sweet, sour, bitter, and salty. However, five taste modalities have been identified in humans: sweet, sour, bitter, salty and umami [5]. However, over the course of human evolution, two of the modalities have been proven to be the most significant to the survival of the species: sweetness provides a mean to search an energy source in the form of carbohydrates and the bitterness gives an ability to avert to alcohols and potentially deadly toxins found in nature [6]. Classically, there are four kinds of taste buds in the mouth of human beings viz. sweet, salt, sour and bitter. The response of each taste bud may be equal to any of the four types of tastes, but some respond quicker to a particular taste than the others. The functioning of taste buds is not so clear till now, beyond the fact that water borne chemicals of these four taste buds can initiate action potential within the neurons of the taste buds and the response conveyed to brain.

Several theories have been proposed to explain the mechanism of sweet taste in man. The original AH-B theories have been supplanted by detailed structural models [7]. Recent identification of human sweet receptor as a dimeric G-protein coupled receptor comprising $T_1R_2$ and $T_1R_3$ subunits [8] has greatly increased our knowledge of the mechanism involved in sweet molecule and binding sweet taste transduction. Stimulation of the receptor could provide the food scientists with the knowledge required to develop new sweeteners, low calorie food products and potentially, a healthier population [7].

**2.1.1.5 Olfactory epithelium/smell:** The mechanism of odour detection is still not completely explained. There are a large numbers of cells in the upper reaches of nasal passages of human beings, which are responsive to the different types of odours. Such cells possess different types of receptive sites responding to different types of smell based on the chemical structure of the molecules to be identified. When the volatile organic compounds bind to one or more of around 1000 protein receptors on the surface of the cells in the nasal passage, the sense of smell is activated. Both aroma and odour are the sensations triggered by the olfactory and trigeminal nerves. Besides this, a third nerve, the terminal nerve, in the nasal cavity of human is also involved in this perception of food. The olfactory mucosa in humans is located high-up in the dorso-posterior region of nasal cavity, on the roof away from the major flow of the inspired air. The message is carried from these receptors to the higher centre through the olfactory bulbs. During normal breathing, only 5% or so of the respired air reaches the receptors, but during sniffing it increases to about 20%.

The flavour of foods comes from its aroma (Plate.1), which watts up the nostrils to sensory cells through a passage way in the back of the mouth. Anatomical studies showed that signals received from olfactory cells reach the olfactory area of the cortex (in brain) after only a single relay in the olfactory bulb. The average human being, it is said, can recognize around 10,000 separate odours. Neurons that sense these odour molecules lie high up in nose, i.e. olfactory epithelium (OE) consisting of around five million olfactory neurons. OE is topped by at least ten hairs like cilia that protrude into a thin bath of mucus at the cell surface. Cilia are supposed to consist of a receptor protein that recognizes and binds odorant molecules, there by stimulating the cells to send signals to brain. The family of olfactory receptor proteins (ORs) is currently the best candidate as a site for the odour transduction in the vertebrate olfactory system. The identification of ORs as member of the G protein coupled receptor ((G PCR) super family was the first evidence for this [9]. Mammals are known to possess more than 1000 different OR subtypes, which is in consistence with the capacity to detect many thousands of odours. Computation analysis of the OR has revealed that 5-10 amino acid residues like to interact with odour molecules.

### *2.1.2 Methods of sensory evaluation*

These are many sensory evaluation methods which are followed for evaluating the sensory quality of a product (Fig. 3).

**2.1.2.1 Analytical tests:** Laboratory analysis with trained or semi-trained panelists is called analytical tests. These tests treat the panelists as discriminative or descriptive measuring instruments and include sensitivity assessment. The interest is in the differences among the treatments or products or samples, the panels being the means of measurements. The quality attributes of the products are assessed, keeping the panel constant.

**2.1.2.2 Affective tests:** Analysis with untrained consumer panels is called affective tests. Affective tests are based on hedonicity and are entirely different from the analytical tests. The importance shifts from the products/samples to the target population [10]. The products are fixed, one or two only, and the segment of the target population that would accept the product is determined. Thus, one is a product study and the other is a population study. There are certain new approaches in the sensory evaluation which are shown in Table 1.

There are a number of food products aimed directly at children considering their increased participation in purchase decisions. Sensory evaluation with children requires a specifically designed introduction to the test methodology and more extensive training. Young children are subject to limitations in cognition, including the concepts of quantity, time/frequency/periodicity and serialization/ progression, abstraction vs. concreteness, and cause/effect relationships [11]. Usually three discrimination tests are used - paired comparison, duo trio and intensity ranking and three consumer tests paired preference, preference ranking and hedonic scaling. It is observed that children can be a valuable measuring tool for sensory testing if treated as a special population and if the appropriate testing environment and protocol are used. Children over the age of two were found to perform reliably a

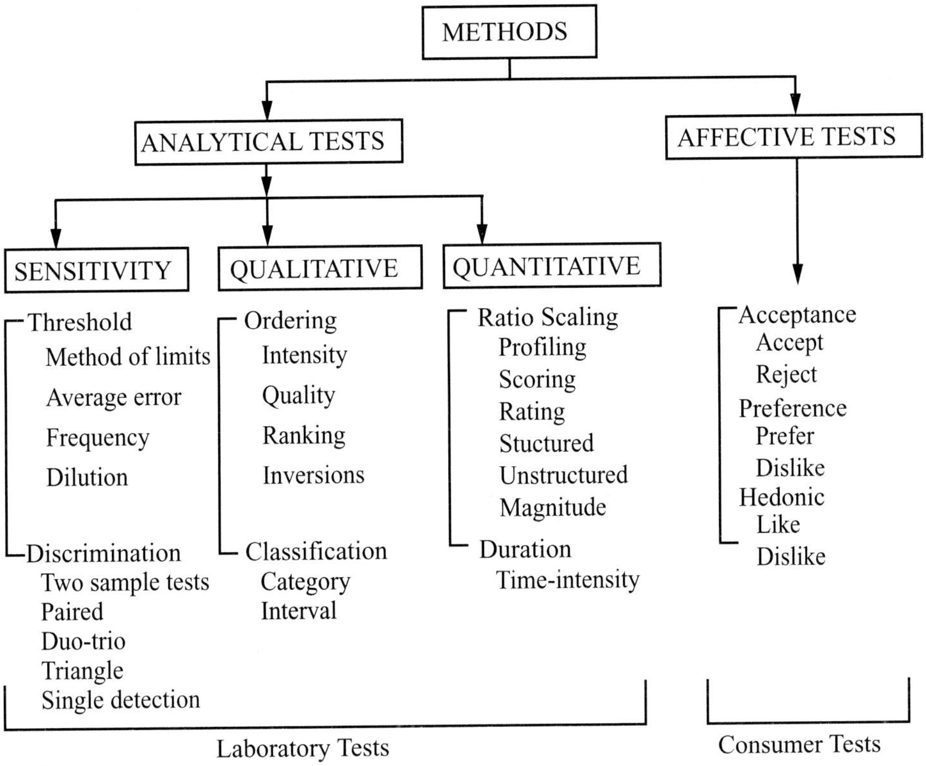


**Fig. 3:** Classification of sensory analysis test methods (Stimulus, colour, appearance, odour, texture, aroma, after taste).

**Table 1:** New approaches in sensory analysis.

**Multiple Regressions**
- Is a popular, extensively used technique
- Develops a working relationship between the dependable variable and several independent variables, ultimately into the model which is based on the correlation between the independent variable and coefficient of variability appearing in the final model

**Factor Analysis/Principal Component Analysis (PCA)**
- Is a powerful set of techniques especially to eliminate unwanted data of a set of samples of large size
- The assumption is the population belonging to the same population whose large variable are studied
- Identifies the subsets of all the variables
- The method creates linear combination which are orthogonal to the rest of combinations
- Characterize the group, find out association or differences

**Multivariate Analysis of Variance**
- Is a logical extension of univariate analysis of variance

**Canonical Discriminate Analysis (CDA)**
- It works on multiple population
- Variable not useful in discriminating are removed
- Works like PCA in orthogonal combinations
- Data on same commodity, found in different regions with same set of attributes

Contd...

Contd...

| |
|---|
| **Multiple Correspondence Analyses** |
| • Is more generalized, working on rectangular matrix through single value decomposition (SVD) or Echert-Young decomposition |
| • Offers mapping of samples in the space of attributes and can be linked to PCA or CDA |
| **Generalized Procrustes Analysis** |
| • It is consensus making technique |
| • Samples measured by different agencies are considered and evaluated for consensus |
| • It makes use of translation, rotation and scaling |
| • Applicable to the data of free choice profile |

paired preference test. More complex tests, such as hedonic scaling discrimination tasks such as the paired comparison, the duo-trio test and ranking for intensity can be performed.

### *2.1.3 The laboratory*

The sensory evaluation programme requires a laboratory equipment and staff. The size of the laboratory is determined by the scale of sensory evaluation carried out by the unit. For a large factory, a separate laboratory with different booths/chambers is required, while a small processing unit could suffice with portable chambers/booths/partitions [12]. Irrespective of these, any sensory evaluation laboratory should be equipped with tables with booths/chambers, balances for weighing of ingredients, heating or cooling systems for the product preparation, calibrated thermometers, graduated cylinders, serving dishes, spoons, ladles, small glasses, beakers, conical flasks, knives, refractometers, pressure testers, etc. The booths for sensory evaluation are prepared in such a manner that one judge should not see the other. It should have proper seat, a light arrangement, a sink and a serving window through which the product could be served to the judge. Each booth should also be provided with a bell, to provide anything required by the judge.

Staffing pattern also varies with the size of the factory. A small staff of 1-3 people is sufficient for a small scale processing unit, while the large plant shall need 2-3 times more persons for assistance in carrying out the sensory evaluation work. This usually consists of a sensory evaluation coordinator and technicians. These employees have to work whole time. Their duties include the test preparation, product preparation, distribution of samples to panelists, house keeping and record keeping.

### *2.1.4. Objectives of sensory evaluation programme*

Sensory testing programmes can be built to serve many objectives [13, 14]. These include; quality control and assurance-for maintenance of quality, process/equipment change, ingredient substitutions, alternate supplier evaluation (Fig. 4).

- Product development or improvement to match competitive product and obtain or retain a share of the market
- Product matching
- Define packaging and storage profiles
- Provide product information for marketing/management groups
- Quality grading/profiling

### *2.1.5 Consumer sensory evaluation*

It should follow the sensory analysis and can take up one or two optimized products and may involve the study of one or more of the following [15]:

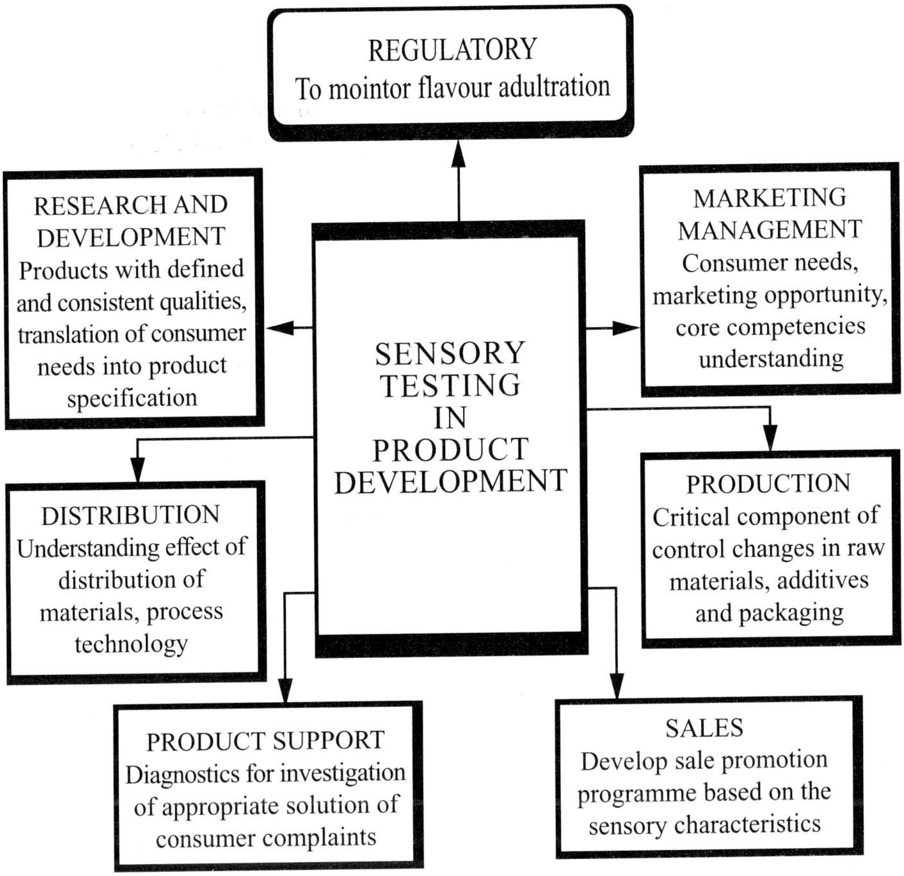


**Fig. 4:** Illustration of sensory analysis in product development.

- Pattern of acceptance/preference of the product by various segments of population
- Effect of package, price, tradition and such other parameters on the acceptance
- Effect of ethnic groups, geographical locations, family size and composition, food habits, etc. on the acceptance of food
- Prediction of market potential

### *2.1.6. Panel screening, selection and training*

2.1.6.1 Panel screening and selection: General information regarding sex, age, food habits, likes and dislikes of foods, willingness to participate and knowledge about the product, and above all their availability are the first few criteria for panel screening and selection [16, 17]. In large processing units, it is easier to choose the panel liberally than smaller groups where they could get few in-house panelists normally. The tests for sensitivity to basic taste, group of odours, ability to identify textural differences can help to identify agusics or anosmics and also group them into broad classes. The sensitivity of panelists will decide the screening limits. The panel performance however, is very important as it is affected by their selection and screening.

Motivation of panel is an important aspect of panel screening and training. The panel member needs a basic sense of interest or involvement in product rather than short-lived interest out of curiosity, or change from routine job or even for a financial remuneration. However, to keep the interest alive, the test method can be made different occasionally and discussion held with panelists along with their performance [18]. For a meaningful sensory evaluation, the panel size is the next criteria to be satisfied as identified in the codified standards of Bureau of Standards of India (BSI) on sensory evaluation. With the use of proper design of experimentation, the number and type of tests can be suitably selected.

2.1.6.2. Panel training: Training ensures the systematic utilization of sensory responses to the product. The training of panel is desirable to give the panelists a definite exposure and opinion about the test product, to perform test without bias and with certain homogeneity [16, 17]. The ability to communicate the perceived sensation needs constant recalling from memory. The training for individual is kept as rigorous exercise for each class and type of food and has to be specific for the attributes to be examined. Besides, the training for fresh panel, subsequent training to keep the panel alert and experienced is also desired. The first training for any product should provide the complete possible range in product type. Use of methodology and scale to be used in the evaluation needs training. One may use basically same methods but anchorage should be specific for a test character. Use of numbers in sensory analysis need to be seen as points to anchor the perception intensity only without associating with any grading or quality analysis of the products.

## 3. SENSORY ANALYSIS

### 3.1 Analysis of Colour and Appearance

Colour and appearance gives the first impression of a food and is associated with quality [19]. Acceptance or rejection depends almost entirely on what the eye picks up quickly. If the colour deviates from the expected, the other attributes of a food product, viz. aroma, texture and taste, etc. are rarely judged [20]. By experience, human eye has learnt to identify certain colours in certain foods. Specifications (CIE or otherwise) should be based on the recommendation of what eye perceives as optimum. Sensory evaluation can play a vital role in defining, optimization and correlating the colour with instrumental analysis. To measure colour difference is another problem and for valid results 'just noticeable difference' (JND) should be established through sensory analysis. In the case of tomatoes and its products, visual grading has successfully been predicted based on instrumental measures [21]. Several methods could be used to evaluate the colour difference, viz., matching with standard colour/sample; interval scaling, descriptive methodology, and ratio scaling.

### 3.2 Analysis of Texture

The texture of food is one of the most important food characteristics. Colour and shape can be described by physical measurements, taste and smell may be approached chemically, whereas, sound can be analyzed by an appropriate audio instrument. Touch, however, has not been characterized at all. It has been defined as the response of the tactile sense to physical stimuli that result from contact between some part of the body and food [22]. It has also been defined "as the attribute of a substance resulting from a combination of physical properties and perceived by the senses of touch (including kinaesthesis and mouth-feel), sight, and hearing. Physical properties may include size, shape, number, nature and conformation of constituents structural elements [23]. The texture is therefore, a sensory attribute and is measurable directly by sensory means. Three senses; touch, sight and hearing may be involved in sensory assessment of texture [24]. The texture characteristics have been categorized into: (i) mechanical characteristics; (ii) geometrical and other miscellaneous characteristics, related to reaction of food to stress, size, shape and orientation of particles or those relating to the moisture and fat contents [25]. Essentially, a sensory feeling is evoked when a piece of food material is put into mouth and chewed. Then, all of its attributes viz., firmness, toughness, chewness, stickiness, brittleness, grittiness, flakiness, etc. may be perceived [26].

Among three senses (touch, sight and hearing) involved in perception of textural and mouthfeel properties of foods, sense of touch plays the most important role [19]. Normally, a food is bitten, masticated and swallowed in the mouth and the sense organs involved in texture evaluation may be grouped into: (i) those in the superficial structure of the mouth, (ii) those around the roots of the teeth, and (iii) those in the muscles and tendon. Methods for measuring texture of the food by sensory testing are finger feel, hedonic rating, descriptive tests, profiling testing. There is no instrument which can perceive, analyze, integrate and interpret a large number of textural sensations at the same time as the mouth does. Even if the human mouth could be duplicated in an instrument, one would still have to duplicate the nervous system which has the important function of receiving and translating the signals from the receptors in the mouth and interpreting the same. The best way is to measure a part of the components of the complex texture using suitable instruments and then, these instrumental measures can be correlated to the sensory perceptions of the texture [27].

Both large untrained panels and small trained panels can be used in the texture evaluation. The consumer panel is preferred for over-all quality evaluation rather than texture only [28]. The small trained panel can have 3 to 30 members. Panelists are educated in the terminology and techniques of sensory evaluation. The type of information needed determine both the selection and degree of training. Discriminatory as well as descriptive tests could be used to evaluate the texture. The discriminate tests include paired comparison, triangle or duo-trio tests while the descriptive terms are used to denote the techniques of ranking and scaling which describe and/or quantify the differences between the samples. Ranking is used only to classify the samples initially or to select the test samples. The scaling tests are frequently employed to indicate the intensity of an attribute on a scale, such as numerical scale from 0 to 3 or 0 to 100 and one or more points may be anchored by a word, description or reference sample. Structured scale for apple has also been developed, which could correlate the sensory characteristic with maturity indices.

### 3.3 Analysis of Aroma

Aroma analysis is an important component of sensory evaluation of food. It is distinguishable from odour. The term 'aroma' is frequently referred to the perception of smelling, while the term 'odour' refers to perception by nose, the term 'aroma' refers to perception by mouth through nasopharyngeal region. It is difficult to classify the odours since the number of odours seems to be large and to make distinctions among several odours is not easy. However, following classification has been made by Henning:

- Spicy e.g., cloves, fennel, anise
- Flowery e.g., heliotrope, coumarin, geranium
- Fruity e.g., oil of orange, oil of bergamot, citronellal
- Resinous or balsamic e.g., turpentine eucalyptus oil, Canadian balsam
- Burnt e.g., pyridine, tar
- Foul e.g., sulphured hydrogen, carbon bisulfide

It is known that the chemicals having sulphur in their molecules have unique odour. Some such chemicals from mango, guava, grapefruit, and tomato such as dimethylsulphide, 2 methyl-1-propanethiol, 1 propanethiol have been identified. The butter flavour with thiol lactones included had a mellow, natural and long lasting odour and taste. The white peach flavour with thiol lacetones included showed a mellow, natural and gave the impression of freshly squeezed juice from a well ripened peach. The different fragrances of flavouring compounds are isolated from natural sieves. The chemical composition and sensory properties of natural isolate are strongly dependent on the production method used [29]. Flavours from the muguet (Lily of the Valley) plants are very small and difficult to extract. It is impossible to produce commercially viable blossom oil from this plant. The compound like 3-alkoxypropane-1-ol and *trans*-alkenol have pronounced muguet note with floral odour of *cis*-alkenol and fully saturated analogue shifting more in the direction of sore [30]. Design and synthesis of novel Muguet material has been made.

There is variation in the sensitivity of aroma in the human beings. 'Anosmic' people are those who have lost the sense of odour/aroma due to tearing-off of fine olfactory nerve or due to some congenital defects. Aroma impact in food follows closely the primary impact from vision. Many decisions on food/beverage quality is based on the nasopharyngeal impact from initial examination for odour by sniffing, continued with and followed by the first impact when the product is put into the oral cavity, wherein the vapour phase of aromatic principals are carried to the nose and till the end of chewing when it is ready for swallowing. These are the minute yet important aspects which decide the consumer reaction and the possibility of repeated purchase of a product. So, aroma plays an important role in determining the acceptability of a product.

It is believed that a fragrance is perceived by the relative volatility of its components. It has a top note, middle note and a bottom note. A new technology has been developed that proved the fragrance is perceived by the diffusion of molecules is an inherent property of the compounds and is independent of their molecular weight, boiling point and odour threshold or odour profile [31]. With the awareness that aroma responses while tasting are due to many compounds some of which are of very low threshold, attempts to correlate the instrumental analysis to sensory responses have been made. Though, the presently available computerized multivariate analysis can provide a measure of the complex relationship, care is necessary to obtain meaningful correlations. Odour-sensing systems are needed in different fields like the food, drink and cosmetic industries. Development of a rapid and reliable odour sensing method has been made [32]. An odour sensing system, using QCM assay and neural network pattern recognition has been applied to whiskey aromas, perfumes and flavours. With all the above inputs, the food production units can achieve very precise estimations of specific compounds and routine analysis for selected components. But this type of analysis will not provide any clue to possible perceivable change and consumer's opinion. Therefore, sensory analysis for aroma is necessary as it provides the important information along with instrumental analysis. The technique cannot be replaced as the total aroma sensation is a perceived sensation and has to be described. Sensory analysis for aroma, has always been considered to be simple and straight forward, but if carried out without due care, it becomes difficult to explain and can be misguiding too. However, with the advancement in sensory evaluation techniques much of the subjectivity of tests is eliminated and reproducible, objective results are obtained. It is important that the techniques should be adopted with care for panel selection and training as already discussed. A significant move was made by Arthur D'Little group in developing flavour profile [33]. Flavour profiling involves the description of a product in user's vernacular and popular language.

In sensory analysis procedure with few discriminative/communicative panel, description of the perceived sensation needs to be understood unambiguously by the panel. Attempts were made to use the chemical terms which could not be defined in the popular language such as diacetyl. To describe a flavour sensation involved a large number of terms, therefore, a multivariate analysis is used to reduce their numbers. An extensive survey of terms has been carried out and 53 different terms have been described and further classified in 4 different groups [34]. Sample of the food material like ginger can be fractionated into sub components which could be identified by the TLC, HPLC or column chromatography. As an illustration, TLC chromograph using different solvents systems and identification of the component, product profile was developed [35]. The selected terms can be used to give the product profile of samples of different types such as garam masala comprising black pepper, fennel, mace, cardamom, cinnamon, etc. The mean intensity can be given in the form of spider web diagram (Fig. 5). The flavour profile of ginger has similarly been made. Use of terms in a free manner has also been advocated instead of pre-tested one. The computer application is made extensively to analyze the data. To obtain the statistical authenticity of the data and appropriate conclusions, recently developed techniques can be utilized.

The ultimate objective of most of the investigations is the commercial utilization of analysis in the food products. Therefore, to make more desirable aroma or flavour or provide equally satisfactory substitute, the sensory examinations become imperative. For this, three distinct groups emerge; the expert observers, the trained observers and the untrained observers. The liquor industry is a classical example using these groups [36]. They need the experts with many years of experience with built-in expertise to recognize minute changes in finished product or even on-line during intermediate stage of production. The ability to recall from memory the built-in knowledge of their product is helpful for the producers. The expert flavourist may be searching to specific odour notes and the knowledge stays as in-house records. In contrast to this, the untrained panel can at best give a hedonic response only. Since odour memory is difficult to be retained, it can be stored for 6-10 seconds; the untrained panel should never be used for intricate test involving odour or aroma evaluation.

Earlier, it was less amenable to statistical analysis though more informative. The data could now easily be analyzed and interpreted using multivariate analytical techniques like Quantitative Descriptive Analysis (Fig. 6), Principal Components Analysis, (Fig. 7). Cluster Analysis, and Canonical Variate Analysis Descriptors (Fig. 8), which have been developed for a number of products including fermented beverages like beer, wine, gin [37, 38]. The specific role of flavour profile technique is (a) to support or interpret instrumental methods of quality analysis-quality to be built, not examined at the

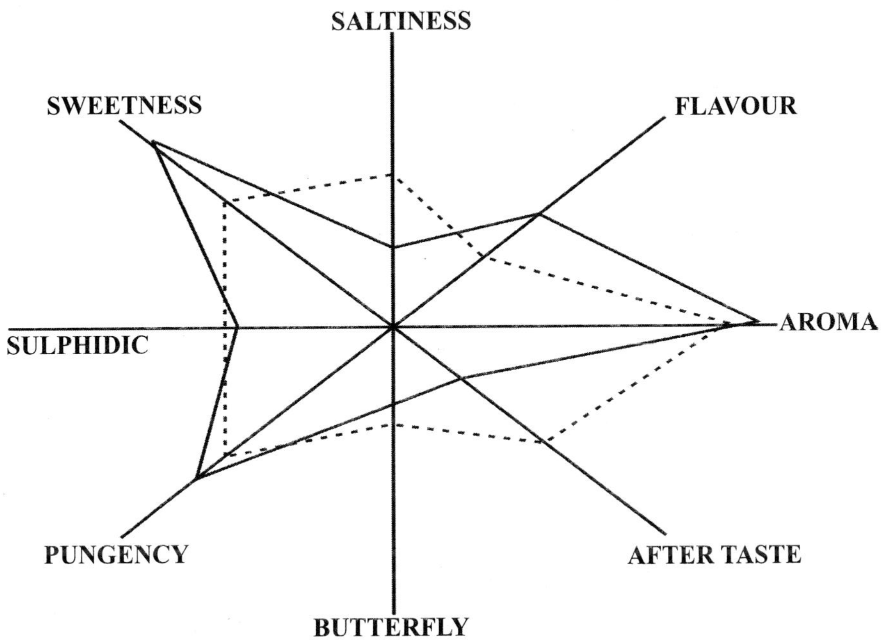


**Fig. 5:** Spider web diagram.

end; (b) to support or interpret other sensory tests where it answers the 'why' or the 'how' of the result; (c) research guidance tool — in product development with descriptive analysis; (d) quality control/ quality assurance — trace quality and intensity differences; (e) easy understanding by visualization; (f) fruit maturity evaluation; and (g) profiling the product as effected by treatment for improvement.

## 3.4 Analysis of Taste

In terms of classic sensory modalities, olfaction and taste represent together with cutanous modalities of machano, thermo and nocicaption, the group of 'lower senses'. The sense of smell or taste is categorized as 'chemical senses' because the primary sensory process initiating a taste or smell sensation, is chemical binding of substances with taste or smell properties, to the surfaces of membranes of the respective receptor cells or their processes. But tasting involves far greater contact of the sense organ and the material being tasted than smelling. The fundamental concept of perception of sense of taste is the perception of the stimulation of gustatory nerves. Four basic tastes which are well recognized are sweet, salt, bitter and sour sensations. The newer ones added are of pungency and umami (from

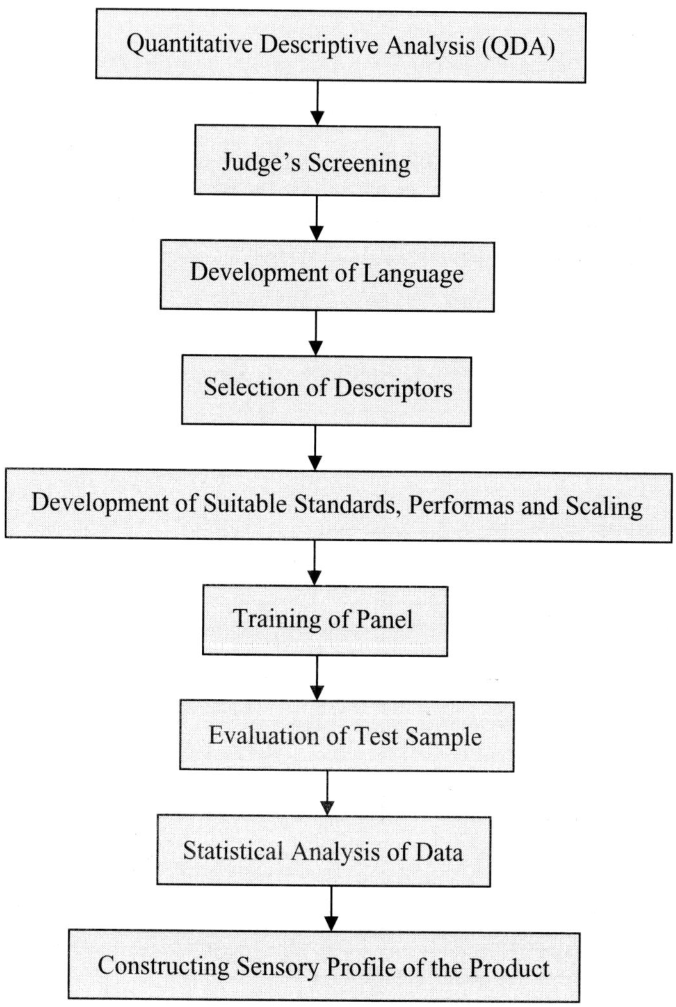


**Fig. 6:** Flow diagram of procedure of quantitative descriptive analysis of a food product.

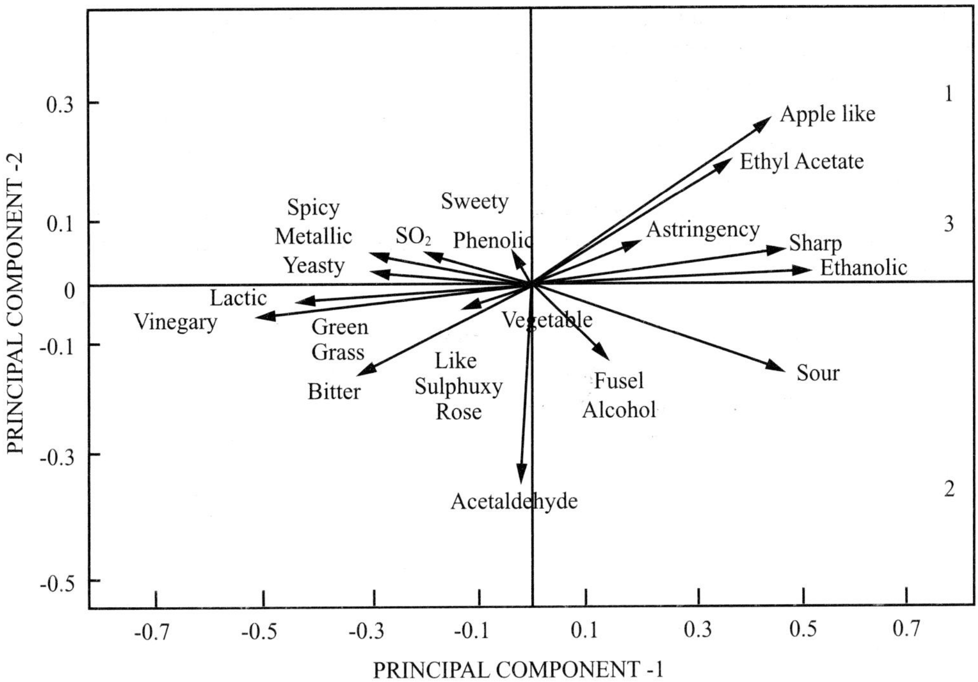


**Fig. 7:** Project of flavour profiling data of apple wines of various treatments in planes defined by the principal component 1 and 2 (1= apple wine from apple juice concentrate, 2= wine from apple juice and sugar, 3= wine from apple juice and apple juice concentrate).

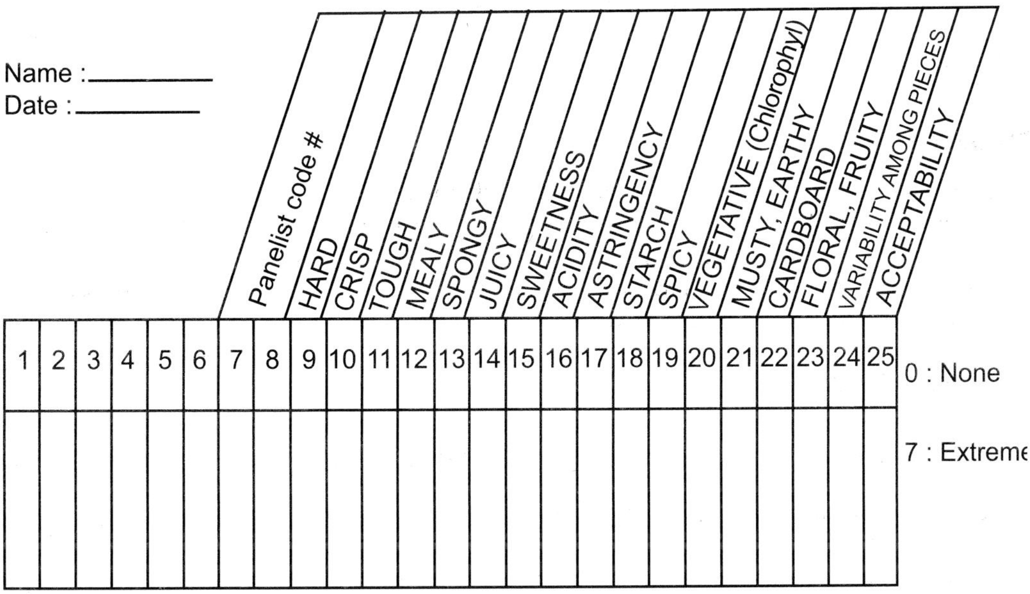


**Fig. 8:** Form used to record intensities of attributes. Column 1 through 6 were used to identify cultivars, harvest data, storage period and sample number. Data were key presented directly from the form.

nucleotides). The combined stimulation of taste buds and olfactory receptors is called the 'Flavour', where the non-volatile components contribute to taste.

The process and nature of receptive and transductive process in human taste is yet to be understood. Most of the information on the tasting phenomenon is from vertebrates rather than man. It is widely accepted that all of gestation is initiated by adsorption and chemical binding of target molecules to quality specific acceptor molecules, which seems to be located on the surface apical taste cell membranes. There, it could possibly be a change in the ultra structural features of acceptor molecules as a result of binding of stimulating molecules, changing thereby, the permeability of the membrane which in turn could stimulate or excite the taste cells. However, different workers have also suggested changes in the phase boundary potential associated with fixed charges at their apical membranes, release of specific proteins especially for sweet and bitter besides release of secondary messenger from permeable membrane. Interference of acetylcholine-esterase in the excitation of taste cells could also be thought of, as indicated by their presence in the taste cells.

For the food manufacturers, analysis of food for the basic taste is easier, both by instrumental analysis and sensory parameters. Definite methods of estimation for acid content of food, pungent principles and to some extent sugar and salt is available. But it should be kept in mind that these estimations should be meaningful for the sensory perception. This is understood clearly when we examine the degree of sweetness imparted by different sweeteners and find a variation. Attempts have been made by psycho-physical studies to establish a relationship between concentration of stimulus and response. Steven's power function law first applied in auditory studies is also applicable here. The formula $S = KC^n$, where S indicates the taste intensity, C the concentration of substance, K and n are experimentally determined constants. The exponent 'n' shows how rapidly sensory perception increases with increase in concentration. In sweetness and saltiness $n > 1$ which indicates that change in concentration of 10:1 will be perceived as more than 10: 1 sensorily. Bitterness on the other hand is governed by a function $n < 1$. These findings are applicable to taste of stimulants singly. However, the situation will be different if factors and their interactions also occur in the products which are tasted. There have been some attempts to quantify sensory responses and measure taste intensity both in simple and in beverage system including the use of descriptors for evaluation of such beverages.

### 3.5 Over-all Quality

Depending upon the objective of the experiment, either individual quality attributes, over-all quality or both could be evaluated. For simple screening type of experiments, ranking is done conveniently. However, for determining the attributes and over-all quality composite scoring or weighed scoring can also be used. Since over-all quality measurements tests for the separate characteristics can also be included, therefore, the methods used and discussion pertaining to these tests shall be applicable to these parameters also. Hedonic tests, though more suitable for consumers research, are also used in the overall quality determination.

## 4. SENSORY ANALYSIS AND CONSUMER REACTION

Consumer reaction about the food product forms the basis for the success or failure of any such product launched in the market. Consumer sensory analysis through survey is normally employed by the food companies before commercial production and marketing of the food products. Normally, one or two products, based on the research are tried for consumer survey. After a product has been optimized, it is always desirable to study the target population preferences using one or two formulations. This can be achieved by correctly identifying the target population and studying the preferences. The sample size required should be around 1000 for correct segregations [39, 40]. Here the sensory analysis conducted with the help of consumer differs from sensory analysis conducted in the laboratory.

In consumer analysis, the food habits play a very significant role. The chronological outline of food habits includes food attitude, food selection, food acceptance/sensory evaluation, food intake and food waste. Food attitude signifies the opinion about food made by the consumer and is associated with the growth of person from childhood. The food selection refers to the act of taking an offered food item. Sensory evaluation indicates the various attributes of food contributing to its acceptance. Food intake signifies the consumption of selected food while waste is the part of food which is not consumed after selection [2]. Either a written questionnaire is supplied or the spoken interview is conducted to determine the acceptability of foods. The term 'survey' is used to refer to either of the method. Out of the two methods, the interview is a labour intensive, time consuming in data collection and data reduction. Analysis need trained persons, provides only one opportunity without follow-up and can influence the person in a big way. The second method where written questionnaire is provided requires more knowledge about the subject matter than the interview. A questionnaire needs substantial time for its development, can be sent by mail, given to individual or small group, chances of loosing data are more due to lack of survey returns, needs less trained persons for carrying out the survey and it can be used as a tool to collect a lot of information at comparatively less costs. The tests are to be very simple with added questions on food habits and the demographic and economic details depending upon the requirement of experiment.

It is better if the samples are prepared under controlled conditions and the test conducted under unbiased conditions at a central location. This will avoid the individual variations in preparations and the sample identity need not be disclosed. With this, it will be possible to predict the consumer acceptance of the product. While conducting such research trial, it should be borne in mind that a consumer oriented optimization study should never be conducted in isolation [41]. If competitive products are not included at this stage, the optimal formulation must subsequently, be tested against such products. The data after survey have to be analyzed to get the clear picture of results. Before attempting to analyze the data statistically, it is equally important to examine the data for linearity or non-linearity. Cost related data could also be included to get better insight for optimized products. Models and optimal formulations for the total samples and sample segments are compared to determine that the total sample formulation represents some composite of opinion differences. This will depend upon the target population. Consumer oriented optimization survey is a powerful tool and the successful companies use it as a routine procedure.

## 5. SENSORY EVALUATION AND RECOMBINANT DNA TECHNOLOGY

Recombinant DNA technology is an important sub-discipline of biotechnology which is traditionally known to be associated the with the production of food and flavours. The recombinant DNA technology has become an important part of the research on flavour. Production of *cis*-3 hexenol, also referred to as leaf alcohol and its esters having high demand being used in various fruit flavours have been introduced in the recombinant yeast cells [42]. The gene has been transferred into the plant. Similarly, tarty peptides such as those found in various meat, cheese, fish, yoghurts have been synthesized either chemically or enzymatically. The recombinant DNA technology approach has been applied to produce octa peptide known as beefy meat peptide which is suggested to enhance the taste of beef gravy. Transgenic plants with ability to produce geraniol, citronelliol have also been developed. Enzymes such as protease, used to generate food protein hydrolysates and lipases for production of natural esters, can be produced using DNA technology as reviewed [42]. Enzymes play a key role in the generation of taste. A number of enzymes are involved in production of molecular having sensory significance [43]. The biotechnical approaches make use of one or more enzymes from plants microbes as such or in isolation form. For example, a multifunctional ingredient is made by fermenting either tomato solids, meat extract or whey by lactic acid bacteria. In all these applications, sensory evaluation techniques are employed to determine the effect of the technology. Another application of DNA recombinant technology in the development of screening system is based on the gustatory and olfactory receptors. It has also been tried in the research on olfactory system and is essential to advance our knowledge of understanding of olfactory and taste organs. Needless to say that it can lead to discovery of novel odour and taste modifying compounds in future.

## REFERENCES

1. Anon (1987) *Basic Techniques in Sensory Analysis of Food,* A course material, Department of Sensory Analysis, Central Food Technological Research Institute, Mysore, India.
2. Piggot JR (1988) *Sensory Analysis of Foods*, Elsevier Applied Science, London, UK.
3. Robert L (1992) Food crops sensory evaluation. In: *Encyclopedia of Food Science and Technology*, Vol 3, Hui YH (ed), John Wiley & Sons, New York, USA, pp 1023-1030.
4. Joshi VK (2006) Sensory Science: Principles and Application in Food Evaluation. Agrotech Publishing Academy, Udaipur, pp 1-800.
5. Temussi P (2006) Natural sweet taste: not exacting a piece of cake. *J Mol Recognit* 19: 189-199.
6. Cui Liang P, Malliet M, Max M, Margolskee RF & Osman R (2006) The heterodimeric sweet taste receptor was multiple potential legend binding sites. *Curr Pharma Cent Des* 12: 4391-4400.
7. Meyer B & Brewer MS (2008) Sweet taste in mini-review. *J Food Sci* 73: 81-90.
8. Fredholm BB, Hokfelt T & Miligam G (2007) G-protein coupled receptors: an update. *Acta Physiol* 190: 3-7.
9. Singer MS & Shepherd GM (1997) Toward a rational structure-function analysis of odour molecules: The olfactory receptor TM 4 domain. In: *Flavours and Fragrances,* Swift KAD (ed), The Royal Society of Chemistry, Cambridge, UK, pp 3-10.
10. IFT (1981) Sensory evaluation guide for testing food and beverage products. *Food Technol* 35: 50-59.
11. Stachy A, Kimmol MSG & Guinard JX (1994) Sensory testing with young children. *Food Technol* 48: 92-99.
12. Mastrain LK (1985) The sensory evaluation programme within a small processing operation. *Food Technol* 39: 127-29.
13. Amerine MA, Pangborn RM & Roessler EB (1965) *Principles of Sensory Evaluation of Foods,* Academic press, New York, USA.
14. ISI (1971) Guide for sensory evaluation of foods: Part 2. Methods and Evaluation Cards. IS 6273 Part 2.
15. Govindarajan VS (1980) Optimizing sensory evaluation. *ISI Bull* 32: 58-63.
16. ASTM Committee E-18 (1981) Guidelines for the selection and training of sensory panel members, STP 758. American Society for Testing and Materials, Philadelphia, USA.
17. ISI (1976) Guide for selection and training of panel for sensory evaluation of foods and beverages. IS 8140.
18. Govindarajan VS (1981) *Problems and possibilities of sensory evaluation,* Proceedings of II$^{nd}$ Indian Convention of Food Scientists & Technologists, AFST, Mysore, India.
19. Mackinney G & Little AC (1962) *Colour of Foods,* AVI Publishing Co, Westport, USA.

20. Francis FJ (1977) Colour and appearance as dominating sensory properties of foods. In: *Sensory Properties of Foods,* Birch CG, Brennan JG & Parker KJ (eds), Applied Science Publishers, London, UK, pp 26-44.
21. Hutchings JB (1977) The importance of visual appearance of foods to the food processor and consumer. In: *Sensory Properties of Foods,* Birch GG, Brennan JG & Parker KJ (eds), Applied Science Publishers, London, UK, pp 45-56.
22. Halmos AL (1997) Food texture and sensory properties of dairy ingredients. *Food Australia* 49: 169-173.
23. BSI (1975) *Glossary of terms relating to the sensory analysis of food.* BS 5098, British Standard Institution, London, UK.
24. Jowitt R (1974) The terminology of food texture. *J Texture Studies* 5: 351-358.
25. Szczesniak AS (1963) Classification of textural characteristics. *J Food Sci* 28: 385-389.
26. Oldfield RC (1960) Perception in the mouth. In: *Texture in Foods*, Monograph No.7, Society of Chemical Industry, London, UK, pp 3-35.
27. Brandt MA, Skinnir EZ & Coleman JA (1963) Texture profile methods. *J Food Sci* 28: 404-410.
28. Civille GV & Szczesniak AS (1973) Guidelines to training a textural profile panel. *J Texture Studies* 4: 204-223.
29. Boelens MH (1997) Production, chemistry and sensory properties of natural isolates. In: *Flavours and Fragrances,* Swift KAD (ed), The Royal Society of Chemistry, Cambridge, UK, pp 70-86.
30. Rossiter KJ (1997) The design and synthesis of novel muguet materials. In: *Flavours and Fragrances,* Swift KAD (ed), The Royal Society of Chemistry. Cambridge, UK, pp 21-35.
31. Mookerjee BD, Palee SM, Trenkle RW & Wilson AR (1997). Aura of aroma: A novel rcchnology to study the emmission from the skin. In: *Flavours and Fragrances,* Swift KAD (ed), The Royal Society of Chemistry, Cambridge, UK, pp 36-47.
32. Ide J, Nakamoto T & Moriizumi T (1997) An odour sensing system for use in measuring volatiles in flavours and fragrances using OCM. In: *Flavours and Fragrances,* Swift KAD (ed), The Royal Society of Chemistry, Cambridge, UK, pp 87-95.
33. Cairncross SE & Sojostrom LB (1950) Flavour profiles: A new approach to flavour problems. *Food Technol* 4: 308-319.
34. Harper R, Bate-Smith ECL, Land DG & Griffiets NH (1968) Odour description and odour classification. *Perfumery Essent Oil Record* 59: 22-37.
35. Govindarajan VS (1982) Ginger: Chemistry, technology and quality evaluation Part 2. *Crit Rev Food Sci Nutr* 17: 189-258.
36. Williams AA (1975) The development of vocabulary and profile assessment method for evaluating flavour contribution of cider and berry aroma constituents. *J Sci Food Agric* 26: 567-582.
37. Clapperton JF (1979) Sensory characterization of the flavour of beer. In: *Progress in Flavour Research,* Land DG & Nursten HF (eds), Applied Science Publishers, London, UK, pp 1-14.
38. Wu LS, Bargamann RE & Powers JJ (1977) Factor analysis applied to wine descriptors. *J Food Sci* 42: 944-952.
39. Box GEP, Hunter WG & Hunter JS (1978) *Statistics for Experiments: An Introduction to Design, Data Analysis and Model Building,* John Wiley and Sons, Inc, New York, USA.
40. Cochran WG & Cox GM (1957) *Experimental Designs,* John Wiley and Sons, Inc, New York, USA.
41. Fishken D (1983) Consumer oriented product optimization. *Food Technol* 37: 49-52.

42. Muheim A, Hausler A, Schilling B & Lerch K (1997) The impact of recombinant DNA technology on flavour and fragrance industry. In: *Flavours and Fragrances,* Swift KAD (ed), The Royal Society of Chemistry, Cambridge, UK, pp 11-20.
43. Laane C, Rietjens I, Haaker H & William V (1997) Generation of taste through (Redox) biocatalysts. In: *Flavours and Fragrances,* Swift KAD (ed), The Royal Society of Chemistry, Cambridge, UK, pp 138-152.

# Biotechnology in Waste Utilization

26

V.K. Joshi and Dev Raj

*Department of Food Science and Technology, Dr. Y.S. Parmar University of Horticulture and Forestry, Nauni, Solan 173 230 (H.P.)*

## 1. INTRODUCTION

The growing, processing and preparation of food result in the production of at least some degree of waste material. The waste material may be in the form of leaf/straw, waste during harvesting, food preparation waste, after processing waste, live excretory waste during rearing, unused material as waste, food processing operation waste like waste water during washing operation [1]. So, food processing is an important sector, the scope of which has increased manifold by which wastage can be avoided on a large scale, particularly in the area of perishable commodities. The changing food habits, depleting joint families and increasing number of working women in towns and cities generated increasing demands for processed foods in recent times. At present, the growth of processed food and packaged food is put at 20 per cent per annum. The changing scenario in rural and urban sectors provides a good market for food industries. The huge wastage of the food material is due to lack of control on such agri-economy practices. So there is a strong need for a sound policy framework with enough fiscal and technical inputs to match India's food status in the global context.

The food processing industries generate a large quantity and variety of food products, provides employment to a large number of people, and uplift their economic status [2]. At the same time, food processing industries generate a huge quantity of biodegradable waste, causing health hazards due to environmental pollution [3]. So, the disposal of waste material has become a challenge to the processors, as different agencies are pressurizing for an environment friendly treatment of waste material. The types of waste from food industry, their pollution load and composition, their utilization for the production of value added products by biotechnological processes are the aspects that are very important in the management of food processing waste. Status of industrial exploitation of the processes and the future strategies for minimizing pollution load coupled with recovery of useful products or transformation of waste into various products have been discussed in this chapter.

## 2. MAGNITUDE OF THE PROBLEM

Food processing industries produce a large quantity of solid wastes in India. Besides this, the waste water is also generated by the food industry, the quantity of which is many times more (Table 1) than the solid waste [3]. Total availability of agro-industrial, agricultural and municipal solid wastes

*Food Biotechnology: Principles and Practices*
*V.K. Joshi & R.S. Singh (Eds.)*
*I.K. International Publishing House Pvt. Ltd., India*
*pp 881–914*

in India is presented in Table 2. Different types of microorganisms associated with solid wastes of food industries result in the production of metabolic end products causing chemical and biological environmental pollution. The extent of waste produced as a result of processing is shown in Tables 3, 4 and 5. The major sources of pollution load in fruit and vegetable processing industry are illustrated in Fig. 1.

**Table 1:** Pollution load of waste water from various food industries effluent COD (mg/l).

| Industry | Effluent COD (mg/l × 1000) |
|---|---|
| Starch gluten manufacture | 9.0 |
| Enzyme production | 4.9 |
| Vegetable processing | - |
| Sugar beet refinery | 10.4 |
| Wine distillery | 29.5 |
| Cane molasses distilleries | - |
| Vegetable canning | 30.0 |
| Potato processing | 3.0 |
| Starch production | 11.0 |
| Yeast | 30.5 |

*Source:* 3

**Table 2:** Total availability of agro-industrial, agricultural and municipal solid wastes ($10^6$ metric tonnes/year).

| Waste | India | Brazil | Sudan | USA | Sweden |
|---|---|---|---|---|---|
| Municipal solid wastes | 135.5 | 44.0 | 2.3 | 148.0 | 5.3 |
| Sewage | 44.9 | 8.02 | 1.4 | 16.0 | 0.6 |
| Manure | 635.0 | 470.4 | 68.0 | 306.0 | 13.2 |
| Agricultural residues | 200.0 | 47.0 | 8.1 | 573.0 | 12.6 |
| Biomass | 140.0 | 496.8 | 192.3 | 427.0 | 14.0 |

*Source:* 4

**Table 3:** Quantities of waste generated from various fruit and vegetable processing industries.

| Commodity | Waste (%, w/w) |
|---|---|
| Apple | 12-47 |
| Apricot | 8-25 |
| Grape fruit | 3-58 |
| Orange | 3 |
| Peach | 11-40 |

Contd...

Contd...

| Commodity | Waste (%, w/w) |
|---|---|
| Pear | 12-46 |
| Asparagus | 3.2-30 |
| Bean, green | 5-20 |
| Beet | 7-4 |
| Broccoli | 20 |
| Cabbage | 5-25 |
| Carrot | 18-52 |
| Peas | 6-79 |
| Potatoes | 5 |
| Spinach | 10-40 |
| Sweet potato | 15 |
| Tomato | 5-25 |

*Source:* 5

**Table 4:** Quantity of fruits and vegetable processing wastes available in India per annum.

| Fruit/ Vegetable | Nature of waste | Production (Tonnes) | Approx. waste (%) | Waste (Tonnes) |
|---|---|---|---|---|
| Mango | Peel, stones | 6987.7 | 45 | 3144.4 |
| Banana | Peel | 2378.0 | 35 | 832.3 |
| Citrus | Peel, rag and seed | 1211.9 | 50 | 6-06.0 |
| Pineapple | Skin, core | 75.7 | 33 | 24.7 |
| Grapes | Stem, skin and seeds | 565 | 20 | - |
| Guavas | Peel and core and seeds | 565 | 10 | - |
| Peas | Shell | 107.7 | 40 | 68.3 |
| Tomato | Skin, core and seeds | 464.5 | 20 | 90.3 |
| Potato | Peel | 2769.0 | 15 | 415.3 |
| Onion | Outer leaves | 1102.0 | - | - |
| Apple | Peel, pomace and seeds | 1376.0 | - | 412.0 |

*Source:* 6

**Table 5:** Average quantity of some of the by-products made for every 100 MT sugarcane ground.

| By-products | Quantity |
|---|---|
| Raw sugar (98.8° Pol) | 11.2 |
| Surplus bagasse (49% moisture) | 5.0 |
| Molasses (89° B) | 27 |
| Filter mud (80% moisture) | 30 |
| Furnace ash | 0.3 |

*Source:* 7

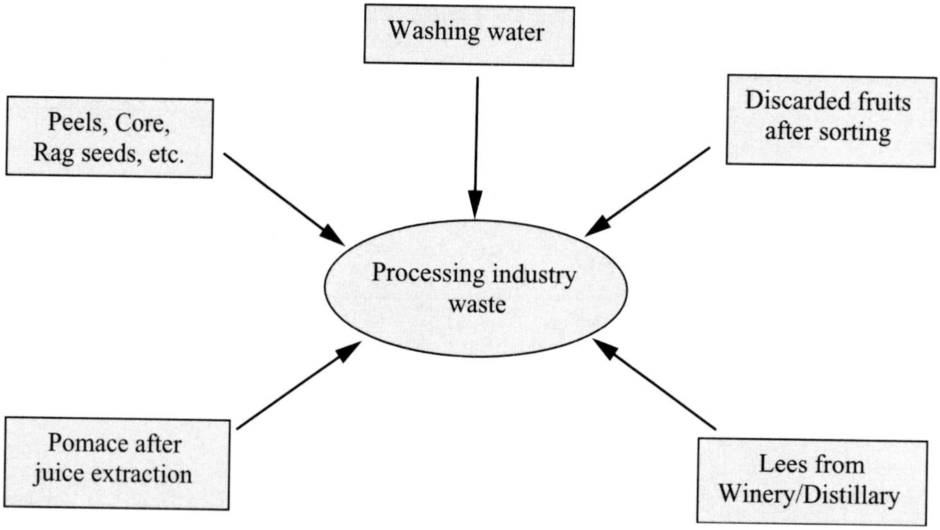


**Fig. 1:** Major sources of pollution load in fruit and vegetable processing industry. *Source:* **8**.

Different types of food industries wastes are: (i) fermentation industry waste, (ii) fruit and vegetable industry waste, (iii) Sugar and starch industry waste, (iv) diary industry waste, (v) meat processing industry waste, (vi) coffee industry waste and (vii) palm oil industry waste [9].

## 3. QUALITY ATTRIBUTES OF THE WASTE MATERIALS

The physico-chemical constituents of different solid wastes from various food processing industries is presented in Tables 6 and 7. Some of the fruit and vegetable wastes are a rich source of vital constituents like carbohydrates, proteins, fats, minerals, fibres, etc. Mango seed kernel is rich in carbohydrates, fats, proteins, minerals; orange, melon and pumpkin seeds can provide fats and mineral matters. Most of these wastes are nutritionally rich.

Conventional aerobic and anaerobic treatment processes are applied to aqueous food processing waste [1]. Aerobic treatment is generally suitable for highly diluted effluent and is designed to yield a final outflow containing minimal BOD and suspended solids levels. A drawback with aerobic treatment is the problem of sludge disposal, which is often achieved only at additional cost. Anaerobic treatment of food waste generally has an application where an aqueous effluent stream has a very high level of BOD. In this treatment, BOD is reduced to such a value which can be handled in an aerobic system. The reaction rates in anaerobic digesters are much slower than aerobic digesters. Anaerobic treatment results in the production of methane gas, which can be burnt as fuel [1]. So, the strength of waste can be defined in terms of biological oxygen demand (BOD) and chemical oxygen demand (COD), and characterized on the basis of their BOD and COD. The BOD and COD of the food industry waste are presented in Table 8. The amount of pollution load and characteristics of the waste depend on the food being processed.

### 3.1 Food Processing Waste and Their Legal Standards

Earlier environmental pollution control board/agencies/legislation usually compelled the generator of waste to pay for the disposal of waste. The cost of its disposal varied depending upon the quantity

**Table 6:** Composition (%) of food processing wastes.

| Constituents | Mango peel | Tomato peel and seeds | Lemon wastes | Deoiled orange waste | Orange peel | Pineapple waste |
|---|---|---|---|---|---|---|
| Moisture | 73.6 | 70.5 | 11.5 | 12.0 | 11.7 | 6.6 |
| Total solids | 26.4 | 29.5 | 88.5 | 88.0 | 88.3 | 93.4 |
| Volatile solids | 96.5 | 95.7 | 95.8 | 93.6 | 94.2 | 92.2 |
| Ash | 3.5 | 4.3 | 4.2 | 6.4 | 5.8 | 7.8 |
| Total carbon | 39.8 | 54.9 | 42.2 | 43.1 | 42.5 | 38.3 |
| Ether extract | 8.7 | 21.0 | 7.8 | 0.7 | 4.3 | - |
| Total nitrogen | 0.56 | 4.2 | 2.46 | 2.22 | 1.09 | 0.85 |
| Total protein | 3.5 | 23.3 | 15.37 | 13.88 | 6.80 | 5.31 |

*Source:* 6

**Table 7:** Composition (%) of different fruit wastes.

| Waste | Moisture | Protein | Fat | Minerals | Fibre | Carbohydrate | Reference |
|---|---|---|---|---|---|---|---|
| Apple pomace | - | 2.99 | 1.71 | 1.65 | 16.16 | 17.35 | 10 |
| Mango seed kernel | 8.2 | 8.50 | 8.85 | 3.66 | - | 74.49 | 11 |
| Jack fruit (inner and outer portion) | 8.5 | 7.50 | 11.82 | 6.50 | 30.77 | 14.16 | 12 |
| Jack fruit seeds | 64.5 | 6.60 | 0.40 | 1.20 | 1.50 | 25.80 | 13 |
| Jack seed flour | 77.0 | 2.64 | 0.28 | 0.71 | 1.02 | 18.12 | 14 |
| Passion fruit peel | 81.9 | 2.56 | 0.12 | 1.47 | 5.01 | - | 15 |
| Banana peel | 79.2 | 0.83 | 0.78 | 2.11 | 1.72 | 5.00 | 16 |
| Sweet orange seeds | 4.00 | 15.80 | 36.90 | 4.00 | 14.00 | - | 17 |
| Watermelon seeds | 4.3 | 34.10 | 52.60 | 3.70 | 0.80 | 4.50 | 13 |
| Muskmelon seeds | 6.8 | 21.00 | 33.00 | 4.00 | 30.00 | - | 18 |
| Pumpkin seeds | 6.0 | 29.50 | 35.40 | 4.55 | 12.00 | 12.53 | 19 |
| Banana stem | | | | | | | |
| Central core | 93.1 | 0.30 | 0.03 | 1.04 | 0.68 | 1.20 | 20 |
| Outer hard | 91.9 | 0.12 | 0.06 | 0.98 | 1.81 | 2.44 | 20 |
| Fibrous sheath | | | | | | | |
| Press juice from stem | 98.6 | 0.05 | - | 0.63 | - | 0.41 | 20 |

and strength of the effluent flowing from the food processing factory. Now a days, it has become mandatory by law to install effluent treatment plant in each food processing factory, the lack of which enforces environmental regulation agencies to take strict actions against the owners of the factory. However, to install an effluent treatment plant adds an increase in both capital and operating costs. So there is a great need to find out the options which have positive values of food waste. The options having positive values are: (i) minimization of the waste generation, (ii) development of value added products [1]. So, stringent standards have been laid down by the regulatory agencies for the discharge of industrial effluent under the Water Prevention and Control of Pollution Act, 1974 and Environmental Protection Act, 1986 (Table 9). These standards are mandatory for the processing industries, necessitating adoption of suitable pollution control measures before the waste can be discharged. To look upon the vast microbial capabilities to transform this waste into economically useful products would be a judicious and appropriate approach.

## 4. OPTIONS FOR UTILIZATION OF FOOD INDUSTRY WASTE

It is clear from the physico-chemical constituents of the food industries waste that these wastes are rich source of vital constituents (Tables 6 and 7). So there is an urgent need to find out suitable options for utilization of the waste materials. Waste material should be utilized in such a manner that after its treatment, best possible benefits can be extracted without any monitory loss or environmental hazards. To get positive results, waste material obtained from food industries can be utilized for composting and landfill by aerobic or anaerobic treatment, biogas production and production of value added products. In this way, recycling of the waste material into useful products can solve the problem of waste disposal and pollution hazards. So, utilization of the waste material into a useful end product can be done by following one of the methods mentioned below:

### 4.1 Composting and Landfill

Waste material procured from the food processing industries can be used as a suitable substrate for composting. The multifold increase in production of processed products during past few decades has also enhanced the quantity of waste material after processing. So, proper disposal of waste material is

**Table 8:** Characteristics of food industry wastes.

| Product | COD (mg/l) | BOD (mg/l) |
|---|---|---|
| Apple | 395-37,000 | 240-19,000 |
| Beets | 445-13,240 | 530-6,400 |
| Carrots | 1,750-2,910 | 817-1,927 |
| Cherries | 1,200-3,795 | 660-1,900 |
| Green beans | 78-2,200 | 43-1,400 |
| Peas | 729-2,284 | 337-1,350 |
| Sauerkraut | 470-65,000 | 300-41,000 |
| Tomatoes | 652-2,305 | 454-1,575 |
| Wine | 4,95-12,200 | 363-7,665 |

*Source:* 3, 21

**Table 9:** Standards for waste water disposal set up by environment protection agency.

| Parameters | Limits |
|---|---|
| BOD (mg/l) | |
| Inland | 30.00 |
| On land | 100.0 |
| Sewage | 350.0 |
| pH | 5.5 |
| Total dissolved solids (TDS) | 9.0 |
| Total soluble solids | 210.0 |
| Electronic conductivity | 100.0 |
| Sulphate (mg/l) | 1000.0 |
| Chloride (mg/l) | 1000.0 |
| Chemical oxygen demand (COD) | 2000.0 |

*Source:* 22

becoming an issue which all industrialists confront, especially those settled in cities, where the land cost is very high. Moreover, over the time, the environment protection agencies have also started questioning the unorganized disposal of waste material, as it affects the surrounding environment and causes nuisance if not handled properly. So, during the recent years, the disposal of waste material has become a challenge to the processors, as different agencies are pressurizing for an environment-friendly treatment of waste material. So composting is one of the best methods for solving the problems of waste material disposal. This can be done using different methods like natural weathering, aerobic and anaerobic processes in pits (5×5×5 feet$^3$ size) for a period of 12-18 months. In natural weathering process, the pits are filled up to the top with waste material and natural recomposting process are allowed to proceed, while in aerobic process, unforced air was introduced in the perforated bottom of the pit with the help of both-end-open 2 inch plastic pipes. In anaerobic process, the pit of 5×5×5 feet$^3$ size after filling with waste material was covered with soil up to 1 feet height and left as such for 12-18 months. The manure prepared after composting is ready to use for crop cultivation (Plate 1). The earlier reports have revealed that how spent mushroom substrate (SMS) can be utilized as organic manure, which includes its utilization for nutritionally poor soil, neutralizing acidic soils and in other cases for improving polluted sites [23-25]. In addition, the SMS also possesses good biocontrol activity against certain foliar and soil-borne diseases [26]. The SMS, after subjecting it to suitable treatments, can completely or partially substitute the growing media for cultivation of horticultural crops of economic importance [27, 28]. This will help in bringing the thousands' hectors of polluted land under cultivation for agricultural and horticultural crops, thereby increasing the net area under crops and food production in the country, and above all it has a positive impact on socio-economic status of the farming community. Vermicomposting is also a viable alternative for utilization of the waste material. Leaves of the fruits, vegetables and floricultural crops can also be utilized in a similar way for purpose of composting and vermicomposting. The process of composting helps in the prevention of spread of the insects/diseases to the nearby crop or areas. So after composting or vermicomposting process, the resultant material can be used for landfilling in such a place where the soil/land is completely barren. The 12-18 months composting of the waste material results in an increase in NPK content, water holding capacity, organic carbon, etc. and thereby can increase the soil fertility. Apple pomace has been used for mushroom cultivation in SSF [29-31].

**Plate 1:** Wheat crop grown by using spent mushroom substrate (SMS).

## 4.2 Biogas Production

Biomass consisting of agricultural, forest, crop residues, solid and liquid wastes from industries, sewage and sludge can be utilized for production of biogas through microbial technology. Similarly, the waste from fruit and vegetable processing industries being rich in biodegradable substances has been used for production of biogas. Biogas is produced by anaerobic digestion of fruit and vegetable wastes. Apple pomace was used for biogas production [32]. Methanotropic bacteria like *Methanobacterium* and *Methanococus* sp. can utilize $CO_2$ from food waste materials to produce methane. During this process, the complex polymers are first hydrolyzed into simple substances by acid forming bacteria, and finally these are digested anaerobically by the methanotropic bacteria and methane gas is liberated. Pretreatment of the waste with selected strains of *Sporotrichum, Aspergillus, Penicillium* and *Fusarium,* etc. resulted in increased methane production, as they are capable of breaking complex polymers. The waste material from various processing industries like food, agriculture, fruits and vegetable processing industry can also be subjected to anaerobic treatment process (AnTP) for biogas production. A process is described for the treatment of waste water from industries. By this process when sugar beet pulp waste was loaded @ 1 kg volatile solids $m^{-3}d^{-1}$, it yielded 0.74 $m^3$ of biogas per kg of loading. Waste water from food industry under anaerobic conditions can also be utilized for biogas production using immobilized microbial cells in a bioreactor. This technique has several benefits like continuous reactor operation at any desired input, no fear of cell wash out, increased productivity, easy cell/liquid separation, coarse screening of rapid gravity settling, repeated batch operation using the same cells and enhanced gas-liquid mass transfer. Fluidized bed bioreactor (FBR) is a common immobilized bioreactor used in the anaerobic treatment of food industry waste. Various types of reactors used for the production of biogas are illustrated in Fig 2. Orange peel can be digested anaerobically to yield biogas containing 50-55% methane as fuel gas [34].

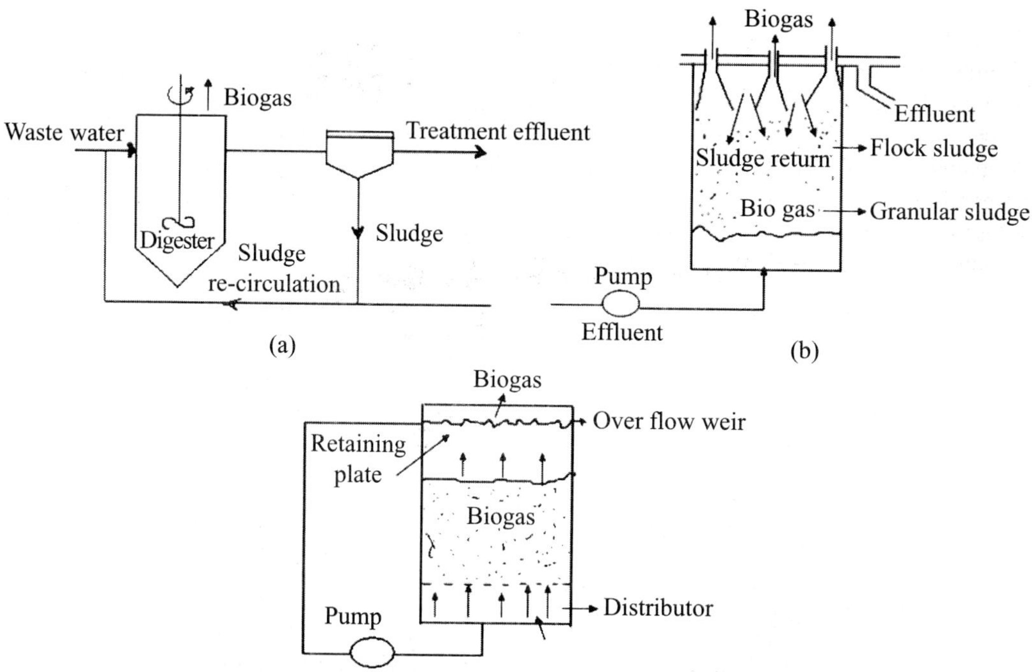


**Fig. 2:** (a) Stirred tank reactor of anaerobic contact process; (b) Upflow anaerobic sludge blanket bioreactor (USAB); (c) Fluidized bed bioreactor (FBR).
*Source:* **33**

### 4.3 Production of Value Added Products

Depending upon the composition of the waste material, they can be exploited as a source material for production of different high value food products (Table 10). The important value added products generated from the agricultural commodity based waste material are discussed ahead.

**Table 10:** Possible by-products from solid wastes in fruit processing units.

| Fruit/ Vegetable | Waste (%) | Nature of waste | By-products |
|---|---|---|---|
| Apple | 20-30 | Pomace | Juice, wine, vinegar, pectin, cattle feed |
| Citrus | | | |
| Orange | 50 | Peels, seeds & pulp | Essential oil, pectin, cattle feed, peel candy, etc. |
| Lime | 60 | Peels, seeds & pulp | Essential oil, pectin, cattle feed, peel candy, etc. |
| Mango | 40-69 | Peel and pulp | Pectin, cattle, feed, alcohol |
| Peels | 12-15 | Fibre | Wine, vinegar, juice |
| Pulper waste | 5-10 | Hull & kernel | Fat, tannins, starch |
| Kernels | 15-20 | Peel, core, trimming, shreds | Juice, wine, syrup, bromeline, cattle feed, biogas |
| Pineapple | 30-60 | Core, peel & seeds | Animal feed, seed oil and meal |
| Tomato | 20-30 | Core, peel & seeds | Animal feed, single cell protein |

*Source:* 35-37

#### *4.3.1 Fat*

The industrial processing of mangoes ends up with a considerable proportion of stones and peels, which are discarded and end up in wastes. The stone content of mangoes ranges from 9-23% with an average of 15% [38]. According to one estimate, 0.3 million tonnes of dry mango kernels are available annually (Table 11) from which 30,000 tonnes of mango fat valued at Rs. 20 crores could be obtained [39]. In 1985, 1035 tonnes of mango fat was produced, although its potential is reckoned at not less than 48,000 tonnes [40]. Peoples are extracting fats from the carcasse of animals and using for adulteration in ghee, which is it a bad and illegal practice.

**Table 11:** Present use and future potential of mango waste.

| | |
|---|---|
| Total production | 10.0 million tonnes |
| Estimated waste | 6.0-36.7% |
| Peels | 12-15% |
| Pulper waste | 5-10% |
| Stones | 15-20% |
| *Uses of immature fallen fruits and held wastes* | |
| Pickles (1 kg per family) | 1,75,000 tonnes |
| Amchur from 92,500 tonnes green mangoes | 37,000 tonnes |
| Estimated availability of dry mango seeds | 3,00,000 tonnes |
| Estimated fat content | 30,000 tonnes |
| Estimated value | Rs. 20 crores |
| Actual production of fat | 1035 tonnes |
| Estimated availability of starch from seed kernel | 1,40,000 tonnes |

*Source:* 38-40

### 4.3.2 Essential oils

India is producing 3 million tonnes of citrus fruits like mandarins, lime, lemon, sweet orange accordingly generating large quantities of wastes (flavedo of peel) around the processing units [35, 41]. Citrus wastes are a rich source of essential oils. Mandarin essential oil is extracted in small quantity at different processing units at Bangalore, Nagpur, Abohar (Punjab) and Sikkim [2, 35]. Mandarin oil is also extracted in small scale unit based on orange peels collected from fresh juice vending stands. Lime oil is extracted at Uttrayan, Jalgaon and Kodur (Table 12). Essential oil of citrus can be concentrated to the degree of 1:5 to 1:20. Such concentrated oils are less sensitive to oxidation and have a pleasant flavour. The oil is used in cosmetics, pharmaceuticals, perfumes and for scenting soap [42]. Physico-chemical characteristics of a large number of Indian mandarins and lime oils, both cold pressed and steam distilled have been analysed [43, 44], and these data have helped to formulate their standards and their suitability for deterpenation. D-Limonine is a major by-product from citrus industry, and at current level of production of citrus fruits in India, its potential availability is about 15,000 tonnes. In citrus, essential oil develops in receptacles on the flavedo of the peel, which is covered by protective wax. Before the juice is extracted, the fruit is washed and the skin oil is removed by passing the fruit over an oil extraction machine, which inflicts hundreds of small cuts on the skin. The oil sacs in the surface of the skin are ruptured and the oil is removed by washing. The oil and wash water are centrifuged so that the emulsion is removed from the water, which is returned to the machine. The emulsion can be removed from the juice after juice extraction, during the finishing operation. The quality and quantity of the oil depend on the ripeness of the fruit. The main steps in the extraction of essential oil are as follows:

- Extraction of the oil by mechanical pressure on the oil vesicles.
- Separation of the oil from the peel by water spray.
- Centrifugation of the water oil emulsion.
- Clarification of the crude oil in a centrifuge.
- Settling of the waxy crystals present in the oil by holding at low temperature for several weeks (winterizing).
- Separating the wax from the oil by centrifugation before packing into drums.

The essential oil can be distilled to produce different aroma fractions for use in the drink industry. An improved method for oil extraction from orange peel is discussed by Sharma *et al.* [45]. Stones fruit kernels can also be a very good source of oil [46].

**Table 12:** Utilization of citrus wastes.

| | |
|---|---|
| Total production | 3 million tonnes |
| Orange waste | 50% approx. |
| Lime waste | 60% approx. |
| Mandarin oil extracted | Bangalore, Nagpur, Abohar, Delhi |
| Lime oil extracted | Uttarayan (MS), Jalgaon and Kodur |
| Dried orange peels, rags, seed as livestock feed | Low in protein, fibre and fat but rich in carbohydrate, classed as carbohydrate concentrate |
| Liquid concentrate from above used for dairy and beef cattle | Molasses |

*Source:* 2, 35

### 4.3.3 Animal feed

Citrus peel is the main processing waste from juice industry. Orange waste in the form of peels, rags and seeds are used as a livestock feed in wet as well as dried form [35]. It is low in crude protein, fibre and fat, but rich in carbohydrate [47, 48]. Often it is classed as a carbonate concentrate [49]. The liquid concentrate made from crude citrus waste is called molasses. Its use is similar to cane molasses and is widely used for feeding dairy and beef cattle. Growing yeast on it and using it for animal feed is further helpful. Ripe banana peels (constituting about 22-30%) are found littered all around the streets, causing pollution and skidding risks. In Vietnam, these peels are sun dried for use as poultry feed (Table 13).

**Table 13:** Utilization of banana, apple and tomato waste.

| | |
|---|---|
| **Banana waste** | |
| Total production | 8 million tonnes |
| Peels | 22-30% (poultry and cattle feed) |
| Pseudostem yield | 20-25 tonnes from 1000 plants |
| Starch in pseudostem | 5% (in food products) |
| Each banana stem | Gives 20-25 leaves |
| Dried leaves | Cups and trays |
| Green leaves | To serve meals |
| **Apple waste** | |
| Total production | One million tonnes |
| Cull apples | 2,00,000 tonnes |
| No. of juice concentration unit | 3 (Parwanoo, Sopore, Jarole) |
| Total crushing capacity | 32,000 tonnes |
| Apple pomace | 10,000 tonnes approx. |
| **Tomato waste** | |
| Total production | 5 million tonnes |
| Tomato used for seeds (0.2%) | 9000 tonnes |
| Total seed production | 125-150 tonnes |
| Tomato juice wasted | 7000-8000 |

*Source:* 50

There is always an acute shortage of fodder in hilly areas during winter months. Apple drops and culls can be partly used to meet these shortages by the use of appropriate technology like slicing fruits, fumigating them and partially drying them and making them into compressed blocks. These blocks can be coated with molten wax to avoid microbial or insect attack. At the time of use, wax can be scrapped and apple block is used as animal feed. About 20,000 tonnes of apple pomace obtained in processing units is a course of great pollution in the factory surroundings. The method of conversion of pomace into ensilage has been developed in 20 tonnes batch. It can also be used to make edible products [2].

Pineapple bran has a great demand as a cattle feed in the Hawaii Island. Cannary wastes are also being fed to the dairy cattle in Hawaii after ensiling [51]. Proximate composition of the silage made out of pineapple tops and leaves with molasses and urea is almost comparable to any of the commonly used grass silage [52]. Major waste like mango seed kernel is a good source of carbohydrate and can replace around 20% of corn in growing and egg laying chicken rations [53]. Drying of mango kernels and peel can make high class energy food for animal feed.

The waste obtained from processing of fruits and vegetables is rich in fibre that includes cellulose, hemicellulose, lignin and silica with poor quantity of protein. Owing to their low protein content and low digestibility of fibre, these are poor sources of animal feed. Various methods have been tried to increase the feed value of these waste materials. With the help of SSF, it is possible to increase the crude protein content of this waste and some degradation of the fibres also takes place, so that these can be used as animal feed.

Fermented potato waste has been successfully tried as an animal feed. When sweet potato waste was fermented with fungi *imperfecti,* their protein content increased up to 31.6%. Fresh or dried apple pomace can be used as an animal feed [54]. Milching cows have been fed with apple pomace in combination with other fodders. Apple pomace, being a poor source of nitrogen is not a complete animal feed in itself. A novel approach of bioconversion of non-protein nitrogen by fermentation into proteinacious nitrogen has been advocated. Apple pomace after fermentation with different yeasts and drying becomes enriched with proteins, vitamins, minerals and fats, and it can be used for feeding animals [55-57]. When apple pomace is subjected to SSF with *Candida utilis,* there is 2.5 fold increase in protein content, 3.4 fold increase in niacin, 2.0 fold increase in pantothenic acid and 1.5 fold increase in thiamin. Coculture of apple pomace with *Candida utilis* and *Aspergillus niger* results in protein enrichment. An increase in true protein from 3.4 to 7.5% took place when apple pomace was fermented with *Kloeckera apiculata* and *Candida utilis* [58]. The apple pomace after ethanol recovery had enough nutrients to be used as an animal feed [59, 60]. Furthermore, protein obtained was rich in limiting amino acid, lysine. Waste from wineries, breweries and distilleries after fermentation can also be used for feeding livestock. Animal feed can also be obtained from grape pomace and wine lees after growing microbes on them. It was reported that dry brewer's grains after addition of molasses become a very good cattle feed. Apple pomace after fermentation with *Torula utilis, Candida utilis, Saccharomyces cerevisiae, Schizo. pombe* and *Kloeckera* spp. and distillation of ethanol, was fed to the broilers. The fermented apple pomace when mixed with standard broiler feed in the ratio of 1:1 was acceptable to broilers with no adverse effect on body weight gain and various biochemical parameters. The results indicated that reformulation of the feed could make apple poamce as one of the constituents of the broiler feed [61, 62].

During the production of cheese, a lot of whey is wasted. However the solids in the whey are potentially valuable, although in very dilute form. Whey can be used for feeding the animals, piggery, etc. and so proved a very convenient method of whey disposal [1]. Whey powder after spray drying also has its applications in animal feed. Bones left after the preparation of boneless meat can be used to feed pet dogs.

### *4.3.4 Pectin*

Most of the pectin produced in the world is extracted from citrus peels. Pectin is used mainly for jam making, pharmaceutical and several other industries. Mango peels, apple pomace (Plate 2 and Fig. 3) which are also available in plenty from processing factories have also been shown to be a good source of pectin. Apple pomace is the processing waste generated after apple juice manufacturing and represents upto 30% of the original fruit. The quality of pectin obtained from different fruit sources is presented in Table 14. In citrus, pectin is generally manufactured from the peel. The essential oil is extracted from the peel prior to pectin processing. The peel is then shredded and washed with water to remove carbohydrates and then is extracted several times with dilute hot acid to dissolve the protopectin in the peel into a water-soluble form. The extractant is filtered and concentrated to about 3-4% pectin. Alcohol is then added to precipitate the pectin as a gelatinous mass before separation using a hydraulic press. After several alcohol and water washing, the pectin is warm air dried. The dried pectin is pulverized and packed as a yellowish white powder. Rinsing of the pectin with dilute acid prior to drying produces rapid set pectin. Schematic diagram showing the manufacturing process of pectin from citrus peel is presented in Fig. 4. Because inorganic acids are used in the manufacture of pectin, heavy metal contaminants in the acids become quality control issues in the food grade pectin product. Another important quality control parameter is the degree of esterification. The number of methoxy groups of the pectin determine its gelling ability. Jelly grade is related to this parameter. As 75% of the pectin manufactured is utilized in the preparation of jam and jelly, jelly grade becomes a

**Plate 2:** Dried apple pomace and powder.

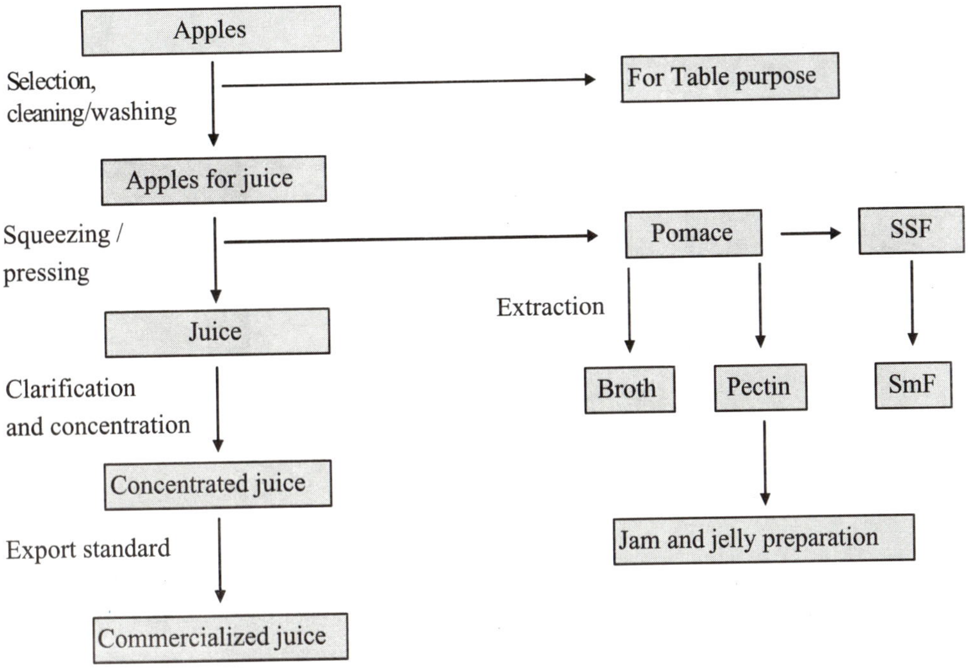


**Fig. 3:** Flow sheet for apple processing and waste utilization.

major factor in pectin quality. The jelly grade is measured by using a Ridgelimeter/gel meter. Pectin, a heteropolysaccharide is mostly prepared by a chemical method, which is tedious and a corrosive process. Pectin can also be produced by growing microorganisms on waste of processing industries [67]. Pectin has been produced from citrus peel by using *Trichosporon penicilliatum,* which produces proto-pectin solubilizing enzyme.

**Table 14:** Quality of pectin obtained from different fruit sources.

| Fruits | Yield % (dry wt) | Methoxyl content (%) | Anhydrouronic acid (%) | Jelly grade |
|---|---|---|---|---|
| Mango peels | | | | |
| Dushehari | 18-19 | 8.99 | 58.40 | 155 |
| Langra | 15-16 | 8.07 | 54.60 | 175 |
| Alphanso | 13.0 | 9.20 | 61.12 | 200 |
| Totapari | 15.7 | 8.31 | - | 200 |
| Assam lemon | 20.2 | 9.20 | 71.20 | 277 |
| Mandarin | - | 9.86 | 84.73 | 250 |
| Pumello | - | 9.31 | 79.89 | 200 |
| Galgal | 26.5 | 6.12 | 47.05 | 350 |
| Apple pomace | 15-18 | 8.40 | 59.40 | 200 |
| Raw papaya | 9.0 | 7.50 | 66.80 | 227 |
| Guava | 9.4 | 8.90 | 79.90 | - |
| Cashew apple | - | 3.50 | 45.10 | 75 |
| Fig | - | 9.28 | 78.36 | 150 |

*Source:* 2, 15, 63-66

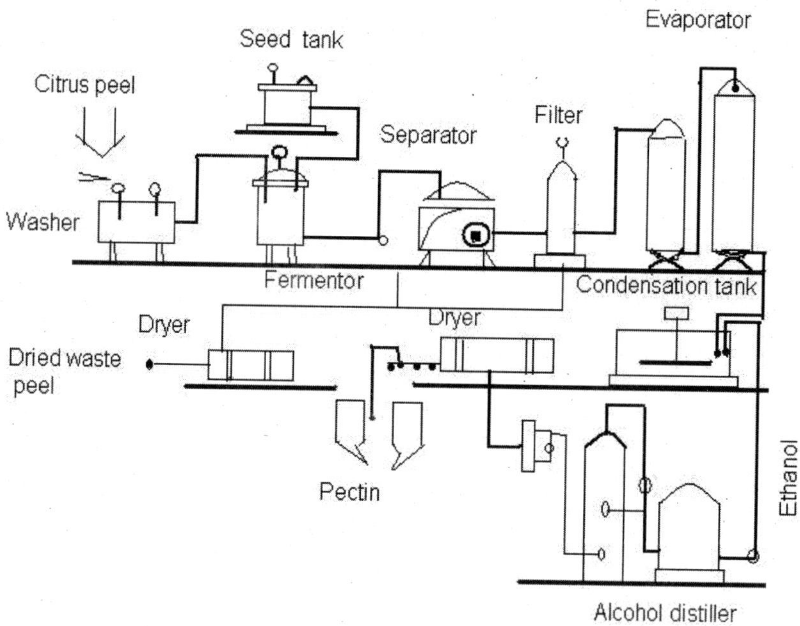


**Fig. 4:** Schematic diagram showing the manufacturing process of pectin from citrus peel.

### *4.3.5 Starch*

Banana pseudostem having a yield of 20-25 tonnes from about 1000 banana plants, when extracted, gives about 5% edible starch. The method for strarch extraction from pseudostem has been standardized and physico-chemical properties of starch have been studied. Likewise, 1, 40,000 tonnes of starch may also be available from mango seed kernels [2, 68].

### 4.3.6 Natural colours

Natural colors have been added from ancient times in our food, but with increase in demand, synthetic colours like coal tar dyes were developed which dominated the natural colours. The growing public concern over the adverse health effects of addition of synthetic colours has compelled the scientists to develop alternative to these colours. The natural colours have the action as pro-vitamin, antioxidant or possible tumor inhibiting agents. A number of microorganisms are also known to produce pigments such as *Rhodotorula, Sarcina, Cryptococcus, Monascus purpureus, Phaffia rhodozyma, Bacillus* sp. Apple pomace, due to the presence of several nutrients, proved to be a good component of medium to produce microbial pigments. Apple pomace has been utilized for the production of microbial colours using solid state fermentation [69]. It has been used successfully for the cultivation of *Rhodotorula* for pigment production (Plate 3). The effect of carbon and nitrogen sources on yield and carotenoids production by *Chromobacter* sp., a dark red colour producing bacteria showed, that apple pomace at a concentration of 20 g/l gave maximum yield of biomass and carotenoids in the basic medium [70, 71]. The use of maltose (0.3%) in the apple pomace based medium produced the maximum yield of biomass (6.65 g/l) and carotenoids (46.33 mg/100g). Amongst the different nitrogen sources tried, potassium nitrate (0.1%) gave the maximum production of biomass (6.50 g/l), carotenoids (46.60 mg/100g), and beyond this concentration, both the yield and carotenoids were reduced. A pH value 6.0 was found optimum, whereas a temperature of 35°C produced the highest carotenoids with incubation period of 48 h. The use of 50 g/l of apple pomace in the medium gave the highest yield of biomass and carotenoids. Maltose (0.2%) was the best carbon source. A pH range of 5.5 to 6.0 was optimized, whereas a temperature of 30°C and incubation period of 72 h produced the highest amount of carotenoids. 10-50 g/l of apple pomace have been incorporated to the basic medium for the producton of *Rhodotorula* (pink colour), *Sarcina* sp. (dark yellow), *Chromobacter* sp. (dark red) and *Micrococcus* sp. (light yellow) as shown in Plate 4 [70-74]. The production of pigment in apple pomace based medium using SSF gives the better yield of biomass and carotenoids. Pigment production from *Sarcina* and *Micrococchus* has also been optimized. About 70% of the carotenoids of the citrus fruit are concentrated in the flavedo [35]. The main components are xanthophylls along with other carotenoids. The extraction of the colour from peel is not easy, as the carotenoids are not water-soluble. Therefore, only a few methods of extraction have been developed at industrial level. Several research institutes have suggested different ways of colour extraction from wet peel. Flavedo flakes are liquefied at pH 4.5 with the aid of pectolytic enzymes and cellulose at 40°C for 6 hours. The terpene D-limonene is added to the mixture in order to absorb the readily extracted carotenoids. After first centrifugation of the solid particles, the liquid is again recentrifuged to remove the highly coloured D-limonene which contains 75% of the carotenoids present in the flavedo. The oil phase is dewaxed, concentrated and purified. About 3-3.5% carotenoid solution can be achieved by this method.

An extraction process for colour from dried citrus peel has been invented by the late M. Koffler on industrial scale in Israel. Peel, rag and flakes were shredded and treated with lime to a pH of 8.0-9.0. They were then pressed and dried in a rotating dryer with a directly heated drum. Temperatures were carefully controlled to avoid overheating the peel. The dried peels were stored in a cool silo and extracted as soon as possible to prevent the loss of carotenoids. Extraction was carried out in a commercial extractor, as used in the oil industry, with light petroleum ether mixed with alcohol. The extract was concentrated under vacuum and D-limonene was added to the highly concentrated colour solution. The petroleum ether was removed by evaporation at 50°C and a concentrated carotenoid solution in *D*-limonene of 4000 ppm was obtained.

Natural colours of blue grapes skin, kokum (*Garcinia indica*), phalsa (*Grewia subinaequalis*), plum (Plate 5) and Jamun (*Syzygium cumini*) have been thoroughly investigated for their nature, concentration, extractability, stability and suitability as food colours. Methods for making colour concentrates and powders from them have also been developed [75]. A new fat-soluble yellow pigment, garcinol was isolated from the rind of kokum [76]. Central committee on Food Standards is actively considering permitting the use of kokum and safflower pigments in the food industry. Attri and Joshi [71] used apple pomace based medium for carotenoids production using *Micrococcus* sp. Addition of 0.3% (v/w) ferrous ammonium sulphate gave the highest pigmentation [71, 72].

**Plate 3:** *Rhodotorula* grown on apple pomace based medium as biocolour.

**Plate 4:** Biocolour from *Micrococcus*.

**Plate 5:** Biocolour from Plum anthocyanin.

### *4.3.7 Ethanol*

The waste from fruits and vegetable processing industries being rich in polysaccharides (cellulose, hemicellulose and lignin) can be subjected to SSF for the production of ethanol, which has several uses [77, 78]. It can be used as a liquid fuel or liquid fuel supplement and as a solvent in many industries.

Traditionally, alcohol is produced from liquid or liquid mash via submerged microbial fermentation. In recent years, there has been a considerable interest in the production of alcohol from food processing wastes such as apple pomace because of (i) the rising energy costs of molasses (ii) the negative cost of values of wastes as substrates. Apple pomace is not readily amenable to submerged microbial fermentation due to its nature. But solid state fermentation of apple pomace offers several advantages for ethanol production such as higher yield, but has one difficulty of ethanol extraction from the solid materials. Different microorganisms have been used for the production of ethanol.

Predominantly, the yeast belonging to *Saccharomyces cerevisiae* has been a microorganism of choice [79]. A detailed process for the production of ethanol from apple pomace has been illustrated in Fig. 5 by Joshi *et al.* [9]. Natural fermentation of apple pomace is inferior to the yeast inoculated fermentation for ethanol, crude and soluble proteins. The production of ethanol in natural fermentation is almost half than that of *Saccharomyces cerevisiae* fermented apple pomace. Partial aseptic and anaerobic conditions were provided to the solid state fermentation of apple pomace by the addition of $SO_2$ and it was found that the addition of $SO_2$ upto 200 ppm increased the ethanol content by *Saccharomyces cerevisiae,* while it was 150 ppm for *Candida utilis* and *Torula utilis* [80]. The original pH and the initial moisture content of apple pomace was found to be suitable for ethanol production; decreasing the pH or increasing the moisture content reduced the ethanol content. Fermentation time increased the ethanol production upto 96 h at 30°C, and among the different nitrogen sources tried, ammonium sulphate gave the highest ethanol production and *Saccharomyces cerevisiae* giving better response to it than *Candida utilis* and *Torula utilis* [60]. Addition of 0.4% of ammonium sulphate increased the ethanol yield. The combined effect of AMS and $ZnSO_4$, however, was detrimental to ethanol production, but AMS alone gave better ethanol yield. It was found that all the yeast fermented apple pomace

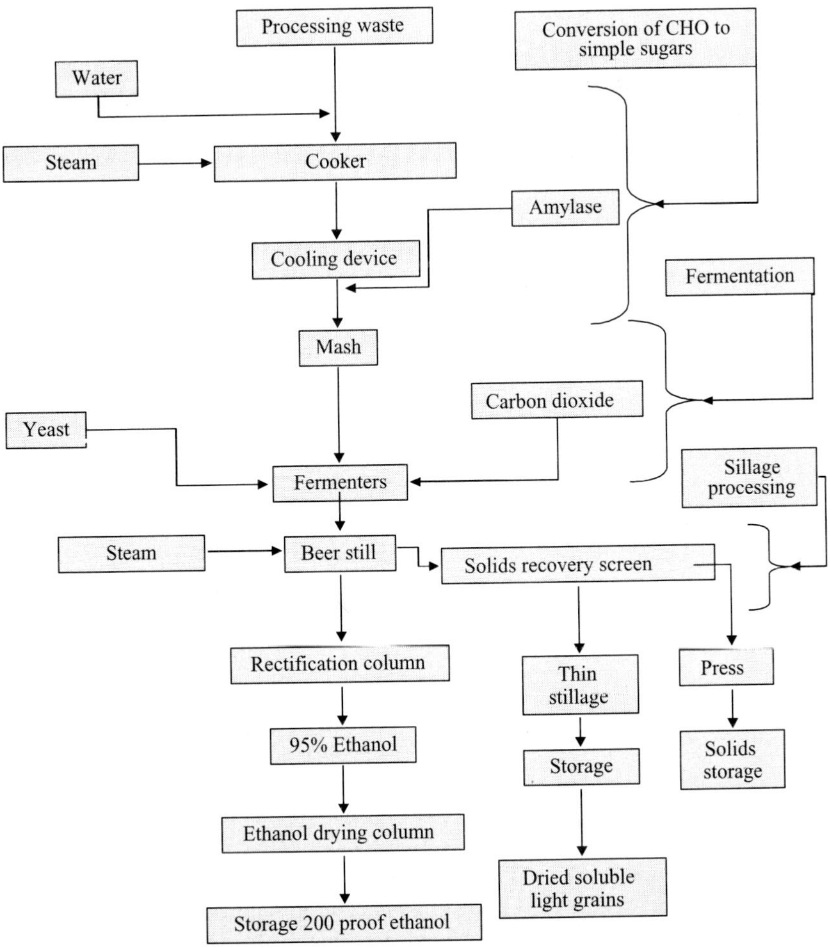


**Fig. 5:** Flow sheet for production of ethanol from processing waste.
*Source:* 9

distillates contained methyl and butyl alcohols, and aldehyde. *Saccharomyces cerevisiae* fermented distillate had more desirable characteristics than those obtained from fermentation with other yeasts, and thus had potential for conversion into potable alcohol. The distillate obtained from *Saccharomyces* fermented apple pomace had more desirable characteristics. After fermenting apple pomace with *Saccharomyces cerevisiae* for ethanol production, four methods i.e. hydraulic pressing, direct distillation, steam distillation and vacuum distillation of fermented apple pomace were applied for separation of ethanol. Out of these, the steam distillation was found to be the best as it induced minimum alteration in the fermented apple pomace [59, 60]. Apple, pear and cherry wastes have also been utilized for the production of ethanol. Waste from processing of orange can be employed for production of ethanol. Orange peels after enzymatic hydrolysis were found suitable to SSF by using *Saccharomyces cerevisiae* for ethanol production [81]. Recombinant bacterial strains of *E. coli* and *Klebsiella oxyloca* have been utilized for production of ethanol from lignocellulosic waste. Bacteria i.e. *Clostridium thermocellum* produces enzymes and is also capable of producing ethanol from lignocellulose waste. The major limitations in the utilization of *C. thermocellum* for ethanol production from waste materials are its low substrate conversion efficiencies and marked intolerance to ethanol. This limitation can be overcome to a certain extent by isolating and using ethanol tolerant strains or improving the tolerance of natural strains by gradual adaptation and mutation.

#### *4.3.8 Fermented edible products*

A number of beverages such as cider, beer, wine and brandy and vinegar can be obtained from the fermentation of fruit waste. An acceptable low alcoholic beverage (3-5% alcohol) may be prepared from apple pomace by adding colour, citric acid, $SO_2$ and raising the TSS of pomace extract from 4 to 12° B by adding sugar [82]. Citrus processing waste can also be used for the manufacture of press liquor being used in unleaded gasoline as an efficient fuel [35].

4.3.8.1 Cider: Cider is a product obtained after fermentation of apple juice containing 1 to 8% of alcohol. Apple pomace has been utilized for the production of cider and the quality of drink was improved by carbonating it. Good quality apple cider and brandy by fermenting milled apple pulp has been produced. The possibility of making brandy from dried, culled and surplus apples, grapes, oranges and other fruits has also been explored. Cider production by combining the distillate from fermented apple pomace with fermented apple juice has successfully been attempted [83].

4.3.8.2 Vinegar: Vinegar can also be prepared from fruit wastes [84]. The fruit waste is initially subjected to alcoholic fermentation, which is followed by acetic acid fermentation by *Acetobacter* bacteria which produce acetic acid. The vinegar can be prepared by Orleans Slow process or German Quick process. After the acetic acid fermentation, the product is kept for some time/days to allow acetic acid to react with alcohol to produce ethyl acetate, which imparts fruity flavour to the vinegar. Finally, the product is bottled and pasteurized at 175°F for 20 min. Vinegar production by fermenting waste from pineapple juice has been reported [85, 86]. Vinegar production by fermenting orange peel juice has also been attempted successfully. It can also be produced by mixing apple pomace extract and molasses in the ratio of 2:1. To this mixture, 0.5 g of ammonium sulphate per litre was added and it was subjected to alcoholic fermentation for 4 days, prior to acetic acid fermentation. Alcohol produced from solid state fermentation of apple pomace could be used as a base for the vinegar production. Apple pomace was mixed with molasses in the ratio of 2:1 and fortified with 0.05% ammonium sulphate for alcoholic fermentation that was carried out for four days followed by acetic acid fermentation by acetic acid bacteria. Vinegar produced by this method was found to be physico-chemically and sensorily acceptable.

### 4.3.9 Citrus juice cells

Juice cells are separated from the juice by sieving with variable pressure through a rotating drum finisher. Embryonic pips and small pieces of peel are separated by a cyclone installed in the juice production line. Cells are pasteurized and cooled immediately after sieving in order to destroy pectic enzyme activity and to prevent fermentation. The cells contain natural juice and have the same Brix as that for juice. This product is used mainly to improve flavour and mouth feel in single strength juice made from reconstituted concentrate. Cells may be frozen or aseptically packed. An alternative industrial possibility of utilizing the juice retained in the kinnow pulp is by extraction with water. This is usually done by a multistage countercurrent extraction system, using juice extractors for pressing the washed cells. By this process 3-6% °brix pulpwash can be prepared. The washed pulp is pasteurized, cooled and packed in the same way as premium pulp. Pulpwash has a low °brix, high percentage of broken cells and less flavour. This product is mainly used in drinks to increase 'body' and improve the appearance. Due to its high pectin content, the liquid recovered from washed cells is very viscous and difficult to concentrate. So this liquid can be treated with pectolytic enzymes at 45°C. After pasteurization and centrifugation, the pulpwash extract can be concentrated under vacuum to 58-64° B and stored frozen in drums. The market for juice cells is limited. About 10-15% juice cells are produced during juice extraction process. These can be dried by drum drying. Juice cells are rich source of dietary fiber and pectin, which makes them a valuable dietary product. Juice cells help in preventing diseases of colon, lowering cholesterol and preventing heart diseases. Dried juice cells have high water and fat absorption ability. One part of the dried cells can absorb 10-15 parts of water and 3-5 parts of fat.

### 4.3.10 Single cell proteins

In developing countries, including India, there is a deficiency of proteins and people suffer from malnutrition. The major cause is explosive increase in population and damage to agricultural crop leading to less availability of protein to a common man. This has given rise to explore new non-conventional resources of protein production [87]. Amongst the various processes used to supply protein, those based on the microbial growth and microbial biomass has attracted the attention of the scientific community. These proteins are called single cell proteins (SCP) and can be used as protein supplement for feed or food [21]. A number of microorganisms like yeast, fungi, algae and bacteria can be employed for the production of SCP, and each of them has its own advantages and disadvantages. For SCP production, the substances which are considered as waste but can support the growth of microorganism are used [88]. The microorganisms in turn use these substances as starting materials for fermentation and SCP production by assimilation. This aspect would be illustrated by taking specific examples of waste. Processes for production of SCP from baked bean and sauerkraut waste are presented in Figures 6 and 7. Saurkraut brine with high BOD and salt content, a waste from fermentation industry, is successfully used for the growth of a number of yeasts, but *Candida utilis* was preferred because of higher yield per unit time (Fig. 6). The process needs neither pH adjustment nor any nutrient addition, but can remove 90% BOD within 24 h. Apple pomace is the residue left after extraction of juice from apples. It is rich in sugars and has a high moisture content (80%), but it is poor in proteins. Due to its poor protein content, apple pomace is not suitable for feeding animals. SCP can be produced from dried and pectin extracted apple pomace by using *Trichoderma viride* and *Aspergillus niger* under SSF and LSF conditions. Apple pomace has a high content of lignocellulose, and a co-culture system with a mold having cellulolytic activity and yeast fermentation in SSF has been used to increase the protein content of apple poamce [90]. Different yeasts viz., *Saccharomyces cerevisiae, Candida utilis* and *Torula utilis* were grown on apple pomace, and it was found that the addition of 1.8 per cent ammonium sulphate along with mineral nutrients to the diluted apple pomace (1:6) gave higher biomass yield. The grape waste and pressed apple pulp have also been employed as a subtrate for *Aspergillus niger* to generate 35% crude protein and vitamins. Pineapple waste for SCP production has been utilized. Citrus peel juice has also been used to generate SCP by using *Fusarium, Candida utilis, Aspergillus niger* and *Trichoderma koningii* [91-93]. Cassava is a carbohydrate rich substance, but it is poor in protein. The increase in the protein content of cassava upto 20% has been achieved by using *Aspergillus niger* [94]. It has also been treated concomitantly for producing SCP using *Endomycopsis fibuliger,* and *Candida utilis. Aspergillus tamari* has been employed to produce SCP from sugar beet pulp on a pilot scale with the protein content of about 22.4% in 48 h. Mixed cultures of *Trichoderma* and yeasts, and a combination of *Kluyveromyces marxianus* efficiently converted beet pulp into protein and offered a maximum yield of 51%. The feasibility of production of SCP from baked bean waste using *Aspergillus foetidus* has been explored and flow sheet is presented in Fig. 7. These by products of oil industries can be used as an animal feed. Tomato pulp and pomace from tomato processing waste have been made as animal feed. Carob or locust tree is found in Mediterranean countries. Its pods are rich in carbohydrates, but they also have high tannin content and, thus, cannot be used for feeding animals. The tannins are broken down by certain fungi including *Aspergillus niger*. The growth of fungi on liquid carob sugar media for microbial protein production has been reported. *Rhizopus oligosporus* and *Monascus* rubber can be grown on chopped carob pods. Similarly, the growth of fungus on carob pods and protein production using two strains of *Aspergillus niger,* which depends-upon the particle size of carob pods, has also been tried. Waste from fermentation industry e.g. Sauerkraut brine with high BOD and salt content has successfully been used for the growth of a number of yeasts even in non-sterile conditions, but *Candida utilis* was preferred most as it gave a higher yield in a short time. This process does not need pH adjustment and nutrient addition, and can remove 90% BOD within 24 h. *Saccharomyces cerevisiae* and *Torulopsis utilis* have been

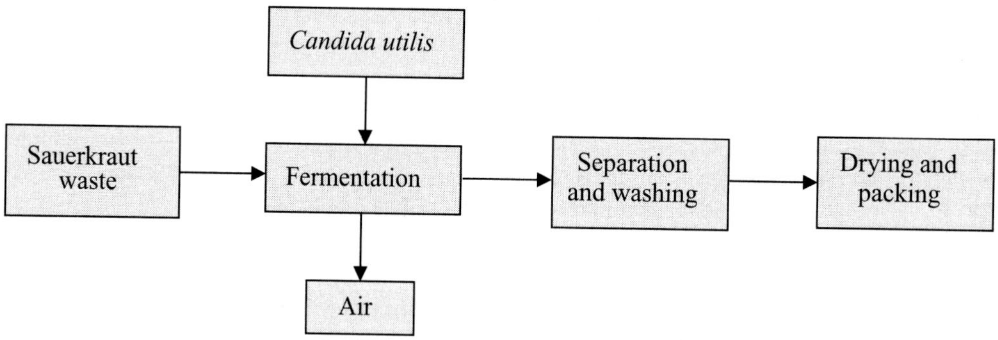


**Fig. 6:** Flow sheet for production of yeast SCP from sauerkraut brine.
*Source:* 89

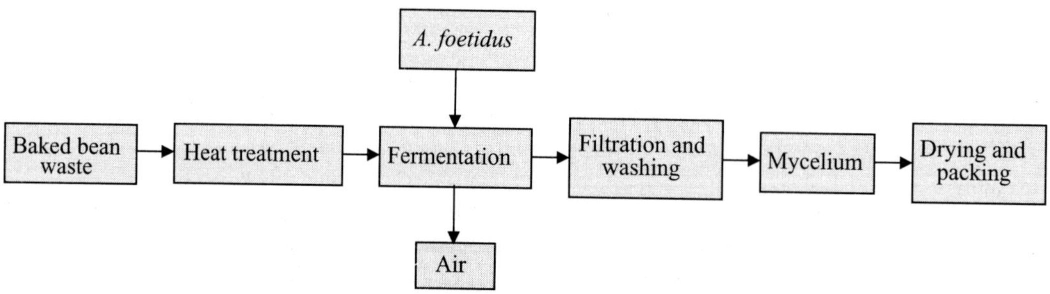


**Fig. 7:** Flow sheet for production of fungal SCP from baked bean waste.
*Source:* 95

successfully grown on the molasses with protein yield of 42-47 and 36-38%, respectively. With the addition of corn steep liquor, the protein yield increased to 60 and 52.6%, respectively. *Saccharomyces cerevisiae* had 50% protein, 2.5% fat and 13% minerals, while *Torulopsis utilis* had 42% protein, 5.6% fat and 15.8% minerals.

The waste from brewery and distilleries also supported the production of SCP. SCP can also be produced from baked bean waste using *Aspergillus foetidus*. Symba-yeast on potato waste has been successfully propagated. When two yeasts i.e. *Saccharomcyes fibuliger* and *C. utilis* were grown symbiotically, the BOD was reduced to 90% with a yeast yield of 300 kg/hl with 48% protein. Potato peels supplemented with 0.04% ammonium chloride have also been used for the production of protein by using a non-toxic fungi *Pleorotus ostreatus* [96]. Similarly, waste from orange, sugarcane and grape processing industry have also been utilized for the production of SCP [97, 98]. Nutrition of microbial cells compared with the foods of animal and plant origin is presented in Table 15.

**Table 15:** Nitrogen and protein content of microbial cells compared with selected foods of animals and plant origin.

| Source | Nitrogen (%) | Crude proteins (%) |
|---|---|---|
| Filamentous fungi | 5-8 | 31-50 |
| Algae | 7.5-10 | 47-63 |
| Yeast | 7.5-8.5 | 47-63 |
| Bacteria | 11.5-12.5 | 72-78 |
| Milk | 3.5-4.0 | 22-25 |
| Beef | 13.0-14.0 | 81-90 |
| Beef | 5.6 | 35.00 |
| Rice | 1.2-1.4 | 7.5-9.0 |
| Wheat flour | 1.6-2.2 | 9.8-13.5 |
| Corn meal | 1.1-1.5 | 7.0-9.4 |

*Source:* 99

**Table 16:** Food processing waste used and enzymes produced through microbial fermentation.

| Waste | Microorganisms | Enzyme produced |
|---|---|---|
| Apple pomace | *Trichoderma viride* | Cellulase |
| | *Aspergillus* sp. | Xylanase |
| Agro-waste | *T. reesei* | Cellulase |
| | *A. niger* | Cellulase & β-glucosidase |
| | *Pleurotus* sp. | Cellulase & xylanase |
| Bagasse | *T. reesei* | Cellulase |
| | *A. phoenicis* | |
| Grape wine trimming dust | *Cerrena unicolor* | Cellulase, xylanase and ligninase |
| Sauerkraut waste | *Candida utilis* | Invertase |
| | *A. phoenicis* | |
| Sugar beet pulp | *Cerrena unicolour* | β-glucosidase |
| Tea production waste | *Coriolus hirsutus* | CMCase, xylanase and laccase |
| | *Pleurotus ostreatus* | |

### 4.3.11 Enzymes

Both submerged fermentation (SF) and solid state fermentation (SSF) are generally employed for the production of enzymes. But SSF is a better method than SF for the production of enzymes. Various enzymes have been produced by fermenting food processing waste (Table 16). Invertase enzyme has been produced by fermenting Sauerkraut waste using *Canidida utilis*. This enzyme is widely used in the food processing industry. Subsequently, fungal amylase has been produced by using baked bean waste. Enzymes cellulase and xylanase have been produced from dried apple pomace, using *Trichoderma viridae* and *Aspergillus* sp. [100]. Native microflora has to be utilized for the generation of enzymes from cabbage waste. The enzymes produced are amylase, protease and cellulase. Bacterium *Pseudomonas* is the predominant microorganism involved in the fermentation. Apple pomace has proved to be a good substrate for the production of many enzymes [101]. Enzymes like amylase, xylanase and cellulase have been produced from apple pomace as a substrate. Among pectinases, pectin esterase is the first enzyme of pectin hydrolysis that catalyses the hydrolysis of methylated carboxylic ester group in pectin into pectic acid and methanol [101-103].

Production of pectinase from apple pomace is promising due to several advantages like easy availability of cheaper raw material and easier processing of the substrate. In solid state fermentation of apple pomace with *Aspergillus niger* among different dilutions tried, 1:3 dilution (w/v) at pH 4.0, incubation temperature of 25°C for 96 h was found optimal for higher PME production. Out of the different nitrogen sources and additives evaluated, ammonium sulphate @ 0.2 per cent and sodium chloride @ 0.5 per cent gave the highest PME production. The pectin esterase production in SSF using *Aspergillus niger* using apple pomace as a substrate is described in detail by Parmar [103] (Plate 6). The SSF gave 2.3 times higher PME activity than SmF using optimized parameters of fermentation [101]. Citrus waste can also be used for the production of pectinase in solid state culture [104].

Hydrolysed whey syrup is prepared by enzymatically hydrolyzing the lactose into its component monosaccharides, glucose and galactose. The hydrolysates along with residual proteins is concentrated to syrup by evaporation and so used in confectionery and ice cream.

### 4.3.12 Other value added products

**4.3.12.1 Citric acid:** Citric acid has been produced by fermenting brewery waste with *Aspergillus niger* [105]. Citric acid can also be produced on commercial scale by solid state fermentation (SSF). SSF with *Aspergillus niger* on various substrates like pineapple juice, molasses, sweet potato residue, sugar-cane bagasse impregnated with pineapple juice, mandarine orange waste, apple pomace, grape pomace was carried out to evaluate their potential for citric acid production [106-108].

**4.3.12.2 Microbial flavours and gums:** Microbial gums like xanthan can be produced from processing of waste. *Xanthamonas campestris* has been employed to produce xanthan from fermentation of cabbage waste and citrus waste. Various substances including sucrose, hydrolysed cereal grains, molasses and hydrolysed lactic acid from cheese whey have been used as subtrates for *X. campestris* to produce xanthan employed extensively as a food additive. D-limonine, a by product from citrus processing industry, is used as a flavouring compound [35].

**4.3.12.3 Amino acids:** Amino acids have also been produced by microbial fermentation of waste. These amino acids are used as additives in food, feed and also as flavouring agents. Glutamic acid is one of the amino acids, which is produced by fermenting hydrolyzed starch solutions, cane molasses and beet molasses by using bacateria like *Corynebacterium glutamicum, Arthrobacter* and *Brevibacterium flavum*. Glutamic acid in the form of monosodium glutamate or Ajinomoto is used as a flavouring agent in foods [109].

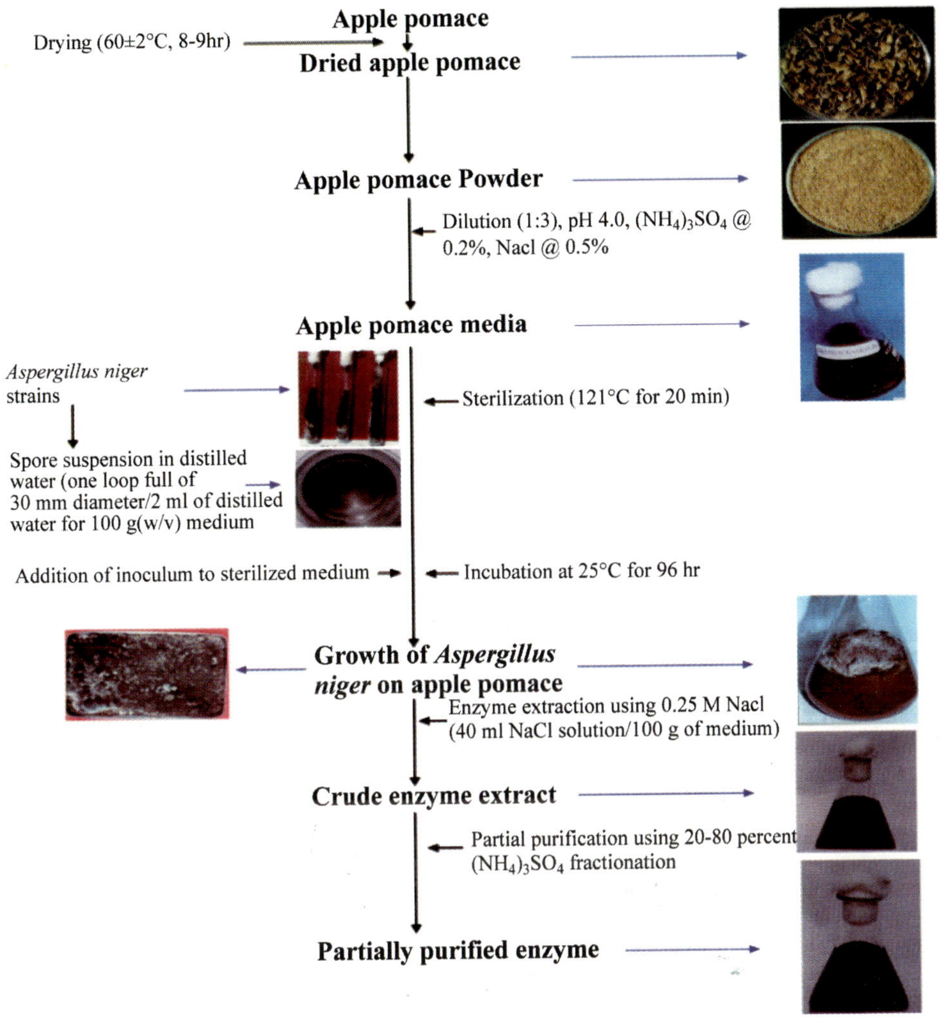


**Plate 6:** Flow-sheet of optimized process for production of pectin estrases in SSF.

4.3.12.4 Vitamins: Microbial fermentation can also be used for the production of vitamins, which are very important biomolecules and their deficiency can cause several diseases in human beings and animals. Bateria *Pseudomonas denitrificans, Propionibacterium shermanii* and *Streptomycetes* have been used for commercial production of vitamin $B_{12}$ from soyabean meal, corn steep liquor, fish meal, meat extract, etc. Corn steep liquor has also been used as a substrate for a mold *Ashbya gossypii* to produce vitamin $B_2$ or riboflavin. β-carotene, which is a precursor of vitamin A, and also used as a pigment in food industry can be produced by microbial fermentation using fungus *Blackeslea trispora* and algae *Dunaliella.*

4.3.12.5 Baker's yeast: Apple pomace has been found suitable for the production of baker's yeast [110, 111]. Comparison of different substrates for the production of baker's yeast *Saccharomyces diastaticus* was done, and it was found that among the different substrates, apple pomace extract supported higher biomass yield though at par with jaggery [112]. Propagation of baker's yeast starting from a slant culture to a commercial fermenter is conducted under limited aeration [113]. Apple pomace was successfully employed as a substrate to produce baker's yeast in a medium by a fed batch process. The highest biomass production was observed in apple pomace extract having pH 4.5. An optimum temperature of 30°C or below has been recommended for baker's yeast growth [114, 115]. The performance of the experimentally produced yeast from apple pomace was also judged by preparing different bakery products such as bread, bun and rusk, and comparing with commercial baker's yeast produced (Plate 7). The quantitative descriptive analysis (QDA) profiling, triangle tests as well as composite scoring methods were not able to differentiate the products. Effect of addition of apple pomace on characteristic of bread and leavening activity of yeast has also been studied [116]. Bhushan and Joshi [111] used apple pomace extract as a carbon source in an anaerobic fed batch culture for the production of baker's yeast. The fermentable sugar concentration in the bioreactor was regulated at 1 to 2%, and a biomass yield of 0.48 g $g^{-1}$ of sugar was obtained. Interestingly, the dough raising capacity of the baker's yeast grown on the apple pomace extract was apparently the same as that of commercial yeast. The use of apple pomace extract as a substrate is an interesting alternative to molasses, traditionally used as a carbon source for baker's yeast production.

**Plate 7:** Bakery products prepared from experimental yeast grown on apple pomace.
a: Bread, b: Bun, c: Rusk

4.3.12.6 Organic compounds: SSF technique has been used to produce organic compounds like butanol, acetone, citric acid, acetic acid and lactic acid from molasses. 2, 3 butylene glycol can be produced by fermenting citrus peel juice and citrus waste after adding molasses.

4.3.12.7 Proteins: The constituents present in the whey are given in Table 17. The whey protein is concentrated by ultrafiltration, and may then be dried to give a high protein powder [1]. Cashew apple pomace is utilized for the production of proteins [118]. A flow sheet for the bioconversion of cashew apple pomace to a protein rich product is given in Fig. 8. The whey protein left after Press cake left after oil extraction from apricot kernels can be a very good souce of essential oil and protein isolate [119]. Cattle

**Table 17:** Some constituents present in the whey.

| Constituent | Amount (%) |
|---|---|
| Dry matter | 5.8-6.8 |
| Lactose | 4.2-4.7 |
| Albumin | 0.8-1.0 |
| Fat | 0.2-0.4 |
| Biotin (μg/kg)* | 325-599 |
| Iron (mg/kg)* | 9.16 |

* Dry weight basis
*Source:* 117

blood after slaughtering can be a very good source of proteins, which otherwise goes waste. This blood is dried and used as a protein source.

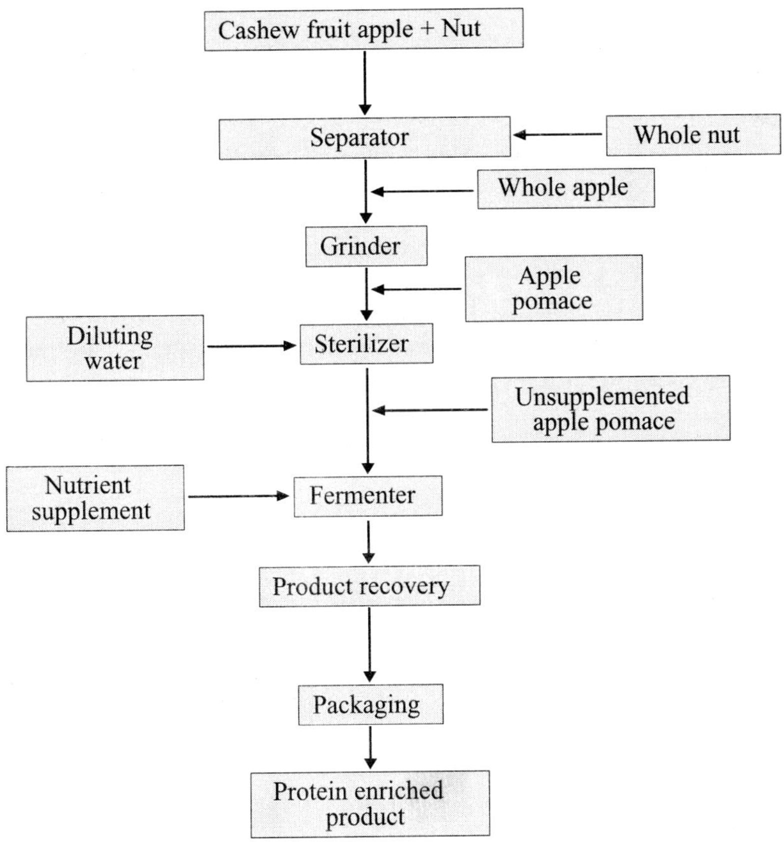


**Fig. 8:** Flow sheet for the bioconversion of cashew apple pomace to a protein rich product.
*Source:* **118**

## 5. COMPREHENSIVE STRATEGIES FOR BIOTECHNOLOGY OF WASTE UTILIZATION

It is apparent that waste from fruit and vegetable processing is a rich source of several constituents and has a great potential for value addition. In view of these, the adoption of right approach or strategy would be of utmost importance. Some of the strategies, which could be adopted, are summarized as ahead.

- Use such techniques which generate no or less quantity of waste (dry casting technology/ enzyme technology for juice processing).
- Recover maximum useful materials (e.g., pectin from apple pomace, lactose from whey).
- Convert the waste into useful products (ethanol from fruit waste, citric acid from pomace, enzyme from wheat bran, animal feed from waste).
- Use the waste as substrate for production of valuable substances (SSF of waste for mushroom production, single cell proteins, pigment production).
- The waste material should get first preference for food/feed related substances, followed by biogas.

Waste water treatment should be the last strategy to meet the requirement of pollution control agencies.

## 6. WASTE TREATMENT TECHNOLOGY

Effluent treatment plant (ETP) in the food processing industries is an important component for treatment of polluted water, which otherwise awakened the environment protection agencies, as this polluted water/effluent affects the surrounding environment and causes nuisance if not handled properly. For this, the pollution control boards of the states are playing an important role in installation of the effluent treatment plants in food processing industries. Effluent is a liquid/water of industrial waste, which may be the outflow from sewage during purification. A large part of it is used in distilleries, which generate around $30m^3$ distillery effluents/waste/spent wash every day. This effluent is a rich source of organic carbon and contains plant nutrients in varying amounts. Its use in agriculture is very limited because of its high BOD (Biochemical Oxygen Demand). Many distilleries have now installed biogas plants to produce methane which in turn reduces BOD and can be used for pre-sowing irrigations with beneficial effects on crops. The effluent can also be used for irrigating crops with appropriate dilution.

Waste water i.e., effluent generated from various points at the plant such as outlet of Westfeling separator and Treatment tanks, etc. is polluted and it must be given a treatment before draining so that pollution control board norms are met. It is achieved through a specific treatment in a plant called effluent treatment plant. ETP treats the waste/polluted water coming from the plant and makes it able to dispose off in the river by neutralizing its toxicity. Effluent treatment plant (ETP) consists of lime doser, oil grease trap, sump tank, flash mixer tank, flocculation tank, primary settling tank, aeration tank and secondary settling tank as discussed ahead [119]. Flow sheet for effluent treatment used in the food processing industries is presented in Fig. 9.

(a) Lime Doser: In this unit alum, calcium, lime, etc are added to maintain the pH of the effluent waste water.

(b) Oil-Grease Trap: In this unit operation, in tank oil and grease are separated from the waste water.

(c) Sump and Flash Mixer Tank: Alum and lime are mixed properly with the help of agitator and mixers.

(d) Flocculation Tank: Flocculates of microorganisms are allowed to be formed for the degradation of organic waste present.

(e) Primary Settling Tank: It provides a smooth transition as effluent flows from flocculation tank to aeration tank. The sludge settled out into sludge drying beds built adjacently. The tank is provided with clarifier and scraper.

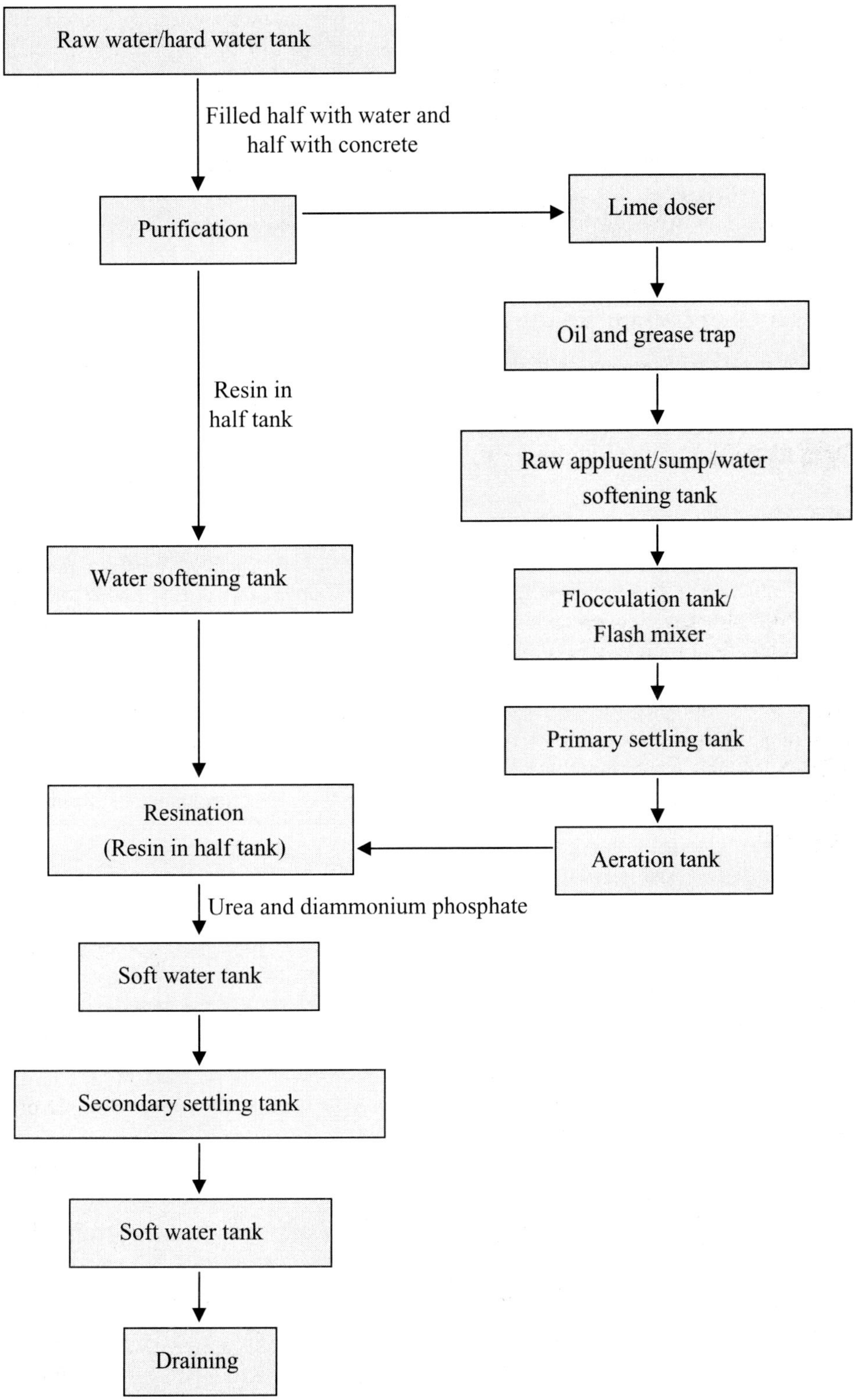


**Fig. 9:** Flow sheet for effluent treatment plant used in the food processing industries.

(f) Aeration Tank: Aeration tank is provided for the biological treatment of organic matter, which is easily biodegradable. The essential features of this process are development of microbial floccules held in suspension in an aeration chamber. Prior to the biological treatment, nutrients like DAP and urea are added to over flow of the primary clarifier. After the retention time, the water from aeration tank is sent to a secondary clarifier.

(g) Secondary Settling Tank: Secondary settling tank is used for removal of excess biomass from aeration basin. The design of the unit is similar to primary clarifier. The over flow from this tank is out drained and the excess sludge is then dried in sludge drying beds built adjacently.

### 6.1 Advantages of Effluent Treatment

- Prevention of environmental pollution
- Good supplement for irrigation of the agriculture crop
- Good source of nutrients and can increase fertility of the soil
- Solve the problems of salinity hazards.

## 7. CONCLUSIONS

The waste of food industry is a rich source of several nutrients, and disposing it off is a sheer wastage of natural resources. So the waste from food processing industries is valuable and have large economic potential. A number of value added products can be prepared from food processing industries waste. However, lack of pilot testing of the technologies developed, luck-warm attitude of the industrialists and perhaps, less helping hand from the government sector have led to virtually no adoption of the technologies developed for the waste utilization. Besides this, another hurdle is the economy behind setting up of waste processing units and their operation. The food processing industries should be encouraged at global level for setting up waste utilization plants in the form of some incentives and concession. Nevertheless, when industrialists will come to know about the economic potential of such units, they will themselves come forward.

## 8. FUTURE THRUSTS

An integrated approach comprising stripping off the useful recoverable products followed by waste treatment technology is needed. Waste can be converted into a number of useful products by suitable technologies especially by the deployment of microorganisms. A number of techniques (genetic engineering, reverse osmosis, enzyme technology, etc.) can be used to improve the production of useful products from wastes. It is an area where government could intervene by giving specific incentives to the industries employing waste management technology and imposing heavy penalties to defaulters.

## REFERENCES

1. Angold R, Beech G & Taggart J (1989) *Food Biotechnology*, Cambridge University Press, Cambridge, USA, pp 1-167.
2. Maini SB & Sethi V (2000) Utilization of fruits and vegetables processing waste. In: *Post Harvest Technology of Fruits and Vegetables*, Verma LR & Joshi VK (eds), Indus Publishing Co, New Delhi, India, pp 1006-1018.

3. Joshi C & Joshi VK (1990) Food processing waste management - Need for an integrated approach. *Indan Food Packer* 44: 56-67.
4. Nand K, Manjrekar SP & Kadam PS (1993) Processing of agriculture and agro-industrial wastes for the economical production of important products. In: *Proceedings of 3rd International Food Convention*, CFTRI, Mysore, India, pp 35-40.
5. Gera IB & Kramer A (1969) The utilization of food industries wastes. In: *Advances in Food Research*, Chichester CO, Mkak EM & Stewart GF (eds), Academic Press, London, UK, pp 28-42.
6. Nand K (1999) Biomethanation from agro-industrial and food processing wastes. In: *Biotechnology: Food Fermentation,* Vol 2, Joshi VK & Pandey A (eds), Educational Publishers and Distributors, New Delhi, India, pp 1349-1372.
7. Nigam P & Pandey A (1986) Utilization of by-products of cane sugar industry. *Cooperative Sugar* 17: 529-534.
8. Joshi VK, Sandhu DK, Sharma N & Kaushal BBL (1996) Microbial utilization of fruit and vegetable processing industry waste. *Front Appl Environ Microbiol* 2: 15-21.
9. Joshi VK, Pandey A & Sandhu DK (1999) Fermentation technology for food industry waste utilization. In: *Biotechnology: Food Fermentation - Microbiology, Biochemistry and Technology*, Vol 1, Joshi VK & Pandey A (eds), Educational Publishers and Distributors, New Delhi, India, pp 1291-1348.
10. Johar DS, Krishnamurthy GV & Bhatia BS (1960) Utilization of apple pomace. *Food Sci* 9: 82-84.
11. Kehar CL & Chandra R (1946) Mango seed kernel - a new source of food. *Curr Sci* 15: 8-50.
12. Siddappa GS and Bhatia BS (1952) Preparation of jelly from jackfruit rind. *Bull Cent Food Technol Res Inst, Mysore*, 2: 70-72.
13. Gopalan C, Ramasastri BV & Balasubramanian SC (1985) *Nutritive Value of Indian Foods*, Narasinga Rao BS, Deosthale YG & Pant KC (eds), National Institute of Nutrition, Indian Council of Medical Research, Hyderabad, India, pp 46-103.
14. Siddappa GS & Bhatia BS (1955) Preparation and use of Jack seed floor. *Bull Cent Food Technol Res Inst, Mysore* 5: 3-6.
15. Pruthi JS (1960) Nutritive value and utilization of passion fruit waste (skin and rind). *Food Sci* 9: 397-399.
16. Maini SB (1963) Some studies on banana powder, banana juice powder and utilization of banana peels, Associateship Thesis, Central Food Technological Research Institute, Mysore, India.
17. Khan RYA, Vishwanathan PK & Rao SDT (1970) Processing of Indian sweet orange seed. *La Rivista Italiana Delle Sostanzl Grasse* 47: 262-264.
18. Ramakrishna G, Vishwanathan RK & Rao STD (1970) Pilot plant oil production from muskmelon. *Oil Mill Gaz* 75: 8-11.
19. Kour S, Teotaia MS & Berry SK (1989) Sorption behaviour of pumpkin seeds and kernels. *Indian Food Packer* 43: 25-30.
20. Nanjundaswamy AM & Siddappa GS (1964) Preservation of the central edible core of the banana stem in the form of candy, canned, curried product and dehydrated slices. *Indian Food Packer* 18: 1-3.
21. Hang YD (1986) Production of single cell proteins from food processing wastes. In: *Food Processing Waste,* Green JH & Krammer A (eds), AVI Publishing Co Inc, Westport, USA, pp 442-455.

22. Hang YD & Woodams EE (1982) A solid state fermentation system for production of ethanol from apple pomace. *J Food Sci* 47: 1851-1852.
23. Pannier W (1993) Spent mushroom compost: A natural resource that provides solution to environmental problems. *Mushroom News* 41: 10-11.
24. Ahlawat OP & Rai RD (2002) Recycling of spent mushroom substrate. In: *Recent Advances in the Cultivation Technology of Edible Mushrooms*, Verma RN & Vijay B (eds), National Research Centre for Mushroom, Solan, India, pp 261-282.
25. Ahlawat OP, Indu Rani C & Sagar MP (2005) Spent mushroom substrate - properties and recycling for beneficial purposes. In: *Frontiers in Mushroom Biotechnology*, Rai RD, Upadhaya RC & Sharma SR (eds), National Research Centre for Mushrooms, Solan, India, pp 314-334.
26. Devrajan A, Joshi VK, Gupta K, Shekhar C & Lal BB (2004) Evaluation of apple based reconstituted feed in rats after solid state fermentation and ethanol recovery. *Braz Arch Biol Technol* 47: 93-106.
27. Beyer M (1996) The impact of mushroom industry on the environment. *Mush News* 44: 6-13.
28. Beyer D & Noble R (2006) Opportunities for using spent mushroom substrate (SMS). *Mush Intl* 103: 19-21.
29. Upadhyay RC & Sohi HS (1988) Apple pomace a good substrate for the cultivation of edible mushrooms. *Curr Sci* 57: 1189-1190.
30. Worrall JJ & Yang CS (1992) Shiitake and oyster mushroom production on apple pomace and sawdust. *Hort Sci* 27: 1131-1133.
31. Vendruscolo F, Albuquerque PM & Streit F (2008) Apple pomace: A versatile substrate for biotechnological applications. *Crit Rev Biotechnol* 28: 1-12.
32. Sharma RC (1983) Waste utilization of apple pomace for vinegar generation and briquettes and biogas production, M.Sc. Thesis, HP Agricultural University, Palampur, India.
33. Madhukara K, Nand K, Raju NR & Srilatha HR (1993) Ensilage of mango peel for methane generation. *Process Biochem* 28: 119-123.
34. Lane AG (1985) Anaerobic digestion of orange peel. *Indian Food Packer* 39: 27-33.
35. Maini SB & Ladaniya MS (2004) Citrus processing waste utilization and value addition. In: *Advances in Citricultutre,* Singh S, Shivankar VJ, Srivastava AK & Singh IP (eds), Jagmander bood Agency, Delhi, India, pp 657-671.
36. Bhalerao SD, Mulsmuley GV, Ananthakrishna SM & Potty VH (1989) Effluent management in food industry. *Indian Food Packer* 43: 5-7.
37. Srirangrajan AN & Shrikhande AJ (1977) Characterization of mango peel pectin. *J Food Sci* 42: 279-280.
38. Palaniswamy KP, Muthukrishanan CR & Shanmugavelu KG (1974) Physico-chemical characteristics of some varieties of mango. *Indian Food Packer* 28: 12-19.
39. Narsimhachar BL, Reddy BR & Thirmula RSD (1977) Processing of mango stones for fat. *J Am Oil Chem Soc* 54: 494-495.
40. Anon (1985) Annual Report, Central Food Technological Research Institute, Mysore, India.
41. Parekh CM, Pruthi JS & Lal G (1959) Utilization of citrus waste. *Indian Food Packer* 14: 7-12.
42. Ranade GS (1987) Future of natural essential oils in India. *Indian Perfumers* 31: 13-16.
43. Pruthi JS, Parekh CM & Lal G (1961a) Physico-chemical examination of of Indian mandarin oils and their quality standards. *Food Science* 10: 345-350.

44. Pruthi JS, Parekh CM & Lal G (1961b) An integrated process for recovery of essential oil and pectin from mandarin orange waste. *Food Science* 10: 372-378.
45. Sharma A, Joseph R & Sreenivasa MV (1980) Improved method of extraction of oil from orange peels using fungal xylanase. In: *Proceedings of the Symposium on Wastes from Food Industries: Utilization and Disposal,* Association of Food Scientists and Technologists, CFTRI, Mysore, India, pp 18-25.
46. Sharma PC, Prasant Kamboj, Rakesh Sharma & Dev Raj (2006) Storage behavior of stone fruit kernel oils in different packages. *J Food Sci Tech* 43: 297-300.
47. Kimball D (1991) *Citrus Processing: Quality Control and Technology,* Van Nostrand reinhold, New York, USA.
48. Rao VS (1989) Studies on processing and waste utilization of 'Nagpur' mandarin. National Reaserch centre for Vitrus, Nagpur, Annual report of 1988-89, pp 18-19.
49. Anon (1962) *Agricultural Hand Book: Citrus Products and Byproducts,* Department of Agriculture, Chemistry and Technology of Citrus, USA, No. 98
50. Anon (1987) World Bank Review of Horticulture in South West Hill Region, India.
51. Dev DK & Ingle UM (1982) Utilization of pineapple by products and wastes - A review. *Indian Food Packer* 36: 15-22.
52. Alibes X, Munoz F & Rodriquez J (1984) Feeding value of apple pomace silage for sheep. *Animal Feed Sci Technol* 11: 189-198.
53. Bose S, Thukral BM & Narayana S (1952) Utilization of mango seed kernel and Jamun seed meal in a simplified poultry ration for growing chickens. *Indian J Vet Sci* 22: 247-251.
54. Joshi VK (1998) Apple pomace utilization - Present status and future strategies. In: *Advances in Biotechnology*, Pandey A (ed), Educational Publishers & Distributors, New Delhi, India, pp 141-155.
55. Arora M, Wadhwa M, Thapar VK & Sehgal VK (1995) *In vivo* evaluation of fermented processing waste as ruminant feed. *Indian J Microbiol* 35: 259-261.
56. Joshi VK & Sandhu DK (1994) Solid state fermentation of apple pomace for production of ethanol and animal feed. In: *Solid State Fermentation*, Pandey A (ed), New Age Publisher, New Delhi, India, pp 93-98.
57. Joshi VK & Sandhu DK (1996b) Preparation and evaluation of animal feed using solid state fermentation of apple pomace. *Biores Technol* 56: 251-255.
58. Rahmat H, Hodge RA, Manderson GJ & Yu PL (1995) Solid substrate fermentation of *Kloeckera apiculata* and *Candida utilis* on apple pomace to produce an improved stock feed. *World J Microbiol Biotechnol* 11: 168-170.
59. Devrajan A (1997) Solid state fermentation of apple pomace: methods of ethanol recovery, physico-chemical evaluation and acceptability of animal feed produced, MSc Thesis, Dr. YS Parmar University of Horticulture and Forestry, Nauni, Solan, India.
60. Joshi VK & Devrajan A (2008) Ethanol recovery from solid state fermentated apple pomace and evaluation of physico-chemical characteristics of residue. *Nat Prod Rad* 7: 127-132.
61. Sandhu DK & Joshi VK (1997) Solid state fermentation of apple pomace for concomitant production of ethanol and animal feed. *J Sci Ind Res* 56: 86-90.
62. Joshi VK, Gupta K, Devarajan A, Lal BB & Arya SP (2000) Production and evaluation of fermented apple pomace in the feed of broilers. *J Food Sci Technol* 37: 609-612.
63. Beer OP, Raghuramaiah B & Krishnamurthy GV (1976) Utilization of mango waste peel as a source of pectin. *J Food Sci Technol* 13: 96-97.

64. Bhatia BS, Krishnamurthy GV & Lal G (1959) Preparation of pectin from raw papaya (*Carica papaya*) by aluminium chloride precipitation method. *Food Sci* 8: 314-320.
65. Chaliha BP, Barua AD & Siddappa GS (1963) Assam lemon as a source of pectin. Part I. Effect of method of extraction, drying and storage of peel and pomace on the recovery of the pectin during storage. *Indian Food Packer* 17: 8-14.
66. Srivas SR & Pruthi JS (1976) Studies on the quantitative estimation of chemical components of cashew apple, citrus and other fruit pectins. *J Inst Chemists* 48: 95-99.
67. Sakai T & Okushima M (1980) Microbial production of pectin from citrus peel. *Appl Environ Microbiol* 39: 908-912.
68. Das NB & Benerjee RM (1952) Utilization of mango seed kernels as a source of starch. *Sci Culture* 17: 33-36.
69. Attri D (2004) Production and evaluation of microbial colours using apple pomace, Ph.D. thesis, Dr. YS Parmar University of Horticulture and Forestry, Nauni, Solan, India.
70. Attri D & Joshi VK (2005a) Optimization of apple pomace based medium and fermentation conditions for pigment production by *Chromobacter* species. *J Food Sci Technol* 43: 515-520.
71. Attri D & Joshi VK (2005b) Optimization of apple pomace based medium and fermentation conditions for pigment production by *Micrococcus* species. *J Sci Ind Res* 64: 598-601
72. Sandhu DK & Joshi VK (1997b) Development of apple pomace based medium, optimizing pigment production by *Rhodotorula* and its characterization. *Adv Food Sci* 19: 1-34.
73. Joshi VK, Attri D, Anju B & Shashi B (2003) Microbial pigments. *Ind J Biotechnol* 2: 362-369.
74. Joshi VK & Attri, D (2005) Optimization of apple pomace based medium and fermentation conditions for pigment production by *Rhodotorula* species. *Proc Nat Acad Sci* 76: 171-176.
75. Krishnamurthy N, Shivashankar S, Sampathu SR, Ravinderanath B, Shanakaranarayan R, Lewis YS & Netrajan CP (1978) *Proceedings of First Indian Convention of Food Scientists and Technologists*, Mysore, India p 74.
76. Krishnamurthy N, Lewis YS & Ravinderanath B (1981) On the structure of garcinol, isogarcinol and camboginol. *Tetrahed Lett* 22: 793-796.
77. Badger PC & Broder DJ (1989) Ethanol production from food processing wastes. *Hort Sci* 24: 227-233.
78. Jarosz K (1988) Solid state alcoholic fermentation of apple pomace. *Acta Aliment Pol* 14: 139-144.
79. Joshi VK & Sandhu DK (1996) Effect on type of alcohols in the distillate from the solid state fermentation of apple pomace by different yeasts. *Nat Acad Sci Lett* 49: 219-224.
80. Hang YD, Lee CY, Woodams BE & Cooley HJ (1981) Production of alcohol from apple pomace. *Appl Environ Microbiol* 42: 1128-1129.
81. Converti A, Perego P, Borghi MD & Ferraiolo G (1989) Pretreatment operations and alcohol fermentation of orange wastes. *J Ferm Bioengg* 68: 277-281.
82. Barwal VS (1990) Low alcoholic beverages from culled apples. *J Food Sci Technol* 28: 257-258.
83. Masuda H, Naritaka S & Maraki H (1964) Studies on fermentative processing of apple fruit. IV Fermentation of milled apple pulp in cider making. *J Ferment Technol* 42: 379-384.
84. Ethiraj S & Suresh ER (1992) Studies on the utilization of mango processing waste for production of vinegar. *J Food Sci Technol* 29: 48-50.

85. Richardson KC (1967) Submerged acetification of a vinegar base produced from waste pineapple juice. *Biotech Bioeng* 9: 171-186.
86. Spurgin MM (1964) Vinegar base production from waste pineapple juice. *Queensland J Agr Sci* 21: 213-232.
87. Chen SL & Peppler HJ (1979) Single cell proteins in food applications. *Dev Ind Microbiol* 19: 79-94.
88. Nigam P & Singh D (1996) Processing of agricultural wastes in solid state fermentation for microbial protein production. *J Sci Ind Res* 55: 373-380.
89. Hang YD, Splittstoesser OF & Landschoot RL (1972) Sauerkraut waste: a favourable medium for cultivating yeasts. *Appl Microbiol* 24: 1007-1008.
90. Bhalla TC & Joshi M (1994) Protein enrichment of apple pomace by coculture of cellulolytic molds and yeasts. *World J Microbiol Biotechnol* 10: 116-117.
91. Gai SK, Kahlon SS & Sood SM (1979) SCP from canning industry waste 1. Citrus fruit pulp as substrate. *J Res* 16: 64-68.
92. CFTRI (1989) Annual Report of 1988-89, Central Food Technological Research Institute, Mysore, India.
93. Bahar S & Azuaze JT (1984) Studies on the growth of *Fusarium* sp. on citrus waste for the production of single cell proteins. *J Food Sci Technol* 21: 63-67.
94. Mani Lal VB, Narayanan CS & Balagopalan C (1991) Cassava starch effluent, treatment with concomitant SCP production. *World J Microbiol Biotechnol* 7: 185-190.
95. Forney LJ & Reddy CA (1977) Fermentative conversion of potato processing waste into a crude protein feed supplement by *Lactobacilli*. *Dev Ind Microbiol* 18: 135-143.
96. Green JH & Kramer A (1979) *Food Processing Waste Management,* AVI Publishing Co Inc, Westport, USA.
97. Lines KK (1980) Production of SCP from potato processing waste, United States Patent-4, 144, 132, 1976.
98. Kahlon SS & Arora M (1986) Utilization of potato peels by fungi for protein production. *J Food Sci Technol* 23: 264-267.
99. Kharatyan GS (1978) Microbes as food for humans. *Ann Rev Microbiol* 32: 301-327.
100. Bhalla TC & Joshi M (1993) Production of cellulase and xylanase by *Trichoderma viride* and *Aspergillus* sp. on apple pomace. *Ind J Microbiol* 33: 253-255.
101. Joshi VK, Parmar M & Neerja SR (2006) Pectin esterase production from apple pomaceinsolid state and submerged fermentations. *Food Technol Biotechnol* 44: 253-256.
102. Hours RA, Voget CE & Ertola RJ (1988) Apple pomace as a raw material for petinase production in solid state cultures. *Biol Wastes* 24: 147-157.
103. Parmar M & Joshi VK, Lal BB & Neerja R (2004) Effect of fermentation parameters for pectin methyl esterase production in solid state fermentation of apple pomace. In: *Proceedings of National symposium on Advances in Industrial Biotechnology,* Punjabi University, Patiala, Punjab, India, p 35.
104. Garzon GG & Hours RA (1992) Citrus waste: an alternative substrate for pectinase production in solid state culture. *Biores Technol* 39: 93-95.
105. Crunele CN, Cimerman A & Perdih A (1986) *Aspergillus niger* mutants for bioconversion of apple distillery wastes. *Enzyme Microbial Technol* 8: 166-170.
106. Kapoor KK, Chaudhary K & Tauro P (1982) Citric acid production. In: *Prescott and Dunn's Industrial Microbiology*, Reed G (ed), AVI Publ Co, West Port, Connecticut, pp 709-747.
107. Hang YD (1985) Utilization of apple pomace for microbial production of citric acid. Special Report, New York State Agricultural Experiment Station, Geneva, Switzerland.

108. de Lima VAG, Stamford TM & Salgueiro AA (1995) Citric acid production from pineapple waste by solid state fermentation using *Aspergillus niger. Arch Biol Technol* 38: 777-783.
109. Nakayama K (1982) Production of amino acids. In: *Presscott and Dunn's Industrial Microbiology*, Reed G (ed), AVI Publishing Co Inc, Westport, USA, pp 748-80.
110. Bhushan S & Joshi VK (2006) Baker's yeast production under fed batch culture from apple pomace. *J Sci Ind Res* 65: 72-76.
111. Bhushan S & Joshi VK (2007) Apple pomace utilization for the production of baker's yeast. Performance in preparation of bakery products. *Proc Nat Acad Sci India* 77: 303-312.
112. Joshi VK, Shashi B, Attri D & Kaushal BBL (2005) Utilization of apple pomace in comparison to molasses and Jaggery under fed-batch cultivation for the production of Baker's yeast. In: *Acta Horticulture* II, Chauhan JS, Sharma SD, Sharma RC, Rehalia AS & Kumar K (eds), Dr. Y.S. Parmar University of Agriculture & Forestry, Nauni, Solan, India, pp 479-484.
113. Joshi VK & Bhushan S (2003) Apple pomace utilization for the production of Baker's yeast: Effect of different substrate concentration and growth stimulators. *Ind J Biotechnol* 2: 220-226.
114. Joshi VK, Jaiswal S & Lal BB (1997) Apple pomace: Effect of sulphur dioxide and temperature of storage on its preservation and medium optimization for baker's yeast produced from it. *Front Biotechnol* S57: 692-697.
115. Joshi VK, Jaswal S & Lal B (1998) Apple pomace: Effect of sulphur dioxide and temperature on its preservation and medium optimization for yeast biomass production. *J Sci Ind Res* 57: 692-697.
116. Joshi VK & Kaushal NK (2001) Effect of apple pomace on characteristics of and leavening activity of yeast. *Natl Acad Sci Lett* 2: 1-7.
117. Viz S & Gandhi DN (1993) Whey - an alternative substrate for the production of Bakers yeast. *Indian Food Ind* 12: 41-45.
118. Kaur K (1989) Microbial transformation of apple pomace to recover industrial products, MSc Thesis, Punjab University, Chandigarh, India.
119. Tilakaratne BMKS (2007) Studies on utilization of kernel oil and press cake of apricot for value addition, Ph.D. Thesis, Department of Post Harvest Technology, Dr YS Parmar University of Agriculture and Forestry, Nauni, Solan, India.
120. Raj D & Gupta A (2008) Water/effluent treatment plant in the food processing industries: An important process. *Beverages Food World* 35: 37-39.

# Index